Birkhäuser

Trends in Mathematics

Trends in Mathematics is a series devoted to the publication of volumes arising from conferences and lecture series focusing on a particular topic from any area of mathematics. Its aim is to make current developments available to the community as rapidly as possible without compromise to quality and to archive these for reference.

Proposals for volumes can be submitted using the Online Book Project Submission Form at our website www.birkhauser-science.com.

Material submitted for publication must be screened and prepared as follows:

All contributions should undergo a reviewing process similar to that carried out by journals and be checked for correct use of language which, as a rule, is English. Articles without proofs, or which do not contain any significantly new results, should be rejected. High quality survey papers, however, are welcome.

We expect the organizers to deliver manuscripts in a form that is essentially ready for direct reproduction. Any version of TEX is acceptable, but the entire collection of files must be in one particular dialect of TEX and unified according to simple instructions available from Birkhäuser.

Furthermore, in order to guarantee the timely appearance of the proceedings it is essential that the final version of the entire material be submitted no later than one year after the conference.

Firuz Kamalov • R. Sivaraj • Ho-Hon Leung
Editors

Advances in Mathematical Modeling and Scientific Computing

International Conference on Recent Developments in Mathematics, Dubai, 2022 – Volume 2

Birkhäuser

Editors
Firuz Kamalov
Department of Electrical Engineering
Canadian University Dubai
Dubai, United Arab Emirates

R. Sivaraj
Department of Mathematics
Dr B R Ambedkar National Institute of Technology
Jalandhar, Punjab, India

Ho-Hon Leung
Department of Mathematical Sciences
United Arab Emirates University
Al Ain, United Arab Emirates

ISSN 2297-0215 ISSN 2297-024X (electronic)
Trends in Mathematics
ISBN 978-3-031-41419-0 ISBN 978-3-031-41420-6 (eBook)
https://doi.org/10.1007/978-3-031-41420-6

This book is published under the imprint Birkhäuser, www.birkhauser-science.com by the registered company Springer Nature Switzerland AG
The registered company address is: Gewerbestrasse 11, 6330 Cham, Switzerland

Paper in this product is recyclable.

Preface

The Canadian University Dubai, UAE, and United Arab Emirates University, UAE, jointly organized the International Conference on Recent Developments in Mathematics (ICRDM 2022) during August 24–26, 2022, in hybrid mode at Canadian University Dubai, UAE. The major objective of ICRDM 2022 is to promote scientific and educational activities toward the advancement of common man's life by improving the theory and practice of various disciplines of Mathematics. The conference was a grand success and more than 500 participants (Professors/Scholars/Students) enriched their knowledge in the wings of mathematics through ICRDM 2022. Over 200 leading researchers worldwide served in various capacities to organize ICRDM 2022. Thirty-one eminent speakers worldwide delivered the keynote address and invited talks in this conference. Three hundred and seventy-six researchers submitted their quality research articles to ICRDM 2022 through EasyChair. We shortlisted more than 300 research articles for oral presentations authored by dynamic researchers around the world. After peer review, 119 manuscripts were shortlisted for publication in the Springer book series: Trends in Mathematics. We hope that ICRDM 2022 inspired several researchers in mathematics: shared research interest and information, and created a forum of collaboration to build a trust relationship. We feel honored and privileged to serve the best recent developments in the field of mathematics to the readers in two volumes: Volume I, Recent Developments in Algebra and Analysis, and Volume II, Advances in Mathematical Modeling and Scientific Computing.

This book comprises the advances in mathematical modeling and scientific computing. A basic premise of this book is that the quality assurance is effectively achieved through the selection of quality research articles by the scientific committee that consists of several potential reviewers worldwide. This book comprises the contribution of several dynamic researchers in 82 chapters. Each chapter identifies the existing challenges in the areas of differential equations, mechanics, operation research, statistics, graph theory, and mathematical education and emphasizes the importance of establishing new methods and algorithms to addresses the challenges. Each chapter presents a selection of research problem, the technique suitable for solving the problem with sufficient mathematical background, and discussions on the obtained results with physical interruptions to understand the domain of applicability. This book also provides a comprehensive literature survey which reveals the challenges, outcomes, and developments of higher-level mathematics in this decade. The theoretical coverage of this book is relatively at a higher level to meet the global orientation of applied mathematics.

The target audience of this book is postgraduate students, researchers, and industrialists. This book promotes a vision of applied mathematics as integral to modern science and engineering. Each chapter contains important information emphasizing applied mathematics, intended for the professionals who already possess a basic understanding. In this book, theoretically oriented readers will find an overview of applied mathematics and applications. Industrialists will find a variety of techniques with sufficient discussion in terms of physical point of view to adapt for solving the particular application-based mathematical models. The readers can make use of the literature survey of this book to identify the current trends in applied mathematics. It is our hope and expectation that this book will provide an effective learning experience and referenced resource for all young mathematicians in the areas of applied mathematics.

As editors, we would like to express our sincere thanks to all the administrative authorities of Canadian University Dubai, UAE, and United Arab Emirates University, UAE, for their motivation and support. We also extend our profound thanks to all faculty members and staff members of the institutes. We especially thank all the members of the organizing committee of ICRDM 2022 who worked as a team by investing their time to make the conference a grand success. We express our sincere gratitude to all the referees for spending their valuable time to review the manuscripts which led to substantial improvements and sort out the quality research papers for publication. We thank EasyChair platform for providing the manuscript submission and review service. We are thankful to the project coordinator and team members from Springer Nature for their commitment and dedication toward the publication of this book. The organizing committee is grateful to Dr. Chris Eder, Associate Editor, Mathematics, Birkhäuser, Springer Nature, for his continuous encouragement and support toward the publication of this book.

Dubai, United Arab Emirates — Firuz Kamalov
Jalandhar, Punjab, India — R. Sivaraj
Al Ain, United Arab Emirates — Ho-Hon Leung

Contents

Part I
Differential Equations

High-Precision Algorithms for Fredholm Integral Equations

Fadi Awawdeh and Linda Smail

1 Introduction

In this paper, we present numerical schemes for second kind linear Fredholm integral equations (FIEs) of the form,

$$(\lambda - \mathcal{K})\, y(t) = f(t), \qquad t \in \Gamma, \tag{1}$$

where

$$\mathcal{K}y(t) = \int_{\Gamma} K(t,s)y(s)ds \tag{2}$$

and Γ can be a closed interval $[a, b]$ or an infinite interval like $(0, \infty)$ or $(-\infty, \infty)$.

Integral equation formulations of a wide variety of real-world problems are used in many branches of engineering and sciences. For example, boundary integral equation methods arise widely in potential flow calculations, crack problems in elasticity, electrostatic and elastostatic calculations, and perturbation theory in quantum mechanics.

There have been recent efforts to construct accurate numerical integrators for the Fredholm integral equation (1); see [1–6] and references therein. While such methods of good order of convergence exist, they require huge computations in

F. Awawdeh (✉)
Faculty of Science, Department of Mathematics, The Hashemite University, Zarqa, Jordan
e-mail: awawdeh@hu.edu.jo

L. Smail
College of Interdisciplinary Studies, Zayed University, Dubai, United Arab Emirates
e-mail: linda.smail@zu.ac.ae

F. Kamalov et al. (eds.), *Advances in Mathematical Modeling and Scientific Computing*, Trends in Mathematics, https://doi.org/10.1007/978-3-031-41420-6_1

order to achieve high accuracy. In applications such as high-degree polynomial integration and spectral methods, the Gauss quadrature rules suffer occasionally because they require the use of a large number of nodes. Computing Chebyshev, Legendre, Jacobi, and Gauss-Jacobi nodes and weights implies locating the zeros of polynomials and evaluating their derivatives. Existing approaches for computing such nodes and weights have complexity $O(n^2)$, and the errors grow with n, which can be limiting when n is large. The algorithms described in this paper utilize exact and direct formulae for both nodes and weights. The new algorithms allow the computation of the nodes and the weights of the presented quadrature rules in just $O(n)$ operations to about full double precision for any value of n.

In this work, the integral in (2) is replaced by a high-precision quadrature scheme. After collocating at the quadrature nodes, this yields a system of linear algebraic equations. To obtain high accuracy methods, the discrete values of the solution at the nodes are replaced by a Nyström interpolation formula. We show that we can achieve a high accuracy for problems with smooth kernels on finite and infinite intervals.

After introducing the double exponential formulae in Sect. 2, we propose high-precision quadrature schemes to approximate the integral in (2). In Sect. 3, we describe a connection of these methods with the integral equation and propose numerical schemes for certain FIEs of the second kind. Finally, in Sect. 4, we show the excellent rate of convergence of the proposed methods applied to our equation and discuss some examples.

2 High-Precision Quadrature Schemes

Consider the integral

$$I = \int_a^b f(x)dx, \tag{3}$$

where the integrand $f(x)$ is assumed to be analytic on (a, b) except possibly at the endpoints.

Let $\phi : (-\infty, \infty) \to (a, b)$ be an analytic function. We also assume that ϕ is monotone increasing. By employing the transformation $x = \phi(t)$, we can express the integral (3) as

$$I = \int_{-\infty}^{\infty} f(\phi(t))\phi'(t)dt. \tag{4}$$

We employ the uniformly divided trapezoidal formula to numerically approximate the integral (4). To this end, we write, for $h > 0$,

$$I = I_h + E,$$

where E is the discretization error and

$$I_h = h \sum_{j=-\infty}^{\infty} w_j f(x_j) \tag{5}$$

with $x_j = \phi(hj)$ and $w_j = \phi'(hj)$ [7]. A suitable choice of $\phi(t)$ is when $|f(\phi(hj))\phi'(hj)|$ decays rapidly as $|j| \to \infty$. To truncate the infinite summation (5), we replace (5) by

$$I_h^{(N)} = h \sum_{j=-N^-}^{N^+} w_j f(x_j).$$

A formula that is constructed such that

$$|f(\phi(t))\phi'(t)| \approx \exp(-C\exp(|t|)), \qquad t \to \pm\infty,$$

where C is a positive constant, is called a double exponential formula. In this way, we can obtain the highest precision with the minimum number of function's evaluations [8]. For example, if we deal with the integral

$$I = \int_{-1}^{1} f(x)dx,$$

then the transformation

$$x = \phi(t) = \tanh\left(\frac{\pi}{2}\sinh t\right)$$

gives a truncated double exponential formula, usually called the tanh – sinh quadrature formula,

$$I_h^{(N)} = h \sum_{j=-N}^{N} w_{j,N} f\left(x_{j,N}\right), \tag{6}$$

with

$$x_{j,N} = \tanh\left(\frac{\pi}{2}\sinh hj\right), \quad w_{j,N} = \frac{\pi}{2}\frac{\cosh hj}{\cosh^2\left(\frac{\pi}{2}\sinh hj\right)}. \tag{7}$$

The tanh – sinh rule is generally unsuitable for a direct use over large integration intervals, while the following transformations still display a double exponential decay rate:

$$t = \exp\left(\frac{\pi}{2}\sinh t\right) \quad \text{for} \quad \int_0^{\infty} g(x)dx \tag{8}$$

and

$$t = \sinh\left(\frac{\pi}{2}\sinh t\right) \quad \text{for} \quad \int_{-\infty}^{\infty} g(x)dx. \tag{9}$$

Using the mapping functions (8) and (9), we can construct the double exponential formulae:

$$\int_0^{\infty} g(x)dx \approx \frac{\pi h}{2}\sum_{j=-\infty}^{\infty} g\left(\exp\left(\frac{\pi}{2}\sinh hj\right)\right)(\cosh hj)\exp\left(\frac{\pi}{2}\sinh hj\right) \tag{10}$$

and

$$\int_{-\infty}^{\infty} g(x)dx \approx \frac{\pi h}{2}\sum_{j=-\infty}^{\infty} g\left(\sinh\left(\frac{\pi}{2}\sinh hj\right)\right)(\cosh hj)\cosh\left(\frac{\pi}{2}\sinh hj\right). \tag{11}$$

To better understand double exponential formulae and their error terms, we refer the reader to [7–9].

3 Computational Algorithms

3.1 *Integral Equation with Smooth Kernel Function on* $[a, b]$

In this section, we restrict our attention to the FIE (1)–(2) where $f(t) \in \mathcal{C}^{\infty}([a, b])$ and $K(t, s)$ is smooth on $[a, b] \times [a, b]$. We first extend the tanh – sinh quadrature formula [10–13] to be applied on a general interval $[a, b]$ as

$$\int_a^b g(t)dt = \frac{b-a}{2}\int_{-1}^{1} g\left(\frac{1}{2}((b-a)t + b + a)\right)dt \approx \frac{(b-a)}{4}\pi h \sum_{j=-\infty}^{\infty} w_{j,n} g(s_{j,n}),$$

where

$$s_{j,n} = \frac{1}{2}\left((b-a)\tanh\left(\frac{\pi}{2}\sinh hj\right) + b + a\right), \quad w_{j,n} = \frac{\cosh hj}{\cosh^2\left(\frac{\pi}{2}\sinh hj\right)}. \tag{12}$$

Using this formula, we approximate the integral $\mathcal{K}y(t)$ in (2), for $a \le t \le b$, as

$$\mathcal{K}_n y(t) = \frac{(b-a)}{4}\pi h \sum_{j=-n}^{n} w_{j,n} K(t, s_{j,n}) y(s_{j,n}). \tag{13}$$

Applying the double exponential formula (13) to the integral part of (2) yields

$$\lambda y_n(t) - \frac{(b-a)}{4}\pi h \sum_{j=-n}^{n} w_{j,n} K(t, s_{j,n}) y(s_{j,n}) = f(t), \qquad a \leq t \leq b, \tag{14}$$

where $w_{j,n}$ and $s_{j,n}$ are given by (12).

To solve (14), we first find the values of y_n at the abscissas $\{s_{j,n}\}$. To this end, we collocate (14) at the abscissas to have the linear system:

$$\lambda y_n(s_{i,n}) - \frac{(b-a)}{4}\pi h \sum_{j=-n}^{n} w_{j,n} K(s_{i,n}, s_{j,n}) y_n(s_{j,n}) = f(s_{i,n}). \tag{15}$$

We can obtain the solution to (1), for any $t \in [a, b]$, by the Nyström interpolation formula:

$$y_n(t) = \frac{1}{\lambda}\left(f(t) + \frac{(b-a)}{4}\pi h \sum_{j=-n}^{n} w_{j,n} K(t, s_{j,n}) y_n(s_{j,n}) \right), \qquad a \leq t \leq b. \tag{16}$$

More details about the presentation and the analysis of the Nyström method can be found in [2].

3.2 Convergence Analysis

In this part, we will discuss the convergence and error analysis of the approximate solutions of the proposed method in Sect. 3.1. We will primarily use the well-known results regarding the Nyström methods from [2] and the errors of the double exponential formulae from [13].

It has been proved in [2] that, for sufficiently large n, the approximating equation (14) has a unique solution. Moreover, we guarantee the existence of a constant $c > 0$ such that

$$\|y - y_n\|_\infty \leq c \,\|\mathcal{K}y - \mathcal{K}_n y\|_\infty . \tag{17}$$

The error formula (17) is considered to be the error of the approximation (13).

The Euler-Maclaurin summation formula [11, 12] gives a very quick rate of convergence of order $O(h^{2m+2})$ if $K(t,\cdot) \in \mathcal{C}^{2m+2}[a,b]$ for $a \leq t \leq b$. In this case, the bound (17) implies

$$\|y - y_n\|_\infty = O(h^{2m+2}). \tag{18}$$

The parameter h is set to be equal to 2^{-k} and n is chosen to be large enough so that

$$\left|s_{n,n} - b\right| < eps,$$

where eps is the arithmetic precision being used. Moving from one level 2^{-k} to the larger level $2^{-(k+1)}$ doubles the accuracy in the approximation.

Another error estimate of the proposed method can be obtained by following the methods presented in [1]. To accomplish this, the new scheme is assumed to predict the error in a geometric behavior:

$$\|y - y_n\|_\infty = O(h^{-q})$$

for some $q \geq 1$. Using three successive values of h, say h_k, $h_{k+1} = 2h_k$, $h_{k+2} = 2h_{k+1}$, we can estimate h using

$$2^{-q} \approx \frac{\left\|y_{n_{k+2}} - y_{n_{k+1}}\right\|_\infty}{\left\|y_{n_{k+1}} - y_{n_k}\right\|_\infty} \equiv R,$$

where all the differences are calculated at the common nodes. In this case, we have the estimate:

$$\left\|y - y_{n_{k+2}}\right\|_\infty \approx \frac{R}{1-R}\left\|y_{n_{k+2}} - y_{n_{k+1}}\right\|_\infty.$$

We now look into establishing the stability of the Nyström method we have proposed. Let $X = \mathcal{C}([a, b])$ with the max norm and let D be an open ball in $\mathcal{C}([a, b])$ of radius 1 about the origin. Assume $K(t, s)$ is continuous for $t, s \in [a, b]$. Following the procedure in [2], we can easily arrive to the following stability result.

Theorem 1 *Let $X = \mathcal{C}([a, b])$ with the max norm. Assume $K(t, s)$ and $K_t(t, s)$ are both continuous for $t, s \in [a, b]$. Then*

1. *for any compact operator $\phi : X \to X$,*

$$\|(\mathcal{K} - \mathcal{K}_n)\,\phi\| \to 0 \quad \text{as } n \to \infty,$$

2. $\|(\mathcal{K} - \mathcal{K}_n)\,\mathcal{K}_n\| \to 0$ *as $n \to \infty$.*

3.3 Integral Equations over Infinite Intervals

We also study integral equations over unbounded intervals of the form

$$\lambda y(t) - \int_0^\infty K(t, s) y(s) ds = f(t), \qquad 0 \leq t < \infty \tag{19}$$

and

$$\lambda y(t) - \int_{-\infty}^{\infty} K(t,s)y(s)ds = f(t), \qquad -\infty \leq t < \infty. \tag{20}$$

Using the abscissa-weight pairs [11, 13],

$$\tilde{w}_{j,n} = \frac{\pi h}{2}\cosh hj \exp\left(\frac{\pi}{2}\sinh hj\right), \qquad \tilde{s}_{j,n} = \exp\left(\frac{\pi}{2}\sinh hj\right), \tag{21}$$

we have the approximation

$$\mathcal{K}y(t) = \int_{0}^{\infty} K(t,s)y(s)ds \approx \sum_{j=-n}^{n} \tilde{w}_{j,n}K(t,\tilde{s}_{j,n})y(\tilde{s}_{j,n}) = \mathcal{K}_n y(t).$$

As in Sect. 3.1, this approximation leads to

$$\lambda y_n(t) - \sum_{j=-n}^{n} \tilde{w}_{j,n}K(t,\tilde{s}_{j,n})y_n(\tilde{s}_{j,n}) = f(t), \qquad 0 \leq t \leq \infty. \tag{22}$$

Once the solution, $y_n(\tilde{s}_{i,n})$, of the linear system

$$\lambda y_n(\tilde{s}_{i,n}) - \sum_{j=-n}^{n} \tilde{w}_{j,n}K(\tilde{s}_{i,n},\tilde{s}_{j,n})y_n(\tilde{s}_{j,n}) = f(\tilde{s}_{i,n}),$$

is found, we can solve (22) for $y_n(t)$ to get the Nyström interpolant:

$$y_n(t) = \frac{1}{\lambda}\left(f(t) + \sum_{j=-n}^{n} \tilde{w}_{j,n}K(t,\tilde{s}_{j,n})y_n(\tilde{s}_{j,n})\right). \tag{23}$$

Using the sinh − cosh double exponential formula [11, 13] and the methods presented in Sect. 3.1, the solution of (20) can be approximated using the Nyström interpolant formula:

$$y_n(t) = \frac{1}{\lambda}\left(f(t) + \sum_{j=-n}^{n} \tilde{w}_{j,n}K(t,\tilde{s}_{j,n})y_n(\tilde{s}_{j,n})\right), \tag{24}$$

where $\tilde{w}_{j,n}$ and $\tilde{s}_{j,n}$ are given by (21).

4 Computational Details and Numerical Tests

In this section, we present the results of some numerical tests, to demonstrate the efficiency and accuracy of the approaches developed in Sects. 3.1 and 3.3. In what follows, we will denote the approximations (16), (23), and (24) by HPNS.

4.1 *Implementation of the Method*

Our implementations require a selection of the parameter h that controls the quadrature calculations and the estimation of the solution of the linear systems that results from the collocation process. The proposed methods were implemented as MATLAB m-files.

For testing the efficiency of the algorithms, we present the maximum true error at the nodes, e_{nod},

$$e_{nod} = \left\| y(s_{i,n}) - y_n(s_{i,n}) \right\|_\infty .$$

The maximum true errors, reported as e_I, estimated by sampling 200 regularly spaced points in the interval $[a, b]$, will be helpful to assess the error of the interpolant. When the interval is infinite, we choose an argument section value, b, and measure the errors in $[0, b]$ for intervals $(0, \infty)$ and in $[-b, b]$ for intervals $(-\infty, \infty)$. We solved the linear system (15) using the backslash operator in MATLAB.

In our implementation, we set $h = 2^{-m}$, where m is the "level" of the quadrature calculations and n is chosen large enough to satisfy

$$\left| s_{k,n} - b \right| < eps \qquad \text{for } k \geq n.$$

The HPNS begins with $h = 2^{-1}$ and reduces h in half until either a solution y_n is found with sufficient accuracy or the final level $m = 10$ has been reached. For a given h, the abscissas $t_{j,n}$ and the weights $w_{j,n}$ can be computed once and then used for numerous problems.

4.2 *Numerical Experiments*

We will study, in this first example, Love's equation [14]:

$$\lambda y(t) - \int_0^1 \frac{cy(s)}{c^2 + (t-s)^2} ds = f(t), \qquad 0 \leq t \leq 1, \tag{25}$$

Table 1 Numerical results for Equation (25) with $c = 0.01$ and $\lambda = -0.5$

Level m	n	Running time	e_{nod}	e_I
5	102	0.012279	$1.3e-02$	$1.5e-01$
6	203	0.039353	$1.0e-03$	$1.3e-02$
7	405	0.25533	$6.1e-06$	$7.4e-05$
8	810	2.1349	$2.2e-10$	$2.7e-09$
9	1619	16.1349	$1.0e-14$	$6.2e-15$

where $f(t)$ is defined so that the exact solution is $y(t) = 0.06 - 0.8s + s^2$. This integral equation has a Runge kernel and arises in many models such as in electrostatics [1]. Results for the increasing values of the level m are shown in Table 1. The errors decrease rapidly when the number of nodes is sufficient.

In the results shown in Table 1, we find that roughly $3.2 \cdot 2^k$ abscissa-weight pairs are generated at level k, so that the total required for m levels is approximately $6.4\,(2^m - 1)$. It should be mentioned that, as c approaches 0, the kernel function becomes very peaked along $s = t$; hence, we need to increase the level m to approximate the integral accurately.

We also studied the FIE:

$$y(t) = f(t) + \int_0^\infty \frac{1}{\left(s^2 + t^2 + 1\right)^2} y(s)ds, \tag{26}$$

where the function $f(t)$ is defined so that the integral equation will have $y(t) = e^{-0.2t} \cos t$ as its solution. The level of abscissa-weight pairs required to achieve the target accuracy 10^{-12} is $m = 8$. For this problem, 8 levels were generated for HPNS. Numerical results for this example, with $n = 512$ and running time 4.530171 seconds, are shown in Fig. 1. Note, however, that the estimated bound on the error decreases with n until about machine precision and we get that

$$e_{nod} = 1.7 \times 10^{-12} \quad \text{and} \quad e_I = 1.6 \times 10^{-12}.$$

It is mentioned in [15] that direct Nyström method is not at all appropriate for this type of integrals, because the integrals do not have exponentially decaying characters. In this case, a rule obtained by mapping to a finite interval and then using an ordinary Gauss quadrature is far more appropriate.

5 Conclusion and Future Work

This work proposed and also compared different approaches to the numerical solution of the Fredholm integral equations of the second kind. All these approaches are based on the quadrature methods with piecewise linear, polynomial, or rational approximations. Numerical experiments demonstrate that the method based on

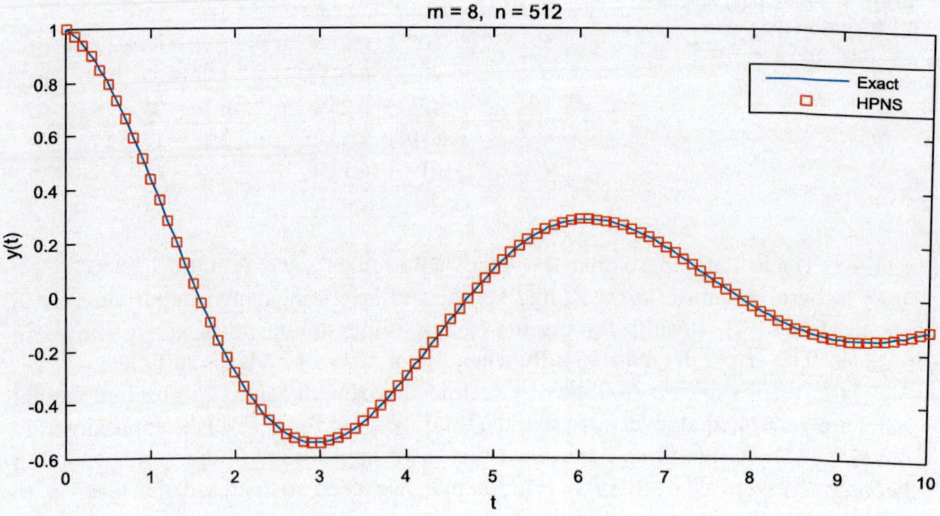

Fig. 1 Numerical results for Equation (26)

double exponential formulae with the Nyström interpolation is the most accurate for a variety of problems including problems with smooth kernels as well as kernels ranging from well-behaved functions on finite intervals, except possibly across the diagonal $s = t$, to kernels on an infinite interval. The HPNS algorithms are vectorized, making it very efficient in a variety of architectures and languages. Numerical results for a couple of examples have been reported. Those results showed the efficiency of the new methods for solving Fredholm integral equations with the low CPU time. We remark that all results of this paper extend without any additional difficulty to various types of integral equations. The method has flexibility to be expanded on integral equations with singular kernels. Looking into such extensions will be our future work.

References

1. Atkinson, K.E., Shampine, L.F.: Solving Fredholm integral equations of the second kind in Matlab. ACM Trans. Math. Softw. **34** (2008)
2. Atkinson, K.E.: The Numerical Solution of Integral Equations of the Second Kind. Cambridge University Press (1997)
3. Atkinson, K.E.: Elementary Numerical Analysis. Wiley, New York (1993)
4. Atkinson, K.E., Bogomolny, A.: The Discrete Galerkin Method for Integral Equations. Math. Comput. **48**(178), 595–616 (1987)
5. Atkinson, K.E., Potra, F.: Projection and iterated projection methods for nonlinear integral equations. SIAM J. Num. Anal. **24**, 1352–1373 (1987)

6. Amosov, A., Ahues, M., Largillier, A.: Superconvergence of some projection approximations for weakly singular integral equations using general grids. SIAM J. Numer. Anal. **47**(1), 646–674 (2008)
7. Takahasi, H., Mori, M.: Double exponential formulas for numerical integration. Publ. RIMS, Kyoto Univ. **9**, 721–741 (1974)
8. Takahasi, H., Mori, M.: Error estimation in the numerical integration of analytic functions. Rep. Comput. Centre Univ. Tokyo **3**, 41–108 (1970)
9. Mori, M.: Developments in the double exponential formula for numerical integration. In: Proceedings of the International Congress of Mathematicians (Springer, Berlin, 1991)
10. Bailey, D.H., Jeyabalan, K., Li, X.S.: A comparison of three high-precision quadrature schemes. Exp. Math. **14**, 317–329 (2005)
11. Bailey, D.H., Borwein, J.M.: Effective error bounds in Euler-Maclaurin-based quadrature schemes. In: Proceedings of the 2006 Conference on High-Performance Computing Systems. IEEE Computer Society (2006)
12. Borwein, J.M., Ye, L.: Quadratic Convergence of the tanh-sinh Quadrature Rule, D-Drive preprint #342 (2006). http://locutus.cs.dal.ca:8088/archive/00000342
13. Borwein, J.M., Bailey, D.H., Girgensohn, R.: Experimentation in Mathematics, Computational Paths to Discovery. A.K. Peters, Ltd (2004)
14. Pastore, P.: The numerical treatment of Love's integral equation having very small parameter. J. Comput. Appl. Math. **236**, 1267–1281 (2011)
15. Sloan, I.: Quadrature methods for integral equations of the second kind over infinite intervals. Math. Comput. **36**, 511–523 (1981)

General Decay Estimate for a Weakly Dissipative Viscoelastic Suspension Bridge

Salim A. Messaoudi, Soh Edwin Mukiawa, and Mohammad M. Al-Gharabli

AMS 2010: 35B40, 74F05, 93C20, 93D20, 93D15

1 Introduction

Let $\Omega = (0, \pi) \times (-\ell, \ell) \subset \mathbb{R}^2$ be a thin rectangular plate as modeling for a suspension bridge, where the plate is partially hinged on the vertical edges:

$$u(0, y) = u_{xx}(0, y) = u(\pi, y) = u_{xx}(\pi, y) = 0$$

and free on the horizontal edges; see [1, 2]:

$$u_{yy}(x, \pm\ell) + \sigma u_{xx}(x, \pm\ell) = u_{yyy}(x, \pm\ell) + (2 - \sigma)u_{xxy}(x, \pm\ell) = 0. \tag{1}$$

The work of Ferrero and Gazzola [3] suggested the plate equation:

$$u_{tt} + \Delta^2 u + \delta u_t + h(x, y, u) = f \tag{2}$$

S. A. Messaoudi
Department of Mathematics and Statistics, University of Sharjah, Sharjah, United Arab Emirates
e-mail: smessaoudi@sharjah.ac.ae

S. E. Mukiawa
Department of Mathematics and Statistics, University of Hafr Al Batin, Hafar Al Batin, Saudi Arabia
e-mail: mukiawa@ubh.edu.sa

M. M. Al-Gharabli (✉)
The Preparatory Year Program, King Fahd University of Petroleum and Minerals, Dhahran, Saudi Arabia

The Interdisciplinary Research Center in Construction and Building Materials, King Fahd University of Petroleum and Minerals, Dhahran, Saudi Arabia
e-mail: mahfouz@kfupm.edu.sa

F. Kamalov et al. (eds.), *Advances in Mathematical Modeling and Scientific Computing*, Trends in Mathematics, https://doi.org/10.1007/978-3-031-41420-6_2

as the downward displacement of a suspension bridge. The authors in [3] established a global existence result for (2) and also discussed several stationary problems. Berchio et al. [4] established the structural instability of nonlinear plates modelling suspension bridges. Gazzalo and Wang [5] modelled a suspension bridges through the von Karman plate equation. Messaoudi and co-authors [6–8] have investigated extensively the suspension bridge plate model (2) and established existence, decay and global attractor results. **In this work**, we consider

$$u_{tt}(x,y,t) + \Delta^2 u(x,y,t) + \int_0^t g(t-s)u_{xx}(x,y,s)ds - u_{xxt} = 0, \quad \Omega \times (0,T), \tag{3}$$

coupled with the partially hinged boundary conditions

$$\begin{cases} u(0,y,t) = u_{xx}(0,y,t) = 0, & \text{for } (y,t) \in (-\ell,\ell)\times(0,T), \\ u(\pi,y,t) = u_{xx}(\pi,y,t) = 0, & \text{for } (y,t) \in (-\ell,\ell)\times(0,T), \\ u_{yy}(x,\pm\ell,t) + \sigma u_{xx}(x,\pm\ell,t) = 0, & \text{for } (x,t) \in (0,\pi)\times(0,T), \\ u_{yyy}(x,\pm\ell,t) + (2-\sigma)u_{xxy}(x,\pm\ell,t) = 0, & \text{for } (x,t) \in (0,\pi)\times(0,T) \end{cases} \tag{4}$$

and initial conditions

$$u(x,y,0) = u_0(x,y), \;\; u_t(x,y,0) = u_1(x,y), \;\text{in } \Omega. \tag{5}$$

2 Preliminaries

In this section, we recall some useful materials and assumptions. Throughout this work, we use C to denote a generic positive constant and consider the Hilbert space $H_*^2(\Omega)$, defined in [3] by

$$H_*^2(\Omega) = \left\{ w \in H^2(\Omega) : w = 0 \text{ on } \{0,\pi\}\times(-\ell,\ell) \right\}$$

and endowed with the inner product

$$(u,v)_{H_*^2} = \int_\Omega [(\Delta u \Delta v + (1-\sigma)(2u_{xy}v_{xy} - u_{xx}v_{yy} - u_{yy}v_{xx})]dxdy.$$

We denote by $\mathcal{H}(\Omega)$ the dual of $H_*^2(\Omega)$ and introduce, the following space:

$$H_*^1(\Omega) = \left\{ w \in H^1(\Omega) : w = 0 \text{ on } \{0,\pi\}\times(-\ell,\ell) \right\},$$

equipped with the norm $\|w\|_{H^1_*(\Omega)} = \left(\int_\Omega |\nabla u|^2 dxdy\right)^{\frac{1}{2}}$ .Using the embedding of $H^2_*(\Omega)$ in $H^1_*(\Omega)$, we let $\Lambda := \min_{w \in H^2_*(\Omega), w \neq 0} \frac{\|w\|_{H^2_*(\Omega)}}{\|w\|_{H^1_*(\Omega)}}$.

Lemma 1 (embedding) *Suppose $1 < p < +\infty$. Then for any $v \in H^2_*(\Omega)$, there exists an embedding constant $C_e = C_e(\Omega, p) > 0$ such that*

$$\|v\|_p \leq C_e \|v\|_{H^2_*(\Omega)}.$$

Lemma 2 *For any $v \in H^2_*(\Omega)$, we have*

$$\|v_x\|^2_{L^2(\Omega)} \leq \Lambda^2 \|v\|^2_{H^2_*(\Omega)}. \tag{6}$$

Proof It is straightforward to see that

$$\|v_x\|^2_{L^2(\Omega)} \leq \|v\|^2_{H^1_*(\Omega)} \leq \Lambda^2 \|v\|^2_{H^2_*(\Omega)}.$$

□

We assume the relaxation function g admits the following assumptions:

(C1) $g : [0, +\infty) \longrightarrow (0, +\infty)$ is a decreasing C^1- function such that

$$g(0) > 0, \quad 1 - \Lambda^2 \int_0^\infty g(s)ds = l_0 > 0. \tag{7}$$

(C2) We assume that there exist a C^1 function $G : [0, +\infty) \to [0, +\infty)$ which is linear or strictly convex C^2- function on $(0, r], r \leq g(0)$, with $G(0) = G'(0) = 0$ and a positive non-increasing differentiable function $\xi : [0, +\infty) \to (0, +\infty)$ such that

$$g'(t) \leq -\xi(t) G\left(g(t)\right), \quad \forall\, t \geq 0. \tag{8}$$

Remark 1 [9]

1. There exist $t_0 \geq 0$ such that $g(t_0) = r$. and

$$g'(t) \leq -\xi(t) G(g(t)) \leq -\xi(t) \frac{b_1 g(0)}{g(0)} \leq -\frac{b_1}{g(0)} \xi(t) g(t), \tag{9}$$

which gives

$$\xi(t) g(t) \leq -\frac{g(0)}{b_1} g'(t), \quad \forall\, t \in [0, t_0]. \tag{10}$$

2. We can find an extension of G say $\bar{G} : [0, +\infty) \to (0, +\infty)$ which is also strictly increasing and strictly convex C^2-function.

For completeness, we state without proof the existence uniqueness result. The result can be established using the Galerkin approximation method as in [10].

Proposition 1 *Let $(u_0, u_1) \in H_*^2(\Omega) \times L^2(\Omega)$ be given. Assume g satisfies $(C1)$ and $(C2)$. Then, problem (3) has a unique global weak solution which satisfies*

$$u \in C([0, T), H_*^2(\Omega)) \cap C^1([0, T), L^2(\Omega)). \tag{11}$$

Furthermore, if $(u_0, u_1) \in \left(H^4(\Omega) \cap H_*^2(\Omega)\right) \times H_*^2(\Omega)$, then

$$u \in C([0, T), H^4(\Omega) \cap H_*^2(\Omega)) \cap C^1([0, T), H_*^2(\Omega)). \tag{12}$$

Next, we define the following "modified" energy functionals associated with problem (3) which is defined by

$$E(t) = \frac{1}{2}\|u_t\|^2_{L^2(\Omega)} + \frac{1}{2}\|u\|^2_{H_*^2(\Omega)} - \frac{1}{2}\left(\int_0^t g(s)ds\right)\|u_x\|^2_{L^2(\Omega)} + \frac{1}{2}(g \diamond u_x)(t), \tag{13}$$

where

$$(g \diamond u_x)(t) = \int_0^t g(t-s)\|u_x(t) - u_x(s)\|^2_{L^2(\Omega)}ds.$$

Lemma 3 *Let u be the solution to problem (3) and assume g satisfies $(C1)$ and $(C2)$. Then, the energy functionals (13) satisfy*

$$E'(t) = \frac{1}{2}(g' \diamond u_x)(t) - \|u_{xt}\|^2_{L^2(\Omega)} - \frac{1}{2}g(t)\|u_x\|^2_{L^2(\Omega)} \leq 0, \ \forall\, t \geq 0. \tag{14}$$

Proof Multiplying (3) by u_t and integrating over Ω, using assumptions $(C1)$ and $(C2)$, we obtain (14). □

Lemma 4 *Under condition (7), the energy function (13) satisfies*

$$0 \leq E(t) \leq E(0), \ \ \forall\, t \geq 0. \tag{15}$$

Proof Using Lemma 2, (7) and (13), we have

$$\begin{aligned} E(t) &\geq \frac{1}{2}\|u_t\|^2_{L^2(\Omega)} + \frac{1}{2}\|u\|^2_{H_*^2(\Omega)} - \frac{\Lambda^2}{2}\int_0^t g(s)ds\|u\|^2_{H_*^2(\Omega)} + \frac{1}{2}(g \diamond u_x)(t) \\ &= \frac{1}{2}\|u_t\|^2_{L^2(\Omega)} + \frac{1}{2}\left(1 - \Lambda^2\int_0^t g(s)ds\right)\|u\|^2_{H_*^2(\Omega)} + \frac{1}{2}(g \diamond u_x)(t) \geq 0. \end{aligned} \tag{16}$$

From Lemma 3 and (16), we get that the energy is decreasing and

$$0 \leq E(t) \leq E(0), \quad \forall\, t \geq 0.$$

□

The following lemmas will be applied frequently in our calculations.

Lemma 5 (Lemma 4.2 [10]) *Letting* $w \in L^2_{loc}([0, +\infty), L^2(\Omega))$, *we have*

$$\int_\Omega \left(\int_0^t g(t-s)(w(t) - w(s))ds \right)^2 dxdy \leq \left(\frac{1 - l_0}{\Lambda^2} \right) \int_\Omega \int_0^t g(t-s)(w(t) - w(s))^2 dsdxdy. \tag{17}$$

As in [11], for any $0 < \epsilon < 1$, let

$$h_\epsilon(t) = \epsilon g(t) - g'(t) \quad \text{and} \quad A_\epsilon = \int_0^{+\infty} \frac{g^2(s)}{\epsilon g(s) - g'(s)} ds.$$

We have the following lemma.

Lemma 6 *[9] Let u be the solution of problem* (3). *Then, for any* $0 < \epsilon < 1$, *we have*

$$\int_\Omega \left(\int_0^t g(t-s) \left((u_x(x, y, t) - u_x(x, y, s) \right) ds \right)^2 dxdy \leq A_\epsilon \left(h_\epsilon \diamond u_x \right)(t), \forall\, t \geq 0. \tag{18}$$

3 Essential Lemmas

In this section, we introduce some functionals and prove some related lemmas that are essential to establish our main result.

Lemma 1 *Under the conditions* $(C1)$ *and* $(C2)$, *the functional*

$$F_1(t) = \int_\Omega u u_t dxdy$$

satisfies, along the solution of (3),

$$F_1'(t) \leq \|u_t\|^2_{L^2(\Omega)} - \frac{l_0}{2} \|u\|^2_{H^2_*(\Omega)} + C A_\epsilon (h_\epsilon \diamond u_x)(t), \forall\, t \geq 0. \tag{19}$$

Proof Using (3) and the boundary conditions (4), we have

$$\begin{aligned}
F_1'(t) &= \|u_t\|^2_{L^2(\Omega)} - \|u\|^2_{H^2_*(\Omega)} - \int_\Omega u \int_0^t g(t-s)u_{xx}(x,y,s)ds \\
&= \|u_t\|^2_{L^2(\Omega)} - \|u\|^2_{H^2_*(\Omega)} \\
&\quad - \int_{-\ell}^{\ell} \left[\overbrace{u \int_0^t g(t-s)u_x(s)ds\Big|_{x=0}^{x=\pi}}^{\to 0} - \int_0^\pi u_x \int_0^t g(t-s)u_x(x,y,s)dsdx \right] dy \\
&= \|u_t\|^2_{L^2(\Omega)} - \|u\|^2_{H^2_*(\Omega)} + \int_0^t g(s)ds\|u_x\|^2_{L^2(\Omega)} \\
&\quad - \int_\Omega u_x \int_0^t g(t-s)(u_x(x,y,t) - u_x(x,y,s))dsdxdy. \qquad (20)
\end{aligned}$$

Applying Cauchy-Schwarz' and Young's inequalities and Lemmas 2, 3, 4, 5, and 6, we obtain (19). □

Lemma 2 *Let u be the solution of problem* (3)*. Then, the functional*

$$F_2(t) = \int_0^t J(t-s)\|u_x(s)\|^2_{L^2(\Omega)}ds$$

satisfies

$$F_2'(t) \le 3(1-l_0)\|u\|^2_{H^2_*(\Omega)} - \frac{1}{2}(g \diamond u_x)(t), \qquad (21)$$

where $J(t) = \int_t^{+\infty} g(s)ds$.

Proof Direct differentiation, using the fact that $J'(t) = -g(t)$ and recalling that $J(t) \le J(0) = \frac{(1-l_0)}{\Lambda^2}$ and $\int_0^t g(s)ds \le \frac{(1-l_0)}{\Lambda^2}$, we obtain (21). □

Now, let

$$F(t) = NE(t) + N_1F_1(t),$$

where N, N_1 are positive constants to be specified later.

Lemma 3 *Assume conditions* (C1) *and* (C2) *hold. Then, for suitable choices of* N, N_1 *the functional* $F(t)$ *satisfies for* $\beta_1, \beta_2 > 0$

$$\beta_1 E(t) \le F(t) \le \beta_2 E(t), \quad \text{that is } F \sim E \qquad (22)$$

and

$$F'(t) \leq -c\|u_t\|^2_{L^2(\Omega)} - 4(1-l_0)\|u(t)\|^2_{H^2_*(\Omega)} + \frac{1}{4}(g \diamond u_x)(t), \quad \forall\, t \geq t_0. \tag{23}$$

Proof Routine computations, using Cauchy-Schwarz' and Young's inequalities with the embedding Lemma 2, yield (22). To establish (23), we set $g_0 = \int_0^{t_0} g(s)ds > 0$, recall $g'(t) = \epsilon g(t) - h_\epsilon(t)$ and use Lemmas 3, 1 and Poincaré's inequality ($\|u_t\|_2^2 \leq c_p\|u_{xxt}\|_2^2$), and we obtain for all $t \geq t_0$

$$\begin{aligned} F'(t) \leq & -\left(\frac{N}{c_p} - N_1\right)\|u_t\|^2_{L^2(\Omega)} - \frac{N_1 l_0}{2}\|u\|^2_{H^2_*(\Omega)} \\ & + \frac{N\epsilon}{2}(g \diamond u_x)(t) - \left(\frac{N}{2} - CA_\epsilon N_1\right)(h_\epsilon \diamond u_x)(t). \end{aligned}$$

We choose $N_1 = \frac{8(1-l_0)}{l_0}$, and then select N so large such that (22) remains valid and

$$N > \max\left\{c_p N_1, 2CA_\epsilon N_1\right\}. \tag{24}$$

Thus, with $\epsilon = \frac{1}{2N}$, we obtain (23).

□

4 Main Decay Result

Now, we are ready to state our main decay result.

Theorem 1 *Assume conditions (C1) and (C2) hold. Then, there exist positive constants m_1 and m_2 such that the energy functional (13) satisfies*

$$E(t) \leq m_2 G_0^{-1}\left(\frac{m_1}{\displaystyle\int_{g^{-1}(r)}^{t} \xi(s)ds}\right), \ \forall\, t \geq t_0, \quad \text{where } G_0(s) = sG'(s). \tag{25}$$

Proof By using (9) and (14), we see that

$$\begin{aligned} \int_0^{t_0} g(s)\|u_x(t)-u_x(t-s)\|^2_{L^2(\Omega)}ds &\leq -\frac{g(0)}{b_1}\int_0^{t_0} g'(s)\|u_x(t)-u_x(t-s)\|^2_{L^2(\Omega)}ds \\ &\leq -CE'(t). \end{aligned} \tag{26}$$

From (26) and (23), we get

$$F'(t) \leq -\mu E(t) - CE'(t) + C \int_{t_0}^{t} g(s) \|u_x(t) - u_x(t-s)\|_{L^2(\Omega)}^2 ds, \quad \forall\, t \geq t_0,$$

for some positive constant μ. Now, we define the functional L by

$$L(t) = F(t) + F_2(t).$$

Using Lemmas 2 and 3, we see that L is non-negative and, for some $d > 0$ and $\forall\, t \geq t_0$,

$$L'(t) \leq -c\|u_t\|_{L^2(\Omega)}^2 - (1 - l_0)\|u(t)\|_{H_*^2(\Omega)}^2 - \frac{1}{4}(g \diamond u_x)(t) \leq -dE(t). \tag{27}$$

Thus, we have

$$d \int_{t_0}^{t} E(s)ds \leq L(t_0) - L(t) \leq L(t_0).$$

Hence, we obtain

$$\int_0^{+\infty} E(s)ds < +\infty. \tag{28}$$

Next, we define

$$\varphi(t) = p \int_{t_0}^{t} \|u_x(t) - u_x(t-s)\|_{L^2(\Omega)}^2 ds, \quad \forall\, t \geq t_0. \tag{29}$$

From (6) and Lemma 4, we deduce that

$$E(t) \geq \frac{l_0}{2\Lambda^2} \|u_x(t)\|_{L^2(\Omega)}^2, \quad \forall\, t \geq 0. \tag{30}$$

So, using (13), (28) and (30), we have

$$\begin{aligned}
\int_{t_0}^{t} \|u_x(t) - u_x(t-s)\|_{L^2(\Omega)}^2 ds &\leq 2 \int_{t_0}^{t} \left(\|u_x(t)\|_{L^2(\Omega)}^2 + \|u_x(t-s)\|_{L^2(\Omega)}^2 \right) ds \\
&\leq \frac{2}{l_0} \Lambda^2 \int_{t_0}^{t} (E(t) + E(t-s))\, ds \\
&\leq \frac{4}{l_0} \Lambda^2 \int_{t_0}^{+\infty} E(s)ds < +\infty, \ \forall\, t \geq t_0.
\end{aligned} \tag{31}$$

Consequently, we can choose $0 < p < 1$ such that

$$\varphi(t) < 1, \ \forall \ t \geq t_0. \tag{32}$$

We also define the functional

$$\omega(t) = -\int_{t_0}^{t} g'(s)\|u_x(t) - u_x(t-s)\|_{L^2(\Omega)}^2 ds$$

and note that, from (26), we get $\omega(t) \leq -CE'(t)$. Using the fact that G is strictly convex on $(0, r]$ and $G(0) = 0$, we have, for any $0 \leq \theta \leq 1$ and $s \in (0, r]$, $G(\theta s) \leq \theta G(s)$. Therefore, using this fact, (32) and Jensen's inequality, we arrive at

$$\begin{aligned}
\omega(t) &= \frac{1}{p\varphi(t)} \int_{t_0}^{t} \varphi(t)(-g'(s))p\|u_x(t) - u_x(t-s)\|_{L^2(\Omega)}^2 ds \\
&\geq \frac{\xi(t)}{p} \bar{G}\left(\int_{t_0}^{t} pg(s)\|u_x(t) - u_x(t-s)\|_{L^2(\Omega)}^2 ds\right),
\end{aligned}$$

where $\bar{G}$ is the extension of G, introduced in Remark 9. This implies

$$\int_{t_0}^{t} g(s)\|u_x(t) - u_x(t-s)\|_{L^2(\Omega)}^2 ds \leq \frac{1}{p}\bar{G}^{-1}\left(\frac{p\omega(t)}{\xi(t)}\right)$$

and (4) becomes

$$K'(t) \leq -\mu E(t) + \frac{C}{p}\bar{G}^{-1}\left(\frac{p\omega(t)}{\xi(t)}\right), \ \forall t \geq t_0, \tag{33}$$

where $K(t) = F(t) + CE(t)$. Letting $r_0 < r$, we define the functional

$$L_1(t) = \bar{G}'\left(r_0 \frac{E(t)}{E(0)}\right) K(t). \tag{34}$$

Using the fact that $L_1 \sim E$ (since $F \sim E$), (33) and the fact that $E'(t) \leq 0, \ G'(t) > 0, \ G''(t) > 0$ yield

$$\begin{aligned}
L_1'(t) &= r_0 \frac{E'(t)}{E(0)} \bar{G}''\left(r_0 \frac{E(t)}{E(0)}\right) K(t) + \bar{G}'\left(r_0 \frac{E(t)}{E(0)}\right) K'(t) \\
&\leq -\mu E(t)\bar{G}'\left(r_0 \frac{E(t)}{E(0)}\right) + C\bar{G}'\left(r_0 \frac{E(t)}{E(0)}\right) \bar{G}^{-1}\left(\frac{p\omega(t)}{\xi(t)}\right).
\end{aligned} \tag{35}$$

Let $\bar{G}^*$ be the convex conjugate of G in the sense of Young (see [12], pp. 61–64). Then

$$\bar{G}^*(s) = s(\bar{G}')^{-1}(s) - \bar{G}\left((\bar{G}')^{-1}(s)\right) \tag{36}$$

and $\bar{G}^*$ satisfies the generalized Young's inequality:

$$XY \leq \bar{G}^*(X) + \bar{G}(Y). \tag{37}$$

By setting

$$X = \bar{G}'\left(r_0\frac{E(t)}{E(0)}\right) \text{ and } Y = \bar{G}^{-1}\left(\frac{p\omega(t)}{\xi(t)}\right)$$

and using (14) and (35), (36), and (37), we obtain for all $t \geq t_0$

$$\begin{aligned} L_1'(t) &\leq -\mu E(t)\bar{G}'\left(r_0\frac{E(t)}{E(0)}\right) + C\bar{G}^*\left(\bar{G}'\left(r_0\frac{E(t)}{E(0)}\right)\right) + \frac{Cp\omega(t)}{\xi(t)} \\ &\leq -(\mu E(0) - Cr_0)\frac{E(t)}{E(0)}\bar{G}'\left(r_0\frac{E(t)}{E(0)}\right) + \frac{Cp\omega(t)}{\xi(t)}. \end{aligned}$$

We then choose r_0 small enough such that $\mu E(0) - Cr_0 > 0$ and get

$$L_1'(t) \leq -\lambda_1\frac{E(t)}{E(0)}\bar{G}'\left(r_0\frac{E(t)}{E(0)}\right) + C\frac{\omega(t)}{\xi(t)}, \quad \forall\, t \geq t_0. \tag{38}$$

Multiplying (38) by $\xi(t)$ and observing that

$$r_0\frac{E(t)}{E(0)} < r,\ \omega(t) \leq -CE'(t) \text{ and } \bar{G}'\left(r_0\frac{E(t)}{E(0)}\right) = G'\left(r_0\frac{E(t)}{E(0)}\right),$$

we arrive at

$$\xi(t)L_1'(t) \leq -\lambda_1\xi(t)\frac{E(t)}{E(0)}G'\left(r_0\frac{E(t)}{E(0)}\right) - CE'(t), \quad \forall t \geq t_0.$$

Letting $L_2(t) = \xi(t)L_1(t) + CE(t)$, then

$$\lambda_1\xi(t)\frac{E(t)}{E(0)}G'\left(r_0\frac{E(t)}{E(0)}\right) \leq -L_2'(t), \quad \forall\, t \geq t_0, \tag{39}$$

since ξ is decreasing. Now, using the fact that $G'' > 0$ and E is decreasing, we conclude that the function $t \longmapsto \frac{E(t)}{E(0)}G'\left(r_0\frac{E(t)}{E(0)}\right)$ is decreasing. Thus, integrating (39) over (t_0, t), we obtain

$$\begin{aligned}\lambda_1 \frac{E(t)}{E(0)} G'\left(r_0 \frac{E(t)}{E(0)}\right) \int_{t_0}^{t} \xi(s)ds \leq & \lambda_1 \int_{t_0}^{t} \frac{E(s)}{E(0)} G'\left(r_0 \frac{E(s)}{E(0)}\right) \xi(s)ds \\ \leq & L_2(t_0) - L_2(t) \leq L_2(t_0).\end{aligned} \tag{40}$$

Letting $G_0(s) = sG'(r_0 s)$, then from assumption (C_2), G_0 is strictly increasing. It follows from (40) that

$$\lambda_1 G_0 \left(r_0 \frac{E(t)}{E(0)}\right) \int_{t_0}^{t} \xi(s)ds \leq L_2(t_0), \quad \forall\, t \geq t_0.$$

This implies

$$E(t) \leq m_2 G_0^{-1} \left(\frac{m_1}{\int_{t_0}^{t} \xi(s)ds} \right), \; \forall\, t \geq t_0 \tag{41}$$

for some positive constants m_1 and m_2. This completes the proof. □

4.1 Examples

(1) Let $g_1(t) = k_0 e^{-k_1 t}$, $t \geq 0$, $k_0, k_1 > 0$ be chosen in a way that (C_1) holds. Then, the solution energy (13) satisfies, for a constant $K > 0$,

$$E(t) \leq \frac{K}{t - t_0}, \quad \forall\, t > t_0.$$

(2) Let $g_2(t) = \frac{k_0}{(1+t)^{k_1}}$, $t \geq 0$, $k_0 > 0$, $k_1 > 1$ be chosen in such a way that (C_1) holds. Then, the solution energy (13) satisfies for $t > t_0 > 0$, large enough and for some $K > 0$,

$$E(t) \leq \frac{K}{(1+t)^{\frac{k_1}{k_1+1}}}.$$

5 Conclusion

In this work, we considered a problem modelling a viscoelastic suspension bridge, where the viscoelastic dampings are effective on the $x-$direction only. We showed that the decay rate of the energy is weaker than that of the relaxation function. An open question is: Can we obtain a similar or even weaker decay rate if the damping term u_{xxt} is absent?

Acknowledgments The authors thank the University of Sharjah, University of Hafr Al Batin and King Fahd University of Petroleum and Minerals. The first and third authors are supported by KFUPM, project #SB201003.

References

1. Brownjohn, J.M.W.: Observations on non-linear dynamic characteristics of suspension bridges. Earthquake Eng. Struct. Dyn. **23**, 1351–1367 (1994)
2. Ventsel, E., Krauthammer, T.: Thin Plate Sand Shells: Theory, Analysis and Applications. Marcel Dekker Inc., New York (2001)
3. Ferrero, A., Gazzola, F.: A partially hinged rectangular plate as a model for suspension bridges. Discrete Contin. Dyn. Syst. **35**(12), 5879–5908 (2015)
4. Berchio, E., Ferrero, A., Gazzola, F.: Structural instability of nonlinear plates modelling suspension bridges: mathematical answers to some long-standing questions. Nonlin. Anal. Real World Appl. **28**, 91–125 (2016)
5. Gazzola, F., Wang, Y.: Modeling suspension bridges through the Von Karman quasilinear plate equations. Progress in Nonlinear Differential Equations and Their Applications. In: Contributions to Nonlinear Differential Equations and Systems, a tribute to Djairo Guedes de Figueiredo on occasion of his 80th birthday (2015), pp. 269–297
6. Messaoudi, S.A., Mukiawa, S.E., Enyi, C.D.: Finite dimensional global attractor for a suspension bridge problem with delay. C.R. Acad. Sci. Paris, Ser. I **354**, 808–824 (2016)
7. Messaoudi, S.A., Bonfoh, A., Mukiawa, S.E., Enyi, C.D.: The global attractor for a suspension bridge with memory and partially hinged boundary conditions. Z. Angew. Math. Mech. (ZAMM) **97**(2), 1–14 (2016)
8. Messaoudi, S.A., Mukiawa, S.E.: A Suspension Bridge Problem: Existence and Stability. Mathematics Across Contemporary Sciences. Springer International Publishing, Switzerland (2016)
9. Mustafa, M.I.: General decay result for nonlinear viscoelastic equations. J. Math. Anal. Appl. **457**, 134–152 (2018)
10. Messaoudi, S.A., Mukiawa, S.E.: Existence and decay of solutions to a viscoelastic plate equation. Electron. J. Differ. Equ. **2016**(22), 1–14 (2016)
11. Jin, K.-P., Liang, J., Xiao, T.-J.: Coupled second order evolution equations with fading memory: optimal energy decay rate. J. Differ. Equ. **257**(5), 1501–1528 (2014)
12. Arnold, V.I.: Mathematical Methods of Classical Mechanics. Graduate Texts in Mathematics, vol. 60. Springer, New York (1989)

Existence and Uniqueness of Renormalized Solution to Noncoercive Elliptic Problem with Measure Data

Tiyamba Valea, W. Basile Yaméogo, and Arouna Ouédraogo

1 Introduction

Let Ω be an open bounded polygonal subset of $\mathbb{R}^N$, $N \geq 2$. The purpose of this work is to establish the existence and the uniqueness of a renormalized solution to noncoercive convection-diffusion elliptic equation:

$$\begin{aligned} -\Delta u + \operatorname{div}(\mathbf{v} u) + bu &= \mu \text{ in } \Omega, \\ u &= 0 \qquad \text{on } \partial\Omega, \end{aligned} \tag{1}$$

where the vector fields $\mathbf{v} \in (L^p(\Omega))^N$, $2 < p < +\infty$ if $N = 2$, $p = N$, if $N \geq 3$ and $b \in L^2(\Omega)$ is a nonnegative function, $\mu = f - \operatorname{div} F$ with $f \in L^1(\Omega)$ and $F \in (L^p(\Omega))^d$, is a diffuse measure.

In the sequel, we define the functions T_t and d_j, respectively, by

$$T_t(s) = \min(t, \max(-t, s))$$

and

$$d_j(s) = 1 - \left| \frac{T_{2j}(s) - T_j(s)}{j} \right|, \quad \forall s \in \mathbb{R}. \tag{2}$$

Now, we give the following.

T. Valea · W. B. Yaméogo · A. Ouédraogo (✉)
Laboratoire de Mathématiques, Informatique et Applications, Université Norbert ZONGO, Koudougou, Burkina Faso

F. Kamalov et al. (eds.), *Advances in Mathematical Modeling and Scientific Computing*, Trends in Mathematics, https://doi.org/10.1007/978-3-031-41420-6_3

Definition 1 A measurable function u defined from Ω into $\overline{\mathbb{R}}$ and finite a.e. in Ω is a renormalized solution of (1) if

$$\forall t > 0,\ T_t(u) \in H_0^1(\Omega), \tag{3}$$

$$\lim_{j\to+\infty} \int_{\{j<|T_t(u)|<j+1\}} |\nabla T_t(u)|^2\, dx = 0, \tag{4}$$

$$\forall d \in C_c^1(\mathbb{R}), \forall I \in H_0^1(\Omega) \cap L^\infty(\Omega),$$

$$\int_\Omega d(u)\, \nabla u \cdot \nabla I\, dx + \int_\Omega I\, \nabla u \cdot \nabla u\, d'(u)\, dx - \int_\Omega u\, d(u)\, \mathbf{v} \cdot \nabla I\, dx$$
$$- \int_\Omega u\, d'(u)\, I \mathbf{v} \cdot \nabla u\, dx + \int_\Omega b\, u\, d(u)\, I\, dx = \int_\Omega I\, d(u)\, d\mu. \tag{5}$$

2 Existence and Uniqueness Results

Theorem 1 *Let* $\mathbf{v} \in (L^p(\Omega))^d$ *and* μ *a diffuse measure. Then there exists at least one renormalized solution of the problem* (1).

Proof We will divide the proof of Theorem 1 into several steps.

2.1 *Approximate Problem*

By density, we can find a sequence of functions $\{f_\epsilon\} \subset L^\infty(\Omega)$ strongly converging to f in $L^1(\Omega)$ such that $\|f_\epsilon\|_{L^1(\Omega)} \le \|f\|_{L^1(\Omega)}$. Similarly, $C_c^\infty(\Omega)$ is dense in $L^2(\Omega)$ and then we can also find a sequence $\{F_\epsilon\} \in C_c^\infty(\Omega)$ strongly converging to F in $L^2(\Omega)^N$ such that $\|F_\epsilon\|_{L^2(\Omega)} \le \|F\|_{L^2(\Omega)}$.

Then, we set $\mu_\epsilon = f_\epsilon -$ div F_ϵ and consider the approximate problems of (1):

$$\begin{aligned} -\Delta u_\epsilon + \operatorname{div}(\mathbf{v}\, u_\epsilon) + b\, u_\epsilon &= \mu_\epsilon \text{ in } \Omega, \\ u_\epsilon &= 0 \quad \text{on } \partial\Omega. \end{aligned} \tag{6}$$

2.2 *A Priori Estimates*

Lemma 1 *For every* $t > 0$*, there exists a positive constant* C *such that*

$$\|\nabla T_t(u_\epsilon)\|^2_{L^2(\Omega)} \leq C.$$

Proof Taking $\phi = T_t(u_\epsilon)$ as test function in the formulation (6), we arrive at

$$\int_\Omega |\nabla T_t(u_\epsilon)|^2\, dx + \int_\Omega b\, u_\epsilon\, T_t(u_\epsilon)\, dx$$
$$= \int_\Omega T_t(u_\epsilon)\, \mathbf{v} \cdot \nabla T_t(u_\epsilon)\, dx + \int_\Omega T_t(u_\epsilon)\, d\mu_\epsilon. \tag{7}$$

Thanks to the Sobolev embedding, the Young inequality, and the fact that $bu_\epsilon T_t(u_\epsilon) \geq 0$, we deduce from (7)

$$\|\nabla T_t(u_\epsilon)\|^2_{L^2(\Omega)} \leq C, \tag{8}$$

with $C = 2k|\mu|(\Omega) + S(\Omega, t)t^2\|\mathbf{v}\|^p_{L^p(\Omega)}$.

Lemma 2 *Let $u_\epsilon \in H^1_0(\Omega)$ be a weak solution of* (6)*. Then there exists $C > 0$, such that*

$$\|\ln(1 + |u_\epsilon|)\|^2_{H^1_0(\Omega)} \leq C, \tag{9}$$

$$meas\{|u_\epsilon| \geq t\} \leq \frac{C}{(\ln(1+t))^2},\ \textit{for all } t \textit{ large enough} \tag{10}$$

and

$$meas\{|\nabla u_\epsilon| \geq t\} \leq \frac{C}{t} + \frac{C}{(\ln(1+t))^2},\ \textit{for all } t \textit{ large enough.} \tag{11}$$

Proof of (9): Let's define a Lipschitz function $\phi(u_\epsilon) = \int_0^{u_\epsilon} \frac{1}{(1+|s|)^2} dt$ with $\phi(0) = 0$. Since $u_\epsilon \in H^1_0(\Omega)$, then by the Stampacchia lemma, we deduce that $\phi(u_\epsilon) \in H^1_0(\Omega)$. Using $\phi(u_\epsilon)$ as a test function in (6), and with Sobolev embedding, we deduce that

$$\|\ln(1 + |u_\epsilon|)\|^2_{H^1_0(\Omega)} \leq C,$$

where $C = C(\mu, \Omega, \mathbf{v})$ is a positive constant, and then (9) is proved.

Proof of (10): Using again Poincaré's inequality and (9), we get

$$\int_{\{|u_\epsilon| \geq t\}} (\ln(1+t))^2\, dx \leq C(\mu, \Omega, \mathbf{v}),$$

which implies

$$meas\{|u_\epsilon| \geq t\} \leq \frac{C(\mu, \Omega, \mathbf{v})}{(\ln(1+t))^2}.$$

Proof of (11): Given $t, \lambda \geq 0$, set

$$I(t, \lambda) = \text{meas}\{|\nabla u_\epsilon|^2 > \lambda,\ |u_\epsilon| > t\}.$$

According to (10), we have

$$I(t, 0) \leq \frac{C}{(\ln(1+t))^2}, \quad \text{for any } t > 0 \text{ large enough.} \tag{12}$$

Thanks to (8), we obtain

$$\int_0^\infty \big(I(0, s) - I(t, s)\big) ds \leq C. \tag{13}$$

We deduce from (12) and (13) that

$$I(0, \lambda) \leq \frac{C}{\lambda} + \frac{C}{(\ln(1+t))^2}, \quad \text{for all } t \geq 1, \lambda > 0. \tag{14}$$

Setting $\lambda = t$ in (14) gives (11).

2.3 *Convergence Results*

Lemma 3

(i) For all $t > 0$, $T_t(u_\epsilon) \to T_t(u)$ in $L^2(\Omega)$ and a.e. in Ω, as $\epsilon \to 0$.
(ii) There exists a measurable function u such that $u_\epsilon \to u$ in measure and a.e. in Ω, as $\epsilon \to 0$.

Proof The proof of Lemma 3 follows the same lines as the proof of Lemma 4.8 -(i) in [7].

Lemma 4 *For all $j > 0$,*

$$\lim_{j \to +\infty} \limsup_{\epsilon \to 0} \int_{\{j < |T_t(u_\epsilon)| < j+1\}} |\nabla T_t(u_\epsilon)|^2 \, dx = 0. \tag{15}$$

Proof Taking $\phi = T_1\big(T_t(u_\epsilon) - T_j(T_t(u_\epsilon))\big)$ as test function in (6) and using the fact that b is nonnegative, T_1 is nondecreasing, and $\nabla\big[T_1\big(T_t(u_\epsilon) - T_j(T_t(u_\epsilon))\big)\big] = \nabla T_t(u_\epsilon)\, \chi_{\{j < |T_t(u_\epsilon)| < j+1\}}$, we obtain

$$\int_{\{j<|T_t(u_\epsilon)|<j+1\}} |\nabla T_t(u_\epsilon)|^2\,dx \le \int_{\{j<|T_t(u_\epsilon)|<j+1\}} |T_t(u_\epsilon)\mathbf{v}\cdot\nabla T_t(u_\epsilon)|\,dx$$
$$+\int_\Omega |f_\epsilon\, T_1\big(T_t(u_\epsilon)-T_j(T_t(u_\epsilon))\big)|\,dx+\int_{\{j<|T_t(u_\epsilon)|<j+1\}} |F_\epsilon\cdot\nabla T_t(u_\epsilon)|\,dx. \quad (16)$$

For the third integral of (16), we can use the Young inequality and the Sobolev embedding, $L^p(\Omega)^N \hookrightarrow L^2(\Omega)^N$, to obtain

$$\int_{\{j<|T_t(u_\epsilon)|<j+1\}} |F_\epsilon\cdot\nabla T_t(u_\epsilon)|\,dx \le \int_{\{j<|T_t(u_\epsilon)|<j+1\}} \frac{1}{2}|\nabla T_t(u_\epsilon)|^2\,dx$$
$$+\int_{\{j<|T_t(u_\epsilon)|<j+1\}} \frac{1}{2}|F_\epsilon|^2\,dx. \quad (17)$$

Therefore, combining (16) and (17), we have

$$\frac{1}{2}\int_{\{j<|T_t(u_\epsilon)|<j+1\}} |\nabla T_t(u_\epsilon)|^2\,dx \le \int_{\{j<|T_t(u_\epsilon)|<j+1\}} |T_t(u_\epsilon)\mathbf{v}\cdot\nabla T_t(u_\epsilon)|\,dx \quad (18)$$
$$+\int_\Omega |f_\epsilon\, T_1\big(T_t(u_\epsilon)-T_n(T_t(u_\epsilon))\big)|\,dx+\int_{\{j<|T_t(u_\epsilon)|<j+1\}} \frac{1}{2}|F_\epsilon|^2\,dx. \quad (19)$$

We evaluate each term of (18).

Using the Sobolev embedding, $L^p(\Omega)^N \hookrightarrow L^2(\Omega)^N$, we have $F_\epsilon\,\chi_{\{j<|T_t(u_\epsilon)|<j+1\}} \to F\,\chi_{\{j<|T_t(u_\epsilon)|<j+1\}}$ strongly in $L^2(\Omega)^N$. Then, we can pass to the limit in the third term in the right-hand side of (18), as $\epsilon\to 0$ to obtain

$$\lim_{\epsilon\to 0}\int_{\{j<|T_t(u_\epsilon)|<j+1\}} |F_\epsilon|^2\,dx = \int_{\{j<|T_t(u_\epsilon)|<j+1\}} |F|^2\,dx. \quad (20)$$

Using the fact that meas $\{T_t(u)\ge j\}\to 0$ as $j\to+\infty$ (see Lemma 2) and $F\in L^2(\Omega)^N$, we can pass to the limit as $j\to+\infty$ in (20) to obtain

$$\lim_{j\to+\infty}\int_{\{j<|T_t(u_\epsilon)|<j+1\}} |F|^2\,dx \le \lim_{j\to+\infty}\int_{\{|T_t(u)|>j\}} |F|^2\,dx = 0,$$

therefore,

$$\lim_{j\to+\infty}\lim_{\epsilon\to 0}\int_{\{j<|T_t(u_\epsilon)|<j+1\}} |F_\epsilon|^2\,dx = 0. \quad (21)$$

Regarding the second integral of the right side of (18), we use the fact that $f_\epsilon\, T_1\big(T_t(u_\epsilon)-T_j(T_t(u_\epsilon))\big)\to f\, T_1\big(u-T_j(u)\big)$ a.e. in Ω as $\epsilon\to 0$, $|f_\epsilon\, T_1\big(T_t(u_\epsilon)-T_j(T_t(u_\epsilon))\big)|\le |f_\epsilon|\in L^1(\Omega)$ and $f_\epsilon\to f$ strongly in $L^1(\Omega)$. Then,

$$\lim_{\epsilon\to 0}\int_{\Omega}|f_\epsilon\, T_1\big(T_t(u_\epsilon)-T_j(T_t(u_\epsilon))\big)|\,dx=\int_{\Omega}|f\, T_1\big(u-T_j(u)\big)|\,dx.$$

On the other hand, $f\, T_1\big(u-T_j(u)\big)\to 0$ a.e. in Ω as $j\to+\infty$, $|f\, T_1\big(u-T_n(u)\big)|\le |f|\in L^1(\Omega)$. Thus, Lebesgue's dominated convergence theorem gives

$$\lim_{j\to\infty}\lim_{\epsilon\to 0}\int_{\Omega}|f_\epsilon\, T_1\big(T_t(u_\epsilon)-T_j(T_t(u_\epsilon))\big)|\,dx=0. \tag{22}$$

Next, for the first term of (18), we serve Lemma 3 and the Sobolev embedding $L^p(\Omega)^N\hookrightarrow L^2(\Omega)^N$, to conclude that $T_t(u_\epsilon)\,\mathbf{v}\,\chi_{\{j<|T_t(u_\epsilon)|<j+1\}}\to T_t(u)\,\mathbf{v}\,\chi_{\{j<|T_t(u)|<j+1\}}$ in $L^2(\Omega)^N$ as $\epsilon\to 0$. Combining with the weak convergence of

$$\nabla T_t(u_\epsilon)\,\chi_{\{j<|T_t(u_\epsilon)|<j+1\}} \text{ to } \nabla T_t(u)\,\chi_{\{j<|T_t(u)|<j+1\}} \text{ in } L^2(\Omega)^N,$$

we deduce that

$$\lim_{\epsilon\to 0}\int_{\{j<|T_t(u_\epsilon)|<j+1\}}|T_t(u_\epsilon)\mathbf{v}.\nabla T_t(u_\epsilon)|\,dx=\int_{\{j<|T_t(u)|<j+1\}}|T_t(u)\,\mathbf{v}\cdot\nabla T_t(u)|\,dx. \tag{23}$$

We claim that

$$\lim_{j\to+\infty}\int_{\{j<|T_t(u)|<j+1\}}|T_t(u)\,\mathbf{v}\cdot\nabla T_t(u)|\,dx=0. \tag{24}$$

To prove (24), we need the following lemma.

Lemma 5 *For $t>0$, if $\nabla T_t(u)\in H_0^1(\Omega)$, then $T_t(u)\,\nabla T_t(u)\in L^s(\Omega)^N$ with*

$$\begin{cases}1<s<2 & \textit{for } N=2,\\ 1<s\le \frac{N}{N-1} & \textit{for } N\ge 3.\end{cases}$$

Proof ***(Lemma 5)*** From the Sobolev embedding, we have

$$\begin{cases}H_0^1(\Omega)\hookrightarrow L^q \text{ for all } q<+\infty, & \text{if } N=2,\\ H_0^1(\Omega)\hookrightarrow L^{q*} \text{ with } q*=\frac{2N}{N-2}, & \text{if } N\ge 3.\end{cases}$$

By the Hölder inequality, for all pair of positive numbers (θ,θ') such that $\dfrac{1}{\theta}+\dfrac{1}{\theta'}=1$, we have

$$\int_{\Omega}|T_t(u)\,\nabla T_t(u)|^s\le\left(\int_{\Omega}|T_t(u)|^{s\theta}\,dx\right)^{\frac{1}{\theta}}\left(\int_{\Omega}|\nabla T_t(u)|^{s\theta'}\,dx\right)^{\frac{1}{\theta'}}.$$

For $N \geq 3$, we must conclude if $1 < s\theta \leq q*$ and $1 < s\theta' \leq 2$, i.e., $s(\frac{1}{q*}+\frac{1}{2}) < 1$ and $s > 1$.

As $(\frac{1}{q*}+\frac{1}{2}) = \frac{N-1}{N}$, we obtain the desired result.

For $N = 2$, a similar computation gives $s(\frac{1}{q}+\frac{1}{2}) < 1$ for any $q < +\infty$, then $s < 2$.

Next, using Lemma 5 and the fact that $\mathbf{v} \in (L^p(\Omega))^N$, with $2 < p < +\infty$ if $N = 2$ and $p = N$, if $N \geq 3$, we conclude that $T_t(u)\,\mathbf{v}\cdot\nabla T_t(u) \in L^1(\Omega)$. Moreover, from Lemma 2, we have meas $\{T_t(u) \geq j\} \to 0$ as $j \to +\infty$.

Therefore, we can pass to the limit in the second integral of (23) to obtain

$$\lim_{j\to+\infty}\int_{\{j<|T_t(u)|<j+1\}} |T_t(u)\,\mathbf{v}\cdot\nabla T_t(u)|\,dx \leq \lim_{j\to+\infty}\int_{\{|T_t(u)|>j\}} |T_t(u)\,\mathbf{v}\cdot\nabla T_t(u)|\,dx = 0,$$

which proves (24). Combining (23) and (24), we obtain

$$\lim_{n\to\infty}\lim_{\epsilon\to 0}\int_{\{j<|T_t(u_\epsilon)|<j+1\}} |T_t(u_\epsilon)\,\mathbf{v}\cdot\nabla T_t(u_\epsilon)|\,dx = 0. \tag{25}$$

Combining (18), (20), (21), (22), (23), (24), and (25), we obtained the desired result (15).

2.4 Passage to the Limit

In order to show that the limit function u is a renormalized solution of problem (1), we need the following result.

Lemma 6 *For all $d \in C_c^1(\mathbb{R})$ and $I \in H_0^1(\Omega)\cap L^\infty(\Omega)$,*

$$\nabla[d(u_\epsilon)I] \to \nabla[d(u)I] \quad \textit{strongly in} \quad (L^2(\Omega))^2, \quad \textit{as} \quad \epsilon \to 0.$$

Proof For any $d \in C_c^1(\mathbb{R})$ and $I \in H_0^1(\Omega)\cap L^\infty(\Omega)$, we have

$$\begin{aligned}\int_\Omega \left|\nabla[d(u_\epsilon)I] - \nabla[d(u)I]\right|^2 dx &\leq \int_\Omega \left|(d(u_\epsilon)-d(u))\nabla I\right|^2 dx \\ &\quad + \int_\Omega \left|d'(u_\epsilon)I[\nabla u_\epsilon - \nabla u]\right|^2 dx \\ &\quad + \int_\Omega \left|(d'(u_\epsilon)-d'(u))I\nabla u\right|^2 dx \\ &:= \Phi_1^\epsilon + \Phi_2^\epsilon + \Phi_3^\epsilon.\end{aligned} \tag{26}$$

(a): $\Phi_1^\epsilon = \int_\Omega |(d(u_\epsilon) - d(u))\nabla I|^2 \, dx$.
Set $\omega_1^\epsilon(x) = |(d(u_\epsilon) - d(u))\nabla I|^2$. By Lebesgue's dominated convergence theorem, we obtain that

$$\|\Phi_1^\epsilon\|_{L^2(\Omega)} \to 0 \text{ as } \epsilon \to 0. \tag{27}$$

(b): Let $l > 0$ so that $\mathrm{supp}(d) \subset [-l, l]$ and $\Phi_2^\epsilon = \int_\Omega |d'(u_\epsilon)I(\nabla T_l(u_\epsilon) - \nabla T_l(u))|^2 \, dx$.
Set $\omega_2^\epsilon(x) = |d'(u_\epsilon)I(\nabla T_l(u_\epsilon) - \nabla T_l(u))|^2$. We have $\omega_2^\epsilon(x) \to 0$ a.e. $x \in \Omega$ as $\epsilon \to 0$ and

$$|\omega_2^\epsilon(x)| \leq C(d, \|I\|_\infty)|\nabla T_l(u_\epsilon) - \nabla T_l(u)|^2.$$

By the generalized convergence theorem of Lebesgue, we get

$$\|\Phi_2^\epsilon\|_{L^2(\Omega)} \to 0 \text{ as } \epsilon \to 0. \tag{28}$$

(c): $\Phi_3^\epsilon = \int_\Omega |(d'(u_\epsilon) - d'(u))I\nabla u|^2 \, dx$.
Set $\omega_3^\epsilon(x) = |(d'(u_\epsilon) - d'(u))I\nabla u|^2$. We have $\omega_3^\epsilon(x) \to 0$ a.e. $x \in \Omega$ as $\epsilon \to 0$ and

$$|\omega_3^\epsilon(x)| \leq C(d, \|I\|_\infty)|\nabla T_l(u)|^2 \in L^1(\Omega),$$

given $l > 0$ such that $\mathrm{supp}(d) \subset [-l, l]$.

So, by Lebesgue's dominated convergence theorem, we have $\lim\limits_{\epsilon \to 0} \Phi_3^\epsilon = 0$.
Hence,

$$\|\Phi_3^\epsilon\|_{L^2(\Omega)} \to 0 \text{ as } \epsilon \to 0. \tag{29}$$

Thanks to (27), (28), and (29), we get

$$\left\|\Phi_1^\epsilon + \Phi_2^\epsilon + \Phi_3^\epsilon\right\|_{L^2(\Omega)^2} \to 0 \text{ as } \epsilon \to 0.$$

This ends the proof of Lemma 6.

Next, by testing (6) with the function $d(u_\epsilon)I$, we obtain

$$\begin{aligned} \int_\Omega \nabla u_\epsilon \cdot \nabla[d(u_\epsilon)I]dx - \int_\Omega u_\epsilon \mathbf{v} \cdot \nabla[d(u_\epsilon)I]dx + \int_\Omega b\, u_\epsilon \, d(u_\epsilon)\, I dx \\ = \int_\Omega f_\epsilon d(u_\epsilon) I dx + \int_\Omega F_\epsilon \cdot \nabla[d(u_\epsilon)I]dx. \end{aligned} \tag{30}$$

Since d has compact support, then there exists a positive real number t such that $supp(d) \subset [-t, t]$, so u_ϵ can be replaced by its truncation $T_t(u_\epsilon)$. We start with the first integral of (30). By Lemma 6, $\nabla[d(u_\epsilon)I]$ converges strongly to $\nabla[d(u)I]$ in $(L^2(\Omega))^2$ and, since $\nabla T_t(u_\epsilon)$ converges weakly to $\nabla T_t(u)$ in $(L^2(\Omega))^2$, we deduce that

$$\lim_{\epsilon\to 0}\int_\Omega \nabla T_t(u_\epsilon)\cdot\nabla[d(u_\epsilon)I]dx = \int_\Omega \nabla T_t(u)\cdot\nabla[d(u)I]dx. \tag{31}$$

In the same way, thanks to Lemma 6 and the fact that F_ϵ converges to F strongly in $L^2(\Omega)^N$ when ϵ goes to zero, we can pass to the limit in the second integral on the right-hand side of (30) to obtain

$$\lim_{\epsilon\to 0}\int_\Omega F_\epsilon\cdot\nabla[d(u_\epsilon)I]dx = \int_\Omega F\cdot\nabla[d(u)I]dx. \tag{32}$$

As far as the second integral on the left-hand side of (30) is concerned, we use Lemma A.1 of [4] to deduce that the sequence $(T_t(u_\epsilon)\mathbf{v})_{\epsilon>0}$ converges weakly to $T_t(u)\mathbf{v}$ in $L^2(\Omega)^N$. Furthermore, using Lemma 6, we get, when $\epsilon\to 0$,

$$\lim_{\epsilon\to 0}\int_\Omega T_t(u_\epsilon)\mathbf{v}\cdot\nabla[d(u_\epsilon)I]dx = \int_\Omega T_t(u)\mathbf{v}\cdot\nabla[d(u)I]dx. \tag{33}$$

Since $d(u_\epsilon)I$ converges weakly-$\star$ to $d(u)I$ in $L^\infty(\Omega)$, using the fact that $b\in L^2(\Omega)$ and $T_t(u_\epsilon)$ converges strongly to $T_t(u)$ in $L^2(\Omega)$ and a.e. in Ω as $\epsilon\to 0$, we deduce that $bT_t(u_\epsilon)\longrightarrow bT_t(u)$ strongly in $L^1(\Omega)$. We also have $f_\epsilon\to f$ strongly in $L^1(\Omega)$. Hence, when ϵ goes to zero, we can pass to the limit in the third integral on the left-hand side of (30) and in the first integral on the right-hand side of (30), respectively, to obtain

$$\lim_{\epsilon\to 0}\int_\Omega bT_t(u_\epsilon)d(u_\epsilon)Idx = \int_\Omega bT_t(u)d(u)Idx, \tag{34}$$

$$\lim_{\epsilon\to 0}\int_\Omega f_\epsilon d(u_\epsilon)Idx = \int_\Omega fd(u)Idx. \tag{35}$$

Combining (31), (32), (33), (34), and (35), we obtain (5).

2.5 *Uniqueness of Renormalized Solution*

We follow the idea of [1]. Let u_1 and u_2 be two renormalized solutions to problem (1) in the sense of Definition 1. Let $d_j\in C_c^1(\mathbb{R})$ with compact support $supp\,(d_j) = [-2j, 2j]$ and for all $\sigma, t>0$, $T_\sigma\big(T_t(u_1)-T_t(u_2)\big)\in H_0^1(\Omega)\cap L^\infty(\Omega)$. For u_1, we take $\frac{1}{\sigma}T_\sigma\big(T_t(u_1)-T_t(u_2)\big)d_j(u_1)$ as test function in (5) and we let t go to $+\infty$. Using such a process in (5) written for u_2 gives by subtraction

$$A_{\sigma,j} + B_{\sigma,j} + C_{\sigma,j} + D_{\sigma,j} + E_{\sigma,j} = F_{\sigma,j} + G_{\sigma,j} + H_{\sigma,j}, \tag{36}$$

where

$$\begin{aligned}
A_{\sigma,j} &= \tfrac{1}{\sigma}\int_\Omega \left[\nabla u_1\, d_j(u_1) - \nabla u_2\, d_j(u_2)\right]\nabla T_\sigma(u_1-u_2)dx,\\
B_{\sigma,j} &= \tfrac{1}{\sigma}\int_\Omega \left[\nabla u_1.\nabla u_1\, d_j'(u_1) - \nabla u_2.\nabla u_2\, d_j'(u_2)\right] T_\sigma(u_1-u_2)dx,\\
C_{\sigma,j} &= -\tfrac{1}{\sigma}\int_\Omega \left[u_1\, d_j(u_1) - u_2\, d_j(u_2)\right]\mathbf{v}.\nabla T_\sigma(u_1-u_2)dx,\\
D_{\sigma,j} &= -\tfrac{1}{\sigma}\int_\Omega \left[u_1 d_j'(u_1)\nabla u_1 - u_2 d_j'(u_2)\nabla u_2\right].\mathbf{v}\; T_\sigma(u_1-u_2)dx,\\
E_{\sigma,j} &= \tfrac{1}{\sigma}\int_\Omega \left[u_1 d_j(u_1) - u_2 d_j(u_2)\right] b\; T_\sigma(u_1-u_2)dx,\\
F_{\sigma,j} &= \tfrac{1}{\sigma}\int_\Omega \left[d_j(u_1) - d_j(u_2)\right] f\; T_\sigma(u_1-u_2)dx,\\
G_{\sigma,j} &= \tfrac{1}{\sigma}\int_\Omega \nabla\big(d_j(u_1) - d_j(u_2)\big) F\; T_\sigma(u_1-u_2)dx,\\
H_{\sigma,j} &= \tfrac{1}{\sigma}\int_\Omega \left[d_j(u_1) - d_j(u_2)\right] F.\nabla\; T_\sigma(u_1-u_2)dx.
\end{aligned}$$

In the following, we will examine the limit of each term in (36) when σ tends to 0 and j tends to $+\infty$, respectively. The functions $d_j(u_1)$ and $d_j(u_2)$ converge to 1 almost everywhere in Ω and in $L^\infty(\Omega)$ weak-$*$. Consequently, the quantity $(d_j(u_1) - d_j(u_2))$ converges to 0 as j tends to $+\infty$. Then, thanks to Lebesgue's dominated convergence theorem, we obtain

$$\lim_{j\to+\infty}\lim_{\sigma\to 0} F_{\sigma,j} = \lim_{j\to+\infty}\lim_{\sigma\to 0}\frac{1}{\sigma}\int_\Omega \big(d_j(u_1) - d_j(u_2)\big) f\; T_\sigma(u_1-u_2)dx = 0 \tag{37}$$

and

$$\begin{aligned}
\lim_{j\to+\infty}\lim_{\sigma\to 0} H_{\sigma,j} = &\lim_{j\to+\infty}\limsup_{\sigma\to 0}\frac{1}{\sigma}\int_\Omega \big(d_j(u_1) - d_j(u_2)\big)\, F\cdot\nabla\, T_\sigma(u_1-u_2)\, dx\\
&= 0.
\end{aligned}\tag{38}$$

On the other hand,

$$\begin{aligned}
\lim_{j\to+\infty}\lim_{\sigma\to 0} E_{\sigma,j} &= \lim_{j\to+\infty}\lim_{\sigma\to 0}\frac{1}{\sigma}\int_\Omega \left[u_1 d_j(u_1) - u_2 d_j(u_2)\right] b\; T_\sigma(u_1-u_2)\, dx\\
&= \int_\Omega b\,|u_1-u_2|\, dx.
\end{aligned}\tag{39}$$

By virtue of the definition of d_j, we have $|d_j'(s)| = \frac{1}{j}\chi_{\{j\le |s|\le 2j\}}$, and then we deduce that

$$\left|\frac{1}{\sigma}\int_{\Omega}\left(d_j'(u_1)\nabla u_1 - d_j'(u_2)\nabla u_2\right)F\ T_\sigma(u_1-u_2)dx\right|$$
$$\leq \int_{\Omega}|d_j'(u_1)\nabla u_1 - d_j'(u_2)\nabla u_2|\,|F|dx$$
$$\leq \frac{1}{j}\int_{\{j\leq|u_1|\leq 2j\}}|\nabla u_1||F|dx + \frac{1}{j}\int_{\{j\leq|u_2|\leq 2j\}}|\nabla u_2||F|dx. \quad (40)$$

From (40), we deduce that

$$\lim_{j\to+\infty}\ \limsup_{\sigma\to 0}|G_{\sigma,j}| = 0. \quad (41)$$

Using the previous arguments, we also deduce that

$$\lim_{j\to+\infty}\ \lim_{\sigma\to 0}\sup|B_{\sigma,j}| = 0. \quad (42)$$

From the same way, we obtain that

$$\lim_{j\to+\infty}\ \lim_{\sigma\to 0}\sup|D_{\sigma,j}| = 0. \quad (43)$$

It remains to pass to the limit in $A_{\sigma,j}$. It can be rewritten as

$$\begin{aligned}A_{\sigma,j} &= \frac{1}{\sigma}\int_{\Omega}\left[\nabla u_1\ d_j(u_1) - \nabla u_2\ d_j(u_2)\right]\nabla T_\sigma(u_1-u_2)dx\\ &= \frac{1}{\sigma}\int_{\Omega}\left(\nabla u_1 - \nabla u_2\right)d_j(u_1)\,\nabla T_\sigma(u_1-u_2)dx\\ &\quad + \frac{1}{\sigma}\int_{\Omega}(d_j(u_1)-d_j(u_2))\nabla u_2\,\nabla T_\sigma(u_1-u_2).\end{aligned} \quad (44)$$

Let's use the previous results and pass to the limit in the two integrals of the right-hand side of (44). We have

$$\lim_{j\to+\infty}\ \lim_{\sigma\to 0}\frac{1}{\sigma}\int_{\Omega}(d_j(u_1)-d_j(u_2))\nabla u_2\,\nabla T_\sigma(u_1-u_2) = 0 \quad (45)$$

and

$$\lim_{j\to+\infty}\ \lim_{\sigma\to 0}\frac{1}{\sigma}\int_{\Omega}\left(\nabla u_1 - \nabla u_2\right)d_j(u_1)\,\nabla T_\sigma(u_1-u_2)dx = \int_{\Omega}|\nabla u_1 - \nabla u_2|\,dx. \quad (46)$$

From (45) and (46), we deduce that

$$\lim_{j\to+\infty}\lim_{\sigma\to 0} A_{\sigma,j} = \int_{\Omega} |\nabla u_1 - \nabla u_2|\, dx. \tag{47}$$

Putting (37), (38), (39), (40), (41), (42), (43) and (47) together, we can pass to the limit in (36) as σ tends to 0 and n tends to $+\infty$, respectively, to obtain

$$\int_{\Omega} b\,|u_1 - u_2|\, dx + \int_{\Omega} |\nabla u_1 - \nabla u_2|\, dx = 0. \tag{48}$$

From (48), we deduce that $\int_{\Omega} b\,|u_1 - u_2|\, dx = 0$ and then, as $b \geq 0$, $u_1 = u_2$ a.e. in Ω.

Notes and Comments The concept of renormalized solution is appropriate for studying problem with measure and L^1-data. This concept of solutions was initiated in [2] for Boltzmann equations and has been taken up in [1] and [6] for elliptic problems with L^1-data. A generalization to coercive elliptic problem with measure data is addressed in [3]. It is well-known that the renormalized solutions are a convenient framework for parabolic and elliptic equations with L^1-data which provides in general existence, stability, and uniqueness results (see [5]).

References

1. Ben Cheick, M., Guibé, O.: Nonlinear and non-coercive elliptic problems with integrable data. Adv. Math. Sci. Appl. **16**, $n^0 1$ 275–297 (2006)
2. DiPerna, R.-J., Lions, P.-L.: On the Cauchy problem for Boltzmann equations: global existence and weak stability. Ann. Math. **130**, 321–366 (1989)
3. Droniou, J.: Solving convection-diffusion equations with mixed, Neumann and Fourier boundary conditions and measures as data, by a duality method. Adv. Differ. Equ. **5**(10–12), 1341–1396 (2000)
4. Droniou, J.: Non-coercive linear elliptic problems, Potential Anal. **17**(2), 181–203 (2002)
5. Guibé, O., Mercaldo, A.: Uniqueness results for noncoercive nonlinear elliptic equations with two lower order terms. Commun. Pure Appl. Anal. **7**, 163–192 (2008)
6. Murat, F.: Soluciones renormalizadas de EDP elipticas no lineales. Technical Report R93023, Laboratoire d'Analyse Numérique, Paris VI, 1993
7. Ouédraogo, A., Valea, T.: Existence and uniqueness of renormalized solution to multivalued homogeneous Neumann problem with L^1-*data*. Gulf J. Math. **13**(2), 42–66 (2022)

Fixed-Point Theorems Based Evaluation of Analytical Solution in Fractional Diffusion Equations

B. Malathi and S. Chelliah

1 Introduction

Nowadays, fractional calculus is a fantastic tool for illustrating the memory and developmental behaviors of numerous elements and processes. Interest in area mathematics and its applications is expanding in numerous study fields. In fractional analyses [1], we generally find the arrangement of the different states of the fragmented demand. Large-scale fragmentation conditions have accumulated in various areas of design and physics, especially signal design, electromagnetism, biology, fluid dynamics, electrochemistry, scattering operations, dynamic, continuous and real dynamics of viscoelastic materials, and the design of circular shapes. To find the answer to fragmented difference conditions [2], we need mathematical methods to obtain a difficult arrangement. In most cases, there are no careful arrangements for partial differentiation conditions. There is no technique to find the exact arrangement [3] with the basic/limiting conditions of the fractional difference condition. Thus, mathematical methods have been devised to estimate fractional demand variance. Pade approximation, homotopy analysis, homotopy perturbation, variation iteration, standard Adomian decomposition method (SADM), Fourier and Laplace transforms, and others have been used to address fragmentary circumstances [4]. Mathematically differentiating fragmentation circumstances are common [5]. Mathematical methods let us organize fragmentary differential conditions [6]. SADM excels in all mathematical strategies. The typical Adomian decomposition method (ADM) has solved several difficulties [7, 11].

B. Malathi (✉) · S. Chelliah
PG Research Scholar (Reg. no 921072092017) and Research Department of Mathematics, The M.D.T. Hindu College, Affiliated to Manonmaniam Sundaranar University, Tirunelveli, Tamilnadu, India

F. Kamalov et al. (eds.), *Advances in Mathematical Modeling and Scientific Computing*, Trends in Mathematics, https://doi.org/10.1007/978-3-031-41420-6_4

Most mathematical methods use uniqueness, which makes it harder to fix mistakes and takes more time and calculations. Due to the absence of customization, the standard Adomian decomposition method avoids error correction and reduces calculation time. The two-step Adomian decomposition method (TSADM) and improved Adomian decomposition techniques (NIADM) are also used to solve SADM difficulties [8]. The two-ventricular Adomian decomposition technique (DSATM) and many other mathematical methods were used for two-dimensional problems, but for a partial diffraction wave rather than a two-dimensional time interval, which becomes increasingly complex compared to expanding the components of a condition [9, 10]. Thus, the mathematical system faces hurdles in solving such problems. In the Middle Ages, we wanted to solve such problems simply, without the mathematical system's constraints [12].

2 Literature Review

The authors' devised fractional diffusion equations can be solved using a wide variety of techniques. This section reviews a few of the techniques.

Manoj Kumar et al. [13] have shown that the multi-dimensional time-space tempered fractional diffusion wave (TSTFDW) state exists and is unique, utilizing the static-point hypothesis. The principal advantage of this technique is that it provides precise arrangements individually, without aligning and without changing the problem of regulating conditions. However, there were some restrictions on this technique that should be used when it involves the initial time period of the series (zero period) problem completion period and the corresponding start/limit conditions. The two-step Adomian decomposition approach (TSADM) has been developed by Dumitru Baleanu [14] to solve the multi-dimensional Riesz space distributed-order advection diffusion (RSDOAD) equation. The analytical solution was produced satisfactorily utilizing the TSADM [20]. Mariam Abushammala [15] has introduced a fractional-order epidemic model for youth diseases with a new piece of supplementary system proposed by Kaputo and Fabrizio. To deal with the problem, arrangements were made in unlimited series using the Laplace Adomian decomposition method (LADM). Here, we demonstrate the existence, uniqueness, and sound of the arrangement using the static-point hypothesis. In addition, we present certain mathematical impacts to highlight the prospective partner's credibility.

Khuri et al. [16] introduced a selective integrated action plan that covers all of the direct and non linear third demand limit evaluation issues that come up in real applications. Significant fixed-point emphasis, including the projects of Picard's and Krasnoselskii–Mann's schemes, depends on developing green capabilities. Dumitru Baleanu [17]. has introduced the novel method based on Ishikawa's application of the green ability for mathematical rendering of a wide range of rating problems in the third requirement into a fixed-point reactivation technique. The proposed technique would operate at a pace that would add to other existing strategies based on Picard and Mann's action plans. The point of our electoral system is to defeat the significant deficit of other action plans, which in general brings about the decay of error when the area increases. A direct required administrator was contacted until the green capacity was configured, and then, at that time, the notable Ishikawa Fixed-Point Action Plan was used to obtain another action plan.

3 Background Study of Fractional Derivatives

Initially, the basic study of fractional derivatives and fractional integrals is presented in this section. Based on the properties, the fractional integral and derivatives are described. In this proposed methodology, Caputo's derivatives are utilized, which are mainly used in the mathematical applications.

Definition 1 The real function of the derivatives is presented as $G(Y)$, $Y > 0$, which considered as the space C^{θ}. The condition checks this real function, $\theta \epsilon \mathbb{R}$, and $\exists$ is a real number with the condition of $v(>\theta)$, which is presented in Eq. (1):

$$G(Y) = Y^{V} G1(Y) \tag{1}$$

Where, $G1(Y) \in C[0, \infty]$ and is defined as the space Eq. (2):

$$C_{\theta}^{n} \text{ if } G^{N} \varepsilon \mathbb{N} \cup \{0\} \tag{2}$$

Definition 2 "The operator known as the Riemann–Liouville fractional order integral $\sigma \geq 0$ with the consideration of function $G \in C_{\theta}$, $\theta \geq -1$, which is defined as Eqs. (3) and (4):

$$J^{\sigma} YG(Y) = \frac{1}{\Gamma(\sigma)} \int_{0}^{Y} (Y - \zeta)^{\sigma - 1} G(d\zeta)\, d\zeta, \quad \sigma > 0, Y > 0 \tag{3}$$

$$J^{\sigma} YG(Y) = G(Y) \tag{4}$$

Based on the properties of the $J^{\sigma} Y$ operator, it is formulated as Eqs. (5, 6, 7, and 8):

$$G \varepsilon C_{\theta}, \theta \geq -1, \sigma, \omega \geq 0 \text{ and } \zeta > -1 \tag{5}$$

$$J_{Y}^{\sigma} J_{Y}^{\omega} G(Y) = J_{Y}^{\sigma+\omega} G(Y) \tag{6}$$

$$J_{Y}^{\sigma} J_{Y}^{\omega} G(Y) = J_{Y}^{\sigma} J_{Y}^{\omega} G(Y) \tag{7}$$

$$J_{Y}^{\sigma} Y^{\zeta} = \frac{\Gamma(\zeta + 1)}{\Gamma(\sigma + \zeta + 1)} Y^{\sigma+\zeta} \tag{8}$$

The Riemann–Liouville derivative has some disadvantages that are removed with the combination of Caputo's derivative. This derivative is designed for fractional derivatives.

Definition 3 The Caputo's derivative of fractional derivative is formulated as Eqs. (9) and (10):

$$d_Y^{\sigma} G(Y) = J_Y^{N-\sigma} D_Y^N G(Y) \tag{9}$$

$$= \frac{1}{\Gamma(N-\sigma)} \int_0^Y (Y-\zeta)^{N-\sigma-1} G^N(\zeta)\, d\zeta, \quad \sigma > 0, Y > 0 \tag{10}$$

The equation is as follows, Eq. (11).

$$\text{For } N-1 < \sigma \leq N, \quad N \in \mathbb{N}, Y > 0, G \in C_{-1}^N \tag{11}$$

And,

$$D_Y^{\sigma} A = 0 \tag{12}$$

The linear operators of Caputo's derivatives are presented as Eq. (13):

$$D_Y^{\sigma}(AS(Y) + BH(Y)) = AD_Y^{\sigma} S(Y) + BD_Y^{\sigma} H(Y) \tag{13}$$

Where, B and A are constant.

Definition 4 In this definition, the Caputo time fractional derivation [21, 22] operator is checked with the condition of the smallest integer (N) and increases, $\sigma > 0$, which are formulated as Eqs. (14) and (15):

$$D_{\zeta}^{\sigma} W(Y,\zeta) = \frac{\partial^{\sigma} w(Y,\zeta)}{\partial \xi^{\sigma}} \tag{14}$$

$$= \begin{cases} \frac{1}{\Gamma N-\sigma} \int_0^{\xi} (\zeta-\tau)^{N-\sigma-1} \frac{\partial^N W(Y,\tau)}{\partial \tau^N} d\tau, N-1 < \sigma < N \\ \frac{\partial^N W(Y,\tau)}{\partial \tau^N}, \sigma = N \in \mathbb{N} \end{cases} \tag{15}$$

To identify the solution for the considered problems, the lemma functions are utilized, which are presented as Eqs. (16, 17, and 18).

Lemma 1

$$\text{If } N-1 < \sigma \leq N, N \in \mathbb{N} \text{ and } G \in C_{\theta}^N, \theta \geq -1 \tag{16}$$

$$D_Y^{\sigma} J_Y^{\sigma} G(Y) = G(Y) \tag{17}$$

$$D_Y^{\sigma} J_Y^{\sigma} G(Y) = G(Y) - \sum_{Q=0}^{N-1} G^{Q}\left(0^{+}\right) \frac{Y^{Q}}{Q!}, \quad Y > 0 \tag{18}$$

Definition 5 The Mittag-Leffler function is formulated as the special function, denoted as Eq. (19):

$$E^{\sigma}(Y) = \sum_{P=0}^{\infty} \frac{Y_P}{\Gamma(1+\sigma P)}, \quad \sigma \in C, \mathbb{R}(\sigma) > 0, Y \in C \tag{19}$$

If $\sigma = 1$, the exponential function of Mittag-Leffler can be described as Eq. (20):

$$\sum_{P=0}^{\infty} \frac{Y_P}{\Gamma(1+\sigma P)} = \sum_{P=0}^{\infty} \frac{Y_P}{P!} = e_Y \tag{20}$$

3.1 *Background Study About Fixed-Point Theorems*

In this section, the lemmas and theorems of fixed-point theorems are explained.

Definition 6 In this definition, a continuous function is defined as the operator. However, if a function transforms constrained sets from a set before and is continuous.

Definition 7 The mapping function (Y) with normal space $(Y, \|.\|)$ satisfies the condition $P : Y : \rightarrow Y$ at the condition of $Y1, Y2 \in Y$. This condition is formulated as Eq. (21):

$$||P(Y1) - P(Y2)|| \leq \delta \, ||Y1 - Y2|| \tag{21}$$

These operators are presented in the values of $0 \leq \delta < 1$.

In this proposed methodology, Caputo's derivative function is utilized to solve problems, and it plays a major role in the proposed analysis. The Caputo's derivative based on lemma functions are presented as follows:

Lemma 2 The normal solution of the homogenous equation with the consideration of $\sigma > 0$. This condition is presented as Eq. (22):

$$D_{0+}^{\sigma} \varphi(Y) = 0 \tag{22}$$

It is explained as Eqs. (23) and (24):

$$\varphi(Y) = A0 + A1Y + A2Y^2 + A3Y^3 + \cdots + AM - 1Y^{M-1} \tag{23}$$

$$\varphi \in \mathbb{R}, J = 1, 2, \ldots, M - 1\,(M = [\sigma] + 1) \tag{24}$$

Lemma 3 In this function, $\sigma > 0$ is formulated as Eqs. (25) and (26):

$$J_{0+}^{\sigma} D_{0+}^{\sigma} \varphi(Y) = \varphi(Y) + A0 + A1Y + A2Y^2 + A3Y^3 + \cdots + AM - 1Y^{M-1} \tag{25}$$

$$\varphi \in \mathbb{R}, J = 1, 2, \ldots, M - 1\,(M = [\sigma] + 1) \tag{26}$$

Theorem 1 Schaefer's Krasnoselskii fixed-point theorem

In this theorem [23], the Y is a bounded and closed convex with the consideration of Banach space X,and $P : Y \to Y$ is described as the continuous function. Here, P has a fixed point in terms of P.

Theorem 2 Arzela–Ascoli theorem

The sub norm metric [18, 24] can be taken as $C(Y, \mathbb{R})$,and the set $P \subset C(Y)$ is considered as compact with the condition of P is equicontinous, closed, and bounded.

Theorem 3 Banach fixed-point theorem

Every contraction mapping [19, 25] on a whole metric space consists of a unique fixed point.

3.2 *Evaluation of Banach Space in Differential Equation*

In this section, the Banach fixed-point theorem is applied to differential equations and evaluated. Here, $C([0, \theta] \times \mathbb{R}^3, \mathbb{R})$ is considered as a Banach space of the continuous function with the basis of $[0, \theta] \times \mathbb{R}^3$ into $\mathbb{R}$ with the function of norm $\|.\|_\infty$ defined by $||\xi||_\infty :- \sup\{|\xi(Y)| : Y \in [0, \theta]\}$.

Consider the nonlinear beginning value problem for a multi-order fractional differential equation, as seen in Eqs. (27) and (28):

$$D^{\alpha}\xi + H\left(Y, \xi(Y), D^{\alpha 1}\xi, D^{\alpha 2}\xi\right) = G(Y) \tag{27}$$

$$Y \in [0, \theta]\,, \mathrm{N} - 1 < \alpha, \alpha 1, \alpha 2 \leq N \tag{28}$$

With the initial Eq. (29),

$$\xi^{K}(0) = C^{K}, K = 0, 1, 2, \ldots, M-1 \tag{29}$$

Where, Y can be described as a variable, $D^{\alpha} = \frac{\partial^{\alpha}}{\partial Y^{\alpha}}$ can be described as the Caputo fractional derivative of the order α, the continuous function is denoted as $[0, \theta] \times \mathbb{R}^{3} \rightarrow \mathbb{R}$, and the source term is formulated as $G(Y)$.

Definition 8 The function of $\xi(Y) \in C([0, \theta] \times \mathbb{R}^{3}, \mathbb{R})$ derivatives, as it exists on $[0, \theta]$, is denoted here as α. The issue with Eqs. (27) and (29) has a solution like this. The variable o$\xi(Y)$ is regarded as the beginning conditions, since it solves the issue in Eqs. (27) and (29) linked to $[0, \theta]$. The Banach fixed-point theorems are the first results.

To solve the problems, the initial conditions are considered, which are presented as follows:

P1: The constant value is $\omega^{I} > 0, I = 0, 1, 2,$which is formulated as Eqs. (30) and (31):

$$\left|H\left(Y, \xi 1(Y), D^{\alpha 1}\xi 1(Y), D^{\alpha 2}\xi 1(Y) - H\left(Y, \xi 2(Y), D^{\alpha 1}\xi 2(Y), D^{\alpha 2}\xi 2(Y)\right|\right.\right. \tag{30}$$

$$\leq \omega^{0}\left|\xi 1 - \xi 2\right| + \sum_{J=1}^{2} \omega^{J}\left|D^{\alpha J}\xi 1(Y) - D^{\alpha J}\xi 2(Y)\right| \tag{31}$$

in its entirety $Y \in [0, \theta], \xi 1, \xi 2 \epsilon \mathbb{R}$

P2: The constant value is $\zeta^{\alpha I} > 0, I = 1, 2,$ which is formulated as Eqs. (32) and (33):

$$\left|D^{\alpha I}\xi 1(Y) - D^{\alpha I}\xi 2(Y)\right| \leq \zeta^{\alpha I}\left|\zeta^{I} - \zeta^{2}\right|, I = 1, 2, \tag{32}$$

For complete $Y \in [0, \theta], \xi 1, \xi 2 \epsilon \mathbb{R}$.

If

$$\left(\frac{\theta^{\alpha}}{\Gamma(\alpha+1)}\left(\omega 0 + \omega 1\zeta^{\alpha I} + \omega 2\zeta^{\alpha 2}\right)\right) < 1 \tag{33}$$

To solve the problem of Eqs. (27) and (28), it has a specific solution on $[0, \theta]$.

Proof The problem is changed into a fixed-point problem. Here, consider the operator $\phi : C([0, \theta] \times \mathbb{R}^{3}, \mathbb{R}) \rightarrow C([0, \theta] \times \mathbb{R}^{3}, \mathbb{R})$, which is defined as Eq. (34):

$$\phi\xi(Y) = \sum_{K=0}^{M-1} \frac{\partial^{K}\xi\left(0^{+}\right)}{\partial \xi_{K}} \frac{Y^{K}}{K!} + j^{\alpha} = \left(G(Y) - H\left(y, \xi(Y) D^{\alpha I}\xi, D^{\alpha 2}\xi\right),\right.$$

$$\xi = \xi(Y), N - 1 \leq \alpha, \alpha 1, \alpha 2 \leq N \tag{34}$$

In the Banach fixed-point theorem, the fixed point of the operator ϕ is the solution of the problem in Eqs. (26) and (27). In the Banach theorem, the operator ϕ is considered the fixed point. At last, the operator ϕ is defined as the contraction. Here, the assumptions are $\xi 1, \xi 2 \in, C([0, \theta] \times \mathbb{R}^3, \mathbb{R})$, after that $Y \in [0, \theta]$ is described as Eq. (35):

$$\begin{aligned}
&| \phi\ (\xi 1(Y)) \\
&-\phi\left(\xi 2(Y) \right| = \left| \left(\sum_{K=0}^{M-1} \frac{\partial^K \xi(0^+)}{\partial \xi_K} \frac{Y^K}{K!} + j^\alpha \left(G(Y) - H\left(Y, \xi 1(Y), D^{\alpha 1}\xi 1(Y), D^{\alpha 2}\xi 1(Y)\right)\right)\right.\right. \\
&\left.\left. - \left(\sum_{K=0}^{M-1} \frac{\partial^K \xi(0^+)}{\partial \xi_K} \frac{Y^K}{K!} + j^\alpha \left(G(Y) - H\left(Y, \xi 2(Y), D^{\alpha 1}\xi 2(Y), D^{\alpha 2}\xi 1(Y)\right)\right)\right)\right| \\
&\leq \left| j^\alpha H\left(Y, \xi 1(Y), D^{\alpha 1}\xi 1(Y), D^{\alpha 2}\xi 1(Y)\right.\right. \\
&\left. -\left(Y, \xi 2(Y), D^{\alpha 1}\xi 2(Y), D^{\alpha 2}\xi 1(Y)\right| , \leq j^\alpha \left(\omega^0 \left|\xi 1 - \xi 2\right| + \omega^1 \left| D^{\alpha 1}\xi 1 \right.\right.\right. \\
&\left. -D^{\alpha 1}\xi 2\right| + \omega^0 \left| D^{\alpha 1}\xi 1 - D^{\alpha 1}\xi 2 \right|
\end{aligned}$$

$$\begin{aligned}
&\leq j^\alpha \left(\omega^0 \left|\xi 1 - \xi 2\right| + \omega^1 \zeta^{\alpha I} \left|\xi 1 - \xi 2\right| + \omega^2 \zeta^{\alpha 2} \left|\xi 1 - \xi 2\right| \right. \\
&\leq \left(\frac{\theta^\alpha}{\Gamma(\alpha+1)} \left(\omega^0 + \omega^1 \xi 1 + \omega^2 \xi 2\right)\right) + \sup \left|\xi 1 - \xi 2\right|, \\
&\leq j^\alpha \left(\omega^0 \left|\xi 1 - \xi 2\right| + \omega^1 \zeta^{\alpha I} \left|\xi 1 - \xi 2\right| + \omega^2 \zeta^{\alpha 2} \left|\xi 1 - \xi 2\right| \right. \\
&\leq \left(\frac{\theta^\alpha}{\Gamma(\alpha+1)} \left(\omega^0 + \omega^1 \xi 1 + \omega^2 \xi 2\right)\right) + \left|\xi 1 - \xi 2\right|_\infty,
\end{aligned} \tag{35}$$

Hence Eq. (36):

$$\left\| \phi\ (\xi 1(Y)) - \phi\left(\xi 2(Y)\right\|_\infty \leq \gamma \alpha, \theta, \omega^0, \omega^1, \omega^2, \zeta^{\alpha I}, \zeta^{\alpha 2} \|\xi 1 - \xi 2\|_\infty, \tag{36}$$

Where, $\gamma \alpha, \theta, \omega^0, \omega^1, \omega^2, \zeta^{\alpha I}, \zeta^{\alpha 2} = \left(\frac{\theta^\alpha}{\Gamma(\alpha+1)} \left(\omega^0 + \omega^1 \xi 1 + \omega^2 \xi 2\right)\right) < 1$
Based on Eq. (36), the operator ϕ is defined as the contraction. Related with the Banach fixed-point theorem, we can deduce that operator ϕ has a specific fixed point that is problem-solving Eqs. (26) and (27). This is the complete proof of the theorem.

4 Description of Caputo Derivative Iterative Adomian Decomposition Method

The Caputo-derivative based Iterative Adomian Decomposition (CIADM) technique [26] is utilized to solve the problem in the fixed-point theorems. In the proposed methodology, the iterative method provides efficient results for solving fixed-point theorems. This same iterative method and the step-by-step iteration procedure are very similar. The operator form of Eqs. (27) and (28) is formulated as Eq. (37):

$$D^{\alpha}\xi + L\xi + N\xi = G(Y), N - 1 \leq \alpha \leq N \tag{37}$$

Where,

- $G(Y)$ can be described as the source term.
- N can be described as the nonlinear term of order equal or less than α.
- L can be regarded as a linear term with a value lower than α.

By utilizing J_Y^{α} presented applied in Eq. (37), we achieve, as has been derivated by Eqs. (38 and 39),

$$\xi = \sum_{K=0}^{M-1} \frac{\partial^K \xi\left(0^+\right)}{\partial \xi_K} \frac{Y^K}{K!} + j^{\alpha}\left(G(Y) - (L\xi + N\xi)\right) \tag{38}$$

$$\xi = \xi(Y), N - 1 \leq \alpha \leq N \tag{39}$$

Here, ξ can be described as an unknown parameter. This unidentified solution is shown in the definition of an infinite series, which is given as Eq. (40):

$$\xi(Y) = \sum_{N=0}^{\infty} \xi_N(Y) \tag{40}$$

The infinite series of Adomian polynomial function in nonlinear term is formulated as Eq. (41):

$$N\xi = \sum_{N=0}^{\infty} B_N \tag{41}$$

The Adomian polynomial function is formulated as Eq. (42):

$$B_N\left(\zeta^0, \zeta^1, \zeta^2, \ldots\right) = \frac{1}{N!}\frac{dN}{dXN}\left(N\left(\sum_{I=0}^{N} XI\zeta^1\right)\right)_{X=0} \tag{42}$$

Eqs. (40) and (41) are substituted in (38), which has been denoted by Eq. (43):

$$)\xi = \sum_{K=0}^{M-1} \frac{\partial^K \xi\left(0^+\right)}{\partial \xi_K} \frac{Y^K}{K!} + j^\alpha \left(G(Y) - \left(L\left(\sum_{N=0}^{\infty} \xi_N(Y) \right) + \left(\sum_{N=0}^{\infty} B_N \right) \right) \right) \tag{43}$$

The recursion formula for the solution of the above equation is presented as Eq. (44):

$$\xi = \sum_{K=0}^{M-1} \frac{\partial^K \xi\left(0^+\right)}{\partial \xi_K} \frac{Y^K}{K!} + j^\alpha \left(G(Y)\right) \tag{44}$$

And,

$$\xi_{N+1} = -J^\alpha \left(L\left(\xi_N(Y) + B_N \right) \right) \tag{45}$$

The approximate solution of the problem is presented as Eqs. (46 and 47):

$$\eta N = \sum_{N=0}^{N-1} \xi_M(Y) \tag{46}$$

And,

$$\lim_{M \to \infty} \eta N = \xi(Y) \tag{47}$$

ξ (Y) is the arrangement of the problem in fixed-point hypotheses, which is addressed using the proposed atomic technique. The main cycle of the proposed Adomian strategy is introduced as follows.

In this section, we introduce Eq. (27) to show the proposed strategy. All cycles are recursive connections [27], such as Eq. (42), and change occurs after this progression. To use the proposed strategy, consider the primary cycle time of the series arrangement, that is ξ_0 Eq. (41) from the condition. At present, we agree that the word ξ_0 can be divided into a number of categories, which has been derived by Eq. (48):

$$\xi_0 = P^0 + P^1 + P^2 + P^3 + \ldots \tag{48}$$

We are currently illustrating the way to obtain the specific arrangement of Eq. (42) by using the proposed technique. Start with the main section of ξ_0, that is ξ_0, and check if the section meets Eq. (27) and if the basic condition is met; ξ_0 is the specific provision of Eq. (27). In case the main part of Eq. (27) or the basic condition is not met, we go to the following section and continue with a similar contact. The equation for satisfaction Eq. (27), and any condition becomes the specific arrangement of Eq. (27) with the basic conditions. In the event that each part does not meet Condition (27), we proceed to the next stage, that is, using regular ADM operations.

5 Numerical Analysis of the Proposed Method

In the section about nonlinear initial value fractional-order differential equations, the proposed approach is shown in numbers.

Example 1 In this numerical analysis, the nonlinear fractional differential equation is shown as Eqs. (49 and 50):

$$D^3\xi + D^{\frac{5}{2}}\zeta + \xi^2 = Y^4 \tag{49}$$

With origin,

$$\zeta(0) = 0, \xi^{`}(0) = 0, \xi^{``}(0) = 2 \tag{50}$$

Following that, the precise solution can be determined. The operator representation of Eq. 40 is denoted by Eq. (53):

$$D^3\xi + L\zeta + N\zeta = G(Y) \tag{51}$$

Where, the source term is $G(Y) = Y^4$, the nonlinear term is $N\zeta = \xi^2$, and the linear term is denoted as $L\zeta = D^{\frac{5}{2}}\zeta$. Multiplying the inverse operator [28] $J^{\frac{5}{2}}$ of $D^{\frac{5}{2}}$ on both sides of the equation results in Eq. (52):

$$Y = \zeta(0) + Y\zeta^{`}(0) + \frac{Y^2}{2}Y\zeta^{``}(0) + J^3\left(Y^4 - (L\zeta + N\zeta)\right) \tag{52}$$

The nonlinear and linear terms are computed by an infinite series, which are presented as Eqs. (53 and 54):

$$\xi(Y) = \sum_{N=0}^{\infty} \xi_N(Y) \tag{53}$$

$$N\xi = \sum_{N=0}^{\infty} B_N \tag{54}$$

Where, B_N can be described as the Adomian polynomial, and the formula of computing the polynomials is presented as Eq. (55):

$$B_N = \frac{1}{N!}\frac{dN}{dXN}\left(N\left(\sum_{I=0}^{\infty} XI\zeta^1\right)\right)_{X=0} \tag{55}$$

The series formulation of Eq. (52) is presented as Eq. (56):

$$\zeta = \zeta(0) + Y\zeta^{'}(0) + \frac{Y^2}{2}Y\zeta^{''}(0) + J^3\left(Y^4 - \left(L\left(\sum_{N=0}^{\infty}\xi_N(Y)\right) + N\left(\sum_{N=0}^{\infty}B_N\right)\right)\right) \tag{56}$$

The recursive relation is achieved from Eqs. (57 and 58):

$$\xi_0 = Y^2 + J^3\left(Y^4\right) \tag{57}$$

And,

$$\xi_{N+1} = Y^2 + J^3\left((L\zeta + N\zeta)\right) \tag{58}$$

Now, this specified example is solved with the help of the proposed methodology. The proposed method produced the best results while resolving the problem of initial values of a multi-order fractional differential equation. Based on Eq. 57, ξ_0 can be divided into different components, which are presented as Eq. (59):

$$\xi_0 = P0 + P1 \tag{59}$$

From the above equation, $P0$, we can conclude, Eq. (60):

$$P0 = Y^2 \tag{60}$$

To validate $\xi = P0$ is a resolve to Eq. (39), substitute $\xi = P1$ in the left side of Eq. (39), and by using different definitions, we achieve Eq. (61):

$$D^3 Po + D^{5/2} Po + PO^2 = Y^4 \tag{61}$$

The first component $\xi_0 = Y^2$ from Eq. (60) satisfies Eq. (49), furthermore providing the initial condition (49). According to the recommended technique, the component Y^2 becomes the correct answer of Eq. (49) with the initial Eq. (50). On multiplying the inverse operator $D^{3/2}$ of $J^{5/2}$ two sides of Eq. (62):

$$\xi = \xi(o) + Y\xi^{'}(0) + J^{3/2}\left(\frac{4\sqrt{Y}}{\sqrt{\pi}} + \frac{8}{3}\frac{Y^{3/2}}{\sqrt{\pi}} + 2y + Y^2 + e^{Y^2} - (L\xi + N\xi)\right) \tag{62}$$

The linear and nonlinear terms are estimated by infinite series as Eqs. (53, 54, and 55):

$$(Y) = \sum_{N=0}^{\infty} \xi_N(Y) \tag{53}$$

$$N\xi = \sum_{N=0}^{\infty} B_N \tag{55}$$

Example 2 In this condition, the non-homogenous Bagly Torvik equation is considered with the linear fractional differential equation, which is presented by Eq. (56):

$$D^2\xi(y) + D^{\frac{3}{2}}\zeta(y) + \xi(y) = 1 + y,\ y \in [0, 1] \tag{56}$$

The initial condition is formulated as Eq. (57):

$$\xi(0) = \xi^{`}(0) = 1 \tag{57}$$

The recursion formula is achieved with the help of the proposed methodology presented in Eqs. (58, 59, and 60),

$$\xi(0) = 1 + y + \frac{1}{2}y^2 + \frac{1}{6}y^6 \tag{58}$$

$$p0 + p1 + p2 \tag{59}$$

And,

$$\xi_{N+1} = -J^2\,(l\xi_N) \tag{60}$$

Example 3

$$D^2\xi(y) - 2D\xi(y) + D^{\frac{1}{2}}\zeta(y) + \zeta(y) = y^3 - 6y^2 + 6y + \frac{16}{5\sqrt{\pi}}y^{\frac{5}{2}},\ y\epsilon\,[0, 1] \tag{61}$$

The initial condition is formulated as Eq. (61):

$$\xi(0) = \xi^{`}(0) = 0 \tag{62}$$

The recursion formula is achieved with the help of the proposed methodology in Eqs. (63, 64, and 65),

$$\xi(0) = \frac{1}{20}y^5 - \frac{1}{2}y^4 + y^3 + \frac{64}{310\sqrt{\pi}}y^{9/2} \tag{63}$$

$$p0 + p1 + p2 + p3 \tag{64}$$

And,

$$\xi_{N+1} = -J^2 (l\xi_N) \tag{65}$$

Based on the formulation, the solutions are achieved with the help of the proposed methodology.

6 Conclusion

In this study, "a fractional order nonlinear differential equation's solution was calculated using CIADM and the Caputo derivative." In this chapter, two main constraints have been considered, such as without changing the nonlinear fractional differential equation to the linear algebraic fractional equation and limited number of iterations. This method has been investigated taking into consideration some fixed-point theorems. Additionally, the uniqueness and existence solution has been evaluated by different examples. This proposed method has been evaluated with the consideration of conventional ADM and modified Adomian decomposition method.

References

1. Kishore, G.N.V.: Caristi type cyclic contraction and common fixed-point theorems in bipolar metric spaces with applications. J. Fixed Point Theory Appl. **1**, 1–13 (2018)
2. Ansari, A.H.: New approach for common fixed-point theorems via-class functions in-metric spaces. J. Funct. Spaces. **2017**, 1–9 (2017)
3. Ozturk, V.: Fixed point theorems for (F, ψ, φ)-contractions on ordered S-Complete Hausdorff uniform spaces. New Trends Math. Sci. **5**, 243–249 (2017)
4. Taş, N.: Common fixed points of continuous mappings on S-metric spaces. Math. Notes. **104**, 587–600 (2018)
5. Mutlu, A.: Coupled fixed point theorems on bipolar metric spaces. Eur. J. Pure Appl. Math. **10**, 655–667 (2017)
6. Nazam, D.: A common fixed-point theorem for a pair of generalized contraction mappings with applications. J. Comput. Anal. Appl. **25**, 552–564 (2018)
7. Yao, Y.: Mann-type iteration method for solving the split common fixed-point problem. J. Nonlinear Convex Anal. **18**, 875–882 (2017)
8. Miculescu, R.: Fixed point theorems for set-valued contractions in b-metric spaces. J. Fixed-Point Theory Appl. **19**, 2153–2163 (2017)

9. Chen, L.: Fixed point theorems of mean nonexpansive set-valued mappings in Banach spaces. J. Fixed-Point Theory Appl. **19**, 2129–2143 (2017)
10. Ahmed, A.: Modified Adomian decomposition method for solving fuzzy Volterra-Fredholm integral equation. J. Indian Math. Soc. **85**, 53–69 (2018)
11. Moradweysi, P.: Application of modified Adomian decomposition method to pull-in instability of nano-switches using nonlocal Timoshenko beam theory. Appl. Math. Model. **54**, 594–604 (2018)
12. Alderremy, A.A.: Modified Adomian decomposition method to solve generalized Emden–Fowler systems for singular IVP. Math. Probl. Eng. **12**, 1–6 (2019)
13. Kumar, M.: Existence and uniqueness results and analytical solution of the multi-dimensional Riesz space distributed-order advection–diffusion equation via two-step Adomian decomposition method. Eng. Comput. **38**, 2051–2066 (2020)
14. Baleanu, D.: On modelling of epidemic childhood diseases with the Caputo-Fabrizio derivative by using the Laplace Adomian decomposition method. Alex. Eng. J. **59**, 3029–3039 (2020)
15. Abushammala, M.: A novel fixed point iteration method for the solution of third order boundary value problems. Appl. Math. Comput. **271**, 131–141 (2015)
16. Khuri, S.A.: A new fixed point iteration method for nonlinear third-order BVPs. Int. J. Comput. Math. **98**, 2220–2232 (2021)
17. Garcia-Falset: Schaefer–Krasnoselskii fixed point theorems using a usual measure of weak noncompactness. J. Differ. Equ. **252**, 3436–3452 (2012)
18. Li, R.: An improvement of the Arzela–Ascoli theorem. Topol. Appl. **159**, 2058–2061 (2012)
19. Shukla, S.: A generalized Banach fixed point theorem. Bull. Malays. Math. Sci. Soc. **39**, 1529–1539 (2016)
20. Khan Marwat, D.N.: Solution of the heat equation with variable properties by two-step Adomian decomposition method. Math. Comput. Model. **48**, 83–90 (2008)
21. Akgül, A.: Approximate solutions to the conformable Rosenau-Hyman equation using the two-step Adomian decomposition method with Padé approximation. Math. Methods Appl. Sci. **43**, 7632–7639 (2020)
22. Verma, P.: Analytical solution with existence and uniqueness conditions of non-linear initial value multi-order fractional differential equations using Caputo derivative. Eng. Comput. **38**, 661–678 (2020)
23. Garcia-Falset, J., Latrach, K., Moreno-Galvez, E., Taoudi, M.-A.: Schaefer–Krasnoselskii fixed point theorems using a usual measure of weak noncompactness. J. Differ. Equ. **252**(5), 3436–3452 (2012)
24. Li, R., Zhong, S., Swartz, C.: An improvement of the Arzela–Ascoli theorem. Topol. Appl. **159**(8), 2058–2061 (2012)
25. Shukla, S., Balasubramanian, S., Pavlović, M.: A generalized Banach fixed point theorem. Bull. Malays. Math. Sci. Soc. **39**(4), 1529–1539 (2016)
26. Khan Marwat, D.N., Asghar, S.: Solution of the heat equation with variable properties by two-step Adomian decomposition method. Math. Comput. Model. **48**(1–2), 83–90 (2008)
27. Akgül, A., Aliyu, A.I., Inc, M., Yusuf, A., Baleanu, D.: Approximate solutions to the conformable Rosenau-Hyman equation using the two-step Adomian decomposition method with Pad é approximation. Math. Methods Appl. Sci. **43**(13), 7632–7639 (2020)
28. Verma, P., Kumar, M.: Analytical solution with existence and uniqueness conditions of non-linear initial value multi-order fractional differential equations using Caputo derivative. Eng. Comput. 1–18 (2020)

Control and Synchronization of a Modified Chaotic Finance System with Integer and Non-integer Orders

Khaled Moaddy and Talal Al Mutairi

1 Introduction

The latest financial crises have had an impact on the world economy. Therefore, there is a great need for financial mathematical modeling to examine the financial systems' internal dynamics in order to identify and control the current state of chaos. Since the majority of chaotic systems with integer and non-integer orders without exact solutions are extremely sensitive to time step sizes, it is necessary to employ a trustworthy numerical approach to test the accuracy over the long term. Modern economies are built on financial systems. A balanced economy and a developed society are produced when the financial system is kept secure and stable [1]. Nonlinear economics, sometimes known as chaos economics, is a discipline that dates back to the nineties of the last century [2], which uses nonlinearity chaotic theory to explain actual economic occurrences. It is well recognized that the appearance of irregular order behavior in a chaotic system's dynamics is one of its most significant characteristics.

Since its discovery, chaotic synchronization has been used in a wide range of disciplines, including ecological models, biological and physical systems, and structural engineering [3]. Financial crises have recently had an impact on the world economy [4]. As a result, there was a pressing need for financial mathematical modeling to investigate the underlying workings of the financial systems in order to identify and manage the current state of economic chaos [5]. The Routh-Hurwitz criterion served as the foundation for the creation of linear feedback controllers, which are used to direct the financial system's chaotic trajectory toward one of its

K. Moaddy (✉) · T. Al Mutairi
Department of Mathematics, College of Science and Humanities, Shaqra University, Shaqra, Saudi Arabia

F. Kamalov et al. (eds.), *Advances in Mathematical Modeling and Scientific Computing*, Trends in Mathematics, https://doi.org/10.1007/978-3-031-41420-6_5

unstable equilibrium points [6]. The advantage of a linear feedback controller is that it stabilizes the modified system's state variables.

The major objective of this study is to develop a class of modified finance chaotic systems with integer and non-integer orders. Different integer and non-integer orders are investigated in a variety of cases. For each parameter of the system responses with various order ranges, i.e., stable, periodic, and chaos, are illustrated. The second goal is to describe the stability analysis and the condition of oscillation. The last goal is to create a feedback control law to stabilize the state variables of the changed system. In order to study the synchronization of an identical changed system using adaptive control, the Adams-Bashforth-Moulton predictor corrector technique is proposed for the numerical solutions of the master, slave, and error systems. To solve these issues, numerical methods and programs are created.

2 The Modified System

A modified finance chaotic system is proposed as follows:

$$\begin{aligned} \frac{dx_1(t)}{dt} &= x_3 + (x_2 - a_1)x_1 + d_1x_1, \\ \frac{dx_2(t)}{dt} &= 2 - b_1x_2 - \mid x_1 \mid, \\ \frac{dx_3(t)}{dt} &= x_1x_2 - x_1 - c_1x_3, \end{aligned} \tag{1}$$

where x_1 denotes the interest rate, x_2 the demand for investments, and x_3 the price index and $\beta_1 = \beta_2 = \beta_3 = 1$. The parameters (a_1, b_1, c_1) with positive values indicate the amount saved, the cost-per-investment ratio, and the elasticity of demand in the commercial market, respectively, and the value of parameter d_1 can be varied within a specific range [7]. The modified system (1) has two modifications from system [8], and the term ${x_1}^2$ in the second equation in system [8] has been substituted with $|x|$ in the second equation in system (1) to maintain a low and positive interest rate [7]. Adding the d_1x_1 item to the first equation in system (1) means that the changes of x_1 are proportional to the rate of average profit margin [7]. These modifications have a simple algebraic structure and can display complex dynamics: chaotic and quasiperiodic phenomena occur alternately.

2.1 Non-integer Order Case

The mathematical model of the modified finance system with fractional order offered in this work can be expressed by the state equations listed below:

$$
\begin{aligned}
D^{\beta_1}x_1 &= x_3 + (x_2 - a_1)x_1 + d_1x_1,\\
D^{\beta_2}x_2 &= 2 - b_1x_2 - |x_1|,\\
D^{\beta_3}x_3 &= x_1x_2 - x_1 - c_1x_3,
\end{aligned} \tag{2}
$$

where β_i, $(i = 1, 2, 3)$ are parameters describing the non-integer order of the non-integer time derivative. In this case, nonstandard finite difference scheme is applied to give numerical solutions for system (1).

3 Stability Analysis

When $(a_1 = 0.8, b_1, = 0.2, c_1 = 1.0)$ and $d_1 = 1.5$, the modified system (1) has three unstable steady states:

$$
\begin{aligned}
E_1 &= (0, 10, 0),\\
E_2 &= (1.97, 0.15, -1.6745),\\
E_3 &= (-1.97, 0.15, 1.6745).
\end{aligned}
$$

The Lyapunov exponent is an important indicator for identifying chaotic attractor, and a mathematical description can help clarify how sensitive beginning values are. When the parameters of system (1) are selected as $a_1 = 0.8$, $b_1 = 0.2$, $c_1 = 1.0$, and $d_1 = 1.5$, with initial conditions $x_1(0) = 1.0$, $x_2(0) = 0.1$, and $x_3(0) = 2.0$. Using the Wolf algorithm [9], we calculate Lyapunov exponents as follows: $L_1 = 0.566142$, $L_2 \approx 0$, and $L_3 = -1.35197$ to define the Kaplan-Yorke dimension [10] of the modified finance chaotic system (1) which illustrates the complexity of the attractor may be numerically produced as

$$
D_{ky} = 2 + \frac{L_1 + L_2}{|L_3|} \approx 2.41875, \tag{3}
$$

while the modified finance chaotic system's maximum Lyapunov exponent (1) is $L_1 = 0.566142$ and $\sum L_i, i : 1, 2, 3 < 0$ which shows that the new system is chaotic. When $\beta = \beta_i$, $(i = 1, 2, 3)$, the phase portraits are shown in Fig. 1.

Figure 1 shows that Lyapunov exponents' values agree with bifurcation diagrams; when $\beta < 0.835$ the modified system (2) displays stable behavior. Near $0.985 < \beta < 0.989$, the system (2) displays periodic behavior. Also, when $0.835 < \beta < 0.989$, the system (2) displays chaotic behavior.

Choosing $b_1 = 0.2$, $c_1 = 1.0$ and $d_1 = 1.5$, Figs. 2a and 3 show the system bifurcation and phase diagrams, respectively, when a_1 takes different values $\in [0.1, 4]$. The effect of different factors $(a_1 = 0.8, c_1 = 1.0, d_1 = 1.5)$ on investment costs and bifurcation diagram of variable x_3 versus b_1 is shown in Fig. 2b. Figure 2c

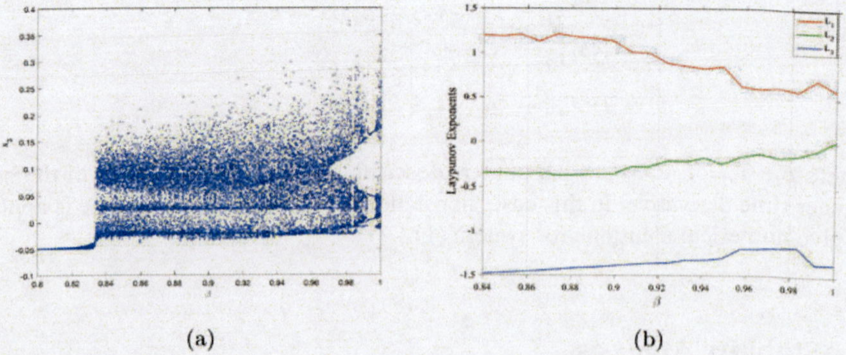

Fig. 1 Bifurcation diagram of $\beta \in [0.8, 1]$. (**a**) Lyapunov exponents with respect to fractional order β (**b**) where $a_1 = 0.8, b_1 = 0.2, c_1 = 1.0, d_1 = 1.5$ and initial values $x_1(0) = 1.0, x_2(0) = 0.1, x_3(0) = 2.0$

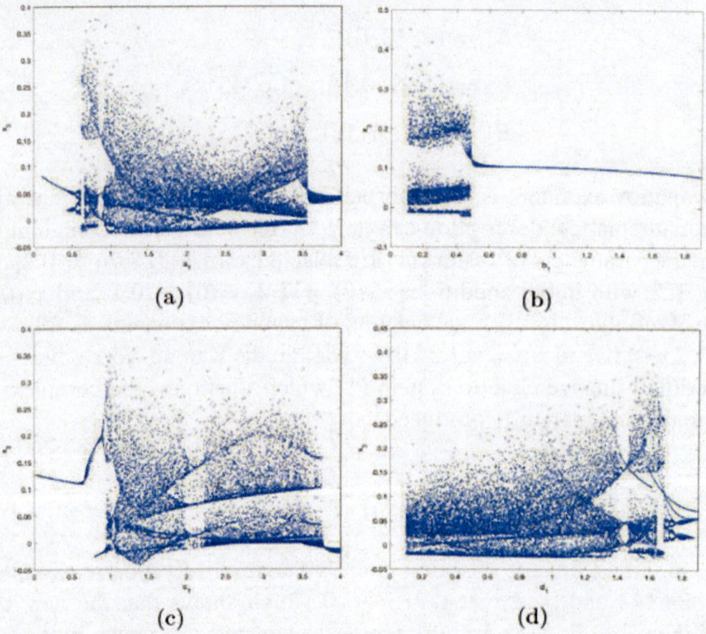

Fig. 2 Bifurcation diagram where $\beta_1 = \beta_2 = \beta_3 = 1$ with initial values $x_1(0) = 1.0, x_2(0) = 0.1, x_3(0) = 2.0$. (**a**) $d_1 = 1.5, b_1 = 0.2, c_1 = 1.0$, and $a_1 \in [0.1, 4]$. (**b**) $a_1 = 0.8, c_1 = 1.0, d_1 = 1.5$, and $b_1 \in [0.1, 1.9]$. (**c**) $a_1 = 0.8, b_1 = 0.2, d_1 = 1.5$, and $c_1 \in [0, 4]$. (**d**) $a_1 = 0.8, b_1 = 0.2, c_1 = 1.0$, and $d_1 \in [0.1, 1.9]$

and d show the system phase diagrams when b_1, d_1 take different values $c_1 \in [0, 4]$ and $d_1 \in [0.1, 1.9]$, respectively.

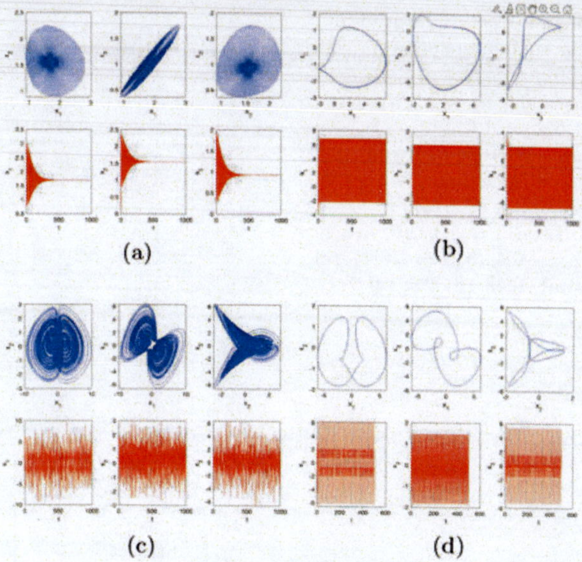

Fig. 3 Phase diagrams and time histories of system where $b_1 = 0.2, d_1 = 1.5, c_1 = 1.0, \beta_1 = \beta_2 = \beta_3 = 1$ with initial values $x_1(0) = 1.0, x_2(0) = 0.1, x_3(0) = 2.0$. **(a)** $a_1 = 3.6$. **(b)** $a_1 = 0.4$. **(c)** $a_1 = 1.4$. **(d)** $a_1 = 0.85$

4 Control of the Modified Chaotic Finance System

In this section, chaotic system (1) trajectories are controlled by linear feedback controllers to one of the system's unstable equilibrium points, E_2. The controlled system is provided by

$$\begin{aligned}
\frac{dx_1(t)}{dt} &= x_3 + (x_2 - a_1)x_1 + d_1x_1 + \beta_1(x_1 - x_1{}^*),\\
\frac{dx_2(t)}{dt} &= 2 - b_1x_2 - |x_1| + \beta_2(x_2 - x_2{}^*),\\
\frac{dx_3(t)}{dt} &= x_1x_2 - x_1 - c_1x_3 + \beta_3(x_3 - x_3{}^*),
\end{aligned} \tag{4}$$

where $\beta_i, (i = 1, 2, 3)$ must be defined in order to describe the feedback gains and $x_1{}^*, x_2{}^*, x_3{}^*$ are any equilibrium point of system (1). The Jacobian matrix (5) of the system (4) evaluated at the equilibrium point $E^* = (1.97, 0.15, -1.6745)$ is given by

$$J(E^*) = \begin{pmatrix} \beta_1 + 0.85 & 1.97 & 1 \\ \pm 1 & \beta_2 - 0.2 & 0 \\ -0.85 & 1.97 & \beta_3 - 1 \end{pmatrix}. \tag{5}$$

To simplify our task, let $\beta_1 = \beta_2 = \beta_3 = \beta$ and then, the characteristic equation of the Jacobian matrix (5) is

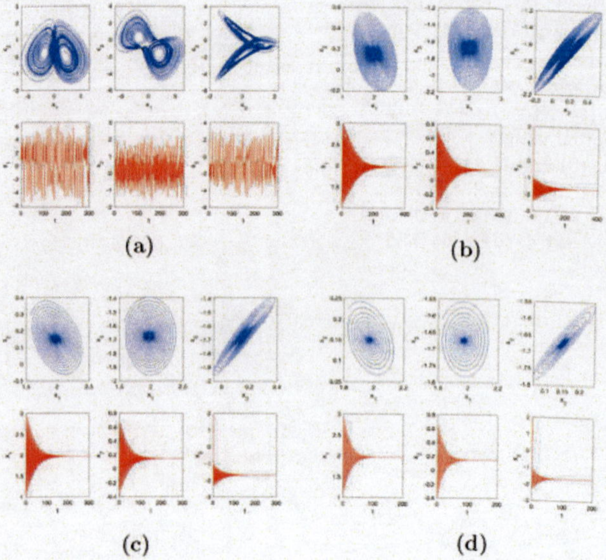

Fig. 4 Controlling chaos to the equilibrium point E_2 with different values of k when $a_1 = 0.8, b_1 = 0.2, d_1 = 1.5, c_1 = 1.0$, $\beta_1 = \beta_2 = \beta_3 = 1$ with initial values $x_1(0) = 1.0, x_2(0) = 0.1, x_3(0) = 2.0$. **(a)** $k = -0.4$. **(b)** $k = -0.47$. **(c)** $k = -0.48$. **(d)** $k = -0.5$

$$\lambda^3 + \eta_1\lambda^2 + \eta_2\lambda + \eta_3 = 0, \tag{6}$$

where

$$\begin{cases} \eta_1 = -3\beta + 0.35, \\ \eta_2 = 3\beta^2 - 0.7\beta + 2, \\ \eta_3 = -2\beta + 0.35\beta^2 - \beta^3 + 3.94. \end{cases} \tag{7}$$

The system (4) is asymptotically stable at the equilibrium point E^* according to the Routh-Hurwitz criterion [6], which we established.

1. $\eta_1 > 0$,
2. $\eta_1\eta_2 > \eta_3, \eta_3 > 0$,

After some calculations, we determine that the control gain must meet the following inequality (4) in order to satisfy those two conditions. Thus, when β is chosen such as the condition $\beta < -0.4529$ is satisfied, the characteristic equation (6) has three eigenvalues with negative real parts [5, 11]. For simulations, we choose $\beta = -0.46, -0.47, -0.48$ and initial conditions as $(1.0, 0.1, 2.0)$ followed by the stabilization and convergence of the controlled system toward the equilibrium point E^*. Figure 4 shows the stabilization of chaos with different values of k. As a result, the chaos can be easily controlled to reach the equilibrium point E_1 with this choice. The control inputs are switched on at $t = 200$ in every simulation.

5 Adaptive Synchronization of the Identical Modified System

Assume that we have two chaotic systems: one is the master system and the other is the slave system and that we need to use adaptive control functions to modify the slave system's reaction so that it synchronizes with the master chaotic system. We will define the drive master Eqs. (1) and response slave systems as follows:

$$\begin{aligned}\frac{dy_1(t)}{dt} &= y_3 + (y_2 - a_1)y_1 + d_1y_1 + u_1,\\ \frac{dy_2(t)}{dt} &= 2 - b_1y_2 - |y_1| + u_2,\\ \frac{dy_3(t)}{dt} &= y_1y_2 - y_1 - c_1y_3 + u_3.\end{aligned} \tag{8}$$

We specify the error functions among the systems' state variables (1) and (8) as follows:

$$\begin{aligned}e_1 &= y_1 - x_1,\\ e_2 &= y_2 - x_2,\\ e_3 &= y_3 - x_3,\end{aligned} \tag{9}$$

System (9) together with system (1) and (8) yields the error system:

$$\begin{aligned}\frac{de_1(t)}{dt} &= e_3 + y_2y_1 - x_2x_1 - a_1e_1 + d_1e_1 + u_1,\\ \frac{de_2(t)}{dt} &= b_1e_2 - |y_1| + |x_1| + u_2,\\ \frac{de_3(t)}{dt} &= y_1y_2 - x_1x_2 - e_1 - c_1e_3 + u_3.\end{aligned} \tag{10}$$

Define the adaptive control functions u_1, u_2, and u_3 as follows:

$$\begin{aligned}u_1 &= -e_3 + a_1{}^*e_1 - d_1{}^*e_1 - y_2y_1 + x_2x_1 - \delta_1e_1,\\ u_2 &= b_1{}^*e_2 + |y_1| - |x_1| - \delta_2e_2,\\ u_3 &= e_1 + c_1{}^*e_3 - y_1y_2 + x_1x_2 - \delta_3e_3,\end{aligned} \tag{11}$$

in which $a_1{}^*, b_1{}^*, c_1{}^*$, and $d_1{}^*$ are the estimates of the parameters a_1, b_1, c_1, and d_1, respectively, and $\delta_i, (i = 1, 2, 3)$ are positive constants:

$$\frac{da_1{}^*}{dt} = -e_1{}^2 - \delta_4e_a,$$

$$\frac{db_1{}^*}{dt} = -e_2{}^2 - \delta_5 e_b,$$
$$\frac{dc_1{}^*}{dt} = -e_3{}^2 - \delta_6 e_c, \tag{12}$$
$$\frac{dd_1{}^*}{dt} = e_1{}^2 - \delta_7 e_d.$$

The following are the definitions of the estimation error of unknown parameters:

$$\begin{aligned} e_a &= a_1 - a_1{}^*, \\ e_b &= b_1 - b_1{}^*, \\ e_c &= c_1 - c_1{}^*, \\ e_d &= d_1 - d_1{}^*. \end{aligned} \tag{13}$$

System (13) together with Eq. (12) yields the error system:

$$\begin{aligned} \frac{de_1(t)}{dt} &= (e_d - e_a)e_1 - \delta_1 e_1, \\ \frac{de_2(t)}{dt} &= -e_b e_2 - \delta_2 e_2, \\ \frac{de_3(t)}{dt} &= -e_c e_3 - \delta_3 e_3. \end{aligned} \tag{14}$$

Consider the Lyapunov function defined as follows:

$$V = \frac{1}{2}\left(e_1{}^2 + e_2{}^2 + e_3{}^2 + e_a{}^2 + e_b{}^2 + e_c{}^2 + e_d{}^2\right). \tag{15}$$

Using (14) and (13), the derivative of equation (15) gives

$$\frac{dV}{dt} = -\delta_1 e_1{}^2 - \delta_2 e_2{}^2 - \delta_3 e_3{}^2 - \delta_4 e_a{}^2 - \delta_5 e_b{}^2 - \delta_6 e_c{}^2 - \delta_7 e_d{}^2 < 0. \tag{16}$$

As a result, for all initial conditions, the error functions converge globally and asymptotically toward zero [5]. So the master system and the slave system can reach synchronization.

The initial conditions for the master system are $(x_1(0), x_2(0), x_3(0)) = (1.0, 0.1, 2.0)$ and for the slave system, $(y_1(0), y_2(0), y_3(0)) = (3.0, 1.0, 0.4)$, and the parameters of the modified system are chosen as follows: $a_1 = 0.8, b_1 = 0.2, c_1 = 1.0, d_1 = 1.5$ and we choose $\delta_i = 2, i : 1, \cdots, 7$ for the adaptive and update laws. The initial parameter estimation values of the system are $a_1{}^* =$

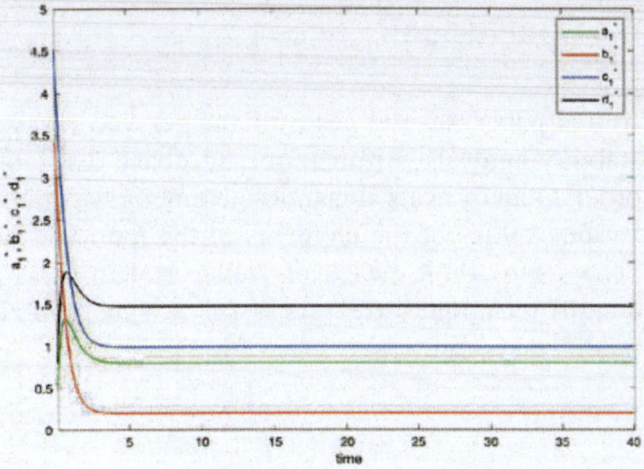

Fig. 5 Parameter estimates for adaptive control when $\beta = 1$

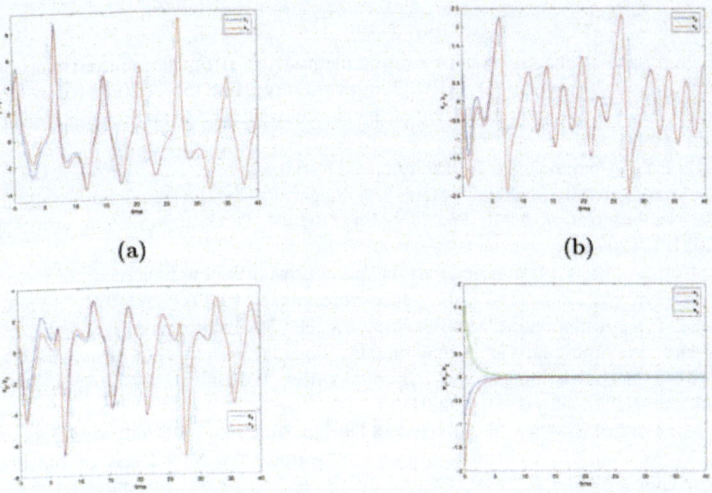

(a) (b)

(c) (d)

Fig. 6 Time evolutions of the master and slave system state variables (x_1, y_1), (x_2, y_2), (x_3, y_3), respectively, when $a_1 = 0.8, b_1 = 0.2, d_1 = 1.5, c_1 = 1.0$ in fractional order case $\beta_1 = \beta_2 = \beta_3 = 0.95$

$0.1, b_1{}^* = 4.0, c_1{}^* = 5.0$, and $d_1{}^* = 0.3$. Figure 5 shows the estimated parameters $a_1{}^*, b_1{}^*, c_1{}^*$, and $d_1{}^*$ converge toward the real values of the parameters of the system (8). By applying the adaptive control law (11) and the parameter update law (12) to the modified controlled financial system (1), Fig. 6 displays the time evolutions of two identical systems that are synchronized with the adaptive controller with fractional and variable orders.

6 Conclusion

In this paper, several cases of integer and non-integer orders are introduced with stability analysis. Bifurcation and phase diagrams of the system are employed in order to analyze its dynamics. Numerical simulations observed that the effect of various values of the elements in the modified finance system change the chaotic behavior to limit cycles or stable system. The proposed feedback and adaptive control techniques are successful at synchronizing and controlling chaos in the modified system.

References

1. Lu, X.: A financial chaotic system control method based on intermittent controller. Math. Problems Eng. **2020**, Article ID 5810707. https://doi.org/10.1155/2020/5810707
2. Huang, D., Li, H.: Theory and Method of the Nonlinear Economics. Sichuan University Press, Chengdu (1993)
3. Dousseh, P.Y., Ainamon, C., Miwadinou, C.H., Monwanou, A.V., Orou, J.B.C.: Adaptive control of a new chaotic financial system with integer order and fractional order and its identical adaptive synchronization. Math. Problems Eng. Article ID 5512094 (2021). https://doi.org/10.1155/2021/5512094
4. Yu, H., Cai, G. , Li, Y.: Dynamic analysis and control of a new hyperchaotic finance system. Nonlinear Dyn. **67**, 2171–2182 (2012). https://doi.org/10.1007/s11071-011-0137-9
5. Dousseha, P.Y., Ainamona, C., Miwadinoua, C.H., Monwanoua, A.V., Chabi, J.B.: Chaos control and synchronization of a new chaotic financial system with https://doi.org/10.1142/S021812742150187Xh integer and fractional order. J. Nonlinear Sci. Appl. **14**(6) (2021). https://doi.org/10.22436/jnsa.014.06.01
6. Gopal, M.: Control Systems: Principles and Design. McGraw-Hill Education, India (2002)
7. Zheng, W., Zhang, G.: Control and Synchronization of a New Class of Finance Chaotic Systems. CONFERENCE 2019, CCDC. IEEE, Nanchang (2019). https://doi.org/10.1109/CCDC.2019.8833169
8. Liping, C., Khan, M.A., Atangana, A., Kumar, S.: A new financial chaotic model in Atangana-Baleanu stochastic fractional differential equations. Alexandria Eng. J. **60**(6), 5193–5204 (2022). https://doi.org/10.1016/j.aej.2021.04.023
9. Danca, M.: Matlab code for Lyapunov exponents of fractional-order systems, Part II: The non-commensurate case. Int. J. Bifurcation Chaos **31**(12), 2150187 (2021). https://doi.org/10.1142/S021812742150187X
10. Kaplan, J.L., Yorke, J.A.: Preturbulence: a regime observed in a fluid flow model of Lorenz. Commun. Math. Phys. **67**, 93–108 (1979)
11. Xin, B., Zhang, J.: Finite-time stabilizing a fractional-order chaotic financial system with market confidence. Nonlinear Dyn. **79**, 1399–1409 (2015). https://doi.org/10.1007/s11071-014-1749-7

Dutch Book Methods for Difference and Differential Equations

Alberto Gandolfi and Jianhan Hu

MSC Classification: 34B99, 39A05, 12D99, 60A99

1 Introduction

The term Dutch Book possibly originated in the English language due to seventeenth-century rivalry with the Dutch to mean some mechanism by which an agent would gain no matter what the outcome of a certain set of events is [10]. Nowadays, it is used on gambling, to indicate a set of odds set up by a bookie which would guarantee a sure win to the bookie and a sure loss to the collective set of bettors; when the outcome is random, as in most cases in gambling, the sure win is to be interpreted in the sense of zero loss in all possible cases, and some positive probability of a strictly positive gain. The point that elicits some theoretical consideration is that in this context, it is irrational for the bettors to bet, as the bookie's odds implied probabilities are incoherent [16].

The fact that the exploitation of an incoherent assessment can be turned into the sure gain of a Dutch Book has then been used as a basis of mathematical methods, named Dutch Book Arguments, which help explore consistency of assumptions; the arguments proceed by purporting an agent, the player, who believes in the consistency of some inconsistent set of assumptions, and another agent, the bookie, who can trap the player into a game which the player sees as profitable, and instead it turns out to provide a sure loss to the player and a profit to the bookie. In this way, the inconsistency of the player's assumptions is witnessed by the Dutch Book [6]. Sometimes, a second, equivalent definition of Dutch Book is used, in which it is the agent with inconsistent beliefs that is forced to set up a book of odds, after which (s)he is forced to accept any bets; we adopt the first interpretation since the second one seems to contain some impositions on agent's behaviors.

A. Gandolfi (✉) · J. Hu
New York University Abu Dhabi, Abu Dhabi, UAE
e-mail: ag189@nyu.edu

F. Kamalov et al. (eds.), *Advances in Mathematical Modeling and Scientific Computing*, Trends in Mathematics, https://doi.org/10.1007/978-3-031-41420-6_6

Dutch Book Arguments have been mainly exploited in foundations of Probability Theory: De Finetti [14] and Ramsey [13] have shown how an agent whose beliefs violate the basic axioms of probability can be subjected to a Dutch Book. Other works have explored the role of Dutch Books for other probabilistic problems [4, 6, 7, 11]

In economics, an agent, such as a consumer, can be Dutch-booked if they have incoherent preferences, such as being intransitive [8].

In philosophy, Dutch Book Arguments are used to explore rational behavior and degrees of certainty of certain common beliefs for agents taking independent decisions [2].

In this paper, we develop the concepts of Dutch Book and Dutch Book Arguments for difference and differential equations. A Dutch Book Argument in these contexts is meant to identify the absence of solutions of difference or differential equations with specific boundary conditions, and consists of the determination of a Dutch Book, as described before, against a purported incorrect believer that a difference equation has a solution $x_i, i = 1, \ldots$ or that a differential equation with mixed boundary conditions has a solution $x(t)$. The formal definition, given in Sect. 2, is based on a vector $\mathbf{y}(x)$ or a function $y(t, x)$ which witnesses the absence of a solution x and allows to determine the Dutch Book setup.

Various mathematical tools can be used to form Dutch Books in our context.

For linear difference equations, the theorem of the alternative in linear algebra, also called Fredholm alternative, can be used to form Dutch Books. For difference equations containing only polynomial expressions in the unknowns, the Positivstellensatz in algebraic geometry can be used.

For linear differential equations, one can use a function $y(t)$ which is the solution of the adjoint equation as determined in Fredholm alternative; see Sect. 4. To show that the Dutch Book Argument is more general than Fredholm alternative, even for linear differential equations, we present several examples in Sect. 4.2, in which one can take $y = y(t, x)$ depending on the unknown solution x as well. Although this last choice of y might sound counterintuitive, in the sense that the function x does not exist when y exists, one can concoct a strategy to get rid of the unknown parts in the needed integration (see below).

The importance of defining Dutch Books for differential equations can be first realized by their being a generalization of the Fredholm alternative theorem, whose relevance in linear algebra and in the analysis of linear differential equations and, more generally, linear operators are well established [12]: in fact, Fredholm alternative is used to prove existence of solutions, variational principles, and also numerical approximations; applications to economics, for instance, can be found in [18].

Secondly, absence of a Dutch Book can then be used to establish existence results for solutions of suitable equations, which, even independently from uniqueness theorems, are of extreme importance and interest. As an example, the existence of solutions for Navier-Stokes equations is a Millennium Prize problem [5], and it involves existence problems for nonlinear differential equations. In fact, the results for difference equations and those more limited for differential equations presented in this paper suggest a potentially novel strategy to address the existence of solutions of some differential equations: if the differential equations have only polynomial terms in x and its derivatives (as it happens in Navier-Stokes equations, for instance), one could hope to follow a path analogous to the one we use here for difference equations. For these, in fact, we go from the linear algebra Fredholm alternative to Positivstellensatz for polynomial equations. More explicitly, one could try to express the existence of solutions of differential equations in terms of absence of a Dutch Book based on some $y = y(t, x)$; the function $y = y(t, x)$ could possibly be found in analogy with the what insured by the Positivstellensatz for difference equations.

For differential equations, there exists, in fact, a similar development in the theory of differential closed fields [9], which in fact applies to differential equations which are polynomial in x and its derivatives; unfortunately, the theory of differential closed fields does not include boundary conditions.

In the direction of potential extensions of existence methods for polynomial differential equations, notice that in that case, functions $y = y(t)$ dependent on t only cannot be sufficient: this highlights the relevance of our extension of Fredholm alternative in Sect. 4.2 in which we use functions of the form $y = y(t, x)$. Our example constitutes then the very first step toward the potential extension.

Although many more general cases could be treated, such as systems, complex valued functions, PDEs, etc., we restrict to single difference and differential equations in the real case for simplicity of exposition.

2 Dutch Books Differential Equations: Definition

Given a differential equation with mixed boundary conditions

$$Dx = f, \quad Ux = \gamma, \quad t \in [a, b] \tag{1}$$

where D is any differential operator, f is a continuous function, U is a boundary operator depending only on x and on its derivatives computed at a and b, $\gamma \in \mathbb{R}^\nu$ for a suitable ν, and $a, b \in \mathbb{R}$, a Dutch Book can be constructed under the following conditions.

- One agent, hereafter called **player**, believes in the existence of a solution of (1).
- A second agent, hereafter called **bookie**, is able to determine a function $y = y(t, x, \gamma)$ such that

  (I) $\alpha := \int_a^b (Dx - f) y dt \in \mathbb{R}$ can be computed regardless of x, for all the x satisfying $Ux = \gamma$,

  (II) $\alpha \neq 0$.

  The Dutch Book is then constructed by the bookie as follows:

(i) Each player receives the amount $|\alpha|/2 > 0$ while entering the game and can play as many simultaneous games as desired.

(ii) The player will then pay the amount $\beta := \min_{x:Ux=\gamma} | \int_a^b (Dx - f)ydt|$ at the end of the game (no need for the player to come up with an explicit x; an abstract calculation of β is sufficient).

Any agent who incorrectly believes in the existence of a solution to (1) finds the game strictly advantageous since $|\alpha|/2 > 0$ and he/she thinks that $\beta = 0$ when the solution x of (1) is used; such agent is then lead by rational behavior to be a player in as many, even simultaneous, games as possible. On the other hand, $\beta = |\alpha| > |\alpha|/2$ and hence each player incurs in a secure strictly positive loss in each game.

Example 1 Let $a = 0, b = 1, Dx = x' - x, f = 0, Ux = \gamma$ s.t. $x(0) = 1, x(1) = 2$. The bookie can then take $y = y(t) = ce^{-t}$ since the integral

$$\int_a^b (Dx - f)ydt = \int_0^1 (x' - x)ce^{-t}dt = c[xe^{-t}]_0^1 = c(2/e - 1)$$

can be computed regardless of x, and it is nonzero for some choices of c. Setting $c = (2/e - 1)^{-1}$, for instance, the player would receive $1/2$ upon entry, and pay 1, with a net loss of $1/2$ in each game.

The whole construction of the Dutch Book hinges then upon finding a suitable y; in the next sections, we see some methods for finding y in the linear case.

3 Dutch Book Arguments for Linear and Polynomial Difference Equations

If (1) is replaced by a system of difference equations

$$Fx = b \qquad Ux = \gamma, \tag{2}$$

where $x \in \mathbb{R}^n$, $b \in \mathbb{R}^m$, F is any map from $\mathbb{R}^n$ to $\mathbb{R}^m$, and U is a boundary operator assigning ℓ conditions depending only on x_1 and x_n, a Dutch Book can be constructed under the conditions analogous to the one of Sect. 2, in which the player believes in the existence of a solution of (2), and the bookie is able to determine a vector $y = y(x, b) \in \mathbb{R}^m$ such that the $\alpha := \sum_{j=1}^m ((Fx)_j - b_j)y_j + \sum_{j=m+1}^{\ell} ((Ux)_j - \gamma_j)y_j$ does not depend on x, and $\alpha \neq 0$. The Dutch Book is then constructed by the bookie as before with $\beta := \min_{x \in \mathbb{R}^n} |\alpha|$.

For a linear difference equation of the form (2), let

$$A = \begin{bmatrix} F & 0 \\ 0 & U \end{bmatrix}, \qquad \tilde{\gamma} = \begin{bmatrix} 0 \\ \gamma \end{bmatrix}$$

where A is the matrix of coefficients of the linear system augmented by the boundary conditions, so that there is a solution to (2) if and only if there exists a vector $\tilde{x}$ such that $A\tilde{x} = \tilde{\gamma}$. A Dutch Book can then be constructed from applying Fredholm alternative to A: there is a solution of $A\tilde{x} = \tilde{\gamma}$ if and only if all solutions $\tilde{y}$ of the adjoint problem $\tilde{y}A = 0$ are such that $\tilde{y} \cdot \tilde{\gamma} = 0$.

More generally, if the maps F and U in (2) are polynomial in the variables x_i's, then one can exploit Stengle's Positivstellensatz [1, 17] to construct a Dutch Book. In this case, (2) can be written as $G_j(x) = 0$, $j = 1, \ldots, n$ for a suitable n, where G_j's are polynomials in the variables x_i's. By the Positivstellensatz, such system has no solution if and only if there exists a polynomial P in the ideal generated by the G_j's in $\mathbb{R}[\mathbf{x}]$ such that P is independent of the x_i's and $P = -1$. More explicitly, there are polynomials $t_j \in \mathbb{R}[\mathbf{x}]$, $j = 1, \ldots, n$ such that $\sum_{j=1}^{n} t_j G_j(x) \equiv -1$. [1]. Let then $y_j := t_j$, which are functions of x, b and γ; notice that, nonetheless, $\alpha = \sum_{j=1}^{n} y_j G_j(x)$ does not depend on the x_i's, and that $\alpha \neq 0$. Having satisfied (I) and (II), we can construct the Dutch Book as described in Sect. 3. This proves

Lemma 1 *If F and U are polynomials, then there is a solution x to (2) if and only if there is no Dutch Book against the believer of the existence of such solution.*

In the nonlinear case, the determination of the polynomials G_j's required in the Dutch Book is, however, not straightforward as in the linear case [15]. Here is a simple example of polynomial difference equations mimicking $x' = x^2$.

Example 2 Let

$$\begin{cases} p_1 := x_1 - x_0 - x_0^2 = 0 \\ p_2 := x_2 - x_1 - x_1^2 = 0 \\ p_3 := x_0 - a = 0 \quad p_4 := x_2 - b = 0 \end{cases} \tag{3}$$

which does not have a solution if $b \neq a + 2a^2 + 2a^3 + a^4$. This is witnessed by

$$\alpha = p_2 + p_1 + p_3 + p_3^2 + 2p_3 + (p_1 + p_3 + p_3^2 + 2p_3)^2$$
$$+2(a + a^2)(p_1 + p_3 + p_3^2 + 2ap_3) - p_4.$$

Notice that $\alpha = b - (a + 2a^2 + 2a^3 + a^4)$, although it would be zero based on (3), so that it can be used to create a Dutch Book when no solution x exists.

4 Dutch Book Arguments for Linear OdEs

4.1 *Dutch Book from Fredholm Alternative*

For linear OdEs, one can exploit Fredholm alternative to construct Dutch Books. Expressing (1) as

$$Lx = p_0 x^n + p_1 x^{n-1} + p_2 x^{n-2} + \ldots p_n x = 0,$$

$$U_i x = \sum_{j=1}^{n} M_{ij} x^{j-1}(a) + N_{ij} x^{j-1}(b) = \gamma_i \quad i = 1, \ldots, m, \tag{4}$$

where we use the notation $D = L$ since we now have a linear operator, the matrix U can be completed by a matrix U_c to have full rank $2n$, and two further matrices U_c^+ of dimension $m \times 2n$ and U^+ of dimension $2n - m \times 2n$ are uniquely determined by $[xy](b) - [xy](a) = Ux \cdot U_c^+ y + U_c x \cdot U^+ y$ where for $t \in \mathbb{R}$

$$[xy](t) = \sum_{m=1}^{n} \sum_{j+k=m-1} (-1)^j x^{(k)}(t)(p_{n-m}y)^{(j)}(t). \tag{5}$$

Then one form of Fredholm alternative is that there exists a solution of (1), with $D = L$, if and only if $\alpha = \int_a^b f(t)y(t)dt - \gamma \cdot U_c^+ y = 0$ holds for every solution y of the adjoint homogeneous problem $L^+ y = (-1)^n p_0 y^{(n)} + (-1)^{n-1} p_1 y^{(n-1)} + \ldots p_n y = 0, \quad U^+ y = 0$, [3]. We then have

Corollary 1 *For the Problem* (1), *with* $D = L$, *there is a solution if and only if there is not a Dutch Book based on* $\alpha = \int_a^b f(t)y(t)dt - \gamma \cdot U_c^+ y \neq 0$ *for a suitable function* $y = y(t)$.

Example 3 For the self-adjoint problem

$$x'' + x = 0, \quad x(0) = 1, \quad x(2\pi) = \overline{x} \tag{6}$$

we have

$$y'' + y = 0, \quad y(0) = 0, \quad y(2\pi) = 0, \tag{7}$$

and $\alpha = [xy']_0^{2\pi} = \overline{x}y'(2\pi) - y'(0)$. This implies $y = u_1 \sin(t) + u_2 \cos(t)$ for any choice of u_i's. If $\overline{x} = 1$ we have $\alpha = 0$ by periodicity, and else one can take u_i's so that $\alpha \neq 0$, confirming Corollary 1.

Example 4 For

$$x' - x = f, \quad x(a) = x_a, \quad x(b) = x_b, \tag{8}$$

the adjoint equation is $y' + y = 0$ with no boundary conditions, so Corollary 1 implies that there is no solution to (8) if and only if

$$\alpha = \int_a^b fydt - [x\ y]_a^b = c_0 \left(\int_a^b fe^{-t}dt - x_b e^{-b} + x_a e^{-a} \right) \neq 0 \tag{9}$$

for some c_0.

4.2 Beyond Fredholm Alternative

We present alternative possibilities for the last two examples illustrating the fact that the concept of Dutch Book goes beyond Fredholm alternative even for linear differential equations.

Let's denote by $y_0(t)$ the function provided by the Fredholm alternative. Then, in each case, to go beyond FA, one needs to explore what are the possible forms of $y(t,x)$ beyond $y(t,x) = y(t) = y_0(t)$.

For Problem (8) in Example 4, it is enough to add a linear term to $y_0(t)$, obtaining $y(t,x) = y_0(t) + y_1(t)x$; but for Problem (6) in Example 3, this would not be sufficient to determine a viable $y(t,x)$: recall that $y(t,x)$ should make the integral in Part (I) of the definition of Dutch Book computable regardless of x as long as the functions y_ℓ's satisfy certain supplementary differential equations which are determined along the calculation. If we take $y(t,x) = y_0(t) + y_1(t)x$ in Example 3, then the only solution to the supplementary differential equations has $y_1(t) = 0$, hence bringing us back to FA. It turns out, however, that using $y(t,x) = y_0(t) + y_1(t)x + y_2(t)x'$, one finds nontrivial solutions of the differential equations for the y_ℓ's, as shown below.

Which forms of $y(t)$ produce extensions to the FA cannot be immediately predicted at this stage of the research and has to be determined by trial and error.

An Alternative Dutch Book for $x'' + x = 0$ with Dirichlet Boundary Conditions In the first case, which is a continuation of Example (3), we can take $y = y_0(t) + y_1(t)x + y_2(t)x'$.

Theorem 1 *There exists a solution of (6) if and only if there exists no $y = y(t,x) = y_0(t) + y_1(t)x + y_2(t)x'$ such that*

$$\begin{cases} y_0'' + y_0 = 0 \\ y_1 + \frac{1}{2}y_2' = 0 \\ y_1'' + 4y_1 = 0 \\ y_0(0) = y_0(2\pi) = y_1(0) = y_1(2\pi) = y_2(0) = y_2(2\pi) = 0 \\ [\frac{1}{2}x^2\ y_1']_0^{2\pi} + [x\ y_0']_0^{2\pi} \neq 0 \end{cases} \tag{10}$$

Proof Integrating by parts several times, we get

$$\int_0^{2\pi} (x'' + x)(y_0 + y_1 x + y_2 x')dt \tag{11}$$

$$= [x'\ y_0]_0^{2\pi} + [x'\ y_1\ x]_0^{2\pi} + [\frac{1}{2}(x')^2\ y_2]_0^{2\pi} + [\frac{1}{2}x^2\ y_2]_0^{2\pi} - [\frac{1}{2}x^2\ y_1']_0^{2\pi} - [x\ y_0']_0^{2\pi}$$

$$+ \int_0^{2\pi} x(y_0'' + y_0) - (x')^2(y_1 + \frac{1}{2}y_2') + x^2(\frac{1}{2}y_1'' + y_1 - \frac{1}{2}y_2')dt$$

Equating to zero the coefficients of the various expressions of x under the integral, we obtain the system:

$$\begin{cases} y_0'' + y_0 = 0 \\ y_1 + \frac{1}{2}y_2' = 0 \\ \frac{1}{2}y_1'' + y_1 - \frac{1}{2}y_2' = 0, \end{cases} \tag{12}$$

which is easily seen to be equivalent to the first three equations in (10).

The various boundary terms of x different from $x(2\pi) = x(0) = 0$ are not known; expanding the first terms in (11), one gets

$$[x'\ y_0]_0^{2\pi} + [x'\ y_1\ x]_0^{2\pi} + [\frac{1}{2}(x')^2\ y_2]_0^{2\pi} = x'(2\pi)y_0(2\pi) - x'(0)y_0(0) + \ldots + \frac{1}{2}(x'(0))^2 y_2(0);$$

we can obtain an expression independent from x by equating to zero the multiplicative factors of each term involving boundary terms of x different from $x(2\pi) = x(0) = 0$; each of these factors is expressed in terms of a boundary value of y. This gives boundary conditions for the y_i's, namely,

$$y_0(0) = y_0(2\pi) = y_1(0) = y_1(2\pi) = y_2(0) = y_2(2\pi) = 0$$

which coincides with the fourth equation in (10).

If there exists a solution x of (6), then the l.h.s. of (11) is zero; since a solution of the first three equations in (10) must also satisfy (12), and the fourth relation in (10) eliminates all other terms of the r.h.s. of (11) except two, we must have $[\frac{1}{2}x^2\ y_1']_0^{2\pi} + [x\ y_0']_0^{2\pi} = 0$, violating the last equation of (10).

On the other hand, to discuss the case in which $x(2\pi) \neq 1$ so that there is no solution x of (6), observe that all solutions of the first four equations in (10), and hence (12), have the form

$$y_0 = c_0 \sin t, \quad y_1 = c_1 \sin 2t, \quad y_2 = c_1 \cos 2t \tag{13}$$

for any value of c_0, c_1. We have then

$$[\frac{1}{2}x^2\ y_1']_0^{2\pi} + [x\ y_0']_0^{2\pi} = c_1(x^2(2\pi) - 1) + c_0(x(2\pi) - 1),$$

and we can take the functions in (13) with $c_0 \neq -c_1(x(2\pi) + 1)$ as witness of absence of solutions. This proves the theorem.

We can now construct an alternative Dutch Book: let $y = y_0 + y_1 x + y_2 x'$ with y_0, y_1, y_2 satisfying (10). Since $y = y_0 + y_1 x + y_2 x'$ satisfies the first three equations in (10), then $\alpha = \int_0^{2\pi} (x'' + x)(y_0 + y_1 x + y_2 x')dt$ can be computed regardless of x, and it is nonzero by the inequality in (10). So, $y = y_0 + y_1 x + y_2 x'$ is the function required to set up a Dutch Book.

An Alternative Dutch Book for $x' - x = f$ with Dirichlet Boundary Conditions
This result expands on Example 4 and provides a function y which depends on x, next to t.

Theorem 2 *Given f continuous in $[a, b]$, $x_a, x_b \in \mathbb{R}$, there exists a solution of (8) if and only if there exists no $y = y(t, x) = y_0(t) + y_1(t)x$ such that*

$$\begin{cases} y_1'/2 + y_1 = 0 \\ y_0' + y_0 + f y_1 = 0 \\ \int_a^b f\ y_0\ dt \neq [x\ y_0]_a^b + [\frac{1}{2}\ x^2\ y_1]_a^b \end{cases} \tag{14}$$

Proof Integrating by parts, we get

$$\int_a^b (x' - x - f)(y_0 + y_1 x) dt =$$

$$= \int_a^b (x' y_0 + x' x y_1 - x y_0 - x^2 y_1 - f y_0 - f y_1 x) dt$$

$$= [x y_0]_a^b - \int_a^b x y_0' dt - \int_a^b x y_0$$

$$+ [\frac{x^2}{2} y_1]_a^b - \int_a^b (x^2/2) y_1' dt - \int_a^b x^2 y_1 - \int_a^b f y_0 - \int_a^b f y_1 x$$

$$= \quad [x\ y_0]_a^b + [\frac{1}{2}\ x^2\ y_1]_a^b$$

$$- \int_a^b x(y_0' + y_0 + f y_1) dt - \int_a^b x^2 (y_1'/2 + y_1) dt - \int_a^b f y_0 dt$$

If there exists a solution x of (8), then the l.h.s. is zero and hence any solution of the first two equations in (14) must satisfy $\int_a^b f\ y_0\ dt = [x\ y_0]_a^b + [\frac{1}{2}\ x^2\ y_1]_a^b$.

On the other hand, if there is no solution x of (8), then observe that all solutions of the first two equations in (14) must satisfy

$$y_1 = c_1 e^{-2t}, \quad y_0 = c_0 e^{-t} + c_1 \tilde{y}_0 \tag{15}$$

where $\tilde{y}_0$ is a particular solution of $y_0' + y_0 + f(t) e^{-2t} = 0$. We then have

$$\int_a^b f\ y_0\ dt - [x\ y_0]_a^b - [\frac{1}{2}\ x^2\ y_1]_a^b$$

$$= \int_a^b f(t)\ (c_0 e^{-t} + c_1 \tilde{y}_0)\ dt - x(b)(c_0 e^{-b} + c_1 \tilde{y}_0)$$

$$+ x(a)(c_0 e^{-a} + c_1 \tilde{y}_0) - \frac{1}{2} x(b)^2 c_1 e^{-2b} + \frac{1}{2} x(a)^2 c_1 e^{-2a}$$

$$= c_0 \left(\int_a^b f e^{-t} dt - x(b)e^{-b} + x(a)e^{-a} \right)$$

$$+ c_1 \left(\int_a^b f \tilde{y}_0 dt - x(b)\tilde{y}_0 + x(a)\tilde{y}_0 - \frac{1}{2}x(b)^2 e^{-2b} + \frac{1}{2}x(a)^2 e^{-2a} \right)$$

If the coefficient of c_1 is nonzero, then c_1 can be taken such that the inequality in (14) holds. If, instead, the coefficient of c_1 is zero, then one can take $c_1 = 0$. In this case, the second equation in (14) reduces to $y_0' + y_0 = 0$, which is the adjoint equation of the problem (8). Hence, by Fredholm alternative, there is a solution $y = c_0 e^{-t}$ such that $\int_a^b f\ y_0\ dt \neq [x\ y_0]_a^b$, which is exactly the inequality in (14) when c_1, and hence y_1, are zero. This proves the theorem.

We can now construct an alternative Dutch Book as indicated at the end of the previous theorem by means of $y = y_0 + y_1 x$ with y_0, y_1 satisfying (14).

From the point of view of the Dutch Book, using $y = y(t, x)$ is better than a simple $y = y(t)$ as it makes the integration in (I) less immediate.

5 Conclusions

We have introduced Dutch Books Methods for difference and differential equations as tools to discuss existence or absence of solutions. We have seen how these can be reduced to various known methods: to the Positivstellensatz for polynomial difference equations, and to Fredholm alternative for linear differential equations. We have shown by examples that Dutch Book Methods are more general than Fredholm alternative.

We restricted here to single ordinary differential equations for brevity, but the extension to other linear differential operators does not seem to pose any difficulty.

On the other hand, we hope that, in analogy with what done here for difference equations, the Dutch Book Method can be extended to polynomial differential equations or operators to tackle some open existence issues.

Acknowledgments We would like to thank Andrea Aspri for several relevant discussions and an anonymous referee for carefully reading the paper.

References

1. Bochnak, J., Coste, M., Roy, M.-F.: Real Algebraic Geometry. Springer, Berlin (2010)
2. Bovens, L., Rabinowicz, W.: Bets on hats: on Dutch books against groups, degrees of belief as betting rates, and group-reflection. Episteme **8**, 281–300 (2011)

3. Coddington, E.A., Levinson, N.: Theory of Ordinary Differential Equations. Tata Mcgraw-Hill, New Delhi (2012)
4. Elga, A.: Subjective probabilities should be sharp. Philosophers **10**, 1–11 (2010)
5. Fefferman, C.: Existence and smoothness of the Navier-Stokes equation the Euler and Navier-Stokes equations describe the motion of a fluid in $\mathbb{R}^n$ (2000)
6. Hájek, A.: Dutch book arguments. In Paul Anand, Prasanta Pattanaik & Clemens Puppe (eds.), *The Oxford Handbook of Rational and Social Choice*. Oxford University Press, Oxford (2008)
7. Hitchcock, C.: Beauty and the bets. Synthese **139**(3), 405–420 (2004)
8. Kelsey, D.: Maxmin expected utility and weight of evidence. Oxford Econ. Pap. **46**(3), 425–444 (1994)
9. Marker, D.: Model theory of differential fields. Model Theory Algebra Geom. **39**, 53–63 (2000)
10. Wakker, P.P.: History of the term Dutch book, 07 (2011)
11. Pettigrew, R.: Dutch Book Arguments. Cambridge University Press, Cambridge (2020)
12. Ramm, A.G.: A simple proof of the Fredholm alternative and a characterization of the Fredholm operators. Am. Math. Mon. **108**(9), 855–860 (2001)
13. Ramsey, F.P.: Truth and Probability the Foundations of Mathematics and Other Logical Essays. Kegan, Paul, Trench, Trubner & Co., London (1931)
14. de Finetti Roma, B.: Sul significato soggettivo della probabilità. «Fundamenta Mathematicae **17**, 298–329 (1931)
15. Scheiderer, C.: A positivstellensatz for projective real varieties. Manuscripta Mathematica **138**(1), 73–88 (2012)
16. Sidney, C.: *The Art of Legging: [The History, Theory and Practice of Bookmaking on the English Turf]*. Rotex Publications, Cincinnati (1976)
17. Stengle, G.: A nullstellensatz and a positivstellensatz in semialgebraic geometry. Mathematische Annalen **207**, 87–97 (1974)
18. Tsekrekos, A.E., Yannacopoulos, A.N.: Approximation of optimal stopping problems and variational inequalities involving multiple scales in economics and finance. PhD thesis, Cham (2016)

Fourier Modes in Fluid Flow and Energy Cascade

Irina Afanasyeva, Bailey Downing, and Rathinam Panneer Selvam

1 Introduction

Devastating extreme wind events like hurricanes, thunderstorms, and tornadoes can affect and destroy large geographical areas, causing significant economic losses, injury, or even death. To avoid structural collapse, buildings and structures must be designed for design-level winds to be able to resist applied wind loads. Understanding wind loads and their effects on buildings and structures can be extremely important to avoid catastrophic failures. Field measurements can give us information about realistic wind velocities and pressures, but because they are expensive and unpredictable, alternative methods like wind tunnel (WT) testing and computational fluid dynamics (CFD) are in use.

The ASCE 7–16 wind load provisions set the standards for calculating wind loads on buildings and structures. Despite the fact that CFD analysis is not a method used in the ASCE to predict wind loads, it is still a powerful tool for gaining insight into wind flow behavior around the building. Using an experimentally validated parametrized CFD model, an engineer can perform simulations of different conditions to get a better insight into the flow structure and development, and consequently, changes in transient wind loads, within a shorter timeframe and at a lower cost.

The main challenge that engineers and scientists face nowadays, in terms of wind loading, is a precise prediction of peak wind pressures on buildings and structures, which are determined by the surrounding turbulent airflow. Despite extensive research done [1, 2], still there is no clear understanding of the turbulence phenomena.

I. Afanasyeva (✉) · B. Downing · R. P. Selvam
4190 BELL, University of Arkansas, Fayetteville, AR, USA
e-mail: irinaa@uark.edu; bgdownin@uark.edu; rps@uark.edu

F. Kamalov et al. (eds.), *Advances in Mathematical Modeling and Scientific Computing*, Trends in Mathematics, https://doi.org/10.1007/978-3-031-41420-6_7

For the flow with an isotropic turbulence, the universal theory of fully developed turbulent flow exists. The key feature of this theory is the presence of a phenomenon called an *energy cascade* in the flow. Energy is transferred from the low wavenumber ($k = 2\pi/L$, where L is the wavelength of the eddy) to the high wavenumber in the inertial subrange. The transfer of energy in the energy spectrum follows the $k^{-5/3}$ Kolmogorov law. For civil engineering wind-related problems, where the turbulent flows are homogeneous or inhomogeneous with mean velocity, there are not many theoretical works, like methods used for isotropic turbulence, available in the literature. Therefore, CFD is one of the most attractive tools that can give answers if used properly. For engineering applications with a very high Reynold's number, for now, large eddy simulation (LES) becomes the only feasible approach to model the turbulent flow reasonably accurately.

The major challenge faced in applying LES is to have proper inflow turbulence, as reviewed by [3–5]. Current inflow turbulence generators, which exist and are available in literature as reported in [3, 5], lose their energy approaching the building location compared to what is used as an input. Patruno and Ricci [4] showed that the implementation of the divergence-free condition can be done easily in several random Fourier methods (RFM), which makes this class of methods more advantageous. When several Fourier functions are considered in the RFM method, they interact with each other, and because of that the energy provided at the inlet is lost and only less energy is available at the target location, as reported by [6, 7].

To understand the Fourier mode interaction further from the fundamental point of view, Fourier modes as two types of analytical solutions for the Euler and Navier-Stokes equations were proposed in [8]. The first solution represents modes that satisfy only n-constraints and are called required constrained (RC) modes. When n^2 constraints (required and all the additional) are satisfied, these modes represent the second type of solution and are called fully constrained (FC) modes.

1.1 Objectives

This work is an extension of the work presented in [8, 9]. The main objective is to study the interaction of two and three modes (sine waves) with different combinations of wavelengths and amplitudes that will represent not only cases of RC and FC modes but partially constrained (PC) modes when some of the additional constraints are satisfied in addition to the required (divergence-free) constraints. This should give us more clarity on the initiation and development of energy cascade phenomena for multiple modes of interaction, depending on the number of constraints satisfied. That will help shed light on the turbulence generation issue when using LES for peak wind loads prediction as well as help engineers and scientists solve turbulence problems to mitigate damages. As CFD programs advance, researchers, engineers, and designers will have the ability to conduct a more reliable detailed analysis of wind loads and effects on buildings and structures.

2 Formulation of the Problem

2.1 Numerical Simulation Setup

As governing equations, the incompressible Navier-Stokes (NS) equations reported in [8, 10] are solved using the finite difference method (FDM) implemented in the in-house CFD code "pb3d2.exe." The NS equations are non-dimensionalized with a reference length L_{ref} and a reference velocity U_{ref}. Here the reference length is half of the domain length, and the reference velocity is the mean velocity U_m (in the x-direction). The NS equations are approximated by the 2nd-order finite difference method in space and the Crank-Nicolson method in time. The momentum equations and the pressure equation are solved by the line iteration. The solution is sought on a non-staggered grid. For each time step, a maximum of 10 sub-iterations are allowed to reduce the error to the required tolerance. A periodic domain of length $L_{di} = 2L_{ref}$ is considered $(L_{dx}, L_{dy}, L_{dz}) = (2, 2, 2)$. The mean flow is along the x-direction. In CFD modeling, the Smagorinsky LES model is used to diffuse the wavenumbers higher than the k_{grid}, where k_{grid} is considered as $2\pi/4h$ for the FDM. Here h is the grid spacing. The computational domain is discretized with an equal grid spacing, where the number of grid points of $51 \times 51 \times 51$ is used. Different initial conditions are considered (Table 1), and the computation is done for a non-dimensional time of 50-time units. The time step used for all the runs is 0.01 time units, which satisfies the CFL condition: $\text{CFL} = \text{U}_{max} \cdot \frac{\Delta t}{\Delta x} = 0.3 < 1$. The mean velocity U_m is kept as 1, and the kinematic viscosity is kept as a small number of 10^{-5} to have a minimal loss of energy.

Initial Conditions For initial conditions, the velocities in all three directions (U, V, W) are expressed as Fourier series in space with fixed amplitudes. Here for simplification, we consider only sine terms (Eqs. 1, 2, and 3):

$$\text{U} = U_m + \sum a_{xi} \cdot \sin\left(k_{xi} \cdot (x - U_m t) + k_{yj} \cdot y + k_{zk} \cdot z\right), \tag{1}$$

$$\text{V} = \sum a_{yj} \cdot \sin\left(k_{xi} \cdot (x - U_m t) + k_{yj} \cdot y + k_{zk} \cdot z\right), \tag{2}$$

$$\text{W} = \sum a_{zk} \cdot \sin\left(k_{zi} \cdot (x - U_m t) + k_{yj} \cdot y + k_{zk} \cdot z\right), \tag{3}$$

where U_m is the mean velocity, a_{xi}, a_{yj}, a_{zk} are the amplitudes, k_{xi}, k_{yj}, k_{zk} are the wavenumbers, and indexes *i, j, k* represent x-, y-, z- direction, respectively.

When the velocities (Eqs. 1, 2, and 3) are substituted into the continuity and momentum equations, we get the following expression after simplification (detailed derivation can be found in [8]):

Table 1 Studied test cases with input parameters varied

Test case name	Mode type	Constraints type – # of constraints	Input parameters (L_x, L_y, L_z)	(a_x, a_y, a_z)
2 modes				
TC21	E	FC – 4 ($C_{ij} = 0$)	(2, 2, 2)	(0.1, −0.05, −0.05)
	E		(1, 1, 1)	(0.1, −0.05, −0.05)
TC22	E	RC – 2 ($C_{ij} = 0, i = j$)	(2, 2, 2)	(0.1, −0.05, −0.05)
	UE		(2, 1, 0.5)	(0.1, −0.025, −0.0125)
TC23	E	PC – 3 ($C_{ij} = 0, i = j$ & $C_{12} = 0$)	(2, 2, 2)	(0.1, −0.15, −0.05)
	UE		(2, 1, 0.5)	(0.1, −0.025, −0.0125)
3 modes				
TC31	E	FC – 9 ($C_{ij} = 0$)	(2, 2, 2)	(0.1, −0.05, −0.05)
	E		(1, 1, 1)	(0.1, −0.05, −0.05)
	E		(0.5, 0.5, 0.5)	(0.1, −0.05, −0.05)
TC32	UE	RC – 3 ($C_{ij} = 0, i = j$)	(2, 1, 0.5)	(0.1, −0.025, −0.0125)
	UE		(1, 0.5, 2)	(0.05, −0.0125, −0.05)
	UE		(0.5, 2, 1)	(0.025, −0.05, −0.025)
TC33	UE	PC – 4 ($C_{ij} = 0, i = j$ & $C_{12} = 0$)	(2, 1, 0.5)	(0.1, −0.05, 0)
	UE		(1, 0.5, 2)	(0.05, −0.0125, −0.05)
	UE		(0.5, 2, 1)	(0.025, −0.05, −0.025)
TC34	UE	PC – 5 ($C_{ij} = 0, i = j$ & $C_{12} = C_{21} = 0$)	(2, 1, 1)	(0.1, −0.025, −0.025)
	UE		(1, 0.5, 0.5)	(0.1, −0.025, −0.025)
	E		(0.5, 0.5, 0.5)	(0.05, −0.025, −0.025)
TC35	UE	PC – 6 ($C_{23} \neq 0, C_{31} \neq 0, C_{32} \neq 0$)	(2, 1, 1)	(0.1, 0, −0.05)
	UE		(1, 0.5, 0.5)	(0.05, −0.025, 0)
	UE		(0.5, 0.5, 0.25)	(0.1, −0.05, −0.025)
TC36	UE	PC – 7 ($C_{31} \neq 0, C_{32} \neq 0$)	(2, 1, 1)	(0.1, 0, −0.05)
	UE		(1, 0.5, 0.5)	(0.1, 0, −0.05)
	UE		(0.5, 0.5, 0.25)	(0.1, −0.05, −0.025)

$$C = A \cdot K^T, \tag{4}$$

where A is an $n \times 3$ matrix of modes' amplitudes in three directions x, y, and z; K is an $n \times 3$ matrix of wavenumbers in three directions x, y, and z; symbol "T" means matrix transpose; C is an $n \times n$ matrix of constraints; n – a number of modes considered.

In theory, one can have an infinite length in all three directions, and hence the sine wave is *consistent*. Here, only two units of length are considered; hence, the largest consistent wavelength L will be two units. Beyond that length, it will be considered inconsistent for the domain. For the given domain with lengths of L_d, a consistent wavelength will be $L = L_d/m$ (or wavenumbers, $k = 2\pi m/L_d$), where m is an integer. In our discussions, if m is not an integer, then we call those modes as *inconsistent*. When generating inflow turbulence using RFM, many researchers used inconsistent waves to satisfy the continuity equation. This produces energy cascade or energy decay and spurious pressures [11]. To have reasonable accuracy, we will have minimum five grid points for the smallest sine wave used in this work. The amplitude of the sine wave is kept 0.15 unit or less. If more than one wave is used, the total amplitude of all the waves is around 0.15 unit. This will make sure that the Taylor hypothesis is applicable. The wavelengths and amplitudes are given as input to calculate the initial conditions for the CFD programs, and this way it is easy to calculate the frequency as $1/L$.

Full Constrained Multiple Modes (FC) For the case of homogenous turbulence and the flow with mean velocity when the computational domain dimensions are equal in all three directions, if the wavenumbers (or wavelengths) are taken equally for all three directions for all considered modes (Eq. 5) and the amplitudes of each mode are proportional for all three directions, respectively (Eq. 6), then the matrix of constraints C will be zero (Eq. 7), which means that the number of constraints satisfied is equal to n^2. Then there is no energy cascade for a long time.

$$K = \frac{2\pi}{L_d}\begin{bmatrix} m_1 & m_1 & m_1 \\ \vdots & \ddots & \vdots \\ m_n & m_n & m_n \end{bmatrix}, \tag{5}$$

$$A = A_1 \circ \begin{bmatrix} 1 & 1 & 1 \\ b_2 & b_2 & b_2 \\ \vdots & \ddots & \vdots \\ b_n & b_n & b_n \end{bmatrix}, \tag{6}$$

where L_d is the domain length; m_i is an integer; $i = 1...n$ is a mode number; $A_1 = (a_{x1}, a_{y1}, a_{z1})$ is a vector of the 1st mode amplitudes in three directions representing the first row of a matrix A; b_i are the constants; $i = 1...n$ is a mode number; and symbol "∘" is the elementwise product.

$$C_{ij} = 0, \qquad \text{for i, j} = 1..\text{n}, \tag{7}$$

where i and j are matrix indices representing the row and column numbers, respectively; n is the number of considered modes.

Required Constrained Multiple Modes (RC) If the amplitudes and wavenumbers satisfy only the continuity equation, then all n diagonal terms of matrix C will be zero (Eq. 8).

$$C_{ij} = 0, \text{if i} = \text{j} \qquad \text{for i, j} = 1..\text{n}, \tag{8}$$

where i and j are matrix indices representing the row and column numbers, respectively; n is the number of considered modes.

Partially Constrained Multiple Modes (PC) In addition to n zero diagonal terms of the martix C, cross-terms (off-diagonal terms) could be equal to zero as well. Such cases will represent PC multiple modes with additional constraints satisfied (the number of constraints in these cases will be more than n and less than n^2).

For a single mode that has an unequal wavelength (or wavenumber) in all three directions and amplitudes that satisfy the criteria above (Eq. 8), an energy cascade happens in some time, while for a single mode with equal wavelengths, there is no energy cascade. For multiple modes, if the criteria developed in Eq. 7 are not satisfied but in Eq. 8 are satisfied, then an energy cascade happens from the beginning or after some short period of time.

2.2 *Test Cases*

The current study consists of 11 test cases described in Table 1 that represent the interaction of two and three consistent modes with different combinations of wavelengths (L_x, L_y, L_z) and amplitudes (a_x, a_y, a_z). The name of each test case is constructed in such a way: "[TC][# of modes considered][# of the test case for the particular # of modes]." All test cases consist of modes that can be characterized as "E" – mode with equal wavelengths in all three directions and "UE" – mode with unequal wavelengths in three directions. Depending on the input parameters – different combinations of wavelengths (L_x, L_y, L_z) and amplitudes ($\mathrm{a_x}$, $\mathrm{a_y}$, $\mathrm{a_z}$), all the test cases represent the following cases of constraints: (1) fully constrained multiple modes (FC), (2) required constrained multiple modes (RC), and (3) partially constrained multiple modes (PC). In Table 1 one can identify the number of satisfied constraints for the test case by the number following the constraints type. For instance, in the TC31 three modes with equal wavelengths are considered; these modes are FC modes with all nine constraints satisfied (all nine terms of the matrix C equal to zero).

3 Results and Discussion

This section discusses the results of CFD simulations based on the velocities and pressure time series (Figs. 1 and 2, left) obtained in the middle of the computational domain (1,1,1) (see Fig. 1) and U-velocity wavelet scalograms (Figs. 1 and 2, right). For the wavelet analysis, continuous wavelet transform (CWT) implemented in MATLAB software is obtained using the analytic Morse wavelet. Below, Figs. 1 and 2 demonstrate an energy cascade initiation for two- and three-modes interaction with different modes and constraints types. All computations are done for the fluid flow domain without a building.

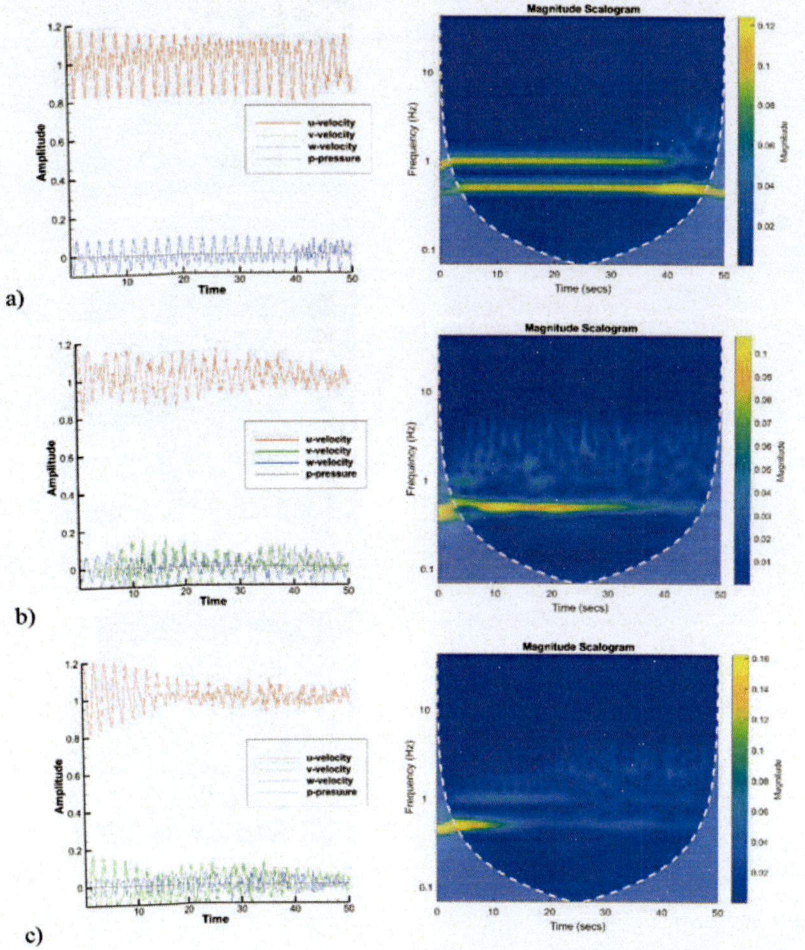

Fig. 1 Energy cascade initiation for two-modes interaction: (a)–(c) TC21-TC23

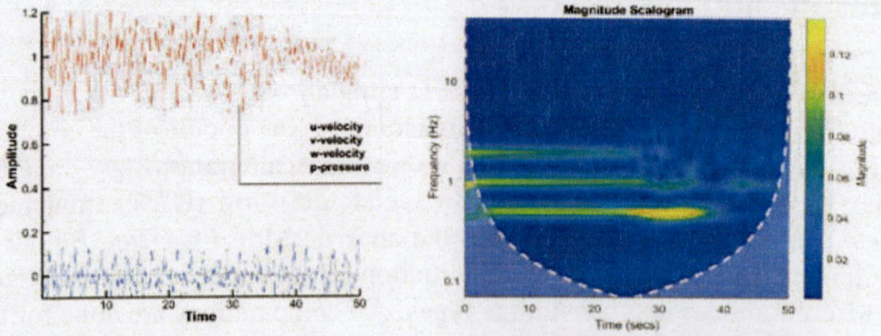

a)

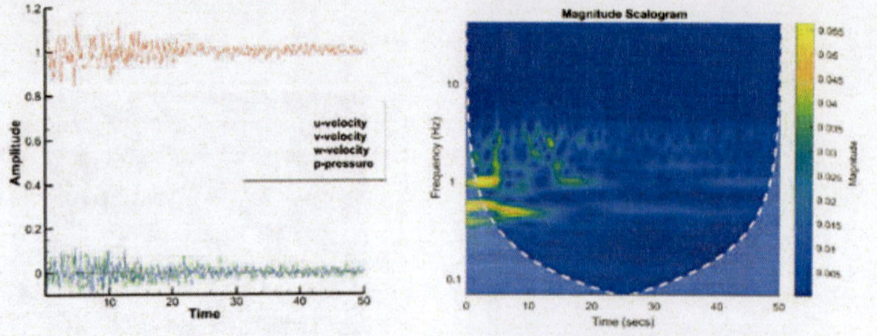

b)

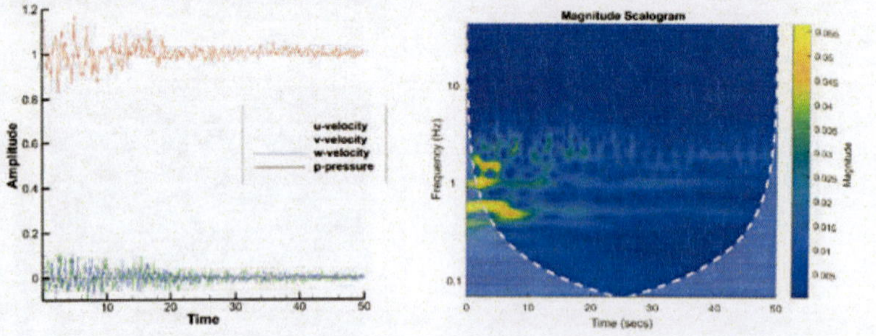

c)

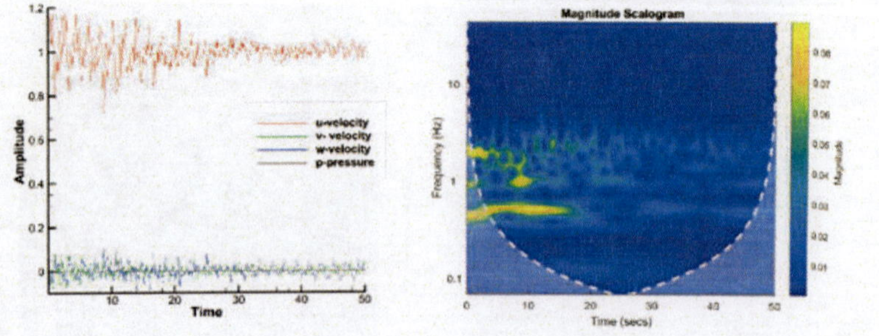

d)

Fig. 2 (continued)

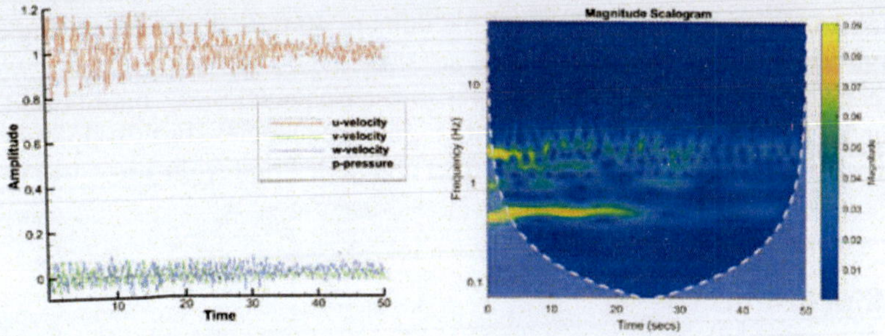

e)

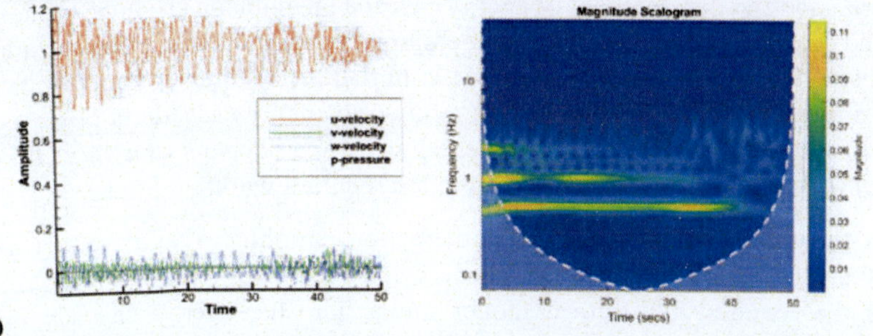

f)

Fig. 2 Energy Cascade initiation for three-modes interaction: **(a)–(d)** TC31-TC34. Energy Cascade initiation for three-modes interaction: **(e, f)** TC35-TC36

Two-Modes Interaction For the two-modes interaction, three test cases have been studied. Figure 1a shows the development of flow characteristics for the FC case, where one can notice a slight reduction and then growth of amplitude in time for the velocity U, which can happen due to the viscosity effect. The energy cascade starts to happen in a long time (after 40 time units), where the mode with the highest frequency starts dissipating, while the one with the lowest frequency is still maintained. Ideally, we should not observe an energy cascade for FC modes. Besides the viscosity effect, dissipation of the mode with the highest frequency may be caused by not enough grid resolution. Further investigations on this issue need to be done. For the RC and PC cases (Fig. 1b, c) amplitudes change differently depending on what constraints and how many of them are satisfied. For all three cases, an energy cascade happens right away from the beginning. As in the FC case, the mode with the highest frequency starts dissipating first, while the one with the lowest frequency is still maintained for some time.

Three-Modes Interaction For the three-modes interaction, six test cases have been studied. Figure 2a illustrates the FC case where three distinct modes and maintained amplitudes are observed up to 30 time units of simulation. For this case, we can say that an energy cascade happens in a long time. For the RC and PC cases (Fig. 2b–f), three distinct modes can be distinguished at the very beginning, but the dissipation of the modes with the highest frequencies starts very quickly, while the mode with the lowest one is maintained from 5 time units to 40 time units depending on what constraints and how many of them are satisfied. One can notice a trend where the more constraints are satisfied, the longer lower frequencies are maintained. As for two-modes interaction cases, modes with higher frequencies start dissipating earlier, which can occur due to the viscosity effect and poor grid resolution.

4 Conclusions

The presence and development of energy cascade phenomena are investigated when two and three modes with varied input parameters interact. The effect of a number of constraints satisfied on energy cascade initiation is studied. The results of the CFD simulations show that for certain modes, no energy cascade is observed for a long time, while for other modes, an energy cascade happens in a short time. The following conclusions can be drawn from the reported work:

- For the fully constrained modes interaction, amplitudes and frequencies are maintained for a long time, and an energy cascade takes over in a long time.
- For the required constrained modes interaction, the energy cascade happens quickly at the beginning of the simulation.
- For cases representing partially constrained modes, the energy cascade is observed in the beginning or in a short time depending on the number of constraints satisfied. The more constraints are satisfied, the more distinct becomes the mode with the lowest frequency, while others start dissipating right away.
- For both full constrained cases for two- and three-modes interaction, the energy cascade presents and happens in a long time, while in theory there should not be an energy cascade. Also for all nine cases, modes with higher frequencies start dissipating earlier. All these effects can occur due to the viscosity effect and not enough grid resolution. So further investigation on this issue needs to be done.
- In previous research done by [8, 9], it was noticed that the more modes with equal wavelengths are considered, the later energy cascade is initiated. In current work almost all cases consist of modes with unequal wavelengths; therefore, for considered cases the energy cascade phenomena happen quicker. So, further investigation needs to be done.

The results of this study can help develop more efficient inflow turbulence generators without much loss of energy. To confirm the conclusions stated here, more investigations of modes interactions need to be conducted. Also, further research needs to be done to investigate the effect of different boundary conditions and domain size and shape on the initiation of the energy cascade. Since the current work considers only consistent modes, the study of inconsistent modes interaction needs to be performed as well. Besides that, numerical error in the input for wavelength or amplitude may trigger the numerical stability issue, which needs to be studied.

Acknowledgments We would like to thank Dr. M. Hale for the generous donation to the *Civil Engineering Department Scholarship* and Mr. R.D. Thomas and Mr. W.F. Thomas for funding the *W.R. Thomas Endowed Doctoral Fellowship Fund*, which supported this work.

References

1. Ecke, R.: The turbulence problem: an experimentalist perspective. Los Alamos Sci. **29**, 124–141 (2005)
2. Eckert, M.: The Turbulence Problem: A Persistent Riddle in Historical Perspective. Springer (2019)
3. Wu, X.: Inflow turbulence generation methods. Annu. Rev. Fluid Mech. **49**, 23–49 (2017)
4. Patruno, L., Ricci, M.: On the generation of synthetic divergence-free homogeneous anisotropic turbulence. Comput. Methods Appl. Mech. Eng. **315**, 396–417 (2017)
5. Dhamankar, N.S., Blaisdell, G.A., Lyrintzis, A.S.: Overview of turbulent inflow boundary conditions for large-Eddy simulations. AIAA J. **56**(4), 1317–1334 (2018)
6. Atencio Mojica, Z.D.: Frequency effect on peak pressure coefficients using the narrowband synthesis random flow generator (NSRFG) method. Graduate Theses and Dissertations (2021). Retrieved from https://scholarworks.uark.edu/etd/4204
7. Mansouri, Z., Selvam, R.P., Chowdhury, A.G.: Maximum grid spacing effect on peak pressure computation using inflow turbulence generators. Results Eng. **15** (2022). https://doi.org/10.1016/j.rineng.2022.100491
8. Selvam, R.P.: Fourier modes as an analytical solution for the Euler and Navier-Stokes equations and energy cascade. Annual report. Department of Civil Engineering, University of Arkansas (2022)
9. Downing, B.: Wave interactions and energy cascade. Civil Engineering Undergraduate Honors Theses (2022). Retrieved from https://scholarworks.uark.edu/cveguht/70
10. Selvam, R.P.: Computational Fluid Dynamics for Wind Engineering. Wiley (2022)
11. Mansouri, M., Selvam, R.P., Chowdhury, A.G.: Performance of different inflow turbulence methods for wind engineering applications. J. Wind Eng. Ind. Aerodyn. **229** (2022). https://doi.org/10.1016/j.jweia.2022.105141

Approximate Solutions of Third-Order Time Fractional Dispersive Equations with Singular and Nonsingular Kernel Derivatives

K. Pavani, K. Raghavendar, and K. Aruna

1 Introduction

In recent decades, fractional calculus is a developing field in numerous disciplines of science and engineering. Fractional calculus involves differentiation and integration of arbitrary orders. In the investigation of physics, chemistry, applied mathematics, and several other branches of engineering and science, nonlinear phenomena are always visible. Dynamical events have been seen in a variety of scientific domains, including electrodynamics, chaos theory, biotechnology, and many other fields [1–3]. Numerous approaches for studying the computational and exact solutions of fractional differential equations (FDEs) have also been developed due to their importance in diverse domains. Convergence and divergence of the outcomes, in addition to the modeling, are equally essential. The fractional generalization of a physical process requires an adequate definition. In the previous few centuries, numerous fractional derivative (FD) definitions have appeared. Riemann-Liouville (RL), Caputo (C), Caputo-Fabrizio (CF), Atangana-Baleanu derivative in the Caputo manner (ABC), Grunwald-Letnikov, and Riesz are some well-known FD definitions that have been utilized in this research. More information can be found in [4, 5] and the references therein. Numerous researchers have concentrated their attention in recent years on studying partial differential equation (PDE) solutions using various methodologies, such as homotopy perturbation transform method [6], q-homotopy analysis transform method [7], differential transform method (DTM) [8], variational iteration method (VIM) [9], parametric cubic spline [10], tension spline [11, 12], and references therein. Several researchers have recently investigated the time fractional third-order dispersive equation (TFTDE) using

K. Pavani · K. Raghavendar (✉) · K. Aruna
Department of Mathematics, School of Advanced Sciences, Vellore Institute of Technology, Vellore, India

F. Kamalov et al. (eds.), *Advances in Mathematical Modeling and Scientific Computing*, Trends in Mathematics, https://doi.org/10.1007/978-3-031-41420-6_8

various methods and techniques, such as VIM [13], DTM [14], and so on. Recently, R. Shah et al. [15] studied TFTDE by using the Laplace-Adomian decomposition method.

This study is focused on using natural transform decomposition method (NTDM) [16] to resolve the TFTDE. The NTDM eliminates round-off errors by not requiring linearization, assumptions, discretization, or perturbation. This innovative approach is regarded as an outstanding tool for quickly and easily solving particular classes of coupled nonlinear PDEs. This method offers a solution with the aid of a fast convergence series that can be exact or approximate. As a result, the NTDM is becoming an important mathematical technique for a variety of linear and nonlinear PDEs [17]. NTDM was recently used for the Klein-Gordon equation [18], time fractional Fisher's equation [19], Zakharov-Kuznetsov equation [20], time fractional Sawada-Kotera-Ito equation [21], and Kawahara and modified Kawahara equations [22]. The paper is formatted as follows. In Sect. 2, we present primary definitions which are useful in this study. In Sect. 3, the basic concept of NTDM is presented. In Sect. 4, the solutions uniqueness and convergence were investigated. In Sect. 5, we present few examples of TFTDE. In Sect. 6, we presented conclusions.

2 Definitions

In this section, we have presented the basic definitions [23, 24].

Definition 1 ([25]) The Caputo FD of $\mathtt{f}(\omega)$is given by

$$ {}_0^C D_\omega^\mu \mathtt{f}(\omega) = \frac{1}{\mathtt{m}-\mu}\int_0^\omega (\omega-\varsigma)^{\mathtt{m}-\mu-1}\mathtt{f}^{\mathtt{m}}(\varsigma)d\varsigma,\ \mathtt{m}\in\mathbb{N},\ \mathtt{m}-1<\mu\le\mathtt{m},\ \omega>0, \quad \mathtt{f}\in C_\nu^{\mathtt{m}},\ \nu\ge -1. \tag{1} $$

Definition 2 ([26]) The FD of $\mathtt{f}(\omega)$ in CF is given by

$$ {}_0^{CF} D_\omega^\mu (\mathtt{f}(\omega)) = \frac{\mathtt{B}(\mu)}{1-\mu}\int_0^\omega \exp\left(\frac{-\mu(\omega-\varsigma)}{1-\mu}\right) D(\mathtt{f}(\varsigma))d\varsigma,\ 0<\mu<1, \quad \mathtt{B}(0)=\mathtt{B}(1)=1. \tag{2} $$

Definition 3 ([27]) The FD of $\mathtt{f}(\omega)$ in ABC is presented as

$$ {}_0^{ABC} D_\omega^\mu \mathtt{f}(\omega) = \frac{\mathtt{B}(\mu)}{1-\mu}\int_0^\omega \mathrm{E}_\mu\left(\frac{(\omega-\varsigma)^\mu}{1-\mu}\right) D(\mathtt{f}(\varsigma))d\varsigma,\ 0<\mu<1. \tag{3} $$

Definition 4 ([28]) The NT of $\mathtt{u}(\mathtt{t})$ is given by

$$\mathbf{NT}(\mathtt{u}(\mathtt{t})) = \mathfrak{U}(s, v) = \int_{-\infty}^{\infty} e^{-s\mathtt{t}}\mathtt{u}(v\mathtt{t})d\mathtt{t}, \quad s, v \in (-\infty, \infty). \tag{4}$$

Definition 5 ([28]) The inverse NT of $\mathfrak{U}(s, v)$is defined by

$$\mathbf{NT}^{-1}\left[\mathfrak{U}(s, v)\right] = \mathtt{u}(\mathtt{t}),\ \forall \mathtt{t} \geq 0 \tag{5}$$

Definition 6 ([29]) NT of $D_{\mathtt{t}}^{\mu}\mathtt{u}(\mathtt{t})$ by means of Caputo sense is given as

$$\mathbf{NT}\left[{}_{0}^{C}D_{\mathtt{t}}^{\mu}\mathtt{u}(\mathtt{t})\right] = \left(\frac{s}{\mathtt{v}}\right)^{\mu}\left(\mathbf{NT}[\mathtt{u}(\mathtt{t})] - \left(\frac{1}{s}\right)\mathtt{u}(0)\right) \tag{6}$$

Definition 7 ([30]) NT of $D_{\mathtt{t}}^{\mu}\mathtt{u}(\mathtt{t})$ by means of CF is given by

$$\mathbf{NT}\left[{}_{0}^{CF}D_{\mathtt{t}}^{\mu}\mathtt{u}(\mathtt{t})\right] = \frac{1}{1-\mu+\mu\left(\frac{\mathtt{v}}{s}\right)}\left(\mathbf{NT}[\mathtt{u}(\mathtt{t})] - \frac{1}{s}\mathtt{u}(0)\right) \tag{7}$$

Definition 8 ([21]) NT of $D_{\mathtt{t}}^{\mu}\mathtt{u}(\mathtt{t})$ by means of ABC derivative is given as

$$\mathbf{NT}\left[{}_{0}^{ABC}D_{\mathtt{t}}^{\mu}\mathtt{u}(\mathtt{t})\right] = \frac{M[\mu]}{1-\mu+\mu\left(\frac{\mathtt{v}}{s}\right)^{\mu}}\left(\mathbf{NT}[\mathtt{u}(\mathtt{t})] - \left(\frac{1}{s}\right)\mathtt{u}(0)\right) \tag{8}$$

3 Methodology

In this section, we present NTDM [31] to the following equation:

$$D_{\mathtt{t}}^{\mu}\mathtt{u}(\zeta, \mathtt{t}) = \mathscr{L}(\mathtt{u}(\zeta, \mathtt{t})) + \mathscr{N}(\mathtt{u}(\zeta, \mathtt{t})) + \mathtt{h}(\zeta, \mathtt{t}), \tag{9}$$

with initial condition

$$\mathtt{u}(\zeta, 0) = \phi(\zeta), \tag{10}$$

where $\mathscr{L}$, $\mathscr{N}$, and $\mathtt{h}(\zeta, \mathtt{t})$ are linear, nonlinear, and source terms, respectively.
Case I (NTDM$_C$):
By taking NT of equation (9), by means of Caputo FD, we obtain

$$\left(\frac{s}{\mathtt{v}}\right)^{\mu}\left(\mathbf{NT}\left[\mathtt{u}(\zeta, \mathtt{t})\right] - \frac{\phi(\zeta)}{s}\right) = \mathbf{NT}\left[\mathtt{M}(\zeta, \mathtt{t})\right], \tag{11}$$

where $\mathtt{M}(\zeta, \mathtt{t}) = \mathscr{L}(\mathtt{u}(\zeta, \mathtt{t})) + \mathscr{N}(\mathtt{u}(\zeta, \mathtt{t})) + \mathtt{h}(\zeta, \mathtt{t})$. By applying inverse NT using (5), we rewrite (11) as

$$\mathrm{u}(\zeta,\mathrm{t}) = \mathbf{NT}^{-1}\left[\frac{\phi(\zeta)}{s} + \left(\frac{\mathrm{v}}{s}\right.^{\mu} \mathbf{NT}\left[\mathrm{M}(\zeta,\mathrm{t})\right]\right. . \tag{12}$$

$\mathscr{N}(\mathrm{u}(\zeta,\mathrm{t}))$ can be decomposed into

$$\mathscr{N}\left[\mathrm{u}(\zeta,\mathrm{t})\right] = \sum_{l=0}^{\infty} A_l, \tag{13}$$

where A_l is the Adomian polynomials. We assume that the analytical expansion of equation (9) is

$$\mathrm{u}(\zeta,\mathrm{t}) = \sum_{l=0}^{\infty} \mathrm{u}_l(\zeta,\mathrm{t}). \tag{14}$$

By substituting equations (13)–(14) into (12), we obtain

$$\begin{aligned}\sum_{l=0}^{\infty} \mathrm{u}_l(\zeta,\mathrm{t}) =& \mathbf{NT}^{-1}\left[\frac{\phi(\zeta)}{s}\right. + \mathbf{NT}^{-1}\left[\left(\frac{\mathrm{v}}{s}\right.^{\mu} \mathbf{NT}\left[\mathrm{h}(\zeta,\mathrm{t})\right]\right] \\ &+ \mathbf{NT}^{-1}\left[\left(\frac{\mathrm{v}}{s}\right.^{\mu} \mathbf{NT}\left[\sum_{l=0}^{\infty} \mathscr{L}(\mathrm{u}_n(\zeta,\mathrm{t})) + A_l\right]\right].\end{aligned} \tag{15}$$

From (15), we get

$$\begin{aligned}\mathrm{u}_0^C(\zeta,\mathrm{t}) &= \mathbf{NT}^{-1}\left[\frac{\phi(\zeta)}{s}\right. + \mathbf{NT}^{-1}\left[\left(\frac{\mathrm{v}}{s}\right.^{\mu} \mathbf{NT}\left[\mathrm{h}(\zeta,\mathrm{t})\right]\right] \\ \mathrm{u}_1^C(\zeta,\mathrm{t}) &= \mathbf{NT}^{-1}\left[\left(\frac{\mathrm{v}}{s}\right.^{\mu} \mathbf{NT}\left[\mathscr{L}(\mathrm{u}_0(\zeta,\mathrm{t})) + A_0\right]\right] \\ &\vdots \\ \mathrm{u}_{l+1}^C(\zeta,\mathrm{t}) &= \mathbf{NT}^{-1}\left[\left(\frac{\mathrm{v}}{s}\right.^{\mu} \mathbf{NT}\left[\mathscr{L}(\mathrm{u}_l(\zeta,\mathrm{t})) + A_l\right]\right], \; l = 1, 2, \ldots.\end{aligned} \tag{16}$$

By substituting (16) into (14), we get the NTDM$_C$ solution of (9)–(10) as

$$\mathrm{u}^C(\zeta,\mathrm{t}) = \mathrm{u}_0^C(\zeta,\mathrm{t}) + \mathrm{u}_1^C(\zeta,\mathrm{t}) + \mathrm{u}_2^C(\zeta,\mathrm{t}) + \ldots. \tag{17}$$

In a similar manner, we can get the solution of (9)–(10) for CF and ABC derivatives.
Case II (NTDM$_{CF}$):

$$\mathrm{u}^{CF}(\zeta,\mathrm{t}) = \mathrm{u}_0^{CF}(\zeta,\mathrm{t}) + \mathrm{u}_1^{CF}(\zeta,\mathrm{t}) + \mathrm{u}_2^{CF}(\zeta,\mathrm{t}) + \ldots. \tag{18}$$

Case III (NTDM$_{ABC}$):

$$u^{ABC}(\zeta, t) = u_0^{ABC}(\zeta, t) + u_1^{ABC}(\zeta, t) + u_2^{ABC}(\zeta, t) + \dots. \tag{19}$$

4 Convergence Analysis

We have presented convergence and uniqueness of the NTDM$_C$, NTDM$_{CF}$, and NTDM$_{ABC}$ in this section [21].

Theorem 1 *The NTDM$_C$ solution of* (9) *is unique when* $0 < (\delta_1 + \delta_2)\frac{t^\mu}{\Gamma(1+\mu)} < 1$.

Proof Let $F = (C[J], ||.||)$ be the Banach space with $||\phi(t|| = \max_{t \in J} |\phi(t)|$, $\forall$ continuous functions on J. Let $G : F \to F$ be a nonlinear mapping, where

$$u_{l+1}^C(\zeta, t) = u_0^C + \mathbf{NT}^{-1}\left[\left(\frac{v}{s}\right)^{\mu} \mathbf{NT}\left[\mathscr{L}(u_l(\zeta, t)) + \mathscr{N}(u_l(\zeta, t))\right]\right], \quad l \geq 0.$$

Suppose that $|\mathscr{L}(u) - \mathscr{L}(u^*)| < \delta_1 |u - u^*|$ and $|\mathscr{N}(u) - \mathscr{N}(u^*)| < \delta_2 |u - u^*|$, where δ_1 and δ_2 are Lipschitz constants and $u := u(\zeta, t)$ and $u^* := u^*(\zeta, t)$ are two different function values:

$$\begin{aligned}
||Gu - Gu^*|| &= \max_{t \in J} \Big|\mathbf{NT}^{-1}\left[\left(\frac{v}{s}\right)^{\mu} \mathbf{NT}\left[\mathscr{L}(u) + \mathscr{N}(u)\right]\right] \\
&\quad - \mathbf{NT}^{-1}\left[\left(\frac{v}{s}\right)^{\mu} \mathbf{NT}\left[\mathscr{L}(u^*) + \mathscr{N}(u^*)\right]\right]\Big| \\
&\leq \max_{t \in J}\Big[\delta_1 \mathbf{NT}^{-1}\left[\left(\frac{v}{s}\right)^{\mu} \mathbf{NT}\left|u - u^*\right|\right] \\
&\quad + \delta_2 \mathbf{NT}^{-1}\left[\left(\frac{v}{s}\right)^{\mu} \mathbf{NT}\left|u - u^*\right|\right]\Big] \\
&\leq (\delta_1 + \delta_2)\Big[\mathbf{NT}^{-1}\left[\left(\frac{v}{s}\right)^{\mu} \mathbf{NT}\|u - u^*\|\right]\Big] \\
&= (\delta_1 + \delta_2)\frac{t^\mu}{\Gamma(1+\mu)}\,||u - u^*||,
\end{aligned}$$

G is contraction as $0 < (\delta_1 + \delta_2)\frac{t^\mu}{\Gamma(1+\mu)} < 1$. The solution of (9) is unique by using Banach fixed-point theorem. □

Theorem 2 *NTDM$_C$ solution of* (9) *is convergent.*

Proof Let $u_m = \sum_{r=0}^{m} u_r(\zeta, t)$. To prove that u_m is a Cauchy sequence in F, consider

$$\|\mathrm{u}_m - \mathrm{u}_n\| = \max_{\mathrm{t}\in J} |\mathrm{u}_m - \mathrm{u}_n| = \max_{\mathrm{t}\in J} \left| \sum_{r=n+1}^{m} \mathrm{u}_r \right|, \quad n = 1, 2, 3, \ldots.$$

$$\leq \max_{\mathrm{t}\in J} \left| \mathbf{NT}^{-1} \left[\left(\frac{v}{s}^{\mu} \mathbf{NT} \left[\sum_{r=n+1}^{m} (\mathscr{L}(\mathrm{u}_{r-1}) + \mathscr{N}(\mathrm{u}_{r-1})) \right]\right]\right|$$

$$\leq \max_{\mathrm{t}\in J} \left| \mathbf{NT}^{-1} \Big[\Big(\frac{v}{s}^{\mu} \mathbf{NT} \Big[\Big(\mathscr{L}(\mathrm{u}_{m-1}) - \mathscr{L}(\mathrm{u}_{n-1}) \right.$$

$$\left. + \mathscr{N}(\mathrm{u}_{m-1}) - \mathscr{N}(\mathrm{u}_{n-1}) \Big) \Big] \Big] \right|$$

$$\leq \delta_1 \max_{\mathrm{t}\in J} \left| \mathbf{NT}^{-1} \Big[\Big(\frac{v}{s}^{\mu} \mathbf{NT} \left[(\mathscr{L}(\mathrm{u}_{m-1}) - \mathscr{L}(\mathrm{u}_{n-1})) \right] \Big] \right|$$

$$+ \delta_2 \max_{\mathrm{t}\in J} \left| \mathbf{NT}^{-1} \left[\left(\frac{v}{s}^{\mu} \mathbf{NT} \left[(\mathscr{N}(\mathrm{u}_{m-1}) - \mathscr{N}(\mathrm{u}_{n-1})) \right] \right] \right|$$

$$= (\delta_1 + \delta_2) \frac{\mathrm{t}^{\mu}}{\Gamma(\mu + 1)} \|\mathrm{u}_{m-1} - \mathrm{u}_{n-1}\|$$

Let $m = n + 1$, and then $\|\mathrm{u}_{n+1} - \mathrm{u}_n\| \leq \delta \|\mathrm{u}_n - \mathrm{u}_{n-1}\| \leq \ldots \leq \delta^n \|\mathrm{u}_1 - \mathrm{u}_0\|$, where $\delta = (\delta_1 + \delta_2) \frac{\mathrm{t}^{\mu}}{\Gamma(\mu+1)}$. Similarly, we have

$$\|\mathrm{u}_m - \mathrm{u}_n\| \leq \|\mathrm{u}_{n+1} - \mathrm{u}_n\| + \|\mathrm{u}_{n+2} - \mathrm{u}_{n+1}\| + \ldots + \|\mathrm{u}_m - \mathrm{u}_{m-1}\|,$$

$$\leq (\delta^n + \delta^{n+1} + \ldots + \delta^{m-1}) \|\mathrm{u}_1 - \mathrm{u}_0\| \leq \delta^n \left(\frac{1 - \delta^{m-n}}{1 - \delta} \right) \|\mathrm{u}_1\|,$$

As $0 < \delta < 1$, we get $1 - \delta^{m-n} < 1$. Therefore, $\|\mathrm{u}_m - \mathrm{u}_n\| \leq \frac{\delta^n}{1-\delta} \max_{\mathrm{t}\in J} \|\mathrm{u}_1\|$. Since $\|\mathrm{u}_1\| < \infty$. $\|\mathrm{u}_m - \mathrm{u}_n\| \to 0$ when $n \to \infty$. Hence, the series u_m is convergent as u_m is a Cauchy sequence in F. □

In simlilar lines of Theorems 1 and 2, we can prove that NTDM_{CF} and NTDM_{ABC} solutions of (9) is unique and convergent.

5 Numerical Examples

This section provides approximate analytical solutions for a few TFTDE examples. These equations were chosen due to the availability of closed-form solutions and well-established methods for studying the solutions in the literature.

Example 5 Consider TFD-Kdv equation: $\mathrm{u}_{\mathrm{t}}^{\mu} = -2\mathrm{u}_{\zeta} - \mathrm{u}_{\zeta\zeta\zeta}$ with initial condition $\mathrm{u}(\zeta, 0) = \sin\zeta$. When $\mu = 1$, the exact solution is $\mathrm{u}(\zeta, \mathrm{t}) = \sin(\zeta - \mathrm{t})$ [32]. □

NTDM$_C$ With the aid of NTDM_C, we get the successive solution as

$$u_0^C(\zeta,t)=\sin(\zeta),\, u_1^C(\zeta,t)=-\frac{\cos(\zeta)t^{\mu}}{\Gamma(\mu+1)},\, u_2^C(\zeta,t)=-\frac{\sin(\zeta)t^{2\mu}}{\Gamma(2\mu+1)},\ldots.$$

Substituting $u_0^C(\zeta,t), u_1^C(\zeta,t),\ldots$ in(17), we get the NTDM$_C$ series solution as

$$u^C(\zeta,t)\approx\sin(\zeta)-\frac{\cos(\zeta)t^{\mu}}{\Gamma(\mu+1)}-\frac{\sin(\zeta)t^{2\mu}}{\Gamma(2\mu+1)}+\ldots. \tag{20}$$

NTDM$_{CF}$ With the aid of NTDM$_{CF}$, we get the successive solution as

$$u_0^{CF}(\zeta,t)=\sin(\zeta),\, u_1^{CF}(\zeta,t)=\cos(\zeta)(\mu+\mu(-t)-1),$$
$$u_2^{CF}(\zeta,t)=-\frac{\sin(\zeta)}{2}\left(\mu^2\left(t^2-4t+2\right)+4\mu(t-1)+2\right),\ldots$$

Substituting $u_0^{CF}(\zeta,t), u_1^{CF}(\zeta,t),\ldots$ in(18), we get the NTDM$_{CF}$ series solution.

NTDM$_{ABC}$ With the aid of NTDM$_{ABC}$, we get the successive solution as

$$u_0^{ABC}(\zeta,t)=\sin(\zeta),\, u_1^{ABC}(\zeta,t)=\cos(\zeta)\left(\mu-\frac{\mu t^{\mu}}{\Gamma(\mu+1)}-1\right),$$
$$u_2^{ABC}(\zeta,t)=-\sin(\zeta)\left((\mu-1)^2+\frac{\mu^2t^{2\mu}}{\Gamma(2\mu+1)}-\frac{2(\mu-1)\mu t^{\mu}}{\Gamma(\mu+1)}\right),\ldots$$

Substituting $u_0^{ABC}(\zeta,t), u_1^{ABC}(\zeta,t),\ldots$ in (19), we obtain NTDM$_{ABC}$ solution.

For $\mu=1$, $u^C(\zeta,t)$,$u^{CF}(\zeta,t)$ and $u^{ABC}(\zeta,t)$ approximations converge to the exact solution. Table 1 presents absolute error.

Table 1 Absolute errors of NTDM$_C$, NTDM$_{CF}$, and NTDM$_{ABC}$ of Example 5 for $\mu=1$

ζ	t	NTDM$_C$	NTDM$_{CF}$	NTDM$_{ABC}$
−5	0.25	1.7764e − 15	1.7764e − 15	1.7764e − 15
	0.5	3.9524e − 12	3.9524e − 12	3.9524e − 12
	0.75	3.6228e − 10	3.6228e − 10	3.6228e − 10
	1.	9.05197e − 9	9.05197e − 9	9.05197e − 9
0	0.25	5.9674e − 15	5.9674e − 15	5.9674e − 15
	0.5	1.2213e − 11	1.2213e − 11	1.2213e − 11
	0.75	1.0543e − 09	1.0543e − 09	1.0543e − 09
	1.	2.4892e − 08	2.4892e − 08	2.4892e − 08
5	0.25	1.5543e − 15	1.5543e − 15	1.5543e − 15
	0.5	2.9763e − 12	2.9763e − 12	2.9763e − 12
	0.75	2.3584e − 10	2.3584e − 10	2.3584e − 10
	1.	5.0700e − 09	5.0700e − 09	5.0700e − 09

Example 6 Consider TFD-Kdv equation:

$$u_t^\mu = -u_{\zeta\zeta\zeta} - u_{yyy} \tag{21}$$

with initial condition $u(\zeta, y, 0) = \cos(\zeta + y)$. When $\mu = 1$, the exact solution is $u(\zeta, y, t) = \cos(\zeta + 2t + y)$[32].

NTDM$_C$ With the aid of NTDM$_C$, we get the successive solution as

$$u_0^C(\zeta, y, t) = \cos(\zeta + y), u_1^C(\zeta, y, t) = -\frac{2t^\mu \sin(\zeta + y)}{\Gamma(\mu + 1)}, u_2^C(\zeta, y, t)$$
$$= -\frac{4t^{2\mu} \cos(\zeta + y)}{\Gamma(2\mu + 1)}, \ldots.$$

Substituting $u_0^C, u_1^C, \ldots$ in (17), we obtain the NTDM$_C$ series solution.

NTDM$_{CF}$ With the aid of NTDM$_{CF}$, we get the successive solution as

$$u_0^{CF}(\zeta, y, t) = \cos(\zeta + y), u_1^{CF}(\zeta, y, t) = 2(\mu + \mu(-t) - 1) \sin(\zeta + y),$$
$$u_2^{CF}(\zeta, y, t) = -2\left(\mu^2\left(t^2 - 4t + 2 \right. \right. + 4\mu(t - 1) + 2 \ \cos(\zeta + y), \ldots.$$

Substituting $u_0^{CF}, u_1^{CF} \ldots$ in(18), we obtain the NTDM$_{CF}$ series solution.

NTDM$_{ABC}$ With the aid of NTDM$_{ABC}$, we get the successive solution as

$$u_0^{ABC}(\zeta, y, t) = \cos(\zeta + y),$$
$$u_1^{ABC}(\zeta, y, t) = 2\left(\mu - \frac{\mu t^\mu}{\Gamma(\mu + 1)} - 1\right) \sin(\zeta + y),$$
$$u_2^{ABC}(\zeta, y, t) = -4\left(\frac{\mu^2 t^{2\mu}}{\Gamma(2\mu + 1)} - \frac{2(\mu - 1)\mu t^\mu}{\Gamma(\mu + 1)} + (\mu - 1)^2\right) \cos(\zeta + y), \ldots.$$

Substituting $u_0^{ABC}, u_1^{ABC}, \ldots$ in (19), we obtain NTDM$_{ABC}$ solution. For $\mu = 1$, $u^C(\zeta, y, t)$,$u^{CF}(\zeta, y, t)$ and $u^{ABC}(\zeta, y, t)$ approximations converge to the exact solution. Table 2 presents NTDM$_C$, NTDM$_{CF}$, and NTDM$_{ABC}$ absolute error.

Example 7 Consider nonhomogeneous TFD-Kdv equation:

$$u_t^\mu = -u_{\zeta\zeta\zeta} - \sin(\pi\zeta) \sin(t) - \cos(\pi\zeta) \cos(t) \tag{22}$$

with $u(\zeta, 0) = \sin(\pi\zeta)$. When $\mu = 1$, the exact solution is $u(\zeta, t) = \cos(t) \sin(\pi\zeta)$ [32].

Table 2 Absolute errors of $NTDM_C$, $NTDM_{CF}$, and $NTDM_{ABC}$ of Example 6 for $\mu = 1$

ζ	t	$NTDM_C$	$NTDM_{CF}$	$NTDM_{ABC}$
−5	0.25	8.9101e − 12	8.9101e − 12	8.9101e − 12
	0.5	1.7481e − 08	1.7481e − 08	1.7481e − 08
	0.75	1.4417e − 06	1.4417e − 06	1.4417e − 06
	1.	3.2384e − 05	3.2384e − 05	3.2384e − 05
0	0.25	1.0552e − 11	1.0552e − 11	1.0552e − 11
	0.5	2.2068e − 08	2.2068e − 08	2.2068e − 08
	0.75	1.9420e − 06	1.9420e − 06	1.9420e − 06
	1.	4.6607e − 05	4.6607e − 05	4.6607e − 05
5	0.25	2.9237e − 12	2.9237e − 12	2.9237e − 12
	0.5	4.9617e − 09	4.9617e − 09	4.9617e − 09
	0.75	3.3994e − 07	3.3994e − 07	3.3994e − 07
	1.	5.9422e − 06	5.9422e − 06	5.9422e − 06

$NTDM_C$ With the aid of $NTDM_C$, we get the successive solution as

$$u_0^C(\zeta,t) = -\frac{\pi^3 t^{\mu}\cos(\pi\zeta)}{\Gamma(1+\mu)} + \frac{\pi^3 t^{2+\mu}\cos(\pi\zeta)}{\Gamma(3+\mu)} - \frac{\pi^3 t^{4+\mu}\cos(\pi\zeta)}{\Gamma(5+\mu)} + \sin(\pi\zeta) - \frac{t^{1+\mu}\sin(\pi\zeta)}{\Gamma(2+\mu)} + \frac{t^{3+\mu}\sin(\pi\zeta)}{\Gamma(4+\mu)} - \frac{t^{5+\mu}\sin(\pi\zeta)}{\Gamma(6+\mu)},$$

$$u_1^C(\zeta,t) = -\pi^3 t^{\mu}\Big(-\frac{\cos(\pi\zeta)}{\Gamma(1+\mu)} + t^{\mu}\Big(\frac{t\cos[\pi\zeta]}{\Gamma(2+2\mu)} - \frac{t^3\cos(\pi\zeta)}{\Gamma(4+2\mu)} + \frac{t^5\cos(\pi\zeta)}{\Gamma(6+2\mu)} - \frac{\pi^3\sin(\pi\zeta)}{\Gamma(1+2\mu)}\Big)\Big) - \pi^3 t^{2\mu}\left(\frac{\pi^3 t^2\sin(\pi\zeta)}{\Gamma(3+2\mu)} - \frac{\pi^3 t^4\sin(\pi\zeta)}{\Gamma(5+2\mu)}\right), \ldots.$$

Substituting $u_0^C(\zeta,t)$, $u_1^C(\zeta,t)$, ... in(17), we obtain the $NTDM_C$ series solution. Similarly, we will get $NTDM_{CF}$ and $NTDM_{ABC}$ solutions. For $\mu = 1$, $u^C(\zeta,t)$, $u^{CF}(\zeta,t)$ and $u^{ABC}(\zeta,t)$ approximations converge to the exact solution.

Figure 1 demonstrates $NTDM_C$, $NTDM_{CF}$, and $NTDM_{ABC}$ solutions for non-integer μ values.

6 Conclusions

In this article, we examined the approximate solutions of the TFTDEs based on the C, CF, and ABC fractional derivative operators using the NTDM. The behavior of the derived series solution has been represented in terms of 3D graphs for various fractional orders. The derived solutions converge extremely fast to the actual solutions, indicating that approximate solutions are quite close to exact solutions. The NTDM technique is a good solution for a wide range of nonlinear fractional

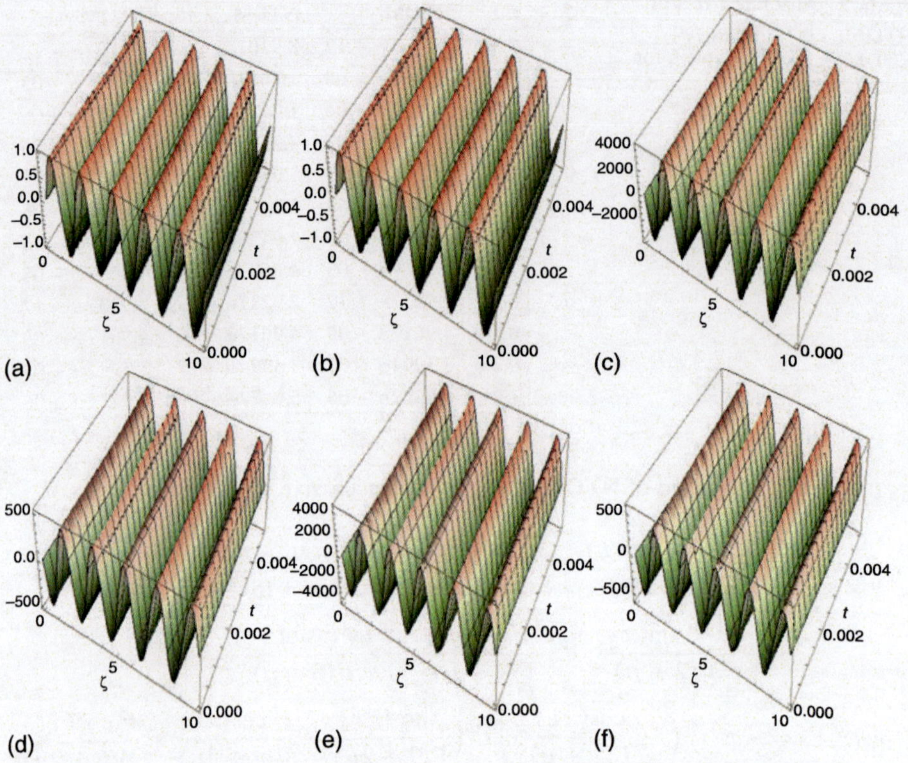

Fig. 1 Approximate solutions of Example 7. (**a**) NTDM$_C$ for $\mu = 0.75$. (**b**) NTDM$_C$ for $\mu = 0.85$. (**c**) NTDM$_{CF}$ for $\mu = 0.75$. (**d**) NTDM$_{CF}$ for $\mu = 0.85$. (**e**) NTDM$_{ABC}$ for $\mu = 0.75$. (**f**) NTDM$_{ABC}$ for $\mu = 0.85$

differential equations that arise in mathematical physics since its principles are straightforward and it is efficient in solving such equations.

References

1. Agarwal, P., El-Sayed, A.A.: Non-standard finite difference and Chebyshev collocation methods for solving fractional diffusion equation. Phys. A: Stat. Mech. Appl. **500**, 40–49 (2018)
2. Baleanu, D., Wu, G.C., Zeng, S.D.: Chaos analysis and asymptotic stability of generalized Caputo fractional differential equations. Chaos Solitons Fractals **102**, 99–105 (2017)
3. Kumar, D., Seadawy, A.R., Joardar, A.K.: Modified Kudryashov method via new exact solutions for some conformable fractional differential equations arising in mathematical biology. Chin. J. Phys. **56**(1), 75–85 (2018)
4. Imtiaz, A., Foong., O.M., Aamina, A., Khan, N., Ali, F., Khan, I.: Generalized model of blood flow in a vertical tube with suspension of gold nanomaterials: applications in the cancertherapy. Comput. Mater. Continua. **65**(1), 171–192 (2020)

5. Srivastava, H.M., Kumar, D., Singh, J.: An efficient analytical technique for fractional model of vibration equation. Appl. Math. Model. **45**, 192–204 (2017)
6. Jleli, M., Kumar, S., Kumar, R., Samet, B.: Analytical approach for time fractional wave equations in the sense of Yang-Abdel-Aty-Cattani via the homotopy perturbation transform method. Alexandria Eng. J. **59**(5), 2859–2863 (2020)
7. Prakash, A., Goyal, M., Gupta, S.: q-homotopy analysis method for fractional Bloch model arising in nuclear magnetic resonance via the Laplace transform. Indian J. Phys. **94**(4), 507–520 (2020)
8. Kanth, A.S.V.R., Aruna, K.: Comparison of two dimensional DTM and PDTM for solving time-dependent Emden-Fowler type equations. Int. J. Nonlinear Sci. **13**, 228–239 (2012)
9. Wu, G.C., Baleanu, D.: Variational iteration method for fractional calculus-a universal approach by Laplace transform. Adv. Differ. Equ. **2013**(1), 1–9 (2013)
10. Ravi Kanth, A.S.V., Deepika, S.: Application and analysis of spline approximation for time fractional mobile–immobile advection dispersion equation. Numer. Methods Part. Differ. Equ. **34**(5), 1799–1819 (2018)
11. Kanth, A.S.V., Sirswal, D.: Analysis and numerical simulation for a class of time fractional diffusion equation via tension spline. Numer. Algorithm. **79**(2), 479–497 (2018)
12. Kumar, P., Ravi Kanth, A.S.V.: Computational study for a class of time-dependent singularly perturbed parabolic partial differential equation through tension spline. Comput. Appl. Math. **39**(3), 1–19 (2020)
13. Prakash, A., Kumar, M.: Numerical method for fractional dispersive partial differential equations. Commun. Numer. Anal. **1**, 1–18 (2017)
14. Kanth, A.R., Aruna, K.: Solution of fractional third-order dispersive partial differential equations. Egyptian J. Basic Appl. Sci. **2**(3), 190–199 (2015)
15. Shah, R., Khan, H., Arif, M., Kumam, P.: Application of Laplace–Adomian decomposition method for the analytical solution of third-order dispersive fractional partial differential equations. Entropy **21**(4), 335 (2019)
16. Rawashdeh, M., Maitama, S.: Finding exact solutions of nonlinear PDEs using the natural decomposition method. Math. Methods Appl. Sci. **40**(1), 223–236 (2017)
17. Rawashdeh, M.S., Maitama, S.: Solving coupled system of nonlinear PDE's using the natural decomposition method. Int. J. Pure Appl. Math. **92**(5), 757–776 (2014)
18. Kanth, A.R., Aruna, K., Raghavendar, K., Rezazadeh, H., Inc, M.: Numerical solutions of nonlinear time fractional Klein-Gordon equation via natural transform decomposition method and iterative Shehu transform method. J. Ocean Eng. Sci. (2021). https://doi.org/10.1016/j.joes.2021.12.002
19. Veeresha, P., Prakasha, D.G., Baskonus, H.M.: Novel simulations to the time-fractional Fisher's equation. Math. Sci. (Springer) **13**(1), 33–42 (2019)
20. Zhou, M.X., Kanth, A.S.V., Aruna, K., Raghavendar, K., Rezazadeh, H., Inc, M., Aly, A.A.: Numerical solutions of time fractional Zakharov-Kuznetsov equation via natural transform decomposition method with nonsingular kernel derivatives. J. Funct. Spaces **2021**, 9884027 (2021)
21. Adivi Sri Venkata, R.K., Kirubanandam, A., Kondooru, R.: Numerical solutions of time fractional Sawada Kotera Ito equation via natural transform decomposition method with singular and nonsingular kernel derivatives. Math. Methods Appl. Sci. **44**(18), 14025–14040 (2021)
22. Koppala, P., Kondooru, R.: An efficient technique to solve time-fractional Kawahara and modified Kawahara equations. Symmetry **14**(9), 1777 (2022)
23. Diethelm, K.: The Analysis of Fractional Differential Equations. Lecture Notes in Mathematics, vol. 2004. Springer, Berlin (2010)
24. Kilbas, A.A., Srivastava, H.M., Trujillo, J.J.: Theory and Applications of Fractional Differential Equations. North-Holland Mathematics Studies, vol. 204. Elsevier Science B.V., Amsterdam (2006)
25. Caputo, M.: Linear models of dissipation whose Q is almost frequency independent-II. Geophys. J. Int. **13**(5), 529–539 (1967)

26. Caputo, M., Fabrizio, M.: A new definition of fractional derivative without singular kernel. Program. Fract. Diff. Appl. **1**(2), 73–85 (2015)
27. Atangana, A., Baleanu, D.: New fractional derivatives with nonlocal and non-singular kernel: theory and application to heat transfer model. Thermal Sci. **20**(2), 763–769 (2016)
28. Khan, Z.H., Khan, W.A.: N-transform properties and applications. NUST J. Eng. Sci. **1**(1), 127–133 (2008)
29. Loonker, D., Banerji, P.K.: Solution of fractional ordinary differential equations by natural transform. Int. J. Math. Eng. Sci. **12**(2), 1–7 (2013)
30. Khalouta, A., Kadem, A.: A new numerical technique for solving fractional Bratu's initial value problems in the Caputo and Caputo-Fabrizio sense. J. Appl. Math. Comput. Mech. **19**(1), 43–56 (2020)
31. Rawashdeh, M.S.: The fractional natural decomposition method: theories and applications. Math. Methods Appl. Sci. **40**(7), 2362–2376 (2017)
32. Wazwaz, A.M.: An analytic study on the third-order dispersive partial differential equations. Appl. Math. Comput. **142**(2–3), 511–520 (2003)

Choosing Between Vaccine Efficacy and Vaccine Price: A Mathematical Model for COVID-19

Mo'tassem Al-arydah

1 Introduction

COVID-19, a new coronavirus, is a disease first detected in Wuhan, China. The infection that causes COVID-19 is spread primarily through droplets when an infected person coughs, gasps, or exhales. These droplets can remain airborne and eventually land on floors or surfaces. As of February 17, 2021, there are more than 110.5 million confirmed cases of COVID-19 around the world, with approximately 2.5 million people who have lost their lives [1].COVID-19 has high transmission and death rates, which is why it has become a major concern for researchers, governments, and health organizations. Scientists have started exploring superior way to control the transmission of the disease. Mathematicians have started looking for realistic models to understand the spread of the disease and the optimal ways to combat it.

There are several studies on COVID-19 based on reported data [2–6]. For example, a simple mathematical model for susceptible, exposed, infected, and recovered (SEIR) for transmission of COVID-19 is proposed and analyzed in [7], in which the authors calculate the basic reproductive number based on data estimated from Wuhan, China. More complicated models that consider more compartments are found in literature, for example, see [8]. Social distancing has been shown to definitely control the epidemic; however, it may harm the country economically. In [9] we use optimal control theory to search for optimal vaccine distribution that controls the COVID-19 disease with the lowest possible cost.

M. Al-arydah (✉)
Department of Mathematics, Khalifa University, Abu Dhabi, UAE
e-mail: motassem.alarydah@ku.ac.ae

F. Kamalov et al. (eds.), *Advances in Mathematical Modeling and Scientific Computing*, Trends in Mathematics, https://doi.org/10.1007/978-3-031-41420-6_9

In this work, we seek the answers for the following questions:

1. Should the priority be given to the vaccine price or efficacy when choosing between two vaccines? Is the price the only key factor when determining the total cost?
2. What is the optimal way to introduce combination of the two vaccines for COVID-19? Do the optimal vaccines control the disease within a 90-day period?

We will present the work as follows: In Sect. 2, we create a mathematical model for COVID-2019. In Sect. 3, we derive formulas for the disease-free equilibrium (DFE) and the basic reproduction number (threshold). We show that the DFE is LAS when the threshold is less than or equal to one and unstable otherwise. We present an optimal control problem and show the existence and uniqueness of the optimal vaccine solution. Then, we estimate the optimal vaccination for the optimal problem. Finally, we finish with a summary in Sect. 4.

2 Background

For our model, we implemented the following assumptions: (1) this model assumes homogeneous mixing among individuals, that is, all people have the same chance of getting the disease whenever exposed to COVID-19; (2) birth rate is ignored as we study the disease in a short period; (3) recovered individuals clean the virus without everlasting immunity; (4) we consider two types of vaccine with different efficacies and prices; (5) recovery rate for infected vaccinated individuals is higher when efficacy rate is higher; and (6) vaccines protect against serious symptoms and prevent death. To form the model, we divide the population into six categories depending on disease status and immunization condition. The categories at time t are susceptible (unvaccinated) ($S(t)$), infected ($I(t)$), vaccinated with vaccine type 1 ($V_1(t)$),vaccinated with vaccine type 2 ($V_2(t)$), recovered with permanent immunity $R(t)$, and passed away from the disease $D(t)$. Susceptibles become infected when coming into contact with infected individuals, and β is the transmission rate. Susceptibles are vaccinated at percentages of vaccinated per unit time $u_1(t)$ and $u_2(t)$ are t. $u_1(t) + u_2(t) \leq 1$ and enter the vaccinated classes V_1 and V_2, respectively. Vaccinated individuals may get infected with the reduced transmission rates $(1 - \epsilon_i)\beta$, $i = 1, 2$. Here $0 < \epsilon_i < 1$ is the efficacy of the vaccine. The parameter η $(0 < \eta < 1)$ means a decrease in the infectivity of vaccinated persons. Infected individuals recover either at rate r_1 for unvaccinated or r_2 for vaccinated. Unvaccinated infected may die from the disease at the rate d and enter the induced death class $D(t)$. Λ and μ are the birth rate and the natural death rate, respectively.

The dynamics of this model is represented by the following ODEs:

$$
\begin{aligned}
\frac{dS}{dt} =& \Lambda + r_1 I + r_2 I_v - \beta S(I + \eta I_v) - (u_1(t) + u_2(t) + \mu)S \\
\frac{dI}{dt} =& \beta S(I + \eta I_v) - (r_1 + d + \mu)I \\
\frac{dI_v}{dt} =& (1 - \epsilon_1)\beta V_1(I + \eta I_v) + (1 - \epsilon_2)\beta V_2(I + \eta I_v) - (r_2 + \mu)I_v \\
\frac{dV_1}{dt} =& u_1(t)S - (1 - \epsilon_1)\beta V_1(I + \eta I_v) - \mu V_1 \\
\frac{dV_2}{dt} =& u_2(t)S - (1 - \epsilon_2)\beta V_2(I + \eta I_v) - \mu V_2 \\
\frac{dD}{dt} =& dI
\end{aligned}
\tag{1}
$$

Note that the last equation is giving no extra information to the model and for this reason will be dropped from the analysis.

3 Results and Findings

3.1 *Stability of the DFE*

Here we assume u_1 and u_2 constants. The DFE for the dynamical system (1) is $(\overline{S}, 0, 0, \overline{V}_1, \overline{V}_2)$, with

$$
\begin{aligned}
\overline{S} &= \frac{\Lambda}{u_1 + u_2 + \mu}, \\
\overline{V}_1 &= \frac{u_1}{\mu}\overline{S}, \\
\overline{V}_2 &= \frac{u_2}{\mu}\overline{S}
\end{aligned}
\tag{2}
$$

Using the next-generation matrix with

$$
F = \begin{bmatrix} \beta\overline{S} & \eta\beta\overline{S} \\ \beta((1 - \epsilon_1)\overline{V}_1 + (1 - \epsilon_2)\overline{V}_2) & \eta\beta((1 - \epsilon_1)\overline{V}_1 + (1 - \epsilon_2)\overline{V}_2) \end{bmatrix}
$$

$$
V = \begin{bmatrix} r_1 + d + \mu & 0 \\ 0 & r_2 + \mu \end{bmatrix}
$$

Note that $\det(FV^{-1}) = 0$; therefore, $R_u := \rho(FV^{-1})$, which is the trace of this matrix. Therefore,

$$R_u = \frac{\Lambda\beta}{r_1 + d + \mu}\Big(\frac{1}{r_1 + d + \mu} + \frac{\eta((1-\epsilon_1)u_1 + (1-\epsilon_2)u_2)}{\mu(r_2 + \mu_2)}\Big) \tag{3}$$

R_u represents the COVID-19 disease threshold value which measures the mean number of secondary infections created by one infected individual when placed in a population of susceptibles that are partially vaccinated at a demographic steady state. The following theorem is established using [10].

Theorem 1 *The DFE of the system* (1)*, given by* (2)*, is LAS if* $R_u \leq 1$ *and unstable if* $R_u > 1$.

3.2 *Optimal Control Model and Analysis*

Assume that $u_1(t)$ and $u_2(t)$ are Lebesgue-measurable functions of time (integrable). Over the time period $[0, t_f]$, we introduce the following total cost objective functional:

$$J(u_1, u_2) = \int_0^{t_f} (AI(t) + A_v I_v(t) + \frac{A_1}{2}u_1^2(t) + \frac{A_2}{2}u_2^2(t))dt \tag{4}$$

A, A_v, A_1, and A_2 are positive constants representing the relative costs of the implemented controls. $AI(t)$ and $A_v I_v(t)$ are the costs of infected vaccinated and unvaccinated. $A_1 u_1^2(t)$ and $A_2 u_2^2(t)$ are the costs of vaccines which are assumed nonlinear and estimated by quadratic simple function based on literature on optimal control of epidemics.

J will be minimized over all $(u_1, u_2) \in U$, with

$$U = \{(u_1(t), u_2(t))) : 0 \leq u_1(t), 0 \leq u_2(t),\ u_1(t) + u_2(t) \leq 1 \ and\ t \in [0, t_f]\} \tag{5}$$

Our aim is to identify (u_1^*, u_2^*) that minimize (4) over the control set U given in (5). The existence of an optimal solution can be proved using standard optimal control theory [11], and a similar proof can be found in [12].

Pontryagin's maximum principle [13] is now applied to determine the necessary conditions on the optimal control (u_1^*, u_2^*), with

$$J(u_1^*, u_2^*) = \min_{(u_1, u_2) \in U} J(u_1, u_2). \tag{6}$$

We need to minimize pointwise the following Hamiltonian function:

$$H \equiv AI + A_v I_v + \frac{A_1}{2}u_1^2 + \frac{A_2}{2}u_2^2 + \sum_{i=1}^{5} \lambda_i(t) g_i,$$

where $\lambda_i, g_i, i = 1, \ldots, 5$ are the adjoint functions and, respectively, the right-hand sides of the equations in the model (1). The following result includes the necessary conditions formulated by Pontryagin's maximum principle [13].

Theorem 2 *An optimal control (u_1^*, u_2^*) for (6) and a solution (S, I, I_v, V_1, V_2) for the system (1) exist. Furthermore, there exist λ_i, $i = 1, 2, \ldots, 5$, adjoint functions, that satisfy*

$$\frac{d\lambda_1}{dt} = (\lambda_1 - \lambda_2)\beta(I + \eta I_v) + \lambda_1(u_1 + u_2 + \mu) - \lambda_4 u_1 - \lambda_5 u_2)$$

$$\frac{d\lambda_2}{dt} = -A + \lambda_1(\beta S - r_1) - \lambda_2(\beta S - r_1 - d - mu) - \lambda_3((1-\epsilon_1)\beta V_1 + (1-\epsilon_2)\beta V_2)$$
$$+ \lambda_4(1-\epsilon_1)\beta V_1 + \lambda_5(1-\epsilon_2)\beta V_2);$$

$$\frac{d\lambda_3}{dt} = -A_v + \lambda_1(\beta\eta S - r_2) - \lambda_2(\beta\eta S(t)) - \lambda_3((1-\epsilon_1)\beta\eta V_1$$
$$+ (1-\epsilon_2)\beta\eta V_2 - r_2 - mu) + \lambda_4(1-\epsilon_1)\beta\eta V_1 + \lambda_5(1-\epsilon_2)\beta\eta V_2);$$

$$\frac{d\lambda_4}{dt} = (\lambda_4 - \lambda_3)(1-\epsilon_1)\beta(I + \eta I_v) + \lambda_4 \mu$$

$$\frac{d\lambda_5}{dt} = (\lambda_5 - \lambda_3)(1-\epsilon_2)\beta(I + \eta I_v) + \lambda_5 \mu$$

with $\lambda_i(T) = 0, i = 1, 2, 3, 4, 5$ that are the transversality conditions. Moreover, the controls fulfill the following:

$$u_1^* = min\left\{max\left\{\frac{(\lambda_1 - \lambda_4)S}{A_1}, 0\right\}, 1\right\},$$

$$u_2^* = min\left\{max\left\{\frac{(\lambda_1 - \lambda_5)S}{A_2}, 0\right\}, 1 - u_1^*\right\},$$

Proof The proof is similar to that carried out for Theorem 2 [14, 15]; for the sake of completeness, we repeat the proof here. The adjoint equations follow from $\frac{d\lambda}{dt} = -\frac{\partial H(x,u,\lambda)}{\partial x}$ that is the partial derivative of the Hamiltonian (4) with respect to the state $x = S, \ldots, V_2$. The transversality conditions $\lambda_i(t_f) = 0$, for $i = 1, 2, 3, 4, 5$, follow from considering the final values for the states $S(t_f), \ldots, V_2(t_f)$ which are free. The optimal controls u_1^* and u_2^* are determined as follows: within the interior of the control set U, we have

$$\frac{\partial H}{\partial u_1} = A_1 u_1 - \lambda_1 S + \lambda_4 S = 0 \Rightarrow u^* = \frac{(\lambda_1 - \lambda_4)S}{A_1}.$$

and

$$\frac{\partial H}{\partial u_2} = A_2 u_2 - \lambda_1 S + \lambda_5 S = 0 \Rightarrow u^* = \frac{(\lambda_1 - \lambda_5)S}{A_2}.$$

Now when considering the bounds in the set of admissible controls U, we have

$$u_1^* = \begin{cases} 0, & if \;\; A_1 u_1 - \lambda_1 S + \lambda_4 S < 0, \\ u_1^*, & if \;\; 0 \leq A_1 u_1 - \lambda_1 S + \lambda_4 S = A_1, \\ 1, & if \;\; A_1 u_1 - \lambda_1 S + \lambda_4 S > A_1. \end{cases}$$

$$u_2^* = \begin{cases} 0, & if \;\; A_2 u_2 - \lambda_1 S + \lambda_5 S < 0, \\ u_2^*, & if \;\; 0 \leq A_2 u_2 - \lambda_1 S + \lambda_5 S = A_2(1 - u_1^*), \\ 1 - u_1^*, & if \;\; A_2 u_2 - \lambda_1 S + \lambda_5 S > A_2(1 - u_1^*). \end{cases}$$

which may be written on the compact form (7). Also, from the second derivative of the Hamiltonian function, we have

$$\frac{\partial^2 H}{\partial u} = \begin{bmatrix} A_1 & 0 \\ 0 & A2 \end{bmatrix} \tag{7}$$

which is a positive definite matrix. As a result, the optimal control found is a minimizer for J. □

3.3 *Estimating Optimal Solution*

In this section, we estimate the optimal vaccine that minimizes the cost function within finite time $[0, t_f]$. As explained in [16], we use Runge-Kutta's fourth-order (RK4) method [17] to solve (6) subject to (1) in which education parameter is considered function of time. Starting with the initial control guess $(u_1, u_2) = (0, 0)$, parameters' values and the initial conditions are shown in Tables 1 and 2, respectively. We will consider the weight constants $A_1 = 1000$ and $A_2 = 10$ USD.

The code is carried out in MATLAB, as follows: The state variable equations (1) are initially solved by the RK4, with the initial conditions given in Table 2, with $t_f = 90$ days length and the step length is 0.01. Resulting state values are located in adjoint equations (7). Then, by using the backward RK4, we solved the adjoint equations (7) with the transversality conditions. The controls (7) are updated using the estimated states and adjoint solutions. This procedure will be repeated continuously until the current state, and adjoint and optimal control values converge in the consecutive process. That means the final iteration value has a sufficiently close value to the iteration before it in this process.

Table 1 Sample values

Parameter	Value	Unit	Reference
A	1000	USD	Assumed
A_v	300	USD	Assumed
A_1	400	USD	Assumed
A_2	200	USD	Assumed
Λ	5	Days$^-$1	Estimated
μ	1/(36500)	Days$^-$1	Estimated
r_1	1/14	Days$^-$1	Estimated
r_2	1/7	Days$^-$1	Estimated
ϵ_1	0.9	–	Assumed
ϵ_2	0.6	–	Assumed
d	0.03	Days$^-$1	[18]
β	0.0009	days$^-$1	Estimated from [18]

Table 2 Initial conditions for the model

$S(0)$	10000
$I(0)$	20
$I_v(0)$	0
$V_1(0)$	0
$V_2(0)$	0

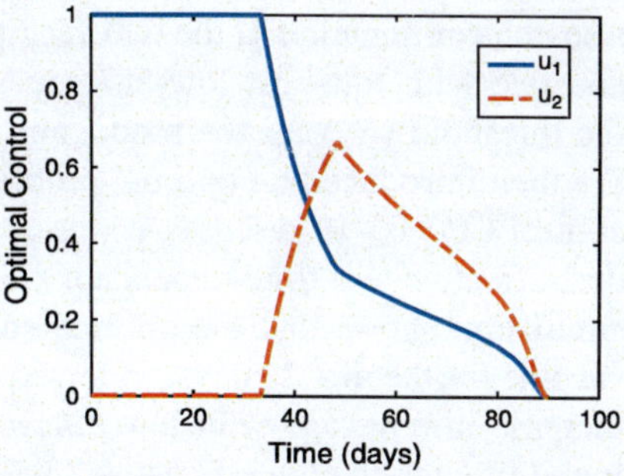

Fig. 1 Optimal vaccines

Figure 1 shows the optimal vaccines: an expensive one (400$ per individual) with high efficacy (90%) and a cheap one (200$ per individual) with low efficacy (60%). In the 90-day time period, the time-dependent optimal vaccines are: 100% high-efficacy coverage for 33 days which can be reduced gradually to 35% coverage at day 45 at the same time increasing the low-efficacy cheap vaccine starts from zero coverage to 65% coverage. Then both can be gradually decreased to reach zero coverage at day 90.

Figure 2 is evaluating the cost function and number of infected individuals for the optimal vaccines and constant vaccines. The figure indicates that optimal vaccines have the least cost and the least number of infected. In summary, we have found that the most effective way is to introduce vaccines that not only have the lowest cost but also can manage the COVID-19 disease in a 90-day period.

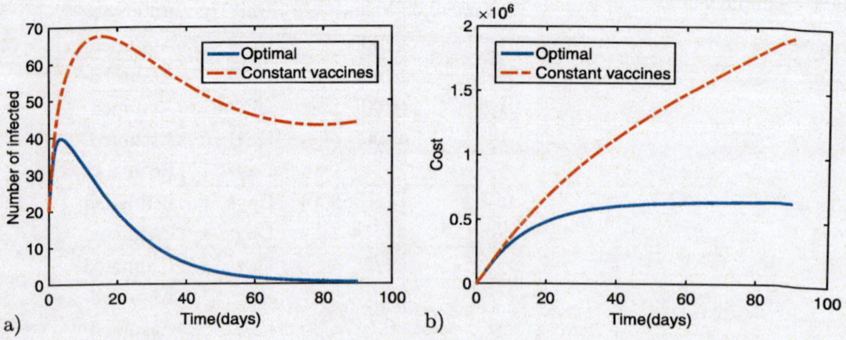

Fig. 2 (**a**) The number of infected, (**b**) the cost for optimal vaccines and constant vaccines $u_1 = u_2 = 0.5$ scenarios

4 Conclusion and Future Scope

We have implemented the SVIS (susceptible-vaccinated-infectious-susceptible) compartment disease model for COVID-19 to find the optimal way to introduce the vaccine to the population at the lowest cost and with the best control. We then have a search combination of the two vaccines, one with high-cost high-efficacy vaccines and one with low-cost low-efficacy vaccines. We have established a formula for the threshold (R_0) for the model and then used it to discuss the LAS of the DFE. We then introduced an optimal problem to find the best allocation of vaccines that control COVID-19 diseases within a 90-day period using two vaccines. The ideal cost-effective and time-dependent way to introduce vaccines has been found. The simulation shows that the optimal solution has full coverage of the high-potency vaccine for the first 33 days, followed by a gradual reduction and an equally gradual increase in coverage with low efficacy. We have shown that the optimal vaccines found provide the lowest cost and best control, by comparing its cost with the cost of constant 50–50 high- and low-efficacy vaccine coverage. We should point out that the results found are sensitive to parameters' values; for this reason, one needs to be careful before making use of them.

References

1. WHO, Coronavirus disease, Situation Report – 207. https://www.who.int/emergencies/diseases/novel-coronavirus-2019/situation-reports
2. Zhao, S., Lin, Q., Ran et al.: Preliminary estimation of the basic reproduction number of novel coronavirus (2019-nCoV) in China, from 2019 to 2020: a data-driven analysis in the early phase of the outbreak. Int. J. Infectious Dis. **92**, 214–217 (2020)
3. Zhao, S., Musa, S.S., Lin Q., et al.: Estimating the unreported number of novel coronavirus (2019-nCoV) cases in China in the first half of January 2020: a data-driven modelling analysis of the early outbreak. J. Clin. Med. **9**(2), 388 (2020)

4. Hu, Z., Cui, Q., et al.: Evaluation and prediction of the COVID-19 variations at different input population and quarantine strategies, a case study in Guangdong province, China. Int. J. Infectious Dis. **95**(2020), 231–240
5. Tang, B., Wang, X., Li, Q., et al.: Estimation of the transmission risk of the 2019-nCoV and its implication for public health interventions. J. Clin. Med. **9**, 462 (2020)
6. Wu, J., Tang, B., Bragazzi, N.L., Nah, K., McCarthy, Z.: Quantifying the role of social distancing, personal protection and case detection in mitigating COVID-19 outbreak in Ontario, Canada. J. Math. Ind. **10**(1), 15 2020. https://doi.org/10.1186/s13362-020-00083-3
7. Lina, Q., Zhaob, S., Gaod, D., Loue, Y., Yangf, S., Musae, S.S., Wangb, M.H., Caig, Y., Wangg, W., Yangh, L., Hee, D.: A conceptual model for the coronavirus disease 2019 (COVID-19) outbreak in Wuhan, China with individual reaction and governmental action. Int. J. Infectious Dis. **95**, 211–216 (2020)
8. DarAssi, M., Shatnawi, T., Safi, M.: Mathematical analysis of a MERS-Cov coronavirus model. Demonstratio Math. **55**(1), 265–276 (2022)
9. Olaniyi, S., Obabiyi, O.S., Okosun, K.O., Oladipo, A.T., Adewale, S.O.: Mathematical modelling and optimal cost-effective control of COVID-19 transmission dynamics. Eur. Phys. J. Plus **135**, 938 (2020). https://doi.org/10.1140/epjp/s13360-020-00954-z
10. Van Den Driessche, P., Watmough, J.: Reproduction numbers and sub-threshold endemic equilibria for compartmental models of disease transmission. Math. Biosci. **180**, 29–48 (2002)
11. Lenhart, S., Workman, J.T.: Optimal control applied to biological models. Chapman and Hall/CRC, New York (2007)
12. Alrabaiah, H., Safi, M. , DarAssi, M., Al-Hdaibat, B., Ullah, S., Khan, M., Shah, S.: Optimal control analysis of hepatitis B virus with treatment and vaccination. Results Phys. **19**, 103599 (2020)
13. Pontryagin's, L., Boltyanskii, V.: The Mathematical Theory of Optimal Processes. Golden and Breach Science Publishers, London (1980)
14. Madhu, K., Al-arydah, M.: Optimal vaccine for human papillomavirus and age-difference between partners. Math. Comput. Simul. **185**, 325–346 (2021)
15. Al-Arydah, M.: Two-sex logistic model for human papillomavirus and optimal vaccine. Int. J. Biomath. **14**(3), 2150011 (2021)
16. Al-Arydah, M., Malik, T.: An age-structured model of the human papillomavirus dynamics and optimal vaccine control. Int. J. Biomath. **10**(6), 1750083 (2017)
17. Devipriya, G., Kalaivani, K.: Optimal control of multiple transmission of water-borne diseases. Int. J. Math. Math. Sci. **2012**, Article ID 421419 (2012)
18. Al-arydah, M., Berhe, H., Dib, K., Madhu, K.: Mathematical modeling of the spread of the coronavirus under strict social restrictions. Math. Methods Appl. Sci. 1–11 (2021). https://doi.org/10.1002/mma.7965

[illegible]

[illegible]

[illegible]

[illegible]

[illegible]

[illegible]

[illegible]

[illegible]

[illegible]

[illegible]

[illegible]

[illegible]

[illegible]

Classification of Cosmological Wormhole Solutions in the Framework of General Relativity

Usamah S. Al-Ali

1 Introduction

The cosmic speed limit imposed by special relativity makes the human dream of interstellar travel far beyond our present capabilities. If the speed restriction is taken into account, straightforward calculations show that the exploration of nearby stars would take a few human lifetimes no matter how technologically advanced our starship industry. Nevertheless, general relativity allows the existence of shortcuts in the spacetime fabric connecting two distant regions admitting unlimited possibilities of interstellar travel. This hypothetical shortcut is called wormhole, and it was introduced in literature as early as 1916 by Flamm [1], and then by Einstein and Rosen in 1935 [2], and was later stimulated by Morris and Thorne [3]. These efforts culminated in the construction of a theoretical time machine using a static, spherically symmetric metric [4].

However, exotic matter is needed in order to support the geometry of wormhole as the energy conditions are inevitably violated in GR [3, 4]. This violation raises skepticism about the existence of wormhole in the real world. Recall that such violation could be avoided if alternative or modified theories of gravity are used where successful attempts were made to construct wormhole solutions that fulfill at least the null energy conditions; see, for instance, [5–8].

It turned out that the violation of energy conditions could be avoided even in Einstein's theory if the wormhole is allowed to evolve. In this context, a non-stable evolving wormhole geometry was constructed by Kar where the weak energy conditions could be respected for arbitrarily large intervals of time [9]. Furthermore,

U. S. Al-Ali (✉)
Basic Science Department, College of Science and Theoritical Studies, Saudi Electronic University, Riyadh, Saudi Arabia
e-mail: U.ALALI@seu.edu.sa

F. Kamalov et al. (eds.), *Advances in Mathematical Modeling and Scientific Computing*, Trends in Mathematics, https://doi.org/10.1007/978-3-031-41420-6_10

a cosmological model with traversable wormhole was considered by Kim where the need for exotic matter was minimized [10].

Motivated by the fact that the expansion of space could be exploited to minimize or even eliminate the need of exotic matter for supporting wormhole geometry, we explore in this project the effect of the scale factor on the exoticity of matter supporting a cosmological traversable wormhole metric. Recall that we can construct a traversable wormhole metric by embedding a static wormhole in the Friedman universe. Such model has been already addressed in [11] where Chaplygin gas EoS was implemented to derive wormhole solutions supported by dark matter and dark energy. It was also studied in [12] in the context of Einstein-Cartan theory. In this work, we adopt different approach where we classify the possible closed-form solutions corresponding to the considered wormhole metric that may arise in GR and investigate their corresponding energy conditions. The classification is performed on the basis of a variable EoS given as $p = \omega(r)\rho$ where p and ρ represent respectively the radial pressure and the energy density.

The organization of this paper is given as follows. In the next section, we introduce the studied wormhole metric and the energy conditions. Then, EFEs are derived in Sect. 3 and classification of their solutions is performed in Sect. 4. Finally, concluding remarks and discussion are presented in the last section.

2 Cosmological Wormhole Metric and Energy Conditions

We consider the wormhole that is obtained by embedding the following static traversable wormhole metric:

$$ds^2 = e^{\Psi(r)}dt^2 - \frac{dr^2}{1 - \frac{b(r)}{r}} - r^2 d\Omega^2, \tag{1}$$

in the homogeneous expanding universe with the following line element:

$$ds^2 = dt^2 - a^2(t)[\frac{dr^2}{1 - kr^2} + r^2 d\Omega^2], \tag{2}$$

where Ψ is the red shift function, b is the shape function , $d\Omega^2 = d\theta^2 + sin^2\theta d\phi^2$ is the 2-sphere, $a(t)$ is the scale factor of the universe, and k is a constant representing the curvature of space. Recall that $k \in \{-1, 0, +1\}$ where $k = -1$, $k = 0$, and $k = +1$ for respectively open, flat, and closed universes.

Combining metrics (1) with (2) yields the following wormhole metric:

$$ds^2 = e^{\Psi(r)}dt^2 - a^2(t)[\frac{dr^2}{1 - kr^2 - \frac{b(r)}{r}} + r^2 d\Omega^2]. \tag{3}$$

In order to allow for time travel, we need to avoid the horizon which occurs if the red shift function Ψ is infinite. We restrict our attention to the zero radial tide case ($\Psi = 0$). As a result, our metric (3) is reduced to the following metric:

$$ds^2 = dt^2 - a^2(t)[\frac{dr^2}{1 - kr^2 - \frac{b(r)}{r}} + r^2 d\Omega^2]. \tag{4}$$

The shape function b has to satisfy the throat and the flare-out conditions given respectively as $b(r_0) = r_0$ and $\frac{b - rb'}{b^2} > 0$, where r_0 is the throat radius and $r \in (r_0, \infty)$. Further restrictions include $1 - \frac{b(r)}{r} > 0$ and $b'(r_0) < 1$.

On the other hand, the energy conditions were constructed to ensure the fulfillment of the conservation of energy-momentum tensor, i.e., $T^{ab}_{;} = 0$, and the energy densities for a matter field are almost always nonnegative. The null, weak, strong, and dominant energy conditions are given respectively as follows:

$$\rho + p_a \geq 0, \tag{NEC}$$

$$\rho \geq 0, \quad \rho + p_a \geq 0. \tag{WEC}$$

$$\rho + \sum_a p_a \geq 0, \quad \rho + p_a \geq 0, \tag{SEC}$$

$$\rho \geq 0, \quad -\rho \leq p_a \leq \rho. \tag{DEC}$$

3 Einstein Field Equations for Metric (4)

Recall that EFEs are given by the following tensor equation:

$$R_{\mu\nu} - \frac{1}{2} R g_{\mu\nu} = T_{\mu\nu}, \tag{5}$$

where $R_{\mu\nu}$, $g_{\mu\nu}$, and $T_{\mu\nu}$ are respectively the Ricci, the metric, and the energy-momentum tensors, while R is the Ricci scalar. We apply the field equations (5) to metric (4), and for the sake of simplicity, we use a proper reference frame obtained by rotating the basis to get orthonormal vectors which can be expressed as $e_t = e_{\hat{t}}$, $e_r = \frac{1}{\sqrt{1 - kr^2 - \frac{b(r)}{r}}} e_{\hat{r}}$; $e_\theta = r e_{\hat{\theta}}$ and $e_\phi = r sin\theta e_{\hat{\phi}}$.

Thus, the EFEs in the above given reference frame can be written as

$$\rho = 3\frac{\dot{a}^2}{a^2} + 3\frac{k}{a^2} + \frac{b'}{r^2 a^2}, \tag{6}$$

$$\tau = -2\frac{\ddot{a}}{a} - \frac{\dot{a}^2}{a^2} - \frac{k}{a^2} - \frac{b}{r^3 a^2}, \tag{7}$$

$$p = -2\frac{\ddot{a}}{a} - \frac{\dot{a}^2}{a^2} - \frac{k}{2a^2} - \frac{b'}{2r^2a^2} + \frac{b}{2r^3a^2}, \tag{8}$$

where the dot and the dash denote differentiation with respect to t and r, respectively, while the energy-momentum tensor is given by $T_{\mu\nu} = diag(\rho, -\tau, -p, -p)$, where τ and p represent respectively tangential and radial pressures.

4 Classification of Solutions for Variable EoS

We consider the relation: $p = \omega(r)\rho$, where ω is the EoS parameter. Applying this relation to Eqs. (6), (7), and (8) yields the following equation:

$$2a\ddot{a} + (1 + 3\omega)\dot{a}^2 + (\frac{1}{2} + 3w)k + (\frac{1}{2} + \omega)\frac{b'}{r^2} - \frac{b}{2r^3} = 0. \tag{9}$$

Differentiating Eq. (9) wrt to t and the resulting equation wrt r yields

$$\dot{a}(t)\dddot{a}(t)\omega'(r) = 0 \tag{10}$$

In the following analysis, we consider three possibilities arising from Eq. (10), namely, (1) $\omega'(r) = 0$, $\dddot{a}(t) \neq 0$, (2) $\dddot{a}(t) = 0$, $\omega'(r) \neq 0$, and 3) $\dddot{a}(t) = \omega'(r) = 0$.
Case I: $\omega'(r) = 0$ **and** $\dddot{a}(t) \neq 0$
In this case, $\omega = \omega_0$ and consequently $p = \omega_0\rho$. Therefore, the solution of Eq. (9) depends on the value of the parameter ω_0.
Case I-A ($\omega_0 = 0$): Recall that $\omega = 0$ corresponds to a pressureless matter such as dust and galaxies. In this case, we can write Eq. (9) in a separable form as follows:

$$2a\ddot{a} + \dot{a}^2 = \frac{b}{2r^3} - \frac{b'}{2r^2} - \frac{1}{2}k = \lambda, \tag{11}$$

where λ is the separation constant that can be zero or nonzero.
Case I-A-1 Zero Separation Constant ($\lambda = 0$): This case leads to the following shape function: $b(r) = (1 + \frac{k}{2}r_0^2)r - \frac{k}{2}r^3$ and the following scale factor: $a(t) = (c_1t + c_2)^{\frac{2}{3}}$, where c_1 and c_2 are constants. The wormhole conditions are satisfied only if the universe is flat or closed and generally violated otherwise. The solution of EFEs for this case is given as follows:

$$\rho = [\frac{4}{3}c_1^2(c_1t + c_2)^{-\frac{2}{3}} + \frac{1 + 0.5kr_0^2}{r^2} + 1.5k][c_1t + c_2]^{-\frac{4}{3}},$$

$$\tau = [-0.5k - \frac{1 + 0.5kr_0^2}{r^2}][c_1t + c_2]^{-\frac{4}{3}}, \quad p = 0. \tag{12}$$

Note that all the energy conditions are satisfied for flat and closed universes.

Case I-A-2 Nonzero Separation Constant ($\lambda \neq 0$): The scale factor in this case is any nonlinear solution of the equation:

$$2a\ddot{a} + \dot{a}^2 = \lambda, \tag{13}$$

while the shape function is given by $b(r) = (1 + 0.5r_0^2 + \lambda r_0^2)r - (\lambda + 0.5k)r^3$. Thus, the solution of EFEs is given by

$$\rho = (3\dot{a}^2 + 1.5k + \frac{1 + 0.5kr_0^2 + \lambda r_0^2}{r^2} - 3\lambda)a^{-2}(t),$$
$$\tau = (-0.5k - \frac{1 + 0.5kr_0^2 + \lambda r_0^2}{r^2})a^{-2}(t), \qquad p = 0. \tag{14}$$

A sufficient condition for all the energy conditions to be satisfied is $k > 3\lambda$.
Case I-B: $\omega_0 \neq 0$: First, we consider the case of zero separation constant.
Case I-B-1: Zero Separation Constant ($\lambda = 0$): In this case, Eq. (9) can be written as

$$(3\omega_0 + 1)\dot{a}^2 + 2a\ddot{a} = \frac{b}{2r^3} - (\omega_0 + 0.5)\frac{b'}{r^2} - (3\omega_0 + 0.5)k = 0, \tag{15}$$

leading to the following shape function and scale factor:

$$b(r) = (r_0^{\frac{2\omega_0}{2\omega_0+1}} + \frac{6\omega_0 + 1}{6\omega_0 + 2}kr_0^{3-\frac{1}{2\omega_0+1}})r^{\frac{1}{2\omega_0+1}} - \frac{6\omega_0 + 1}{6\omega_0 + 2}kr^3, \tag{16}$$

$$a(t) = (c_1t + c_2)^{\frac{2}{3(\omega_0+1)}}. \tag{17}$$

In fact, interesting cases can be considered. For instance, for relativistic matter ($\omega_0 = \frac{1}{3}$), the shape function (16) and the scale factor (17) respectively become

$$b(r) = (r_0^{\frac{2}{5}} + 0.75kr_0^{\frac{12}{5}})r^{\frac{3}{5}} - 0.75kr^3, \qquad a(t) = \sqrt{c_1t + c_2}. \tag{18}$$

In this case, the flare-out condition is satisfied without any restriction for a flat and a closed universe and the solution of EFEs is given as

$$\rho = \frac{3c_1^2}{4(c_1t + c_2)^2} + \frac{0.6(r_0^{\frac{2}{5}} + 0.75kr_0^{\frac{12}{5}})r^{\frac{-12}{5}} + 0.75k}{(c_1t + c_2)}$$
$$\tau = \frac{c_1^2}{4(c_1t + c_2)^2} - \frac{(r_0^{\frac{2}{5}} + 0.75kr_0^{\frac{12}{5}})r^{\frac{-12}{5}} - 0.25k}{(c_1t + c_2)}, \quad p = \frac{1}{3}\rho. \tag{19}$$

Notice that the null, weak, and strong energy conditions could be satisfied for a closed universe. Furthermore, for dark energy ($\omega_0 = -1$), the shape function is $b(r) = \frac{r_0^2+1.25kr_0^4}{r} - 1.25kr^3$, while the scale factor is given by $a(t) = c_1 e^{c_2 t}$. Therefore, the solution of EFEs is given by

$$\rho = 3c_2^2 - (\frac{3k}{4c_1^2} + \frac{r_0^2 + 1.25kr_0^4}{c_1^2 r^4})e^{-2c_2 t},$$

$$\tau = -3c_2^2 + (\frac{k}{4c_1^2} + \frac{r_0^2 + 1.25kr_0^4}{c_1^2 r^4})e^{-2c_2 t}, \quad p = -\rho. \tag{20}$$

It is found that the null, the weak, and the dominant energy conditions are satisfied for a flat universe if the inequality $3c_2^2 \geq \frac{r_0^2}{c_1^2 r^4} e^{-2c_2 t}$ holds.

Case I-B-2: Nonzero Separation Constant ($\lambda \neq 0$): This case gives rise to the following shape function:

$$b(r) = (r_0^{\frac{2\omega_0}{2\omega_0+1}} + \frac{(6\omega_0+1)k+2\lambda}{6\omega_0+2} r_0^{3-\frac{1}{2\omega_0+1}})r^{\frac{1}{2\omega_0+1}} - \frac{(6\omega_0+1)k+2\lambda}{6\omega_0+2} r^3, \tag{21}$$

while the scale factor is a nonlinear solution of the equation:

$$(3\omega_0 + 1)\dot{a}^2 + 2a\ddot{a} = \lambda. \tag{22}$$

In particular, if the wormhole is supported by relativistic matter ($\omega_0 = \frac{1}{3}$), then the solution of Eq. (22) becomes $a(t) = \sqrt{\frac{\lambda}{2}t^2 + \lambda t + c_1}$. Furthermore, the shape function is obtained from Eq. (21) as $b(r) = (r_0^{\frac{2}{5}} + \frac{3k+2\lambda}{4} r_0^{\frac{12}{5}})r^{\frac{3}{5}} - \frac{3k+2\lambda}{4} r^3$. A sufficient condition for the above shape function to describe a wormhole is to have $2\lambda + 3k > 0$. As a result, the solution of EFEs can be written as

$$\rho = (\frac{3(\lambda t+\lambda)^2}{4(\frac{\lambda}{2}t^2 + \lambda t + c_1)} + 0.6(r_0^{\frac{2}{5}} + \frac{2\lambda+3k}{4} r_0^{\frac{12}{5}})r^{-\frac{12}{5}} + 0.75k - 1.5\lambda)(\frac{\lambda}{2}t^2 + \lambda t + c_1)^{-1},$$

$$\tau = (\frac{(\lambda t+\lambda)^2}{4(\frac{\lambda}{2}t^2 + \lambda t + c_1)} - (r_0^{\frac{2}{5}} + \frac{2\lambda+3k}{4} r_0^{\frac{12}{5}})r^{-\frac{12}{5}} - 0.2k - 0.5\lambda)(\frac{\lambda}{2}t^2 + \lambda t + c_1)^{-1},$$

$$p = \frac{1}{3}\rho.$$

A sufficient condition for the null, weak, and strong energy conditions to be satisfied is to have $k \geq 4\lambda$ and $\frac{(\lambda t+\lambda)^2}{4(\lambda t^2+\lambda t+c_1)} \geq \frac{2}{5}(r_0^{\frac{2}{5}} + \frac{3k+2\lambda}{4} r_0^{\frac{12}{5}})r^{-\frac{12}{5}}$. On the other hand, if the wormhole is supported by dark energy ($\omega = -1$), then the shape function is given by $b(r) = \frac{4r_0^2-(2\lambda-5k)r_0^4}{4r} + \frac{2\lambda-5k}{4} r^3$, while the scale factor is $a(t) = \sqrt{\frac{\lambda}{2}} \cosh t$, where $\lambda > 0$. Thus, the solution of EFEs is

$$\rho = \frac{2}{\lambda}(3\frac{\lambda}{2}\cosh^2 t + \frac{(2\lambda - 5k)r_0^4 - 4r_0^2}{4r^4} - 0.75k)sech^2 t,$$

$$\tau = \frac{2}{\lambda}(\frac{-3\lambda}{2}\cosh^2 t + \frac{(2\lambda - 5k)r_0^4 - 4r_0^2}{4r^4} + 0.25k)sech^2 t, \qquad p = \rho.$$

It is a sufficient condition to satisfy the inequality $\frac{(2\lambda-5k)r_0^4-4r_0^2}{4r^4} \geq 0.75k$ in order for the null and weak energy conditions to hold. However, the strong energy condition is inevitably violated if the universe is either flat or closed.

Case II: $\ddot{a}(t) = 0$ **and** $\omega'(r) \neq 0$

This leads to a linear scale factor $a(t) = c_1 t + c_2$, while the shape function cannot be determined explicitly. The solution of EFEs in this case can be written as

$$\rho = (3c_1^2 + 3k + \frac{b'}{r^2})(c_1 t + c_2)^{-2}, \tag{23}$$

$$\tau = (-c_1^2 - k - \frac{b}{r^3})(c_1 t + c_2)^{-2}, \tag{24}$$

$$p = (-c_1^2 - 0.5k - \frac{b'}{2r^2} + \frac{b}{2r^3})(c_1 t + c_2)^{-2}, \tag{25}$$

Solving the above field equation for ω using the EoS $p = \omega(r)\rho$ yields

$$\omega(r) = \frac{b - rb' - kr^3 - 2c_1^2 r^2}{6c_1^2 r^3 + 6kr^3 + 2rb'}. \tag{26}$$

At this stage, we can construct explicit wormhole solutions by assuming specific shape functions and substituting in Eq. (26) to derive the value of ω. In order to be consistent with the actual universe, we restrict ourselves to asymptotically flat wormholes. This requires the shape function to satisfy $\frac{b}{r} \to 0$. Furthermore, since a linear EoS is more consistent with our actual universe, we assume that it is approximately linear far away from the wormhole throat. Keeping in mind these two restrictions , we consider the following two shape functions. First, the shape function $b(r) = r_0(\frac{r_0}{r})^n$, with $n \geq 1$, leads to the following solution:

$$\rho = (3c_1^2 + 3k - \frac{n}{r^2}(\frac{r_0}{r})^{n+1})(c_1 t + c_2)^{-2}, \tag{27}$$

$$\tau = (-c_1^2 - k - \frac{1}{r^2}(\frac{r_0}{r})^{n+1})(c_1 t + c_2)^{-2}, \tag{28}$$

$$p = (-c_1^2 - 0.5k + \frac{n+1}{2r^2}(\frac{r_0}{r})^{n+1})(c_1 t + c_2)^{-2}. \tag{29}$$

If r is large enough and $c_1^2 > 2.5$, then the null and weak energy conditions are satisfied for all types of universes. For closed or flat universes, all the energy conditions are satisfied and the EoS becomes $p = \frac{r_0(\frac{r_0}{r})^n[n+1]-(2c_1^2+k)r^3}{(6c_1^2+6k)r^3-2nr(\frac{r_0}{r})^{n+1}}\rho$. On the other hand, if we consider an exponential shape function of the form $b(r) = r_0e^{r_0-r}$, the solution of the EFEs is given by

$$\rho = (3c_1^2 + 3k - \frac{r_0}{r^2}e^{r_0-r})(c_1t + c_2)^{-2}, \tag{30}$$

$$\tau = (-c_1^2 - k - \frac{r_0}{r^3}e^{r_0-r})(c_1t + c_2)^{-2}, \tag{31}$$

$$p = (-c_1^2 - 0.5k + \frac{r_0}{2r^2}e^{r_0-r} + \frac{r_0}{2r^3}e^{r_0-r})(c_1t + c_2)^{-2}. \tag{32}$$

Similarly, all energy conditions are satisfied for flat and closed universes if $c_1^2 > 2.5$, while only the null and weak conditions hold for an open universe.

Case III: $\ddot{a}(t) = \omega'(r) = 0$,

This is a straightforward case leading to $\omega = \omega_0$ and $a(t) = c_1t + c_2$. The EoS $p = \omega(r)\rho$ in this case gives rise to the following shape function:

$$b(r) = (r_0^{\frac{2\omega_0}{2\omega_0+1}} + \frac{6c_1^2\omega_0 + 6k\omega_0 + 2c_1^2 + k}{6\omega_0 + 2}r_0^{3-\frac{1}{2\omega_0+1}})r^{\frac{1}{2\omega_0+1}}$$
$$- \frac{(6c_1^2\omega_0 + 6k\omega_0 + 2c_1^2 + k}{6\omega_0 + 2}r^3. \tag{33}$$

At this stage, we can consider the following subcases.

Case III-A: ($\omega_0 = 0$): The shape function becomes $b(r) = (1 + \frac{2c_1^2+k}{2}r_0^2)r - \frac{2c_1^2+k}{2}r^3$, giving rise to the following solution of EFEs:

$$\rho = (\frac{1 + (c_1^2 + 0.5k)r_0^2}{r^2} + 1.5k)(c_1t + c_2)^{-2},$$

$$\tau = (\frac{1 + (c_1^2 + 0.5k)r_0^2}{r^2} - 0.5k)(c_1t + c_2)^{-2}, \qquad p = 0. \tag{34}$$

In this case, all energy conditions are satisfied for open and closed universes.

Case III-B: ($\omega_0 \neq 0$): We can solve for interesting cases as we did before. In particular, for relativistic matter ($\omega_0 = \frac{1}{3}$), the shape function is $b(r) = (r_0^{\frac{2}{5}} + (c_1^2 + 0.75k)r_0^{\frac{12}{5}})r^{\frac{3}{5}} - (c_1^2 + 0.75k)r^3$ leading to the following solution:

$$\rho = (0.6r_0^{\frac{2}{5}} + 0.6(c_1^2 + 0.75k)r_0^{\frac{12}{5}})r^{-\frac{12}{5}} + 0.75k)(c_1t + c_2)^{-2},$$

$$\tau = (-r_0^{\frac{2}{5}} - (c_1^2 + 0.75k)r_0^{\frac{12}{5}})r^{-\frac{12}{5}} - 0.25k)(c_1 t + c_2)^{-2}, \quad p = \frac{1}{3}\rho. \tag{35}$$

Thus, all energy conditions are satisfied for a closed universe if $r^{\frac{12}{5}} \geq 0.8r_0^{\frac{2}{5}} + 0.8(c_1^2 + 0.75k)r_0^{\frac{12}{5}}$ holds. An interesting case arises if we consider a wormhole supported by dark energy ($\omega_0 = -1$) which leads to the shape function: $b(r) = \frac{r_0^2+(c_1^2+1.25k)r_0^4}{r} - (c_1^2 + 1.25k)r^3$ giving rise to the following solution:

$$\rho = (-\frac{r_0^2 + (c_1^2 + 1.25k)r_0^4}{r^4} - 0.75k)(c_1 t + c_2)^{-2},$$

$$\tau = (-\frac{r_0^2 + (c_1^2 + 1.25k)r_0^4}{r^4} + 0.25k)(c_1 t + c_2)^{-2}, \quad p = -\rho. \tag{36}$$

Evidently, a universe with negative curvature satisfies the null, weak, and dominant energy conditions if $c_1^2 \geq 1.25$ and $r_0^2 + (c_1^2 - 1.25)r_0^4 \leq \frac{r^4}{4}$. The solutions of this section with their corresponding energy conditions are presented in Table 1.

5 Conclusion

A classification of cosmological wormhole solutions in an expanding Friedman universe was performed in the framework of GR by assuming an EoS of the form $p = \omega(r)\rho$ where exact solutions describing interesting wormhole geometries were obtained. While energy conditions are inevitably violated for static wormholes, our obtained solutions reveal that the expansion of space has a considerable effect on the exoticity of matter supporting the wormhole. In this context, the violation of energy conditions could be reduced and, in some interesting cases, could be eliminated leading to attractive geometries. Although many of the obtained solutions were not asymptotically flat, they are still interesting from theoretical point of view as they show that the exoticity of matter threading the wormhole could be minimized even in Einstein theory if the scale factor of the universe is incorporated in our wormhole models.

Our investigation has been extended to nonlinear EoS where the consistency of this equation with the actual universe was maintained by assuming the local behavior of the fluid near the throat tends to a global EoS in the large scale. This case gave rise to a linear scale factor and an EoS parameter expressed as a function depending on the wormhole shape as shown by Eq. (26). It has also been shown that the considered wormhole could be supported by ordinary matter if certain given inequalities hold.

It is also worth mentioning that the scale factor could play a role in enlarging the wormhole size which means that even a microscopic wormhole could, in principle, be suitable for time travel as the expansion of space could stretch it to larger size.

Table 1 Cosmological wormhole solutions for the EoS $p = \omega(r)\rho$

Shape function $b(r)$	Scale factor $a(t)$	Constraints	Type of universe	Satisfied energy conditions
$(1+\frac{k}{2}r_0^2)r - \frac{k}{2}r^3$	$(c_1 t + c_2)^{2/3}$	–	$k=0$ and $k=1$	All E.Cs
$(1+\frac{k}{2}r_0^2+\lambda r_0^2)r - (\lambda+\frac{k}{2})r^3$	Nonlin sol of Eq. (13)	$k > 3\lambda$	$k=0, k=-1$ and $k=1$	All E.Cs
$(r_0^{2/5} + (0.75k+0.5\lambda)r_0^{12/5})r^{3/5}$ $-(0.75k+0.5\lambda)r^3$	$\sqrt{0.5\lambda t^2 + \lambda t + +c_1}$	$k > max(12\lambda, -2\lambda)$	$k=0, k=1$	NEC+WEC+SEC
$\frac{4r_0^2+(5k-2\lambda)r_0^4}{4r} - \frac{5k-2\lambda}{4}r^3$	$\sqrt{0.5\lambda}\cosh t$	$\frac{(2\lambda-5k)r_0^4-4r_0^2}{4r^4} \geq 0.75k$ and $\lambda > 0$	$k=0, k=-1$ and $k=1$	NEC+WEC
$(r_0^{2/5} + 0.75kr_0^{12/5})r^{3/5}$ $-0.75kr^3$	$\sqrt{c_1 t + c_2}$	$(\frac{c_1}{c_1 t + c_2} + 0.5k)r^{12/5} \geq$ $0.4(r_0^{2/5} + 0.75kr_0^{12/5})$	$k=0$ and $k=1$	All E.Cs
$\frac{r_0^2+1.25kr_0^4}{r} - 1.25kr^3$	$c_1 e^{c_2 t}$	$3c_1^2c_2^2e^{2c_2t} \geq \frac{r_0^2}{r^4}$	$k=0$	NEC+WEC+DEC
$r_0(\frac{r_0}{r})^n$	$c_1 t + c_2$	$c_1^2 \geq 2.5$	$k=0$ and $k=1$	All E.Cs
$r_0 e^{r_0 - r}$	$c_1 t + c_2$	–	$k=0, k=1$	All E,Cs
$[1+(c_1^2+0.5k)r_0^2]r - (c_1^2+0.5k)r^3$	$c_1 t + c_2$	–	$k=0$ and $k=1$	All E,Cs
$[r_0^{2/5} + (c_1^2+0.75k)r_0^{12/5}]r^{3/5}$ $-(c_1^2+0.75k)r^3$	$c_1 t + c_2$	$r^{12/5} \geq 0, 8(r_0^{2/5}+$ $(c_1^2+0.75k)r_0^{12/5})$	$k=1$	All E.Cs
$\frac{r_0^2+(c_1^2+1.25k)r_0^4}{r} - (c_1^2+1.25k)r^3$	$c_1 t + c_2$	$r_0^2 + (c_1^2-1.25)r_0^4 \leq \frac{r^4}{4}$ and $c_1^2 \geq 1.25$	$k=-1$	NEC+WEC+DEC

Finally, it will be worthwhile to generalize this study by considering cosmological wormholes of nonzero radial tide as well.

References

1. Flam, L.: Bitrage zur Einsteinschen Gravitationstheorie. Phys. Z. **17**, 448 (1916)
2. Einstein, A., Rozen, N.: The particle problem in the general theory of relativity. Phys. Rev. **48**, 73–77 (1935)
3. Morris, M.S., Thorne, K.S.: Wormholes in space-time and their use for interstellar travel: a tool for teaching general relativity. K.S. Am. J. Phys. **56**, 395 (1988)
4. Morris, M.S., Thorne, K.S., Yurtsever, U.: Wormholes, time machines, and the weak energy condition. Phys. Rev. Lett. **61**, 1466 (1988)
5. Bhawal, B., Kar, S.: Lorentzian wormholes in Einstein-Gauss-Bonnet theory. Phys. Rev. D **46**, 2464 (1992)
6. Bohmer, C.G., Harko, T., Lobo, F.S.N.: Wormhole geometries in modified teleparralel gravity. Phys. Rev. D **85**, 044033 (2012)
7. Furey, N., DeBenedictis, A.: Wormhole throats in R^m gravity. Class. Quantum Grav. **22**, 313 (2004)
8. Banerjee, A., Singh, K., Jasim, M., Rahaman, F.: Traversible wormholes in f(R,T) gravity with conformal motions, arXiv:1908.04754v1 [gr-qc] (2019)
9. Kar, S.: Evolving wormholes and the weak energy condition. Phys. Rev. D **49**, 862–865 (1994)
10. Kim, S.W.: Cosmological model with a traversable wormhole. Phys. Rev. D **53**, 6889 (1996)
11. Ovgun, A.: Evolving wormhole supported by dark matter and dark energy, arXiv:1803.04256 [physics.gen-ph] (2018)
12. Mehdizadeh, M., Ziaie, A.: Dynamic wormhole solutions in Einstein-Cartan gravity. Phys. Rev. D **96**, 124017 (2017)

Use of Software and Technology in Math Education

Sonal Jain, Ho-Hon Leung, and Firuz Kamalov

1 Introduction

Technology offers a lot of opportunities to help expand, improve, and enrich math in the education system. However, the use of computers to, well, calculate has drastically changed the applications of mathematics and the knowledge of applying mathematics outside of the classroom. At the moment, it appears that the majority of its use in schools is focused on improving the efficiency of exposition, worked examples, and repeated practice which are examples of conventional pedagogical strategies. Although attempts to incorporate technology into the nature of mathematics instruction in university have a long history, nowadays this is only the case in a small number of institutions. Wolfram [31] and others have called for a fundamental shift in the way that mathematics is learned, emphasizing the description, generalization, and explanation aspects of problem-solving far more while leaving the laborious, physical computing to machines. Will this approach be the norm for mathematical futures in universities, or will technology's function be expanded to encompass more difficult mathematical activities, or will it remain limited to only replacing professors in their setup and breakdown duties?

The major question is what technology can do in math education? There are many different technologies being employed in the field of education today, but

S. Jain (✉)
School of Technology, Woxsen University, Hyderabad, Telangana, India

H.-H. Leung
Department of Mathematical Sciences, UAE University, Al Ain, UAE
e-mail: hohon.leung@uaeu.ac.ae

F. Kamalov
Department of Electrical Engineering, Canadian University Dubai, Dubai, UAE
e-mail: firuz@cud.ac.ae

F. Kamalov et al. (eds.), *Advances in Mathematical Modeling and Scientific Computing*, Trends in Mathematics, https://doi.org/10.1007/978-3-031-41420-6_11

who benefits from them and how does it improve learning? Technology possesses six traits, according to Loveless [14]: "provisionality, interaction, capability, range, velocity, and automatic functions." Users were able to perform tasks "that could not be performed as efficiently, or at all, using alternative technologies," according to the argument. Each of these traits directly relates to the teaching of mathematics. Provisionality, for instance, enables students to test out concepts by investigating how altering a variable in a computer graphics program impacts the result, adjusting a quadratic's coefficients to alter its graph and solutions, or even practicing possible answers for an online test. In dynamic geometry software, interaction allows students to explore the features of geometric creations that are invariant, and automatic functions allow students to receive instant feedback on their exam answers. Given the rapid advancements in machine learning, this list may now be expanded to include adaptability. These systems are able to quickly alter their operations in reaction to the data that is supplied. For instance, the questions that students are asked in a quiz or game may be automatically changed to provide additional practice or a greater challenge, simulating the way that people adjust to new circumstances, based on a student's prior performance or the performance of other students on comparable exercises.

In particular, Lewin et al. [18] state that the biggest impact on arithmetic performance comes from simulations, structure, and clever tutoring systems, despite the fact that there is now only a limited amount of software used in universities for these purposes. Other benefits of technology in math teaching may include assisting investigative reports, finding novel concepts outside the parameters of exam specifications, cross-curricular activities, and cooperative methods. However, it seems like not much is being done right now to focus on these areas.

At the classroom level, mathematics education may consist of three components: role of technology in teaching mathematics, technology impact on learning mathematics, and practicing mathematics with technology. The three components are interconnected because students learn mathematics as a consequence of their teachers' instruction and as a result of their real participation in mathematics. The goal of mathematics instruction and learning should, at least in part, be to help students become better mathematicians. All three of these factors are susceptible to the influence of technology.

2 Technology's Place in Math Education

Today, schooling now includes a significant amount of technology. In educational environments, various forms of technology are being used everywhere we look. Computers, the Internet, calculators, email, iPads, mobile phones, digital games, electronic books, televisions, and interactive whiteboards are just a few examples of the technology used in classrooms today.

Teachers themselves employ digital tools in mathematics lessons that are similar to those used in other areas. As a result, mathematics is taught:

1. By utilizing common presentation techniques and software
2. By giving an exam to the entire class utilizing a general platform like Kahoot, Blackboard Learn, Quizizz, Canvas LMS, Poll Everywhere, Quizlet, Schoology, TalentLMS, and Mindtickle
3. By employing teaching aids or modeling computations on interactive whiteboards or visualizers
4. By giving activities to students, occasionally through teacher-created digital worksheets or electronic textbooks, but also through online forms
5. By using online learning settings which frequently contain tools for online assessment such as Google Classroom, Microsoft Teams, etc.
6. Via specialized tutoring software or video conferencing tools like Zoom or Meet
7. Through the use of multimedia, such as narrated presentations, worked example films, or instructional videos

   Additionally, Math teachers can use interactive visualizations and virtual manipulatives or can demonstrate how students might use some of these tools. These digital resources were made specifically for teaching mathematics. Teachers use dynamic geometry software on interactive whiteboards and display screens to depict geometric structure, shape characteristics, symmetry, and transformations. The automatic translation of lessons or the use of automatic captioning during classes content into students first languages are just two examples of how digital technology can enhance inclusion in the mathematics classroom. Adaptive learning approaches and, more recently, the use of machine learning can be utilized to create personalized practice material based on a student's past success and current objectives. Since answers are either correct or incorrect, it is easy for computers to generate questions automatically for school mathematics, and more advanced methodologies allow for some ways for the computer to check mathematical reasoning; it appears that school mathematics is particularly well suited to this type of methodology. By offering multimedia content, rankings, or badges in addition to the actual questions, some software tries to "gamify" learning. In their latest evaluation of studies in mathematics instruction, Ofsted only identified one use of technology: repetition and practice.

3 Technology Impact on Learning Mathematics

As a result, students will interact with online learning environments, complete in-class or assigned homework quizzes, take part in video courses or tutorials, and use other online resources as directed by their teachers. Technology use by students to learn mathematics typically mirrors that of their teachers. While most of the students' work is still done on paper, they can answer questions on screens and often get feedback right away. Practice questions can be created randomly using a predetermined pattern or selected from a bank of prepared materials, providing limitless practice. By giving students enough repetitions and honest feedback in a manner they find enjoyable, computing technology can also assist students in learning numerical facts [25].

It is useful to make a distinction between quick, narrowly focused online exams that build on direct instruction and worked examples, the electronic equivalent of the worksheet, and more in-depth, adaptable intelligent tutoring systems that offer a variety of practice questions combining spaced repetition of previous topics with carefully structured new content. The former is better suited to more individualized approaches to mathematics instruction, such as those that might be used in the context of homework or even homeschooling. The latter may be especially beneficial for kids whose arithmetic aptitude is far below or above the class average. The first strategy is suggested by Ofsted: "Students make more progress and enjoy computer math games when core subject is provided as different learning components that are carefully followed by 'mini-games'" (2021, Ofsted).

Beyond this limited method, current technology might also allow pupils to research issues independently. This could be achieved through reading about interesting mathematical subjects, utilizing virtual manipulatives, locating intriguing puzzles and problems, watching video lessons, and interacting with other students. Again, technology may support inclusiveness by, for example, allowing for comments in other formats and providing jobs in other formats, such as text to speech and automatic translation.

4 Practicing Math with Technology

Certain technologies are employed to help students experience math at school, and these technologies help students construct their internal models of math [9]. Since pencil (or pen) and paper are the primary tools used when teaching arithmetic to students, it is understandable but incorrect to assume that math is primarily concerned with how to conduct manual calculations. In recent years, there has been an upsurge in emphasis on using physical tools to think about mathematics in an embodied rather than abstract approach.

Calculators are no longer a component of the core curriculum or a prerequisite for the "SATs." Beginning in secondary school, students begin using scientific or graphing calculators, with A Level courses specifically encouraging their use [3].

The bulk of students' school-level interactions with mathematics seem to be extremely dissimilar from the academic discipline's wide range of applications for comprehending systems and solving issues, where sophisticated, automated computing is more common than not. All too frequently, it appears that the emphasis at school is on repetitive repetition of computations that machines can accomplish accurately and effectively.

In numerous studies over the years, technology has served as a medium for students' original mathematical work. By utilizing turtle visuals and other "microworlds" in the Logo programming language, Papert [26] and others produced the most notable and distinctive work. Scratch was developed as a result of Papert's ambitious predictions about the possibilities of a constructionist method for teaching mathematics through programming [27], and turtle graphics packages were also introduced in Python and other programming languages. But his desire was rarely realized because the depth of Logo programming was typically limited to learning the outside angles of regular polygons in math lectures [1].

Some signs point to a resurgence of interest in teaching students' own maths using programming. Students can now use their programming knowledge in mathematics because it is part of their computing education. This is simpler in primary schools, because the same instructor often teaches both computing and mathematics subjects. There have been effective methods used in these settings, such as Scratch Maths [23]. Students study a text-based language in secondary schools, typically Python.

Although some computer science teachers encourage their students to use their programming skills to solve relatively complex mathematical problems (such as Project Euler1), there doesn't seem to be much programming being used in math lessons at the moment. It can be the result of the absence of such methods in the national curriculum or test requirements. A notable exception to this general trend is MEI's [22] data science section, which uses Pandas, Python, and Jupyter Notebooks on Kaggle for visualizing and analyzing extremely large data sets.

Spreadsheet programs can be utilized as a teaching tool for mathematics, frequently by formulating the relationship between variables, but they can also be used for numerical methods and linear programming. Even with enormous data volumes, spreadsheet software is commonly utilized for data visualization. Most teachers appear to favor a spreadsheet-based approach when it comes to the A Level math big data set subject.

There is some use of interactive data visualization software. The National Curriculum makes non-statutory guidelines about how pupils should use "coordinate-plotting ICT technologies" [7, p. 134]. Students may also create charts from large data sets or survey findings using visualization tools. Many of these programs have components for statistical analysis.

Students are allowed to use digital manipulatives, such as dynamic geometric doodles that their teachers share with the class. For a generation of children who have grown up with touchscreen interfaces, such virtual manipulatives have many of the same advantages as their physical equivalents. However, in comparison to their physical counterparts, virtual manipulatives can be used to investigate a significantly greater variety and depth of mathematics. Even more options are provided by virtual and augmented reality, but the existing educational applications are pricey and constrained.

Many of the aforementioned ways that students could use digital technology to do mathematics are covered and extended by computer algebra systems like Mathematica and the related Wolfram Language. Most of the math, algebra, and calculus tasks needed for school can be completed using these tools, which also offer an interactive interface for exploring what occurs when parameters are changed. Teachers may worry that the software performs too much math, like calculators in elementary schools, but proponents like Wolfram [31] contend that this software shifts the emphasis to more practical rather than mechanical parts of such calculating, solving equations, or integration, posing issues and evaluating solutions.

The research done by Crompton is one instance of how mobile devices may be used to promote the learning of mathematical concepts [6]. Crompton proposed a design-based research study as part of her work that would employ iPads as a tool to help young children learn angles.

5 Education for Mathematics Via Mobile Learning

Academics and education professionals are paying increased attention to how teachers and students in the discipline of mathematics are using mobile technologies, such as tablets and smartphones. Mobile technologies are swiftly becoming an integral part of society as they are incorporated into people's daily lives on a global scale. Mobile technologies like cell phones, smartphones, and tablet computers, as well as the affordances they permit, have already revolutionized how information is accessible and shared, how business transactions are conducted, and how people interact with one another, according to Liu et al. [17].

Researchers commonly state that mobile technology devices and their use are becoming "increasingly prevalent in society, particularly with young." [12]. Mobile learning, according to Traxler, is "any service where the single or dominant technology is a portable or palmtop device" from 2005 (or m-learning). Since technology is changing the very landscape of teaching and learning, it "is of great importance." According to Black [2], learning using mobile devices is no longer seen as a lonely pursuit but as a complex activity that deserves research into how people, including teachers and students, work together and interact in groups.

According to Drijvers [8], stakeholders have been emphasizing the promise of digital technologies for mathematics for more than 20 years as having a wide range of potential applications. There are still many questions among educators, researchers, and teachers concerning how technology should be included into the teaching and learning process. Mobile learning is a technology that is "ubiquitous in nature, wireless, highly portable, and equipped with multimedia capabilities, introducing a new dimension to curriculum delivery," according to researchers [16, 19]. Two study investigations into the integration of information and communication technology (ICT) in mathematics education confirm these conclusions with the remark that "digital technologies were becoming ever more widespread and their effect affecting most, if not all, education institutions." The distinctive features of mobile devices, such as their capacity for data collection and communication, user collaboration, and media consumption, as well as their ability to create and produce distinctive forms of expression and representation, can be easily adapted to the scientific, mathematical, and engineering practices emphasized by the Common Core Math and Next Generation Standards, as claimed by [32, p. 64] (NGSS Lead States [24]).

Researchers are paying more attention to the potential applications and opportunities of mobile devices, yet this area of mathematics education research is still understudied. However, some research has been done on the potential use of this technology in the teaching and learning of mathematics (e.g., [4, 15]).

The earliest studies on the application of mobile learning in mathematics date from the latter part of the 2000s (e.g., [11]), and since then, there has been a significant expansion in this type of research in terms of both international conferences and industry-specific journals. The majority of the research papers analyzed in this study can be grouped into three major categories: (a) studies on potential mobile device applications for mathematics teaching and learning, (b) affective studies on the use of mobile devices, and (c) the use of mobile devices for mathematics teacher preparation.

Numerous studies have concentrated on using the benefits of mobile devices, including their portability, mobility, and ability to capture images and videos of actual events that can later be researched and mathematically evaluated. In one such study, Wijers et al. [30] employed a GPS-enabled mobile game called Mobile Math based on location that made it simpler to construct and investigate quadrilateral equations and their properties in a real-world environment outside of the classroom. The perceptions and feelings that math teachers and students experience when learning or teaching mathematics using mobile devices have been the subject of other studies. For example, the "Bring Your Own Device" (BYOD) program, which encouraged the use of the Internet and mobile devices during math lectures, drew criticism from both instructors and students, according to Holubz [13], who gathered their thoughts.

Finally, it can be demonstrated that numerous research have looked into the usage of mobile devices for instructing math teachers. These studies have looked at the design of inquiry assignments for preservice and in-service learners in mobile environments. For instance, Yerushalmy and Botzer [33] talked about teaching's theoretical foundations as well as its problems and potential advantages.

5.1 Touchscreen and Pen Tablets in Math Teaching

Researchers have proposed that the input devices people use when engaging in activities or tasks assisted by computers may have an impact on their attention spans [5, 10, 21]. Using touchscreens and pen tablets for problem-solving activities in the subject of mathematics with virtual manipulatives, Chen et al. [5] conducted a study to compare and evaluate students' attention spans in connection with the amount of time spent on a work and the quantity of distractions. The results showed that students using touchscreens had greater attention spans than those using pen tablets, enabling them to concentrate on their work for longer periods of time with fewer interruptions.

Mangen [21] asserted that using a mouse to view information on a computer screen could distract the user. The use of virtual manipulatives on touchscreen devices to encourage the learning of mathematics is one of the most current methods of integrating technology into educational practice made possible by recently produced technologies like touchscreens, which offer simple and sharing interfaces (e.g., [20, 29]). Studies reveal that while using touchscreens, there is a higher correlation between the user's hand movements and the outcomes displayed on the screen than when using a mouse or physical keyboard [28].

6 Future Research Direction

Whether we should use digital technology to give students ever-more efficient ways to perform calculations manually and to reflect the ways that real mathematics has already altered, we should alter the way that classroom mathematics is taught. As a result of the automation of those computations is the fundamental question that must be answered in order to determine the role of digital technology in the future of mathematics.

1. Are the existing, mostly didactic methods of teaching and learning mathematics still the most effective? Or do we need a change?
2. What adjustments to evaluation procedures would encourage greater genuine use of digital technologies in mathematics?
3. What strategies for integrating digital technology into mathematics instruction have been successful in other countries?
4. Which schools are already effective at employing digital technologies for math?
5. What prevents students from utilizing their coding expertise more in mathematics lessons?

7 Recommendations

The researchers advise the following in light of their findings:

1. At different educational levels, these digital technologies must be taught in mathematical curriculum.
2. The study findings should be used by the stakeholders to motivate instructors to keep implementing these technologies in mathematics instruction.
3. To address the issues raised in the discussion section, more investigation is required.

8 Summary

The research revealed that a variety of digital tools were used in mathematics education, such as computer algebra systems (CAS) like Mathematical, Maple, MuPAD, MathCAD, Derive, and Maxima, mobile devices, touchscreens, and pen tablets, as well as digital libraries and the creation of learning objects. Depending on the sort of educational technology used, the results differed. However, it is evident that the initial success in the teaching of mathematics is due to digital learning.

Acknowledgments The United Arab Emirates Ministry of Education (MOE) provides financial support for the project. The assistance is based on the 2019 Collaborative Research Program Grant being awarded (CRPG 2019). We appreciate the opportunity the MOE has provided for us.

References

1. Agalianos, A., Whitty, G., Noss, R.: The social shaping of Logo. Soc. Stud. Sci. **36**(2), 241–267 (2006)
2. Black, A.: Gen Y: who they are and how they learn. Educ. Horizons **88**(2), 92–101 (2010)
3. Button, T.: Ten calculations A Level students should do on a calculator (2020). Available at: https://www.ncetm.org.uk/features/ten-calculations-a-level-students-should-do-on-a-calculator/
4. Crompton, H., Traxler, J. (eds.): Mobile Learning and Mathematics. Foundations, Design and Case Studies. Taylor & Francis, Routledge (2015)
5. Chen, C.-H., Chiu, C.-H., Lin, C.-P., Chou, Y.-C.: Students attention when using touchscreens and pen tablets in a mathematics classroom. J. Inf. Technol. Educ.: Innov. Pract. **16**, 91–106 (2017). https://doi.org/10.28945/3691
6. Crompton, H.: Understanding angle and angle measure: A design-based research study using context aware ubiquitous learning. Int. J. Technol. Math. Educ. **22**(1), 19–30 (2015). https://doi.org/10.1564/tmev22.1.02
7. Department for Education: The National Curriculum in England: Framework Document. Department for Education, London (2013)
8. Drijvers, P.: Digital technology in mathematics education: why it works (ordoesn't) (2012). Retrieved from http://www.icme12.org/upload/submission/2017f.pdf
9. Drijvers, P., Kieran, C., Mariotti, M.-A., et al.: Integrating technology into mathematics education: theoretical perspectives. In: Mathematics Education and Technology-Rethinking the Terrain, pp. 89–132. Springer (2009)
10. Evans, M.A., Feenstra, E., Ryon, E., McNeill, D.: A multimodal approach to coding discourse: collaboration, distributed cognition, and geometric reasoning. International J. Comput-Supported Collab. Learn. **6**(2), 253–278 (2011). https://doi.org/10.1007/s11412-011-9113-0
11. Franklin, T., Peng, L.-W.: Mobile math: math educators and students engage in mobile learning. J. Comput. Higher Educ. **20**(2), 69–80 (2008). https://doi.org/10.1007/s12528-008-9005-0
12. Grant, M.M., Barbour, M.K.: Mobile teaching and learning in the classroom and online: case studies in K-12. In: Berge, Z., Muilenburg, L. (eds.) Handbook of Mobile Learning, pp. 285–292. Routledge, New York (2013)
13. Holubz, B.J.: Mobilizing mathematics. Participants' perspectives on bring your own device (2015)
14. Loveless, A.: Literature Review in Creativity, New Technologies and Learning. Futurelab, Bristol (2002)
15. Larkin, K., Calder, N.: Mathematics education and mobile technologies. Math. Educ. Res. J. (2015). https://doi.org/10.1007/s13394-015-0167-6
16. Littlejohn, A., Pegler, C.: Preparing for Blended e-Learning. Routledge, London (2007)
17. Liu, M., Scordino, R., Geurtz, R., Navarrete, C., Ko, Y., Lim, M.: A look at research on mobile learning in K-12 education from 2007 to the present. J. Res. Technol. Educ. **46**(4), 325–372 (2014)

18. Lewin, C., Smith, A., Morris, S., et al.: Using Digital Technology to Improve Learning: Evidence Review. Education Endowment Foundation, London (2019)
19. Maxfield, M.B., Romano, D.: Chapter 8 – Videos with Mobile Technologies: pre-service teacher observations during the first day of school. In: Keengwe, J. (ed.) Pedagogical Applications and Social Effects of Mobile Technology Integration, pp. 133–155. IGI Global, Hershey (2013)
20. Moyer-Packenham, P.S., Bullock, E.K., Shumway, J.F., Tucker, S.I., Watts, C.M., Westenskow, A., ..., Jordan, K.: The role of affordances in children's learning performance and efficiency when using virtual manipulative mathematics touch-screen apps. Math. Educ. Res. J. **28**(1), 79–105 (2016). https://doi.org/10.1007/s13394-015-0161-z
21. Mangen, A.: Hypertext fiction reading: haptics and immersion. J. Res. Read. **31**(4), 404–419 (2008). https://doi.org/10.1111/j.1467-9817.2008.00380.x
22. MEI: Introduction to data science (2021). Available at: https://mei.org.uk/introduction-to-data-science/
23. Noss, R., Hoyles, C., Saunders, P., et al.: Making constructionism work at scale: the story of Scratchmaths. In: Holbert, N., Berland, M., Kafai, Y. (eds.) Designing Constructionist Futures: The Art, Theory and Practice of Learning Designs, pp. 39–52. MIT Press, Cambridge, MA (2020)
24. NGSS Lead States: Next generation science standards: For states, by states (2013). http://www.nextg. enscience.org. Accessed 15 May 2020
25. Ofsted: Research Review Series: Mathematics. Ofsted, London (2021). Available at: https://www.gov.uk/government/publications/research-review-series-mathematics/research-review-series-mathematics
26. Papert, S.: Mindstorms: Children, Computers and Powerful Ideas. Basic Books, New York (1980)
27. Resnick, M.: The seeds that Seymour sowed. Int. J. Child-Comput. (2017)
28. Romeo, G., Edwards, S., McNamara, S., Walker, I., Ziguras, C.: Touching the screen: issues related to the use of touchscreen technology in early childhood education. Br. J. Educ. Technol. **34**(3), 329–339 (2003). https://doi.org/10.1111/1467-8535.00330
29. Watts, C.M., Moyer-Packenham, P.S., Tucker, S.I., Bullock, E.P., Shumway, J.F., Westenskow, A., ..., Jordan, K.: An examination of children's learning progression shifts while using touch screen virtual manipulative mathematics apps. Comput. Hum. Behav. **64**, 814–828 (2016). https://doi.org/10.1016/j.chb.2016.07.029
30. Wijers, M., Jonker, V., Drijvers, P.: MobileMath: exploring mathematics outside the classroom. ZDM–Int. J. Math. Educ. **42**(7), 789–799 (2010). https://doi.org/10.1007/s11858-010-0276-3
31. Wolfram, C.: The Math(s) Fix: An Education Blueprint for the AI Age. Wolfram Media, Inc (2020)
32. White, T., Martin, L.: Mathematics and mobile learning. TechTrends **58**(1), 64–70 (2014). https://doi.org/10.1007/s11528-013-0722-5
33. Yerushalmy, M., Botzer, G.: Guiding mathematical inquiry in mobile settings. In: Zaslavsky, O., Sullivan, P. (eds.) Constructing Knowledge for Teaching Secondary Mathematics, pp. 191–207. Springer (2011). https://doi.org/10.1007/978-0-387-09812-812

Modified VIM for the Solutions of Gas Dynamics

Nahid Fatima

1 Introduction

Numerous scientists and technologists have demonstrated the significance of non-linear partial differential gas dynamics-type equations, which are the same as gas kinetics, which is a study of the fluid mechanical properties of gases based on fluid mechanics and kinetics, with its various applications in quantum mechanics, quantum field theory, relativistic physics, and dispersive wave phenomena. Fluid Mechanical properties is used in particular for the study of flight for aircraft and propulsion system design, though the principles can also be applied to manufacturing processes.

For the last two decades, various methods were proposed to solve the gas dynamics type equations numerically and analytically. However, they analyzed and compared the solution of gas dynamics equations with integer-order only.

Various numerical techniques were studied and applied for approximate and numerical analytical solutions of nonlinear PDE equations such as gas dynamics equations. In this chapter, we propose a numerical and computation method, modified variation iteration method (MVIM), never applied to gas dynamics type equations. However, this chapter presents a novel solution for gas equations among the various analytical numerical solutions with better approximate and convergence properties and specific applications. Further, in literature, it is found that there are various methods to solve this type of equations [1, 7]. Also, in recent decades, Laplace transform method, Adomian decomposition method (ADM), variation iteration method (VIM), computational approach, homotopy perturbation method (HPM), etc., have been proposed and applied to gas equations [6,7,14,20]. In

N. Fatima (✉)
Prince Sultan University, Riyadh, KSA
e-mail: lncs@springer.com

F. Kamalov et al. (eds.), *Advances in Mathematical Modeling and Scientific Computing*, Trends in Mathematics, https://doi.org/10.1007/978-3-031-41420-6_12

some other literatures [10, 11], authors suggested and applied MVIM on linear and nonlinear PDE, ODE, fractional partial, or ordinary differential equations to obtain more accurate results compared to other methods like HAM, ADM, and VIM. Among the methods presented in these references, modified variation iteration method (MVIM) outperforms in terms of convergence; however, as per our best knowledge, none of the open literature presents a solution for the above equation along with various examples. MVIM is already a numerical method for solving PDE with differential operators. Some papers on gas dynamics equations [7] have also investigated the analysis of the solution of nonlinear partial differential equations. There are various methods for finding analytical or approximate solutions to these kinds of equations [1, 3, 14]. Many scientists and researchers have shown interest in the equations encountered in science and technology.

In this research, we present a strategy for dealing with nonlinear issues by using a homotopy technique in tandem with a perturbation methodology [5–15]. In the past decades, the problem of a complex system has been expressed in ODE or PDE of fractional order. Finding the exact solutions to many PDEs may be complicated. So, using vivid numerical techniques, approximate solutions of these PDEs could be found [4–7, 12, 15]. Researchers are focusing on analytical and approximation solutions of PDEs in boundary value and initial value problems. Many mathematicians have used different derivative operators in various proposed PDEs of a nonlinear system [9, 13]. Physical phenomena are usually time and space dependent.

Many numerical methods for solving PDEs have been developed in the past two decades. These methods include the Adomian decomposition method (ADM) [2] for obtaining a more meaningful solution. In the references [5, 8, 15], the authors applied the homotopy method for PDE. Some authors employed the fractional wave equation, the Burgers equation, and the gas dynamics equation. They proved and compared various numerical methods such as ADM, HAM, VIM, HPM as effective and promising methods.

We have introduced a modified variational iteration method (MVIM). The lack of rounding mistakes and the modest resource requirements of this method make it the preferred numerical method. J.H. He [4, 8] has used the variation iteration method to derive analytical solutions to nonlinear partial differential equations with variable coefficients and autonomous ordinary differential equations.

It is noteworthy to mention that none of the open literature presents the solution of the gas equation employing MVIM. Also, the examples considered in this chapter are missing from the literature. Furthermore, this chapter is organized in different sections. In Sect. 1.1 the analytical expressions for equations using MVIM are proposed. In Sects. 2.1, 2.2, and 2.3 different examples are given to prove the efficacy of the proposed method. Section 3 concluded this chapter.

A closer look at how the VIM's problems are solved reveals a few drawbacks. These include the time and energy spent on unnecessary computations and computations that are performed many times. To address these shortcomings, we provide the modified variational iteration method (MVIM).

1.1 Modified Variational Iteration Method

The modified variational iteration method is a slight tweak to the original variational iteration approach. Let us understand the following partial differential equation to better understand the basic concept of the modified variational iteration method (MVIM.

$$Lu\,(x,\tau) + Ru\,(x,\tau) + Nu\,(x,\tau) = g\,(x,\tau)$$

$$u\,(x,0) = f\,(x),$$

Where, $L = \frac{\partial}{\partial \tau}$, R is a linear operator that has partial derivatives with respect to x, Nu(x, τ) is a nonlinear term, and g(x, t) is an inhomogeneous term.

The iteration formula obtained by the variational iteration approach is as follows.

$$U_{n+1} = U_n - \int_0^t \{L\,(U_n) + R\,(U_n) + NU_n - g\}\,d\tau$$

We can rewrite the equation as

$$U_{n+1} = U_n - \int_0^t \{R\,(U_n - U_{n-1}) + (G_n - G_{n-1})\}\,d\tau,$$

where

$$U_{-1}=0,\ U_0=f\,(x),\ U_1=U_0-\int_0^t \{R\,(U_0-U_{-1}) + (G_0-G_{-1}) - g\}\,d\tau \textit{ and } G_0\,(x,t)$$

is obtained from

$$NU_n\,(x,\tau) = G_n\,(x,\tau) + O\left(\tau^{n+1}\right)$$

The equation can be solved iteratively to obtain an approximate solution that takes the form

$$u\,(x,\tau) \cong U_n\,(x,\tau)$$

Where, n is the final iteration step. This is the required modified variational iteration method (MVIM.

2 K-dV Equation

The Kd-V equation is

$$u_t + \alpha u u_x + \beta\left(u^2\right)_x + u_{xxx} = 0$$

In this case, the coefficient β has a positive or negative value, and the appropriate sign has critical implications for the resulting structures.

Using a few case studies, we can now analyze the variational iteration methods with the modified variational iteration methods.

2.1 Example 1

Let us consider the K-dV equation

$$u_t - 5\left(u^2\right)_x + u_{xxx} = 0$$

$$u\,(x, 0) = 6x$$

$$u_1 = u_0 - \int_0^t \left[(u_0)_{xxx} + G_0\right]\, d\tau$$

Where, R= u_{xxx}

$$G_0\,(x, t) = -Nu_0\,(x, t)$$

$$G_0\,(x, t) = -5\left(u_0^2\right)_x$$

$$G_0\,(x, t) = -5\left(36x^2\right)_x$$

$$G_0\,(x, t) = -360x$$

$$u_1 = u_0 - \int_0^t (6x)_{xxx} - (360x) d\tau$$

$$= 6x + 360xt$$

$$u_1 = 6x\,(1 + 60t)$$

$$u_2 = u_1 - \int_0^t R\,(u_1 - u_0) + (G_1 - G_0)\,d\tau$$

$$Ru_0 = (6x)\,xxx = 0$$

$$Ru_1 = (u_1)\,xxx = (6x + 360xt)\,xxx = 0$$

$$\begin{aligned} G_1\,(x, t) &= -Nu_1\,(x, t) \\ &= -5\left(u_1^2\right)_x \end{aligned}$$

After substituting all values in the above formula, we have

$$u_2 = 6x\left(1 + 60t + 3600t^2\right)$$

and

$$u_3 = 6x\left(1 + 60t + 3600t^2 + 15552\,t^3\right)$$

.

.

.

.

And so on.

$$u = \lim_{n \overset{yields}{\rightarrow} \infty} u_n$$

$$u = \frac{6x}{1 - 6t^2}$$

This is the closed-form solution.

2.2 Example 2

Consider another form of Kd-V equation:

$$u_t + \left(u^2\right)_x + \left(u^2\right)_{xxx} = 0$$

$$u(x, 0) = 5x$$

$$Nu_n = \left(u_n^2\right)_x + \left(u_n^2\right)_{xxx}$$

By using the above formula, we can calculate

$$G(u_0) = N(u_0) = \left(u_0^2\right)_{xxx} + \left(u_0^2\right)_x$$

$$= \left(25x^2\right)_{xxx} + \left(25x^2\right)_x = 0 + 50x$$

$$G_0 = 50x$$

$$u_0 = 5x$$

$$u_1 = u_0 - \int_0^t R(u_0 - u_{-1}) + (G_0 - G_{-1})\, d\tau$$

$$u_1 = 5x - 50xt$$

- $G(u_1) = N(u_1) = \left(u_1^2\right)_{xxx} + \left(u_1^2\right)_x$

$$= 0 + \left(u_1^2\right)_x$$

$$u_2 = u_1 - \int_0^t R(u_1 - u_0) + (G_1 - G_0)\,d\tau$$

$$u_2 = 5x - 100xt + 500xt^2$$

And so on.

$$u_n = 5x\left(1 - 10t + 100t^2 - \dots\right)$$

$$= \frac{5x}{1 - 10t}\,u$$

The exact solution to the problem is being approached.

2.3 *Example 3*

$$u_t + uu_x - u(1 - u) = 0$$

With the boundary conditions $u(x, 0) = e^{-x}$

$$Nu_n = u_n(u_n)_x + u_n^2 - u_n$$

$$Nu_0 = u_0(u_0)_x + u_0^2 - u_0$$

$$= e^{-x}\left(e^{-x}\right)_x + e^{-2x} - e^{-x} = -e^{-x}$$

$$G_n = Nu_n = u_n(u_n)_x - u_n^2$$

$$G_0 = -e^{-x}$$

$$u_1 = u_0 - \int_0^t 0 + (G_0 - G_{-1})\,d\tau$$

$$= e^{-x} + te^{-x}$$

$$u_2 = u_1 - \int_0^t 0 + (G_1 - G_0)\,d\tau$$

$$u_2 = e^{-x} \quad 1 + t + \frac{t^2}{2}\Big)$$

Similarly, we have obtained

$$u_3 = e^{-x} \quad 1 + t + \frac{t^2}{2} + \frac{t^3}{3!}\Big)$$

And so on.

$$u_n = e^{-x}\left(1 + t + \frac{t^2}{2!} + \frac{t^3}{3!} + \ldots\ldots\ldots \frac{t^n}{n!}\right)$$

The above modified VIM results are obtained.

3 Conclusions

In this work, we have examined a few MVIM-related issues in depth, and we have drawn the conclusion that the method can be used to mitigate some of VIM's drawbacks. By removing unnecessary terminology and redundant calculations, MVIM streamlines the VIM language. When it comes to reducing workload and speeding up computations, MVIM dominates. Engineers and non-experts alike can benefit from MVIM since it provides a closed-form approximation of a solution to a nonlinear condition. Utilizing variational iteration theory, the technique for finding a solution is straightforward, and only a small number of iterations are required to obtain precise results. Specifically, variational iteration method (VIM) includes repetitive calculations and the computation of unnecessary terms. Therefore, by utilizing the modified variational iteration method (MVIM), we are able to overcome the shortcomings of VIM, which in turn prevents the repetition of old computations and thus minimizes extra effort and calculations.

References

1. Abassy, T.A., El-Tawil, M., Kamel, H.: The solution of KdV and mKdV equations using Adomian Padé approximation. Int. J. Nonlinear Sci. Numer. Simul. **5**(4), 327–339 (2004)
2. Wazwaz, A.-M.: A comparison between the variational iteration method and Adomian decomposition method. J. Comput. Appl. Math. **207**, 129–136 (2007)
3. He, J.H.: A new approach to nonlinear partial differential equations. Commun. Nonlinear Sci. Numer. Simul. **2**(4), 230–235 (1997)
4. He, J.H.: Variational iteration method—a kind of non-linear analytical technique: some examples. Int. J. Non-Linear Mech. **34**(4), 699–708 (1999)
5. Fatima, N., Daniel, S.: Solution of wave equations and heat equations using HPM. Appl. Math. Sci. Comput. Trends Math., 367–374 (2019)
6. He, J.H.: Approximate analytical solution of Blasius' equation. Commun. Nonlinear Sci. Numer. Simul. **4**(1), 75–78 (1999)
7. Fatima, N.: © Solution of gas dynamic and wave equations with VIM springer nature Singapore Pte Ltd. In: Advances in Fluid Dynamics Lecture Notes in Mechanical Engineering, pp. 81–91 (2021)
8. He, J.H.: Variational iteration method for autonomous ordinary differential systems. Appl. Math. Comput. **114**(2–3), 115–123 (2000)
9. He, J.H.: Some asymptotic methods for strongly nonlinear equations. Int. J. Mod. Phys. B. **20**(10), 1141–1199 (2006)
10. Tamer, A., et al.: Toward a modified variational iteration method. J. Comput. Appl. Math. **207**, 137–147 (2007)
11. Tari, H.: Modified variational iteration method. Phys. Lett. A. **369**, 290–293 (2007)
12. He, J.H., Wu, X.H.: Construction of solitary solution and compacton-like solution by variational iteration method. Chaos, Solitons Fractals. **29**(1), 108–113 (2006)
13. Hopf, E.: The partial differential equation ut + uux = μuxx. Commun. Pure Appl. Math. **3**, 201–230 (1950)
14. Soliman, A.A.: A numerical simulation and explicit solutions of KdV-Burgers' and Lax's seventh-order KdV equations. Chaos, Solitons Fractals. **29**(2), 294–302 (2006)
15. Fatima, N.: The study of heat conduction equation by homotopy perturbation method. SN Comput. Sci. **3**, 65 (2022)

Part II
Mechanics

Nanofluid Containing Motile Gyrotactic Microorganisms Squeezed Between Parallel Disks

J. Prathap Kumar, J. C. Umavathi, and A. S. Dhone

1 Introduction

Many academics have looked at the squeezing flow of Newtonian and non-Newtonian fluids in two parallel disks due to its importance in engineering and technological uses. Several studies exist to clarify and show the characteristics and conduct of nanofluid suction for industrial use such as nuclear reactors, materials, electronics, biomechanics, transportation, and food. Flow formation is caused by the relative motion of parallel adjacent surfaces. Stefan [1] first came up with the notion of squeezing flow in 1874 in the context of lubrication estimates. After Stefan's final work, several researchers examined such flows for various geometries. There are several precedents of squeezing flow, but the most notable are injection, compression, polymer forming and such type of relative motion is mostly used in modelling the oil flow in the bearings, to determine the load bearing capacity. After this foundational work, a number of scientists have shown their keen interest towards squeezing flow, and many studies have been carried out to understand these flows. Haq et al. [2] inquired into the occurrence of a squeezing flow for water-efficient carbon nanotubes enclosed by two parallel discs in the occupancy of a magnetic field dependent on time. The movement of heat and fluid flow mechanisms for nanofluid flow in parallel disks in the presence of a variable magnetic field is explored by Hatami et al. [3]. Domairry et al. [4] studied MHD squeezing flow of sticky liquid surrounded by two parallel disks.

Nanofluid is a modern class of nanotechnology defined as the colloidal suspension of 1-100 nm sized ultrafine particles in ordinary fluids. The term "nanofluid"

J. Prathap Kumar · J. C. Umavathi (✉)
Department of Mathematics, Gulbarga University, Kalaburgi, Karnataka, India

A. S. Dhone
Department of Mathematics, Sangameshwar College (Autonomous), Solapur, Maharashtra, India

F. Kamalov et al. (eds.), *Advances in Mathematical Modeling and Scientific Computing*, Trends in Mathematics, https://doi.org/10.1007/978-3-031-41420-6_13

was initially coined by Choi and Eastman [5] in ANI (1995). This is a very interdisciplinary and fast growing field. Recently, modern investigators are fascinated to nanofluids because of their significant industrial applications such as heat exchangers, power production, nanodrug delivery, cancer therapy, coolants, nuclear reactors, sensing and imaging, nanocryosurgery, nanoelectronics, energy production, and consumer products. Khaled et al. [6] has analysed the hydromagnetic effects on the dynamic and the characteristics of thermal boundary layer on a sensor surface placed in the designated volatile external squeezing flows. Acharya et al. [7] inspected the impacts of magnetic field over a squeezing flow of copper-water and copper-kerosene nanofluids in parallel plates.

The movement of particles in a fluid causes heat transfer inside the fluid, called convection, which is caused by the inequality between viscous and buoyancy forces. The convective flow of nanofluids and microorganisms is a significant field in recent studies of bio-nano-material processing applications [8]. The impact of Stefan blowing and various slips on the nonlinear gyrotactic bioconvection 2-D flow of nanofluid on a horizontal plate pointing upwards is examined by Uddin et al. [9]. Microorganisms are widely employed in the chemical and food sectors to make pharmaceuticals/medications, ordinary foods, and pesticides, as well as in the decomposition of organic matter. By swimming upwards, microorganisms are concentrated in the upper layer of fluid. As a result, the surface becomes densely unstable and the microorganisms fall down, initiating bioconvection. Bioconvection in nanofluids is now an impending field for investigative work, as it is widely used to measure the toxity of nanoparticles in biomicrosystem [10], biosensors [11], and microdevices [12].

Findings from the above literature show that limited studies are available in the field of nanofluid squeezing flow involving microorganisms in two parallel disks. The purpose of this research work is to study how various parameters affect a bioconvective nanofluid flow containing microorganisms. Squeeze parameters and Hartman numbers show significant effects on concentration profile and microorganisms number. Thermophoresis, Brownian motion parameters, Lewis numbers, Peclet numbers, bioconvective Schmidt numbers, and Prandtl number effects were observed and graphical results were drawn.

2 Mathematical Formulation

We consider a 2-D incompressible squeezing nanofluid flow containing motile microorganism between two parallel disks placed at a distance $h(t) = H(1 - at)^{1/2}$. The lower disk and upper disks are solids. The magnetic intensity $B(t) = B_0(1 - at)^{-1/2}$ is acting vertically on the disks. In real situation, it is obvious that $1 - at$ must be non-negative, i.e. $1 - at > 0$, so that $a < \frac{1}{t}$. The plates are fixed when $a = 0$. Further, the top disk is squeezed to the bottom disk when $0 < a < \frac{1}{t}$ (until they come into contact $t = \frac{1}{a}$) whereas $a < 0$, the disks retard from each other.

Fig. 1 Physical configuration

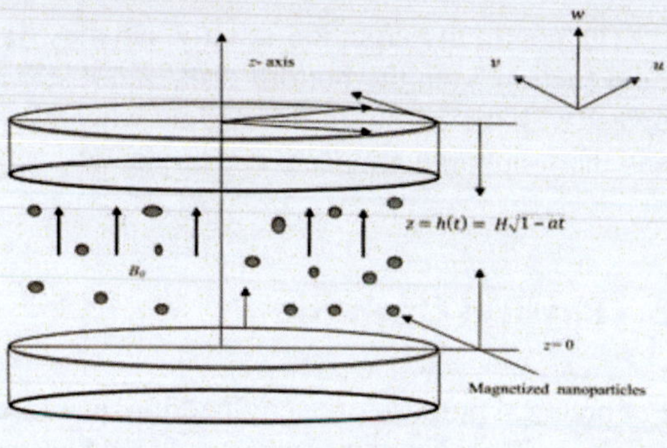

The lower disk is immobile, i.e. $z = 0$, while the top disk at a distance $z = h(t)$ moves with velocity $\frac{dh}{dt}$ in the direction of or apart from the bottom disk. Figure 1 depicts the flow geometry of the problem. The equations describing the present physical model in dimensionless form [4, 13] are as follows:

$$f'''' - S\left(\eta f''' + 3f'' - 2f\, f'''\right) - M^2 f'' = 0 \tag{1}$$

$$\theta' + \Pr Nb\, \theta'\phi' + \Pr Nt\, \theta'^2 + S \Pr\, (2f - \eta)\, \theta' = 0 \tag{2}$$

$$\phi'' + \left(\frac{Nt}{Nb}\right)\theta'' + S\, Le\left(2f\phi' - \eta\, \phi'\right) = 0 \tag{3}$$

$$\chi'' + S\, Lb\, \left(2f\chi' - \eta\chi'\right) - Pe\, Lb\, \left(\chi'\phi' + \chi\, \phi''\right) = 0 \tag{4}$$

with relevant non-dimensional boundary conditions

$$f(0) = A,\;\; f'(0) = 0,\;\; \theta(0) = 1,\;\; \phi(0) = 1 \text{ and } \chi(0) = 1 \text{ at } \eta = 0$$

$$f(1) = \frac{1}{2},\;\; f'(1) = 0,\;\; \theta(1) = 0,\;\; \phi(1) = 0 \text{ and } \chi(1) = 0 \text{ at } \eta = 1 \tag{5}$$

The various parameters used above are defined below

$$S = \frac{a\,H^2}{2\nu},\;\; M^2 = \frac{H^2 B_0^2 \sigma}{\mu},\;\; Nb = \frac{(\rho\, C)_p D_B\, (C_w - C_h)}{(\rho\, C)_f \nu},\;\; \Pr = \frac{\nu}{\alpha},\;\; A = \frac{w_0}{aH},$$

$$Le = \frac{\nu}{D_B},\;\; Lb = \frac{\nu}{D_n},\;\; Pe = \frac{b\, W_c}{\nu},\;\; Nt = \frac{(\rho\, C)_p D_T\, (T_w - T_h)}{(\rho\, C)_f \nu\, T_m} \tag{6}$$

The continuity equation strictly satisfies. It is worth noting that $A > 0$ denotes fluid suction from the bottom disk, and $A < 0$ denotes injection flow. The problem agrees with the case of ordinary fluid when $Nb = Nt = 0$, i.e. both Brownian motion and thermophoretic impacts are very small.

3 Physical Quantities

For practical purposes, the engineering quantities of interest—the skin friction coefficient, reduced Nusselt number, reduced Sherwood number, and motile microorganism number—in dimensionless form can be given as follows:

$$\left(\frac{H^2}{r^2}\right)\mathrm{Re}_r C_{f\,r} = f''\,(1),\ \mathrm{Re}_r = \frac{r\ a\ H(1-a\ t)}{2\ \nu}^{1/2},\quad Nur = \sqrt{1-a\ t}Nu = -\theta'\,(1),$$

$$Shr = \sqrt{1-a\ t}\ Sh = -\phi'\,(1),\quad Qnr = \sqrt{1-a\ t}\ Qn = -\chi'\,(1) \tag{7}$$

4 Results and Discussions

The RK fourth-order approach, which is incorporated in Matlab's bvp4c software's collocation formula, can be utilized to solve the highly nonlinear coupled system of O-D-Es that executes the current problem quantitatively (following [14–16]). For this purpose, the set of PDEs is turned into nonlinear ODEs. The impact of pertinent parameters namely squeeze number, blowing/suction parameter, Brownian motion, and thermophoresis parameters are discussed in detail. Different implications of different numbers, Prandtal, bioconvective Lewis, Lewis, Peclet, and Hartman numbers (*M*), were observed in either cases of suction and blowing. The values of non-dimensional parameters for plotting graphs are chosen as follows: $Lb = Le = \mathrm{Pe} = \ \mathrm{Pr}\ = 1,\ M = S = 1,\ Nb = 0.1(or1.5),\ Nt = 0.1$. Throughout the analysis, these values were kept constant, except for the varied values that are displayed in the respective graphs.

Figures 2, 3, 4, 5, 6, and 7 present the graphical results for momentum ($f'(\eta)$), heat ($\theta\ (\eta)$), nanoparticle species concentration ($\phi(\eta)$), and motile microorganisms ($\chi(\eta)$). The figures are depicted for two cases of suction ($A > 0$) and injection ($A < 0$). Figure 2 portrays the momentum distributions for both suction and injection. It is important to notice that perforations in the disk surface do not cause suction or injection. The rate of mass flow from the potential flow to the surface of the solid disk is related to the influence of suction (opposite of Stefan blowing). Likewise, the Stefan blowing is created by the rate of mass flow from the revolving disk surface to the potential flows. The velocity profile for the blowing situation ($A < 0$)

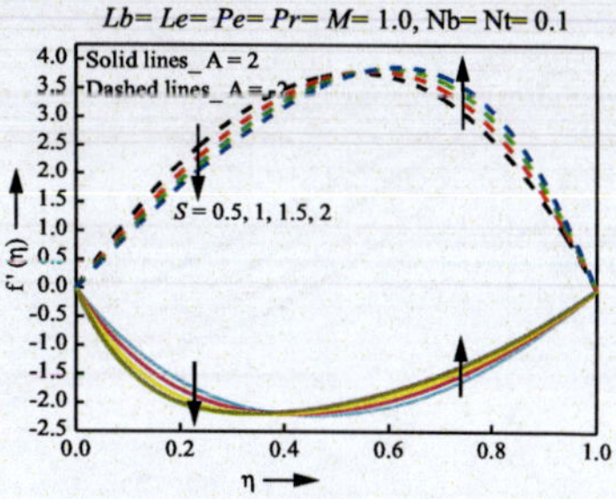

Fig. 2 Influence of S on $f(\eta)$

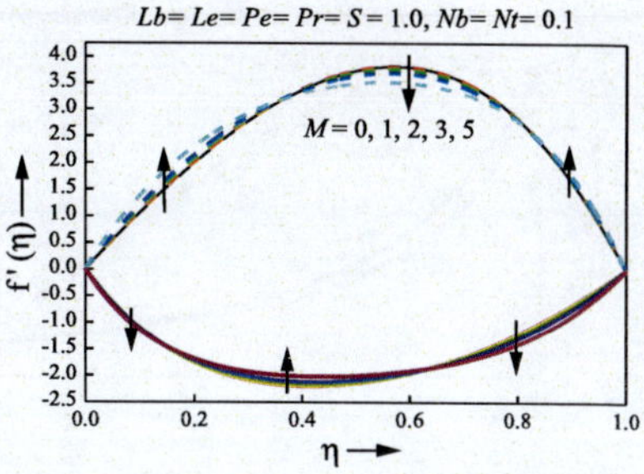

Fig. 3 Influence of M on $f(\eta)$

reduces initially with an increasing value of S in the bottom half space $0 \le \eta \le 0.56$; however, for $\eta \sim 0.56$ in the top half space of the gap, a critical point is achieved at which velocity begins to increase with S. This is because the intense blowing at the bottom disk occurs beyond the gap's midline ($\eta = 0.5$) effectively, causing augmentation in the top half space for the no-slip requirement imposed at both disk surfaces ($\eta = 0, 1$). For suction at the lower disk ($A > 0$) for $S > 0$ (enhanced pace/rate of rise of the bottom disk in the direction of the top disk), the flow is again reduced. The adhesion between the bottom disk surfaces and the fluid closely above the disk surface causes bigger negative values in velocity. Acceleration starts with S in the top half region $0.4 \le \eta \le 1.0$. A similar distribution of momentum in the absence of nanoparticles and motile microorganisms was observed by Chandrashekhar et al. [17] and Verma [18]. The symmetry of velocity profiles may be achieved by setting $A = 0$, which makes both discs equivalent.

Figure 3 demonstrates the upshot of M on the momentum distribution. The Lorentz force owing to a magnetic field is described by the Hartman number. Taking $M = 0$, or the electrically non-conducting nanofluid situation, yields the regulating equations for hydrodynamic flow. Physically, an increment in the Hartman number declines the contribution of the Lorentz force through the term $-M^2 f''(\eta)$ in the momentum Eq. (1). The effect of rising Hartman number is to increase the velocity near the bottom disk for the blowing scenario since ($A < 0$) lateral mass influx into the regime firstly overcomes the decelerating Lorentzian magnetic drag force impact. The resistive magnetic body force, on the other hand, dominates in the core zone $0.38 \le \eta \le 0.8$, resulting in a considerable slowing in the flow with bigger M values. There is a weak revival of acceleration in the neighbourhood of the upper

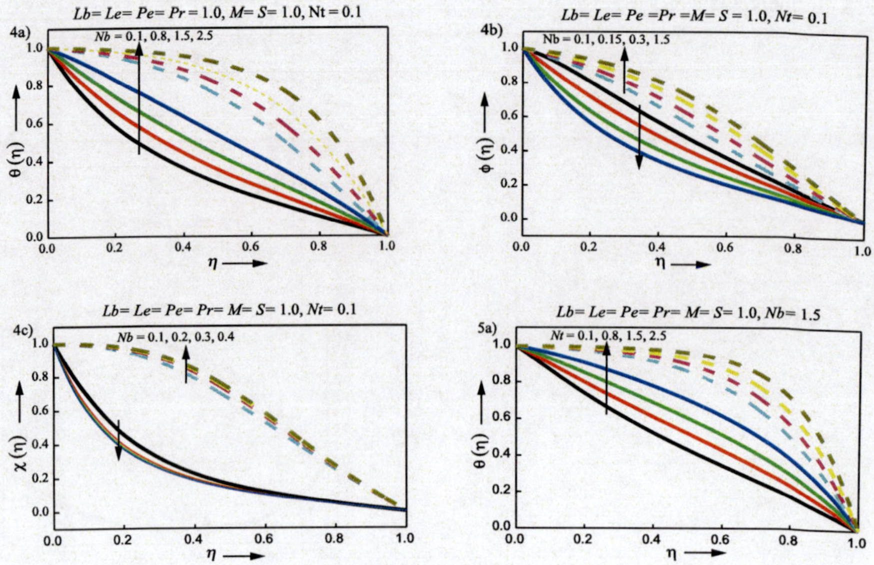

Fig. 4 Influence of Nb on **(a)** $\theta\ (\eta)$, **(b)** $\phi\ (\eta)$, and **(c)** $\chi\ (\eta)$

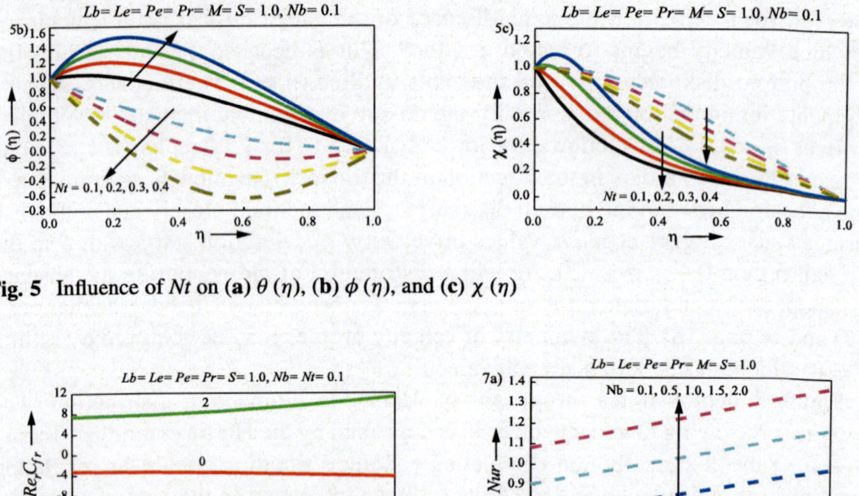

Fig. 5 Influence of Nt on **(a)** $\theta\ (\eta)$, **(b)** $\phi\ (\eta)$, and **(c)** $\chi\ (\eta)$

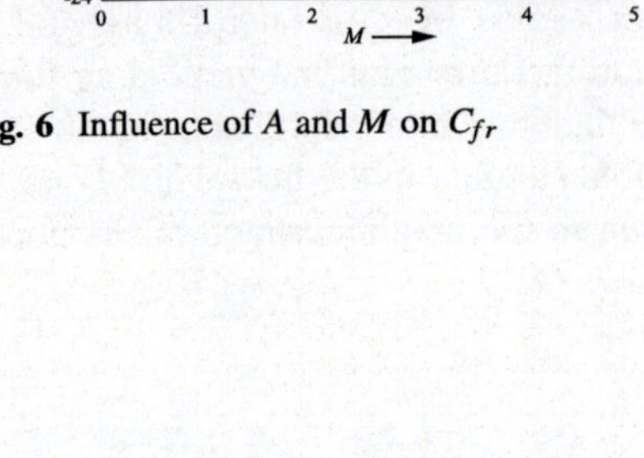

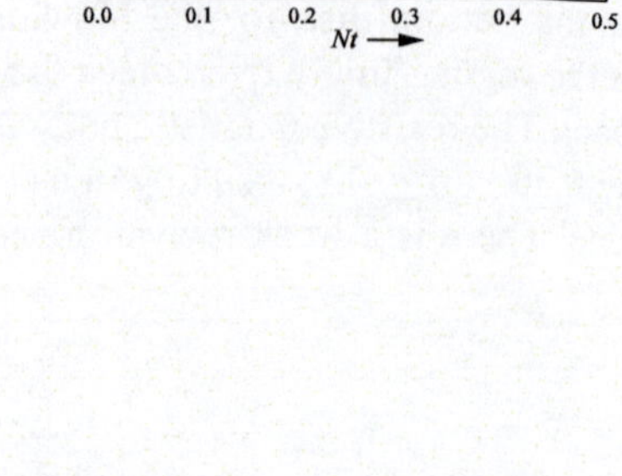

Fig. 6 Influence of A and M on C_{fr}

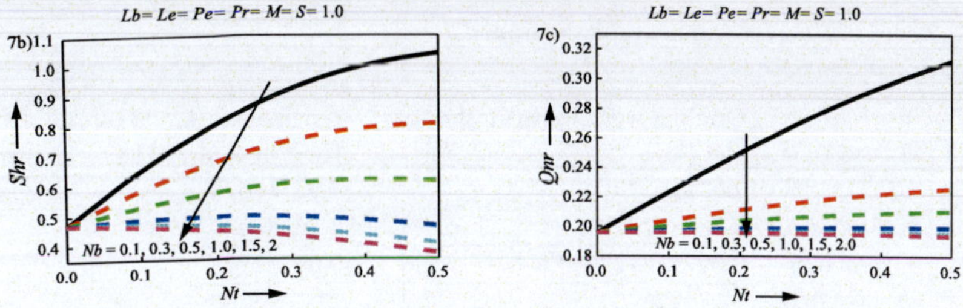

Fig. 7 Influence of *Nb* and *Nt* on **(a)** *Nur*, **(b)** *Shr*, **(c)** *Qnr*

disk, which is because of the decreasing impact of the magnetic field near the disk surface "$0.8 \leq \eta \leq 1.0$" and the redistribution of velocity further from the core region. In case of rotating disk, similar results are observed by Zueco and Beg [19].

The significance of *Nb* on heat, nanoparticle concentration, and density of motile microorganisms is revealed in Fig. 4a–c respectively. Figure 4a reveals that the thermal energy distribution ($\theta(\eta)$) enhances with *Nb* across the gap from bottom disk ($\eta = 0$) to the top disk ($\eta = 1$) for both suction and blowing. Physically haphazard motions of nanoparticles induce ballistic collisions between nanoparticles, as observed by Buongiorno [20] that lead to extra heating in the magnetic nanofluid saturated with motile microorganisms, culminating in an augmentation in the temperature profiles with greater *Nb* values.

In the blowing situation, however, rising *Nb* values cause the temperature profile topology to warp into a more parabolic shape. The characteristics of *Nb* occur in Eq. (2) in the terms $\Pr Nb\, \theta' \phi'$ and also in Eq. (3) ($\phi(\eta)$) in the terms $+\frac{Nt}{Nb}\theta''$. Evidently, nano-scale Brownian motion has a significant effect on thermal energy profiles. Figure 4b reflects that the nanoparticles concentration profile $\phi(\eta)$ rises with *Nb* because of the haphazard transition of nanoparticles from the top disk to the nanofluid. Clearly, nanoparticle diffusion from $\eta = 0$ to $\eta = 1$ is generally enhanced with blowing ($A = -2$) and is suppressed with suction ($A = 2$). Figure 4c exhibits the effect of *Nb* on microorganism profiles $\chi(\eta)$ in both suction and blowing cases. The motile microorganism is at its maximum at the lowest disk, while at the topmost disk, it is at its lowest. In blowing ($A < 0$) case, the profile $\chi(\eta)$ enhances with rise in *Nb*. The reverse impact is observed when$A > 0$.

Figure 5a–c illustrated the effect of *Nt* on $\theta(\eta)$, $\phi(\eta)$, and $\chi(\eta)$, respectively. There is a similar response of *Nt* on temperature to that calculated for *Nb* for both suction and blowing; see Fig. 5a. The magnitudes of temperature are clearly higher, and profiles are extensively restricted to the top half space of the gap, boosting thermophoretic body force and heating the regime/system greatly. Figure 5b shows the upshot of *Nt* on $\phi(\eta)$. The nanoparticle concentration profiles are skewed/tilted towards the lower disk for $A = 2$ and skewed/leaned towards the upper disk for $A = -2$. In the suction case, there is a noticeable rise in the concentration $\phi(\eta)$ as the thermophoretic action strengthens. From a physical standpoint, an increment in the thermophoretic impact causes the temperature gradient to generate a higher mass

Table 1 Values of $f''(1)$ for various values of M and S

M	S	$f''(1)$ Present work	Original work HAM Hashmi et al. [21]	Original work ND solver Hashmi et al. [21]
0	1	7.53313483	7.53316579	7.53316579
2		8.26385368	8.26387231	8.26387230
3		9.09730026	9.09732572	9.09732573
5		11.34927822	11.3492890	11.3492890
1	0.1	8.97524706	8.97552394	8.97552394
	0.5	8.34918684	8.34924578	8.34924578
	1	7.72191589	7.72194601	7.72194601
	2	6.94077588	6.94077326	6.94077334

flux, which elevates the concentration. Reversal effect is attained for the effects of Nt for blowing in comparison with suction. Figure 5c illustrates that near the lower disk (for weaker η), when $A > 0$ the microorganism profile $\chi(\eta)$ is significantly greater than when $A < 0$ at small values of parameters $Nb = Nt = 0.1$. The profiles of $\chi(\eta)$ get more pronounced as the thermophoresis parameter Nt is increased, and the profiles overlap near the upper disk when $A > 0$. When $A < 0$ occurs, a decrease in the microorganism profile is indicated and the distance between the profiles is found to be much less near the bottom disk and converge on the topmost disk.

The skin friction at the upper disk with Stefan blowing and suction parameter for the variations of Hartman number is depicted in Fig. 6. The skin friction reduces as M increases for blowing. The friction component for suction is amplified by a strong magnetic Lorentz force compared to viscous force ($M > 1$). The effects of skin friction on the skin are insignificant if there is no suction or injection.

Figure 7a–c displayed the combined impact of Nb and Nt on the Nusselt number, Sherwood number, and local wall motile microorganism number. In the case of suction, increasing the parameters Nb and Nt continually increases the magnitude of the Nusselt number (Fig. 7a). However, enhancing the values of Nt has the reverse effect on Sherwood number and local wall motile microorganisms' number, lowering the mass transfer rate to the disk surface. As shown in Fig. 7b, magnetic nano-lubricants significantly alter heat transfer, mass transfer rate, and the quantity of local wall motile microorganisms' number on the disk surface.

Table 1 indicates that the present solutions obtained using bvp4c MATLAB agreed very well with the numerical solutions obtained by Homotopy analysis method for all the values of the parameters in the absence of $\chi(\eta)$ which justify the current analysis. The comparison at various values of M, S, Nb, and Nt with that of $f''(1), -\theta'(1), -\phi'(1)$, and $-\chi'(1)$ are equivalent to the $\left(\frac{H^2}{r^2}\right) \mathrm{Re}_r C_{f\,r}, Nur, Shr$, and Qnr, respectively, was established when $A = 2, M = S = \mathrm{Pr} = Pe = Le = Lb = 1.0$, and $Nb = Nt = 0$.

5 Final Remarks

The impact of suction/blowing and squeezing parameters on the time-dependent squeezing flow of nanofluids in the presence of a magnetic field in two parallel disks, one impermeable and the other porous, is investigated in this work. Results of Brownian and thermophoresis factors are also shown. The following are the key findings of this study:

1. The investigation shows that the impacts of the traditional Brownian motion (Nb) and thermophoresis parameters (Nt) on temperature distribution are similar and increasing in both suction ($A > 0$) and blowing ($A < 0$) flow cases, while nanoparticle concentration and microorganisms' concentration are reverse in both $A > 0$ and $A < 0$ flow cases.
2. We can deduce from Fig. 3 that the magnetic field is critical for controlling the velocity of various conducting fluids. In both suction and blowing, it has a similar effect on velocity. It is also seen that the Hartman number has no impact on $\theta(\eta)$, $\phi(\eta)$, and $\chi(\eta)$.
3. The momentum distribution f in the suction case is a rising function of S in the lower half, towards the lower disk, and a declining function of S in the top half, approaching the upper disk. Injection flow, on the other hand, exhibits the reverse behaviour.
4. In both suction and injection, the variation in squeezing parameter S boosts the heat transfer rate at the lower disk half, whereas it is suppressed in the topmost disk half.
5. For $A > 0$, the increase of S causes the increase in nanoparticle concentration while decrease in case of injection $A < 0$. The opposite effects are observed in case of motile microorganisms.
6. The increase in Nb and Nt tends to boost the Nusselt number, while diminishing the Shr and Qnr in a similar way.

References

1. Stefan, M.J.: Versuch Uber die scheinbare adhesion. Sitzungsber. Akad. Wiss. Wien Math. Naturwiss. **69**, 713–721 (1874)
2. Haq R.U., Hammouch Z., Khan W.A.: Water-based squeezing flow in the presence of carbon nanotubes between two parallel disks. Therm. Sci. **20**(6), 1973–1981 (2016)
3. Hatami, M., Ganji, D.D.: Heat transfer and nanofluid flow in suction and blowing process between parallel disks in presence of variable magnetic field. J. Mol. Liq. **190**, 159–168 (2014)
4. Domairry, G., Aziz, A.: Approximate analysis of MHD squeeze flow between two parallel disks with suction or injection by homotopy perturbation method. Math. Probl. Eng. **2009**, 603916, 19 pages (2009)
5. Choi, S.U.S., Eastman, J.A.: Enhancing thermal conductivity of fluids with nanoparticles. In: ASME International Mechanical Engineering Congress & Exposition. American Society of Mechanical Engineers, San Francisco (1995)

6. Khaled, A.R.A., Vafai, K.: Hydromagnetic squeezed flow and heat transfer over a sensor surface. Int. J. Eng. Sci. **42**, 509–519 (2004)
7. Acharya, N., Das, K., Kundu, P.K.: The squeezing flow of Cu-water and Cu-kerosene nanofluids between two parallel plates. Alex. Eng. J. **55**, 1177–1186 (2016)
8. Das, K., Duari, P.R., Kundu, P.K.: Nanofluid bioconvection in presence of gyrotactic microorganisms and chemical reaction in a porous medium. J. Mech. Sci. Technol. **29**(11), 4841–4849 (2015)
9. Uddin, M.J., Alginahi, Y., Beg, O.A., Kabir, M.N.: Numerical solutions for nonlinear gyrotactic bioconvection in nanofluid saturated porous media with Stefan blowing and multiple slip effects. Comput. Math. Appl. **72**, 2562–2581 (2016) ISSN: 0898-1221
10. Tsai, T.H., Liou, D.S., Kuo, L.S., Chen, P.H.: Rapid mixing between ferro-nanofluid and water in a semi-active Y-type micromixer. Sens. Actuators A Phys. **153**, 267–273 (2009)
11. Li, H., Liu, S., Dai, Z., Bao, J., Yang, X.: Applications of nanomaterials in electrochemical enzyme biosensors. Sensors. **9**, 8547–8561; ISSN 1424-8220 (2009)
12. Munir, A., Wang, J., Zhou, H.S.: Dynamics of capturing process of multiple magnetic nanoparticles in a flow through microfluidic bioseparation system. IET Nanobiotechnol. **3**(3), 55–64 (2009)
13. Joneidi, A.A., Domairry, G., Babaelahi, M.: Effect of mass transfer on the flow in the magnetohydrodynamic squeeze film between two parallel disks with one porous disk. Chem. Eng. Commun. **198**, 299–311 (2011)
14. Bandar, B.M., Naveed, A., Adnan, K.U., Syed, T.M.: A bioconvection model for a squeezing flow of nanofluid between parallel plates in the presence of gyrotactic microorganisms. Eur. Phys. J. Plus. **132**, 187 (2017)
15. Hussain, T., Xu, H., Raees, A., Zhao, Q.K.: Unsteady three-dimensional MHD flow and heat transfer in porous medium suspended with both microorganisms and nanoparticles due to rotating disks. J. Therm. Anal. Calorim. **147**, 1607 (2021). https://doi.org/10.1007/s10973-020-10528-x
16. Latiff, N.A., Uddin, M.J., Md. Ismail A. I.: Stefan blowing effect on bioconvective flow of nanofluid over a solid rotating stretchable disk. Propuls. Power Res. **5**(4), 267–278 (2016)
17. Chandrashekhara, B.C., Kantha, R., Rudraiah, N.: Effect of slip on porous-walled squeeze films in the presence of a transverse magnetic field. Appl. Sci. Res. **1978**(34), 393–411 (1978)
18. Verma, P.D.S.: Magnetic fluid-based squeeze film. Int. J. Eng. Sci. **24**, 395–401 (1986)
19. Zueco, J., Beg, O.A.: Network numerical analysis of hydromagnetic squeeze film flow dynamics between two parallel rotating disks with induced magnetic field effects. Tribol. Int. **43**, 532–543 (2010)
20. Buongiorno, J.: Convective transport in nanofluids. J. Heat Transf. **128**, 240–250 (2010)
21. Hashmi, M.M., Hayat, T., Alsaedi, A.: On the analytic solutions for squeezing flow of nanofluid between parallel disks. Nonlinear Anal. Model. Control. **17**(4), 418–430 (2012)

Linear Model for Two-Layer Porous Bed Suspended with Nano-Sized Particles

J. C. Umavathi

1 Introduction

The modeling of convective flows is a vital downside in both applied and theoretical terms. The convective flows contacting through the interface play a necessary role, as an example, in technology, the nuclear business, and additionally to cooling gadgets in electronics. Some of the works on this topic can be seen in [1–4]. Umavathi and her group have examined the flow tendency of two immiscible fluids filled in various geometries, such as vertical and horizontal channels, wavy vertical channels, and vertical ducts, since 1997. Some of the important papers published are [5–8]. For more details, researchers can visit the scholar account.

The problem of researching the multi-fluid flow state of affairs specializes in the interplay among specific sorts of fluids, specifically inside the same geometry. Some pioneers have, to this point, made a few progress. Laminar flow in a horizontal channel filled with non-mixing fluids was measured by Packham and Shail [9]. Ground water remediation and oil extraction require the knowledge of flow through porous media. Ghassemi and Pak [10] published the work related to immiscible fluids including porous matrix. Researchers have been working on this topic for more than five decades. Initially, research was carried out by sprinkling micro- and millimeter-sized particles, since Maxwell [11] regulated the analytical model. But this type of model created many obstacles, such as jamming the small channels and settling very fast. In 1995, Choi and Eastman [12] first doped small nano-sized particles in natural fluids, which are applicable in technology and industries for its unique chemical and physical characteristics. For the first time in the literature, Umavathi [13] explored the thermal modulation of a nanofluid saturated with porous media. Natural convection through a porous wavy cavity filled with nanofluid was

J. C. Umavathi (✉)
Department of Mathematics, Gulbarga University, Gulbarga, Karnataka, India

F. Kamalov et al. (eds.), *Advances in Mathematical Modeling and Scientific Computing*, Trends in Mathematics, https://doi.org/10.1007/978-3-031-41420-6_14

investigated by Sheremet et al. [14], including the impact of thermal dispersion, by adopting Forchheimer-Buongiorno approach. Further, Sheikholeslami and Ganji [15] furnished diversified semi-analytical methods and nanotechnology applications to open doors for researchers. Zeinali Heris et al. [16] designed an experiment to calculate the rate of heat transfer for the Al_2O_3/water nanofluid flowing through a square duct. They proved that there was a reinforcement of the coefficient of heat transfer up to 27.6% (2.5% volume fraction) when compared with only base fluid (in the absence of nanoparticles). Also, they observed that heat transfer was augmented by increasing the nanoparticle concentration at high flow rates.

Botong et al. [17] explored the flow characteristics of not-miscible fluids stuffed inside a two-dimensional horizontal annulus. They found that the initial distribution using denser fluid was fussy to resolve the final fluid distribution. They further continued the analysis by adding nanoparticles and discussing some aspects on the control of two-fluid gravity-driven flow, which has applications in heating power generation technologies and also in solar cooling. Botong et al. [18] also explored convective heat transfer of double immiscible fluids in a vertical channel. They concluded that the nanofluid with a lower volume fraction was a better model to achieve higher fluid flow and increasing the volume fraction of nanoparticles in one layer also boosted heat transfer in both the layers. Farooq and Liao et al. [19] also explicated on the nanofluid properties of a duct. Similar work was carried out by Khan et al. [20] using Eyring-Powell fluid inside a vertical channel.

By citing the prior literature on two immiscible fluid layers, this work is spotlighted on reckoning distinct naoparticles in the upper fluid layer and saturating two not-miscible layers by porous beds having different permeabilities. This work is focused on reporting the impacts of nanoparticle volume fraction, ratio of permeabilities to resolve the effects of interfacial drag in two-layer fluids. This study has applications in petroleum reservoir simulations and in energy conservation.

2 Mathematical Formulation

Figure 1 narrates the geometry of the problem, which exhibits that the duct is filled with two Newtonian not-miscible fluids that are marked as layer-1 (upper layer) and layer-2 (lower layer). The upper layer is occupied with a porous matrix with permeability κ_1 saturated with nanofluid, and the lower layer is filled with a porous bed with permeability κ_2. The Darcy and viscous dissipations are also considered. The porous beds and nanofluids are modeled mathematically with Forchhhiemer effects and Tiwari–Das model, respectively. The boundary conditions on temperature are flux at the top and bottom and T_{w1} and T_{w2} at the vertical left and right boundaries in such a way that $T_{w2} > T_{w1}$. Velocity is zero on all the boundaries. The width of the duct is b, whereas height of layer-1 is $a_1/2$ and layer-2 is $a_2/2$. The physical aspects such as viscosity and thermal conductivity are assumed to be constant. Further, the interface conditions are simulated to have continuity of shear stress, heat flux, temperature, and velocity. The nanofluid in

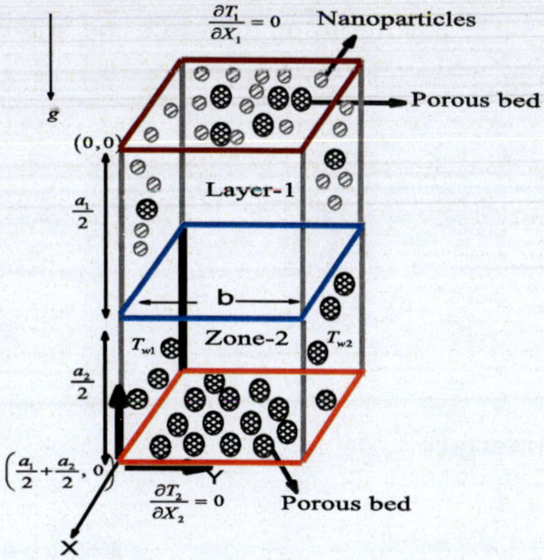

Fig. 1 Physical configuration

layer-1 is detailed to have viscosity μ_{nf}, density ρ_{nf}, thermal expansion coefficient β_{nf}, thermal conductivity K_{nf}, and permeability κ_1, and layer-2 is said to have viscosity μ_2, density ρ_2, thermal expansion coefficient β_2, thermal conductivity K_2, and permeability κ_2. Three combinations of two layers, such as engine oil-mineral oil, ethylene glycol-mineral oil, and ethylene oil-kerosene oil, are acknowledged for base fluids along with copper as a nanoparticle. Further, using water as the base fluid in both the layers, the prospects of copper, diamond, and titanium oxide nanoparticles are prepared. Following Choi and Eastman [12], the Nervier-Stokes equations are defined as follows:

Layer-1

$$\frac{\partial^2 w_1}{\partial {x_1}^2} + \frac{\partial^2 w_1}{\partial y^2} + GR\ \theta_1 - \frac{1}{Da}\ w_1 - I_F\ w_1^2 \ = p_1 \tag{1}$$

$$\frac{\partial^2 \theta_1}{\partial {x_1}^2} + \frac{\partial^2 \theta_1}{\partial y^2} + BR\left[\left(\frac{\partial\ w_1}{\partial x_1}\right)^2 + \left(\frac{\partial\ w_1}{\partial y}\right)^2\right] + \frac{BR}{Da}\ w_1^2 \ = 0 \tag{2}$$

Layer-2

$$\frac{\partial^2 w_2}{\partial {x_1}^2} + \frac{\partial^2 w_2}{\partial y^2} + \frac{Gr\ n\ \beta}{\lambda}\ \theta_2 - \frac{\kappa}{Da}\ w_2 - \frac{I\ n\ \sqrt{\kappa}}{\lambda}\ {w_2}^2 \ = \ \frac{p}{\lambda} \tag{3}$$

$$\frac{\partial^2 \theta_2}{\partial {x_2}^2} + \frac{\partial^2 \theta_2}{\partial y^2} + \frac{Br\ \lambda}{K}\left[\left(\frac{\partial\ w_2}{\partial x_2}\right)^2 + \left(\frac{\partial\ w_2}{\partial y}\right)^2\right] + \frac{Br\ \lambda\ \kappa}{Da\ K}\ {w_2}^2 \ = 0 \tag{4}$$

The conditions on the walls and at the interface are as follows:

Layer-1

$$w_1 = 0,\ \ \theta_1 = -\frac{1}{2}\ \text{at}\ Y = 0;\ \ w_1 = 0,\ \ \theta_1 = \frac{1}{2}\ \ \text{for}\ \ 0 \le x_1 < A_1;\ \ w_1 = 0,\ \ \frac{\partial \theta_1}{\partial x_1} = 0\ \text{at}\ x_1 = 0\ \text{for}\ 0 \le y \le 1 \tag{5}$$

Interface

$$w_1 = w_2,\quad \frac{\partial w_1}{\partial x_1} = \frac{\lambda}{(1-\phi)^{2.5}}\frac{\partial w_2}{\partial x_2},\quad \theta_1 = \theta_2,\quad \frac{\partial \theta_1}{\partial x_1} = \frac{K\,K_f}{K_{nf}}\frac{\partial \theta_2}{\partial x_2}\ \ \text{at}\ \ x = A_1\ \ \text{for}\ \ 0 \le y \le 1 \tag{6}$$

Layer-2

$$\left.\begin{array}{l} w_2 = 0,\ \theta_2 = -\frac{1}{2},\ \ w_2 = 0,\ \ \theta_2 = \frac{1}{2}\ \text{at}\ \ y = 0\ \ \text{for}\ \ A_1 < x_2 \le A_2 \\ w_2 = 0,\ \ \frac{\partial \theta_2}{\partial x} = 0\ \text{at}\ \ x = A_1 + A_2\ \ \ \text{for}\ 0 \le y \le 1 \end{array}\right\} \tag{7}$$

Here

$$\left.\begin{array}{l} Gr = \frac{g\ \beta_f b^3 (T_{w2}-T_{w1})}{{\nu_f}^2},\ \ Da\ =\ \frac{\kappa}{b^2},\ \ I = \frac{C_F\ b}{\sqrt{\kappa}},\ \ \ p = \frac{\rho_f\ b^3}{{\mu_f}^2}\ \frac{\partial P}{\partial Z}, p_1 = p\ (1-\phi)^{2.5}, \\ Br = \frac{{\mu_f}^3}{K_f\ {\rho_f}^2\ b^2\ (T_{w2}-T_{w1})},\ \ GR = Gr\ \left((1-\phi) + \phi\ \frac{(\rho\beta)_s}{(\rho\beta)_f}\right)\ (1-\phi)^{2.5}, \beta = \frac{\beta_2}{\beta_f}, n = \frac{\rho_2}{\rho_{nf}}, \kappa = \frac{\kappa_1}{\kappa_2}, \\ BR = \frac{Br}{(1-\phi)^{2.5}}\ \left(\frac{K_s + 2\ K_f + \phi\ \ (K_f - K_s)}{K_s + 2\ K_f - 2\ \ \phi\ \ (K_f - K_s)}\right), \lambda = \frac{\mu_2}{\mu_f},\ \ K = \frac{K_2}{K_f},\ \ \ A_1 = \frac{a_1}{2b},\ \ \ A_2 = \frac{a_2}{2b} \end{array}\right\} \tag{8}$$

$$\frac{\rho_{nf}}{\rho_f} = (1-\phi) + \phi\ \frac{\rho_s}{\rho_f},\ \frac{\beta_{nf}}{\beta_f} = (1-\phi) + \phi\ \frac{\beta_s}{\beta_f},\ \ \frac{\mu_{nf}}{\mu_f} = (1-\phi)^{-2.5},\ \frac{K_{nf}}{K_f} = \left(\frac{K_s + 2K_f - 2\phi\,(K_f - K_s)}{K_s + 2K_f + \phi\,(K_f - K_s)}\right) \tag{9}$$

3 Numerical Solutions

A well-known finite difference, along with Southwell-Over-Relaxation technique, is invited to find numerical solutions of governing Eqs. (9)–(12) along with prescribed boundary and interface conditions as mentioned in Eqs. (13)–(15). Layer-1 is divided into $Nx1$ and layer-2 is divided into $Nx2$ grids along x-axis with Δx as the length of one step and Ny grids along y-axis with Δy step length. Applying central finite differences of the first and second kinds, the finite difference equations become

Layer-1

$$\left(\frac{w_{1\,(i+1,j)} - 2w_{1\,(i,j)} + w_{1\,(i-1,j)}}{(\Delta x_1)^2}\right) + \left(\frac{w_{1\,(i,j+1)} - 2w_{1\,(i,j)} + w_{1\,(i,j-1)}}{(\Delta y)^2}\right) + GR\,\theta_{1\,(i,j)} - \frac{1}{Da} w_{1\,(i,j)} - I\, w^2_{1(i,j)} - p_1 = 0 \tag{10}$$

$$\left(\frac{\theta_{1\,(i+1,j)} - 2\theta_{1\,(i,j)} + \theta_{1\,(i-1,j)}}{(\Delta x_1)^2}\right) + \left(\frac{\theta_{1\,(i,j+1)} - 2\theta_{1\,(i,j)} + \theta_{1\,(i,j-1)}}{(\Delta y)^2}\right) + BR\left[\left(\frac{w_{1\,(i+1,j)} - w_{1\,(i-1,j)}}{2\,\Delta x_1}\right)^2 + \left(\frac{w_{1\,(i,j+1)} - w_{1\,(i,j-1)}}{2\,\Delta y}\right)^2\right] + \frac{BR}{Da} w^2_{1\,(i,j)} = 0 \tag{11}$$

Layer-2

$$\left(\frac{w_{2\,(i+1,j)} - 2w_{2\,(i,j)} + w_{2\,(i-1,j)}}{(\Delta x_2)^2}\right) + \left(\frac{w_{2\,(i,j+1)} - 2w_{2\,(i,j)} + w_{2\,(i,j-1)}}{(\Delta y)^2}\right) + \frac{Gr\,\beta\,n}{\lambda}\,\theta_{2\,(i,j)} - \frac{\kappa}{Da} w_{2\,(i,j)} - \frac{I\,n\,\sqrt{\kappa}}{\lambda} w^2_{2\,(i,j)} - \frac{p}{\lambda} = 0 \tag{12}$$

$$\left(\frac{\theta_{2\,(i+1,j)} - 2\theta_{2\,(i,j)} + \theta_{2\,(i-1,j)}}{(\Delta x_2)^2}\right) + \left(\frac{\theta_{2\,(i,j+1)} - 2\theta_{2\,(i,j)} + \theta_{2\,(i,j-1)}}{(\Delta y)^2}\right) + \frac{Br\,\lambda}{K}\left[\left(\frac{w_{2\,(i+1,j)} - w_{2\,(i-1,j)}}{2\Delta x_2}\right)^2 + \left(\frac{w_{2\,(i,j+1)} - w_{2\,(i,j-1)}}{2\Delta y}\right)^2\right] + \frac{Br\,\lambda\,\kappa}{Da\,K} w^2_{2\,(i,j)} = 0 \tag{13}$$

The discretized boundary and interface wall conditions take the form

Layer-1

$$\left.\begin{aligned} & w_{1\,(i,0)} = -w_{1\,(i,1)},\ \theta_{1\,(i,0)} = -1 - \theta_{1\,(i,1)},\ w_{1\,(i,Ny+1)} = -\,w_{1\,(i,Ny)},\ \theta_{1\,(i,Ny+1)} \\ & \quad = 1 - \theta_{1\,(i,Ny)} \text{ at } Y = 0 \text{ for } 0 \le x_1 < A_1 \\ & w_{1\,(0,j)} = -w_{1\,(1,j)},\ \theta_{1\,(1,j)} = \theta_{1\,(0,j)} \text{ at } x_1 = 0 \text{ for } 0 \le y \le 1 \end{aligned}\right\} \tag{14}$$

Layer-2

$$\left.\begin{aligned}
&w_{2\ (i,0)} = -w_{2\ (i,1)},\ \ \theta_{2\ (i,0)} = -1 - \theta_{2\ (i,1)}, \\
&\text{at } y = 0 \quad \text{for} \quad A_1 < x_2 \le A_2 \\
&w_{2\ (i,Ny+1)} = -w_{2\ (i,Ny)}, \qquad \theta_{2\ (i,Ny+1)} = 1 - \theta_{2\ (i,Ny)} \qquad \text{at} \\
&y = 0 \quad \text{for} \quad A_1 < x_2 \le A_2 \\
&w_{2\ (0,j)} = -w_{2\ (1,j)},\ \ \theta_{2\ (1,j)} = \theta_{2\ (0,j)} \ \text{ at } x = A_1 + A_2 \\
&\text{for } 0 \le y \le 1
\end{aligned}\right\} \tag{15}$$

Interface

$$\left.\begin{aligned}
&\left.\begin{aligned}
&w_{2\ (Nx2+1,j)} = w_{1\ (Nx+1,j)} + w_{1\ (Nx1,j)} - w_{2\ (Nx2,j)}, \\
&w_{1\ (Nx+1,j)} = \lambda \frac{\Delta x_1}{\Delta x_2}\left(w_{2\ (Nx2,j)} - w_{2\ (Nx2+1,j)}\right) + w_{1\ (Nx1,j)},
\end{aligned}\right\} \\
&\text{at } x = A_1 \quad \text{for } 0 \le y \le 1 \\
&\left.\begin{aligned}
&\theta_{2\ (Nx2+1,j)} = \theta_{1\ (Nx+1,j)} + \theta_{1\ (Nx1,j)} - \theta_{2\ (Nx2,j)}, \\
&\theta_{1\ (Nx+1,j)} = K \frac{\Delta x_1}{\Delta x_2}\left(\theta_{2\ (Nx2,j)} - \theta_{2\ (Nx2+1,j)}\right) + \theta_{1\ (Nx1,j)},
\end{aligned}\right\} \\
&\text{at } x = A_1 \quad \text{for } 0 \le Y \le 1
\end{aligned}\right\} \tag{16}$$

The values of average Nusselt values using different grids are tabulated in Table 1. This table confirms that 101 × 101 grids are sufficient to attain the accuracy. In the absence of nanoparticles and porous bed, the solutions tally with Umavathi and Bég [21], which is shown in Table 2. To further validate the code, the solutions obtained by Oztop et al. [22], Moshkin [23] (two-layer fluid flows), and De Vahl Davis [24, 25] were tallied in Umavathi and Bég [21].

Table 1 Grid independence test using water as base fluid and copper as nanoparticle

Grid size	Zone-I $\left[\frac{d\theta}{dy}\right]_{y=0}$	Zone-II $\left[\frac{d\theta}{dy}\right]_{y=0}$
11 × 11	1.03189193105140	1.00173000007397
51 × 51	1.03187749001005	1.00171580368600
101 × 101	1.03187845234035	1.00171672743727
151 × 151	1.03187868136563	1.00171692777853
201 × 201	1.03187877486040	1.00171700365681

Table 2 Similarity of $\left[\frac{d\theta}{dy}\right]_{y=0}$

Present $\phi = 0.01$, $Da = 0.01$, $I = 4.0$, $\kappa = 1.0$		Present $\phi = 0$, $Da = 0$, $I = 0$, $\kappa = 0$		Umavathi and Bég [21] $\phi = 0$, $Da = 0$, $I = 0$	
Layer-1	Layer-2	Layer-1	Layer-2	Layer-1	Layer-2
1.03187845	1.001716727	1.00455555	1.00455555	1.00455555	1.00455555

4 Results and Discussion

Figure 2a, b details the velocity distributions using the mixture of engine oil (layer-1) – mineral oil (layer-2), ethylene glycol (layer-1) – mineral oil (layer-2), and ethylene glycol (layer-1) – kerosene oil (layer-2). The 3D visualizes that there is almost no flow in layer-1 and the curvature is convexly bent on the top and bottom plates of the duct for engine oil-mineral oil, and there is almost equal distribution of fluid in both the layers and the contours are flat at the top and bottom plates for ethylene glycol-mineral oil and for ethylene glycol-kerosene oil. Flow occurs in both the layers, but convexity is drenched in layer-2 compared with layer-1. The flow is still better viewed in 2D picture which parade that there are no contours in layer-1 and two cells are formed in layer 2 which are almost symmetric for the combination of engine oil-mineral oil. Flow contours appear in both the layers and are proportional for ethylene glycol-mineral oil and for the mixture of ethylene glycol-kerosene. Contours are formed in both the layers, and the distributions are not symmetric. The 1D chart clearly manifests that at the interface point, engine oil-mineral oil takes the minimum at the left half and maximum at the right half of the duct, whereas the other two mixtures are not much reformed. That is to say that the flow patterns are almost same for the other two mixtures. Figure 3 presents the repercussion of the presence of nanoparticles. Copper, diamond, and titanium oxide are utilized as nanoparticles with water as base fluid in both the layers for equal permeabilities. Both the 3D and 2D visualize that there are no differences on the flow by using different nanoparticles. The number of contours is balanced in both the fluid layers. The 1D graph also did not reflect any changes. Hence, values of velocity are provided in Table 3, which exhibit that copper nanoparticle produces slightly more flow in comparison with the other two.

Figure 4 pageants the influence of solid volume fraction. For regular fluid ($\phi = 0$), the flow structure is flat in the upper and lower fluid layers and symmetric contours are captured. Almost similar 3D and 2D view is anticipated for $\phi = 0.0$ and $\phi = 0.01$ but by appending more nanoparticles ($\phi = 0.5$) the flow is almost nil in layer-1 and the flow is convexly symmetric in both the upper and lower plates of the duct (3D) and the number of contours are five in layer-2 and is one in layer-1. Hence, the occurrence of extensive nanoparticles reduces the flow. This is further justified by Fig. 4b (1D), which exposes that as ϕ increases velocity reduces. The reason for reduction in velocity by increasing ϕ is due to the fact the viscosity of the base fluid increases as ϕ increases.

The leverage of Da on the momentum and energy is manifested in Fig. 5a–c. For $Da = 0.001$, the 3D plot shows that the flow is flat in both the layers, and the contours (2D) are linear and symmetric in both the layers indicating that flow is almost stagnant. For higher values of Da, the shape of the flow is bulging outside in both the layers (3D). The flow contours (2D) are circular at the core region and become flat at the boundaries. Figure 5b (1D) also exposes that the profile is almost linear for $Da = 0.001$ and no much difference in the flow for values of $Da = 1, 2$. The presence of porous bed does not pressurize the energy distributions as can be

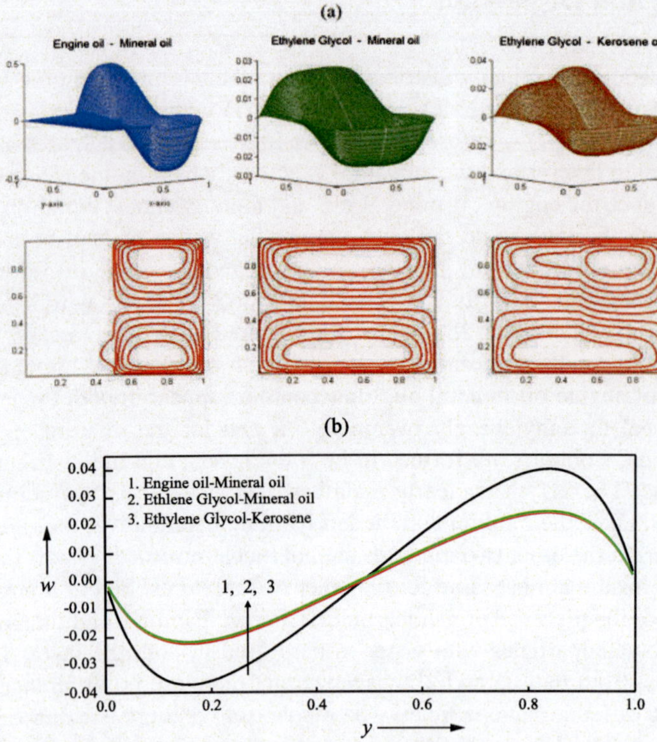

Fig. 2 Velocity **(a)** contours **(b)** profiles for altered base fluids

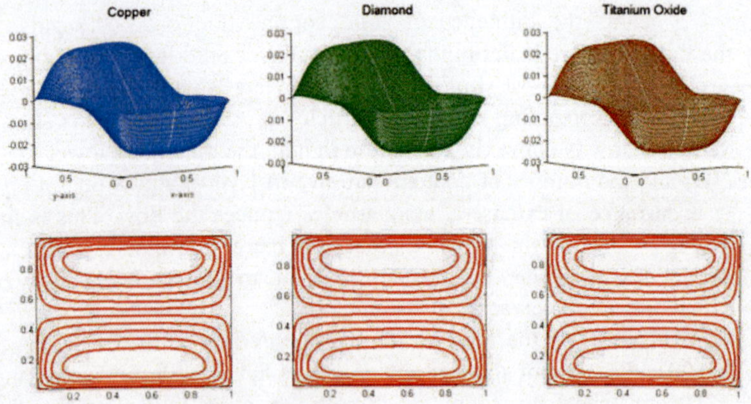

Fig. 3 Velocity contours for altered nanoparticles

seen in Fig. 5c. However when enlarged a portion of the profiles, the temperature is low for $Da = 0.001$. The suppression in the velocity and temperature for low values of Da physically indicates that the porous bed is packed thickly and hence

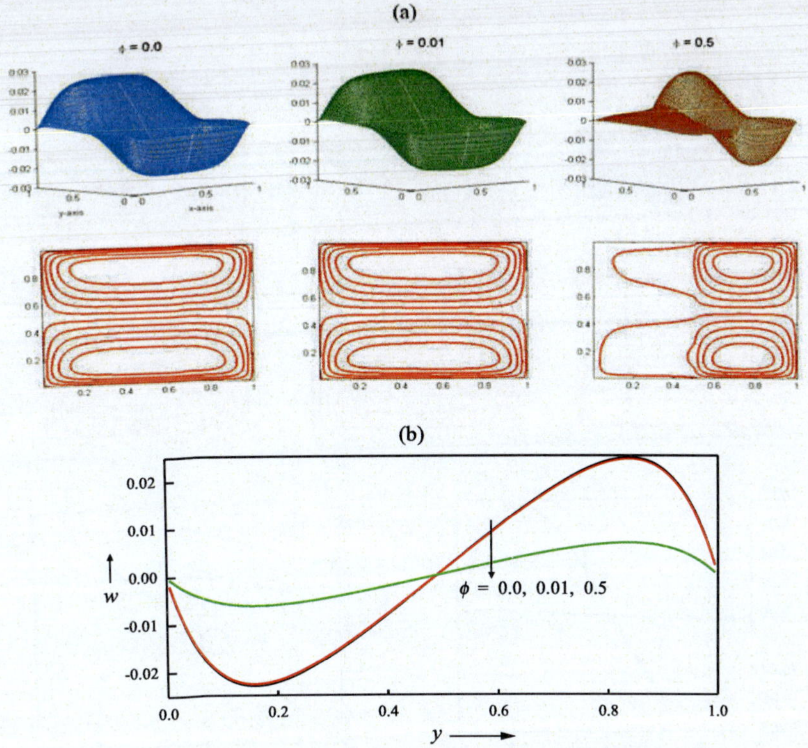

Fig. 4 Velocity (**a**) contours and (**b**) profiles for altered solid volume fraction

fluid cannot flow fast, which also reduces the transfer of heat. Figure 7b (1D) also reflects that as κ increases, velocity decreases in the range $0.5 \le y \le 1$ (right wall) and increases in the range $0.0 \le y \le 0.5$ (left wall). The low values of permeability represent densely packed porous bed and high values stand for sparsely packed porous bed.

The reaction of viscosity ratio λ is seen in Fig. 6a, b. The contours are observed when $\lambda = 1.0$, whereas they are dense at the right plate when $\lambda = 0.1,\ 0.5$. Velocity decreases at the right plate as the λ values are heightened (Fig. 6b). The influence K can be viewed in Fig. 7. The number of contours is symmetric in both the layers for all values of K.

5 Conclusions

Two fluids, which are not miscible, are filled in a duct. The upper portion is filled with nanofluid saturated with porous medium, and the lower portion is filled with only viscous fluid saturated with porous medium whose permeability is

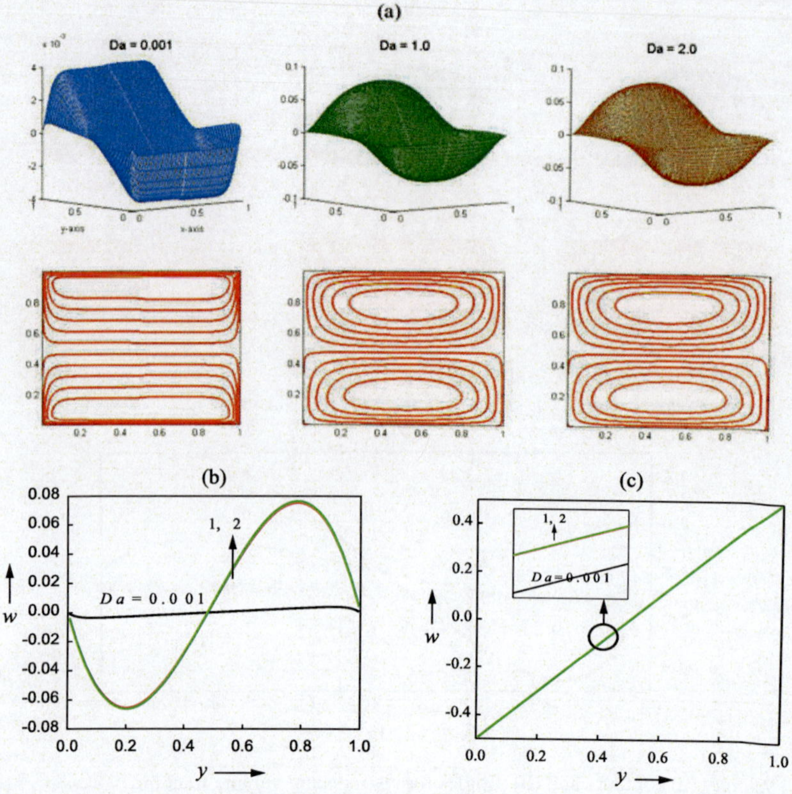

Fig. 5 Velocity (**a**) contours, (**b**) profiles, and (**c**) temperature profiles for altered Darcy number

different from the upper portion. The Tiwari-Das model and Forchheimer effects are incorporated. The numerical values are computed using difference equations along with Southwell-Over-Relaxation technique. The consequences are as follows: (a) The engine oil (layer-1) and mineral oil (layer-2) do not show any flow in layer-1, and two cell formations are attained in layer-2, whereas ethylene glycol (layer-1) – mineral oil (layer-2) and ethylene glycol (layer-1) – kerosene oil (layer-2) cause the fluid to flow in both the layers. (b) Using copper, diamond, and titanium oxide nanoparticles in water does not develop noticeable variations on the velocity and temperature. (c) Da expands but ϕ, λ shrinks the velocity and K do not alter the flow. (d) The solutions agree with Umavathi and Beg [21] in the absence of nanoparticles and porous bed.

Darcy number

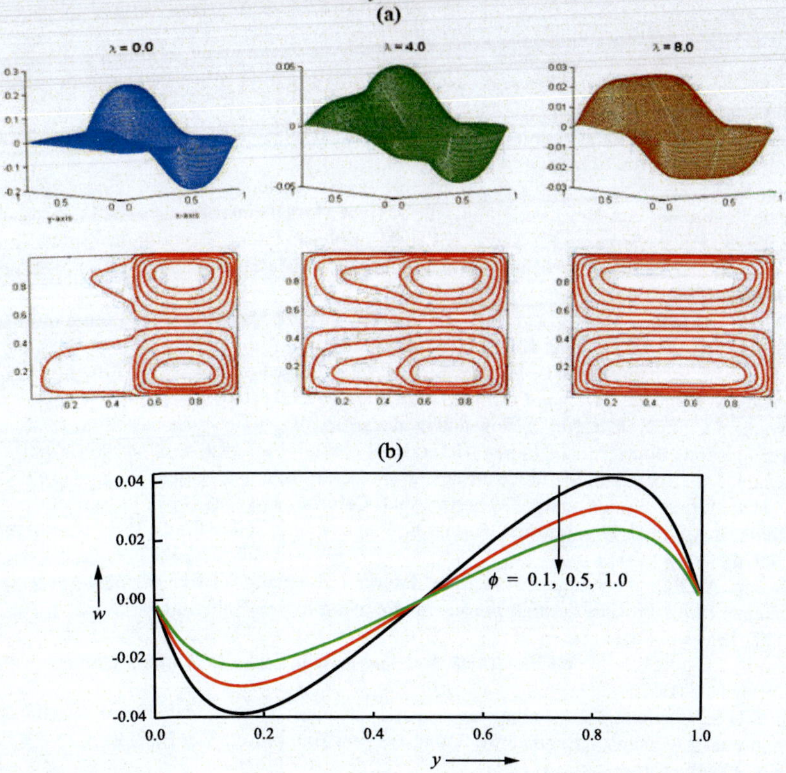

Fig. 6 Velocity (**a**) contours and (**b**) profiles for altered ratio of viscosities

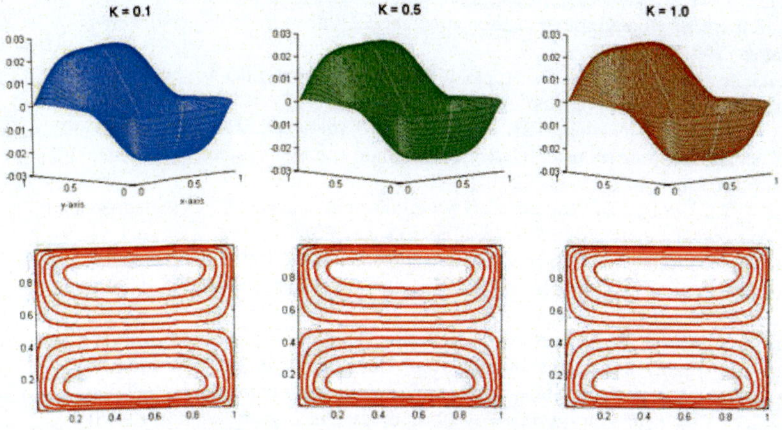

Fig. 7 Velocity contours for altered ratio of thermal conductivities

References

1. Li, H.Y., Yap, Y.F., Lou, J., Shang, Z.: Numerical investigation of heat transfer in three-fluid stratified flows. Int. J. Heat Mass Transf. **89**, 576–587 (2015)
2. Khaled, A.R.A.: Heat transfer enhancement in a vertical tube confining two immiscible falling co-flows. Int. J. Therm. Sci. **85**, 138–150 (2014)
3. Huang, Y., Li, H., Wong, T.N.: Two immiscible layers of electro-osmotic driven flow with a layer of conducting non-Newtonian fluid. Int. J. Heat Mass Transf. **74**, 368–375 (2014)
4. Redapangu, P.R., Vanka, S.P., Sahu, K.C.: Multiphase lattice Boltzmann simulations of buoyancy-induced flow of two immiscible fluids with different viscosities. Eur. J. Mech. B Fluids. **34**, 105–114 (2012)
5. Umavathi, J.C., Shaik Meera, D., Liu, I.C.: Unsteady flow and heat transfer of three immiscible fluids. Int. J. Appl. Mech. Eng. **13**, 1079–1100 (2008)
6. Umavathi, J.C., Chamkha, A.J., Sridhar, K.S.R.: Generalised plain Couette flow heat transfer in a composite channel. Transp. Porous Media. **85**, 157–169 (2010)
7. Umavathi, J.C., Sheremet, M.A.: Flow and heat transfer of couple stress nanofluid sandwiched between viscous fluids. Int. J. Numer. Methods Heat Fluid Flow. **29**, 4262–4276 (2019)
8. Umavathi, J.C., Sheremet, M.A.: Heat transfer of viscous fluid in a vertical channel sandwiched between nanofluid porous zones. J. Therm. Anal. Calorim. **144**, 1389–1399 (2021)
9. Packham, B.A., Shail, R.: Stratified laminar flow of two immiscible fluids. Proc. Camb. Philos. Soc. **69**, 443–448 (1971)
10. Ghassemi, A., Pak, A.: Numerical study of factors influencing relative permeability's of two immiscible fluids flowing through porous media using lattice Boltzmann method. J. Pet. Sci. Eng. **77**, 135–145 (2011)
11. Maxwell, J.C.: A Treatise on Electricity and Magnetism, 2nd edn. Oxford University Press, Cambridge (1904)
12. Choi, S.U.S., Eastman, J.A.: Enhancing thermal conductivity of fluids with nanoparticles. In: International Mechanical Engineering Congress and Exhibition, San Francisco, CA (United States). ASME, San Francisco (1995)
13. Umavathi, J.C.: Rayleigh-Benard convection subject to time dependent wall temperature in a porous medium layer saturated by a nanofluid. Meccanica. **50**, 981–994 (2015)
14. Sheremet, M.A., Revnic, C., Pop, I.: Free convection in a porous wavy cavity filled with a nanofluid using Buongiorno's mathematical model with thermal dispersion effect. Appl. Math. Comput. **299**, 1–15 (2017)
15. Sheikholeslami, M., Ganji, D.: Applications of Semi Analytical Methods for Nanofluid Flow and Heat Transfer. Elsevier Inc, Amsterdam (2018)
16. Zeinali Heris, S., Nassan, T.H., Noie, S.H., Sardarabadi, H., Sardarabadi, M.: Laminar convective heat transfer of Al_2O_3/water nanofluid through square cross-sectional duct. Int. J. Heat Fluid Flow. **44**, 375–382 (2013)
17. Botong, L., Yiming, D., Xuehui, C.: Two immiscible stratified fluids with one nanofluid layer in a horizontal annulus. Eur. Phys. J. Plus. **135**, 337 (2020)
18. Botong, L., Yiming, D., Liancun, Z., Xiaochuan, L., Xinxin Zhang, Z.: Mixed convection heat transfer of double immiscible fluids in functional gradient material preparation. Int. J. Heat Mass Transf. **121**, 812–818 (2018)
19. Farooq, U., Hayat, T., Alsaedi, A., Liao, S.: Heat and mass transfer of two-layer flows of third-grade nano-fluids in a vertical channel. Appl. Math. Comput. **242**, 528–540 (2014)
20. Khan, N.A., Sultan, F., Rubbab, Q.: Optimal solution of nonlinear heat and mass transfer in a two-layer flow with nano-Eyring–Powell fluid. Results Phys. **5**, 199–205 (2015)
21. Umavathi, J.C., Bég, O.A.: Effects of thermophysical properties on heat transfer at the interface of two immiscible fluids in a vertical duct: numerical study. Int. J. Heat Mass Transf. **154**, 1–18 (2020)
22. Oztop, H.F., Yasin, V., Ahmet, K.: Natural convection in a vertically divided square enclosure by a solid partition into air and water regions. Int. J. Heat Mass Transf. **52**, 5909–5921 (2009)

23. Moshkin, N.P.: Numerical model to study natural convection in a rectangular enclosure filled with two immiscible fluids. Int. J. Heat Fluid Flow. **23**, 373–379 (2002)
24. De Vahl Davis, G.: Laminar natural convection in an enclosed rectangular cavity. Int. J. Heat Mass Transf. **1**, 1167–1693 (1968)
25. De Vahl Davis, G.: Natural convection of air in a square cavity. Int. J. Numer. Methods Fluids. **3**, 249–264 (1983)

Effect of Variable Viscosity on Magnetohydrodynamics Mixed Convection Flow from a Vertical Flat Plate

M. Ajaykumar, C. K. Ajay, and A. H. Srinivasa

Nomenclature

x	Streamwise coordinate
G	Dimensionless temperature
f, F	Dimensionless stream functions
u, v,	Components of velocity in the x and y direction
T	Temperature of the fluid
Pr	Prandtl number
g	Gravitational acceleration
τ_x	Local shear stress
y	Transverse coordinate
Gr_x	Local Grashof number
q_x	Surface heat flux

Greek Symbols

γ, ω	Constants
ψ	Dimensional stream functions
μ	The dynamic viscosity
ν	Kinematic viscosity
η	Similarity variable
β	The thermal expansion factor
θ	Dimensionless temperature
α	Thermal diffusivity
M	Magnetic field parameter

M. Ajaykumar (✉) · C. K. Ajay · A. H. Srinivasa
Department of Mathematics, MIT Mysore, Belawadi, Karnataka, India

F. Kamalov et al. (eds.), *Advances in Mathematical Modeling and Scientific Computing*, Trends in Mathematics, https://doi.org/10.1007/978-3-031-41420-6_15

ξ	Dimensionless distance, along the surface
ε	Viscosity variation parameter
ρ	Fluid density

Superscript

$'$	Differentiation with respect to η

Subscripts

∞	Free stream
w, e	Conditions at the surface and at the edge of the boundary layer

1 Introduction

The forced and free convection flow composition is an example of mixed convection, which is one of the significant transport occurrences. At the same time, these flow patterns are detected by volumetric forces and an external forcing mechanism within. It is utilized for various environmental, geophysical, and energy-related designing purposes, containing twist-free solar boards, electronic devices known as multi-stage emergency-cooled nuclear fans, and fixed heat exchangers in a low-slung velocity climate.

The impacts of both natural and forced convection are found in equivalent order in a mixed convection flow. In several realistic fields, we find major temperature changes between the hot body surface and the natural flow. Such temperature variation creates density gradients within fluid media and becomes important in the existence of gravitational mixed convection impacts. The two-dimensional forced and natural convection laminar flow across a vertical flat plate, on which in-depth investigations [1–8] have been concentrated, is the least physical representation of such a stream. It has typically been accepted that $\xi = Gr_x/Re_x^2$, where Re_x is the Reynolds number and Gr_x is the Grashof number, is the leading parameter for the laminar boundary mixed convection flow, which signifies the percentage of buoyancy forces to the inertial forces within the boundary layer.

The effects of viscosity variations on the natural convection laminar boundary layer flow past a vertical isothermal flat plate are examined by Kafoussius N.G. et al. [9]. The authors Hossain M.A. et al. [10] calculated free convective flow from a vertical wavy surface with varying viscosity. Hady F.M. et al. [11] observed the varying viscosity mixed convective boundary layer flow on a continuous flat plate. On a continuously moving flat plate, both the authors Gorla and Rashidi

[12] investigated the outcomes of temperature-dependent viscosity on the flow and transmission of heat. The majority of the analytical studies that are now available for the mixed convection issue rely on the persistent physical characteristics of the ambient fluid. Thus far, Herwig H. [13] makes it clear that certain characteristics, like fluid viscosity, can fluctuate significantly. To correctly anticipate the transfer of heat rates, it is vital to account for the viscosity relationship to temperature. Recently, many authors studied the outcomes of changing viscosity over a mixed convection flow [14–16].

The use of magnetohydrodynamics has a significant impact on a variety of technical and industrial operations. Magnetohydrodynamics flows and heat transfer were examined in the field of astrophysics, geophysics, electrical transformers, coils for cooling, pumps, meters, bearings, and petroleum production. The magnetic field has a propensity to get faster the motion of the fluid, since the Lorentz force as a consequence of the scale of the boundary layer is condensed. Researchers [17–24] have looked into how magnetic fields affect boundary layer flow and heat transmission.

This study examines how varying viscosity affects heat transmission from a vertical flat plate and magneto hydrodynamic-mixed convection boundary layer flow.

2 Mathematical Formulation

We analyze a semi-infinite vertical plate with a continuous laminar mixed convection flow of a viscous, electrically conducting, incompressible fluid. The plate is situated in the x and y planes. In the positive direction, the induced magnetic field (B_0) is accepted to be uniform and normal to the surface. It is accepted that the magnetic number of Reynolds is minute, so it is feasible to disregard the induced magnetic field. T_w represents the temperature of the plate, which is consistent and higher than that of the free stream (T_∞). Additionally, it is identified that U_∞ is a constant free stream velocity that is parallel to a vertical plate. Furthermore, we accept that temperature-related property discrepancies are limited to viscosity and density, with density being treated separately so that the buoyancy element solely influences its outcomes in the momentum equation (Fig. 1).

Beneath the stated hypotheses, the two-dimensional boundary layer equations for a fluid-mixed convection flow over the semi-infinite vertical plate is shown.

$$\frac{\partial u}{\partial x} + \frac{\partial v}{\partial y} = 0 \tag{1}$$

$$u\frac{\partial u}{\partial x} + v\frac{\partial u}{\partial y} = g\beta\,(T - T_\infty) - \frac{\sigma B_0^2}{\rho}u + \frac{1}{\rho}\frac{\partial}{\partial y}\left(\mu\frac{\partial u}{\partial y}\right) \tag{2}$$

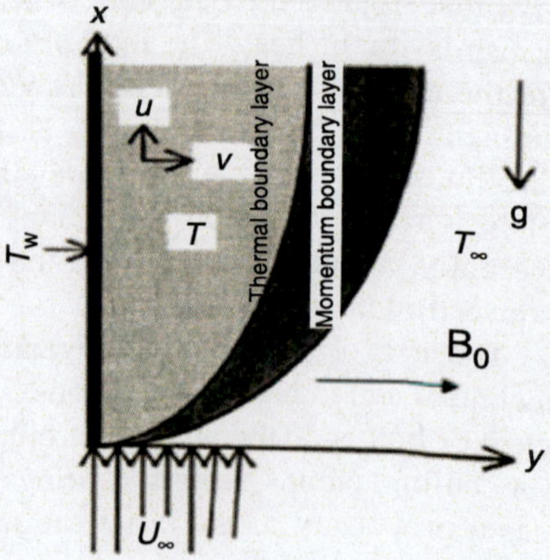

Fig. 1 Domain of flow and coordinate system

$$u\frac{\partial T}{\partial x} + v\frac{\partial T}{\partial y} = \alpha\frac{\partial^2 T}{\partial y^2} \tag{3}$$

The overhead equations and necessary boundary conditions are.

$$u = 0,\ \ v = 0,\ \ T = T_w\ \text{ at }\ y = 0$$

$$u = U_\infty,\ \ T = T_\infty\ \text{ as }\ y \to \infty \tag{4}$$

In this work, a semi-empirical formula for viscosity is provided by

$$\frac{\mu}{\mu_\infty} = \frac{1}{1 + \gamma\ (T - T_\infty)} \tag{5}$$

And as well-known authors Ling and Dybbs [25], have recognized, here μ_∞ is the viscosity.

Further,

$$\psi\,(x, y) = \upsilon_\infty \mathrm{Re}_x^{1/2}(1+\xi)^{1/4} f\,(\xi, \eta)\,,\ \ \eta = \frac{y}{x}\mathrm{Re}_x^{1/2}(1+\xi)^{1/4},\ \ \xi = \frac{Gr_x}{\mathrm{Re}_x^2}$$

$$\mathrm{Re}_x = \frac{U_\infty x}{\upsilon_\infty},\ \ Gr_x = \frac{g\beta\,(T_w - T_\infty)\,x^3}{\upsilon_\infty^2},\ \ \theta\,(\xi, \eta) = \frac{T - T_\infty}{T_w - T_\infty} \tag{6}$$

The equation of continuity (1) is precise by

$$u = \frac{\partial \psi}{\partial y}\ \text{ and }\ v = -\frac{\partial \psi}{\partial x} \tag{7}$$

By replacing out the aforementioned changes in Eqs. (1) and (4), we get

$$F''-\left(\frac{\varepsilon}{1+\varepsilon G}\right)G'F'+(1+\varepsilon G)\,F'f\left(\frac{2+3\xi}{4(1+\xi)}\right)+(1+\varepsilon G)\,G\left(\frac{\xi}{1+\xi}\right)-$$
$$(1+\varepsilon G)\,(1+\xi)^{-1/2}MF-(1+\varepsilon G)\,F^2\left(\frac{\xi}{2(1+\xi)}\right)-\xi\,(1+\varepsilon G)\left[FF_{\xi}-F'f_{\xi}\right]=0 \tag{8}$$

$$G''-\Pr\xi\left[FG_{\xi}-G'f_{\xi}\right]+\left(\frac{2+3\xi}{4\,(1+\xi)}\right)G'f\,\Pr=0 \tag{9}$$

The transformed boundary conditions are

$$G\,(\xi,0)=1,\;F\,(\xi,0)=f\,(\xi,\mathbf{0})=0$$

$$F\,(\xi,\infty)=(1+\xi)^{-1/2},\quad G\,(\xi,\infty)=0 \tag{10}$$

In accordance, the local shear stress and surface heat flux can be imparted as

$$\tau_x=\left(\xi^{-1/2}(1+\xi)^{3/4}f''\,(\xi,0)\right)/\,(1+\varepsilon)$$

$$q_x=\xi^{-1/2}(1+\xi)^{1/4}G'\,(\xi,0) \tag{11}$$

3 Results and Discussion

The related nonlinear partial differential Eqs. (8) and (9) as well as the boundary conditions (10) are solved by means of an implicit finite difference approach. Collected by means of a quasilinearization system, this is agreed upon. The technique's depiction is removed here for the sake of conciseness because it is discussed in Inoue and Tate [26], Ajaykumar M. et al. [27, 28], and A. H. Srinivasa et al. [29]. The outcomes of the present study were equated with those from Raju et al. [6], when $\varepsilon = 0.0$ and $M = 0.0$ (ε is the variable viscosity and M is the magnetic field), in order to assess dependability. For $Pr = 0.7$, various values of ξ, transfer of heat (q_x), and skin friction (τ_x) are presented in Table 1. Our findings are discovered to be in good correlation with Raju et al.'s [6] results to the exact fourth decimal point.

Figure 2 shows that given a fixed Prandtl number ($Pr = 0.72$) and magnetic field ($M = 0.5$), an upsurge in the variable viscosity (ε) is observed, the local shear stress [τ_x] factor decreases along the downward movement of the plate, and the transfer

Table 1 Numerical values of $F^{'}(\xi, 0)$ and $G^{'}(\xi, 0)$ taking $Pr = 0.7$ while $M = 0.0$ and $\varepsilon = 0.0$

ξ	$G^{'}(\xi, 0)$		$F^{'}(\xi, 0)$	
	Current	Raju et al. [6]	Current	Raju et al. [6]
0.0000	0.2926	0.2928	0.3321	0.3323
0.2901	0.3371	0.3373	0.5740	0.5742
0.4211	0.3503	0.3505	0.6707	0.6709
0.5299	0.3582	0.3584	0.7346	0.7348
0.6292	0.3637	0.3639	0.7895	0.7897
0.7239	0.3678	0.3680	0.8229	0.8230
1.2005	0.3821	0.3823	0.9354	0.9357
1.7790	0.3825	0.3828	0.9886	0.9888
2.5915	0.3791	0.3794	1.0097	1.0099
3.8729	0.3750	0.3751	1.0142	1.0140
6.1694	0.3703	0.3706	1.0078	1.0079
11.0660	0.3661	0.3662	0.9956	0.9959
24.9799	0.3617	0.3619	0.9824	0.9826
99.9950	0.3574	0.3576	0.9691	0.9693
∞	0.3530	0.3531	0.9565	0.9567

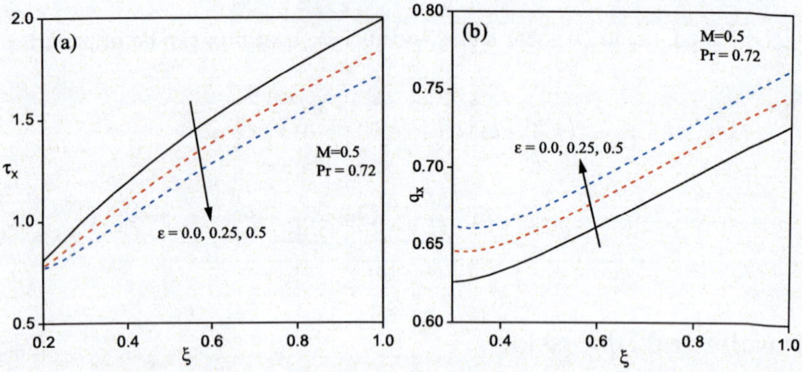

Fig. 2 (**a**) The local shear stress $[\tau_x]$ and (**b**) local surface heat flux $[q_x]$ factor for changed data of ε with $M = 0.5$

of heat $[q_x]$ rates increase. In this case, it is determined that a highly viscous fluid has a sluggish shear stress and a high rate of transfer of heat in the same proportion. The percentage decrease in $[\tau_x]$ is 19.88% and increase $[q_x]$ is 2.94% at $\xi = 0.8$, justifying the important role of Prandtl number and temperature-dependent viscosity across the boundary layer.

The outcome of the variable viscosity (ε) on the equivalent velocity and temperature outlines with a stable MHD field $(M = 0.5)$, and $Pr = 0.72$ as shown in Fig. 3a, b. It displays that collected velocity increases and temperature declines with an upsurge of ε. The variations in velocity profile in the η direction show the usual mixed convection boundary layer flow velocity profiles. That is, the velocity is null at the boundary wall; however, as η rises, the velocity grows to the maximum values.

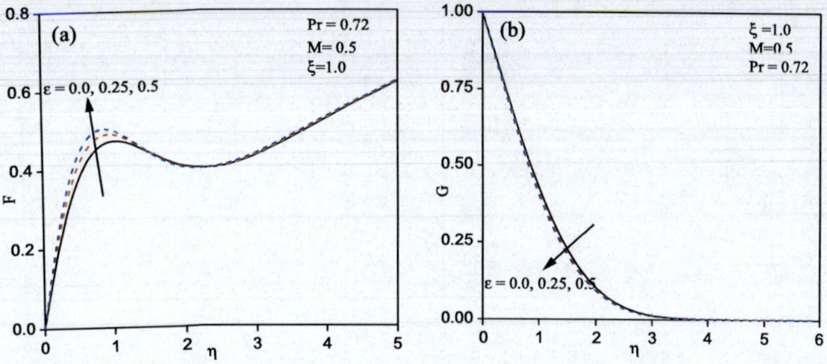

Fig. 3 The velocity (**a**) and temperature factors (**b**) for some data of ε with $M = 0.5$

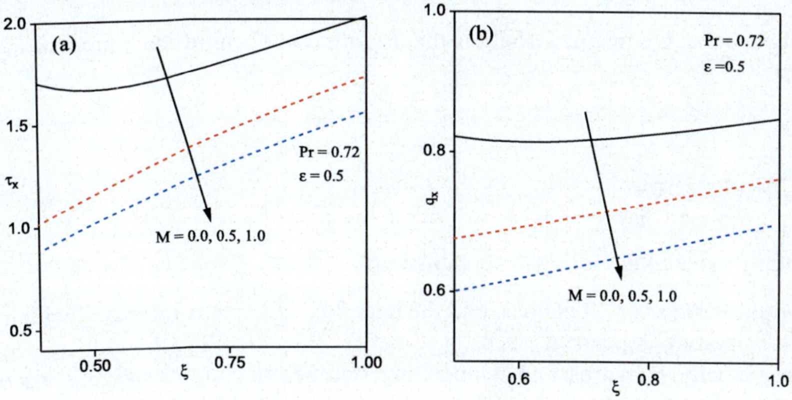

Fig. 4 (**a**) Represents the local shear stress $[\tau_x]$ and (**b**) local surface heat flux $[q_x]$ coefficient for different values of M with $\varepsilon = 0.5$

As a result, the numerical outcomes in these figures show that at increasing distances from the foremost edge, in terms of η, the thermal boundary layer thickness drops while the momentum boundary layer thickness rises.

The shear stress $[\tau_x]$ and transfer of heat $[q_x]$ factors are highlighted in Fig. 4a, b, which together explain the rising data of the magnetic parameter M for constant data of $Pr = 0.72$ and $\varepsilon = 0.5$. As a result, it appears that in the range of $0 \leq M \leq 1.0$, heat transfer $[q_x]$ increases by 21.83% while $[\tau_x]$ decreases by 63.92%.

Figure 5 highlights the velocity (F) and temperature (G) contours for several M data. It very well may be seen that the ascent in the magnetic parameter boundary prompts crumbles in the velocity. It does so because the Lorentz force, a type of resistive force, is produced when a transverse magnetic field is applied to a fluid that conducts electricity. This force has an inclination to slow the fluid's axial movement. Also, the temperature profile increases along η direction. Due to overheating, the magnetic field (M) raises the fluid temperature within the boundary layer. Thus,

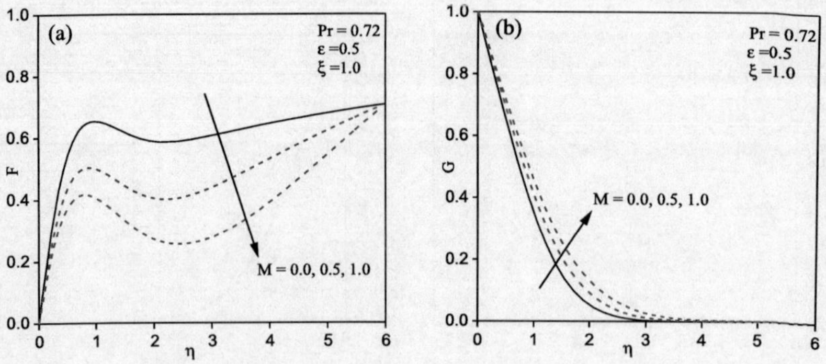

Fig. 5 (**a**) The velocity and (**b**) temperature factors for different values of M with $\varepsilon = 0.5$

we propose that the magnetic field is the driving force behind the boundary layer's decline.

4 Conclusions

This study came to the following conclusions.

1. The shear stress (τ_x) declines, and the heat flux (q_x) rate is increased with a rise in temperature-dependent viscosity values for the fixed Prandtl number.
2. The velocity increases and temperature outlines decline as varying viscosity upsurges.
3. The fixed Prandtl number and varying viscosity cause the transfer of heat and shear stress rate to decrease as values of the magnetic parameter upsurge.
4. The velocity outlines declines and the temperature outlines rises as the values of the magnetic field upsurge with fixed Prandtl number and temperature-dependent viscosity.

Conflict of Interest The writers affirm that there are no conflicting interests between the subjects of this study.

References

1. Sparrow, E.M., Eichorn, R., Gregg, J.L.: Combined forced and free convection in boundary layer flow. Phys. Fluids. **2**(3), 319–328 (1959)
2. Merkin, J.H.: The effects of buoyancy forces on the boundary layer flow over semi-infinite vertical flat plate in a uniform free stream. J. Fluid Mech. **35**(3), 4398–4450 (1969)

3. Lloyd, J.R., Sparrow, E.M.: Combined forced and free convection flow on vertical surfaces. Int. J. Heat Mass Transf. **13**(2), 434–438 (1970)
4. Wilks, G.: Combined forced and free convective flow on vertical surfaces. Int. J. Heat Mass Transf. **16**(10), 1958–1964 (1973)
5. Tingwei, G., Bachrum, R., Dagguent, M.: Influence de la Convective Natural le Sur la Convection Force Andes-sus D'Une Surface Plane Vertical Voumise a un Flux de Rayonnement. Int. J. Heat Mass Transf. **25**(7), 1061–1065 (1982)
6. Raju, M.S., Liu, X.R., Law, C.K.: A formulation of combined forced and free convection past horizontal and vertical surfaces. Int. J. Heat Mass Transf. **27**(12), 2215–2224 (1984)
7. Srinivasa, A.H., Eswara, A.T., Jayakumar, K.R.: Unsteady MHD mixed convection boundary layer flow and heat transfer in the stagnation region of a vertical plate due to impulsive motion. Indian J. Math. Math. Sci. **5**, 17–26 (2009)
8. Siddiqa, S., Hossain, M.A.: Mixed convection boundary layer flow over a vertical flat plate with radiative heat transfer. Appl. Math. **3**, 705–716 (2012)
9. Kafoussius, N.G., Williams, E.W.: The effect of temperature-dependent viscosity on the free convective laminar boundary layer flow past a vertical isothermal flat plate. ActaMechanica. **110**, 123–137 (1995)
10. Hossain, M.A., Kabir, S., Rees, D.A.S.: Natural convection flow from vertical wavy surface with variable viscosity. Z. Angew. Math. Phys. (1999)
11. Hady, F.M., Bakier, A.Y., Gorla, R.S.R.: Mixed convection boundary layer flow on a continuous flat plate with variable viscosity. Heat Mass Transf. **31**, 169–172 (1996) Springer-Verlag
12. Gorla, R.S.R., Rashidi, M.: Effects of variable viscosity on the flow and heat transfer on a continuous moving flat plate. Int. J. Eng. Sci. **30**, 1–6 (1992)
13. Herwig, H., Wicken, G.: The effect of variable properties on laminar boundary layer flow. Warme-und Stoffubertrag. **20**, 47–57 (1986)
14. Pal, D., Mondal, H.: Effects of temperature-dependent viscosity and variable thermal conductivity on MHD non-Darcy mixed convective diffusion of species over a stretching sheet. J. Egypt. Math. Soc. **22**(1), 123–133 (2014)
15. Jha, B.K., Aina, B.: Investigation of temperature dependent viscosity on steady fully developed mixed convection flow in a vertical microchannel. Int. J. Fluid Mech. Res. **44**(4), 357–373 (2017)
16. Pandey, A.K., Upreti, H.: Mixed convective flow of Ag–H2O magnetic nanofluid over a curved surface with volumetric heat generation and temperature dependent viscosity. Heat Transfer. **50**(7), 7251–7270 (2021)
17. Chamka A, J., Takhar H. S., Nath G: Mixed convection flow over a vertical plate with localized heating (cooling), magnetic field and suction (injection) , 835–841,(2004)
18. Nandkeolyar, R., Narayana, M.S., Motsa, S., Sibanda, P.: Magnetohydrodynamic mixed convective flow due to a vertical plate with induced magnetic field. J. Therm. Sci. Eng. Appl. **10**(6), 061005 (2018)
19. Makinde, O.D., Sibanda, P.: Magnetohydrodynamic mixed connective flow and heat and mass transfer past a vertical plate in a porous medium with constant wall suction. J. Heat Transfer. **130**(11), 112602 (2018)
20. Bhattacharyya, K., Mukhopadhyay, S., Layek, G.C.: MHD boundary layer slip fiow and heat transfer over a fiat plate Chin. Phys. Leo. **28**(2), 024701 (2011)
21. Nandkeolyar, R., Narayana, M., Motsa, S.S., Sibanda, P.: Magnetohydrodynamic mixed connective flow due to a vertical plate with induced magnetic field. J. Therm. Sci. Eng. Appl. **10**(6), 061005 (2018)
22. Ajay, C.K., Srinivasa, A.H.: Internal heat generation effects on MHD mixed convection flow from a vertical flat plate. AIP Conf. Proc. **2277**, 030023 (2020)
23. Sankar, M., Girish, N., Siri, Z.: Fully developed magnetoconvective heat transfer in vertical double-passage porous annuli. In: Narayanan, N., Mohanadhas, B., Mangottiri, V. (eds.) Flow and Transport in Subsurface Environment Springer Transactions in Civil and Environmental Engineering. Springer, Singapore (2018)

24. Girish, N., Sankar, M.: Fully developed mixed convection in vertical double passage porous annuli. J. Phys. Conf. Ser. **1597**(1), 012050). IOP Publishing (2020)
25. Ling, X., Dybbs, A.: Force convection over a flat plate submersed in a porous medium: variable viscosity case. In: Conference: ASME Winter Meeting, Boston, MA, USA., Technical Paper 87-WA/HT-23 (1987)
26. Inouye, K., Tate, A.: Finite difference version of quasilinearization applied to boundary layer equations. AIAAJ. **12**, 558–560 (1974)
27. Ajaykumar, M., Srinivasa, A.H.: Variable viscosity effects on unsteady MHD laminar boundary layer flow and heat transfer over a stretching sheet. In: AIP Conference Proceedings, vol. 2277. No. 1. AIP Publishing LLC (2020)
28. Ajaykumar, M., Srinivasa, A.H.: Effect of pressure work on free convection flow from an isothermal truncated cone with temperature dependent viscosity. J. Math. Comput. Sci. **10**(6), 2422–2433 (2020)
29. Srinivasa, A.H., Eswara, A.T., Jayakumar, K.R.: Unsteady MHD mixed convection boundary layer flow and heat transfer in the stagnation region of a vertical plate due to impulsive motion. Indian J. Math. Math. Sci. (IJMMS). **5**, 17–26 (2009)

LTNE Effects in the Darcy-Bénard Instability in a Rotating Anisotropic Porous Layer Saturated with a Fluid of Variable Viscosity

Om P. Suthar, B.S. Bhadauria, and Aiyub Khan

1 Introduction

The stability of rotating porous media convective flow is of great importance in a variety of fields related to food packaging industry, geothermal reservoirs, chemical engineering, rotating turbo machinery, etc. Also, the study of the Coriolis effect on the thermal convection in porous domains makes the investigation applicable to many fields of geophysical as well as astrophysical sciences. These important applicability aspects have attracted many researchers to study the convective instability in rotating porous media. Excellent review of most of the available studies on rotating porous media convection can be found in the monographs due to Bejan [1], Vadasz [2], and Nield and Bejan [3].

In most of the studies on natural convection, the fluids are assumed to have a constant viscosity, which, in general, is not true for most of the fluids. It is a common phenomenon that the viscosity of the fluid varies with the physical factors such as temperature. As we know that for liquids (e.g., oils, mercury, etc.) the increasing temperature decreases viscosity, whereas the opposite happens in the case of gases. Although this change in the viscosity of the fluid may be very small in some fluids, the importance of this effect cannot be ignored, in particular, when we deal with porous media flows where the viscosity has a destabilizing effect on the convective

O. P. Suthar (✉)
Department of Mathematics, Malaviya National Institute of Technology, Jaipur, India

B. S. Bhadauria
Department of Mathematics, School of Physical and Decision Sciences, Babasaheb Bhimrao Ambedkar University, Lucknow, India

A. Khan
Faculty of Science, Department of Mathematics and Statistics, Jai Narain Vyas University, Jodhpur, India

F. Kamalov et al. (eds.), *Advances in Mathematical Modeling and Scientific Computing*, Trends in Mathematics, https://doi.org/10.1007/978-3-031-41420-6_16

flow. Among the few available studies on temperature-dependent viscosity, few studies [4–6] are of particular interest in the context of the present work. Patil and Vaidyanathan [4] studied the effect of variable viscosity on the onset of thermal convection in a Darcy porous medium, whereas Richardson and Straughan [5] reported the effect of temperature-dependent viscosity on the nonlinear stability in a Brinkman porous medium. Ramanathan and Muchikel [6] investigated the threshold of ferroconvection in a sparse porous medium using the Brinkman model when the fluid's viscosity is temperature-dependent. Later, Vanishree and Siddheswar [7] studied the effects of rotation as well as temperature-dependent viscosity on natural convection in an anisotropic porous medium.

A vital aspect of porous media convection is the local thermodynamic equilibrium (LTE) between the fluid and the solid phases of the porous domain, which is a consequence of equal thermal conductivities of both the phases. Conversely, when the thermal conductivities of fluid and solid phases are not the same, then the porous domain is said to be in local thermal nonequilibrium (LTNE). For such porous domains, a two-field model for energy equation must be used, in which the single energy equation, describing the common temperature of the saturated porous domain, is replaced by two equations used for the fluid and the solid phases separately. Most of the studies on LTNE have been carried out only during the last few decades. In the most significant studies, one can reckon studies due to Banu and Rees [8] and Malashetty et al. [9]. Straughan [10] investigated the effect of LTNE on the nonlinear stability of Darcy-Bénard convection. Following these studies, several authors reported studies concerning the applications of LTNE effects (see [11–16], and [17] and references therein).

In this article, we investigate the onset of thermal convection in an infinitely extended anisotropic porous layer subjected to rotation. The fluid, saturating the porous layer, is considered to be of variable viscosity, i.e., the viscosity depends on the temperature. The porous layer is also assumed to lack local thermal equilibrium. Thus, the present work considers the effects of four different factors on Darcy-Bénard convection: anisotropy, rotation, temperature-dependent viscosity, and LTNE.

2 Mathematical Formulation

A temperature-dependent Boussinesq fluid is thought to rotate about the z-axis in an infinitely extended horizontal porous layer between two rigid parallel horizontal planes. Assuming the origin to be on the lower plane, we may assume the planes to be $z = 0$ and $z = d$. To maintain an adverse temperature gradient ΔT across the layer, we keep the lower and upper plates at $T_0 + \Delta T$ and T_0 temperatures, respectively. The porous medium is considered to be highly porous, and as a consequence, the Darcy-Brinkman momentum equation is used to describe the fluid motion. The Coriolis force is taken into account by adding it as an external force to the momentum equation. The centrifugal force is considered to be absorbed by the pressure term. Furthermore, the absence of local thermal equilibrium is taken into account, and the heat flow is described using a two-equation model as in [8]. Under the abovementioned assumptions, the equations that govern the flow are

$$\nabla \cdot \mathbf{V} = 0 \tag{1}$$

$$\frac{\rho_{f0}}{\delta}\frac{\partial \mathbf{V}}{\partial t} = -\nabla p + \rho_f \mathbf{g} - \mu_f \mathbf{K} \cdot \mathbf{V} + 2\frac{\rho_{f0}}{\delta}(\boldsymbol{\Omega} \times \mathbf{V}) + \nabla \cdot \left[\mu_p \left(\nabla \mathbf{V} + \nabla \mathbf{V}^{tr}\right)\right] \tag{2}$$

$$\delta \rho_{f0} c_f \frac{\partial T_f}{\partial t} + \rho_{f0} c_f \mathbf{V} \cdot \nabla T_f = \nabla \cdot \left(\delta \boldsymbol{\kappa}_\mathbf{f} \cdot \nabla T_f\right) + h(T_s - T_f) \tag{3}$$

$$(1-\delta)\rho_{s0} c_s \frac{\partial T_s}{\partial t} = \nabla \cdot ((1-\delta)\boldsymbol{\kappa}_\mathbf{s} \cdot \nabla T_s) - h(T_s - T_f) \tag{4}$$

$$\rho_f = \rho_{f0}\left[1 - \alpha(T_f - T_0)\right], \tag{5}$$

where $\mathbf{V}$, p, T_f, and T_s are the fluid velocity, fluid pressure (including the centrifugal force term), fluid temperature, and solid phase temperature, respectively, at any point (x, y, z) inside the porous layer, at time t. The vector $\boldsymbol{\Omega} = (0, 0, \Omega)$ is the rotation speed, $\mathbf{g} = g_0\mathbf{k}$ is the gravitational acceleration, and ρ_f is the fluid density. The symbols μ_f and μ_p denote the fluid viscosity and the effective viscosity, respectively, at some reference temperature. The permeability and thermal conductivity vectors are assumed as

$$\mathbf{K} = K_z\,(\xi,\ \xi,\ 1)\,; \quad \boldsymbol{\kappa}_\mathbf{f} = \kappa_{fz}\left(\eta_f,\ \eta_f,\ 1\right); \quad \boldsymbol{\kappa}_\mathbf{s} = \kappa_{sz}\,(\eta_s,\ \eta_s,\ 1)$$

where $\xi = K_x/K_z = K_y/K_z$ and K_z is the permeability in the z-direction. Also, $\eta_f = \kappa_{fx}/\kappa_{fz} = \kappa_{fy}/\kappa_{fz}$, $\eta_s = \kappa_{sx}/\kappa_{sz} = \kappa_{sy}/\kappa_{sz}$, and κ_{fz} and κ_{sz} are the thermal conductivities of fluid and solid in z-direction, respectively. The symbol α is the coefficient of volume expansion in Eq. (5). The fluid viscosity is assumed to be temperature sensitive so as to depend on temperatures as

$$\mu_f = \frac{\mu_1}{\mathcal{F}(T_f, T_s)}; \qquad \mu_p = \frac{\mu_2}{\mathcal{F}(T_f, T_s)} \tag{6}$$

where $\mathcal{F}(T_f, T_s) = 1 + \Gamma_0\left[\delta T_f + (1-\delta)T_s - T_0\right]$ and μ_1, μ_2 represent reference fluid and the effective viscosity, respectively. The parameter Γ_0 is the variable viscosity parameter. The conditions on the temperatures at the boundary are

$$T_f(0) = T_s(0) = T_0 + \Delta T, \qquad T_f(d) = T_s(d) = T_0.$$

A motionless basic state ($\mathbf{V}_H = 0$) implies the other basic state quantities:

$$\rho_f = \rho_{fb}(z),\ p = p_b(z),\ T_f = T_{fb}(z),\ T_s = T_{sb}(z),\ \mu_f = \mu_{fb}(z),\ \mu_p = \mu_{pb}(z),$$

to satisfy the following equations:

$$Dp_b = -\rho_b g, \quad D^2 T_{fb} = 0, \quad D^2 T_{sb} = 0. \tag{7}$$

where $D \equiv d/dz$. Solving (7) we get $T_{fb} = T_{sb} = T_0 + \Delta T(1 - z/d)$. Moreover,

$$\rho_{fb} = \rho_{f0}\,[1 - \alpha \Delta T(1 - z/d)]; \quad \mu_{fb} = \mu_1 F(z); \quad \mu_{pb} = \mu_2 F(z), \tag{8}$$

where $F(z) = 1/[1 + \Gamma_0 \Delta T(1 - z/d)]$.

3 Linear Stability Analysis

Following the standard analysis, we superpose small perturbations to the basic state and obtain the linearized governing equations as

$$\nabla \cdot \mathbf{V}' = 0 \tag{9}$$

$$\frac{\rho_{f0}}{\delta}\frac{\partial \mathbf{V}'}{\partial t} = -\nabla p' + \alpha\rho_{f0} g T'_f \mathbf{k} - \mu_{fb}\mathbf{K}\cdot\mathbf{V}' + 2\frac{\rho_{f0}}{\delta}\Omega\,(\mathbf{k}\times\mathbf{V}') + \mu_{pH}\nabla^2\mathbf{V}' + \nabla\mu_{pH}\left[\nabla\mathbf{V} + \nabla\mathbf{V}^{tr}\right] \tag{10}$$

$$\frac{\partial T'_f}{\partial t} + \delta^{-1}\mathbf{V}'\cdot\nabla T_{fb} = \frac{1}{\delta\rho_{f0}c_f}\nabla\cdot\left(\delta\boldsymbol{\kappa}_{\mathbf{f}}\cdot\nabla T'_f\right) + \frac{h}{\delta\rho_{f0}c_f}(T'_s - T'_f) \tag{11}$$

$$\frac{\partial T'_s}{\partial t} = \frac{1}{\rho_{s0}c_s}\nabla\cdot\left(\boldsymbol{\kappa}_{\mathbf{s}}\cdot\nabla T'_s\right) - \frac{h}{(1-\delta)\rho_{s0}c_s}(T'_s - T'_f) \tag{12}$$

The above equations are made dimensionless using the transformations $\mathbf{r}' = d\mathbf{r}'_*$, $\mathbf{V}' = \frac{\delta\kappa_{fz}}{\rho_{f0}c_f d}\mathbf{V}'_*$, $p' = \frac{\delta\kappa_{fz}\mu_1}{\rho_f c_f K_z}p_*$, $T'_f = T_0 + \Delta T T_{f*}$, $T'_s = T_0 + \Delta T T_{s*}$, and $t = \frac{\rho_{f0}c_f d^2}{\kappa_{fz}}t_*$. Thus, Eqs. (10), (11), and (12) transform to

$$\frac{1}{\mathrm{Va}K_z}\frac{\partial\mathbf{V}_*}{\partial t_*} = -\nabla p_* + \mathrm{Ra}K_z^{-1}T_{f*}\mathbf{k} + \sqrt{\mathrm{Ta}}\,K_z^{-1}(\mathbf{k}\times\mathbf{V}_*) - F(z_*)\mathbf{K}\cdot\mathbf{V}_* + \frac{\Lambda}{d^2}F(z_*)\nabla_*^2\mathbf{V}_* + \frac{\Lambda}{d^2}\nabla F(z_*)\cdot\left[\nabla_*\mathbf{V}_* + \nabla_*\mathbf{V}_*^{tr}\right] \tag{13}$$

$$\frac{\partial T_{f*}}{\partial t_*} - w_* = \eta_f\nabla_*^2 T_{f*} + \lambda(T_{s*} - T_{f*}) \tag{14}$$

$$\chi\frac{\partial T_{s*}}{\partial t_*} = \eta_s\nabla_*^2 T_{s*} - \lambda\gamma(T_{s*} - T_{f*}) \tag{15}$$

where $F(z_*) = [1+\Gamma(1-z_*)]^{-1}$. For the sake of convenience, we drop the asterisks in further analysis. We take curl of Eq. (13) twice to eliminate the pressure term and obtain the following equations for the z-component of the fluid velocity (w), and the vorticity vector (ζ)

$$\frac{1}{\mathrm{Va}}\frac{\partial}{\partial t}\left(\nabla^2 w\right) = \mathrm{Ra}\nabla_1{}^2 T_f + F(z)\left[\mathrm{Br}\nabla^4 w - \left(\nabla_1{}^2 w + \frac{1}{\xi}\frac{\partial^2 w}{\partial z^2}\right)\right]$$

$$+\Gamma\,[F(z)]^2\left[2\,\mathrm{Br}\nabla^2 - \frac{1}{\xi}\right]\frac{\partial w}{\partial z} - \sqrt{Ta}\frac{\partial \zeta}{\partial z} - 2\,\mathrm{Br}\,\Gamma^2\,[F(z)]^3\left(\nabla_1{}^2 w - \frac{\partial^2 w}{\partial z^2}\right) \quad (16)$$

$$\frac{1}{\mathrm{Va}}\frac{\partial \zeta}{\partial t} = \sqrt{\mathrm{Ta}}\frac{\partial w}{\partial z} + F(z)\left(\mathrm{Br}\nabla^2 - \frac{1}{\xi}\right) + \Gamma\,\mathrm{Br}\,[F(z)]^2\,\frac{\partial \zeta}{\partial z} \quad (17)$$

In the above equations (14), (15), (16), and (17), the parameters Va, Ra, Ta, Br, Γ, λ, and γ are the Vadasz number ($\mathrm{Va} = \delta\,\mathrm{Pr}/\mathrm{Da}$), Rayleigh number ($\mathrm{Ra} = \alpha\, g\, \Delta T\, K_z d/\delta\nu\kappa_{fz}$), Taylor number ($\mathrm{Ta} = (4\Omega^2 K_z^2)/(\delta^2\nu^2)$), Darcy-Brinkman number ($\mathrm{Br} = \Lambda\mathrm{Da}$), nondimensional variable viscosity parameter ($\Gamma = \Gamma_0\,\Delta T$), nondimensional inter-phase heat transfer coefficient ($\lambda = (h\, d^2)/(\delta\,\kappa_{fz})$), and porosity modified conductivity ratio ($\gamma = \delta\kappa_{fz}/((1-\delta)\kappa_{sz})$), respectively, with Pr being the Prandtl number, $\mathrm{Da} = K_z/d^2$ the Darcy number, and $\Lambda = \mu_2/\mu_1$ the Brinkman number and $\chi = (\rho_s c_s \kappa_{fz})/(\rho_f c_f \kappa_{sz})$. We need to solve the modified dimensionless governing equations (14), (15), (16), and (17), subject to the boundary conditions:

$$w = Dw = D\zeta = T = 0 \quad at \quad z = 0, 1. \quad (18)$$

4 Method of Solution

We express the dependent variables using normal mode technique as

$$\left[w, \zeta, T_f, T_s\right] = \left[W(t,z), Z(t,z), T_f(t,z), T_s(t,z)\right]\exp[i\,(a_x x + a_y y) + \sigma t]$$

where a_x, a_y are the wave numbers in x- and y-directions, respectively, and σ is the frequency of perturbation. Substituting above in equations (15), (16), (17) and (18), we obtain the following equations:

$$\mathrm{Ra}a^2 T_f + (D^2-a^2)\sigma\, W - F(z)\left[Br(D^2-a^2)^2 - (\xi^{-1}d^2 - a^2)\right] W + \sqrt{Ta}DZ$$

$$= \Gamma\,[F(z)]^2\left[2\,Br(D^2-a^2) - \tfrac{1}{\xi}\right] DW + 2\,Br\,\Gamma^2\,[F(z)]^3\,(D^2+a^2)W, \quad (19)$$

$$\sqrt{Ta}DW+\left[F(z)\left\{Br(D^2-a^2)-\frac{1}{\xi}\right\}-\sigma\right]Z+\Gamma Br\,[F(z)]^2\,DZ=0 \tag{20}$$

$$\eta_f(D^2-a^2)T_f-\sigma\,T_f\,Va+W+\lambda\,(T_s-T_f)=0 \tag{21}$$

$$\eta_s(D^2-a^2)T_s-\sigma\,T_f\,Va-\lambda\,\gamma\,(T_s-T_f)=0 \tag{22}$$

In deriving above equations, we have rescaled the time variable using Vadasz number. The Galerkin method is used to solve the above system by assuming

$$\left[W,\,Z,\,T_f,\,T_s\right]=\sum_{n=1}^{N}\left[A_n(t)\phi_n(z),\;B_n(t)\psi_n(z),\;C_n(t)\theta_{f,n}(z),\;D_n(t)\theta_{s,n}(z)\right] \tag{23}$$

where $\phi_n(z)$, $\psi_n(z)$, $\theta_{f,n}(z)$, and $\theta_{s,n}(z)$ are the basis functions, chosen such that they satisfy the hydrodynamic boundary conditions given by Eq. (18). However, most of the useful stability results can be drawn easily using first-order Galerkin method. Thus, considering $N = 1$ in Eq. (23) and substituting in Eqs. (19), (20), (21), and (22), we get the following equations in the unknown coefficients A_1, B_1, C_1, and D_1 by minimizing the residual over the interval [0, 1]:

$$\begin{aligned}&\left[Br(I_6-2\,a^2I_7+a^4I_8)-(\xi^{-1}I_7-a^2I_8)-\sigma(I_9-a^2I_{10})+2\,\Gamma\;Br I_{11}\right.\\&\left.-(2\,Br\,a^2+\xi^{-1})I_{12}+2\,\Gamma^2\;Br\;(I_{13}+a^2I_{14})\right]A_1=\sqrt{Ta}I_{15}B_1+\mathrm{Ra}a^2I_{16}C_1\end{aligned} \tag{24}$$

$$\sqrt{Ta}I_1A_1+\left[BrI_2-(Bra^2+\xi^{-1})I_3+\Gamma\;BrI_4-\sigma I_5\right]B_1=0 \tag{25}$$

$$I_{17}A_1+\left[\left(\eta_f a^2+\lambda+\sigma Va\right)I_{18}-\eta_f I_{19}\right]C_1+\lambda I_{20}D_1=0 \tag{26}$$

$$\lambda\,\gamma\,I_{21}C_1+\left[\eta_s I_{22}-\left(\eta_s a^2+\sigma Va+\lambda\,\gamma\right)I_{23}\right]D_1=0 \tag{27}$$

The values of I_k's have been given in the appendix. For nontrivial solution of the above system of equations, we obtain a relation in the physical parameters that yields Ra as a function of other parameters characterizing the flow. The stationary convection mode ($\sigma = 0$) requires

$$\mathrm{Ra}^{st}=\frac{[G_1G_2+\mathrm{Ta}\,I_1I_{15}]\left[\lambda^2\gamma I_{20}I_{21}-G_3G_4\right]}{a^2I_{16}I_{17}G_2G_4} \tag{28}$$

The values of G_i's are given in appendix A. The critical value of Ra_{st} is obtained by minimizing it with respect to the wave number a. Thus, $Ra_c = Ra_{st}\,(a_c)$. Various physical parameters' effect on Ra_c is reported in the next section.

5 Result and Discussion

In this section, we shall discuss the effect of these physical parameters on the critical Rayleigh number ($\mathrm{Ra_c}$) as well as on critical wave number a_c, which has been depicted in Figs. 1, 2, 3, and 4. In all the configurations considered, it was found that the effect of inter-phase heat transfer coefficient λ is to stabilize the system, i.e., as the value of λ increases, the critical Rayleigh number also increases, particularly for moderated values of λ. Also, the increase in smaller values of λ results in increase in the value of the critical wave number a_c. This increase continues till λ approaches to a certain value, say λ_0, in general, $15 \leq \lambda_0 \leq 20$. For further increase in λ, i.e., for $\lambda > \lambda_0$, the value of a_c starts decreasing, and this trend continues for all greater values of λ till, $\lambda \approx 10^4$, after which the value of a_c is almost the same. These effects of λ, on Ra_c and a_c, persist for all values of various physical parameters, as shown in the figures. Now, we shall discuss the effect of other physical parameters for different values of λ. The reference constant values set for the physical parameters to plot the graphs are $\Gamma = 0.2$, $Ta = 10$, $Br = 0.12$, $\gamma = 1.0$, $\xi = \eta_f = \eta_s = 1.2$. Figure 1a demonstrates the critical Rayleigh number, $\mathrm{Ra_c}$, as a function of the inter-phase heat transfer coefficient λ for different values of the variable

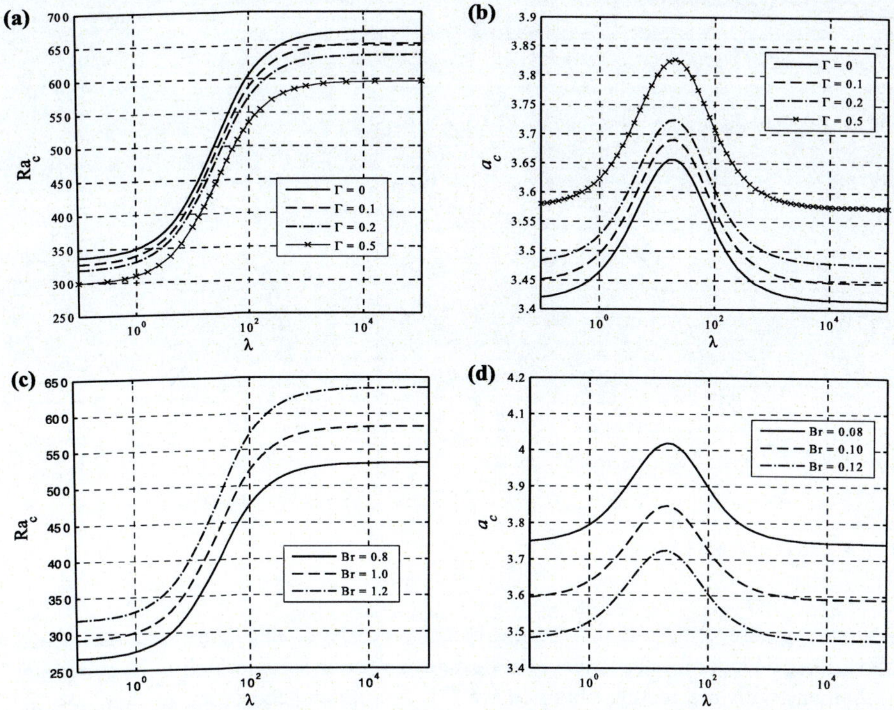

Fig. 1 Effect of Γ and Br on Ra_c and a_c

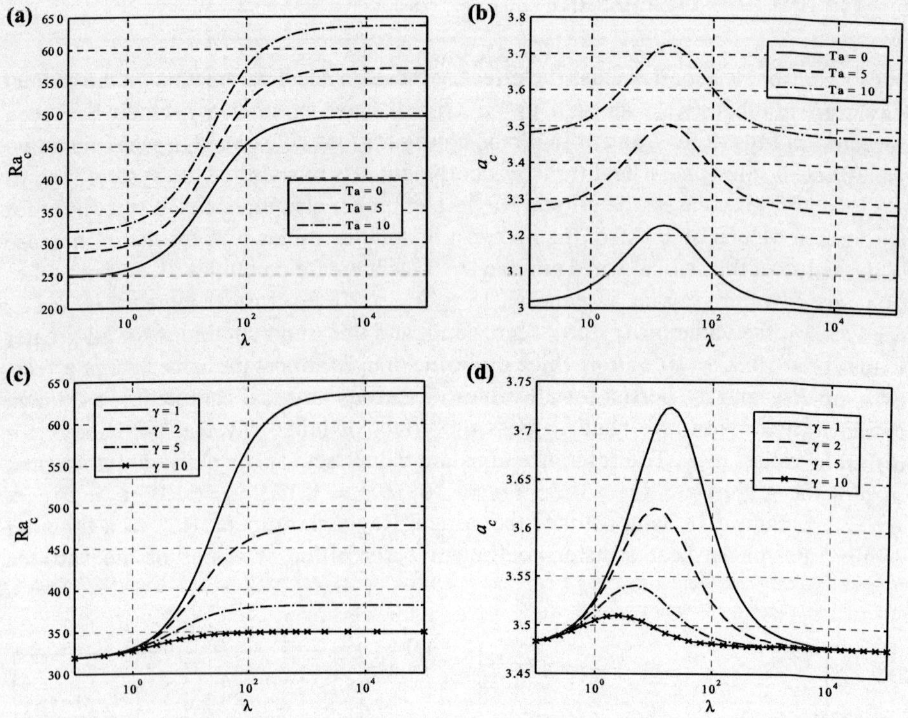

Fig. 2 Effect of Ta and γ on Ra_c and a_c

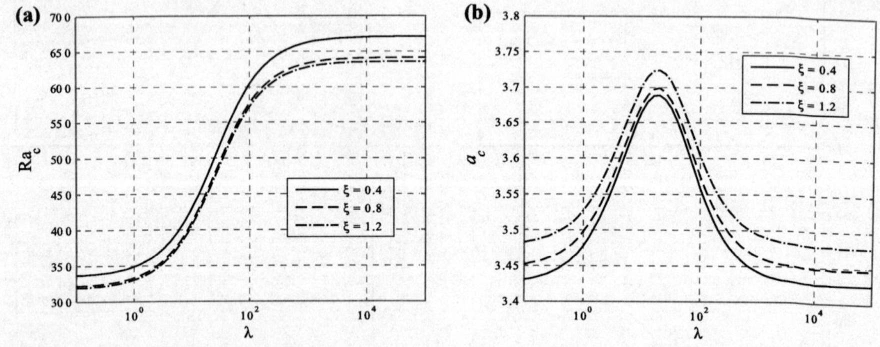

Fig. 3 Effect of ξ on Ra_c and a_c

viscosity parameter Γ. It can be easily observed from the figure that the effect of variable viscosity parameter is to destabilize the system, which is in complete agreement with the results obtained by [7]. A similar accordance to their result, related to the cell size, can also be found by observing Fig. 1b, that increasing values of Γ decrease the cell size, at the onset of convection, i.e., increase the

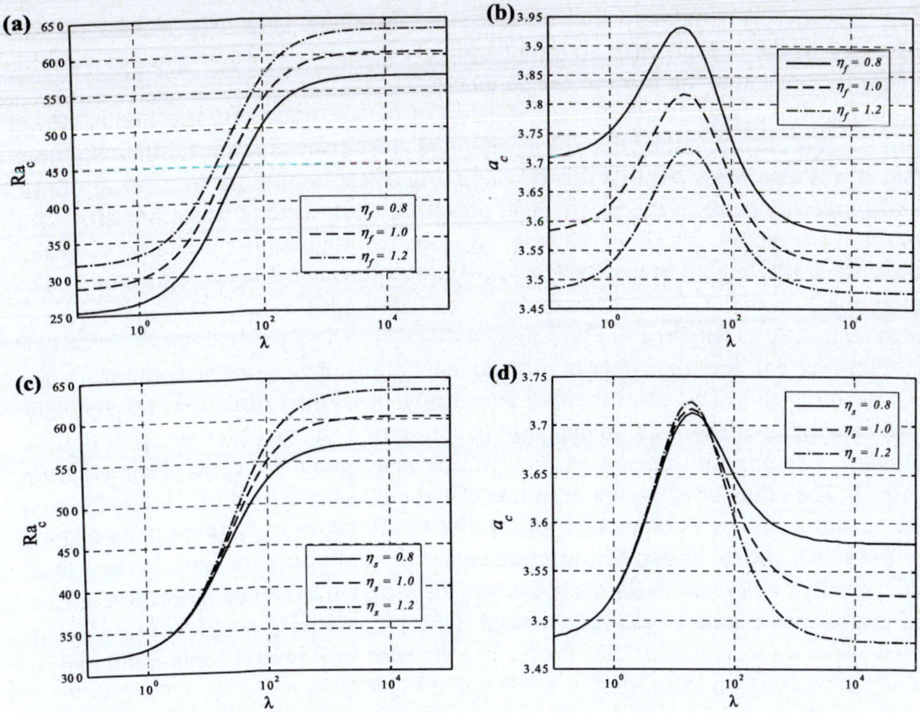

Fig. 4 Effect of η_f and η_s on Ra_c and a_c

critical wave number a_c. The Darcy-Brinkman number Br, consisting of the ratio of both the viscosities, stabilize the present system, as depicted in Fig. 1c. From the previous studies on porous media convection, we know that viscosity has a destabilizing effect, whereas, the Darcy number (Da) has a stabilizing effect on thermal convection. Both parameters are included in Br [$= (\mu_2/\mu_1)$Da] and also register a similar impact in the present case. The critical wave number (a_c) behaves in a very similar way, as Ra_c behaves, for different values of Br, as shown in Fig. 1d.

The effect of the Taylor number (Ta) on Ra_c and a_c is displayed in Fig. 2, respectively. Vadasz [18], in his comprehensive work, has shown that rotation has stabilizing impact on porous media convection. A very similar conclusion can be made by analyzing Fig. 2a, that as the value of the Taylor number increases, the system becomes more stable for all values of the inter-phase heat transfer coefficient λ. Also, the effect of the Taylor number on the critical wave number is same as on Ra_c, i.e., increase in Ta results in increase in a_c (see Fig. 2b). An exactly opposite impact on Ra_c and a_c is produced by the modified conductivity ratio γ, as shown in Fig. 2c and d. The parameter γ is basically the porosity modified ratio of fluid and solid conductivity in the z-direction. The conclusions, which can be drawn from

Fig. 2c, are very similar to that obtained by Govender [11]. The values of Ra_c obtained in the present case are comparatively less due to the impact of variable viscosity, present in the form of the parameter Γ.

In Fig. 3, we have demonstrated the effect of the anisotropy on $\mathrm{Ra_c}$ and a_c. From Fig. 3a, it is clear that the effect of mechanical anisotropy is to destabilize the flow. But it was also observed that this destabilizing effect is only up to a certain limit, and further increase in the mechanical anisotropy parameter ξ does not affect the onset of convection for all values of λ, e.g., one can find almost the same values of the critical Rayleigh number for $\xi = 1.5$ and $\xi = 2.0$ at different values of λ. The effect of ξ on the critical wave number is exactly opposite, i.e., as ξ increases, a_c also increases, as depicted in Fig. 3b.

Now, we consider the effect of thermal anisotropy, present in the form of η_f and η_s, denoting fluid and solid thermal anisotropy in vertical direction, respectively. The stabilizing effect of η_f on the onset of convection can be observed from Fig. 4a. However, its opposite impact on the critical wave number can be observed from Fig. 4b. The effect of η_s is similar to that of η_f, but rather complicated. Its effect on $\mathrm{Ra_c}$ is quite similar as obtained by [11]. For small values of λ, there is no change in the values of $\mathrm{Ra_c}$ and a_c for different values of η_s. Further increase in the values of λ results as increase in $\mathrm{Ra_c}$ as usual (see Fig. 4c), but a dual phenomenon in case of critical wave number can be observed. For comparatively small values of λ, an increase in η_s increases a_c, but the further increase in λ reverses this trend and a_c starts decreasing on increasing η_s as shown in Fig. 4d.

6 Conclusion

The conclusions drawn from the present study are the following:

1. The effect of λ is to stabilize the thermal convection and to increase the critical wave number for moderated values.
2. The Darcy-Brinkman number (Br) and the Taylor number (Ta) increase the stability of the system.
3. The conductivity ratio's effect on both $\mathrm{Ra_c}$ and a_c is exactly opposite to that of the Taylor number, i.e., it decreases both $\mathrm{Ra_c}$ and a_c.
4. The mechanical anisotropy destabilizes the system to a certain limit and increases the critical wave number a_c.
5. Fluid's thermal anisotropy parameter η_f decreases the wave number, whereas solid's thermal anisotropy parameter η_s shows a twofold impact, which switches at a certain value of λ.

Appendix A

The G_i's used in Eq. (28) are given by

$$G_1 = \Big[Br(I_6 - 2a^2 I_7 + a^4 I_8) - (2\,Br\,a^2 + \xi^{-1}) I_{12} - (\xi^{-1} I_7 - a^2 I_8) + 2\,\Gamma^2\,Br\,(I_{13} + a^2 I_{14} + \Gamma^{-1} I_{11}) \Big],$$

$$G_2 = \Big[Br I_2 - (Bra^2 + \xi^{-1}) I_3 + \Gamma\,Br I_4 \Big],$$

$$G_3 = \Big[\left(\eta_f a^2 + \lambda\right) I_{18} - \eta_f I_{19} \Big],$$

$$G_4 = \Big[\eta_s I_{22} - \left(\eta_s a^2 + \lambda\,\gamma\right) I_{23} \Big].$$

In Eqs. (24), (25), (26), and (27), the symbols I_k's denote various integrals as given below:

$$\begin{array}{llll}
I_1 = \langle D\phi, \psi \rangle, & I_2 = \langle F\,D^2\phi, \phi \rangle, & I_3 = \langle F\,\psi, \psi \rangle, & I_4 = \langle F^2\,D\psi, \psi \rangle, \\
I_5 = \langle \psi, \psi \rangle, & I_6 = \langle F\,D^4\phi, \phi \rangle, & I_7 = \langle F\,D^2\phi, \phi \rangle, & I_8 = \langle F\,\phi, \phi \rangle, \\
I_9 = \langle D^2\phi, \phi \rangle, & I_{10} = \langle \phi, \phi \rangle, & I_{11} = \langle F^2\,D^3\phi, \phi \rangle, & I_{12} = \langle F^2\,D\phi, \phi \rangle, \\
I_{13} = \langle F^3\,D^2\phi, \phi \rangle, & I_{14} = \langle F^3\,\phi, \phi \rangle, & I_{15} = \langle D\psi, \phi \rangle, & I_{16} = \langle \theta_f, \phi \rangle, \\
I_{17} = \langle \phi, \theta_f \rangle, & I_{18} = \langle \theta_f, \theta_f \rangle, & I_{19} = \langle D^2\theta_f, \theta_f \rangle, & I_{20} = \langle \theta_s, \theta_f \rangle, \\
I_{21} = \langle \theta_f, \theta_s \rangle, & I_{22} = \langle D^2\theta_s, \theta_s \rangle, & I_{23} = \langle \theta_s, \theta_s \rangle, &
\end{array}$$

where $\langle L, M \rangle = \int_0^1 L\,M\,dz$. While writing the above values of I_k's, we have omitted the subscript "1" for convenience.

References

1. Bejan, A.: Convective Heat Transfer, 2nd edn. Wiley, New York (1995)
2. Vadasz, P.: Flow in Rotating Porous Media, Chapter 4. In: Du Plessis, P. (ed.) Fluid Transport in Porous Media. Computational Mechanics Publications, Southampton (1997)
3. Nield, D.A., Bejan, A.: Convection in Porous Media, 3rd edn. Springer, New York (2006)
4. Patil, P.R., Vaidyanathan, G.: Effect of variable viscosity on the setting up of convection currents in a porous medium. Int. J. Eng. Sci. **19**, 421–426 (1981)
5. Richardson, L., Straughan, B.: Convection with temperature dependent viscosity in a porous medium: nonlinear stability and the Brinkman effect. Atti. Accad. Naz.Lincei-Ci-Sci-Fis. Mat. **4**, 223–232 (1993)
6. Ramanathan, A., Muchikel, N.: Effect of temperature-dependent viscosity on ferroconvection in a porous medium. Int. J. Appl. Mech. Eng. **11**, 93–104 (2006)
7. Vanishree, R., Siddheswar, P.G.: Effect of rotation on thermal convection in an anisotropic porous medium with temperature-dependent viscosity. Trans. Porous Med. **81**(1), 73–87 (2010)
8. Banu, N., Rees, D.A.S.: Onset of Darcy-Bénard convection using a thermal non-equilibrium model. Int. J. Heat Mass Transfer **45**, 2221–2228 (2002)

9. Malashetty, M.S., Shivakumara, I.S., Kulkarni, S.: The onset of Lapwood-Brinkman convection using a thermal non-equilibrium model. Int. J. Heat Mass Transfer **48**, 1155–1163 (2005)
10. Straughan, B.: Global nonlinear stability in porous convection with a thermal non-equilibrium model. Proc. R. Soc. A **462**, 409–418 (2006)
11. Govender, S., Vadasz, P.: The effect of mechanical and thermal anisotropy on the stability of gravity driven convection in rotating porous media in the presence of thermal non-equilibrium. Trans. Porous Media **69**, 55–56 (2007)
12. Malashetty, M.S., Swamy, M., Kulkarni, S.: Thermal convection in a rotating porous layer using a thermal non-equilibrium model. Phys. Fluids **19**, 054102 (2007)
13. Malashetty, M.S., Heera, R.: Linear and non-linear double diffusive convection in a rotating porous layer using a thermal non-equilibrium model. Int. J. Non-Linear Mech. **43**, 600–621 (2008)
14. Malashetty, M.S., Kulkarni, S.: The convective instability of Maxwell fluid-saturated porous layer using a thermal non-equilibrium model. J. Non-Newtonian Fluid Mech. **162**, 29–37 (2009)
15. Siddheshwar, P.G., Sakshath, T.N., Siddabasappa, C.: Effect of rotation on Brinkman-Bénard convection of a Newtonian nanoliquid using local thermal non-equilibrium model. Thermal Sci. Eng. Progress **25**, 100994 (2021)
16. Capone, F., Gentile, M., Gianfrani, J.A.: Optimal stability thresholds in rotating fully anisotropic porous medium with LTNE. Transp. Porous Media **139**(2), 185–201 (2021)
17. Capone, F., Gianfrani, J.A.: Natural convection in a fluid saturating an anisotropic porous medium in LTNE: effect of depth-dependent viscosity. Acta Mechanica **233**(11), 4535–4548 (2022)
18. Vadasz, P.: Coriolis effect on gravity-driven convection in a rotating porous layer heated from below. J. Fluid Mech. **376**, 351–375 (1998)

Soret and Dufour Effects on Radiative MHD Thermosolutal Viscoplastic Nanofluid Mixed Convective Flow Past a Bidirectional Stretching Sheet

K. Venkatadri, N. Vedavathi, G. Dharmaiah, C. H. Suresh Babu, R. Sivaraj, Ho-Hon Leung, Firuz Kamalov, and Mariam AlShamsi

1 Introduction

Due to its significance in the applications of engineering, such as crystallization in liquids, purifying molten metals, electromagnetic and mechatronic devices, nuclear reactor cooling, and solar concentrators, the magnetic field correlated with electrically conducting fluids has received much interest among several researchers. Fluid transport mechanisms continue to find wide applications in energy systems, chemical process engineering, transportation, industry, geosciences, and microelectronics. The primary indicators of electrolyzed liquid flows are electrolytes, plasmas, and liquid metals. Magnetohydrodynamics (MHD) finds a variety of applications in the fields of engineering and technology, including contemporary metallurgy, fusion reactors, MHD generators, and geophysics. Many investigators who work in the

K. Venkatadri (✉)
Department of Mathematics, Indian Institute of Information Technology, Sri City, Chittoor, India

N. Vedavathi
Department of Mathematics, Koneru Lakshmaiah Education Foundation, Vaddeswaram, Andhra Pradesh, India

G. Dharmaiah
Department of Mathematics, Narasaraopeta Engineering College, Narasaraopet, Andhra Pradesh, India

C. H. Suresh Babu
Department of Instrumentation Engineering, MIT Campus, Anna University, Chennai, India

R. Sivaraj · H.-H. Leung · M. AlShamsi
Department of Mathematical Sciences, United Arab Emirates University, Al Ain, United Arab Emirates

F. Kamalov
Faculty of Engineering, Canadian University Dubai, Dubai, United Arab Emirates

F. Kamalov et al. (eds.), *Advances in Mathematical Modeling and Scientific Computing*, Trends in Mathematics, https://doi.org/10.1007/978-3-031-41420-6_17

flow field were inspired by Pavlov's [1] pioneering research on characterizing the influence of magnetic strength on flow across a stretching plane. The influences of the Hall effect and ion slip on second-grade fluid flow over the penetrable moving vertical plane (semi-infinite) with the account of heat generation/absorption properties were analyzed by Krishna et al. [2]. In the past few decades, thermosolutal (double-diffusive) convection has gained significant implications in some specific applications, including thermal management technologies, geoscience, soil science, food processing, petroleum reservoir simulation, marine engineering, limnology, thermal storage, nanodrug delivery, and cancer thermal therapy [3]. Lewis et al. [4] addressed the features of thermal and solutal transport in detail.

Non-Newtonian fluids are proven to be commercially and industrially significant. Most engineering and industrial applications are associated with the heat transport features of non-Newtonian flows. In comparison to Newtonian fluids, the governing equations of fluids with non-Newtonian nature are very nonlinear and complex; hence, closed form solutions are not achievable. In nature non-Newtonian flows are found in extensive industrial applications, including polymers, poultry, cement, chemicals, geothermal pools, fermentation cycles, pore drying, oil regeneration, heat insulation, metallurgy, flow of molten steel, science of planetary and astronomy, and reactor fusions. Fluids like cosmetic products, custard, clay coating, grease, paints, slurries, and shampoo are some examples of non-Newtonian fluids [5]. The Casson fluid flows over different geometries with different effects and different boundary conditions have been studied by several researchers for various applications [6].

The advanced way of cooling using fluids includes the suspension of nanometer-sized (1–100 nm) particles into ordinary fluids. Due to the excessive heat transport features of the normal fluids, the nanoparticles are suspended in the base fluids to improve the cooling process. This mixture is called nanofluid. At Agronne national laboratory in the United States, Choi and Bestman [7] pioneered the concept of nanofluids. Nanotechnology is one of today's most fascinating fields because of its numerous applications in electronics, healthcare, food, solar cells, batteries, fuel cells, etc. Metals such as nickel, copper, and aluminum; other elements such as silicon carbide, graphene, titanium, carbon nanotubes, and calcium carbonate; and oxides such as silicone, alumina, silicon carbide, titanium, and silicone carbonate are a few examples of nanoparticles. In the aftermath of these models, Buongiorno [8] introduced a two-phase nanofluid model. Recently, many researchers have been involved in exploring the characteristics of nanofluids.

The Navier–Stokes expressions for energy and concentration enforced by the constraints at the boundary effects on their relevant fields have traditionally been used to model the slip-flows [9]. In the presence of velocity and thermal slips, as well as a nonlinear heat reservoir, Zeeshan et al. [10] demonstrated the stagnate point of magnetic flow and heat transport on a nonlinear oscillating surface. They observed that by enhancing the power index, the velocity is enhanced. Due to widespread applications in radiation therapy, ceramics technology, powder metallurgy technology, and other fields, several studies have been conducted on boundary layer flow with thermal properties by considering the impact of thermal radiative flux.

Krishna et al. [11] investigated the laminar convection flows of radiative micropolar fluid above an infinite vertical permeable plane with fixed suction velocity and Hall current in the presence of an inclined magnetic field. Dogonchi et al. [12] analyzed the performance of hybrid nanofluid's thermal properties in a porous enclosure embedded with three spheres.

This study examines the thermosolutal double-diffusion MHD mixed convective flow of Casson nanofluid across the bidirectional stretching surface with Soret, Dufour, velocity slip, and radiative effects. The bvp4c approach [13] is implemented to solve the governing PDEs. The bvp4c approach produces a convergent simulation outcome for well-posed boundary layer ODEs. The graphical trends of various regulating parameters are plotted and tabulated. As for the literature survey, there has not been any research done on the collective impacts mentioned above. As a result, the current simulation is a worthwhile endeavor, and the present results are useful to understand the features of MHD Casson nanofluid flow across the bidirectional stretching surface with velocity slip, Soret, Dufour, and thermal radiation effects.

2 Mathematical Model and Governing Equations

The flow of a steady, incompressible Casson nanofluid over a bidirectional domain is considered. According to the coordinate system, the flow occurs in the domain $z > 0$, and the sheet is considered without z-plane (i.e., xy-plane and $z = 0$). Figure 1 shows the domain of interest.

The stretching sheet with velocity $u = U_w(x) = ax$ in the x-axis direction and $v = V_w(y) = by$ in the y-axis direction is considered. The stretching plane is saturated by the Casson nanofluid with dissolved solutal. An external magnetic field B_0 is considered to be uniform. Let T_w be the sheet's temperature, C_w be the sheet's nanoparticles concentration, and γ_w be the solutal concentration of the sheet. Here T_w, C_w, and γ_w are constants, and T_w, C_w, and γ_w are all taken to be greater than T_∞, C_∞, and γ_∞. In the energy equation, Brownian motion, thermal radiation, thermophoresis, and Dufour effects are incorporated, whereas the Soret effect is incorporated in the solutal equation. The following governing equations in line with Buongiorno [8], Casson [14], Gupta et al. [15], and Nield and Kuznetsov [16] are presented as a result of the assumptions described above:

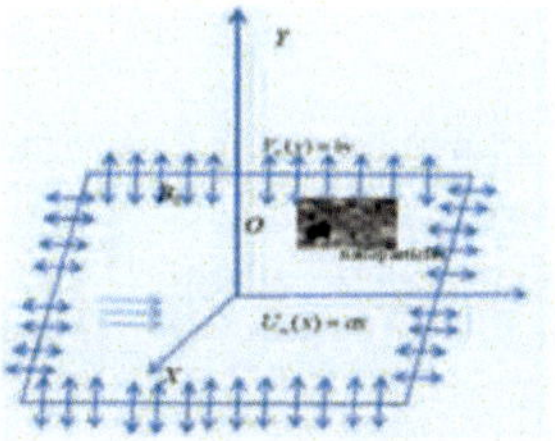

Fig. 1 Flow configuration of the problem

$$\frac{\partial u}{\partial x}+\frac{\partial v}{\partial y}+\frac{\partial w}{\partial z}=0 \tag{1}$$

$$u\frac{\partial u}{\partial x}+v\frac{\partial u}{\partial y}+w\frac{\partial u}{\partial z}=\upsilon\left(1+\frac{1}{\beta}\right)\frac{\partial^2 u}{\partial z^2}-\frac{\sigma B_0^2}{\rho_f}u$$
$$+g^*\left[(1-C_\infty)\rho_{f\infty}\left\{\beta_T(T-T_\infty)+\beta_\gamma(\gamma-\gamma_\infty)\right\}-(\rho_p-\rho_{f\infty})(C-C_\infty)\right] \tag{2}$$

$$u\frac{\partial v}{\partial x}+v\frac{\partial v}{\partial y}+w\frac{\partial v}{\partial z}=\upsilon\left(1+\frac{1}{\beta}\right)\frac{\partial^2 v}{\partial z^2}-\frac{\sigma B_0^2}{\rho_f}v \tag{3}$$

$$u\frac{\partial T}{\partial x}+v\frac{\partial T}{\partial y}+w\frac{\partial T}{\partial z}=\alpha\frac{\partial^2 T}{\partial y^2}+\tau_N\left[D_B\frac{\partial C}{\partial z}\frac{\partial T}{\partial z}+\left(\frac{D_T}{T_\infty}\right)\left(\frac{\partial T}{\partial z}\right)^2\right]$$
$$-\frac{1}{\rho_f c_p}\frac{\partial q_r}{\partial z}+D_{TC}\frac{\partial^2\gamma}{\partial z^2} \tag{4}$$

$$u\frac{\partial C}{\partial x}+v\frac{\partial C}{\partial y}+w\frac{\partial C}{\partial z}=D_B\frac{\partial^2 C}{\partial y^2}+\left(\frac{D_T}{T_\infty}\right)\frac{\partial^2 T}{\partial z^2} \tag{5}$$

$$u\frac{\partial \gamma}{\partial x}+v\frac{\partial \gamma}{\partial y}+w\frac{\partial \gamma}{\partial z}=D_S\frac{\partial^2 \gamma}{\partial z^2}+D_{CT}\frac{\partial^2 T}{\partial z^2} \tag{6}$$

The boundary constraints for the current flow investigation are

$$u=U_w(x)+u_{slip},\ v=V_w(y)+v_{slip},\ w=0,\ T=T_w,\ C=C_w,\ \gamma=\gamma_w,\ at\ z=0$$
$$u\to 0,\ \ v\to 0,\ \ T\to T_\infty,\ \ C\to C_\infty,\ \ \gamma\to\gamma_\infty\ as\ \ z\to\infty \tag{7}$$

Where $u_{slip}=\frac{2-\sigma_v}{\sigma_v}\lambda_0\frac{\partial u}{\partial z},\quad U_w(x)=ax,\quad v_{slip}=\frac{2-\sigma_v}{\sigma_v}\lambda_0\frac{\partial v}{\partial z},$

$$v_w(y)=by\ \ and\ \ \tau_N=\frac{(\rho c)_p}{(\rho c)_f}.$$

Using Rosseland approximation [17], the radiative heat flux is written as

$$q_r=-\frac{4\sigma^*}{3k^*}\frac{\partial T^4}{\partial y} \tag{8}$$

T^4 can be written as a linear function of temperature because the thermal gradient between the fluid controlled by the boundary layer and the free stream is diminutive. Expanding T^4 about T_∞ by means of the expansions in the Taylor series, without considering higher-order terms, T^4 can be represented as

$$T^4 \cong 4T_\infty^3 T - 3T_\infty^4 \tag{9}$$

From Eqs. (4), (8), and (9), one can get

$$\begin{aligned} & u\frac{\partial T}{\partial x} + v\frac{\partial T}{\partial y} + w\frac{\partial T}{\partial z} = \alpha\frac{\partial^2 T}{\partial z^2} + \tau_N\left[D_B\frac{\partial T}{\partial z}\frac{\partial C}{\partial z} + \frac{D_T}{T_\infty}\left(\frac{\partial T}{\partial z}\right)^2\right] \\ & + \frac{1}{(\rho c)_f}\frac{16\sigma^* T_\infty^3}{3k^*}\frac{\partial^2 T}{\partial z^2} + D_{TC}\frac{\partial^2 \gamma}{\partial z^2} \end{aligned} \tag{10}$$

The following similarity transformations are introduced:

$$\begin{aligned} & \eta = z\sqrt{\tfrac{q}{v}}, \quad u = axf'(\eta), \quad v = aug'(\eta), \\ & w = -\sqrt{av}\,(f(\eta) + g(\eta)), \quad T = T_\infty\,(1 + (\theta_w - 1)\,\theta(\eta)), \\ & \phi(\eta) = \tfrac{C - C_\infty}{C_w - C_\infty}, \quad \gamma(\eta) = \tfrac{h - h_\infty}{h_w - h_\infty}, \quad \theta_w = \tfrac{T_f}{T_\infty} > 1 \end{aligned} \tag{11}$$

In view of Eqs. (1) and (11), the mass conservation equation is satisfied, and Eqs. (2), (3), (10), (5), and (6) respectively are transformed into following ordinary differential equations:

$$\left(1 + \frac{1}{\beta}\right) f''' + (f + g)\,f'' - \left(f'\right)^2 - M\,f' + \lambda\,(\theta - Nr\phi + Nc\,h) = 0 \tag{12}$$

$$\left(1 + \frac{1}{\beta}\right) g''' + (f + g)\,g'' - \left(g'\right)^2 - M\,g' = 0 \tag{13}$$

$$\begin{aligned} & \left(\left(1 + Rd(1 + (\theta_w - 1)\,\theta)^3\right)\theta'\right)' \\ & + \Pr\left((f + g)\,\theta' + Nb\,\theta'\,\phi' + Nt\,\left(\theta'\right)^2 + Nd\,h''\right) = 0 \end{aligned} \tag{14}$$

$$\phi'' + Le\,(f + g)\,\phi' + \frac{Nt}{Nb}\theta'' = 0 \tag{15}$$

$$h'' + Ln\,(f + g)\,h' + Ld\,\theta'' = 0 \tag{16}$$

Respective boundary conditions are

$$\begin{aligned} & f = 0,\, f' = 1 + Af'',\, g = 0,\, g' = c + Ag'',\, \theta = 1,\, \phi = 1,\, h = 1\ at\ \eta = 0 \\ & f' \to 0, \quad g' \to 0, \quad \theta \to 0, \quad \phi \to 0, \quad h \to 0 \quad as \quad \eta \to \infty \end{aligned} \tag{17}$$

where

$\lambda = \frac{Gr_x}{{\mathrm{Re}_x}^2}$, $Gr_x = \frac{x^3(1-C_\infty)\rho_f g\beta_T(T_w-T_\infty)}{\upsilon^2}$, $Nc = \frac{\beta_\gamma(\gamma_w-\gamma_\infty)}{\beta_T(T_w-T_\infty)}$,
$Nr = \frac{(\rho_p-\rho_f)(C_w-C_\infty)}{(1-C_\infty)\rho_f\beta_T(T_w-T_\infty)}$, $\mathrm{Pr} = \frac{\upsilon}{\alpha}$, $Rd = \frac{16\sigma * {T_\infty}^3}{3kk^*}$,
$Nb = \frac{\tau_N D_B(C_w-C_\infty)}{\upsilon}$, $Nt = \frac{\tau_N D_T(T_w-T_\infty)}{\upsilon T_\infty}$, $Nd = \frac{D_{TC}(\gamma_w-\gamma_\infty)}{\upsilon(T_w-T_\infty)}$
$M = \frac{\sigma B_0^2}{\rho_f}$, $Le = \frac{\upsilon}{D_B}$, $Ln = \frac{\upsilon}{D_S}$, $Ld = \frac{D_{CT}(T_w-T_\infty)}{D_s(\gamma_w-\gamma_\infty)}$,
$c = \frac{b}{a}$, $\mathrm{Re}_x = \frac{{U_w}^2}{a\,\upsilon}$, $A = \frac{2-\sigma_\upsilon}{\sigma_\upsilon}\lambda_0\sqrt{\frac{a}{\upsilon}}$

The local Nusselt number is used to calculate the heat transfer rate along the surface and is defined as

$$Nu = \frac{x}{k\,(T_w - T_\infty)}\left(-k\frac{\partial T}{\partial z} + (q_r)_w\right) \tag{18}$$

In line with Eq. (11), Eq. (18) is written as

$$(\mathrm{Re}_x)^{-1/2} Nu = -\left(1 + Rd\,\theta_w^3\right)\theta'(0) \tag{19}$$

The friction factor coefficients along x and y directions are written as

$$C_{fx}(\mathrm{Re}_x)^{1/2} = \quad 1 + \frac{1}{\beta}\Big)\; f''(0) \tag{20}$$

$$C_{fy}(\mathrm{Re}_x)^{1/2} = \quad 1 + \frac{1}{\beta}\Big)\; g''(0) \tag{21}$$

where A—velocity slip parameter, a, b—constants, c—stretching sheet ratio parameter, B_0—uniform magnetic field in z-direction, C—nanoparticle concentration, C_{fx} and C_{fy}—skin friction coefficients in x and y directions, D_B—Brownian diffusion coefficient, D_T—thermophoretic diffusion coefficient, f, g—dimensionless stream functions, k—thermal conductivity K—thermal diffusivity of the fluid, Gr_x—Grashof number, g^*—gravitational acceleration, M—magnetic parameter, Nr—nanofluid buoyancy ratio parameter, Nt—thermophoresis parameter, Nb—Brownian motion parameter, q_r—radiative heat flux, Pr—Prandtl number, Nu—local Nusselt number, D_{TC}—Dufour diffusivity, D_{CT}—Soret diffusivity, Nc—buoyancy ratio parameter, Rd—thermal radiation parameter, D_s—solutal diffusivity, k^*—mean absorption coefficient, T—fluid temperature, Nd—modified Dufour number, Le—nanofluid Lewis number, Ln—Lewis number, Ld—modified Soret number, Re_x—local Reynolds number, u, v, w—dimensionless velocity components in the directions of x, y and z respectively, α—thermal diffusivity, η—non-dimensional radial coordinate, β—Casson fluid parameter, β_T—thermal expansion coefficient, β_γ—solutal expansion coefficient, μ—dynamic viscosity, ϕ—dimensionless concentration, σ—electric conductivity, τ_N—ratio of effective heat capacity of nanoparticle to fluid, υ—kinematic viscosity, θ—dimensionless temperature, h—solute concentration distributions, ρ_f—fluid density, ρ_p—particle density, $\sigma*$—Stefan-Boltzmann constant, θ_w—temperature ratio parameter, σ_v—tangential momentum accommodation coefficient, λ_0—molecular mean free path, γ_w—sheet mass concentration, and λ—Richardson number.

3 Numerical Procedure and Validation

The fluid flow nonlinear ODEs (12)–(16) with relevant boundary conditions (17) are numerically computed via the bvp4c technique. The adequate information about the numerical technique can be found in Vedavathi et al. [18].

In order to verify the current numerical code, the results of reduced Nusselt number (rate of local heat transport) *Nu* are mapped with the outcomes of Wang [19], Gorla and Sidawi [20], and Khan and Pop [21] for distinct values of *Pr* and are presented in Table 1. An outstanding agreement is obtained, and therefore the significance of bvp4c code is high.

4 Results and Discussions

Rigorous mathematical computations are studied to investigate the MHD-mixed convection double-diffusion Casson nanofluid flow past a bidirectional stretching plane with velocity slip along with Dufour, radiation, and Soret effects. The system of coupled nonlinear boundary layer Eqs. (2), (3), (10), (5), and (6) are converted into a well-posed system of coupled nonlinear ODEs (12, 13, 14, 15, and 16). Equations (12), (13), (14), (15), and (16) are simplified by using the bvp4c solver. The default values of emerging parameters in the governing equations are frozen with $\beta = 0.2$, $\lambda = 0.01$, $Ld = 0.5$, $Nc = 0.5$, $Nr = 0.2$, $\theta_w = 1.2$, $Rd = 0.6$, $\mathrm{Pr} = 5$, $Nd = 0.5$, $Nt = 0.1$, $Nb = 0.2$, $Ln = 2$, $Le = 4$, $C = 0.3$, $A = 0.1$, and $M = 0.2$, until otherwise mentioned particularly.

Increasing the M values diminishes the skin friction and Nusselt number along *the* x and y axes, respectively, as shown in Table 2. Further, an augmentation in β values is observed to decrease the Nusselt number but increases skin friction. Magnifications in Rd values are seen to increase the skin friction and Nusselt

Table 1 Comparison values of *Nu* for various values of *Pr*

Pr	Wang [19]	Gorla and Sidawi [20]	Khan and Pop [21]	Present
0.2	0.1691	0.1691	0.1691	0.1689
2	0.9114	0.9114	0.9113	0.9112
7	1.8954	1.8905	1.8954	1.8950

Table 2 The values of $Re_x^{-1/2}Nu$, $Re_x^{1/2}C_{fx}$, $Re_x^{1/2}C_{fy}$, for several values of M, β, and Rd

M	β	Rd	$Re_x^{-1/2}Nu$	$Re_x^{1/2}C_{fx}$	$Re_x^{1/2}C_{fy}$
1			0.2222	−3.3044	−0.8865
2			0.2032	−3.9750	−1.1114
3			0.1871	−4.5328	−1.2925
	0.3		0.2297	−2.2093	−0.5428
	0.5		0.2143	−1.8111	−0.4464
	1.0		0.1937	−1.4504	−0.3589
		0.5	0.1064	−2.6297	−0.6447
		1.0	0.6312	−2.6294	−0.6447
		1.5	0.9834	−2.6291	−0.6448

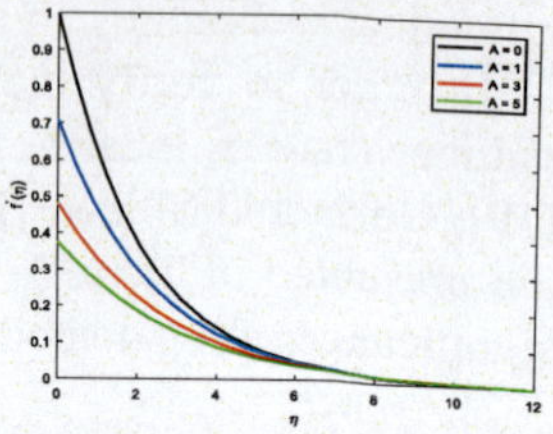

Fig. 2 Variability for f' with A

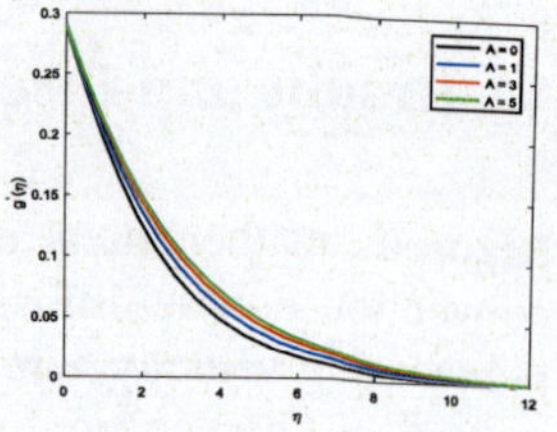

Fig. 3 Variability for g' with A

number in the direction of x-axis but decrease the skin friction in the direction of y-axis.

Figures 2 and 3 depict the velocity-slip parameter's (A) impact on velocity along x-axis and y-axis ($f'(\eta)$ and $g'(\eta)$). It can be noticed that with an augmentation in velocity-slip parameter, the fluid velocity is decreased along the x-axis. However, as A increases, the velocity along y-axis increases. Figure 4 represents the effect of the Casson fluid parameter β on velocity. It is observed that the velocity decreases with escalating β values, implying that the viscoplastic fluid acts like a non-Newtonian fluid as β increases. The viscoplastic fluid parameter appears in the shear term in the equation of momentum boundary layer. It is seen that increasing β reduces the yield stress, and consequently the thickness of the momentum boundary layer is reduced. Tensile tension causes the thickness of the boundary layer to contract. The effect of the magnetic parameter (M) on velocity along the x-axis is illustrated in Fig. 5. The influence of M on the conducting fluid causes the resistive force due to the Lorentz force. The magnetic Lorentzian drag force in the flow is expressed by the parameter M, which is defined by the inverse ratio of viscous hydrodynamic force to the magnetic Lorentzian drag force in the flow. As a result, the velocity decreases

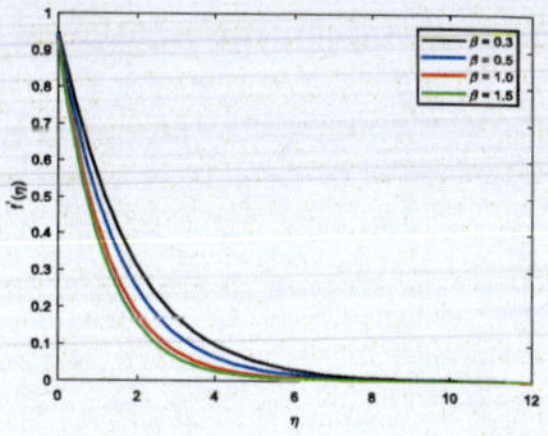

Fig. 4 Variability for f' with β

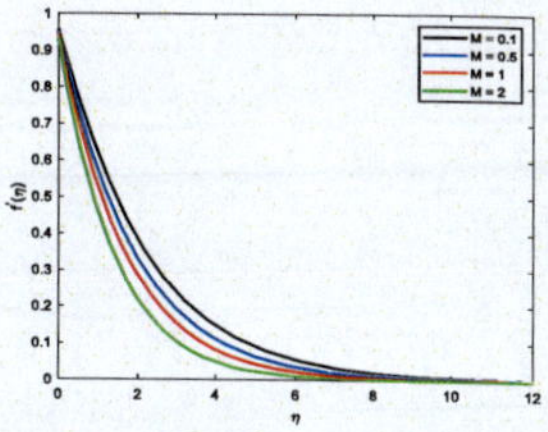

Fig. 5 Variability for f' with M

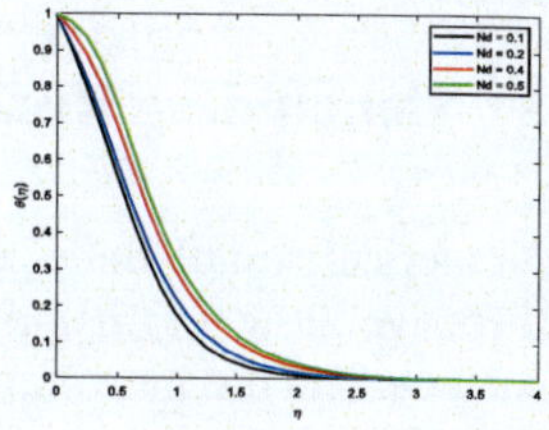

Fig. 6 Variability for θ with Nd

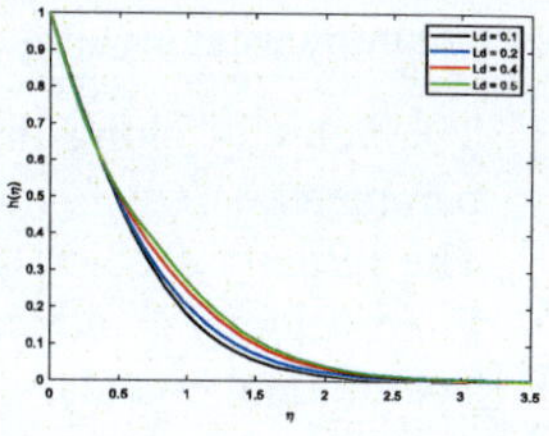

Fig. 7 Variability for h with Ld

when M is strengthened. This indicates that the magnetic force controls fluid transport. Further, augmenting the M value induces a Lorentz force that opposes the flow of fluid. The radial magnetic field operates to induce a drag force acting perpendicularly, which is consistent with several earlier findings of magnetized nanofluid convection. Figure 6 presents the effects of modified Dufour parameter (*Nd*) on temperature distribution. Increasing values of *Nd* increases the temperature profiles. Figure 7 presents the effects of modified Soret number (*Ld*) on Solute concentration profiles. A slight increase in the solute concentration distributions ($h(\eta)$) is observed with an increase in *Ld* values. Figures 8 and 9 depict the profiles for heat transport rate (*Nu*) for distinct values of thermal radiation *Rd and* magnetic parameter *M*, respectively. These graphs reveal that the heat transportation rate increases for amplifying the radiation parameter but decreases for augmenting the magnetic parameter.

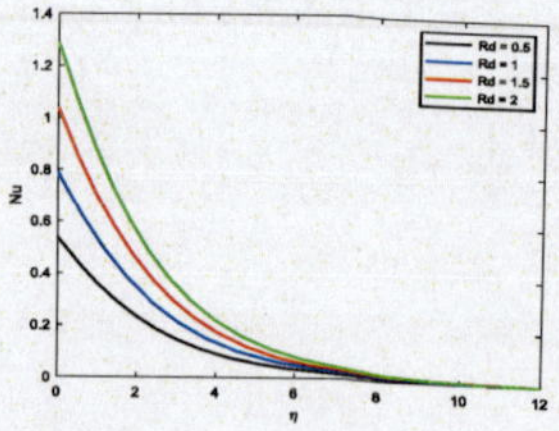

Fig. 8 Variability for *Nu* with *Rd*

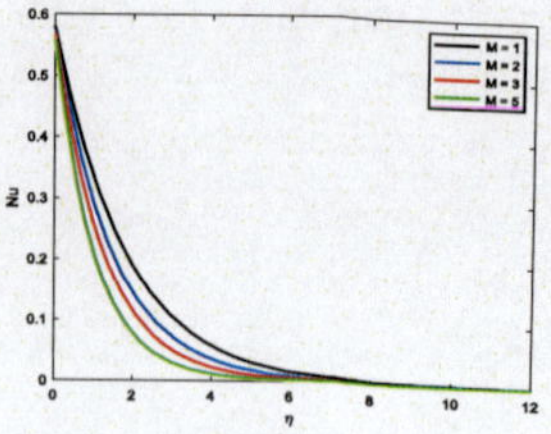

Fig. 9 Variability for *Nu* with *M*

5 Concluding Results

Numerical simulations are executed to investigate the boundary layer double-diffusive flow of magnetized Casson nanofluid over a bidirectional stretching surface. The thermosolutal transport mechanism of Casson nanofluid is investigated in terms of the existence of velocity slip, Soret, Dufour, and radiation effects. A bvp4c numerical technique is adopted to numerically solve the ODEs.

- The velocity along the x-axis is diminished by raising the values of the slip parameter.
- The velocity decelerates with escalating Casson fluid parameter values.
- Fluid velocity is reduced due to the strengthening of the magnetic field.
- The thermal boundary layer is thicker for augmenting Dufour parameter.
- An increase in the modified Soret number causes a magnification in the solute concentration distributions.
- The heat transportation rate augments for magnifying the radiation parameter.

References

1. Pavlov, K.B.: Magnetohydrodynamic flow of an incompressible viscous fluid caused by deformation of a plane surface. Magnitnaya Gidrodinamika. **4**, 146–147 (1974)
2. Krishna, M.V., Ahamad, N.A., Chamkha, A.J.: Hall and ion slip impacts on unsteady MHD convective rotating flow of heat generating/absorbing second grade fluid. Alex. Eng. J. **60**(1), 845–858 (2021)
3. Patra, J.K., Das, G., Fraceto, L.F., et al.: Nano based drug delivery systems: recent developments and future prospects. J. Nanobiotechnol. **16**, 71 (2018)
4. Lewis, R.W., Nithiarasu, P., Seetharamu, K.N.: Fundamentals of the Finite Element Method for Heat and Fluid Flow. Wiley (2004)

5. Prasad, V.R., Gaffar, S.A., Kumar, B.R.: Non-similar computational solutions for double-diffusive MHD transport phenomena for non-Newtonian nanofluid from a horizontal circular cylinder. Non-Linear Eng. **8**(1), 470–485 (2018)
6. Ramesh, K., Riaz, A., Dar, Z.A.: Simultaneous effects of MHD and Joule heating on the fundamental flows of a Casson liquid with slip boundaries. Propul. Power Res. **10**(2), 118–129 (2021)
7. Choi, S.U., Eastman, J.A.: Enhancing Thermal Conductivity of Fluids with Nanoparticles. Argonne National Laboratory, United States (1995)
8. Buongiorno, J.: Convective transport in nanofluids. J. Heat Transf. **128**, 240–250 (2006)
9. Yu, S., Ameel, T.A.: Slip-flow heat transfer in rectangular microchannels. Int. J. Heat Mass Transf. **44**(22), 4225–4234 (2001)
10. Khan, Z., Rasheed, H.U., Islam, S., Noor, S., Khan, W., Abbas, T., Khan, I., Kadry, S., Nam, Y., Nisar, K.S.: Impact of magnetohydrodynamics on stagnation point slip flow due to nonlinearly propagating sheet with nonuniform thermal reservoir. Math. Probl. Eng. **2020**, 1–10 (2020)
11. Krishna, M.V., Anand, P.V.S., Chamkha, A.J.: Heat and mass transfer on free convective flow of a micropolar fluid through a porous surface with inclined magnetic field and hall effects. Spec. Top. Rev. Porous Media. **10**(3), 203–223 (2019)
12. Dogonchi, A.S., Tayebi, T., Karimi, N., Chamkha, A.J., Alhumade, H.: Thermal-natural convection and entropy production behavior of hybrid nanofluid flow under the effects of magnetic field through a porous wavy cavity embodies three circular cylinders. J. Taiwan Inst. Chem. Eng. **124**, 162–173 (2021)
13. Ibrahim, W., Zemedu, C.: MHD nonlinear mixed convection flow of micropolar nanofluid over non-isothermal sphere. Math. Probl. Eng. **2020**, 3596368 (2020)
14. Casson, N.: A flow equation for pigment oil-suspensions of the printing ink type. In: Mill, C.C. (ed.) Rheology of Disperse Systems, pp. 84–102. Pergamon Press, London (1959)
15. Gupta, U., Sharma, J., Sharma, V.: Instability of binary nanofluid with magnetic field. Appl. Math. Mech. **36**(6), 693–706 (2015)
16. Nield, D.A., Kuznetsov, A.V.: The onset of convection in a horizontal nanofluid layer of finite depth: a revised model. Int. J. Heat Mass Transf. **77**, 915–918 (2014)
17. Rosseland, S.: Astrophysik und Atom-Theoretische Grundlagen. Springer, Berlin (1931)
18. Vedavathi, N., Dharmaiah, G., Venkatadri, K., Gaffar, S.A.: Numerical study of radiative non-Darcy nanofluid flow over a stretching sheet with convective Nield conditions and energy activation. Nonlinear Eng. **10**(1), 159–176 (2021)
19. Wang, C.Y.: Free convection on a vertical stretching surface. J. Appl. Math. Mech. **69**, 418–420 (1989)
20. Gorla, R.S.R., Sidawi, I.: Free convection on a vertical stretching surface with suction and blowing. Appl. Sci. Res. **52**, 247–257 (1994)
21. Khan, W.A., Pop, I.: Boundary-layer flow of a nanofluid past a stretching sheet. Int. J. Heat Mass Transf. **53**, 2477–2483 (2010)

Influence of Slip Velocity on an Infinite Cylinder and Rough-Flat Plate Lubricated with Couple Stress Fluid

Hanumagowda B.N, Sreekala C.K, Vishu Kumar M, and Neha Yadav

1 Introduction

A lot of importance has been attained by the lubrication technology due to the advancement of technology and massive development in the machine industry. Every industry's motto is to decrease friction and increase the competence of its machines This has piqued the interest of many scientists, resulting in a huge demand for lubrication. The continuum theory was the basic lubrication theory used for Newtonian fluids, and it was successful in explaining fluids without additives but was unsuccessful in modeling the behavior of non-Newtonian fluids. Hence, a more general theory was introduced for couple stress fluids —Stokes theory [1] has been used by various researchers to study hydrodynamic lubrication [2, 3]. The wear and tear of materials due to surface roughness gained the attention of researchers. Hence, Christensen [4] stochastic theory gained much attention. Many researchers have exclusively studied the roughness effects on different bearings [5–7]. The effect of slip velocity on a porous fluid interface has been studied by Patel and Dehri [8] and Kalavathi et al. [9]. The slip velocity effects were studied by Oladeinde and Akpobi [10] on a slider bearing, Santhosh et al. [11] studied on inclined porous slider bearing, and Sangeetha [12] on porous triangular plates.

A further study was inspired by this way, as the authors had no idea how the collective effect of slip velocity and roughness affect the cylinder-plate system in the existence of couple stress fluid.

Hanumagowda B. N. · Vishu Kumar M. · N. Yadav
Department of Mathematics, School of Applied Sciences, REVA University, Bengaluru, Karnataka, India

Sreekala C. K. (✉)
Department of Mathematics, Nitte Meenakshi Institute of Technology, Yelahanka, Karnataka, India

F. Kamalov et al. (eds.), *Advances in Mathematical Modeling and Scientific Computing*, Trends in Mathematics, https://doi.org/10.1007/978-3-031-41420-6_18

2 Mathematical Formulation

The physical configuration (Fig. 1) represents a cylinder-rough flat-plate system. The radius of the cylinder is R, and it is the plate that is having slip velocity. The thickness of the film is $h = h_m + \frac{x^2}{2R}$, and the fluid used for lubrication is a couple stress fluid.

By Stokes' theory, the basic equations of motion subject to normal assumptions are

$$\mu \frac{\partial^2 u}{\partial y^2} - \eta \frac{\partial^4 u}{\partial y^4} = \frac{\partial p}{\partial x} \tag{1}$$

$$\frac{\partial p}{\partial y} = 0 \tag{2}$$

$$\frac{\partial u}{\partial x} + \frac{\partial v}{\partial y} = 0 \tag{3}$$

The boundary limits for slip velocity are

$$u = \frac{1}{s} \frac{\partial u}{\partial y}\bigg|_{y=0}, \quad \frac{\partial^2 u}{\partial y^2} = 0, \; v = 0 \; at \; y = 0, u = 0, \; \frac{\partial^2 u}{\partial y^2} = 0, \quad v = \frac{\partial h}{\partial t} \; at \; y = h \tag{4}$$

Fig. 1 Physical configuration of cylinder-plate system

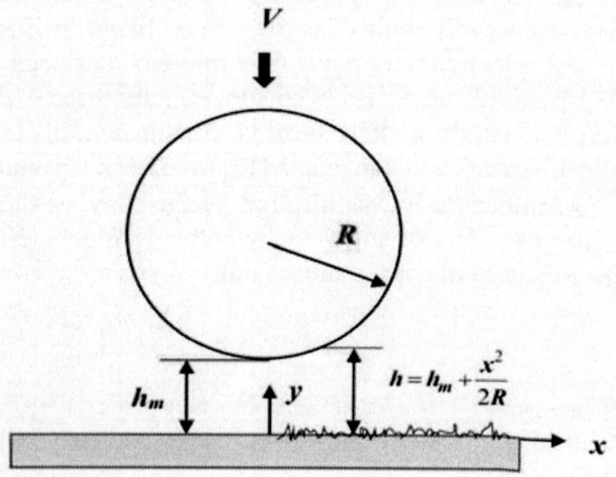

Solving (1) and using slip velocity conditions (4), we get

$$u = \left\{ y^2 - h^2 (\xi_1 + \xi_2 y) \right\} \frac{1}{2\mu} \frac{\partial p}{\partial x} + (h - y)\, \xi_1 l \frac{1}{\mu} \frac{\partial p}{\partial x} \tanh \frac{h}{2l} + \frac{l^2}{\mu} \frac{\partial p}{\partial x} \left\{ 1 - \frac{Cosh\left(\frac{2y-h}{2l}\right)}{\text{Cosh} \frac{h}{2l}} \right\} \tag{5}$$

where, $\xi_1 = \frac{\sigma_1}{(h+\sigma_1)}$, $\xi_2 = \frac{1}{(h+\sigma_1)}$, $\sigma_1 = \frac{1}{s}$

Plugging (5) in (3) and integrating we arrived at Reynolds equation

$$\frac{\partial}{\partial x} \left[j\,(h, l, \sigma_1) \frac{\partial p}{\partial x} \right] = -\mu V \tag{6}$$

Where $j\,(h, l, \sigma_1) = (1 - 3\xi_1) \frac{h^3}{6} + \left(4l^2 - \xi_2 h^3\right) \frac{h}{4} + \left(\xi_1 h^2 l - 4l^3\right) \frac{1}{2} \tan h \left(\frac{h}{2l}\right)$.

By Christensen's approach, to express the surface roughness, the mathematical expressions used are

$$H = h + h_s\,(x, y, \xi) \tag{7}$$

Taking expectation over Eq. (6), the averaged stochastic Reynolds equation reduces to

$$\frac{\partial}{\partial x} \left\{ E\,\{ j\,(H, l, \sigma_1)\} \frac{\partial E(p)}{\partial x} \right\} = -\mu V \tag{8}$$

where $E\,(*) = \int_{-\infty}^{\infty} (*)\, q\,(h_s)\, dh_s$

The stochastic film thickness h_s is expected to have a distribution function of probability $q(h_s)$ given by

$$q\,(h_s) = \begin{cases} \frac{35}{32c^7}\left(c^2 - h_s^2\right)^3 & |h_s| < c, \\ 0 & \text{elsewhere,} \end{cases} \tag{9}$$

where the variance $\overline{\sigma}^2 = c^2/9$.

The two types of roughness forms studied in the current work are an interpretation of the stochastic theory, where striations of roughness are expected to take the form of long narrow ridges and valleys.

Longitudinal roughness: In this form, they move in the direction of the x-axis, and the thickness of film in this case is

$$H = h + h_s\,(y, \xi) \tag{10}$$

And the modified Reynold's equation for expression (8) is

$$\frac{\partial}{\partial x}\left\{E\left\{j\left(H, l, \sigma_1\right)\right\} \frac{\partial E(p)}{\partial x}\right\}=-\mu V \tag{11}$$

Transverse roughness: In this form, they move in the direction of the y-axis. And the film thickness considered is $H = h + h_s(x, \xi)$.

And the modified Reynolds equation for expression (8) is

$$\frac{\partial}{\partial x}\left\{\frac{1}{E\left(\frac{1}{j(H,l,\sigma_1)}\right)} \frac{\partial E(p)}{\partial x}\right\}=-\mu V \tag{12}$$

Combining (11) and (12), we get

$$\frac{\partial}{\partial x}\left\{J\left(H, l, \sigma_1, c\right) \frac{\partial E(p)}{\partial x}\right\}=-\mu V \tag{13}$$

For longitudinal and transverse roughness they are respectively

$$J\left(H, l, \sigma_1, c\right)=\begin{cases}E\left(j\left(H, l, \sigma_1\right)\right) & for\ longitudinal\ roughness \\ \left\{E\left(1 /\left(j\left(H, l, \sigma_1\right)\right)\right)\right\}^{-1} & for\ transverse\ roughness\end{cases} \tag{14}$$

For longitudinal roughness,

$$E\left(j\left(H, l, \sigma_1\right)\right)=\frac{35}{32 c^7} \int_{-c}^c j\left(H, l, \sigma_1\right)\left(c^2-h_s^2\right)^3 d h_s \tag{15}$$

For transverse roughness,

$$E\left(\frac{1}{j\left(H, l, \sigma_1\right)}\right)=\frac{35}{32 c^7} \int_{-c}^c \frac{\left(c^2-h_s^2\right)^3}{j\left(H, l, \sigma_1\right)} d h_s \tag{16}$$

The non-dimensional terms introduced are

$$x^*=\frac{x}{R}, l^*=\frac{l}{h_{m0}}, H^*=\frac{h}{h_{m0}}, C=\frac{c}{h_{m0}}, h_m^*=\frac{h_m}{h_{m0}}, \sigma=\frac{\sigma_1}{h_{m0}},$$

$$P^*=\frac{E(p) h_{m0}^2}{\mu R(-\partial h / \partial t)}, \beta=\frac{h_{m0}}{R}$$

Non-dimensionalizing Equation (13) becomes

$$\frac{\partial}{\partial x^*}\left\{J^*\left(H^*, l^*, \sigma, C\right) \frac{\partial P^*}{\partial x^*}\right\}=\frac{1}{\beta} \tag{17}$$

The boundary condition for pressure is

$$P^* = 0 \; at \; x^* = \pm 1 \; and \; \frac{dP^*}{dx^*} = 0 \; at \; x^* = 0 \tag{18}$$

Integrating (17) and using the boundary values (18) we get

$$P^* = \frac{E(p)h_{m0}^2}{\mu R\left(-\partial h/\partial t\right)} - \frac{1}{\beta}\int_{x^*}^{1} \frac{x^*}{J^*(H^*, l^*, \sigma, C)} dx^* \tag{19}$$

Load-supporting capability

$$W^* = \frac{E(W)h_{m0}^2}{\mu R\left(-\partial h/\partial t\right)} = -\frac{1}{\beta}\int_{x^*=-1}^{x^*=1} \left\{\int_{x^*}^{1} \frac{x^*}{J^*(H^*, l^*, \sigma, C)} dx^*\right\} dx^* \tag{20}$$

Squeezing time

$$\begin{aligned} T^* &= \frac{th_{m0}^2}{\mu R\left(-\partial h/\partial t\right)} \\ &= -\frac{1}{\beta}\int_{h_m^*}^{1} \left[\int_{x^*=-1}^{x^*=1} \left\{\int_{x^*}^{1} \frac{x^*}{J^*(H^*, l^*, \sigma, C)} dx^*\right\} dx^*\right] dh_m^* \end{aligned} \tag{21}$$

3 Results and Discussion

The impact of roughness of the surface C and slip velocity σ are investigated for the geometry of a cylinder-plate system in the existence of couple stress fluids (CSF) l^*. The non-dimensional bearing characteristics like pressure P^*, load W^*, and squeezing time T^* have been derived. The results are discussed graphically. The graphs for each characteristic are plotted for longitudinal (dotted lines) and transverse (solid lines) roughness.

3.1 Film Pressure

In Fig. 2, plots of pressure $P*$ against $x*$ with $\beta = 0.04, l^* = 0.2, h_m^* = 0.6, \sigma = 0.2$ by incrementing the values of parameter C, which accounts for roughness, are made. We can see an enhancement in film pressure for increasing values of C for the transverse roughness, whereas a reverse trend is observed in longitudinal roughness. Also when C approaches 0, both the roughness coincides, which corresponds to a

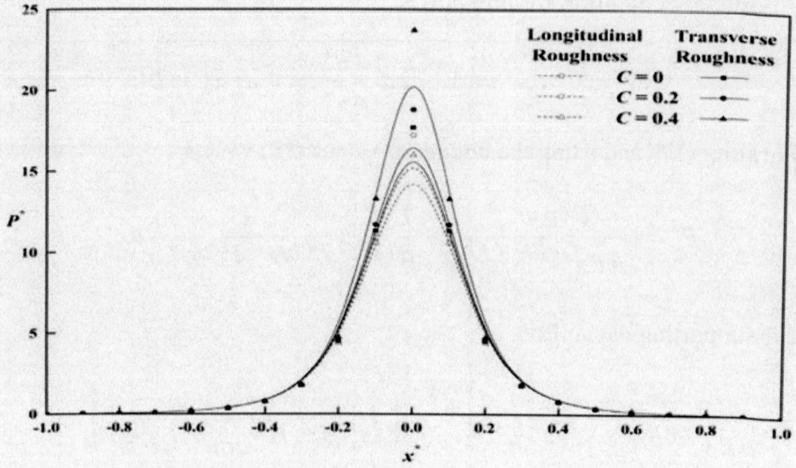

Fig. 2 Variation of $P*$ with $x*$ for distinct values of C where $l^* = 0.2$, $\beta = 0.04$, $h_m^* = 0.6$, $\sigma = 0.2$

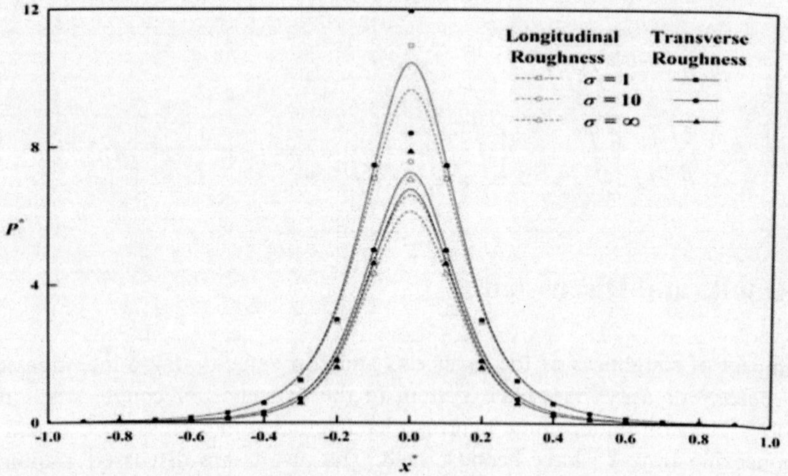

Fig. 3 Variation of $P*$ with $x*$ for distinct values of σ where $l^* = 0.2$, $\beta = 0.04$, $h_m^* = 0.6$, $C = 0.2$

smooth case. In Fig. 3, the graphs of fluid pressure are plotted by changing the slip velocity parameter $\sigma(1{\sim}\infty)$ and by fixing $\beta = 0.04$, $l^* = 0.2$, $h_m^* = 0.6$, $C = 0.2$. It is observed that pressure decreases with increasing values of slip parameter.

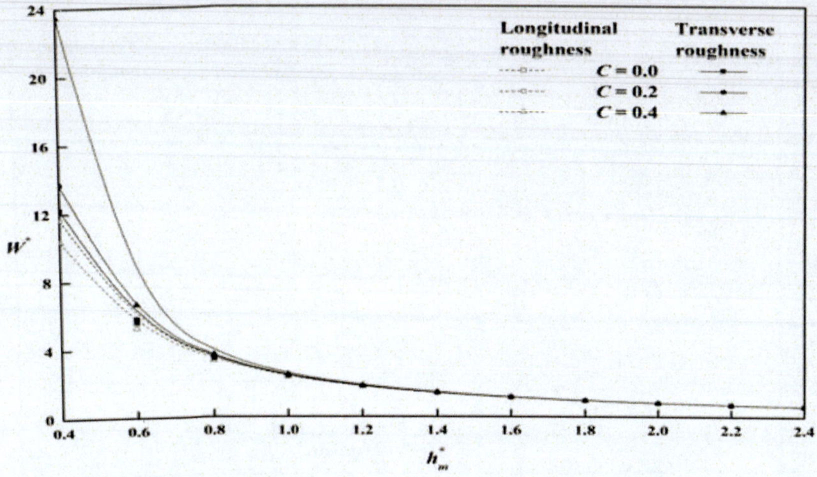

Fig. 4 Variation in $W*$ for distinct values of C, where $l* = 0.2$, $\beta = 0.04$, $\sigma = 0.2$

3.2 Dimensionless Load

In Fig. 4, W^* (the dimensionless load) is graphed against h_m^* (the least film thickness) for different values of C by keeping $\beta = 0.04$, $l^* = 0.2$, $h_m^* = 0.6$, $\sigma = 0.2$. $C = 0$ represents the smooth case, whereas $C(0.2 \sim 0.4)$ represents roughness. It is viewed incrementing C value leads W^* to increases (decreases) for transverse (longitudinal) roughness. Figure 5 illustrates that the capacity of carrying load W^* increases for the decreasing value of slip parameter σ.

3.3 Squeezing Time

Figures 6 and 7 represent plots of squeezing time T^* again dimensionless minimum film thickness h_m^* by varying C, σ. In Fig. 6, it is viewed that the squeezing time increases for transverse roughness and decreases for longitudinal roughness as C values increment from smooth to rough surface. In Fig. 7, T^* decreases with the increasing values of σ, which accounts for slip velocity. The CSF will provide more fluid pressure in the film region than the classical Newtonian lubricant $l^* = 0$.

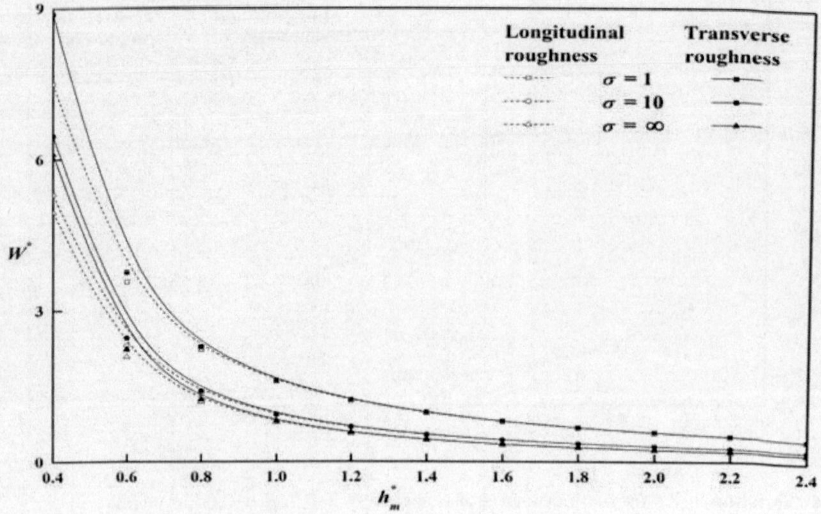

Fig. 5 Variation of $W*$ for distinct values of σ, where $l* = 0.2$, $\beta = 0.04$, $C = 0.2$

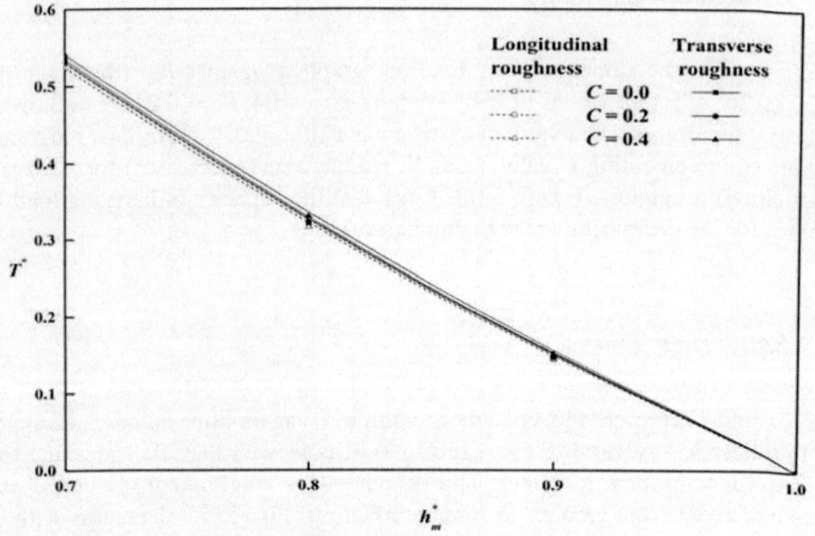

Fig. 6 Variation of $T*$ for distinct values of C, where $l* = 0.2$, $\beta = 0.04$, $\sigma = 0.2$

4 Conclusion

Based on this study, the results are summarized as follows.

The impact of slip velocity, CSF, and roughness of surface is analyzed in this chapter. As the value of slip velocity increases, the fluid pressure, load-supporting

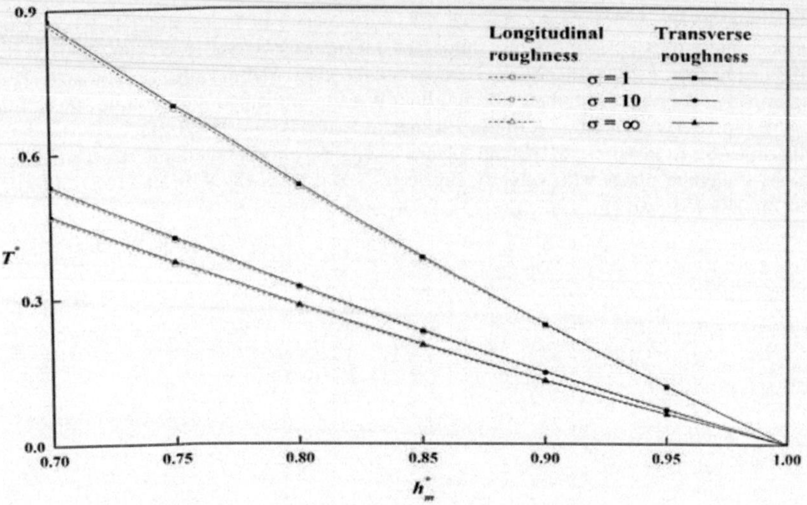

Fig. 7 Variation of $T*$ for distinct values of σ, where $l* = 0.2$, $\beta = 0.04$, $C = 0.2$

capability, and squeezing time decrease. When the value of the roughness parameter increases, the transverse/longitudinal film pressure, dimensionless load, and non-dimensional squeezing time increase/decrease.

References

1. Stokes, V.K.: Couple Stresses in Fluids. Phys. Fluids. **9**, 1709–1715 (1966)
2. Lin, J.R.: Squeeze film characteristics between a sphere and a flat plate: couple stress fluid model. Comput. Struct. **75**, 73–80 (2000)
3. Ramanaiha, G.: Squeeze films between finite plates lubricated by fluids with couple stress. Wear. **54**, 315–320 (1979)
4. Christensen, H.: Stochastic models for hydrodynamic lubrication of rough surfaces. Proc. Inst. Mech. Eng. Part I. **184**, 1013–1026 (1969)
5. Naduvinamani, N.B., Hanumagowda, B.N., Fathima, S.T.: Combined effects of MHD and surface roughness on couple stress squeeze film lubrication between porous circular stepped plates. Tribol. Int. **56**, 19–29 (2012)
6. Hanumagowda, B.N., Salma, A., Nagarajappa, C.: Effects of surface roughness, MHD and couple stress on squeeze film characteristics between curved circular plates. J. Phys. Conf. Ser. **1000**, 012075 (2018)
7. Hanumagowda, B.N., Raju, B.T., Santhosh Kumar, J., Vasanth, K.R.: Combined effect of surface roughness and pressure-dependent viscosity over couple-stress squeeze film lubrication between circular stepped plates. Proc. Inst. Mech. Eng. Part J: J. Eng. Tribol. **232**, 525–534 (2018)
8. Patel, N.D., Dehri, G.M.: Hydromagnetic lubrication of a rough porous parabolic slider bearing with slip velocity. J. Appl. Mech. Eng. **3**, 1–8 (2014)
9. Kalavathi, G.K., Dinesh, P.A., Gururajan, K.: Influence of roughness on porous finite journal bearing with heterogeneous slip/no-slip surface. Tribol. Int. **102**, 174–181 (2016)

10. Oladeinde, M.H., Akpobi, J.A.: A study of load capacity of finite slider bearings with slip surfaces and Stokesian couple stress fluids. Int. J. Eng. Res. Africa. **1**, 57–66 (2010)
11. Santhosh Kumar, J., Hanumagowda, B.N., Sreekala, C.K., Padmavathi, R.: An analysis on the performance of an inclined plane porous slider bearing lubricated with couple stress fluid and having slip velocity. Palestine J. Math. **10**(Special Issue I), 93–102 (2021)
12. Sangeetha, S., Kesavan, S.: MHD and surface roughness effects on couple stress fluid between porous triangular plates with velocity slip. Eur. J. Sci. Res.. ISSN 1450-216X / 1450-202X. **150**(2), 240–251 (2018)

Simulation of MHD Quadratic Natural Convective Flow of Nanofluid Inside a Square Enclosure with Thermal Radiation Effect

K. Venkatadri, V. Raja Rajeswari, A. Shobha, C. Venkata Lakshmi, R. Sivaraj, Firuz Kamalov, Ho-Hon Leung, and Mariam AlShamsi

1 Introduction

Heat transfer by buoyancy-driven flow inside the closed cavity has received considerable attention since it finds applications in several engineering and industrial processes. The flow induced by thermal difference is one of the most interesting mechanisms of heat transfer in the enclosures, and its importance increases interest among the researchers. As a result, several research efforts are made to investigate the energy transmission mechanism inside the cavity with various thermal effects and other effects. The literature contains ample works on natural convection flows inside cavities of square shape with several aspects of heat transfer investigations.

Buongiorno's mathematical model was adopted by Ahmed et al. [1] to examine the MHD free convection flow and energy transmission within a wavy enclosure consisting of a heat-generating porous medium. They observed that the heat transmission rate is a rising function of the wavy contraction ratio and

K. Venkatadri (✉)
Department of Mathematics, Indian Institute of Information Technology, Sri City, Chittoor, India

V. Raja Rajeswari
Department of Electronics and Communication Engineering, School of Engineering and Technology, Sri Padmavati Mahila Visvavidyalayam, Tirupati, India

A. Shobha · C. Venkata Lakshmi
Department of Applied Mathematics, Sri Padmavathi Mahila Visva Vidyalyam, Tirupati, AP, India

R. Sivaraj · H.-H. Leung · M. AlShamsi
Department of Mathematical Sciences, United Arab Emirates University, Al Ain, United Arab Emirates

F. Kamalov
Faculty of Engineering, Canadian University Dubai, Dubai, United Arab Emirates

F. Kamalov et al. (eds.), *Advances in Mathematical Modeling and Scientific Computing*, Trends in Mathematics, https://doi.org/10.1007/978-3-031-41420-6_19

the undulation number. Elshehabey et al. [2] computationally investigated the magnetohydrodynamic free convective circulations of Fe_3O_4-water nanoliquid in a tilted partially open wavy wall cavity by employing the Boussinesq approximation. They considered that the left wall of the cavity is uniformly heated, and the rest of the walls remain adiabatic. They discussed entropy generation tools like local entropy generation and its impact on heat transfer, the influence of the average Bejan number, and fluid friction on local entropy generation. It is noticed that the best position of the aperture in the cavity is its middle portion on the right wall for higher heat distribution and fluid flow. The rise of the aperture length replicates the respective enhancement of natural convection and heat transport. Sheikholeslami et al. [3] investigated the double diffusive nanofluid flow with an induced magnetic field by using the Runge-Kutta approach. Their results elucidate that an increase in the thermophoretic parameter enhances the shear stress. The temperature gradient enhances with the enhancement in suction parameter and decreases with an increase in Schmidt number, thermophoretic, and Brownian motion parameters. Alsabery et al. [4] examined the MHD mixed convective flow of Al_2O_3 nanofluid inside a lid-driven enclosure with conductive inner block. They identified that the huge size of a solid obstacle rises the energy transmission for higher Reynolds and Richardson numbers. Bondareva et al. [5] analyzed the unsteady laminar free convection flow and thermal transport of H_2O-based nanoliquid inside a trapezoidal enclosure with the impact of an inclined magnetic field. Based on their analysis, their conclusion states that Sherwood and Nusselt numbers are rising functions of Rayleigh numbers as well as magnetic field inclination angle and show quite opposite behavior in the occurrence of Hartmann numbers, aspect ratios, and Lewis numbers. Ho et al. [6] experimentally examined the natural convective movement and energy transport of alumina-water nanofluid inside three different vertical square cavities of distinct sizes. Based on a correlation analysis of the nanofluid's thermophysical properties, they reported that the impact of the nanofluid on free convection and energy transmission within the enclosure is generally infeasible and consistent with their experimental results. The heat transport rate across three different cavities appeared to be consistent, and systematic heat transfer decrements were found for nanofluids whose nanoparticles volume fraction is greater than or equal to 2% for the entire range of considered Rayleigh numbers. Wakif et al. [7] demonstrated the impact of radiative flux on MHD unsteady free convective CuO nanofluid Couette flow through vertical parallel plates. In their study, they considered a nanofluid with rheological behavior, and the impact of distinct dimensionless parameters like volumetric fraction of nanomaterials is investigated.

The convective thermal energy transport of a water-based nanoliquid in a two-dimensional square porous cavity with sinusoidal energy distribution on vertical and adiabatic horizontal walls has been analyzed by Sheremet et al. [8]. They exposed the numerical results within the limit of 10–10^4 for Rayleigh numbers and found that the findings are consistent with those found in the literature. Shehzad et al. [9] examined the impact of nanoliquid's convective energy transmission by implementing Buongiorno's two-phase model within a wavy channel. Their analysis revealed that the mass transfer rate and energy transmission rate diminish with the rise of the Prandtl number at the bottom of the wavy channel, whereas the Sherwood number is raised with the rise of the Prandtl number at the top of the wavy channel. Sheremet et al. [10] studied the impact of phase deviation, undulation number, and amplitude ratio on free convective nanoliquid flow inside a cavity with a wavy-wall. It was found in their study that the undulation number has a control over the rate of mass transmission and rate of heat transmission. Moreover, rise of the undulation number causes an expansion of the non-homogenous zones.

To the best of the authors' knowledge, there is no documentation on free convection of Buongiorno's nanofluid model under the influence of thermal radiation with a nonlinear Boussinesq approximation. Therefore, this investigation analyzes the free convective nanofluid flow inside the square domain with a nonlinear Boussinesq approximation and thermal radiation effects. It is worth noting that this study deals with a numerical simulation of passive cooling complex for controlling governing parameters in nanofluid flow.

2 Basic Equations

The two-dimensional graphical view of the current analysis is exhibited in Fig. 1a. The cavity has a cold right wall, a hot left wall, and thermally insulated horizontal walls. The present computations are performed with working fluid as water-based nanofluid and fixed with Pr = 6.8. The left surface of the square enclosure is sustained at high temperature T_h, and the right wall is preserved at cold temperature T_c. The fluid circulations are incompressible and laminar. Additionally, the standard Boussinesq approximation is adopted for adding the energy equation. The heat generation, chemical reaction, and viscous dissipation are ignored.

Using the assumptions discussed earlier, the governing partial differential equations are provided in dimensional forms as

$$\nabla . \boldsymbol{V} = 0 \tag{1}$$

$$\rho_f \left[\frac{\partial \boldsymbol{V}}{\partial t} + (\boldsymbol{V}.\nabla)\, \boldsymbol{V} \right] = -\nabla p + \mu \nabla^2 \boldsymbol{V} \text{-} \sigma B_0^2 \boldsymbol{V} + \left[(1 - C_c)\, \rho_f \left((T - T_c)\, \beta_0 + (T - T_c)^2 \beta_1 \right) - (C - C_c) \left(\rho_p - \rho_f \right) \right] g_y \tag{2}$$

Fig. 1 (a) Schematic representation of the problem. (b) Uniform mesh grid

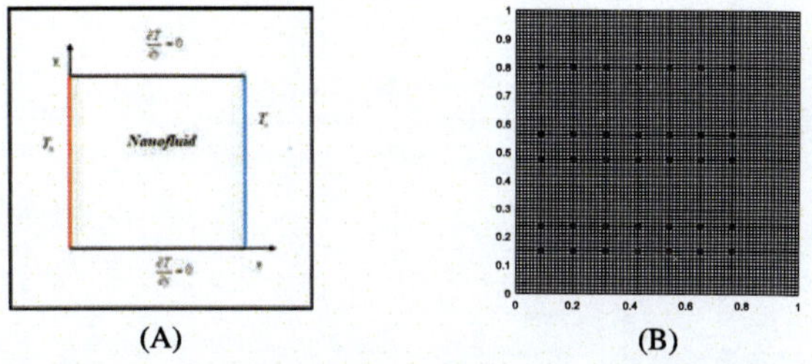

$$\frac{\partial T}{\partial t} + (V.\nabla)\, T = \alpha\, \nabla^2 T - \frac{1}{(\rho c_p)_f}\left(\frac{\partial q_{rx}}{\partial x} + \frac{\partial q_{ry}}{\partial y}\right) + \frac{(\rho c_p)_p}{(\rho c_p)_f}\left\{D_B\left(\frac{\partial C}{\partial x}\cdot\frac{\partial T}{\partial x} + \frac{\partial C}{\partial y}\cdot\frac{\partial T}{\partial y}\right) + \frac{D_T}{T_c}\left(\left(\frac{\partial T}{\partial x}\right)^2 + \left(\frac{\partial T}{\partial y}\right)^2\right)\right\} \tag{3}$$

$$\frac{\partial C}{\partial t} + u\frac{\partial C}{\partial x} + v\frac{\partial C}{\partial y} = D_B\left(\frac{\partial^2 C}{\partial x^2} + \frac{\partial^2 C}{\partial y^2}\right) + \frac{D_T}{T_c}\left(\frac{\partial^2 T}{\partial x^2} + \frac{\partial^2 T}{\partial y^2}\right) \tag{4}$$

Using the Rosseland approximation, the radiative flux is modeled:

$$q_{rx} = -\frac{4\sigma}{3\beta}\frac{\partial T^4}{\partial x},\quad q_{ry} = -\frac{4\sigma}{3\beta}\frac{\partial T^4}{\partial y}$$

The Taylor series expansion of T^4 can be expressed in terms of T_c and remove higher-order T_c terms:

$$T^4 \approx 4TT_c^3 - 3T_c^4 \tag{5}$$

Based on Eq. (5), the radiative flux terms are expressed as follows:

$$q_{rx} = -\frac{4\sigma T_c^3}{3\beta}\frac{\partial T}{\partial x},\quad q_{ry} = -\frac{4\sigma T_c^3}{3\beta}\frac{\partial T}{\partial y}$$

The following dimensional quantities are used

$$\tau=\frac{tu_0}{L},\, X=\frac{x}{L},\, U=\frac{u}{u_0},\, Y=\frac{y}{L},\, V=\frac{v}{u_0},\, \phi=\frac{C-C_c}{C_h-C_c},\, \theta=\frac{T-T_c}{T_h-T_c},\, P=\frac{p}{u_0^2\rho_f},\, u_0=\frac{\alpha}{L} \tag{6}$$

The nonlinear partial governing differential equations in nondimensional forms are

$$\frac{\partial V}{\partial Y} + \frac{\partial U}{\partial X} = 0 \tag{7}$$

$$U\frac{\partial U}{\partial X} + V\frac{\partial U}{\partial Y} + \frac{\partial U}{\partial \tau} = \Pr\left(\frac{\partial^2 U}{\partial X^2} + \frac{\partial^2 U}{\partial Y^2}\right) - \frac{\partial P}{\partial X} \tag{8}$$

$$U\frac{\partial V}{\partial X}+V\frac{\partial V}{\partial Y}+\frac{\partial V}{\partial \tau}=\Pr\left(\frac{\partial^2 V}{\partial X^2}+\frac{\partial^2 V}{\partial Y^2}\right)-\frac{\partial P}{\partial Y}-Ha^2\Pr V+Ra.\Pr\left(\theta\,(1+\chi\theta)-Nr\phi\right) \tag{9}$$

$$\frac{\partial\theta}{\partial\tau} + U\frac{\partial\theta}{\partial X} + V\frac{\partial\theta}{\partial Y} = \left(1 + \frac{4}{3}\mathrm{Rd}\right)\left(\frac{\partial^2\theta}{\partial X^2} + \frac{\partial^2\theta}{\partial Y^2}\right) + Nb\left(\frac{\partial\phi}{\partial X}\cdot\frac{\partial\theta}{\partial X} + \frac{\partial\phi}{\partial Y}\cdot\frac{\partial\theta}{\partial Y}\right) + Nt\left[\left(\frac{\partial\theta}{\partial X}\right)^2 + \left(\frac{\partial\theta}{\partial Y}\right)^2\right] \tag{10}$$

$$\frac{\partial\phi}{\partial\tau} + U\frac{\partial\phi}{\partial X} + V\frac{\partial\phi}{\partial Y} = \frac{1}{\mathrm{Le}}\left(\frac{\partial^2\phi}{\partial X^2} + \frac{\partial^2\phi}{\partial Y^2}\right) + \frac{Nt}{\mathrm{Nb.Le}}\left(\frac{\partial^2\theta}{\partial X^2} + \frac{\partial^2\theta}{\partial Y^2}\right) \tag{11}$$

Here

$$Nr = \frac{(C_h - C_c)(\rho_p - \rho_{f_0})}{\beta(1-C_c)\rho_{f_0}(T_h - T_c)},\ Nb = D_B\frac{(\rho c)_p}{(\rho c)_f}\frac{(C_h - C_c)}{\alpha_f},\ \mathrm{Pr} = \frac{\nu_f}{\alpha_f}$$
$$Nt = \frac{D_T}{T_c}\frac{(\rho c)_p}{(\rho c)_f}\frac{(T_h - T_c)}{\alpha_f},\ Ra = \frac{g\beta(T_h - T_c)(1-C_c)L^3}{\nu_f^2},\ Le = \frac{\alpha_f}{D_B}$$

The entire enclosure's walls are subjected to no-slip velocity boundary conditions, that is $U = V = 0$.

$\theta = 1,\ Nb\frac{\partial\phi}{\partial Y} + Nt\frac{\partial\theta}{\partial Y} = 0$ along the hot wall (i.e., left wall, $X = 0$).

$\theta = 0 = Nb\frac{\partial\phi}{\partial Y} + Nt\frac{\partial\theta}{\partial Y}$ along the cold wall (i.e., right wall, $X = 1$).

$\frac{\partial\phi}{\partial n} = \frac{\partial\theta}{\partial n} = 0$ on horizontal wall (i.e., $Y = 0$, 1).

3 Solution Technique and Validation

The MATLAB code is used to solve the nonlinear partial differential equations depending on the initial and boundary conditions of the current situation. The convective terms are approximated by using the forward finite difference upwind scheme, the diffusion terms are approximated by using the second-order central difference scheme, and the set of nonlinear finite difference approximated equations are solved by using the MAC algorithm with a staged grid-based uniform grid system. The detailed procedure of the method can be seen in the literature [11]. The house-computational MATLAB code is tested for the problem of air-filled square enclosure with different numerical methods [12–14]. The present MAC method's results on average Nusselt number (in Table 1) are seems to have a good agreement between the literature results and confirms the exactness on the present investigation.

Table 1 Comparison of average rate of heat transfer (Nu) with Pr = 0.71

Nu	Ra	Ref. [12]	Ref. [13]	Ref. [14] FEM	Present's MAC
Average	10^3	1.074	1.12	1.117	1.0904
Average	10^4	2.084	2.243	2.254	2.2427
Average	10^5	4.3	4.52	4.598	4.5699
Average	10^6	8.743	8.8	8.976	9.0169

4 Results and Discussion

A numerical investigation has been accomplished to examine the heat exchange and fluid flow variations within the enclosure (Table 2). The computed results are depicted for various thermal and fluid control parameters such as Hartmann (Magnetic) number (*Ha*), Rayleigh (buoyancy) number (*Ra*), thermal radiation (*Rd*), and nonlinear Boussinesq parameter (χ). The value of Prandtl number (*Pr*), Lewis number (*Le*), thermophoretic diffusivity (*Nt*), buoyancy-ratio parameter (*Nr*), and Brownian motion parameter (*Nb*) is fixed at $Pr = 6.8$, $Le = 2$, $Nt = 0.5$, $Nr = 0.1$, and $Nb = 0.5$.

Figure 2 demonstrates the influence of buoyancy force through *Ra* on flow fields and isotherms inside the square cavity, which is occupied by nanoliquid with $\chi = 1$, $Ha = 5$, and $Rd = 1$. The flow field is observed from $Ra = 10^4$ to $Ra = 10^6$. As seen, a single enlarged circulation is formed within the closed domain when $Ra = 10^4$. The inner circulation cells of circular shape are transformed to horizontal elliptical shaped circulation cells for increasing the *Ra* to $Ra = 10^5$. The temperature contours are steeper, and the energy transmission is enhanced because of the enhanced fluid

Table 2 Variation of average rate of heat transfer (Nu) for various fluids

Ra	Nu	$Pr = 0.71$ Air	$Pr = 6.8$ Water	$Pr = 6.8$ Nanofluid
10^3	Average	1.0904	1.0905	1.0225
10^4	Average	2.2427	2.2743	1.5513
10^5	Average	4.5699	4.7745	3.3416
10^6	Average	9.0169	9.3045	6.7145

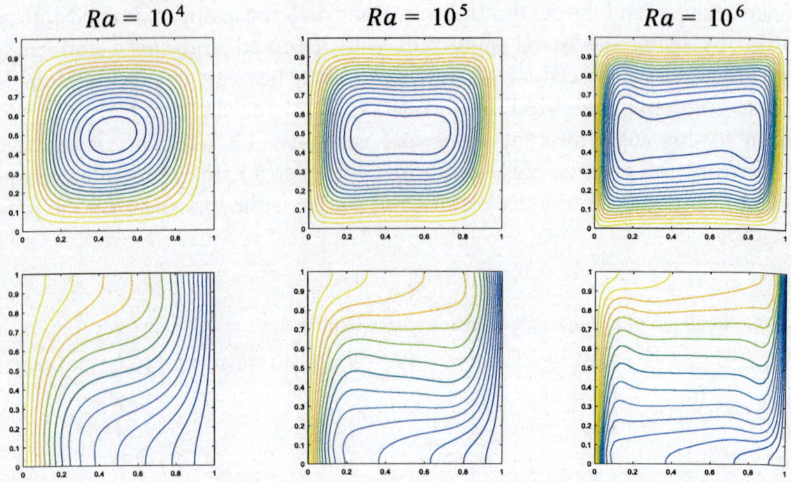

Fig. 2 Streamlines and isotherms for various values of *Ra* with $\chi = 1$, $Rd = 1$, $Ha = 5$

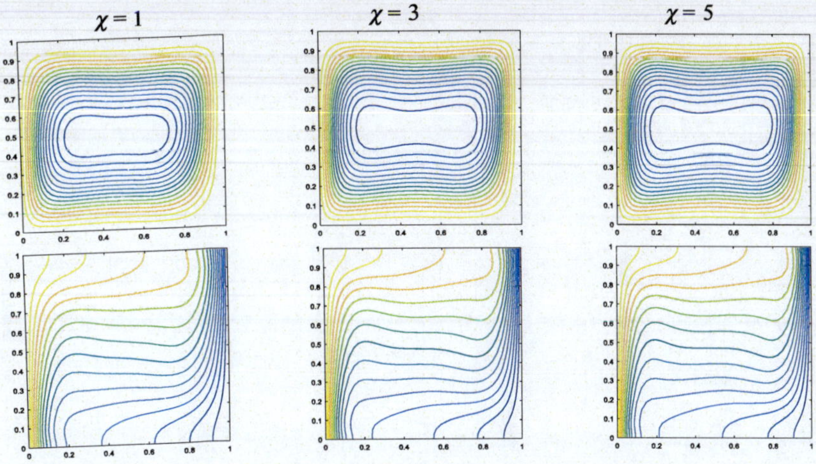

Fig. 3 Variation of nonlinear Boussinesq parameter on streamlines and isotherms with $Ha = 5$, $Rd = 1$, $Ra = 10^5$

velocity along the vertical walls. When $Ra = 10^6$, the fluid velocity is gradually enhanced nearby the vertical walls. Due to enhancement of velocity, and thermal boundary layer along the vertical walls the isotherm lines are synthesized almost parallel to top and bottom walls.

Figure 3 describes the impact of χ on streamlines and temperature contours with $Ha = 5$, $Rd = 1$, and $Ra = 10^5$. The considerable changes are observed with an enhancement in the nonlinear Boussinesq parameter (χ). A single circulation cell is found within the enclosure for $\chi = 1$, and the related temperature contours occupy the entire cavity. The energy transmission is gradually raised with higher values of nonlinear Boussinesq parameter (χ). The thermal contours are pushed to the cold wall and overlapped nearer to the cold wall, and fluid movement is enhanced in the neighborhood of vertical walls. The isothermal contours are stronger along the vertical walls when increasing the nonlinear Boussinesq parameter.

The impact of magnetic parameter Ha on velocity contours and isotherms is exhibited through Fig. 4. The results are analyzed by varying Ha (i.e., Hartmann number) from 0 to 50 with fixed values of $\chi = 1$, $Rd = 1$, and $Ra = 10^5$. The graphical outcomes expose the convective flow and energy transmission behavior. The lower energy transmission is observed for the zero impact of hydromagnetic effect (i.e., $Ha = 0$), whereas the stronger velocity is noticed inside the enclosure in this case. The enlarged horizontal inner circulation cells are transformed into circular shape when a higher magnetic field strength ($Ha = 50$) is imposed. The stronger energy transmission is noticed for growing values of Ha, whereas the fluid motion gradually diminishes at the middle of the enclosure.

The thermal radiation effect on local rate of energy transmission is illustrated through Fig. 5a, b with $Ra = 10^5$, $Ha = 5$, $\chi = 1$. The local rate of energy transmission along both the hot (Fig. 5a) and cold (Fig. 5b) walls is observed

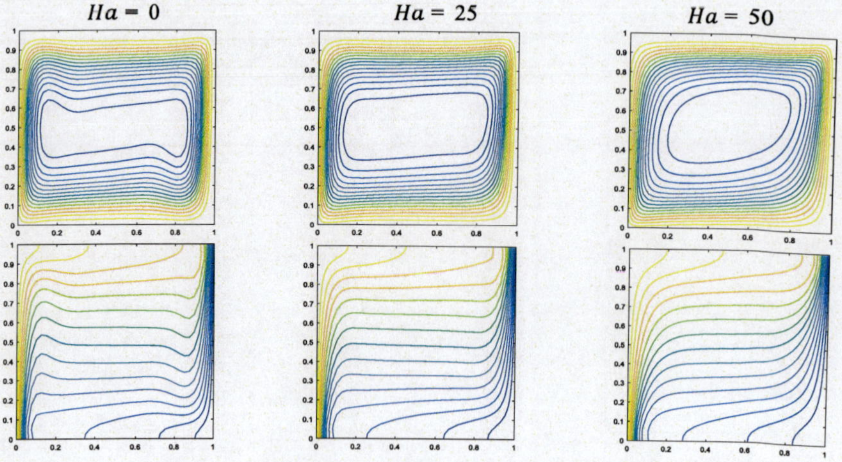

Fig. 4 Influence of *Ha* on streamlines and isotherms for $\chi = 1$, $Rd = 1$, $Ra = 10^5$

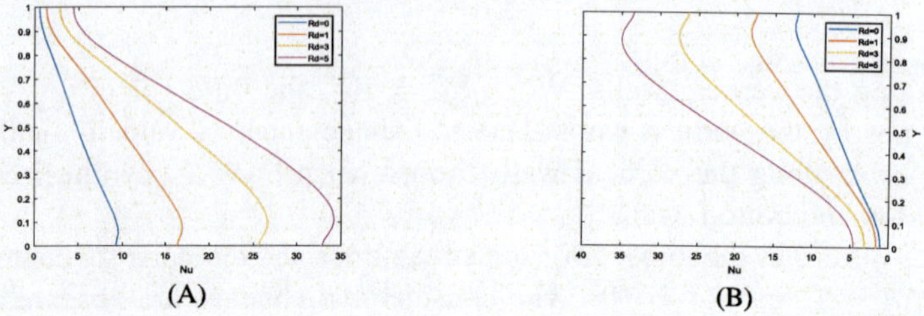

Fig. 5 Local Nusselt number for **(a)** hot wall and **(b)** cold wall for $\chi = 1$, $Ha = 5$, $Ra = 10^5$

with dissimilar values of *Rd*. The minimal heat transmission rate is noticed in the nonappearance of *Rd*, and maximal energy transmission rate is registered in the presence of the thermal radiation parameter (*Rd*) with its highest value. As seen, at both the side walls, the temperature dispersion/absorption increases gradually with the raising of the thermal radiation parameter *Rd*.

5 Conclusions

This study is committed to numerically examine the water-based nanofluid flow and energy transmission within the enclosure of a square shape using Buongiorno's nanofluid model. The studied relevant parameters are nonlinear temperature parameter, Rayleigh number (*Ra*), Hartmann number (*Ha*), thermophoresis, Brownian motion, and thermal radiation parameters. The obtained results reveal the following key remarks:

- Regardless of *Ra* values, when the magnetic parameter (*Ha*) increases, the fluid flow and energy transmission are significantly affected. The presence of *Ha* causes to reduces the flow circulation.
- Growth of the *Ra* and radiative heat flux parameter leads to intensify the heat distribution.
- It is evident that the Rayleigh number has a significant influence on flow patterns and isotherms. As *Ra* upsurges, the influence of conduction increases.
- The local rate of heat transmission at the cold and hot walls increases by raising the radiation parameters inside the enclosure.
- The isothermal lines are densely formed near the side walls for high Rayleigh number values.

References

1. Ahmed, S.E., Rashed, Z.Z.: MHD natural convection in a heat generating porous medium-filled wavy enclosures using Buongiorno's nanofluid model. Case Stud. Therm. Eng. **14**, 100430 (2019)
2. Elshehabey, H.M., Raizah, Z., Oztop, H.F., Ahmed, S.E.: MHD natural convective flow of Fe_3O_4 - H_2O ferrofluids in an inclined partial open complex-wavy-walls ringed enclosures using non-linear Boussinesq approximation. Int. J. Mech. Sci. **170**, 105352 (2020)
3. Sheikholeslami, M., Rokni, H.B.: Nanofluid two phase model analysis in existence of induced magnetic field. Int. J. Heat Mass Transf. **107**, 288–299 (2017)
4. Alsabery, A.I., Ismael, M.A., Chamkha, A.J., Hashim, I.: Effects of two-phase nanofluid model on MHD mixed convection in a lid-driven cavity in the presence of conductive inner block and corner heater. J. Therm. Anal. Calorim. **135**, 729–750 (2019)
5. Bondareva, N.S., Sheremet, M.A., Pop, I.: Magnetic field effect on the unsteady natural convection in a right-angle trapezoidal cavity filled with a nanofluid. Int. J. Numer. Methods Heat Fluid Flow. **25**(8), 1924–1946 (2015)
6. Ho, C.J., Liu, W.K., Chang, Y.S., Lin, C.C.: Natural convection heat transfer of alumina-water nanofluid in vertical square enclosures: an experimental study. Int. J. Therm. Sci. **49**, 1345–1353 (2010)
7. Wakif, A., Boulahia, Z., Ali, F., Eid, M.R., Sehaqui, R.: Numerical analysis of the unsteady natural convection MHD Couette nanofluid flow in the presence of thermal radiation using single and two-phase nanofluid models for Cu–water nanofluids. Int. J. Appl. Comput. Math. **4**, 81 (2018)
8. Sheremet, M.A., Pop, I.: Natural convection in a square porous cavity with sinusoidal temperature distributions on both side walls filled with a nanofluid: Buongiorno's mathematical model. Transp. Porous Media. **105**, 411–429 (2014)
9. Shehzad, N., Zeeshan, A., Ellahi, R., Vafai, K.: Convective heat transfer of nanofluid in a wavy channel: Buongiorno's mathematical model. J. Mol. Liq. **222**, 446–455 (2016)
10. Sheremet, M.A., Pop, I.: Free convection in wavy porous enclosures with non-uniform temperature boundary conditions filled with a nanofluid: Buongiorno's mathematical model. Therm. Sci. **21**(3), 1183–1193 (2017)
11. Venkatadri, K., GouseMohiddin, S., Reddy, S.M.: Numerical analysis of unsteady MHD mixed convection flow in a lid-driven square cavity with central heating on left vertical wall, applications of fluid dynamics. Lect. Notes Mech. Eng., Chapter. **26**, 355–370 (2018)

12. Manzari, M.T.: An explicit finite element algorithm for convective heat transfer problems. Int. J. Numer. Methods Heat Fluid Flow. **9**, 860–877 (1999)
13. Davis de, G.V.: Natural convection of air in a square cavity: a benchmark numerical solution. Int. J. Numer. Methods Fluids. **3**, 249–264 (1983)
14. Wan, D.C., Patnaik, B.S.V., Wei, G.W.: A new benchmark quality solution for the buoyancy-driven cavity by discrete singular convolution. Numer. Heat Transfer, Part B. **40**, 199–228 (2001)

Combined Effects of Magnetic Field and Heat Source on Double-Diffusive Marangoni Convection in Fluid-Porous Structure

N. Manjunatha, Yellamma, and R. Sumithra

1 Introduction

In a variety of conditions, double-diffusive natural convection, or flows caused by buoyancy due to simultaneous temperature and concentration gradients, can occur. Some literature is available on the single and double component layer. Sumithra and Manjunatha [1] conducted an analytical study of magneto convection in a combined layer limited by adiabatic boundaries. Sankar et al. [2] and Jagadeesha et al. [3] examined the two-component convection using the Darcy model for porous enclosure in the presence of heat and solute source. Pushpa et al. [4] explored the thermosolutal convection. The influence of magnetic field on double-diffusive mixed convection in a rectangular inclined domain with an aspect ratio was investigated using a finite volume technique by Shivananda and Satheesh [5]. Shivakumara et al. [6] explored the effect of cross-diffusion on the beginning of convective instability. Sumithra and Arul Selvamary [7] discussed the single component convection for couple stress fluid in combined system for two boundary combinations. In the presence of magnetic field, heat generation or absorption, and chemical reaction, Xiaoli Qiang et al. [8] investigated unstable MHD double-diffusive convection flows between two infinite vertical parallel plates.

N. Manjunatha (✉)
Department of Mathematics, School of Applied Sciences, REVA University, Bengaluru, Karnataka, India
e-mail: manjunatha.n@reva.edu.in

Yellamma
School of Applied Sciences, REVA University, Bengaluru, Karnataka, India

R. Sumithra
Department of Mathematics, Nrupathunga University, Bengaluru, Karnataka, India

F. Kamalov et al. (eds.), *Advances in Mathematical Modeling and Scientific Computing*, Trends in Mathematics, https://doi.org/10.1007/978-3-031-41420-6_20

Peristaltically deformable and double-diffusive characteristics have been examined by Tanveer et al. [9].

Recently, the influence of a power law fluid and an angled magnetic field on a porous material in a staggered cavity is studied by Hussain et al. [10]. The influence of the Soret and Dufour factors is also considered. Meften [11] investigated two models of double-diffusive convection in a fluid layer where viscosity varies quadratically with temperature and nonlinear results are obtained using conditional energy analysis. Manjunatha et al. [12] and Manjunatha and Sumithra [13] studied the effects of three profiles and a heat source on convection in a combined structure in the presence of magnetic field. In the present study, the effect of magnetic field and heat source on onset convection is examined in detail for two types of boundary combinations.

2 Mathematical Formulation

Consider a double component, electrically conducting liquid saturated isotropic, sparsely packed porous layer of thickness d_p with an imposed magnetic field intensity H_0 underlying a triple component liquid layer of thickness d_f and with heat sources Φ_p and Φ_f, respectively. The porous layer's lower surface is hard, while the fluid layer's upper surface is free, with surface tension effects depending on temperature and concentration as shown in Fig. 1. A Cartesian coordinate system (x, y, z) is used with the origin at the interface between the porous and fluid layers, and the z-axis is vertically upward. Let ΔT and ΔC be the temperature and concentration difference between the lower and upper boundaries, $T = T_0$ the reference temperature, and $C = C_0$ the reference salinity. With the Boussinesq

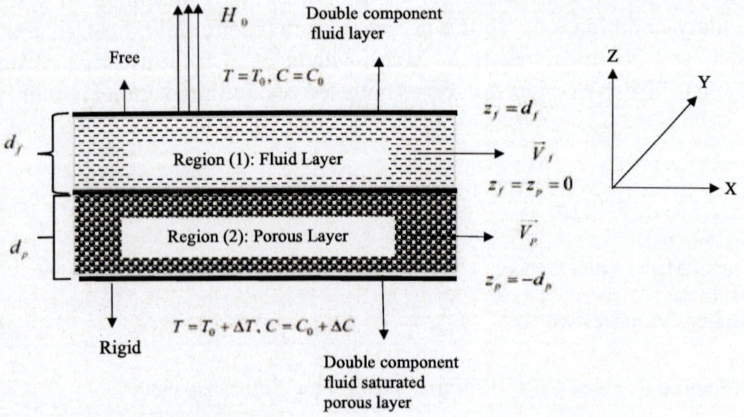

Fig. 1 Geometry of the problem

approximation, the basic equations for region 1 and region 2 (see Sumithra and Manjunatha [1] and Shivakumara et al. [14]).

Fluid layer: Region 1

$$\nabla_f.\overrightarrow{V}_f = 0 \tag{1}$$

$$\nabla_f.\overrightarrow{H} = 0 \tag{2}$$

$$\frac{\partial \overrightarrow{V}_f}{\partial t} + (\overrightarrow{V}_f.\nabla_f)\overrightarrow{V}_f = -\frac{1}{\rho_0}\nabla_f P_f + \frac{\mu_f}{\rho_0}\nabla_f^2 \overrightarrow{V}_f + \frac{\gamma_f}{\rho_0}(\overrightarrow{H}.\nabla_f)\overrightarrow{H} \tag{3}$$

$$\frac{\partial T_f}{\partial t} + (\overrightarrow{V}_f.\nabla_f)T_f = \kappa_f \nabla_f^2 T_f + \Phi_f \tag{4}$$

$$\frac{\partial C_f}{\partial t} + (\overrightarrow{V}_f.\nabla_f)C_f = \kappa_{cf}\nabla_f^2 C_f \tag{5}$$

$$\frac{\partial \overrightarrow{H}}{\partial t} = \nabla_f \times \overrightarrow{V}_f \times \overrightarrow{H} + \nu_f \nabla_f^2 \overrightarrow{H} \tag{6}$$

Porous layer: Region 2

$$\nabla_p.\overrightarrow{V}_p = 0 \tag{7}$$

$$\nabla_p.\overrightarrow{H} = 0 \tag{8}$$

$$\frac{1}{\varepsilon_p}\frac{\partial \overrightarrow{V_p}}{\partial t} + \frac{1}{\varepsilon_p^2}(\overrightarrow{V_p}.\nabla_p)\overrightarrow{V_p} = -\frac{1}{\rho_0}\nabla_p P_p - \frac{\mu_p}{K\rho_0}\overrightarrow{V_p} + \frac{\gamma_p}{\rho_0}(\overrightarrow{H}.\nabla_p)\overrightarrow{H} \tag{9}$$

$$M\frac{\partial T_p}{\partial t} + (\overrightarrow{V_p}.\nabla_p)T_p = \kappa_p \nabla_p^2 T_p + \Phi_p \tag{10}$$

$$\phi_p \frac{\partial C_p}{\partial t} + (\overrightarrow{V_p}.\nabla_p)C_p = \kappa_{cp}\nabla_p^2 C_p \tag{11}$$

$$\phi_p \frac{\partial \overrightarrow{H}}{\partial t} = \nabla_p \times \overrightarrow{V_p} \times \overrightarrow{H} + \nu_p \nabla_p^2 \overrightarrow{H} \tag{12}$$

where the fluid and porous regions are denoted by the subscripts f and p, respectively. $\overrightarrow{V}_f$ is the velocity vector, ρ_0 is the fluid density, μ_f is the fluid viscosity, P_f is the total pressure, $\overrightarrow{H}$ is the magnetic field, T_f is the temperature, ν_f is the magnetic viscosity, γ_f is the magnetic permeability, κ_{cf} is the solute diffusivity of the fluid, and C_f is the salinity field. For region 2, ϕ_p is the porosity, K is the permeability , M is the heat capacity ratio, κ_p is the thermal diffusivity, and ν_p is the effective magnetic viscosity.

The goal of this research is to see if a quiescent state can withstand tiny perturbations superimposed on the basic state; the solutions are as follows:

$$\vec{V}_f = 0,\ P_f = P_{fb}(z_f),\ T_f = T_{fb}(z_f),\ C_f = C_{fb}(z_f),\ \vec{H} = H_0(z_f) \quad (13)$$

$$\vec{V}_p = 0,\ P_p = P_{pb}(z_p),\ T_p = T_{pb}(z_p),\ C_p = C_{pb}(z_p),\ \vec{H} = H_0(z_p) \quad (14)$$

where

$$T_{fb}(z_f) = \frac{-\Phi_f z_f(z_f - d_f)}{2\kappa_f} + \frac{(T_u - T_0)z_f}{d_f} + T_0 \qquad 0 \le z_f \le d_f$$

$$T_{pb}(z_p) = \frac{-\Phi_p z_p(z_p + d_p)}{2\kappa_p} + \frac{(T_0 - T_l)z_p}{d_p} + T_0 \qquad -d_p \le z_p \le 0$$

$$C_{fb}(z_f) = C_0 - \frac{(C_0 - C_u)z_f}{d_f} \qquad 0 \le z_f \le d_f$$

$$C_{pb}(z_p) = C_0 - \frac{(C_l - C_0)z_p}{d_p} \qquad -d_p \le z_p \le 0$$

$T_0 = \frac{\kappa_f d_p T_u + \kappa_p d_f T_l}{\kappa_f d_p + \kappa_p d_f} + \frac{d_f d_p(\Phi_p d_p + \Phi_f d_f)}{2(\kappa_f d_p + \kappa_p d_f)}$, $C_o = \frac{\kappa_{cf} d_p C_u + \kappa_{cp} d_f C_l}{\kappa_{cf} d_p + \kappa_{cp} d_f}$.

To investigate the stability of the basic state, regions 1 and 2 are subjected to infinite perturbations:

$$\vec{V}_f = \vec{V}'_f,\ P_f = P_{fb} + P'_f,\ T_f = T_{fb}(z_f) + \theta'_f,\ C_f = C_{fb}(z_f) + S'_f,$$
$$\vec{H} = H_0(z_f) + \vec{H}' \quad (15)$$

$$\vec{V}_p = \vec{V}_p',\ P_p = P_{pb} + P'_p,\ T_p = T_{pb}(z_p) + \theta'_p,\ C_p = C_{pb}(z_p) + S'_p,$$
$$\vec{H} = H_0(z_p) + \vec{H}' \quad (16)$$

where $\vec{V}'_f, P'_f, \theta'_f, S'_f, \vec{H}'$velocity, pressure, temperature, salinity, and magnetic field are respectively perturbed quantities for region 1 and the similar quantities in region 2 are $\vec{V}_p', P'_p, \theta'_p, S'_p, \vec{H}'$. The variables are nondimensionalized for regions 1 and 2 using $d_f, \frac{d_f^2}{\kappa_f}, \frac{\kappa_f}{d_f}, T_0 - T_u, C_0 - C_u, H_0$, and $d_p, \frac{d_p^2}{\kappa_p}, \frac{\kappa_p}{d_p}, T_l - T_0, C_l - C_0, H_0$.

We arrive at the following stability equations in region 1 and region 2, respectively, using the conventional linear stability analysis approach and assuming that the concept of exchange of stability holds (see Sumithra and Manjunatha [1] and Shivakumara et al. [14]):

$$\left((D_f^2 - a_f^2)^2 - Q_f D_f^2\right) W_f(z_f) = 0 \quad (17)$$

$$(D_f^2 - a_f^2)\theta_f(z_f) + \left(1 + R_{If}^*(2z_f - 1)\right) W_f(z_f) = 0 \quad (18)$$

$$\tau_f(D_f^2 - a_f^2)S_f(z_f) + W_f(z_f) = 0 \quad (19)$$

$$\left((D_p^2 - a_p^2) + Q_p\beta^2 D_p^2\right) W_p(z_p) = 0 \tag{20}$$

$$(D_p^2 - a_p^2)\theta_p(z_p) + \left(1 + R_{Ip}^*(2z_p + 1)\right) W_p(z_p) = 0 \tag{21}$$

$$\tau_p(D_p^2 - a_p^2)S_p(z_p) + W_p(z_p) = 0 \tag{22}$$

In the above equations, a_f, a_p, Q_f, Q_p, $R_{If}^* = \frac{R_{if}}{2(T_0-T_u)}$, $R_{Ip}^* = \frac{R_{ip}}{2(T_l-T_0)}$, $R_{if} = \frac{\Phi_f d_f^2}{\kappa_f}$, $R_{ip} = \frac{\Phi_p d_p^2}{\kappa_p}$, $\tau_f = \frac{\kappa_{cf}}{\kappa_f}$, and $\tau_p = \frac{\kappa_{cp}}{\kappa_p}$ are, namely, the horizontal wave numbers, the Chandrasekhar numbers, the modified internal Rayleigh numbers, the internal Rayleigh numbers, and the diffusivity ratios, $\beta = \sqrt{\frac{K}{d_p^2}}$ is the porous parameter, $W_f(z_f)$ and $W_p(z_p)$ are the vertical velocities, $\theta_f(z_f)$ and $\theta_p(z_p)$ are the temperatures, and $S_f(z_f)$ and $S_p(z_p)$ are the concentration distributions. Because the wave numbers for the combined layers must be the same, so that we have $\frac{a_f}{d_f} = \frac{a_p}{d_p}$ and hence $a_p = \hat{d}a_f$, here $\hat{d} = \frac{d_p}{d_f}$ is the depth ratio.

The boundary conditions are nondimensionalized after:

$$D_f^2 W_f(1) + \left(M_t\theta_f(1) + M_s S_f(1)\right) a_f^2 = 0 \tag{23}$$

The velocity conditions are

$$W_f(1) = 0,\, W_p(-1) = 0,\, D_p W_p(-1) = 0,\, \hat{T} W_f(0) = W_p(0),$$
$$\hat{T}\hat{d}^2(D_f^2 + a_f^2)W_f(0) = \hat{\mu}(D_p^2 + a_p^2)W_p(0),\, \hat{T}\hat{d}D_f W_f(0) = D_p W_p(0),$$
$$\hat{T}\hat{d}^3\beta^2[(D_f^3 - 3a_f^2 D_f)]W_f(0) = [-D_p + \hat{\mu}\beta^2(D_p^3 - 3a_p^2 D_p)]W_p(0) \tag{24}$$

The conditions for adiabatic-adiabatic and adiabatic-isothermal, respectively, are

$$D_f\theta_f(1) = 0,\, \theta_f(0) = \hat{T}\theta_p(0),\, D_f\theta_f(0) = D_p\theta_p(0),\, D_p\theta_p(-1) = 0 \tag{25}$$

$$D_f\theta_f(1) = 0,\, \theta_f(0) = \hat{T}\theta_p(0),\, D_f\theta_f(0) = D_p\theta_p(0),\, \theta_p(-1) = 0 \tag{26}$$

The salinity conditions are

$$D_f S_f(1) = 0,\, S_f(0) = \hat{S}S_p(0),\, D_f S_f(0) = D_p S_p(0),\, D_p S_p(-1) = 0 \tag{27}$$

In the above equations, $\hat{S}$ is the solute diffusivity ratio, $\hat{T}$ is the thermal ratio, $\hat{\mu}$ is the viscosity ratio, $M_t = \frac{\partial\sigma_t}{\partial T_f}\frac{(T_u-T_0)d_f}{\mu_f\kappa_f}$ is the thermal Marangoni number (tMn), $M_s = \frac{\partial\sigma_t}{\partial C_f}\frac{(C_u-C_0)d_f}{\mu_f\kappa_f}$ is the solute Marangoni number (sMn), and σ_t is the surface tension.

3 Methodology

Introducing (24), the velocity profiles are obtained by solving (17) and (20) and appropriately written as follows:

$$W_f(z_f)=A_1[\cosh\psi_f z_f+a_1\sinh\psi_f z_f+a_2\cosh\varphi_f z_f+a_3\sinh\varphi_f z_f] \quad (28)$$

$$W_p(z_p)=A_1[a_4\cosh\delta_p z_p+a_5\sinh\delta_p z_p] \quad (29)$$

where

$$\psi_f=\frac{\sqrt{Q_f}+\sqrt{Q_f+4a_f^2}}{2},\ \varphi_f=\frac{\sqrt{Q_f}-\sqrt{Q_f+4a_f^2}}{2},\ \delta_p=\sqrt{\frac{a_p^2}{1+Q_p\beta^2}},\ a_1=-\frac{a_3\Delta_2}{\Delta_1},$$

$$a_2=\frac{\Delta_5\Delta_7-\Delta_8\Delta_4}{\Delta_3\Delta_7-\Delta_6\Delta_4},\ a_3=\frac{\Delta_5\Delta_6-\Delta_8\Delta_3}{\Delta_4\Delta_6-\Delta_7\Delta_3},\ a_4=\hat{T}(1+a_2),\ a_5=\frac{1}{\delta_p}(\hat{T}\hat{d}a_1\psi_f+a_3\varphi_f)$$

$$\Delta_1=\widehat{d^2}\beta^2(\psi_f^3-3a_f^2\psi_f)+\psi_f,\ \Delta_2=\widehat{d^2}\beta^2(\varphi_f^3-3a_f^2\varphi_f)+\varphi_f,$$

$$\Delta_3=\hat{T}\cosh\delta_p,\ \Delta_4=-\frac{\hat{d}\hat{T}\sinh\delta_p}{\delta_p}(\varphi_f-\frac{\Delta_2\psi_f}{\Delta_1}),\Delta_5=-\Delta_3,\ \Delta_6=\cosh\varphi_f,$$

$$\Delta_7=\sinh\varphi_f-(\frac{\Delta_2}{\Delta_1})\sinh\psi_f,\ \Delta_8=-\cosh\psi_f.$$

Introducing (19) and (22), the salinity profiles are obtained using the condition (27), as follows:

$$S_f(z_f)=A_1[c_{13}\cosh a_f z_f+c_{14}\sinh a_f z_f+\Sigma_{f1}(z_f)] \quad (30)$$

$$S_p(z_p)=A_1[c_{15}\cosh a_p z_p+c_{16}\sinh a_p z_p+\Sigma_{p1}(z_p)] \quad (31)$$

where $\Sigma_{f1}(z_f)=\frac{-1}{\tau_f}\left[\frac{\cosh\psi_f z_f+a_1\sinh\psi_f z_f}{\psi_f^2-a_f^2}+\frac{a_2\cosh\varphi_f z_f+a_3\sinh\varphi_f z_f}{\varphi_f^2-a_f^2}\right]$,

$$\Sigma_{p1}(z_p)=\frac{-1}{\tau_p}\left[\frac{a_4\cosh\delta_p z_p+a_5\sinh\delta_p z_p}{\delta_p^2-a_p^2}\right],\ c_{13}=\hat{S}c_{15}+\Delta_{100}+\Delta_{101},$$

$$c_{14}=\frac{1}{a_f}(c_{16}a_p+\Delta_{102}+\Delta_{103}),\ c_{15}=\frac{\Delta_{108}a_p\cosh a_p-\Delta_{107}\Delta_{105}}{a_p\sinh a_p\Delta_{107}+\Delta_{106}a_p\cosh a_p},$$

$$c_{16}=\frac{\Delta_{105}\Delta_{106}+a_p\sinh a_p\Delta_{108}}{a_p\sinh a_p\Delta_{107}+\Delta_{106}a_p\cosh a_p},\Delta_{100}=\frac{-\hat{S}}{\tau_p}\left(\frac{a_4}{\delta_p^2-a_p^2}\right),$$

$$\Delta_{101}=\frac{1}{\tau_f}\left[\frac{1}{\psi_f^2-a_f^2}+\frac{a_2}{\varphi_f^2-{}_f^2}\right]\Delta_{102}=\frac{-1}{\tau_p}\left(\frac{\delta_p a_5}{\delta_p^2-a_p^2}\right),\Delta_{103}=\frac{1}{\tau_f}(\frac{a_1\psi_f}{\psi_f^2-a_f^2}+\frac{a_3\varphi_f}{\varphi_f^2-{}_f^2})$$

$$\Delta_{104}=\frac{1}{\tau_f}\left[\frac{(\sinh\psi_f+a_1\cosh\psi_f)\psi_f}{\psi_f^2-a_f^2}+\frac{(a_2\sinh\varphi_f+a_3\cosh\varphi_f)\varphi_f}{\varphi_f^2-a_f^2}\right],$$

$$\Delta_{105}=\frac{1}{\tau_p}\left[\frac{\delta_p(-a_4\sinh\delta_p+a_5\cosh\delta_p)}{\delta_p^2-a_p^2}\right],$$

$$\Delta_{106}=\hat{S}a_f\sinh a_f\cosh a_p+a_p\sinh a_p\cosh a_f,$$

$$\Delta_{107}=\hat{S}a_f\sinh a_p\sinh a_f+a_p\cosh a_f\cosh a_p,$$

$$\Delta_{108}=\Delta_{104}-(\Delta_{100}+\Delta_{101})a_f\sinh a_f-(\Delta_{102}+\Delta_{103})\cosh a_f.$$

4 Thermal Marangoni Number

4.1 *Case (i): Adiabatic-Adiabatic Boundary Condition*

The fluid-porous structure is horizontally enclosed by the adiabatic boundaries. Using the boundary conditions for temperature (25), the distributions $\theta_f(z_f)$ and $\theta_p(z_p)$ are produced by solving (18) and (21):

$$\theta_f(z_f) = A_1[c_1 \cosh a_f z_f + c_2 \sinh a_f z_f + \Sigma_{f2}(z_f)] \tag{32}$$

$$\theta_p(z_p) = A_1[c_3 \cosh a_p z_p + c_4 \sinh a_p z_p + \Sigma_{p2}(z_p)] \tag{33}$$

where $\Sigma_{f2}(z_f) = A_1[\delta_1 - \delta_2 + \delta_3 - \delta_4]$, $\Sigma_{p2}(z_p) = A_1[\delta_5 - \delta_6]$

$\delta_1 = \frac{(\alpha_{2f} z_f + \alpha_{1f})}{(\psi_f^2 - a_f^2)}(\cosh \psi_f z_f + a_1 \sinh \psi_f z_f)$,

$\delta_2 = \frac{2\psi_f \alpha_{2f}}{(\psi_f^2 - a_f^2)^2}(a_1 \cosh \psi_f z_f + \sinh \psi_f z_f)$,

$\delta_3 = \frac{(\alpha_{2f} z_f + \alpha_{1f})}{(\varphi_f^2 - a_f^2)}(a_2 \cosh \varphi_f z_f + a_3 \sinh \varphi_f z_f)$,

$\delta_4 = \frac{2\varphi_f \alpha_{2f}}{(\varphi_f^2 - a_f^2)^2}(a_3 \cosh \varphi_f z_f + a_2 \sinh \varphi_f z_f)$,

$\delta_5 = \frac{(\alpha_{1p} + \alpha_{2p} z_p)}{(\delta_p^2 - a_p^2)}(a_4 \cosh \delta_p z_p + a_5 \sinh \delta_p z_p)$,

$\delta_6 = \frac{2\alpha_{2p}\delta_p}{(\delta_p^2 - a_p^2)^2}(a_5 \cosh \delta_p z_p + a_4 \sinh \delta_p z_p)$,

$\alpha_{1f} = R_{If}^* - 1$, $\alpha_{2f} = -2R_{If}^*$, $\alpha_{1p} = -(R_{Ip}^* + 1)$, $\alpha_{2p} = -2R_{Ip}^*$,

$c_1 = c_3 \hat{T} + \Delta_{10} - \Delta_{11}$, $c_2 = \frac{1}{a_f}(c_4 a_p + \Delta_{12} - \Delta_{13})$, $c_3 = \frac{\Delta_{19}\Delta_{14} - \Delta_{16}\Delta_{18}}{\Delta_{15}\Delta_{18} + \Delta_{17}\Delta_{14}}$,

$c_4 = \frac{\Delta_{16}\Delta_{17} + \Delta_{19}\Delta_{15}}{\Delta_{14}\Delta_{17} + \Delta_{18}\Delta_{15}}$, $\Delta_9 = -[\delta_7 + \delta_8 + \delta_9 + \delta_{10}]$,

$\delta_7 = \frac{\psi_f(\alpha_{2f} + \alpha_{1f})}{(\psi_f^2 - a_f^2)}(a_1 \cosh \psi_f + \sinh \psi_f)$,

$\delta_8 = \left[\frac{\alpha_{2f}}{(\psi_f^2 - a_f^2)} - \frac{2\psi_f^2 \alpha_{2f}}{(\psi_f^2 - a_f^2)^2}\right](\cosh \psi_f + a_1 \sinh \psi_f)$,

$\delta_9 = \frac{\varphi_f(\alpha_{2f} + \alpha_{1f})}{(\varphi_f^2 - a_f^2)}(a_3 \cosh \varphi_f + a_2 \sinh \varphi_f)$,

$\delta_{10} = \left[\frac{\alpha_{2f}}{(\varphi_f^2 - a_f^2)} - \frac{2\varphi_f^2 \alpha_{2f}}{(\varphi_f^2 - a_f^2)^2}\right](a_2 \cosh \varphi + a_3 \sinh \varphi)$,

$\Delta_{10} = \hat{T}\left[\frac{\alpha_{1p} a_4}{(\delta_p^2 - a_p^2)} - \frac{2\alpha_{2p}\delta_p a_5}{(\delta_p^2 - a_p^2)^2}\right]$,

$\Delta_{11} = \frac{\alpha_{1f}}{(\psi_f^2 - a_f^2)} - \frac{2\psi_f a_1 \alpha_{2f}}{(\psi_f^2 - a_f^2)^2} + \frac{a_2 \alpha_{1f}}{(\varphi_f^2 - a_f^2)} - \frac{2\varphi_f a_3 \alpha_{2f}}{(\varphi_f^2 - a_f^2)^2}$,

$\Delta_{12} = \left[\frac{\alpha_{2p}}{(\delta_p^2 - a_p^2)} - \frac{2\delta_p^2 \alpha_{2p}}{(\delta_p^2 - a_p^2)^2}\right] a_4 + \frac{a_5 \alpha_{1p}}{(\delta_p^2 - a_p^2)}$,

$\Delta_{13} = \frac{\alpha_{1f}\psi_f a_1 + \alpha_{2f}}{(\psi_f^2 - a_f^2)} - \frac{2\alpha_{2f}\psi_f^2}{(\psi_f^2 - a_f^2)^2} + \frac{\alpha_{1f}\varphi_f a_3 + \alpha_{2f} a_2}{(\varphi_f^2 - a_f^2)} - \frac{2a_2 \alpha_{2f}\varphi_f^2}{(\varphi_f^2 - a_f^2)^2}$,

$\Delta_{14} = a_p \cosh a_p$, $\Delta_{15} = a_p \sinh a_p$,

$$\Delta_{16} = \left. \frac{-\alpha_{2p}}{(\delta_p^2 - a_p^2)} + \frac{2\delta_p^2 \alpha_{2p}}{(\delta_p^2 - a_p^2)^2} \right] (a_4 \cosh \delta_p - a_5 \sinh \delta_p) - \Delta_{160},$$
$$\Delta_{160} = \frac{\delta_p(\alpha_{1p} - \alpha_{2p})}{(\delta_p^2 - a_p^2)} (a_5 \cosh \delta_p - a_4 \sinh \delta_p),\ \Delta_{17} = \hat{T} a_f \sinh a_f,$$
$$\Delta_{18} = a_p \cosh a_f,\ \Delta_{19} = \Delta_9 - a_f(\Delta_{10} - \Delta_{11}) \sinh a_f - (\Delta_{12} - \Delta_{13}) \cosh a_f$$

We get tMn by using the expression (23) as

$$M_{t1} = \frac{-\Lambda_1 - M_s a_f^2 \mho}{a_f^2(c_1 \cosh a_f + c_2 \sinh a_f + \Lambda_2 + \Lambda_3)} \tag{34}$$

where

$$\Lambda_1 = \psi_f^2(\cosh \psi_f + a_1 \sinh \psi_f) + \varphi_f^2(a_2 \cosh \varphi_f + a_3 \sinh \varphi_f),$$
$$\mho = c_{13} \cosh a_f z_f + c_{14} \sinh a_f z_f - \frac{1}{\tau_f} \left[\frac{\cosh \psi_f + a_1 \sinh \psi_f}{\psi_f^2 - a_f^2} + \frac{a_2 \cosh \varphi_f + a_3 \sinh \varphi_f}{\varphi_f^2 - a_f^2} \right],$$
$$\Lambda_2 = \frac{(\alpha_{2f} + \alpha_{1f})}{(\psi_f^2 - a_f^2)} (\cosh \psi_f + a_1 \sinh \psi_f) - \frac{2\psi_f \alpha_{2f}}{(\psi_f^2 - a_f^2)^2} (a_1 \cosh \psi_f + \sinh \psi_f),$$
$$\Lambda_3 = \frac{(\alpha_{2f} + \alpha_{1f})}{(\varphi_f^2 - a_f^2)} (a_2 \cosh \varphi_f + a_3 \sinh \varphi_f) - \frac{2\varphi_f \alpha_{2f}}{(\varphi_f^2 - a_f^2)^2} (a_3 \cosh \varphi_f + a_2 \sinh \varphi_f)$$

4.2 Case (ii): Adiabatic-Isothermal Boundary Condition

The lower porous boundary of the combined layer is adiabatic and the upper fluid boundary of the composite layer is isothermal. Using the temperature boundary conditions (26), the distributions $\theta_f(z_f)$ and $\theta_p(z_p)$are produced by solving (18) and (21):

$$\theta_f(z_f) = A_1[c_5 \cosh a_f z_f + c_6 \sinh a_f z_f + \Sigma_{f3}(z_f)] \tag{35}$$
$$\theta_p(z_p) = A_1[c_7 \cosh a_p z_p + c_8 \sinh a_p z_p + \Sigma_{p3}(z_p)] \tag{36}$$

where $\Sigma_{f3}(z_f) = A_1[\delta_{11} - \delta_{12} + \delta_{13} - \delta_{14}]$, $\Sigma_{p3}(z_p) = A_1[\delta_{15} - \delta_{16}]$,

$$\delta_{11} = \frac{(\alpha_{4f} z_f + \alpha_{3f})}{(\psi_f^2 - a_f^2)} (\cosh \psi_f z_f + a_1 \sinh \psi_f z_f),$$
$$\delta_{12} = \frac{2\psi_f \alpha_{4f}}{(\psi_f^2 - a_f^2)^2} (a_1 \cosh \psi_f z_f + \sinh \psi_f z_f),$$
$$\delta_{13} = \frac{(\alpha_{4f} z_f + \alpha_{3f})}{(\varphi_f^2 - a_f^2)} (a_2 \cosh \varphi_f z_f + a_3 \sinh \varphi_f z_f),$$
$$\delta_{14} = \frac{2\varphi_f \alpha_{4f}}{(\varphi_f^2 - a_f^2)^2} (a_3 \cosh \varphi_f z_f + a_2 \sinh \varphi_f z_f),$$
$$\delta_{15} = \frac{(\alpha_{3p} + \alpha_{4p} z_p)}{(\delta_p^2 - a_p^2)} (a_4 \cosh \delta_p z_p + a_5 \sinh \delta_p z_p),$$
$$\delta_{16} = \frac{2\alpha_{4p} \delta_p}{(\delta_p^2 - a_p^2)^2} (a_5 \cosh \delta_p z_p + a_4 \sinh \delta_p z_p),$$
$$\alpha_{3f} = R_{If}^* - 1,\ \alpha_{4f} = -2R_{If}^*,\ \alpha_{3p} = -(R_{Ip}^* + 1),\ \alpha_{4p} = -2R_{Ip}^*,$$

$c_5 = c_7\hat{T} + \Delta_{21} - \Delta_{22}, c_6 = \frac{1}{a_f}(c_8 a_p + \Delta_{23} - \Delta_{24}), c_7 = \frac{\Delta_{30}\Delta_{25} - \Delta_{27}\Delta_{29}}{\Delta_{28}\Delta_{25} + \Delta_{26}\Delta_{29}},$

$c_8 = \frac{\Delta_{30}\Delta_{26} + \Delta_{27}\Delta_{28}}{\Delta_{29}\Delta_{26} + \Delta_{25}\Delta_{28}}, \Delta_{20} = -[\delta_{17} + \delta_{18} + \delta_{19} + \delta_{20}],$

$\delta_{17} = \frac{\psi_f(\alpha_{4f} + \alpha_{3f})}{(\psi_f^2 - a_f^2)}(a_1 \cosh \psi_f + \sinh \psi_f),$

$\delta_{18} = \left[\frac{\alpha_{4f}}{(\psi_f^2 - a_f^2)} - \frac{2\psi_f^2 \alpha_{4f}}{(\psi_f^2 - a_f^2)^2}\right](\cosh \psi_f + a_1 \sinh \psi_f),$

$\delta_{19} = \frac{\varphi_f(\alpha_{4f} + \alpha_{3f})}{(\varphi_f^2 - a_f^2)}(a_3 \cosh \varphi_f + a_2 \sinh \varphi_f),$

$\delta_{20} = [\frac{\alpha_{4f}}{(\varphi_f^2 - a_f^2)} - \frac{2\varphi_f^2 \alpha_{4f}}{(\varphi_f^2 - a_f^2)^2}](a_2 \cosh \varphi_f + a_3 \sinh \varphi_f),$

$\Delta_{21} = \hat{T}\left[\frac{\alpha_{3p} a_4}{(\delta_p^2 - a_p^2)} - \frac{2\alpha_{4p}\delta_p a_5}{(\delta_p^2 - a_p^2)^2}\right],$

$\Delta_{22} = \frac{\alpha_{3f}}{(\psi_f^2 - a_f^2)} - \frac{2\psi_f a_1 \alpha_{4f}}{(\psi_f^2 - a_f^2)^2} + \frac{a_2 \alpha_{3f}}{(\varphi_f^2 - a_f^2)} - \frac{2\varphi_f a_3 \alpha_{4f}}{(\varphi_f^2 - a_f^2)^2},$

$\Delta_{23} = [\frac{\alpha_{4p}}{(\delta_p^2 - a_p^2)} - \frac{2\delta_p^2 \alpha_{4p}}{(\delta_p^2 - a_p^2)^2}]a_4 + \frac{a_5 \alpha_{3p}}{(\delta_p^2 - a_p^2)},$

$\Delta_{24} = \frac{\alpha_{3f}\psi_f a_1 + \alpha_{4f}}{(\psi_f^2 - a_f^2)} - \frac{2\alpha_{4f}\psi_f^2}{(\psi_f^2 - a_f^2)^2} + \frac{\alpha_{3f}\varphi_f a_3 + \alpha_{4f} a_2}{(\varphi_f^2 - a_f^2)} - \frac{2a_2 \alpha_{4f}\varphi_f^2}{(\varphi_f^2 - a_f^2)^2},$

$\Delta_{25} = \cosh a_p, \Delta_{26} = \sinh a_p,$

$\Delta_{27} = -\frac{\alpha_{3p} - \alpha_{4p}}{(\delta_p^2 - a_p^2)}(a_4 \cosh \delta_p - a_5 \sinh \delta_p) + \frac{2\delta_p \alpha_{4p}}{(\delta_p^2 - a_p^2)^2}(a_5 \cosh \delta_p - a_4 \sinh \delta_p),$

$\Delta_{28} = a_f \hat{T} \sinh a_f, \Delta_{29} = a_p \cosh a_f,$

$\Delta_{30} = \Delta_{20} - a_f(\Delta_{21} - \Delta_{22}) \sinh a_f - (\Delta_{23} - \Delta_{24}) \cosh a_f.$

We get tMn from (23) as

$$M_{t2} = \frac{-\Lambda_1 - M_s a_f^2 \mho}{a_f^2(c_5 \cosh a_f + c_6 \sinh a_f + \Lambda_4 + \Lambda_5)} \tag{37}$$

where $\Lambda_4 = \frac{(\alpha_{4f} + \alpha_{3f})}{(\psi_f^2 - a_f^2)}(\cosh \psi_f + a_1 \sinh \psi_f) - \frac{2\psi_f \alpha_{4f}}{(\psi_f^2 - a_f^2)^2}(a_1 \cosh \psi_f + \sinh \psi_f),$

$\Lambda_5 = \frac{(\alpha_{4f} + \alpha_{3f})}{(\varphi_f^2 - a_f^2)}(a_2 \cosh \varphi_f + a_3 \sinh \varphi_f) - \frac{2\varphi_f \alpha_{4f}}{(\varphi_f^2 - a_f^2)^2}(a_3 \cosh \varphi_f + a_2 \sinh \varphi_f).$

5 Results and Discussion

The tMns for two cases of thermal boundary conditions M_{t1} and M_{t2} are obtained as expressions of the depth ratio $\hat{d}$, the porous parameter β, the thermal ratio $\hat{T}$, the solutal ratio $\hat{S}$, sMn M_s and the Chandrasekhar number Q_f, the horizontal wave numbers a_f and a_p, the diffusivity ratio $\tau_f and \tau_p$, and R_{If}^* and R_{Ip}^* the modified internal Rayleigh numbers for region 1 and region 2. The thermal boundary conditions taken are adiabatic-adiabatic and that the combined layer is horizontally enclosed by the adiabatic boundaries and adiabatic-isothermal, the lower porous

Fig. 2 Comparison of tMn for case (i) adiabatic-adiabatic and case (ii) adiabatic-isothermal

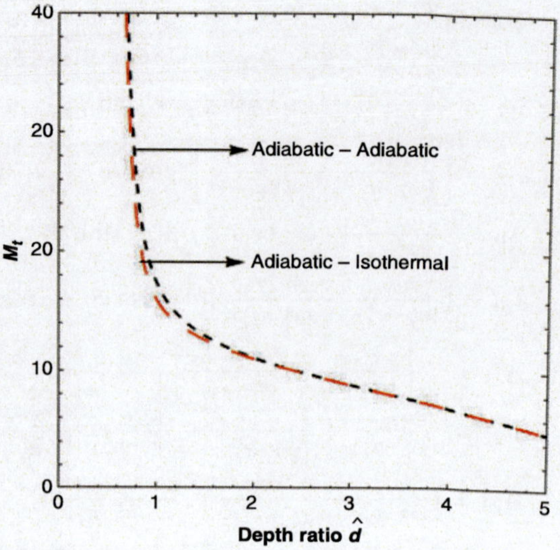

boundary of the composite system is adiabatic, and the upper fluid boundary of the composite system is isothermal. These tMns M_t are drawn versus the depth ratio$\hat{d}$ using Mathematica software. The other parameters are $a_f = 0.5$, $Q_f = 10$, $\beta = 1.0$, $\hat{\mu} = 2$, $\hat{S} = \hat{T} = 1$, $\tau_f = \tau_p = 0.25$, $M_s = 10$, $R^*_{If} = 1$ and $R^*_{Ip} = 1$, the effects of the various parameters are described in detail and represented in the graphs below.

Figure 2 represents the comparison of M_{t1} and M_{t2}, where M_t is the dependent variable and $\hat{d}$, the depth ratio, is the independent variable. The tMn decreases up to some value of the depth ratio, and later it increases as the value of the depth ratio also increases. This behavior is qualitatively the same for both types of thermal boundary combination (TBC). It is interesting to note that for larger values of depth ratios, the Marangoni numbers coincide and no change in them for $\hat{d} \geq 2$, i.e., for porous layer dominant (in depth) systems which is physically impressive as the TBC at the boundary of the porous layer is changed. But for the smaller depth ratio values, the thermal Marangoni number for case (i) is smaller than that for case (ii), indicating that the system with case (i) TBC is more stable.

For both situations of TBCs, the effects of the Chandrasekhar number Q_f on double-diffusive Marangoni convection (DDMC) (DDMC) are shown in Fig. 3. The values of Q_f that were used were 1, 10, and 100. The curves are diverging, showing that Q_f is more prominent at higher depth ratios, i.e., for the porous layer dominant composite system (PDCS). Because an increase in the value of Q_f raises the tMns for a certain depth ratio, the DDC can be advanced by lowering the values of Q_f, and the system can be destabilized. This parameter's effect is analogous to that of TBCs.

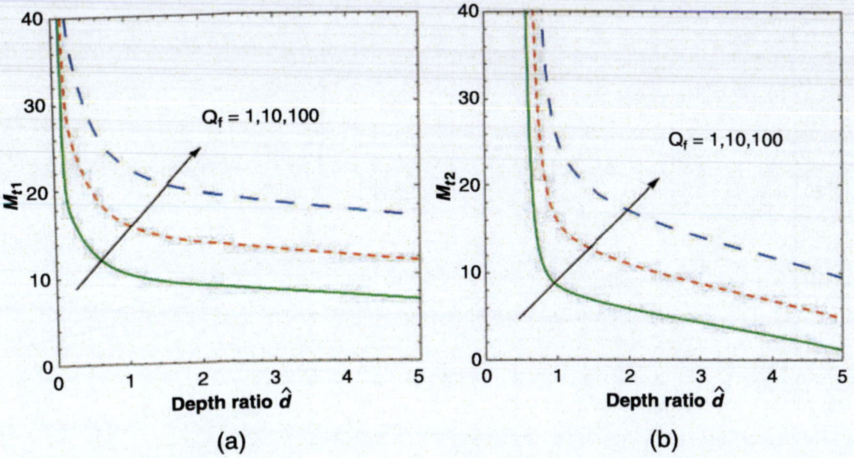

Fig. 3 Variations of $Q_f = 1, 10, 100$ on tMn when $a_f = 0.5$, $\beta = 1.0$, $\hat{\mu} = 2$, $\hat{S} = \hat{T} = 1$, $\tau_f = \tau_p = 0.25$, $M_s = 10$, $R^*_{If} = 1$ and$R^*_{Ip} = 1$

The effect of a modified internal Rayleigh number R^*_{If} on the tMn is shown in Fig. 4 for both the TBCs $R^*_{If} = -1, 0$ and 1. For larger depth ratio values, i.e., for porous layer dominant composite system (PDCS), the curves diverge, reflecting the importance of the modified internal Rayleigh number. The tMn increases when the value of R^*_{If} is increased, that is, from sink to source, for a certain depth ratio. As a result, the DDMC in the presence of a magnetic field can be delayed by raising the value of R^*_{If}. As a result, the heat absorption stabilizes the system. For TBCs, the effect of this parameter is comparable.

The effect of a modified internal Rayleigh number for porous layer R^*_{Ip} on the tMn is shown in Fig. 5 for $R^*_{Ip} = -1, 0$ and 1. The Marangoni number grows as the amount of R^*_{Ip} is increased from sink to source; hence, the DDMC in the presence of a magnetic field can be delayed by raising the value of R^*_{Ip}. As a result, the heat absorption stabilizes the system. This parameter has the same effect in both scenarios of TBCs. Also, as shown in the diagram, this parameter is effective for some modest depth ratios, that is, for PDCS in both scenarios of TBCs.

The effect of solute Marangoni number M_s on the Marangoni number is similar for both the cases of TBCs, which is exhibited in Fig. 6 for $M_s = 5, 10$ and 50. On increasing the values of M_s, the tMn for DDMC in the presence of magnetic field increases; hence, the DDMC can be delayed by increasing the values of M_s. The diverging curves reveal that the effect of sMn is intensive for larger values of depth ratios, that is, for PDCS. The effect of this parameter is comparable to the cases of TBCs.

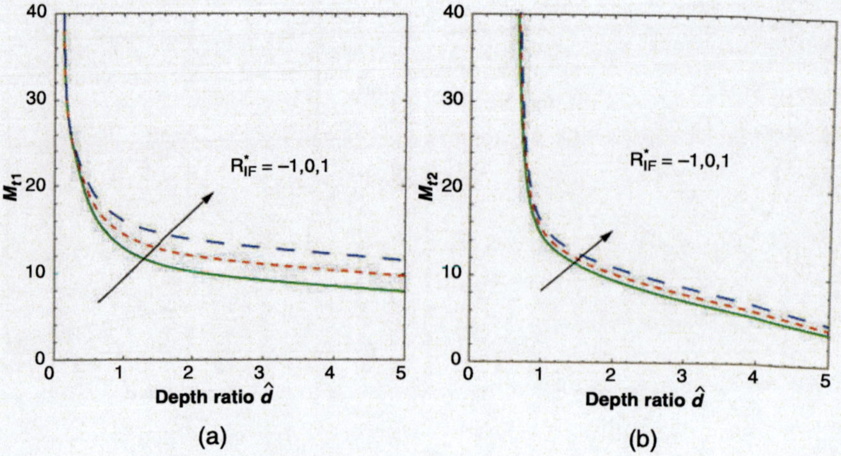

Fig. 4 Variations of $R^*_{If} = -1, 0, 1$ on tMn when $a_f = 0.5$, $Q_f = 10$, $\beta = 1.0$, $\hat{\mu} = 2$, $\hat{S} = \hat{T} = 1$, $\tau_f = \tau_p = 0.25$, $M_s = 10$ and $R^*_{Ip} = 1$

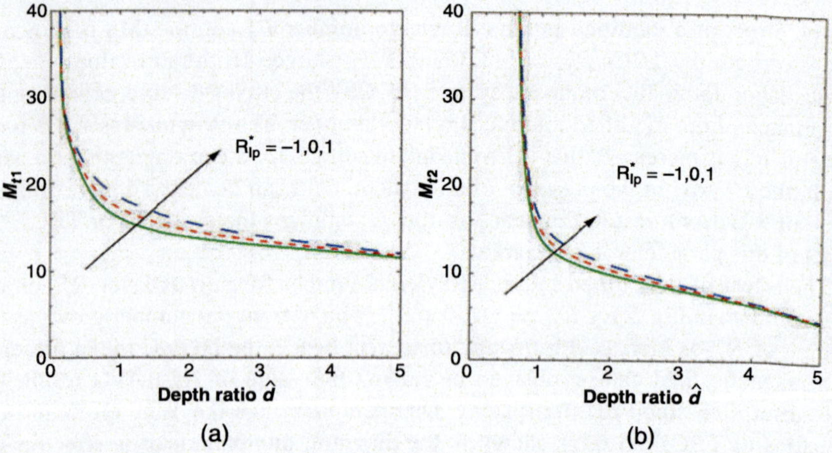

Fig. 5 Variations of $R^*_{Ip} = -1, 0, 1$ for porous region on tMn when $a_f = 0.5$, $Q_f = 10$, $\beta = 1.0$, $\hat{\mu} = 2$, $\hat{S} = \hat{T} = 1$, $\tau_f = \tau_p = 0.25$, $M_s = 10$ and $R^*_{If} = 1$

6 Conclusion

In the current study, the impact of a heat source and magnetic field on onset convection is thoroughly investigated for two different boundary configurations. Because of this, the composite layer system is stable and can be used in adiabatic-adiabatic thermal boundary conditions, where convection needs to be regulated. The following are the key findings from the current study:

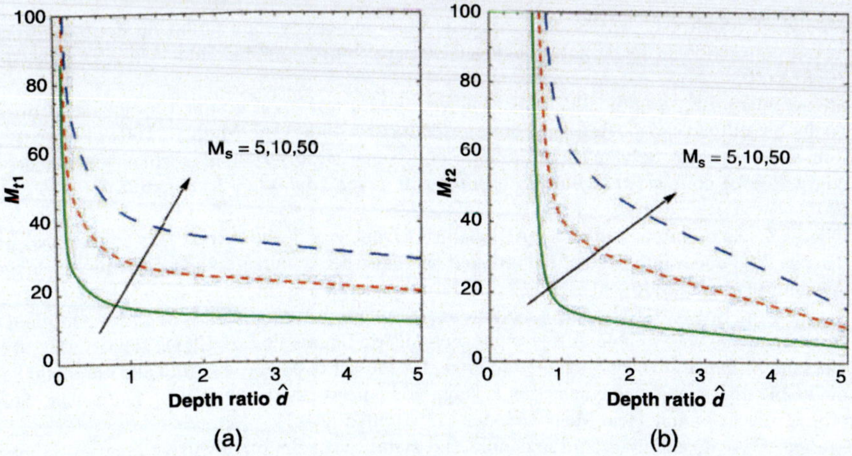

Fig. 6 Variations of $M_s = 5, 10, 50$ on tMn when $a_f = 2.5$, $Q_f = 10$, $\beta = 1.0$, $\hat{\mu} = 2$, $\hat{S} = \hat{T} = 1$, $\tau_f = \tau_p = 0.25$, $R^*_{If} = 1$ and $R^*_{Ip} = 1$

(i) When compared to the case (ii) thermal boundary condition, the tMn for the case (i) thermal boundary condition is large. As a result, in the case of case (i), the fluid-porous system is stable and can be used in circumstances where convection must be controlled.
(ii) All the physical parameters are effective for the larger values for depth ratios that are for the PDCS.
(iii) In the present study, by increasing the values of the Chandrasekhar number Q_f, the modified internal Rayleigh numbers R^*_{If}, R^*_{Ip}, and the solute Marangoni number M_s in the presence of magnetic field, the thermal Marangoni numbers, i.e., to stabilize the system, so the onset of DDMC is delayed.
(iv) The DDMC in the combined system can be controlled by selecting the proper values for the physical parameters. Results are in good accordance with earlier work.

References

1. Sumithra, R., Manjunatha, N.: Analytical study of surface tension driven magneto convection in a composite layer bounded by adiabatic boundaries. Int. J. Eng. Innov. Technol. **1**(6), 249–257 (2012)
2. Sankar, M., Park, Y., Lopez, J.M., Younghae D.: Double-diffusive convection from a discrete heat and solute source in a vertical porous annulus. Transp. Porous Med. **91**, 753–775 (2012)
3. Jagadeesha, R.D., Prasanna, B.M.R., Sankar, M.: Double diffusive convection in an inclined parallelogrammic porous enclosure. Proc. Eng. **127**, 1346–1353 (2015)
4. Pushpa, B.V., Sankar, M., Makinde, O.D.: Optimization of thermosolutal convection in vertical porous annulus with a circular baffle, Thermal Sci. Eng. Progress **20**, 100735 (2020)

5. Moolya, S., Satheesh, A.: Role of magnetic field and cavity inclination on double diffusive mixed convection in rectangular enclosed domain. Int. Commun. Heat Mass Transfer **118**, 104814 (2020)
6. Shivakumara, I.S., Raghunatha, K.R., Pallavi, G.: Intricacies of coupled molecular diffusion on double diffusive viscoelastic porous convection. Results Appl. Math. **7**, 100124 (2020)
7. Sumithra, R., Arul Selvamary, T.: Single component Darcy-Benard surface tension driven convection of couple stress fluid in a composite layer. Malaya J. Matematik **9**(1), 797–804 (2021)
8. Qiang, X., Siddique, I., Sadiq, K., Ali Shah, N.: Double diffusive MHD convective flows of a viscous fluid under influence of the inclined magnetic field, source/sink and chemical reaction. Alexandria Eng. J. **59**(6), 4171–4181 (2021)
9. Tanveer, A., Hayat, T., Alsaedi, A.: Numerical simulation for peristalsis of Sisko nanofluid in curved channel with double diffusive convection. Ain Shams Eng. J. **12**(3), 3195–3207 (2021)
10. Hussain, S., Jamal, M., Pekmen Geridonmez, B.: Impact of power law fluid and magnetic field on double diffusive mixed convection in staggered porous cavity considering Dufour and Soret effects. Int. Commun. Heat Mass Transfer **121**, 105075 (2021)
11. Meften, G.A.: Conditional and unconditional stability for double diffusive convection when the viscosity has a maximum. Appl. Math. Comput. **392**, 125694 (2021)
12. Manjunatha, N., Sumithra, R., Vanishree, R.K.: Influence of constant heat source/sink on non-Darcian-Bnard double diffusive Marangoni convection in a composite layer system. JAMI: J. Appl. Math. Inform. **40**(1–2), 99–115 (2022)
13. Manjunatha, N., Sumithra, R.: Effects of heat source/sink on Darcian-Bnard-Magneto-Marangoni convection in a composite layer subjected to non uniform temperature gradients. TWMS J. Appl. Eng. Math. **12**(3), 669–684 (2022)
14. Shiva kumara, I.S., Suma, S.P., Krishna, B.: Onset of surface tension driven convection in superposed layers of fluid and saturated porous medium. Arch. Mech. **58**(2), 71–92 (2006)

Peristalsis and Taylor Dispersion of Solute in the Flow of Casson Fluid

P. Nagarani and Victor M. Job

Mathematics Subject Classification (2010) 76A05, 76R50, 76Z05

1 Introduction

The study of peristaltic flow phenomena is important for understanding fluid flows in renal, digestive, cardiovascular and lymphatic systems. It is also useful for the design and operation of peristaltic pumps and heart-long machines. Consequently, much research has been conducted on peristaltic flows in recent years [1–5]. In particular, drug transport in physiological systems and the spread of pollutants in the environment can be described using mathematical models for solute dispersion in moving fluids [6]. One such model was constructed by Marbach and Alim [7], who examined Taylor dispersion [8] effects on the transport of solute in a cylindrical tube using the invariant manifold technique. The impact of zero wall mass flux, passive absorption of solute, peristalsis and Navier slip on long-time solute advection and dispersion were considered in the cases of Newtonian and Herschel-Bulkley fluids. The authors determined an optimal axial flow velocity so that Taylor dispersion is maximized. It was also identified that wall slip has no significant influence on dispersion. In the case of peristaltic flow, it was found that the solute dispersion can be adjusted according to the contraction frequency and wavelength of the tube wall. Chakrabarti and Saintillan [9] utilized the generalized Taylor dispersion approach to explore peristaltic flow and solute dispersion in a Newtonian fluid through a flexible channel. An analytical solution was obtained under the long-wavelength assumption using the regular perturbation method, whereas the finite volume method was used to obtain a numerical solution when the channel aspect ratio is finite. It was determined that the solute dispersivity is enhanced by raising the value of the Péclet number Pe. Furthermore, an increase in the wave amplitude

P. Nagarani (✉) · Victor M. Job
Department of Mathematics, The University of the West Indies, Kingston-7, Jamaica
e-mail: victor.job@uwimona.edu.jm

F. Kamalov et al. (eds.), *Advances in Mathematical Modeling and Scientific Computing*, Trends in Mathematics, https://doi.org/10.1007/978-3-031-41420-6_21

causes the long-time dispersion to increase for small values of Pe; this trend is reversed for large values of Pe. Ponalagusamy and Murugan [10] investigated peristaltic electro-magnetohydrodynamic channel flow of a Jeffrey fluid with solute dispersion and first-order chemical reaction. It was determined that the effective dispersion coefficient is reduced with increases in the chemical reaction parameter and Hartmann number and is increased when the electrokinetic and electro-osmotic parameters increase.

One type of non-Newtonian fluid that is frequently used in the mathematical modelling of blood flow in arterioles and capillaries is the Casson fluid [11]. One of the early theoretical studies on peristaltic Casson fluid flows was conducted by Srivastava and Srivastava [12]. In their work, a two-layer blood flow through a tube was considered, where the core fluid is a Casson fluid and the peripheral fluid is a Newtonian fluid. Nagarani and Lewis [1] examined peristaltic Casson fluid flow in an annular region. The long-wavelength approximation was taken for low Reynolds number, and the flow velocity, pressure rise, flow rate, wall friction and stream function were obtained and the impact of pertinent parameters were investigated. Peristaltic Casson fluid flow within a flexible channel under the assumptions of long-wavelength and low Reynolds number was investigated by Devaki et al. [13], and the impact of elastic wall properties, fluid slip and heat transfer were explored. The effects of electro-osmosis on the peristaltic flow of Casson fluid flow through a flexible tube with heated sinusoidal walls were considered by Saleem et al. [3]. The long-wavelength approximation and low Reynolds number were assumed, and an analytical solution to the velocity, streamlines, pressure gradient and temperature was determined.

To the best of the authors' knowledge, there is no existing study in which the Taylor dispersion of solute in a Casson fluid under peristaltic flow is investigated. Consequently, the present manuscript offers an original contribution to this gap in the literature by exploring the effect of peristalsis on Taylor dispersion in the flow of Casson fluid through a tube.

2 Mathematical Formulation

2.1 *Peristaltic Flow of Casson Fluid*

In the present work, the axially symmetric, laminar and incompressible flow of Casson fluid through a horizontal distensible tube is considered as shown in Fig. 1. We use a cylindrical polar coordinate system (R, θ, X), where R and X are the radial and axial coordinates, respectively, and θ is the azimuthal angle. Hence, governing equations for peristaltic Casson fluid flow in the fixed frame are as follows.

Conservation of mass equation:

$$\frac{\partial U}{\partial X} + \frac{1}{R}\frac{\partial}{\partial R}(RV) = 0 \tag{1}$$

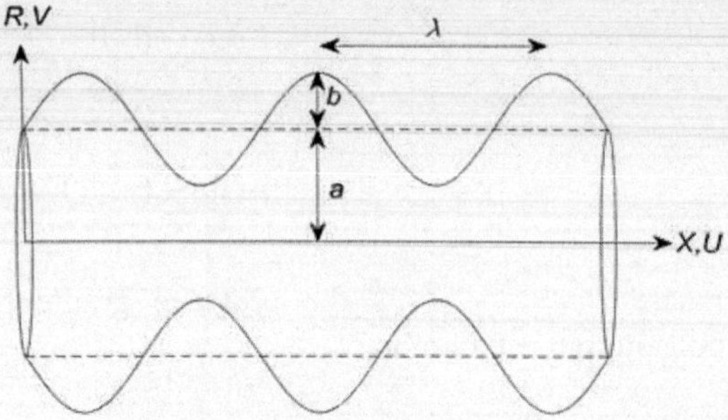

Fig. 1 Schematic diagram of the model

Momentum equation:

$$\rho\frac{\partial U}{\partial t} = -\frac{\partial p}{\partial X} - \frac{1}{R}\frac{\partial}{\partial R}(R\, T_{RX}) \tag{2}$$

Here, the symbols U, V, p, T_{RX} and ρ denote the axial velocity, transverse velocity, pressure, shear stress and density of the fluid, respectively, and t is time. The tube radius is given by

$$H(X,t) = a + b\sin\frac{2\pi}{\lambda}(X - ct), \tag{3}$$

where a is the mean half-width of the passage, λ is the wavelength, b is the amplitude and c is the wave propagation speed.

We define the following transformation between the fixed frame $O(X, R)$ and the wave frame $O(x, r)$:

$$x = X - ct,\ \ r = R,\ \ u(x,r) = U(X, R, t) - c,$$

$$v(x,r) = V(X, R, t),\ \ \tau_{rx}(x,r) = T_{RX}(X, R, t), \tag{4}$$

where u is the axial velocity, v is the transverse velocity and τ_{rx} is the shear stress in the wave frame. Moreover, the governing equations are non-dimensionalized according to the dimensionless quantities below:

$$\hat{x} = \frac{x}{\lambda},\ \hat{r} = \frac{r}{a},\ \hat{t} = \frac{ct}{\lambda}\ \hat{u} = \frac{u}{c},\ \hat{v} = \frac{v}{c\delta},\ \delta = \frac{a}{\lambda},$$

$$\hat{p} = \frac{p}{\mu_\infty c\lambda/a^2},\ \hat{\tau}_{\hat{r}\hat{x}} = \frac{\tau_{rx}}{\mu_\infty c/a},\ \hat{\tau}_y = \frac{\tau_y}{\mu_\infty c/a}, \tag{5}$$

where τ_y is the yield stress and μ_∞ is the Newtonian viscosity. After dropping the 'hats' notation for convenience, and using the long-wavelength and low Reynolds number assumptions, the dimensionless governing equations are

$$\frac{\partial u}{\partial x} + \frac{1}{r}\frac{\partial}{\partial r}(rv) = 0, \tag{6}$$

$$-\frac{1}{r}\frac{\partial}{\partial r}(r\tau_{rx}) + P = 0, \tag{7}$$

where $P = -\frac{\partial p}{\partial x}$ is the pressure gradient. The dimensionless Casson fluid constitutive equation is

$$\left|\tau_{rx}^{1/2}\right| = \tau_y^{1/2} + \left|-\frac{\partial u}{\partial r}\right|^{1/2} \quad \text{if} \ \ \tau_{rx} \geq \tau_y, \tag{8}$$

$$\frac{\partial u}{\partial r} = 0 \text{ if } \ \tau_{rx} \leq \tau_y. \tag{9}$$

The dimensionless boundary conditions are

$$\tau_{rx} \text{ and } v \text{ are finite at } r = 0, \tag{10}$$

$$u|_{r=h} = -1. \tag{11}$$

In Eq. (11), the symbol

$$h = 1 + \phi \sin 2\pi x \tag{12}$$

denotes the non-dimensionalized peristaltic wave function, and $\phi = \frac{b}{a}$ is the amplitude ratio.

2.2 *Solute Dispersion in Casson Fluid*

The following convection-diffusion equation is used to model the solute dispersion in a flowing Casson fluid with concentration χ in the fixed frame:

$$\frac{\partial \chi}{\partial t} + U\frac{\partial \chi}{\partial X} + V\frac{\partial \chi}{\partial R} = D\left[\frac{\partial^2 \chi}{\partial X^2} + \frac{1}{R}\frac{\partial}{\partial R}\left(R\frac{\partial \chi}{\partial R}\right)\right], \tag{13}$$

where D is the solutal mass diffusivity. We use the transformation

$$C(x, r) = \chi(X, R, t)$$

for the concentration C in the wave frame $O(x, r)$, and define the dimensionless solute concentration:

$$\hat{C} = \frac{C}{C_0}, \tag{14}$$

where C_0 is the initial solute concentration. Assume that the wave propagation speed c is equal to the mean velocity U_m. After dropping the 'hat' notation, and using the long-wavelength approximation ($\delta \ll 1$) and Taylor's assumptions, Eq. (13) becomes

$$Pe_m\overline{U}\frac{\partial C}{\partial x} = \frac{1}{r}\frac{\partial}{\partial r}\left(r\frac{\partial C}{\partial r}\right), \tag{15}$$

where $\overline{U} = (U - U_m)$, $Pe_m = \left(\frac{ca}{D}\right)\frac{a}{\lambda}$ is the modified Peclet number. We assume that $\frac{\partial C}{\partial x}$ is constant; hence by taking

$$T_1 = Pe_m\frac{\partial C}{\partial x}, \tag{16}$$

Eq. (15) becomes

$$T_1\overline{U} = \frac{1}{r}\frac{\partial}{\partial r}\left(r\frac{\partial C}{\partial r}\right). \tag{17}$$

3 Method of Solution

3.1 *Flow Velocity of Casson Fluid*

We use the finiteness condition on τ_{rx} from Eq. (10) in the momentum Eq. (7) to obtain

$$\tau_{rx} = \frac{1}{2}Pr. \tag{18}$$

From Eqs. (8)–(9), we have

$$r_p = \frac{2\tau_y}{|P|}. \tag{19}$$

Substituting Eqs. (18) and (19) into Casson's constitutive equations in (8) and (9), and applying the boundary conditions (10) and (11), the axial velocity u is obtained as

$$u(x,r) = \begin{cases} u^-(x,r), \; 0 \le r \le r_p \\ u^+(x,r), \; r_p \le r \le h \end{cases}, \tag{20}$$

where

$$u^-(x,r) = \frac{P}{4}\left[h^2 + 2hr_p - \frac{8}{3}h^{3/2}r_p^{1/2} - \frac{1}{3}r_p^2\right] - 1$$

and

$$u^+(x,r) = \frac{P}{4}\left[h^2 - r^2 + 2r_p(h-r) - \frac{8}{3}r_p^{1/2}\left(h^{3/2} - r^{3/2}\right)\right] - 1,$$

represent the axial velocity in plug and shear regions, respectively, in the wave frame. The mean axial velocity is given by

$$u_m = \frac{1}{4}P\left(\frac{1}{2}h^2 - \frac{8}{7}r_p^{1/2}h^{3/2} + \frac{2}{3}r_p h - \frac{1}{42}\frac{r_p^4}{h^2}\right) - 1. \quad (21)$$

The transverse velocity v is determined using the conservation of mass Eq. (6) and the finiteness condition on v from Eq. (10); this yields

$$v(x,r) = \begin{cases} v^-(x,r),\ 0 \le r \le r_p \\ v^+(x,r),\ r_p \le r \le h \end{cases}, \quad (22)$$

where

$$v^-(x,r) = -\frac{P}{4}\left[h + r_p - 2h^{1/2}r_p^{1/2}\right]h'r - \frac{P'}{8}\left[h^2 - \frac{4}{3}h^{3/2}r_p^{1/2} + \frac{1}{3}r_p^2\right]r$$

and

$$v^+(x,r) = -\frac{P}{4}\left[h + r_p - 2r_p^{1/2}h^{1/2}\right]h'r$$
$$-\frac{P'}{16}\left[(2h^2 - r^2)r - \frac{8}{3}r_p^{1/2}\left(h^{3/2} - \frac{4}{7}r^{3/2}\right)r + \frac{1}{7}\frac{r_p^4}{r}\right].$$

The volume flow rate Q is calculated as

$$Q = \int_0^h u\,r\,dr = \int_0^{r_p} u^-\,r\,dr + \int_{r_p}^h u^+\,r\,dr \quad (23)$$

and is given by

$$\frac{P}{4}\left\{\frac{1}{4}h^4 + \frac{1}{3}h^3 r_p - \frac{4}{7}r_p^{\frac{1}{2}} h^{\frac{7}{2}} - \frac{1}{84}r_p^4\right\} - \frac{1}{2}h^2 = Q. \tag{24}$$

A numerical solution to the pressure gradient P can be computed from the nonlinear Eq. (24) for constant flow rate Q using the fixed point iteration technique.

3.2 Convection-Diffusion

We solve Eq. (17) using Eqs. (20) and (21) with the boundary condition

$$\frac{\partial C}{\partial r} = 0 \text{ at } r = h \tag{25}$$

to obtain the expression

$$C(x,r) = \begin{cases} C^-(x,r), & 0 \le r \le r_p \\ C^+(x,r), & r_p \le r \le h \end{cases}, \tag{26}$$

where

$$C^-(x,r) = \frac{PT_1}{4}\left[\frac{1}{2}h^2 - \frac{32}{21}r_p^{1/2}h^{3/2} + \frac{4}{3}r_p h - \frac{1}{3}r_p^2 + \frac{1}{42}\frac{r_p^4}{h^2}\right]\frac{r^2}{4} + C_{r0}(x)$$

and

$$C^+(x,r) = \frac{PT_1}{4}\left[\frac{1}{8}r^2h^2 - \frac{1}{16}r^4 + 2r_p\left(\frac{1}{6}r^2h - \frac{1}{9}r^3\right) - \frac{8}{3}r_p^{1/2}\left(\frac{1}{7}r^2h^{3/2} - \frac{4}{49}r^{7/2}\right)\right.$$

$$\left. + \frac{1}{168}r^2\frac{r_p^4}{h^2} - \frac{\ln r}{84}r_p^4 + \frac{\ln(r_p)}{84}r_p^4 - \frac{115}{7056}r_p^4\right] + C_{r0}(x).$$

Given that the mean concentration C_m is

$$C_m(x) = \frac{2}{h^2}\int_0^h C\, r\, dr = \frac{2}{h^2}\left[\int_0^{r_p} C^- r dr + \int_{r_p}^h C^+ r dr\right], \tag{27}$$

the function C_{r0} can be expressed as

$$C_{r0}(x) = C_m(x) - \frac{2}{h^2}\frac{PT_1}{4}\left[\frac{1}{48}h^6 - \frac{30}{539}r_p^{\frac{1}{2}}h^{\frac{11}{2}} + \frac{7}{180}r_p h^5\right.$$

$$+\left(\frac{\ln(r_p)}{168}-\frac{1}{336}(2\ln h-1)-\frac{47}{7056}\right)h^2r_p^4$$
$$-\left(\frac{\ln(r_p)}{168}-\frac{1}{336}(2\ln(r_p)-1)-\frac{433}{18480}\right)r_p^6\Bigg]. \quad (28)$$

The dimensional mass flow rate of solute is

$$q(x)=\int_0^h C\overline{U}r dr=\int_0^{r_p} C^-\overline{U}r dr+\int_{r_p}^h C^+\overline{U}r dr=T_1(I_1+I_2), \quad (29)$$

where

$$I_1=\frac{P^2}{16}\left[\frac{r_p^4}{16}\left(\frac{1}{2}h^2-\frac{32}{21}r_p^{1/2}h^{3/2}+\frac{4}{3}r_ph-\frac{1}{3}r_p^2+\frac{1}{42}\frac{r_p^4}{h^2}\right)^2\right]$$

and

$$I_2=\frac{P^2}{451584}\Bigg\{4r_p^8\left(\ln r_p-\frac{213751}{3640}-\ln h\right)+\frac{1512}{55}\frac{r_p^{10}}{h^2}+128\frac{r_p^{17/2}}{h^{1/2}}$$
$$-413h^4r_p^4-\frac{140728}{55}h^2r_p^6-\frac{147}{2}h^8-112\frac{r_p^9}{h}+\frac{84672}{55}hr_p^7-\frac{95936}{15}h^3r_p^5$$
$$+\frac{201216}{77}h^{7/2}r_p^{9/2}+7168h^{5/2}r_p^{11/2}-\frac{96768}{55}h^{3/2}r_p^{13/2}-\frac{r_p^{12}}{h^4}$$
$$+\frac{20608}{55}h^{15/2}r_p^{1/2}+\frac{451584}{715}h^{13/2}r_p^{3/2}-\frac{3136}{15}h^6r_p^2-\frac{279184}{385}h^7r_p\Bigg\}.$$

The effective diffusion coefficient for a Casson fluid is given by

$$D_{\text{eff}}(x)=-q\left(\frac{\partial C}{\partial x}\right)^{-1}=-Pe_m(I_1+I_2). \quad (30)$$

In the case of Newtonian fluid ($\tau_y=0$), the effective diffusion coefficient is

$$D_0(x)=-q\left(\frac{\partial C}{\partial x}\right)^{-1}=\frac{P^2}{6144}Pe_mh^8. \quad (31)$$

4 Results and Discussion

We discuss the peristaltic Taylor dispersion of solute in a Casson fluid for different values of yield stress τ_y, wave amplitude ratio ϕ and volumetric flow rate Q. Unless otherwise stated, the values $\tau_y = 0.05$, $\phi = 0.15$, $Pe_m = 10$ and $Q = 0.05$ are used for the graphical results. The effective diffusion coefficient ratio D_{eff}/D_0 is displayed in Figs. 2, 3, and 4. In these figures, $D_{eff}/D_0 < 1$ for $\tau_y \neq 0$, since the dispersion of solute is lower in Casson fluid than in Newtonian fluid due to a higher viscosity in the Casson fluid case.

Figure 2 displays the periodic axial variation of the effective diffusion coefficient ratio D_{eff}/D_0 for different values of the yield stress τ_y. We found that D_{eff}/D_0 decreases when τ_y is increased; this is due to an increase in the effective Casson

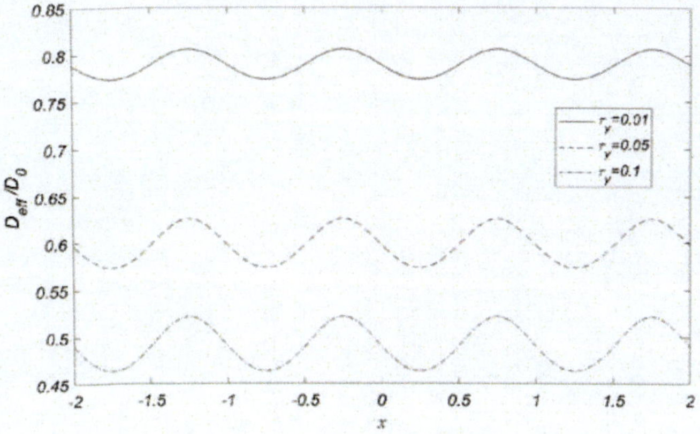

Fig. 2 Plot of D_{eff}/D_0 vs x for different τ_y with $\phi = 0.15$, $Pe_m = 10$ and $Q = 0.05$

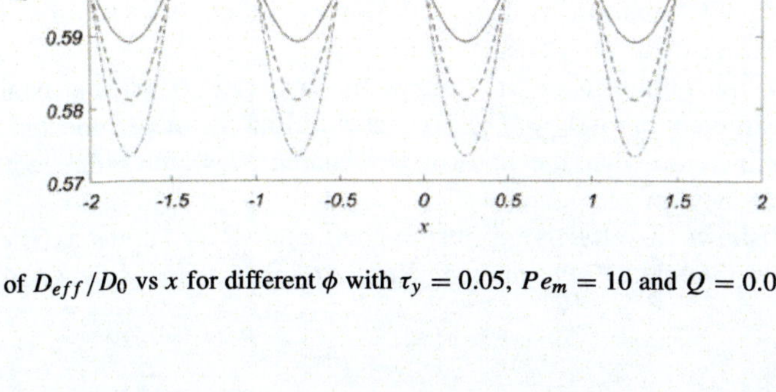

Fig. 3 Plot of D_{eff}/D_0 vs x for different ϕ with $\tau_y = 0.05$, $Pe_m = 10$ and $Q = 0.05$

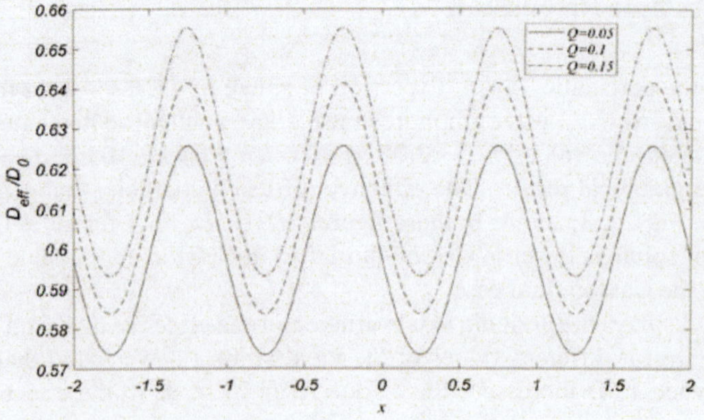

Fig. 4 Plot of D_{eff}/D_0 vs x for different Q with $\tau_y = 0.05$, $\phi = 0.15$ and $Pe_m = 10$

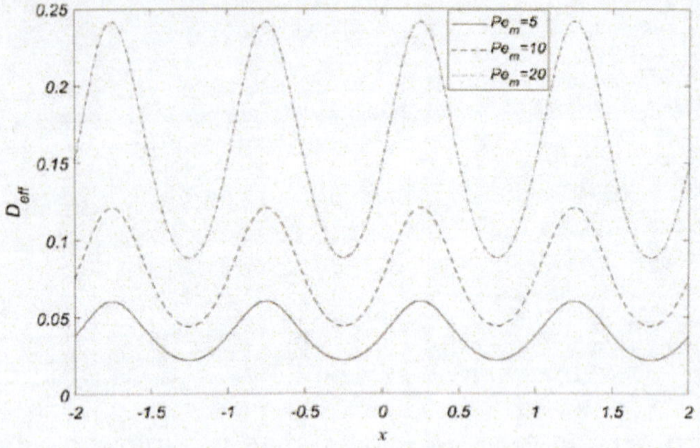

Fig. 5 Plot of D_{eff} vs. x for different Pe_m with $\tau_y = 0.05$, $\phi = 0.15$ and $Q = 0.05$

fluid viscosity and a corresponding reduction in flow velocity with increased yield stress.

The impact of wave amplitude ratio ϕ on the effective diffusion coefficient ratio D_{eff}/D_0 is depicted in Fig. 3. The amplitude of D_{eff}/D_0 is enhanced with increasing values of ϕ as a result of the increased peristaltic effects on the fluid within the flexible tube as ϕ is increased. However, there is no significant effect of ϕ on the mean value of D_{eff}/D_0.

In Fig. 4, the influence of the volumetric flow rate Q on the axial effective diffusion coefficient ratio D_{eff}/D_0 is displayed. Both the magnitude and amplitude of D_{eff}/D_0 increase when the value of Q is increased, which is due to an increase in velocity as the flow rate increases.

Figure 5 shows the effective diffusion coefficient D_{eff} vs. the axial coordinate x for varying modified Péclet number Pe_m. An increase in D_{eff} is noticed when

Pe_m increases, and this occurs as a result of enhanced solutal convection within the Casson fluid. An increase in Pe_m also increases the oscillation amplitude of the effective diffusion coefficient D_{eff}. Moreover, we note that the ratio D_{eff}/D_0 is unchanged since this ratio is independent of Pe_m.

5 Conclusions

In this paper, Taylor dispersion of solute during the peristaltic flow of Casson fluid was explored analytically. The effects of yield stress, peristaltic wave amplitude, modified Péclet number and volumetric flow rate on the effective diffusion coefficient were investigated. It was determined from the graphical results that the effective diffusion coefficient can be enhanced by increasing the yield stress, wave amplitude of peristalsis, modified Péclet number and volumetric flow rate.

References

1. Nagarani, P., Lewis, A.: Peristaltic flow of a Casson fluid in an annulus. Korea-Aust. Rheol. J. **24**(1), 1–9 (2012)
2. Shehzad, S.A., Abbasi, F.M., Hayat, T., Alsaadi, F., Mousa, G.: Peristalsis in a curved channel with slip condition and radial magnetic field. Int. J. Heat Mass Transf. **91**, 562–569 (2015)
3. Saleem, S., Akhtar, S., Nadeem, S., Saleem, A., Ghalambaz, M., Issakhov, A.: Mathematical study of Electroosmotically driven peristaltic flow of Casson fluid inside a tube having systematically contracting and relaxing sinusoidal heated walls. Chin. J. Phys. **71**, 300–311 (2021)
4. Asghar, Z., Waqas, M., Gondal, M.A., Khan, W.A.: Electro-osmotically driven generalized Newtonian blood flow in a divergent microchannel. Alexandria Eng. J. **61**(6), 4519–4528 (2022)
5. Palmada, N., Cater, J.E., Cheng, L.K., Suresh, V.: Experimental and computational studies of peristaltic flow in a duodenal model. Fluids **7**, 40 (2022)
6. Rana, J., Murthy, P.V.S.N.: Unsteady solute dispersion in small blood vessels using a two-phase Casson model. Proc. R. Soc. A **473**, 20170427 (2017)
7. Marbach, S., Alim, K.: Active control of dispersion within a channel with flow and pulsating walls. Phys. Rev. Fluids **4**, 114202 (2019)
8. Taylor, G.: Dispersion of soluble matter in solvent flowing slowly through a tube. Proc. R. Soc. London Ser. A **219**(1137), 186–203 (1953)
9. Chakrabarti, B., Saintillan, D.: Shear-induced dispersion in peristaltic flow. Phys. Fluids **32**, 113102 (2020)
10. Ponalagusamy, R., Murugan, D.: Impact of electro-magnetohydrodynamic nature on dispersion of solute in the peristaltic mechanism. J. Phys.: Conf. Ser. **1850**, 012097 (2021)
11. Fung, Y.C.: Biomechanics: Mechanical Properties of Living Tissues. Springer Science + Business Media, New York (1993)
12. Srivastava, L.M., Srivastava, V.P.: Peristaltic transport of blood: Casson model-II. J. Biomech. **17**(11), 821–829 (1984)
13. Devaki, P., Sreenadh, S., Vajravelu, K., Prasad, K.V., Vaidya, H.: Wall properties and slip consequences on peristaltic transport of a Casson liquid in a flexible channel with heat transfer. Appl. Math. Nonlinear Sci. **3**(1), 277–290 (2018)

An Aligned Magnetic Field Effect on Unsteady Heat and Mass Transfer Flow of Non-Newtonian Fluid Through Porous Medium

S. Rama Mohan, N. Maheshbabu, and M. Eswara Rao

1 Introduction

When a force is applied to non-Newtonian fluids, the viscosity may change, becoming more liquid or perhaps more solid. Fluid viscosity depends linearly on shear stress. Examples of non-Newtonian fluids are blood, honey, ketchup, toothpaste, peanut oil, paints, and inks. The typical viscous fluid model is unable to explain the ductile fracture characteristic of non-Newtonian fluids; however, Jeffrey's liquid model can. The fluid model proposed by Jeffrey is a useful tool for describing the class of non-Newtonian fluids with their distinctive recollection time scales, also referred to as relaxation time. Several authors contributed their analysis of non-Newtonian fluids with the assistance of mathematical modeling. Rundora et al. [1] studied non-Newtonian fluid unstable MHD reactive flow over a porous filled medium with irregular circumferential layer examples. Eddy et al. [2] have discussed the effects of an aligned magnetic field on the Casson liquid flow via a vertical oscillation plate in a porous medium. Jena et al. [3] described that a chemical reaction has an impact on the flow of Jeffery MHD fluid on a stretched surface via a porous medium. An associated irresistible field has an impact on the unstable fluid flow of Casson liquid over a stretch sheet examined by Sailaja et al. [4]. Non-

S. Rama Mohan
Department of Mathematics, PACE Institute of Technology & Sciences (Autonomous), Ongole, A.P, India

N. Maheshbabu (✉)
Department of Mathematics, Dr.S.R.K. Govt Arts College, Yanam, U.T. of Puducherry, India
e-mail: nmahesh@dhtepdy.edu.in

M. Eswara Rao
Department of Mathematics, Saveetha School of Engineering, SIMATS, Chennai, Tamil Nadu, India

F. Kamalov et al. (eds.), *Advances in Mathematical Modeling and Scientific Computing*, Trends in Mathematics, https://doi.org/10.1007/978-3-031-41420-6_22

Newtonian Ferro fluid flow across an unsteady contracted cylinder underneath the effect of an associated irresistible field was studied by Saranya et al. [5]. Santoshi et al. [6] scrutinized non-Newtonian liquid flow characteristics through a parabolic reflector of rotation and discussed them. Thermal effects and an inclined irresistible field on the MHD free convection flow of Casson fluid through a segmentation were investigated by Kumar et al. [7]. An unstable MHD dual diffusional free convection movement of Kuvshinski fluid is affected by an irresistible field across an inclined, movable porosity surface. This has been studied by Rama Prasad et al. [8]. Endalew et al. [9] defined a porous medium as the location of the occurrences of the irresistible field oriented on an unsteady MHD that passes in front of an inclined plate with parabolic acceleration. Abd-Alla et al. [10], in the presence of heat and mass transfer, investigated the impact of an inclined magnetism on the peristaltic blood flow in an asymmetrical porous channel. Kodi et al. [11] studied heat and mass transfer using a vertical inclination porosity surface with Soret thermal diffusion, an associated magnetic field, and the magnetohydrodynamic convective unsteady flow of a Jeffrey fluid. In their study, Nazir et al. [12]. On the heat and mass transmission of non-Newtonian liquids involving interface chemical processes. The analysis of heat and mass transmission above an unstable infinite porous material with a chemical process was discovered by Arshad et al. [13].

The influence of viscous dissipation and buoyancy force on the Jeffrey fluid's inconsistent flow in a vertical porous layer with changing viscosity has not, as far as we are aware, been investigated. To comprehend the alignment of irresistible field influence on the unsteady heat and mass transfer flow of non-Newtonian Jeffrey fluid through porous media under the influence of chemical response and heat absorption, the current problem is modeled. The geological processes in the Earth's mantle, MHD compressors, accelerators, geothermal reservoirs, and subsurface energy transport are all important applications for this model. The chapter is also structured as follows. The plate's lengthwise periodic oscillation induces the flow, and a uniform transversal irresistible field is supplied in the direction of the flow.

Using the Laplace transform approach, the systems of non-dimensional dominating PDEs are analytically solved. Graphs and tables are used to derive and discuss the effects of various flow quantities on velocity, temperature, and concentration, in addition to the frictional pressure coefficient and the rate of mass and heat transfer factors.

2 Mathematical Formulation

The fluid flow is transversely subjected to a uniform magnetic field B while an unstable MHD convective heat and mass transfer of a viscous, immiscible, electrically conducting, radiative, and chemically reactive fluid passes past a vertical plate. Assume that the x^*-axis is taken perpendicular to the surface and that the y^*-axis is parallel to the surface in the direction of the transversal magnetic field that is being provided. The fluid and the surface were originally static, with constant temperatures and concentrations at the time $t^* \leq 0$. The plate starts oscillating on its surface with a velocity of $z = Z_o \cos(w^* t^*)$ against the gravity field whenever $t^* > 0$, where w* denotes the magnitude of the surface oscillation. The tube's concentration and heating are increased to θ_w^* and $\varnothing_w^*$ both at the same time. It is considered that a consistent magnetic field B_0 is applied normally to the flow. So, because the high Reynolds numbers of the flux are thought to be very low, it is also expected that the irresistible field it induces will be minimal for free convection flow. The fluid that is being thought about here is gray, which absorbs/emits radiation, but it is not a dispersion medium. The mathematical model (see References [8, 9]) is listed below.

Momentum Equation:

$$\frac{\partial z^*}{\partial t^*} = v\left(\frac{1}{1+\lambda}\right)\frac{\partial^2 z^*}{\partial y^{*2}} - \frac{\sigma B_0^2 \sin^2\alpha}{\rho} z^* - \frac{v}{K} z^* \tag{1}$$

Energy Equation:

$$\frac{\partial \theta^*}{\partial t^*} = \frac{\kappa}{\rho C_p}\frac{\partial^2 \theta^*}{\partial y^{*2}} - \frac{1}{\rho C_p}\frac{\partial q_r}{\partial y^*} - \frac{Q^*}{\rho C_p}\left(\theta^* - \theta_\infty^*\right) \tag{2}$$

Diffusion Equation:

$$\frac{\partial \phi^*}{\partial t^*} = D_m \frac{\partial^2 \phi^*}{\partial y^{*2}} - Kr\left(\phi^* - \phi_\infty^*\right) \tag{3}$$

According to the original boundary conditions:

$$\begin{aligned} &t^* \leq 0, z^* = 0, \theta^* = \theta_\infty^*, \phi^* = \phi_\infty^* \ for\ all\ y^*, \\ &t^* > 0, z^* = Z_0\ \cos(wt),\ \theta^* = \theta_w^*,\ \phi^* = \phi_w^*\ at\ y^* = o, \\ &z^* \to 0,\ \theta^* \to \theta_\infty^*,\ \phi^* \to \phi_\infty^*\ as\ y^* \to \infty. \end{aligned} \tag{4}$$

Here z^*,λ, B_0, v, q_r, Q, σ, D_m, t, K ,ρ, θ^*, $\varnothing^*$, C_p, C_s, K_T, w, *and* K_r are the velocity of the liquid in the x^*- direction, Jeffrey liquid quantity, external magnetic field, kinematic viscosity, concentration susceptibility, radiative heating absorption, the temperature of the liquid near the surface, heat fluctuation, stereotyped potential, coefficient of mass diffusivity, time, thermal potential, liquid capacity species concentration, specific heat at perpetual pressure, periodicity parameter, and chemical reaction quantity, respectively.

Under the Rosseland estimate, the radiant heat variation of the type shown below is used.

$$q_r = -\frac{4\sigma^*}{3k^*}\frac{\partial \theta^*}{\partial y^*} \tag{5}$$

Here, $\sigma*$ denotes the Stefan–Boltzmann constant, and k* represents the average absorption ratio. The stream's temperature variation θ^{*^4} is thought to be sufficiently minimal to be stated to be a linear function of temperature. The higher-order terms are disregarded to achieve this and expanding θ^{*^4} a Taylor series aboutθ^*_∞.

As a result, we get

$$\theta^{*^4} = 4\theta^{*^3}_\infty \theta^* - 3\theta^{*^4}_\infty. \tag{6}$$

Equation (2) reduces after incorporating Eqs. (5) and (6).

$$\frac{\partial\theta^*}{\partial t^*} = \frac{\kappa}{\rho C_P}\frac{\partial^2\theta^*}{\partial y^{*2}} - \frac{16\sigma^*\theta^*_\infty}{3\rho C_P k^*}\frac{\partial^2\theta^*}{\partial y^{*2}} - \frac{Q^*}{\rho C_p}\left(\theta^* - \theta^*_\infty\right) \tag{7}$$

Adding the aforementioned dimensionless amounts

$$z = \frac{z^*}{Z_0}, y = \frac{Z_0}{\upsilon}y^*, t = \frac{Z_0^2}{\upsilon}t^*, M = \frac{\sigma B_0^2\upsilon}{\rho Z_0^2}, \theta = \frac{(\theta^* - \theta^*_\infty)}{(\theta^*_W - \theta^*_\infty)}, K = \frac{k^* z_0^2}{\upsilon}, \phi = \frac{(\phi^* - \phi^*_\infty)}{(\phi^*_W - \phi^*_\infty)},$$
$$\Pr = \frac{\rho\,\upsilon\,C_P}{\kappa}, Q = \frac{Q^*\upsilon}{\rho C_p Z_0^2}, R = \frac{16\sigma^*\theta^{*^3}_\infty}{3k^*k} w = w^*\frac{\upsilon}{Z_0^2}. \tag{8}$$

Because of Eqs. (1), (3), and (7), transmission constitutes (8)

$$\frac{\partial z}{\partial t} = \left(\frac{1}{1+\lambda}\right)\frac{\partial^2 z}{\partial y^2} - \left(M\sin^2\alpha + \frac{1}{K}\right)z \tag{9}$$

$$\frac{\partial\theta}{\partial t} = \left(\frac{1+R}{\Pr}\right)\frac{\partial^2\theta}{\partial y^2} - Q\,\theta \tag{10}$$

$$\frac{\partial\phi}{\partial t} = -Kr\,\phi + \frac{1}{Sc}\frac{\partial^2\phi}{\partial y^2} \tag{11}$$

The associated boundary conditions become

$$\begin{aligned} &z = 0,\ \theta = 0,\ \phi = 0\ for\ all\ y, t \le 0,\\ &z = \cos(wt),\ \theta = 1, \phi = 1\ at\ y = 0, t > 0,\\ &z \to 0,\ \theta \to 0,\ \phi \to 0\ as\ y \to \infty, t > 0. \end{aligned} \tag{12}$$

3 Solution of the Problem

By employing the Laplace transform method and the boundary condition formulas (12), the equation system (9) to (11) is parameter estimation.

$$z(y,t)=\frac{e^{iwt}}{4}\left[\begin{array}{l}\exp\left(-y\sqrt{H_3\left(H_4+iw\right)}\right)\ \mathrm{erfc}\left(\frac{y\sqrt{H_3}}{2\sqrt{t}}-\sqrt{\left(H_4+iw\right)\,t}\right)\\ +\exp\left(y\sqrt{H_3\left(H_4+iw\right)}\right)\ \mathrm{erfc}\left(\frac{y\sqrt{H_3}}{2\sqrt{t}}+\sqrt{\left(H_4+iw\right)\,t}\right)\end{array}\right]$$
$$+\frac{e^{-iwt}}{4}\left[\begin{array}{l}\exp\left(-y\sqrt{H_3\left(H_4-iw\right)}\right)\ \mathrm{erfc}\left(\frac{y\sqrt{H_3}}{2\sqrt{t}}-\sqrt{\left(H_4-iw\right)\,t}\right)\\ +\exp\left(y\sqrt{H_3\left(H_4-iw\right)}\right)\ \mathrm{erfc}\left(\frac{y\sqrt{H_3}}{2\sqrt{t}}+\sqrt{\left(H_4-iw\right)\,t}\right)\end{array}\right] \tag{13}$$

$$\theta\,(y,t)=\frac{1}{2}\left[\begin{array}{l}\exp\left(-y\sqrt{H_1 Q}\right)\ \mathrm{erfc}\left(\frac{y\sqrt{H_1}}{2\sqrt{t}}-\sqrt{Qt}\right)\\ +\exp\left(y\sqrt{H_1 Q}\right)\ \mathrm{erfc}\left(\frac{y\sqrt{H_1}}{2\sqrt{t}}+\sqrt{Q\,t}\right)\end{array}\right] \tag{14}$$

$$\phi\,(y,t)=\frac{1}{2}\left[\begin{array}{l}\exp\left(-y\sqrt{Sc\,kr}\right)\ \mathrm{erfc}\left(\frac{y\sqrt{Sc}}{2\sqrt{t}}-\sqrt{Krt}\right)\\ +\exp\left(y\sqrt{Sc\,Kr}\right)\ \mathrm{erfc}\left(\frac{y\sqrt{Sc}}{2\sqrt{t}}+\sqrt{Kr\,t}\right)\end{array}\right] \tag{15}$$

3.1 Skin-Friction Coefficient

The formula for the skin-graft coefficient at y = 0 is

$$\tau=-\left(\frac{1}{1+\lambda}\right)\left[\frac{\partial z}{\partial y}\right]_{y=0} \tag{16}$$

We derive the following skin-graft coefficient from Eqs. (13) and (16):

$$\tau=\left(\frac{1}{1+\lambda}\right)\left[\begin{array}{l}\frac{e^{iwt}}{2}\left(\sqrt{H_3\,(H_4+iw})\mathrm{erf}\left(\sqrt{(H_4+iw)\ t}\right)+\sqrt{\frac{H_3}{\pi t}}e^{-(H_4+iw)t}\right)\\ +\frac{e^{-iwt}}{2}\left(\sqrt{H_3\,(H_4-iw})\mathrm{erf}\left(\sqrt{(H_4-iw)\ t}\right)+\sqrt{\frac{H_3}{\pi t}}e^{-(H_4-iw)t}\right)\end{array}\right]$$

3.2 Nusselt Number

Nusselt number at y = 0 is given by

$$Nu=-\left[\frac{\partial\theta}{\partial y}\right]_{y=0} \tag{17}$$

The following is how we derive the Nusselt number from Eqs. (14) and (17).

$$Nu=\sqrt{H_1 Q}\ \mathrm{erf}\left(\sqrt{Q\,t}\right)+\sqrt{\frac{H_1}{\pi t}}e^{-Qt}$$

3.3 Sherwood Number

When y = 0, the Sherwood number is determined by

$$Sh = -\left[\frac{\partial \phi}{\partial y}\right]_{y=0} \tag{18}$$

The following is how we derive the Sherwood number using Eqs. (15) and (18).$Sh = \sqrt{Sc\ Kr}\mathrm{erf}\left(\sqrt{Kr\ t}\right) + \sqrt{\frac{Sc}{\pi t}} e^{-Kr\ t}$

Here

$$H_1 = \frac{\mathrm{Pr}}{1+R},\ H_2 = \frac{1}{1+\lambda},\ H_3 = \frac{1}{H_2},\ H_4 = M\sin^2\alpha + \frac{1}{K}.$$

4 Results and Discussion

In relation to this physical problem, graphs and tables have been used to discuss the frequency of velocity, the rate of heat transfer, temperature, the rate of mass transfer, and velocity by providing numerical ranges to the quantities *M*, K, *t*, λ, *α*, *Pr, Sc, Kr,* Q > 0, *and* w. The results of the comparison show sufficient agreement to support the validity of this numerical approach.

In Fig. 1, with an increase in the magnetic parameter M, it appears that the flow's velocity is greatly lowered. The Lorentz force, which attempts to impede the flow, was produced by the magnetic parameter M's rising value. As a result, the velocity curve flattens out as the magnetic parameter increases. The relationships between the values of (K) and (t) and the liquid velocity curve are shown in Figs. 2 and 3. The liquid velocity curve is seen to increase along with the fluid flow's (K) and (t) parameters. Figures 4 and 5 depict how the Jeffrey liquid quantity (λ) and oriented angle (*α*) affect the fluid velocity. It is noticeable that as the fluid flow's Jeffrey liquid quantity (λ) and oriented angle (*α*) increase, the velocity decreases. The liquid velocity curve remained constant, as shown in Fig. 6, as the time interval quantity (w) in the fluid flow increased. Figures 7, 8, and 9 exhibit how various flow variables influence fluid temperature. The Prandtl number is stated as a function of kinematic viscosity and thermal diffusivity. The relationship between the liquid temperature and the Prandtl number (Pr) and heat emission quantity (Q > 0) is shown in Figs. 7 and 8. The temperature drops as the Prandtl number (Pr), or the quantity of the heat source (Q > 0), increases in the liquid flow. The relationship between amount (R) and liquid temperature appears in graph (Fig. 9). Figures 10 and 11 illustrate how the *Sc* number and chemical quantity (Kr) have an impact on the fluid concentration. The concentration was found to fall as the Sc number or the (Kr) increased.

The Nu, Sh, and skin-graft coefficients' shifts are shown in Table 1. As *M* or α grows, the skin-graft coefficient falls, and as λ, *K,* or *t* increases, the skin-graft

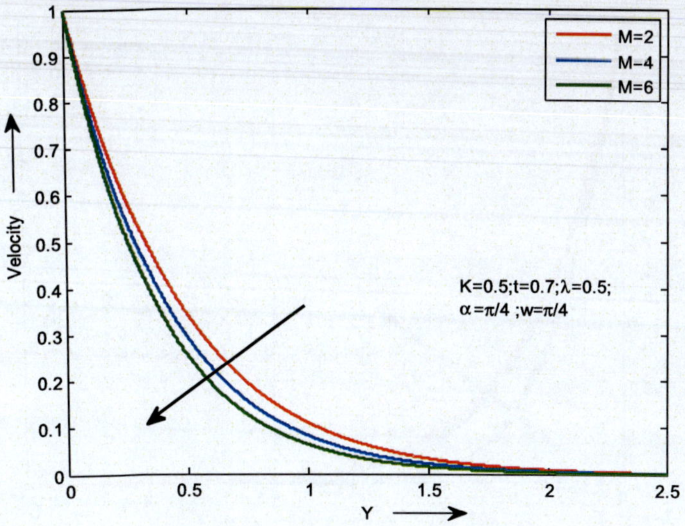

Fig. 1 Velocity lineation for variegated ranges of *M*

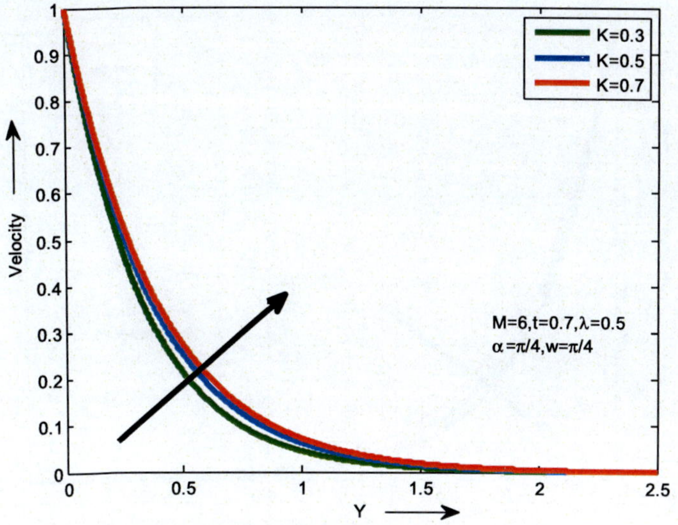

Fig. 2 Velocity lineation for variegated ranges of *K*

coefficient grows. As *R* rises, the Nu number falls, and as *Pr* or *Q* rises, the Nusselt number rises. As *Sc* or *Kr* rises, *Sh* number rises.

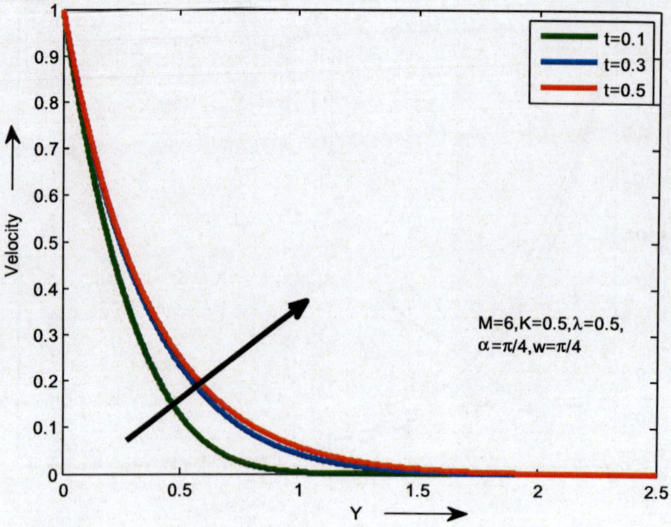

Fig. 3 Velocity lineation for variegated ranges of *t*

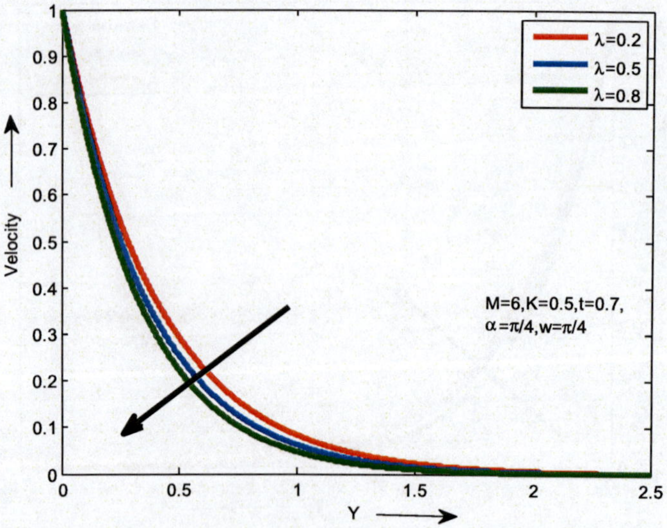

Fig. 4 Velocity lineation for variegated ranges of λ

5 Conclusion

In this work, consideration was given to the non-Newtonian Jeffrey fluid's steady flow, which is necessary for heat and mass transmission in porous media. The emerging parameters' effects are listed.

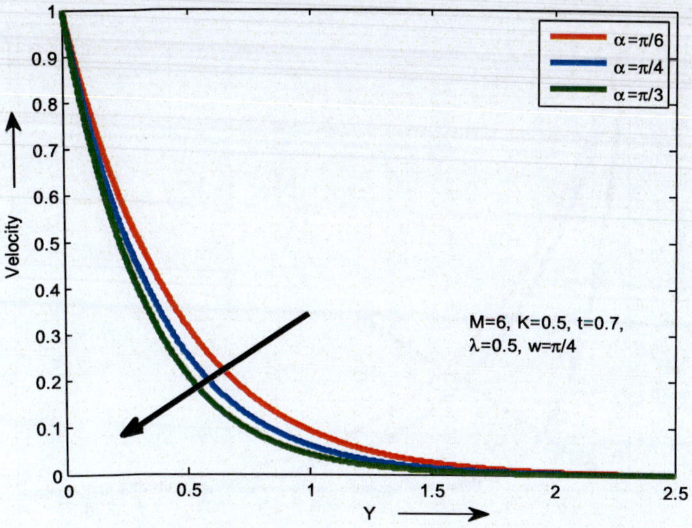

Fig. 5 Velocity lineation for variegated ranges of α

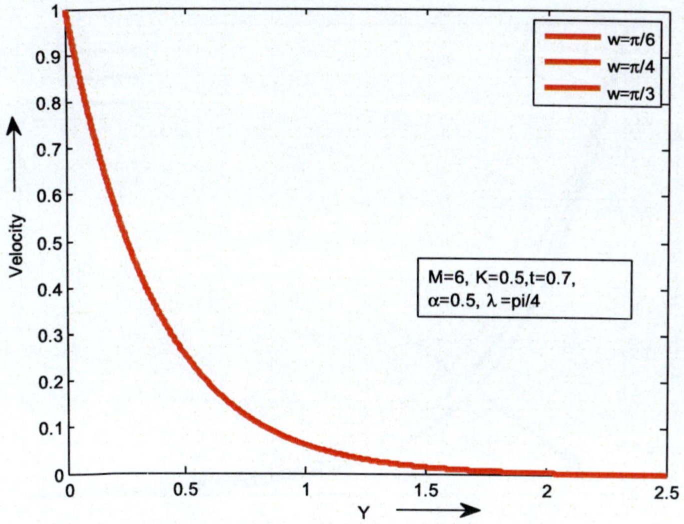

Fig. 6 Velocity lineation for variegated ranges of w

1. The process of heat and mass transport is significantly impacted by the magnetic amount (M).
2. Reduced heat- and mass-transfer rates are caused by increases in M, λ, α, Pr, or Q.
3. The heat and mass transfer, however, increase as K, t, R, Sc, or Kr grows.

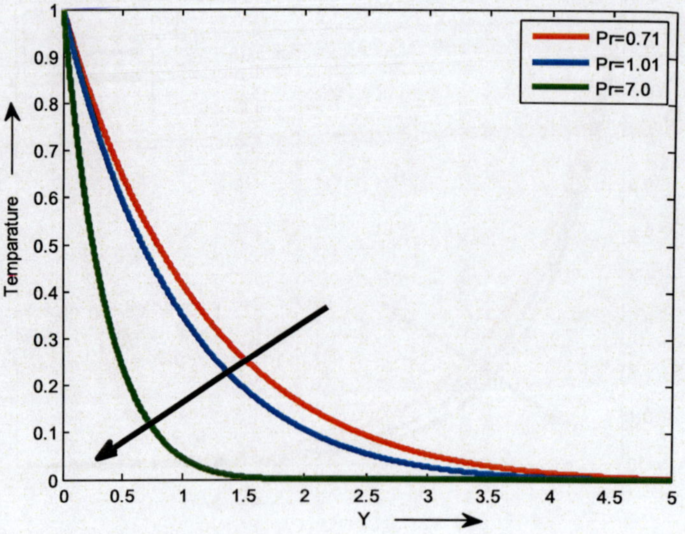

Fig. 7 Temperature lineation for variegated ranges of *Pr*

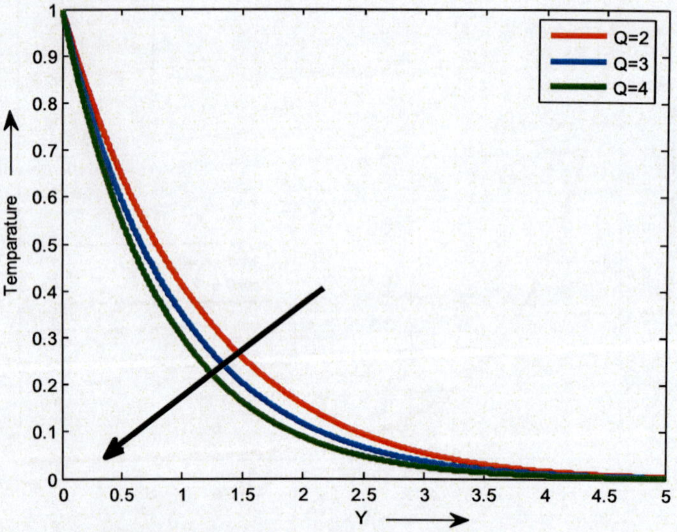

Fig. 8 Temperature lineation for variegated ranges of *Q*

4. The amount of skin graft decreases as the Jeffrey fluid flows, yet mass and heat transmission efficiency at the surface is found to slightly rise.

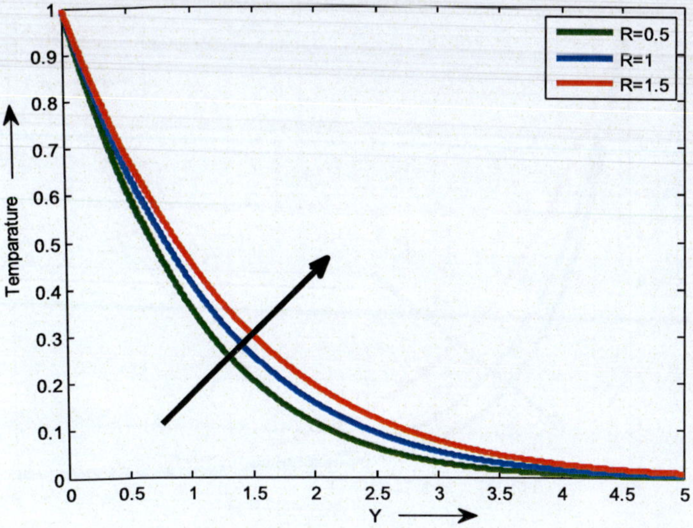

Fig. 9 Temperature lineation for variegated ranges of *R*

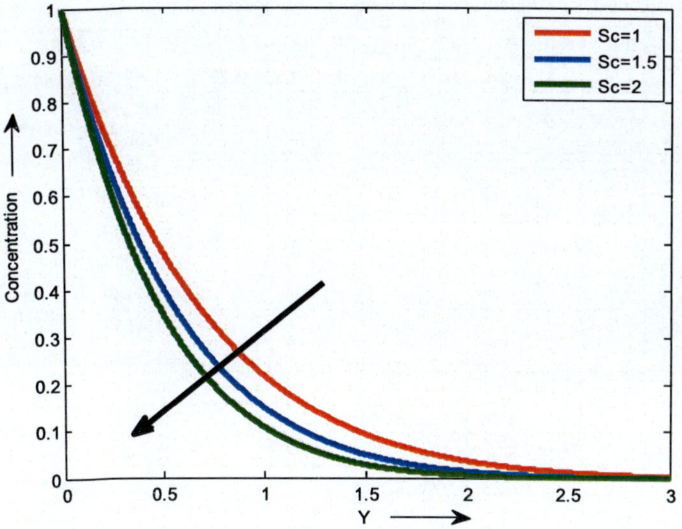

Fig. 10 Concentration lineation for variegated ranges of *Sc*

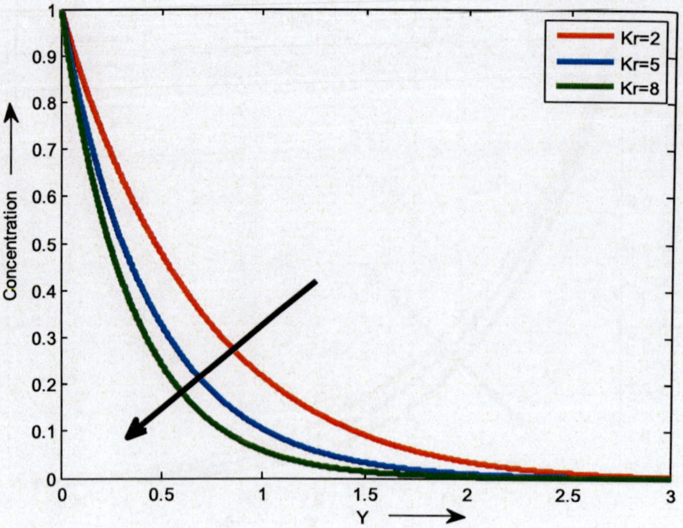

Fig. 11 Concentration lineation for variegated ranges of *K*

Table 1 Skin-friction coefficient, Nusselt number, and Sherwood number

M	λ	*t*	*w*	K	A	*Pr*	*Q*	*R*	*Kr*	*Sc*	τ	*Nu*	*Sh*
2	0.5	1	π/4	0.5	π/4	0.71	2	1	0.5	0.22	−1.4169	0.8498	0.3869
3											−1.5290		
4											−1.6338		
	0.6										−1.3719		
	0.7										−1.3310		
		2									−1.4297		
		3									−1.4274		
			π/6								−1.4169		
			π/3								−1.4169		
				0.6							−1.3374		
				0.7							−1.2780		
					π/6						−1.2961		
					π/3						−1.5290		
						1.01						1.0135	
						7						2.6682	
							3					1.0340	
							4					1.1922	
								2				0.6938	
								3				0.6009	
									0.7				0.4309
									0.9				0.4726
										0.62			0.6496
										0.96			0.8083

References

1. Rundora, L., Makinde, O.D.: Analysis of unsteady MHD reactive flow of non-Newtonian fluid through a porous saturated medium with asymmetric boundary conditions. Iran. J. Sci. Technol. Trans. Mech. Eng. **40**(3), 189–201 (2016)
2. Reddy, J.V., Sugunamma, V., Sandeep, N.: Effect of aligned magnetic field on Casson fluid flow past a vertical oscillating plate in porous medium. J. Adv. Phys. **5**(4), 295–301 (2016)
3. Jena, S., Mishra, S.R., Dash, G.: Chemical reaction effect on MHD Jeffery fluid flow over a stretching sheet through porous media with heat generation/absorption. Int. J. Appl. Comput. Math. **3**(2), 1225–1238 (2017)
4. Sailaja, M., Reddy, R.H., Saravana, R., Avinash, K.: Aligned magnetic field effect on unsteady liquid film flow of Casson fluid over a stretching surface. In: IOP Conference Series: Materials Science and Engineering, vol. 263, No. 6, p. 062008. IOP Publishing (2017). https://dx.doi.org/10.1088/1757-899X/263/6/062008
5. Saranya, S., Al-Mdallal, Q.M.: Non-Newtonian ferrofluid flow over an unsteady contracting cylinder under the influence of aligned magnetic field. Case Stud. Therm. Eng. **21**, 100679 (2020)
6. Santoshi, P.N., Reddy, G.V.R., Padma, P.: Flow features of non-Newtonian fluid through a parabolic of revolution. Int. J. Appl. Comput. Math. **6**(3), 1–22 (2020)
7. Kumar, C.P.: Thermal diffusion and inclined magnetic field effects on MHD free convection flow of Casson fluid past an inclined plate in conducting field. Turk. J. Comput. Math. Educ. **12**(13), 960–977 (2021)
8. Rama Prasad, J.L., Balamurugan, K.S., Varma, S.V.K.: Aligned magnetic field effect on unsteady MHD double diffusive free convection flow of Kuvshinski fluid past an inclined moving porous plate. In: Advances in Fluid Dynamics, pp. 255–262. Springer, Singapore (2021)
9. Endalew, M.F., Sarkar, S.: Incidences of aligned magnetic field on unsteady MHD flow past a parabolic accelerated inclined plate in a porous medium. Heat Tran. **50**(6), 5865–5884 (2021)
10. Abd-Alla, A.M., Abo-Dahab, S.M., Thabet, E.N., Abdelhafez, M.A.: Impact of inclined magnetic field on peristaltic flow of blood fluid in an inclined asymmetric channel in the presence of heat and mass transfer. Waves Random Complex Media, **32**, 1–25 (2022)
11. Kodi, R., Konduru, V.: Heat and mass transfer on MHD convective unsteady flow of a Jeffrey fluid past an inclined vertical porous plate with thermal diffusion Soret and aligned magnetic field. Mater. Today Proc. **50**, 2128–2134 (2022)
12. Nazir, S., Kashif, M., Zeeshan, A., Alsulami, H., Ghamkhar, M.: A study of heat and mass transfer of non-Newtonian fluid with surface chemical reaction. J. Indian Chem. Soc. **99**(5), 100434 (2022)
13. Arshad, M., Hussain, A., Hassan, A., Shah, S.A.G.A., Elkotb, M.A., Gouadria, S., Galal, A.M.: Heat and mass transfer analysis above an unsteady infinite porous surface with chemical reaction. Case Stud. Therm. Eng. **36**, 102140 (2022)

Effect of Viscosity Variation and Slip Velocity on the Squeeze-Film Characteristics Between a Cylinder and a Plane Plate with Couple Stress Fluid

K. Arshiya Kousar, A. Salma, and B. N. Hanumagowda

1 Introduction

In several applications, including the lubrication of machine parts, automatic transmissions, and artificial joints, squeeze-film properties are essential. In general, analysis of the performance of squeeze films assumes that the lubricant behaves essentially like a viscous fluid that satisfies Newton's laws; however, efforts have been made to improve flow properties and enhance lubricating qualities by using a variety of additives. As a result, adding a minimal number of additives to the lubricant has piqued attention. Classical continuum theory is unable to properly describe the rheological flow properties of Newtonian lubricants mixed with different kinds of additives. As a result, a plethora of micro-continuum models were proposed [1–3]. Stokes micro-continuum model [1] is the most basic of them all, and it permits polar impacts like couple stresses and body couples. This framework has been applied to several numerical calculations and analytical studies on the performance of squeeze films such as finite parallel plates [4], a sphere on a flat plate, as well as long partial journal bearings by Lin [5, 6], cylinder and plane surfaces [7], and circular stepped plates [8]. In the preceding investigation, lubricant viscosity was presumed to be a fixed value. Barus and collaborators [9, 10] reported the PVD with the equation $\mu = \mu_0 e^{\alpha p}$, where μ_0 is the viscosity at medium pressure at room temperature and $\propto$ represents the viscosity pressure coefficient. The previous relationship demonstrates that the viscosity of the lubricant increases exponentially, which might affect the expected squeeze-film bearing execution. Lin et al. [11, 12] investigated the effect of viscosity changes and non-Newtonian fluid on parallel circular plates and wide parallel plates and concluded that load-carrying

K. Arshiya Kousar · A. Salma (✉) · B. N. Hanumagowda
Research Scholar, Department of Mathematics, SOAS, REVA University, Bengaluru, Karnataka, India

F. Kamalov et al. (eds.), *Advances in Mathematical Modeling and Scientific Computing*, Trends in Mathematics, https://doi.org/10.1007/978-3-031-41420-6_23

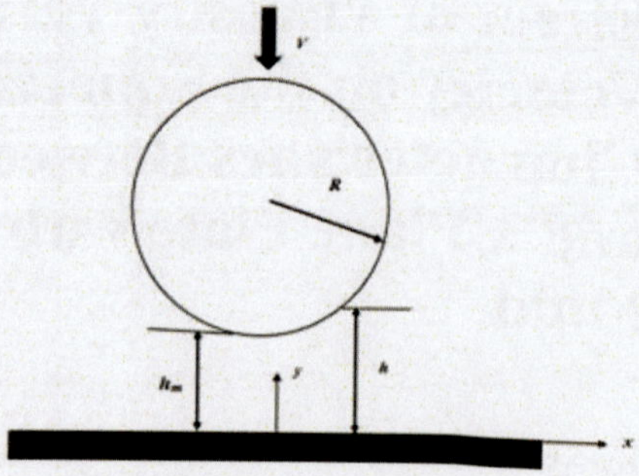

Fig. 1 Geometry of cylinder and plane surface

capacity and squeezing time have increased due to an increase in the viscosity variation parameter. As a result, several research investigations are being conducted that incorporate the variation of viscosity with pressure on distinct bearings [13, 14]. Under the no-slip boundary condition, most Newtonian and non-Newtonian fluid flows have been analyzed. This assumption, however, may not always hold, and fluid leakage may exist at solid boundaries [15, 16]. The effect of slip velocity was studied by many authors for several geometries; a few of them are porous rectangular plates by Hai [17], porous-walled squeeze films by Sparrow et al. [18], rotating porous annular disks by Prakash and Vij [19], and porous circular disks by Patel [20]. From the obtained results, it is found that the presence of slip velocity decreases load-carrying capacity and response time. The purpose of this work is to investigate the influence of slip velocity and viscosity variation on squeeze-film lubrication of a cylinder and a plane plate with non-Newtonian fluid and compare it with Jaw-Ren et al. [7].

1.1 Theoretical Solution

The geometry under study is displayed in Fig. 1 in which the cylinder surface is moving with velocity $V = \left(-\frac{\partial h}{\partial t}\right)$ toward the fixed plane surface.

In this system, the couple stress fluid flow in the film region is laminar, the body forces and body couples are negligible, and the viscosity varies with pressure are considered. Based upon these assumptions, the basic equation of motion and continuity are:

$$\mu \frac{\partial^2 u}{\partial y^2} - \eta \frac{\partial^4 u}{\partial y^4} = \frac{\partial p}{\partial x} \tag{1}$$

$$\frac{\partial p}{\partial y} = 0 \tag{2}$$

$$\frac{\partial u}{\partial x} + \frac{\partial v}{\partial y} = 0 \tag{3}$$

The u and v are considered as components of fluid velocity in the direction x and y respectively, p is the film pressure, μ is the material constant of viscosity, and η is the constant that is accountable for non-Newtonian fluids.

The boundary conditions for the velocity components are:

For lower surface $y = 0$,

$$"u = \frac{1}{s}\frac{\partial u}{\partial y}\bigg|_{y=0}, \quad \frac{\partial^2 u}{\partial y^2} = 0, \;\; v = 0" \tag{4}$$

For upper surface $y = h$,

$$"u = 0, \;\; \frac{\partial^2 u}{\partial y^2} = 0, \;\; v = -\frac{\partial h}{\partial t}" \tag{5}$$

The film thickness h, as in the film region $x \ll R$

$$h = h_m + \frac{x^2}{2R} \tag{6}$$

Where, h_m represents the initial minimum film thickness.

Solving Eq. (1) using boundary conditions (4) and (5), we can get the expression of u

$$\begin{aligned} u = &\left\{y^2 - h^2\left(\xi_1 + \xi_2 y\right)\right\}\frac{1}{2\mu}\frac{\partial p}{\partial x} + (h - y)\,\xi_1 l\frac{1}{\mu}\frac{\partial p}{\partial x}\tanh\frac{h}{2l} \\ &+ \frac{l^2}{\mu}\frac{\partial p}{\partial x}\left[1 - \frac{\mathrm{Cosh}\left\{(2y - h)/2l\right\}}{\mathrm{Cosh}\,(h/2l)}\right] \end{aligned} \tag{7}$$

Here $\xi_1 = \frac{\sigma_1}{(h+\sigma_1)}$, $\xi_2 = \frac{1}{(h+\sigma_1)}$.

Substituting u in Eq. (3) and using Eqs. (4) and (5), the Reynolds equation in modified form is

$$\frac{\partial}{\partial x}\left[\zeta\left(h, l, \sigma_1, \alpha, p\right)\frac{\partial p}{\partial x}\right] = -\mu\frac{\partial h}{\partial t} \tag{8}$$

Where,

$$\begin{aligned}\zeta\left(h, l, \sigma_1, \alpha, p\right) = &\left(1 - 3\xi_1\right)\tfrac{h^3}{6}e^{-\alpha p} + \left(4m^2 h e^{-2\alpha p} - \xi_2 h^4 e^{-\alpha p}\right)\tfrac{1}{4} + \\ &\left(\xi_1 h^2 m e^{-3\alpha p/2} - 4m^3 e^{-5\alpha p/2}\right)\tfrac{1}{2}\tanh\left(\tfrac{h e^{\alpha p/2}}{2m}\right)\end{aligned}$$

The following non-dimensional quantities are substituted in Eq. (8):

$$x^* = \frac{x}{R}, l^* = \frac{m}{h_{m0}}, h^* = \frac{h}{h_{m0}}, h_m^* = \frac{h_m}{h_{m0}}, \sigma = \frac{\sigma_1}{h_{m0}}, P^* = \frac{ph_{m0}^3}{\mu_0 R^2(-dh/dt)}, \beta = \frac{h_{m0}}{R},$$
$$G = \frac{\alpha\mu_0 R^2(-dh/dt)}{h_{m0}^3}, \xi_1{}^* = \frac{\sigma}{(h^*+\sigma)}, \xi_2{}^* = \frac{1}{(h^*+\sigma)}$$

The non-dimensional Reynolds equation in modified form is

$$\frac{\partial}{\partial x^*}\left\{S\left(h^*, l^*, \sigma, G, P^*\right)\frac{\partial P^*}{\partial x^*}\right\} = \frac{1}{\beta} \tag{9}$$

Where,

$$S\left(h^*, l^*, \sigma, G, P^*\right) = \left(1 - 3\xi_1^*\right)\frac{h^{*3}}{6}e^{-GP^*} + \left(4h^* l^{*2} e^{-2GP^*} - \xi_2^* h^{*4} e^{-GP^*}\right)\frac{1}{4}$$
$$+ \left(\xi_1^* h^{*2} l^* e^{-1.5GP^*} - 4l^{*3} e^{-2.5GP^*}\right)\frac{1}{2}\tanh\left(\frac{h^* e^{0.5GP^*}}{2l^*}\right)$$

Equation (9) is highly non-linear and cannot be solved. Hence, to get the analytical solution in first order for viscosity parameter small values $0 \leq G \ll 1$, the film pressure is calculated using a small perturbation approach.

$$P^* = P_0^* + GP_1^* \tag{10}$$

Using Eq. (10) in (9) and ignoring second and higher order of G, the equation for film pressure so obtained is

$$\frac{\partial}{\partial x^*}\left\{S_0\left(h^*, l^*, \sigma\right)\frac{\partial P_1^*}{\partial x^*} + S_1\left(h^*, l^*, \sigma\right) P_0^* \frac{\partial P_0^*}{\partial x^*}\right\} = 0 \tag{11}$$

Where,

$$S_0\left(h^*, l^*, \sigma\right) = \tfrac{1}{6}\left(1 - 3\xi_1^*\right)h^{*3} + \tfrac{1}{4}\left(4h^* l^{*2} - \xi_2^* h^{*4}\right) + \tfrac{1}{2}\left(\xi_1^* h^{*2} l^* - 4l^{*3}\right)\tanh\left(\tfrac{h^*}{2l^*}\right)$$
$$S_1\left(h^*, l^*, \sigma\right) = -\tfrac{h^{*3}}{6}\left(1 - 3\xi_1^*\right) - \tfrac{1}{4}\left(8h^* l^{*2} - \xi_2^* h^{*4}\right) - \tfrac{1}{4}\left(3\xi_1^* h^{*2} l^* - 20l^{*3}\right)\tanh\left(\tfrac{h^*}{2l^*}\right)$$
$$+ \tfrac{1}{8}\left(\xi_1^* h^{*3} - 4h^* l^{*2}\right)\operatorname{sech}^2\left(\tfrac{h^*}{2l^*}\right)$$

For film pressure, the boundary conditions are

$$P^* = 0 \text{ at } x^* = \pm 1 \tag{12}$$

$$\frac{dP^*}{dx^*} = 0 \text{ at } x^* = 0 \tag{13}$$

Simplifying Eq. (11) using boundary conditions (12) and (13), the non-dimensional pressure is obtained as follows

$$P^*=-\frac{1}{\beta}\int_{x^*}^{1}\frac{x^*}{S_0\left(h^*,l^*,\sigma\right)}dx^*-\frac{G}{\beta^2}\int_{x^*}^{1}\left[\frac{S_1\left(h^*,l^*,\sigma\right)x^*}{\left\{S_0\left(h^*,l^*,\sigma\right)\right\}^2}\int_{x^*}^{1}\frac{x^*}{S_0\left(h^*,l^*,\sigma\right)}dx^*\right]dx^* \tag{14}$$

The non-dimensional load-carrying capacity W^* is

$$W^*=\frac{Wh_{m0}^2}{\mu R^2 B\left(-dh\Big/dt\right)}=-\frac{1}{\beta}\int_{x^*=-1}^{x^*=1}\left\{\int_{x^*}^{1}\frac{x^*}{S_0(h^*,l^*,\sigma)}dx^*\right\}dx^*-\frac{G}{\beta}\int_{x^*=-1}^{x^*=1}\left\langle\int_{x^*}^{1}\left[\frac{S_1(h^*,l^*,\sigma)x^*}{\{S_0(h^*,l^*,\sigma)\}^2}\int_{x^*}^{1}\frac{x^*}{S_0(h^*,l^*,\sigma)}dx^*\right]dx^*\right\rangle dx^* \tag{16}$$

Integrating Eq. (16), the non-dimensional Squeeze film time is

$$T^*=\frac{Wh_{m0}^2}{\mu R^3 B}t=-\frac{1}{\beta}\int_{h_m^*}^{1}\left[\int_{x^*=-1}^{x^*=1}\left\{\int_{x^*}^{1}\frac{x^*}{S_0(h^*,l^*,\sigma)}dx^*\right\}dx^*\right]dh_m^*-\frac{G}{\beta^2}\int_{h_m^*}^{1}\left\langle\int_{x^*=-1}^{x^*=1}\left[\int_{x^*}^{1}\left\{\frac{S_1(h^*,l^*,\sigma)x^*}{\{S_0(h^*,l^*,\sigma)\}^2}\int_{x^*}^{1}\frac{x^*}{S_0(h^*,l^*,\sigma)}dx^*\right\}dx^*\right]dx^*\right\rangle dh_m^* \tag{17}$$

2 Results and Discussion

This chapter investigates the impact of viscosity variation, non-Newtonian fluid, and velocity slip on the squeeze-film features of a cylinder and a plane plate. Using the Barus formula for viscosity and Stoke's model for non-Newtonian fluid, the modified Reynold's equation is found, and the non-dimensional expressions for film pressure, load-carrying capacity, and Squeeze film time are subsequently derived using the perturbation technique. These squeeze-film features are examined in terms of non-dimensional factors, such as non-Newtonian parameter l^*, viscosity parameter G, and slip velocity parameter σ.

Squeeze-film pressure The dimensionless film pressure P^* along x^* is explained in Figs. 2, 3, 4, and 5. It is noted that by increasing l^* values, the film pressure P^* is also increased (Fig. 2). Furthermore, it is found that higher pressure is obtained for larger values of the curvature parameter β (Fig. 3). Figure 4 shows that when slip velocity is zero, the film pressure reaches its maximum value, and by increasing slip velocity, P^* decreases. The effect of the viscosity parameter is explained in Fig. 5, and it is seen that, as compared to the non-viscous case ($G = 0$), pressure P^* is significant for larger values of G.

Load-Carrying Capacity The deviation of dimensionless load W^* versus film thickness is depicted in Figs. 6, 7, 8, and 9. In Fig. 6, it is seen that the impact of the non-Newtonian fluid enhances the load as compared to the Newtonian fluid, whereas in Fig. 7, the load is significant for larger values of β. For increasing values of the

Fig. 2 Variation of P^* versus x^* for various values of l^*

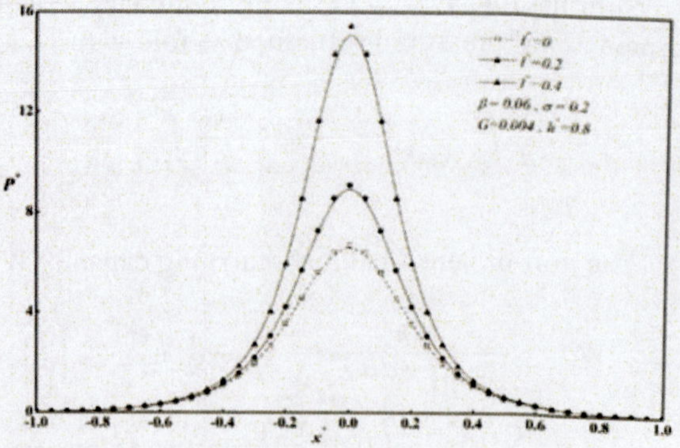

Fig. 3 Variation of P^* versus x^* for various values of β

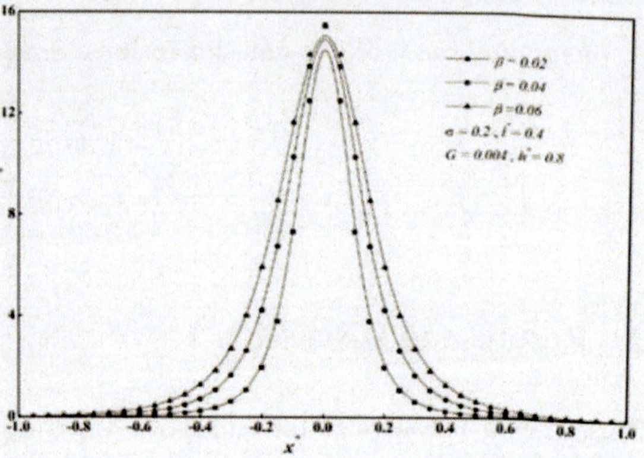

Fig. 4 Variation of P^* versus x^* for various values of σ

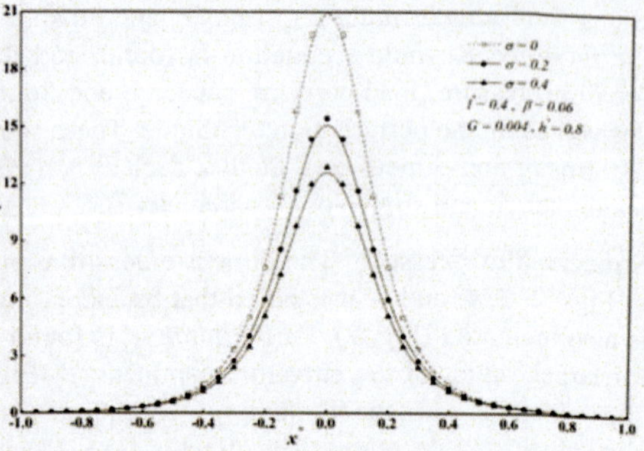

slip velocity parameter, load W^* decreases when compared to the no-slip velocity, as depicted in Fig. 8, but W^* rises for increasing values of viscosity variation G, as seen in Fig. 9. From all these figures, it is illustrated that for lower values of h^*, the load W^* is more.

Fig. 5 Variation of P^* versus x^* for various values of G

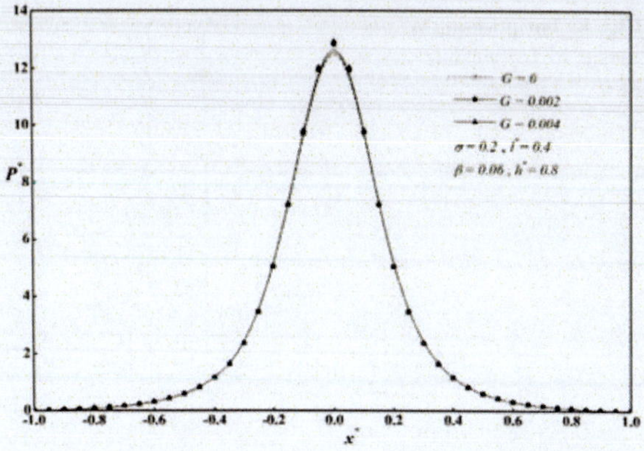

Fig. 6 Variation of W^* versus h^* for various values of l^*

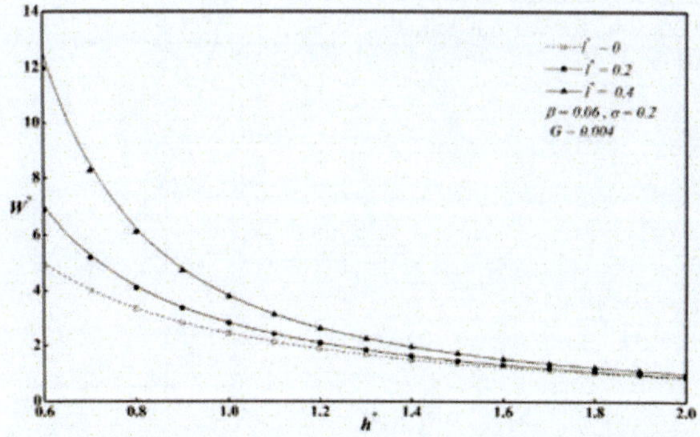

Fig. 7 Variation of W^* versus h^* for various values of β

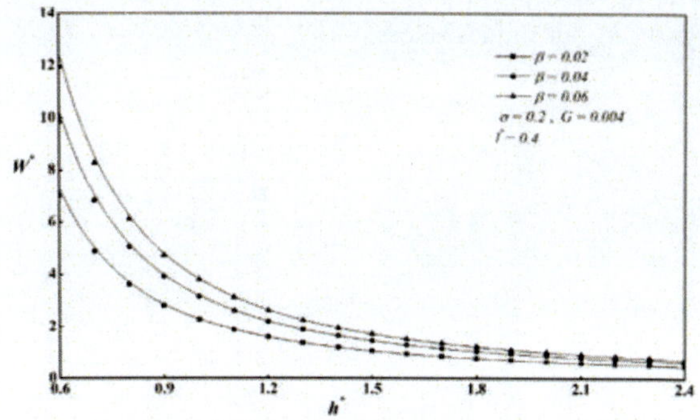

Squeeze Film Time Figures 10, 11, 12, and 13 explain the response time T^* versus film height h_m^*.

The impact of l^* increases the film time, as shown in Fig. 10. T^* enhances for larger values of curvature parameter β, as explained in Fig. 11. Figure 12 describes the Squeeze film time versus h_m^* as a function of slip velocity σ, and it is seen that T^* is decreased by increasing the values of slip velocity parameter. The deviation

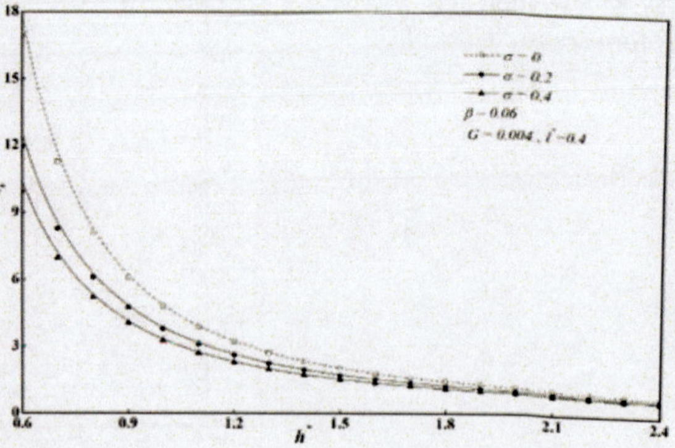

Fig. 8 Variation of W^* versus h^* for various values of σ

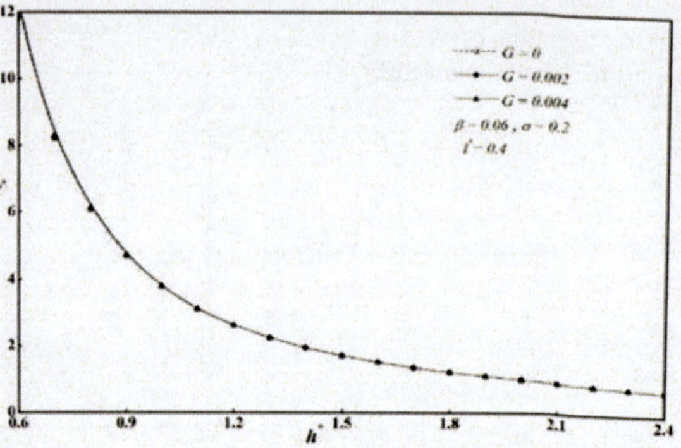

Fig. 9 Variation of W^* versus h^* for various values of G

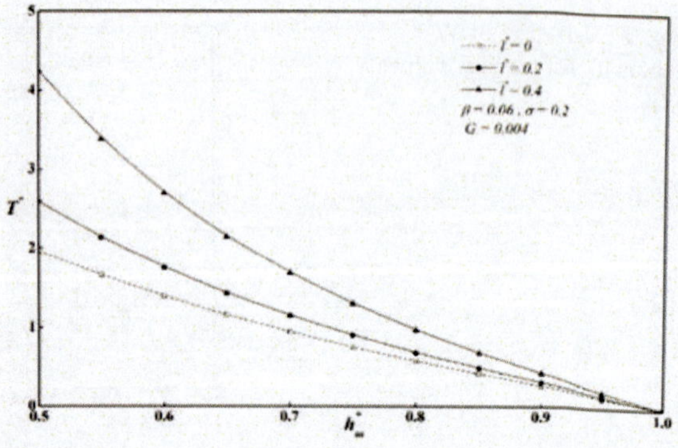

Fig. 10 Variation of T^* versus h_m^* for various values of l^*

of T^* along h_m^* is elaborated in Fig. 13 for distinct values of viscosity parameter G. As compared to the non-viscous case ($G = 0$), the Squeeze film time increases for larger values of viscosity parameter G.

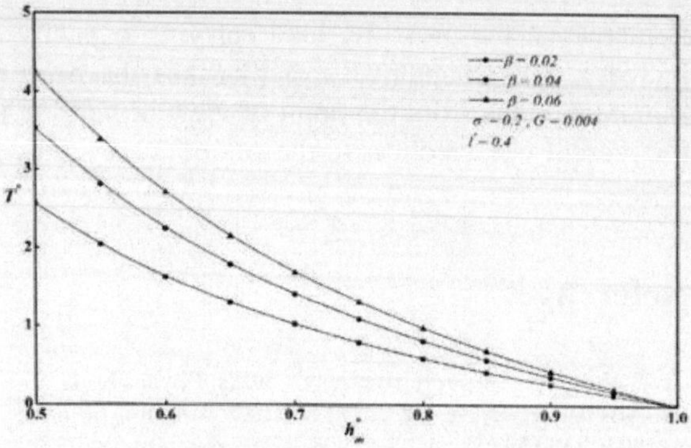

Fig. 11 Variation of T* versus h_m^* for various values of β

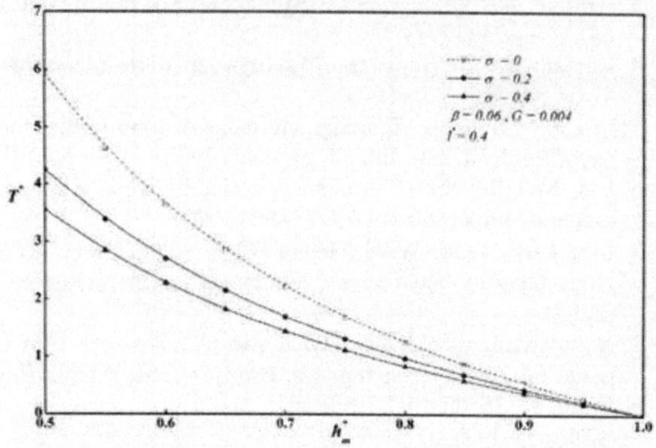

Fig. 12 Variation of T* versus h_m^* for various values of σ

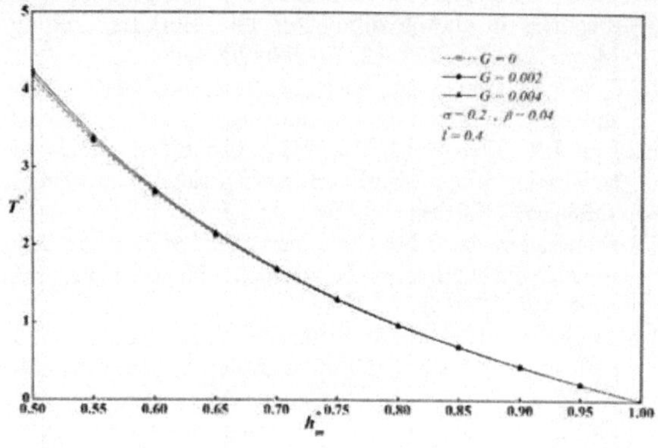

Fig. 13 Variation of T* versus h_m^* for various values of G

3 Conclusion

Based on Stoke's theory and the Barus analysis for viscosity variation, the following analysis is made: the bearing features such as $P*$, $W*$, and $T*$ increase steadily for increasing values of non-Newtonian parameter ($l*$) and viscosity parameter (G) as compared to the Newtonian and non-viscous case. Furthermore, as the value of

β increases, the pressure, load carrying capacity and Squeeze film time steadily increase. This chapter also emphasizes the need to keep the slip parameter to a minimum value from the point of view of bearing lifetime. When $G = 0$ & $\sigma = 0$, the present work reduces to the smooth case studied by Jaw-Ren.et al. [7].

References

1. Stokes, V.K.: Couple stresses in fluids. Phys. Fluids. **9**, 1709–1715 (1966)
2. Ariman, T., Sylvester, N.D.: Micro continuum fluid mechanics, a review. Int. J. Eng. Sci. **11**, 905–930 (1973)
3. Ariman, T., Sylvester, N.D.: Applications of micro continuum fluid mechanics. Int. J. Eng. Sci. **12**, 273–293 (1974)
4. Ramanaish, G.: Squeeze films between finite plates lubricated by fluids with couple stress. Wear. **54**, 315–320 (1979)
5. Lin, J.R.: Squeeze film characteristics of long partial journal bearings lubricated with couple stress fluids. Tribol. Int. **30**, 53–58 (1997)
6. Lin, J.R.: Squeeze-film characteristics between a sphere and a flat plate: couple stress fluid model. Comput. Struct. **75**, 73–80 (2000)
7. Lin, J.-R., Liao, W.-H., Hung, C.-R.: The effects of couple stresses in the squeeze film characteristics between a cylinder and a plane surface. J. Mar. Sci. Technol. **12**, 119–123 (2004)
8. Naduvinamani, N.B., Siddangouda, A.: Squeeze film lubrication between circular stepped plates of couple stress fluids. J. Braz. Soc. Mech. Sci. Eng. **31**, 21–26 (2008)
9. Barus, C.: Isothermal, isopiestics, and isometrics relative to viscosity. Am. J. Sci. **45**(6), 87–96 (1893)
10. Bartz, W.J., Ether, J.: Influence of pressure viscosity oils on pressure, temperature, and film thickness in elastohydrodynamically lubricated rolling contacts. Proc. IMechE, Part C: J. Mech. Eng. Sci. **222**, 1271–1280 (2008)
11. Lin, J.R., Chu, L.M., Liang, L.J.: Effects of viscosity dependency on the non-Newtonian squeeze film of parallel circular plates. Lubr. Sci. **25**, 1–9 (2012)
12. Lin, J.R., Chu, L.M., Li, W.L.: Combined effects of piezo-viscus dependency and non-Newtonian couple stresses in wide parallel-plate squeeze-film characteristics. Tribol. Int. **44**, 1598–1602 (2011)
13. Hanumagowda, B.N.: Combined effect of pressure-dependent viscosity and couple stress on squeeze film lubrication between circular step plates. Proc. IMechE, Part J: J Eng. Tribol. **229**, 1056–1064 (2015)
14. Aminkhani, H., Dali, M.: Effects of piezo-viscous–coupled stress lubricant on the squeeze film performance of parallel triangular plates. Proc. IMechE Part J: J. Eng. Tribol., 1–8 (2019)
15. Ashmawy, E.A.: Unsteady Couette flow of a micropolar fluid with slip. Meccanica. **47**, 85–94 (2012)
16. Thompson, P.A., Troian, S.M.: A general boundary condition for liquid flow at solid surfaces. Nature. **389**, 360–362 (1997)
17. Hai, W.: Effect of velocity-slip on the squeeze film between porous rectangular plates. Wear. **20**(1), 67–71 (1972)
18. Sparrow, E.M., Beavers, G.S., Hwang, I.T.: Effect of velocity slip on porous-walled squeeze films. J. Tribol. **94**(3), 260–264 (1972)
19. Prakash, J., Vij, S.K.: Effect of velocity slip on the squeeze film between rotating porous annular discs. Wear. **38**(1), 73–85 (1976)
20. Patel, K.C.: The hydromagnetic squeeze film between porous circular disks with velocity slip. Wear. **58**(2), 275–281 (1980)

Turbulence Generators and Turbulence Structure

R. Panneer Selvam

1 Introduction

Most of the flows one faces in everyday life are turbulent flow. Understanding these flows helps to design many devices and structures. Experimental methods to understand turbulent flow are preferable but very expensive. Many times, experimental modeling like the wind tunnel may not represent the flow in the actual situation like atmospheric flow. Computer modeling is an alternative tool emerging due to computer speed and storage availability. At the same time, computer modeling of turbulent flow is a challenging area. Many of the industrial applications are modeled using Reynolds-averaged Navier-Stokes (RANS) equations as detailed in [1–3]. But the RANS equations do not provide time-varying quantities like peak pressure on a building. They only provide mean velocities and pressure as well as statistical quantities like turbulent kinetic energy.

For several important applications, time-varying quantities are important. This can be accomplished using direct simulation (DS) or large eddy simulation (LES). The direct simulation can be applied only for a limited situation, but in general practical applications, LES is preferred due to developments in computer storage and speed. For improved accuracy in modeling turbulence using LES, one needs to use inflow turbulence at the inlet [2] and [4, 5] or turbulence in the flow [6]. Different inflow turbulence methods are reviewed by Dhamankar et al. [7] and Wu [8].

Mansouri et al. [4, 5] used several inflow turbulence generators (six different methods) to evaluate their performance in both predicting flow and peak pressure on a building. For flow, evaluation, or understanding the transport of turbulence,

R. Panneer Selvam (✉)
Univeristy of Arkansas, Fayetteville, AR, USA
e-mail: rps@uark.edu

F. Kamalov et al. (eds.), *Advances in Mathematical Modeling and Scientific Computing*, Trends in Mathematics, https://doi.org/10.1007/978-3-031-41420-6_24

they compared wind spectrum at the inflow and in the interior of the computational domain. They found that turbulence in the interior of the domain where the building is located loses its energy for several reasons. To understand further turbulence in the flow, the flow structure of turbulence using three different inflow turbulence generators are considered here. The three different methods considered in this work are two types of Random Fourier method (RFM) and precursor method. The RFM-based NSRFG method is discussed in detail by Yu et al. in [9], and the other RFM-based method called CSDW is reported in Selvam et al. [10]. The precursor method discussed in detail by Shi and Yeo [11] and Yeo and Shi [12] is followed in the CFD implementation.

2 Identification of the Turbulent Flow Structure

During the turbulent flow, eddies or coherent structure (CS) are formed. Sometimes, they are also called vortex. The CS regions are high vorticity regions. To identify these regions, several researchers proposed several ways. Methods of coherent structure identification as per Sengupta and Sharma [13] are as follows:

1. Q-criteria by Hunt et al. [14]
2. $\lambda 2$-criteria by Jeong and Hussain [15]
3. Disturbance enstrophy transport equation (DETE) by Sengupta et al. [16]
4. Disturbance mechanical energy (DME) equation by Sengupta et al. [17]

The first two methods are independent of time, and the last two methods depend on time. This is an ongoing research, and several researchers propose more new methods to identify the CS much better. Q can be calculated by taking the divergence of the NS equation as shown below:

$$\rho \frac{\partial^2 \mathrm{p}}{\partial \mathrm{x}_i \partial \mathrm{x}_i} = -\frac{\partial u_i}{\partial x_j}\frac{\partial u_j}{\partial x_i} = 2Q \tag{1}$$

where:

$$Q = \frac{1}{2}\left(\left(\frac{\partial u}{\partial x}\right)^2 + \left(\frac{\partial v}{\partial y}\right)^2 + \left(\frac{\partial w}{\partial z}\right)^2 + 2\left(\left(\frac{\partial u}{\partial y}\right)\left(\frac{\partial v}{\partial x}\right) + \left(\frac{\partial u}{\partial z}\right)\left(\frac{\partial w}{\partial x}\right) + \left(\frac{\partial v}{\partial z}\right)\left(\frac{\partial w}{\partial y}\right)\right)\right) \tag{2}$$

In the above equations, u, v, and w are velocities and p is the pressure. The Q criterion, magnitude of vorticity (ω_{mag}), and ω_z will be used in this work for turbulence structure comparisons. The Q criterion can be calculated directly from any of the visualization tool like tecplot and paraview.

3 Computer Modeling

The flow is modeled as 3D incompressible flow. The turbulence is modeled using large eddy simulation (LES). The governing equations are approximated by finite difference method. The details of the solution solver and flow computation are reported in [2, 4]. The boundary conditions for RFM as inflow turbulence are inlet and outlet in the streamwise direction (x) and symmetric boundary conditions in the spanwise (y) direction. In the z-direction, the boundary conditions are for bottom wall no-slip condition with a law of the wall wind profile and for top boundary symmetric boundary condition. For the precursor method, in the z-direction same as the boundary conditions used in RFM as inflow turbulence and in the x- and y-directions periodic boundary conditions.

For the precursor method, the initial turbulence is calculated using random numbers and then the flow is computed with a pressure source term in the x-direction of the Navier-Stokes equations. For the RFM as inflow turbulence; the initial velocities are calculated using the inflow generator equations for the whole domain. The inflow boundary conditions for the RFM at every time step are computed using the inflow turbulence generator. For all the methods, the mean velocities are considered to have logarithmic profile and the details are provided in [2, 4].

The computational region for RFM using the NSRFG and the CSDW as inflow turbulence are 13.25H × 9.375H × 5H and for precursor method 10H × 10H × 5H, where H is the reference height. To have higher numerical accuracy, equal spacing grid is considered. The nodes in each direction for both domains are (213 ×151 × 81) and (101 × 101 × 50), respectively. For flow computation, the governing equations are non-dimensionalized by the mean velocity (U_{ave}) at the reference height H and reference length H.

4 Results

Mansouri et al. in [4, 5] investigated several inflow turbulence generators and compared the velocity spectrum at the inflow and at the interior point where structure may be introduced. Some methods compared well with Von Karman spectrum at least on the low frequency region and some methods did not compare well at all. All methods have difficulty in transporting the high frequency side comparing to the inflow turbulence provided. This was shown to be due to grid resolution. A grid spacing h can transport maximum frequency of $f_{Nq} = 1/(2h)$, where f_{Nq} is the Nyquist frequency. For the second-order FDM, the grid frequency $f_{grid} = 1/4h$ or less only can be transported. In addition, Mansouri et al. [4] observed spurious pressure at the inlet and in the interior region because the inflow turbulence does not satisfy the NS equations.

The wind spectrum is a statistical quantity, and hence it does not provide a good picture of the turbulence structure. Also, they are continuous functions. Hence, we

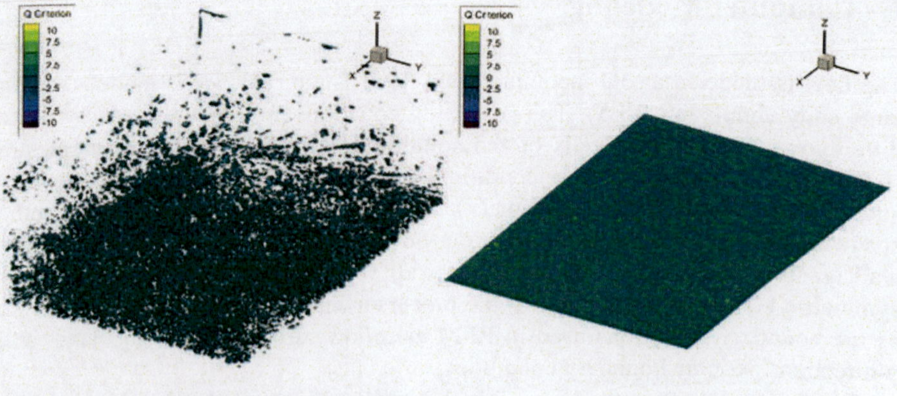

Fig. 1 Iso-surface (Q = −1) and contour plot at the plane H from the ground of Q-criterion using NSRFG inflow turbulence generator

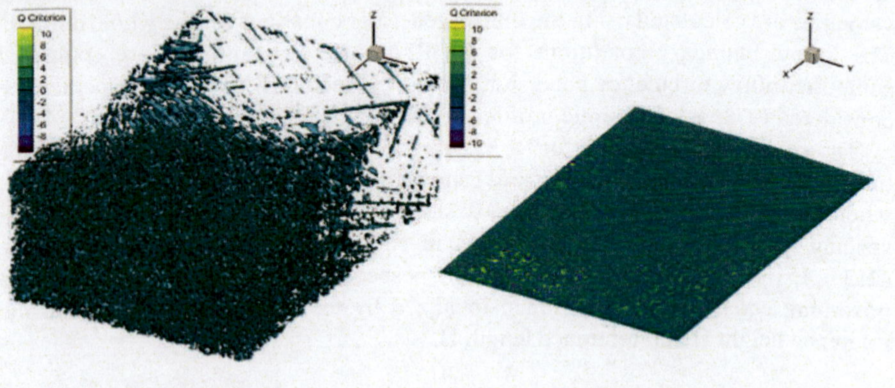

Fig. 2 Iso-surface (Q = −1) and contour plot at the plane H from the ground of Q-criterion using CSDW inflow turbulence generator

will try to visualize the flow structure with different quantities like Q-criterion, vorticity, and magnitude of the vorticity.

4.1 Visualization of Turbulent Structure Using Q-Criterion

The Q-criterion values were calculated for all the three methods stated above using tecplot for visualization. The peak values for each method are as follows: NSRFG (−25 to 23), CSDW (−47 to 59), and precursor (−7 to 5). The iso-surface for $Q = -1$ is shown in Figs. 1, 2, and 3 for the three methods on the left side. The iso-surface of the CSDW method is much denser than other two methods. The precursor method has lots of blank space. The Q contours in the z-plane at the reference height H from the wall is shown in Figs. 1, 2, and 3 on the right side. The CSDW method

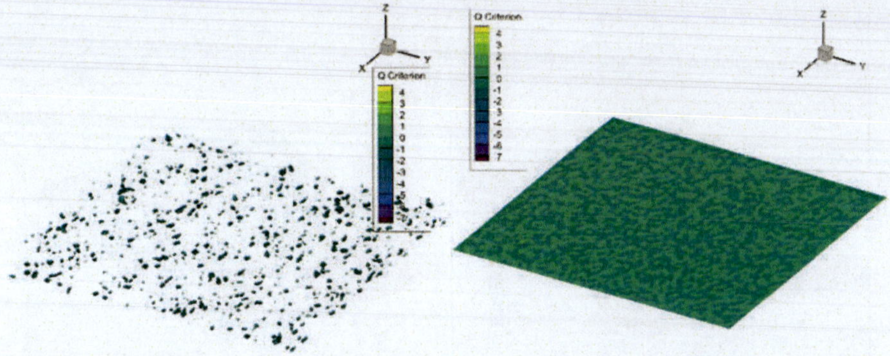

Fig. 3 Iso-surface ($Q = -1$) and contour plot at the plane H from the ground of Q-criterion using precursor turbulence generator

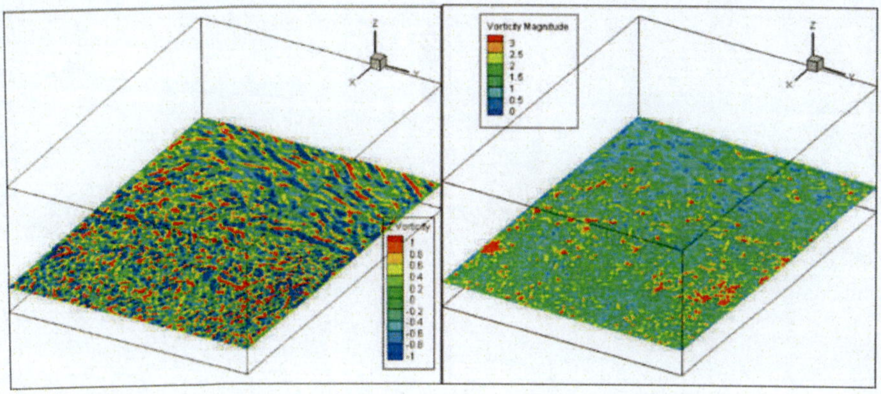

Fig. 4 Comparison of ω_z and ω_{mag} for NSRFG inflow turbulence generator

has a longer structure until the mid-portion in the x-direction, and later, due to the energy cascade, the structure becomes smaller. The precursor method has a smaller uniform structure, and the Q values are much smaller than other methods.

4.2 *Visualization of Turbulent Structure Using Vorticity and Magnitude of Vorticity*

For further understanding of the turbulence structure, ω_z in the z-plane and ω_{mag} in the z-plane are plotted as shown in Figs. 4, 5, and 6. The ω_z are plotted on the left side and the ω_{mag} are plotted on the right side. The contours are plotted at the reference height H as before. From the figure, one can observe that vorticity contours have much more information. The maximum ω_{mag} in the computational domain

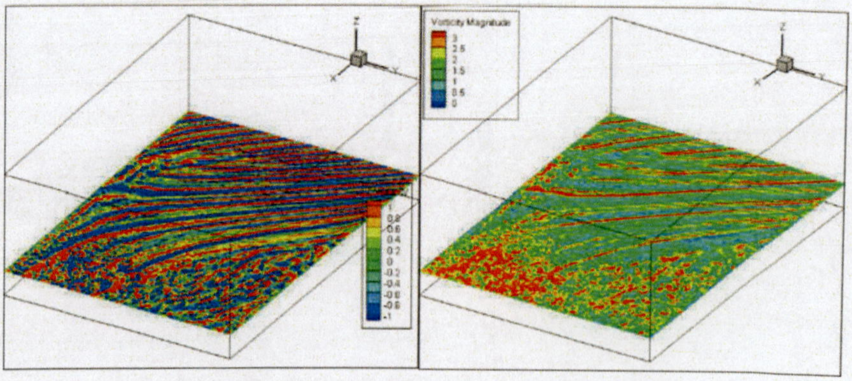

Fig. 5 Comparison of ω_z and ω_{mag} for CSDW inflow turbulence generator

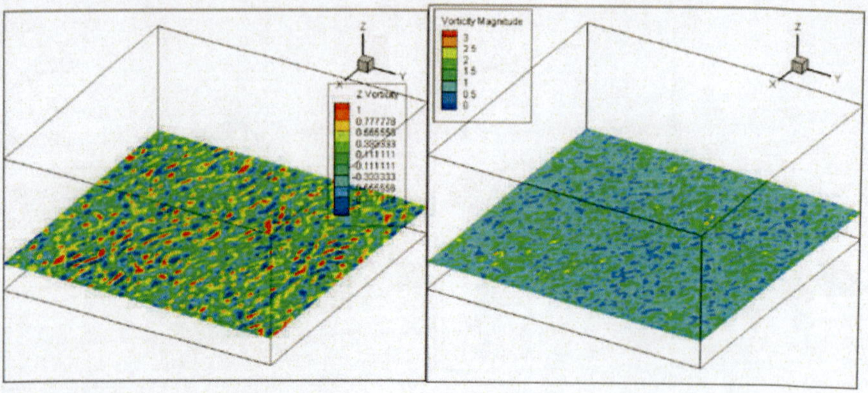

Fig. 6 Comparison of ω_z and ω_{mag} for precursor turbulence generator

for NSRFG, CSDW, and precursor method are 14, 17.5, and 6.3, respectively. Hence, CSDW has maximum vortex strength. From the contour plots, one can see that CSDW has long vorticity structure from the inflow compared to NSRFG and precursor method. The NSRFG has elongated structure in the y-direction closer to the inflow. The NSRFG and CSDW both have smaller vorticity structure in the outflow region. The precursor method has smaller uniform structure over the entire region.

Further investigation with other inflow turbulence generators is needed to know more about the turbulence structure and the one that needs to be used for computing peak pressure around buildings. Mansouri et al. [5] considered seven different inflow turbulence methods for wind engineering, and they will be investigated for turbulence structure in the future work.

5 Conclusions

Three inflow turbulence generators are considered to compare the turbulence structure. Methods used to identify coherent structure are reviewed, and the methods selected to compare the inflow turbulence generators are the Q-criterion and vorticity magnitude. Each inflow turbulence generator has a different turbulent structure. The CSDW and NSRFG methods have a longer structure at the entrance compared to the precursor method. As the flow develops with the inflow and outflow boundary conditions, due to the energy cascade, the inflow turbulence structure changes as it progresses into the domain. The precursor method has a smaller uniform structure all over the region due to periodic boundary conditions, and the flow gets recirculated several times. From the study, it is concluded that each inflow turbulence generator is producing a different turbulence structure, and hence the expected computation of pressures around a building should be different. More understanding is needed to model a turbulent flow around buildings. In addition, the turbulence structure in the wind tunnel and field measurements need to be measured.

References

1. Kajishima, T., Taira, K.: Computational Fluid Dynamics: Incompressible Turbulent Flows. Springer (2017)
2. Selvam, R.P.: Computational Fluid Dynamics for Wind Engineering. Wiley-Blackwell, Chichester (2022)
3. Versteeg, H., Malalasekera, W.: An Introduction to Computational Fluid Dynamics: The Finite Volume Method, 2nd edn. Pearson (2007)
4. Mansouri, Z., Selvam, R.P., Chowdhury, A.G.: Maximum grid spacing effect on peak pressure computation using inflow turbulence generators. Results Eng. **15**, 100491 (2022)
5. Mansouri, Z., Selvam, R.P., Chowdhury, A.G.: Performance of different inflow turbulence methods for wind engineering applications. J. Wind Eng. Ind. Aerodyn. **226**, 105141 (2022)
6. Orlandi, P.: Fluid Flow Phenomena: A Numerical Tool Kit. Kluwer Academic Publishers (2001)
7. Dhamankar, N., Blaisdell, G., Lyrintzis, A.: Overview of turbulent inflow boundary conditions for large-eddy simulations. AIAA J. **56**, 1–18 (2017)
8. Wu, X.: Inflow turbulence generation methods. Annu. Rev. Fluid Mech. **49**, 23–49 (2017)
9. Yu, Y., Yi Yang, Y., Xie, Z.: A new inflow turbulence generator for large eddy simulation evaluation of wind effects on a standard high-rise building. Build. Environ. **138**, 300–313 (2018)
10. Selvam, R.P., Chowdhury, A., Irwin, P., Mansouri, Z., Moravej, M.: CFD peak pressures on TTU building using continuity satisfied dominant waves (CSDW) method as inflow turbulence generator. Report, Department of Civil Engineering, University of Arkansas (2020)
11. Shi, L., Yeo, D.: Large eddy simulations of model-scale turbulent atmospheric boundary layer flows. J. Eng. Mech. **143**(9), 06017011 (2017)
12. Yeo, D., Shi, L.: Chapter 9: Computational versus wind tunnel simulation of atmospheric boundary layer flow for structural engineering applications. In: Wind Engineering for Natural Hazards: Modeling, Simulation and Mitigation of Windstorm Impact on Critical Infrastructure, pp. 169–191. ASCE (2018)

13. Sengupta, T., Sharma, P.K.: Chapter 2: Space-time resolution for transitional and turbulent flows. In: Pirozzoli, S., Sengupta, T.K. (eds.) High-Performance Computing of Big Data for Turbulence and Combustion CISM International Centre for Mechanical Sciences, vol. 592, pp. 31–54. Springer, Cham (2019)
14. Hunt, J.C.R., Wray, A.A., Moin, P.: Eddies, streams, and convergence zones in turbulent flows. In: Proceedings of the Summer Program 1988, CTR Report Stanford University, pp. 193–208. Center for Turbulence Research, Stanford (1988)
15. Jeong, J., Hussain, F.: On the identification of a vortex. J. Fluid Mech. **332**, 339–363 (1995)
16. Sengupta, A., Suman, V.K., Sengupta, T.K., Bhaumik, S.: An enstrophy-based linear and nonlinear receptivity theory. Phys. Fluids. **30**(5), 054105 (2018)
17. Sengupta, T.K., De, S., Sarkar, S.: Vortex-induced instability of an incompressible wallbounded shear layer. J. Fluid Mech. **493**, 277–286 (2003)

Conjugate Buoyant Convection of Nanoliquids in a Porous Saturated Annulus

B. V. Pushpa, A. Rex Macedo Arokiaraj, Geetha Baskaran, and R. D. Jagadeesha

1 Introduction

Buoyant induced convection (BIC) and associated thermal transport (TT) in finite-sized geometries, in particular an annulus geometry structured between two coaxial cylindrical tubes filled with different fluids, have been widely investigated due to the relevance of this geometry to the physical structure of many industrial applications. In this direction, Khan and Kumar [1] made a detailed analysis on BIC in an upright annular domain to investigate the wide range of parameters on convective flow and related thermal transport processes and proposed heat transfer correlations. In the same geometry, the significant impacts of isolated heating on BIC have been thoroughly investigated with an emphasis on the cooling of electronic equipment having discrete thermal sources [2, 3]. In the nanofluid-filled annular region, the pioneering work of Abouali and Falahatpisheh [4] investigates the detailed impacts of nanoparticle (NP) concentrations, radius ratios, and derived thermal transport correlations from their simulations. By utilizing oil-based nanofluids

B. V. Pushpa
Department of Information Technology, University of Technology and Applied Sciences, Nizwa, Oman

A. R. M. Arokiaraj (✉)
Department of Information Technology, University of Technology and Applied Sciences, Ibri, Oman
e-mail: rex.ibr@cas.edu.om

G. Baskaran
Department of Mechanical Engineering, University of Technology and Applied Sciences, Ibri, Oman

R. D. Jagadeesha
Department of Mathematics, Government Science College, Hassan, Karnataka, India

F. Kamalov et al. (eds.), *Advances in Mathematical Modeling and Scientific Computing*, Trends in Mathematics, https://doi.org/10.1007/978-3-031-41420-6_25

(NFs), Cadena-de la Peña et al. [5] carried out an experimental investigation to explore the thermal transport enhancement of two NPs in an upright annular domain. The BIC and TT of nanofluids with various base fluids (BFs) in an annular region have been analyzed by Oudina [6] to understand the effect of isolated heaters and concluded that heat dissipation rates are greatly influenced by the choice of base fluids. Recently, the impacts of a thin baffle on BIC and TT rate in the annular domain have been investigated with an aim of controlling the buoyant nanofluid flow with the help of a thin baffle [7].

The flow movement and associated TT behavior in a finite-sized geometries filled with nanofluid-saturated porous medium is being investigated in great details. Reddy and Sankar [8] made a detailed investigation on the BIC and TC of nanofluids with two NPs inside a porous annulus by accounting three models. The detailed flow and thermal structure as well TT rates have been discussed based on different porous media models and brought out the influences of each model. In an inclined porous annulus, Swamy et al. [9] analyzed the TT and entropy generation rates for various aspect ratios and cavity inclinations by utilizing the Darcy model. BIC in a confined geometry with one or more of the boundaries having finite thickness have sound applications, such as thermal bridge, gas turbine blade cooling, and heat barrier where the wall conductivity needs to be taken into account. The pioneering investigation in a square domain with thick boundaries has been performed by Kaminski and Prakash [10] by choosing three wall conductivity models. For higher magnitudes of Ra, they noticed that the symmetric flow structure is disturbed and the temperature at the interface is nonuniform. Few important and prominent studies pertaining to conjugate TT of NFs in a square porous region and nonuniform thermal conditions are due to [11–13]. An annular domain with a thick inner wall containing three different NFs has been numerically analyzed by Sankar et al. [14], and they found that the TT at the wall and interface strongly depends on the conductivity ratio. A comprehensive review of various studies involving different NFs, geometries, models, and thermal boundary conditions can be found in [15, 16]. Some of the recent studies in porous geometries could be found in [17, 18]. The systematic and comprehensive inspection of literature vividly indicates that the combined impacts of wall thickness and nonuniform thermal condition in a porous annulus have not been attempted.

2 Mathematical Formulation

In this analysis, we chose a cylindrical annular domain constructed between two coaxial cylinders as shown in Fig. 1. Here the inner cylinder with radius r_i has finite thickness ϵ and is uniformly heated with a higher temperature. However, the outer cylinder is imposed with a sinusoidally varying thermal condition. Also, the upper and lower boundaries are thermally insulated. The annular gap is filled with $Cu - H_2O$ nanofluid saturated with porous medium. In addition, the fluid is chosen as incompressible and Newtonian, and the fluid and porous material are in thermal

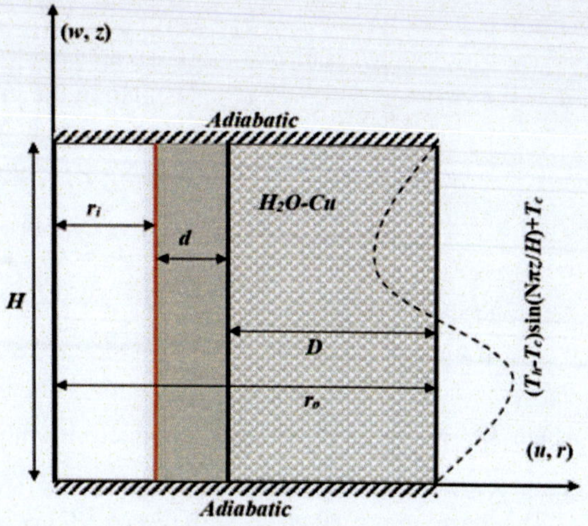

Fig. 1 The axisymmetric structure of annulus along with coordinates

Table 1 Thermophysical characteristics of H_2O and Cu NP [9]

Property	ρ(kg/m^3)	C_p(J/kg K)	k(W/mK)	β(K^{-1})
H_2O	997.1	4179	0.613	21×10^{-5}
Cu	8933	385	400	1.67

equilibrium condition. The momentum equations are modeled using the Darcy's law, and the standard Boussinesq approximation has been applied to the body force term of the momentum equation. The thermophysical characteristics of base fluid and NP are provided in Table 1 [9]. Based on these assumptions, the dimensionless PDEs governing the buoyant convection flow are [9, 14]:

$$\frac{\partial T_{nf}}{\partial t} + U\frac{\partial T_{nf}}{\partial R} + W\frac{\partial T_{nf}}{\partial Z} = \frac{\alpha_{nf}}{\alpha_f}\nabla^2 T_{nf}, \tag{1}$$

$$\frac{\partial T_w}{\partial t} = \frac{\alpha_w}{\alpha_f}\nabla^2 T_w, \tag{2}$$

$$\frac{1}{R}\left[\frac{\partial^2\psi}{\partial R^2} - \frac{1}{R}\frac{\partial\psi}{\partial R} + \frac{\partial^2\psi}{\partial Z^2}\right] = -R(1-\phi)^{2.5} Ra_D \frac{(\rho\beta)_{nf}}{\rho_{nf}\beta_f}\frac{\partial T_{nf}}{\partial R}, \tag{3}$$

$$U = \frac{1}{R}\frac{\partial\psi}{\partial Z}, \quad W = -\frac{1}{R}\frac{\partial\psi}{\partial R}. \tag{4}$$

where $\nabla^2 = \frac{\partial^2}{\partial R^2} + \frac{1}{R}\frac{\partial}{\partial R} + \frac{\partial^2}{\partial Z^2}$ and $Ra_D = \frac{gK\beta_f \Delta\theta D}{\nu_f \alpha_f}$ is the Darcy-Rayleigh number. Also, the subscripts nf, w, and f respectively indicate the NF, wall, and base fluid.

The auxiliary conditions in nondimensional form are:

$$t = 0: \quad U = W = T_w = T_{nf} = 0, \ \psi = 0; \qquad \text{all over the region.}$$

$$t > 0: \quad \psi = \frac{\partial \psi}{\partial R} = 0, \ T_w = 1 \qquad \text{along the inner boundary}$$

$$\psi = \frac{\partial \psi}{\partial R} = 0, \ T_{nf} = sin(N\pi Z) \quad \text{along the outer boundary}$$

$$\psi = \frac{\partial \psi}{\partial Z} = 0, \ \frac{\partial T_w}{\partial Z} = \frac{\partial T_{nf}}{\partial Z} = 0 \quad \text{at the upper and lower boundaries}$$

$$\frac{\partial T_{nf}}{\partial R} = Kr \frac{\partial T_w}{\partial R} \ \text{at the interface,}$$

where $Kr = \frac{k_w}{k_{nf}}$ is the thermal conductivity ratio, N is the periodicity parameter, A and λ are respectively the aspect and radius ratios.

The thermophysical characteristics of NF are based on the following expressions [15, 16]:

$$\rho_{nf} = (1-\phi)\rho_f + \phi\rho_p,$$

$$(\rho\beta)_{nf} = (1-\phi)(\rho\beta)_f + \phi(\rho\beta)_p,$$

$$(\rho C_p)_{nf} = (1-\phi)(\rho C_p)_f + \phi(\rho C_p)_p,$$

$$\frac{k_{nf}}{k_f} = \frac{k_p + 2k_f - 2\phi(k_f - k_p)}{k_p + 2k_f + \phi(k_f - k_p)},$$

$$\mu_{nf} = \frac{\mu_f}{(1-\phi)^{2.5}},$$

$$\alpha_{nf} = \frac{k_{nf}}{(\rho C_p)_{nf}}.$$

Here, the subscripts nf, f, and p denoted the NF, BF, and NP, respectively.

The thermal transport efficiency is estimated using the average Nusselt numbers at solid interior boundary ($\overline{Nu_W}$) and along the interface ($\overline{Nu_I}$) using the following expressions:

$$\overline{Nu_W} = \frac{1}{A}\int_0^A Nu_W \, dZ \ \text{ and } \ \overline{Nu_I} = \frac{1}{A}\int_0^A Nu_I \, dZ,$$

where $Nu_W = -\left(\frac{\partial T}{\partial R}\right)_{R=\frac{1}{\lambda-1}}$ and $Nu_I = -\frac{k_{nf}}{k_f}\left(\frac{\partial T}{\partial R}\right)_{R=\varepsilon}.$

3 Solution Technique

In this analysis, though the Darcy model has been utilized, the model PDEs are nonlinear and coupled; thus as a result, the closed-form solutions could not be obtained. Hence, an implicit finite difference methodology has been chosen to obtain the solution of the governing PDEs. In particular, we adopt the time splitting and line over relaxation methods for energy and momentum equations, respectively. A systematic grid independence resolution has been performed, and after careful inspection, a grid size of 161×161 has been adopted for all simulations. We have developed an in-house code to solve the finite difference equations, and the simulations are validated before the actual results are obtained. The more detailed description of the methodology and validation can be found in our recent works [8, 9] and the same is not given for brevity.

4 Results and Discussion

In this section, the simulation results for flow and thermal distribution as well as TT rates are discussed for various critical parameters chosen in this analysis. Firstly, the influences of pertinent parameters on the flow and temperature distributions have been predicted, and the influences of these parameters on TT rates have been also estimated.

Figure 2 portrays the impacts of *Ra* on streamline and isotherm contours by fixing other parameters. Due to the presence of porosity, the flow rate at low as well as high magnitudes of *Ra* is moderate; however, the impact of higher *Ra* is noticeable compared to lower magnitude. Also, as the periodicity parameter is

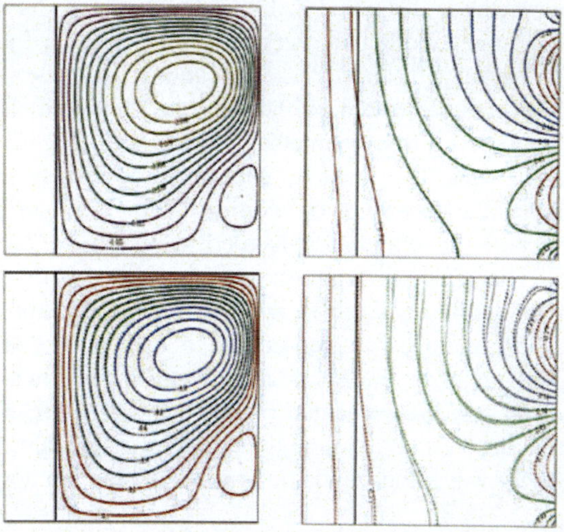

Fig. 2 Impact of Ra_D and ϕ on flow and thermal contours for $Kr = 1, N = 2$ and $\epsilon = 0.2$. $Ra_D = 10^2$ (top) and $Ra_D = 10^3$ (bottom). Solid line (H_2O) and dotted line (NF)

Fig. 3 Impacts of wall thickness (ε) on flow and thermal contours $Kr = 1, N = 2$ and $Ra_D = 10^3$. $\epsilon = 0.1$ (left), $\epsilon = 0.3$ (middle) and $\epsilon = 0.5$ (right). Solid line (H_2O) and dotted line (NF)

kept at $N = 2$, an additional eddy is visible, and thermal contours reveal hot and cold regions along the outer boundary. The wall thickness is another important parameter in conduction-convection analysis, and its significance on flow and thermal distribution is identified through Fig. 3 by choosing 10%, 30% and 50% of the inner wall thickness. An increase in wall width leads to a significant drop in NF buoyant flow rate due to nonavailability of annular space, and also an enhancement in ϵ reveals less availability of thermal energy at the interface. In conduction-convection thermal transport investigation, the conductivity of solid wall and NF along the interface is not necessarily the same, and this leads to another important parameter, known as thermal conductivity ratio, which is the ratio between solid and fluid thermal conductivity. The importance of this parameter on streamlines and isopleths is discussed in Fig. 4 by choosing the ratio of conductivities as meager, moderate, and higher. A considerable increase in the flow circulation rate has been observed with an increase in conductivity ratio, and bicellular flow structure has been reduced to unicellular pattern. The significance of conductivity ratio can also be observed from isopleths, indicating the higher to lower thermal gradients in the solid wall. The impacts of periodicity parameter N on flow and thermal pattern reveal the generation of additional eddies with an enhancement of N and associated variation in temperature isopleths (Fig. 5).

The TT rate from the solid wall and along the interface is an important quantitative measure in any thermal design, and this quantity has been estimated for different parameter ranges in Figs. 6, 7, 8, and 9. The collective impacts of Ra and ϕ on the boundary as well as interface average Nu number have been estimated and presented in Fig. 6 by fixing other parameters. It should be noted that the TT rate is found to be higher for smaller magnitudes of Ra as compared to larger value and this is due to the conduction dominant mode. Also, the wall heat

Fig. 4 Impact of Kr on flow and thermal contours $Ra_D = 10^3$, $N = 2$ and $\epsilon = 0.2$. $Kr = 0.1$ (left), $Kr = 1$ (middle) and $Kr = 10$ (right). Solid line (H_2O) and dotted line (NF)

Fig. 5 Influence of N on flow and thermal contours for $Ra_D = 10^3$, $\epsilon = 0.2$ and $Kr = 1$. $N = 0$, $N = 2$, $N = 4$ and $N = 6$ (Left to right). Solid line (H_2O) and dotted line (NF)

transport rate is on higher magnitude than the transport rates at the interface. Further, the NP concentration has no impact on Nu_W; however, it has significantly altered the Nu_I. Figure 7 depicts the combined influences of Rayleigh number and wall thickness (ϵ) on the total TT rate at the interior boundary and interface. From the predictions, it can be noticed that the dissipation rate at the wall increases, while Nu_I decreases with ϵ. The impacts of conductivity ratio on the global TT rate along the inner boundary and interface, shown in Fig. 8, indicate that Kr is the most significant parameter to control the transport rates along both the wall and interface. Figure 9 displays the impacts of periodicity parameter N for three values of Ra. It is interesting to note that the variation of Nu along the wall and interface appears almost similar. This reveals an important fact that the uniform heating of the outer wall ($N = 0$) produces higher TT rate as compared to nonuniform heating ($N > 0$).

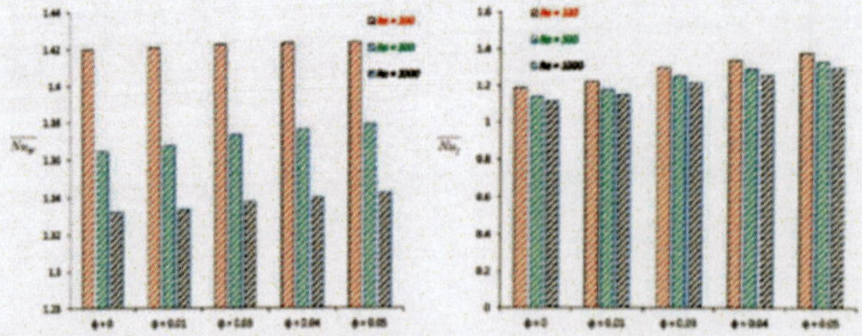

Fig. 6 Collective influences of Ra_D and ϕ on $\overline{Nu}$ for $Kr = 1$, $N = 2$ and $\epsilon = 0.2$

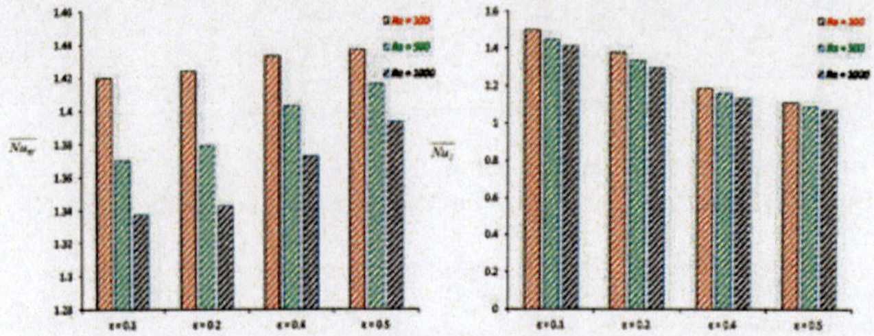

Fig. 7 Collective influences of Ra_D and ε on $\overline{Nu}$ for $Kr = 1$, $N = 2$ and $\phi = 0.05$

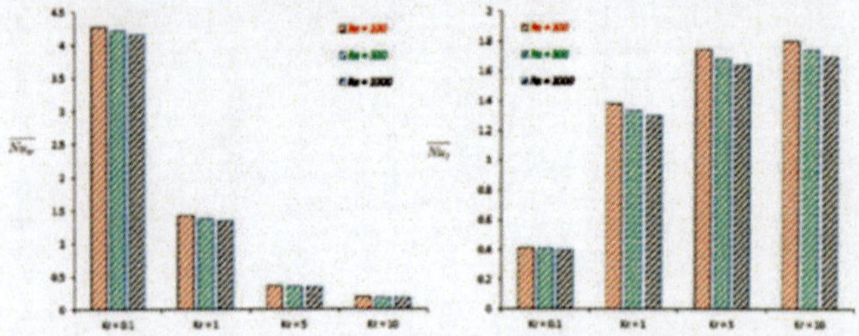

Fig. 8 Collective influences of Ra_D and Kr on $\overline{Nu}$ for $\epsilon = 0.2$, $N = 2$ and $\phi = 0.05$

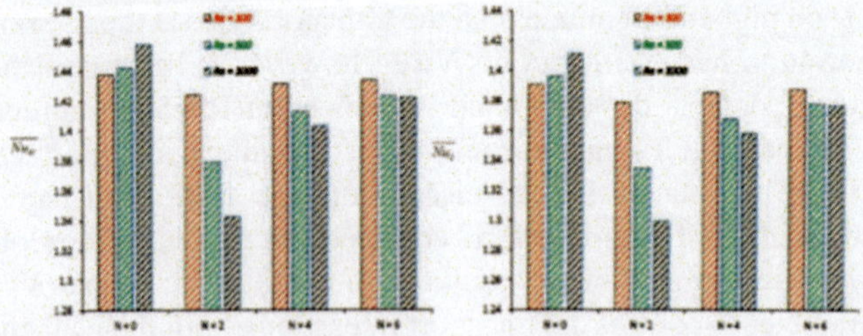

Fig. 9 Collective influences of Ra_D and N on $\overline{Nu}$ for $Kr = 1$, $\epsilon = 0.2$ and $\phi = 0.05$

5 Conclusions

The collective effects of inner wall thickness and nonuniform heating of the outer wall on Cu-based NF buoyant flow and TT in a porous annular region has been numerically investigated and the following predictions are observed:

1. The periodicity parameter (N) induces multicellular flow and hence reduces TT rate. Nu_W increases with Kr, but Nu_I decreases with Kr. An increase in wall thickness produces higher TT rates along the wall, but lower TT rates along the interface.
2. The current study utilizes the Darcy model which is valid for low-velocity flows. This limitation could be improved by considering a generalized porous medium model to predict the results.

References

1. Khan, J., Kumar, R.: Natural convection in vertical annuli: a numerical study for constant heat flux on the inner wall. ASME J. Heat Transfer-Trans. ASME **111**(4), 909–915 (1989)
2. Sankar, M., Kim, B., Lopez, J.M., Do, Y.: Thermosolutal convection from a discrete heat and solute source in a vertical porous annulus. Int. J. Heat Mass Transf. **55**(15–16), 4116–4128 (2012)
3. Lopez, J.M., Sankar, M., Do, Y.: Constant-flux discrete heating in a unit aspect-ratio annulus. Fluid Dyn. Res. **44**, 0655077 (2012)
4. Abouali, O., Falahatpisheh, A.: Numerical investigation of natural convection of Al_2O_3 nanofluid in vertical annuli. Heat Mass Transf. **46**(1), 15–23 (2009)
5. Cadena-de la Peña, N.L., Rivera-Solorio, C.I., Payán-Rodríguez, L.A., García-Cuéllar, A.J., López-Salinas, J.L.: Experimental analysis of natural convection in vertical annuli filled with AlN and TiO_2/mineral oil-based nanofluids. Int. J. Therm. Sci. **111**, 138–145 (2017).
6. Mebarek-Oudina, F.: Convective heat transfer of Titania nanofluids of different base fluids in cylindrical annulus with discrete heat source. Heat Transf.-Asian Res. **48**(1), 135–147 (2019)
7. Pushpa, B.V., Sankar, M., Mebarek-Oudina, F.: Buoyant convective flow and heat dissipation of Cu–H_2O nanoliquids in an annulus through a thin baffle. J. Nanofluids **10**, 292–304 (2021)
8. Reddy, N.K., Sankar, M.: Buoyant heat transfer of nanofluids in a vertical porous annulus: a comparative study of different models. Int. J. Numer. Methods Heat Fluid Flow **33**(2), 477–509 (2023)
9. Swamy, H.A.K., Sankar, M., Reddy, N.K.: Analysis of entropy generation and energy transport of Cu-water nanoliquid in a tilted vertical porous annulus. Int. J. Appl. Comput. Math. **8**(1), 10 (2022)
10. Kaminski, D.A., Prakash, C.: Conjugate natural convection in a square enclosure: effect of conduction in one of the vertical walls. Int. J. Heat Mass Transf. **29**(12), 1979–1988 (1986)
11. Sheremet, M.A., Pop, I.: Conjugate natural convection in a square porous cavity filled by a nanofluid using Buongiorno's mathematical model. Int. J. Heat Mass Transf. **79**, 137–145 (2014)
12. Alsabery, A.I., Chamkha, A.J., Saleh, H., Hashim, I.: Heatline visualization of conjugate natural convection in a square cavity filled with nanofluid with sinusoidal temperature variations on both horizontal walls. Int. J. Heat Mass Transf. **100**, 835–850 (2016)
13. Zahan, I., Alim, M.A.: Effect of conjugate heat transfer on flow of nanofluid in a rectangular enclosure. Int. J. Heat Technol. **36**(2), 397–405 (2018)

14. Sankar, M., Reddy, N.K., Do, Y.: Conjugate buoyant convective transport of nanofluids in an enclosed annular geometry. Sci. Rep. **11**(1), 1–22 (2021)
15. Guo, Z.: A review on heat transfer enhancement with nanofluids. J. Enhanc. Heat Transf. **27**(1), 1–70 (2020)
16. Mebarek Oudina, F., Chabani, I.: Review on nano-fluids applications and heat transfer enhancement techniques in different enclosures. J. Nanofluids **11**(2), 155–168 (2022)
17. N. Manjunatha, R. Sumithra, Effects of heat source/sink on Darcian-Benard-magneto-Marangoni convective instability in a composite layer subjected to nonuniform temperature gradients. TWMS J. App. Eng. Math. **12**(3), 969–984 (2022)
18. Manjunatha, N., Sumithra, R., Vanishree, R.K., Influence of constant heat source/sink on non-Darcian-Benard double diffusive Marangoni convection in a composite layer system. J. Appl. Math. Inf. **40**(1–2), 99–115 (2022)

Study of MHD with Couple Stress Fluid on Squeeze-Film Characteristics of Curved Annular Circular Plates

B. N. Hanumagowda, Swapna S. Nair, and A. Salma

1 Introduction

Magnetohydrodynamics is defined by the movement of electrically conducting fluids when electric and magnetic fields exist. For some decades, in the area of lubrication related to hydrodynamics, the characteristics of couple stress fluid (CSF) for various bearings have been studied by assuming that the lubricant has constant viscosity, though it is dependent on both temperature and pressure. In the past few years, Stokes micro continuum model [1] is the most fundamental one, and it allows polar effects for both, couple stress and body couples.

Researchers in the field of tribology are more interested in the usage of some additives as lubricants when a magnetic field is applied. Current research experiments [2, 3] based on oil combined with long-chain additives have shown that they enhance lubrication by reducing friction and surface damage. A modified form of the Reynolds equation was solved using a finite difference method by Elsharkawy [4], who attempted to examine the effects of geometry, pressure distribution, load carrying capacity, side leakage flow, and friction factor. The results showed that additives increase load carrying capacity while decreasing friction and side leakage coefficients. The authors formulated the modified Reynolds equation and found that lubricants with additives efficiently increase load supporting capacity and decrease the friction coefficient in comparison to Newtonian fluids. Naduvinamani et al. [5] and Ramesh Kudenatti et al. [6] detected that the impact of magnetohydrodynamics

B. N. Hanumagowda · A. Salma
Department of Mathematics, REVA University, Bengaluru, India

S. S. Nair (✉)
Department of Management and Commerce, Amity University, Dubai, United Arab Emirates
e-mail: snair@amityuniversity.ae

F. Kamalov et al. (eds.), *Advances in Mathematical Modeling and Scientific Computing*, Trends in Mathematics, https://doi.org/10.1007/978-3-031-41420-6_26

is essential for improving the time of squeezing film and load carrying. Syeda et al. [7] analyzed the squeeze-film character on various finite plates and noticed the enhanced performance of bearings in the presence of couple stress and MHD. Hanumagowda et al. [8] found that in conical bearings the magnetic field along with CSF enhances the properties of squeeze film. The study of the squeezing behavior of couple stress, between a flat plate and a sphere is analyzed by Lin [9], and it is noticed that the squeezing features of the system are enhanced. Naduvinamani and Siddangouda [10] analyzed theoretically that the couple stress variable impacts the performance of the squeeze film linking the circular stepped plates. It was observed that the CSF increases load support capacity and pressure and then decreases the squeezing time compared to the Newtonian fluids. Naduvinamani et al. [11] investigate the rheological effects of CSF on the squeezing behavior of porous journal bearings and indicate an increase in loading capacity. The combined impact of MHD and couple stress in anisotropic plates has been studied by Fathima et al. [12]. They found that the modified Reyonld's equation obtained with the combination of MHD and CSF is very beneficial for industrial applications. Hanumagowda et al. [13] examined the impact of CSF and magnetohydrodynamics on the characteristics of curved circular plates between squeeze film. With these results, it can be observed that there is an improvement in squeeze film. In this chapter, the consequences of magnetohydrodynamics and CFS on squeeze-film behavior between the curved annular circular plates are examined.

1.1 Theoretical Solution

Figure 1 shows the lubrication of squeeze film between two curved annular circular plates in the presence of a transverse magnetic field. The two curved annular circular plates are separated by a fluid film of central thickness h_m, and 'a' and 'b' are the inner and outer radii of the circle of the annular plate, respectively. A magnetic field B_0 is enforced perpendicular to the plates. β and γ are the curvature parameters for the upper and lower plates, respectively.

The thickness of fluid film is given as $h(r)$

$$h(r) = h_m \left\{ \exp\left(-\beta . r^2\right) - \frac{1}{1+\gamma r} + 1 \right\}; \quad a \le r \le b \tag{1}$$

The derived modified Reynolds equation by Hanumagowda et al. [13] for MHD couple-stress squeeze film between curved circular plates is

$$\frac{1}{r\mu} . \frac{\partial}{\partial r}\left[r . S\left(h, l, M_0\right) \frac{\partial p}{\partial r} \right] = V \tag{2}$$

Where,

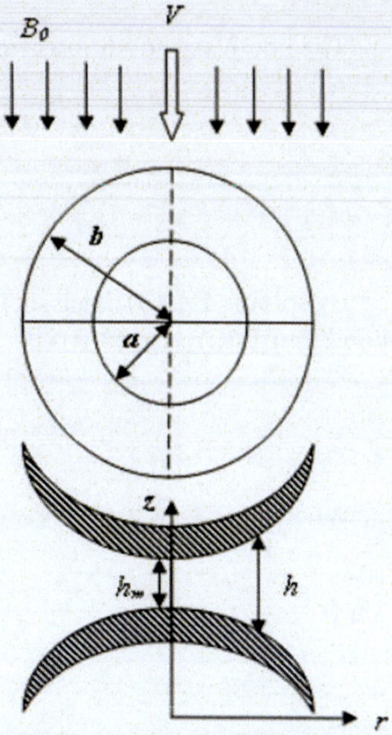

Fig. 1 Geometrical arrangement of curved annular circular plates

$$S(h, l, M_0) = \begin{cases} \frac{h_m^2}{M_0^2}\left\{\frac{2l}{(A^2-B^2)}\left(\frac{B^2}{A}\tanh\frac{A.h}{2l} - \frac{A^2}{B}\tanh\frac{Bh}{2l}\right) + h\right\}, & \text{for } M_0^2 l^2/h_m^2 < 1 \\ \frac{h_m^2}{M_0^2}\left\{\frac{h}{2}\sec h^2\left(\frac{h}{2\sqrt{2l}}\right) - 3\sqrt{2l}\tanh\left(\frac{h}{2\sqrt{2l}}\right) + h\right\}, & \text{for } M_0^2 l^2/h_m^2 = 1 \\ \frac{h_m^2}{M_0^2}\left\{\frac{2lh_0}{M}\left(\frac{(A_2\cot\theta - B_2)\sin B_2 h - (B_2\cot\theta + A_2)\sin A_2 h}{\cos B_2 h + \cosh A_2 h}\right) + h\right\}, & \text{for } M_0^2 l^2/h_m^2 > 1 \end{cases}$$

The non-dimensional quantities given below are used in (2)

$$r^* = \frac{r}{a},\ h^* = \frac{h}{h_{m0}},\ h_m^* = \frac{h_m}{h_{m0}},\ l^* = \frac{2l}{h_{m0}},\ C = \beta a^2,\ K = \gamma a,\ P = -\frac{h_{m0}^3 p}{\mu a^2 V}$$

The Reynolds equation in the modified form is

$$\frac{1}{r^*}\frac{\partial}{\partial r^*}\left\{r^* F\left(h^*, l^*, M_0\right)\frac{\partial P}{\partial r^*}\right\} = -1 \tag{3}$$

Where,

$$F(h^*, l^*, M_0) = \begin{cases} \frac{1}{M_0^2}\left\{\frac{l^*}{(A^{*2}-B^{*2})}\left(\frac{B^{*2}}{A^*}\tanh\frac{A^*h^*}{l^*} - \frac{A^{*2}}{B^*}\tan\text{h}\frac{B^*h^*}{l^*}\right) + h^*\right\} & \text{for } 4M_0^2 l^{*2} < 1 \\ \frac{1}{M_0^2}\left\{\frac{h^*}{2}Sech^2\left(\frac{h^*}{\sqrt{2}l^*}\right) - \frac{3l^*}{\sqrt{2}}\tanh\left(\frac{h^*}{\sqrt{2}l^*}\right) + h^*\right\} & \text{for } 4M_0^2 l^{*2} = 1 \\ \frac{1}{M_0^2}\left\{\frac{l^*(A_2^* Cot\theta^* - B_2^*)Sin B_2^* h^* - l^*(B_2^* Cot\theta^* + A_2^*)Sinh A_2^* h^*}{M_0(Cos B_2^* h^* + Cosh A_2^* h^*)} + h^*\right\} & \text{for } 4M_0^2 l^{*2} > 1 \end{cases}$$

The boundary conditions for pressure is

$$P = 0 \;\; at \;\; r^* = a^* = a/b \tag{4(i)}$$

$$P = 0 \;\; at \;\; r^* = 1 \tag{4(ii)}$$

By solving Eq. (3) the association for film pressure P using the pressure boundary conditions (4(i)) and (4(ii)) is

$$P = \frac{f_2(r^*) f_1(1) - f_1(r^*) f_2(1)}{2 f_2(1)} \tag{5}$$

Where,

$$f_1(r^*) = \int_{r^*=a^*}^{r^*} \frac{r^*}{F(h^*, l^*, M_0)} dr^* \text{ and } f_2(r^*) = \int_{r^*=a^*}^{r^*} \frac{1}{r^* F(h^*, l^*, M_0)} dr^* \tag{6}$$

The equation for pressure is integrated over the film region to get the load supporting capacity

$$W = \int_{r=a}^{b} 2\pi r p \, dr \tag{7}$$

The load-carrying capacity W^* is given by

$$W^* = \frac{W h_{m0}^3}{2\pi \mu b^4 (-dh_m/dt)} = \frac{-1}{2} \int_{r^*=a^*}^{1} f_1(r^*) \, r^* \, dr^* + \frac{1}{2} \frac{f_1(1)}{f_2(1)} \int_{r^*=a^*}^{1} f_2(r^*) \, r^* \, dr^* \tag{8}$$

The squeeze-film time is

$$T^* = \frac{W h_{m0}^2}{\pi \mu b^4} t = \int_{h_m^*}^{1} \left(\frac{2 f_2(1)}{f_2(1) \int_{r^*=a^*}^{1} f_1(r^*) \, r^* \, dr^* - f_1(1) \int_{r^*=a^*}^{1} f_2(r^*) \, r^* \, dr^*} \right) dh_m^* \tag{9}$$

2 Interpretation of Results

In this work, the characteristic behavior of squeeze-film lubrication in the presence of MHD and CSF on curved annular circular plates is analyzed. Numerical and graphical interpretation was carried out on parameters like Hartmann number M_0,

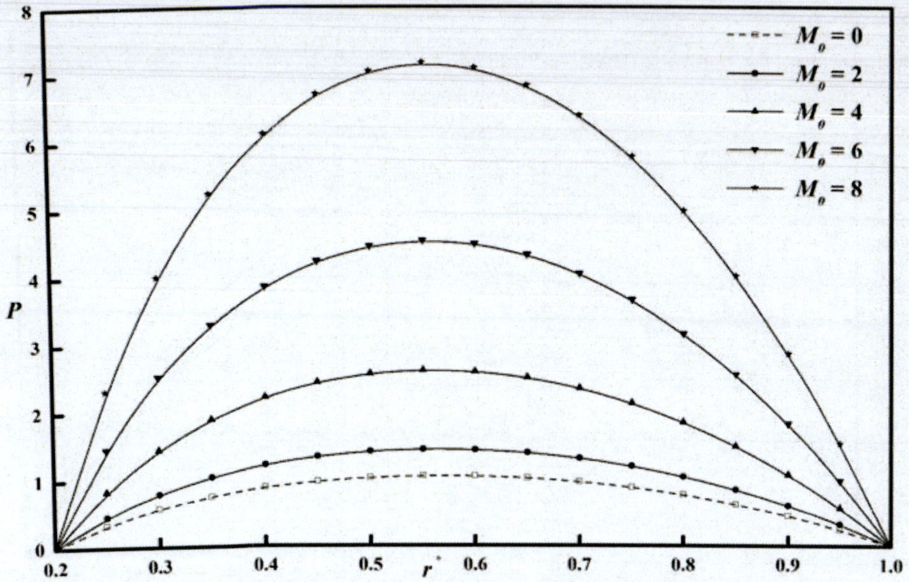

Fig. 2 Non-dimensional pressure *P* variation with r* for different values of M_o with $t = 0.3$, $\beta = 0.5$, $\gamma = 0.6$, $\alpha^* = 0.2$, and $h_w^* = 1$

couple stress parameters l^*, β and γ. For a detailed analysis of the above quantities, we have chosen the following range: $M_0 = 0, 2, 4, 6, 8$, $l^* = 0, 0.2, 0.4, 0.6, 0.8$, $\beta = 0.5$, $\gamma = 0.6$, $a^* = 0.2$

2.1 *Squeeze-Film Pressure*

Figures 2 and 3 show the deviation of pressure *P* with r^* with distinct M_0 and l^* values with $\beta = 0.5$, $\gamma = 0.6$, and $h_m{}^* = 1$. It was noticed that pressure *P* increases with increasing M_0 and l^* values. The variation in *P* with r^* with varying values of β and γ is shown in Figs. 4 and 5. It is noted that pressure *P* is enhanced with higher β values, while *P* reduces with higher γ values.

2.2 *Load-Conducting Capacity*

The deviation of load W^* with respect to β for varying M_0 and l^* values is shown in Figs. 6 and 7, respectively. It is understood that the impact of magnetohydrodynamics increases the load-conducting capacity than that of the non-magnetic field. Similarly, the couple stress parameter l^* also increases load than the Newtonian

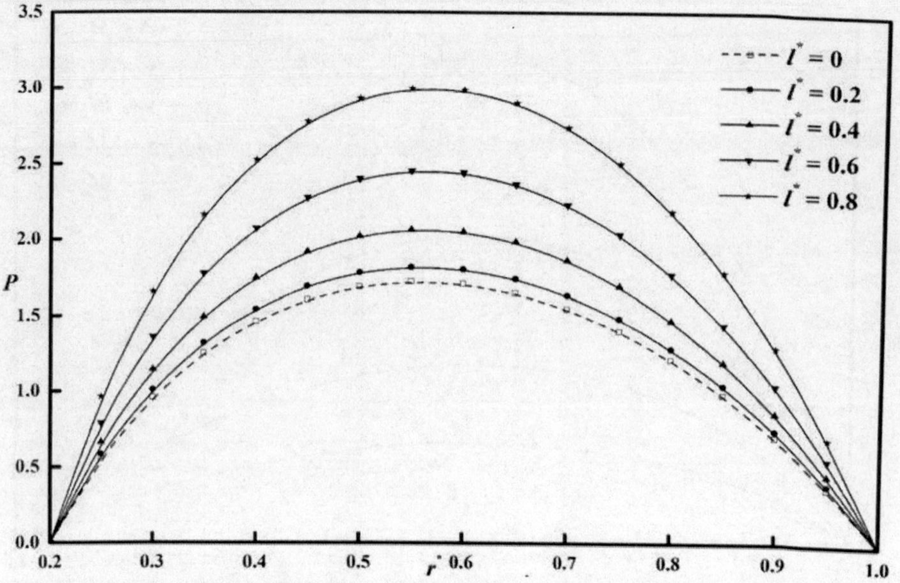

Fig. 3 Non-dimensional pressure P variation with r^* for variation values of t^* with $M_0 = 3$, $\beta = 0.5$, $\gamma = 0.6$, $\alpha^* = 0.2$, and $h_m^* = 1$

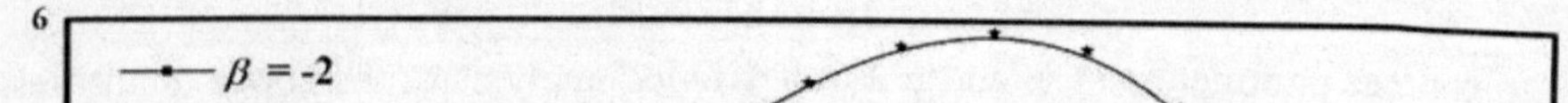

Fig. 4 Non-dimensional pressure P variation with r^* for variation values of β with $M_0 = 3$, $l^* = 0.3$, $\gamma = 0.6$, $a* = 0.2$, and $h_m^* = 1$

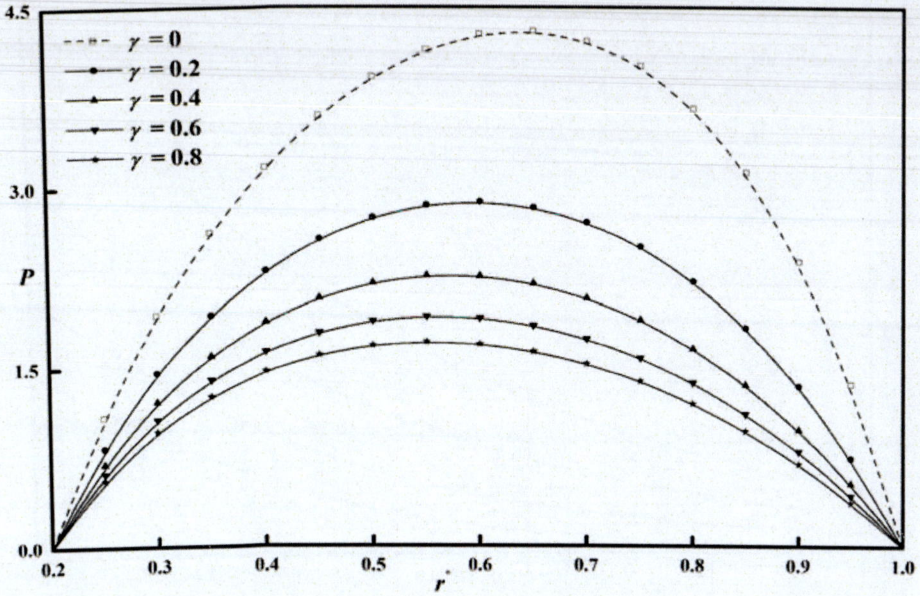

Fig. 5 Non-dimensional pressure P variation with r^* for variation values of γ with $l^* = 0.3$, $\beta = 0.5$, $M_0 = 3$, $a^* = 0.2$, and $h_w^* = 1$

case. Figure 8 displays the change in load W^* with respect to β as a function of γ with $M_0 = 3, a^* = 0.2, l^* = 0.3$ and $h_m{}^* = 1$ and notes that the impact of γ is that it reduces load W^*. The graph of W^* with respect to β for distinct a^* values with $M_0 = 3$, $\gamma = 0.6$, $l^* = 0.3$, and $h_m{}^* = 1$ is displayed in Fig. 9. It is found that for higher $a^* = a/b$ values, the load W^* decreases.

2.3 *Squeeze-Film Time*

The change in time T^* versus h_m^* for various M_0 and l^* values is noted in Figs. 10 and 11, respectively. It was observed that the impact of Hartmann number and CSF increases the squeeze-film time in comparison to the non-magnetic and Newtonian case. Figure 12 displays the deviation T^* with respect to h_m^* for the function of γ with $M_0 = 3$, $a^* = 0.2$, $l^* = 0.3$, and $\beta = 0.5$. We note that with higher γ values, the squeezing time decreases. The profile of T^* with respect to h_m^* for definite a^* values with $M_0 = 3$, $\gamma = 0.6$, $l^* = 0.3$ and $\beta = 0.5$ is shown in Fig. 13. It is found that T^* reduces as $a^* = a/b$ increases.

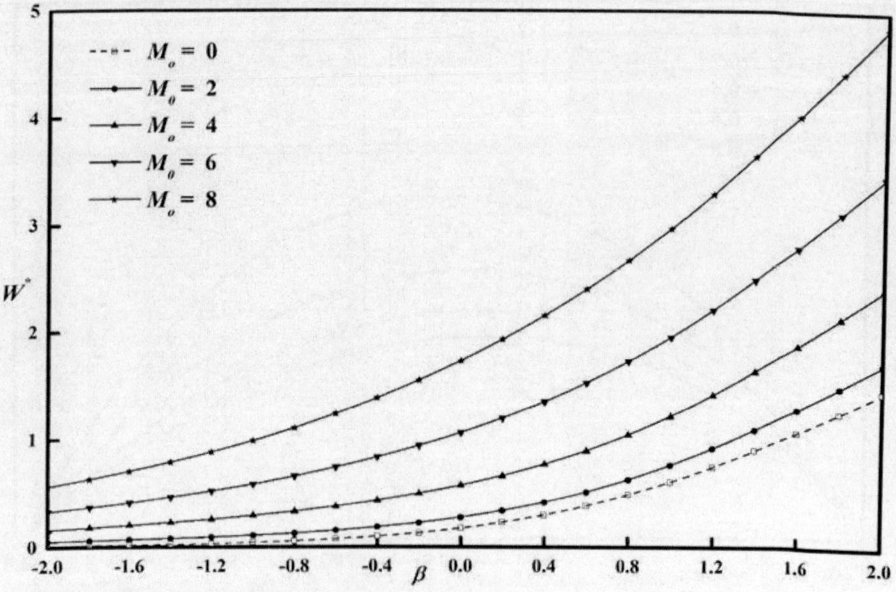

Fig. 6 Non-dimensional load W^* variation with β for variation values of M_O with $l^* = 0.3$, $a^* = 0.2$, $\gamma = 0.6$, and $h_w^* = 1$

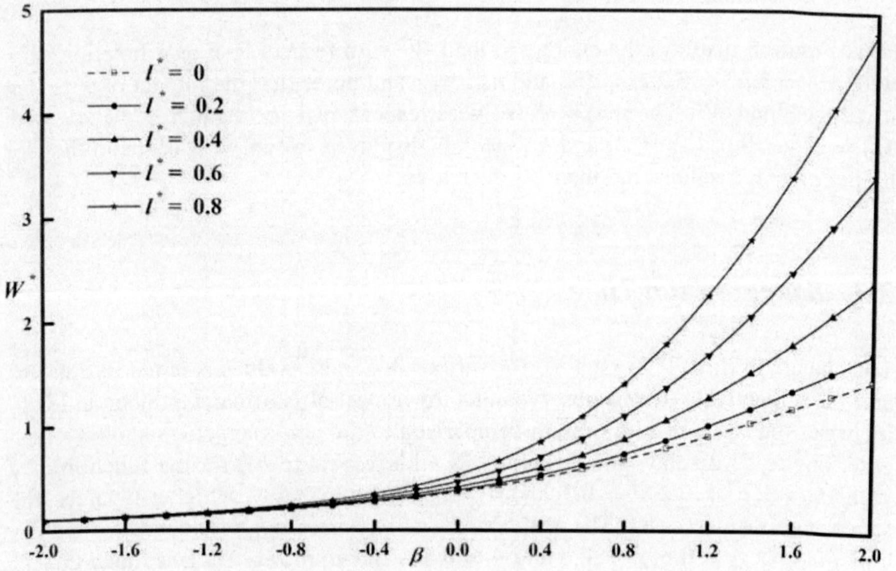

Fig. 7 Non-dimensional load W^* variation with β for variation values of l^* with $M_o = 3$, $a^* = 0.2$, $\gamma = 0.6$, and $h_w^* = 1$

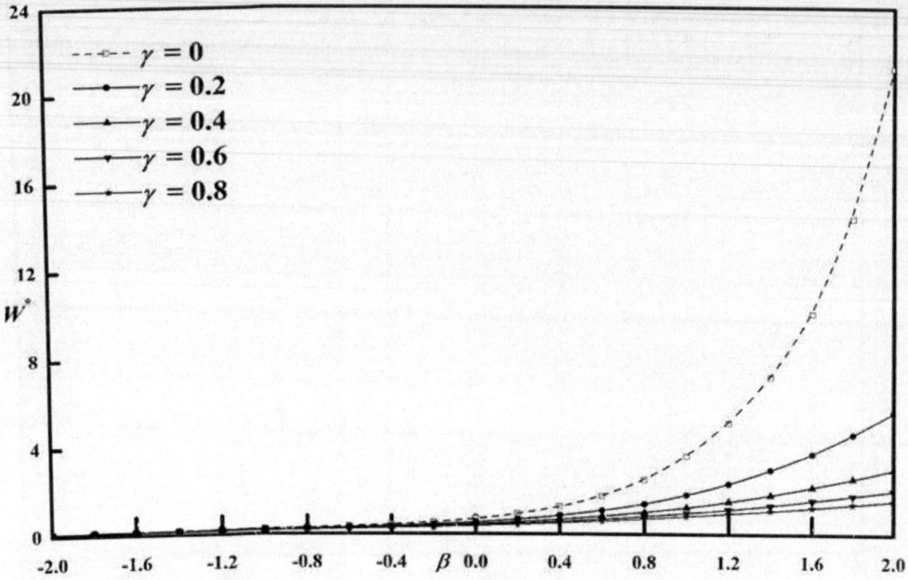

Fig. 8 Variation of non-dimensional load W^* with β for variation values of γ with $M_o = 3$, $a^* = 0.2$, $l^* = 0.3$, and $h_w^* = 1$

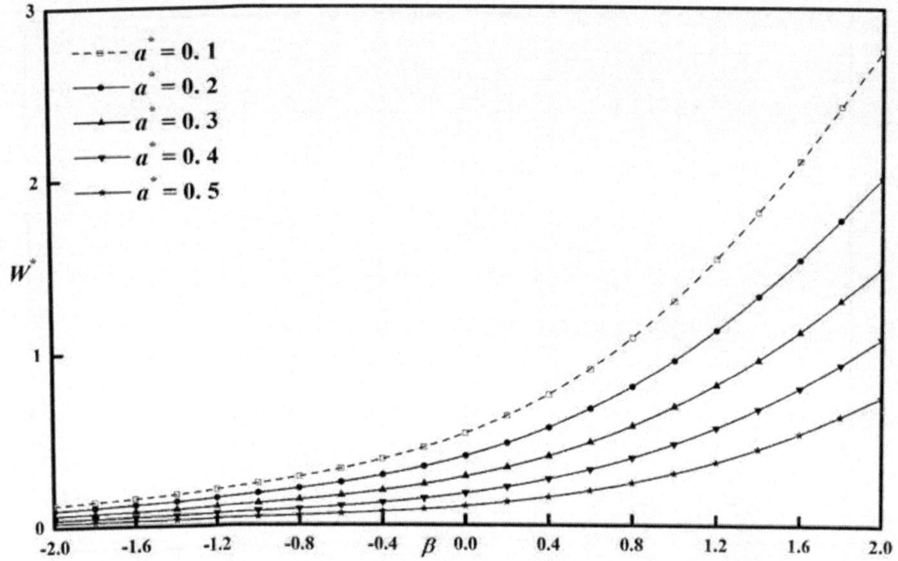

Fig. 9 Non-dimensional load W^* varriation with β for variation values of a^* with $M_s = 3$, $\gamma = 0.6$, $l^* = 0.3$, and $h_w^* = 1$

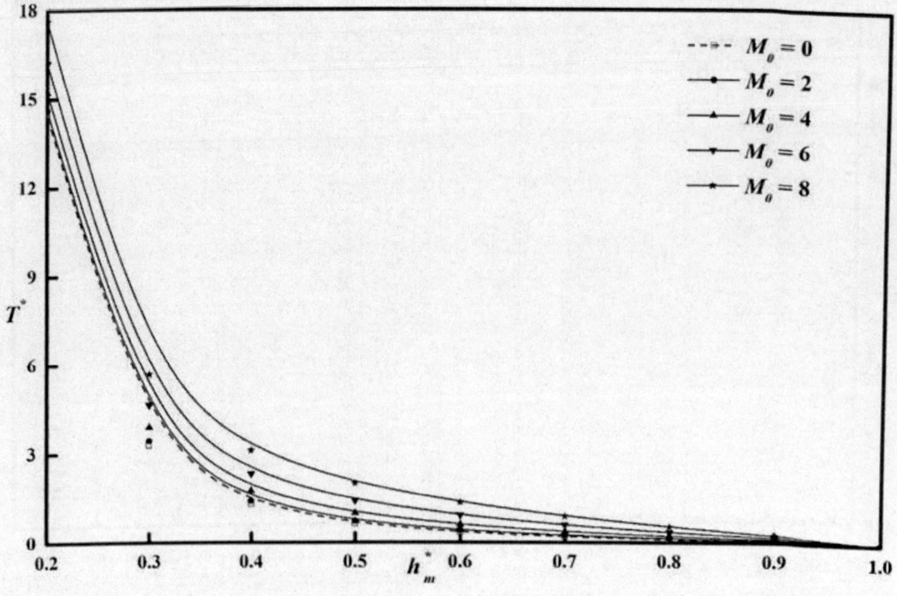

Fig. 10 Non-dimensional squeezing T^* variation with h_m^* for variation values of M_o with $l^* = 0.3$, $a^* = 0.2$, $\gamma = 0.6$, and $\beta = 0.5$

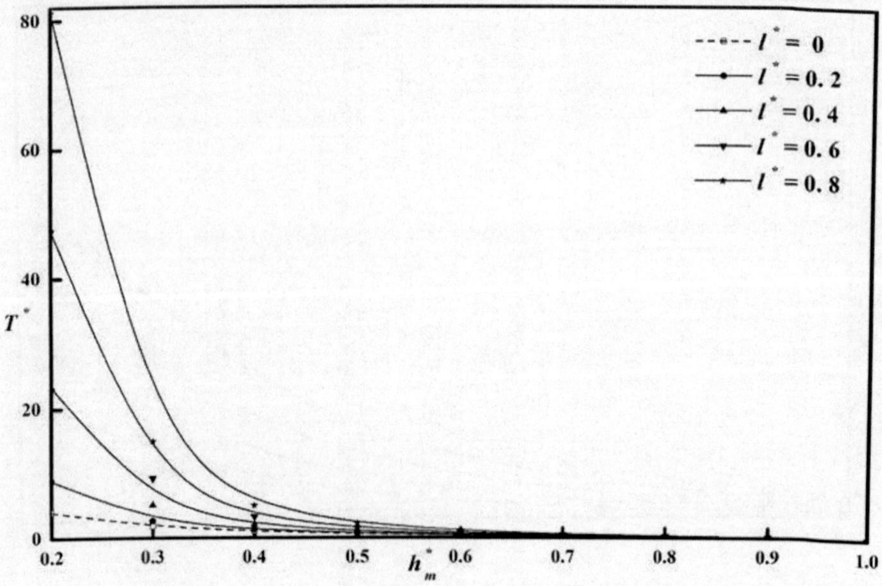

Fig. 11 Non-dimensional squeezing T^* variation with h_m^* for variation values of l^* with $M_o = 3$, $a^* = 0.2$, $\gamma = 0.6$, and $\beta = 0.5$

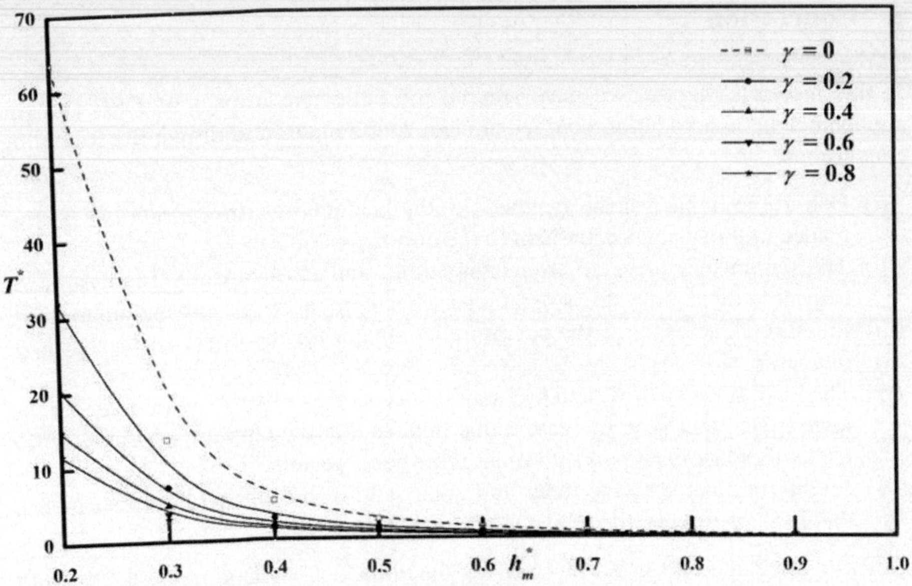

Fig. 12 Non-dimensional squeezing T^* variation with h_m^* for variation values of γ with $M_o = 3$, $a^* = 0.2$, $l^* = 0.3$, and $\beta = 0.5$

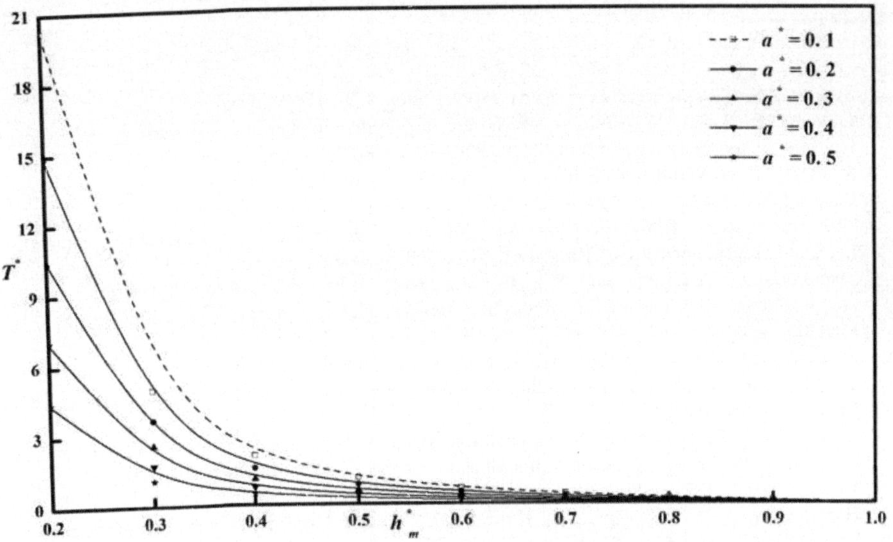

Fig. 13 Non-dimensional squeezing T^* variation with h_j^* for variation values of a^* with $M_o = 3$, $\gamma = 0.6$, $l^* = 0.3$, and $\beta = 0.5$

3 Conclusion

In this research chapter, we have studied the collective impact of CSF and the applied magnetic field among curved annular round plates and drawn the following conclusions:

(i) Due to the impact of the magnetic field, the squeeze-film lubrication performance improves in comparison to the non-magnetic case.
(ii) The parameters pressure, squeezing time, and load-capacity increase with respect to the couple stress parameter when compared to the Newtonian fluid.
(iii) Pressure P increases as the curvature parameter for the upper plate, that is β, increases.
(iv) The load, squeeze time, and pressure reduce as the curvature parameter for the lower plate, that is γ, increases; this implies that the curvature parameter for the lower plate must be kept minimal for better results.
(v) As the ratio between the inner and outer radii of the circle of annular plates, that is a^*, increases, the load and squeezing time reduce.

This work can be extended by studying the impact of MHD along with CFS and surface roughness on the squeezing behavior of curved annular circular plates.

References

1. Stokes, V.K.: Couple stresses in fluids. Phys. Fluids. **9**(9), 1709 (1966)
2. Oliver, D.R.: Load enhancement effects due to polymer thickening in a short model journal bearing. J. Non-Newtonian Fluid Mech. **30**(2–3), 185–196 (1988)
3. Scott, W., Sunti Wttana, P.: Effect of oil additives on the performance of a wet friction clutch material. Wear. **181–183**, 850–855 (1995)
4. Elsharkawy, A.A.: Effects of lubricant additives on the performance of hydro dynamically lubricated journal bearings. Tribol. Lett. **18**(1), 63–73 (2005)
5. Naduvinamani, N.B., Fathima, S.T., Hanumagowda, B.N.: Magneto-hydro dynamics couple stress squeeze film lubrication of circular stepped plates. J. Eng. Tribol. **225**(3), 111–119 (2011)
6. Ramesh Kudenati, B., Marulidhara, N., Patil, H.P.: Numerical solution of the MHD Reynolds equation for squeeze film lubrication between porous and rough rectangular plates. ISRN Tribol., 1–10 (2013)
7. Syeda, T.F., Naduvinamani, N.B., Hanumagowda, B.N., Kumar, S.J.: Modified Reynolds equation for different types of finite plates with the combined effect of MHD and couple stresses. Tribol. Trans. **58**(4), 660–667 (2015). https://doi.org/10.1080/10402004.2014.981906
8. Hanumagowda, B.N., Swapna, N., Kumar, V.: Effect of MHD and couple stress on conical bearing. Int. J. Pure Appl. Math. **113**(6), 316–324 (2017)
9. Lin, J.R.: Squeeze film characteristics between a sphere and a flat plate: couple stress fluid model. Comp. Struct. **75**(1), 73–80 (2000)
10. Naduvinamani, N.B., Siddagouda, A.: Squeeze film lubrication between circular stepped plates of couple stress fluids. J Braz. Soc. Mech. Sci. Eng. **31**(1), 21–26 (2009)
11. Naduvinamani, N.B., Hiremath, P.S., Gurubasavaraj, G.: Squeeze film lubrication of short porous journal bearing with couple stress fluids. Tribol. Int. **34**(11), 739–747 (2001)

12. Fathima, S.T., Naduvinamani, N.B., Shivakumar, H.M., Hanumagowda, B.N.: A study on the performance of hydro magnetic squeeze film between anisotropic porous rectangular plates with couple stress fluids. Tribol. Online. **9**(1), 1–9 (2014)
13. Hanumagowda, B.N., Salma, A., Raju, B.T., Nagarajappa, C.S.: The magneto-hydrodynamic lubrication of curved circular plates with couple stress fluid. Int. J. Pure Appl. Math. **113**(6), 307–314 (2017)

Heat and Mass Transfer of Carbon Nanotubes with Marangoni Convection in the Porous Medium with the Presence of Heat Source/Sink and Chemical Reaction

G. P. Vanitha, U. S. Mahabaleshwar, and M. Hatami

1 Introduction

A carbon tube with a nanometer-sized diameter is commonly referred to as a carbon nanotube (CNT). Engine oil is considered as the base fluid with two types of walled CNTs. Single-walled carbon nanotubes (SWCNTs) are the carbon allotropes that are interposed between bucky balls aviary and flat graphene, while multi-walled carbon nanotubes (MWCNTs) are the CNTs embedded with SWCNT. Carbon nanotubes can have excellent electrical conductivity [1], extensile strength [2], and thermal conductivity [3] due to their nano-structure and the strong bonding between the carbon atoms. They could potentially be chemically altered [4]. It is believed that these properties will be helpful in a number of technological domains, such as electronics, environmental monitoring systems, composite accoutrements (replacing or enhancing carbon fibers), nanotechnology, tissue engineering where CNTs are used as revetment for bone development, and other material science-related applications.

In the past few decades, an analysis of boundary layer flows with CNTs has been extensively investigated by many researchers. Raja et al. [5] analyzed water-based CNTs (single and multi-wall) heat transfer and flow across a porous inclined plate.

G. P. Vanitha (✉)
Department of Mathematics, Siddaganga Institute of Technology, Tumku, India

Department of Studies in Mathematics, Davangere University, Shivagangotri, India

U. S. Mahabaleshwar
Department of Studies in Mathematics, Davangere University, Shivagangotri, India

M. Hatami
Department of Mechanical Engineering, Esfarayen University of Technology, Esfarayen, North Khorasan, Iran
e-mail: m-hatami@um.ac.ir

F. Kamalov et al. (eds.), *Advances in Mathematical Modeling and Scientific Computing*, Trends in Mathematics, https://doi.org/10.1007/978-3-031-41420-6_27

The problem of heat transfer and CNT flow over a flat surface with slip conditions was explained by Khan et al. [6]. Estelle et al. [7] explained the performance and thermal properties of water-based CNT nanofluid flow. Yazid et al. [8] explained the heat transfer characteristics of CNT nanofluids. The thermal efficiency of engine oil-based CNTs is investigated by Choi et al. [9].

Marangoni convection is caused due to the deviation of the surface tension gradients; it is quite useful in the disciplines of welding and crystal expansion [10]. The Marangoni effect [11] is particularly important in the disciplines of metal melting by electron beam, crystal formation, coatings and welding, and space technology. CNT Marangoni boundary layer flow problems are studied by many researchers [12–15].

In view of all the above studies, the current investigation is carried out to analyze the impact of radiation and the chemically reactive Marangoni convection flow of water, kerosene, and engine oil-based CNT with heat source/sink and mass transpiration. Two types of CNTs like SWCNTs and MWCNTs are employed. The various parameters involved in the flow field are analyzed and discussed through plots.

2 Model Description and Solution

We consider the two-dimensional Marangoni convection flow of CNTs past a porous boundary layer. Figure 1 depicts the physical representation of the boundary layer problem. We choose the Cartesian coordinates such that T (surface temperature) and C (surface concentration) are considered quadratic in the x-axis direction, which is perpendicular to the y-axis.

The following equations can be used to represent the flow problem under consideration:

$$\frac{\partial u}{\partial x} + \frac{\partial \mathrm{v}}{\partial y} = 0, \tag{1}$$

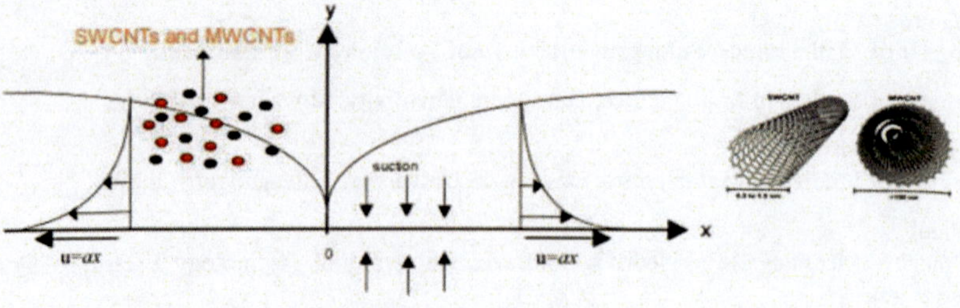

Fig. 1 Schematic diagram of the model

$$u\frac{\partial u}{\partial x}+\mathrm{v}\frac{\partial u}{\partial y}=\frac{\mu_{nf}}{\rho_{nf}}\frac{\partial^2 u}{\partial y^2}-\frac{\mu_{nf}}{\rho_{nf}k}u, \tag{2}$$

$$u\frac{\partial T}{\partial x}+\mathrm{v}\frac{\partial T}{\partial y}=\alpha_{nf}\frac{\partial^2 T}{\partial y^2}+\frac{Q_0}{\left(\rho c_p\right)_{nf}}\left(T-T_\infty\right)+\frac{1}{\left(\rho c_p\right)_{nf}}\frac{16\sigma^* T_\infty^3}{3\rho c_p k^* \kappa_f}\frac{\partial^2 T}{\partial y^2}, \tag{3}$$

$$u\frac{\partial C}{\partial x}+\mathrm{v}\frac{\partial C}{\partial y}=D\frac{\partial^2 C}{\partial y^2}-k_r\left(C-C_\infty\right), \tag{4}$$

where (u, v) are the tangential and transverse velocities of the flow. 'D' is the concentration diffusion. Suffix 'nf' and 'bf' refer to nanofluid and base fluid, respectively. The surface tension and its coefficients are as below:

$$\sigma=\sigma_0\left[1-\left(T-T_\infty\right)\xi_T-\left(C-C_\infty\right)\xi_C\right], \tag{5}$$

$$\xi_T=-\frac{1}{\sigma_0}\frac{\partial \sigma}{\partial T}\bigg|_T \quad and \quad \xi_C=-\frac{1}{\sigma_0}\frac{\partial \sigma}{\partial C}\bigg|_C \tag{6}$$

The imposed Marangoni convective boundary constraints are as below:

$$\mu\frac{\partial u}{\partial y}\bigg|_{y=0}=-\frac{\partial \sigma}{\partial x}\bigg|_{y=0}=\sigma_0\left(\xi_T\frac{\partial T}{\partial x}\bigg|_{y=0}+\xi_C\frac{\partial C}{\partial x}\bigg|_{y=0}\right). \tag{7}$$

$$\mathrm{v}\left(\mathrm{x},0\right)=\mathrm{v}_0,\ T\left(x,0\right)=T_\infty+T_0X^2,\ C\left(x,0\right)=C_\infty+C_0X^2, \tag{8}$$

$$u\left(x,\infty\right)=0,\ T\left(x,\infty\right)=T_\infty,\ C\left(x,\infty\right)=C_\infty \tag{9}$$

The dimensionless variables are

$$\left.\begin{array}{l}\psi\left(x,y\right)=\nu X\, f\left(\eta\right), \quad \eta=\frac{y}{l}, \quad X=\frac{x}{l}, \\ T\left(x,y\right)=\ T_\infty+T_0\ X^2\ \Theta\left(\eta\right),\ \mathrm{C}\left(x,y\right)=C_\infty+C_0\ X^2\ \Phi\left(\eta\right)\end{array}\right\}. \tag{10}$$

The dimensional velocity components are given below:

$$u\left(x,y\right)=\frac{\partial \psi}{\partial y}=\frac{\nu}{l}X\ f_\eta\left(\eta\right)\ \text{and}\ \mathrm{v}\left(x,y\right)=-\frac{\partial \psi}{\partial x}=-\frac{\nu}{l}f\left(\eta\right). \tag{11}$$

The dimensional parameters of nanofluid are given as below:

$$\left.\begin{array}{l}\rho_{nf} = (1-\phi)\,\rho_{bf} + \phi\,\rho_{CNT},\quad (\rho C_p)_{nf} = (1-\phi)\,(\rho C_p)_{bf} + \phi(\rho C_p)_{CNT}\\ \mu_{nf} = \frac{1}{(1-\phi)^{2.5}},\quad \kappa_{nf} = \kappa_f\left[\frac{1-\phi+2\phi\frac{\kappa_{CNT}}{\kappa_{CNT}-\kappa_{bf}}\ln\frac{\kappa_{CNT}+\kappa_{bf}}{2\kappa_{bf}}}{1-\phi+2\phi\frac{\kappa_{bf}}{\kappa_{CNT}-\kappa_{bf}}\ln\frac{\kappa_{CNT}+\kappa_{bf}}{2\kappa_{bf}}}\right]\end{array}\right\}$$

The modeled PDEs Eqs. (2)–(4) are reduced to the following ODEs:

$$\frac{d^3 f}{d\eta^3} - (Da)^{-1}\frac{df}{d\eta} - A_1 A_2\left\{\left(\frac{df}{d\eta}\right)^2 - f\frac{d^2 f}{d\eta^2}\right\} = 0, \tag{12}$$

$$(A_3 + Nr)\frac{d^2\Theta}{d\eta^2} + \Pr A_4\, f\frac{d\Theta}{d\eta} + \Pr\left(I - 2A_4\frac{df}{d\eta}\right)\Theta = 0, \tag{13}$$

$$\frac{d^2\Phi}{d\eta^2} + Sc\, f\frac{d\Phi}{d\eta} - \left(K + 2\frac{df}{d\eta}\right)\Phi = 0. \tag{14}$$

With the corresponding boundary conditions,

$$\left.\begin{array}{l} f(0) = V_c,\quad \frac{d^2 f(0)}{d\eta^2} = -2\,(1 + Ma),\quad \Theta(0) = 1,\ \ \Phi(0) = 1,\\ \frac{df(\infty)}{d\eta} = 0,\quad \Theta\,(\infty) = 0,\quad \Phi\,(\infty) = 0,\end{array}\right\} \tag{15}$$

where V_c is the mass transpiration parameter, $Ma = \frac{Ma_C}{Ma_T}$ the Marangoni number, $Ma_C = \frac{\xi_C\sigma_0 C_0 l}{\delta\mu}$ the solutal Marangoni number, $Ma_T = \frac{\xi_T\sigma_0 T_0 l}{\delta\mu}$ the thermal Marangoni number,

$$K = \frac{k_r l^2}{D},\ Sc = \frac{\nu}{D},\ \Pr = \frac{\nu}{\alpha},\ Nr = \frac{16\sigma^* T^3}{3k^*\kappa\ \rho\ C_p},\ (Da)^{-1} = \frac{l^2}{k},\ \ I = \frac{Q_0 l^2}{\rho\nu c_p},$$
$$A_1 = (1-\phi)^{2.5},\ A_2 = 1 - \phi + \phi\frac{\rho_{CNT}}{\rho_{bf}},\ A_3 = \frac{\kappa_{nf}}{\kappa_{bf}},\ \text{and}\ A_4 = \frac{(\rho C_p)_{nf}}{(\rho C_p)_{bf}}.$$

3 Solution for Velocity

The analytical solution is determined for the velocity equation using function f with the corresponding boundary constraints in Eq. (12) as below:

$$f\,(\eta) = f_\infty + (V_c - f_\infty)\,Exp\,[-a\,\eta], \tag{16}$$

where

$$f_\infty = \frac{a}{A_1 A_2} - \frac{(Da)^{-1}}{a\,A_1 A_2} \tag{17}$$

f_∞ is known as the drawing velocity of the fluid.

Here, $\frac{df(\infty)}{d\eta} = 0$ and $\frac{d^2 f(0)}{d\eta^2} = -2\,(Ma + 1)$ are contented for $a > 0$, and

$$f_\infty = \frac{V_c a^2 + 2\,(1 + Ma)}{a^2}. \tag{18}$$

Now, solving Eqs. (17) and (18) to yield the below cubical equation:

$$a^3 - V_c A_1 A_2 a^2 - (Da)^{-1} a - 2\,(Ma + 1)\,A_1 A_2 = 0. \tag{19}$$

Therefore, following roots of the above cubic equation using Cardano's formula are listed below:

$$a_1 = S + T + \frac{V_c A_1 A_2}{3},\ a_{2,3} = -\frac{(S+T)}{2} + \frac{V_c A_1 A_2}{3} \pm \frac{i\sqrt{3}}{2}\,(S - T).$$

where

$$S = \sqrt[3]{R + \sqrt{Q^3 + R^2}},\ T = \sqrt[3]{R - \sqrt{Q^3 + R^2}},\ Q = -\left[\frac{Da^{-1}}{3} + \frac{V_c^2 A_1^2 A_2^2}{9}\right]$$

and $R = \frac{V_c\, Da^{-1} A_1 A_2}{6} + (1 + Ma)\,A_1 A_2 + \frac{V_c^3 A_1^3 A_2^3}{27}$.

4 Solution for Temperature and Concentration

Equations (13) and (14) represent the hydrodynamic boundary layer problem associated with Θ and Φ boundary conditions in Eq. (15). By substituting $z_1 = \left[\frac{\Pr\, A_4 (V_c - f_\infty)}{a(A_3 + Nr)}\right] Exp\,[-a\,\eta]$ in Eq. (13) and $z_2 = \left[\frac{Sc\,(V_c - f_\infty)}{a}\right] Exp\,[-a\eta]$ in Eq. (14) respectively, we get

$$z_1 \frac{d^2\Theta}{d{z_1}^2} + (1 - p_\Theta - z_1)\,\frac{d\Theta}{dz_1} + \left(\frac{q_\Theta}{z_1} + 2\right) = 0, \tag{20}$$

$$z_2 \frac{d^2\Phi}{d{z_2}^2} + (1 - p_\Phi - z_2)\,\frac{d\Phi}{dz_2} + \left(\frac{q_\Phi}{z_2} + 2\right) = 0. \tag{21}$$

where

$$p_\Theta = \frac{\Pr A_4 f_\infty}{a\,(A_4 + Nr)},\ q_\Theta = \frac{\Pr\ I}{a^2\,(A_4 + Nr)},\ p_\Phi = \frac{Sc\ f_\infty}{a} \text{ and } q_\Phi = -\frac{Sc\ K}{a^2}. \tag{22}$$

Using Kummer's confluent hypergeometric functions (${}_1F_1$) to solve Eqs. (20) and (21), we obtain

$$\Theta(\eta) = \Phi(\eta) = \left(\frac{z}{(z)_0}\right)^{\frac{k_1+k_2}{2}} \frac{{}_1F_1\left[\frac{k_1+k_2-4}{2}, k_1+1,\ z\right]}{{}_1F_1\left[\frac{k_1+k_2-4}{2}, k_1+1,\ (z)_0\right]}, \tag{23}$$

where

$$(z)_0 = (z_1)_0 = \left(\frac{\Pr\ A_4\ (V_c - f_\infty)}{a\,(A_3 - Nr)}\right)\ \left(\text{thermal case } z = z_1\right)$$

and

$$(z)_0 = (z_2)_0 = \left(\frac{Sc\ (V_c - f_\infty)}{a}\right)\ \left(\text{concentration case } z = z_2\right).$$

Equation (23) serves as the solution for thermal and mass equations.

5 Result Analysis

The flow field of three distinct base fluids of SWCNTs and MWCNTs has been investigated. The analytical solutions are represented as Kummer's confluent hypergeometric functions. The thermal characteristics of the base fluids and CNTs are listed in Table 1.

The impact of physical parameters like *Ma* and Da^{-1} on the axial and transverse velocities of the CNT flow are investigated and shown through graphs. Figure 2a–f illustrates the impact of *Ma* on the flow velocity for both injection and suction case, respectively. It is observed that more induced flows are produced due to increase in Marangoni convection (*Ma* value). Therefore, the velocity is increased with the volume fraction of CNTs (both SWCNT and MWCNT), as a result of the fluid flow spreading more widely within the boundary. It is also observed that the higher velocities are achieved in engine oil-based CNTs compared to kerosene or water-based CNTs.

Table 1 Thermal characteristics of the base fluid and CNTs

	Base fluids			Carbon nanotubes	
Physical characteristics	Kerosene	Water	Engine oil	SWCNT	MWCNT
$\rho(kg\,m^{-3})$	783	997	884	2600	1600
$c_p(J\,kg^{-1}K^{-1})$	2090	4179	1910	425	796
$\kappa(W\,m^{-1}K^{-1})$	0.145	0.613	0.144	6600	3000

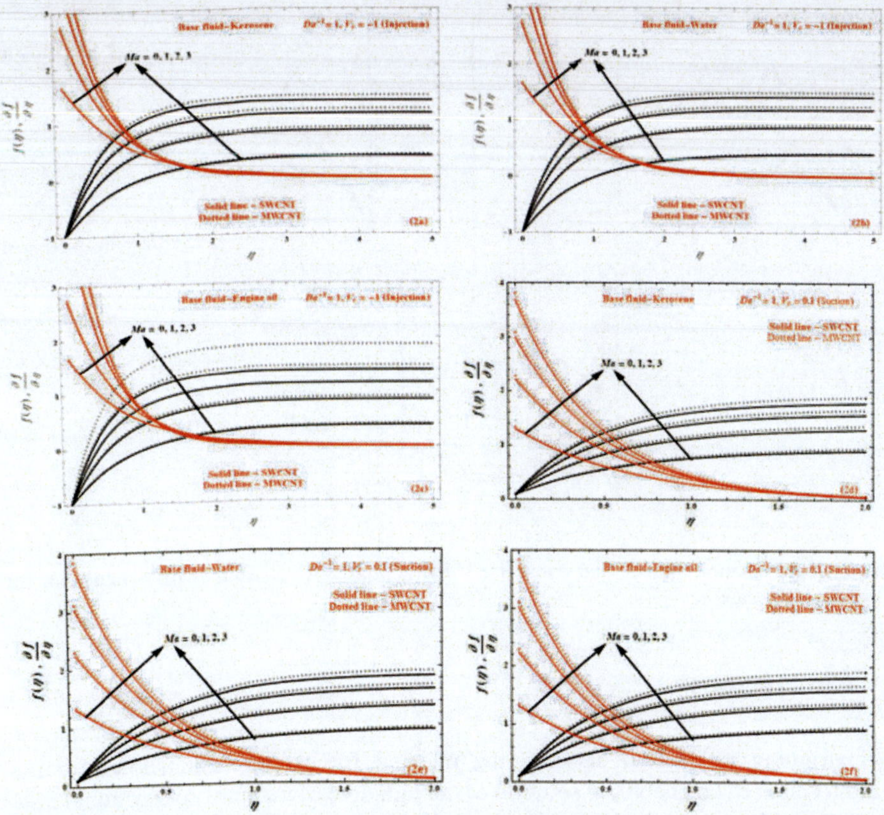

Fig. 2 Representation of axial and transverse velocity profiles for various *Ma* values for suction and injection cases

Figure 3a–c depicts the significance of porosity in the media on the velocity profile. The plots signify the comparison of velocity of the fluid for SWCNT and MWCNT. It is evident from this figure that the velocity is decreased with increased porosity.

The effects of *Nr* and *Sc* are depicted in Fig. 4a–f and have a substantial impact on temperature or concentration profiles. Increasing values of the *Nr* cause the fluid flow closer to the surface, which reduces the thickness of the temperature boundary. As a result, the thermal boundary thicknesses decrease. But the reverse trend is observed in the concentration profile for various values of *Sc*.

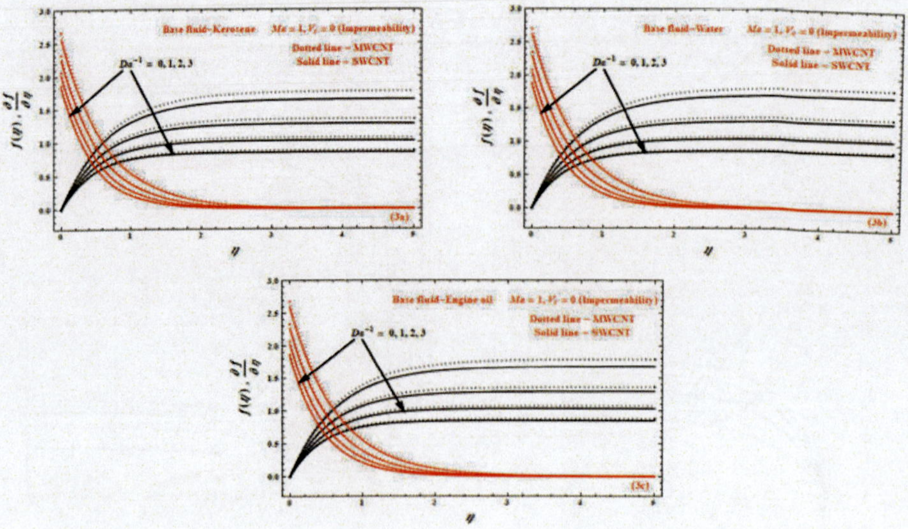

Fig. 3 Representation of axial and transverse velocity profiles for various $(Da)^{-1}$ values for the impermeability case

6 Conclusions

An intriguing interaction between the relevant forces has been shown by the establishment of an analytical solution of the radiative Marangoni convectional flow of CNTs with heat source/sink and chemical reaction. Various physical parameters that play a significant role in the flow field are discussed through plots. From this study, we conclude the following:

- Induced flows are produced due to the increasing Marangoni convection. This is significant since additional flow propagation within the boundary causes the velocity to increase.
- The addition of porosity significantly reduces heat gain/loss and speeds up the process of heat generation/absorption, depending on the situation.

The boundary layer thickness of the thermal increases with the increasing Nr values.

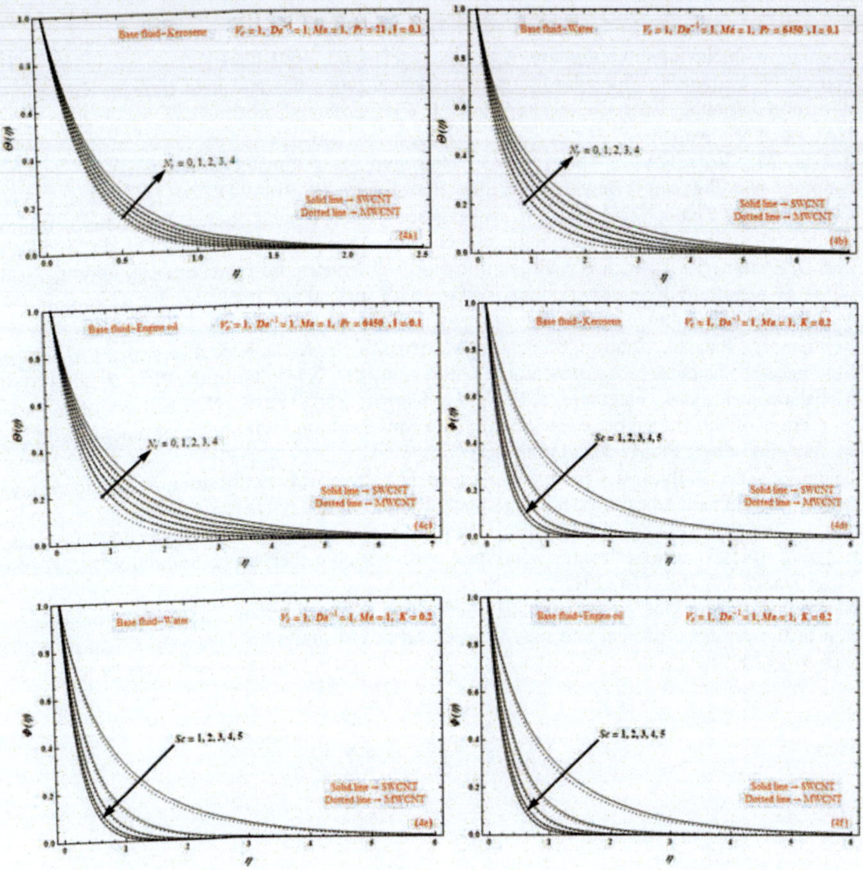

Fig. 4 Representation of thermal and concentration profiles for various *Nr* and *Sc* values

References

1. Sander, J., Devoret, M.H., Hongjie, D., Andreas, T., Richard, S.E., Geerligs, L.J., Dekker, C.: Individual single-wall carbon nanotubes as quantum wires. Nature. **386**(6624), 474–477 (1997)
2. Yu, M., Lourie, O., Dyer, M.J., Moloni, K., Kelly, T.F., Ruoff, R.S.: Strength and breaking mechanism of multi-walled carbon nanotubes under tensile load. Science. **287**(5453), 637–640 (2000)
3. Berber, S., Young-Kyun, K., David, T.: Unusually high thermal conductivity of carbon nanotubes. Phys. Rev. Lett. **84**(20), 4613–4616 (2000)
4. Nikolaos, K., Tagmatarchis, N., Dimitrios, T.: Current progress on the chemical modification of carbon nanotubes. Chem. Rev. **110**(9), 5366–5397 (2010)
5. Raja, O.R., Suryanarayana, M.R.: Heat and mass transfer analysis of single walled carbon nanotubes and multi walled carbon nanotubes-water nanofluid flow over porous inclined plate with heat generation/absorption. Journal of Nanofluids. **8**, 1147–1157 (2019)

6. Khan, W.A., Khan, Z.H., Rahi, M.: Fluid flow and heat transfer of carbon nanotubes along a flat plate with Navier slip boundary. Appl. Nanosci. **4**, 633–641 (2014)
7. Patrice, E., Salma, H., Thierry, M.: Thermophysical properties and heat transfer performance of carbon nanotubes water-based nanofluids. J. Therm. Anal. Calorim. **127**, 2075–2081 (2017) HAL open science
8. Yazida, M., Sidika, A.C., Yahyab, W.J.: Heat and mass transfer characteristics of carbon nanotube nanofluids: a review. Renew. Sust. Energ. Rev. **80**, 914–941 (2017)
9. Choi, S.U.S., Zhang, Z.G., Yu, W., Lockwood, F.E., Grulke, E.A.: Anomalous thermal conductivity enhancement in nanotube suspensions. Appl. Phys. Lett. **79**, 2252–2254 (2001)
10. Zari, I., Anum, S., Tahir, S.K.: Simulation study of Marangoni convective flow of kerosene oil based nanofluid driven by a porous surface with suction and injection. Int. Commun. Heat Mass Transfer. **105493**, 127 (2021)
11. Rehman, A., Taza, G., Zabidin, S., Safyan, M., Hussain, F., Nisar, K.S., Poom, K.: Effect of the Marangoni convection in the unsteady thin film spray of CNT Nanofluids. PRO. **7**, 392 (2019)
12. Mahabaleshwar, U.S., Nagaraju, K.R., Vinay Kumar, P.N., Martin, N.A.: Effect of radiation on thermosolutal Marangoni convection in a porous medium with chemical reaction and heat source/ sink. Phys. Fluids. **32**(11), 113602 (2020)
13. Anum, S., Zari, I., Ilyas, K., Tahir, S.K., Asiful, H.S., Sherif, E.M.: Marangoni boundary layer flow of carbon nanotubes toward a Riga plate. Front. Phys. **7** (2015)
14. Lin, Y., Zheng, L., Zhang, X.: Radiation effects on Marangoni convection flow and heat transfer in pseudo-plastic non-Newtonian nanofluids with variable thermal conductivity. Int. J. Heat Mass Transf. **77**, 708–716 (2014)
15. Magyari, E., Chamkha, A.J.: Exact analytical solutions for thermosolutal Marangoni convection in the presence of heat and mass generation or consumption. Heat Mass Transf. **43**(9), 965–974 (2007)

Hybrid Nanofluid Flow and Thermal Transport Analysis in a Linearly Heated Cylindrical Annulus

N. Keerthi Reddy, Nagaraj Harthikote, M. Sankar, and H. A. Kumara Swamy

1 Introduction

Buoyant convection in a vertical annular geometry is a standard physical configuration representing various technological and industrial applications. The convectional fluids such as water, ethylene glycol, etc., which are used in the industrial sectors for the cooling requirement in heat transfer applications, possess poor thermal conductivity. In view of this drawback of conventional fluids, Choi [1] obtained the new concept of nanofluids which has gained popularity in industries. Nanofluid is a novel heat transfer fluid, designed by dispersing solid nanometer-sized particles in conventional fluids. Study of nanofluids has become an important research area due to its enhanced thermal conductivity. In view of the applications of annular geometry and nanofluids, many researchers have focused on the buoyant convective heat transfer in a cylindrical annular geometry filled with nanofluids [2–5]. Abouali and Falahatpisheh [2] performed a pioneering detailed investigation addressing natural convection in an annular gap filled with water-based alumina nanofluids. They predicted several correlations to estimate the total heat dissipation rate. Considering the impact of discrete heating in the same geometry, Oudina [3] carried

N. K. Reddy
Department of Mechanical Engineering, Ulsan National Institute of Science and Technology, Ulsan, South Korea

N. Harthikote
Department of Electrical Engineering, UTAS-Higher College of Technology, Muscat, Oman

M. Sankar (✉)
Department of Information Technology, University of Technology and Applied Sciences, Ibri, Oman

H. A. K. Swamy
Department of Mathematics, Kyungpook National University, Daegu, Republic of Korea

F. Kamalov et al. (eds.), *Advances in Mathematical Modeling and Scientific Computing*, Trends in Mathematics, https://doi.org/10.1007/978-3-031-41420-6_28

out a numerical analysis to examine the impact of various base fluids with titania nanoparticles on the thermal dissipation in the enclosure. In the annulus containing nanofluids, two different non-uniform bottom wall heating effects on overall Nusselt number have been reported by Reddy and Sankar [4]. The combined conduction-convection phenomena on thermal transfer rate of nanofluids in conjugate annular geometry have been recently discussed by Sankar et al. [5] to analyze wall thickness and conductivity ratio effects. Further, Berrahil et al. [6] numerically analyzed buoyancy-driven convective flow of nanofluids in a differentially heated annulus considering magnetic field effects. Recently, entropy generation along with heat dissipation rate in a vertical annulus under different constraints has been numerically studied [7–9].

Hybrid nanofluids are a novel kind of heat transfer fluids which possess enhanced thermophysical and chemical properties. With the idea of heat transport enhancement, these hybrid nanofluids were developed by dispersing two or more nanoparticles in conventional fluids. Many studies on convection of hybrid nanofluids are reported in the literature [10–13]. Tayebi and Chamkha [10] considered hybrid nanofluid as the working medium in a sinusoidally heated square cavity and reported that the utilization of hybrid nanofluids is more pronounced at higher magnitudes of Rayleigh number. In the study of conjugate heat transport analysis inside a square enclosure, the porous medium saturated with Ag-MgO hybrid nanofluid has been considered by Ghalambaz et al. [11]. Mehryan et al. [12] reported the convective heat transport of hybrid nanofluids in porous cavity and found heat transport deterioration with the dispersion of Ag-MgO hybrid nanoparticles in water. Recently, Reddy et al. [13] investigated the influence of different nanoparticle proportions utilized in the hybrid nanofluid to find the optimum proportion and combination of nanoparticles for higher heat transport enhancement in annular enclosure.

Buoyant convection in enclosed geometries aptly portrays many industrial applications such as solar collectors, electronic cooling devices, and heat exchangers. Many studies are available in the literature which deal with buoyant convection in enclosures having side walls maintained at uniform temperatures. In many industrial applications, the walls of the enclosure need not be maintained at uniform temperatures. Instead, the walls can be maintained at non-uniform temperatures. To consider the impact of such non-uniform heating profiles, linear temperature distribution on the enclosure surfaces has been considered in the literature [14–19]. The impact of two different linear heating cases at the side wall(s) of square geometry has been analyzed by Sathiyamoorthy et al. [14] by considering top adiabatic wall and isothermally cold bottom wall. With the same thermal boundary conditions at the side walls of porous annulus, Sankar et al. [15] investigated the heat transport rates and obtained higher thermal transport for non-uniform heating compared to the case of uniform heating. The same work has been extended in a square cavity by Sathiyamoorthy and Chamkha [16] by considering magnetic field effects. The numerical analysis performed by Kefayati [17] on buoyant convection of ferrofluid in a square enclosure subjected to external magnetic source effect showed that the thermal dissipation rate declines with the addition of ferromagnetic nanoparticles. For five different center locations of linear heating case considered by Sahin [18], enhancement in the thermal transport has been observed as the center of linear temperature profile shifts upwards. Three different cases of linear heating of side walls of a square cavity have been considered in a numerical study carried out by Mahmoodi et al. [19].

It has been noticed from the extensive literature survey that buoyant flow phenomena in an annulus with linear heating at the left wall has not been analyzed in the literature though this geometry finds immense applications in the industrial field. The current analysis thus focuses on the buoyant convection of hybrid nanofluid in a vertical annulus by comparing linear and uniform heating conditions imposed at the inner cylindrical surface.

2 Mathematical Formulation

The current analysis is based on buoyant convection of water-based Ag-MgO hybrid nanofluid-filled vertical annulus having top adiabatic wall, isothermally cold outer cylinder, and linearly or uniformly heated inner cylinder as portrayed in Fig. 1. Two different temperature boundary conditions are imposed on the left wall; linear heating (Case-I) and uniform heating (Case-II). The thermophysical properties of base fluid (water), Ag, and MgO nanoparticles are taken from the literature [11, 12, 20]. Also, the thermophysical properties of hybrid nanoparticles and hybrid nanofluid are mentioned in our earlier studies [13]. The following assumptions are made in the current investigation: fluid is incompressible, flow is laminar, axisymmetric and unsteady, Boussinesq approximation is applied, water and hybrid nanofluid are in thermal equilibrium. Under the above assumptions, the vorticity-stream function form of governing equations can be written as follows:

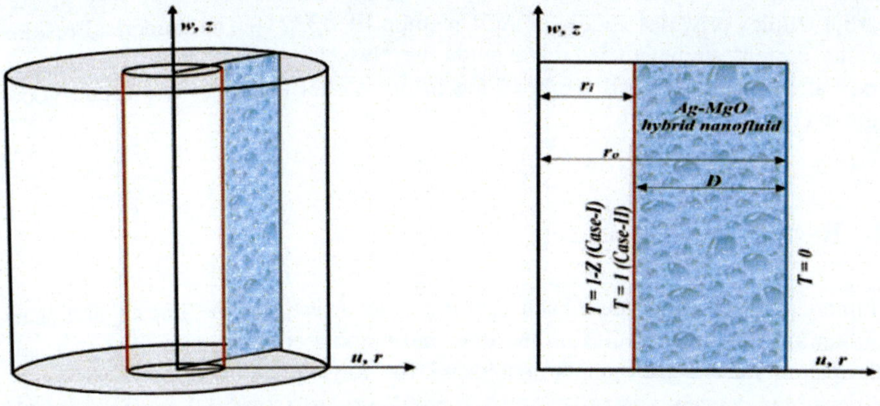

Fig. 1 Physical configuration and its axisymmetric view

$$\frac{\partial \zeta}{\partial t} + U\frac{\partial \zeta}{\partial R} + W\frac{\partial \zeta}{\partial Z} - \frac{U\zeta}{R} = -\frac{\mu_{hnf}}{\rho_{hnf}\alpha_f}\left[\frac{\partial^2 \zeta}{\partial R^2} + \frac{1}{R}\frac{\partial \zeta}{\partial R} + \frac{\partial^2 \zeta}{\partial Z^2} - \frac{\zeta}{R^2}\right] - \frac{(\rho\beta)_{hnf}}{\rho_{hnf}\beta_f}\mathrm{Ra}\,\mathrm{Pr}\frac{\partial T}{\partial R}, \quad (1)$$

$$\frac{\partial T}{\partial t} + U\frac{\partial T}{\partial R} + W\frac{\partial T}{\partial Z} = \frac{\alpha_{hnf}}{\alpha_f}\left(\frac{\partial^2 T}{\partial R^2} + \frac{1}{R}\frac{\partial T}{\partial R} + \frac{\partial^2 T}{\partial Z^2}\right), \quad (2)$$

$$\zeta = \frac{1}{R}\left[\frac{\partial^2 \psi}{\partial R^2} - \frac{1}{R}\frac{\partial \psi}{\partial R} + \frac{\partial^2 \psi}{\partial Z^2}\right], \quad (3)$$

$$U = \frac{1}{R}\frac{\partial \psi}{\partial Z} \text{ and } W = -\frac{1}{R}\frac{\partial \psi}{\partial R}. \quad (4)$$

The total heat transport rate, calculated in terms of average Nusselt number, is given by

$$\overline{Nu} = \frac{1}{A}\left(-\frac{k_{hnf}}{k_f}\right)\int_0^A \left(\frac{\partial T}{\partial R}\right)_{R=\frac{1}{\lambda-1}} dZ \quad (5)$$

3 Solution Procedure

Finite difference methodology is adopted to numerically solve the governing equations. Alternating Direction Implicit method is employed to convert the transient governing equations to a system of algebraic equations. The obtained equations are tri-diagonal in nature, which are solved using Thomas algorithm. One of our earlier studies provides details of ADI method [9, 13, 21]. The numerical results of the current study are validated with the Nusselt number values of Abouali and Falahatpisheh [2] and are provided in our earlier study [5] and found good agreement.

4 Results and Discussion

Numerical simulations have been performed to investigate the impact of linear and uniform heating profiles on the flow and thermal contours as well as thermal dissipation rate in the annular enclosure for Rayleigh number (10^3–10^6) and nanoparticle concentration (0–5%) by fixing λ (radius ratio) = 2 and Pr (Prandtl number) = 6.2. Also, the variation of heat transport rate with different proportions of nanoparticle combinations present in the hybrid nanofluid is analyzed. The obtained

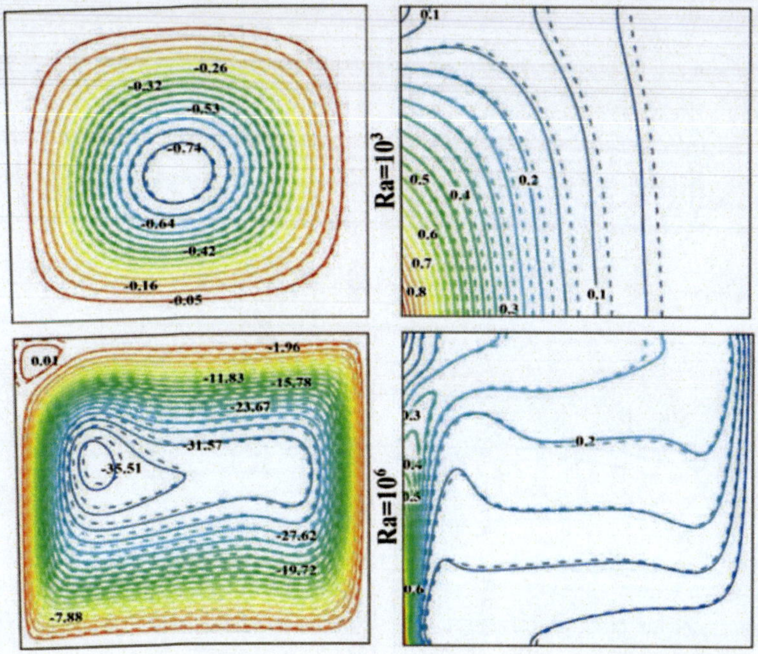

Fig. 2 Effect of Ra on streamline and isotherm contours for Case-I. $\phi = 0$ (solid line) and 5% Ag-MgO hybrid nanofluid (dotted line)

numerical results are graphically represented through flow and thermal contours, and local and average Nusselt numbers.

The influence of Ra on the flow field and isothermal lines for linear heating case is displayed in Fig. 2 for base fluid and 5% Ag-MgO nanofluid. At lower magnitude of Ra (10^3), the buoyancy-induced flow in the enclosure is minimal and dominance of conduction is prevailed. As a result, the fluid in the enclosure behaves similar to a stagnant medium and thus fluid flow strength is minimal which can be seen from $|\psi_{max}| = 0.74$ at Ra $= 10^3$. The isothermal structure also reveals the conduction dominance at lower Ra. Also, the thermal contours are mainly concentrated at the lower portion of inner cylinder. This is due to the hot temperature profile generated at the bottom portion of the left wall for the linearly heating case. This isothermal pattern confirms the non-uniform thermal profile imposed along the left wall. An enhancement in Ra from 10^3 to 10^6 greatly enhances the flow rate $|\psi_{max}| = 35.51$. This is due to a change in heat transport mechanism from conduction to convection dominant. At higher Ra, higher buoyancy forces are induced inside the enclosure which significantly changes the fluid flow and thermal contour pattern. Due to non-uniform heating profile, a smaller circulation cell with very minimal flow strength is noticed at the top left corner, the portion having cold temperature when the wall is linearly heated. Also, the contours for base fluid and hybrid nanofluid show considerable variation at higher magnitude of Ra.

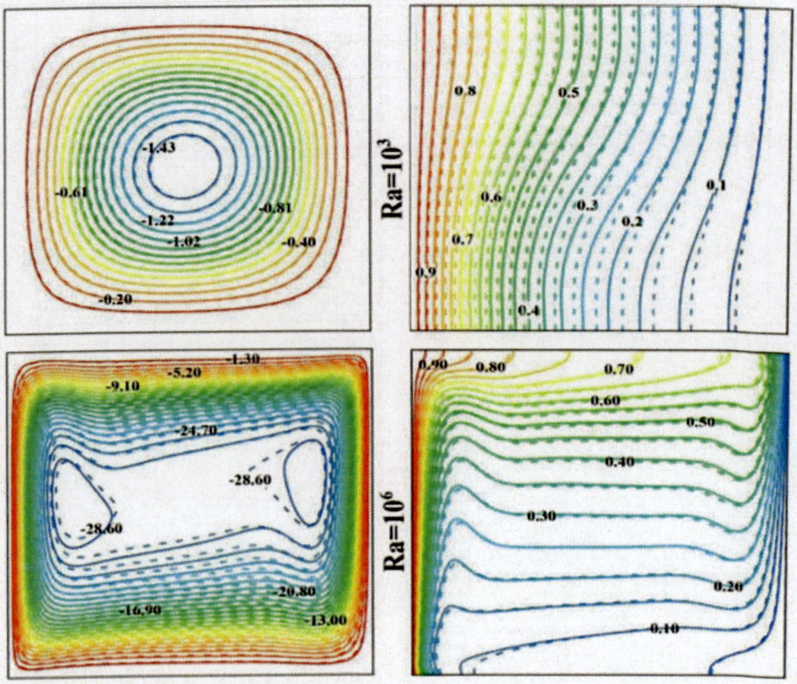

Fig. 3 Effect of Rayleigh number on streamlines and isotherms for Case-II. $\phi = 0$ (solid line) and 5% Ag-MgO hybrid nanofluid (dotted line)

The fluid flow and thermal variation structure for uniform heating case illustrated in Fig. 3 is varied from that of linear heating case (Fig. 2) as the heating rate is significantly varied with the type of thermal profile imposed. Thus, the induced buoyancy forces are also altered for Case-II as compared to Case-I. The fluid flow strength also shows remarkable change for both the cases. The stratification of isotherms is more for uniform heating compared to linear heating. An increase in Ra leads to profound enhancement in flow strength due to the dominance of convection at higher Ra. This convection dominance can be noticed from the stratified isothermal pattern, whereas the conduction dominance is revealed from the vertical isothermal lines at lower Ra.

The heat transport rate at various points along the left wall is evaluated through local Nu which is displayed in Fig. 4 for linear as well as uniform heating profiles. Linear heating profile produces hot temperature at the bottom left and temperature decreases along the height of the left wall. The decrease in temperature leads to deterioration in the thermal dissipation rate. This decreasing trend is clearly observed at greater magnitude of Ra. However, at Ra $= 10^3$, the least thermal transport rate is noticed due to conduction dominance and a marginal decrease in Nu is noticed along the height of the wall. For the uniform heating case, the thermal transport rate along the height of the left wall is almost uniform and consistent with

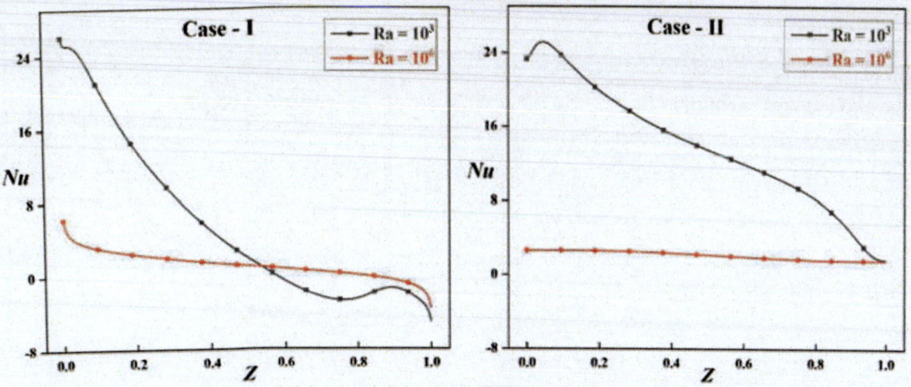

Fig. 4 Local Nusselt number profiles for Case-I and Case-II at Ra $= 10^3$ and Ra $= 10^6$

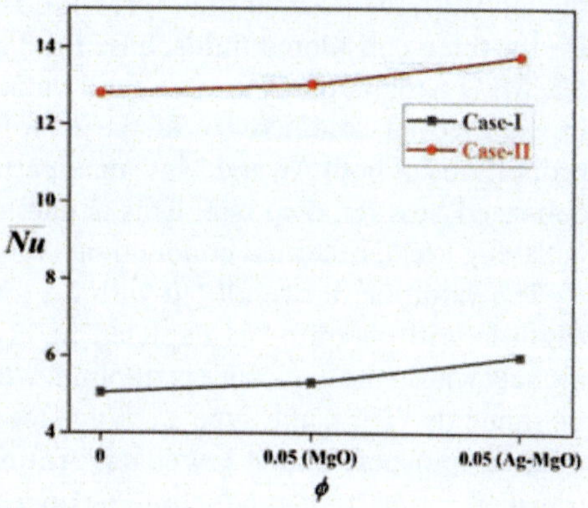

Fig. 5 Variation of average Nu with thermal boundary condition at Ra $= 10^6$ for $\phi = 0$, 5% MgO nanofluid and 5% Ag-MgO hybrid nanofluid

the thermal boundary condition imposed. At higher Ra, Nu decreases with increase in distance from the bottom wall. This is because the interaction between the cold fluid and the hot left wall is more near the bottom left of the wall. The fluid then gets heated up and rises upward and thus less heat transport occurs as the distance from the bottom wall increases. The decrease in Nu along the axial direction is more significant for linear heating case as there exists cold temperature profile near the top left region. It can be concluded that the local Nu alongside left wall is profoundly altered with the temperature profile.

The impact of linear and uniform thermal boundary conditions on the overall thermal dissipation rate for base fluid, 5% MgO nanofluid, and Ag-MgO hybrid nanofluid with equal proportion is presented in Fig. 5. As illustrated in the figure, the enhanced thermal transport rate is obtained for the uniform heating case and a remarkably lower thermal dissipation rate is noticed for linear heating case for all the three fluids considered in this analysis. This could be due to the linear heating for Case-I in which the temperature decreases along the left wall, whereas for uniform heating condition, since hot temperature is imposed at all points along

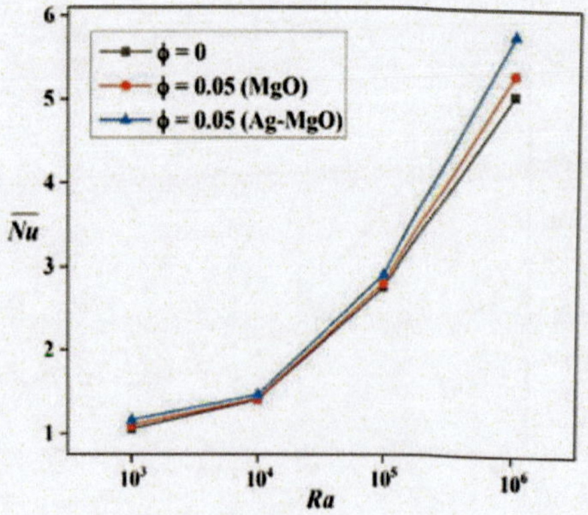

Fig. 6 Effect of Ra on average Nu for water, 5% MgO nanofluid, and 5% Ag-MgO hybrid nanofluid (Case-I)

the left wall, the thermal transport for Case-II is higher. Comparing the average Nu for the three considered fluids, base fluid gives least heat transport rate. Addition of 5% MgO nanoparticles to base fluid enhances the average Nu due to enhancement in the thermal conductivity of the nanofluid. Furthermore, the addition of equal proportions of both Ag and MgO nanoparticles to base fluid produces comparatively enhanced heat removal rate. This is due to the addition of Ag nanoparticles which is having greater thermal conductivity value compared to MgO nanoparticle.

The variation in overall Nu with Ra for water, 5% MgO nanofluid, and 5% Ag-MgO hybrid nanofluid is depicted in Fig. 6. At lower Ra values, the buoyancy inducement in the enclosure is minimal which leads to conduction dominance inside the annulus. Thus, the type of fluid does not produce any significant impact on thermal transport rate at lower Ra. An increment in Ra changes the primary heat transport mode from conduction to convection. When convection starts dominating in the annulus, the impact of type of fluid on average Nu is clearly noticed. Higher thermal transport rate is obtained for hybrid nanofluid compared to single nanofluid and base fluid. The variation in the average Nu for three different fluids is more pronounced at higher magnitudes of Ra.

The variation in the global Nu values with different proportions of nanoparticles present in Ag-MgO hybrid nanofluid for linear (Case-I) and uniform (Case-II) heating is displayed in Fig. 7 for lower and higher Ra values. As discussed earlier, lower magnitude of Ra possesses conduction dominant mechanism in the enclosure and thus, at lower Ra, the variation in the proportions of nanoparticle present in the hybrid nanofluid does not show significant change in the average Nu value. However, for the convection dominant regime at Ra $= 10^6$, an increase in the proportion of Ag nanoparticle greatly enhances the heat dissipation rate. This is due to greater thermal conducting capacity of Ag nanoparticle than that of MgO nanoparticle. It can be concluded that an increase in the proportion of nanoparticle which is having higher thermal conductivity leads to greater enhancement in the thermal dissipation rate. Also, compared to Case-I, Case-II helps greatly in achieving enhanced thermal transport rate for all the nanoparticle proportions and Ra values due to lower heating rate near the wall for linear heating compared to that of uniform heating.

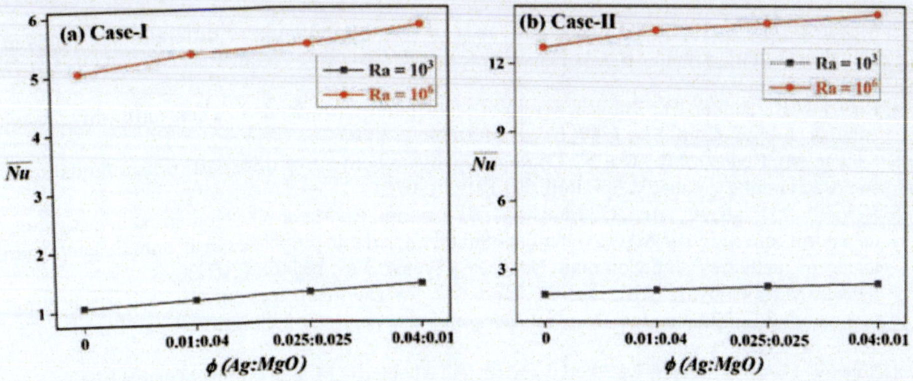

Fig. 7 Effect of different nanoparticles proportions of 5% Ag-MgO hybrid nanofluid and Ra on Avg. Nu for (**a**) Case-I and (**b**) Case-II

5 Conclusion

In this chapter, natural convection of Ag-MgO hybrid nanofluid filled in a vertical annulus formed by two concentric cylinders has been investigated by finite difference methodology. From the extensive numerical computations, it has been noticed that the type of temperature profile imposed at the side wall significantly affects the flow and thermal contours. The important findings are summarized as follows: Rayleigh number has profound influence on flow and heat removal rate. Greater thermal transfer is observed for higher Ra irrespective of the heating condition. Due to lower heating rate along the inner cylinder for linear heating compared to that of uniform heating, the thermal dissipation rate is higher for Case-II. The addition of nanoparticles significantly enhances the heat dissipation rate. In particular, higher the proportion of nanoparticle which is having higher thermal conductivity, greater is the thermal dissipation rate. Hybrid nanofluid greatly helps in achieving enhanced thermal transport rate as compared to single nanofluid.

Acknowledgments NKR acknowledges the support from Ulsan National Institute of Science and Technology, Republic of Korea. HAKS sincerely acknowledges Kyungpook National University, Republic of Korea for the support and encouragement. MS acknowledges the financial support of UTAS, Ibri, Oman, under the Internal Research Funding via Project No. DSR-IRPS-2021-22-PROP-1.

References

1. Choi, S.U.S.: Enhancing thermal conductivity of fluids with nanoparticles. ASME FED. **231**, 99–105 (1995)
2. Abouali, O., Falahatpisheh, A.: Numerical investigation of natural convection of Al_2O_3 nanofluid in vertical annuli. Heat Mass Transf. **46**, 15–23 (2009)

3. Mebarek-Oudina, F.: Convective heat transfer of titania nanofluids of different base fluids in cylindrical annulus with discrete heat source. Heat Transfer – Asian Research. **48**, 135–147 (2019)
4. Reddy, N.K., Sankar, M.: Buoyant convective transport of nanofluids in a non-uniformly heated annulus. J. Phys. Conf. Ser. **1597**, 012055 (2020)
5. Sankar, M., Reddy, N.K., Do, Y.: Conjugate buoyant convective transport of nanofluids in an enclosed annular geometry. Sci. Rep. **11**, 17122 (2021)
6. Berrahil, F., Filali, A., Abid, C., Benissaad, S., Bessaih, R., Matar, O.: Numerical investigation on natural convection of Al_2O_3/water nanofluid with variable properties in an annular enclosure under magnetic field. Int. Commun. Heat Mass Transf. **126**, 105408 (2021)
7. Sankar, M., Swamy, H.A.K., Do, Y., Altemeyer, S.: Thermal effects of non-uniform heating in a nanofluid-filled annulus: Buoyant transport versus entropy generation. Heat Transfer. **51**, 1062–1091 (2022)
8. Swamy, H.A.K., Sankar, M., Reddy, N.K.: Al Manthari, M.S, Double diffusive convective transport and entropy generation in an annular space filled with alumina-water nanoliquid. Eur. Phys. J. Spec. Top. **231**, 13–14
9. Swamy, H.A.K., Sankar, M., Do, Y.: Entropy and energy analysis of MHD nanofluid thermal transport in a non-uniformly heated annulus. Waves in Random and Complex Media (2022). https://doi.org/10.1080/17455030.2022.2145522
10. Tayebi, T., Chamkha, A.J.: Buoyancy-driven heat transfer enhancement in a sinusoidally heated enclosure utilizing hybrid nanofluid. Comput. Therm. Sci. **9**(5), 405–421 (2017)
11. Ghalambaz, M., Sheremet, M.A., Mehryan, S.A.M., Kashkooli, F.M.: Local thermal non-equilibrium analysis of conjugate free convection within a porous enclosure occupied with Ag–MgO hybrid nanofluid. J. Therm. Anal. Calorim. **135**, 1381–1398 (2019)
12. Mehryan, S.A.M., Ghalambaz, M., Chamkha, A.J., Izadi, M.: Numerical study on natural convection of Ag–MgO hybrid/water nanofluid inside a porous enclosure: a local thermal non-equilibrium model. Powder Technol. **367**, 443–455 (2020)
13. Reddy, N.K., Swamy, H.A.K., Sankar, M.: Buoyant convective flow of different hybrid nanoliquids in a non-uniformly heated annulus. Eur. Phys. J. Spec. Top. **230**(5), 1213–1225 (2021)
14. Sathiyamoorthy, M., Basak, T., Roy, S., Pop, I.: Steady natural convection flows in a square cavity with linearly heated side wall(s). Int. J. Heat Mass Transf. **50**, 766–775 (2007)
15. Sankar, M., Kiran, S., Sivasankaran, S.: Natural convection in a linearly heated vertical porous annulus. J. Phys. Conf. Ser. **1139**, 012018 (2018)
16. Sathiyamoorthy, M., Chamkha, A.J.: Effect of magnetic field on natural convection flow in a liquid gallium filled square cavity for linearly heated side wall(s). Int. J. Therm. Sci. **49**, 1856–1865 (2010)
17. Kefayati, G.H.R.: Natural convection of ferrofluid in a linearly heated cavity utilizing LBM. J. Mol. Liq. **191**, 1–9 (2014)
18. Sahin, B.: Effects of the center of linear heating position on natural convection and entropy generation in a linearly heated square cavity. Int. Commun. Heat Mass Transf. **117**, 104675 (2020)
19. Mahmoodi, M., Arani, A.A.A., Sebdani, S.M., Tajik, P.: Natural convection in nanofluid-filled square chambers subjected to linear heating on both sides: a numerical study. Heat Transf. Res. **48**(9), 771–785 (2017)
20. Reddy, N.K. and Sankar, M., Buoyant heat transfer of nanofluids in a vertical porous annulus: a comparative study of different models, Int. J. Numer. Meth. Heat Fluid Flow. **33**(2), 477–509 (2023)
21. Reddy, N.K., Sankar, M., Jang, B.: Impact of thermal source-sink arrangements on buoyant convection in a nanofluid-filled annular enclosure. J. Heat Transf. **144**(11), 112601 (2022)

Influence of Non-similar Heating on Nanofluid Buoyant Convection in a Tilted Porous Parallelogrammic Geometry

S. Vishwanatha, C. V. Vinay, M. Sankar, and N. Keerthi Reddy

1 Introduction

Thermal convection induced by buoyancy or buoyancy-induced convection (BIC) of traditional fluids or nanofluids (NFs) and corresponding heat transport (HT) in non-regularly shaped domains with a particular case of oblique- or parallelogram-shaped geometries have a specific application related to thermal diodes. In this geometry, the buoyant motions and associated HT rates could be effectively monitored through a proper choice of tilting wall angles of the geometry. Hence, several experimental and numerical investigations are performed in this geometry to understand the implications of different geometrical and physical factors involved in this geometry. A detailed flow analysis and thermal characteristics inside a parallelogram-shaped domain has been numerically studied by converting to a regularly shaped region using an appropriate coordinate transformation by Hyun and Choi [1]. Through the predictions from the vast parametric range of pertinent parameters, an optimum angle of side tilting walls has been proposed for higher HT rates. In the same geometry, Costa [2] studied thermosolutal convective flow and the related thermal as well as solutal rates by considering different aspect ratios of the geometry.

S. Vishwanatha
Department of Mathematics, Faculty of Engineering & Technology, Jain Deemed-to-be-University, Kanakapura Taluk, Ramanagaram District, Karnataka, India

C. V. Vinay
Department of Mathematics, JSS Academy of Technical Education, Bangalore, India

M. Sankar (✉)
Department of Information Technology, University of Technology and Applied Sciences, Ibri, Oman

N. K. Reddy
Department of Mechanical Engineering, Ulsan National Institute of Science and Technology, Ulsan, South Korea

F. Kamalov et al. (eds.), *Advances in Mathematical Modeling and Scientific Computing*, Trends in Mathematics, https://doi.org/10.1007/978-3-031-41420-6_29

Later, Bairi and co-researchers [3, 4] developed new correlations for HT rates in this geometry and also made a detailed review on potential applications of this enclosure. Jagadeesha et al. [5, 6] considered the tilting impact of this cavity by adopting the Darcy model for flow motion generated by the collective buoyancies of thermal and solutal forces by taking account of magnetic force as well. The impacts of various non-similar thermal boundary conditions on BIC and HT rates in different geometries have also been studied to understand the flow behavior and heat dissipation rates under these conditions [7, 8].

Lately, due to the high demand in rapid cooling techniques of various electronic equipment, a novel fluid known as "nanofluid," has been invented and is found to be more advantageous than the traditional fluids, which has also been utilized as one of the cooling agents in square- and parallelogram-shaped geometries. Sheremet et al. [9], utilizing Tiwari and Das model, studied BIC and HT rates in a square region containing NF-saturated porous material of two different kinds. By considering the same NF model and porous materials, Ghalambaz et al. [10] studied the flow and associated thermal dissipation rates in a parallelogram geometry by varying the aspect ratio to predict the demerits of NF in porous domains. A similar analysis in parallelogram-shaped geometry containing NF-saturated porous material with LTNE model has been analyzed by Alsabery et al. [11] and analyzed the impacts of nonequilibrium thermal conditions. Later, Alsabery et al. [12] took up a similar analysis in a tilted square geometry partially occupied with porous material. They considered sinusoidal temperature variation along the vertical boundaries with NF formed by four different NPs and found that the porous structure width plays a vital role in controlling thermal dissipation across the non-similarly heated boundary surfaces. Some of the prominent investigations in a parallelogram-structured enclosure considering additional constraints, such as magnetic force and open boundary, in recent times could be found in [13, 14]. In the comprehensive review studies of Guo [15], and Oudina and Chabani [16], the various impacts affecting the BIC flow of NFs and associated HT rates are discussed at length by considering different nanoparticles and base fluids.

The above detailed review of literature identifies the vivid research gap between the present analysis and existing studies. It could be realized that the BIC flows and related thermal dissipation characteristics of NFs in an inclined parallelogram-structured geometry having sinusoidal thermal profile have not been attempted. In this paper, therefore, detailed numerical experiments are undertaken to analyze the impacts of sinusoidal heating on NF buoyant movement and HT rates in this geometry.

2 Mathematical Formulation

In this analysis, we choose an oblique-shaped structure whose sidewalls are inclined at an angle φ with the positive $y-$axis and flat horizontal boundaries as displayed in Fig. 1. The oblique structured geometry is tilted at an angle α with positive abscissa

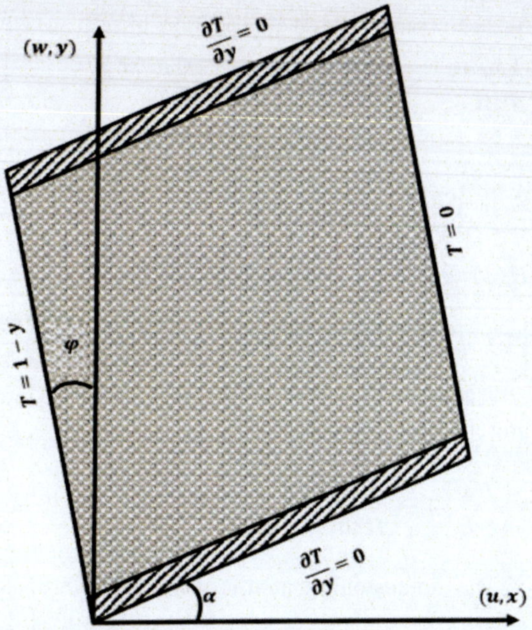

Fig. 1 Geometrical structure with thermal conditions

and contains porous saturated $Cu - H_2O$ nanoliquid. A variable temperature in the form of linear profile has been assigned to the left inclined surface, while a uniform cold thermal condition is maintained at right tilted surface. However, the upper and lower boundary regions are treated with adiabatic thermal condition. The fluid is supposed to be Boussinesq and incompressible as well as the flow is two-dimensional and axisymmetric. The thermophysical aspects of H_2O and Cu NP are given in Table 1 [16, 17]. Further, to restructure oblique-shaped geometry to a square domain, the coordinate transformations $X = x - y tan\varphi$ and $Y = y$ adopted by Jagadeesha et al. [5, 6] have been utilized in the present investigation. Implementing the Darcy flow assumptions and eliminating the pressure components, the constitutive equations for the momentum and energy conservation principles are [10, 12]:

$$\frac{\partial^2 \psi}{\partial \xi^2} - 2\frac{sin\varphi}{A}\frac{\partial^2 \psi}{\partial \xi \partial \eta} + \frac{1}{A^2}\frac{\partial^2 \psi}{\partial \eta^2} = RaH(\phi)\frac{sin\alpha}{A^2}\left[\frac{\partial \theta}{\partial \eta} - tan\varphi\frac{\partial \theta}{\partial \xi}\right] - RaH(\phi)cos^2\varphi cos\alpha\frac{\partial \theta}{\partial \xi}, \quad (1)$$

$$\frac{\partial \theta}{\partial t} + \frac{\partial \psi}{\partial \eta}\frac{\partial \theta}{\partial \xi} - \frac{\partial \psi}{\partial \xi}\frac{\partial \theta}{\partial \eta} = \frac{A}{cos\varphi}\left[\frac{\partial^2 \theta}{\partial \xi^2} - 2\frac{sin\varphi}{A}\frac{\partial^2 \theta}{\partial \xi \partial \eta} + \frac{1}{A^2}\frac{\partial^2 \theta}{\partial \eta^2}\right]. \quad (2)$$

Table 1 Thermophysical characteristics of H_2O and NP (Cu) [10, 17]

Property	ρ(kg/m^3)	C_p(J/kg K)	k(W/mK)	β(K^{-1})
H_2O	997.1	4179	0.613	21×10^{-5}
Cu	8933	385	400	1.67×10^{-5}

In the above equations, $Ra = \dfrac{gK(\rho\beta)_f \Delta\theta L}{\mu_f \alpha_m}$ is the Rayleigh number and $H(\phi) = \dfrac{(H_1 \times H_2)}{H_3}(1-\phi)^{2.5}$, where $H_1 = \left[1 - \phi + \dfrac{\phi(\rho\beta)_p}{(\rho\beta)_f}\right]$, $H_2 = \left[1 - \phi + \dfrac{\phi(\rho C_p)_p}{(\rho C_p)_f}\right]$ and $H_3 = \left[1 - \dfrac{3\epsilon\phi k_f(k_f - k_p)}{k_m[k_p + 2k_f + \phi(k_f - k_p)]}\right]$.

The following dimensionless quantities are introduced to non-dimensionalize the present model equations:

$$\xi = \frac{X}{L}, \eta = \frac{Y}{H\cos\varphi}, t = t^*\left(\frac{\alpha_f}{LH\cos\varphi}\right), \psi = \frac{\psi^*}{\alpha_f}, \theta = \frac{(T - T_r)}{\Delta T}, T_r = \frac{T_h + T_c}{2}.$$

The dimensionless auxiliary conditions are:

$$\begin{aligned}
t = 0: \ & \psi = \theta = 0; \quad \text{over the entire geometry.} \\
t > 0: \ & \psi = \frac{\partial\psi}{\partial\eta} = 0, \ \theta = 1 - \eta \ \text{at left boundary,} \\
& \psi = \frac{\partial\psi}{\partial\eta} = 0, \ \theta = 0 \ \text{at right boundary,} \\
& \psi = \frac{\partial\theta}{\partial\eta} - A\sin\varphi\frac{\partial\theta}{\partial\xi} = 0, \ \text{at } \eta = 0 \text{ and } 1
\end{aligned}$$

The thermophysical characteristics of NF are based on the following expressions [10, 17, 18]:

$$\begin{aligned}
(\rho C_p)_{nfp} &= (1 - \epsilon)(\rho C_p)_s + \epsilon(\rho C_p)_{nf}, \\
k_{nfp} &= (1 - \epsilon)k_s + \epsilon k_{nf}, \\
\alpha_{nfp} &= \frac{k_{nfp}}{(\rho C_p)_{nfp}}.
\end{aligned}$$

Here, the subscripts nfp, s, and nf are identified as the NF-saturated porous material, solid matrix of porous material, and NF, respectively.

The global HT rates are estimated from the average Nusselt number using the below equation:

$$\overline{Nu} = -\frac{k_{nfp}}{k_m}\int_0^1 \frac{1}{\cos\varphi}\left(\frac{\sin\varphi}{A}\frac{\partial\theta}{\partial\eta} - \frac{\partial\theta}{\partial\xi}\right) d\eta$$

3 Solution Technique

In this analysis, the model PDEs are nonlinear and coupled; thus as a result, the closed-form solutions could not be obtained. Hence, an implicit finite difference methodology has been chosen to obtain the solution of governing PDEs. In particular, the ADI and SLOR methods are utilized for transient and steady-state equations, respectively. A systematic grid independence resolutions have been performed, and after careful inspection, a grid size of 101×101 has been adopted for all simulations. We have developed an in-house code to invert the finite difference equations and all calculations are validated before the actual results are obtained. The more detailed description of the methodology and validation can be found in our recent works [5, 6] and the same is not given for brevity.

4 Discussion of Simulation Results

This segment of the analysis is mainly focused on interpreting the outcomes of simulation results for the vast range of parametric values chosen in this study. The results are discussed broadly in two categories, namely, qualitative analysis and quantitative analysis. In qualitative discussion, the flow and temperature variations are predicted in terms of their isopleths for various parametric ranges. However, the quantitative analysis deals with the estimation of average heat dissipation with respect to various parameters affecting the dissipation rates. First, the impacts of buoyancy-induced parameter, namely, Rayleigh number, tilt angle of the sidewall and geometry on isopleths of stream function, and thermal lines are shown in Figs. 2, 3, and 4 for different magnitudes of these control parameters with other pertinent factors being fixed. Figure 2 illustrates the importance of Ra on flow and thermal isopleths; a single vortex flow pattern with moderate strength could be witnessed along with meagerly varied isotherms at $Ra = 10^2$. However, for $Ra = 10^3$, the streamline structure reveals a strong flow with a weak secondary vortex at the top and several order magnitude of increase in extreme stream function value. The contours of thermal distribution are consistent with streamline pattern for higher magnitudes of Ra. Next, the impacts of sidewall tilt angle, φ, on the flow and temperature distributions are portrayed in Fig. 3. It can be noticed that the flow strength increases as the sidewall tilt angle is raised. With regard to the variation of geometry inclination, shown in Fig. 4, the isopleths of flow and thermal lines undergone a notable variation with respect to flow strength as well as temperature stratification. Figures 5, 6, 7, and 8 depict the quantitative variation in terms of $\overline{Nu}$ for different pertinent factors affecting the heat dissipation process by setting other key parameters to fixed value. Figure 5 portrays the collective influence of Ra and φ on the average Nu. Among the various magnitudes of φ, it has been witnessed that higher Ra with $+15^o$ produces an enhanced HT. The variations of HT rate with α have also been estimated and are shown in Fig. 6 for three different

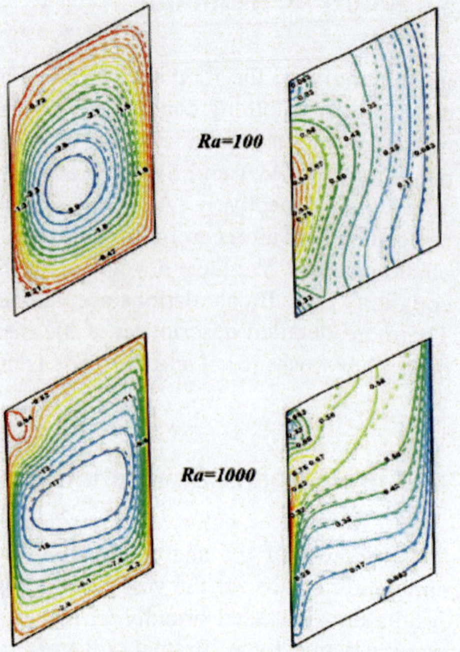

Fig. 2 Effect of Ra on flow and temperature contours (top) $Ra = 10^2$ and (bottom) $Ra = 10^3$. Dotted and continuous contours respectively correspond to $\phi = 0.0$ and $\phi = 0.05$

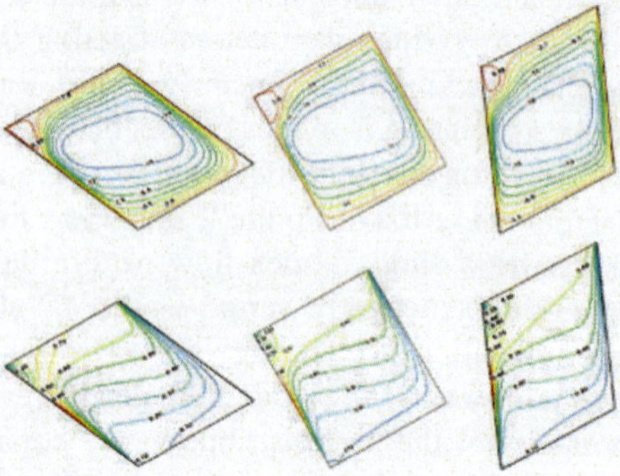

Fig. 3 Impact of φ on flow and temperature contours at $Ra = 10^3$, $\alpha = 30^o$ and $\phi = 0.05$. (Left) $\varphi = -30^o$, (middle) $\varphi = 0^o$ and (right) $\varphi = +30^o$

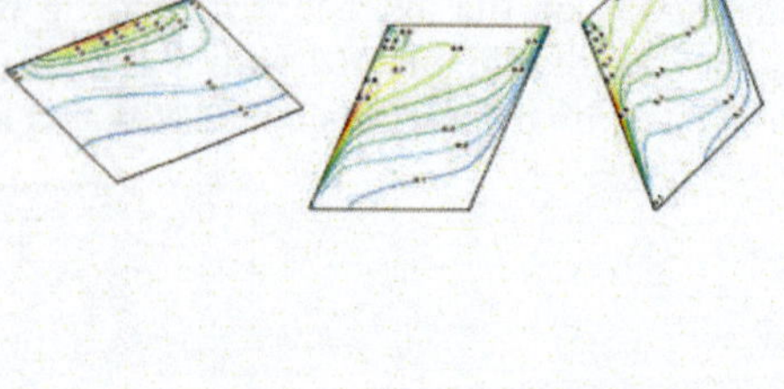

Fig. 4 Impact of α on flow and temperature contours at $Ra = 10^3$, $\varphi = 30^o$ and $\phi = 0.05$. (Left) $\alpha = -45^o$, (middle) $\alpha = 0^o$ and (right) $\alpha = +45^o$

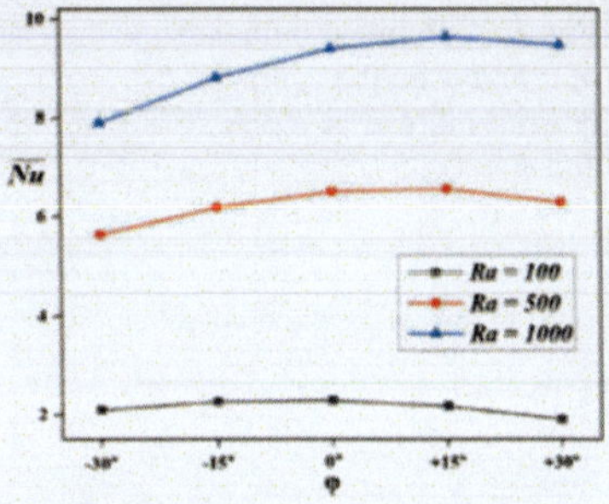

Fig. 5 Combined influence of φ and Ra on the global Nu at $\alpha = 30^o$ and $\phi = 0.03$

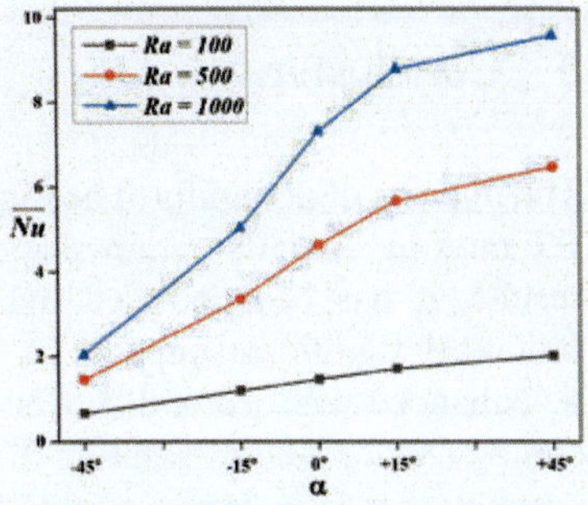

Fig. 6 Combined influence of α and Ra on global Nu $\varphi = 30^o$ and $\phi = 0.03$

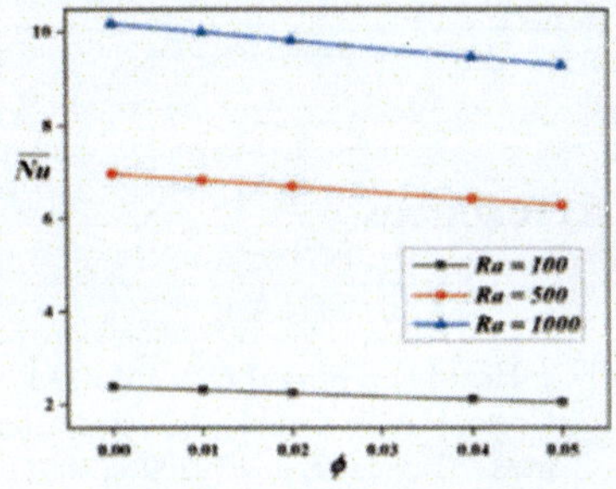

Fig. 7 Combined influence of ϕ and Ra on the average Nusselt number at $\alpha = 30^o$ and $\varphi = 15^o$

values of Ra. From the predictions, it has been witnessed that the HT rates could be enhanced by choosing a higher positive angle of α ($\alpha = +45^o$). In addition, the variation of $\overline{Nu}$ with nanoparticle volume fraction is depicted in Fig. 7 for different magnitudes of Ra. Interestingly, it has been noticed that the addition of nanoparticle produces adverse impacts on heat dissipation rates, and this prediction is consistent with those of Ghalambaz et al.[10]. Finally, the combined effects of φ and α on the HT rates have also been predicted to identify the optimum pair angles to produce maximum heat dissipation and the same has been presented in Fig. 8 by considering three combinations of each parameter. It can be noticed from the predictions that a choice of $\alpha = +45^0$ and $\varphi = +30^o$ increases the heat dissipation rate.

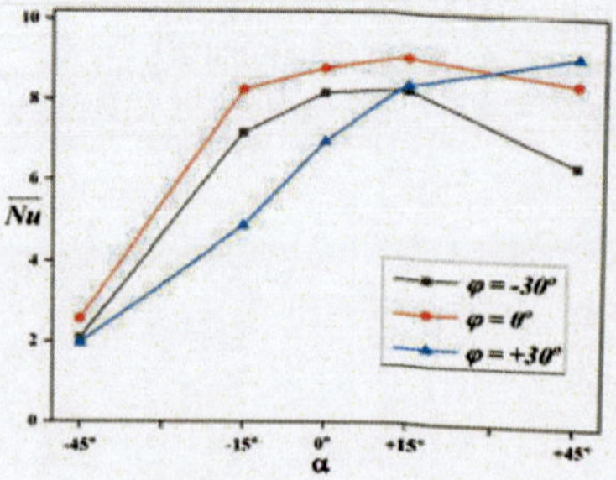

Fig. 8 Combined influence of φ and α on the average Nusselt number at $Ra = 10^3$ and $\phi = 0.05$

5 Conclusions

The impacts of nonuniform heating and geometry tilting on BIC flow and associated HT rates in a parallelogram-shaped domain are predicted. Based on the simulation results, it has been noticed that two tilting angles of geometry and sidewalls produced significant impacts on BIC and HT of NF. Heat dissipation rate could be enhanced with Ra and φ. Also, an optimum choice with $\alpha = 45^o$ and $\varphi = 30^o$ could enhance heat dissipation in the geometry. It is worth to mention here that the current study is mainly applicable for low-velocity flows as governed by the Darcy model.

References

1. Hyun, H.M., Choi, B.S.: Transient natural convection in a parallelogram-shaped enclosure. Int. J. Heat Fluid Flow **11**(2), 129–134 (1990)
2. Costa, V.A.F.: Double-diffusive natural convection in parallelogrammic enclosures. Int. J. Heat Mass Transf. **47**, 2913–2026 (2004)
3. Bairi, A.: On the Nusselt number definition adapted to natural convection in parallelogrammic cavities. Appl. Therm. Eng. **28**, 1267–1271 (2008)
4. Baïri, A., Zarco-Pernia, E., Garcia de Maria, J.: A review on natural convection in enclosures for engineering applications. The particular case of the parallelogrammic diode cavity. Appl. Therm. Eng. **63**, 304–322 (2014)
5. Jagadeesha, R.D., Prasanna, B.M.R., Sankar, M.: Double diffusive convection in an inclined parallelogrammic porous enclosure. Proc. Eng. **127**, 1346–1353 (2015)
6. Jagadeesha, R.D., Prasanna, B.M.R., Sankar, M.: Numerical simulation of double diffusive magnetoconvection in an inclined parallelogrammic porous enclosure with an internal heat source. Mater. Today: Proc. **4**, 10544–10548 (2017)
7. Deng, Q.-H., Chang, J.-J.: Natural convection in a rectangular enclosure with sinusoidal temperature distributions on both side walls. Numer. Heat Transf. Part A: Appl.: Int. J. Comput. Methodol. **54**(5), 507–524 (2008)
8. Sankar, M., Kemparaju, S., Prasanna, B.M.R., Eswaramoorthi, S.: Buoyant convection in porous annulus with discrete sources-sink pairs and internal heat generation. J. Phys.: Conf. Ser. **1139**, 012026 (2018)
9. Sheremet, M.A., Grosan, T., Pop, I.: Free convection in a square cavity filled with a porous medium saturated by nanofluid using Tiwari and Das' nanofluid model. Transp. Porous Med. **106**. 595–610 (2015)

10. Ghalambaz, M., Sheremet, M.A., Pop, I.: Free convection in a parallelogrammic porous cavity filled with a nanofluid using Tiwari and Das' nanofluid model. PLoS One **10**(5), e0126486 (2015)
11. Alsabery, A.I., Saleh, H., Hashim, I., Siddheshwar, P.G.: Transient natural convection heat transfer in nanoliquid-saturated porous oblique cavity using thermal non-equilibrium model. Int. J. Mech. Sci. **114**, 233–245 (2016)
12. Alsabery, A.I., Chamkha, A.J., Saleh, H., et al.: Natural convection flow of a nanofluid in an inclined square enclosure partially filled with a porous medium. Sci. Rep. **7**, 2357 (2017)
13. Hussein, A.K., Mustafa, A.W.: Natural convection in fully open parallelogrammic cavity filled with Cu–water nanofluid and heated locally from its bottom wall. Therm. Sci. Eng. Prog. **1**, 66–77 (2017)
14. Mallick, H., Mondal, H., Biswas, N., Manna, N.K.: Buoyancy driven flow in a parallelogrammic enclosure with an obstructive block and magnetic field. Mater. Today: Proc. **44**, 3164–3171 (2021)
15. Guo, Z.: A review on heat transfer enhancement with nanofluids. J. Enhanc. Heat Transf. **27**(1), 1–70 (2020)
16. Mebarek Oudina, F., Chabani, I.: Review on nano-fluids applications and heat transfer enhancement techniques in different enclosures. J. Nanofluids **11**(2), 155–168 (2022)
17. Reddy, N.K., Sankar, M.: Buoyant heat transfer of nanofluids in a vertical porous annulus: a comparative study of different models. Int. J. Numer. Methods Heat Fluid Flow **33**(2), 477–509 (2023)
18. Reddy, N.K., Swamy, H.A.K., Sankar, M.: Buoyant convective flow of different hybrid nanoliquids in a non-uniformly heated annulus. Eur. Phys. J. Spec. Top. **230**, 1213–1225 (2021)
19. Sankar, M, Swamy, H.A.K., Do, Y., Altmeyer, S.: Thermal effects of nonuniform heating in a nanofluid filled annulus: buoyant transport versus entropy generation. Heat Transf. **51**, 1062–1091 (2022)

Buoyant Convection of Nanofluid in an Annular Domain with Linear Heating

S. Kiran, M. Sankar, N. Girish, and H. A. Kumara Swamy

1 Introduction

Buoyant-induced convection (BIC) of nanofluids (NF) and corresponding thermal transport (TT) is a frequent phenomenon in numerous industrial and engineering applications, which could be represented using finite-sized enclosures. Among those, an annular chamber with a thermal gradient perpendicular to the gravity and adiabatic top-bottom walls is a typical and largely used model, and therefore several experimental and numerical investigations are performed in such geometries [1–3]. Further, the high demand for effective cooling of electronic equipment in high-temperature applications leads to the utilization of NFs as cooling agents replacing the traditional fluids/liquids. Hence, the associated BIC and TT rates in finite annular geometries are analyzed in detail by considering various constraints. In this direction, Abouali and Falahatpisheh [4] made the first attempt in analyzing the BIC and corresponding TT in an annular domain by considering wide parametric ranges and proposed thermal correlations. In the same geometry, an experimental visualization using oil-based NFs has been carried out by conducting for a vast number of cases and was considered to the pioneering experimental study [5].

S. Kiran
Department of Mathematics, Nitte Meenakshi Institute of Technology, Bengaluru, Karnataka, India

M. Sankar
Department of Information Technology, University of Technology and Applied Sciences, Ibri, Oman

N. Girish (✉)
Department of Mathematics, JSS Academy of Technical Education, Bengaluru, India

H. A. Kumara Swamy
Department of Mathematics, Nonlinear Dynamics and Mathematical Application Center, Kyungpook National University, South Korea

F. Kamalov et al. (eds.), *Advances in Mathematical Modeling and Scientific Computing*, Trends in Mathematics, https://doi.org/10.1007/978-3-031-41420-6_30

Oudina [6] analyzed the impacts of isolated thermal source, placed along the interior boundaries of annular geometry, on BIC and TT rates and predicted that a small source has high thermal dissipation as compared to larger source. Bouzerzour et al. [7] investigated a similar setup by focusing on the different thermal source-sink arrangements. Utilizing a thin baffle, the BIC and associated TT rates in an annular region are effectively controlled with the length and position of the baffle placed along the hot cylinder [8]. Swamy et al. [9] performed a numerical study to understand the buoyant convection transport and entropy minimization strategies in a tilted porous annular domain and found an optimum tilt angle to maximize TT with minimum production of entropy.

For numerous thermal transport applications, the geometry of equipment could also be a rectangular-shaped structure, and as a result, BIC and TT rates in such geometries have been investigated in detail. Basak et al. [10] numerically studied the convective HT rate of NFs in a square-shaped enclosure heated with different temperature profiles covering a wide range of parameters and provided thermal dissipation correlations. Later, Roy [11] studied the BIC and TT rates of water-based NFs in a square geometry to understand the importance of various shapes of inner geometry. The influence of different thermal conditions, such as sinusoidal and linear heating of active walls on BIC and heat transport rate in square-shaped geometries, has also been analyzed [12, 13]. The augmentation or reduction of BIC flow and associated TT rates is subjected to various constraints, and these aspects have been systematically discussed by including a full-blown survey of NF heat transfer literature and can be found in [14, 15]. Sathiyamoorthy and co-workers [16, 17] analyzed the impacts of linear thermal heating of square enclosure boundaries on BIC and thermal dissipation rates by considering a vast range of critical parameters with and without magnetic field.

The above detailed review of literature reveals the obvious research gap that the BIC flow motion and related HT characteristics of NFs in an annular configuration having linear thermal profiles have not been attempted. In this chapter, therefore, detailed numerical experiments are undertaken to analyze the impacts of linear heating on NF buoyant movement and HT rates in the annular geometry.

2 Mathematical Formulation

The geometrical structure chosen in this analysis is an upright annular geometry between two coaxial cylinders filled with $Cu - H_2O$ NF and whose side boundaries are supplied with axially varying temperature in the form of linear heating, while the upper and lower portions are perfectly insulated. The fluid is supposed to be Boussinesq and incompressible, and the flow is two-dimensional and axisymmetric. The thermophysical aspects of H_2O and NP are given in Table 1 [15, 18]. Implementing the aforementioned assumptions, and eliminating the pressure components through

Table 1 Thermophysical characteristics of H_2O and NP (Cu) [14, 15, 18]

Property	ρ(kg/m^3)	C_p(J/kg K)	k(W/mK)	β(K^{-1})
H_2O	997.1	4179	0.613	21×10^{-5}
Cu	8933	385	400	1.67

cross-differentiation, the constitutive equations for the momentum and energy in the vorticity-stream function formulation are [6, 9]:

$$\frac{\partial T}{\partial t} + U\frac{\partial T}{\partial R} + W\frac{\partial T}{\partial Z} = \frac{\alpha_{nf}}{\alpha_f}\nabla^2 T, \tag{1}$$

$$\frac{\partial \zeta}{\partial t} + U\frac{\partial \zeta}{\partial R} + W\frac{\partial \zeta}{\partial Z} - \frac{U\zeta}{R} = \frac{\mu_{nf}}{\rho_{nf}\alpha_f}\left[\nabla^2\zeta - \frac{\zeta}{R^2}\right] - \frac{(\rho\beta)_{nf}}{\rho_{nf}\beta_f} Ra\ Pr\ \frac{\partial T}{\partial R}, \tag{2}$$

$$\zeta = \frac{1}{R}\left[\frac{\partial^2\psi}{\partial R^2} - \frac{1}{R}\frac{\partial\psi}{\partial R} + \frac{\partial^2\psi}{\partial Z^2}\right], \tag{3}$$

$$U = \frac{1}{R}\frac{\partial\psi}{\partial Z}, \quad W = -\frac{1}{R}\frac{\partial\psi}{\partial R}, \tag{4}$$

where $Ra = \dfrac{g\beta_f \Delta\theta D^3}{\nu_f \alpha_f}$ and $Pr = \dfrac{\nu_f}{\alpha_f}$ are the Rayleigh and Prandtl numbers, respectively. Also, $\nabla^2 = \dfrac{\partial^2}{\partial R^2} + \dfrac{1}{R}\dfrac{\partial}{\partial R} + \dfrac{\partial^2}{\partial Z^2}$. The following dimensionless quantities are introduced to non-dimensionalize the present model equations:

$$(R, Z) = \frac{(r, z)}{D}, U = \frac{uD}{\alpha_f}, W = \frac{wD}{\alpha_f}$$

$$t = \frac{t^*}{\left(D^2/\alpha_f\right)}, T = \frac{(\theta - \theta_c)}{(\theta_h - \theta_c)}, P = \frac{p}{\rho_{nf}\left(\alpha_f/D\right)^2}$$

The auxiliary conditions in nondimensional form are:

$t = 0:\ U = W = T = 0,\ \psi = 0;$ all over the region.

$t > 0:\ \psi = \dfrac{\partial\psi}{\partial R} = 0,\ T_i = T_o = 1 - Z$ along inner and outer boundaries

$\psi = \dfrac{\partial\psi}{\partial Z} = 0,\ T = 1$ along lower boundary

$\psi = \dfrac{\partial\psi}{\partial Z} = 0,\ \dfrac{\partial T}{\partial Z} = 0$ along upper region

It is worth to mention that the boundary conditions for vorticity have been derived by Taylor's series expansion of the known stream function quantities.

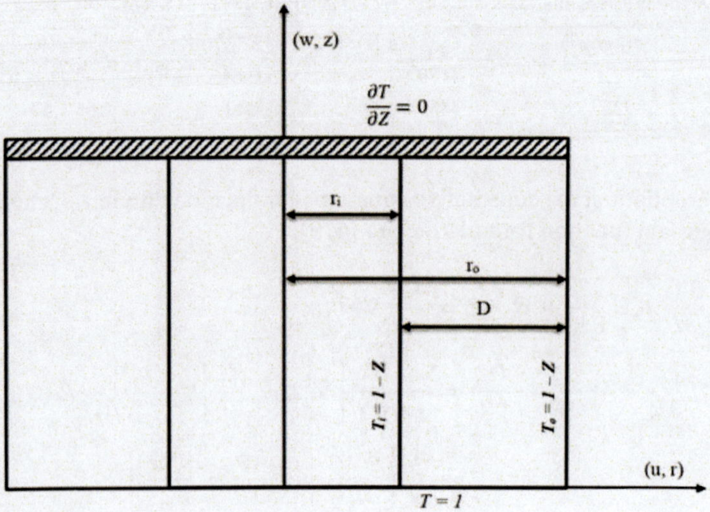

Fig. 1 Geometrical structure and coordinates

The thermophysical characteristics of NF are based on the following expressions [14, 15]:

$$\begin{aligned}
\rho_{nf} &= (1-\phi)\rho_f + \phi\rho_p, \\
(\rho\beta)_{nf} &= (1-\phi)(\rho\beta)_f + \phi(\rho\beta)_p, \\
(\rho C_p)_{nf} &= (1-\phi)(\rho C_p)_f + \phi(\rho C_p)_p, \\
\frac{k_{nf}}{k_f} &= \frac{k_p + 2k_f - 2\phi(k_f - k_p)}{k_p + 2k_f + \phi(k_f - k_p)}, \\
\mu_{nf} &= \frac{\mu_f}{(1-\phi)^{2.5}}, \\
\alpha_{nf} &= \frac{k_{nf}}{(\rho C_p)_{nf}}.
\end{aligned}$$

Here, the subscripts nf, f, and p denoted the NF, BF, and NP, respectively.

The global HT rates are estimated using the below equation:

$$\overline{Nu} = -\frac{k_{nf}}{k_f}\frac{1}{A}\int_0^A \frac{\partial T}{\partial R}\, dZ$$

3 Solution Technique

In this analysis, the model PDEs are nonlinear and coupled; thus as a result, the closed-form solutions could not be obtained. Hence, an implicit finite difference methodology has been chosen to obtain the solution of governing PDEs. In particular, we adopt the ADI and SLOR methods for transient and steady-state equations, respectively. A systematic grid independence resolutions have been performed, and after careful inspection, a grid size of 161×161 has been adopted for all simulations. We have developed an in-house code to invert the finite difference equations and all calculations are validated before the actual results are obtained. The more detailed description of the methodology and validation can be found in our recent works [2, 7] and the same is not given for brevity.

4 Discussion of Simulations

This segment is mainly devoted to discuss the outcomes of simulation results for the vast range of parametric values chosen in this analysis. The influence of various parameters on the buoyant flow and thermal structure has been presented for different magnitudes of the respective parameters.

First, the impacts of buoyancy-induced parameter, namely, Rayleigh number, aspect and radius ratios on isopleths of stream function and thermal lines are shown in Figs. 2, 3, and 4 for different magnitudes of these control parameters with other pertinent factors being fixed. Figure 2 illustrates the importance of Ra on flow

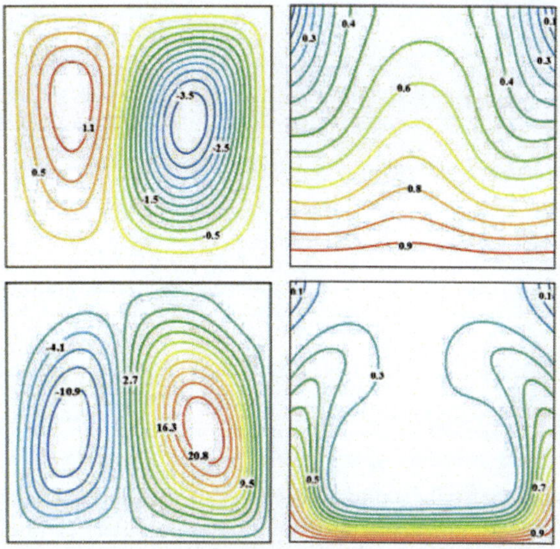

Fig. 2 Effect of Ra and ϕ on flow and temperature contours (top) $Ra = 10^4$ and (bottom) $Ra = 10^6$

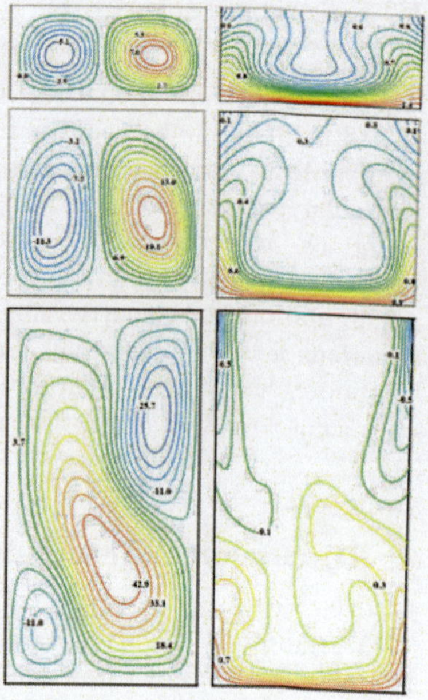

Fig. 3 Impact of A flow and temperature contours (top) $A = 0.5$, (middle) $A = 1.0$, and (bottom) $A = 2.0$

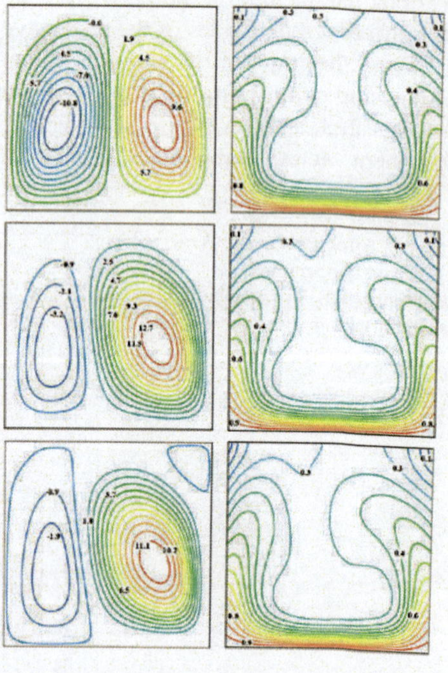

Fig. 4 Impact of λ on flow and temperature contours (top) $\lambda = 1$, (middle) $\lambda = 5$, and (bottom) $\lambda = 10$

and thermal isopleths; a two-eddy regularly structured flow pattern with an eddy appearing near the inner and outer cylinders can be witnessed, while the isotherms reveal similar variations. The streamline structure reveals a strong two-eddy flow

Fig. 5 Combined influence of λ and Ra on average Nusselt number. Continuous lines for right wall and dotted lines for bottom wall

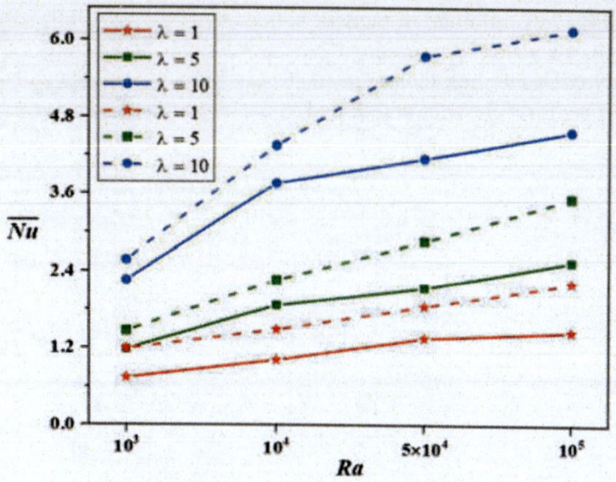

Fig. 6 Combined influence of ϕ and Ra on average Nusselt number. Continuous lines for right wall and dotted lines for bottom wall

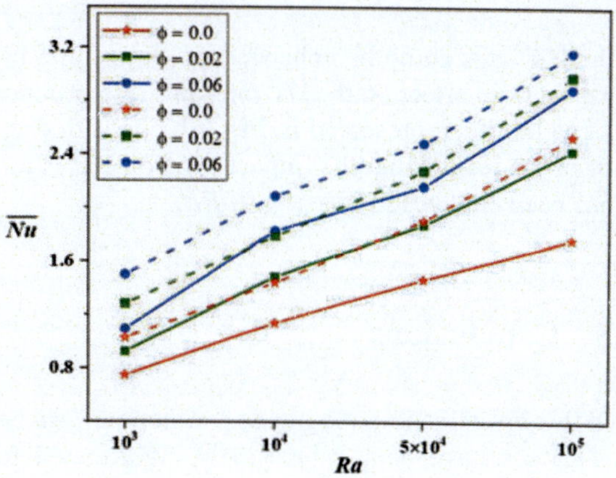

with several order magnitude of increase in extreme stream function value for higher magnitude of $Ra(10^6)$. The contours of thermal distribution are consistent with streamline pattern for both magnitudes of Ra. Next, the impacts of aspect ratio on the flow and temperature distributions are portrayed in Fig. 3. It can be noticed that the flow strength has increased to multifold as the height of the annular structure is raised. With regard to the variation of radius ratio, a critical parameter of this geometry, the isopleths of flow and thermal lines undergone a moderate variation (Fig. 4).

The collective impacts of Ra and other critical parameters on the HT rates have been illustrated in Figs. 5, 6, and 7 by setting remaining key parameters to a constant value. A general observation made from these predictions is that among inner, outer, and bottom walls, higher thermal dissipation has been noticed from bottom boundary. Figure 5 portrays the collective influence of Ra and radius ratio (λ) on the average Nu rate. Among these three magnitudes of λ, it has been witnessed that the larger radius ratio produces a higher HT due to the availability of larger annular area. The variations of HT rate with ϕ have also been estimated and are shown in Fig. 6 for three different volume fractions. From the predictions, it has been witnessed that

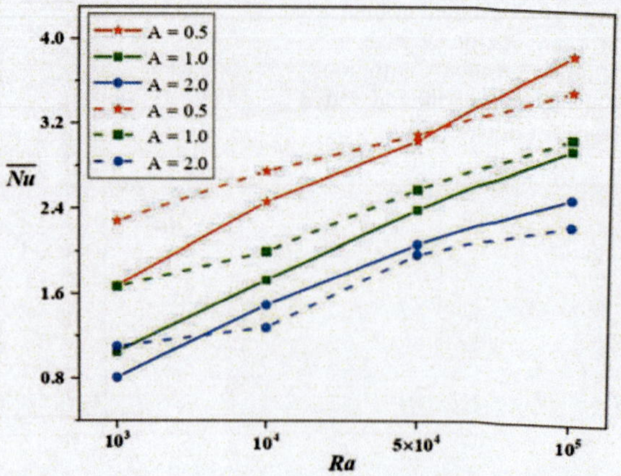

Fig. 7 Combined influence of A and Ra on average Nusselt number. Continuous lines for right wall and dotted lines for bottom wall

the HT rates could be enhanced by choosing a higher magnitude of ϕ. Finally, the aspect ratio effect on the HT rates has also produced interesting predictions, and the same has been presented in Fig. 7 by considering three values of A. It can be seen from the results that the impact of A on $\overline{Nu}$ is to increase heat dissipation rate and has been enhanced further with Ra.

5 Conclusions

In this investigation, the impacts of nonuniform heating on BIC flow and associated HT rates have been numerically investigated for broad ranges of key parameters. Based on the simulation results, the following predictions are observed and recorded:

1. Nonuniform heating and taller annular geometries produce multicellular flows.
2. Heat dissipation rate could be enhanced with radius ratio and NP volume concentration.

References

1. Sankar, M., Kiran, S., Ramesh, G.K., Makinde, O.D.: Natural convection in a non-uniformly heated vertical annular cavity. Defect Diffus. Forum **377**, 189–199 (2017)
2. Kiran, S., Sankar, M., Sivasankaran, S.: Numerical study on conjugate convective thermal transport in an annular porous geometry. Kuwait J. Sci. **49**(2), 1–23 (2022)
3. Kemparaju, S., Kumara Swamy, H.A., Sankar, M., Mebarek-Oudina, F.: Impact of thermal and solute source-sink combination on thermosolutal convection in a partially active porous annulus. Phys. Scr. **97**, 055206 (2022)
4. Abouali, O., Falahatpisheh, A.: Numerical investigation of natural convection of Al_2O_3 nanofluid in vertical annuli. Heat Mass Transf. **46**(1), 15–23 (2009)

5. Cadena-de la Peña, N.L., Rivera-Solorio, C.I., Payán-Rodríguez, L.A., García-Cuéllar, A.J., López-Salinas, J.L.: Experimental analysis of natural convection in vertical annuli filled with AlN and TiO_2/mineral oil-based nanofluids. Int. J. Therm. Sci. **111**, 138–145 (2017).
6. Mebarek-Oudina, F.: Convective heat transfer of Titania nanofluids of different base fluids in cylindrical annulus with discrete heat source. Heat Transf.-Asian Res. **48**(1), 135–147 (2019)
7. Bouzerzour, A., Djezzar, M., Oztop, H.F., Tayebi, T., and Abu-Hamdeh, N.: Natural convection in nanofluid filled and partially heated annulus: effect of different arrangements of heaters. Phys. A **538**, 122479 (2020)
8. Pushpa, B.V., Sankar, M., Mebarek-Oudina, F.: Buoyant convective flow and heat dissipation of Cu–H_2O nanoliquids in an annulus through a thin baffle. J. Nanofluids **10**, 292–304 (2021)
9. Swamy, H.A.K., Sankar, M., Reddy, N.K.: Analysis of entropy generation and energy transport of Cu-water nanoliquid in a tilted vertical porous annulus. Int. J. Appl. Comput. Math. **8**(1), 10 (2022)
10. Basak, T., Chamkha, A.J.: Heatline analysis on natural convection for nanofluids confined within square cavities with various thermal boundary conditions. Int. J. Heat Mass Transf. **55**, 5526–5543 (2012)
11. Roy, N.C.: Natural convection of nanofluids in a square enclosure with different shapes of inner geometry. Phys. Fluids **30**, 113605 (2018)
12. Alsabery, A.I., Chamkha, A.J., Saleh, H., Hashim, I.: Heatline visualization of conjugate natural convection in a square cavity filled with nanofluid with sinusoidal temperature variations on both horizontal walls. Int. J. Heat Mass Transf. **100**, 835–850 (2016)
13. Mahmoudi, A., Mejri, I., Abbassi, M.A., Omri, A.: Lattice Boltzmann simulation of MHD natural convection in a nanofluid-filled cavity with linear temperature distribution. Powder Technol. **256**, 257–271 (2014)
14. Guo, Z.: A review on heat transfer enhancement with nanofluids. J. Enhanc. Heat Transf. **27**(1), 1–70 (2020)
15. Mebarek Oudina, F., Chabani, I.: Review on nano-fluids applications and heat transfer enhancement techniques in different enclosures. J. Nanofluids **11**(2), 155–168 (2022)
16. Sathiyamoorthy, M., Basak, T., Roy, S., Pop, I.: Steady natural convection flows in a square cavity with linearly heated side wall(s). Int. J. Heat Mass Transf. **50**, 766–775 (2007)
17. Sathiyamoorthy, M., Chamkha, A.: Effect of magnetic field on natural convection flow in a liquid gallium filled square cavity for linearly heated side wall(s). Int. J. Therm. Sci. **49**, 1856–1865 (2010)
18. Reddy, N.K., Sankar, M.: Buoyant heat transfer of nanofluids in a vertical porous annulus: a comparative study of different models. Int. J. Numer. Methods Heat Fluid Flow **33**(2), 477–509 (2023)

Thermal and Entropy Management of Nanoliquid in a Discretely Heated Inclined Square Geometry

B. M. R. Prasanna, H. A. Kumara Swamy, M. Sankar, and S. R. Sudheendra

Nomenclature

Be	Bejan number	α	Thermal diffusivity
g	Acceleration due to gravity	β	Thermal expansion coefficient
k	Thermal conductivity	γ	Inclination angle
L	Length and height of the cavity	δ	Source-sink length
$\overline{Nu}$	Average Nusselt number	θ	Dimensional temperature
Pr	Prandtl number	ν	Kinematic viscosity
Ra	Rayleigh number	ρ	Density
S_{tot}	Total entropy generation	ϕ	Nanoparticle volume fraction
T	Dimensionless temperature		

B. M. R. Prasanna
Department of Mathematics, Siddaganga Institute of Technology, Tumkuru, India

H. A. Kumara Swamy (✉)
Department of Mathematics, Nonlinear Dynamics and Mathematical Application Center, Kyungpook National University, Daegu, Republic of Korea

M. Sankar
Department of Information Technology, University of Technology and Applied Sciences – Ibri, Ibri, Oman

S. R. Sudheendra
Department of Mathematics, School of Engineering, Presidency University, Bengaluru, India

F. Kamalov et al. (eds.), *Advances in Mathematical Modeling and Scientific Computing*, Trends in Mathematics, https://doi.org/10.1007/978-3-031-41420-6_31

1 Introduction

Due to its popularity in areas of industrial and engineering applications, such as cooling of electronic devices, extraction of oil, nuclear reactors, solar ponds and collectors, and many other, buoyant thermal transport in an enclosure has been widely investigated. In view of enhanced heat transfer rate (HTR), several researchers addressed the influence of nanoliquid on fluidity and HTR in differently shaped closed enclosures [1–3]. In several engineering applications, the supply of thermal energy may be constrained to particular positions of the walls which is known as the discrete heating. Because of its important applications, the impact of discrete thermal boundary conditions on nanofluid movement and energy dissipation in an enclosure has been numerically investigated under different constraints. Sheikhzadeh et al. [4] studied the impact of source-sink positions on nanoliquid flow and HTR in a cavity. They reported that greater thermal transport has been gained with bottom-middle and middle-middle locations for the high and low Rayleigh numbers, respectively. The influence of three heaters placed at the left wall of a tall cavity on fluidity and HTR has been numerically analyzed by Ben-Mansour and Habib [5]. Bhuiyana et al. [6] performed numerical simulations to analyze the bottom wall discrete heating effect on thermal transport rate of different nanofluids and concluded that silver-water nanofluid dissipates maximum thermal energy compared to other considered nanofluids. Recently, the impact of single and dual source-sink positions on fluid movement and energy transport rate has been numerically studied, and the best source-sink position has been proposed to enhance the energy transport rate in an annular geometry [7, 8].

In several industries, along with enhancement in heat dissipation rate, minimization of entropy generation (EG) is the challenging factor to enhance the thermal system efficiency. This is because generation of entropy leads to degradation of device efficiency. In this prospect, minimization of EG with maximizing HTR under different constraints has become a trending area of research from past few decades [9, 10]. In addition to minimization of EG, the impact of enclosure inclination angle on nanofluid flow and HTR has become an emerging field of research due to the appearance of gravitational force in dual direction. In this regard, Armaghani et al. [11] studied the influence of Cu-water nanoliquid on HTR and EG in an inclined partially porous layered cavity and found that the highest tilt angle and largest porous layer thickness enhance the system performance at lower Rayleigh number. By adopting finite difference lattice Boltzmann method, Kefayati and Che-Sidik [12] investigated the impact of cavity tilt angle on nanofluid thermal transport and irreversibility rates. Computational study on HTR and EG of Cu-water nanofluid filled in an inclined partially active porous cavity under the influence of magnetic field has been performed by Rashad et al. [13]. The influence of magnetic force and porous medium on HTR and EG has been numerically studied in a nanofluid-filled cavity [14–16]. Recently, numerical experiments have been carried out on the influence of nonuniform thermal profile, magnetic force, geometric tilt angle, and dual buoyancy effect on HTR and EG of nanoliquid in an annulus and provided the appropriate set of parameters to increase the system efficiency [17–20]. The physical representation of many thermal industries may be non-regular, such as C-shaped, Γ-shaped, L-shaped, wavy enclosure, quadrantal, and many other odd shapes. It is worth to mention that the shape of the geometry shows significant impact on fluid flow, HTR, and EG. By considering this, a wide range of investigation are performed under various constraints [21–23].

The above state of the art shows that the influence of a thermal source-sink length and cavity inclination angle on nanoliquid convective flow, energy transport, and EG in a tilted square geometry has not yet investigated. Hence, the current work aims to examine the impact of cavity tilt angle, length of source and sink, Rayleigh number, and nanoparticle volume fraction on HTR along with EG. It has been believed that the outcomes of the current study contributes in manufacturing thermal devices that enhance system efficiency.

2 Mathematical Modeling

The physical domain considered in this analysis is an inclined square cavity occupied with silver-water nanoliquid subjected to partial heating/cooling. Except at source-sink positions, the remaining boundaries are maintained at adiabatic and are illustrated in Fig. 1. The source (red in color) is kept at higher temperature than the sink (blue in color) throughout the study. The thermophysical properties and expression for density, thermal expansion coefficient, heat capacitance, thermal conductivity, dynamic viscosity, and diffusivity of nanoliquid are taken from [8, 19]. In this study, the Newtonian, incompressible fluid with unsteady, laminar flow has been assumed along with Boussinesq approximation. By adopting the nondimensional variables used in [17, 20] and abovementioned postulates, the nondimensional equations are as below [11, 13]:

$$\frac{\partial T}{\partial t} + U\frac{\partial T}{\partial X} + V\frac{\partial T}{\partial Y} = \frac{\alpha_{nf}}{\alpha_f}\nabla^2 T \tag{1}$$

$$\left[\frac{\partial \zeta}{\partial t} + U\frac{\partial \zeta}{\partial X} + V\frac{\partial \zeta}{\partial Y}\right] = \frac{\nu_{nf}}{\alpha_f}\nabla^2\zeta + \frac{(\rho\beta)_{nf}}{\rho_{nf}\beta_f} Ra\,Pr\left(sin\gamma\frac{\partial T}{\partial Y} - cos\gamma\frac{\partial T}{\partial X}\right) \tag{2}$$

$$\zeta = \left[\frac{\partial^2\psi}{\partial X^2} + \frac{\partial^2\psi}{\partial Y^2}\right] \tag{3}$$

The heat dissipation rate at sources is represented in terms of average Nusselt number which is defined and given by

Fig. 1 Physical domain

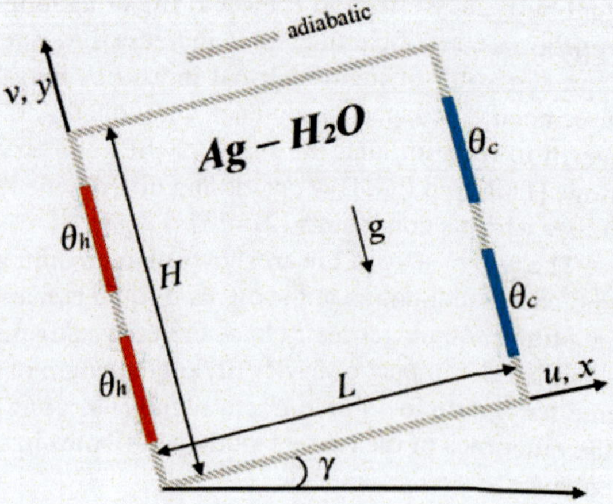

$$\overline{Nu} = \frac{1}{\delta}\int_{0.25-\frac{\delta}{2}}^{0.25+\frac{\delta}{2}} -\frac{k_{nf}}{k_f}\left(\frac{\partial T}{\partial X}\right) dY + \frac{1}{\delta}\int_{0.75-\frac{\delta}{2}}^{0.75+\frac{\delta}{2}} -\frac{k_{nf}}{k_f}\left(\frac{\partial T}{\partial X}\right) dY \quad (4)$$

2.1 Entropy Production Equation

According to the second law of thermodynamics, postulates made, the dimensionless forms of HTE ($S_{l.T}$) and FFE ($S_{l.\Psi}$) are as follows [16]:

$$S_{l.T} = \frac{k_{nf}}{k_f}\left[\left(\frac{\partial T}{\partial X}\right)^2 + \left(\frac{\partial T}{\partial Y}\right)^2\right. \quad (5)$$

$$S_{l.\Psi} = \Phi\frac{\mu_{nf}}{\mu_f}\left[2\left\{\left(\frac{\partial U}{\partial X}\right)^2 + \left(\frac{\partial V}{\partial Y}\right)^2\right\} + \left(\frac{\partial U}{\partial Y} + \frac{\partial V}{\partial X}\right)^2\right. \quad (6)$$

The global entropy production is given by

$$S_{tot} = \int_{Y=0}^{1}\int_{X=0}^{1} (S_{l.T} + S_{l.\Psi})\, dX\, dY \quad (7)$$

The relative dominance of EG due to heat transfer and fluid friction is given by the Bejan number and is defined as

$$Be = \int\limits_{Y}\int\limits_{X} \frac{S_{l,T}}{S_{l,T} + S_{l,\Psi}}\, dX\, dY \tag{8}$$

It is worth to mention that FFE and HTE are dominant respectively for $Be < 0.5$ and $Be > 0.5$, and in other case, the contribution of HTE and FFE in an enclosure is the same.

3 Numerical Technique and Validation

The PDE's (1)–(2) are solved by adopting time-splitting technique and Eq. (3) has been solved by adopting SLOR iterative method. The tri-diagonal system of equations is solved by utilizing the TDMA method. Local EG is estimated by solving Eqs. (5) and (6) using second-order central difference approximation. Finally, Simpson's and trapezoidal rules are utilized to compute the overall HTR and total EG, respectively. The complete details and step-by-step procedure along with comparison of streamlines, isotherms, and entropy generation between our result and benchmark problems can be found in our earlier works [17, 18, 20] and are not repeated for brevity. After performing grid independence study, a mesh size of 161×161 is adopted for all computations.

4 Results and Discussion

The steady-state streamlines, isotherms, and entropy generation contours for different source-sink pair lengths are depicted in Fig. 2 by maintaining $Ra = 10^6$, $\gamma = 45°$ for base fluid (dotted line) and nanoliquid (solid line). Though there is no significant variation in contour pattern of flow and thermal and entropy generation, change in heater and cooler length causes a noticeable variation in the fluid flow strength. Enhancement in length of source-sink pair increases the thermally active portion of the annulus which leads to enhancement in flow strength. The isothermal lines are observed to be more packed near the active portion of the cavity. The core values of entropy generation contours reveal that the strength of EG greatly enhances with an increment in length of sources and sinks. This could be due to the dependence of EG on the velocity gradient of the fluid flow which increases with increment in length of active locations.

Figure 3 illustrates the impact of cavity tilt angles on state streamlines, isotherms, and entropy generation contours by fixing $Ra = 10^6$, $\delta = 0.3$, and $\phi = 0.04$. The profound change in strength and flow pattern of nanoliquid has been noticed with a change in cavity tilt angle. This is due to variation in gravity field direction. It has been observed that enhancement of cavity tilt angle from $\gamma = -60°$ to $60°$ increases the fluidity. Isothermal lines are found to be almost parallel to vertical

Fig. 2 Streamlines, isotherms, and entropy generation for $\delta = 0.2$ (top), $\delta = 0.3$ (middle), and $\delta = 0.4$ (bottom) at $Ra = 10^6$, $\gamma = 45°$

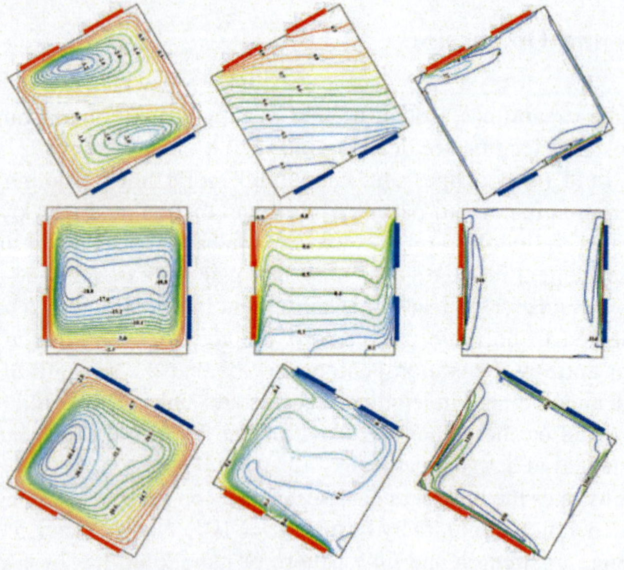

Fig. 3 Streamlines, isotherms, and entropy generation for $\gamma = -60°$ (top), $\gamma = 0°$ (middle), and $\gamma = 60°$ (bottom) at $Ra = 10^6$, $\delta = 0.3$, $\phi = 0.04$

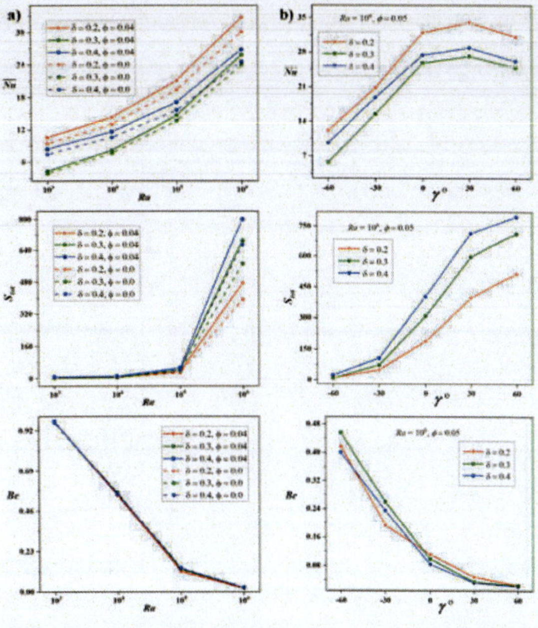

Fig. 4 Impact of δ on different (a) Ra and (b) γ on $\overline{Nu}$, S_{tot}, and Be

walls of the cavity at $\gamma = -60°$ indicating that the thermal transfer takes place due to conduction mode. This can also be justified through the strength of fluid flow. However, stratification in isopleths of thermal field has been noticed at $\gamma = 0°$ and $60°$. Due to the fact that the EG is the function of thermal and velocity gradients, the structure of entropy generation contours varied along the cavity angle due to variation in flow and thermal lines. Through the core value of entropy contours, it has been observed that the minimal EG is achieved with $\gamma = -60°$.

The influence of Rayleigh number and source-sink length on average Nu, S_{tot}, and Be has been displayed in Fig. 4a with $\gamma = 45°$ for both water (dotted line) and nanoliquid (solid line). Irrespective of δ, an increase in HTR and S_{tot} has been noticed with enhancement of Ra. This is due to the fact that enhancement of Ra enhances the buoyancy force. Among the considered three different source-sink lengths, maximum HTR with minimum EG has been found with smaller source-sink pair length. Since the Be is inversely proportional to S_{tot}, decrease in Be has been noticed against Ra. It has also been found that $Be > 0.5$ for $Ra \leq 10^4$ indicates that heat transfer entropy dominates friction entropy, while the reverse has been noticed for $Ra \geq 10^5$. Figure 4b depicts the impact of cavity tilt angle and source-sink length on $\overline{Nu}$, S_{tot} and Be with $Ra = 10^6$, $\phi = 0.05$. As discussed before, minimum EG with greater HTR has been found at $\delta = 0.2$, and this obeys for all γ values. It is interesting to note that among five different tilt angles, maximum $\overline{Nu}$ has been observed at $\gamma = 30°$, while the minimum S_{tot} has been noticed at $\gamma = -60°$. Because of the same mechanism as explained in Fig. 4, Be decreases against γ.

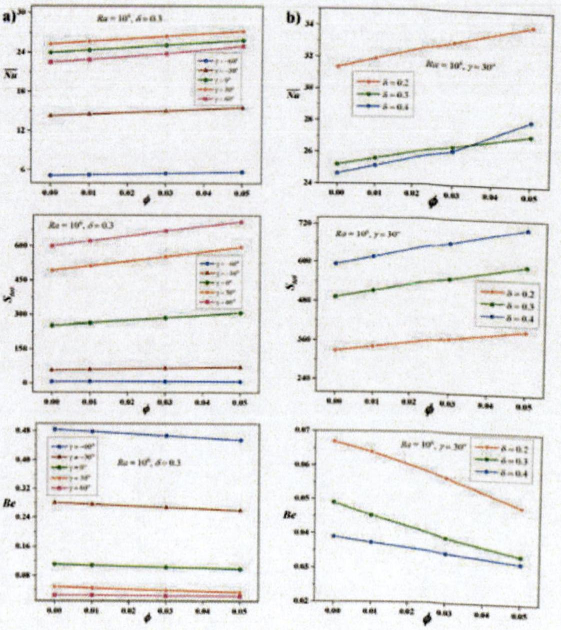

Fig. 5 Impact of ϕ on different **(a)** γ and **(b)** δ on $\overline{Nu}$, S_{tot}, and Be

The effects of ϕ on the average Nu, S_{tot}, and Be for various inclination angles and source-sink lengths are examined in Figs. 5a,b respectively. Irrespective of cavity tilt angle, increase in nanoparticle volume fraction enhances the average Nu due to increase in thermal conductivity of nanoliquid. The same mechanism has been found with respect to source-sink length. Among the considered set of parametric values, it has been found that maximum HTR has been found at $\gamma = 30°$ and $\delta = 0.2$. It is quite interesting to note that as δ value is decreased, enhancement in HTR and decrease in S_{tot} are noticed for all ϕ values. Since Ra is fixed at 10^6 in both Figs. 5a,b regardless of γ, δ and ϕ, $Be \ll 0.5$, this shows that the contribution of fluid friction is greater compared to heat transfer in producing the entropy.

5 Conclusions

The influence of Rayleigh number, source-sink length, cavity inclination angle and nanoparticle concentration on fluid movement, heat transport rate, and irreversibility is numerically investigated in the current analysis. From the obtained results, the following conclusions can be drawn:

- Increase in fluidity has been found with an enhancement of source-sink pair length.
- Increase in nanoparticle volume fraction and Rayleigh number enhances the heat transfer rate.

- The main objective of the current investigation i.e., maximum heat transfer rate with minimum entropy generation, has been achieved with $\gamma = 30°$, $\delta = 0.2$ and $\phi = 0.05$.

The current work deals with the thermal and entropy analysis in a partially active nanoliquid-filled tilted cavity. This work can be extended for three-dimensional geometry filled with hybrid nanoliquid subjected to nonuniform thermal boundary conditions under the influence of magnetic force applied in various directions. The impact of nanoparticle shape and size by adopting a two-phase model can also be studied in the future as an extension of the current investigation.

References

1. Bendaraa, A., Charafi, M.M., Hasnaoui A.: Numerical study of natural convection in a differentially heated square cavity filled with nanofluid in the presence of fins attached to walls in different locations. Phys. Fluids. **31**, 052003 (2019). https://doi.org/10.1063/1.5091709
2. Mohebbi, R., Khalilabad, S.H., Ma, Y.: Effect of γ-Al2O3/water nanofluid on natural convection heat transfer of corrugated Γ shaped cavity: study the different aspect ratio of grooves. J. Appl. Fluid Mech. **12**, 1151–1160 (2019). https://doi.org/10.29252/JAFM.12.04.29455
3. Raizah, Z.A.S., Aly, A.M., Ahmed, S.E.: Natural convection flow of a nanofluid-filled V-shaped cavity saturated with a heterogeneous porous medium: incompressible smoothed particle hydrodynamics analysis. Ain Shams Eng. J. **12**, 2033–2046 (2021). https://doi.org/10.1016/j.asej.2020.09.026
4. Sheikhzadeh, G.A., Arefmanesh, A., Kheirkhah, M.H., Abdollahi, R.: Natural convection of Cu-water nanofluid in a cavity with partially active side walls. Eur. J. Mech. B Fluids **30**, 166–176 (2011). https://doi.org/10.1016/j.euromechflu.2010.10.003
5. Ben-Mansour, R., Habib, M.A.: Use of nanofluids for enhanced natural cooling of discretely heated enclosures. Appl. Mech. Mater. **302**, 422–428 (2013). https://doi.org/10.4028/www.scientific.net/amm.302.422
6. Bhuiyanaa, A.H., Alamb, M.S., Alima M.A.: Natural convection of water-based nanofluids in a square cavity with partially heated of the bottom wall. Proc. Eng. **194**, 435–441 (2017). https://doi.org/10.1016/j.proeng.2017.08.168
7. Kemparaju, S., Swamy, H.A.K., Sankar, M., Mebarek-Oudina, F.: Impact of thermal and solute source-sink combination on thermosolutal convection in a partially active porous annulus. Phys. Scr. **97**, 055206 (2022). https://doi.org/10.1088/1402-4896/ac6383
8. Reddy, N.K., Sankar, M., Jang, B.: Impact of thermal source-sink arrangements on buoyant convection in a nanofluid-filled annular enclosure. J. Heat Transf. (2022). https://doi.org/10.1115/1.4055146
9. Kaluri, R.S., Basak, T.: Entropy generation minimization versus thermal mixing due to natural convection in differentially and discretely heated square cavities. Numer. Heat Transf. A: Appl. 58, 475–504 (2010). https://doi.org/10.1080/10407782.2010.511982
10. Oueslati, F., Ben-Beya, B., Lili, T.: Double-diffusive natural convection and entropy generation in an enclosure of aspect ratio 4 with partial vertical heating and salting sources. Alex. Eng. J. **52**, 605–625 (2013). https://doi.org/10.1016/j.aej.2013.09.006
11. T. Armaghani, M.A. Ismael, A.J. Chamkha, Analysis of entropy generation and natural convection in an inclined partially porous layered cavity filled with a nanofluid. Can. J. Phys. **95**, 238–252 (2017). https://doi.org/10.1139/cjp-2016-0570

12. Kefayati, G.H.R., Che-Sidik, N.A.: Simulation of natural convection and entropy generation of non-Newtonian nanofluid in an inclined cavity using Buongiorno's mathematical model (Part II, entropy generation). Powder Technol. **305**, 679–703 (2017). https://doi.org/10.1016/j.powtec.2016.10.077
13. Rashad, A.M., Armaghani, T., Chamkha, A.J., Mansour, M.A.: Entropy generation and MHD natural convection of a nanofluid in an inclined square porous cavity: effects of a heat sink and source size and location. Chin. J. Phys. **56**, 193–211 (2018). https://doi.org/10.1016/j.cjph.2017.11.026
14. Jino, L., Kumar, A.V.: Mathematical modelling of MHD natural convection in a linearly heated porous cavity. Math. Model. Eng. Probl. **8**, 149–157 (2021). https://doi.org/10.18280/mmep.080119
15. Jino, L., Kumar, A.V.: Cu-water nanofluid MHD quadratic natural convection on square porous cavity. Int. J. Appl. Comput. Math. **7**, 164 (2021). https://doi.org/10.1007/s40819-021-01103-5
16. Kumar, A.V., Lawrence, J., Saravanakumar, G.: Fluid friction/heat transfer irreversibility and heat function study on MHD free convection within the MWCNT-water nanofluid-filled porous cavity. Heat Transf. **51**, 4247–4267 (2022). https://doi.org/10.1002/htj.22498
17. Kiran, S., Sankar, M., Swamy, H.A.K., Makinde, O.D.: Unsteady buoyant convective flow and thermal transport analysis in a nonuniformly heated annular geometry. Comput. Therm. Sci. **14**, 1–17 (2022). https://doi.org/10.1615/ComputThermalScien.2021039723
18. Swamy, H.A.K., Sankar, M., Reddy, N.K.: Analysis of entropy generation and energy transport of Cu-water nanoliquid in a tilted vertical porous annulus. Int. J. Appl. Comput. Math. **8**, 10 (2022). https://doi.org/10.1007/s40819-021-01207-y
19. Swamy, H.A.K., Sankar, M., Reddy, N.K., Al-Manthari, M.S.: Double diffusive convective transport and entropy generation in an annular space filled with alumina-water nanoliquid. Eur. Phys. J. Spec. Top. **231**, 2781–2800 (2022). https://doi.org/10.1140/epjs/s11734-022-00591-w
20. Swamy, H.A.K., Sankar, M., Do, Y.: Entropy and energy analysis of MHD nanofluid thermal transport in a non-uniformly heated annulus. Waves Random Complex Media (2022). https://doi.org/10.1080/17455030.2022.2145522
21. Parvin, S., Chamkha, A.J.: An analysis on free convection flow, heat transfer and entropy generation in an odd-shaped cavity filled with nanofluid. Int. Commun. Heat Mass Transf. **54**, 8–17 (2014). https://doi.org/10.1016/j.icheatmasstransfer.2014.02.031
22. Armaghani, T., Rashad, A.M., Vahidifar, O., Mishra, S.R., Chamkha, A.J.: Effects of discrete heat source location on heat transfer and entropy generation of nanofluid in an open inclined L-shaped cavity. Int. J. Numer. Methods Heat Fluid Flow. **29**, 1363–1377 (2019). https://doi.org/10.1108/HFF-07-2018-0412
23. Dogonchi, A.S., Sadeghi, M.S., Ghodrat, M., Chamkha, A.J., Elmasry, Y., Alsulami, R.: Natural convection and entropy generation of a nanoliquid in a crown wavy cavity: effect of thermo-physical parameters and cavity shape. Case Stud. Therm. Eng. **27**, 101208 (2021). https://doi.org/10.1016/j.csite.2021.101208

Linear and Nonlinear Analysis of Unicellular Rayleigh-Bénard Magneto-convection in a Micropolar Fluid Occupying Enclosures

Sandra Jestine and S. Pranesh

1 Introduction

Fluids have always been an important subject of research due to their broad range of applications and diversity. The classical Navier-Stokes model has the limitation of being unable to explain fluids with microstructure. Eringen (see [1, 2]) simplified the theory of microfluids and proposed micropolar fluids (MPF) as a subclass of these fluids. This model became helpful, in describing a fluid system that does not satisfy the Navier-Stokes equation and conservative laws are derived for such fluids. MPF are classified as non-Newtonian due to the rotation (on its own axis) of microelements contained in the fluid. The early papers on the study of RBC in a MPF started with [3–6].

When magnetic field is applied across the fluid layer, the Lorentz force is generated and will affect the flow, and such a study is termed as "magneto-convection." Siddheshwar and Pranesh (see [7]) were the first to investigate magneto-convection in a MPF. The study reveals that electrically conducting fluids with microelements are more stable than electrically conducting Newtonian fluids. Motivated by this work, the authors (see [8]) extended their previous paper to study temperature and gravity modulation in a MPF. Pranesh and Kiran (see [9]) investigated the stability of MPF in the presence of nonclassical Fourier law with magnetic field and observed that the Cattaneo number due to this nonclassical law decreases the Rayleigh number. Over a saturated porous medium, Anncy et al. (see [10]) performed linear/nonlinear analysis with cross-diffusion effects on two-component convection in MPF. They observed that Dufour-Soret parameters exhibit opposing effects on this system.

S. Jestine (✉) · S. Pranesh
Department of Mathematics, Centre for Mathematical Needs, CHRIST(Deemed to be University), Bangalore, India
e-mail: sandra.jestine@res.christuniversity.in; pranesh.s@christuniversity.in

F. Kamalov et al. (eds.), *Advances in Mathematical Modeling and Scientific Computing*, Trends in Mathematics, https://doi.org/10.1007/978-3-031-41420-6_32

Enclosures are finite spaces that are bounded and filled with liquids or gases. Natural convection in an enclosed system has become the most active field in current-day research due to several applications in the field of engineering. The early insight on flow inside a rectangular cavity is explored by authors (see [11, 12]). RBC in Newtonian liquids and nanoliquids, occupying enclosures, was studied by Siddheshwar and Kanchana (see [13]). They reduced the analytically intractable Lorenz model into Ginzburg-Landau equation (GLE), using multi-scale approach. An analytical study of RBC inside four different enclosures in a NIF (non-inertial frame) was carried out by Kanchana et al. (see [14]). It is observed that in NIF, the transfer of heat is reduced and thereby the system becomes more stable.

From the above survey of literature, it is noticed that not much work has been conducted to investigate the study of Unicellular Rayleigh-Bénard magneto-convection (URBMC) in a MPF occupying enclosures. Thus, the main objective of the present study is to investigate linear and nonlinear analysis of URBMC in a MPF occupying shallow, square, and tall enclosures.

2 Mathematical Formulation

Two-dimensional URBMC in a MPF occupying enclosures of width b and height h, as shown in Fig. 1, is considered. The temperature difference ΔT is maintained between the two horizontal boundaries. We constrain ourselves to the $xz-$ plane for mathematical manageability with gravity acting vertically downward.

The equation of continuity, linear momentum, angular momentum, and energy along with the equation of density variation with temperature, magnetic induction, and magnetic continuity for Rayleigh-Bénard situation in a MPF is given by Siddheshwar and Pranesh [7]:

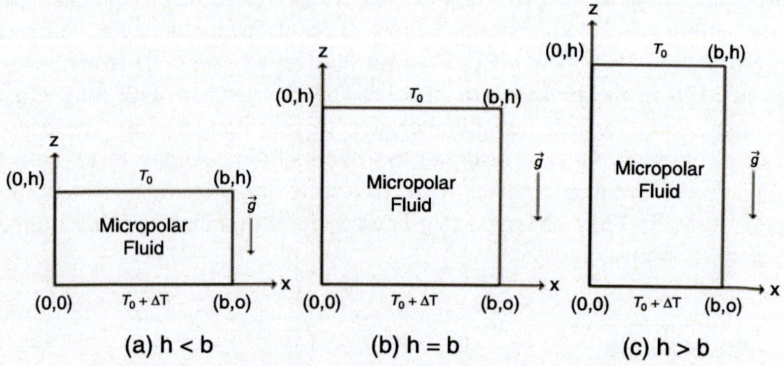

Fig. 1 MPF occupying (**a**) shallow, (**b**) square, (**c**) tall enclosures – schematic diagram

$$\nabla.\vec{\mathfrak{q}} = 0, \tag{1}$$

$$\varrho_o\left(\frac{\partial\vec{\mathfrak{q}}}{\partial t} + [\vec{\mathfrak{q}}.\nabla]\vec{\mathfrak{q}}\right) = -\nabla\mathfrak{p} - \varrho g\hat{k} + 2\xi[\nabla\times\vec{\omega}] + [2\xi+\eta]\nabla^2\vec{\mathfrak{q}} + \mu_m[\vec{\mathcal{H}}.\nabla]\vec{\mathcal{H}}, \tag{2}$$

$$\varrho_o\mathcal{I}\left(\frac{\partial\vec{\omega}}{\partial t} + [\vec{\mathfrak{q}}.\nabla]\vec{\omega}\right) = \eta'\nabla^2\vec{\omega} + [\lambda' + \eta']\nabla[\nabla.\vec{\omega}] + 2\xi[\nabla\times\vec{\mathfrak{q}} - 2\vec{\omega}], \tag{3}$$

$$\frac{\partial\mathcal{T}}{\partial t} + [\vec{\mathfrak{q}}.\nabla]\mathcal{T} = \chi\nabla^2\mathcal{T} + \frac{\beta}{\varrho_0 C_v}[\nabla\times\vec{\omega}].\nabla\mathcal{T}, \tag{4}$$

$$\varrho = \varrho_0\left(1 - \alpha[\mathcal{T} - \mathcal{T}_0]\right), \tag{5}$$

$$\frac{\partial\vec{\mathcal{H}}}{\partial t} + [\vec{\mathfrak{q}}.\nabla]\vec{\mathcal{H}} = \nu_m\nabla^2\vec{\mathcal{H}} + [\vec{\mathcal{H}}.\nabla]\vec{\mathfrak{q}}, \tag{6}$$

$$\nabla.\vec{\mathcal{H}} = 0. \tag{7}$$

The quantities in (1)–(7) are $\vec{\mathfrak{q}}$, velocity; ϱ_0, density at reference temperature; $\mathfrak{p}$, pressure; g, acceleration; ξ, coupling viscosity coefficient; $\vec{\omega}$, angular momentum; t, time; η, shear kinematic viscosity coefficient; μ_m, magnetic permeability; C_v, specific heat; $\mathcal{I}$, moment of inertia; $\mathcal{H}$, magnetic field; λ' and η', bulk and shear spin viscosity coefficients; β, micropolar heat conduction coefficient; $\mathcal{T}$, temperature; and α, thermal expansion coefficient.

(1)–(7) are solved for the following stress-free, isothermal, no-spin, and magnetic boundary conditions:

$$\left.\begin{aligned} w = \frac{\partial^2 w}{\partial z^2} = \mathcal{T} = \omega_y = \frac{\partial\mathcal{H}_z}{\partial z} = 0 \quad \text{at} \quad z = 0 \\ w = \frac{\partial^2 w}{\partial z^2} = \mathcal{T} = \omega_y = \frac{\partial\mathcal{H}_z}{\partial z} = 0 \quad \text{at} \quad z = h \end{aligned}\right\} 0 < x < b \tag{8}$$

Initially, in the motionless state, the quantities are defined by:

$$\vec{\mathfrak{q}_b} = (0,0), \quad \mathcal{T} = \mathcal{T}_b(z), \quad \vec{\omega_b} = (0,0), \quad \varrho = \varrho_b(z), \quad \mathfrak{p} = \mathfrak{p}_b(z), \quad \vec{\mathcal{H}_b} = \mathcal{H}_0\hat{k}. \tag{9}$$

The temperature distribution in the motionless state is given by:

$$\mathcal{T}_b = \left[1 - \frac{z}{h}\right]\Delta\mathcal{T} + \mathcal{T}_0 \tag{10}$$

The motionless state solutions are superimposed by an infinitesimal perturbations given by:

$$\vec{\mathfrak{q}} = \vec{\mathfrak{q}_b} + \vec{\mathfrak{q}}', \; \mathcal{T} = \mathcal{T}_b + \mathcal{T}', \; \vec{\omega} = \vec{\omega_b} + \vec{\omega}', \; \varrho = \varrho_b + \varrho', \; \mathfrak{p} = \mathfrak{p}_b + \mathfrak{p}', \; \vec{\mathcal{H}} = \mathcal{H}_0\hat{k} + \vec{\mathcal{H}}'. \tag{11}$$

Substituting (11) in (1)–(7), following the general procedure of stability analysis, introducing the stream functions ϕ (magnetic field) and ψ (velocity) in the form:

$$\mathcal{H}'_x = \frac{\partial \phi'}{\partial z}, \quad \mathcal{H}'_z = -\frac{\partial \phi'}{\partial x} \quad \text{and} \quad u' = \frac{\partial \psi'}{\partial z}, \quad w' = -\frac{\partial \psi'}{\partial x},$$

and non-dimensionalizing the resultant equations using:

$$(x^*, z^*) = \left(\frac{x}{b}, \frac{z}{h}\right), w^* = \frac{w'}{(\gamma/h^2)}, \mathcal{T}^* = \frac{\mathcal{T}'}{\Delta\mathcal{T}}, \psi^* = \frac{\psi'}{\gamma}, \phi^* = \frac{\phi'}{\mathcal{H}_0},$$

we get the dimensionless equations for URBMC:

$$\frac{1}{\mathcal{P}r}\frac{\partial}{\partial t}\nabla^2_{\mathrm{A_r}}\psi = -\mathcal{R}a_{\mathrm{A_r}}{\mathrm{A_r}}^4\frac{\partial \mathcal{T}}{\partial x} + (1+\mathcal{N}_1)\nabla^4_{\mathrm{A_r}}\psi - \mathcal{N}_1\nabla^2_{\mathrm{A_r}}\omega_y \tag{12}$$
$$-\frac{\mathrm{A_r}}{\mathcal{P}r}\mathcal{J}(\psi, \nabla^2_{\mathrm{A_r}}\psi) + \mathcal{Q}\frac{\mathcal{P}r}{\mathcal{P}m}\nabla^2_{\mathrm{A_r}}\left(\frac{\partial \phi}{\partial z}\right) - \mathcal{Q}\frac{\mathcal{P}r}{\mathcal{P}m}\mathrm{A_r}\mathcal{J}(\phi, \nabla^2_{\mathrm{A_r}}\phi),$$

$$\frac{\mathcal{N}_2}{\mathcal{P}r}\frac{\partial \omega_y}{\partial t} = \mathcal{N}_3\nabla^2_{\mathrm{A_r}}\omega_y - 2\mathcal{N}_1\omega_y + \mathcal{N}_1\nabla^2_{\mathrm{A_r}}\psi + \frac{\mathcal{N}_2}{\mathcal{P}r}\mathrm{A_r}\mathcal{J}(\psi, \omega_y), \tag{13}$$

$$\frac{\partial \mathcal{T}}{\partial t} = -\mathrm{A_r}\frac{\partial \psi}{\partial x} + \nabla^2_{\mathrm{A_r}}\mathcal{T} - \mathrm{A_r}\mathcal{N}_5\frac{\partial \omega_y}{\partial x} \tag{14}$$
$$+ \mathrm{A_r}\mathcal{J}(\psi, \mathcal{T}) + \mathrm{A_r}\mathcal{N}_5\mathcal{J}(\omega_y, \mathcal{T}),$$

$$\frac{\partial \phi}{\partial t} = \frac{\partial \psi}{\partial z} + \frac{\mathcal{P}r}{\mathcal{P}m}\nabla^2_{\mathrm{A_r}}\phi + \mathrm{A_r}\mathcal{J}(\psi, \phi). \tag{15}$$

$$\text{where,} \quad \mathrm{A_r} = \frac{h}{b}, \quad \nabla_{\mathrm{A_r}} = \mathrm{A_r}\frac{\partial}{\partial x}\hat{i} + \frac{\partial}{\partial z}\hat{k}, \quad \mathcal{J}(\ ,\ ) = \text{Jacobian}$$

The dimensionless groups in (12)–(15) are:

$\mathcal{N}_1$: Coupling parameter – $-\frac{\xi}{\xi+\eta}$, $\mathcal{P}m$: Magnetic Prandtl number – $-\frac{\xi+\eta}{\nu_m}$,
$\mathcal{N}_2$: Inertial parameter – $-\frac{1}{h^2}$, $\mathcal{Q}$: Chandrasekhar number: – $-\frac{\mu_m \mathcal{H}_0^2 h^2}{(\xi+\eta)\nu_m}$,
$\mathcal{N}_3$: Couple stress parameter – $-\frac{\eta'}{(\xi+\eta)h^2}$, $\mathcal{R}a_{\mathrm{A_r}}$: Rayleigh number: – $-\frac{\alpha g \Delta T b^3 \varrho_0}{(\xi+\eta)\gamma}$,
$\mathcal{N}_5$: Heat conduction parameter – $-\frac{\beta}{\varrho_0 C_v h^2}$, $\mathcal{P}r$: Prandtl number – $-\frac{\xi+\eta}{\gamma \varrho_0}$.

Boundary condition (8) in dimensionless form is:

$$\psi = \nabla^2_{\mathrm{A_r}}\psi = \mathcal{T} = \phi = 0 \quad \text{at} \quad z = 0, 1. \tag{16}$$

2.1 *Linear Stability Theory*

Now, we discuss about the linear stability analysis of the problem, by neglecting the Jacobians in (12)–(15), to get the condition for onset of URBMC.

To find the condition for onset of convection in URBMC, we represent the ψ, ω_y, $\mathcal{T}$, and ϕ by Fourier series, as follows by considering steady state (see Siddheshwar and Pranesh [7]):

$$\left.\begin{array}{ll} \psi = \psi_0 \, sin(\pi x)\, sin(\pi z), & \omega_y = \omega_0 \, sin(\pi x)\, sin(\pi z), \\ \mathcal{T} = \mathcal{T}_0 \, cos(\pi x)\, sin(\pi z), & \phi = \phi_0 \, sin(\pi x)\, cos(\pi z). \end{array}\right\} \tag{17}$$

Substituting (17) in the linear version of (12)–(15), we get:

$$\begin{bmatrix} (1+\mathcal{N}_1)\kappa_{A_r}^4 & \mathcal{N}_1\kappa_{A_r}^2 & \mathcal{R}a_{A_r}{A_r}^4\pi & \frac{\mathcal{Q}\mathcal{P}r\kappa_{A_r}^2\pi}{\mathcal{P}m} \\ \mathcal{N}_1\kappa_{A_r}^2 & \mathcal{N}_3\kappa_{A_r}^2 + 2\mathcal{N}_1 & 0 & 0 \\ A_r\pi & A_r\mathcal{N}_5\pi & \kappa_{A_r}^2 & 0 \\ \pi & 0 & 0 & -\frac{\mathcal{P}r\kappa_{A_r}^2}{\mathcal{P}m} \end{bmatrix} \begin{bmatrix} \psi_0 \\ \omega_0 \\ \mathcal{T}_0 \\ \phi_o \end{bmatrix} = \begin{bmatrix} 0 \\ 0 \\ 0 \\ 0 \end{bmatrix},$$

where $\kappa_{A_r} = \pi^2(1 + {A_r}^2)$.

To get the nontrivial solution, the determinant of the coefficient matrix must be zero, which yields:

$$\mathcal{R}a_{A_r} = \frac{\kappa_{A_r}^8\mathcal{N}_3(1+\mathcal{N}_1) + \kappa_{A_r}^6\mathcal{N}_1(2+\mathcal{N}_1) + \pi^2\kappa_{A_r}^2(2\mathcal{N}_1+\mathcal{N}_3)\mathcal{Q}}{\pi^2{A_r}^5\left[\kappa_{A_r}^2(\mathcal{N}_3 - \mathcal{N}_1\mathcal{N}_5) + 2\mathcal{N}_1\right]}. \tag{18}$$

Keeping $\mathcal{N}_3$ arbitrary and setting $\mathcal{N}_1 = \mathcal{N}_5 = \mathcal{Q} = 0$ in (18), we get:

$$\mathcal{R}a_{A_r} = \frac{\kappa_{A_r}^6}{\pi^2{A_r}^5} \tag{19}$$

which is the classical Rayleigh number for RBC occupying enclosures in Newtonian fluid.

The linear theory fails to predict the amount of heat transfer in URBMC system as it gives only the condition for onset of convection. So, we perform local nonlinear analysis to estimate the amount of heat transfer in URBMC by deriving GLE, using the multi-scale approach in the next section.

3 Derivation of GLE for Local Nonlinear Stability Analysis

We assume ψ, ω_y, $\mathcal{T}$, and ϕ in the form:

$$\left.\begin{array}{rl} \psi &= \mathscr{A}(t)\, sin(\pi x)\, sin(\pi z), \\ \omega_y &= \mathscr{B}(t)\, sin(\pi x)\, sin(\pi z), \\ \mathcal{T} &= \mathscr{C}(t)\, cos(\pi x)\, sin(\pi z) + \mathscr{E}(t)\, sin(2\pi z), \\ \phi &= \mathscr{F}(t)\, sin(\pi x)\, cos(\pi z) + \mathscr{G}(t)\, sin(2\pi x). \end{array}\right\} \tag{20}$$

where $\mathscr{A}$, $\mathscr{B}$, $\mathscr{C}$, $\mathscr{E}$, $\mathscr{F}$, and $\mathscr{G}$ are the amplitudes that have to be determined.

On substituting (20) into (12)–(15), we get the six-mode Lorenz system as:

$$\frac{d\mathscr{A}}{dt} = -[(1+\mathcal{N}_1)\kappa_{A_r}^2 \mathcal{P}r]\mathscr{A} - [\mathcal{N}_1 \mathcal{P}r]\mathscr{B} - \left[\frac{\mathcal{R}a_{A_r}\pi {A_r}^4 \mathcal{P}r}{\kappa_{A_r}^2}\right]\mathscr{C} - \left[\frac{\mathcal{Q}\mathcal{P}r^2\pi}{\mathcal{P}m}\right]\mathscr{F} - \left[\frac{\mathcal{Q}\mathcal{P}r^2 A_r(\pi^2\kappa_{A_r}^2 - 4\pi^4 {A_r}^2)}{\kappa_{A_r}^2 \mathcal{P}m}\right]\mathscr{F}\mathscr{G}, \tag{21}$$

$$\frac{d\mathscr{B}}{dt} = -\left[\frac{\mathcal{N}_1\kappa_{A_r}^2 \mathcal{P}r}{\mathcal{N}_2}\right]\mathscr{A} - \left[\frac{(2\mathcal{N}_1 + \mathcal{N}_3\kappa_{A_r}^2)\mathcal{P}r}{\mathcal{N}_2}\right]\mathscr{B}, \tag{22}$$

$$\frac{d\mathscr{C}}{dt} = -[A_r\pi]\mathscr{A} - [A_r\mathcal{N}_5\pi]\mathscr{B} - [\kappa_{A_r}^2]\mathscr{C} - [A_r\pi^2]\mathscr{A}\mathscr{E} - [A_r\mathcal{N}_5\pi^2]\mathscr{B}\mathscr{E}, \tag{23}$$

$$\frac{d\mathscr{E}}{dt} = -[4\pi^2]\mathscr{E} + \left[\frac{A_r\pi^2}{2}\right]\mathscr{A}\mathscr{C} + \left[\frac{A_r\mathcal{N}_5\pi^2}{2}\right]\mathscr{B}\mathscr{C}, \tag{24}$$

$$\frac{d\mathscr{F}}{dt} = [\pi]\mathscr{A} - \left[\frac{\mathcal{P}r\kappa_{A_r}^2}{\mathcal{P}m}\right]\mathscr{F} + [\pi^2 A_r]\mathscr{A}\mathscr{G}, \tag{25}$$

$$\frac{d\mathscr{G}}{dt} = -\left[\frac{4\pi^2 {A_r}^2 \mathcal{P}r}{\mathcal{P}m}\right]\mathscr{G} - \left[\frac{A_r\pi^2}{2}\right]\mathscr{A}\mathscr{F}. \tag{26}$$

The generalized six-mode Lorenz systems (21)–(26) are used to derive GLE for URBMC by expanding the amplitudes and Rayleigh number, by regular perturbation as follows:

$$\left.\begin{aligned} \mathcal{S} &= \epsilon\mathcal{S}_{11} + \epsilon^2\mathcal{S}_{12} + \epsilon^3\mathcal{S}_{13} + \ldots \\ \mathcal{R}a_{A_r} &= \mathcal{R}a_{A0} + \epsilon\mathcal{R}a_{A1} + \epsilon^2\mathcal{R}a_{A2} + \epsilon^3\mathcal{R}a_{A3} + \ldots \end{aligned}\right\}, \tag{27}$$

where $\mathcal{S} = \mathcal{S}(\mathscr{A}, \mathscr{B}, \mathscr{C}, \mathscr{E}, \mathscr{F}, \mathscr{G})$.

On assuming small timescale $\tau^* = \epsilon^2 t$, substituting (27) into (21)–(26) and equating coefficients of similar powers of ϵ, we get the following system of equations:

$$o(\epsilon) : \mathcal{L}_A U_{11} = 0, \tag{28}$$

$$o(\epsilon^2) : \mathcal{L}_A U_{22} = [\mathcal{M}_{21}, \quad \mathcal{M}_{22}, \quad \mathcal{M}_{23}, \quad \mathcal{M}_{24}, \quad \mathcal{M}_{25}, \quad \mathcal{M}_{26}]^{Tr}, \tag{29}$$

$$o(\epsilon^3) : \mathcal{L}_A U_{33} = [\mathcal{M}_{31}, \quad \mathcal{M}_{32}, \quad \mathcal{M}_{33}, \quad \mathcal{M}_{34}, \quad \mathcal{M}_{35}, \quad \mathcal{M}_{36}]^{Tr}. \tag{30}$$

where $\mathcal{L}_A = \begin{bmatrix} -(1+\mathcal{N}_1)\kappa_{\mathrm{A_r}}^2 \mathcal{P}r & -\mathcal{N}_1 \mathcal{P}r & -\frac{\mathcal{R}a_{A0}\mathrm{A_r}^4 \mathcal{P}r\pi}{\kappa_{\mathrm{A_r}}^2} & 0 & -\frac{\mathcal{Q}\mathcal{P}r^2\pi}{\mathcal{P}m} & 0 \\ -\frac{\mathcal{N}_1\kappa_{\mathrm{A_r}}^2 \mathcal{P}r}{\mathcal{N}_2} & -\frac{(2\mathcal{N}_1+\mathcal{N}_3\kappa_{\mathrm{A_r}}^2)\mathcal{P}r}{\mathcal{N}_2} & 0 & 0 & 0 & 0 \\ -\mathrm{A_r}\pi & -\mathrm{A_r}\mathcal{N}_5\pi & -\kappa_{\mathrm{A_r}}^2 & 0 & 0 & 0 \\ 0 & 0 & 0 & -4\pi^2 & 0 & 0 \\ 0 & 0 & 0 & 0 & 0 & -\frac{4\pi^2\mathrm{A_r}^2\mathcal{P}r}{\mathcal{P}m} \end{bmatrix}$,

$$U_{ii} = [\,\mathscr{A}_{1i}, \mathscr{B}_{1i}, \mathscr{C}_{1i}, \mathscr{E}_{1i}, \mathscr{F}_{1i}, \mathscr{G}_{1i}\,]^{Tr}, \qquad i = 1, 2, 3$$

$$\left.\begin{aligned} \mathcal{M}_{21} &= \frac{\mathcal{R}a_{A1}\pi^2\mathrm{A_r}^5\mathcal{P}r}{\kappa_{\mathrm{A_r}}^4}\left[\frac{\mathcal{N}_1\mathcal{N}_5\kappa_{\mathrm{A_r}}^2}{2\mathcal{N}_1+\mathcal{N}_3\kappa_{\mathrm{A_r}}^2} - 1\right]\mathscr{A}_{11}, \quad \mathcal{M}_{22} = \mathcal{M}_{23} = 0, \\ \mathcal{M}_{24} &= \frac{\mathrm{A_r}^2\pi^3}{2\kappa_{\mathrm{A_r}}^2}\left[\frac{\mathcal{N}_1\mathcal{N}_5\kappa_{\mathrm{A_r}}^2}{2\mathcal{N}_1+\mathcal{N}_3\kappa_{\mathrm{A_r}}^2} - 1\right]^2 \mathscr{A}_{11}^2, \quad \mathcal{M}_{25} = 0, \quad \mathcal{M}_{26} = \frac{\mathrm{A_r}\pi^3}{2\kappa_{\mathrm{A_r}}^2}\frac{\mathcal{P}m}{\mathcal{P}r}\mathscr{A}_{11}^2 \end{aligned}\right\} \tag{31}$$

and

$$\left.\begin{aligned} \mathcal{M}_{31} &= \frac{d\mathscr{A}_{11}}{d\tau} + \frac{\mathcal{R}a_{A2}\pi^2\mathrm{A_r}^5\mathcal{P}r}{\kappa_{\mathrm{A_r}}^4}\left[\frac{\mathcal{N}_1\mathcal{N}_5\kappa_{\mathrm{A_r}}^2}{2\mathcal{N}_1+\mathcal{N}_3\kappa_{\mathrm{A_r}}^2} - 1\right]\mathscr{A}_{11} - \frac{\mathcal{Q}\pi^2\mathcal{P}m^2(\pi^2\kappa_{\mathrm{A_r}}^2 - 4\pi^4\mathrm{A_r}^2)}{8\kappa_{\mathrm{A_r}}^6\mathcal{P}r}\mathscr{A}_{11}^3, \\ \mathcal{M}_{32} &= -\left[\frac{\mathcal{N}_1\mathcal{N}_5\kappa_{\mathrm{A_r}}^2}{2\mathcal{N}_1+\mathcal{N}_3\kappa_{\mathrm{A_r}}^2}\right]\frac{d\mathscr{A}_{11}}{d\tau}, \\ \mathcal{M}_{33} &= \frac{\mathrm{A_r}\pi}{\kappa_{\mathrm{A_r}}^2}\left[\frac{\mathcal{N}_1\mathcal{N}_5\kappa_{\mathrm{A_r}}^2}{2\mathcal{N}_1+\mathcal{N}_3\kappa_{\mathrm{A_r}}^2} - 1\right]\frac{d\mathscr{A}_{11}}{d\tau} + \frac{\mathrm{A_r}^3\pi^3}{8\kappa_{\mathrm{A_r}}^2}\left[\frac{\mathcal{N}_1\mathcal{N}_5\kappa_{\mathrm{A_r}}^2}{2\mathcal{N}_1+\mathcal{N}_3\kappa_{\mathrm{A_r}}^2} - 1\right]^3\mathscr{A}_{11}^3, \\ \mathcal{M}_{34} &= \frac{\mathrm{A_r}^2\pi^3}{\kappa_{\mathrm{A_r}}^2}\left[\frac{\mathcal{N}_1\mathcal{N}_5\kappa_{\mathrm{A_r}}^2}{2\mathcal{N}_1+\mathcal{N}_3\kappa_{\mathrm{A_r}}^2} - 1\right]^2\mathscr{A}_{11}\mathscr{A}_{12}, \quad \mathcal{M}_{35} = \frac{\pi}{\kappa_{\mathrm{A_r}}^2}\frac{\mathcal{P}m}{\mathcal{P}r}\frac{d\mathscr{A}_{11}}{d\tau} + \frac{\pi^3}{8\kappa_{\mathrm{A_r}}^2}\frac{\mathcal{P}m^2}{\mathcal{P}r^2}\mathscr{A}_{11}^3, \\ \mathcal{M}_{36} &= \frac{\mathrm{A_r}\pi^3}{\kappa_{\mathrm{A_r}}^2}\frac{\mathcal{P}m}{\mathcal{P}r}\mathscr{A}_{11}\mathscr{A}_{12}. \end{aligned}\right\} \tag{32}$$

By solving (28), we get the solution as:

$$U_{11} = \mathscr{A}_{11}\left[1, \left[-\frac{\mathcal{N}_1\kappa_{\mathrm{A_r}}^2}{2\mathcal{N}_1+\mathcal{N}_3\kappa_{\mathrm{A_r}}^2}\right], \frac{\mathrm{A_r}\pi}{\kappa_{\mathrm{A_r}}^2}\left[\frac{\mathcal{N}_1\mathcal{N}_5\kappa_{\mathrm{A_r}}^2}{2\mathcal{N}_1+\mathcal{N}_3\kappa_{\mathrm{A_r}}^2} - 1\right], 0, \frac{\pi}{\kappa_{\mathrm{A_r}}^2}\frac{\mathcal{P}m}{\mathcal{P}r}, 0\right]^{Tr}. \tag{33}$$

Using the solvability condition given by:

$$\sum_{j=1}^{6}\mathcal{M}_{ij}U_{11} = 0. \tag{34}$$

Taking $i = 2$ in (34) and using (29), we get $\mathcal{R}a_{A1} = 0$. Now, taking $i = 3$ in (34) and using (32), we get GLE for URBMC in the form:

$$\frac{d\mathscr{A}_{11}}{d\tau} = \mathcal{P}_1\mathscr{A}_{11}(\tau) - \mathcal{P}_2\mathscr{A}_{11}^3(\tau). \tag{35}$$

where

$$\mathcal{P}_1 = \frac{-\pi^2 \mathrm{A_r}^5}{\kappa_{\mathrm{A_r}}^4}\left[\frac{\mathcal{R}a_{A2}\mathcal{P}r x_2}{1+x_1^2+\frac{\mathrm{A_r}^2\pi^2 x_2^2}{\kappa_{\mathrm{A_r}}^4}+\frac{\pi^2\mathcal{P}m^2}{\kappa_{\mathrm{A_r}}^4\mathcal{P}r^2}}\right],\ \mathcal{P}_2 = \frac{\frac{\mathrm{A_r}^4\pi^4 x_2^4}{8\kappa_{\mathrm{A_r}}^4}+\frac{\pi^4\mathcal{P}m^3}{8\kappa_{\mathrm{A_r}}^4\mathcal{P}r^3}-\frac{\mathcal{Q}\pi^2\mathcal{P}m^2(\pi^2\kappa_{\mathrm{A_r}}^2-4\pi^4\mathrm{A_r}^2)}{8\kappa_{\mathrm{A_r}}^6\mathcal{P}r}}{1+x_1^2+\frac{\mathrm{A_r}^2\pi^2 x_2^2}{\kappa_{\mathrm{A_r}}^4}+\frac{\pi^2\mathcal{P}m^2}{\kappa_{\mathrm{A_r}}^4\mathcal{P}r^2}},$$

$$x_1 = \frac{\mathcal{N}_1\kappa_{\mathrm{A_r}}^2}{2\mathcal{N}_1+\mathcal{N}_3\kappa_{\mathrm{A_r}}^2}, \qquad x_2 = \frac{\mathcal{N}_1\kappa_{\mathrm{A_r}}^2}{2\mathcal{N}_1+\mathcal{N}_3\kappa_{\mathrm{A_r}}^2} - 1.$$

The amount of heat transport in URBMC is estimated in the next section by deriving the Nusselt number expression.

4 Estimation of Heat Transport in URBMC

The Nusselt number, ($\mathcal{N}u(\tau)$), which quantifies the amount of heat transport in URBMC is given by:

$$\begin{aligned}\mathcal{N}u(\tau) &= 1 - 2\pi\mathscr{E}\\ &= 1 + \epsilon^2\frac{\mathrm{A_r}^2\pi^2 x_2^2}{4\kappa_{\mathrm{A_r}}^2}\mathscr{A}_{11}^2.\end{aligned} \tag{36}$$

5 Results and Discussions

The linear and nonlinear analysis of URBMC in an enclosed boundary is considered. The horizontal boundaries are assumed to be stress-free and isothermal, and no-spin conditions are applied on the boundaries. The stability analysis of the problem is studied by performing linear analysis, and $\mathcal{R}a_{\mathrm{A_r}}$ is expressed as a function of $\mathcal{N}_1$, $\mathcal{N}_3$, $\mathcal{N}_5$, $\mathcal{Q}$, and $\mathrm{A_r}$. In the nonlinear analysis part, the amount of heat transfer is calculated through $\mathcal{N}u(\tau)$.

In this investigation, because of the presence of microelement, the value of $\mathcal{P}r$ is taken more than that of fluids without microelements. Also, fixed values are taken for aspect ratio, i.e., $\mathrm{A_r} = 0.8$, $\mathrm{A_r} = 1$, and $\mathrm{A_r} = 1.2$, respectively, for shallow, square, and tall enclosures. In accordance with Clausius-Duhem inequality, the parameters $\mathcal{N}_1$, $\mathcal{N}_3$, and $\mathcal{N}_5$ are taken within the limit, $0 \le \mathcal{N}_1 \le 1$, $0 \le \mathcal{N}_3 \le m_1$, and $0 \le \mathcal{N}_5 \le m_2$, where m_1 and m_2 are real numbers.

Figures 2, 3, 4, and 5 represent the effects of $\mathcal{N}_1$, $\mathcal{N}_3$, $\mathcal{N}_5$, and $\mathcal{Q}$ on the onset of convection in URBMC for different enclosures. The impact of $\mathcal{N}_1$ on $\mathcal{R}a_{\mathrm{A_r}}$ for different enclosures is plotted in Fig. 2. It is observed that $\mathcal{R}a_{\mathrm{A_r}}$ increases with $\mathcal{N}_1$. This is due to the increased concentration of microelements with $\mathcal{N}_1$, which consumes the majority of the energy to form gyrational energy. Thus, the increase in microelements delays the onset of convection and stabilizes the system. The plot of $\mathcal{N}_3$ on $\mathcal{R}a_{\mathrm{A_r}}$ for different enclosures is shown in Fig. 3. $\mathcal{N}_3$ is having destabilizing

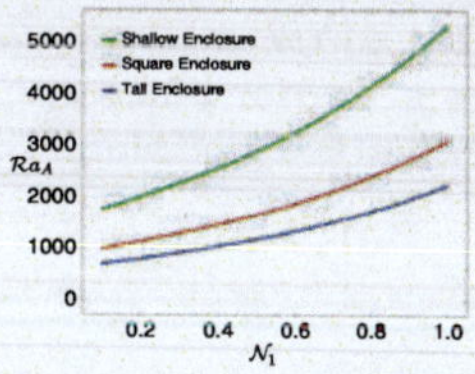

Fig. 2 The impact of $\mathcal{N}_1$ on $\mathcal{R}a_{A_r}$ for different enclosures

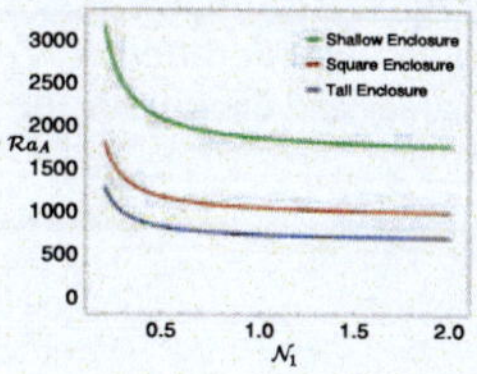

Fig. 3 The impact of $\mathcal{N}_3$ on $\mathcal{R}a_{A_r}$ for different enclosures

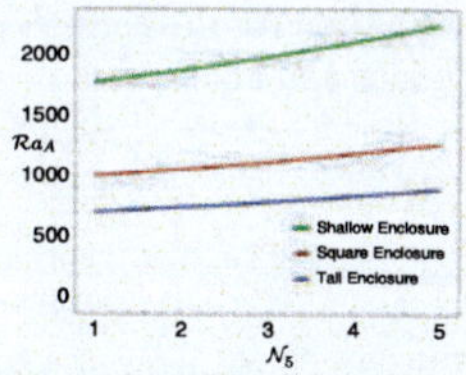

Fig. 4 The impact of $\mathcal{N}_5$ on $\mathcal{R}a_{A_r}$ for different enclosures

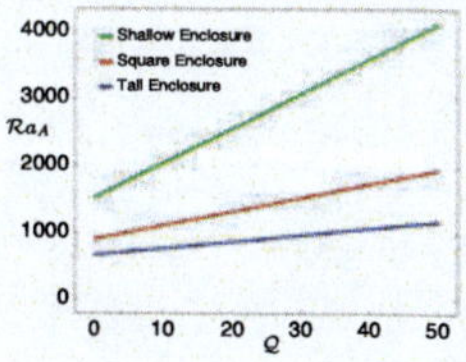

Fig. 5 The impact of $\mathcal{Q}$ on $\mathcal{R}a_{A_r}$ on for different enclosures

effect on the system. This is because when $\mathcal{N}_3$ increases, micropolar diffusion decreases as couple stress of the fluid increases, which in turn advances the onset of convection. Therefore, as $\mathcal{N}_3$ increases $\mathcal{R}a_{A_r}$ decreases. The effect of $\mathcal{N}_5$ on $\mathcal{R}a_{A_r}$ for different enclosures is shown in Fig. 4. The increase in $\mathcal{N}_5$ enhances the heat induced into the system by microelements, which cause delay in convection and makes the system stable. Figure 5 illustrates the plot of $\mathcal{Q}$ on $\mathcal{R}a_{A_r}$ for different enclosures. Increase in $\mathcal{Q}$ increases $\mathcal{R}a_{A_r}$ and is having stabilizing effect on the system. This is because the viscosity is induced into the fluid when the strength of magnetic field increases, which hampers the growth of disturbance and leads to delay in onset of convection. From the figures plotted, the results w.r.t enclosures can be summarized as:

$$\mathcal{R}a_{A_r}|_{[A_r=0.8]} > \mathcal{R}a_{A_r}|_{[A_r=1]} > \mathcal{R}a_{A_r}|_{[A_r=1.2]},$$

i.e., the above inequalities are due to the fact that the extent of instability is based on the change in the width of the lower wall, as energy transport in the vertical direction

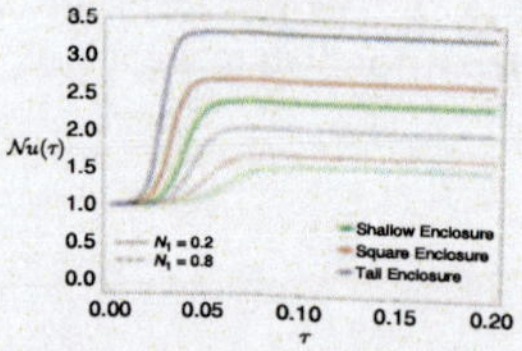

Fig. 6 The impact of $\mathcal{N}u(\tau)$ with time for different $\mathcal{N}_1$ values and enclosures

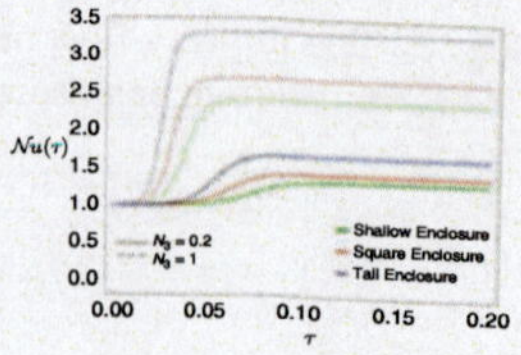

Fig. 7 The impact of $\mathcal{N}u(\tau)$ with time for different $\mathcal{N}_3$ values and enclosures

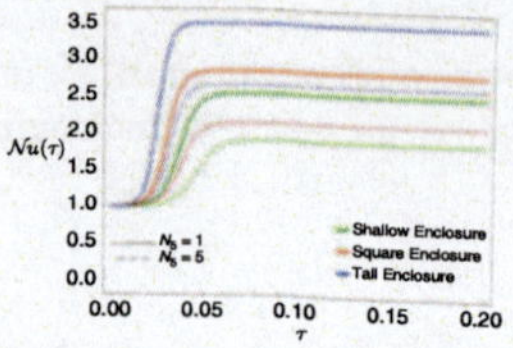

Fig. 8 The impact of $\mathcal{N}u(\tau)$ with time for different $\mathcal{N}_5$ values and enclosures

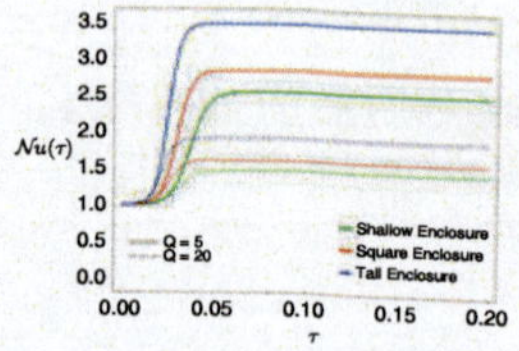

Fig. 9 The impact of $\mathcal{N}u(\tau)$ with time for different $\mathcal{Q}$ values and enclosures

remains the same. Thus, it is observed that the onset of convection is enhanced in tall enclosure and then square enclosure and delayed in shallow enclosure.

In the following part, the results obtained from nonlinear analysis of the problem are discussed to study the rate of heat transport in the system. GLE is derived from Lorenz equations using multi-scale approach. The variation of $\mathcal{N}u(\tau)$ with time for different enclosures and different values of $\mathcal{N}_1$, $\mathcal{N}_3$, $\mathcal{N}_5$, and $\mathcal{Q}$ are presented in Figs. 6, 7, 8, and 9. The following are observed from these figures:

(i) $\mathcal{N}u(\tau)\,|\,_{[\mathcal{N}_1=0.2]} > \mathcal{N}u(\tau)\,|\,_{[\mathcal{N}_1=0.8]}$
(ii) $\mathcal{N}u(\tau)\,|\,_{[\mathcal{N}_3=0.2]} < \mathcal{N}u(\tau)\,|\,_{[\mathcal{N}_3=1]}$
(iii) $\mathcal{N}u(\tau)\,|\,_{[\mathcal{N}_5=1]} > \mathcal{N}u(\tau)\,|\,_{[\mathcal{N}_1=5]}$
(iv) $\mathcal{N}u(\tau)\,|\,_{[\mathcal{Q}=5]} > \mathcal{N}u(\tau)\,|\,_{[\mathcal{Q}=20]}$,

i.e., the rate of heat transfer is reduced by the increase in concentration of microelements, heat induced by microelements, and strength of magnetic field as they have stabilizing effect. The spin diffusion coefficient enhances heat transfer as it is destabilizing the system. This reconfirms the results obtained in linear case.

On comparison with square and shallow enclosures, the tall enclosure exhibits more heat transfer as it requires small amount of energy in forming convection cell:

$$\text{i.e}\quad \mathcal{N}u(\tau)\,|\,_{[A_r=1.2]} > \mathcal{N}u(\tau)\,|\,_{[A_r=1]} > \mathcal{N}u(\tau)\,|\,_{[A_r=0.8]}$$

6 Conclusion

1. On the one hand, $\mathcal{N}_1$, $\mathcal{N}_5$, and $\mathcal{Q}$ stabilize the system and reduce heat transfer. On the other hand $\mathcal{N}_3$ destabilizes the system and enhances heat transfer.
2. Shallow enclosure is more stable than square and tall enclosures. Thus, the rate of heat transfer is more in tall enclosure than other enclosures.
3. The increase in aspect ratio advances the onset of convection.

References

1. Eringen, A.C.: Simple microfluids. Int. J. Eng. Sci. **2**, 205–217 (1964). https://doi.org/10.1016/0020-7225(64)90005-9
2. Eringen, A.C.: Theory of MPF. J. Math. Mech. 1–18 (1966). https://doi.org/10.1512/iumj.1967.16.16001
3. Datta, A.B., Sastry, V.U.K.: Thermal instability of a horizontal layer of MPF heated from below. Int. J. Eng. Sci. **14**(7), 631–637 (1976). https://doi.org/10.1016/0020-7225(76)90005-7
4. Ahmadi, G.: Stability of a MPF layer heated from below. Int. J. Eng. Sci. **14**, 81–89 (1976). https://doi.org/10.1016/0020-7225(76)90058-6
5. Rao, K.R.: Thermal instability in a MPF layer subject to a magnetic field. Int. J. Eng. Sci. **18**, 741–750 (1980). https://doi.org/10.1016/0020-7225(80)90107-X
6. Qin, Y., Kaloni, P.: A thermal instability problem in a rotating MPF. Int. J. Eng. Sci. **30**, 1117–1126 (1992). https://doi.org/10.1016/0020-7225(92)90061-K
7. Siddheshwar, P.G., Pranesh, S.: Magnetoconvection in a MPF. Int. J. Eng. Sci. **36**, 1173–1181 (1998). https://doi.org/10.1016/S0020-7225(98)00013-5
8. Siddheshwar, P.G., Pranesh, S.: Effect of temperature/gravity modulation on the onset of magneto-convection in electrically conducting fluids with internal angular momentum. J. Magn. Magn. Mater. **192**, 159–176 (1999). https://doi.org/10.1016/S0304-8853(98)00384-9
9. Pranesh, S., Kiran, R.V.: Study of Rayleigh-Bénard Magneto convection in a micropolar fluid with Maxwell-Cattaneo law. Appl. Math. **1**(06), 470 (2010)
10. Anncy, M., Joseph, T.V., Pranesh, S.: Linear and non-linear analyses of double-diffusive-chandrasekhar convection coupled with cross-diffusion in micropolar fluid over saturated porous medium. Multidiscip. Model. Mater. Struct. **17**(1), 211–236 (2020)
11. Ostrach, S.: Natural convection in enclosures. Adv. Heat Transf. **8**, 161–227 (1972). https://doi.org/10.1016/S0065-2717(08)70039-X
12. Gelfgat, A.Y.: Different modes of Rayleigh-Bénard instability in two-and three-dimensional rectangular enclosures. J. Comput. Phys. **156**, 300–324 (1999). https://doi.org/10.1006/jcph.1999.6363
13. Siddheshwar, P.G., Kanchana, C.: Unicellular unsteady Rayleigh-Bénard convection in Newtonian liquids and Newtonian nanoliquids occupying enclosures: new findings. Int. J. Mech. Sci. **131**, 1061–1072 (2017). https://doi.org/10.1016/j.ijmecsci.2017.07.050
14. Kanchana, C., Suthar, O.P., Siddheshwar, P.G.: A study of Rayleigh-Bénard-Taylor convection in very-shallow, shallow, square and tall enclosures. Int. J. Appl. Comput. Math. **6**, 1–17 (2020). https://doi.org/10.1007/s40819-020-00833-2

Study of Bénard-Marangoni Convection in a Microfluid with Coriolis Force

Riya Baby

1 Introduction

Microfluids are a subcategory of generalized fluids that arose from the classical fluid of hydrodynamics. It would be explained by the fact that fluid particles have no internal structure. The result is reflected in the well-known Navier–Stokes equations, which are nothing more than a description of numerous hydrodynamical phenomena. The primary variance between microfluids and classical fluids is that they favour couple stresses caused by rotation and the force stresses caused by translatory motion. Microfluids are fluids with microstructure, and these fluids contain rigid, randomly oriented particles with their spin inertia.

Eringen [1] initiated the theory of microfluids in 1966. In his research, he explained the presence and structure of microparticles in a given fluid. Micropolar fluid is derived from microfluids and exhibits micro-rotational inertia. The micropolar fluid applications are more appropriate to applied mathematicians as they are more readily amenable to solutions in various mathematical formulations. Micropolar fluids are fluids that include suspended molecules, such as magnetic fluids, animal blood, polymeric fluids, liquid crystals with complex molecules, and so on.

Micropolar fluid, on which many studies have taken place, is based on the books of Eringen [2], Lukasazewicz [3], and Power [4]. Thermo-convection on micropolar fluid has been researched in extension to these authors' books, and to name them, a few are Datta and Sastry [5], Ahmadi [6], and Bhattacharya and Jena [7]. The study of micropolar fluids is continued by the authors Siddheshwar and Pranesh [8] and Pranesh and Riya [9].

R. Baby (✉)
Department of Computational Sciences, CHRIST (Deemed to be University), Bangalore, India
e-mail: riya.baby@christuniversity.in

F. Kamalov et al. (eds.), *Advances in Mathematical Modeling and Scientific Computing*, Trends in Mathematics, https://doi.org/10.1007/978-3-031-41420-6_33

The effect of rotation in the fluid layer was not considered in any of the previous studies. The rotation or Coriolis Effect occurs when moving objects are observed in a rotating reference. Chandrasekhar [10] was the first to examine the impact of rotation on Benard convection. Based on this theory, many people have researched the subject matter, and to name them, a few are Vidal and Acrivos [11], McConaghy and Finlayson [12], and Shivakumara and Nanjundappa [13]. The importance of Coriolis force is mainly in meteorology and oceanography, long-range artillery, and sight modifications for the rotation of the Earth. Chemical engineering and biochemical conditions are vital applications of the impact of Coriolis force on heat convection in a fluid with suspended particulates. The topic of stability analysis in a spinning non-Newtonian fluid has been researched by many authors like Sastry and Rao [14], Reena and Rana [15], Joseph et al. [16], and Anncy et al. [17].

Marangoni convection generally refers to the convection induced by surface tension variation with temperature without buoyancy. The Marangoni effect results from a surface tension gradient that causes mass transfer at a fluid-fluid contact. This phenomenon might be referred to as thermo-capillary convection in the case of temperature dependency. Temperature or concentration gradients can both contribute to surface tension gradients. According to Block, Bénard cells in shallow pools are caused by more surface tension fluctuations than buoyancy. The Marangoni effect is the name given to this phenomenon. According to Pearson's theory, if the non-dimensional Marangoni number is sufficiently high and positive, the surface tension over temperature variability would result in consistent Marangoni convection in a fluid layer. Hashim [18] proposed a theory for the influence of uniform rotation on the onset of oscillatory and steady surface tension-driven convection in a horizontal fluid layer heated from below. Gupta and Shandil [19] theorized about the effect of the Coriolis force on the beginning of surface tension-driven convection in a substantially hotter or cooler liquid layer confined between two horizontal limits in the presence of uniform rotation. Ahmed et al. [20] recently studied the Marangoni effects and tilted magnetic fields in three-dimensional non-Newtonian Nanofluids.

This chapter aims to understand the impact of rotation on the emergence of convection in a microfluid with upper free velocity and adiabatic temperature boundary conditions and lower rigid velocity and isothermal temperature boundary conditions.

2 Mathematical Framework

It is considering the micropolar fluid enclosed among two infinite horizontal surfaces distanced by d. With a constant angular velocity, the fluid is rotated. The origin is considered in the lower border, and the z-axis is acted in the upward direction. The rotation and the vertical coordinate axis coincide in a Cartesian system (Fig. 1).

The following are the required basic equations considered in this chapter.

The continuity equation:

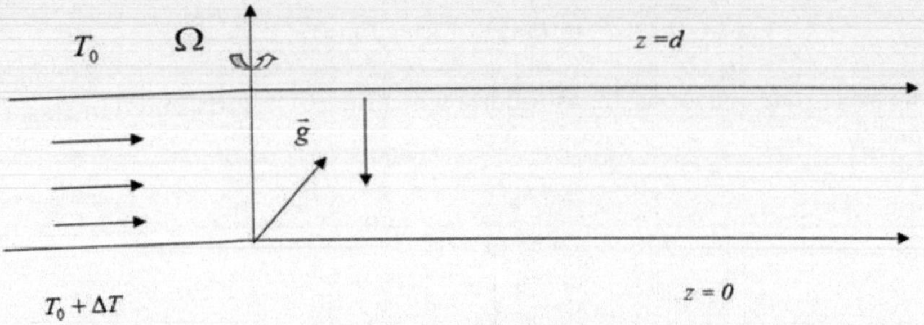

Fig. 1 Basic diagram

$$\nabla.\vec{V} = 0. \tag{1}$$

The conservation of linear momentum:

$$\rho_o\left[\frac{\partial \vec{V}}{\partial t}+\left(\vec{V}.\nabla\right)\vec{V}+2\vec{\Omega}\times\vec{V}\right] = -\nabla P-\rho g\hat{k}+(2\xi+\eta)\nabla^2\vec{V}+\xi\nabla\times\vec{\omega}. \tag{2}$$

The conservation of angular momentum:

$$\rho_o I\left[\frac{\partial \vec{\omega}}{\partial t}+\left(\vec{V}.\nabla\right)\vec{\omega}-\vec{\omega}\times\vec{\Omega}\right] = (\lambda'+\eta')\nabla\left(\nabla.\vec{\omega}\right)+\left(\eta'\nabla^2\vec{\omega}\right) +\zeta\left(\nabla\times\vec{V}-2\vec{\omega}\right). \tag{3}$$

The conservation of energy:

$$\frac{\partial T}{\partial t}+\left(\vec{V}-\frac{\beta}{\rho_o C_v}\nabla\times\vec{\omega}\right).\nabla T\Big] = \chi_e\nabla^2 T. \tag{4}$$

The equation of state:

$$\rho = \rho_o\left[1-\alpha_t\left(T-T_o\right)\right]. \tag{5}$$

where $\vec{V}$ is velocity, ρ_0 is density of the fluid at temperature T = T_0, P is the pressure, ρ is density, ξ is the coupling viscosity coefficient, η is the shear spin-viscosity coefficients, ω is the angular velocity, I is the moment of inertia, η' and λ' are shear and bulk spin-viscosity coefficients, C_v is the specific heat, χ_e is the thermal conductivity, β is micropolar heat conduction coefficient, and α_t is coefficient of thermal expansion.

The fluid's primary condition of quiescence is characterized as follows:

$$\vec{V}_b = \vec{0}, \quad P = P_b(z), \quad T = T_b(z), \quad \rho = \rho_b(z) \tag{6}$$

The basic state equations are as follows when Eq. (6) is substituted in Eqs. (1, 2, 3, 4 and 5).

$$\frac{-\partial P_b}{\partial z} - \rho_b g = 0. \tag{7}$$

$$\frac{\partial^2 T_b}{\partial z^2} = 0. \tag{8}$$

$$\rho_b = \rho_0 \left[1 - \alpha_t \left(T_b - T_o\right)\right]. \tag{9}$$

Consider the instability by introducing slight perturbations on the quiescent state. Using the given equations, allow an infinitesimal thermal disturbance to disrupt the quiescent state.

$$\left.\begin{array}{l} \vec{V} = \vec{V}_b + \vec{V}', P = P_b + P', \rho = \rho_b + \rho', \\ T = T_b + T', \Omega = \Omega_b + \Omega'. \end{array}\right\}. \tag{10}$$

The prime denotes tiny fluctuations in the quantities.

Substituting Eq. (10) in the governing equations and using Eqs. (7, 8, and 9) in the resulting equations. Since we are considering the linear stability analysis, we are removing all the non-linear expressions,

$$\nabla.\vec{V}' = 0. \tag{11}$$

$$\rho_o \left[\frac{\partial \vec{V}'}{\partial t} + 2\Omega_b \times \vec{V}'\right] = -\nabla P' - \rho' g\hat{k} + (2\xi + \eta)\nabla^2 \vec{V}' + \left(\xi \nabla \times \vec{\omega}'\right) \tag{12}$$

$$\rho_o I \left[\frac{\partial \vec{\omega}'}{\partial t} - \vec{\omega}' \times \Omega_b\right] = (\lambda' + \eta')\nabla\left(\nabla \vec{\omega}'\right) + \left(\eta' \nabla^2 \vec{\omega}'\right) + \xi\left(\nabla \times \vec{V}' - 2\vec{\omega}'\right) \tag{13}$$

$$\frac{\partial T'}{\partial t} = \frac{\Delta T}{d}\left[\vec{V}' - \frac{\beta}{\rho_o C_v \varphi} \nabla \times \overleftarrow{\omega}'\right] + \chi_e \nabla^2 T' \tag{14}$$

$$\rho' = -\alpha_t \rho_0 T' \tag{15}$$

Substitute (15) in (12) we get,

$$\rho_o\left[\frac{\partial \overrightarrow{V}'}{\partial t}+2\Omega_b\times\overrightarrow{V}'\right]=-\nabla P'+\alpha_t\rho_0 g T'\hat{k}+(2\xi+\eta)\nabla^2\overrightarrow{V}'+\left(\xi\nabla\times\overrightarrow{\omega}'\right). \tag{16}$$

Removing the physical quantities of the perturbation expressions (11, 12, 13, 14, 15 and 16) and operating curl twice in Eqs. (13) and (16). For the resulting expressions, we are employing the rules of Eq. (17):

$$\left.\begin{array}{l}(x^*,y^*,z^*)=\frac{(x',y',z')}{d},\ w^*=\frac{w'}{\left(\chi_e/d\right)},\ \overrightarrow{\omega}^*=\frac{\omega'}{\left(\chi_e/d^2\right)},t^*=\frac{t'}{\left(d^2/\chi_e\right)},\\ T^*=\frac{T'}{\Delta T},\ \rho^*=\frac{\rho'}{\rho_0},\xi^*=\frac{\xi'}{\chi_e/d^2}\phi^*=\frac{\phi'}{\chi_e/d^3},\varphi^*=\frac{\varphi'}{\chi_e/d^2}.\end{array}\right\} \tag{17}$$

Substituting Eq. (17) in the resultant expression, we get the below non-dimensional equations:

$$(1+N_1)\nabla^2\varphi+N_1\left(D\phi+a^2\omega_3\right)+TaDw=0 \tag{18}$$

$$(1+N_1)\nabla^4 w+R_o\nabla_1^2 T+N_1\nabla^2\Omega_w-TaD\varphi=0 \tag{19}$$

$$N_3\nabla^2\Omega_w-N_1\nabla^2 w-2N_1\Omega_w-\frac{1}{2}TaN_2\phi=0 \tag{20}$$

$$N_3\nabla^2\left(D\phi+a^2\omega_3\right)-2N_1\left(D\phi+a^2\omega_3\right)-N_1\nabla^2\varphi+\frac{1}{2}TaN_2D\Omega_w=0 \tag{21}$$

$$N_4D^2\omega_3+N_3\nabla^2\omega_3-2N_1\omega_3+N_4D\phi+N_1\varphi=0 \tag{22}$$

$$\nabla^2T+w-N_5\Omega_w=0 \tag{23}$$

We removed the asterisks for legibility. The micropolar and rotation parameters are specified as: N_1 Coupling parameter, N_2Inertia Parameter, N_3 Couple Stress Parameter, N_4 Stress parameter, N_5 Micropolar Heat Conduction Parameter, R_0 Rayleigh number, *Ta* Taylor's number.

Using normal mode analysis, we convert the Eqs. (18)–(23) as follows:

$$(1+N_1)\left(D^2-a^2\right)\varphi+N_1\left(D\phi+a^2\omega_3\right)+TaDw=0. \quad (24)$$

$$(1+N_1)\left(D^2-a^2\right)^2 w-R_o a^2 T+N_1\left(D^2-a^2\right)\Omega_w-TaD\varphi=0 \quad (25)$$

$$N_3\left(D^2-a^2\right)\Omega_w-N_1\left(D^2-a^2\right)w-2N_1\Omega_w-\frac{1}{2}TaN_2\phi=0. \quad (26)$$

$$N_3\left(D^2-a^2\right)\left(D\phi+a^2\omega_3\right)-2N_1\left(D\phi+a^2\omega_3\right)-N_1\left(D^2-a^2\right)\varphi$$
$$+\frac{1}{2}TaN_2D\Omega_w=0 \quad (27)$$

$$N_4D^2\omega_3+N_3\left(D^2-a^2\right)\omega_3-2N_1\omega_3+N_4D\phi+N_1\varphi=0 \quad (28)$$

$$\left(D^2-a^2\right)T+w-N_5\Omega_w=0 \quad (29)$$

where

$$D=\frac{d}{dz}.$$

The set of Eqs. (24)–(29) we are solving by the Galerkin method. Multiply expression (24) by ϕ, (25) by w, (26) by Ω_w, (27) by ψ, (28) by ω_3, and (29) by T, respectively, and use integration by parts to solve these equations. Consider $\phi=B\phi_1$, $w=Dw_1$, $\Omega_z=E\Omega_{z1}$, $\psi=F\psi_1$, $\omega_3=G\omega_{31}$, and $T=HT_1$ in which B, D, E, F, G, and H are considered as constants and ϕ_1, w_1, Ωw_1, ψ_1, ω_{31}, and T_1 are considered as trial functions. The matrix form of the above Eigen value problem is

$$AY_1=0 \quad (30)$$

$$A=\begin{bmatrix} B \\ D \\ E \\ F \\ G \\ H \end{bmatrix}$$

$$Y_1 = \begin{bmatrix} (1+N_1)\,X_1 & Ta^{1/2}X_4 & 0 & N_1X_2 & N_1a^2X_3 & 0 \\ -Ta^{1/2}X_8 & (1+N_1)\,X_5 & N_1X_7 & 0 & 0 & -R_oa^2X_6 \\ 0 & -N_1X_{11} & \begin{pmatrix} N_3X_9 \\ -2N_1 \end{pmatrix} & -\frac{1}{2}N_2Ta^{1/2}X_{12} & 0 & 0 \\ -N_1X_{17} & 0 & \frac{1}{2}N_2Ta^{1/2}X_{18} & \begin{pmatrix} N_3X_{13} \\ -2N_1X_{15} \end{pmatrix} & \begin{pmatrix} N_3a^2X_{14} \\ -2N_1a^2X_{16} \end{pmatrix} & 0 \\ N_1X_{23} & 0 & 0 & N_4X_{22} & \begin{pmatrix} N_4X_{19} \\ +N_3X_{20} \\ -2N_1X_{21} \end{pmatrix} & 0 \\ 0 & X_{25} & -N_5X_{26} & 0 & 0 & X_{24} \end{bmatrix}$$

$$\begin{aligned}
&X_1 = \langle \varphi_1 \left(D^2 - a^2\right) \varphi_1 \rangle, && X_{13} = \langle \phi_1 \left(D^2 - a^2\right) D\phi_1 \rangle, \\
&X_2 = \langle \varphi_1 D\phi_1 \rangle, && X_{14} = \langle \phi_1 \left(D^2 - a^2\right) w_{31} \rangle, \\
&X_3 = \langle \varphi_1 \omega_{31} \rangle, && X_{15} = \langle \phi_1 D\phi_1 \rangle, \\
&X_4 = \langle \varphi_1 w_1 \rangle, && X_{16} = \langle \phi_1 \omega_{31} \rangle \\
&X_5 = \left\langle w_1 \left(D^2 - a^2\right)^2 w_1 \right\rangle, && X_{17} = \langle \phi_1 \left(D^2 - a^2\right) \varphi_1 \rangle, \\
&X_6 = \langle w_1 T_1 \rangle, && X_{19} = \langle \omega_{31} D^2 \omega_{31} \rangle, \\
&X_7 = \langle w_1 \left(D^2 - a^2\right) w_1 \rangle, && X_{20} = \langle \omega_{31} \left(D^2 - a^2\right) \omega_{31} \rangle, \\
&X_8 = \langle w_1 D\varphi_1 \rangle, && X_{21} = \langle w_{31}^2 \rangle, \\
&X_9 = \langle \Omega w_1 \left(D^2 - a^2\right) \Omega w_1 \rangle && X_{22} = \langle \omega_{31} D\phi_1 \rangle, \\
&X_{10} = \langle \Omega_{w1}^2 \rangle, && X_{23} = \langle \omega_{31} \varphi_1 \rangle \\
&X_{11} = \langle \Omega_{w1} \left(D^2 - a^2\right) w_1 \rangle, && X_{24} = \langle T_1 \left(D^2 - a^2\right) T_1 \rangle, \\
&X_{12} = \langle \Omega_{w1} \phi_1 \rangle, && X_{25} = \langle w_1 T_1 \rangle, \\
& && X_{26} = \langle T_1 \Omega_{z1} \rangle,
\end{aligned}$$

For R_o, in the given matrix (36) has a non-trivial solution.

In the above equations,

$\langle - - - \rangle$ represents that the equations X_1 to X_{26} are integrating with respect to $z = 0$ and $z = 1$.

The critical Rayleigh number has a value that is determined by the boundaries, and it can be calculated by using the boundary conditions which are as follows:

$$w = Dw = DT = \Omega_z = \varphi_r = D\phi_r = D\omega_{31} = 0, \quad at \quad z = 0,$$

$$w = D^2w + +a^2M_{at} = DT = \Omega_z = \varphi_r = D\phi_r = D\omega_{31} = 0, \quad at \quad z = 1. \tag{31}$$

Here, we considered M_{at} as the Marangoni number.

3 Results and Discussion

In this chapter, the author has investigated the stability study of the impact of rotation on the surface tension convection in a fluid with small micro-sized particles. Different velocities and various temperature boundary conditions are presented on the graph, along with different micropolar and rotational characteristics.

The graph of critical Rayleigh number, M_c vs. N_1, for upper free isothermal temperature boundary and lower rigid isothermal boundaries is shown in Fig. 2. It has been noticed that as N_1 rises, M_c rises as well. A rise in N_1 implies that the concentration of microelements has increased. Because these microelements spend most of their energy creating gyrational velocity, the convection is getting delayed. As a result, we may conclude that increasing N_1 will help to stabilize the overall system. The graph of M_c vs. N_2 is shown in Fig. 3. A rise in N_2 raises M_c, which stabilizes the system. The plots of M_c vs. N_3 for upper free and lower rigid velocity boundary combinations with adiabatic/isothermal limits are shown in Fig. 4. Couple stress takes part in the same effect in the conservation of the linear momentum equation as shear stress does in the conservation of angular momentum. In this problem, it has been noticed that as N_3 levels rise, M_c falls. Because as N_3 levels rise, the fluid's couple stress rises, causing the micro-rotation to decrease. As a result, a rise in N_3 causes the system to become unstable. The reduction in M_c is considerable for lower levels of N_3, but it is inconsequential for larger values, as seen in the graph. As a result, we may deduce that couple stress is effective for only a small level of N_3. The plots of Mc vs. N_4 for upper free and lower rigid velocity boundary combinations with adiabatic/isothermal limits are shown in Fig. 5. A rise

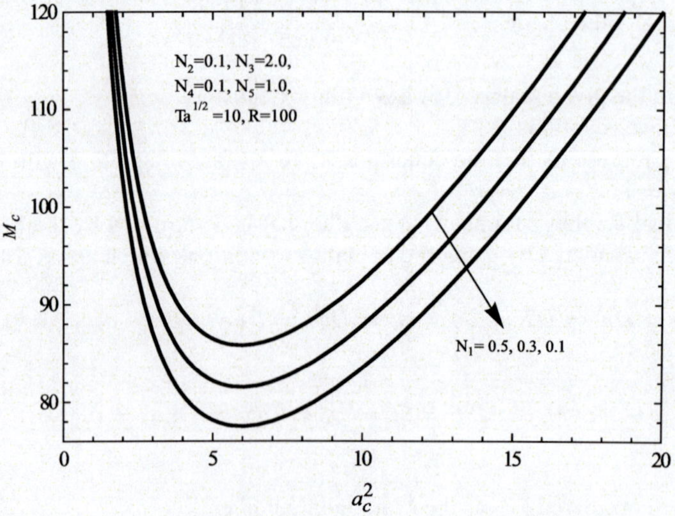

Fig. 2 Graph of M_c versus a_c^2 for various values of N_1

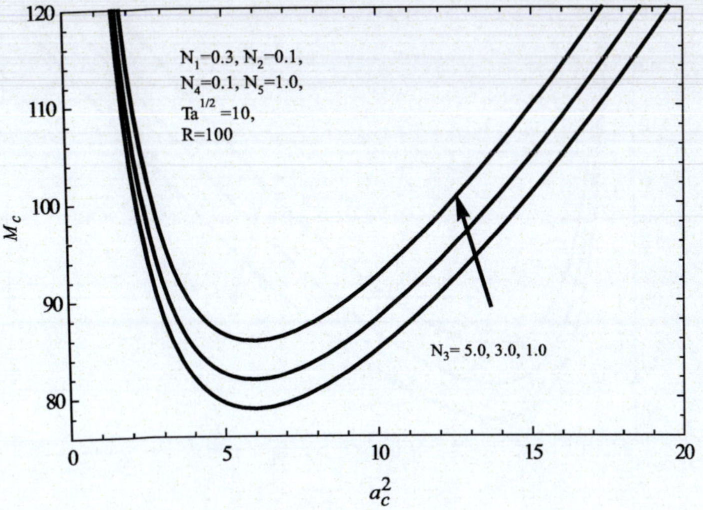

Fig. 3 Graph of M_c versus $a_c{}^2$ for various values N_2

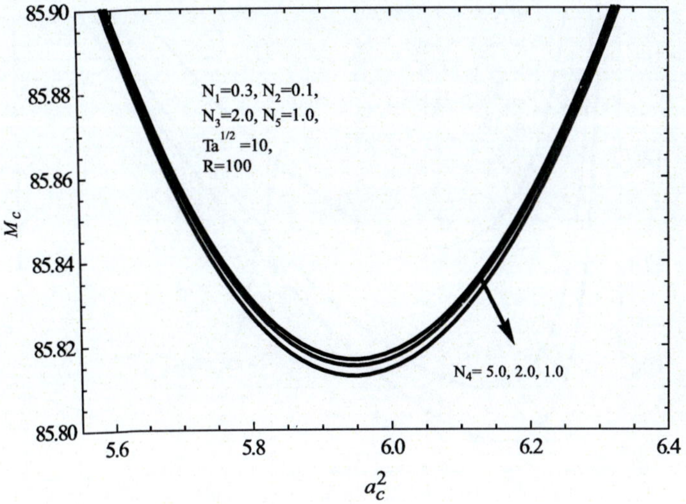

Fig. 4 Graph of M_c versus $a_c{}^2$ for various values, N_3

in N_4 raises M_c, which stabilizes the system. The graphs of M_c vs. N_5 for upper free and lower rigid velocity boundary combinations with adiabatic/isothermal limits are shown in Fig. 6. When N_5 levels rise, the heat produced in the fluid by these microelements also rises, which reduces heat transmission from the lower plate to the upper plate. The slow-down on the convection process is due to a reduction in heat transmission. As a result, we infer that increasing N_5 raises M_c and therefore

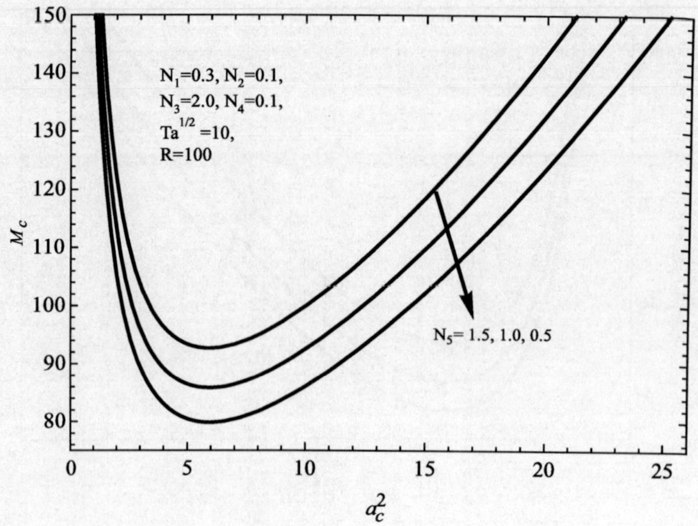

Fig. 5 Graph of M_c versus a_c^2 for various values, N_4

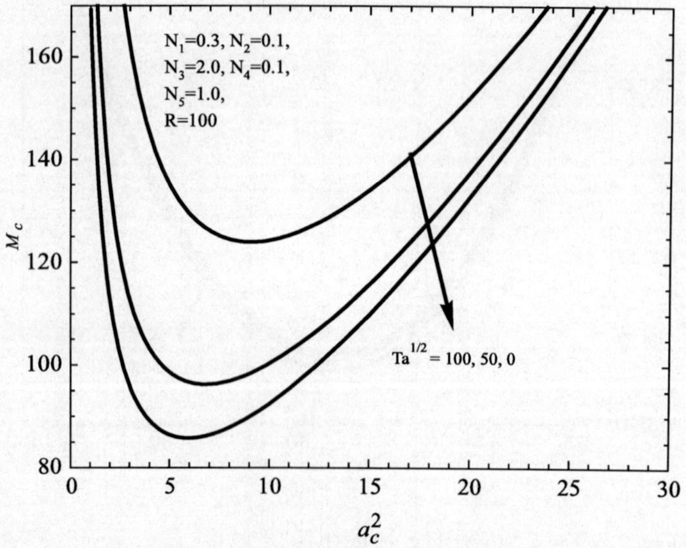

Fig. 6 Graph of M_c versus a_c^2 for various values N_5

stabilizes the system. The graphs of M_c vs. Taylor number Ta for three various velocity boundary combinations with isothermal limits are shown in Fig. 7. M_c rises when Ta rises, showing that Coriolis force has a stabilizing effect. In stationary convection, the impact of rotation stabilizes the system, and with the large values

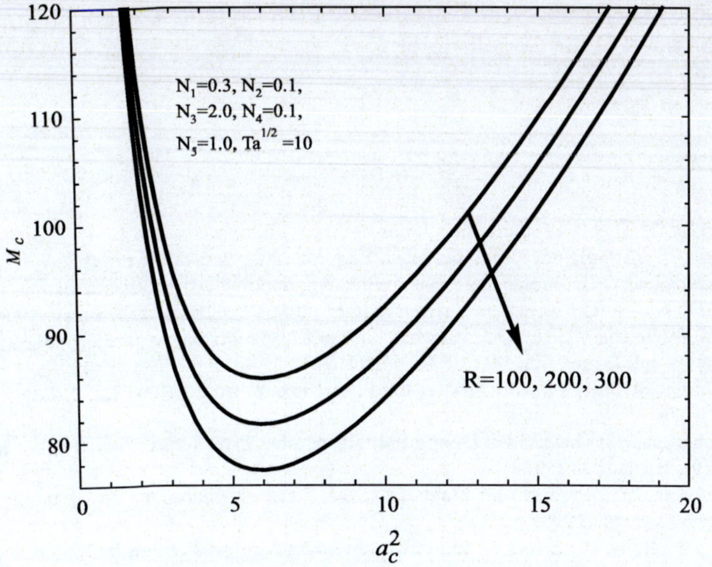

Fig. 7 Graph of M_c versus a_c^2 for various values Ta

of the micropolar parameters, stabilizing effect is highest than that compared to small values. Vortices are introduced into a Newtonian fluid when it rotates. Due to this reason, the fluid then moves at a faster rate in horizontal planes. The velocity of the fluid perpendicular towards the planes decreases as a result of this motion. Convection is therefore being extended. Since the micropolar fluid also has a micro-rotation effect, both the micro-rotation and rotation effects of the entire model reinforce each other, which has the net consequence of reducing the vertical velocity component. Because the presence of micropolar additions reduces the motion for large rotations, the systems are susceptible to instability.

As a limiting situation, we get the following results from the current study:

1. Chandrashekhar [10] result we get when considering $N_1 = 0$ and $Ta = 0$.
2. Siddheshwar and Pranesh [8] findings we get when considering $N_1 \neq 0$ and $Ta = 0$.

4 Conclusion

This research work's significant findings are that stationary convection is the favoured form of instability in a Newtonian fluid with micron-sized suspended particles. By incorporating micron-sized suspended particles with Newtonian fluids, it is possible to defer the system's convection. The system is becoming more stable due to the impact of the Coriolis force.

References

1. Eringen, A.C.: Theory of micropolar fluids. J. Appl. Math. Mech. **16**, 1–18 (1966)
2. Eringen, A.C.: Simple microfluidics. Int. J. Eng. Sci. **2**(2), 205–217 (1964)
3. Lukaszewicz, G.: Micropolar Fluid Theory and Applications. Birkhauser, Boston (1999)
4. Power, H.: Bio-Fluid Mechanics: Advances in Fluid Mechanics. W. I. T. Press (1995)
5. Datta, A.B., Sastry, V.U.K.: Thermal instability of a horizontal layer of micropolar fluid heated from below. Int. J. Eng. Sci. **14**(7), 631–637 (1976)
6. Ahmadi, G.: Stability of a micropolar fluid layer heated from below. Int. J. Eng. Sci. **14**(1), 81–89 (1976)
7. Bhattacharyya, S.P., Jena, S.K.: On the stability of a hot layer of micropolar fluid. Int. J. Eng. Sci. **21**(9), 1019–1024 (1983)
8. Siddheshwar, P.G., Pranesh, S.: Magneto convection in a micropolar fluid. Int. J. Eng. Sci. **36**(10), 1173–1181 (1998)
9. Pranesh, S., Baby, R.: Effect of non-uniform temperature gradient on the onset of Rayleigh-Bénard electro convection in a micropolar fluid. Appl. Math. **3**(5), 442–450 (2012)
10. Chandrasekhar, S.: Hydrodynamic and Hydro Magnetic Stability. Clarendon Press, Oxford (1961)
11. Vidal, A., Acrivos, A.: The influence of Coriolis force on surface-tension-driven convection. J. Fluid Mech. **26**(4), 807–818 (1966)
12. McConaghy, G.A., Finlayson, B.A.: Surface tension driven oscillatory instability in a rotating fluid layer. J. Fluid Mech. **39**(1), 49–55 (1969)
13. Shivakumara, I.S., Nanjundappa, C.E.: Effects of Coriolis force and different basic temperature gradients on Marangoni ferroconvection. Acta. Mechanica. **182**(1), 113–124 (2006)
14. Sastry, V.U.K., Rao, V.R.: Numerical solution of thermal instability of a rotating micropolar fluid layer. Int. J. Eng. Sci. **21**(5), 449–461 (1983)
15. Reena, Rana, U.S.: Thermosolutal convection of micropolar rotating fluids saturating a porous medium. Int. J. Eng. Iran. **4**, 189–217 (2008)
16. Joseph, T.V., Manjunath, S.R., Pranesh, S.: Effect of non-uniform basic temperature gradient on the onset of Rayleigh-Bénard-Marangoni electro-convection in a micropolar fluid. Appl. Math. **4**, 1–9 (2013)
17. Anncy, M., Joseph, T.V., Pranesh, S.: Linear and non-linear analyses of double-diffusive-Chandrasekhar convection coupled with cross-diffusion in micropolar fluid over saturated porous medium. Multidiscip. Model. Mater. Struct. **17**, 211–236 (2020)
18. Hashim, I.: On competition between modes at the onset of Bénard-Marangoni convection in a layer of fluid. ANZIAM J. **43**(3), 387–395 (2002)
19. Gupta, A.K., Shandil, R.G.: Marangoni convection in a relatively hotter or cooler liquid layer. Proc. Natl. Acad. Sci. **82**(2), 103–106 (2012)
20. Ahmed, S.E., Elshehabey, H.M., Oztop, H.F.: Natural convection of three-dimensional non-Newtonian nanofluids with Marangoni effects and inclined magnetic fields. Int. Commun. Heat Mass Transf. **137**, 106288 (2022)

Performance of Magnetic Dipole Contribution on Ferromagnetic Heat and Mass Transfer Flow with the Influence of Nonlinear Radiative Heat Flux

K. Venkatadri, V. Raja Rajeswari, G. Dharmaiah, R. Sivaraj, Firuz Kamalov, Ho-Hon Leung, and Mariam AlShamsi

1 Introduction

Ferrofluids are made by suspending the ferromagnetic nanoparticles in a base liquid and are attracted under the regulation of a magnetic region [1]. Ferrofluids have several applications in various engineering and technological processes. Ferrofluids are utilized in a variety of appurtenances, including lasers, avionics, nuclear plants, X-rays, crystal processing, fiber optics, robotics, textile tools, refrigeration, loudspeakers, and peripherals of computers [2, 3]. Various researchers and scientists have accelerated their examination of ferrofluids based on their immeasurable value. Albrecht et al. [4] investigated the features of ferromagnetic effects in liquids and metals. Anderson and Valnes [5] explored the characteristics of ferromagnetic flow over a stretching sheet. The fundamental analytical specimens for the ferrofluid's motion were developed by Shliomis [6]. Neuringer and Rosensweig [7] explored the characteristics of magnetomechanical and thermomagnetic effects on heat transmission and flow features of fluid. Due to the dependence of magnetization

K. Venkatadri (✉)
Department of Mathematics, Indian Institute of Information Technology, Sri City, India

V. R. Rajeswari
Department of Electronics and Communication Engineering, School of Engineering and Technology, Sri Padmavati Mahila Visvavidyalayam, Tirupati, India

G. Dharmaiah
Department of Mathematics, Narasaraopeta Engineering College, Narasaraopet, Andhra Pradesh, India

R. Sivaraj · H.-H. Leung · M. AlShamsi
Department of Mathematical Sciences, United Arab Emirates University, Al Ain, UAE

F. Kamalov
Faculty of Engineering, Canadian University Dubai, Dubai, UAE

F. Kamalov et al. (eds.), *Advances in Mathematical Modeling and Scientific Computing*, Trends in Mathematics, https://doi.org/10.1007/978-3-031-41420-6_34

on temperature, a coupled thermomagnetic effect may result in a variety of practical application for ferrofluids [8].

The liquid flow over a stretched membrane has become a classic fluid dynamical problem due to the fact that it possesses closed structure solution. In view of many industrial applications, several researchers analyzed the liquid flows across stretched sheets. Gupta and Gupta [9] studied the mass and heat transmission features on a stretched sheet with suction/blow. They offered the numerical solutions to the boundary layer equations with various factors. Rasool et al. [10] explored the mass and thermal transport on a nonlinearly stretched surface with the Darcy–Forchheimer connection. Shahid et al. [11] used the Cattaneo-Christov model to investigate the impact of chemically reactive and thermally radiative flow over a permeable stretched surface.

According to the existing literature, investigations on ferromagnetic steady fluid flows are rare. None of the published work separately addressed the impact of nonlinear radiative heat flux and quadric thermal radiation effect on a stretched surface with a magnetic dipole. With the intention of overcoming this research gap, the current work focuses on the efficacious factor's impact on the flow characteristics of ferromagnetic fluids at the interface. To produce the nonlinear system of ODEs, a suitable similarity transformation is employed. The influence of distinct factors on velocity, temperature, and concentration is examined. In addition, the temperature and mass gradients are investigated numerically.

2 Problem Formulation

The incompressible ferromagnetic flow accompanied by nonlinear radiative heat flux and quadric thermal radiation is investigated. Figure 1 depicts the flow diagram and initial stretching rate, denoted by the letter "*a*." The magnetic dipole's centers lie down along the y-axis. A magnetic dipole directed toward the x-axis enhances the influence of magnetic dipoles and exposes to ferrofluid saturation. Due to the existence of ferromagnetic activity, it is reasonable to assume that Curie temperature (Tc) is higher than the temperature of the liquid and magnetic region is homogeneous. However, when the ferrofluid achieves Curie temperature, further magnetization is not possible. This characteristic is significant for physical applications since the Curie temperature is quite large; for instance, it is 1043 K in the case of iron.

The boundary layer governing equations are considered as follows (Kumar et al. [12]):

$$u\frac{\partial u}{\partial x} + v\frac{\partial u}{\partial y} = 0 \tag{1}$$

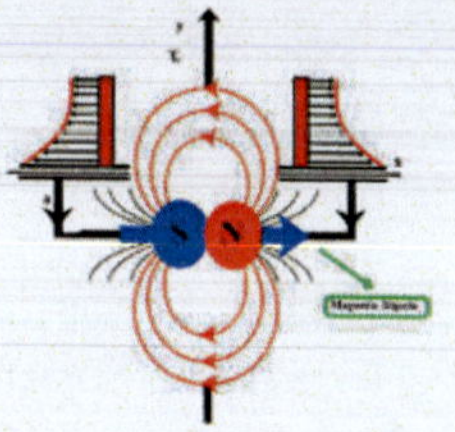

Fig. 1 Flow diagram

$$u\frac{\partial u}{\partial x} + v\frac{\partial u}{\partial y} = \nu\frac{\partial^2 u}{\partial y^2} + \frac{\lambda_0 M}{\rho}\frac{\partial H}{\partial x} \tag{2}$$

$$u\frac{\partial T}{\partial x} + v\frac{\partial T}{\partial y} = \alpha\frac{\partial^2 T}{\partial y^2} - \frac{1}{\rho C_p}\frac{\partial q_r}{\partial y} + q''' - \mu_0 T\frac{\partial M}{\partial T}\left(u\frac{\partial H}{\partial x} + v\frac{\partial H}{\partial y}\right) \tag{3}$$

$$u\frac{\partial C}{\partial x} + v\frac{\partial C}{\partial y} = D_m\frac{\partial^2 C}{\partial y^2} - k\left(C - C_\infty\right) \tag{4}$$

The boundary conditions (Kumar et al. [12]) of the problem is considered as:

$$u = u_w + L\frac{\partial u}{\partial y}, v = 0, C = C_w, -k\frac{\partial T}{\partial y} = h_f\left(T_w - T\right) \quad at \quad y = 0 \tag{5}$$

$$u \to 0, C \to C_\infty, T \to T_\infty \qquad as \qquad y \to \infty \tag{6}$$

where

$$q''' = \frac{k u_w}{x\upsilon}\left(f'\left(T_w - T_\infty\right)A + \left(T - T_\infty\right)B\right) \tag{7}$$

Here $T_w - T_\infty = bx$, where T_∞ and T_w are the far away and nearby temperatures. Moreover, $B < 0$ and $A < 0$ resemble the internal heat absorption of the system while $B > 0$ and $A > 0$ describe the heat generation impact.

$$q_r = -\frac{16}{3}\frac{\sigma^*}{k^*}\frac{\partial T}{\partial y}T^3 \tag{8}$$

The scalar magnetic potential, its components, and magnitude of magnetic field H are considered based on the references [13–17].

$$\Phi = \frac{\gamma}{2\pi}\frac{x}{x^2 + (y + a)^2} \tag{9}$$

$$H_x = -\frac{\partial \Phi}{\partial x} = \frac{\gamma}{2\pi}\ \frac{x^2 - (y+a)^2}{\left[x^2 + (y+a)^2\right]^2} \tag{10}$$

$$H_y = -\frac{\partial \Phi}{\partial y} = \frac{\gamma}{2\pi}\ \frac{2x\,(y+a)}{\left[x^2 + (y+a)^2\right]^2} \tag{11}$$

$$H = \left(\frac{\partial \Phi}{\partial x}\right)^2 + \left(\frac{\partial \Phi}{\partial y}\right)^2 \Bigg\}^{\frac{1}{2}} \tag{12}$$

From Eqns. (10) and (11), we have

$$\frac{\partial H}{\partial x} = -\frac{\gamma}{2\pi}\ \frac{2x}{(y+a)^4} \tag{13}$$

$$\frac{\partial H}{\partial y} = \frac{\gamma}{2\pi}\ \left\{-\frac{2}{(y+a)^3} + \frac{4x^2}{(y+a)^5}\right\} \tag{14}$$

where T – fluid temperature (K), v – vertical velocity (m/s), u – horizontal velocity (m/s), y – vertical coordinate (m), x – horizontal coordinate (m), H – magnetic field of intensity (A/m), Cp – specific heat at constant pressure [J/ (kg K)], Dm – mass diffusivity, a – the distance between origin and center of magnetic dipole/distance parameter, τw – wall shear stress, γ – the strength of magnetic field.

The similarity transformations are

$$\eta = \sqrt{\frac{a}{\upsilon}}\,y;\ \ \psi = \sqrt{a\upsilon}\,x f(\eta);\ \ u = \frac{\partial \psi}{\partial y} = a x f'(\eta);\ \ u = -\frac{\partial \psi}{\partial x} = -\sqrt{a\upsilon}\, f(\eta);$$

$$\phi(\eta) = \frac{C - C_\infty}{C_w - C_\infty};\ \ T = T_\infty\left[1 + (\theta_w - 1)\,\theta\right];\ \ \Pr = \frac{\upsilon \rho C_p}{k} = \frac{\mu C_p}{k};\ \ Sc = \frac{\upsilon}{D_m};$$

$$M = \frac{\sigma B_0^2}{a\rho};\ \ Kr = \frac{K}{a};\ \ Rd = \frac{16\sigma^* T_\infty^3}{3kk^*};\ \ \gamma = L\sqrt{\frac{a}{\upsilon}};\ \ Bi = \sqrt{\frac{\upsilon}{a}}\,\frac{h_f}{k};$$

$$\lambda = \frac{\gamma}{2\pi}\ \frac{\lambda_0}{\rho}\ \frac{1}{\upsilon^2} K\,(T_w - T_c);\ \ \varepsilon = \frac{T_\infty}{T_w - T_\infty};\ \ \beta = H^2 a K\,(T_c - T_\infty);\ \ \theta_w = \frac{T_w}{T_\infty};$$

$$M = K\,(T_c - T);\ \ \theta(\eta) = \frac{T - T_\infty}{T_w - T_\infty}. \tag{15}$$

The dimensionless equations become

$$f''' - f'^2 + f\ f'' - \frac{2\lambda\theta}{(\eta + A)^4} = 0 \tag{16}$$

$$\left\{1 + Rd\left[1 + 3\left(\theta_w - 1\right)\theta + 3(\theta_w - 1)^2\theta^2 + (\theta_w - 1)^3\theta^3\right]\right\}\theta'' + \Pr f\theta' - \Pr f'\theta + Rd\left[3\left(\theta_w - 1\right)\theta'^2 + 6\left(\theta_w - 1\right)\theta\theta'^2 + 3(\theta_w - 1)^3\theta^2\theta'^2\right] + Af' + B\theta + \frac{2\lambda\beta}{(\eta+a)^3}f\theta - \lambda\beta\left(\theta - \varepsilon\right)\left[\frac{4f}{(\eta+a)^5} + \frac{2f'}{(\eta+a)^4}\right] = 0 \tag{17}$$

$$\phi'' + Scf\phi' - KrSc\phi = 0 \tag{18}$$

The boundary conditions are transformed into

$$f' = 1 + f_w\ f'',\ f = 0,\ \theta'(0) = -Bi\left(1 - \theta(0)\right),\ \phi = 1 \quad at \quad \eta = 0 \tag{19}$$

$$f'(\eta) \to 0,\ \theta(\eta) \to 0,\ \phi(\eta) \to 0 \qquad as \qquad \eta \to \infty \tag{20}$$

The expressions for *Sk*, *Nu*, and *Sh* are stated as:

$$C_f \mathrm{Re}_x^{\frac{1}{2}} = f''(0) \tag{21}$$

$$Nu\mathrm{Re}_x^{-\frac{1}{2}} = -\left(1 + \frac{4}{3}Rd\right)\theta'(0) \tag{22}$$

$$Sh\mathrm{Re}_x^{-\frac{1}{2}} = -\phi'(0) \tag{23}$$

where, λ_0 – magnetic permeability ($\mathrm{m}^{-2}\ \mathrm{s}^2$), $k*$ – the mean absorption coefficient, $\sigma*$ – the Stefen-Boltzmann constant, $\mathrm{Re}_x = \frac{xu_w}{\upsilon}$ is the Reynolds number, *Rd* – Radiation parameter, *Bi* – Biot number, *Pr* – Prandtl number, *Sc* – Schmidt number, *fw* – the first order velocity slip parameter, M – magnetization ($\sqrt{m}\sqrt{kg}s^{-2}$), *Cw* – the concentration of the fluid near the surface, *hf* – the convective heat transport, *L* – the molecular free mean path, Θw – the temperature ratio parameter, $\theta(\eta)$ – the dimensionless temperature, $f'(\eta)$ – the dimensionless velocity, *A, B* – nonlinear heat flux constants, *K* – gyromagnetic coefficient constant, λ – ferromagnetic interaction parameter, ε – dimensionless curie temperature, C_f – skin-friction coefficient, β – Eckart number, *Sh* – the local Sherwood number, *Nu* – the local Nusselt number.

3 Methodology

The Eqs. (16, 17, and 18) with appropriate boundary conditions (19 and 20) are nonlinear equations and have no analytical solution. In this investigation, these equations are numerically solved by applying the RK technique using the MATLAB software. The shooting approach is used to transform the boundary conditions (BCs) into appropriate initial conditions. This approach operates by evaluating the BCs as a function of the initial condition's multivariate at certain locations and hence transform the boundary value problem into the initial value problem. The estimated initial values must correspond with the actual boundary values.

4 Results and Discussion

In this study, computations are carried out to investigate the changes in transport characteristics against the variations of the ferrohydrodynamic parameter (λ), distance parameter (a), first order velocity slip parameter (fw), chemical reaction parameter (Kr), Schmidt number (Sc), temperature ratio parameter (θw), radiation parameter (Rd), Prandtl number (Pr), temperature-dependent heat sink/source parameter (B), Biot number (Bi), and space-dependent internal heat sink/source parameter (A). For the graphical results, the variables are assumed to be constant with vales: $\lambda = 0.5$; $\beta = 0.01$; $a = 1.0$; $\varepsilon = 2.0$; $A = 0.1$; $B = 0.1$; $\theta w = 1.2$; $Kr = 0.2$; $Bi = 2$; $fw = 0.1$; $Pr = 6.2$; Rd. = 2, unless specified. Figures 2, 3, 4, 5, 6, 7, 8, 9, 10, 11, 12, 13, 14 and 15 show the key parameters' impact on velocity, temperature, and concentration distributions.

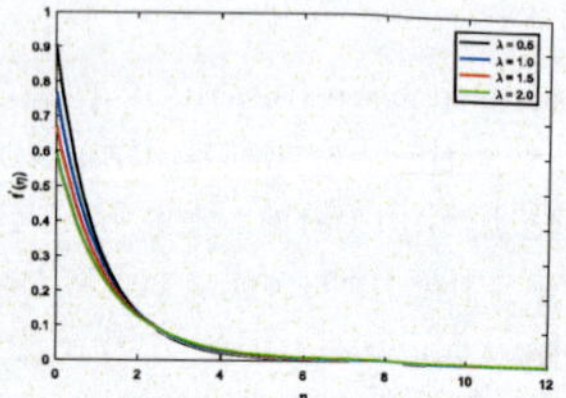

Fig. 2 Preeminence of λ on $f'(\eta)$

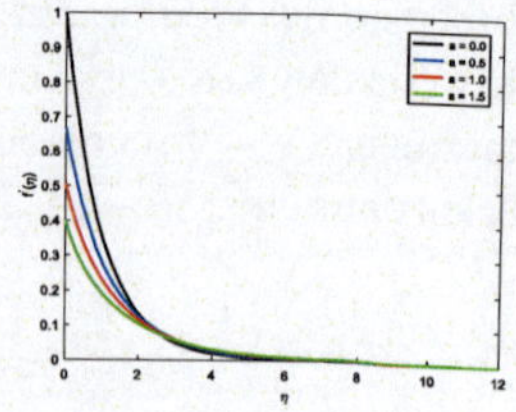

Fig. 3 Preeminence of a on $f'(\eta)$

Fig. 4 Preeminence of *fw* on $f'(\eta)$

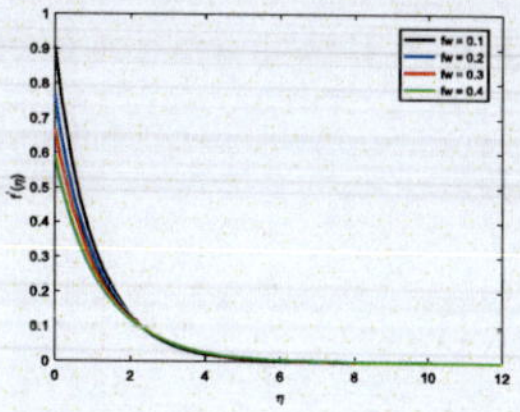

Fig. 5 Preeminence of a on $\theta(\eta)$

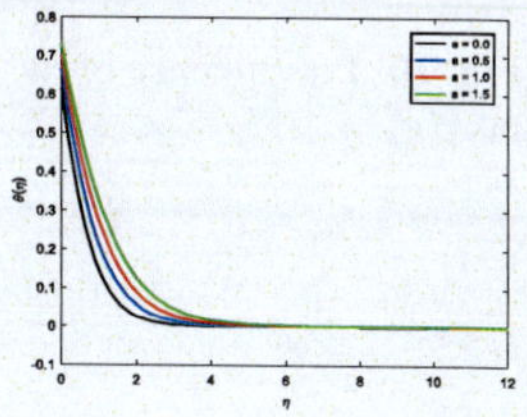

Fig. 6 Preeminence of A on $\theta(\eta)$

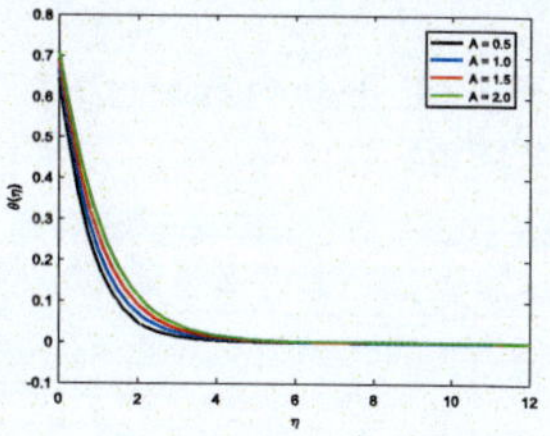

Fig. 7 Preeminence of B on $\theta(\eta)$

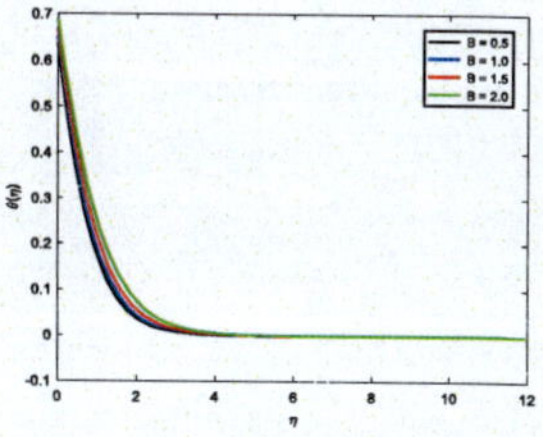

Fig. 8 Preeminence of β on $\theta(\eta)$

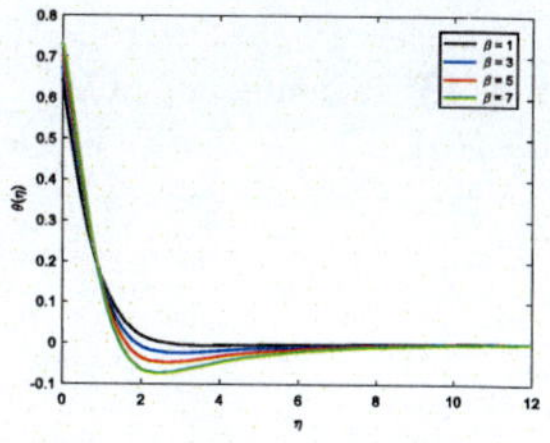

The velocity curves versus the ferromagnetic interaction variable are shown in Fig. 2. Here, velocity decreases as λ increases. Physically, greater values of λ provide greater fluid resistance. Consequently, the velocity falls. Figures 3, 5, and 15 explore the velocity, temperature, and mass transfer sketches against the variations of a. It has been deduced that increasing this parameter reduces the velocity while increasing the concentration and temperature distributions. Figures 4 and 10, in order, illustrate an improvement of *fw* on fluid flow and heat transmission

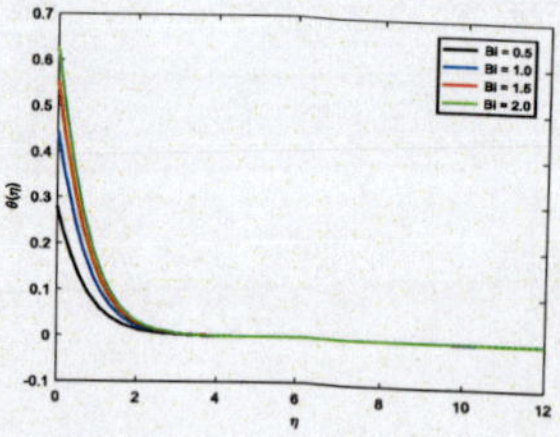

Fig. 9 Preeminence of *Bi* on $\theta(\eta)$

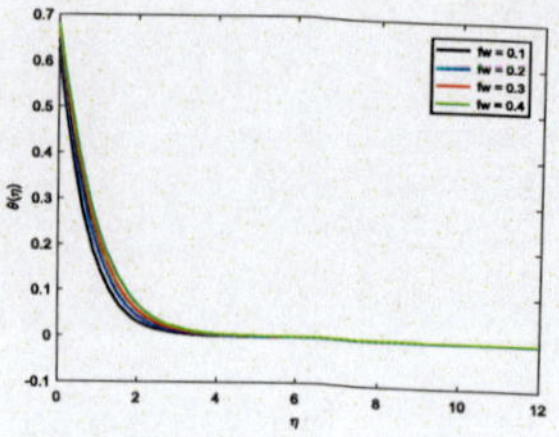

Fig. 10 Preeminence of *fw* on $\theta(\eta)$

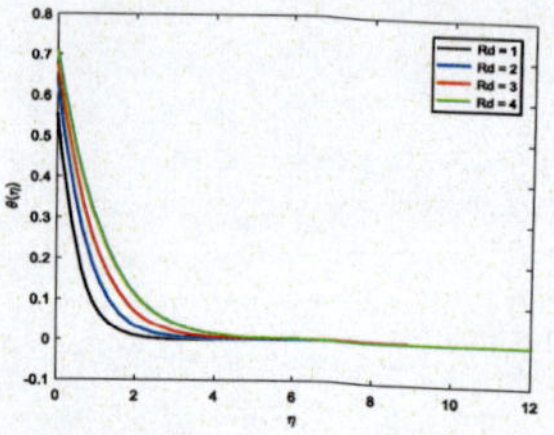

Fig. 11 Preeminence of *Rd* on $\theta(\eta)$

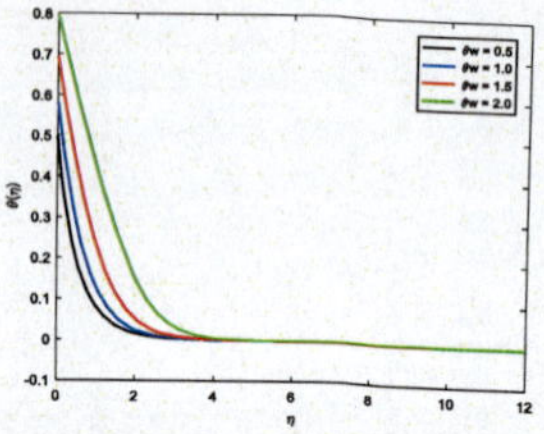

Fig. 12 Preeminence of θw on $\theta(\eta)$

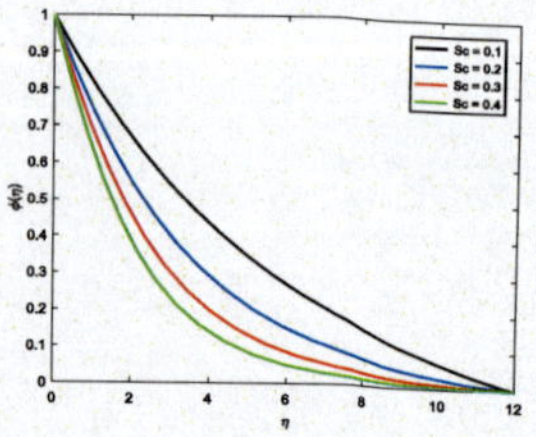

Fig. 13 Preeminence of *Sc* on $\phi(\eta)$

profiles. By raising *fw*, the fluid flow decelerates whereas the heat transmission improves. Figures 6 and 7 illustrate the consequences of heat sources. Clearly, the temperature increases for increasing the A and B. According to the temperature curve displayed in Fig. 8, the fluid temperature increases for enhancing the Eckart number close to the sheet. However, it declines away from the sheet ($\eta > 1$). Figure 9 depicts the changes in heat transmission with varying Biot numbers.

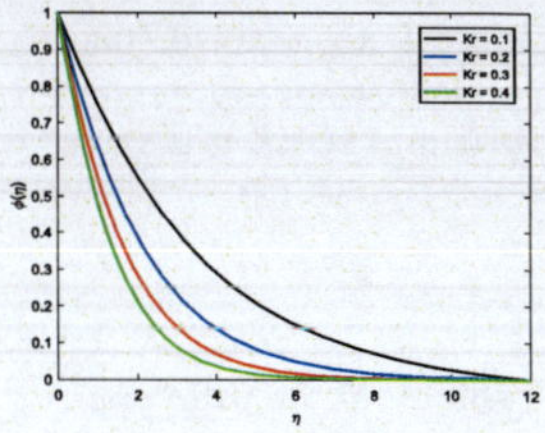

Fig. 14 Preeminence of Kr on $\phi(\eta)$

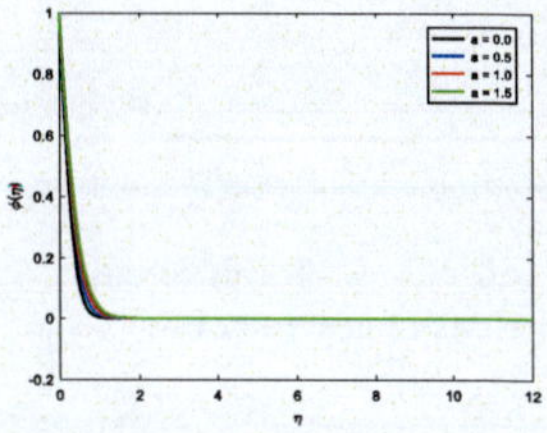

Fig. 15 Preeminence of a on $\phi(\eta)$

Magnifying the Biot number accelerates the heat transport. Figure 11 illustrates the relationship between Rd and heat transmission characteristics. Clearly, the energy transmission grows as Rd upgrades. Radiation irradiates energy which enriches the heat transmission. Figure 12 depicts the energy transmission variations against the temperature ratio parameter. Enhancements in the temperature ratio parameter increases the temperature field. Magnification in the Schmidt number (Sc) diminishes the diffusion coefficient. Therefore, an enhancement in Sc decreases the concentration field, as seen in Fig. 13. Figure 14 investigates the deviation of mass transfer for various Kr values. As Kr grows, the concentration decreases.

The outcomes of a comparison study are illustrated in Table 1. For this, the present model is changed according to Chen [18]. For some selected Pr values, the Nusselt number values are obtained using the present numerical scheme and are compared with the results of Chen [18]. The comparison analysis reveals a strong correlation. Tables 2, 3 and 4 present the computational results of Skin-friction, temperature gradient (Nusselt number), and Sherwood number in relation to certain physical factors. For rising values of fw, the Skin friction coefficient increases. For rising values of A, θw, and B, the Nusselt number exhibits a decreasing trend but it rises for increasing Bi. For rising values of, λ, Rd, and fw, the Sherwood number exhibits a decreasing trend but it rises for increasing Kr.

5 Final Remarks

The effect of magnetic dipole on radiative ferromagnetic fluid flow across a stretched surface is investigated numerically using a broad range of key parameters. The prime findings of the current computational study are listed below:

Table 1 Comparison of Nusselt number for preferred Pr with $\beta = 1.5$, $\lambda = 0.01$, $\varepsilon = 2.0$, $\gamma = 1.0$

Pr	Chen [18]	Present
1.0	1.3333	1.333373
2.0	2.0210	2.021059
3.0	2.5097	2.509748
10.0	4.7968	4.796829

Table 2 *Sk* for different values of *fw*

fw	C_f
0.1	−0.964324
0.3	−0.782763
0.5	−0.665369
0.7	−0.582281

Table 3 *Nu* for different values of key factors

A	θw	*B*	*Bi*	*Nu*
0.2				2.704063
0.4				2.638647
0.6				2.573490
0.8				2.508592
	0.5			3.673746
	1.0			3.048390
	1.5			2.239583
	2.0			1.489647
		0.1		2.736868
		0.3		2.700278
		0.5		2.661980
		0.7		−5.985819
			1	2.036610
			2	2.736868
			3	3.066917
			4	3.255490

- Velocity is a decreasing function of the ferromagnetic interaction parameter and distance parameter.
- Temperature increases with the distance parameter, the first order velocity slip parameter, and the heat sources.
- Concentration profiles increase with increasing dipole magnetic source distance and decrease with increasing *Sc* and *Kr*.
- The skin friction coefficient increases with increasing *fw*.
- The Nusselt number boosts with improving Biot number whereas drops with increasing heat sources.
- The mass transfer rate augments as the chemical reaction parameter grows while decreases as the radiation parameter enhances.

Table 4 *Sh* for different values of key factors

λ	*Rd*	*fw*	*Kr*	*Sh*
0.1				2.905898
0.2				2.901339
0.3				2.896740
0.4				2.892099
	0.5			3.049660
	1.0			2.892160
	1.5			2.889588
	2.0			2.887416
		0.1		2.887416
		0.3		2.720565
		0.5		2.596014
		0.7		2.496379
			0.1	2.676426
			0.3	3.086029
			0.5	3.452936
			0.7	3.787553

References

1. Hayat, T., Ahmad, S., Khan, M.I., Alsaedi, A.: Simulation of ferromagnetic nanomaterial flow of Maxwell fluid. Results Phys. **8**, 34–40 (2018)
2. Godson, L., Raja, B., Lal, D.M., Wongwises, S.E.A.: Enhancement of heat transfer using nanofluids—an overview. Renew. Sust. Energ. Rev. **14**, 629–641 (2010)
3. Ali, L., Liu, X., Ali, B., Mujeed, S., Abdal, S., Khan, S.A.: Analysis of magnetic properties of Nano-particles due to a magnetic dipole in micropolar fluid flow over a stretching sheet. Coatings. **10**, 170 (2020)
4. Albrecht, T., Bührer, C., Fähnle, M., Maier, K., Platzek, D., Reske, J.: First observation of ferromagnetism and ferromagnetic domains in a liquid metal. Appl. Phys. A Mater. Sci. Process. **65**, 215–220 (1997)
5. Andersson, H., Valnes, O.: Flow of a heated ferrofluid over a stretching sheet in the presence of a magnetic dipole. Acta Mech. **128**, 39–47 (1998)
6. Shliomis, M.: Comment on "ferrofluids as thermal ratchets". Phys. Rev. Lett. **92**, 1–6 (2004)
7. Neuringer, J.L., Rosensweig, R.E.: Ferrohydrodynamics. Phys. Fluids. **7**, 1927–1937 (1964)
8. Bailey, R.: Lesser known applications of ferrofluids. J. Magn. Magn. Mater. **39**, 178–182 (1983)
9. Gupta, P.S., Gupta, A.S.: Heat and mass transfer on a stretching sheet with suction or blowing. Can. J. Chem. Eng. **55**(6), 744–746 (1977)
10. Rasool, G., Shafiq, A., Alqarni, M.S., Wakif, A., Khan, I., Bhutta, M.S.: Numerical scrutinization of Darcy-forchheimer relation in convective magnetohydrodynamic nanofluid flow bounded by nonlinear stretching surface in the perspective of heat and mass transfer. Micromachines. **12**(4), 374 (2021)
11. Shahid, A., Bhatti, M.M., Bég, O.A., Kadir, A.: Numerical study of radiative Maxwell viscoelastic magnetized flow from a stretching permeable sheet with the Cattaneo–Christov heat flux model. Neural Compu. Appl. **30**(11), 3467–3478 (2018)
12. Anantha Kumar, K., Venkata Ramudu, A.C., Sugunamma, V., Sandeep, N.: Effect of non-linear thermal radiation on MHD Casson fluid flow past a stretching surface with chemical reaction. Int. J. Ambient Energy. (2022). https://doi.org/10.1080/01430750.2022.2097947
13. Devi, T.S., Lakshmi, C.V., Venkatadri, K., Reddy, M.S.: Influence of external magnetic wire on natural convection of non-Newtonian fluid in a square cavity. Partial Differ. Equ. Appl. Math. **4**, 100041 (2021)

14. Venkatadri, K., Shobha, A., Venkata Lakshmi, C., Ramachandra Prasad, V., Hidayathulla Khan, B.M.: Influence of magnetic wire positions on free convection of Fe_3O_4-water nanofluid in a square enclosure utilizing with MAC algorithm. J. Comput. Appl. Mech. **51**(2), 323–331 (2020)
15. Kandelousi, M.S.: Effect of spatially variable magnetic field on ferrofluid flow and heat transfer considering constant heat flux boundary condition. Eu. Phys. J. Plus. **129**(11), 1–12 (2014)
16. Senapati, M., Parida, S.K., Swain, K., Ibrahim, S.M.: Analysis of variable magnetic field on chemically dissipative MHD boundary layer flow of Casson fluid over a nonlinearly stretching sheet with slip conditions. Int. J. Ambient Energy. (2020). https://doi.org/10.1080/01430750.2020.1831601
17. Gowda, R.P., Kumar, R.N., Prasannakumara, B.C., Nagaraja, B., Gireesha, B.J.: Exploring magnetic dipole contribution on ferromagnetic nanofluid flow over a stretching sheet: an application of Stefan blowing. J. Mol. Liq. **335**, 116215 (2021)
18. Chen, C.H.: Laminar mixed convection adjacent to vertical, continuously stretching sheets. Heat Mass Transf. **33**, 471–476 (1998)

Thermogravitational Convective Flow Inside a Cavity with a Heated Circular Cylinder: A Finite Difference Analysis via Vorticity Stream Function Approach

K. Venkatadri, Veena Chandanam, C. Venkata Lakshmi, R. Sivaraj, Ho-Hon Leung, Firuz Kamalov, and Mariam AlShamsi

1 Introduction

Heat transfer mechanism through natural convection (heat transport without any external forces) is the trending investigation area in the computational fluid dynamics. Fire control, ventilation heat transfer has ample applications, which includes building constructions, cooling mechanisms in electronic devices, nuclear reactors, aerospace technologies and chemical apparatus, natural convection's impact is remarkable. Recently, the energy transmission mechanism within the cavities in the presence of various thermal boundary conditions are analyzed by many researchers.

Number of numerical investigations examined the energy transmission mechanism inside the cavities without any obstacle while few investigations analyzed the heat transfer and flow mechanism within the enclosures with obstacles inside the domain. Fluid flow with obstacles in various shapes is one of the interesting areas of research. Sheikhzadeh et al. [1] theoretically analyzed the natural convective flow of Cu-H_2O nanoliquid around a square cold obstacle placed within the square enclosure. The obstacle size and thermal Rayleigh number impact on heat transmission, entropy generation, and fluid flow characteristics was studied several authors. Zadeh et al. [2] studied the influence of aspect ratio obstacle on natural convective

K. Venkatadri (✉)
Department of Mathematics, Indian Institute of Information Technology, Sri City, Andhra Pradesh, India

V. Chandanam · C. V. Lakshmi
Department of Applied Mathematics, Sri Padmavathi Mahila Visvavidyalayam, Tirupati, India

R. Sivaraj · H.-H. Leung · M. AlShamsi
Department of Mathematical Sciences, United Arab Emirates University, Al Ain, UAE

F. Kamalov
Faculty of Engineering, Canadian University Dubai, Dubai, UAE

F. Kamalov et al. (eds.), *Advances in Mathematical Modeling and Scientific Computing*, Trends in Mathematics, https://doi.org/10.1007/978-3-031-41420-6_35

heat and flow transport in a Cu-H_2O nanoliquid. With vertical and horizontal rectangular obstacles, they studied the flow pattern and energy transmission with distinct Rayleigh numbers. Results indicate that for low Rayleigh numbers, notice a very low entropy generation while Rayleigh number is increased entropy generation is increases. Zoubair et al. [3] scrutinized the natural convective heat transport and flow of Cu-H_2O filled enclosure of square shape with a cold rectangular block. The flow pattern and heat transport for distinct Rayleigh numbers by considering different heights of the obstacle were investigated. In their observation, rate of energy transmission is increased for augmenting the Rayleigh number and the obstacle's height. Mohsen et al. [4] placed a hot square obstacle within the cavity to investigate the double diffusion natural convective heat and flow transport of the fluid. They analyzed the flow behavior and heat transport for distinct values of aspect ratio parameter, Lewis number, and Rayleigh number. A rise in Rayleigh number or Lewis number generates multi-cell flow inside the cavity and the circulation cells are vanished when buoyancy number is increased. Faith et al. [5] explored the features of hydromagnetic free convective heat and flow transport inside an enclosure of a square shape filled with nano liquid by implanting different obstacles like a square, diamond, and sphere within the cavity. By considering linearly heated vertical walls, an adiabatic top wall, and a heated bottom wall with distinct adiabatic obstacles, they compared the results with a non-obstacle cavity in the magnetic field. Compared to the non-obstacle cavity, the energy transmission within the cavity is lower when the obstacle of any considered shapes is present. Parvin et al. [6] analyzed the free hydromagnetic convective heat transfer and fluid flow characteristics within a square cavity containing a heated solid circular block. They presented the numerical simulations for a variety of parameters, like the diameter of the heated circular block, Rayleigh number, and Hartmann number. Their outcomes exposed that the thermal layer around the heated circular surface becomes thick for rising the Rayleigh number and thin for the elevation of the Hartmann number.

Rasul et al. [7] investigated the $Al_2O_3 - H_2O$ nanoliquid flow inside a cavity of L-shape with a heated rectangular solid block to test the free convection heat transport. This analysis specifically focused on the impact of distinct key parameters like aspect ratio of flow channel, height and position of the heated block, nanoparticle volume fraction, Rayleigh number, and nanoparticle's diameter. A rise in nanoparticle's diameter and Rayleigh number causes a rise in the heat transfer rate. The impact of the location of the heated block shows a considerable influence on temperature and fluid flow distributions. When the obstacle is placed at the L-shaped cavity's lower left wall, high rate of heat transfer is observed for each aspect ratio. By choosing U-shaped cavity, Yuan et al. [8] studied the free convective heat transport of MWCNT-H_2O nanoliquid having a hot rectangular block by implementing the LBM method. They studied the heat transport and fluid flow characteristics for 15 distinct positions of heated block which is placed inside the cavity. Better outcomes are noticed in the rate of heat transfer when the hot obstacle is positioned at the enclosure's bottom with a high Rayleigh number ($Ra = 10^6$). Ahmad et al. [9] investigated the magnetic field's influence on free convective $Al_2O_3 - H_2O$ nanoliquid flow within an enclosure of square shape with a pair of isothermal square obstacles by applying sinusoidal wall temperature. Both the isothermal heat sources are placed inside the cavity at an equal distance from the center. As per the outcomes, it is noticed that fluid with low velocity takes additional time to move upward for smaller values of aspect ratio because of the available wide space among the walls of the enclosure and heat sources positioned. Abchouyeh et al. [10] considered a two-dimensional horizontal channel to analyze the free convective energy transmission of nanoliquid with CuO nanoparticles having sinusoidal obstacle. Their study has been done for various Reynolds numbers and nanoparticle's volume fraction parameter. Augmenting the Reynolds numbers and solid volume fraction parameter rises the average nusselt number. An electro-thermo-convective flow within the square cavity containing a hot solid block was theoretically examined by Hassen et al. [11]. For the elimination of dead zone generated at the enclosure's bottom, they implemented a potential difference between horizontal walls and hot obstacle. The rate of heat transfer is improved upto 43% for altering the shape factor of the obstacle. A theoretical investigation on mixed convective fluid flow inside an open cavity contained with hot obstacle is performed by Miansari et al. [12]. The influence of positions of the obstacles and fluid flow outlet on heat transport and flow field within the cavity are investigated by them.

Motivated by the developments in heat transport in enclosures, the present chapter is intended to investigate the free convective flow inside a cavity featuring obstacles located at the middle part of the cavity. In the scientific literature, such a configuration has not found much attention. Studies on hot obstacles within enclosures play a vital role in various applications. A numerical solution is established with a vorticity and stream function approach, and a fine grid difference scheme is utilized. The graphical representation of streamlines and temperature contours is elucidated to examine the consequence of buoyancy number.

2 Mathematical Formulation

Figure 1 displays the schematic diagram for the current investigation. Cavity's middle regime, a hot, circular cylinder, is considered. The enclosure's side walls are maintained at cold temperatures, and its bottom and top walls are thermally insulated. The flow is considered laminar and incompressible without considering the slip conditions. All the external forces are ignored except the buoyancy force due to gravity. The flow configuration is depicted by the system of equations, which are written in a cartesian form as follows:

$$\frac{\partial u}{\partial x} + \frac{\partial v}{\partial y} = 0 \tag{1}$$

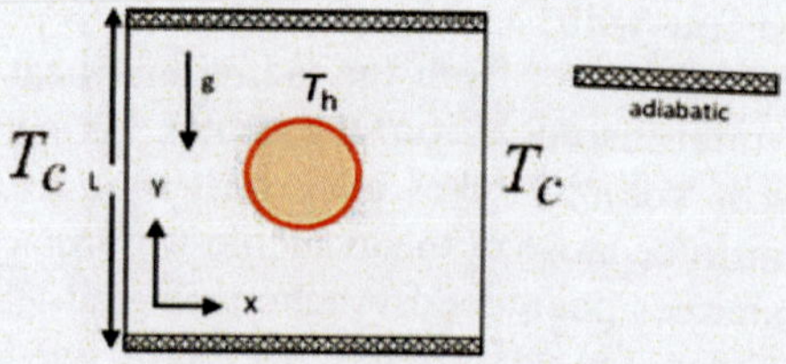

Fig. 1 The present model's schematic view

$$\frac{\partial u}{\partial t}+u\frac{\partial u}{\partial x}+v\frac{\partial u}{\partial y}=-\frac{1}{\rho}\frac{\partial p}{\partial x}+\upsilon\left(\frac{\partial^2 u}{\partial x^2}+\frac{\partial^2 u}{\partial y^2}\right) \tag{2}$$

$$\frac{\partial v}{\partial t}+u\frac{\partial v}{\partial x}+v\frac{\partial v}{\partial y}=-\frac{1}{\rho}\frac{\partial p}{\partial y}+\upsilon\left(\frac{\partial^2 v}{\partial x^2}+\frac{\partial^2 v}{\partial y^2}\right)+g\beta_T\,(T-T_c) \tag{3}$$

$$\frac{\partial T}{\partial t}+u\frac{\partial T}{\partial x}+v\frac{\partial T}{\partial y}=\alpha\left(\frac{\partial^2 T}{\partial x^2}+\frac{\partial^2 T}{\partial y^2}\right) \tag{4}$$

The enclosure's boundary conditions are considered as:

$$\begin{aligned} u=v=0, T=T_c\ at\ x=0, L \\ u=v=0,\ \tfrac{\partial T}{\partial y}=0\ at\ y=0, L \end{aligned} \tag{5}$$

The dimensionless variables are introduced as follows:

$$\tau=\frac{t\alpha}{L^2}, X=\frac{x}{L}, Y=\frac{y}{L}, U=\frac{uL}{\alpha}, V=\frac{vL}{\alpha}, \theta=\frac{T-T_c}{T_h-T_c},\ \ P=\frac{pL^2}{\rho\alpha^2} \tag{6}$$

The dimensionless constitutive equations are presented as follows:

$$\frac{\partial U}{\partial X}+\frac{\partial V}{\partial Y}=0 \tag{7}$$

$$\frac{\partial U}{\partial \tau}+U\frac{\partial U}{\partial X}+V\frac{\partial U}{\partial Y}=-\frac{\partial P}{\partial X}+\Pr\left(\frac{\partial^2 U}{\partial X^2}+\frac{\partial^2 U}{\partial Y^2}\right) \tag{8}$$

$$\frac{\partial V}{\partial \tau}+U\frac{\partial V}{\partial X}+V\frac{\partial V}{\partial Y}=-\frac{\partial P}{\partial Y}+\Pr\left(\frac{\partial^2 V}{\partial X^2}+\frac{\partial^2 V}{\partial Y^2}\right)+Ra\Pr\theta \tag{9}$$

$$\frac{\partial \theta}{\partial \tau}+U\frac{\partial \theta}{\partial X}+V\frac{\partial \theta}{\partial Y}=\left(\frac{\partial^2 \theta}{\partial X^2}+\frac{\partial^2 \theta}{\partial Y^2}\right) \tag{10}$$

Dimensionless vorticity (ω) and stream function (ψ) are determined as:

$$V = -\frac{\partial \psi}{\partial X}, \quad U = \frac{\partial \psi}{\partial Y} \tag{11}$$

$$\omega = \frac{\partial V}{\partial X} - \frac{\partial U}{\partial Y} = -\frac{\partial^2 \psi}{\partial X^2} - \frac{\partial^2 \psi}{\partial Y^2} \tag{12}$$

With the help of vorticity expressions and stream function, the vorticity transport equation leads to:

$$\frac{\partial \omega}{\partial \tau} + U\frac{\partial \omega}{\partial X} + V\frac{\partial \omega}{\partial Y} = \Pr\left(\frac{\partial^2 \omega}{\partial X^2} + \frac{\partial^2 \omega}{\partial Y^2}\right) + Ra \Pr \frac{\partial \theta}{\partial X} \tag{13}$$

$$\frac{\partial \theta}{\partial \tau} + U\frac{\partial \theta}{\partial X} + V\frac{\partial \theta}{\partial Y} = \left(\frac{\partial^2 \theta}{\partial X^2} + \frac{\partial^2 \theta}{\partial Y^2}\right) \tag{14}$$

The corresponding boundary conditions become:

$$\begin{array}{l} U = V = \psi = 0, \theta = 0 \; at \; X = 0, 1 \\ U = V = \psi = 0, \frac{\partial \theta}{\partial Y} = 0 \; at \; Y = 0, 1 \end{array} \tag{15}$$

where $Ra = \frac{g\beta(T_h - T_c)L^3}{\alpha\nu}$, $\Pr = \frac{\nu}{\alpha}$, Ψ – dimensionless stream function, μ – viscosity of the fluid [Nsm^{-2}], θ – non-dimensional temperature, β – thermal expansion coefficient [K^{-1}], α – thermal diffusivity, h – convective heat transfer coefficient [Wm^{-2} K^{-1}], Cp – specific heat [J kg^{-1} K^{-1}], ρ – density [kgm^{-3}], T_c – cold wall's temperature [K], T_h – hot wall's temperature [K], P – pressure, Ra – Rayleigh number, (X, Y) – dimensionless co-ordinates, (x, y) – Cartesian co-ordinates, (U, V) – dimensionless velocity components, ψ – stream function, (u, v) – velocity components, T – temperature, Nu – local Nusselt number, g – acceleration due to gravity [m/s^2], Pr – Prandtl number, L – length of walls [m], k – thermal conductivity [Wm^{-1} K^{-1}].

3 Method of Solution and Validation

The reduced governing partial differential equations (Eqs. (12)–(14)) in dimensionless form are discretized with finite difference scheme. The diffusive terms are approximated with central difference approximation. The vorticity-stream function approach adopted by Venkatadri [16] is utilized to solve the energy and momentum equations. The numerical scheme is based on collocated grid system. Venkatadri et al. [17] presented the adequate details about the present scheme. The numerical computation is ended when the continuity residual and the energy residual reach 10^{-8} and momentum residual reaches 10^{-6}. Comparison of the present outcomes

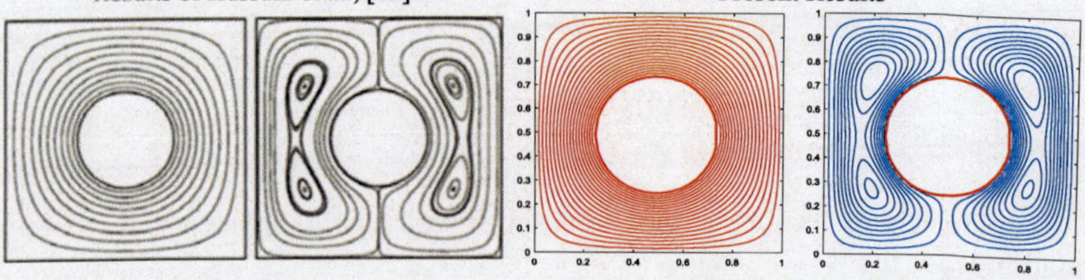

Fig. 2 Comparison of the outcomes of previous study (Hussain et al. [13]) with the present outcomes when $Ra = 10^3$

Table 1 Validation of *Nu* with the benchmark results

	Nu			Error percentage	
Ra	Kim et al. [14]	Kim et al. [15]	Present study	Kim et al. [14] and Present study	Kim et al. [15] and Present study
10^3	5.02	5.093	5.039	1.06	1.0600
10^4	5.113	5.108	5.131	0.35	0.3500
10^5	7.75	7.767	7.807	0.73	0.5120
10^6	14.2	14.11	14.13	0.49	0.1415

and the outcomes of Hussain et al. [13] when $Ra = 10^3$ are depicted in Fig. 2. As seen, the present results match closely with the outcomes of Hussain et al. [13]. In addition, the free convective flow inside the cavity of square shape with hot circular cylinder is studied and compared with Kim et al. [14, 15]. The outcomes of the comparative study are presented in Table 1. The present results give good accordance with the outcomes of Kim et al. [14, 15].

4 Results and Discussion

In this chapter, thermogravitational convective flow inside a cavity of a square shape is investigated. The enclosure's side walls are considered cold walls whereas the horizontal walls are assumed to be adiabatic. At the cavity's center, a hot circular cylinder is fixed and the respective convective flow behavior within the cavity is analyzed. For all the figures discussed in this chapter, the respective results are obtained by fixing $Pr = 0.71$.

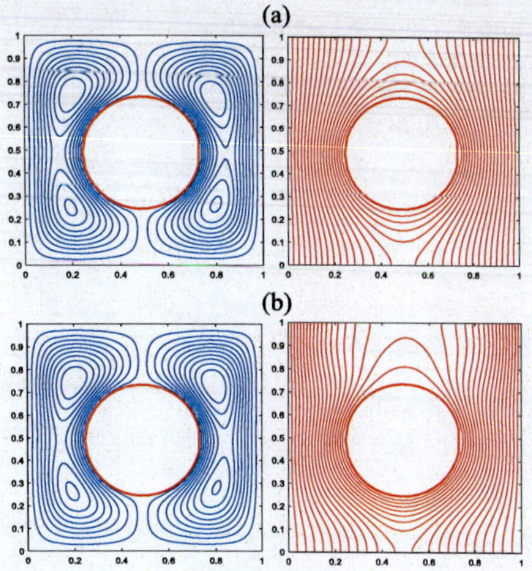

Fig. 3 Streamlines and isotherms with $Pr = 0.71$ for **(a)** $Ra = 10^3$ and **(b)** $Ra = 10^4$

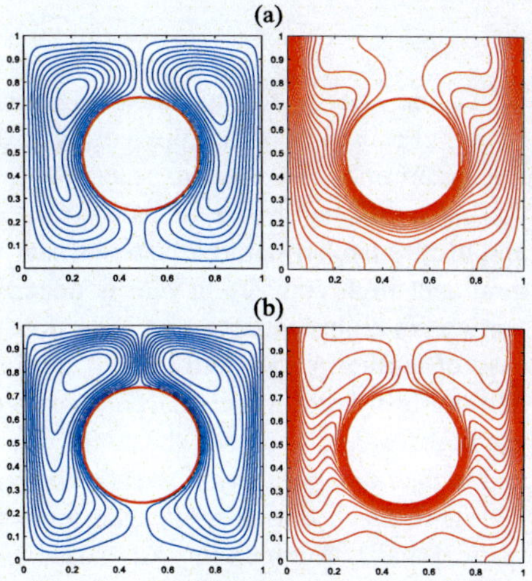

Fig. 4 Streamlines and isotherms for working fluid with $Pr = 0.71$ for **(a)** $Ra = 10^5$ and **(b)** $Ra = 10^6$

Figure 3 demonstrates the isotherms (right) and streamlines (left) for working fluid with $Pr = 0.71$ with lower Ra values. Fig. 3a represents the streamlines and isotherms for $Ra = 10^3$. A symmetric streamlines structure is observed for $Ra = 10^3$ which shows that the flow pattern is uniform and almost symmetric for low Rayleigh number. In case of isotherms, a uniform temperature distribution from the hot cylinder toward the cold side walls is observed. When Ra value is increased to 10^4, a slight enlargement of vortices and a rise in the flow of fluid is noticed in the streamlines near the left wall. In the respective isotherms, a slight heat transmission augmentation is identified from the hot cylinder toward the vertical cold walls.

In Fig. 4, isotherms (right) and streamlines (left) patterns for higher Ra values are observed. When Ra is risen from 10^4 to 10^5, a full transformation in streamlines and temperature contours is noticed. In this scenario, enhancement of convection

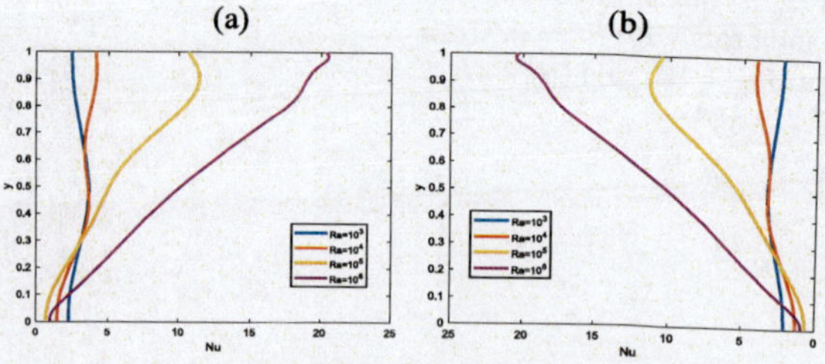

Fig. 5 Local Nusselt number along the side walls with $Pr = 0.71$ for (**a**) left wall and (**b**) right wall

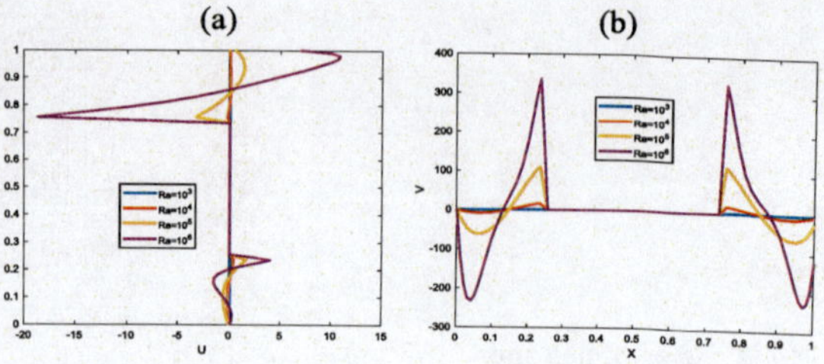

Fig. 6 Centerline velocity profiles of (**a**) U and (**b**) V velocities with $Pr = 0.71$

effect and energy transmission augmentation is observed. With this convection effect, elimination of bottom vortex and enlargement of top vortex is noticed in the streamlines. This means, the flow is boasted toward top adiabatic walls from the hot cylinder. In case of isotherms, a non-uniform thermal bloom is observed and all the isotherms are clustered near the top walls. When Ra is risen to 10^6, a dominant high convection rate is occurred in the fluid flow. As seen, the size of vortices is highly enlarged and the flow is highly boosted toward the top wall. In case of isotherms, smooth curved uniform pattern is entirely changed into wavy non-uniform pattern and sharper heat plume is observed around the hot obstacle. For high Rayleigh number, the isotherms are densely formed around the hot obstacle and in the neighborhood of the top adiabatic wall.

Figure 5a, b demonstrates the local rate of heat transfer variations along the vertical walls. Augmenting Ra causes a similar Nusselt number variation on both left and right vertical walls. Distribution of Nusselt number is gradually increased near the top wall and a massive change is observed for high Ra values.

Figure 6a represents the centerline velocity profiles of U-velocity for different Rayleigh numbers. Due to the buoyancy effect, high fluid flow is detected at the cavity's top portion, and a low convection effect is noticed in the cavity's bottom portion. Figure 6b depicts the centerline velocity profiles of V-velocity for different Rayleigh numbers. On both left and right sides of the wall, symmetry in flow is observed and flow is boasted symmetrically toward both the side walls. High fluid flow is observed at the cavity's top portion and low convection effect is deducted in the cavity's bottom portion.

5 Conclusion

In the present study, thermogravitational convective flow within the square cavity including a heated circular obstacle with the variation of *Ra* is examined numerically. A fine grid finite diffcrence analysis through vorticity stream function (pressure elimination) technique is employed to compute the flow behavior and heat transfer features within the cavity. The effects of key parameters are discussed and the respective results are depicted by means of graphs. Some of the following key results were obtained from the current investigations:

- When the Rayleigh number (*Ra*) values are upraised gradually, the innermost bottom vortices merge with top vortex in the streamlines which indicates the rise in convection flow within the cavity near the top wall.
- The incremental values of *Ra* convert the smooth uniform isotherms into wavy non-uniform pattern and expose the sharper heat plume around the hot obstacle. This indicates a noticeable enhancement in the fluid temperature around the hot obstacle.
- Rising values of *Ra* increase the denseness of streamlines and isotherms in the neighborhood of top walls.
- Distribution of Nusselt number is gradually increased and a massive change is observed for high values of *Ra*.
- Through the centerline velocity profiles of *U* and *V* velocities, the impact of buoyancy on the fluid flow inside the cavity is explored.

References

1. Sheikhzadeh, G.A., Nikfar, M., Fattahi, A.: Numerical study of natural convection and entropy generation of Cu-water nanofluid around an obstacle in a cavity. J. Mech. Sci. Technol. **26**(10), 3347–3356 (2012)
2. Sheikhzadeh, G.A., Nikfar, M.: Aspect ratio effects of an adiabatic rectangular obstacle on natural convection and entropy generation of a nanofluid in an enclosure. J. Mech. Sci. Technol. **27**(11), 3495–3504 (2013)
3. Boulahia, Z., Wakif, A., Sehaqui, R.: Natural convection heat transfer of the nanofluids in a square enclosure with an inside cold obstacle. Int. J. Innov. Sci. Res. **21**(2), 367–375 (2016)
4. Nazari, M., Louhghalam, L., Kayhani, M.H.: Lattice Boltzmann simulation of double diffusive natural convection in a square cavity with a hot square obstacle. Chin. J. Chem. Eng. **23**(1), 22–30 (2015)
5. Selimefendigila, F., Öztop, H.F.: Natural convection and entropy generation of nanofluid filled cavity having different shaped obstacles under the influence of magnetic field and internal heat generation. J. Taiwan Inst. Chem. Eng. **56**, 42–56 (2015)
6. Parvin, S., Nasrin, R.: Analysis of the flow and heat transfer characteristics for MHD free convection in an enclosure with a heated obstacle. Nonlinear Anal. Model. Contr. **16**(1), 89–99 (2011)
7. Mohebbi, R., Rashidi, M.M.: Numerical simulation of natural convection heat transfer of a nanofluid in an L-shaped enclosure with a heating obstacle. J. Taiwan Inst. Chem. Eng. **72**, 70–84 (2017)

8. Yuan, M., Mohebbi, R., Rashidi, M.M., Yang, Z.: Effect of hot obstacle position on natural convection heat transfer of MWCNTs-water nanofluid in U-shaped enclosure using lattice Boltzmann method. Int. J. Numer. Method. Heat Fluid Flow. **29**(1), 223–250 (2019)
9. Pordanjani, A.H., Jahanbakhshi, A., Nadooshan, A.A., Afrand, M.: Effect of two isothermal obstacles on the natural convection of nanofluid in the presence of magnetic field inside an enclosure with sinusoidal wall temperature distribution. Int. J. Heat Mass Transf. **121**, 565–578 (2018)
10. Abchouyeh, M.A., Mohebbi, R., Fard, O.S.: Lattice Boltzmann simulation of nanofluid natural convection heat transfer in a channel with a sinusoidal obstacle. Int. J. Modern Phy. C. **29**(9), 1850079 (2018)
11. Hassen, W., Kolsi, L., Ghachem, K., Almeshaal, M.A., Maatki, C., Borjini, M.N.: Numerical investigation of electro-thermo-convection in a square enclosure with incorporated hot solid body. J. Therm. Anal. Calorim. **143**, 2647–2661 (2021)
12. Miansari, M., Halimi, A., Miansari, M., Toghraie, D.: Mixed convection in a ventilated enclosure by considering both geometrical parameters and thermo-physical properties of water/Cu nanofluid. J. Therm. Sci. **30**, 950–961 (2021)
13. Hussain, S.H., Hussein, A.K.: Numerical investigation of natural convection phenomena in a uniformly heated circular cylinder immersed in square enclosure filled with air at different vertical locations. Int. Commun. Heat Mass Transf. **37**(8), 1115–1126 (2010)
14. Kim, B.S., Lee, D.S., Ha, M.Y., Yoon, H.S.: A numerical study of natural convection in a square enclosure with a circular cylinder at different vertical locations. Int. J. Heat Mass Transf. **51**(7), 1888–1906 (2008)
15. Kim, M., Doo, J.H., Park, Y.G., Yoon, H.S., Ha, M.Y.: Natural convection in a square enclosure with a circular cylinder according to the bottom wall temperature variation. J. Mech. Sci. Technol. **28**(12), 5013–5025 (2014)
16. Venkatadri, K., Bég, O.A., Rajarajeswari, P., Prasad, V.R.: Numerical simulation of thermal radiation influence on natural convection in a trapezoidal enclosure: heat flow visualization through energy flux vectors. Int. J. Mech. Sci. **171**, 105391 (2020)
17. Venkatadri, K., Bég, O.A., Kuharat, S.: Magneto-convective flow through a porous enclosure with Hall current and thermal radiation effects: numerical study. Eur. Phys. J. Spec. Top. **231**, 2555–2568 (2022)

An Unsteady Flow of Fluid Velocity, Temperature, and Heat Emission on MHD Free Convection Flow of Some Nanofluids

K. Ramesh Babu and J. Buggaramulu

1 Introduction

An example of a fluid is a liquid or gas that can be easily manipulated. Because of its high relative compressibility, the fluid stands out among other substances. Fluids are regarded as continuous media, and their properties, such as velocity, pressure, density, and so on, can be expressed in terms of their spatial and temporal distributions.

A subdivision of fluid mechanics, i.e. fluid dynamics, is concerned with the movement of liquids and gases. There are many sub-disciplines within this field, including aerodynamics, which is the study of the properties of a solid object as they relate to the flow of air around it. Fluid dynamics can be used to calculate the stresses and moments of aeroplanes, determine the mass flow rate of petroleum through pipelines, explain nebulas in interstellar space, and model fission weapon explosion. Even in traffic engineering, where traffic is viewed as a continuous fluid, some of its ideas are employed.

It is possible to study the dynamics of magnetic fields in electrically conducting media, such as plasmas and liquid metals, using a physical-mathematical framework known as magneto-hydrodynamics. Magnetic and hydrodynamics are combined in the term "magneto-hydrodynamics", for example (meaning water or liquid).

K. R. Babu (✉)
Department of Mathematics, University College of Engineering, Osmania University, Hyderabad, Telangana, India

J. Buggaramulu
Research Scholar, Department of Mathematics, Osmania University, Hyderabad, India

F. Kamalov et al. (eds.), *Advances in Mathematical Modeling and Scientific Computing*, Trends in Mathematics, https://doi.org/10.1007/978-3-031-41420-6_36

Hannes Alfven, a Swedish physicist who earned the Nobel Prize in Physics in 1970 for his revolutionary work and discoveries in MHD, was the first to discover the concept.

Magnetic fields can be supported by conductive fluids, according to magneto-hydrodynamics (MHD) theory. Fluid (usually plasma) undergoes stresses due to magnetic fields, which can change the magnetic fields' geometrical properties as well as their strength. Adverting motions in a particular conducting fluid are critical, as they are more powerful than electrical resistivity's diffusive effects. The foundations of magneto-hydrodynamics are Alfven waves, magneto-convection, MHD reconnection, and hydro-magnetic dynamo theory.

To further emphasise the relevance of researching the thermal flow characteristics of these viscous fluids, solid nanoparticle addition can be employed to enhance the natural convective flow of these fluids in heat transfer instruments. Nanofluids offer a wide range of practical uses because of their increased convective heat transfer rates and thermal conductivity, including enhanced nuclear power plants and space aircraft.

The amount of thermal radiation and the injection/consumption of heat have a significant impact on the heat transfer rate. Heating, ventilation, and air conditioning are just a few of the industries where these components are employed. In order to improve cooling and heating processes, heat sources and sinks are added to free and forced convection MHD flows.

Using a vertical plate to induce the flow of MHD spontaneous convective nanofluid, Chamkha and Aly discovered heat absorption and generation. Researchers Turkyilmazoglu and Pop hypothesised about how radiation influences nanofluid MHD natural convection flow when it crosses a vertically extended sheet. Heat radiation flux has an impact on MHD flow and heat transfer in nanofluids, according to a two-phase model developed by Sheikholeslami et.al. According to Li and colleagues, solitons based on the energy comparison of non-linear equations were found to be stable globally.

Nanofluid stagnation point flow is affected by heat injection/consumption, as Soomro et al. found out when they investigated the topic. A porous fat vertical wall was examined by Hamad and Pop in a rotating frame of reference with continuous heat production to determine the natural convective nanofluid mobility that occurs. Reddy investigated the effects of thermal radiation and heat generation on micropolar fluid flow over a stretched surface. In the presence of thermal radiation and heat generation/consumption, Khan et al. explored Casson-type nanofluid flow. It is possible to compare and contrast similar studies.

2 Literature Review

Convective flow and heat transfer engineering techniques are widely used in a wide variety of disciplines of study and practise. Numerous studies have been conducted on water, oil, and ethylene glycol with the goal of optimising their thermal properties as common heat transfer fluids. [1] Choi invented the term 'nanofluids' to describe the method of suspending nanoparticles in these fluids. Because nanofluids have a higher thermal conductivity than ordinary fluids, they are the subject of numerous studies.

Water with 5% CuO nanoparticles added to it enhanced its thermal conductivity by 60%, as reported by Eastman et al. [2]. In addition, Choi et al. [3] discovered that adding nanotubes at a concentration of less than one percent increased the thermal conductivity of oil-based nanofluids by twofold (less than 1 percent). An increase in thermal conductivity of nanofluids was seen when Akmal et al. [4] investigated the effect of temperature on thermal conductivity.

Further reading can be found in the review articles by Kumar et al. [5], Das and Choi [6], and Kumar [7]. Nanofluid flow's characteristics and applications have been uncovered over the course of several decades of study. A natural convection flow of nanofluid over a vertical plate is influenced by Brownian motion and thermophoresis according to Kumar et al. [8]. A decrease in the local Nusselt number was found to be caused by increasing the buoyancy ratio, Brownian motion, and thermophoresis factors.

The convective heat transport of water-based nanofluids was studied by Shah et al. [9] using the Keller box method and a stretched sheet. The Ag-water nanofluid was shown to have a higher heat transfer rate than the Cu-water nanofluid. Tiwari [10] studied the temperature-dependent heat source/sink impact in nanofluid mixed convection flow using the finite element approach. Ag nanoparticles were shown to be the most effective in cooling, whereas TiO2 nanoparticles were found to be the least effective.

Non-linear flow and heat transmission were studied by Mahdy and El-Shehaby [11] using nanofluids. Nanofluids were studied in an unstable mixed convection flow with indeterminate thermal conductivity and dynamic viscosity in another work [12]. Nanofluids, he found, boosted the temperature boundary layer while reducing the momentum boundary layer in convection.

The Crank-Nicolson implicit finite-difference method was used by Rajesh et al. [13] to investigate nanofluid flow on a semi-infinite vertical plate with changing surface temperature. Kumaresan et al. [14] investigated the flow of a nanofluid boundary layer across a moving semi-continuous porous plate using the ideal collocation technique (OCM). Cao et al. [15] studied the flow of an unstable Maxwell viscoelastic nanofluid across a moving plate using finite difference and the L1 technique.

Using a non-linear bidirectional surface, Mohanthesh and colleagues [16] investigated the three-dimensional MHD flow of a viscous nanofluid. Flow of nanofluids across a semi-infinite vertical flat plate was investigated by Hamad et al. [17]. Nanofluids naturally convect through permeable vertical plates in Chamkha and Khan et al.'s work [18].

Researchers who explored MHD free convective nanofluid flow across stretched surfaces found that the heat transfer rate in the Cu-water nanofluid decreased when volume percent of nanoparticles was increased [19]. Turkyilmazoglu and Pop [20] employed nanofluids with a radiation effect to investigate transient free convection flow over a vertical flat plate. An isothermal and isoflux boundary condition was taken into account when solving these equations using a similarity variable analysis.

Turkyilmazoglu [21] explored natural convection flow in nanofluids with isothermal and isoflux boundary conditions. He discovered that the physical properties of nanofluids were enhanced when exposed to radiation [22]. MHD CuO-water nanofluid in an enclosure was studied using the lattice Boltzmann method and the Koo-Kleinstreuer-Li correlation. Ahmed et al. [23] studied both MHD unsteady nanofluid flow and thermal radiation heat transfer.

Over a vertical plate, a nanofluid with a higher volume concentration of nanoparticles had a higher rate of heat transfer and a higher temperature [24]. A sheet of elastic/shrinking nanoparticles was used to model nanofluid flow in a porous environment. [25] Ehsan and Pedram are the two protagonists in this story. A magnetic field can drastically affect the flow of nanofluids. According to Nadjia and Nadji [26], homotopic analysis was used to investigate the melting heat transfer in stagnation point flow across a stretching sheet.

Using a vertical plate, Ali and his colleagues [27] studied the transient MHD motion of water-based nanofluids in a porous medium. They utilised a Prandtl number fixed at $Pr = 0.71$ (air) for water-based nanofluids, which is wrong. Seeing what they've gotten to, it's hard to believe. Kataria and Mittal [28] recently explored MHD gravity-driven convective flow of nanofluids across an oscillating vertical plate to see how mass transfer affects the flow. A mistake was made in the presentation of these results, however.

3 Mathematical Modelling

Thermal radiation flux, incompressible MHD flow, and energy transfer equations are all affected by fluid flow through an endlessly long vertical plate nestled in a porous medium that is susceptible to the Boussinesq approximation:

$$\nabla . V = 0 \tag{1}$$

$$\rho \left[\frac{\partial V}{\partial t} + (V.\nabla)\, V \right] = r + \mu \nabla^2 V + J \times B + \rho g \beta \,(\Theta - \Theta_\infty) \tag{2}$$

$$\rho C_p \left[\frac{\partial \Theta}{\partial t} + (V.\nabla)\, \Theta \right] = K_f \nabla^2 \Theta + \gamma_1 - \frac{\partial Q_r}{\partial n} - Q_o \,(\Theta - \Theta_\infty) \tag{3}$$

An endlessly long vertical plate imbedded in a porous medium can demonstrate stable natural convection by flowing and transferring energy. Initially, both the plate and fluid are static at the same temperature Θ_∞. At time $t > 0$, the plate starts an impulsive motion with velocity $U_o \frac{t}{t_o}$, and the temperature of vertical plate is raised to $\Theta_\infty + (\Theta_w - \Theta_\infty)\,\frac{t}{t_o}$ for $0 < t \leq t0$. Later on, a uniform velocity U0 and constant temperature Θ_w are maintained for $t > t_0$. Flow in this diagram is assumed to be one-dimensional and unidirectional; therefore, the x- and y-axes are perpendicular to the vertical plate. Only in y values greater than zero can the nanofluid flow, because the plate is assumed to be in position at $y = 0$. Also included in this idealisation of the concept are these assumptions:

- The nanofluid contains nanoparticles of copper, titanium dioxide, and aluminium oxide in addition to water as the fundamental fluid.
- Nanoparticles and the underlying fluid are both kept at a fixed temperature.
- With respect to the velocity equation, density has an effect on the temperature buoyancy force.
- Thermal radiative flux (Qr) is supposed to have a negligible impact in the direction parallel to the plate.
- Viscous dissipation is ignored in energy equations.
- When compared to the magnetic field, the impact of nanofluid flow is underestimated.
- An external electric field isn't used because of the nanofluid polarisation effect.
- Nanoparticles are thought to be homogeneous in size and shape.

The temperature and velocity of nanofluid are solely affected by changes in t and y since the vertical plate is believed to be infinite in length. Figure 1 depicts the physical model's construction, which provides a geometric explanation of the concept.

Here's how the Newtonian fluid meets Darcy's law:

$$r = -\frac{\mu \gamma_2}{k^*} u \tag{4}$$

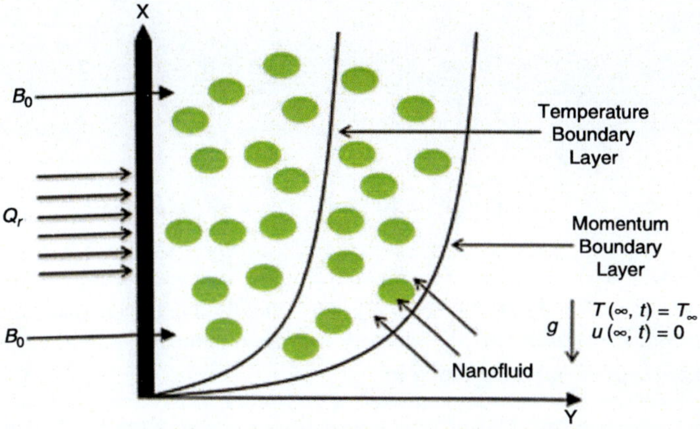

Fig. 1 Specified model's geometry

Maxwell's equations for dealing with the magnetic field can be found here:

$$\text{div}B = 0,\;\; \text{curl}B = \mu_m J,\;\; \text{Curl}E = -\frac{\partial B}{\partial t} \tag{5}$$

along with using Ohm's law, which in turn leads to

$$J \times B = -\left(\sigma B_o^2 u,\; 0, 0\right) \tag{6}$$

For a nanofluid, Eqs. (2) and (3) can be expressed as follows in light of the assumptions in Eqs. (4)–(6):

$$\rho nf\frac{\partial u}{\partial t} = \mu_{nf}\frac{\partial^2 u}{\partial y^2} - \frac{\mu_{nf}\gamma_2}{k^*}u + g(\rho\beta)_{nf}\;(\Theta-\Theta_\infty) - \sigma_{nf}B_o^2 u \tag{7}$$

$$\left(\rho C_p\right)_{nf}\frac{\partial\Theta}{\partial t} = K_{nf}\frac{\partial^2\Theta}{\partial y^2} - -\frac{\partial Q_r}{\partial y} - Q_0\,(\Theta-\Theta_\infty)\;\;(8) \tag{8}$$

The momentum and energy equations have their initial and boundary conditions as follows:

$$u\,(y, 0) = 0,\;\; \Theta\,(y, 0) = \Theta_\infty \text{ for } \; y \geq 0 \tag{9}$$

$$u\,(0, t) = \begin{matrix} U_o\frac{t}{t_o} & 0 < t \leq t_o \\ U_0 & t > t_o \end{matrix} \tag{10}$$

$$\Theta\,(0, t) = \begin{matrix} \Theta_\infty + (\Theta_w - \Theta_\infty)\frac{t}{t_0} & 0 < t \leq t_o \\ \Theta_w & t > t_o \end{matrix} \tag{11}$$

$$u\,(y, t) \to 0,\;\; \Theta\,(y, t) \to \Theta_\infty \tag{12}$$

When $y \to \infty \; for \; t > 0$

The expressions for dynamic viscosity μnf, heat capacitance $(\rho Cp)nf$, coefficient of thermal expansion $(\rho\beta)nf$, density ρnf, and electrical conductivity σnf of nanofluid are respectively calculated as

$$\mu_{nf} = \frac{\mu_w}{(1-\emptyset)^{2.5}},\ (\rho C_p)_{nf} = (\rho C_p)_w \left[1 - \emptyset + \emptyset \frac{(\rho C_p)_{np}}{(\rho C_p)_w}\right],$$

$$(\rho\beta)_{nf} = (\rho\beta)_w \left[1 - \emptyset + \emptyset \frac{(\rho\beta)_{np}}{(\rho\beta)_w}\right],\ \rho_{nf} = \rho_w \left[1 - \emptyset + \emptyset \frac{\rho_{np}}{\rho_w}\right] \tag{13}$$

$$\sigma_{nf} = \sigma_w \left[1 + \frac{3\emptyset(\sigma - 1)}{(\sigma + 2) - \emptyset(\sigma - 1)}\right],\ \sigma = \frac{\sigma_{np}}{\sigma_w}$$

Nanoparticles' thermal conductivity may be accurately predicted using the Hamilton and Crosser model:

$$\frac{K_{nf}}{K_w} = \frac{K_{np} + 2K_w - 2\left(K_w - K_{np}\right)\emptyset}{K_{np} + 2K_w + \left(K_w - K_{np}\right)\emptyset} \tag{14}$$

All of the subscripts w, np, and nf in Eqs. 7 and 8, as well as in 13 and 14, all relate to water's properties as the basis for the equations. It is also important to note that the relationships (13) only apply to nanoparticles with a spherical form. The Rosseland approximation and a thick, optically thick fluid lead to the following radiation heat flux formula:

$$Q_r = -\frac{4\sigma_r}{3k_r}\frac{\partial \Theta^4}{\partial y} \tag{15}$$

The non-linearity of heat radiation flux with respect to temperature can be seen from the expression above. Linearisation can be performed in nanofluid flow because the temperature differences are so small.

4 Results and Analysis

With a volume fraction (Nr) equal to 5, we can see in Fig. 2 how temperature changes in different nanofluids with varying nanoparticle volume fractions. A free stream temperature of 0 is reached after a slow drop in temperature along the ray. The fact that the nanofluid's temperature is higher than that of the normal fluid suggests that it has a better capacity for heat transfer than the regular fluid (water).

Heat absorption for Cu-water is shown in Fig. 3 as a transient temperature profile that changes over time. The temperature rises over time until it reaches a constant state. Increases in plate-to-fluid heat conduction cause a rise in fluid temperature. Gradually, the temperature drops until η meets the free stream requirement.

It is demonstrated in Fig. 4 that heat generation causes the nanofluid to rise in temperature over time. It demonstrates that as the elapsed time grows, so does the temperature profile. Increased time values result in temperatures that peak near the plate and progressively decline as time increases.

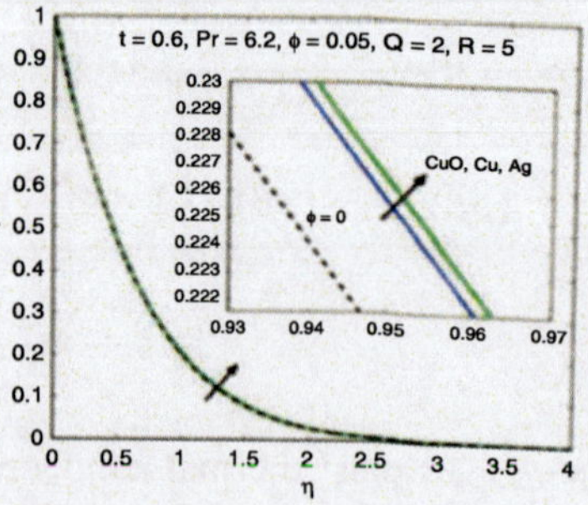

Fig. 2 Nanofluid temperature distributions

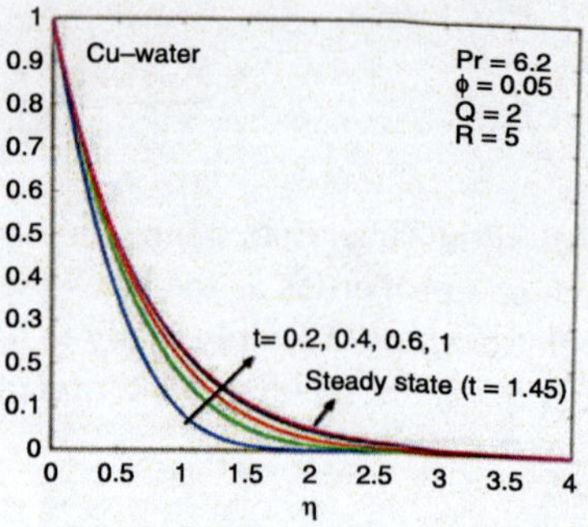

Fig. 3 Heat absorption temperature profiles at various times

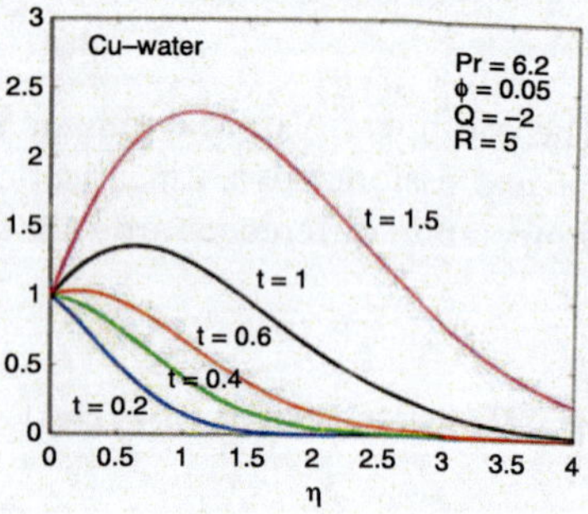

Fig. 4 In the case of heat generation, temperature distributions at separate moments in time

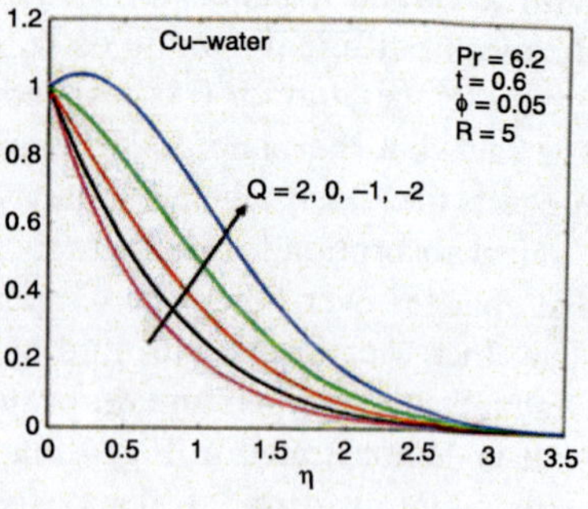

Fig. 5 Temperature graphs at various points in time Q

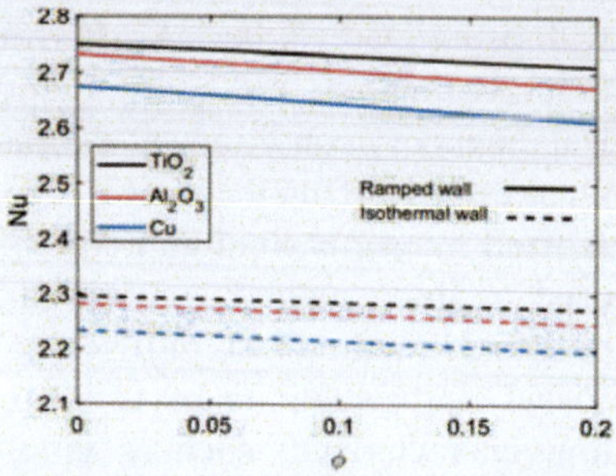

Fig. 6 Temperature graphs at various points in time Q

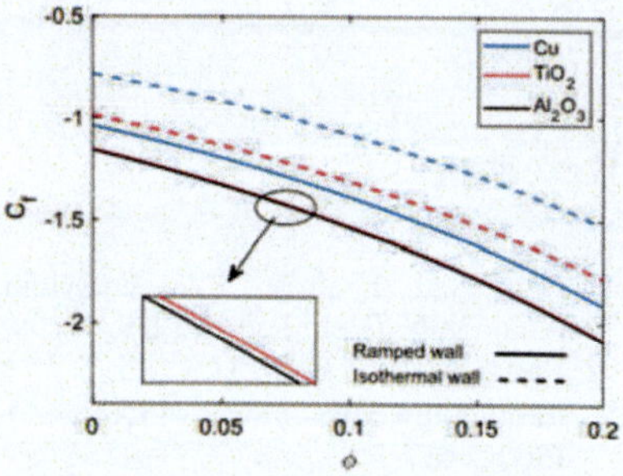

Fig. 7 When M = 2.0, K = 0.5, Gr = 5.0, Nr = 0.5, and Q = 0.5, the skin friction of different nanofluids can be compared

It is shown in Fig. 5 how Cu-water nanofluid temperature distributions change as heat generation/absorption parameters change. However, increasing the amount of heat that is absorbed causes a drop in the temperature profile. When heat absorption is taken into consideration, a temperature peak forms near the plate before cooling to a free stream temperature; however, as the distance between the plate and the sample increases, the temperature decreases constantly.

Figure 6 shows the heat transfer rates of various nanofluids at a wall $\xi = 0$. Cu-water nanofluid has the lowest heat transmission rate of the three nanofluids tested. By virtue of Cu nanoparticles having a higher thermal conductivity, this behaviour may be explained in terms of how hotter Cu-water nanofluid is and how much heat is absorbed by the plates.

For a range of nanofluids, as illustrated in Fig. 7, the skin friction in both ramping wall and isothermal wall situations is quite high, as indicated. The increased density of Cu nanoparticles increases the $\xi = 0$ shear stress in Cu-wall water. As a result, the shear stresses of Al2O3 and TiO2 are nearly identical since their densities are so similar. Isothermal plates have greater shear stress curves than ramping plates.

5 Conclusion

The primary purpose of this research is to investigate the physical consequences of simultaneously ramping the wall velocity and the wall temperature on unstable, MHD convection flow across an infinite vertical plate. Medium porosity is dealt with using Darcy's law. Aside from that, the model takes into account heat injection/consumption and heat radiation. Physically applying ramping and isothermal wall boundary conditions at the same time is important; nevertheless, the analytical management of mathematical expressions that follow from doing so is onerous. A fixed magnetic field and an exponentially accelerating vertical plate have been used to study natural convection heat transfer and nanofluid flow. The Laplace transform method was utilised to solve the dimensionless governing equations. Visual representations are used to explain the relationship between heat and fluid flow characteristics such as time, volume, heat generation/absorption, radiation, magnetic field, and acceleration.

References

1. Choi, S.U., Eastman, J.A.: Enhancing Thermal Conductivity of Fluids with Nanoparticles. Argonne National Lab, IL, USA (1995)
2. Eastman, J.A., Choi, U.S., Li, S., Thompson, L.J., Lee, S.: Enhanced thermal conductivity through the development of nanofluids. MRS Proc. **457**, 3–11 (1996). https://doi.org/10.1557/PROC-457-3
3. Choi, S.U.S., Zhang, Z.G., Yu, W., Lockwood, F.E., Grulke, E.A.: Anomalous thermal conductivity enhancement in nanotube suspensions. Appl. Phys. Lett. **79**, 2252–2254 (2001). https://doi.org/10.1063/1.1408272
4. Akmal, N., Sagheer, M., Hussain, S., Kamran, A.: Investigation of free convection in micropolar nanofluid with induced magnetic field. Eur. Phys. J. Plus **134**, 235 (2019). https://doi.org/10.1140/epjp/i2019-12512-7
5. Kumar, T.S., Kumar, B.R., Makinde, O.D., Vijaya Kumar, A.G.: Magneto-convective heat transfer in micropolar nanofluid over a stretching sheet with non-uniform heat source/sink. Defect Diffus. Forum **387**, 78–90 (2018). https://doi.org/10.4028/www.scientific.net/ddf.387.78
6. Das, S.K., Choi, S.U.S.: A review of heat transfer in nanofluids. In: Irvine, T.F., Hartnett, J.P. (eds.) Advances in Heat Transfer, vol. 41, pp. 81–197. Elsevier (2009). https://doi.org/10.1016/s0065-2717(08)41002-x
7. Sravan kumar, T.: Impact of Lorentz force on free convection flow of a viscous fluid past an infinite vertical plate. SN Appl. Sci. **1**, 1257 (2019). https://doi.org/10.1007/s42452-019-1292-8
8. Sravan Kumar, T., Dinesh, P.A., Makinde, O.D.: Impact of Lorentz force and viscous dissipation on unsteady nanofluid convection flow over an exponentially moving vertical plate. Math. Models Comput. Simul. **12**, 631–646 (2020). https://doi.org/10.1134/s2070048220040110
9. Shah, Z., Alzahrani, E.O., Dawar, A., Ullah, A., Khan, I.: Influence of Cattaneo-Christov model on DarcyForchheimer flow of Micropolar Ferrofluid over a stretching/shrinking sheet. Int. Commun. Heat Mass Transf. **110**, 104385 (2020). https://doi.org/10.1016/j.icheatmasstransfer.2019.104385
10. Tiwari, A., Shah, P.D., Chauhan, S.S.: Analytical study of micropolar fluid flow through porous layered microvessels with heat transfer approach. Eur. Phys. J. Plus **135**, 209 (2020). https://doi.org/10.1140/epjp/s13360-020-00128-x
11. Mahdy, A., ElShehabey, H.M.: Uncertainties in physical property effects on viscous flow and heat transfer over a nonlinearly stretching sheet with nanofluids. Int. Commun. Heat Mass Transf. **39**, 713–719 (2012). https://doi.org/10.1016/j.icheatmasstransfer.2012.03.019

12. Mahdy, A.: Unsteady mixed convection boundary layer flow and heat transfer of nanofluids due to stretching sheet. Nucl. Eng. Des. **249**, 248–255 (2012). https://doi.org/10.1016/j.nucengdes.2012.03.025
13. Vemula, R., Chamkha, A.J., Mallesh, M.P.: Nanofluid flow past an impulsively started vertical plate with variable surface temperature. Int. J. Numer. Methods Heat Fluid Flow **26**, 328–347 (2016). https://doi.org/10.1108/hff-07-2014-0209
14. Kumaresan, E., Sravan Kumar, T., Suresh Babu, R.: MHD slip flow and heat transfer of Cu-Fe3O4/ethylene glycol-based hybrid nanofluid over a stretching surface. Biointerf. Res. Appl. Chem. **11**, 11956–11968 (2020). https://doi.org/10.33263/briac114.1195611968
15. Cao, Z., Zhao, J., Wang, Z., Liu, F., Zheng, L.: MHD flow and heat transfer of fractional Maxwell viscoelastic nanofluid over a moving plate. J. Mol. Liq. **222**, 1121–1127 (2016). https://doi.org/10.1016/j.molliq.2016.08.012
16. Mahanthesh, B., Gireesha, B.J., Gorla, R.S.R., Abbasi, F.M., Shehzad, S.A.: Numerical solutions for magnetohydrodynamic flow of nanofluid over a bidirectional non-linear stretching surface with prescribed surface heat flux boundary. J. Magn. Magn. Mater. **417**, 189–196 (2016). https://doi.org/10.1016/j.jmmm.2016.05.051
17. Hamad, M.A.A., Pop, I., Md Ismail, A.I.: Magnetic field effects on free convection flow of a nanofluid past a vertical semi-infinite flat plate. Non-linear Anal.: Real World Appl. **12**, 1338–1346 (2011). https://doi.org/10.1016/j.nonrwa.2010.09.014
18. Khan, W.A., Rashad, A.M., El-Kabeir, S.M.M., El-Hakiem, A.M.A.: Framing the MHD micropolarnanofluid flow in natural convection heat transfer over a radiative truncated cone. Processes **8** (2020). https://doi.org/10.3390/pr8040379
19. Ibrahim, W., Gadisa, G.: Non-linear convective boundary layer flow of micropolar-couple stress nanofluids past permeable stretching sheet using Cattaneo-Christov heat and mass flux model. Heat Transf. **49**, 2521–2550 (2020). https://doi.org/10.1002/htj.21733
20. Turkyilmazoglu, M., Pop, I.: Heat and mass transfer of unsteady natural convection flow of some nanofluids past a vertical infinite flat plate with radiation effect. Int. J. Heat Mass Transf. **59**, 167–171 (2013). https://doi.org/10.1016/j.ijheatmasstransfer.2012.12.009
21. Turkyilmazoglu, M.: Unsteady convection flow of some nanofluids past a moving vertical flat plate with heat transfer. J. Heat Transf. **136** (2013). https://doi.org/10.1115/1.4025730
22. Sheikholeslami, M., Bandpy, M.G., Ellahi, R., Zeeshan, A.: Simulation of MHD CuO–water nanofluid flow and convective heat transfer considering Lorentz forces. J. Magn. Magn. Mater. **369**, 69–80 (2014). https://doi.org/10.1016/j.jmmm.2014.06.017
23. Ahmed, M.B., Hanan, E., Osama, O., Kholmirzo, T.K., Tarek, H., Medhat, A.I.: Effect of nano metal oxides on heme molecule: molecular and biomolecular approaches. Biointerf. Res. Appl. Chem. **10**, 4837–4845 (2019). https://doi.org/10.33263/briac101.837845
24. Khan, M.N., Nadeem, S., Muhammad, N.: Micropolar fluid flow with temperature-dependent transport properties. Heat Transf. **49**, 2375–2389 (2020). https://doi.org/10.1002/htj.21726
25. Pedram, E., Ehsan, K.: Experimental investigation of rheological properties and formation damage of waterbased drilling fluids in the presence of Al2O3, Fe3O4, and TiO2 nanoparticles. Biointerf. Res. Appl. Chem. **10**, 5886–5894 (2020). https://doi.org/10.33263/briac104.886894
26. Sabri, N., Moulai-Mostefa, N.: Formulation and characterization of oil-in-water emulsions stabilized by Saponins extracted from Hedera Helix Algeriensis using response surface method. Biointerf. Res. Appl. Chem. **10**, 6282–6292 (2020). https://doi.org/10.33263/briac105.62826292
27. Ali, F., Gohar, M., Khan, I.: MHD flow of water-based Brinkman type nanofluid over a vertical plate embedded in a porous medium with variable surface velocity, temperature and concentration. J. Mol. Liq. **223**, 412–419 (2016). https://doi.org/10.1016/j.molliq.2016.08.068
28. Kataria, H.R., Mittal, A.S.: Velocity, mass and temperature analysis of gravity-driven convection nanofluid flow past an oscillating vertical plate in the presence of magnetic field in a porous medium. Appl. Therm. Eng. **110**, 864–874 (2017). https://doi.org/10.1016/j.applthermaleng.2016.08.129

Micropolar Fluid Enfolded with Viscous Nanofluid: Three-Layer Flow

J. C. Umavathi and P. Sutkar

1 Introduction

Transport models for nanofluids [1] are studied by many researchers. A single-phase or two-phase approach is used to model the heat transfer in nanofluids. Microcontinuum fluid mechanics was presented by Eringen [2] as a strong mathematical foundation for simulating such events. The micropolar model is the simple example of a microfluid model, which was customized by Eringen [2] to introduce angular momentum issues that have been shown accurately to simulate [3] correct medical, industrial, and environmental flows, which consist of foodstuffs, polymer melts, liquid crystals, air-borne pollutants, hemodynamics, and sediment transport in river beds. Micropolar fluid dynamics has been a research topic for nearly five decades, and novel applications are still being explored. Prasad et al. [4] used a Keller box numerical approach to calculate the impact of Brownian motion, Grashof number, micro-inertial density, thermophoresis, and angular velocity distributions in a steady axisymmetric double-diffusive flow of a micropolar nanofluid from a cylinder. Umavathi and Sheremet [5] investigated the onset of convection in a porous medium layer saturated by a nanofluid of a sparsely packed micropolar fluid. Umavathi et al. [6, 7] studied multi-fluid flows using Newtonian/non-Newtonian fluids past channels/ducts.

J. C. Umavathi (✉)
Department of Mathematics, Gulbarga University, Kalaburgi, Karnataka, India

P. Sutkar
Department of Mathematics, NES Science College, Nanded, Maharashtra, India

F. Kamalov et al. (eds.), *Advances in Mathematical Modeling and Scientific Computing*, Trends in Mathematics, https://doi.org/10.1007/978-3-031-41420-6_37

2 Mathematical Formulation

We consider a steady, fully developed, micropolar fluid flowing inside a rigid horizontal channel as given by Ariman et al. [8] and laboratory conclusions of Bugliarello and Sevilla [9]. This flow consists of a two-fluid model with micropolar fluid filled in the region $0 \leq y \leq h$, and the peripheral layers $h \leq y \leq 2h$ and $-h \leq y \leq 0$ are viscous Newtonian nanofluid of width h as displayed in Fig. 1. The appropriate equations for one-dimensional, steady, linear velocity, temperature, and microrotation velocity in the dimensionless form are expressed as follows (following Umavathi et al. [10]).

Region-I

$$\frac{d^2u_1}{dy^2} = mp \tag{1}$$

$$\frac{d^2\theta_1}{dy^2} + \frac{Ec\,\mathrm{Pr}\,Cr}{m}\left(\frac{du_1}{dy}\right)^2 = 0 \tag{2}$$

Region-II

$$(1+K)\frac{d^2u_2}{dy^2} + K\frac{dN}{dy} - p = 0 \tag{3}$$

$$\left(1+\frac{K}{2}\right)\frac{d^2N}{dy^2} - 2KN - K\frac{du_2}{dy} = 0 \tag{4}$$

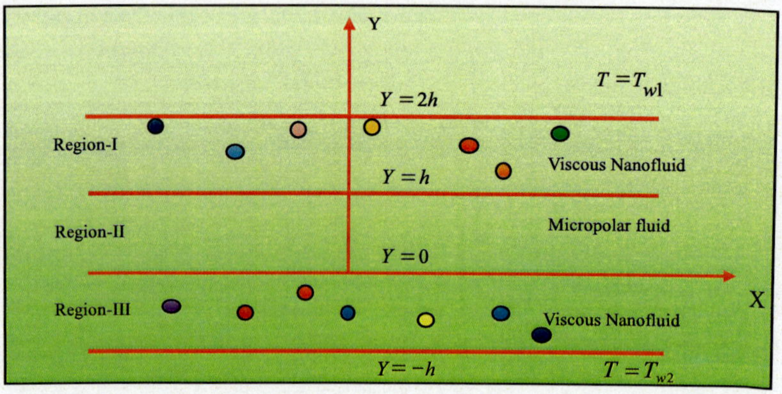

Fig. 1 Physical configuration

$$\frac{d^2\theta_2}{dy^2} + Ec\,\mathrm{Pr}\left(\frac{du_2}{dy}\right)^2 = 0 \tag{5}$$

Region-III

$$\frac{d^2u_3}{dy^2} = mp \tag{6}$$

$$\frac{d^2\theta_3}{dy^2} + \frac{Ec\,\mathrm{Pr}\,Cr}{m}\left(\frac{du_3}{dy}\right)^2 = 0 \tag{7}$$

where

$$m = \frac{\mu_f}{\mu_{nf}} = \frac{1}{1 + 39.11\,\phi + 533.9\,\phi^2},\quad Cr = \frac{k_f}{k_{nf}} = \frac{1}{1 + 7.47\,\phi},$$

$$Ec = \frac{u_1^2}{C_{p_{nf}}\left(T_{w1} - T_{w2}\right)},\quad \mathrm{Pr} = \frac{C_p\mu_f}{k_f},$$

$$Ec = \frac{u_1^2}{C_{p_{nf}}\left(T_{w1} - T_{w2}\right)},\quad p = \frac{h^2}{\mu_f\overline{u_1}}\frac{\partial p}{\partial x},\quad \mu_{nf} = \frac{\mu_f}{(1-\phi)^{2.5}},$$

$$k_{nf} = k_f\left(\frac{k_s + 2k_f + 2\phi\left(k_s - k_f\right)}{2k_f + k_s + \phi\left(k_f - k_s\right)}\right)$$

Here Pr is the Prandtl number, Ec stands for Eckert number, K is micropolar parameter, Cr stands for the thermal conductivity ratio, p represents the non-dimensional pressure, N is microrotation component, and m is viscosity ratio. The values for the viscosity ratio and conductivity ratio have been determined by Bourantas and Loukopoulos [11]. Temperature and linear velocity's boundary and interface conditions are

$$u_1(2) = 0,\;\; u_1(1) = u_2(1),\;\; u_2(0) = u_3(0),\;\; u_3(-1) = 0,$$

$$\frac{du_1}{dy} = m\left(1+\kappa\right)\frac{du_2}{dy} + m\kappa N \;\text{ at }\; y = 1$$

$$m\left(1+\kappa\right)\frac{du_2}{dy} + m\kappa N = \frac{du_3}{dy} \;\text{ at }\; y = 0$$

$$\frac{dN}{dy} = 0 \text{ at } y = 1$$

$$\frac{dN}{dy} = 0 \text{ at } y = 0 \tag{8}$$

$$\theta_1(2) = 1, \ \theta_1(1) = \theta_2(1), \ \theta_2(0) = \theta_3(0), \ \theta_3(-1) = 0,$$

$$\frac{d\theta_1}{dy} = Cr\frac{d\theta_2}{dy} \text{ at } y = 1$$

$$\frac{d\theta_2}{dy} = \left(\frac{1}{Cr}\right)\frac{d\theta_3}{dy} \text{ at } y = 0 \tag{9}$$

3 Closed from Solutions

The solution of Eqs. (1)–(7) for the non-dimensional linear velocities, microrotation velocity, and temperature for all three regions can be derived with the use of boundary and interface conditions described by Eqs. (8) and (9) for all three regions and are not presented.

4 Determination of Skin Friction, Mass Flow Rate, and Nusselt Number

(i) Skin Friction

The non-dimensional skin friction at the upper plate τ_T and lower plates τ_B is as follows:

$$\tau = \left(\frac{du}{dy}\right)_{y=-1,2}$$

$$\tau_T = 2mp + d_1$$

$$\tau_B = -mp + d_7$$

(ii) Mass Flow Rate

Evaluation and determination of the mass flow rate through the channel is

$$\int_{-1}^{2} u\,dy = \frac{l_1}{a}\sinh a + \frac{l_2}{a}\cosh a + \frac{l_3}{3} + \frac{d_5}{2} + d_6 - \frac{l_2}{a} + \frac{4p}{3} + \frac{3d_1}{2} + d_2 - \frac{d_7}{2} + d_6$$

(iii) Nusselt Number

The Nusselt number in non-dimensional form is defined as

$$Nu\Big]_{y=-1} = -\frac{k_{nf}}{k_f}\left(\frac{d\theta_3}{dy}\right)_{y=-1} = \left(\frac{1}{Cr}\right)(4l_{16} - 3l_{17} + 2l_{18} - e_5)$$

$$Nu\Big]_{y=2} = -\frac{k_{nf}}{k_f}\left(\frac{d\theta_1}{dy}\right)_{y=2} = \left(\frac{-1}{Cr}\right)(32l_4 + 12l_5 + 4l_6 + e_1)$$

5 Results and Discussion

The non-dimensional governing Eqs. (1) through (7) are coupled ordinary differential equations, and hence an exact solution can be found. The calculations are performed for alumina nanoparticles and clean water as the base fluid. Figure 2a informs that as K is escalated the velocity is reduced in the core region and is unaltered in the upper and lower regions. With increasing material parameter values, the fluid experiences greater acceleration, which suggests that there is stronger micropolarity and greater vortex viscosity. This results in a reduction in momentum diffusion in the field of linear velocity, which slows the linear velocity. The drag-reducing characteristics of micropolar fluids are therefore confirmed, and these have been identified also in Eringen [12], Umavathi et al. [13, 14]. Figure 2b displays that as K increases, the microrotation velocity also increases. This is owing to the fact that hiking K improves the spin gradient viscosity and, hence, causes rotational velocity to improve. The temperature decreases as the values of the micropolar material parameters increase, as seen in Fig. 2c. The highest temperature is produced for the Newtonian case (K = 0).

There is a definite decrease in linear velocity (Fig. 3a) with increasing nanoparticle doping levels. The largest magnitude of velocity is produced when nanoparticles are not present. Figure 3b shows that the temperature drops in all three areas, that is, as the solid volume fraction increases, velocity diminishes, and hence temperature also diminishes. However, because of interfacial dragging effects, the influence can also be seen in the upper and lower regions. As Ec is enhanced, the temperature in the channel is also elevated as seen in Fig. 4. Minimal temperature is attained in the absence of viscous dissipation.

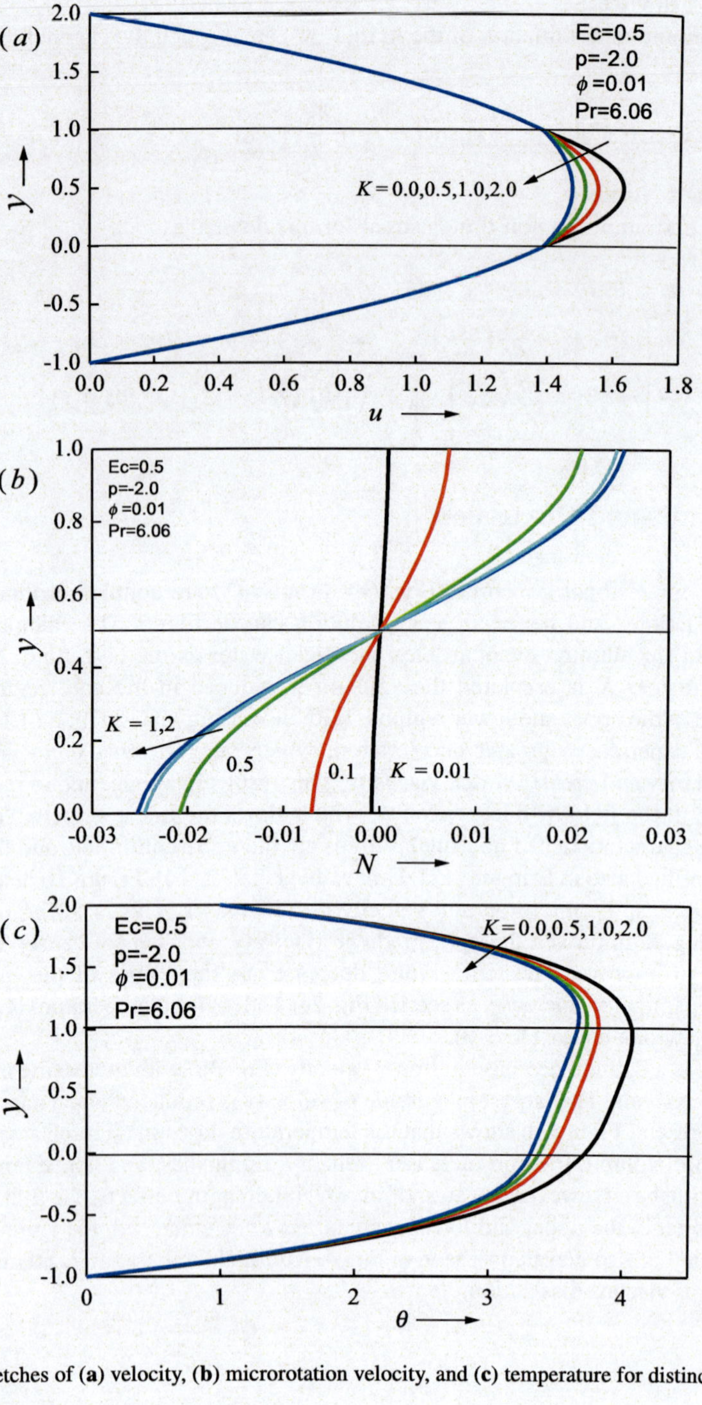

Fig. 2 Sketches of (**a**) velocity, (**b**) microrotation velocity, and (**c**) temperature for distinct K

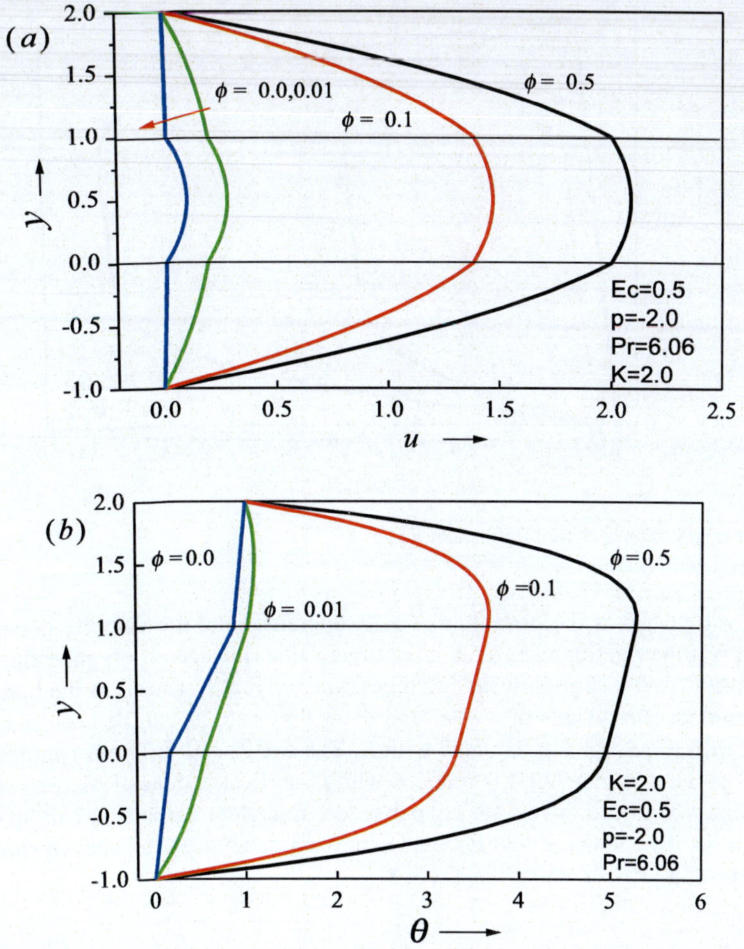

Fig. 3 Sketches of (**a**) linear velocity and (**b**) temperature for distinct ϕ

Adopting the definition of viscosity as defined in Dintenfass [15], the profiles of velocities are shown in Fig. 5a, b for variations of K for concentration of red blood corpuscles to be 20% and 40%. We define μ_R as the cell rotation velocity and the shear viscosity as $\mu = \mu_R/\kappa$ [1]. It is seen from Fig. 5a that the velocity is boosted when $K = 1$, but decreases for values of $K = 2$, which were the similar results obtained by Umavathi et al. [10].

Tables 1 and 2 depict the values of linear velocity and temperature for the effects of using different nanofluids. There are no significant changes on the distributions of either the velocity or the temperature using different nanoparticles. Considering water as base fluid, Tables 3 and 4 illustrate the values of velocity and temperature for different values of the solid volume fraction, whereas Tables 5 and 6 provide the

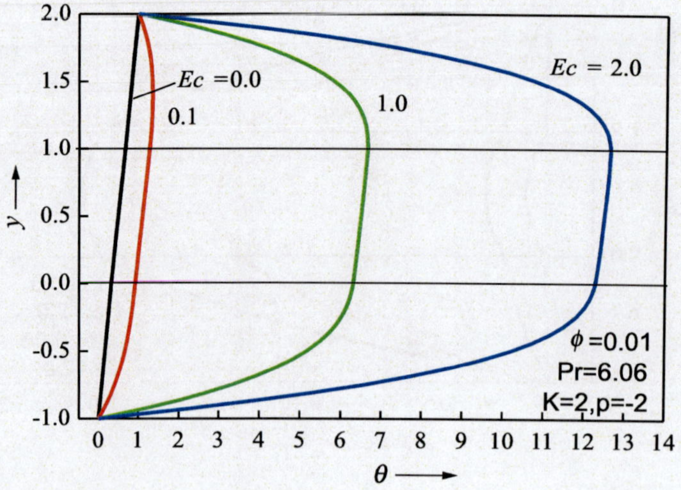

Fig. 4 Sketches of temperature for distinct *Ec*

values for glycerin as the base fluid. The temperature and the velocity decrease as the solid volume fraction rises, and the optimal values are seen for regular fluid ($\phi = 0.0$). Tables 5 and 6 show similar outcomes when glycerin is used as the base fluid. Table 7 reflects the values of τ, Nu, and mass flow rate. It can be seen that Nu is boosted with Ec, Pr , ϕ but depleted with K. The Ec, Pr and different nanofluids do not change the flow rate but it is reduced with K, ϕ. Peak values of Nu are achieved for graphite oxide and lowest for copper oxide. Excellent agreement can be shown when comparing the mass flow rate and skin friction between the current study and with Umavathi [10] as marked in Table 8.

6 Final Remarks

The conclusions drawn are as follows:

1. The material parameter and the volume fraction diminish the magnitude of the linear velocity and temperature. The solid volume fraction does not alter the microrotation velocity, but it was improved with material parameter.
2. The temperature was enhanced with Eckert number.
3. The linear velocity and the microrotational velocity were both decreased by the cell rotational velocity.
4. The linear velocity and temperature are unaffected by the presence of nanoparticles.

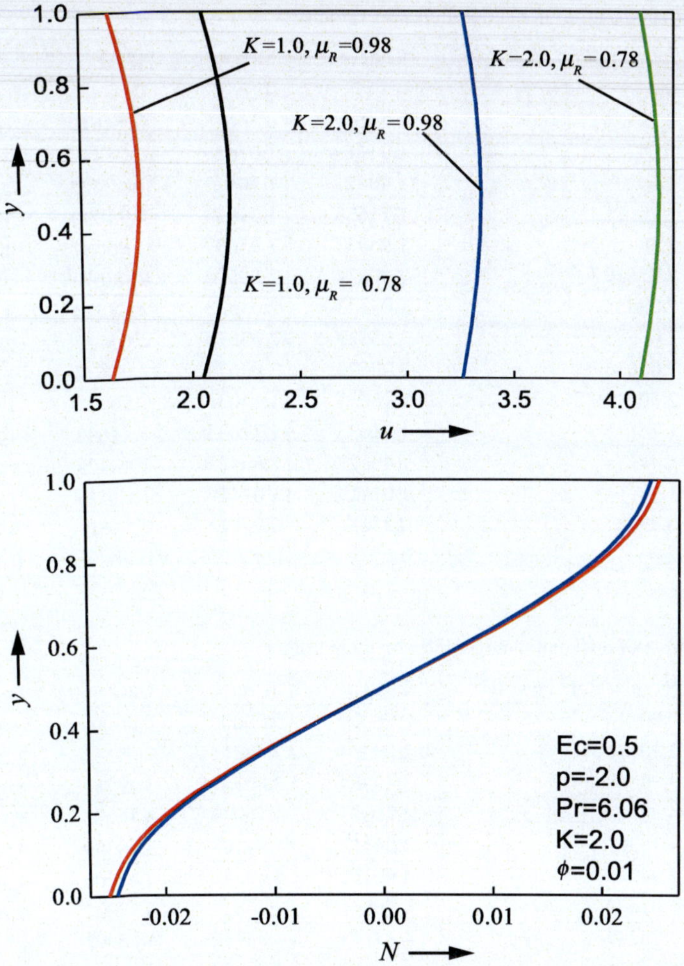

Fig. 5 Sketches of (**a**) velocity and (**b**) microrotation velocity for cell rotational viscosity μ_R

5. The mass flow rate was decreased by incrementing the solid volume fraction and material parameter.
6. The material parameter reduces the values of rate of heat transfer.

Table 1 Effect of different nanoparticles on velocity

Y	Regular fluid ($\phi = 0.0$)	Copper	Silver	CuO	Go
−1	0.0	0	0	0	0
−0.8	0.29	0.5461	0.5461	0.5461	0.5461
−0.6	1.04	1.01419	1.01419	1.01419	1.01419
−0.4	1.44	1.40427	1.40427	1.40427	1.40427
−0.2	1.76	1.71633	1.71633	1.71633	1.71633
0	2.00	1.95037	1.95037	1.95037	1.95037
0.2	2.05636	2.00674	2.00674	2.00674	2.00674
0.4	2.08485	2.03523	2.03523	2.03523	2.03523
0.6	2.08485	2.03523	2.03523	2.03523	2.03523
0.8	2.05636	2.00674	2.00674	2.00674	2.00674
1	2.00	1.95037	1.95037	1.95037	1.95037
1.2	1.76	1.71633	1.71633	1.71633	1.71633
1.4	1.44	1.40427	1.40427	1.40427	1.40427
1.6	1.04	1.01419	1.01419	1.01419	1.01419
1.8	0.29	0.5461	0.5461	0.5461	0.5461
2	0.0	0	0	0	0

Table 2 Effect of different nanoparticles on temperature

Y	Regular fluid ($\phi = 0.0$)	Copper	Silver	CuO	Go
−1	0.00	0.00000	0.00000	0.00000	0.00000
−0.8	2.20671	2.09214	2.09213	2.09331	2.09189
−0.6	3.59088	3.40564	3.40561	3.40752	3.40523
−0.4	4.3852	4.16077	4.16074	4.16306	4.16027
−0.2	4.78361	4.54112	4.54108	4.54359	4.54057
0	4.94123	4.69352	4.69348	4.69606	4.69296
0.2	5.01479	4.76841	4.76837	4.77092	4.76786
0.4	5.08249	4.83743	4.83739	4.83992	4.83688
0.6	5.14915	4.90543	4.90539	4.90789	4.90488
0.8	5.21479	4.97239	4.97235	4.97483	4.97186
1	5.27457	5.03349	5.03346	5.03590	5.03296
1.2	5.25028	5.01309	5.01305	5.01547	5.01257
1.4	4.9852	4.76475	4.76472	4.76696	4.76427
1.6	4.32421	4.14163	4.14160	4.14346	4.14122
1.8	3.07338	2.96014	2.96012	2.96127	2.95989
2	1.000	1.00000	1.00000	1.00000	1.00000

Table 3 Velocity using water as base fluid

Y	$\phi = 0.0$	$\phi = 0.01$	$\phi = 0.1$	$\phi = 0.5$
0	2.00	1.95037	1.53687	0.35355
0.2	2.08234	2.03272	1.61921	0.4359
0.4	2.12375	2.07413	1.66062	0.47731
0.6	2.12375	2.07413	1.66062	0.47731
0.8	2.08234	2.03272	1.61921	0.4359
1	2.00	1.95037	1.53687	0.35355

Table 4 Values of temperature using water as base fluid

Y	$\phi = 0.0$	$\phi = 0.01$	$\phi = 0.1$	$\phi = 0.5$
0	0.34269	0.33888	0.30560	0.16790
0.2	0.40939	0.40690	0.38557	0.30096
0.4	0.47606	0.47490	0.46551	0.43399
0.6	0.54272	0.54290	0.54545	0.56702
0.8	0.60939	0.61089	0.62538	0.70004
1	0.67602	0.67885	0.70529	0.83304

Table 5 Values of velocity using glycerin as base fluid

Y	$\phi = 0.0$	$\phi = 0.01$	$\phi = 0.1$	$\phi = 0.5$
0	2	1.95037	1.53687	0.35355
0.2	2.08234	2.03272	1.61921	0.4359
0.4	2.12375	2.07413	1.66062	0.47731
0.6	2.12375	2.07413	1.66062	0.47731
0.8	2.08234	2.03272	1.61921	0.4359
1	2	1.95037	1.53687	0.35355

Table 6 Values of temperature using glycerin as base fluid

Y	$\phi = 0.0$	$\phi = 0.01$	$\phi = 0.1$	$\phi = 0.5$
0	10.80369	10.24883	6.3943	0.69548
0.2	10.90311	10.34958	6.50699	0.86126
0.4	10.97463	10.42243	6.59178	0.99914
0.6	11.0413	10.49042	6.67172	1.13217
0.8	11.10311	10.55356	6.74681	1.26035
1	11.13703	10.5888	6.794	1.36062

Table 7 The rate of skin friction at the top and bottom plates, and the mass flow rate for different values of K, Ec, and Pr

		τ_T	τ_B	MFR	$Nu]_{y=-1}$	$Nu]_{y=2}$
ϕ	0.0	−3.0000	3.000	4.3922	2.97191	2.30525
	0.01	−2.0769	2.0769	3.0587	2.18005	1.48099
	0.05	−0.6993	0.6993	1.0689	1.03181	0.21752
	0.08	−0.3976	0.3975	0.6331	0.80466	0.08348
κ	0.1	−2.0769	2.0768	3.1516	2.25109	1.55203
	0.5	−2.0769	2.0769	3.1124	2.21349	1.51443
	1.0	−2.0769	2.0769	3.0858	2.19427	1.49521
	2.0	−2.0769	2.0768	3.0587	2.18005	1.48099
Ec	0.01	−2.0769	2.0769	3.0587	0.53258	0.1665
	0.1	−2.0769	2.0769	3.0587	2.18005	1.4810
	0.5	−2.0769	2.0769	3.0587	9.50214	8.8031
	1.0	−2.0769	2.0769	3.0587	18.65476	17.9557
Pr	0.01	−2.0769	2.0769	3.05872	0.35255	0.34651
	0.5	−2.0769	2.0769	3.05872	0.50056	0.19850
	1.0	−2.0769	2.0769	3.05872	0.65160	0.04746
	4.0	−2.0769	2.0769	3.05872	1.55780	0.85874
	Copper	−2.9256	2.9256	4.2846	2.91339	2.23345
	Silver	−2.9256	2.9256	4.2846	2.91339	2.23345
	CuO	−2.9256	2.9256	4.2846	2.91326	2.23358
	Go	−2.9256	2.9256	4.2846	2.91342	2.23342

Table 8 Values of mass flow rate in comparison with Umavathi et al. [10]

		Mass flow rate		Skin friction			
				τ_T		τ_B	
		Present	Umavathi et al. [10]	Present	Umavathi et al. [10]	Present	Umavathi et al. [10]
K	0.01	4.49835	4.49835	−3.00	−3.00	3.00	3.00
	0.5	4.44583	4.44583	−3.00	−3.00	3.00	3.00
	1.0	4.41919	4.41919	−3.00	−3.00	3.00	3.00
	4.0	4.36973	4.36973	−3.00	−3.00	3.00	3.00
Ec	0.01	4.41919	4.41919	−3.00	−3.00	3.00	3.00
	0.5	4.41919	4.41919	−3.00	−3.00	3.00	3.00
	1.0	4.41919	4.41919	−3.00	−3.00	3.00	3.00
	4.0	4.41919	4.41919	−3.00	−3.00	3.00	3.00

References

1. Akbarinia, A., Laur, R.: Investigation the diameter of solid particles effects on a laminar nanofluid flow in a curved tube using a two phase approach. Int. J. Heat Fluid Flow. **30**, 706–714 (2009)
2. Eringen, A.C.: Theory of micropolar fluids. J. Math. Mech. **16**, 1–18 (1996)
3. Beg, O.A., Bhargava, R., Rashidi, M.M.: Numerical Simulation in Micropolar Fluid Dynamics. Lambert Academic Saarbrucken, Germany (2011)
4. Prasad, V.R., Gaffar, S.A., Beg, O.A.: Heat and mass transfer fluid. J. Thermophys. Heat Trans. **29**, 127 (2015)
5. Umavathi, J.C., Mikhail, A.S.: Onset of double-diffusive convection of a sparsely packed micropolar fluid in a porous medium layer saturated with a nanofluid. Microfluid. Nanofluid. **21**, 128 (2017)
6. Umavathi, J.C., Hemavathi, K.: Flow and heat transfer of composite porous medium saturated with nanofluid. Propul. Power Res. **8**, 173 (2019)
7. Umavathi, J.C., Bég, O.A.: Convective fluid flow and heat transfer in a vertical rectangular duct containing a horizontal porous medium and fluid layer. Int. J. Numer. Methods Heat Fluid Flow. **31**, 1320 (2020)
8. Ariman, T., Truk, M.A., Sylvester, N.D.: Microcontinuum fluid mechanic – a review. Int. J. Eng. Sci. **11**, 905–930 (1973)
9. Bugliarello, G., Sevilla, J.: Velocity distribution and other characteristics of steady and pulsatile blood flow in fine glass tubes. Biorheology. **7**, 85 (1970)
10. Umavathi, J.C., Pratap Kumar, J., Chamka, A.J.: Flow and heat transfer of a micropolar fluid sandwiched between viscous fluid layers. Can. J. Phys. **86**, 961–973 (2008)
11. Bourantas, G.C., Loukopoulos, V.C.: MHD natural-convection flow in an inclined square enclosure filled with a micropolar nanofluid. Int. J. Heat Mass Transf. **79**, 930–944 (2014)
12. Eringen, A.C.: Microcontinuum Field Theories. II: Fluent Media. Springer, New York (2001)
13. Umavathi, J.C., Chamkha, A.J., Shekar, M.: Flow and heat transfer of a micropolar fluids separated by viscous fluid layer. Int. J. Microscale Nanoscale Therm. Fluid Trans. Phenom. **5**(1), 23–49 (2014)
14. Umavathi, J.C., Shekar, M.: Flow and heat transfer in a porous medium saturated by a micropolar fluid between parallel permeable disks. J. Porous Media. **8**, 669–684 (2014)
15. Dintensfass, S.: Viscosity factors in hypertensive and cardiovascular disease. Cardiovasc. Med. **2**, 337 (1977)

Finite Element Analysis of Unsteady Dispersion in Casson Fluid Flow

P. Nagarani, Victor M. Job, and Sreedhara Rao Gunakala

Mathematics Subject Classification (2010) 65M60, 76A05, 76R50, 76Z05

1 Introduction

The longitudinal solute dispersion within a solvent that flows through a conduit has abundant applications in environmental engineering, chemical engineering, physiological fluid dynamics and biomedical engineering. Taylor [1] gave one of the first theoretical studies on solutal dispersion and showed that the lateral solute diffusion and the cross-sectional variation of velocity act together to diffuse the solute when it is injected into a fluid as it undergoes steady flow in a tube. The effect of axial diffusion was neglected in this study; however, this study was extended by Aris [2] by incorporating the contribution of axial solute diffusion using the method of moments. It was determined that both of these studies are valid for large values of time after the initial solute injection. Gill and Sankarasubramanian [3] later presented the generalized dispersion model. Later, several authors such as Jiang and Grotberg [4], Kumar and Jayaraman [5] and Wang and Chen [6] analysed the dispersion phenomenon in Newtonian fluids by taking different conduit geometries and flow conditions into consideration using the aforementioned approaches and several other approaches.

The applications of dispersion theory in non-Newtonian fluids are due to developments in the fields such as polymer processing and biomedical engineering. Dash et al. [7] studied the dispersion of solute that is driven by steady Casson

P. Nagarani (✉) · V. M. Job
Department of Mathematics, The University of the West Indies, Mona Campus, Kingston, Jamaica
e-mail: victor.job@uwimona.edu.jm

S. R. Gunakala
Department of Mathematics and Statistics, The University of the West Indies, St. Augustine Campus, St. Augustine, Trinidad and Tobago
e-mail: Sreedhara.Rao@sta.uwi.edu

F. Kamalov et al. (eds.), *Advances in Mathematical Modeling and Scientific Computing*, Trends in Mathematics, https://doi.org/10.1007/978-3-031-41420-6_38

fluid flow through a conduit by employing the generalized dispersion model. Later, Nagarani et al. [8] extended this model for the case of solute absorption through the wall. The influence of oscillatory flow on dispersion in a flowing Casson fluid was analysed by Nagarani and Sebastian [9]. For the unsteady flow case, it was observed that the dispersion coefficients vary cyclically and eventually reach a non-transient state. This study gives the analysis in the case of low Womersley number. Several other studies are available in the existing literature that investigate the flow of non-Newtonian fluids with solute dispersion [10–13].

The particular interest in this study is to explore the impact of pertinent parameters on the dispersion of solute within a Casson fluid-filled tube while the flow is unsteady. In particular, we seek to analyse the solute local and mean concentrations for varying yield stress, fluctuating component of the pressure gradient, initial slug length and Péclet number.

2 Mathematical Formulation

2.1 Convection-Diffusion Equation

Here, consider the axisymmetric, laminar, fully developed and unsteady flow of Casson fluid in a tube having radius a. The time-dependent convection-diffusion equation that describes the local solute concentration $\bar{C}(\bar{t}, \bar{z}, \bar{r})$ within the fluid, is dependent on time $\bar{t}$, axial distance $\bar{z}$ and radial distance $\bar{r}$. Under the above flow assumptions, the dimensionless convection-diffusion equation can be written as

$$\frac{\partial C}{\partial t} + w(r,t)\frac{\partial C}{\partial z} = \left(\frac{1}{r}\frac{\partial}{\partial r}\left(r\frac{\partial}{\partial r}\right) + \frac{1}{Pe^2}\frac{\partial^2}{\partial z^2}\right)C, \tag{1}$$

where w is the non-dimensional axial flow velocity, $Pe = \frac{aw_0}{D_m}$ is the Péclet number, $w_0 = \frac{a^2}{2\mu_\infty}P_0$ is a reference velocity, P_0 is the steady component of pressure gradient, μ_∞ is the Newtonian fluid viscosity and D_m is the molecular diffusivity which is assumed to be constant.

An initial dispersion of a solute bolus with length $\bar{z}_s$ units and constant concentration C_0 is assumed. Hence, the corresponding initial and boundary conditions are provided below:

$$C(0,z,r) = \begin{cases} 1, & \text{if } |z| \le \frac{z_s}{2} \\ 0, & \text{if } |z| > \frac{z_s}{2} \end{cases}, \tag{2}$$

$$\frac{\partial C}{\partial r}(t,z,0) = \frac{\partial C}{\partial r}(t,z,1) = 0, \tag{3}$$

$$C(t,z,r) \to 0 \text{ as } |z| \to \infty. \tag{4}$$

The non-dimensional variables are given by

$$r = \frac{\bar{r}}{a}, \; z = \frac{D_m \bar{z}}{a^2 w_0}, \; t = \frac{D_m \bar{t}}{a^2}, \; C = \frac{\bar{C}}{C_0}, \; w = \frac{\bar{w}}{w_0}, \tag{5}$$

where the variables with overlined '-' refer to the corresponding dimensional variables. The dimensionless axial mean concentration C_m is given by

$$C_m = 2\int_0^1 C r dr. \tag{6}$$

2.2 *Casson Fluid Constitutive Equation and Flow Velocity Distribution*

The dimensionless constitutive equation that relates the shear stress τ and Casson fluid shear rate $\frac{\partial w}{\partial r}$ is given by [14]

$$\tau^{1/2} = \tau_y^{1/2} + \left(-\frac{\partial w}{\partial r}\right)^{1/2}, \text{ if } \tau \geq \tau_y \tag{7a}$$

$$\frac{\partial w}{\partial r} = 0, \text{ if } \tau \leq \tau_y, \tag{7b}$$

where

$$\tau = \frac{\bar{\tau}}{P_0 a/2}; \; \tau_y = \frac{\bar{\tau}_y}{P_0 a/2} \tag{8}$$

are (respectively) the dimensional shear and yield stresses. The analytical solution for the Casson fluid velocity distribution is obtained as [14]

$$w^-(r,t) = \frac{1}{2}P(t)\left\{1 - \frac{8}{3}r_p^{1/2} + 2r_p - \frac{1}{3}r_p^2\right\} - \frac{1}{32}\alpha^2 P'(t)\left\{3 - \frac{1144}{147}r_p^{1/2}\right.$$

$$\left. + \frac{320}{63}r_p + \frac{4}{3}r_p^2 - \frac{16}{9}r_p^{5/2} + \frac{65}{441}r_p^4\right\} \text{ for } 0 \leq r \leq r_p, \tag{9a}$$

$$w^+(r,t) = \frac{1}{2}P(t)\left\{1 - r^2 - \frac{8}{3}r_p^{1/2}(1 - r^{3/2}) + 2r_p(1 - r)\right\}$$

$$-\frac{1}{32}\alpha^2 P'(t)\left\{3-(4-r^2)r^2-r_p^{1/2}\left[\frac{1144}{147}-\frac{16}{3}(r^2+r^{3/2})+\frac{424}{147}r^{7/2}\right]\right.$$

$$\left.+r_p\left[\frac{320}{63}+\frac{128}{63}r^3-\frac{64}{9}r^{3/2}\right]\right\} \tag{9b}$$

where

$$P(t)=1+e\,cos(\alpha^2\,Sc\,t),\ r_p=\frac{\tau_y}{P(t)},\ \alpha^2=\frac{\omega a^2}{\nu},\ e=\frac{P_1}{P_0},\ Sc=\frac{\nu}{D_m}. \tag{10}$$

In Eq. (10), P_1 is the fluctuating pressure component, r_p is the non-dimensional plug flow radius, ν is the kinematic viscosity coefficient, α is the Womersley frequency parameter and Sc is the Schmidt number. The subscripts '-' and '+' in Eq. (9b) are associated with the velocity distribution in the respective plug flow ($0 \le r \le r_p$) and shear flow ($r_p \le r \le 1$) regions.

3 Numerical Solution

We obtain a numerical solution for the convection-diffusion Eq. (1) and the associated initial and boundary conditions (2)–(4) using the finite element technique. First, a spatial semi-discrete Galerkin formulation is constructed based on a suitably chosen triangulation and finite element solution space. Then, time discretization is performed using the Crank-Nicolson scheme. Additional details on the solution methodology are outlined in Sects. 3.1 and 3.2.

3.1 Finite Element Method

Let $\mathcal{T}^h$ be a triangulation of the domain $\Omega = (-l, l) \times (0, 1)$, where each $K \in \mathcal{T}^h$ is a linear triangular element and

$$h = \max\{\text{diam}K : K \in \mathcal{T}^h\}.$$

Let

$$S^h = \left\{q_h \in C(\bar{\Omega}) : q_h|_K \in P_1(K) : \forall K \in \mathcal{T}^h\right\} \tag{11}$$

and suppose that $w_h(t,\cdot,\cdot)$ is the Lagrange interpolant of $w(t,\cdot,\cdot)$ in S^h for each $t > 0$. We identify the finite element space $V^h = S^h \cap H^1(\Omega)$ and construct the Galerkin semi-discrete problem formulation as follows: For any $t > 0$, find $C_h(t, \cdot, \cdot) \in V^h$ such that $\forall \chi_h \in V^h$,

$$\int_\Omega r\left[\left(\frac{\partial C^h}{\partial t} + w_h\frac{\partial C_h}{\partial z}\right)\chi_h + \frac{\partial C_h}{\partial r}\frac{\partial \chi_h}{\partial r} + \frac{1}{Pe^2}\frac{\partial C_h}{\partial z}\frac{\partial \chi_h}{\partial z}\right]dzdr = 0. \tag{12}$$

Let $\{t_n\}_{n=1}^N$ be a uniform partition of the time domain $[0, T]$ with step-size Δt, and suppose that $C_h(t_n) = C_h^n$ and $w_h(t_n) = w_h^n$ for $n \in \{1, \ldots, N\}$. Using the Crank-Nicolson approximation

$$\frac{1}{2}\left(\left.\frac{\partial C_h}{\partial t}\right|_{t=t_{n+1}} + \left.\frac{\partial C_h}{\partial t}\right|_{t=t_n}\right) = \frac{C_h^{n+1} - C_h^n}{\Delta t}, \tag{13}$$

the fully discrete Galerkin formulation is as follows: Given $t_{n+1} > 0$, find $C_h^{n+1} \in V^h$ such that $\forall \chi_h \in V^h$:

$$\int_\Omega r\left[\left(2\frac{C_h^{n+1}}{\Delta t} + w_h^{n+1}\frac{\partial C_h^{n+1}}{\partial z}\right)\chi_h + \frac{\partial C_h^{n+1}}{\partial r}\frac{\partial \chi_h}{\partial r} + \frac{1}{Pe^2}\frac{\partial C_h^{n+1}}{\partial z}\frac{\partial \chi_h}{\partial z}\right]dzdr$$
$$+\int_\Omega r\left[\left(-2\frac{C_h^n}{\Delta t} + w_h^n\frac{\partial C_h^n}{\partial z}\right)\chi_h + \frac{\partial C_h^n}{\partial r}\frac{\partial \chi_h^n}{\partial r} + \frac{1}{Pe^2}\frac{\partial C_h^n}{\partial z}\frac{\partial \chi_h}{\partial z}\right]dzdr = 0. \tag{14}$$

Let $\{\phi_j^h\}_{j=1}^{N_h}$ be the piecewise Lagrange polynomial basis for S^h and assume that

$$C_h^n = \sum_{j=1}^{N_h} c_{j,n}^h \phi_j^h. \tag{15}$$

By substituting this linear combination into the fully discrete formulation, we acquire a linear system of the form

$$\left(2A + \Delta t K^{(n+1)}\right)c^{(n+1)} = \left(2A - \Delta t K^{(n)}\right)c^{(n)}. \tag{16}$$

The finite element solution vector $c^{(n)}$ is computed by solving the above system at each time t_n using MATLAB with $N = 100$ time steps.

3.2 *Grid-Independence Test and Code Validation*

A grid-independence test is conducted by calculating the maximum concentration C_{max} at time $t = 0.1$ on finite element meshes with 363, 531, 814, 1486 and 3365

Table 1 Grid-independence test results for $\alpha = 0.5$, $Sc = Pe = 1000$, $\tau_y = 0.05$ and $\Delta t = 0.001$

No. of elements	C_{max}	Computational time(s)
363	0.1822	11.99
531	0.1688	17.57
814	0.1646	26.42
1486	0.1571	45.53
3365	0.1567	109.45

elements for $\alpha = 0.5$, $Sc = Pe = 1000$, $\tau_y = 0.05$ and $\Delta t = 0.001$. Based on the criterion that the relative error in C_{max} between two successive meshes is less than 1%, a mesh with 1486 elements was chosen for the present investigation.

To validate the present numerical algorithm, we compare the mean concentration computed using our finite element formulation with the previous analytical solution obtained in Nagarani and Sebastian [9] (see Figs. 1 and 2). It was determined from the graphical results that our numerical solution is in good agreement with the analytical solution of Nagarani and Sebastian [9]; hence, the present finite element algorithm is suitable for the analysis of our mathematical model.

4 Results and Discussion

The unsteady dispersion of solute in the case of a non-Newtonian (Casson) fluid with consideration of pulsatile flow is discussed. In this analysis, we considered the yield stress τ_y values from 0 to 0.05 [14]; Womersley frequency parameter α fixed at 0.5; the amplitude of fluctuating component of pressure e fixed at 0.5 [9]; Schmidt number Sc at 1000 [15]; the slug input length z_s of solute with given values 0.004, 0.01 and 0.02; and Peclet number Pe that is taken (1, 10, 1000). The purpose of taking the values here is for the comparison with past studies such as Nagarani and Sebastian [9] with the approach of generalized dispersion and with Dash et al. [7] in steady flow case.

Figure 1 depicts the influence of τ_y on the axial mean concentration; here $\tau_y = 0$ yields the Newtonian fluid case. From Figs. 1a,b in both Newtonian and Casson fluid cases, the generalized dispersion approach and FEM for dispersion in oscillatory flow are in 100% agreement qualitatively and in good agreement quantitatively with values being a little higher in the case of FEM method. The same conclusion can be made from Figs. 2a–b, which illustrate the axial mean concentration distribution for varying dispersion times. As t increases, the peak value of the mean concentration moves forward axially, which is due to a reduction of fluid convection and the peak mean concentration value with time.

Figure 3 shows the surface plot of local concentration, for $t = 0.2$ by fixing the other parameters as in the case of Fig. 2. As in the case of mean concentration, here the peak value is reduced with time, and the width of the mean concentration

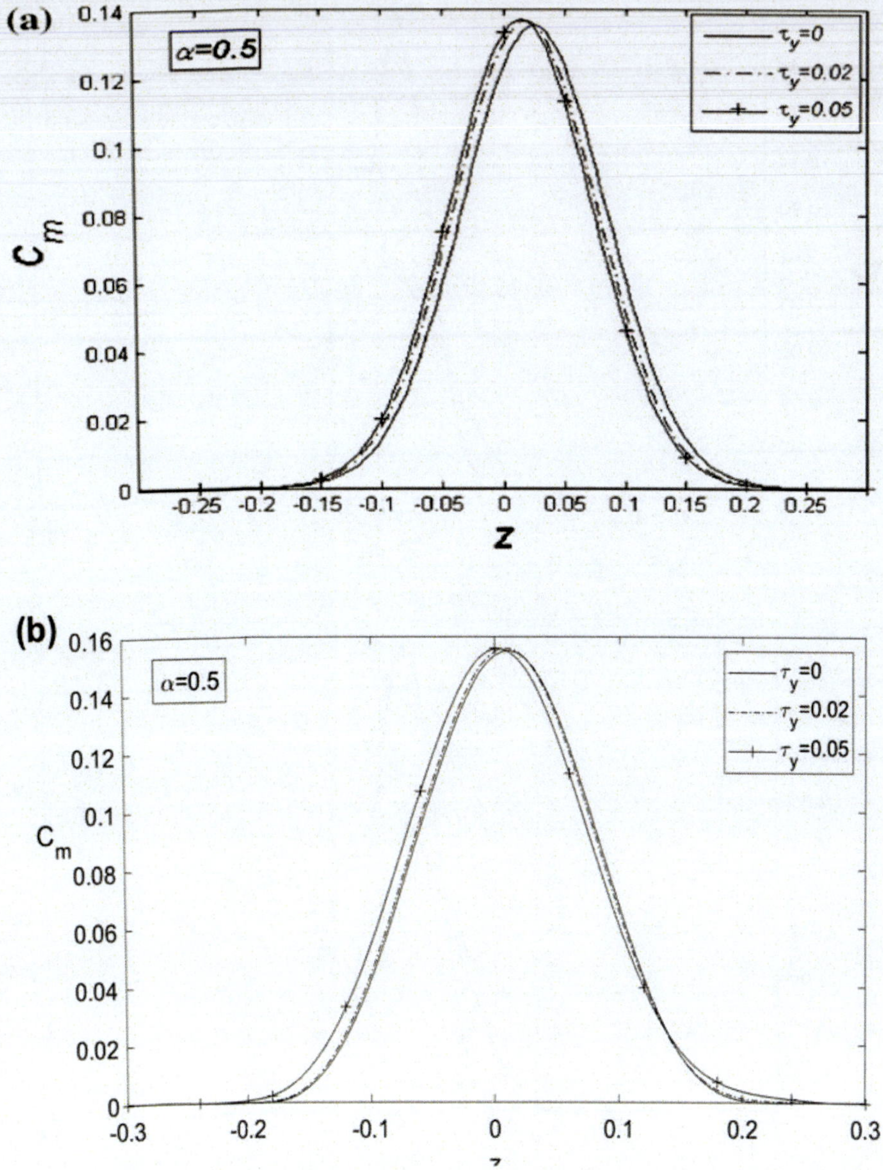

Fig. 1 Comparison of axial distribution of mean concentration obtained by (**a**) Nagarani and Sebastian [9] and (**b**) present code for different values of τ_y when $z_s = 0.02$, $Pe = Sc = 10^3$, $e = 0.5$ and $t = 0.1$.

distribution increases with time. The symmetry of these curves with respect to $z = 0$ is observed.

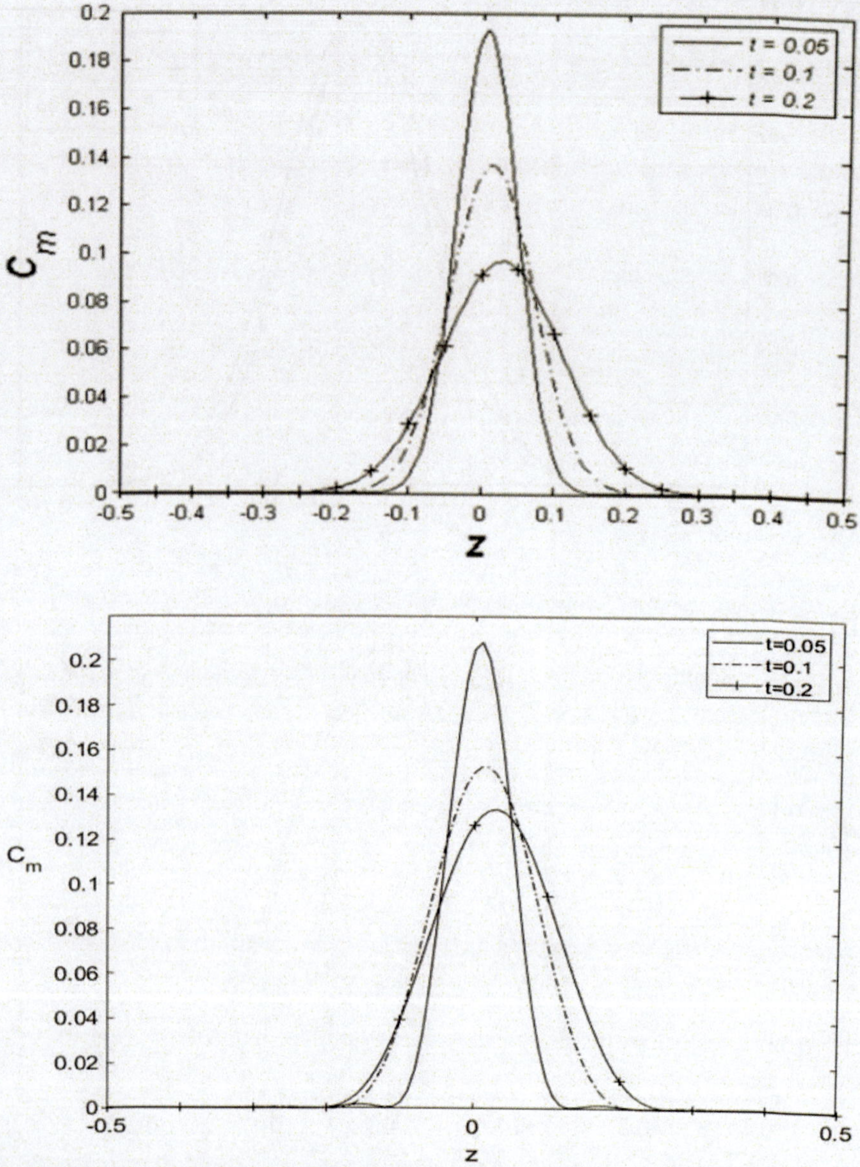

Fig. 2 Comparison of axial distribution of mean concentration obtained by (**a**) Nagarani and Sebastian [9] and (**b**) present code for different values of t when $z_s = 0.02$, $Pe = Sc = 10^3$, $e = 0.5$ and $\tau_y = 0.05$.

Figure 4 indicates that the maximum concentration decays with time for all the values of z_s and shows the trend of reaching a steady state and C_{max} increasing with z_s. Lighthill [16] initially demonstrated the importance of Taylor and Aris analyses

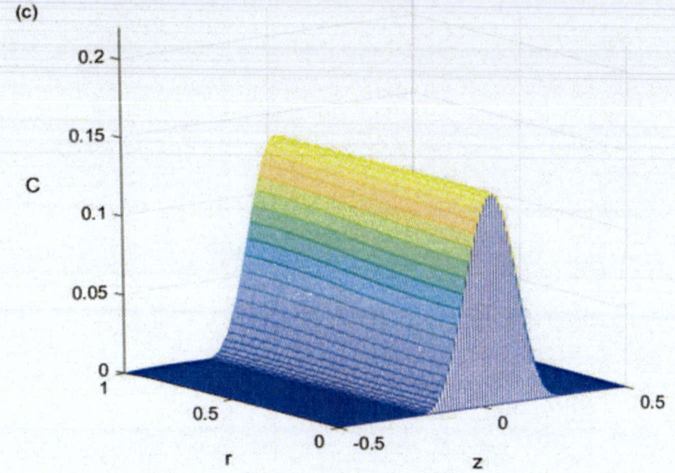

Fig. 3 Surface plot of concentration C for $t = 0.2$ when $Pe = Sc = 10^3$, $e = 0.5$, $z_s = 0.02$, $\tau_y = 0.05$ and $\alpha = 0.5$

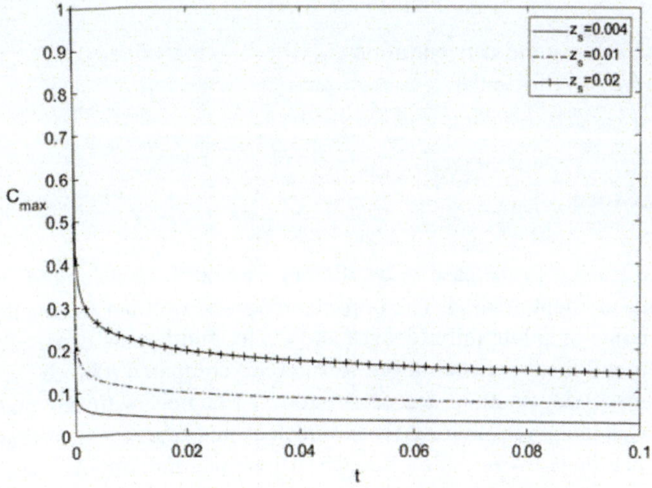

Fig. 4 Time-varying maximum concentration C_{max} for different values of z_s when $Pe = Sc = 10^3$, $e = 0.5$, $\tau_y = 0.05$ and $\alpha = 0.5$

for the study of mass transport in cardiovascular flows. It was later shown that these analyses are best suited for the cases of larger dispersion times [17]. Hence, it is very essential to have a model which discusses the dispersion of solute since after injecting as the measured dispersion time is very small (order of 10^{-5}) in the cardiovascular system [18]. From Fig. 5, it is observed that the C_{max} is higher for larger Pe, and the values of C_{max} corresponding to $Pe = 1$, 10 and 1000 reduce

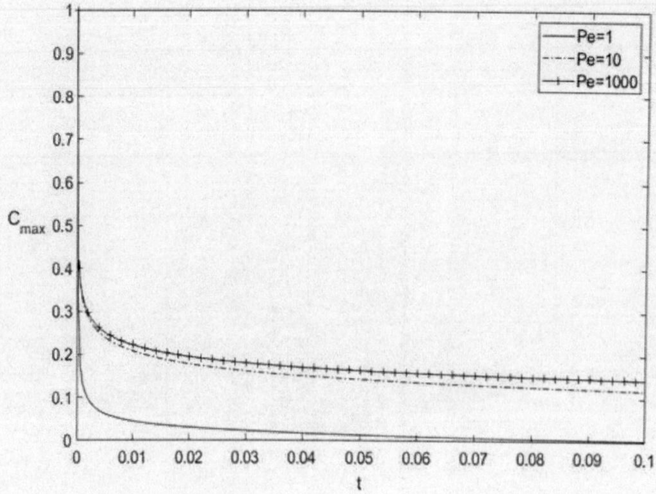

Fig. 5 Time-varying maximum concentration C_{max} for different values of Pe when $Sc = 10^3$, $e = 0.5$, $\tau_y = 0.05$, $z_s = 0.02$ and $\alpha = 0.5$

from the initial maximum concentration $C_{max} = 1$ to the values 0.0034, 0.1185 and 0.1431 (respectively) for t=0.1.

5 Conclusions

Convection-diffusion in the case of oscillatory Casson fluid is studied numerically using the finite element method. The impacts of non-Newtonian and oscillatory flow behaviours, bolus of solute initial length and Peclet number are discussed on mean concentration and local concentration. Results are compared with the approach of generalized dispersion method and are in good agreement with this method. This study has advantages over Taylor-Aris approach as this approach can be applied any time after injection. However, this approach still has its limitations as the Womersley number is assumed to be small in this study. This limitation can be overcome by solving the entire problem numerically for both the flow equations and convection-diffusion equation; this would be essential for applying this model to dispersion problems in cardiovascular flows.

References

1. Taylor, G.: Dispersion of soluble matter in solvent flowing slowly through a tube. Proc. R. Soc. Lond. Ser. A **219**(1137), 186–203 (1953)

2. Aris, R.: On dispersion of a solute in a fluid flowing through a tube. Proc. R. Soc. Lond. Ser. A **235**, 67–77 (1956)
3. Gill, W.N., Sankarasubramanian, R.: Exact analysis of unsteady convective diffusion. Proc. R. Soc. Lond. Ser. A **316**, 341–350 (1970)
4. Jiang, Y., Grotberg, J.B.: Bolus contaminant dispersion in oscillatory tube flow with conductive walls. ASME. J Biomech. Eng. **115**(4A), 424–431 (1993)
5. Kumar, S., Jayaraman, G.: Method of moments for laminar dispersion in an oscillatory flow through curved channels with absorbing walls. Heat Mass Transf. **44**(11), 1323–1336 (2008)
6. Wang, P., Chen, G.Q.: Solute dispersion on open channel flow with bed absorption. J. Hydrol. **543**, 208–217 (2016)
7. Dash, R.K., Jayaraman, G., Mehta K.N.: Shear augmented dispersion of a solute in a Casson fluid flowing in a Conduit. Ann. Biomed. Eng. **28**(4), 373–385 (2000)
8. Nagarani, P., Sarojamma, G., Jayaraman, G.: Effect of boundary absorption in dispersion in Casson fluid flow in a tube. Ann. Biomed. Eng. **32**(5), 706–719 (2004)
9. Nagarani, P., Sebastian, B.T.: Dispersion of a solute in pulsatile non-Newtonian fluid flow through a tube. Acta Mech. **224**, 571–585 (2013)
10. Debnath, S., Saha, A.K., Mazumder, B.S., Roy, A.K.: Dispersion phenomena of reactive solute in a pulsatile flow of three-layer liquids. Phys. Fluids **29**, 097107 (2017)
11. Rana, J., Murthy, P.V.S.N.: Unsteady solute dispersion in non-Newtonian fluid flow in a tube with wall absorption. Proc. R. Soc. Lond. Ser. A. **472**, 20160294 (2016)
12. Rana, J., Liao, S.: A general analytical approach to study solute dispersion in non-Newtonian fluid flow. Eur. J. Mech. – B/Fluids **77**, 183–200 (2019)
13. Chauhan, S.S., Tiwari, A.: Solute dispersion in non-Newtonian fluids flow through small blood vessels: a varying viscosity approach. Eur. J. Mech. – B/Fluids **94**, 200–211 (2022)
14. Aroesty, J., Gross, J.F.: The mathematics of pulsatile flow in small blood vessels 1. Casson Theory. Microvasc. Res. **4**, 1–12 (1972)
15. Caro, C.G., Pedley, T.J., Schroter, R.C., Seed, W.A.: The Mechanics of Circulation. Oxford University Press, New York (1978)
16. Lighthill, M.J.: Initial development of diffusion in Poiseuille flow. IMA J. Appl. Math. **2**(1), 97–108 (1966)
17. Reejhsinghani, N.S., Gill, W.N., Barduhn, A.J.: Laminar dispersion in capillaries: part III. Experiments in horizontal tubes including observations on natural convection effects. AIChE J. **12**, 916–923 (1966)
18. Caro, C.G.: The dispersion of indicator flowing through simplified models of the circulation and its relevance to velocity profiles in blood vessels. J. Physiol. **185**, 501–519 (1966)

Thermosolutal Convection in a Tilted Porous Parallelogrammic Enclosure with Discrete Heating and Salting

P. Ravindra, Mahesha, Maimouna Al Manthari, and M. Sankar

1 Introduction

Buoyant-induced-convection (BIC) processes and related thermal transport (TT) repeatedly appear in several engineering applications which involve heat dissipation in finite-sized enclosures. Among those, the square or annular chamber with thermal gradient perpendicular to the gravity and adiabatic top-bottom walls is a typical and largely used model, and therefore several experimental and numerical investigations are performed in such geometries [1–4]. In addition, the porous medium has a great influence on fluid movement and heat dissipation rate in such geometries and this phenomenon has been analyzed in detail [5, 6]. For numerous thermal and solute transport applications, the geometry of equipment need not have a regular-shaped structure, and some critical structure-based parameters in such geometries considerably affect the BIC and related heat dissipation processes [7]. Hyun and Choi [8] numerically studied the convective HT rate in a parallelogram-shaped enclosure for wide range of parameters and provided thermal dissipation correlations. Later, Baytas and Pop [9] adopted the Darcy model to study the fluid flow and TT rate in an oblique-shaped geometry. The experimental and numerical investigations have been performed by Bairi and co-researchers [10, 11] to analyze heat dissipation in a parallelogram-structured geometry for various sets

P. Ravindra
Department of Mathematics, East Point College of Engineering and Technology, Avalahalli, Bengaluru, Karnataka, India

Mahesha
Department of Mathematics, University BDT College of Engineering, Davanagere, Karnataka, India

M. Al. Manthari · M. Sankar (✉)
Department of Information Technology, University of Technology and Applied Sciences, Ibri, Oman

F. Kamalov et al. (eds.), *Advances in Mathematical Modeling and Scientific Computing*, Trends in Mathematics, https://doi.org/10.1007/978-3-031-41420-6_39

of parametric values and obtained a good agreement between results of numerical study and experimental study. The influence of nanoparticles on fluid flow and heat transport rate in parallelogram-shaped geometries has also been analyzed [12, 13].

In many situations, the convective motions are generated, in addition to thermal gradients, due to the presence of two or more concentrations or species and thus the existence of concentration gradients. Such type of convective flows, where the motion is caused by the combined appearance of thermal and solutal differences, is termed as double diffusive convection (DDC) or thermosolutal convection (TSC). This phenomenon has very important applications in ocean science, materials processing, and crystal growth techniques. A sound knowledge of TSC flows as well as the associated thermal and solute transfer (TST) rates in finite geometries is very important in terms of the abovementioned applications. Costa [14] studied double diffusive convection in a parallelogram-shaped enclosure occupied with fluid-saturated porous medium. The influence of enclosure and sidewall tilt angle on DDC flow and HT rate in a parallelogrammic enclosure is numerically studied by Jagadeesha et al. [15]. The same study has been extended by considering the internal heat generation [16]. Some of the recent studies involving isolated heating and thermosolutal convection in different geometries with different constraints can be found in [17, 18]. The above detailed review of literature obviously reveals the existing research gap that the DDC flow motion and related HMT characteristics in an inclined parallelogram configuration have not been attempted in the presence of source-sink arrangements. In this chapter, therefore, detailed numerical experiments are undertaken to analyze the collective impacts of source-sink pair positional setup on buoyant movement and HMT rates in the tilted parallelogram-shaped geometry.

2 Mathematical Formulation

In this analysis, a parallelogram-shaped porous geometry has been considered and is shown in Fig. 1 alongside the Cartesian coordinates. Here H and L, respectively, denote the elevation and breadth of geometry and are inclined with an angle α to the positive abscissa. A source and sink are, respectively, placed alongside the left and right tilted wall, and the source is being retained at higher thermal and solutal conditions ($T_h \& S_h$), while the sink has been retained at lower thermal and solutal conditions ($T_c \& S_c$). Also, the remaining regions of the geometry are considered as insulated and impermeable. The fluid is supposed to be Boussinesq and incompressible. The impacts from viscous dissipation and porous inertia are taken as insignificant. The DDC flow inside the geometry is caused by a collective buoyancy effect from both thermal and solutal differences. Implementing the Darcy model, and eliminating the pressure components through cross differentiation, the constitutive equations for the momentum, energy, and concentration are [15, 16]:

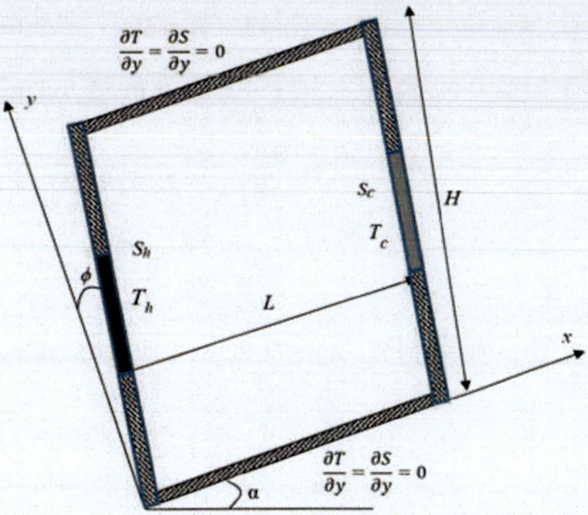

Fig. 1 Geometrical structure and coordinates

$$\frac{\partial^2\psi}{\partial\xi^2} - 2\frac{sin\phi}{A}\frac{\partial^2\psi}{\partial\xi\partial\eta} + \frac{1}{A^2}\frac{\partial^2\psi}{\partial\eta^2} = \frac{Ra_T cos\phi sin\alpha}{A}\left[\frac{\partial\theta}{\partial\eta} + N\frac{\partial S}{\partial\eta}\right] - \frac{Ra_T cos\phi cos(\phi-\alpha)}{A}\left[\frac{\partial\theta}{\partial\xi} + N\frac{\partial S}{\partial\xi}\right], \quad (1)$$

$$\frac{\partial\theta}{\partial t} + \frac{\partial\psi}{\partial\eta}\frac{\partial\theta}{\partial\xi} - \frac{\partial\psi}{\partial\xi}\frac{\partial\theta}{\partial\eta} = \frac{A}{cos\phi}\left[\frac{\partial^2\theta}{\partial\xi^2} - 2\frac{sin\phi}{A}\frac{\partial^2\theta}{\partial\xi\partial\eta} + \frac{1}{A^2}\frac{\partial^2\theta}{\partial\eta^2}\right], \quad (2)$$

$$\frac{\epsilon}{\sigma}\frac{\partial S}{\partial t} + \frac{\partial\psi}{\partial\eta}\frac{\partial S}{\partial\xi} - \frac{\partial\psi}{\partial\xi}\frac{\partial S}{\partial\eta} = \frac{A}{Le\, cos\phi}\left[\frac{\partial^2 S}{\partial\xi^2} - 2\frac{sin\phi}{A}\frac{\partial^2 S}{\partial\xi\partial\eta} + \frac{1}{A^2}\frac{\partial^2 S}{\partial\eta^2}\right] \quad (3)$$

It is worth to mention here that based on standard coordinate transformations used by Baytas and Pop [9], the model equations are converted to regular-shaped region. In the above equations, $Ra_T = \frac{gK\beta_T \Delta T L}{\nu\alpha_m}$ is the thermal Rayleigh number, $N = \frac{\beta_C \Delta S}{\beta_T \Delta T}$ is the ratio of solutal and thermal buoyancies, and $Le = \frac{\alpha_m}{D}$ is the Lewis number. The following dimensionless quantities are introduced to non-dimensionalize the model equations:

$$\xi = \frac{X}{L}, \eta = \frac{Y}{H cos\phi}, t = t^*\left(\frac{\alpha_m}{\sigma L H cos\phi}\right), \psi = \frac{\psi^*}{\alpha_m},$$

$$\theta = \frac{(T - T_r)}{\Delta T}, S = \frac{(C - C_r)}{\Delta C},$$

In the above transformations, $T_r = \frac{T_h + T_c}{2}$ and $C_r = \frac{C_h + C_l}{2}$ are reference temperature and concentration, respectively.

The dimensionless auxiliary conditions are:

$$t = 0: \ \psi = \theta = C = 0, \ \psi = 0; \quad \text{over the entire annulus.}$$

$$t > 0: \ \psi = \frac{\partial\theta}{\partial\eta} = \frac{\partial C}{\partial\eta} = 0 \ \text{at insulated region; } T = C = 1 \ \text{at source.}$$

$$\psi = \frac{\partial\theta}{\partial\eta} = \frac{\partial C}{\partial\eta} = 0 \ \text{at insulated region; } T = C = 0 \ \text{at sink.}$$

$$\psi = \frac{\partial\theta}{\partial\eta} - Asin\phi\frac{\partial\theta}{\partial\xi} = \frac{\partial C}{\partial\eta} - Asin\phi\frac{\partial C}{\partial\xi} = 0, \quad \text{at } \eta = 0 \text{ and } 1$$

The overall HMT rates at the sources are estimated using the below equations:

$$\overline{Nu} = \frac{1}{A}\int_0^A -\frac{1}{cos\phi}\left(\frac{sin\phi}{A}\frac{\partial\theta}{\partial\eta} - \frac{\partial\theta}{\partial\xi}\right)d\eta$$

$$\overline{Sh} = \frac{1}{A}\int_0^A -\frac{1}{cos\phi}\left(\frac{sin\phi}{A}\frac{\partial S}{\partial\eta} - \frac{\partial S}{\partial\xi}\right)d\eta.$$

3 Solution Technique

The partial differential equations which govern the physical processes in this chapter are numerically integrated by utilizing the finite difference methodology with the aid of supplementary conditions. The numerical iteration technique for transient equation is performed using the time-splitting scheme (Alternating Direction Implicit method) whose detailed description, methodology, and validation can be found in our recent works [15, 16], and hence the details are not provided here for brevity. The discretized algebraic equations are formed in tri-diagonal banded structure and are solved through tri-diagonal matrix algorithm. The iteration process used to estimate the field variables is continued until the chosen convergence condition is reached. The grid sizes are varied from 41 × 41 to 161 × 161, and through the systematic grid independency test, it is found that a 101 × 101 mesh is adequate, and the same mesh size is adapted for further calculations in the present analysis.

4 Results and Discussion

This section is mainly devoted to discuss the outcomes of numerical simulations for the vast range of parametric values chosen in this analysis. The influence of various parameters on the buoyant-flow, thermal, and solutal structure has been presented for different magnitudes of the respective parameters.

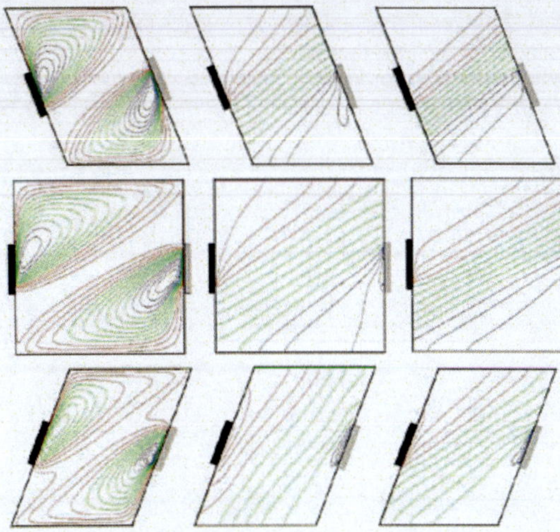

Fig. 2 Influence of sidewall angle (ϕ) on flow, thermal, and concentration contours for $Ra = 10^3, \alpha = -30^o, \epsilon = 0.2, D = 0.5, N = 2, Le = 2 \phi = -30^0$ (top), $\phi = 0^0$ (middle), $\phi = 30^0$ (bottom)

First, the impacts of sidewall tilting on isopleths of stream function, thermal, and solutal lines are shown in Fig. 2 for three different angles (positive, zero, and negative) with other pertinent parameters fixed. For all angles, a two-eddy flow pattern with each eddy appearing near the source and sink can be witnessed, while the isotherms and isoconcentrations reveal similar variations. The streamline structure reveals a moderate single-eddy flow pattern for lower magnitudes of $Ra(100)$, while bi-cellular flow structure can be witnessed for moderate magnitude of $Ra(500)$ and strong two-eddy flow shape for higher magnitude of $Ra(10^3)$ with one eddy nearer to the source and another eddy close to sink along the right boundary. The contours of thermal and solutal distributions are consistent with streamline pattern for all three magnitudes of Ra.

The impact of geometry tilt angle is another prime parameter in this analysis, and the impact of this parameter has been discussed in Fig. 3 for three different angles. It can be clearly observed that the cavity inclination has severe impact on the flow structure as well as thermal and solutal contours. A bi-cellular flow structure could be observed for negative and zero inclination of the geometry, while a multi-cellular flow structure has been noticed inside the cavity for positive inclination of the geometry. The isopleths of thermal and solutal distribution show strong stratified structure for no-inclination and parallel-like structure for negative and positive inclination of the geometry. Figure 4 reveals the significance of Lewis number on the isopleths of flow, temperature, and solutal lines for three different magnitudes of Le. The flow, thermal, and solutal contour structures strongly depend on the magnitude of Le. For lower value of Lewis number, bi-cellular eddies with an encapsulated eddy covering the two eddies near the source and sink could be seen. However, for $Le > 0.1$, a strong bi-cellular flow could be witnessed from the contour, and isopleths of thermal and solutal lines show a similar structure.

The collective impacts of Ra and other critical parameters on the HMT rates have been illustrated in Figs. 5, 6, 7 and 8 by setting other key parameters to a constant

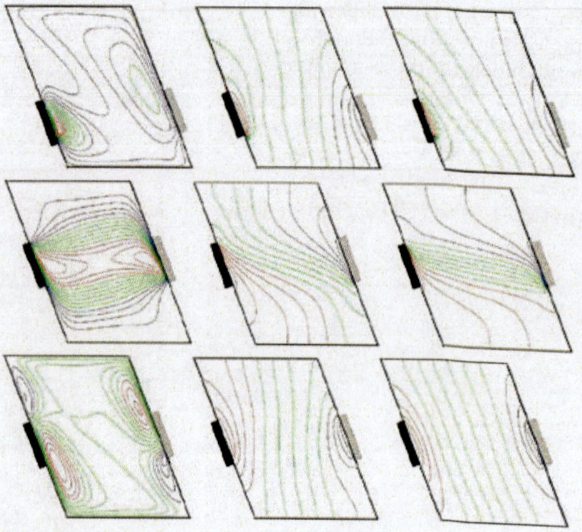

Fig. 3 Influence of geometry tilt angle (α) on flow, thermal, and concentration contour for $Ra = 10^3$, $\phi = -30^o$, $\epsilon = 0.2$, $D = 0.5$, $N = -2$, $Le = 2$, $\alpha = -45^0$ (top), $\alpha = 0^0$ (middle), $\alpha = 45^0$ (bottom)

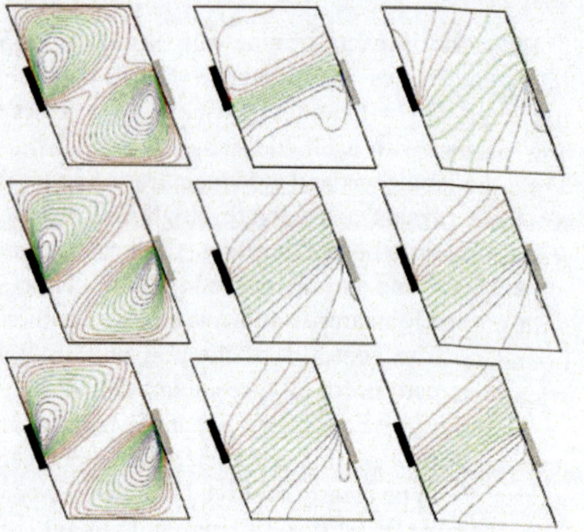

Fig. 4 Influence of Lewis number (Le) on flow, thermal, and concentration contour for $Ra = 10^3$, $\alpha = -30^o$, $\phi = -30^o$, $N = 2$, $\epsilon = 0.2$, $D = 0.5$ $Le = 0.1$ (top), $Le = 1$ (middle), $Le = 2$ (bottom)

value. Figure 5 portrays the combined impacts of Ra and source-sink position (D) on the average Nu and Sh rates. For illustration, three sample locations, such as lower, middle, and top positions on the sidewall, have been chosen. Among these three locations, it has been witnessed that the positioning of source-sink around the lower middle location produces a higher HMT as compared to other locations. The variations of HMT rates with source-sink lengths have also been estimated and are shown in Fig. 6 for three different lengths. From the predictions, it has been witnessed that the HMT rates could be enhanced by choosing a higher length heater as compared to smaller lengths. The wall inclination effect on the HMT rates has also been producing interesting predictions, and the same has been presented in Fig. 7 by considering a wider range of angles. It can be seen from the results that the impact of ϕ on $\overline{Nu}$ and $\overline{Sh}$ strongly depends on the magnitude of Ra. The impacts

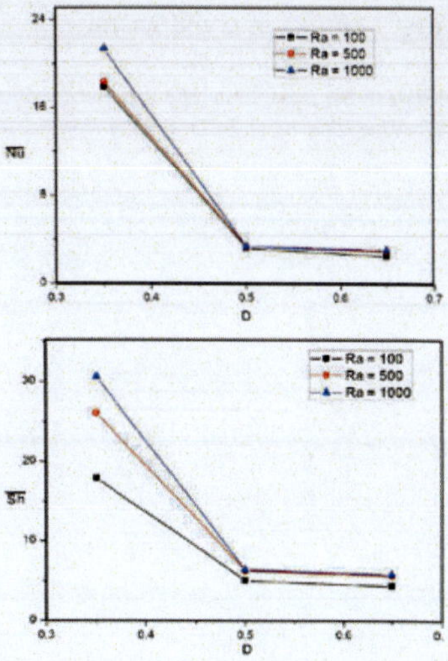

Fig. 5 Effect of source position (D) and Ra on $\overline{Nu}$ (top) and $\overline{Sh}$ (bottom) with $\alpha = -30^o, \phi = 45^o, \epsilon = 0.2, Le = 2, N = 2$

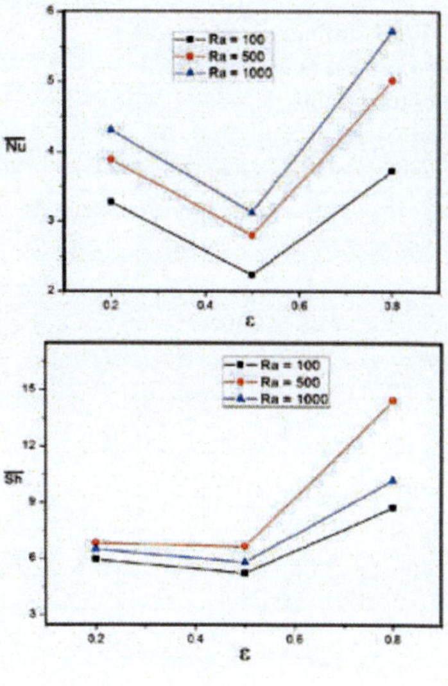

Fig. 6 Influence of source length and Ra on $\overline{Nu}$ (top) and $\overline{SH}$ (bottom) with $\phi = -30^o, \alpha = -30^o, Le = 2, N = 2, D = 0.5$

of buoyancy ratio on the HMT rates for different magnitudes of Ra are important and have been predicted in Fig. 8. The predictions vividly indicate that the HMT rates are maximum for $N = -2$ among other buoyancy ratios.

5 Conclusions

In this investigation, the impacts of source-sink arrangements on DDC flow and associated HMT rates have been numerically investigated for broad ranges of key

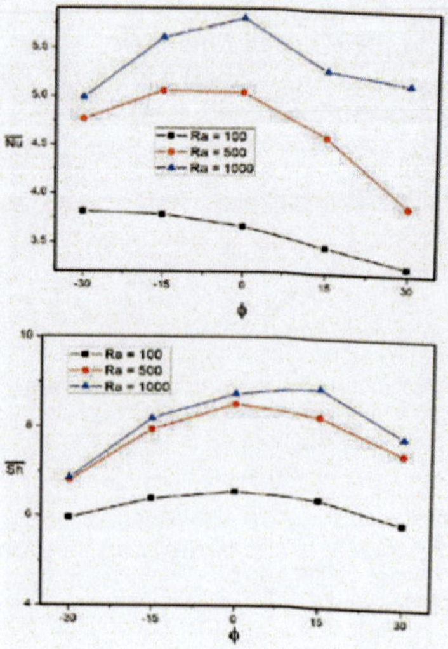

Fig. 7 Effect of ϕ and Ra on $\overline{Nu}$ (top) and $\overline{Sh}$ (bottom) with $\alpha = -30^o$, $Le = 2, N = 2, \epsilon = 0.2, D = 0.5$

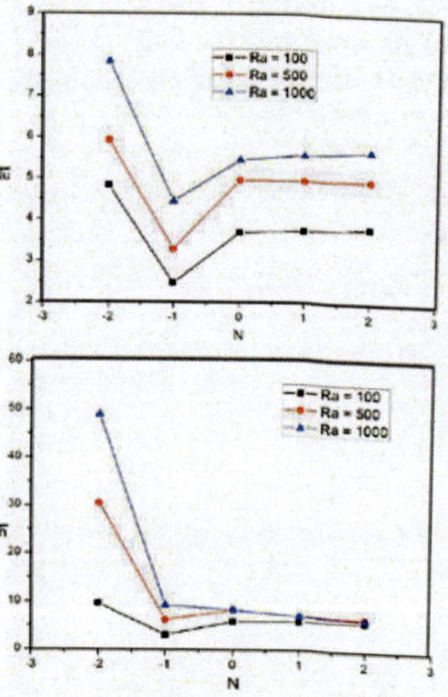

Fig. 8 Influence of (N) and Ra on $\overline{Nu}$ (top) and $\overline{Sh}$ (bottom) with $\alpha = -30^o, Le = 2, \phi = -30^o, \epsilon = 0.2, D = 0.5$

parameters of the problem. Based on the vast simulations, the following predictions are observed and recorded:

1. A large magnitude of Ra produces higher flow circulation strength and thus higher HMT rates could be attained. Bi-cellular flow structure and multi-cellular flow structure are produced by buoyancy ratio and for a particular choice of N, maximum HMT could be produced.
2. The location of source-sink arrangement is another crucial parameter in enhancing the flow strength and HMT rates. The sidewall and geometry inclinations play an important role in altering the DDC flow and resulting HMT rates.

References

1. Deng, Q.H.: Fluid flow and heat transfer characteristics of natural convection in square cavities due to discrete source–sink pairs. Int. J. Heat Mass Transf. **51**, 5949–5957 (2008)
2. Sankar, M., Do, Y.: Numerical simulation of free convection heat transfer in a vertical annular cavity with discrete heating. Int. Commun. Heat Mass Transf. **37**, 600 (2010)
3. Sivasankaran, S., Do, Y., Sankar, M.: Effect of discrete heating on natural convection in a rectangular porous enclosure. Transp. Porous Med. **86**, 29 (2011)
4. Pushpa, B.V., Sankar, M., Mebarek-Oudina, F.: Buoyant convective flow and heat dissipation of Cu–H_2O nanoliquids in an annulus through a thin baffle. J. Nanofluids **10**, 292–304 (2021)
5. Nield, D.A., Bejan, A.: Convection in Porous Media, 2nd edn. Springer, New York, (1999)
6. Vafai, K.: Handbook of Porous Media. Marcel Dekker, New York (2000)
7. Asako, Y., Nakamura, H.: Heat transfer in a parallelogram shaped enclosure. Bull. JSME **27**(228), 1144–1151 (1984)
8. Hyun, J.M., Choi, B.S.: Transient natural convection in a parallelogram-shaped enclosure. Int. J. Heat Fluid Flow **11**(2), 129–134 (1990)
9. Baytas, A.C., Pop, I.: Free convection in oblique enclosures filled with a porous medium. Int. J. Heat Mass Transf. **42**, 1047–1057 (1999)
10. Bairi, A., Garcia de Maria, J.M., Laraqi, N.: Transient natural convection in parallelogrammic enclosures with isothermal hot wall. Experimental and numerical study applied to on-board electronics. Appl. Therm. Eng. **30**, 1115–1125 (2010)
11. Bairi, A., Garcia de Maria, J.M.: Nu–Ra–Fo correlations for transient free convection in 2D convective diode cavities with discrete heat sources. Int. J. Heat Mass Transf. **57**, 623–628 (2013)
12. Nayak, R.K, Bhattacharyya, S., Pop, I.: Numerical study on mixed convection and entropy generation of Cu–water nanofluid in a differentially heated skewed enclosure. Int. J. Heat Mass Transf. **85**, 620–634 (2015)
13. Ghalambaz, M., Sheremet, M.A., Pop, I.: Free convection in a parallelogrammic porous cavity filled with a nanofluid using Tiwari and Das' nanofluid model. PLoS One **10**(5), e0126486 (2015)
14. Costa, V.A.F.: Double-diffusive natural convection in parallelogrammic enclosures filled with fluid saturated porous media. Int. J. Heat Mass Transf. **47**, 2699–2714 (2004)
15. Jagadeesha, R.D., Prasanna, B.M.R., Sankar, M.: Double diffusive convection in an inclined parallelogrammic porous enclosure. Proc. Eng. **127**, 1346–1353 (2015)
16. Jagadeesha, R.D., Prasanna, B.M.R., Sankar, M.: Numerical simulation of double diffusive magnetoconvection in an inclined parallelogrammic porous enclosure with an internal heat source. Mater. Today: Proc. **4**, 10544–10548 (2017)
17. Reddy, N.K., Sankar, M.: Buoyant heat transfer of nanofluids in a vertical porous annulus: a comparative study of different models. Int. J. Numer. Methods Heat Fluid Flow. **33**(2), 477–509 (2023)
18. Swamy, H.A.K., Sankar, M., Reddy, N.K., et al.: Double diffusive convective transport and entropy generation in an annular space filled with alumina-water nanoliquid. Eur. Phys. J. Spec. Top. **231**, 2781–2800 (2022)

An Application of Generalized Fourier and Fick's Law over a Different Non-Newtonian Fluid

R. Padmavathi and A. Revathi

1 Introduction

Applications of mass and heat transfer in design and industrial range widely, from the food industry and pollution control to nuclear reactor cooling and power generation. Parmar and Timol, and Fourier and Darboux, in [1, 2], analysed the passage of a power-law fluid via a cone's surface, which they found to be electrically conductive. During heat transmission, energy is typically transported from a hotter to a colder surface. Predicting the behaviour of heat transport under different situations receives attention because of the relevance of heat transport in a number of industrial fields. The well-known Fourier law of heat transport [2] makes it easier to understand how heat moves in many different situations. The law did have one major flaw, though. It went against the idea of cause and effect by letting even the smallest change in heat transfer be felt immediately throughout the whole system. By making some simple adjustments, Cattaneo [3] was able to conduct a classic research taking into consideration the thermal relaxation time in the Fourier model. Since heat can only be transmitted at a finite rate through the propagation of heat waves, it is evident that a hyperbolic equation is developed for temperature distribution. Abbas et al. [4] used the C-C heat flux on magnetic nanoparticles moving in a curved channel to study the heat transport mechanism. There are more CC-Model findings for other geometries provided in [5–8]. The findings demonstrate an increase in the induced magnetic field in response to a rise in the magnetic parameter and a decrease in

R. Padmavathi (✉)
Nitte Meenakshi Institute of Technology, Bangalore, India
e-mail: padmavathi.r@nmit.ac.in

A. Revathi
Presidency School Bangalore North, Bengaluru, India

F. Kamalov et al. (eds.), *Advances in Mathematical Modeling and Scientific Computing*, Trends in Mathematics, https://doi.org/10.1007/978-3-031-41420-6_40

the same field as a function of surface curvature. In addition, various studies [9–11] have established the concept of an induced magnetic field in a variety of geometries.

The chemical, polymer, and biomedical sectors all rely heavily on non-Newtonian fluids. It's made up of things like blood, mayonnaise, starch suspension, fruit liquids, polymers, toothpaste, etc. Almost all fluids with biological or industrial uses exhibit a non-linear connection between stress and the rate of deformation. All non-Newtonian fluids cannot have their behaviour captured by a single constitutive equation. Given the complexity of non-Newtonian fluids, researchers have created a number of mathematical models that may be found in the current literature. Convective heat transfer in the motion of a non-Newtonian fluid has been the subject of several studies [12–20].

In this work, we focus on flow across a sheet subject to a C-C heat and mass flux. The MHD group is also open to applications. The boundary layer equations are converted into non-linear ODEs along with the necessary boundary conditions, and the problem is numerically solved by employing the RKF-45 solver in Maple software.

2 Mathematical Formulation

We have considered steady non-Newtonian fluid flow over an elaborated surface. C_w and T_w denote concentration and temperature of elaborated surface, respectively. Moreover, C_∞ and T_∞ are the concentration and temperature away from the surface. The surface is moving with velocity u_w where a is constant. Along the positive y-axis, a non-uniform transverse magnetic field with a strength of $B(x) = B_0 x^{-1/3}$ is applied.

$$\frac{\partial u}{\partial y} = \frac{\tau}{\mu_\infty + \frac{\mu_0 - \mu_\infty}{1+\left(\frac{\tau}{\tau_s}\right)^2}}, \quad \text{[Reiner – Philippoff fluid model]} \tag{1}$$

The suggested model's governing equations for continuity, energy, momentum, and mass are [14, 16, 17, 18]:

$$\frac{\partial u}{\partial x} + \frac{\partial u}{\partial x} = 0, \tag{2}$$

$$u\frac{\partial u}{\partial x} + v\frac{\partial u}{\partial y} = \frac{1}{\rho}\frac{\partial \tau}{\partial y} - \frac{\sigma B^2(x)}{\rho}u, \tag{3}$$

$$u\frac{\partial T}{\partial x} + v\frac{\partial T}{\partial y} + \lambda_1 \begin{pmatrix} u^2\frac{\partial^2 T}{\partial x^2} + v^2\frac{\partial^2 T}{\partial y^2} + 2uv\frac{\partial^2 T}{\partial x \partial y} + \\ \left(u\frac{\partial u}{\partial x} + v\frac{\partial u}{\partial y}\right)\frac{\partial T}{\partial x} + \\ \left(u\frac{\partial v}{\partial x} + v\frac{\partial v}{\partial y}\right)\frac{\partial T}{\partial y} \end{pmatrix} = k\frac{\partial^2 T}{\partial y^2} + \frac{\sigma B^2}{\rho c_p}u^2, \tag{4}$$

$$u\frac{\partial C_0}{\partial x} + v\frac{\partial C_0}{\partial y} + \lambda_2 \begin{pmatrix} u^2\frac{\partial^2 T}{\partial x^2} + v^2\frac{\partial^2 T}{\partial y^2} + 2uv\frac{\partial^2 T}{\partial x \partial y} + \\ \left(u\frac{\partial u}{\partial x} + v\frac{\partial u}{\partial y}\right)\frac{\partial T}{\partial x} + \\ \left(u\frac{\partial v}{\partial x} + v\frac{\partial v}{\partial y}\right)\frac{\partial T}{\partial y} \end{pmatrix} = D\frac{\partial^2 C_0}{\partial y^2}. \tag{5}$$

Associative boundary conditions are given by:

$$u = u_w(x) = ax^{1/3},\ v = 0,\ \frac{\partial T}{\partial y} = -h_1 T,\ \frac{\partial C_0}{\partial y} = -h_2 c_0\ at\ y = 0,$$

$$u \to 0,\ T \to T_\infty C_0 \longrightarrow C_\infty \text{ as } y \to \infty, \tag{6}$$

where ρ – the densities of the fluid, B – uniform magnetic field, T, C_0, k – temperature, concentration, and thermal conductivity of the fluid, respectively, D – mass diffusivity coefficient, λ_1 – relaxation time of heat flux, λ_2 – relaxation time of mass flux, h_1 – heat transfer coefficient, and h_2 – mass diffusion coefficient.

The first step is to non-dimensionalize the system we've been provided with. We introduce the dimensionless quantities below for this purpose:

$$\eta = \sqrt{\frac{a}{\nu}}\frac{y}{x^{1/3}},\ \psi = \sqrt{a\nu}x^{2/3},\ \tau = \rho\sqrt{a^3\nu}g(\eta),\ \ \theta(\eta) = \frac{T - T_\infty}{T_w - T_\infty},$$
$$\phi(\eta) = \frac{C_0 - C_\infty}{C_w - C_\infty}. \tag{7}$$

Transforming Eqs. (1)–(5) into their non-dimensional forms, we get:

$$g = f''\frac{g^2 + \lambda\gamma^2}{g^2 + \gamma^2}, \tag{8}$$

$$g' = \frac{1}{3}f'^2 - \frac{2}{3}ff'' + Mf', \tag{9}$$

$$\theta'' + \frac{2}{3}\text{Pr}\ f\theta' + \text{Pr}\Gamma_1\left(ff'\theta' + \eta f^2\theta''\right) + Ec\ M\ f'^2 = 0, \tag{10}$$

$$\phi'' + \frac{2}{3}Sc\ f\phi' + \text{Sc}\Gamma_2\left(ff'\theta' + \eta f^2\phi''\right) = 0. \tag{11}$$

$\lambda_* = 1$, $\lambda_* < 1$, and $\lambda_* > 1$ represent the case of Newtonian fluid, dilatant, and pseudo-plastic, respectively.

Now we have,

$$f(0)=0,\ f'(0)=1,\quad \theta'(0)=-\gamma_1\left(1-\Theta(0)\right),\ \phi'^{(0)}=-\gamma_2\left(1+\phi(0)\right),$$

$$f'(\infty)\to 0,\ \theta(\infty)\to 0,\ \phi(\infty)=0. \tag{12}$$

where $\lambda^* = \frac{\mu_0}{\mu_\infty}$ non-Newtonian fluid parameter, $\gamma = \frac{\tau_s}{\rho\sqrt{a^3\nu}}$ Bingham number, $M = \frac{\sigma B_0^2}{\rho a}$ Magnetic parameter, $\Pr = \frac{\nu}{\alpha}$ Prandtl number, $\varGamma_1 = a\lambda_1$ thermal heat flux, $\varGamma_2 = a\lambda_2$ solutal mass flux, $Ec = \frac{u_w{}^2}{(T_w - T_\infty)c_p}$ Eckert number, $Sc = \frac{\upsilon}{D}$ Schmidt number, $\gamma_1 = \frac{h_1}{k}\sqrt{\frac{\upsilon}{a}}$ conjugate heat flux, $\gamma_2 = \frac{h_2}{k}\sqrt{\frac{\upsilon}{a}}$ conjugate mass flux.

The Nu_x and Sh_x are defined as follows:

$$Nu_x = \frac{xq_w}{k\left(T_w - T_\infty\right)},\ Sh_x = \frac{xJ_w}{k_f\left(C_\infty - C_m\right)}.$$

where

$$q_w = -k\left(\frac{\partial T}{\partial y}\right)_{y=0} \text{ and } J_w = -k_{nf}\left(\frac{\partial C_0}{\partial y}\right)_{y=0}.$$

The reduced Nu_x and Sh_x in dimentionless form are:

$$Nu_x Re_x^{-1/2} = -\theta'(0),\ Sh_x(\mathrm{Re})^{-\frac{1}{2}} = -\phi'(0).$$

3 Method of Solution

By imposing boundary conditions, a non-linear system of ODEs (see Eqs. (8)–(12)) may be solved using RKF-45 in Maple. In this diagrammatic study, we examine the effects of varying the body forces' control variables on the momentum, energy, and mass field. The $Nu_x Re_x^{-1/2}$ and $Sh_x(\mathrm{Re})^{-\frac{1}{2}}$ are tabulated for a variety of values when $\gamma = 0.5$, $M = 0.5$, $Pr = 6$, $\varGamma_1 = 0.5$, $\Gamma_2 = 0.4$, $Ec = 0.5$, $Sc = 3$, $\gamma_1 = 0.5$, and $\gamma_2 = 0.5$. Obtaining first-order forms of higher-order ODEs by reducing them as:

$$f = y_1,\ f' = y_2,\ f'' = y_2',\ g = y_3,\ g' = y_4,$$

$$y_2' = y_3\frac{\left(y_3^2 + \lambda\gamma^2\right)}{\left(y_3^2 + \lambda^2\right)}, \tag{13}$$

$$y_2' = \frac{3}{2}y_1\left[\frac{1}{3}y_2^2 - y_4 + My_2\right], \tag{14}$$

$$\theta = y_5, \theta' = y_6, \theta'' = y_6',$$

$$y_6{}' = -\left[\frac{2}{3}\text{Pr } y_1y_6 + \text{Pr}\Gamma_1\left(y_1y_2y_6 + \eta y_1^2y_6{}'\right) + Ec\ M\ y_1^2\right], \tag{15}$$

$$\phi = y_7, \phi' = y_8, \phi'' = y_8', 0$$

$$\phi'' = -\left[\frac{2}{3}Sc\ y_1y_8 + \text{Sc}\Gamma_2\left(y_1y_2y_6 + \eta y_1y_6'\right)\right], \tag{16}$$

now we have:

$$y_1(0) = 0, y_2(0) = 1,\ \ y_6(0) = -\gamma_1\left(1 - y_5(0)\right),\ \ y_8(0) = -\gamma_2\left(1 + y_7(0)\right),$$

$$y_2(\infty) \to 0, y_5(\infty) \to 0, y_7(0) = 0. \tag{17}$$

4 Result and Discussion

Studying the two-dimensional motion of a non-Newtonian fluid under the influence of Joule heating, MHD, and C-C heat and mass flux via a stretched sheet is the focus of this investigation. Here, we use graphs to examine the set-up of guaranteed parameters. In order to resolve the ODE, we incorporated the Maple function (RKF-45). Initial test calculations for $\theta'(0)$ are performed for viscous fluid for various values of *Pr*, and the results are compared with the available published findings by Gorla and Sidawi [21] in Table 1, where good agreement is obtained.

To explain the observed shift in $f'(\eta)$ distribution as a function of varying *M* parameter, see Fig. 1. The drawing makes clear that a higher estimate of *M* causes a decline in the fluid flow velocity sketch. Furthermore, for highly regarded *M*, the

Table 1 Comparison table for $-\theta'(0)$ Ec = R = M = 0

Pr	Gorla and Sidawi [21]	Present
0.2	0.1691	0.170259788
0.7	0.5349	0.454447258
2	0.9114	0.911352755
7	1.8905	1.895400395

Fig. 1 Sketch of $f'(\eta)$ for M

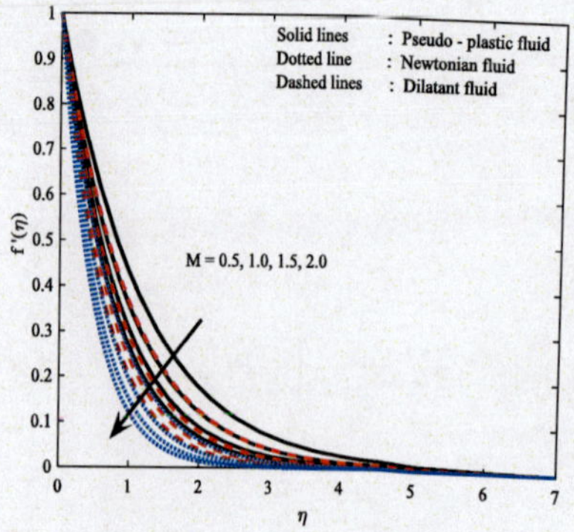

Fig. 2 Sketch of $\theta(\eta)$ for M

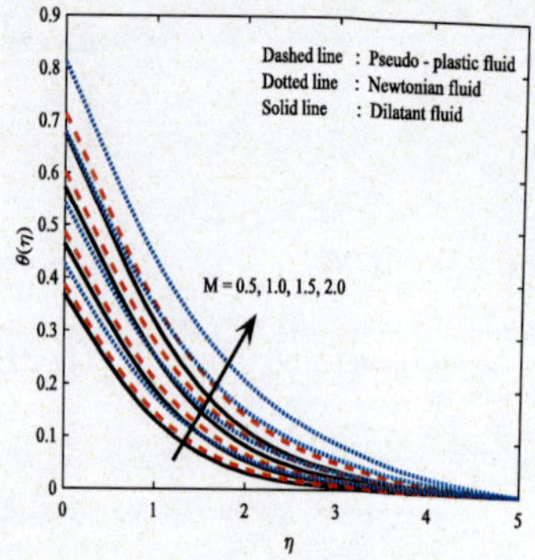

inter-relevance closeness of the momentum layer is reduced. As the intensity of the magnetic field increases, so does the amount of the force operating on us (denoted by M). So the fluid speed drops. The figure also reveals that the fluid travels more slowly in the case of $\lambda_* = 1$ compared to the cases of $\lambda_* < 1$ and $\lambda_* > 1$.

Figure 2 elucidates this facet of the M on $\theta(\eta)$ field. An increase in the M is shown in Fig. 2 by a thickening of the $\theta(\eta)$ field. At the same time, with increased M, the thermal horizon layer thickens and becomes more compact. The instance of $\lambda_* = 1$ sees a larger improvement than the cases of $\lambda_* < 1$ and $\lambda_* > 1$. Figure 3 shows how Γ_1 affects on $\theta(\eta)$ field. For larger values of Γ_1, the thermal field and associated width naturally decrease. Because of thermal relaxation time and the lengthened time necessary for heat to travel to neighbouring nanoparticles, the temperature field gradually fades over time. As a corollary, the rate of heat transfer is greatest when $\lambda_* = 1$, then next highest when $\lambda_* > 1$, and finally lowest when $\lambda_* < 1$.

Figure 4 shows the impact of γ_1 on $\theta(\eta)$. To our surprise, when we increase the size of γ_1, the $\theta(\eta)$ profile begins to exhibit a rising trend. Further, it is evident from this image that the proximity of the matching horizon layer increases rapidly with increasing γ_1. Furthermore, in the situation of $\lambda_* > 1$ and $\lambda_* < 1$, the rate of heat transfer is dominated by the $\lambda_* = 1$ scenario.

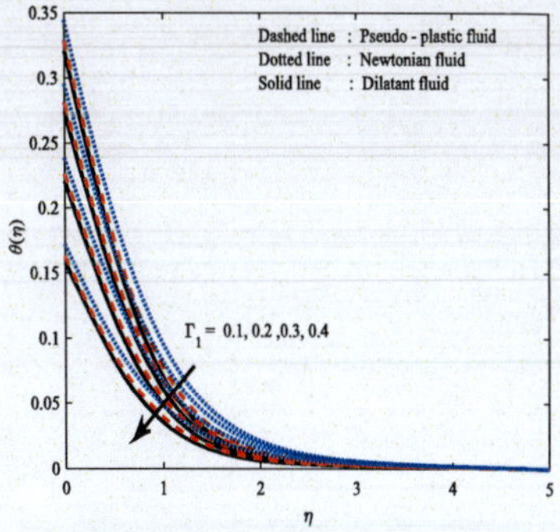

Fig. 3 Sketch of $\theta(\eta)$ for Γ_1

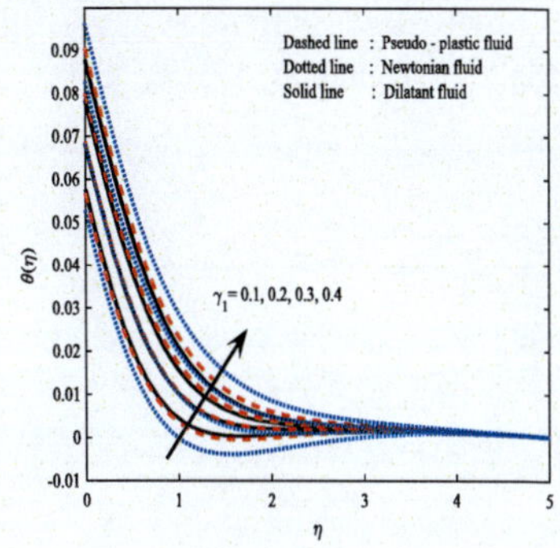

Fig. 4 Sketch of $\theta(\eta)$ for γ_1

Figure 5 shows that when *Pr* rises, the $\theta(\eta)$ structure declines. When comparing thermal diffusion rates to momentum diffusion rates, the generalized rate constant *Pr* is used to characterize the relationship between the two. Since the structure of the thermal layer is degraded, the boundary layer temperature decreases as *Pr* rises. Additionally, as compared to the $\lambda* = 1$ and $\lambda* > 1$ cases, the rate of heat transmission decreases significantly when $\lambda* < 1$. Figure 6 further shows that the $\theta(\eta)$ profile rises with increasing *Ec* parameter values. Furthermore, the effect of increasing *Ec* values is shown by an accompanying increase in layer thickness. Mechanical and electrical energy are converted to heat via the fluid's magnetic resistance and internal friction. That's why an improved *Ec* is good for the thermal horizon layer and the $\theta(\eta)$ field.

The plot of the $\phi(\eta)$ field versus Γ_2 parameter is shown and described in Fig. 7. It indicates that the associated thickness of the solutal layer density will decrease as a function of the increased heat that is supplied to the liquid in proportion to the increase in the Γ_2 value. This is because an increase in Γ_2 leads to a decrease in the corresponding $\phi(\eta)$ field, which explains the observed phenomenon. Figure 8 delves into the nuances of $\phi(\eta)$ for increasing γ_2. As can be seen in the picture, the corresponding thickness and $\phi(\eta)$ field decrease as γ_2 increases. Since nanoparticles take longer to diffuse into neighbouring regions, the solutal relaxation time rises, leading to higher mass concentration and a thicker solutal layer. Figure 9 shows the $\phi(\eta)$ profile, which describes the centre of *Sc* on in detail. *Sc* is a dimensionless

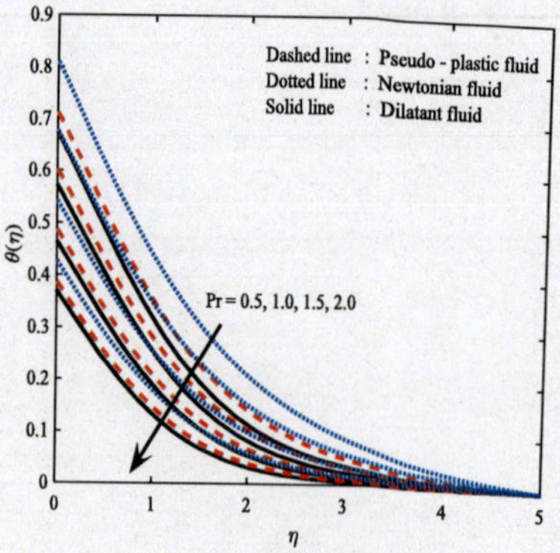

Fig. 5 Sketch of $\theta(\eta)$ for Pr

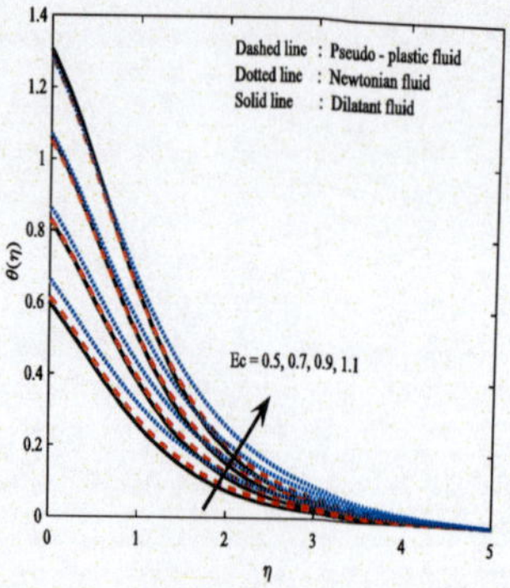

Fig. 6 Sketch of $\theta(\eta)$ for Ec

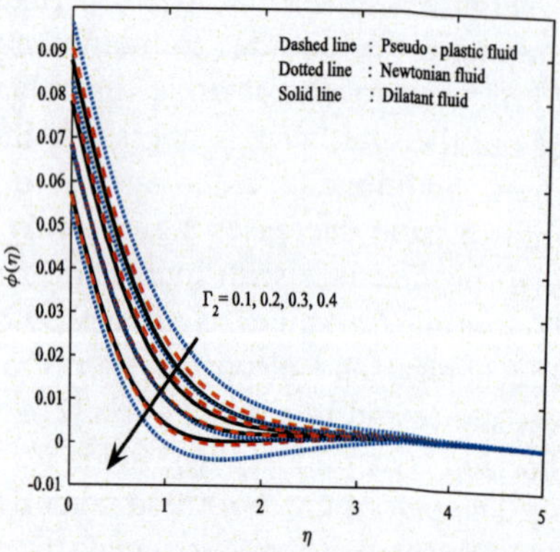

Fig. 7 Sketch of $\phi(\eta)$ for Γ_2

quantity that specifies the thickness ratio between the hydrodynamic boundary layer and the concentration boundary layer. Therefore, as seen in Fig. 9, larger Sc results in flatter $\phi(\eta)$ profiles due to the inverse relationship between Sc and mass diffusivity.

The $Nu_x Re_x^{-1/2}$ as a function of the Pr with contrasting character of Ec and γ_1 parameter have been plotted in the Figs. 10 and 11. It is observed from Fig. 10 that reduction in the $Nu_x Re_x^{-1/2}$ occurs for all $\lambda^* < 1$, $\lambda^* = 1$ and $\lambda^* > 1$ cases by

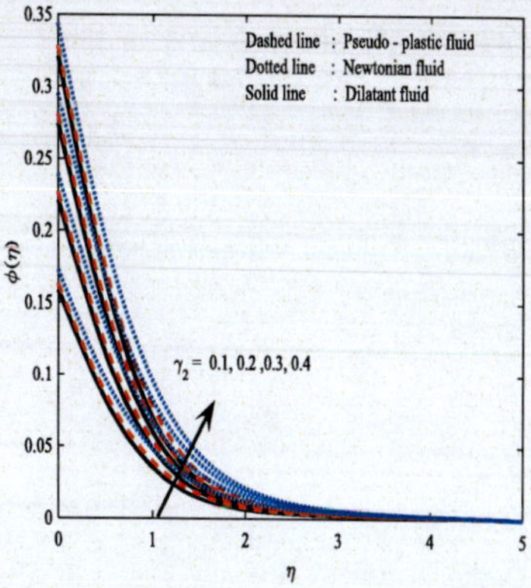

Fig. 8 Sketch of $\phi(\eta)$ for γ_2

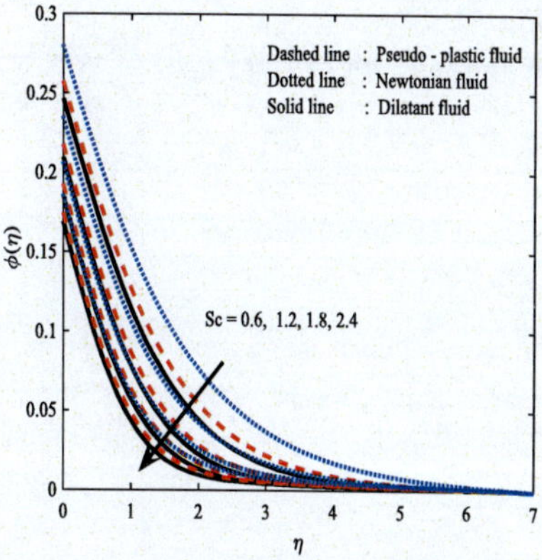

Fig. 9 Sketch of $\phi(\eta)$ for Sc

rising the values of Pr and Ec. Furthermore, heat transfer rate declines faster in all cases. Figure 11 depicts the $Nu_x Re_x^{-1/2}$ as a function of Pr with different range values of γ_1 parameter. It seems that $Nu_x Re_x^{-1/2}$ decreases when Pr and Ec are increasing. Figure 12 is a plot of the $Sh_x(\text{Re})^{-\frac{1}{2}}$ as a function of the Sc and γ_2 parameter. It is checked from Fig. 12 that cutback in the $Sh_x(\text{Re})^{-\frac{1}{2}}$ occurs for all $\lambda_* < 1$, $\lambda_* = 1$, and $\lambda_* > 1$ cases by rising the values of Sc and γ_2. Moreover, in the case of $\lambda_* > 1$, the heat transfer rate decreases at a quicker rate than in the cases of $\lambda_* < 1$ and $\lambda_* = 1$.

5 Conclusion

The Matlab function RKF-45 addresses the flow, heat, and mass transfer model of non-Newtonian fluids with Joule heating, C-C heat, and mass flux.

- A decaying concentration of fluid is noted for Sc.

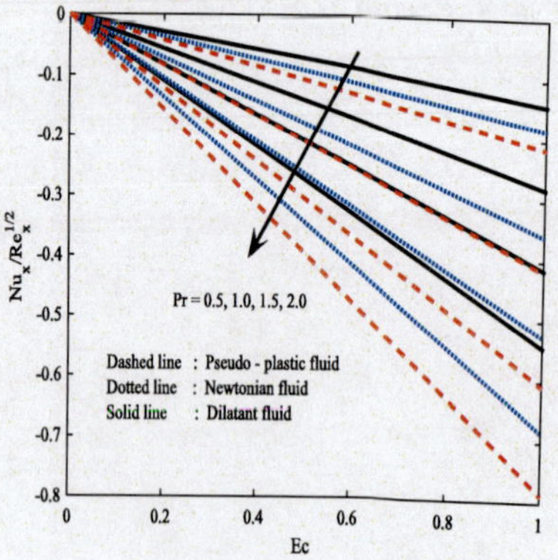

Fig. 10 Sketch of $Nu_x Re_x^{-1/2}$ for Pr and Ec

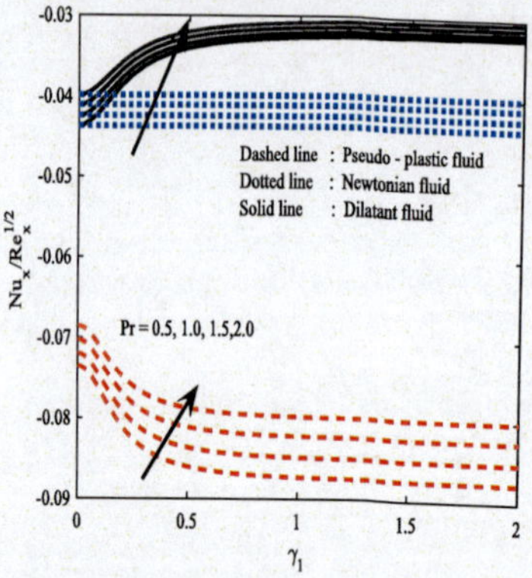

Fig. 11 Sketch of $Nu_x Re_x^{-1/2}$ for Pr and γ_1

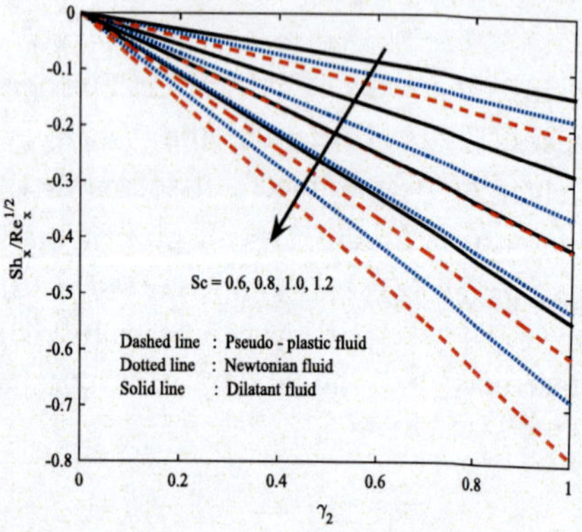

Fig. 12 Sketch of $Sh_x Re_x^{-1/2}$ for Sc and γ_2

- A rapid velocity decrement in velocity is observed when M parameter gets increasing values.
- The temperature fields strengthen with enhanced values of Ec parameter.
- The concentration plot is thinner with higher Γ_2 values.
- The temperature field shows increasing behaviour with γ_1 parameter.
- Intensity of Nusselt number decays for rising $Pr$and $\gamma_1$is noticed.

- Magnitude of $Sh_x(\mathrm{Re})^{-\frac{1}{2}}$ improves for increasing values of Sc and γ_2 parameter.
- Magnitude of $Sh_x(\mathrm{Re})^{-\frac{1}{2}}$ diminished via rising Γ_2.

References

1. Parmar, H., Timol, M.G.: Deductive group technique for MHD coupled heat and mass transfer natural convection flow of non-Newtonian power law fluid over a vertical cone through porous medium. Int. J. Appl. Math Mech. **7**(2), 35–50 (2011)
2. Fourier, J.B.J., Darboux, G.: Théorie analytique de la chaleur, vol. 504. Didot, Paris (1822)
3. Cattaneo, C.: Sulla conduzione del calore. Atti Sem. Mat. Fis. Univ. Modena. **3**, 83–101 (1948)
4. Abbas, Z., Hussain, S., Naveed, M., Nadeem, A., Ali, A.: Analysis of joule heating and generalized slip flow in ferromagnetic nanoparticles in a curved channel using Cattaneo-Christov heat flux theory. Therm. Sci. **26**(1), 437–448 (2022). https://doi.org/10.2298/TSCI200410220A
5. Zhang, Y., Shahmir, N., Ramzan, M., Alotaibi, H., Aljohani, H.M.: Upshot of melting heat transfer in a Von Karman rotating flow of gold-silver/engine oil hybrid nanofluid with cattaneo-christov heat flux. Case Stud. Therm. Eng. **26**, 101149 (2021). https://doi.org/10.1016/j.csite.2021.101149
6. Ramzan, M., Abid, N., Lu, D., Tlili, I.: Impact of melting heat transfer in the timedependent squeezing nanofluid flow containing carbon nanotubes in a Darcy-Forchheimer porous media with Cattaneo-Christov heat flux. Commun. Theor. Phys. **72**(8), 85801 (2020). https://doi.org/10.1088/1572-9494/ab8a2c
7. Bashir, S., Ramzan, M., Ghazwani, H.A.S., Nisar, K.S., Saleel, C.A., Abdelrahman, A.: Magnetic dipole and thermophoretic particle deposition impact on bioconvective Oldroyd-B fluid flow over a stretching surface with Cattaneo–Christov heat flux. Nanomaterials. **12**(13), 2181 (2022). https://doi.org/10.3390/nano12132181
8. Reddy, P.S., Sreedevi, P., Chamkha, A.J.: Heat and mass transfer analysis of nanofluid flow over swirling cylinder with Cattaneo–Christov heat flux. J. Therm. Anal. Calorim. **147**(4), 3453–3468 (2022). https://doi.org/10.1007/s10973-021-10586-9
9. Amjad, M., Zehra, I., Nadeem, S., Abbas, N., Saleem, A., Issakhov, A.: Influence of Lorentz force and induced magnetic field effects on Casson micropolar nanofluid flow over a permeable curved stretching/shrinking surface under the stagnation region. Surf. Interface. **21**, 100766 (2020). https://doi.org/10.1016/j.surfin.2020.100766
10. Al-Hanaya, A.M., Sajid, F., Abbas, N., Nadeem, S.: Effect of SWCNT and MWCNT on the flow of micropolar hybrid nanofluid over a curved stretching surface with inducedmagnetic field. Sci. Rep. **10**(1), 1–18 (2020). https://doi.org/10.1038/s41598-020-65278-5
11. Ghosh, R., Agbaje, T.M., Mondal, S., Shaw, S.: Bio-convective viscoelastic Casson nanofluid flow over a stretching sheet in the presence of induced magnetic field with Cattaneo–Christov double diffusion. Int. J. Biomath. **15**(3), 2150099 (2022). https://doi.org/10.1142/S1793524521500996
12. Bird, R.B., Curtiss, C.F., Armstrong, R.C., Hassager, O.: Dynamics of Polymeric Liquids. Wiley, New York (1987)
13. Bird, R.B.: Useful non-Newtonian models. Annu. Rev. Fluid Mech. **8**, 13–34 (1976)
14. Kumar, K.G., Reddy, M.G., Sudharani, M., Shehzad, S.A., Chamkha, A.J.: Cattaneo–Christov heat diffusion phenomenon in Reiner–Philippoff fluid through a transverse magnetic field. Physica A Stat. Mech. Appl. **541**, 123330 (2020)
15. Reddy, M.G., Sudharani, M., Praveena, M.M., Kumar, K.G.: Effect of thermal conductivity on Blasius–Rayleigh–Stokes flow and heat transfer over a moving plate by considering magnetic dipole moment. Eur. Phys. J. Plus. **137**(1), 1–13

16. Reddy, M.G., Sudharani, M., Ganesh Kumar, K., Chamkha, A.: Physical aspects of Darcy–Forchheimer flow and dissipative heat transfer of Reiner–Philippoff fluid. J. Therm. Anal. Calorimetr. **141**(2), 829–838
17. Reddy, M.G., Rani, S., Kumar, K.G., Seikh, A.H., Rahimi-Gorji, M., Sherif, E.S.M.: Transverse magnetic flow over a Reiner–Philippoff nanofluid by considering solar radiation. Modern Physics Letters B. **33**(36), 1950449
18. Kumar, K.G., Lokesh, H.J., Shehzad, S.A., Ambreen, T.: On analysis of Blasius and Rayleigh–Stokes hybrid nanofluid flow under aligned magnetic field. J. Therm. Anal. Calorimetr. **139**(3), 2119–2127
19. Khan, S.U., Khaled, K., Bhatti, M.M.: Numerical experiment of Reiner–Philippoff nanofluid flow subject to the higher-order slip features, activation energy, and bioconvection. Partial Differ. Equ. Appl. Math. **4**, 100126 (2021)
20. Çolak, A.B.: Investigation of flow parameters of Reiner–Philippoff nanofluid flow with higher-order slip properties, activation energy, and bioconvection by artificial neural networks. Heat Transf. **51** (2022). https://doi.org/10.1002/htj.22529
21. Gorla, R.S.R., Sidawi, I.: Free convection on a vertical stretching surface with suction and blowing. Appl. Sci. Res. **52**, 247–257 (1994)

Part III
Operation Research

Production Inventory Model for Three Levels of Production with Defective Items, Shortages Including Multi-delivery Policy

R. Pavithra and K. Karthikeyan

1 Introduction

Many years ago, Harris develop the economic order quantity (EOQ) model, a mathematical model. To assist practitioners in the production and inventory control domains in determining the economic replenishment batch size that minimizes overall production inventory, Taft (1918) developed the economic production quantity (EPQ) model. Chiu et al. [1] proposed an algebraic technique to determining the optimal common cycle time in the multi-item EPQ model that replaces the usage of differential calculus. Viji and Karthikeyan [2] considered a three-level production model for degradation goods and examined three distinct stages of production. The rate of degradation was determined by two variables Weibull distribution. Jia-Tzer Hsu and Lie-FernHsu [3] developed two optimal production quantity models with imperfect manufacturing processes, inspection mistakes, anticipated backorders, and sales return. Hector River-Gomez et al. [4] proposed a new joint policy that takes into account the relationships between production quality and maintenance. Harun Ozturk [5] presented a deterministic manufacturing inventory model with backorder, damaged items, and an inadequate rework method. A modified inventory model that incorporates defective-quality products was put out by Salameh and Jaber [6]. Essey Kebede Muluneh et al. [7] create and test an inventory model for degrading products, assuming that the production rate is determined by the amount of stock on hand. Zhou et al. [8] presented the inventory replenishment model for a single item with a demand rate that varies with inventory is expanded to the situation of two warehouses (i.e) owned warehouse and rented warehouse. Sabheji das et al. [9] considered the manufactures and consumers trade credit policies are established by

R. Pavithra · K. Karthikeyan (✉)
Department of Mathematics, SAS, VIT Vellore, Vellore, India
e-mail: k.karthikeyan@vit.ac.in

F. Kamalov et al. (eds.), *Advances in Mathematical Modeling and Scientific Computing*, Trends in Mathematics, https://doi.org/10.1007/978-3-031-41420-6_41

taking into account all conceivable scenarios based on permissible credit durations. In a manufacturing inventory model with defective products that considers two variable production rates, Sivashankari and Panayappan [10] incorporate a multi-delivery strategy. Zsuzsanna Hauck et al. [11] created a profit maximization model that took into account price, quality screening, and lot sizing. Allen H. Tai [12] discussed two models one with a single production rework and another with n production sites and rework.

In this chapter, a multi-delivery policy is combined with a manufacturing inventory model with defective items and shortages. Three alternative production rates are examined, and it is possible that manufacturing started at one rate and then switched to another. This situation is preferable in that a high volume of manufacturing goods is avoided at the beginning with the intention of limiting by starting at a low rate of production. By comparing the two models, model II gives a minimum holding cost.

2 Assumptions

The following is an inventory model assumption for the product life cycle:

(a) The demand rate is well-known, constant, and continuous.
(b) Products are manufactured and added to the inventory.
(c) In inventory, three rates of production are analyzed and shortages are allowed.
(d) The item is a single item that has no interactions with any other items in the inventory.
(e) The ratio of the production rate to the demand rate and defective items is always larger than or equal to one.
(f) Consider a production system that has the potential to create x portion of random defective goods at the manufacturing rate of d (i.e.,) $d = Kx$.
(g) A first installment of finished goods is provided to customers during the producer's manufacturing uptime to meet customer demand. A fixed quantity n installment of finished goods will be given to customers at a fixed interval of time at the end of manufacturing.

3 Notations

K – Rate of production in units per unit time
D – Rate of demand in units per unit time
d – Defective items during regular production
Q_1 – On-hand inventory level at time t_1
Q_2 – On-hand inventory level at time t_2

Q_3 – On-hand inventory level at time t_3
Q – Optimum production
c_p – Production cost per unit
c_d – Defective cost per unit
c_h – Holding cost
c_o – Setup cost
c_{FD} – Fixed delivery cost
c_H – Supplier holding cost
c_S – Shortage cost
n – Number of deliveries
T – Cycle duration
t_i – Unit time ($i = 1,2,3,4,5,6$)
TC – Total cost
THC – Total holding cost

4 EPQ Model in Three Levels of Production with Defective Items and Shortages: Model I

At time t_1,the manufacturing rate (K), the demand rate (D), and the defective rate (d). At time $t = 0$, the cycle begins. The inventory is increased at a rate of K-D-d, $t_1 = \frac{Q_1}{K-D-d}$.

At time t_2, the manufacturing rate (aK), the demand rate (aD), and the defective rate (ad), where 'a' is a constant. The inventory is increased at a rate of $a(K - D - d)$, $t_2 = \frac{Q_2-Q_1}{a(K-D-d)}$.

At time t_3, the manufacturing rate (bK), the demand rate (bD), and the defective rate (bd), where 'b' is a constant. The inventory is increased at a rate of $b(K - D - d)$, $t_3 = \frac{Q_3-Q_2}{b(K-D-d)}$.

Customer taste change during the decline time t_4. Care should be made to regulate the number of inventories of the product. The inventory level starts to decline owing to demand at a rate of D $t_4 = \frac{Q_3}{D}$.

Time t_5 required to construct B units of items, $t_5 = \frac{B}{D}$.

The manufacture resumes again at time t_6 at a rate of K-D-d in order to recover both the previous shortages in the period t_6 and to meet demand in the period t_6, $t_6 = \frac{B}{(K-D-d)}$.

Time T needed to consume all unit's Q at demand rate, i.e., $Q = DT$, $T = \frac{Q}{D}$.

The process is repeated. Figure 1 shows the variation of the underlying inventory system for one cycle.

From triangular identities, $\frac{t_1}{t_2} = \frac{Q_1}{Q_2-Q_1} = \frac{K-D-d}{a(K-D-d)}$

$$Q_2 = (1 + a)\, Q_1$$

Fig. 1 On-hand inventory of defective items

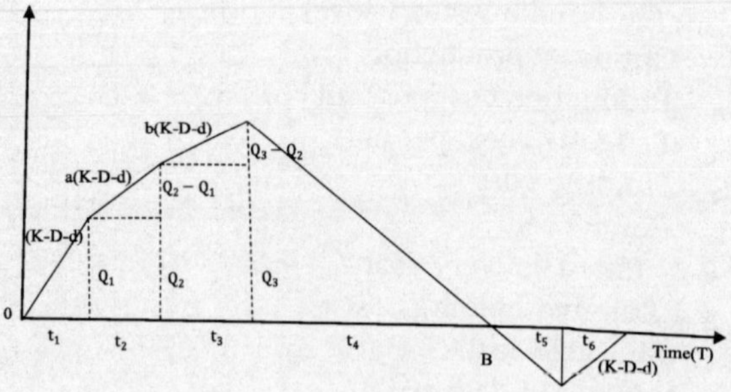

$$\frac{t_1}{t_3} = \frac{Q_1}{Q_3 - Q_2} = \frac{K - D - d}{b\,(K - D - d)}$$

$$Q_3 = (1 + a + b)\,Q_1$$

Inventory level during production cycle, $T = t_1 + t_2 + t_3 + t_4 + t_5 + t_6$

$$Q_1 = \frac{Q\,(K - D - d) - BK\,(1 - x)}{3D + (K - D - d)\,(a + b + 1)}$$

Average inventory (AI) $= \frac{1}{2}Q_1t_1 + Q_1t_2 + \frac{1}{2}t_2\,(Q_2 - Q_1) + Q_2t_3 + \frac{1}{2}t_3\,(Q_3 - Q_2) + \frac{1}{2}Q_3t_4$

$$= \frac{Q_1^2}{2}\left[\frac{(5 + 3a + b)\,D + (1 + a + b)^2\,(K - D - d)}{D\,(K - D - d)}\right]$$

Substituting Q_1 value in the above equation. we get,

$$= \frac{Q^2\,(K - D - d)\,A}{2D} + \frac{B^2K^2(1 - x)^2A}{2D\,(K - D - d)} - \frac{QBK\,(1 - x)\,A}{D}, \text{ where}$$

$$A = \frac{(5 + 3a + b)\,D + (1 + a + b)^2\,(K - D - d)}{(3D + (K - D - d)\,(a + b + 1))^2}$$

The average inventory during shortage period: $I_s = \frac{1}{T}\left[\frac{1}{2}t_5B + \frac{1}{2}t_6B\right]$

$$= \frac{B^2K\,(1 - x)}{2Q\,(K - D - d)}$$

Total cost

Production cost $= \frac{1}{T}Qc_p = Dc_p$

Setup cost $= \frac{1}{T}c_o = \frac{Dc_o}{Q}$
Holding cost $= \frac{c_h}{T}(AI)$

$$= \left[\frac{c_h Q (K-D-d) A}{2} + \frac{c_h B^2 K^2 (1-x)^2 A}{2Q(K-D-d)} - c_h BK(1-x)A\right]$$

Shortage cost $= \frac{c_s B^2 K(1-x)}{2Q(K-D-d)}$
Defective cost $= \frac{1}{T}c_d Qx = Dc_d x$

$$TC = Dc_p + \frac{Dc_o}{Q} + \frac{c_h Q (K-D-d) A}{2} + \frac{c_h B^2 K^2 (1-x)^2 A}{2Q(K-D-d)} - c_h BK(1-x)A + \frac{c_s B^2 K(1-x)}{2Q(K-D-d)} + Dc_d x$$

4.1 Optimality

It can be easily shown that *TC (Q)* is a convex function in Q. Hence, an optimal manufacturing quantity Q^* can be calculated from $\frac{\partial TC}{\partial Q} = 0$, $\frac{\partial^2 TC}{\partial Q^2} > 0$.

From the above equation, $Q^2 = \frac{2Dc_o}{c_h A(K-D-d)} + \frac{B^2 K^2 (1-x)^2}{(K-D-d)^2} + \frac{PB^2 c_s (1-x)}{c_h (K-D-d)^2 A}$.
Partially differentiate with respect to B,

$$\frac{\partial TC}{\partial B} = \frac{c_h K^2 B (1-x)^2 A}{Q(K-D-d)} - c_h K(1-x)A + \frac{c_s KB(1-x)}{Q(K-D-d)} = 0 \text{ and}$$

$$\frac{\partial^2 TC}{\partial B^2} = \frac{c_h K^2 (1-x)^2 A}{Q(K-D-d)} + \frac{c_s K(1-x)}{Q(K-D-d)} > 0.$$

From the above equation, the shortage level B, $B = \frac{c_h Q(K-D-d)A}{Kc_h(1-x)A+c_s}$.
The optimum quantity,

$$Q = \sqrt{\frac{2Dc_o[Kc_h(1-x)A+c_s]^2}{c_h(P-D-d)A\left[[Kc_h(1-x)A+c_s]^2 - K^2(1-x)^2 c_h^2 A^2 - Kc_s(1-x)c_h A\right]}}.$$

4.2 Numerical Illustration

Let us assume some parameters: $K = 5000$, $D = 4000$, $c_o = 80$, $d = 0$, $a = 1$, $c_h = 4$, $c_p = 50$, $c_d = 10$,
$b = 2$, $c_H = 5$, $c_{FD} = 2$, $c_s = 10$, $x = 0.01$.

Solutions

$Q = 1023.8$, $B = 62.51$, $Q_1 = 44.65$, $Q_2 = 89.30$, $Q_3 = 178.60$, T=0.2560

Holding cost = Rs. 218.09, Ordering cost = Rs. 312.55, Shortage cost = Rs. 94.46, Production cost =

Rs. 2,00,000, Defective cost = Rs. 400 and Total cost = Rs. 2,01,030.

5 EPQ Model in Three Levels of Production with Defective Items, Shortages and including Multi-delivery Policy: Model II

At time $t = 0$, the cycle begins. At time t_1, the manufacturing rate (K), the demand rate (D), and the defective rate (d). The inventory is increased at a rate of *K-D-d*, $t_1 = \frac{Q_1}{K-D-d}$.

At time t_2, the manufacturing rate (aK), the demand rate (aD), and the defective rate (ad), where "a" is a constant. The inventory is increased at a rate of $a(K - D - d)$, $t_2 = \frac{Q_2-Q_1}{a(K-D-d)}$.

At time t_3, the manufacturing rate (bK), the demand rate (bD), and the defective rate (bd), where "b" is a constant. The inventory is increased at a rate of $b(K - D - d)$, $t_3 = \frac{Q_3-Q_2}{b(K-D-d)}$.

After the production downtime, the remainder of the manufacturing lot is quality guaranteed, and set amount "n" installments of finished goods are supplied to clients at predetermined intervals of time. The goal of such a delivery policy is to decrease the holding costs. Customer tastes change during the decline time t_4. A multi-delivery time is considered, and it is also believed that final goods may only be supplied to customers if the entire lot is quality guaranteed at the end of the manufacturing process.

During the production downtime t_4, n-instalments of fixed-quantity are provided to customers at regular intervals.

The product becomes technologically obsolete or customer tastes change during the decline time t_4. Care should be made to regulate the number of inventories of the product. The inventory level starts to decline owing to demand at a rate of D $t_4 = \frac{Q_3}{D}$.

Time t_5 required to construct B units of items, $t_5 = \frac{B}{D}$.

The manufacturing resumes again at time t_6 at a rate of *K-D-d* in order to recover both the previous shortages in the period t_6 and to meet demand in the period t_6, $t_6 = \frac{B}{(K-D-d)}$ (Fig. 2).

Time T needed to consume all unit's Q at demand rate, i.e., $Q = DT$, $T = \frac{Q}{D}$.

From triangular identities, $\frac{t_1}{t_2} = \frac{Q_1}{Q_2-Q_1} = \frac{K-D-d}{a(K-D-d)}$

$$Q_2 = (1 + a)\, Q_1$$

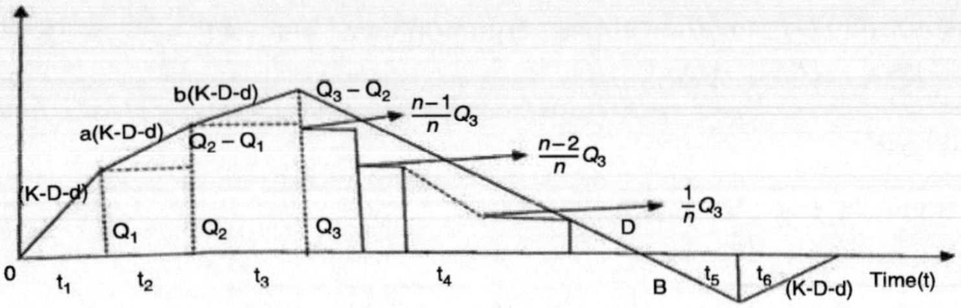

Fig. 2 On-hand inventory level

$$\frac{t_1}{t_3} = \frac{Q_1}{Q_3 - Q_2} = \frac{K - D - d}{b\,(K - D - d)}$$

$$Q_3 = (1 + a + b)\, Q_1.$$

Inventory level during production cycle, $T = t_1 + t_2 + t_3 + t_4 + t_5 + t_6$

$$Q_1 = \frac{Q\,(K - D - d) - BK\,(1 - x)}{3D + (K - D - d)\,(a + b + 1)}.$$

Calculating the holding cost of completed goods held by the producer throughout the distribution period:

(i) If $n = 1$, $THC = 0$ at time t_4.
(ii) If $n = 2$, THC in delivery time are $c_h\left(\frac{Q_3}{2} \bullet \frac{t_4}{2}\right) = c_h\left(\frac{1}{2^2}\right) Q_3 t_4$.
(iii) If $n = 3$, THC in delivery time are $c_h\left(\frac{2Q_3 t_4}{3^2} + \frac{Q_3 t_4}{3^2}\right) = c_h\left(\frac{2+1}{3^2}\right) Q_3 t_4$.
(iv) If $n = 4$, THC in delivery time are $c_h\left(\frac{3Q_3 t_4}{4^2} + \frac{2Q_3 t_4}{4^2} + \frac{Q_3 t_4}{4^2}\right) = c_h\left(\frac{3+2+1}{4^2}\right) Q_3 t_4$.

Therefore, the general term of holding cost is $c_h\left(\frac{1}{n^2}\right)\left(\sum_{i=1}^{n-1} i\right) Q_3 t_4 = c_h\left(\frac{n-1}{2n}\right) Q_3 t_4$.

Average inventory (AI) $= \frac{1}{2}Q_1 t_1 + Q_1 t_2 + \frac{1}{2}t_2\,(Q_2 - Q_1) + Q_2 t_3 + \frac{1}{2}t_3\,(Q_3 - Q_2) + \frac{1}{2}\left(\frac{n-1}{n}\right) Q_3 t_4$

$$= \frac{Q_1^2}{2}\left[\frac{(5 + 3a + b)\,D + (1 + a + b)^2\,(K - D - d)\left(\frac{n-1}{n}\right)}{D\,(K - D - d)}\right].$$

Substituting Q_1 in the above equation we get, AI $= \frac{Q^2(K-D-d)A}{2D} + \frac{B^2K^2(1-x)^2A}{2D(K-D-d)} - \frac{QBK(1-x)A}{D}$.

Total cost

Production cost $= \frac{1}{T}Qc_p = Dc_p$
Setup cost $= \frac{1}{T}c_o = \frac{Dc_o}{Q}$
Holding cost $= \frac{c_h}{T}(AI) = \frac{c_hQ(K-D-d)A}{2} + \frac{c_hB^2K^2(1-x)^2A}{2Q(K-D-d)} - c_hBK\,(1-x)\,A$
Defective cost $= \frac{1}{T}c_dQx = Dc_dx$
Fixed delivery cost $= \frac{1}{T}nc_{FD} = \frac{D}{Q}nc_{FD}$
Stock holding cost for buyer $= \frac{c_HQ_3t_4}{2nT}$

$$= \frac{c_HQ(1+a+b)^2(K-D-d)^2}{2n[3D+(K-D-d)\,(a+b+1)]^2} - \frac{c_HBK\,(K-D-d)\,(a+b+1)^2\,(1-x)}{2n[3D+(K-D-d)\,(a+b+1)]^2} + \frac{c_HB^2K^2(a+b+1)^2(1-x)^2}{2nQ[3D+(K-D-d)\,(a+b+1)]^2}$$

Shortage cost $= \frac{c_sB^2K(1-x)}{2Q(K-D-d)}$
Total cost $(TC\Big) = Dc_p + \frac{Dc_o}{Q} + \frac{c_hQ(K-D-d)A}{2} + \frac{c_hB^2K^2(1-x)^2A}{2Q(K-D-d)} - c_hBK\,(1-x)\,A + Dc_dx + \frac{Dnc_{FD}}{Q} + \frac{c_HQ(1+a+b)^2(K-D-d)^2}{2n[3D+(K-D-d)(a+b+1)]^2} - \frac{c_HBK(K-D-d)(a+b+1)^2(1-x)}{2n[3D+(K-D-d)(a+b+1)]^2} + \frac{c_HB^2K^2(a+b+1)^2(1-x)^2}{2nQ[3D+(K-D-d)(a+b+1)]^2} + \frac{c_sB^2K(1-x)}{2Q(K-D-d)}$.

5.1 Optimality

It is simple to demonstrate that $TC\,(Q)$ is a convex function in Q. Hence, an optimal manufacturing quantity $Q*$ can be reckoned from, $\frac{\partial TC}{\partial Q} = 0$, $\frac{\partial^2 TC}{\partial Q^2} > 0$, $\frac{\partial TC}{\partial B} = 0$, $\frac{\partial^2 TC}{\partial B^2} > 0, \frac{\partial TC}{\partial n} = 0$, and $\frac{\partial^2 TC}{\partial n^2} > 0$.

From the above equation we get, $Q = \sqrt{\frac{Nr}{Dr}}$

$$B = -\frac{\frac{Kc_h\sigma_3(x-1)}{\sigma_2} - \frac{Kc_H(x-1)\sigma_4(D-K+d)}{n\sigma_2}}{\frac{Kc_s(x-1)}{\sigma_1(D-K+d)} + \frac{K^2c_H(x-1)^2\sigma_4}{n\sigma_2\sigma_1} - \frac{K^2c_h\sigma_3(x-1)^2}{\sigma_2\sigma_1(D-K+d)}} \quad \text{where } \sigma_1 = \sqrt{\frac{Nr}{Dr}}$$

$$Nr = N_1 + N_2 + N_3 + N_4 + N_5 + N_6$$

$$\begin{aligned}
N_1 =& B^2K^3c_h - B^2K^3c_H + 8D^4c_{FD}n^2 + 8D^4c_on - 2DK^3c_on - 8D^4ac_on \\
&- 8D^4bc_on + 2Dc_od^3n + B^2DK^2c_H - B^2DK^2c_h - 2B^2K^3ac_H \\
&+ 2B^2K^3ac_h - 2B^2K^3bc_H + 2B^2K^3bc_h + B^2K^2c_Hd - B^2K^2c_hd \\
&- B^2K^3c_hn - B^2K^3c_sn - 2DK^3c_{FD}n^2 - 6D^2K^2c_on + 2B^2K^3c_Hx \\
&- 2B^2K^3c_hx - 8D^4ac_{FD}n^2 + 2D^4a^2c_on - 8D^4bc_{FD}n^2 + 2D^4b^2c_on \\
&+ 2Dc_{FD}d^3n^2 - 6D^2c_od^2n - B^2K^3a^2c_H + B^2K^3a^2c_h \\
&- B^2K^3b^2c_H + B^2K^3b^2c_h - 6D^2K^2c_{FD}n^2 - B^2K^3c_Hx^2 \\
&+ B^2K^3c_hx^2 + 2D^4a^2c_{FD}n^2 + 2D^4b^2c_{FD}n^2 - 6D^2c_{FD}d^2n^2
\end{aligned}$$

$$\begin{aligned}
N_2 =& B^2DK^2a^2c_h + B^2DK^2b^2c_H - B^2DK^2b^2c_h + B^2DK^2c_Hx^2 \\
&- B^2DK^2c_hx^2 + B^2K^2a^2c_Hd - B^2K^2a^2c_hd + B^2K^2b^2c_Hd \\
&- B^2K^2b^2c_hd - B^2K^3a^2c_hn - B^2K^3a^2c_sn - B^2K^3b^2c_hn \\
&- B^2K^3b^2c_sn - 2DK^3a^2c_{FD}n^2 - 6D^3Ka^2c_{FD}n^2 + 6D^2K^2a^2c_on \\
&- 2DK^3b^2c_{FD}n^2 - 6D^3Kb^2c_{FD}n^2 + 6D^2K^2b^2c_on \\
&- 2B^2K^3ac_Hx^2 + 2B^2K^3a^2c_Hx + 2B^2K^3ac_hx^2 - 2B^2K^3a^2c_hx \\
&- 2B^2K^3bc_Hx^2 + 2B^2K^3b^2c_Hx + 2B^2K^3bc_hx^2 - 2B^2K^3b^2c_hx \\
&+ B^2K^2c_Hdx^2 - B^2K^2c_hdx^2 + 2Da^2c_{FD}d^3n^2 + 6D^3a^2c_{FD}dn^2 \\
&+ 6D^2a^2c_od^2n - B^2K^3c_hnx^2 + 2Db^2c_{FD}d^3n^2 + 6D^3b^2c_{FD}dn^2 \\
&+ 6D^2b^2c_od^2n - 4DK^3ac_on + 12D^3Kac_on - 4DK^3bc_on + 12D^3Kbc_on \\
&- 6DKc_od^2n + 6DK^2c_odn + 12D^2Kc_odn + 4D^4abc_on + 4Dac_od^3n \\
&- 12D^3ac_odn + 4Dbc_od^3n - 12D^3bc_odn + 6D^2K^2a^2c_{FD}n^2 \\
&+ 6D^2K^2b^2c_{FD}n^2 - B^2K^3a^2c_Hx^2 + B^2K^3a^2c_hx^2 \\
&- B^2K^3b^2c_Hx^2 + B^2K^3b^2c_hx^2 + 6D^2a^2c_{FD}d^2n^2 \\
&+ 6D^2b^2c_{FD}d^2n^2 + 2B^2DK^2ac_H - 2B^2DK^2ac_h + 2B^2DK^2bc_H \\
&- 2B^2DK^2bc_h - 4B^2DK^2c_hn - 4B^2DK^2c_sn - 4B^2D^2Kc_sn
\end{aligned}$$

$$
\begin{aligned}
N_3 = &-2B^2DK^2c_Hx + 2B^2DK^2c_hx - 2B^2K^3abc_H + 2B^2K^3abc_h \\
&+ 2B^2K^2ac_Hd - 2B^2K^2ac_hd + 2B^2K^2bc_Hd - 2B^2K^2bc_hd \\
&- 2B^2K^3ac_hn - 2B^2K^3ac_sn - 2B^2K^3bc_hn - 2B^2K^3bc_sn \\
&- 4DK^3ac_{FD}n^2 + 12D^3Kac_{FD}n^2 - 2DK^3a^2c_on - 6D^3Ka^2c_on \\
&+ B^2K^2c_hdn - B^2Kc_sd^2n + 2B^2K^2c_sdn - 4DK^3bc_{FD}n^2 \\
&+ 12D^3Kbc_{FD}n^2 - 2DK^3b^2c_on - 6D^3Kb^2c_on - 6DKc_{FD}d^2n^2 \\
&+ 6DK^2c_{FD}dn^2 + 12D^2Kc_{FD}dn^2 + 4B^2K^3ac_Hx - 4B^2K^3ac_hx \\
&+ 4B^2K^3bc_Hx - 4B^2K^3bc_hx - 2B^2K^2c_Hdx + 2B^2K^2c_hdx \\
&+ 4D^4abc_{FD}n^2 + 4Dac_{FD}d^3n^2 - 12D^3ac_{FD}dn^2 + 2Da^2c_od^3n \\
&+ 6D^3a^2c_odn + 2B^2K^3c_hnx + B^2K^3c_snx + 4Dbc_{FD}d^3n^2 \\
&- 12D^3bc_{FD}dn^2 + 2Db^2c_od^3n + 6D^3b^2c_odn + B^2K^2a^2c_Hdx^2 \\
&- B^2K^2a^2c_hdx^2 + B^2K^2b^2c_Hdx^2 - B^2K^2b^2c_hdx^2 \\
&- B^2K^3a^2c_hnx^2 - B^2K^3b^2c_hnx^2 + 2B^2DK^2abc_H \\
&- 2B^2DK^2abc_h - B^2DK^2ac_hn - 2B^2DK^2ac_sn + 4B^2D^2Kac_sn \\
&+ B^2DK^2bc_hn - 2B^2DK^2bc_sn + 4B^2D^2Kbc_sn - 4B^2DK^2ac_Hx \\
&+ 4B^2DK^2ac_hx - 4B^2DK^2bc_Hx + 4B^2DK^2bc_hx + 2B^2K^2abc_Hd \\
&- 2B^2K^2abc_hd + 8B^2DK^2c_hnx + 4B^2DK^2c_snx + 4B^2D^2Kc_snx \\
&- 2B^2K^3abc_hn - 2B^2K^3abc_sn
\end{aligned}
$$

$$
\begin{aligned}
N_4 = &\, 2B^2K^2ac_hdn - 2B^2Kac_sd^2n + 4B^2K^2ac_sdn - 4DK^3abc_{FD}n^2 \\
&- 12D^3Kabc_{FD}n^2 + 12D^2K^2abc_on + 2B^2K^2bc_hdn - 2B^2Kbc_sd^2n \\
&+ 4B^2K^2bc_sdn - 12DKac_{FD}d^2n^2 + 12DK^2ac_{FD}dn^2 - 6DKa^2c_od^2n \\
&+ 6DK^2a^2c_odn - 12D^2Ka^2c_odn - 12DKbc_{FD}d^2n^2 + 12DK^2bc_{FD}dn^2 \\
&- 6DKb^2c_od^2n + 6DK^2b^2c_odn - 12D^2Kb^2c_odn + 4B^2K^3abc_Hx \\
&- 4B^2K^3abc_hx - 4B^2K^2ac_Hdx + 4B^2K^2ac_hdx - 4B^2K^2bc_Hdx \\
&+ 4B^2K^2bc_hdx + 4B^2K^3ac_hnx + 2B^2K^3ac_snx + 4Dabc_{FD}d^3n^2 \\
&+ 12D^3abc_{FD}dn^2 + 12D^2abc_od^2n + 4B^2K^3bc_hnx + 2B^2K^3bc_snx
\end{aligned}
$$

$$- 2B^2K^2c_hdnx + B^2Kc_sd^2nx - 2B^2K^2c_sdnx + B^2DK^2a^2c_hn$$
$$+ 2B^2DK^2a^2c_sn - B^2D^2Ka^2c_sn + B^2DK^2b^2c_hn + 2B^2DK^2b^2c_sn$$
$$- B^2D^2Kb^2c_sn + 2B^2DK^2ac_Hx^2 - 2B^2DK^2a^2c_Hx - 2B^2DK^2ac_hx^2$$
$$+ 2B^2DK^2a^2c_hx + 2B^2DK^2bc_Hx^2 - 2B^2DK^2b^2c_Hx - 2B^2DK^2bc_hx^2$$
$$+ 2B^2DK^2b^2c_hx - 4B^2DK^2c_hnx^2 + B^2K^2a^2c_hdn - B^2Ka^2c_sd^2n$$
$$+ 2B^2K^2a^2c_sdn + 12D^2K^2abc_{FD}n^2 + B^2K^2b^2c_hdn - B^2Kb^2c_sd^2n$$
$$+ 2B^2K^2b^2c_sdn - 6DKa^2c_{FD}d^2n^2 + 6DK^2a^2c_{FD}dn^2$$
$$- 12D^2Ka^2c_{FD}dn^2 - 6DKb^2c_{FD}d^2n^2 + 6DK^2b^2c_{FD}dn^2$$
$$- 12D^2Kb^2c_{FD}dn^2 - 2B^2K^3abc_Hx^2$$

$$N_5 = 2B^2K^3abc_hx^2 + 2B^2K^2ac_Hdx^2 - 2B^2K^2a^2c_Hdx - 2B^2K^2ac_hdx^2$$
$$+ 2B^2K^2a^2c_hdx + 2B^2K^2bc_Hdx^2 - 2B^2K^2b^2c_Hdx - 2B^2K^2bc_hdx^2$$
$$+ 2B^2K^2b^2c_hdx - 2B^2K^3ac_hnx^2 + 2B^2K^3a^2c_hnx + B^2K^3a^2c_snx$$
$$+ 12D^2abc_{FD}d^2n^2 - 2B^2K^3bc_hnx^2 + 2B^2K^3b^2c_hnx + B^2K^3b^2c_snx$$
$$+ B^2K^2c_hdnx^2 + 4B^2DKc_sdn - 4DK^3abc_on - 12D^3Kabc_on$$
$$- 12DKac_od^2n + 12DK^2ac_odn - 12DKbc_od^2n + 12DK^2bc_odn$$
$$+ B^2DK^2a^2c_Hx^2 - B^2DK^2a^2c_hx^2 + B^2DK^2b^2c_Hx^2$$
$$- B^2DK^2b^2c_hx^2 + 4Dabc_od^3n + 12D^3abc_odn + 2B^2K^2abc_Hdx^2$$
$$- 2B^2K^2abc_hdx^2 - 2B^2K^3abc_hnx^2 + 2B^2K^2ac_hdnx^2$$
$$- 2B^2K^2a^2c_hdnx + B^2Ka^2c_sd^2nx - 2B^2K^2a^2c_sdnx$$
$$+ 2B^2K^2bc_hdnx^2 - 2B^2K^2b^2c_hdnx + B^2Kb^2c_sd^2nx$$
$$- 2B^2K^2b^2c_sdnx + 2B^2DKac_sdn + 2B^2DKbc_sdn - 4B^2DKc_sdnx$$
$$- 12DKabc_od^2n + 12DK^2abc_odn - 24D^2Kabc_odn + B^2DK^2a^2c_hnx^2$$
$$+ B^2DK^2b^2c_hnx^2 + B^2K^2a^2c_hdnx^2 + B^2K^2b^2c_hdnx^2$$
$$+ 2B^2DK^2abc_hn + 4B^2DK^2abc_sn - 2B^2D^2Kabc_sn - 2B^2DKa^2c_sdn$$
$$- 2B^2DKb^2c_sdn - 4B^2DK^2abc_Hx + 4B^2DK^2abc_hx + 2B^2DK^2ac_hnx$$
$$+ 2B^2DK^2ac_snx - 4B^2D^2Kac_snx - 2B^2DK^2bc_hnx + 2B^2DK^2bc_snx$$

$$-4B^2D^2Kbc_snx+2B^2K^2abc_hdn$$

$$\begin{aligned}N_6 =& -2B^2Kabc_sd^2n+4B^2K^2abc_sdn-12DKabc_{FD}d^2n^2+12DK^2abc_{FD}dn^2\\
&-24D^2Kabc_{FD}dn^2-4B^2K^2abc_Hdx+4B^2K^2abc_hdx+4B^2K^3abc_hnx\\
&+2B^2K^3abc_snx-4B^2K^2ac_hdnx+2B^2Kac_sd^2nx-4B^2K^2ac_sdnx\\
&-4B^2K^2bc_hdnx+2B^2Kbc_sd^2nx-4B^2K^2bc_sdnx+2B^2DK^2abc_Hx^2\\
&-2B^2DK^2abc_hx^2-B^2DK^2ac_hnx^2-2B^2DK^2a^2c_hnx-2B^2DK^2a^2c_snx\\
&+B^2D^2Ka^2c_snx+B^2DK^2bc_hnx^2-2B^2DK^2b^2c_hnx-2B^2DK^2b^2c_snx\\
&+B^2D^2Kb^2c_snx-4B^2DK^2abc_hnx-4B^2DK^2abc_snx+2B^2D^2Kabc_snx\\
&+2B^2DKa^2c_sdnx+2B^2DKb^2c_sdnx-4B^2K^2abc_hdnx+2B^2Kabc_sd^2nx\\
&-4B^2K^2abc_sdnx+2B^2DK^2abc_hnx^2+2B^2K^2abc_hdnx^2-4B^2DKabc_sdn\\
&-2B^2DKac_sdnx-2B^2DKbc_sdnx+4B^2DKabc_sdnx\end{aligned}$$

$$\begin{aligned}D_r =& (D-K+d)\Big(D^2c_H-D^2c_h+K^2c_H-K^2c_h+c_Hd^2-c_hd^2+D^2a^2c_H\\
&-D^2a^2c_h+D^2b^2c_H-D^2b^2c_h+K^2a^2c_H-K^2a^2c_h+K^2b^2c_H-K^2b^2c_h\\
&+a^2c_Hd^2-a^2c_hd^2+b^2c_Hd^2-b^2c_hd^2-2DKc_H+2DKc_h\\
&+2Dc_Hd-2Dc_hd-2Kc_Hd+2Kc_hd+2D^2ac_H-2D^2ac_h+2D^2bc_H\\
&-2D^2bc_h+2K^2ac_H-2K^2ac_h+2K^2bc_H-2K^2bc_h-4D^2c_hn\\
&+K^2c_hn+2ac_Hd^2-2ac_hd^2+2bc_Hd^2-2bc_hd^2+c_hd^2n\\
&-2DKa^2c_H+2DKa^2c_h-2DKb^2c_H+2DKb^2c_h+2D^2abc_H\\
&-2D^2abc_h+2Da^2c_Hd-2Da^2c_hd+2Db^2c_Hd-2Db^2c_hd\\
&+2K^2abc_H-2K^2abc_h-D^2ac_hn+D^2bc_hn-2Ka^2c_Hd+2Ka^2c_hd\\
&-2Kb^2c_Hd+2Kb^2c_hd+2K^2ac_hn+2K^2bc_hn+2abc_Hd^2\\
&-2abc_hd^2+2ac_hd^2n+2bc_hd^2n+D^2a^2c_hn+D^2b^2c_hn+P^2a^2c_hn\\
&+K^2b^2c_hn+a^2c_hd^2n+b^2c_hd^2n-4DKac_H+4DKac_h-4DKbc_H\\
&+4DKbc_h+3DKc_hn+4Dac_Hd-4Dac_hd+4Dbc_Hd-4Dbc_hd\\
&-4Kac_Hd+4Kac_hd-4Kbc_Hd+4Kbc_hd-3Dc_hdn-2Kc_hdn\end{aligned}$$

$$- 4DKabc_H + 4DKabc_h - DKac_hn - 3DKbc_hn + 4Dabc_Hd$$
$$- 4Dabc_hd - 4Kabc_Hd + 4Kabc_hd + Dac_hdn + 3Dbc_hdn$$
$$- 4Kac_hdn - 4Kbc_hdn - 2DKa^2c_hn - 2DKb^2c_hn$$
$$+ 2D^2abc_hn + 2Da^2c_hdn + 2Db^2c_hdn + 2K^2abc_hn - 2Ka^2c_hdn$$
$$- 2Kb^2c_hdn + 2abc_hd^2n - 4DKabc_hn + 4Dabc_hdn - 4Kabc_hdn)$$

$$\sigma_2 = (3D - (a+b+1)(D-K+d))^2, \sigma_3 = D(3a+b+5) - \frac{(n-1)\sigma_4(D-K+d)}{n} \text{ and } \sigma_4 = (a+b+1)^2$$

$$n \text{ value } n = \sqrt{\frac{Nr_i}{Dr_i}}, Nr_i = Nr_1 + Nr_2 + Nr_3 + Nr_4 + Nr_5 + Nr_6 + Nr_7 + Nr_8 + Nr_9 + Nr_{10} + Nr_{11} + Nr_{12} + Nr_{13}$$

$$Nr_1 = -u_1\left(\frac{K^3c_Hu_3^2}{u_2^2} - \frac{K^3c_hu_3^2}{u_2^2} - \frac{2K^3c_Hxu_3^2}{u_2^2} + \frac{2K^3c_hxu_3^2}{u_2^2} + \frac{K^3a^2c_Hu_3^2}{u_2^2} - \frac{K^3a^2c_hu_3^2}{u_2^2} + \frac{K^3b^2c_Hu_3^2}{u_2^2} - \frac{K^3b^2c_hu_3^2}{u_2^2} - \frac{D^3c_Hu_4}{u_5} + \frac{D^3c_hu_4}{u_5} + \frac{K^3c_Hx^2u_3^2}{u_2^2} - \frac{K^3c_hx^2u_3^2}{u_2^2} + \frac{K^3c_Hu_4}{u_5} - \frac{K^3c_hu_4}{u_5} - \frac{c_Hd^3u_4}{u_5} + \frac{c_hd^3u_4}{u_5} - \frac{DK^2c_Hu_3^2}{u_2^2} + \frac{DK^2c_hu_3^2}{u_2^2}\right)$$

$$Nr_2 = \frac{2K^3c_Hu_3\sqrt{\frac{u_4}{u_5}}}{u_2} - \frac{2K^3c_hu_3\sqrt{\frac{u_4}{u_5}}}{u_2} + \frac{2K^3ac_Hu_3^2}{u_2^2} - \frac{2K^3ac_hu_3^2}{u_2^2} + \frac{2K^3bc_Hu_3^2}{u_2^2} - \frac{2K^3bc_hu_3^2}{u_2^2} - \frac{K^2c_Hdu_3^2}{u_2^2} + \frac{K^2c_hdu_3^2}{u_2^2} - \frac{D^3a^2c_Hu_4}{u_5} + \frac{D^3a^2c_hu_4}{u_5} - \frac{D^3b^2c_Hu_4}{u_5} + \frac{D^3b^2c_hu_4}{u_5} + \frac{K^3a^2c_Hx^2u_3^2}{u_2^2}$$

$$- \frac{K^3a^2c_hx^2u_3^2}{u_2^2} + \frac{K^3b^2c_Hx^2u_3^2}{u_2^2} - \frac{K^3b^2c_hx^2u_3^2}{u_2^2} + \frac{K^3a^2c_Hu_4}{u_5} - \frac{K^3a^2c_hu_4}{u_5}$$

$$+ \frac{K^3b^2c_Hu_4}{u_5} - \frac{K^3b^2c_hu_4}{u_5} - \frac{4DK^2c_Hu_3\sqrt{\frac{u_4}{u_5}}}{u_2} + \frac{2D^2Kc_Hu_3\sqrt{\frac{u_4}{u_5}}}{u_2}$$

$$+ \frac{4DK^2c_hu_3\sqrt{\frac{u_4}{u_5}}}{u_2} - \frac{2D^2Kc_hu_3\sqrt{\frac{u_4}{u_5}}}{u_2} - \frac{2DK^2ac_Hu_3^2}{u_2^2} + \frac{2DK^2ac_hu_3^2}{u_2^2}$$

$$- \frac{2DK^2bc_Hu_3^2}{u_2^2} + \frac{2DK^2bc_hu_3^2}{u_2^2} - \frac{a^2c_Hd^3u_4}{u_5} + \frac{a^2c_hd^3u_4}{u_5} - \frac{b^2c_Hd^3u_4}{u_5}$$

$$+ \frac{b^2c_hd^3u_4}{u_5} + \frac{4K^3ac_Hu_3\sqrt{\frac{u_4}{u_5}}}{u_2}$$

$$Nr_3 = - \frac{4K^3ac_hu_3\sqrt{\frac{u_4}{u_5}}}{u_2} + \frac{2DK^2c_Hxu_3^2}{u_2^2} - \frac{2DK^2c_hxu_3^2}{u_2^2} + \frac{4K^3bc_Hu_3\sqrt{\frac{u_4}{u_5}}}{u_2}$$

$$- \frac{4K^3bc_hu_3\sqrt{\frac{u_4}{u_5}}}{u_2} + \frac{2Kc_Hd^2u_3\sqrt{\frac{u_4}{u_5}}}{u_2} - \frac{4K^2c_Hdu_3\sqrt{\frac{u_4}{u_5}}}{u_2} - \frac{2Kc_hd^2u_3\sqrt{\frac{u_4}{u_5}}}{u_2}$$

$$+ \frac{4K^2c_hdu_3\sqrt{\frac{u_4}{u_5}}}{u_2} + \frac{2K^3abc_Hu_3^2}{u_2^2} - \frac{2K^3abc_hu_3^2}{u_2^2} - \frac{2K^2ac_Hdu_3^2}{u_2^2}$$

$$+ \frac{2K^2ac_hdu_3^2}{u_2^2} - \frac{2K^2bc_Hdu_3^2}{u_2^2} + \frac{2K^2bc_hdu_3^2}{u_2^2} - \frac{2K^3c_Hxu_3\sqrt{\frac{u_4}{u_5}}}{u_2}$$

$$+ \frac{2K^3c_hxu_3\sqrt{\frac{u_4}{u_5}}}{u_2} - \frac{4K^3ac_Hxu_3^2}{u_2^2} + \frac{4K^3ac_hxu_3^2}{u_2^2} - \frac{4K^3bc_Hxu_3^2}{u_2^2}$$

$$+ \frac{4K^3bc_hxu_3^2}{u_2^2} + \frac{2K^2c_Hdxu_3^2}{u_2^2} - \frac{2K^2c_hdxu_3^2}{u_2^2} - \frac{DK^2a^2c_Hu_3^2}{u_2^2}$$

$$+ \frac{DK^2a^2c_hu_3^2}{u_2^2} - \frac{DK^2b^2c_Hu_3^2}{u_2^2} + \frac{DK^2b^2c_hu_3^2}{u_2^2} + \frac{2K^3a^2c_Hu_3\sqrt{\frac{u_4}{u_5}}}{u_2}$$

$$- \frac{2K^3a^2c_hu_3\sqrt{\frac{u_4}{u_5}}}{u_2}$$

$$Nr_4 = -\frac{DK^2c_Hx^2u_3^2}{u_2^2} + \frac{DK^2c_hx^2u_3^2}{u_2^2} + \frac{2K^3b^2c_Hu_3\sqrt{\frac{u_4}{u_5}}}{u_2} - \frac{2K^3b^2c_hu_3\sqrt{\frac{u_4}{u_5}}}{u_2}$$
$$- \frac{3DK^2c_Hu_4}{u_5} + \frac{3D^2Kc_Hu_4}{u_5} + \frac{3DK^2c_hu_4}{u_5} - \frac{3D^2Kc_hu_4}{u_5}$$
$$- \frac{K^2a^2c_Hdu_3^2}{u_2^2} + \frac{K^2a^2c_hdu_3^2}{u_2^2} - \frac{K^2b^2c_Hdu_3^2}{u_2^2} + \frac{K^2b^2c_hdu_3^2}{u_2^2}$$
$$- \frac{2D^3ac_Hu_4}{u_5} + \frac{2D^3ac_hu_4}{u_5} - \frac{2D^3bc_Hu_4}{u_5} + \frac{2D^3bc_hu_4}{u_5} - \frac{3Dc_Hd^2u_4}{u_5}$$
$$- \frac{3D^2c_Hdu_4}{u_5} + \frac{3Dc_hd^2u_4}{u_5} + \frac{3D^2c_hdu_4}{u_5} + \frac{2K^3ac_Hx^2u_3^2}{u_2^2}$$
$$- \frac{2K^3a^2c_Hxu_3^2}{u_2^2} - \frac{2K^3ac_hx^2u_3^2}{u_2^2} + \frac{2K^3a^2c_hxu_3^2}{u_2^2} + \frac{2K^3bc_Hx^2u_3^2}{u_2^2}$$
$$- \frac{2K^3b^2c_Hxu_3^2}{u_2^2} - \frac{2K^3bc_hx^2u_3^2}{u_2^2} + \frac{2K^3b^2c_hxu_3^2}{u_2^2} + \frac{2K^3ac_Hu_4}{u_5}$$
$$- \frac{2K^3ac_hu_4}{u_5} - \frac{K^2c_Hdx^2u_3^2}{u_2^2} + \frac{K^2c_hdx^2u_3^2}{u_2^2} + \frac{2K^3bc_Hu_4}{u_5}$$
$$- \frac{2K^3bc_hu_4}{u_5} + \frac{3Kc_Hd^2u_4}{u_5} - \frac{3K^2c_Hdu_4}{u_5}$$

$$Nr_5 = -\frac{3Kc_hd^2u_4}{u_5} + \frac{3K^2c_hdu_4}{u_5} - \frac{2ac_Hd^3u_4}{u_5} + \frac{2ac_hd^3u_4}{u_5} - \frac{2bc_Hd^3u_4}{u_5}$$
$$+ \frac{2bc_hd^3u_4}{u_5} - \frac{2DK^2ac_Hx^2u_3^2}{u_2^2} + \frac{2DK^2a^2c_Hxu_3^2}{u_2^2} + \frac{2DK^2ac_hx^2u_3^2}{u_2^2}$$
$$- \frac{2DK^2a^2c_hxu_3^2}{u_2^2} - \frac{2DK^2bc_Hx^2u_3^2}{u_2^2} + \frac{2DK^2b^2c_Hxu_3^2}{u_2^2} + \frac{2DK^2bc_hx^2u_3^2}{u_2^2}$$
$$- \frac{2DK^2b^2c_hxu_3^2}{u_2^2} + \frac{2Ka^2c_Hd^2u_3\sqrt{\frac{u_4}{u_5}}}{u_2} - \frac{4K^2a^2c_Hdu_3\sqrt{\frac{u_4}{u_5}}}{u_2}$$
$$- \frac{6DK^2ac_Hu_4}{u_5} + \frac{6D^2Kac_Hu_4}{u_5} - \frac{2Ka^2c_hd^2u_3\sqrt{\frac{u_4}{u_5}}}{u_2} + \frac{4K^2a^2c_hdu_3\sqrt{\frac{u_4}{u_5}}}{u_2}$$
$$+ \frac{6DK^2ac_hu_4}{u_5} - \frac{6D^2Kac_hu_4}{u_5} + \frac{2Kb^2c_Hd^2u_3\sqrt{\frac{u_4}{u_5}}}{u_2} - \frac{4K^2b^2c_Hdu_3\sqrt{\frac{u_4}{u_5}}}{u_2}$$

$$-\frac{6DK^2bc_Hu_4}{u_5}+\frac{6D^2Kbc_Hu_4}{u_5}-\frac{2Kb^2c_hd^2u_3\sqrt{\frac{u_4}{u_5}}}{u_2}+\frac{4K^2b^2c_hdu_3\sqrt{\frac{u_4}{u_5}}}{u_2}$$
$$+\frac{6DK^2bc_hu_4}{u_5}-\frac{6D^2Kbc_hu_4}{u_5}-\frac{2D^3abc_Hu_4}{u_5}+\frac{2D^3abc_hu_4}{u_5}$$

$$Nr_6=-\frac{2K^3a^2c_Hxu_3\sqrt{\frac{u_4}{u_5}}}{u_2}-\frac{6Dac_Hd^2u_4}{u_5}-\frac{6D^2ac_Hdu_4}{u_5}+\frac{2K^3a^2c_hxu_3\sqrt{\frac{u_4}{u_5}}}{u_2}$$
$$+\frac{6Dac_hd^2u_4}{u_5}+\frac{6D^2ac_hdu_4}{u_5}-\frac{2K^3b^2c_Hxu_3\sqrt{\frac{u_4}{u_5}}}{u_2}-\frac{6Dbc_Hd^2u_4}{u_5}$$
$$-\frac{6D^2bc_Hdu_4}{u_5}+\frac{2K^3b^2c_hxu_3\sqrt{\frac{u_4}{u_5}}}{u_2}+\frac{6Dbc_hd^2u_4}{u_5}+\frac{6D^2bc_hdu_4}{u_5}$$
$$+\frac{2K^3abc_Hx^2u_3^2}{u_2^2}-\frac{2K^3abc_hx^2u_3^2}{u_2^2}-\frac{2K^2ac_Hdx^2u_3^2}{u_2^2}+\frac{2K^2a^2c_Hdxu_3^2}{u_2^2}$$
$$+\frac{2K^2ac_hdx^2u_3^2}{u_2^2}-\frac{2K^2a^2c_hdxu_3^2}{u_2^2}+\frac{2K^3abc_Hu_4}{u_5}-\frac{2K^3abc_hu_4}{u_5}$$
$$-\frac{2K^2bc_Hdx^2u_3^2}{u_2^2}+\frac{2K^2b^2c_Hdxu_3^2}{u_2^2}+\frac{2K^2bc_hdx^2u_3^2}{u_2^2}-\frac{2K^2b^2c_hdxu_3^2}{u_2^2}$$
$$+\frac{6Kac_Hd^2u_4}{u_5}-\frac{6K^2ac_Hdu_4}{u_5}-\frac{6Kac_hd^2u_4}{u_5}-\frac{6K^2ac_hdu_4}{u_5}+\frac{6Kbc_Hd^2u_4}{u_5}$$

$$Nr_7=-\frac{6K^2bc_Hdu_4}{u_5}-\frac{6Kbc_hd^2u_4}{u_5}-\frac{6K^2bc_hdu_4}{u_5}+\frac{4DKc_Hdu_3\sqrt{\frac{u_4}{u_5}}}{u_2}$$
$$-\frac{4DKc_hdu_3\sqrt{\frac{u_4}{u_5}}}{u_2}-\frac{2abc_Hd^3u_4}{u_5}+\frac{2abc_hd^3u_4}{u_5}-\frac{DK^2a^2c_Hx^2u_3^2}{u_2^2}$$
$$+\frac{DK^2a^2c_hx^2u_3^2}{u_2^2}-\frac{DK^2b^2c_Hx^2u_3^2}{u_2^2}+\frac{DK^2b^2c_hx^2u_3^2}{u_2^2}-\frac{3DK^2a^2c_Hu_4}{u_5}$$
$$+\frac{3D^2Ka^2c_Hu_4}{u_5}+\frac{3DK^2a^2c_hu_4}{u_5}-\frac{3D^2Ka^2c_hu_4}{u_5}-\frac{3DK^2b^2c_Hu_4}{u_5}$$
$$+\frac{3D^2Kb^2c_Hu_4}{u_5}+\frac{3DK^2b^2c_hu_4}{u_5}-\frac{3D^2Kb^2c_hu_4}{u_5}-\frac{3Da^2c_Hd^2u_4}{u_5}$$

$$-\frac{3D^2a^2c_Hdu_4}{u_5}+\frac{3Da^2c_hd^2u_4}{u_5}+\frac{3D^2a^2c_hdu_4}{u_5}-\frac{3Db^2c_Hd^2u_4}{u_5}$$
$$-\frac{3D^2b^2c_Hdu_4}{u_5}+\frac{3Db^2c_hd^2u_4}{u_5}+\frac{3D^2b^2c_hdu_4}{u_5}-\frac{K^2a^2c_Hdx^2u_3^2}{u_2^2}$$
$$+\frac{K^2a^2c_hdx^2u_3^2}{u_2^2}-\frac{K^2b^2c_Hdx^2u_3^2}{u_2^2}+\frac{K^2b^2c_hdx^2u_3^2}{u_2^2}+\frac{3Ka^2c_Hd^2u_4}{u_5}$$
$$-\frac{3K^2a^2c_Hdu_4}{u_5}-\frac{3Ka^2c_hd^2u_4}{u_5}+\frac{3K^2a^2c_hdu_4}{u_5}+\frac{3Kb^2c_Hd^2u_4}{u_5}$$

$$Nr_8=-\frac{3K^2b^2c_Hdu_4}{u_5}-\frac{3Kb^2c_hd^2u_4}{u_5}+\frac{3K^2b^2c_hdu_4}{u_5}-\frac{8DK^2ac_Hu_3\sqrt{\frac{u_4}{u_5}}}{u_2}$$
$$+\frac{4D^2Kac_Hu_3\sqrt{\frac{u_4}{u_5}}}{u_2}+\frac{8DK^2ac_hu_3\sqrt{\frac{u_4}{u_5}}}{u_2}-\frac{4D^2Kac_hu_3\sqrt{\frac{u_4}{u_5}}}{u_2}$$
$$-\frac{8DK^2bc_Hu_3\sqrt{\frac{u_4}{u_5}}}{u_2}+\frac{4D^2Kbc_Hu_3\sqrt{\frac{u_4}{u_5}}}{u_2}+\frac{8DK^2bc_hu_3\sqrt{\frac{u_4}{u_5}}}{u_2}$$
$$-\frac{4D^2Kbc_hu_3\sqrt{\frac{u_4}{u_5}}}{u_2}-\frac{2DK^2abc_Hu_3^2}{u_2^2}+\frac{2DK^2abc_hu_3^2}{u_2^2}+\frac{4DK^2c_Hxu_3\sqrt{\frac{u_4}{u_5}}}{u_2}$$
$$-\frac{2D^2Kc_Hxu_3\sqrt{\frac{u_4}{u_5}}}{u_2}-\frac{4DK^2c_hxu_3\sqrt{\frac{u_4}{u_5}}}{u_2}+\frac{2D^2Kc_hxu_3\sqrt{\frac{u_4}{u_5}}}{u_2}$$
$$+\frac{4DK^2ac_Hxu_3^2}{u_2^2}-\frac{4DK^2ac_hxu_3^2}{u_2^2}+\frac{4K^3abc_Hu_3\sqrt{\frac{u_4}{u_5}}}{u_2}-\frac{4K^3abc_hu_3\sqrt{\frac{u_4}{u_5}}}{u_2}$$
$$+\frac{4DK^2bc_Hxu_3^2}{u_2^2}-\frac{4DK^2bc_hxu_3^2}{u_2^2}+\frac{4Kac_Hd^2u_3\sqrt{\frac{u_4}{u_5}}}{u_2}-\frac{8K^2ac_Hdu_3\sqrt{\frac{u_4}{u_5}}}{u_2}$$
$$-\frac{4Kac_hd^2u_3\sqrt{\frac{u_4}{u_5}}}{u_2}+\frac{8K^2ac_hdu_3\sqrt{\frac{u_4}{u_5}}}{u_2}+\frac{4Kbc_Hd^2u_3\sqrt{\frac{u_4}{u_5}}}{u_2}$$
$$-\frac{8K^2bc_Hdu_3\sqrt{\frac{u_4}{u_5}}}{u_2}-\frac{4Kbc_hd^2u_3\sqrt{\frac{u_4}{u_5}}}{u_2}$$

$$Nr_9=\frac{8K^2bc_hdu_3\sqrt{\frac{u_4}{u_5}}}{u_2}+\frac{6DKc_Hdu_4}{u_5}-\frac{6DKc_hdu_4}{u_5}-\frac{2K^2abc_Hdu_3^2}{u_2^2}$$

$$+\frac{2K^2abc_hdu_3^2}{u_2^2}-\frac{4K^3ac_Hxu_3\sqrt{\frac{u_4}{u_5}}}{u_2}+\frac{4K^3ac_hxu_3\sqrt{\frac{u_4}{u_5}}}{u_2}$$
$$-\frac{4K^3bc_Hxu_3\sqrt{\frac{u_4}{u_5}}}{u_2}+\frac{4K^3bc_hxu_3\sqrt{\frac{u_4}{u_5}}}{u_2}-\frac{2Kc_Hd^2xu_3\sqrt{\frac{u_4}{u_5}}}{u_2}$$
$$+\frac{4K^2c_Hdxu_3\sqrt{\frac{u_4}{u_5}}}{u_2}+\frac{2Kc_hd^2xu_3\sqrt{\frac{u_4}{u_5}}}{u_2}-\frac{4K^2c_hdxu_3\sqrt{\frac{u_4}{u_5}}}{u_2}$$
$$-\frac{4K^3abc_Hxu_3^2}{u_2^2}+\frac{4K^3abc_hxu_3^2}{u_2^2}+\frac{4K^2ac_Hdxu_3^2}{u_2^2}-\frac{4K^2ac_hdxu_3^2}{u_2^2}$$
$$+\frac{4K^2bc_Hdxu_3^2}{u_2^2}-\frac{4K^2bc_hdxu_3^2}{u_2^2}-\frac{4DK^2a^2c_Hu_3\sqrt{\frac{u_4}{u_5}}}{u_2}$$
$$+\frac{2D^2Ka^2c_Hu_3\sqrt{\frac{u_4}{u_5}}}{u_2}+\frac{4DK^2a^2c_hu_3\sqrt{\frac{u_4}{u_5}}}{u_2}$$
$$-\frac{2D^2Ka^2c_hu_3\sqrt{\frac{u_4}{u_5}}}{u_2}-\frac{4DK^2b^2c_Hu_3\sqrt{\frac{u_4}{u_5}}}{u_2}$$
$$+\frac{2D^2Kb^2c_Hu_3\sqrt{\frac{u_4}{u_5}}}{u_2}+\frac{4DK^2b^2c_hu_3\sqrt{\frac{u_4}{u_5}}}{u_2}-\frac{2D^2Kb^2c_hu_3\sqrt{\frac{u_4}{u_5}}}{u_2}$$

$$Nr_{10}=\frac{8DK^2ac_Hxu_3\sqrt{\frac{u_4}{u_5}}}{u_2}-\frac{4D^2Kac_Hxu_3\sqrt{\frac{u_4}{u_5}}}{u_2}-\frac{8DK^2ac_hxu_3\sqrt{\frac{u_4}{u_5}}}{u_2}$$
$$+\frac{4D^2Kac_hxu_3\sqrt{\frac{u_4}{u_5}}}{u_2}+\frac{8DK^2bc_Hxu_3\sqrt{\frac{u_4}{u_5}}}{u_2}-\frac{4D^2Kbc_Hxu_3\sqrt{\frac{u_4}{u_5}}}{u_2}$$
$$-\frac{8DK^2bc_hxu_3\sqrt{\frac{u_4}{u_5}}}{u_2}+\frac{4D^2Kbc_hxu_3\sqrt{\frac{u_4}{u_5}}}{u_2}+\frac{4DK^2abc_Hxu_3^2}{u_2^2}$$
$$-\frac{4DK^2abc_hxu_3^2}{u_2^2}+\frac{4Kabc_Hd^2u_3\sqrt{\frac{u_4}{u_5}}}{u_2}-\frac{8K^2abc_Hdu_3\sqrt{\frac{u_4}{u_5}}}{u_2}$$
$$-\frac{4Kabc_hd^2u_3\sqrt{\frac{u_4}{u_5}}}{u_2}-\frac{8K^2abc_hdu_3\sqrt{\frac{u_4}{u_5}}}{u_2}+\frac{12DKac_Hdu_4}{u_5}$$

$$- \frac{12DKac_hdu_4}{u_5} + \frac{12DKbc_Hdu_4}{u_5} - \frac{12DKbc_hdu_4}{u_5} - \frac{4K^3abc_Hxu_3\sqrt{\frac{u_4}{u_5}}}{u_2}$$
$$+ \frac{4K^3abc_hxu_3\sqrt{\frac{u_4}{u_5}}}{u_2} - \frac{4Kac_Hd^2xu_3\sqrt{\frac{u_4}{u_5}}}{u_2} + \frac{8K^2ac_Hdxu_3\sqrt{\frac{u_4}{u_5}}}{u_2}$$
$$+ \frac{4Kac_hd^2xu_3\sqrt{\frac{u_4}{u_5}}}{u_2} - \frac{8K^2ac_hdxu_3\sqrt{\frac{u_4}{u_5}}}{u_2}$$

$$Nr_{11} = - \frac{4Kbc_Hd^2xu_3\sqrt{\frac{u_4}{u_5}}}{u_2} + \frac{8K^2bc_Hdxu_3\sqrt{\frac{u_4}{u_5}}}{u_2} + \frac{4Kbc_hd^2xu_3\sqrt{\frac{u_4}{u_5}}}{u_2}$$
$$- \frac{8K^2bc_hdxu_3\sqrt{\frac{u_4}{u_5}}}{u_2} + \frac{4K^2abc_Hdxu_3^2}{u_2^2} - \frac{4K^2abc_hdxu_3^2}{u_2^2}$$
$$+ \frac{4DK^2a^2c_Hxu_3\sqrt{\frac{u_4}{u_5}}}{u_2} - \frac{2D^2Ka^2c_Hxu_3\sqrt{\frac{u_4}{u_5}}}{u_2} - \frac{4DK^2a^2c_hxu_3\sqrt{\frac{u_4}{u_5}}}{u_2}$$
$$+ \frac{2D^2Ka^2c_hxu_3\sqrt{\frac{u_4}{u_5}}}{u_2} + \frac{4DK^2b^2c_Hxu_3\sqrt{\frac{u_4}{u_5}}}{u_2} - \frac{2D^2Kb^2c_Hxu_3\sqrt{\frac{u_4}{u_5}}}{u_2}$$
$$- \frac{4DK^2b^2c_hxu_3\sqrt{\frac{u_4}{u_5}}}{u_2} + \frac{2D^2Kb^2c_hxu_3\sqrt{\frac{u_4}{u_5}}}{u_2} - \frac{2DK^2abc_Hx^2u_3^2}{u_2^2}$$
$$+ \frac{2DK^2abc_hx^2u_3^2}{u_2^2} - \frac{6DK^2abc_Hu_4}{u_5} + \frac{6D^2Kabc_Hu_4}{u_5} + \frac{6DK^2abc_hu_4}{u_5}$$
$$- \frac{6D^2Kabc_hu_4}{u_5} + \frac{6DKa^2c_Hdu_4}{u_5} - \frac{6DKa^2c_hdu_4}{u_5} + \frac{6DKb^2c_Hdu_4}{u_5}$$
$$- \frac{Kb^2c_hdu_4}{6Du_5} - \frac{6Dabc_Hd^2u_4}{u_5} - \frac{6D^2abc_Hdu_4}{u_5} + \frac{6Dabc_hd^2u_4}{u_5}$$
$$+ \frac{6D^2abc_hdu_4}{u_5} - \frac{2Ka^2c_Hd^2xu_3\sqrt{\frac{u_4}{u_5}}}{u_2} + \frac{4K^2a^2c_Hdxu_3\sqrt{\frac{u_4}{u_5}}}{u_2}$$

$$Nr_{12} = \frac{2Ka^2c_hd^2xu_3\sqrt{\frac{u_4}{u_5}}}{u_2} - \frac{4K^2a^2c_hdxu_3\sqrt{\frac{u_4}{u_5}}}{u_2} - \frac{2Kb^2c_Hd^2xu_3\sqrt{\frac{u_4}{u_5}}}{u_2}$$

$$+\frac{4K^2b^2c_Hdxu_3\sqrt{\frac{u_4}{u_5}}}{u_2}+\frac{2Kb^2c_hd^2xu_3\sqrt{\frac{u_4}{u_5}}}{u_2}-\frac{4K^2b^2c_hdxu_3\sqrt{\frac{u_4}{u_5}}}{u_2}$$

$$-\frac{2K^2abc_Hdx^2u_3^2}{u_2^2}+\frac{2K^2abc_hdx^2u_3^2}{u_2^2}+\frac{6Kabc_Hd^2u_4}{u_5}$$

$$-\frac{6K^2abc_Hdu_4}{u_5}-\frac{6Kabc_hd^2u_4}{u_5}+\frac{6K^2abc_hdu_4}{u_5}+\frac{8DKac_Hdu_3\sqrt{\frac{u_4}{u_5}}}{u_2}$$

$$-\frac{8DKac_hdu_3\sqrt{\frac{u_4}{u_5}}}{u_2}+\frac{8DKbc_Hdu_3\sqrt{\frac{u_4}{u_5}}}{u_2}-\frac{8DKbc_hdu_3\sqrt{\frac{u_4}{u_5}}}{u_2}$$

$$-\frac{4DKc_Hdxu_3\sqrt{\frac{u_4}{u_5}}}{u_2}+\frac{4DKc_hdxu_3\sqrt{\frac{u_4}{u_5}}}{u_2}-\frac{8DK^2abc_Hu_3\sqrt{\frac{u_4}{u_5}}}{u_2}$$

$$+\frac{4D^2Kabc_Hu_3\sqrt{\frac{u_4}{u_5}}}{u_2}+\frac{8DK^2abc_hu_3\sqrt{\frac{u_4}{u_5}}}{u_2}-\frac{4D^2Kabc_hu_3\sqrt{\frac{u_4}{u_5}}}{u_2}$$

$$+\frac{4DKa^2c_Hdu_3\sqrt{\frac{u_4}{u_5}}}{u_2}-\frac{4DKa^2c_hdu_3\sqrt{\frac{u_4}{u_5}}}{u_2}+\frac{4DKb^2c_Hdu_3\sqrt{\frac{u_4}{u_5}}}{u_2}$$

$$Nr_{13}=-\frac{4DKb^2c_hdu_3\sqrt{\frac{u_4}{u_5}}}{u_2}-\frac{8DKac_Hdxu_3\sqrt{\frac{u_4}{u_5}}}{u_2}$$

$$+\frac{8DKac_hdxu_3\sqrt{\frac{u_4}{u_5}}}{u_2}-\frac{8DKbc_Hdxu_3\sqrt{\frac{u_4}{u_5}}}{u_2}+\frac{8DKbc_hdxu_3\sqrt{\frac{u_4}{u_5}}}{u_2}$$

$$+\frac{8DK^2abc_Hxu_3\sqrt{\frac{u_4}{u_5}}}{u_2}-\frac{4D^2Kabc_Hxu_3\sqrt{\frac{u_4}{u_5}}}{u_2}-\frac{8DK^2abc_hxu_3\sqrt{\frac{u_4}{u_5}}}{u_2}$$

$$+\frac{4D^2Kabc_hxu_3\sqrt{\frac{u_4}{u_5}}}{u_2}-\frac{4DKa^2c_Hdxu_3\sqrt{\frac{u_4}{u_5}}}{u_2}+\frac{4DKa^2c_hdxu_3\sqrt{\frac{u_4}{u_5}}}{u_2}$$

$$-\frac{4DKb^2c_Hdxu_3\sqrt{\frac{u_4}{u_5}}}{u_2}+\frac{4DKb^2c_hdxu_3\sqrt{\frac{u_4}{u_5}}}{u_2}+\frac{12DKabc_Hdu_4}{u_5}$$

$$-\frac{12DKabc_hdu_4}{u_5}-\frac{4Kabc_Hd^2xu_3\sqrt{\frac{u_4}{u_5}}}{u_2}+\frac{8K^2abc_Hdxu_3\sqrt{\frac{u_4}{u_5}}}{u_2}$$

Table 1 Showing the impact of cost parameters on optimal values

Parameters		Optimum values					
		N	Q	Q_1	Q_2	Q_3	T
c_o	90	1.47	1085.5	46.65	93.31	186.62	0.2714
	100	1.55	1144.3	49.21	98.43	196.87	0.2861
	110	1.63	1200.2	51.65	103.31	206.63	0.3001
	120	1.7	1253.6	53.98	107.96	215.93	0.3134
c_s	20	1.51	942.94	48.03	96.07	192.15	0.2357
	30	1.57	914.76	49.67	99.35	198.7	0.2287
	40	1.59	900.08	50.53	101.07	202.14	0.225
	50	1.61	891.32	51.08	102.17	204.34	0.2228
x	0.02	1.39	1021.9	44.02	88.05	176.1	0.2555
	0.03	1.39	1020.3	44.09	88.19	176.38	0.2551
	0.04	1.39	1018.7	44.16	88.33	176.66	0.2547
	0.05	1.39	1017.1	44.23	88.47	176.94	0.2543
c_p	60	1.38	1023.3	43.94	87.88	175.76	0.2558
	70	1.38	1023.3	43.94	87.88	175.76	0.2558
	80	1.38	1023.3	43.94	87.88	175.76	0.2558
	90	1.38	1023.3	43.94	87.88	175.76	0.2558
a	2	1.42	1247.3	50.42	151.28	252.14	0.3118
	3	1.46	913.5	30.78	123.12	184.68	0.2284
	4	1.48	882	26.81	134.08	187.72	0.2205
	5	1.5	858.41	23.77	142.65	190.2	0.2146

$$+\frac{4Kabc_hd^2xu_3\sqrt{\frac{u_4}{u_5}}}{u_2}-\frac{8K^2abc_hdxu_3\sqrt{\frac{u_4}{u_5}}}{u_2}+\frac{8DKabc_Hdu_3\sqrt{\frac{u_4}{u_5}}}{u_2}$$

$$-\frac{8DKabc_hdu_3\sqrt{\frac{u_4}{u_5}}}{u_2}-\frac{8DKabc_Hdxu_3\sqrt{\frac{u_4}{u_5}}}{u_2}+\frac{8DKabc_hdxu_3\sqrt{\frac{u_4}{u_5}}}{u_2}$$

Sensitivity Analysis

$$Dr_i = u_1$$

where

$$\begin{aligned} u_1 =& 2c_{FD}D^4a^2 + 4c_{FD}D^4ab - 8c_{FD}D^4a + 2c_{FD}D^4b^2 - 8c_{FD}D^4b \\ &+ 8c_{FD}D^4 - 6c_{FD}D^3Ka^2 - 12c_{FD}D^3Kab + 12c_{FD}D^3Ka - 6c_{FD}D^3Kb^2 \\ &+ 12c_{FD}D^3Kb + 6c_{FD}D^3a^2d + 12c_{FD}D^3abd - 12c_{FD}D^3ad + 6c_{FD}D^3b^2d \end{aligned}$$

$$- 12c_{FD}D^3bd + 6c_{FD}D^2K^2a^2 + 12c_{FD}D^2K^2ab + 6c_{FD}D^2K^2b^2$$
$$- 6c_{FD}D^2K^2 - 12c_{FD}D^2Ka^2d - 24c_{FD}D^2Kabd - 12c_{FD}D^2Kb^2d$$
$$+ 12c_{FD}D^2Kd + 6c_{FD}D^2a^2d^2 + 12c_{FD}D^2abd^2 + 6c_{FD}D^2b^2d^2$$
$$- 6c_{FD}D^2d^2 - 2c_{FD}DK^3a^2 - 4c_{FD}DK^3ab - 4c_{FD}DK^3a - 2c_{FD}DK^3b^2$$
$$- 4c_{FD}DK^3b - 2c_{FD}DK^3 + 6c_{FD}DK^2a^2d + 12c_{FD}DK^2abd$$
$$+ 12c_{FD}DK^2ad + 6c_{FD}DK^2b^2d + 12c_{FD}DK^2bd + 6c_{FD}DK^2d$$
$$- 6c_{FD}DKa^2d^2 - 12c_{FD}DKabd^2 - 12c_{FD}DKad^2 - 6c_{FD}DKb^2d^2$$
$$- 12c_{FD}DKbd^2 - 6c_{FD}DKd^2 + 2c_{FD}Da^2d^3 + 4c_{FD}Dabd^3$$
$$+ 4c_{FD}Dad^3 + 2c_{FD}Db^2d^3 + 4c_{FD}Dbd^3 + 2c_{FD}Dd^3$$

$$u_2 = \frac{Kc_s(x-1)}{\sqrt{\frac{u_4}{u_5}}(D-K+d)} + \frac{K^2c_H(x-1)^2u_8}{nu_6\sqrt{\frac{u_4}{u_5}}} - \frac{K^2c_hu_7(x-1)^2}{u_6\sqrt{\frac{u_4}{u_5}}(D-K+d)},$$

$$u_3 = \frac{Pc_hu_7(x-1)}{u_6} - \frac{Pc_H(x-1)u_8(D-K+d)}{nu_6}, u_4 = N_r\ u_5 = D_r$$

$$u_6 = (3D - (a+b+1)(D-K+d))^2,\quad u_7 = D(3a+b+5)$$
$$- \frac{(n-1)u_8(D-K+d)}{n}, u_8 = (a+b+1)^2$$

5.2 Numerical Illustration

Let us assume some parameters: $K = 5000$, $D = 4000$, $c_o = 80$, $d = 0$, $a = 1$, $c_h = 4$, $c_p = 50$, $c_d = 10$,
$b = 2$, $c_H = 5$, $c_{FD} = 2$, $c_s = 10$, $x = 0.01$.

Solutions
$Q = 1023.50$, $B = 64.70$, $Q_1 = 43.95$, $Q_2 = 87.90$, $Q_3 = 175.81$, $T = 0.2560$, $n = 1.389$

Holding cost = Rs. 167.91, Ordering cost = Rs. 312.64, Shortage cost = Rs. 101.22, Production cost = Rs. 2,00,000, Defective cost = Rs. 400, Stock holding cost = Rs. 54.35, Fixed delivery cost = Rs. 10.85, and

Total cost = Rs. 2,01,050

6 Observation

From the Table 1,

1. When setup cost increases, the number of deliveries, optimum quantity and time increases.
2. When shortage cost increases, the number of deliveries increases, optimum quantity and time decreases.
3. When defective cost increases, the number of deliveries have the same value, optimum quantity and time decreases and similarly, other parameters can be observed in Table 1.

7 Conclusion

In this chapter, we proposed the three levels of production with defective items, shortages, and multi-delivery policy in inventory model which reduces the holding cost. We identified that the holding cost is less in model II by comparing with the model I. The sensitivity analysis has been tested with respect to the various parameter and it has been noticed. The suggested model can help manufacturers and retailers determine the best quantity, cycle time, and inventory total cost.

References

1. Chiu, Y.S.P., Wu, M.-F., Chiu, S.W., Chang, H.-H.: A simplified approach to the multi-item economic production quantity model with scrap, rework, and multi-delivery. J. Appl. Res. Technol. **13**, 472–476 (2015)
2. Viji, G., Karthikeyan, K.: An economic production quantity model for three levels of production with Weibull distribution deterioration and shortage. Ain Shams Eng. J. **9**, 1481–1487 (2018)
3. Hsu, J.-T., Hsu, L.-F.: Two EPQ models with imperfect production processes, inspection errors, planned backorders, and sales returns. Comput. Ind. Eng. **64**, 389–402 (2013)
4. Rivera-Gomez, H., Gharbi, A., Kenne, J.-P., Ortiz-Zarco, R., Corona-Armenta, J.R.: Joint production, inspection and maintenance control policies for deteriorating system under quality constraint. J. Manuf. Syst. **60**, 585–607 (2021)
5. Ozturk, H.: A deterministic production inventory model with defective items, imperfect rework process and shortages backordered. Int. J. Oper. Res. **39**, 237–261 (2020)
6. Salameh, M.K., Jaber, M.Y.: Economic production quantity model for items with imperfect quality. Int. J. Prod. Econ. **64**, 59–64 (2000)
7. Muluneh, E.K., Srinivasa Rao, K.: EPQ models for deteriorating items with stock-dependent production rate and time-dependent demand having three-parameter Weibull decay. Int. J. Oper. Res. **14**, 271–300 (2012)
8. Zhou, Y.-W., Yang, S.-L.: A two-warehouse inventory model for items with stock-level-dependent demand rate. Int. J. Prod. Econ. **95**, 215–228 (2005)

9. Das, S., Khan, M.A.-A., Mahmoud, E.E., Abdel-Aty, A.-H., Abualnaja, K.M., Shaikh, A.A.: A production inventory model with partial trade credit policy and reliability. Alex. Eng. J. **60**, 1325–1338 (2021)
10. Sivashankari, C.K., Panayappan, S.: Production inventory model for two levels production with defective items and incorporating multi-delivery policy. Int. J. Oper. Res. **19**, 259–279 (2014)
11. Hauck, Z., Rabta, B., Reiner, G.: Joint quality and pricing decisions in lot sizing models with defective items. Int. J. Prod. Econ. **241** (2021)
12. Allen, H.T.: Economic production quantity models for deteriorating/imperfect products and service with rework. Comput. Ind. Eng. **66**, 879–888 (2013)

Accomplishment Expedients of Batch Arrival Queuing Model by Fuzzy Ordering Approach

K. Sakthivel, N. Paramaguru, R. Ramesh, and P. Syamala

1 Introduction

Nowadays queuing models are taking places in manufacturing/production systems, inventory control, and computer and telecommunication systems [1, 2]. Kaufmann [3, 13] presented for the theory of fuzzy subsets. We transform from the batch arrival fuzzy queue to a crisp queues family by Zadeh's [4] extension principle. In this chapter, we use $FM^{[K]}/FM/1$ model which is developed by Chen [5, 6]. The Parametric Programming method is derived by Lin et al. [7, 8].

Later the bulk arrival queues are analyzed by Jeeva and Rathnakumari [9]. For any queuing system, cost analysis constitutes a very important aspect of its investigation. In the fuzzy batch arrival queues, Ramesh and Kumaraghuru [24] gave the cost measures. Seenivasan and Shiny Epciya [29] surveyed working vacation with feedback queues.

Rue and Roshen shine [10] analyzed the policy of control on arrivals. Seenivasan and Chandiraleka [30] illustrated multiple working vocation queue system. Ritha and Lilly Robert [11] surveyed M/EK/1 fuzzy queuing system. Seenivasan et al. [28], investigated working vacation and catastrophe queueing system. Yager [12] introduced fuzzy ordering in [0,1]. A subjective approach was delivered by Campos

K. Sakthivel
Department of Mathematics, Govt. Arts and Science College, Veppanthattai, Perambalur, Tamilnadu, India

N. Paramaguru
Department of Mathematics, Govt. Arts College for Women, Krishnagiri, Tamilnadu, India

R. Ramesh (✉)
Department of Mathematics, Arignar Anna Government Arts College, Musiri, Tamilnadu, India

P. Syamala
Department of Mathematics, AVVM Sri Pushpam College, Poondi, Thanjavur, Tamilnadu, India

F. Kamalov et al. (eds.), *Advances in Mathematical Modeling and Scientific Computing*, Trends in Mathematics, https://doi.org/10.1007/978-3-031-41420-6_42

et al. [14] and Kwang et al. [17] for sorting fuzzy numerals. The integral value-based ordering procedures were submitted by Liou and Wang [15]. The distance-based method for ranking was given by Cheng [16]. The preference function for ranking was given by Modarres and Nezhad [18]. The area-based method for ordering was delivered by Chu and Tsao [19]. The centroid-based ranking method was given by Deng et al. [20], Wang et al. [22], and Liang et al. [21]. The method for ordering generalized trapezoidal fuzzy numerals was submitted by Chen and Chen [23]. Ramesh and Hari Ganesh [25] used the wingspans fuzzy ranking method in their article. Ramesh and Seenivasan [26, 27] illustrated fuzzy queueing system. Seenivasan et al. [31] analyzed the heterogeneous queueing model. In this chapter, we succeed a technique which can be laid AEC for batch arrival queues.

2 Preliminaries

2.1. **Def:** A set $\tilde{A} = \{(z, \mu_{\tilde{A}}(z)); z \; Z\}$ is said to be fuzzy set, if it is dictated by a membership function $\mu_{\tilde{A}}: Z \rightarrow [0,1]$ and $\mu_{\tilde{A}}(z)$ is the membership grade of z Z, where Z is the universal set.

2.2. **Def:** A number $\tilde{A}\,(a_1, a_2, a_3;1)$ is determined by triangular fuzzy number if it should be dictated by a membership function

$$\mu_{\tilde{A}}(z) = \begin{cases} \frac{z-a_1}{a_2-a_1}, & a_1 \leq z \leq a_2 \\ 1 \quad , & z = a_2 \\ \frac{z-a_3}{a_2-a_3}, & a_2 \leq z \leq a_3 \\ 0, otherwise \end{cases}$$

2.3. **Def:** A number $\tilde{A}\,(a_1, a_2, a_3, a_4;1)$ is determined by trapezoidal fuzzy number if it should be dictated by a membership function

$$\mu_{\tilde{A}}(z) = \begin{cases} \frac{z-a_1}{a_2-a_1}, & a_1 \leq z \leq a_2 \\ 1 \quad , & a_2 \leq z \leq a_3 \\ \frac{z-a_4}{a_3-a_4}, & a_3 \leq z \leq a_4 \\ 0, otherwise \end{cases}$$

3 Accomplishment Expedients of Queuing Model

Let *FM*$^{[K]}$*/FM/1* be our queueing system (Fig. 1). Here customers are arriving with batch arrival rate ($\tilde{\lambda}$) along Poisson process and they serviced by the rate ($\tilde{\mu}$) along exponential stream. The predicted batch size ($\tilde{k}$), the service cost ($\tilde{C}_S$), and the holding cost ($\tilde{C}_h$) are convex fuzzy sets and approximately known. Let $\varphi_{\tilde{\lambda}}(u)$, $\varphi_{\tilde{\mu}}(v)$, $\varphi_{\tilde{k}}(w)$, $\varphi_{\tilde{C}_S}(m)$, and $\varphi_{\tilde{C}_h}(n)$ denote the membership functions of $\tilde{\lambda}$, $\tilde{\mu}$, $\tilde{k}$, $\tilde{C}_S$, and $\tilde{C}_h$.

We treat the universal crisp sets of the arrival, service, batch size, service cost and holding cost rates are U, V, W, M, and N are all at once.

Then

$$\tilde{\lambda} = \left\{ \left(u, \varphi_{\tilde{\lambda}}(u)\right) \middle| \ u \in U \right\}$$

$$\tilde{\mu} = \left\{ \left(v, \varphi_{\tilde{\mu}}(v)\right) \middle| \ v \in V \right\}$$

$$\tilde{k} = \left\{ \left(w, \varphi_{\tilde{k}}(w)\right) \middle| \ w \in W \right\}$$

$$\tilde{C}_S = \left\{ \left(m, \varphi_{\tilde{C}_S}(m)\right) \middle| \ m \in M \right\}$$

$$\tilde{C}_h = \left\{ \left(n, \varphi_{\tilde{C}_h}(n)\right) \middle| \ n \in N \right\}$$

By queueing theory postulates,

The aggregate expected cost (AEC) = $m + wn \left[\frac{E\left(w^2\right)+E(w)}{2[v-uE(w)]}\right]$.

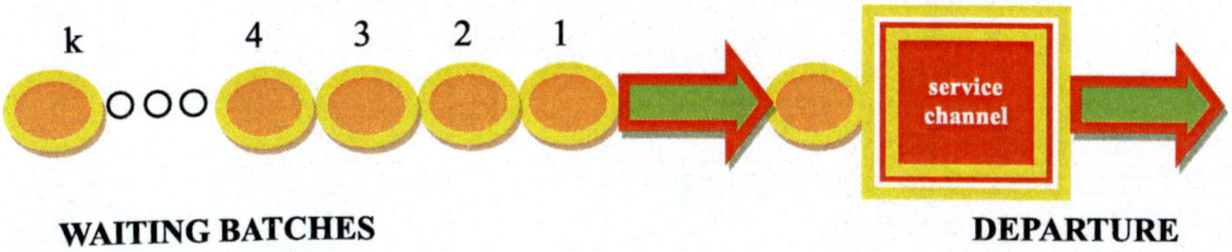

Fig. 1 Batch arrival queue

4 Wingspans Ranking Function Method – Algorithm

Take a fuzzy number $\tilde{A}$ and a_0 be a core point, also its membership function be $\varphi_{\tilde{A}}$.

Now, $W_A = a_0 - \frac{1}{2}\int_{-\infty}^{a_0} \varphi_{\tilde{A}}(x)dx + \frac{1}{2}\int_{a_0}^{\infty} \varphi_{\tilde{A}}(x)dx$ is called the wingspans center of $\tilde{A}$. This wingspans center based on the area between horizontal real axis and the curve of the membership function, also it is symmetric.

The ranking function of the triangular fuzzy number $\tilde{A} = [a_l\ a_0\ a_r]$ is

$$\mathrm{R}\left(\tilde{A}\right) = \frac{1}{2}a_0 + \frac{1}{4}\left(a_l + a_r\right).$$

The ranking function of the trapezoidal fuzzy number $\tilde{A} = [a_l\ a_b\ a_c\ a_r]$ is

$$\mathrm{R}\left(\tilde{A}\right) = \frac{1}{4}\left(a_l + a_b + a_c + a_r\right).$$

5 Illustration

Consider a vehicle washing station system with only one service space. The vehicles are arriving with a Poisson stream by batches. All vehicles other than the servicing one are wanted to stay in that station. Hence the holding charge occurs. The manager wants to evaluate the aggregate expected cost in different varied arrival rates.

5.1 *Triangular Fuzzy Number System*

Illustration 1

Let $\tilde{\lambda}_1 = [1\text{–}3]$, $\tilde{\mu}_1 = [5\text{–}7]$, $\tilde{k} = [7\text{–}9]$, $\tilde{C}_h = [20, 30, 40]$, and $\tilde{C}_S = [1000, 2000, 3000]$

Now

The ranking index of $\tilde{\lambda}$ is $\mathrm{R}(\tilde{\lambda}_1) = \mathrm{R}(1, 2, 3) = \frac{1}{2}(2) + \frac{1}{4}(1 + 3) = 2$

Similarly,

$\mathrm{R}(\tilde{\mu}_1) = \mathrm{R}(5, 6, 7) = 6$
$\mathrm{R}(\tilde{k}) = \mathrm{R}(7, 8, 9) = 8$
$\mathrm{R}(\tilde{C}_S) = \mathrm{R}(1000, 2000, 3000) = 2000$
$\mathrm{R}(\tilde{C}_h) = \mathrm{R}(20, 30, 40) = 30$

The Aggregate expected cost $= m + wn \left[\frac{E(w^2)+E(w)}{2[v-uE(w)]}\right] =$ Rs. 1136.00.

Similarly, we can proceed for some illustrations we got.

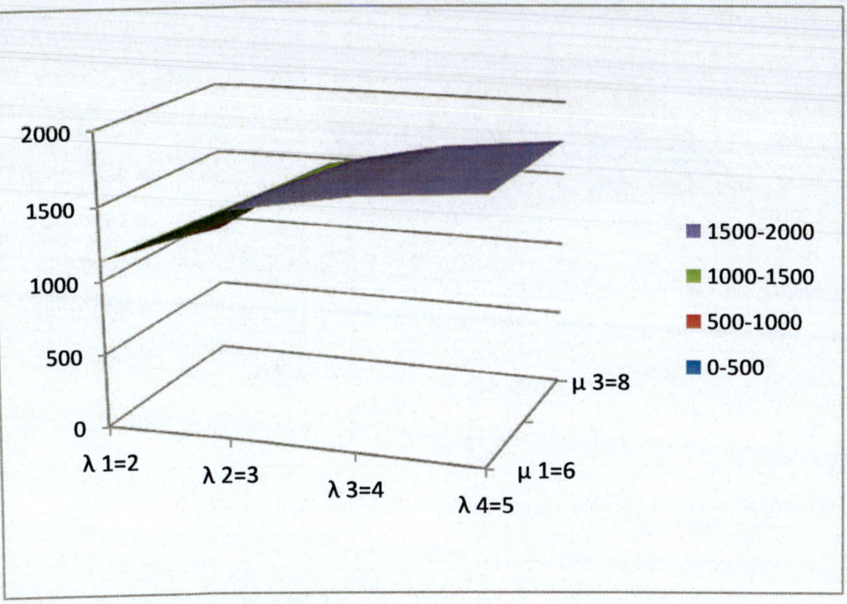

Fig. 2 Arrival rate versus service rate

Table 1 Aggregated expected cost

AEC	$\widetilde{\mu}_1$= [5–7]	$\widetilde{\mu}_2$= [6–8]	$\widetilde{\mu}_4$= [7–9]	$\widetilde{\mu}_4$= [8–10]	$\widetilde{\mu}_5$= [9–11]
$\widetilde{\lambda}_1$= [1–3]	Rs. 1136	Rs. 1040	Rs. 920	Rs. 766	Rs. 560
$\widetilde{\lambda}_2$= [2–4]	Rs. 1520	Rs. 1492	Rs. 1460	Rs. 1424	Rs. 1383
$\widetilde{\lambda}_3$= [3–5]	Rs. 1668	Rs. 1654	Rs. 1640	Rs. 1624	Rs. 1607
$\widetilde{\lambda}_4$= [4–6]	Rs. 1746	Rs. 1738	Rs. 1730	Rs. 1721	Rs. 1712
$\widetilde{\lambda}_5$= [5–7]	Rs. 1794	Rs. 1789	Rs. 1784	Rs. 1778	Rs. 1773

Figure 2 determines that if the defuzzified values of service rate increased, then the aggregate expected cost decreased, and if the defuzzified values of arrival rate increased, then the aggregate expected cost increased with respect to Table 1.

Illustration 2

Let $\widetilde{\lambda}_1 = [6\text{–}8]$, $\widetilde{\mu}_1 = [5\text{–}7]$, $\tilde{k} = [7\text{–}9]$, $\tilde{C}_h = [30, 40, 50]$, and $\tilde{C}_S = [1000, 2000, 3000]$

Now

The ranking index of $\tilde{\lambda}$ is $R(\widetilde{\lambda}_1) = R(6,7,8) = \frac{1}{2}(7) + \frac{1}{4}(6+8) = 7$.

Similarly,

$R(\widetilde{\mu}_1) = R(5,6,7) = 6$
$R(\tilde{k}) = R(7,8,9) = 8$
$R(C_{s1}) = R(1000,2000,3000) = 2000$
$R(C_h) = R(20,30,40) = 30$

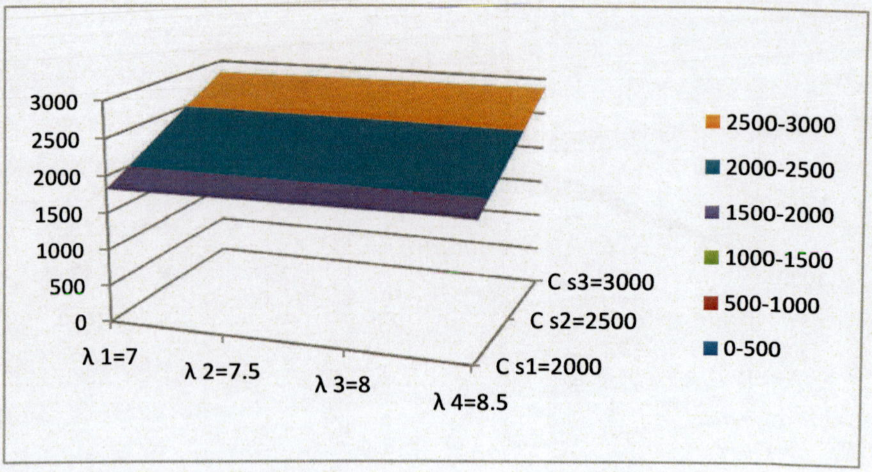

Fig. 3 Arrival rate versus service cost

Table 2 Aggregated expected cost

AEC	$\widetilde{C}_{s1}$= [1000, 2000, 3000]	$\widetilde{C}_{s2}$= [1500, 2500, 3500]	$\widetilde{C}_{s3}$= [2000, 3000, 4000]	$\widetilde{C}_{s4}$= [2500, 3500, 4500]	$\widetilde{C}_{s5}$= [3000, 4000, 5000]
$\widetilde{\lambda}_1$= [6–8]	Rs. 1827	Rs. 2327	Rs. 2827	Rs. 3327	Rs. 3827
$\widetilde{\lambda}_2$= [6.5, 7.5, 8.5]	Rs. 1840	Rs. 2340	Rs. 2804	Rs. 3340	Rs. 3840
$\widetilde{\lambda}_3$= [7–9]	Rs. 1851	Rs. 2351	Rs. 2851	Rs. 3351	Rs. 3851
$\widetilde{\lambda}_4$= [7.5, 8.5, 9.5]	Rs. 1861	Rs. 2361	Rs. 2861	Rs. 3361	Rs. 3861
$\widetilde{\lambda}_5$= [8–10]	Rs. 1869	Rs. 2369	Rs. 2869	Rs. 3369	Rs. 3869

The Aggregate expected cost = $m + wn \left[\frac{E(w^2)+E(w)}{2[v-uE(w)]}\right]$ = Rs. 1827.00.

Similarly, we can proceed for some illustrations we got.

Figure 3 determines that if the defuzzified values of service cost rate increased, then the aggregate expected cost increased, and if the defuzzified values of arrival rate increased, then the aggregate expected cost also increased with respect to Table 2.

5.2 *Trapezoidal Fuzzy Number System*

Illustration 3

Let $\widetilde{\lambda}_1$ = [0, 1, 2, 3], $\widetilde{\mu}_1$ = [5–8], $\tilde{k}$ = [7–10], $\tilde{C}_h$ = [20, 30, 40, 50], and $\tilde{C}_S$ = [4000, 5000, 6000, 7000].

Now

The ranking index of $\tilde{\lambda}$ is R($\widetilde{\lambda}_1$) = R(0, 1, 2, 3) = $\frac{0+1+2+3}{4}$ = 1.5.

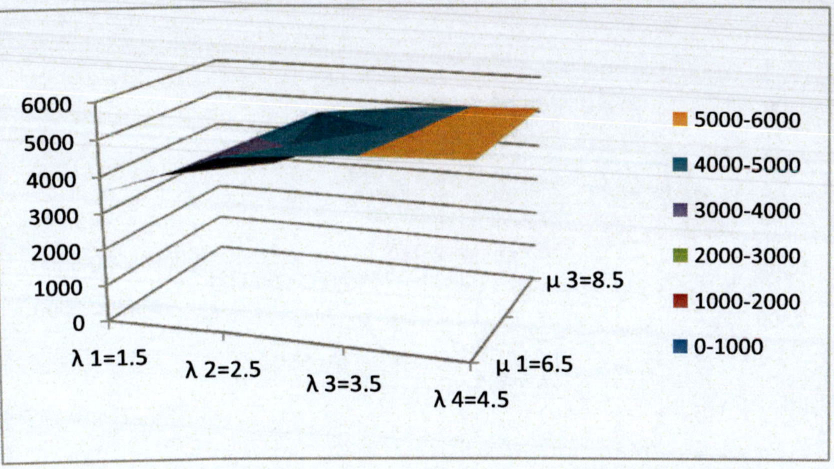

Fig. 4 Arrival rate versus service rate

Table 3 Aggregate expected cost

AEC	$\widetilde{\mu}_1 =$ [5, 6, 7, 8]	$\widetilde{\mu}_2 =$ [6, 7, 8, 9]	$\widetilde{\mu}_3 =$ [7, 8, 9, 10]	$\widetilde{\mu}_4 =$ [8, 9, 10, 11]	$\widetilde{\mu}_5 =$ [9, 10, 11, 12]
$\widetilde{\lambda}_1 = [0, 1, 2, 3]$	Rs. 3578	Rs. 3212	Rs. 2674	Rs. 1804	Rs. 862
$\widetilde{\lambda}_2 = [1, 2, 3, 4]$	Rs. 4686	Rs. 3753	Rs. 4558	Rs. 4478	Rs. 4383
$\widetilde{\lambda}_3 = [2, 3, 4, 5]$	Rs. 4983	Rs. 4960	Rs. 4935	Rs. 4907	Rs. 4876
$\widetilde{\lambda}_4 = [3, 4, 5, 6]$	Rs. 5122	Rs. 5109	Rs. 5096	Rs. 5082	Rs. 5067
$\widetilde{\lambda}_5 = [4, 5, 6, 7]$	Rs. 5202	Rs. 5194	Rs. 5186	Rs. 5178	Rs. 5169

Similarly,

$R(\widetilde{\mu}_1) = R(5, 6, 7, 8) = 6.5$
$R(\tilde{k}) = R(7, 8, 9, 10) = 8.5$
$R(\tilde{C}_S) = R(4000, 5000, 6000, 7000) = 5500$, $R(\tilde{C}_h) = R(20, 30, 40, 50) = 35$.

The aggregate expected cost $= m + wn \left[\frac{E(w^2)+E(w)}{2[v-uE(w)]}\right] =$ Rs. 3578.00.

Similarly, we can proceed for some illustrations we got.

Figure 4 determines that if the defuzzified values of service rate increased, then the aggregate expected cost decreased, and if the defuzzified values of arrival rate increased, then the aggregate expected cost increased with respect to Table 3.

Illustration 4

Let $\widetilde{\lambda}_1 = [6–9]$, $\widetilde{\mu}_1 = [5–8]$, $\tilde{k} = [7–10]$, $\tilde{C}_h = [20, 30, 40, 50]$, and $\widetilde{C}_{s1} = [4000, 5000, 6000, 7000]$.

Now, the ranking index of $\tilde{\lambda}$ is $R(\widetilde{\lambda}_1) = R(6, 7, 8, 9) = \frac{6+7+8+9}{4} = 7.5$.

Similarly,

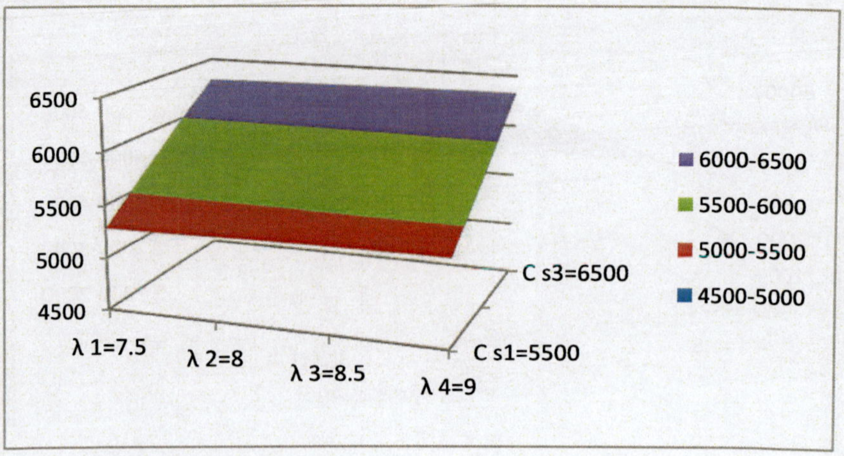

Fig. 5 Arrival rate versus service cost

Table 4 Aggregate expected cost

AEC	$\widetilde{C}_{s1} =$ [4000, 5000, 6000, 7000]	$\widetilde{C}_{s2} =$ [4500, 5500, 6500, 7500]	$\widetilde{C}_{s3} =$ [5000, 6000, 7000, 8000]	$\widetilde{C}_{s4} =$ [5500, 6500, 7500, 8500]	$\widetilde{C}_{s5} =$ [6000, 7000, 8000, 9000]
$\widetilde{\lambda}_1 = [6, 7, 8, 9]$	Rs. 5290	Rs. 5790	Rs. 6290	Rs. 6790	Rs. 7290
$\widetilde{\lambda}_2 = [6.5, 7.5, 8.5, 9.5]$	Rs. 5305	Rs. 5805	Rs. 6305	Rs. 6805	Rs. 7305
$\widetilde{\lambda}_3 = [7, 8, 9, 10]$	Rs. 5317	Rs. 5817	Rs. 6317	Rs. 6817	Rs.7317
$\widetilde{\lambda}_4 = [7.5, 8.5, 9.5, 10.5]$	Rs. 5328	Rs. 5828	Rs. 6328	Rs. 6828	Rs. 7328
$\widetilde{\lambda}_5 = [8, 9, 10, 11]$	Rs. 5338	Rs. 5838	Rs. 6338	Rs. 6838	Rs. 7338

$R(\widetilde{\mu}_1) = R(5, 6, 7, 8) = 6.5$, $R(\tilde{k}) = R(7, 8, 9, 10) = 8.5$
$R(\widetilde{C}_{s1}) = R(4000, 5000, 6000, 7000) = 5500$
$R(\widetilde{C}_h) = R(20, 30, 40, 50) = 35.$

The aggregate expected cost $= m + wn \left[\frac{E(w^2)+E(w)}{2[v-uE(w)]}\right] =$ Rs. 5290.00.

Similarly, we can proceed for some illustrations we got.

Figure 5 determines that if the defuzzified values of service cost rate increased, then the aggregate expected cost increased, and if the defuzzified values of arrival rate increased, then the aggregate expected cost also increased with respect to Table 4.

6 Output

This chapter provides the accomplishment expedients such are the Aggregate expected costs (AEC) with the help of our proposed ordering procedure. The researcher can hold of the supreme verdict. We culminate that our proposed ordering procedure is more beneficial, and this chapter furnishes feasible facts for future inventors.

References

1. Kleinrock, L.: Queuing Systems, Vol. I. Theory. Wiley, New York (1975)
2. Gross, D., Harris, C.M.: Fundamentals of Queueing Theory, 3rd edn. Wiley, New York (1998)
3. Kaufmann, A.: Introduction to the Theory of Fuzzy Subsets, Volume 1. Academic Press, New York (1975)
4. Zadeh, L.A.: Fuzzy sets as a basis for a theory of possibility. Fuzzy Sets Syst. **1**, 3–28 (1978)
5. Chen, S.P.: Parametric nonlinear programming approach to fuzzy queues with bulk service. Eur. J. Oper. Res. **163**, 434–444 (2005)
6. Chen, S.P.: A bulk arrival queuing model with fuzzy parameters and varying batch sizes. Appl. Math. Model. **30**, 920–929 (2006)
7. Lin, C.H., Huang, H.I., Ke, J.C.: Parametric programming approach for batch arrival queues with vacation policies and fuzzy parameters. Appl. Math. Comput. **180**(1), 217–232 (2006)
8. Lin, C.H., Huang, H.I., Ke, J.C.: On a batch arrival queue with set up and uncertain parameters patterns. Int. J. Appl. Sci. Eng. **6**(2), 163–180 (2008)
9. Jeeva, M., Rathnakumari, E.: Bulk arrival single server, Bernoulli feedback queue with fuzzy vacations and fuzzy parameters. ARPN. **2**(5), 492–499 (2012)
10. Rue, R.C., Roshen, S.: Some properties of optimal control policies for entries to an M/M/1 queue. Nav. Res. Logist. Q. **28**, 520–532 (1981)
11. Ritha, W., Robert, L.: Profit analysis of fuzzy M/Ek/1 queuing system non-linear zero one programming approach. IJAIR. **1**(2), 1–14 (2012)
12. Yager, R.R.: A procedure for ordering fuzzy subsets of the unit interval. Inf. Sci. **24**(2), 143 (1981)
13. Kaufmann, A., Gupta, M.M.: Fuzzy Mathematical Models in Engineering and Management Science. Elsevier, Amsterdam (1988)
14. Campos, L., Gonzalez, A.: A subjective approach for ranking fuzzy numbers. Fuzzy Sets Syst. **29**(2), 145 (1989)
15. Liou, T.S., Wang, M.J.: Ranking fuzzy numbers with integral value. Fuzzy Sets Syst. **50**(3), 247 (1992)
16. Cheng, C.H.: A new approach for ranking fuzzy numbers by distance method. Fuzzy Sets Syst. **95**(3), 307 (1998)
17. Kwang, H.C., Lee, J.H.: A method for ranking fuzzy numbers and its application to decision making. IEEE Trans. Fuzzy Syst. **7**(6), 677 (1999)
18. Modarres, M., Sadi Nezhad, S.: Ranking fuzzy numbers by preference ratio. Fuzzy Sets Syst. **118**(3), 429 (2001)
19. Chu, T.C., Tsao, C.T.: Ranking fuzzy numbers with an area between the centroid point and original point. Comput. Math. Appl. **43**(1–2), 111 (2002)
20. Deng, Y., Liu, Q.: A TOPSIS-based centroid-index ranking method of fuzzy numbers and its applications in decision making. Cybern. Syst. **36**(6), 581 (2005)

21. Liang, C., Wu, J., Zhang, J.: Ranking Indices and Rules for Fuzzy Numbers Based on Gravity Center Point, pp. 21–23. Paper Presented at the 6th world Congress on Intelligent Control and Automation, Dalian (2006)
22. Wang, Y.J., Lee, H.S.: The revised method of ranking fuzzy numbers with an area between the centroid and original points. Comput. Math. Appl. **55**(9), 2033 (2008)
23. Chen, S.J., Chen, S.M.: Fuzzy risk analysis based on the ranking of generalized trapezoidal fuzzy numbers. Appl. Intell. **26**(1), 1 (2007)
24. Ramesh, R., Kumaraghuru, S.: Cost measures of fuzzy batch arrival queuing model by ranking function method. Int. J. Sci. Res. **4**(1), 234–238 (2015)
25. Ramesh, R., Hari Ganesh, A.: M/M/1/N fuzzy queueing models with discouraged arrivals under Wingspans Fuzzy ranking method. Int. J. Appl. Eng. Res. **14**, 4 (2019)
26. Ramesh, R., Seenivasan, M.: Performance Calibrations of a Single Server Glycolic Acid Based Beauty Parlour by Fuzzy Retrial Queuing Models, pp. 2422–2426. Materials Today: Proceedings, Volume 51, Part 8 (2022)
27. Ramesh, R., Seenivasan, M.: Achievement expedients of fuzzy queuing models with an unreliable electrical transformer. IEEE Xplore (ICEEICT 2022). **22** (2022) pp. 978-1-6654-3647-2, February 2022
28. Seenivasan, M., Chakravarthy, V.J., Abinaya, R.: Markovian Queueing Model with Server Breakdown, Single Working Vacation and Catastrophe, vol. 881, pp. 409–421. Springer Book Series, (ICAECT 2021) (2022, June 26)
29. Seenivasan, M., Shiny Epciya, J.: M/M/1 queue Server Breakdown, Single Working Vacation, Feedback and State Dependent Customer, pp. 1–5. ICAECT 2022, IEEE Xplore (2022, July 2)
30. Seenivasan, M., Chandiraleka, S.: Single Server Queueing Model with Multiple Working Vacation and with Server Breakdown, pp. 1–5. ICAECT 2022, IEEE Xplore (2022, July 02)
31. Seenivasan, M., Senthilkumar, R., Subasri, K.S.: M/M/2 heterogeneous queueing system having unreliable server with catastrophes and restoration. Mater. Today: Proc. **51**(8), 2332–2338 (December 2021)

Analysis of a Retrial Bulk Arrival Queue with Secondary Optional Servicc Subject to Bernoulli Vacation, Server Breakdown and Customer Balking

R. Harini and K. Indhira

1 Introduction

Due to network complexity and an expanding client base, consumer patterns and the retrial concept may have had a considerable influence on how well computer networks have functioned in recent decades. The main feature of the retrial queue (RQ) entails that a client who enters the service area and realizes the server seems to be loaded has been obliged to evacuate the place and is asked to wait in a virtual location called "orbit" before resubmitting his request. Numerous studies have been conducted on the RQ recently.

Over the past two decades, there has been a substantial surge in research on queueing systems (QS), whereby the server sequentially provides every client with dual levels of service. With timely Bernoulli feedback, Varalakshmi et al. [10] studied an unstable dual-phase RQ. Recently, Nila and Sumitha [6] analysed a batch arrival RQ with second optional service, initial failure, BV, feedback and impatient clients. Furthermore, Abdollahi and Salehi Rad [1] examined how a $M^{[X]}/G/1$ retrial QS behaves in steady state (SS) with dual stages of heterogeneous services and generic retrial time.

Many studies have been conducted in the interim on QS with server vacations. Madhu Jain and Sandeep Kaur [3] analysed an unreliable $M^{[X]}/G/1$ RQ paradigm with BV, Bernoulli feedback, balking and optional service. One may find widespread discussions of the vacation queueing framework in Doshi [2], Takagi [9] and Ke et al.[5]. It is also becoming more crucial to consider a client's conduct because of the intricacy that is emerging in many sectors. Using twin forms of BV, Pavai Madheswari et al. [7] evaluated a single server RQ with secondary optional

R. Harini (✉) · K. Indhira
Department of Mathematics, School of Advanced Sciences, Vellore Institute of Technology, Vellore, TamilNadu, India
e-mail: kindhira@vit.ac.in

F. Kamalov et al. (eds.), *Advances in Mathematical Modeling and Scientific Computing*, Trends in Mathematics, https://doi.org/10.1007/978-3-031-41420-6_43

service and client balking. Service disruptions are an unavoidable issue in several realistic scenarios. Service stations inevitably break down and require fixing. Ke and Choudhury [4] acknowledged the bulk arrival RQ with two serving options with the notions of breakage and deferring repair. The RQ systems have been reviewed by Rajadurai et al.[8] in terms of breakage and restoration.

Thus, the primary objective of this work is to scrutinize the presupposed framework in order to obtain the necessary results, which are also beneficial in several authentic situations.

2 Characterization and Practical Justification of the Model

In this paper, a framework of a $M^{[X]}/G/1$ RQ with dual levels of service, where the latter is optional subjected under BV, server breakage and balking, has been developed. An exhaustive overview of the framework is provided below:

The arrival process: In accordance with a complex "Poisson process", clients congregate in bulk at a rate ϑ. Let $\mathcal{V}_j$ be the no. of clients affiliated with the jth arrival batch, where "$\mathcal{V}_j,\ j = 1, 2, 3, \ldots$" follows an arbitrary distribution "$Pr[\mathcal{V}_j = n] = \gamma_n, n = 1, 2, 3, \ldots$" and $\mathcal{V}(\varsigma)$ refers the PGF of $\mathcal{V}$.

The retrial process: Clients that approach in bulks and discover that the server is occupied, malfunctioning or on vacation either abandon the service area (balks) with a probability h or join in the orbit with a probability $1-h$. The clients in the orbit are resolute in their decision to continue evaluating different services unless they locate the one they desire. The clients are evenly spaced apart by an arbitrary distribution $\mathcal{G}(\acute{\xi})$, Laplace-Stieltjes Transform (LST) $\mathcal{G}^*(\kappa)$ and a conditional completion rate $\varpi(\xi)d\acute{\xi} = \frac{d\mathcal{G}(\acute{\xi})}{1-\mathcal{G}(\xi)}$ and the RQ has a $FCFS$ restriction.

The service process: The server performs an Initial Regular Service (IRS) on the entire approaching bulk of clients. Once the user's IRS is accomplished, they can choose to abandon the system with probability $q(= 1 - p)$ or undertake an Second Optional Service (SOS) with probability p. The ith (i= 1,2) level service is expected to obey a probability distribution function (PDF) $\mathcal{P}_i(\acute{\xi})$, LST $\mathcal{P}_i^*(\kappa)$ and conditional completion rates $\chi_i(\xi)d\acute{\xi} = \frac{d\mathcal{P}_i(\acute{\xi})}{1-\mathcal{P}_i(\acute{\xi})}$.

The vacation process: The server will take a BV, i.e. with probability $1 - d$, the server awaits to handle the next client, whereas with probability d, the server will go on vacation, following each service. The server tends to leave once more for another vacation if there isn't even one client in the orbit. When the vacation is completed, the server is still inactive for users who are orbiting or have just entered. The server's vacation period is of discretionary distance $\mathcal{H}$ with PDF $\mathcal{H}(\acute{\xi})$, LST $\mathcal{H}^*(\kappa)$ and a conditional completion rate $\varphi(\xi)d\acute{\xi} = \frac{d\mathcal{H}(\acute{\xi})}{1-\mathcal{H}(\acute{\xi})}$.

The breakdown process: When the server is operating at any level of service, the service channel could rupture at any time, causing a brief server outage. The malfunctions, or the life expectancy of the server, are caused by "exogenous Poisson processes" with pace μ_1 for IRS and μ_2 for SOS.

The repair process: The server is quickly sent out for repair when it malfunctions. The server stops providing client service throughout that whole period and pauses for the repair to be finished. "Deferring time" is the name given to this waiting interval. The deferring time of the server follows a PDF $\mathcal{A}_i(\dot{\tau})$, LST $\mathcal{A}_i^*(\kappa)$ and conditional completion rates $\iota_i(\dot{\tau})d\dot{\tau} = \frac{d\mathcal{A}_i(\dot{\tau})}{1-\mathcal{A}_i(\dot{\tau})}$, where $i = 1, 2$. Just before the server failure, the client who was really being served awaits for the remaining portion of the service to be accomplished. It is expected that the server's repair time distributions are dispersed randomly for both kinds of services with PDF $\mathcal{K}_i(\dot{\tau})$, LST $\mathcal{K}_i^*(\kappa)$ and conditional completion rates $\pi_i(\dot{\tau})d\dot{\tau} = \frac{d\mathcal{K}_i(\dot{\tau})}{1-\mathcal{K}_i(\dot{\tau})}$.

It is presumed that each stochastic process in the system operates independently of the others.

2.1 *Implementation of the Model in Real Life*

Beyond its theoretical importance, the retrial QS under study has many significant practical applications. An example is the concept of cloud computing, which is fundamentally utilized in Internet banking, a modern electronic payment system. In particular, the IaaS (Infrastructure as a Service) service model, which has been extensively adopted in this industry, has advanced significantly over the previous year. High-level application programming interfaces (APIs) have been received from it, which are utilized to deliver various low-level network infrastructures, such as log access clustering, storage resiliency, etc. (discretionary services). The consumers are instantly routed to a virtual waiting line known as the load balancers (orbit) if the servers are striving to process a demand in an environment where every service has been supplied virtually, or they may choose to give up (balking). IaaS promptly recovers (repairs) the information from another server farm immediately, even if some disaster (breakdown) completely destroys one data centre. Additionally, when the APIs use the virtual machines, breakage might happen and they'll be sent to be repaired. Furthermore, the virtual server goes on what we may call a vacation in order to perform maintenance. As a result, IaaS is a reliable strategy that can be adapted to meet the needs of the clients.

3 System Analysis

We generate the SS differential eqs. for the RQ system and then the PGF for server states, the PGF for system size and orbit are then determined.

3.1 The Steady-State Equations

The elapsed retrial times, elapsed service times, elapsed vacation times, elapsed deferring times and elapsed repair times at time $\widetilde{\zeta}$ are denoted as $\mathcal{G}^0(\widetilde{\zeta})$, $\mathcal{P}_i^0(\widetilde{\zeta})$, $\mathcal{H}^0(\widetilde{\zeta})$ and $\mathcal{A}_i^0(\widetilde{\zeta})$ and $\mathcal{K}_i^0(\widetilde{\zeta})$, respectively.

In this section, we emphasize the usage of the bivariate Markov process to depict the model's state at time $\widetilde{\zeta}$ $\{\Xi(\widetilde{\zeta}),\ \Omega(\widetilde{\zeta});\widetilde{\zeta}\ \geq 0\ \}$. $\Xi(\widetilde{\zeta})$ implies the server state (0, 1, 2, 3, 4, 5, 6 or 7) relying on if the server is inert, actively performing both service levels, on vacation, deferring time on the two levels, or repair on the two levels, $\Omega(\widetilde{\zeta})$ signifies the no. of clients in the orbit. $\mathcal{G}^0(\widetilde{\zeta})$ is proportional to the elapsed retrial time, if $\Xi(\widetilde{\zeta}) = 0$ and $\Omega(\widetilde{\zeta}) > 0$. $\mathcal{P}_1^0(\widetilde{\zeta})(\mathcal{P}_2^0(\widetilde{\zeta}))$ is proportional to the elapsed time of the client under service in IRS (SOS), if $\Xi(\widetilde{\zeta}) = 1(2)$ and $\Omega(\widetilde{\zeta}) \geq 0$. $\mathcal{H}^0(\widetilde{\zeta})$ is proportional to the elapsed vacation time, if $\Xi(\widetilde{\zeta}) = 3$ and $\Omega(\widetilde{\zeta}) \geq 0$. $\mathcal{A}_1^0(\widetilde{\zeta})(\mathcal{A}_2^0(\widetilde{\zeta}))$ is proportional to the elapsed time of the server under deferred repair in IRS (SOS), if $\Xi(\widetilde{\zeta}) = 4(5)$ and $\Omega(\widetilde{\zeta}) \geq 0$. $\mathcal{K}_1^0(\widetilde{\zeta})(\mathcal{K}_2^0(\widetilde{\zeta}))$ is proportional to the elapsed time of the server under repair in IRS (SOS), if $\Xi(\widetilde{\zeta}) = 6(7)$ and $\Omega(\widetilde{\zeta}) \geq 0$.

Let the collection of epochs during a regular service, optional service or repair time termination be termed as $\{\widetilde{\zeta}_n; n = 1, 2, \ldots\}$. Thus, using a series of random vectors $G_n = \{\Xi(\widetilde{\zeta}_n+), \Omega(\widetilde{\zeta})(\widetilde{\zeta}_n+)\}$, an MC is created and incorporated in the retrial QS. Therefore, $G_n; n \in N$ is ergodic iff $[1 - \mathcal{G}^*(\vartheta)]I + \vartheta h\mathcal{I}[d\mathcal{H} + \mathcal{P}_1(1 + \mu_1(\mathcal{A}_1 + \mathcal{K}_1)) + \mathcal{P}_2(1 + \mu_2(\mathcal{A}_2 + \mathcal{K}_2))] < 1$, in order to maintain the stability of our system.

For the process $\{\Omega(\widetilde{\zeta}), \widetilde{\zeta} \geq 0\}$, we mention the probs. $\Lambda_0(\widetilde{\zeta}) = P\{\Xi(\widetilde{\zeta}) = 1, \Omega(\widetilde{\zeta}) = 0\}$ and $\Sigma_0(\widetilde{\zeta}) = P\{\Sigma(\widetilde{\zeta}) = 0, \mathcal{V}(\widetilde{\zeta}) = 0\}$ and the probability densities are

$\Gamma_n(\xi, \widetilde{\zeta})d\xi = P\{\Xi(\widetilde{\zeta}) = 0, \Omega(\widetilde{\zeta}) = n, \xi \leq \mathcal{G}_i^0(\widetilde{\zeta}) < \xi + d\xi\}$,

for $\widetilde{\zeta} \geq 0, \xi \geq 0$ and $n \geq 1$.

$\Lambda_{i,n}(\xi, \widetilde{\zeta})d\xi = P\{\Xi(\widetilde{\zeta}) = 1, 2, \Omega(\widetilde{\zeta}) = n, \xi \leq \mathcal{P}_i^0(\widetilde{\zeta}) < \xi + d\xi\}$,

for $\widetilde{\zeta} \geq 0, \xi \geq 0, n \geq 0, i = 1, 2$.

$\Sigma_n(\xi, \widetilde{\zeta})d\xi = P\{\Xi(\widetilde{\zeta}) = 3, \Omega(\widetilde{\zeta}) = n, \xi \leq \mathcal{H}_i^0(\widetilde{\zeta}) < \xi + d\xi\}$,

for $\widetilde{\zeta} \geq 0, \xi \geq 0$ and $n \geq 1$. $\Upsilon_{i,n}(\xi, \dot{\tau}, \widetilde{\zeta})d\xi = P\{\Xi(\widetilde{\zeta}) = 4, 5, \Omega(\widetilde{\zeta}) = n,\ \dot{\tau} \leq \mathcal{A}_i^0(\widetilde{\zeta}) < \dot{\tau} + d\dot{\tau}/\mathcal{P}_1^0(\widetilde{\zeta}) = \xi\}$,

for $\widetilde{\zeta} \geq 0, (\xi, \dot{\tau}) \geq 0, n \geq 0, i = 1, 2$.

$\Phi_{i,n}(\xi, \dot{\tau}, \widetilde{\zeta})d\xi = P\{\Xi(\widetilde{\zeta}) = 6, 7, \Omega(\widetilde{\zeta}) = n, \dot{\tau} \leq \mathcal{K}_i^0(\widetilde{\zeta}) < \dot{\tau} + d\dot{\tau}/\mathcal{P}_2^0(\widetilde{\zeta}) = \acute{\xi}\}$,

for $\widetilde{\zeta} \geq 0, (\xi, \dot{\tau}) \geq 0, n \geq 0, i = 1, 2$.

The continuation, in our opinion, meets the stability standards; therefore we may assign

$\Lambda_0 = lim_{\widetilde{\zeta}\to\infty}\Lambda_0(\widetilde{\zeta})$, for $\widetilde{\zeta}\ \geq 0$; $\Gamma_n(\acute{\xi}) = lim_{\widetilde{\zeta}\to\infty}\Gamma_n(\xi, \widetilde{\zeta})$;

$\Lambda_{i,n}(\acute{\xi}) = lim_{\widetilde{\zeta}\to\infty}\Lambda_{i,n}(\xi, \widetilde{\zeta})$, $(i = 1, 2)$; $\Sigma_n(\acute{\xi}) = lim_{\widetilde{\zeta}\to\infty}\Sigma_n(\xi, \widetilde{\zeta})$;

$\Upsilon_{i,n}(\acute{\xi}, \dot{\tau}) = lim_{\widetilde{\zeta}\to\infty}\Upsilon_{i,n}(\xi, \dot{\tau}, \widetilde{\zeta})$, $(i = 1, 2)$; $\Pi_{i,n}(\acute{\xi}, \dot{\tau}) = lim_{\widetilde{\zeta}\to\infty}\Pi_{i,n}(\xi, \dot{\tau}, \widetilde{\zeta})$,;

We construct the eqs. below by utilizing the approach of supplementary variable to regulate the dynamics of the system's performance for ($i = 1, 2$).

$$\vartheta o \Lambda_0 = \int_0^\infty \Sigma_0(\acute{\xi})\varphi(\acute{\xi})d\acute{\xi} \tag{1}$$

$$\frac{d\Gamma_n(\acute{\xi})}{d\acute{\xi}} = -(\vartheta + \varpi(\acute{\xi}))\Gamma_n(\acute{\xi}); n \geq 0 \tag{2}$$

$$\frac{d\Lambda_{i,0}(\acute{\xi})}{d\acute{\xi}} + (\vartheta + \mu_i + \chi_i(\acute{\xi}))\Lambda_{i,0}(\acute{\xi}) = \vartheta(1-h)\Lambda_{i,0}(\acute{\xi}) + \int_0^\infty \Pi_{i,n}(\acute{\xi}, \dot{\tau})\varepsilon_i(\dot{\tau})d\dot{\tau} \tag{3}$$

$$\frac{d\Lambda_{i,n}(\acute{\xi})}{d\acute{\xi}} + (\vartheta + \mu_i + \chi_i(\acute{\xi}))\Lambda_{i,n}(\acute{\xi}) = \vartheta h \sum_{k=1}^{n} a_k \Lambda_{i,n-k}(\acute{\xi}) + \vartheta(1-h)\Lambda_{i,n}(\acute{\xi}) + \int_0^\infty \Pi_{i,n}(\acute{\xi}, \dot{\tau})\varepsilon_i(\dot{\tau})d\dot{\tau}; n \geq 1 \tag{4}$$

$$\frac{d\Sigma_0(\acute{\xi})}{d\acute{\xi}} + (\vartheta + \varphi(\acute{\xi}))\Sigma_0(\acute{\xi}) = \vartheta(1-h)\Sigma_0(\acute{\xi}) \tag{5}$$

$$\frac{d\Sigma_n(\acute{\xi})}{d\acute{\xi}} + (\vartheta + \varphi(\acute{\xi}))\Sigma_n(\acute{\xi}) = \vartheta(1-h)\Sigma_n(\acute{\xi}) + \vartheta h \sum_{k=1}^{n} a_k \Sigma_{n-k}(\acute{\xi}); n \geq 1 \tag{6}$$

$$\frac{d\Upsilon_{i,0}(\acute{\xi}, \dot{\tau})}{d\dot{\tau}} + (\vartheta + \iota_i(\dot{\tau}))\Upsilon_{i,0}(\acute{\xi}, \dot{\tau}) = \vartheta(1-h)\Upsilon_{i,0}(\acute{\xi}, \dot{\tau}) \tag{7}$$

$$\frac{d\Upsilon_{i,n}(\acute{\xi}, \dot{\tau})}{d\dot{\tau}} + (\vartheta + \iota_i(\dot{\tau}))\Upsilon_{i,n}(\acute{\xi}, \dot{\tau}) = \vartheta h \sum_{k=1}^{n} a_k \Upsilon_{i,n-k}(\acute{\xi}, \dot{\tau}) + \vartheta(1-h)\Upsilon_{i,n}(\acute{\xi}, \dot{\tau}); n \geq 1 \tag{8}$$

$$\frac{d\Pi_{i,0}(\acute{\xi}, \dot{\tau})}{d\dot{\tau}} + (\vartheta + \varepsilon_i(\dot{\tau}))\Pi_{i,0}(\acute{\xi}, \dot{\tau}) = \vartheta(1-h)\Pi_{i,0}(\acute{\xi}, \dot{\tau}) \tag{9}$$

$$\frac{d\Pi_{i,n}(\acute{\xi}, \dot{\tau})}{d\dot{\tau}} + (\vartheta + \varepsilon_i(\dot{\tau}))\Pi_{i,n}(\acute{\xi}, \dot{\tau}) = \vartheta h \sum_{k=1}^{n} a_k \Pi_{i,n-k}(\acute{\xi}, \dot{\tau}) + \vartheta(1-b)\Pi_{i,n}(\acute{\xi}, \dot{\tau}); n \geq 1 \tag{10}$$

The preceding are the SS boundary conditions at $\acute{\xi} = 0, \dot{\tau} = 0$.

$$\Gamma_n(0) = \int_0^\infty \Sigma_n(\acute{\xi})\varphi(\acute{\xi})d\acute{\xi} + (1-d)q\int_0^\infty \Lambda_{1,n}(\acute{\xi})\chi_1(\acute{\xi})d\acute{\xi} + (1-d)\int_0^\infty \Lambda_{2,n}(\acute{\xi})\chi_2(\acute{\xi})d\acute{\xi} \quad (11)$$

$$\Lambda_{1,n}(0) = \vartheta\int_0^\infty \sum_{k=1}^{n} a_k\Gamma_{n-k-1}(\acute{\xi})d\acute{\xi} + \int_0^\infty \Gamma_{n+1}(\acute{\xi})\varpi(\acute{\xi})d\acute{\xi} + \vartheta h a_{n+1}\Lambda_0 \quad (12)$$

$$\Lambda_{2,n}(0) = p\int_0^\infty \Lambda_{1,n}(\acute{\xi})\chi_1(\acute{\xi})d\acute{\xi} \quad (13)$$

$$\Sigma_n(0) = dq\int_0^\infty \Lambda_{1,n}(\acute{\xi})\chi_1(\acute{\xi})d\acute{\xi} + d\int_0^\infty \Lambda_{2,n}(\acute{\xi})\chi_2(\acute{\xi})d\acute{\xi} \quad (14)$$

$$\Upsilon_{i,n}(\acute{\xi},0) = \mu_i\Lambda_{i,n}(\acute{\xi}) \quad (15)$$

$$\Pi_{i,n}(\acute{\xi},0) = \int_0^\infty \Upsilon_{i,n}(\acute{\xi},0)\iota_i(\dot{\tau})d\dot{\tau} \quad (16)$$

The normalizing condition is

$$\Gamma_0 + \sum_{n=1}^{\infty}\int_0^\infty \Gamma_n(\acute{\xi})d\acute{\xi} + \sum_{n=0}^{\infty}\left(\int_0^\infty \Sigma_n(\acute{\xi})d\acute{\xi} + \sum_{i=1}^{2}\left[\int_0^\infty \Lambda_{i,n}(\acute{\xi})d\acute{\xi} + \int_0^\infty\int_0^\infty \Upsilon_{i,n}(\acute{\xi},\dot{\tau})d\acute{\xi}d\dot{\tau} + \int_0^\infty\int_0^\infty \Pi_{i,n}(\acute{\xi},\dot{\tau})d\acute{\xi}d\dot{\tau}\right]\right) = 1 \quad (17)$$

3.2 *The Steady-State Solution*

The SS soln. of the retrial QS is accomplished with the aid of the generating function method. To resolve the eqs. preceding, the GFs for $|\grave{\varsigma}| < 1$ are specifically defined:

$$\Gamma(\acute{\xi},\grave{\varsigma}) = \sum_{n=1}^{\infty}\Gamma_n(\acute{\xi})\grave{\varsigma}^n;\quad \Gamma(0,\grave{\varsigma}) = \sum_{n=1}^{\infty}\Gamma_n(0)\grave{\varsigma}^n;\quad \Lambda_i(\acute{\xi},\grave{\varsigma}) = \sum_{n=0}^{\infty}\Lambda_{i,n}(\acute{\xi})\grave{\varsigma}^n;$$

$$\Lambda_i(0,\grave{\varsigma}) = \sum_{n=0}^{\infty}\Lambda_{i,0}(\acute{\xi})\grave{\varsigma}^n; i = 1,2;\quad \Sigma(\acute{\xi},\grave{\varsigma}) = \sum_{n=1}^{\infty}\Sigma_n(\acute{\xi})\grave{\varsigma}^n;\quad \Sigma(0,\grave{\varsigma}) = \sum_{n=1}^{\infty}\Sigma_n(0)\grave{\varsigma}^n;$$

$$\Upsilon_i(\acute{\xi}, \dot{\tau}, \grave{\varsigma}) = \sum_{n=0}^{\infty} \Upsilon_{i,n}(\acute{\xi}, \dot{\tau})\grave{\varsigma}^n; \ \ \Upsilon_i(\acute{\xi}, 0, \grave{\varsigma}) = \sum_{n=0}^{\infty} \Upsilon_{i,0}(\acute{\xi}, \dot{\tau})\grave{\varsigma}^n; i = 1, 2;$$

$$\Pi_i(\acute{\xi}, \dot{\tau}, \grave{\varsigma}) = \sum_{n=0}^{\infty} \Pi_{i,n}(\acute{\xi}, \dot{\tau})\grave{\varsigma}^n;$$

$$\Pi_i(\acute{\xi}, 0, \grave{\varsigma}) = \sum_{n=0}^{\infty} \Pi_{i,0}(\acute{\xi}, \dot{\tau})\grave{\varsigma}^n; i = 1, 2; \ \ a(\grave{\varsigma}) = \sum_{n=1}^{\infty} \gamma_n \grave{\varsigma}^n$$

The SS eqs. and SS boundary conditions from (2) to (16) should now be multiplied by $\grave{\varsigma}^n$ and taking summation over n, $(n = 0, 1, 2, \ldots)$

$$\frac{\partial}{\partial \acute{\xi}} \Gamma(\acute{\xi}, \grave{\varsigma}) - (\vartheta + \varpi(\acute{\xi}))\Gamma(\acute{\xi}, \grave{\varsigma}) = 0 \tag{18}$$

$$\frac{\partial}{\partial \acute{\xi}} \Lambda_i(\acute{\xi}, \grave{\varsigma}) + [\vartheta h(1 - a(\grave{\varsigma})) + \mu_i + \chi_i(\acute{\xi})]\Lambda_i(\acute{\xi}, \grave{\varsigma}) = \int_0^{\infty} \Upsilon_i(\acute{\xi}, \dot{\tau}, \grave{\varsigma})\varepsilon_i(\dot{\tau})d\dot{\tau} \tag{19}$$

$$\frac{\partial}{\partial \acute{\xi}} \Sigma(\acute{\xi}, \grave{\varsigma}) + [\vartheta h(1 - a(\grave{\varsigma})) + \varphi(\acute{\xi})]\Sigma(\acute{\xi}, \grave{\varsigma}) = 0 \tag{20}$$

$$\frac{\partial}{\partial \acute{\xi}} \Upsilon_i(\acute{\xi}, \dot{\tau}, \grave{\varsigma}) + [\vartheta h(1 - a(\grave{\varsigma})) + \iota_i(\dot{\tau})]\Upsilon_i(\acute{\xi}, \dot{\tau}, \grave{\varsigma}) = 0 \tag{21}$$

$$\frac{\partial}{\partial \acute{\xi}} \Pi_i(\acute{\xi}, \dot{\tau}, \grave{\varsigma}) + [\vartheta h(1 - a(\grave{\varsigma})) + \varepsilon_i(\dot{\tau})]\Pi_i(\acute{\xi}, \dot{\tau}, \grave{\varsigma}) = 0 \tag{22}$$

$$\Gamma(0, \grave{\varsigma}) = \int_0^{\infty} \Sigma(\acute{\xi}, \grave{\varsigma})\varphi(\acute{\xi})d\acute{\xi} + (1 - d)q \int_0^{\infty} \Lambda_1(\acute{\xi}, \grave{\varsigma})\chi_1(\acute{\xi})d\acute{\xi} + (1-d) \int_0^{\infty} \Lambda_2(\acute{\xi}, \grave{\varsigma})\chi_2(\acute{\xi})d\acute{\xi} - \vartheta h \Gamma_0 \tag{23}$$

$$\Lambda_1(0, \grave{\varsigma}) = \frac{\vartheta}{\grave{\varsigma}} \int_0^{\infty} \Gamma(\acute{\xi}, \grave{\varsigma})a(\grave{\varsigma})d\acute{\xi} + a(\grave{\varsigma})\frac{\vartheta h \Lambda_0}{\grave{\varsigma}} + \frac{1}{\grave{\varsigma}} \int_0^{\infty} \Gamma(\acute{\xi}, \grave{\varsigma})\varpi(\acute{\xi})d\acute{\xi} \tag{24}$$

$$\Lambda_2(0, \grave{\varsigma}) = p \int_0^{\infty} \Lambda_1(\acute{\xi}, \grave{\varsigma})\chi_1(\acute{\xi})d\acute{\xi} \tag{25}$$

$$\Sigma(0, \grave{\varsigma}) = dq \int_0^{\infty} \Lambda_1(\acute{\xi}, \grave{\varsigma})\chi_1(\acute{\xi})d\acute{\xi} + d \int_0^{\infty} \Lambda_2(\acute{\xi}, \grave{\varsigma})\chi_2(\acute{\xi})d\acute{\xi} \tag{26}$$

$$\Upsilon_i(\acute{\xi}, 0, \acute{\varsigma}) = \Lambda_i(\acute{\xi}, \acute{\varsigma})\mu_i \tag{27}$$

$$\Pi_1(\acute{\xi}, 0, \acute{\varsigma}) = \int_0^\infty \Upsilon_{i,n}(\acute{\xi}, \dot{\tau}, \acute{\varsigma})\iota_i(\dot{\tau})d\dot{\tau} \tag{28}$$

By solving the partial differential (Eqs. (18) to (22)), we obtain

$$\Gamma(\acute{\xi}, \grave{\varsigma}) = \Gamma(0, \grave{\varsigma})[1 - \mathcal{G}(\acute{\xi})]e^{-\vartheta\acute{\xi}} \tag{29}$$

$$\Lambda_i(\acute{\xi}, \grave{\varsigma}) = \Lambda_i(0, \grave{\varsigma})[1 - \mathcal{P}_i(\acute{\xi})]e^{-\mathcal{B}_i(\grave{\varsigma})\acute{\xi}}; (i = 1, 2) \tag{30}$$

$$\Sigma(\acute{\xi}, \grave{\varsigma}) = \Sigma(0, \grave{\varsigma})[1 - \mathcal{H}(\acute{\xi})]e^{\mathcal{C}(\grave{\varsigma})\acute{\xi}} \tag{31}$$

$$\Upsilon_i(\acute{\xi}, \dot{\tau}, \grave{\varsigma}) = \Upsilon_i(\acute{\xi}, 0, \grave{\varsigma})[1 - \mathcal{A}_i(\dot{\tau})]e^{-\mathcal{C}(\grave{\varsigma})\dot{\tau}}; (i = 1, 2) \tag{32}$$

$$\Pi_i(\acute{\xi}, \dot{\tau}, \grave{\varsigma}) = \Pi_i(\acute{\xi}, 0, \grave{\varsigma})[1 - \mathcal{K}_i(\dot{\tau})]e^{-\mathcal{C}(\grave{\varsigma})\dot{\tau}}; (i = 1, 2) \tag{33}$$

where $\mathcal{B}_i(\grave{\varsigma}) = \mathcal{C}(\grave{\varsigma}) + \mu_i[1 - \mathcal{A}_i^*(\mathcal{C}(\grave{\varsigma}))\mathcal{K}_i^*(c(\grave{\varsigma}))]$ and $\mathcal{C}(\grave{\varsigma}) = \vartheta h(1 - a(\grave{\varsigma}))$

Theorem 1 *Under the stability criteria, the stationary distribution of the no. of clients in orbit when the server is inactive, in IRS, in SOS and on vacation, during deferring repair and during maintenance, as well as the probability that the server is inactive, is given by* $\nu < 1$.

$$\Gamma(\grave{\varsigma}) = \frac{Nr(\grave{\varsigma})}{Dr(\grave{\varsigma})} \tag{34}$$

$$\begin{aligned} Nr(\grave{\varsigma}) =& b\Lambda_0\left[a(\grave{\varsigma})[d\mathcal{H}^*(\mathcal{C}(\grave{\varsigma})) + (1-d)][\mathcal{P}_1^*(\mathcal{B}_1(\grave{\varsigma}))(q + p\mathcal{P}_2^*(\mathcal{B}_2(\grave{\varsigma})))] - \grave{\varsigma}\right] \\ & [1 - \mathcal{G}^*(\vartheta)] \\ Dr(\grave{\varsigma}) =& \grave{\varsigma} - \left[[1 - \mathcal{G}^*(\vartheta)]a(\grave{\varsigma}) + \mathcal{G}^*(\vartheta)\right]\left[[d\mathcal{H}^*(\mathcal{C}(\grave{\varsigma})) + (1-d)][\mathcal{P}_1^*(\mathcal{B}_1(\grave{\varsigma})) \right. \\ & \left.(q + p\mathcal{P}_2^*(\mathcal{B}_2(\grave{\varsigma})))]\right] \end{aligned}$$

$$\Lambda_1(\grave{\varsigma}) = \frac{\mathcal{B}(\grave{\varsigma})\Lambda_0\mathcal{G}^*(\vartheta)[\mathcal{P}_1^*(\mathcal{B}_1(\grave{\varsigma})) - 1]}{\mathcal{B}_1(\grave{\varsigma})Dg(\grave{\varsigma})} \tag{35}$$

$$\Lambda_2(\grave{\varsigma}) = \frac{\mathcal{B}(\grave{\varsigma})p\Lambda_0\mathcal{G}^*(\vartheta)[\mathcal{P}_2^*(\mathcal{B}_2(\grave{\varsigma})) - 1]\mathcal{P}_1^*(\mathcal{B}_1(\grave{\varsigma}))}{\mathcal{B}_1(\grave{\varsigma})Dg(\grave{\varsigma})} \tag{36}$$

$$\Sigma(\grave{\varsigma}) = \frac{d\Lambda_0\mathcal{G}^*(\vartheta)\mathcal{P}_1^*(\mathcal{B}_1(\grave{\varsigma}))(q + p\mathcal{P}_2^*(\mathcal{B}_2(\grave{\varsigma})))[\mathcal{H}^*(\mathcal{C}(\grave{\varsigma})) - 1]}{Dg(\grave{\varsigma})} \tag{37}$$

$$\Upsilon_1(\grave{\varsigma}) = \frac{\mu_1 \Lambda_0 \mathcal{G}^*(\vartheta)[1-\mathcal{P}_1^*(\mathcal{B}_1(\grave{\varsigma}))][\mathcal{A}_1^*(\mathcal{C}(\grave{\varsigma}))-1]}{\mathcal{B}_1(\grave{\varsigma})Dg(\grave{\varsigma})} \tag{38}$$

$$\Upsilon_2(\grave{\varsigma}) = \frac{\mu_2 p \Lambda_0 \mathcal{G}^*(\vartheta)[1-\mathcal{P}_2^*(\mathcal{B}_2(\grave{\varsigma}))][\mathcal{A}_2^*(\mathcal{C}(\grave{\varsigma}))-1]\mathcal{P}_1^*(\mathcal{B}_1(\grave{\varsigma}))}{\mathcal{B}_2(\grave{\varsigma})Dg(\grave{\varsigma})} \tag{39}$$

$$\Pi_1(\grave{\varsigma}) = \frac{\mu_1 \Lambda_0 \mathcal{G}^*(\vartheta)[1-\mathcal{P}_1^*(\mathcal{B}_1(\grave{\varsigma}))][\mathcal{K}_1^*(\mathcal{C}(\grave{\varsigma}))-1]\mathcal{A}_1^*(\mathcal{C}(\grave{\varsigma}))}{\mathcal{B}_1(\grave{\varsigma})Dg(\grave{\varsigma})} \tag{40}$$

$$\Pi_2(\grave{\varsigma}) = \frac{\mu_2 \Lambda_0 \mathcal{G}^*(\vartheta)[1-\mathcal{P}_2^*(\mathcal{B}_2(\grave{\varsigma}))][\mathcal{K}_2^*(\mathcal{C}(\grave{\varsigma}))-1]\mathcal{P}_1^*(\mathcal{B}_1(\grave{\varsigma}))\mathcal{A}_2^*(\mathcal{C}(\grave{\varsigma}))}{\mathcal{B}_2(\grave{\varsigma})Dg(\grave{\varsigma})} \tag{41}$$

where

$$\Lambda_0 = \frac{1-[1-\mathcal{G}^*(\vartheta)]\mathcal{I}-\vartheta h\mathcal{I}\,[d\mathcal{H}+\mathcal{P}_1(1+\mu_1(\mathcal{A}_1+\mathcal{K}_1))+p\mathcal{P}_2(1+\mu_2(\mathcal{A}_2+\mathcal{K}_2))]}{(1-h)\,[1+\mathcal{I}(1-\mathcal{G}^*(\vartheta))[\vartheta h[d\mathcal{H}+\mathcal{P}_1(1+\mu_1(\mathcal{A}_1+\mathcal{K}_1))+p\mathcal{P}_2(1+\mu_2(\mathcal{A}_2+\mathcal{K}_2))]-1]]+\mathcal{G}^*(\vartheta)h} \tag{42}$$

Proof *By defining the PGF as*
$\Gamma(\grave{\varsigma}) = \int_0^\infty \Gamma(\xi, \grave{\varsigma})d\acute{\xi}$, $\Lambda_i(\grave{\varsigma}) = \int_0^\infty \Lambda_i(\acute{\xi}, \grave{\varsigma})d\acute{\xi}$, $(i = 1,2)$; $\Upsilon_i(\acute{\xi}, \grave{\varsigma}) = \int_0^\infty \Upsilon_i(\acute{\xi}, \dot{\tau}, \grave{\varsigma})d\dot{\tau}$,
$\Upsilon_i(\grave{\varsigma}) = \int_0^\infty \Upsilon_i(\acute{\xi}, \grave{\varsigma})d\acute{\xi}$, $\Pi_i(\acute{\xi}, \grave{\varsigma}) = \int_0^\infty \Pi_i(\acute{\xi}, \dot{\tau}, \grave{\varsigma})d\dot{\tau}$, $\Pi_i(\grave{\varsigma}) = \int_0^\infty \Pi_i(\acute{\xi}, \grave{\varsigma})d\acute{\xi}$, $(i = 1,2)$ *and after some mathematical computations, we get Eqs.* (34) *to* (41).

For Λ_0, *the probability that the server will be inactive when there are no clients in the orbit is the sole unknown, which can be estimated by the normalizing condition. Thus, by making* $\grave{\varsigma} = 1$ *in Eqs.* (34)–(42) *and using the l'Hospitals method wherever needed, we get* $\Lambda_0 + \Gamma(1) + \Sigma_1(1) + \Lambda_1(1) + \Lambda_2(1) + \Upsilon_1(1) + \Upsilon_2(1) + \Pi_2(1) + \Pi_2(1) = 1$.

Corollary 1 *The following eqs. are employed in the SS to derive the PGF for the no. of clients in the system* $K_e(\grave{\varsigma})$ *and the orbit* $R_e(\grave{\varsigma})$.

$$K_e(\grave{\varsigma}) = \Lambda_0+\Gamma(\grave{\varsigma})+\Sigma_1(\grave{\varsigma})+\grave{\varsigma}\{\Lambda_1(\grave{\varsigma})+\Lambda_2(\grave{\varsigma})+\Upsilon_1(\grave{\varsigma})+\Upsilon_2(\grave{\varsigma})+\Pi_2(\grave{\varsigma})+\Pi_2(\grave{\varsigma})\} \tag{43}$$

$$R_e(\grave{\varsigma}) = \Lambda_0+\Gamma(\grave{\varsigma})+\Sigma_1(\grave{\varsigma})+\Lambda_1(\grave{\varsigma})+\Lambda_2(\grave{\varsigma})+\Upsilon_1(\grave{\varsigma})+\Upsilon_2(\grave{\varsigma})+\Pi_2(\grave{\varsigma})+\Pi_2(\grave{\varsigma}) \tag{44}$$

4 System Performance Measures

4.1 Mean Size of the System and the Orbit in SS

(i) By differentiating (43) in regard to $\grave{\varsigma}$ and utilizing $\grave{\varsigma} = 1$, we can get the mean no. of clients in the system (L_s).

$$L_s = K_e^{'}(1) = \lim_{\grave{\varsigma}\to 1} \frac{d}{d\grave{\varsigma}} K_e(\grave{\varsigma}) = \Lambda_0 \left[\frac{Nr_q^{''}(1)Dr_q^{'}(1) - Dr_q^{''}(1)Nr_q^{'}(1)}{2(Dr_q^{'}(1))^2} \right] \tag{45}$$

(ii) By differentiating (44) in regard to $\grave{\varsigma}$ and utilizing $\grave{\varsigma} = 1$, we can get the mean no. of clients in the queue (L_q).

$$L_q = R_e^{'}(1) = \lim_{\grave{\varsigma}\to 1} \frac{d}{d\grave{\varsigma}} R_e(\grave{\varsigma}) = \Lambda_0 \left[\frac{Nr_s^{''}(1)Dr_q^{'}(1) - Dr_q^{''}(1)Nr_q^{'}(1)}{2(Dr_q^{'}(1))^2} \right] \tag{46}$$

(iii) The mean period a client can invest in the system is given by $W_s = \frac{L_s}{\vartheta I}$, and the mean period a client can invest in the queue is given by $W_q = \frac{L_q}{\vartheta I}$

5 Exceptional Cases

Case (1): One Poisson arrival and absence of vacation, breakage and retrial.

Let $Pr[\mathcal{V} = 1] = 1$, $d = 0$, $\mu_1 = \mu_2 = 0$, and $\mathcal{G}^*(\vartheta) \to 1$. Therefore, a single server QS under dual stages of service of which the latter being discretionary and balking has been obtained.

Case (2): Single Poisson arrival and the absence of balking, breakdown and retrial.

Let $Pr[\mathcal{V} = 1] = 1$, $h = 0$, $\mu_1 = \mu_2 = 0$, and $\mathcal{G}^*(\vartheta) \to 1$. Therefore, we obtain a single server QS with second optional service subject to BV.

6 Numerical Results

To demonstrate the many alternatives for the system's dynamic reaction, we have used MATLAB in this segment. We look into exponentially distributed retrial times, two-stage service times with an optional second stage, vacation times and repair times with factors ϑ, ν and μ_i.

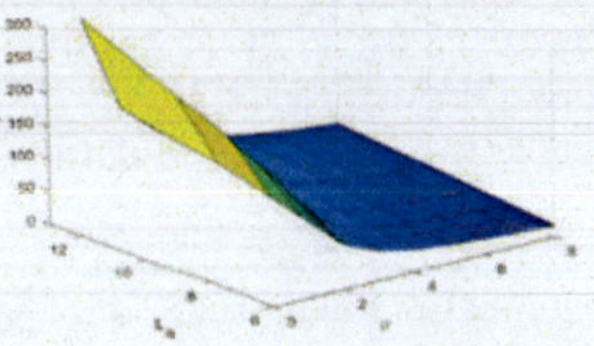

(a) L_q, W_q verses arrival rate ϑ

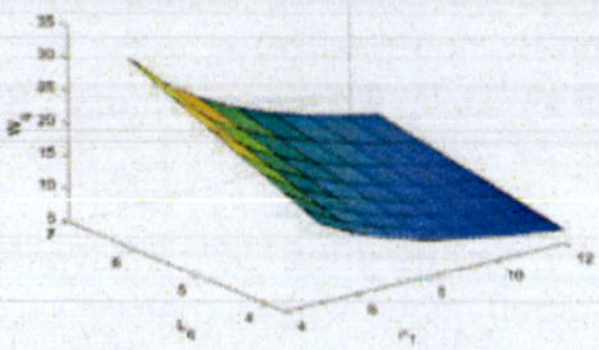

(b) L_q, W_q verses IRS μ_1

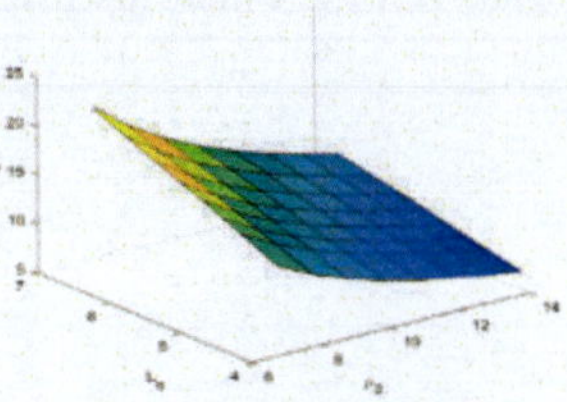

(c) L_q, W_q verses SOS μ_2

7 Conclusion

This paper has scrutinized a $M^{[X]}/G/1$ retrial QS with two segments of service with second discretionary service subjugated to BV, breakdown and repair. The PGF of the no. of consumers when the system is idle, occupied, on vacation or undergoing maintenance is computed by utilizing the supplementary variable approach. Quantitative measurements are gathered, including mean waiting time and the expected queue length. We also took into account certain instances of the model in order to demonstrate that the findings were numerically feasible. Our framework is better equipped to deal with the significant systems of various business units in a variety of real queueing scenarios.

References

1. Abdollahi, S., Salehi Rad, M.R.: Analysis of a batch arrival retrial queue with two-phase services, feedback and admission. Bull. Iranian Math. Soc. **48**(3), 791–804 (2022)
2. Doshi, B.T.: Queueing system with vacations – a survey. Queueing Syst. **1**, 29–66 (1986)
3. Jain, M., Kaur, S.: Bernoulli vacation model for MX/G/1 unreliable server retrial queue with bernoulli feedback, balking and optional service. RAIRO-Oper. Res. **55**, S2027-S2053 (2021)
4. Ke,J.C., Choudhury, G.: A batch arrival retrial queue with general retrial times under Bernoulli vacation schedule for unreliable server and delaying repair. Appl. Math. Model. **36**, 255–69 (2012)
5. Ke, J.C., Wu, C.H., Zhang, Z.G.: Recent developments in vacation queueing models: a short survey. Inter. J. Oper. Res. **7**, 3–8 (2010)

6. Nila, M., Sumitha, D.: Two phase retrial queue with starting failures, customer impatience, feedback and Bernoulli vacation. Adv. Appl. Math. Sci. **21**, 1773–1786 (2022)
7. Pavai Madheswari, P., Krishna Kumar, P., Suganthi, P.: Analysis of $M/G/1$ retrial queues with second optional service and customer balking under two types of Bernoulli vacation schedule. RAIRO-Oper. Res. **53**, 415–443 (2019)
8. Rajadurai, P., Varalakshmi, M., Saravanarajan, M.C., Chandrasekaran, V.M.: Analysis of $M^{[x]}/G/1$ retrial queue with two phase service under Bernoulli vacation schedule and random breakdown. Int. J. Math. Oper. Res. **7**, 19–41 (2015)
9. Takagi, H.: Queueing Analysis, Volume 1: Vacation and Priority Systems. North-Holland, Amsterdan (1991)
10. Varalakshmi, M., Chandrasekaran, V.M., Saravanarajan, M.C.: An unreliable two phase retrial queue with immediate Bernoulli feedbacks. Int. J. Pure Appl. Math. **115**, 447–457. (2017)

Solving Neutrosophic Bi-objective Assignment Problem Using Different Approaches

S. Sandhiya and D. Anuradha

1 Introduction

The linear programming problem (LPP) is a highly well-known mathematical programming problem that has numerous applications in a variety of industries. The assignment problem (AP) is one of the most established areas of optimization. The AP with two conflicting objectives that deals with cost, time, quality, etc., is known as bi-objective AP (BOAP). In 1965, Zadeh [1] was the first to introduce the fuzzy sets (FS), which provides only the degree of membership function (MF). The fuzzy programming approach (FPA) was introduced by Zimmermann [2] for multi-objective optimization problems. Tilva and Dhodiya [3] modified the algorithm with the help of exponential MF to solve the MOAP. Atanassov [4] introduced the intuitionistic fuzzy set, which is an extension of FS. These deal with both the elements of MF and non-MF in the set, whereas FS just addresses the MF. To obtain a POS, Ghosh et al. [5] utilized the FPA, intuitionistic FPA, and goal PA (GPA) to solve the multi-objective transportation problem (MOTP) with a fixed charge three-dimensional problem under an intuitionistic environment. Smarandache [6] was the first to introduce the concept of the neutrosophic set (NS). It takes into account both the degree of truth MF and falsity MF together with the indeterminacy degree when making decisions. The concept of fuzzy decision-making was introduced by Bellman and Zadeh [7]. Khalifa [8] proposed a weighting tchebycheff approach to obtain the neutrosophic solutions for neutrosophic MOAP. Deli and Subas [9] presented the special forms of SVNNs, and for solving the multi-criteria problems, they applied SVTrNNs weighted aggregation operator. Harnpornchai and Wonggattaleekam [10] proposed an NS based on a relative AP.

S. Sandhiya · D. Anuradha (✉)
Department of Mathematics, School of Advanced Sciences, Vellore Institute of Technology, Vellore, Tamil Nadu, India
e-mail: anuradhadhanapal1981@gmail.com

F. Kamalov et al. (eds.), *Advances in Mathematical Modeling and Scientific Computing*, Trends in Mathematics, https://doi.org/10.1007/978-3-031-41420-6_44

Ahmad and Adhami [11] applied the NFPA to solve the nonlinear MOTP with fuzzy parameters. Veeramani et al. [12] established the NGPA to solve the fractional MOTP. In the literature, many researchers have applied FPA and NFPA to solve the MOTP, where the parameters are in crisp and fuzzy nature. According to the latest records, this is the first time that the FPA and NFPA are extended to obtain a POS for BOAP under neutrosophic nature. In general, most of the existing techniques provide POS to the TPs under a crisp and fuzzy environment. Practically, the DM needs to achieve the results not only in the degree of truth and falsity MF but also in the indeterminacy factors. In addition to this, most of the existing techniques provide only deterministic solution to the optimization problems under neutrosophic environment. Practically, the DM may not have specific, reliable, and detailed information regarding these solutions. This motivates us to solve BOAP under neutrosophic environment. The purpose of this chapter is to investigate a POS of NBOAP under FPA and NFPA. Illustrative examples as well as a comparative study have been used to demonstrate the effectiveness and appropriateness of these two approaches. In scientific and engineering contexts, the FPA and NFPA constitute an innovative way to cope with information that is imprecise, ambiguous, partial, and inconsistent. The structure of this research chapter is as follows: Section 2 presents some basic definitions related to NS. The mathematical formulation of NBOAP and the two methods, namely FPA and NFPA, with related formulations are illustrated in Sect. 3. The solution approaches are represented in Sect. 4. In Sect. 5, the numerical illustration is investigated. Section 6 discusses the performance regarding the results. Section 7 outlines the conclusions of the present research and suggests some future directions for research.

2 Preliminaries and Essential Definitions

Some fundamental definitions related to the NS, SVN sets, SVTrNNs, and score function have been discussed in [6, 9].

Definition 1 POS of NBOAP is a feasible solution $x^* = ({x^*}_{ij} : i = 1, 2, \ldots, n; j = 1, 2, \ldots, n)$ such that there exists no other feasible solution $x = (x_{ij} : i = 1, 2, \ldots, n; j = 1, 2, \ldots, n)$ with $Z_a(x) \leq Z_a(x^*)$, $a = 1, 2$ and $Z_a(x) < Z_a(x^*)$ for at least one a.

3 Problem Description and Formulation

The purpose of this section is to introduce assumptions, indices, and formulation for defining a mathematical problem. Here, we develop NBOAP with two resources and two activities—unit cost, time, resources, and activities are expressed as neutrosophic numbers. In NBOAP, the quantity x_{ij}^N must be allocated from resources $i(i = 1, 2, \ldots, n)$ to activities $j(j = 1, 2, \ldots, n)$ with the parameters $c_{ij}^{(1)N}$ and $c_{ij}^{(2)N}$, where $c_{ij}^{(1)N}$ is the production cost and $c_{ij}^{(2)N}$ is the storage cost while assigning the products to plants. The mathematical formulation of BOAP under neutrosophic parameters can be represented as follows:

$$G_1 \text{ Minimize } Z_1^N(x) = \sum_{i=1}^{n}\sum_{j=1}^{n} c_{ij}{}^{(1)N} x_{ij}{}^N; \text{ Minimize } Z_2^N(x) = \sum_{i=1}^{n}\sum_{j=1}^{n} c_{ij}{}^{(2)N} x_{ij}{}^N$$

$$Subject\ to: \sum_{j=1}^{n} x_{ij}{}^N = 1^N, i = 1, 2, \ldots, n$$

$$\sum_{i=1}^{n} x_{ij}{}^N = 1^N, j = 1, 2, \ldots, n$$

$$x^N{}_{ij} = 0^N \ or\ 1^N \ for\ all\ i\ and\ j \tag{1}$$

Here, $c_{ij}{}^{(1)N} = \left(c_{ij}^1, c_{ij}^2, c_{ij}^3, c_{ij}^4, c'^1_{ij}, c'^2_{ij}, c'^3_{ij}\right)$ and $c_{ij}{}^{(2)N} = \left(c_{ij}^1, c_{ij}^2, c_{ij}^3, c_{ij}^4, c'^1_{ij}, c'^2_{ij}, c'^3_{ij}\right)$ denote the first objective and the second objective SVTrN production cost and storage cost, respectively. The SVTrN variable $x_{ij}{}^N = \left(x_{ij}{}^1, x_{ij}{}^2, x_{ij}{}^3, x_{ij}{}^4; x_{ij}{}'^1, x_{ij}{}'^2, x'^3_{ij}\right)$ is assumed to be either 0 or 1 based on the entire assignment of ith resource to j^{th} activity. The above problem is of neutrosophic fuzzy in nature. Therefore, using the score function of the objective function, we can transform the problem G_1 into its equivalent problem. We get

$$G_2 \text{ Minimize } \Re(Z_1(x)) = \Re\left(\sum_{i=1}^{n}\sum_{j=1}^{n}\left(c_{ij}^{1(1)}x_{ij}^1, c_{ij}^{2(1)}x_{ij}^2, c_{ij}^{3(1)}x_{ij}^3, c_{ij}^{4(1)}x_{ij}^4, c'^{1(1)}_{ij}x'^1_{ij}, c'^{2(1)}_{ij}x'^2_{ij}, c'^{3(1)}_{ij}x'^3_{ij}\right)\right)$$

$$\text{Minimize } \Re(Z_2(x)) = \Re\left(\sum_{i=1}^{n}\sum_{j=1}^{n}\left(c_{ij}^{1(2)}x_{ij}^1, c_{ij}^{2(2)}x_{ij}^2, c_{ij}^{3(2)}x_{ij}^3, c_{ij}^{4(2)}x_{ij}^4, c'^{1(2)}_{ij}x'^1_{ij}, c'^{2(2)}_{ij}x'^2_{ij}, c'^{3(2)}_{ij}x'^3_{ij}\right)\right)$$

subject to

$$\sum_{j=1}^{n} x_{ij}^1 = 1^N, \sum_{j=1}^{n} x_{ij}^2 = 1^N, \sum_{j=1}^{n} x_{ij}^3 = 1^N, \sum_{j=1}^{n} x_{ij}^4 = 1^N, i = 1, 2, \ldots, n,$$

$$\sum_{j=1}^{n} x'^1_{ij} = 1^N, \sum_{j=1}^{n} x'^2_{ij} = 1^N, \sum_{j=1}^{n} x'^3_{ij} = 1^N, i = 1, 2, \ldots, n,$$

Table 1 Pay-off matrix

	Z_1	Z_2
X_1	$Z_1(X_1)$	$Z_2(X_1)$
X_2	$Z_1(X_2)$	$Z_2(X_2)$

$$\sum_{i=1}^{n} x_{ij}^{1} = 1^{N}, \sum_{i=1}^{n} x_{ij}^{2} = 1^{N}, \sum_{i=1}^{n} x_{ij}^{3} = 1^{N}, \sum_{i=1}^{n} x_{ij}^{4} = 1^{N}, j = 1, 2, \ldots, n,$$

$$\sum_{i=1}^{n} x'^{1}_{ij} = 1^{N}, \sum_{i=1}^{n} x'^{2}_{ij} = 1^{N}, \sum_{i=1}^{n} x'^{3}_{ij} = 1^{N}, j = 1, 2, \ldots, n, \tag{2}$$

$x'^{1}_{ij}{=}0^{N}$ *or* 1^{N}, $x^{1}_{ij}{-}x'^{1}_{ij}{=}0^{N}$ *or* 1^{N}, $x'^{2}_{ij}{-}x^{1}_{ij}{=}0^{N}$ *or* 1^{N}, $x^{2}_{ij}{-}x'^{2}_{ij}{=}0^{N}$ *or* 1^{N}, $x^{3}_{ij}{-}x^{2}_{ij}{=}0^{N}$ *or* 1^{N}, $x'^{3}_{ij}{-}x^{3}_{ij}{=}0^{N}$ *or* 1^{N}, $x^{4}_{ij} - x'^{3}_{ij} = 0^{N}$ *or* 1^{N}, *for all* $i = 1, 2, \ldots, n$, $j = 1, 2, \ldots, n$

To evaluate the problem G_2, each objective function is considered separately with respect to the constraints. In the pay-off matrix (POM), determine the upper bound (UB) and lower bound (LB) for all the objective functions that are displayed in Table 1. The UB U_a and LB L_a are defined as

$$U_a = \max\left[Z_a\left(X_a\right)\right] \ \textit{and} \ L_a = \min\left[Z_a\left(X_a\right)\right], a = 1, 2 \tag{3}$$

Two solution approaches, namely FPA and NFPA, are used to tackle the problem G_2 so that we may obtain a POS as formulated in the following subsections.

3.1 FPA

There are numerous ways to solve BOAP. FPA is one of these techniques that is effective at determining a POS. Zimmermann [2] introduced FPA, which makes it relatively simple to solve the problems of neutrosophic nature. To solve the NBOAP and to obtain a POS, we utilize FPA. The MF $T_a^N\left(Z_a(x)\right)$ for each objective function is formulated as

$$T_a^N\left(Z_a(x)\right) = \begin{cases} 1, & if \ Z_a(x) \le L_a^T \\ 1 - \frac{Z_a(x) - L_a^T}{U_a^T - L_a^T}, & if \ L_a^T \le Z_a(x) \le U_a^T, (a = 1, 2) \\ 0, & if \ Z_a(x) \ge U_a^T \end{cases} \tag{4}$$

Here, $U_a^T = U_a$ and $L_a^T = L_a$ are the UB and LB for each objective function. Our goal is to maximize the degree of MF of each objective function, where the degree of MF is denoted as λ; then, with the help of FPA, the problem G$_2$ can be formulated as

$$G_3 \text{ Maximize } \lambda$$
$$Subject\ to : T_a^N(Z_a(x)) \geq \lambda, (a = 1, 2)\,;\, \lambda \in [0, 1]$$

along with all the constraints of (2).

Now, the problem G$_3$ can be represented as follows:

$$G_4 \text{ Maximize } \lambda$$
$$Subject\ to : Z_a(x) + \left(U_a^T - L_a^T\right)\lambda \leq U_a^T, (a = 1, 2)\,;\, \lambda \in [0, 1]$$

along with all the constraints of (2).

Theorem 1 If $x^* = (x^*_{ij}, :i = 1, 2, \ldots, n; j = 1, 2, \ldots, n)$ is an optimal solution of the problem G$_4$, then it is also a POS of the problem G$_2$.

Proof
Let x^* be not a POS of the problem G$_2$. Then, by Definition 1 we can consider that there exists at least one x, such that $Z_a(x) \leq Z_a(x^*)$ for $a = 1, 2$ and $Z_a(x) < Z_a(x^*)$ for at least one a. Therefore, the MF $T_a^N(Z_a(x))$ strictly decreases with respect to the objective function $Z_a(x)$ in [0,1]. Hence, $T_a^N(Z_a(x)) \geq T_a^N(Z_a(x^*))$ and $T_a^N(Z_a(x)) > T_a^N(Z_a(x^*))$ for at least one a. Now, $\lambda = \min\left\{T_a^N(Z_a(x))\right\} \geq \min\left\{T_a^N(Z_a(x^*))\right\} = \lambda^*$. This contradicts that x^* is an optimal solution of the problem G$_4$, since λ^* is the value of λ at x^*. Hence, the theorem is proved.

3.2 NFPA

The NFPA is based on NDS to solve the NBOAP, which consists of three MFs—maximizing the degree of truth MF and indeterminacy MF and minimizing the degree of falsity MF, and it is used in specific optimization problems. In numerous real-world decision-making situations under fuzziness, Bellman and Zadeh [7] were the first to propose the idea of a fuzzy decision (P), a fuzzy goal (Q), and a fuzzy constraint (R) in 1970. The truth MF $T_a^N(Z_a(x))$, indeterminacy MF $I_a^N(Z_a(x))$, and falsity MF $F_a^N(Z_a(x))$ for each objective function are formulated as

$$T_a^N(Z_a(x)) = \begin{cases} 1, & if\ Z_a(x) < L_a^T \\ \frac{U_a^T - Z_a(x)}{U_a^T - L_a^T}, & if\ L_a^T \leq Z_a(x) \leq U_a^T, (a = 1, 2) \\ 0, & if\ Z_a(x) > U_a^T \end{cases} \quad (5)$$

$$I_a^N(Z_a(x)) = \begin{cases} 1, & if\ Z_a(x) < L_a^I \\ \frac{U_a^I - Z_a(x)}{U_a^I - L_a^I}, & if\ L_a^I \le Z_a(x) \le U_a^I, (a = 1, 2) \\ 0, & if\ Z_a(x) > U_a^I \end{cases} \tag{6}$$

$$F_a^N(Z_a(x)) = \begin{cases} 1, & if\ Z_a(x) < L_a^F \\ \frac{U_a^F - Z_a(x)}{U_a^F - L_a^F}, & if\ L_a^F \le Z_a(x) \le U_a^F, (a = 1, 2) \\ 0, & if\ Z_a(x) > U_a^F \end{cases} \tag{7}$$

The representation of the bounds for a objective functions under the neutrosophic parameter are $U_a^T = U_a, L_a^T = L_a$ for truth MF,$U_a^I = L_a^T + s_a, L_a^I = L_a^T$ for indeterminacy MF, and $U_a^F = U_a^T, L_a^F = L_a^T + t_a$ for falsity MF, where $s_a\ and\ t_a \in (0, 1)$ are real numbers assigned by DMs. Using (5–7) and the idea of Bellman and Zadeh [7] in 1970, the problem G_2 can be signified as follows:

G_5 Max $\min_{a=1,2} T_a^N(Z_a(x))$; Max $\min_{a=1,2} I_a^N(Z_a(x))$; Min $\max_{a=1,2} F_a^N(Z_a(x))$

Subject to all the constraints of (2).

Our goal is to maximize the truth and indeterminacy degree and to minimize the falsity degree for all the objective functions, where the truth degree is denoted as λ, the indeterminacy degree as δ, and the falsity degree as η; then, with the help of NFPA, the problem G_5 can be formulated as

G_6 Max λ, Max δ, Min η
$Subject\ to: T_a^N(Z_a(x)) \ge \lambda;\ I_a^N(Z_a(x)) \ge \delta;\ F_a^N(Z_a(x)) \ge \eta, (a = 1, 2);$
$\lambda \ge \delta, \lambda \ge \eta, \lambda + \delta + \eta \le 3; \lambda, \delta, \eta \in [0, 1]$

along with all the constraints of (2).

Now, the problem G_6 can also be represented as follows:

G_7 Maximize $\lambda - \eta + \delta$
$Subject\ to: Z_a(x) + (U_a^T - L_a^T)\lambda \le U_a^T;\ Z_a(x) + (U_a^I - L_a^I)\delta \le U_a^I;$
$Z_a(x) - (U_a^F - L_a^F)\eta \le L_a^F;\ \lambda \ge \delta, \lambda \ge \eta, \lambda + \delta + \eta \le 3; \lambda, \delta, \eta \in [0, 1]$

along with all the constraints of (2).

Theorem 2 If $x^* = (x^*_{ij} : i = 1, 2, \ldots, n; j = 1, 2, \ldots, n)$ is an optimal solution of the problem G_7, then it is also a POS of the problem G_2.

Proof
Let x^* be not a POS of the problem G_2. Then by Definition 1, we can consider that there exists at least one x, such that $Z_a(x) \le Z_a(x^*)$ for $a = 1, 2$ and $Z_a(x) < Z_a(x^*)$for at least one a. Therefore, the truth MF $T_a^N(Z_a(x))$ and the

indeterminacy MF $I_a^N(Z_a(x))$ strictly decreases to $Z_a(x)$ in [0,1]. Again the falsity MF $F_a^N(Z_a(x))$ strictly increases to $Z_a(x)$ in [0,1]. Hence, $T_a^N(Z_a(x)) \geq T_a^N(Z_a(x^*)) \forall a$ and $T_a^N(Z_a(x)) > T_a^N(Z_a(x^*))$ for at least one a. Similarly, $I_a^N(Z_a(x)) \geq I_a^N(Z_a(x^*)) \forall a$ and $I_a^N(Z_a(x)) > I_a^N(Z_a(x^*))$ for at least one a. Hence, $F_a^N(Z_a(x)) \leq F_a^N(Z_a(x^*)) \forall a$ and $F_a^N(Z_a(x)) < F_a^N(Z_a(x^*))$ for at least one a. Now, $(\lambda - \eta + \delta) = \min\{T_a^N(Z_a(x)), F_a^N(Z_a(x)), I_a^N(Z_a(x))\} \geq \min\{T_a^N(Z_a(x^*)), F_a^N(Z_a(x^*)), I_a^N(Z_a(x^*))\} = (\lambda^* - \eta^* + \delta^*)$. This contradicts that x^* is an optimal solution of the problem G_7, since $\lambda*$, $\eta*$ *and* $\delta*$ are the values of λ, η *and* δ at x^*. Hence the theorem is proved.

4 Solution Approaches

In this section, we utilize two approaches such as FPA and NFPA to obtain a POS. The procedure of these approaches from problem formulation to final solution are summarized as follows:

Step 1 Formulate the NBOAP as in the problem G_1 for FPA.
Step 2 The problem G_1 is converted into the problem G_2 using the score function.
Step 3 Each objective function with subject to all constraints is solved individually using the LINGO software, and the POM is constructed as in Table 1.
Step 4 The UB and LB are determined for the problem G_2 using (3) from Table 1.
Step 5 The MF for the problem G_2 is defined using (4) for FPA.
Step 6 Formulate the FPA defined as in the problem G_4 and then solve it using the LINGO software to obtain a POS.
Step 7 For NFPA, repeat Steps 1 to 4.
Step 8 The truth MF, indeterminacy MF, and falsity MF for the problem G_2 are defined using (5–7) for NFPA.
Step 9 Formulate the NFPA defined as in the problem G_7 and then solve it using the LINGO software to obtain a POS.

5 Numerical Illustration

Dairies produce three varieties of milk products. They have three plants, one of which is assigned to each product. Let us assume that there are two objectives to be considered: (i) the minimization of total production cost of each product processed at each plant and (ii) the minimization of total storage cost of each product processed at each plant. The production cost $Z_1^N(x)$ and storage cost $Z_2^N(x)$ of the milk products are represented as SVTrNNs and the corresponding data are displayed below.

Minimize $Z_1^N(x)$

$$= \begin{bmatrix} \left((14, 17, 21, 28; 0.8, 0.2, 0.6) \otimes \tilde{x}_{11}^N\right) \oplus \left((13, 18, 20, 24; 0.6, 0.4, 0.5) \otimes \tilde{x}_{12}^N\right) \\ \oplus \left((20, 25, 30, 35; 0.8, 0.4, 0.2) \otimes \tilde{x}_{13}^N\right) \oplus \left((15, 18, 23, 30; 0.9, 0.2, 0.3) \otimes \tilde{x}_{21}^N\right) \\ \oplus \left((11, 16, 25, 28; 0.8, 0.3, 0.2) \otimes \tilde{x}_{22}^N\right) \oplus \left((17, 19, 23, 27; 0.9, 0.3, 0.2) \otimes \tilde{x}_{23}^N\right) \\ \oplus \left((11, 17, 22, 25; 0.6, 0.5, 0.4) \otimes \tilde{x}_{31}^N\right) \oplus \left((10, 14, 26, 30; 0.8, 0.6, 0.2) \otimes \tilde{x}_{32}^N\right) \\ \oplus \left((14, 16, 21, 23; 0.7, 0.5, 0.3) \otimes \tilde{x}_{33}^N\right) \end{bmatrix}$$

Minimize $Z_2^N(x)$

$$= \begin{bmatrix} \left((12, 20, 25, 29; 0.9, 0.3, 0.2) \otimes \tilde{x}_{11}^N\right) \oplus \left((22, 25, 30, 34; 0.8, 0.2, 0.4) \otimes \tilde{x}_{12}^N\right) \\ \oplus \left((12, 18, 21, 24; 0.7, 0.4, 0.5) \otimes \tilde{x}_{13}^N\right) \oplus \left((15, 17, 19, 24; 0.7, 0.2, 0.3) \otimes \tilde{x}_{21}^N\right) \\ \oplus \left((28, 32, 35, 40; 0.9, 0.3, 0.2) \otimes \tilde{x}_{22}^N\right) \oplus \left((12, 15, 27, 29; 0.8, 0.2, 0.3) \otimes \tilde{x}_{23}^N\right) \\ \oplus \left((23, 27, 30, 31; 0.9, 0.3, 0.4) \otimes \tilde{x}_{31}^N\right) \oplus \left((12, 19, 24, 25; 0.8, 0.5, 0.4) \otimes \tilde{x}_{32}^N\right) \\ \oplus \left((13, 18, 23, 25; 0.9, 0.2, 0.2) \otimes \tilde{x}_{33}^N\right) \end{bmatrix}$$

Subject to all the constraints of (1)

Using Step 2, the above problem can be converted based on the score function and is named G_8. As in Step 3, to solve the problem G_8 using the LINGO software, consider the one objective and exclude other objectives with respect to the constraints. The objective function values are shown in Table 2, which are represented as a POM.

6 Results and Discussions

Solving the problem numerically using the two approaches such as FPA and NFPA, we obtain a POS as depicted in Table 3. The POS that we have obtained using the two approaches are provided below.

Table 2 Pay-off matrix

	Z_1	Z_2
X_1	29	28
X_2	38	42

Table 3 Comparison of results in two approaches

Objective function	FPA	NFPA
Z_1^N	(41, 50, 70, 85; 0.8, 0.6, 0.6)	(41, 50, 70, 85; 0.8, 0.6, 0.6)
Z_2^N	(36, 54, 76, 83; 0.8, 0.5, 0.4)	(36, 54, 76, 83; 0.8, 0.5, 0.4)

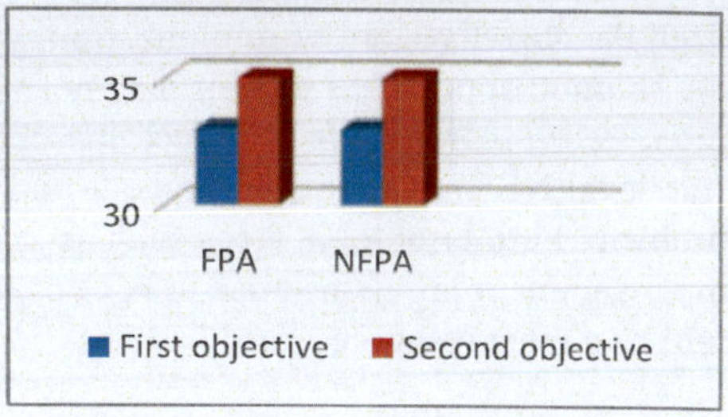

Fig. 1 Comparison of different approaches

For FPA, solving the problem G_8 by considering one objective function at a time with respect to the constraints and then as in Step 4, we obtain the UB and LB from Eq. (3) as $U_1 = 38$, $L_1 = 29$, $U_2 = 42$, and $L_2 = 28$. Using Steps 5 and 6, the POS in FPA with SVTrNNs is $\lambda = 0.5$; $\tilde{x}_{11}^N = (1, 1, 1, 1; 1, 1, 1)$, $\tilde{x}_{12}^N = (0, 0, 0, 0; 0, 0, 0)$, $\tilde{x}_{13}^N = (0, 0, 0, 0; 0, 0, 0)$, $\tilde{x}_{21}^N = (0, 0, 0, 0; 0, 0, 0)$, $\tilde{x}_{22}^N = (0, 0, 0, 0; 0, 0, 0)$, $\tilde{x}_{23}^N = (1, 1, 1, 1; 1, 1, 1)$, $\tilde{x}_{31}^N = (0, 0, 0, 0; 0, 0, 0)$, $\tilde{x}_{32}^N = (1, 1, 1, 1; 1, 1, 1)$, $\tilde{x}_{33}^N = (0, 0, 0, 0; 0, 0, 0)$. Therefore, the neutrosophic values of the objective functions in FPA are$Z_1^N = (41, 50, 70, 85; 0.8, 0.6, 0.6)$, $Z_2^N = (36, 54, 76, 83; 0.8, 0.5, 0.4)$.

For NFPA, using Steps (7–9) the POS with SVTrNNs is $\lambda = 0.5, \eta = 0, \delta = 0.5$; $\tilde{x}_{11}^N = (1, 1, 1, 1; 1, 1, 1)$, $\tilde{x}_{12}^N = (0, 0, 0, 0; 0, 0, 0)$, $\tilde{x}_{13}^N = (0, 0, 0, 0; 0, 0, 0)$, $\tilde{x}_{21}^N = (0, 0, 0, 0; 0, 0, 0)$, $\tilde{x}_{22}^N = (0, 0, 0, 0; 0, 0, 0)$, $\tilde{x}_{23}^N = (1, 1, 1, 1; 1, 1, 1)$, $\tilde{x}_{31}^N = (0, 0, 0, 0; 0, 0, 0)$, $\tilde{x}_{32}^N = (1, 1, 1, 1; 1, 1, 1)$, $\tilde{x}_{33}^N = (0, 0, 0, 0; 0, 0, 0)$. Therefore, the neutrosophic values of the objective functions in NFPA are $Z_1^N = (41, 50, 70, 85; 0.8, 0.6, 0.6)$, $Z_2^N = (36, 54, 76, 83; 0.8, 0.5, 0.4)$.

The solution obtained for two objective functions using the two programming approaches is graphically presented in Fig. 1. A comparison of the POS has been performed using Table 3 and Fig. 1. Since the objectives are of minimization type, we observe that the minimal values of each objective function in FPA and NFPA are the same. Therefore, we conclude that FPA provides the same result as NFPA.

7 Conclusions and Scope for Future Research

In this chapter, we have taken into account NBOAP. We have implemented neutrosophic parameter with the truth, indeterminacy, and falsity MFs to NBOAP in order to investigate these circumstances. The problem is entirely neutrosophic in nature, which gives DMs more flexibility. By introducing the score function, the neutrosophic parameters have been reduced to their equivalent BOAP. We have employed FPA and NFPA to obtain a POS to the formulated BOAP. The applicability of our solution approaches has been illustrated with an example. Based on the comparison of the pareto-optimal values, we conclude that FPA provides the same result as NFPA. These approaches are limited in that they yield the best POS if all the objectives are of minimization type. In the event of conflicting objectives, the solution approaches may or may not yield the best solution. In the future study, interested researchers may reduce the complexity and extend our approaches for fractional NBOAP, multi-item type-2 fuzzy problem, or type-2 neutrosophic fuzzy problem. Furthermore, in areas such as management science and finance, wherever the assignment problems arise in neutrosophic environment, this solution approach will be a great resource.

References

1. Zadeh, L.A.: Fuzzy sets. Inf. Control. **8**(3), 338–353 (1965)
2. Zimmermann, H.J.: Fuzzy programming and linear programming with several objective functions. Fuzzy Sets Syst. **1**(1), 45–55 (1978)
3. Tilva, S., Dhodiya, J.: Multi-objective assignment problem solved by hybrid Jaya algorithm. J. Interdiscip Math. **25**(1), 109–121 (2022)
4. Atanassov, K.T.: Intuitionistic fuzzy sets. Fuzzy Sets Syst. **20**(1), 87–96 (1986)
5. Ghosh, S., Roy, S.K., Ebrahimnejad, A., Verdegay, J.L.: Multi-objective fully intuitionistic fuzzy fixed-charge solid transportation problem. Complex Intell. Syst. **7**(2), 1009–1023 (2021)
6. Smarandache, F.: A Unifying Field in Logics Neutrosophy: Neutrosophic Probability, Set and Logic. American Research Press, Rehoboth, USA (1999)
7. Bellman, R.E., Zadeh, L.A.: Decision-making in a fuzzy environment. Manag. Sci. **17**(4), 141–164 (1970)
8. Khalifa, H.: An approach to the optimization of multi-objective assignment problems with neutrosophic numbers. Int. J. Ind. Eng. Prod. Res. **31**(2), 287–294 (2020)
9. Deli, I., Subas, Y.: Single valued neutrosophic numbers and their applications to multicriteria decision making problem. Neutrosophic Sets and Systems. **2**(1), 1–13 (2014)
10. Harnpornchai, N., Wonggattaleekam, W.: An application of neutrosophic set to relative importance assignment in AHP. Mathematics. **9**(20), 2636 (2021)
11. Ahmad, F., Adhami, A.Y.: Neutrosophic programming approach to multi-objective nonlinear transportation problem with fuzzy parameters. Int. J. Manag. Sci Eng. Manag. **14**(3), 218–229 (2019)
12. Veeramani, C., Edalatpanah, S.A., Sharanya, S.: Solving the multi-objective fractional transportation problem through the neutrosophic goal programming approach. Discret. Dyn. Nat. Soc. **2021**, 1–17 (2021)

Analysis of Attainment Estimates of Loss System Queue

R. Ramesh and M. Seenivasan

1 Introduction

Nowadays, people meet more over-crowding problems in front of reservation centers, ATM points, hospitals, medical shops, ration shops, etc. Stressing the significance of time distribution is the ideal focus of the author. Queueing models play a very prominent role in this environment. The preludes and models of the queueing system [1–4] are essential for our investigative motive. Queues with loss systems [5–7] have applications in computers and telecommunication departments. Recently Ramesh R et al. published [8, 9] articles about retrial queueing models and loss system. More inventors introduced variant strategies for finding the attainments of queues. In that race, the matrix geometry method plays a huge role. Recently Seenivasan et al. presented [10–15] some articles for queueing models using matrix geometry method. Moreover the proposed technique is very concrete.

2 Loss System

In a queue system, if a free server is available, then the customers go in for the service. Suppose all the servers are busy, then the customers can't join the system and terminate the system without service is called Loss system. In a telephone exchange, while a caller acquires one of the "m" outgoing lines only if one such line is available. If every line is engaged, the caller is enforced to quit the system

R. Ramesh
Department of Mathematics, Arignar Anna Government Arts College, Musiri, Tamilnadu, India

M. Seenivasan (✉)
Mathematics Wing, DDE, Annamalai University, Chidambaram, TamilNadu, India

F. Kamalov et al. (eds.), *Advances in Mathematical Modeling and Scientific Computing*, Trends in Mathematics, https://doi.org/10.1007/978-3-031-41420-6_45

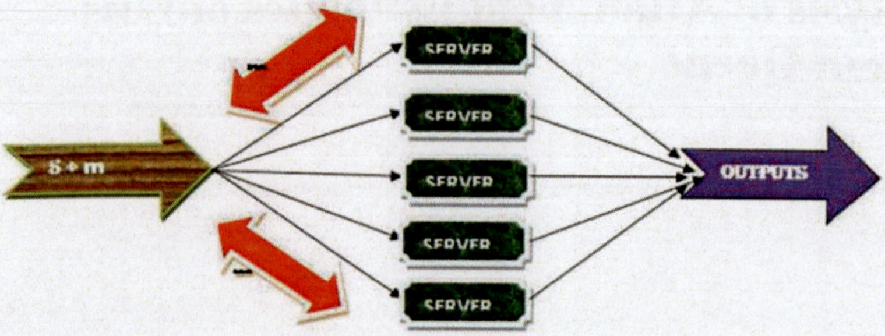

Fig. 1 Loss system with five servers

Fig. 2 Transition diagram

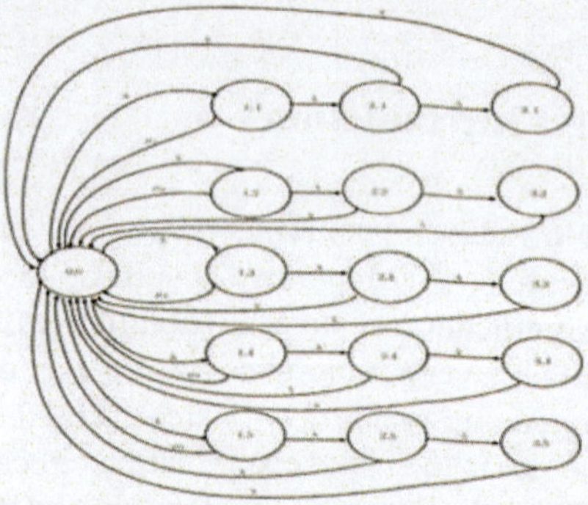

without service. A loss system has 5 + m inputs with five servers (Fig. 1) given below.

Let us consider the arrival rate and service rate as λ and μ, respectively. If the probability of blocking be ϕ, then the mean arrival rate actually entering the queue will be $\lambda(1\ \breve{}\ \phi)$. This system state is limited to (0, 1, ...,m) and will always be ergodic. The rate transition diagram is given in Fig. 2.

The blocking probability ϕ for this queue is determined as the probability that a customer finds each server is engaged and quits the system without service, i.e., $\phi = P_0 \frac{\rho^m}{m!}$, where $P_0 = \frac{1}{\sum_{k=0}^{n} \frac{\rho^k}{k!}}$ and k is the stage of state. Let the rate of rejected calls $\xi = \lambda\phi$.

In the queue, at time t, let $\{(n(t), m(t)); t \geq 0\}$ be a MP, where $n(t)$ and $m(t)$ represent state of process at time t, respectively.

$$m(t) = \begin{cases} 1, & \textit{state of server 1} \\ 2, & \textit{state of server 2} \\ 3, & \textit{state of server 3} \\ 4, & \textit{state of server 4} \\ 5, & \textit{state of server 5} \end{cases}$$

$n(t)$ represents a number of customers.

The quasi-birth-and-death (QBD) process is following the state space

$$\Omega = (0, 0) \bigcup (j, i); i = 1, 2, 3, 4, 5 and j = 1, 2, .., n \geq 1$$

Infinitesimal generator matrix Q is presented below:

$$Q = \begin{pmatrix} B_{00} & B_{01} & 0 & \cdots & \cdots & \cdots \\ B_{10} & A_1 & B & \cdots & \cdots & \cdots \\ 0 & 0 & C & B & \cdots & \cdots \\ 0 & 0 & 0 & C & \cdots & \cdots \\ \vdots & \vdots & \vdots & \vdots & \vdots & \vdots \end{pmatrix}$$

where

$$B_{00} = \begin{pmatrix} -5\lambda \end{pmatrix}; \quad B_{01} = \begin{pmatrix} \lambda \ \lambda \ \lambda \ \lambda \end{pmatrix}; \quad A_0 = \begin{pmatrix} \chi \\ \chi \\ \chi \\ \chi \\ \chi \end{pmatrix} \quad B_{10} = \begin{pmatrix} \mu_1 \\ \mu_2 \\ \mu_3 \\ \mu_4 \\ \mu_5 \end{pmatrix};$$

;

$$B = \begin{pmatrix} \lambda & 0 & 0 & 0 & 0 \\ 0 & \lambda & 0 & 0 & 0 \\ 0 & 0 & \lambda & 0 & 0 \\ 0 & 0 & 0 & \lambda & 0 \\ 0 & 0 & 0 & 0 & \lambda \end{pmatrix}; C = \begin{pmatrix} -(\chi+\lambda) & 0 & 0 & 0 & 0 \\ 0 & -(\chi+\lambda) & 0 & 0 & 0 \\ 0 & 0 & -(\chi+\lambda) & 0 & 0 \\ 0 & 0 & 0 & -(\chi+\lambda) & 0 \\ 0 & 0 & 0 & 0 & -(\chi+\lambda) \end{pmatrix}$$

$$A_1 = \begin{pmatrix} -(\mu_1+\lambda) & 0 & 0 & 0 & 0 \\ 0 & -(\mu_2+\lambda) & 0 & 0 & 0 \\ 0 & 0 & -(\mu_3+\lambda) & 0 & 0 \\ 0 & 0 & 0 & -(\mu_4+\lambda) & 0 \\ 0 & 0 & 0 & 0 & -(\mu_3+\lambda) \end{pmatrix}.$$

We define π_{ij}= $\{r = i, s = j\}$; $\lim_{t\to\infty} r(t) = i$, $\lim_{t\to\infty} s(t) = j$, , where j is the number of arrivals in the queue and i is the state of the server. We define the probability vector as Π= $(\Pi_0, \Pi_1, \Pi_2, \ldots)$, where $\Pi_0 = \pi_{00}$ and $\Pi_j = (\pi_{j1}, \pi_{j2}, \pi_{j3}, \pi_{j4}, \pi_{j5})$,j=1,2,3,4,. . . ,. . . , and we used the row matrix of static probability PQ=O.

$$\Pi_0 B_{00} + \Pi_1 B_{10} + \Pi_2 A_0 + \Pi_3 A_0 + \Pi_4 A_0 + \Pi_5 A_0 + \ldots = 0 \tag{1}$$

$$\Pi_0 B_{01} + \Pi_1 A_1 = 0 \tag{2}$$

$$\Pi_1 B + \Pi_2 C = 0 \tag{3}$$

$$\Pi_2 B + \Pi_3 C = 0 \tag{4}$$

$$\cdots$$

$$\Pi_{i-1} B_1 + \Pi_i A_1 + \Pi_{i+1} C_1 = 0 \tag{5}$$

$$\Pi_i = \Pi_1 R^{i-1} \text{for} \quad i \geq 2 \tag{6}$$

We denote the irreducible nonnegative solution rate matrix to be R. Then we have

$$\Pi_1 R[RC + B] = 0 \tag{7}$$

Using Eqs. (1) and (2), we get

$$(\Pi_0, \Pi_1)(B_{00} B_{01} B_{10} + A_0 R (1 - R)^{-1} A_1) = 0 \tag{8}$$

The balancing mathematical statement is

$$\Pi_0 e + \Pi_1 (1 - R)^{-1} e = 1 \tag{9}$$

Here e is unit matrix.

The static condition takes format $\rho < 1$, where $\rho = \frac{\lambda}{\mu_1+\mu_2+\mu_3+\mu_4+\mu_5}$. Using Eq. (7) we found R matrix by iteration method and from that we found probability vectors using Eqs. (6) and (8).

3 Numerical Example

A telephone switch has only five output lines (m = 5 servers) and a massive number of incoming calls. Suppose a line is available, the arriving call is assigned to an output line, or the call is choked and stray. The output line remains assigned to the call for its entire duration which is of exponentially distributed length. We treat the arrival rate as λ and the service rate as μ concerning Poisson manner. We consider the rate of rejected calls as ξ. Here we take the servers as $m = 5, 6, 7, \ldots$

Now we consider the value of λ differs and the remaining values are fixed.

Illustration 1 We take $\lambda = 0.4$, $\mu_1 = 0.010$, $\mu_2 = 0.014$, $\mu_3 = 0.018$, $\mu_4 = 0.022$, $\mu_5 = 0.026$, $\xi = 0.096$, and

$$R = \begin{pmatrix} 0.8065 & 0 & 0 & 0 & 0 \\ 0 & 0.8065 & 0 & 0 & 0 \\ 0 & 0 & 0.8065 & 0 & 0 \\ 0 & 0 & 0 & 0.8065 & 0 \\ 0 & 0 & 0 & 0 & 0.8065 \end{pmatrix}$$

Table 1 Probability vector

Π_i	π_{j1}	π_{j2}	π_{j3}	π_{j4}	π_{j5}	Total
Π_0	0.0436					0.0436
Π_1	0.0396	0.0382	0.0369	0.0357	0.0346	0.1850
Π_2	0.0319	0.0308	0.0298	0.0288	0.0279	0.1492
Π_3	0.0258	0.0248	0.0240	0.0232	0.0225	0.1203
Π_4	0.0208	0.0200	0.0194	0.0187	0.0182	0.0971
Π_5	0.0168	0.0162	0.0156	0.0151	0.0146	0.0783
Π_6	0.0135	0.0130	0.0126	0.0122	0.0118	0.0631
Π_7	0.0109	0.0105	0.0102	0.0098	0.0095	0.0509
Π_8	0.0088	0.0085	0.0082	0.0079	0.0077	0.0411
Π_9	0.0071	0.0068	0.0066	0.0064	0.0062	0.0331
Π_{10}	0.0057	0.0055	0.0053	0.0052	0.0050	0.0267
Π_{11}	0.0046	0.0044	0.0043	0.0042	0.0040	0.0215
Π_{12}	0.0037	0.0036	0.0035	0.0034	0.0032	0.0174
Π_{13}	0.0030	0.0029	0.0028	0.0027	0.0026	0.0141
Π_{14}	0.0024	0.0023	0.0023	0.0022	0.0021	0.0113
Π_{15}	0.0020	0.0019	0.0018	0.0018	0.0017	0.0092
Π_{16}	0.0016	0.0015	0.0015	0.0014	0.0014	0.0074
Π_{17}	0.0013	0.0012	0.0012	0.0011	0.0011	0.0059
Π_{18}	0.0010	0.0010	0.0010	0.0009	0.0009	0.0048
Π_{19}	0.0008	0.0008	0.0008	0.0007	0.0007	0.0038
Π_{20}	0.0007	0.0006	0.0006	0.0006	0.0006	0.0031
Π_{21}	0.0005	0.0005	0.0005	0.0005	0.0005	0.0025
Π_{22}	0.0004	0.0004	0.0004	0.0004	0.0004	0.0020
Π_{23}	0.0003	0.0003	0.0003	0.0003	0.0003	0.0015
Π_{24}	0.0003	0.0003	0.0003	0.0003	0.0002	0.0014
Π_{25}	0.0002	0.0002	0.0002	0.0002	0.0002	0.0010
Π_{26}	0.0001	0.0001	0.0001	0.0001	0.0001	0.0005
Total						0.9976

Using R in Eq. (8), vector Π_0 and normalization equation $\Pi_0 e+\Pi_1(1-R)^{-1}e = 1$ are obtained. Also Π_0 and Π_1 are found as Π_0 = 0.0436 and Π_1 = (0.0396, 0.0382, 0.0369, 0.0357, 0.0346). The equilibrium vector Π_j is attained from $\Pi_i = \Pi_1 R^{i-1}, i \geq 2$ 2 and tabulated in tabular column 1. Columns 2, 3, 4, 5, and 6 of Table 1 contain the five members of Π_j, j = 0, 1, 2 .... The last column determines the sum of five members. The aggregate probability has been validated as 0.9976 ≈1.

Illustration 2 We take $\lambda = 0.5$, $\mu_1 = 0.010$, $\mu_2 = 0.014$, $\mu_3 = 0.018$, $\mu_4 = 0.022$, $\mu_5 = 0.026$, $\xi = 0.165$, and

$$R = \begin{pmatrix} 0.7519 & 0 & 0 & 0 & 0 \\ 0 & 0.7519 & 0 & 0 & 0 \\ 0 & 0 & 0.7519 & 0 & 0 \\ 0 & 0 & 0 & 0.7519 & 0 \\ 0 & 0 & 0 & 0 & 0.7519 \end{pmatrix}$$

By using R in Eq. (8), vector Π_0 and normalization equation $\Pi_0 e + \Pi_1(1 - R)^{-1}e = 1$ are obtained. Also Π_0 and Π_1 are found as $\Pi_0 = 0.0489$ and $\Pi_1 = (0.0479, 0.0476, 0.0472, 0.0468, 0.0464)$. Now, the equilibrium vector Π_j is attained from $\Pi_i = \Pi_1 R^{i-1}, i \geq 2$ 2 and tabulated in tabular column 1. Columns 2, 3, 4, 5, and 6 of Table 2 contain the five members of Π_j, j = 0, 1, 2 .... The last column determines the sum of five members. The aggregate probability has been validated as 0.9989 ≈1.

Table 2 Probability vector

Π_i	π_{j1}	π_{j2}	π_{j3}	π_{j4}	π_{j5}	Total
Π_0	0.0489					0.0489
Π_1	0.0479	0.0476	0.0472	0.0468	0.0464	0.2359
Π_2	0.0360	0.0358	0.0355	0.0352	0.0349	0.1774
Π_3	0.0271	0.0269	0.0267	0.0265	0.0262	0.1334
Π_4	0.0204	0.0202	0.0201	0.0199	0.0197	0.1003
Π_5	0.0153	0.0152	0.0151	0.0150	0.0148	0.0754
Π_6	0.0115	0.0114	0.0113	0.0112	0.0111	0.0565
Π_7	0.0087	0.0086	0.0085	0.0085	0.0084	0.0427
Π_8	0.0065	0.0065	0.0064	0.0064	0.0063	0.0321
Π_9	0.0049	0.0049	0.0048	0.0048	0.0047	0.0241
Π_{10}	0.0037	0.0037	0.0036	0.0036	0.0036	0.0183
Π_{11}	0.0028	0.0027	0.0027	0.0027	0.0027	0.0136
Π_{12}	0.0021	0.0021	0.0020	0.0020	0.0020	0.0102
Π_{13}	0.0016	0.0015	0.0015	0.0015	0.0015	0.0076
Π_{14}	0.0012	0.0012	0.0012	0.0011	0.0011	0.0058
Π_{15}	0.0009	0.0009	0.0009	0.0009	0.0009	0.0045
Π_{16}	0.0007	0.0007	0.0007	0.0006	0.0006	0.0033
Π_{17}	0.0005	0.0005	0.0005	0.0005	0.0005	0.0025
Π_{18}	0.0004	0.0004	0.0004	0.0004	0.0004	0.0020
Π_{19}	0.0003	0.0003	0.0003	0.0003	0.0003	0.0015
Π_{20}	0.0002	0.0002	0.0002	0.0002	0.0002	0.0010
Π_{21}	0.0002	0.0002	0.0002	0.0002	0.0001	0.0009
Π_{22}	0.0001	0.0001	0.0001	0.0001	0.0001	0.0005
Total						0.9989

Illustration 3 We take $\lambda = 0.6$, $\mu_1 = 0.010$, $\mu_2 = 0.014$, $\mu_3 = 0.018$, $\mu_4 = 0.022$, $\mu_5 = 0.026$, $\xi = 0.240$, and

$$R = \begin{pmatrix} 0.7143 & 0 & 0 & 0 & 0 \\ 0 & 0.7143 & 0 & 0 & 0 \\ 0 & 0 & 0.7143 & 0 & 0 \\ 0 & 0 & 0 & 0.7143 & 0 \\ 0 & 0 & 0 & 0 & 0.7143 \end{pmatrix}$$

By using R in Eq. (8), vector Π_0 and normalization equation $\Pi_0 e + \Pi_1(1 - R)^{-1}e = 1$ are obtained. Also Π_0 and Π_1 are found as $\Pi_0 = 0.0556$ and $\Pi_1 = (0.0547, 0.0543, 0.0540, 0.0536, 0.0533)$. Now, the equilibrium vector Π_j is attained from $\Pi_i = \Pi_1 R^{i-1}, i \geq 2$ 2 and tabulated in tabular column 1. Columns 2, 3, 4, 5, and 6 of Table 3 contain the five members of Π_j, j = 0, 1, 2 .... The last column determines the sum of five members. The aggregate probability has been validated as 0.9989 ≈1.

Table 3 Probability vector

Π_i	π_{j1}	π_{j2}	π_{j3}	π_{j4}	π_{j5}	Total
Π_0	0.0556					0.0556
Π_1	0.0547	0.0543	0.0540	0.0536	0.0533	0.2708
Π_2	0.0391	0.0388	0.0386	0.0383	0.0381	0.1929
Π_3	0.0279	0.0277	0.0275	0.0273	0.0272	0.1376
Π_4	0.0199	0.0198	0.0197	0.0195	0.0194	0.0983
Π_5	0.0142	0.0141	0.0141	0.0139	0.0139	0.0702
Π_6	0.0102	0.0101	0.0100	0.0100	0.0099	0.0502
Π_7	0.0073	0.0072	0.0072	0.0071	0.0071	0.0359
Π_8	0.0052	0.0051	0.0051	0.0051	0.0051	0.0256
Π_9	0.0037	0.0037	0.0037	0.0036	0.0036	0.0183
Π_{10}	0.0026	0.0026	0.0026	0.0026	0.0026	0.0130
Π_{11}	0.0019	0.0019	0.0019	0.0018	0.0018	0.0093
Π_{12}	0.0013	0.0013	0.0013	0.0013	0.0013	0.0065
Π_{13}	0.0010	0.0010	0.0009	0.0009	0.0009	0.0047
Π_{14}	0.0007	0.0007	0.0007	0.0007	0.0007	0.0035
Π_{15}	0.0005	0.0005	0.0005	0.0005	0.0005	0.0025
Π_{16}	0.0003	0.0003	0.0003	0.0003	0.0003	0.0015
Π_{17}	0.0002	0.0002	0.0002	0.0002	0.0002	0.0010
Π_{18}	0.0002	0.0002	0.0002	0.0002	0.0002	0.0010
Π_{19}	0.0001	0.0001	0.0001	0.0001	0.0001	0.0005
Total						0.9989

Illustration 4 We take $\lambda = 0.7$, $\mu_1 = 0.010$, $\mu_2 = 0.014$, $\mu_3 = 0.018$, $\mu_4 = 0.022$, $\mu_5 = 0.026$, $\xi = 0.322$, and

$$R = \begin{pmatrix} 0.6849 & 0 & 0 & 0 & 0 \\ 0 & 0.6849 & 0 & 0 & 0 \\ 0 & 0 & 0.6849 & 0 & 0 \\ 0 & 0 & 0 & 0.6849 & 0 \\ 0 & 0 & 0 & 0 & 0.6849 \end{pmatrix}$$

By using R in Eq. (8), vector Π_0 and normalization equation $\Pi_0 e + \Pi_1(1 - R)^{-1}e = 1$ are obtained. Also Π_0 and Π_1 are found as Π_0 = 0.0607 and Π_1 = (0.0599, 0.0595, 0.0592, 0.0589, 0.0585). Now, the equilibrium vector Π_j is attained from $\Pi_i = \Pi_1 R^{i-1}, i \geq 2$ 2 and tabulated in tabular column 1. Columns 2, 3, 4, 5, and 6 of Table 4 contain the five members of Π_j, j = 0, 1, 2 .... The last column determines the sum of five members. The aggregate probability has been validated as 0.9992 ≈1.

Table 4 Probability vector

Π_i	π_{j1}	π_{j2}	π_{j3}	π_{j4}	π_{j5}	Total
Π_0	0.0607					0.0607
Π_1	0.0599	0.0595	0.0592	0.0589	0.0585	0.2960
Π_2	0.0410	0.0407	0.0405	0.0403	0.0401	0.2026
Π_3	0.0281	0.0279	0.0278	0.0276	0.0274	0.1388
Π_4	0.0192	0.0191	0.0190	0.0189	0.0188	0.0950
Π_5	0.0132	0.0131	0.0130	0.0130	0.0129	0.0652
Π_6	0.0090	0.0090	0.0089	0.0088	0.0088	0.0463
Π_7	0.0062	0.0061	0.0061	0.0061	0.0060	0.0305
Π_8	0.0042	0.0042	0.0042	0.0042	0.0041	0.0209
Π_9	0.0029	0.0029	0.0028	0.0028	0.0028	0.0143
Π_{10}	0.0020	0.0020	0.0019	0.0019	0.0019	0.0098
Π_{11}	0.0014	0.0013	0.0013	0.0013	0.0013	0.0066
Π_{12}	0.0009	0.0009	0.0009	0.0009	0.0009	0.0045
Π_{13}	0.0006	0.0006	0.0006	0.0006	0.0006	0.0030
Π_{14}	0.0004	0.0004	0.0004	0.0004	0.0004	0.0020
Π_{15}	0.0003	0.0003	0.0003	0.0003	0.0003	0.0015
Π_{16}	0.0002	0.0002	0.0002	0.0002	0.0002	0.0010
Π_{17}	0.0001	0.0001	0.0001	0.0001	0.0001	0.0005
Total						0.9992

4 Performance Measures

Using the steady-state probabilities, we got the probability that:

1. System is idle $P(I) = \Pi_0$
2. System is busy $P(B) = 1 - \Pi_0$
3. Customer receives the service from server 1 $P(S_1) = \sum_{j=0}^{\infty} j\pi_{j1}$
4. Customer receives the service from server 2 $P(S_2) = \sum_{j=0}^{\infty} j\pi_{j2}$
5. Customer receives the service from server 3 $P(S_3) = \sum_{j=0}^{\infty} j\pi_{j3}$
6. Customer receives the service from server 4 $P(S_4) = \sum_{j=0}^{\infty} j\pi_{j4}$
7. Customer receives the service from server 5 $P(S_5) = \sum_{j=0}^{\infty} j\pi_{j5}$
8. Average customers in the system $E(N) = P(B) + P(S_1) + P(S_2) + P(S_3) + P(S_4) + P(S_5)$
9. Aggregate yield achieved is $\delta = \mu_1 \sum_{j=0}^{\infty} j\pi_{j1} + \mu_2 \sum_{j=0}^{\infty} j\pi_{j2} + \mu_3 \sum_{j=0}^{\infty} j\pi_{j3} + \mu_4 \sum_{j=0}^{\infty} j\pi_{j4} + \mu_5 \sum_{j=0}^{\infty} j\pi_{j5}$

Table 5 represents the attainment estimates with respect to the arrival rate λ.

The presented graphs elucidated the performance expedients in respect to λ, for instance, the probabilities that the system is ideal; the system is busy; the customer got the service from servers 1, 2, 3, 4, and 5; average customers in the system; and yield achieved from the system all at once. Figures 3, 4, 5, 6, 7, and 8 show that if

Table 5 With the effect of λ

Λ	0.4	0.5	0.6	0.7
P(I)	0.0436	0.0489	0.0556	0.0607
P(B)	0.9564	0.9511	0.9444	0.9393
$P(S_1)$	1.0301	0.7724	0.66097	0.5951
$P(S_2)$	0.9915	0.7661	0.6563	0.5908
$P(S_3)$	0.9646	0.7584	0.6527	0.5864
$P(S_4)$	0.9313	0.7519	0.6466	0.5841
$P(S_5)$	0.9007	0.7431	0.6446	0.5803
E(N)	5.7746	4.7430	4.2055	3.8760
Δ	0.0855	0.0680	0.0585	0.0527

Fig. 3 Arrival versus P(I)

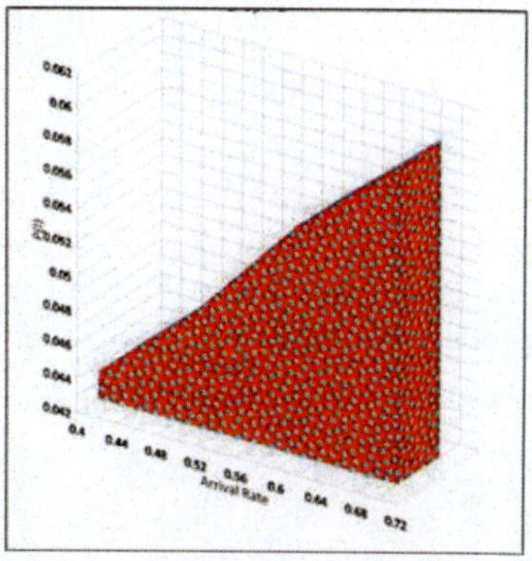

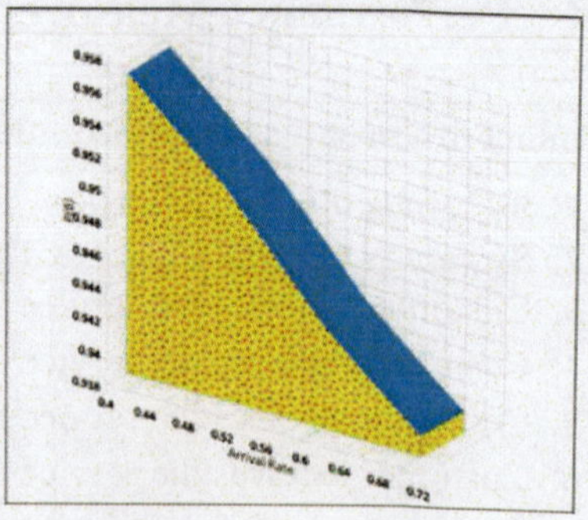

Fig. 4 Arrival versus P(B)

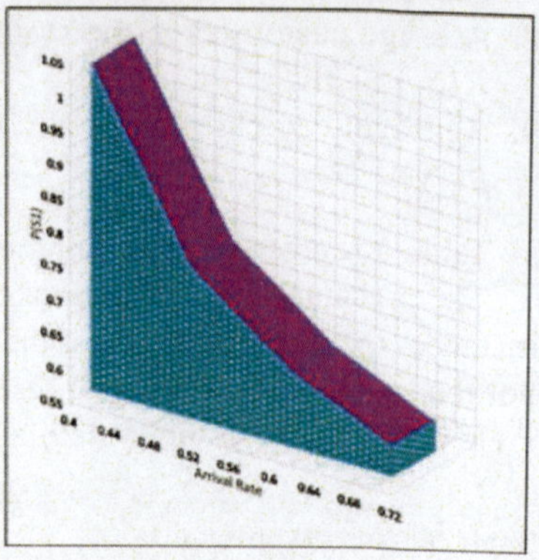

Fig. 5 Arrival versus $P(S_1)$

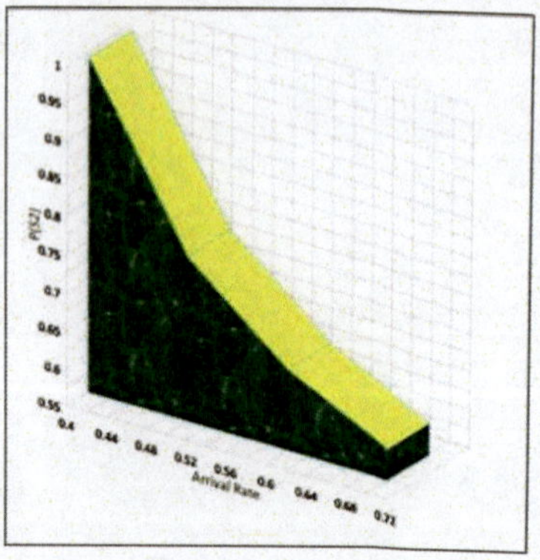

Fig. 6 Arrival versus $P(S_2)$

the arrival rate rises, the probability that the system is busy and the customer gets the service from servers 1, 2, 3, 4, and 5, the average customers throughout the system, and aggregate yield achieved rise, and the Fig. 3 describes that if the λ rises, the probability that the system is idle gets decreased.

Fig. 7 Arrival versus $P(S_3)$

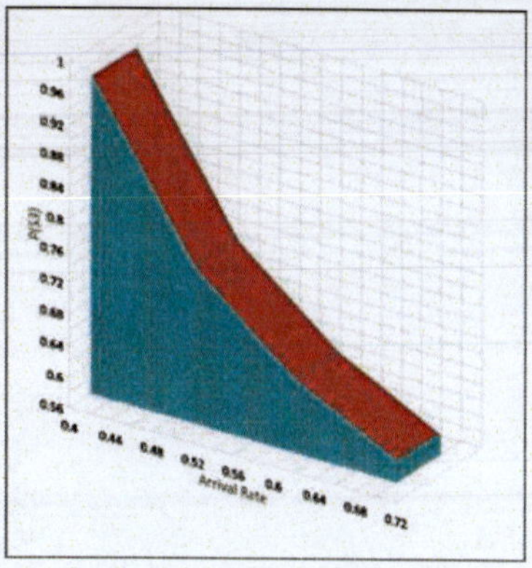

Fig. 8 Arrival versus $P(S_4)$

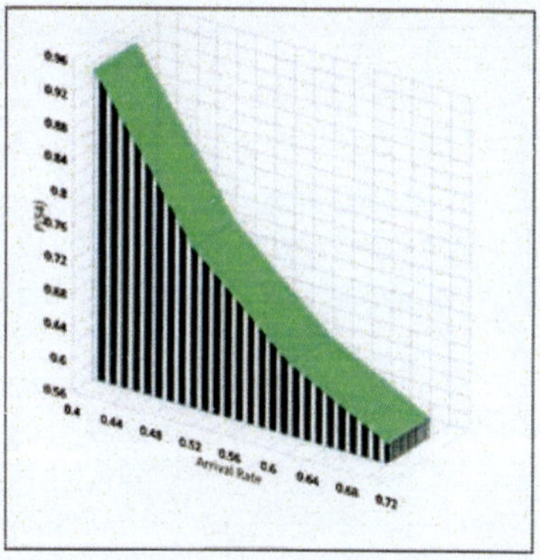

5 Conclusion

This article analyzed with a new procedure for detecting the performance expedients of the loss system under matrix geometry method. We can utilize this methodology for dissimilar queue system. The proposed technique gives more accurate numerals and concrete verdicts. This technique is essential to all the innovators yet to arrive. Figures 9, and 10 show that if the arrival rate rises, the probability that the system is busy and the customer gets the service from servers 1, 2, 3, 4, and 5, the average customers throughout the system, and aggregate yield achieved rise.

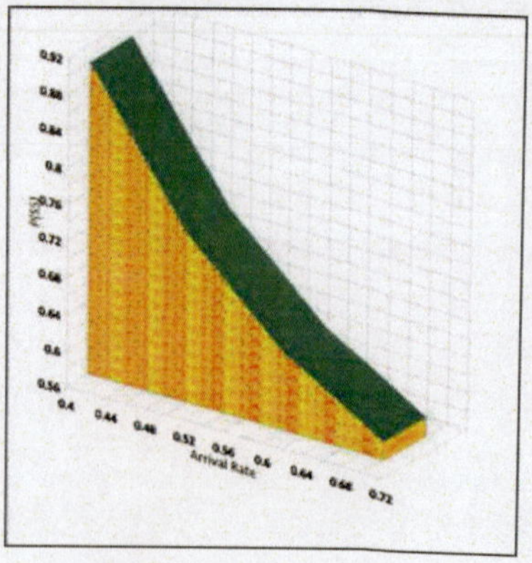

Fig. 9 Arrival versus $P(S_{E(N)})$

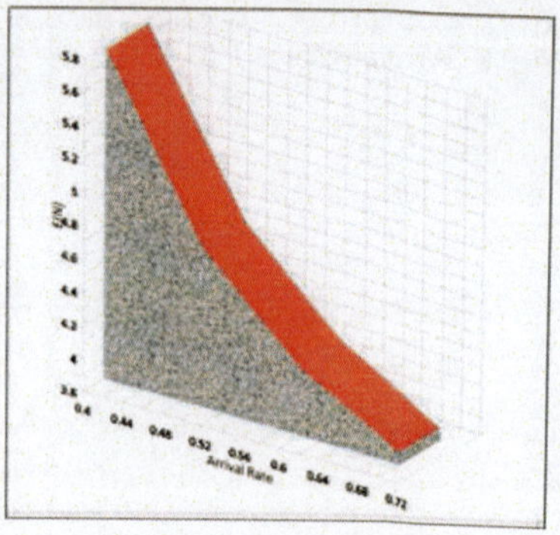

Fig. 10 Arrival versus δ

References

1. Bose, S.: An Introduction to Queueing Systems. Kluvar Academic/Plenum Publishers, New York (2008)
2. Cooper, R.: Introduction to Queueing Theory, 3rd edn. CEE Press, Washington, (1990)
3. Hadidi, N.: Busy periods of queues with state dependent arrival and service rates. J. Appl. Probab. **11**, 842–848 (1974)
4. Janos, S.: Basic Queueing Theory. Globe Edit Publishers, Omniscriptum GMBH, Germany (2016)
5. Klimenok, V.I., Taramin, O.S.: A tandem queueing system with losses and blocking. Appl. Comput. Math. **6**(2), 295–311 (2007)
6. van de Vrugt, M., Litvak, N., Boucherie, R.J.: Blocking probabilities in Erlang loss queues with advance reservation. Stoch. Models **30**(2), 187–196 (2014) (1974)
7. Abdalla, N., Boucherie, R.J.: Blocking probabilities in mobile communications networks with timevarying rates and redialing subscribers. Ann. Oper. Res. **112**(1–4), 15–34 (2002)
8. Ramesh, R., Kumaraghuru, S.: Performance measures of retrial queuing model with fuzzy parameters using robust ranking technique. Int. J. Adv. Innov. Res. **3**(4), 62–66 (2014)
9. Ramesh, R., Hari Ganesh, A.: $M/M/m/m$ Fuzzy loss system under expansion center fuzzy ranking method. J. Appl. Sci. Comput. **VI**(III), 2188–2195 (2019)
10. Seenivasan, M., Indhumathi, M.: Retrial queueing model with two heterogeneous server using matrix geometry method. Mater. Sci. Eng. **1070**, 012028 (2021). IOP Publishing

11. Seenivasan, M., Indhumathi, M.: Performance analysis of two heterogeneous server queueing model with intermittently obtainable server using matrix geometric method. J. Phys. **1724**, 012001 (2021). IOP Publishing
12. Seenivasan, M., Chakravarthy, V.J., Abinaya, R.: Markovian Queueing Model with Server Breakdown, Single Working Vacation and Catastrophe. vol. 881, pp. 409–421. Springer Book Series, Springer, Singapore. (2022)
13. Seenivasan, M., Shiny Epciya, J.: M/M/1 queue Server Breakdown, Single Working Vacation, Feedback and State Dependent Customer. In: IEEE Xplore, pp. 1–5 (2022)
14. Seenivasan, M., Chandiraleka, S.: Single Server Queucing Model with Multiple Working Vacation and with Server Breakdown. In: IEEE Xplore, pp. 1–5 (2022)
15. Seenivasan, M., Senthilkumar, R., Subasri, K.S.: M/M/2 heterogeneous queueing system having unreliable server with catastrophes and restoration. Mater. Today: Proc. **51**(8), 2332–2338 (2021)

Performance Analysis of an $M/G/1$ Retrial Queue with Two-Phase Service and Preemptive Resume Service Under Working Vacations and Working Breakdowns

N. Micheal Mathavavisakan and K. Indhira

1 Introduction

For a long time, there appeared to be a significant contribution to the retrial queueing models. Keilson et al.'s [8] study on the $M/G/1$ retrial queue paved the way to the introduction of the supplementary variable approach. In steady-state approaches, the combined distribution of the channel status and system length was derived. Retrials are a well-known phenomenon in which the server can join the orbital contact of blocked subscribers and retry after an arbitrary period of time if it is occupied. A retrial queue (RQ) is a queueing model that includes the retrial effect. In Artalejo [1] and Artalejo and Corral [2], we may discover generic models in retrial queues. Choi and Park [3] researched a reformed $M/G/1$ RQ with unlimited capacity where the incoming consumers find the service crowded and either join the retrial area or wait indefinitely.

In a working vacation (WV) queueing model, the system provides the services to consumers at a reduced rate but completely discontinues them during the regular vacation period. Chandrasekaran V.M. et al. [4] presented a short survey on WV queueing systems in recent years. Recently, there have been several works on WV [5, 9, 10]. To establish the queue length and orbit length distributions, which will be utilised further to ascertain the system's other performance metrics, is the purpose of this investigation.

N. Micheal Mathavavisakan · K. Indhira (✉)
Department of Mathematics, School of Advanced Sciences, Vellore Institute of Technology, Vellore, Tamil Nadu, India
e-mail: kindhira@vit.ac.in

F. Kamalov et al. (eds.), *Advances in Mathematical Modeling and Scientific Computing*, Trends in Mathematics, https://doi.org/10.1007/978-3-031-41420-6_46

2 Characterisation of the Model

Under WVs and working breakdowns (WB), we construct a $M/G/1$ RQ with two-phase service and preemptive resume service. A proper description of our model is given as follows:

The Arrival Process The arriving consumers of the system enter through a Poisson stream with rate ϑ.

The Retrial Policy The system offers the preliminary service when a consumer arrives and discovers that the server is not in use. Then the server offers the primary service once the preliminary service is done. While providing preliminary service, the server might force the consumer to the orbital area (with probability β) in order to begin preliminary service for an incoming consumer or maintain the current service (with probability $(1-\beta)$) such that the incoming consumer prefers to leave the service area to enter the orbit in conformance under $FCFS$ restraint. That is, the server is only accessible to the consumer at the top of the orbit queue. We presume that such a push-out option is unavailable when the server is performing its primary function. Inter-retrial times are presumed to follow a random variable S with a random distribution $S(t)$ and a density function $S(\tilde{\gamma})$ with a "Laplace-Stieltjes transform" (LST) $S^*(\varpi)$. The model becomes a $M/G/1$ RQ (with no push-out) with $FCFS$ discipline for $\beta = 0$ but, for $\beta = 1$, the system guide to $LCFS$ discipline with preemptive resume service.

The Regular Service Process There is a single server that offers all entering consumers with a "preliminary first phase of regular service" (FRS) indicated by $\mathscr{H}_1$. When a consumer's FRS is over, he or she may depart the model with a probability $\bar{\theta} = 1-\theta$ or perhaps be offered with a "second phase of optional service" (SOS) indicated by $\mathscr{H}_2$ with probability θ. These kinds of processes are known as regular service process. The service times follow general laws with probability distribution function $\mathscr{H}_i(\check{\psi})$ $(i = 1, 2)$ and density function $\mathscr{H}_i(\tilde{\gamma})$ having LST $\mathscr{H}_i^*(\varpi)$ and limited k^{th} moment $\mu_i^{(k)}$, $(i = 1, 2)$ in which the subscript $i = 1$ $(i = 2$, respectively) indicates the FRS (SOS, respectively).

The Working Vacations Process When the orbit is free, the server periodically takes a WV. The vacation period takes an exponential distribution with variable α. If a consumer enters during a vacation time, the server $(\gamma_w < \gamma_b)$ keeps running at a reduced rate. During the WV time, tasks are carried out at a slower pace. If any consumers are in the orbit at a slower service completion moment during the vacation period, the server will end the vacation and return to the normal busy time, interrupting the vacation. If not, the vacation keeps going. When the vacation gets over, the server restores normal operations if there are still customers in the orbit.

The Working Breakdown Process While the server is performing the regular service, it could fail at any point, causing the service route to break for a brief timespan, resulting in the server being inaccessible for a limited period of time. Exogenous Poisson procedure with rates σ_1 for FRS and σ_2 for SOS causes the breakdowns, during FRS and SOS periods, respectively. When calamity strikes, all current consumers are forced to quit the system, and the main server fails. When the main server deteriorates, it is immediately dispatched to mend, and the mending time begins. The repair time has been exponentially distributed with rate ϕ. Then the recovered server is meant to function as if it were brand-new. Nevertheless, when a calamity strikes the regular busy server, the server undergoes a WB. During the WB time, the substitute server serves the arriving clients at a lesser rate ($\gamma_w < \gamma_b$). If there are consumers in the orbital area when the repair is completed, the server returns to regular operation, and a fresh period of high activity begins; or else, it is unoccupied and prepared to serve newcomers. During the WV and WB periods (slower-rate services), the service period is assessed by a random variable $\mathcal{G}_b$ with distribution function $\mathcal{G}_b(t)$ and density function $\mathcal{G}_b(\tilde{\gamma})$ having LST $\mathcal{G}_b^*(\varpi)$.

The system's stochastic processes are considered to be independent of one another.

2.1 Application of the Model in Real Life

Our model might be useful in the field of computer processing systems. The loading size (orbit) utilised to hold texts in a computer processing system is limited, and messages (consumers) arrive one after the other, with the processor (server) which is responsible for handling them. The messages are being processed in two stages where one is primary and the other one is optional (back-up). The processor may toss out the message which is getting the service to buffer in order to process a new arriving message (pop-up add). Viruses may infect the operational mail server (breakdowns), and the computer system's performance may be degraded (WB). The primary mail server is then dispatched to be repaired right away. While the computer system will still be able to do some tasks, it will do so at a much slower pace. If the device is unavailable, information is briefly kept in a load and delivered later (retrial time). When all messages have been received and no fresh ones have been received, the system will do a series of restoration, like a virus detection, to optimise the computer's performance (WVs). To conserve money, the processor may process messages at a slower pace during maintenance (WV period). The processor may choose to perform another maintenance operation if no message is available in the system after each maintenance (multiple WVs). This kind of WV and WB discipline is a proper estimate like a computer processing system. The practical design of stochastic production and inventory models in a multifunctional manufacturing facility and the challenges of machine renewal are examples of where this model might be applied.

3 Overview of Steady-State Probabilities

This division first develops the steady-state (SS) equations for the RQ system by considering the elapsed retrial period, the elapsed service time on providing the preliminary service or the primary service, and the elapsed lower-speed service times as SV. The PGF of the no. of consumers in the orbit and system and the orbit length generating functions for numerous server states are computed.

3.1 *The Steady-State Equations*

It is assumed in SS that $S(0) = 0$, $S(\infty) = 1$, $\mathscr{H}_i(0) = 0$, $\mathscr{H}_i(\infty) = 1$, $(i = 1, 2)$, and $\mathscr{G}_b(0) = 0$, $\mathscr{G}_b(\infty) = 1$ are continuous at $\tilde{\gamma} = 0$. The functions $\varphi(\tilde{\gamma})$, $\zeta_i(\tilde{\gamma})$, and $\eta_b(\tilde{\gamma})$ are the hazard rates for retrial, service on FRS and SOS, and repair, respectively.

$$\varphi(\tilde{\gamma})d\tilde{\gamma} = \frac{dS(\tilde{\gamma})}{1 - S(\tilde{\gamma})}; \zeta_i(\tilde{\gamma})d\tilde{\gamma} = \frac{d\mathscr{H}_i(\tilde{\gamma})}{1 - \mathscr{H}_i(\tilde{\gamma})} (i = 1, 2); \eta_b(\tilde{\gamma})d\tilde{\gamma} = \frac{d\mathscr{G}_b(\tilde{\gamma})}{1 - \mathscr{G}_b(\tilde{\gamma})}$$

Apart from it, let $S^0(\check{\psi})$, $\mathscr{H}_i^0(\check{\psi})$, and $\mathscr{G}_b^0(\check{\psi})$ be the elapsed retrial period, elapsed service period on preliminary or providing the primary, and elapsed lower-speed service times, respectively, at time $\check{\psi}$.

Here, we highlight the usage of bivariate Markov process to describe the system's state at time $\check{\psi}$ $\{\Lambda(\check{\psi}), \mathrm{Y}(\check{\psi}); \check{\psi} \geq 0\}$, where $\Lambda(\check{\psi})$ denote the server state (0, 1, 2 or 3, 4) based on whether the server is empty, preliminary service or providing the primary service, and slower-rate service. If $\Lambda(\check{\psi}) = 1$ and $Y(\check{\psi}) > 0$, then $S^0(\check{\psi})$ is corresponding to the elapsed retrial period. If $\Lambda(\check{\psi}) = 2$ and $Y(\check{\psi}) \geq 0$, then $\mathscr{H}_1^0(\check{\psi})$ is corresponding to the elapsed period of the consumer being provided with preliminary service. If $\Lambda(\check{\psi}) = 3$ and $Y(\check{\psi}) \geq 0$, then $\mathscr{H}_2^0(\check{\psi})$ is equivalent to the elapsed time of the consumer being provided with primary service. If $\Lambda(\check{\psi}) = 4$ and $Y(\check{\psi}) \geq 0$, then $\mathscr{G}_b^0(\check{\psi})$ is equivalent to the elapsed time of the consumer served at a slower-pace service period.

Let $\{\check{\psi}_n; n = 1, 2, \ldots\}$ be the series of epochs where either a time of service completion or a shorter service period happens. $W_n = \{\Lambda(\check{\psi}_n+), Y(\check{\psi}_n+)\}$ sequence of random vectors. The Markov chain (MC) formed by embedded in the RQ system. The embedded MC $\{W_n; n \in N\}$ is ergodic iff $\tau < 1$, to maintain the stability of our system, where $\tau = \mathscr{H}_1^*(\sigma_1 + \beta\vartheta)\{\bar{\theta} + \theta\mathscr{H}_2^*(\sigma_2)\}$.

For the method $\{Y(\check{\psi}), \check{\psi} \geq 0\}$, we specify the probabilities $\Psi_0(\check{\psi}) = P\{\Lambda(\check{\psi}) = 1, Y(\check{\psi}) = 0\}$ and $\Omega_0(\check{\psi}) = P\{\Lambda(t) = 0, Y(\check{\psi}) = 0\}$, and the probability densities are as follows: $\Psi_n(\tilde{\gamma}, \check{\psi})d\tilde{\gamma} = P\{\Lambda(\check{\psi}) = 1, Y(\check{\psi}) = n, \tilde{\gamma} \leq S^0(t) < \tilde{\gamma} + d\tilde{\gamma}\}$,

for $t \geq 0$, $\tilde{\gamma} \geq 0$, and $n \geq 1$.

$\Phi_{i,n}(\tilde{\gamma}, t)d\tilde{\gamma} = P\{\Lambda(\check{\psi}) = 2, 3, Y(\check{\psi}) = n, \tilde{\gamma} \leq \mathscr{H}_i^0(\check{\psi}) < \tilde{\gamma} + d\tilde{\gamma}\}$,

for $t \geq 0, \tilde{\gamma} \geq 0, n \geq 0$, and $i = 1, 2$.
$\Upsilon_{b,n}(\tilde{\gamma}, \check{\psi})d\tilde{\gamma} = P\{\Lambda(\check{\psi}) = 4, Y(\check{\psi}) = n, \tilde{\gamma} \leq \mathscr{G}_b^0(\check{\psi}) < \tilde{\gamma} + d\tilde{\gamma}\}$,
for $t \geq 0, \tilde{\gamma} \geq 0$, and $n \geq 0$.

We assume that the sequel satisfies the stability criteria; we may assign $\Psi_0 = lim_{\check{\psi}\to\infty}\Psi_0(\check{\psi})$, $\Omega_0 = lim_{\check{\psi}\to\infty}\Omega_0(\check{\psi})$ and limiting densities are
$\Psi_n(\tilde{\gamma}) = lim_{\check{\psi}\to\infty}\Psi_n(\tilde{\gamma}, \check{\psi})$; $\Phi_{i,n}(\tilde{\gamma}) = lim_{\check{\psi}\to\infty}\Phi_{i,n}(\tilde{\gamma}, \check{\psi})$; $(i = 1, 2)$
$\Upsilon_{b,n}(\tilde{\gamma}) = lim_{\check{\psi}\to\infty}\Upsilon_{b,n}(\tilde{\gamma}, \check{\psi})$.

We construct the following system of equations for $(i = 1, 2)$, using the supplementary variable approach.

$$\vartheta\Psi_0 = \phi\Omega_0 \tag{1}$$

$$\begin{aligned}(\vartheta + \alpha + \phi)\Omega_0 = \bar{\theta}\int_0^\infty \Phi_{1,n}(\tilde{\gamma})\zeta_1(\tilde{\gamma})d\tilde{\gamma} \\ + \int_0^\infty \Phi_{2,n}(\tilde{\gamma})\zeta_2(\tilde{\gamma})d\tilde{\gamma} \\ + \int_0^\infty \Upsilon_{b,0}(\tilde{\gamma})\eta_b(\tilde{\gamma})d\tilde{\gamma} + \alpha\Omega_0\end{aligned} \tag{2}$$

$$\frac{d}{d\tilde{\gamma}}\Psi_n(\tilde{\gamma}) + (\vartheta + \varphi(\tilde{\gamma}))\Psi_n(\tilde{\gamma}) = 0 \tag{3}$$

$$\frac{d}{d\tilde{\gamma}}\Phi_{1,0}(\tilde{\gamma}) + (\vartheta + \sigma_1 + \zeta_1(\tilde{\gamma}))\Phi_{1,0}(\tilde{\gamma}) = 0, n = 0 \tag{4}$$

$$\frac{d}{d\tilde{\gamma}}\Phi_{1,n}(\tilde{\gamma}) + (\vartheta + \sigma_1 + \zeta_1(\tilde{\gamma}))\Phi_{1,n}(\tilde{\gamma}) = \vartheta(1-\beta)\Phi_{1,n-1}(\tilde{\gamma}), n \geq 1 \tag{5}$$

$$\frac{d}{d\tilde{\gamma}}\Phi_{2,0}(\tilde{\gamma}) + (\vartheta + \sigma_2 + \zeta_2(\tilde{\gamma}))\Phi_{2,0}(\tilde{\gamma}) = 0, n = 0 \tag{6}$$

$$\frac{d}{d\tilde{\gamma}}\Phi_{2,n}(\tilde{\gamma}) + (\vartheta + \sigma_2 + \zeta_2(\tilde{\gamma}))\Phi_{2,n}(\tilde{\gamma}) = \vartheta\Phi_{1,n-1}(\tilde{\gamma}), n \geq 1 \tag{7}$$

$$\frac{d}{d\tilde{\gamma}}\Upsilon_{b,0}(\tilde{\gamma}) + (\vartheta + \alpha + \sigma + \eta_b(\tilde{\gamma}))\Upsilon_{b,0}(\tilde{\gamma}) = 0, n = 0 \tag{8}$$

$$\frac{d}{d\tilde{\gamma}}\Upsilon_{b,n}(\tilde{\gamma}) + (\vartheta + \alpha + \sigma + \eta_b(\tilde{\gamma}))\Upsilon_{b,n}(\tilde{\gamma}) = \vartheta\Upsilon_{b,n-1}(\tilde{\gamma}), n \geq 1 \tag{9}$$

At $\tilde{\gamma} = 0$, the SS boundary conditions are as follows:

$$\Psi_n(0) = \bar{\theta}\int_0^\infty \Phi_{1,n}(\tilde{\gamma})\zeta_1(\tilde{\gamma})d\tilde{\gamma} + \int_0^\infty \Phi_{2,n}(\tilde{\gamma})\zeta_2(\tilde{\gamma})d\tilde{\gamma}$$

$$+\int_0^\infty \Upsilon_{b,n}(\tilde{\gamma})\eta_b(\tilde{\gamma})d\tilde{\gamma}, n \geq 1 \tag{10}$$

$$\Phi_{1,n}(0) = \int_0^\infty \Psi_{n+1}(\tilde{\gamma})\varphi(\tilde{\gamma})d\tilde{\gamma} + \vartheta \int_0^\infty \Psi_n(\tilde{\gamma})d\tilde{\gamma} + (\alpha + \phi)\int_0^\infty \Upsilon_{b,n}(\tilde{\gamma})d\tilde{\gamma} \tag{11}$$

$$+ (\sigma_1 + \beta\vartheta)\int_0^\infty \Phi_{1,n-1}(\tilde{\gamma})d\tilde{\gamma}, n \geq 1$$

$$\Phi_{1,0}(0) = \int_0^\infty \Psi_1(\tilde{\gamma})\varphi(\tilde{\gamma})d\tilde{\gamma} + \vartheta\Psi_0 + (\alpha + \phi)\Upsilon_{b,0}(\tilde{\gamma})d\tilde{\gamma}, n = 0, \tag{12}$$

$$\Phi_{2,0}(0) = \theta \int_0^\infty \Phi_{1,n}(\tilde{\gamma})\zeta_1(\tilde{\gamma})d\tilde{\gamma}, n \geq 1 \tag{13}$$

$$\Upsilon_{b,n}(0) = \begin{cases} \vartheta\Omega_0, & n = 0 \\ 0, & n \geq 1 \end{cases} \tag{14}$$

The normalising condition is

$$\Psi_0+\Omega_0+\sum_{n=1}^{\infty}\int_0^\infty \Psi_n(\tilde{\gamma})d\tilde{\gamma}+\sum_{n=0}^{\infty}\left(\sum_{i=1}^{2}\int_0^\infty \Phi_{i,n}(\tilde{\gamma})d\tilde{\gamma}+\int_0^\infty \Upsilon_{b,n}(\tilde{\gamma})d\tilde{\gamma}\right)=1 \tag{15}$$

3.2 The Steady-State Solutions

The generating functions for $|\breve{\xi}| < 1$ in order to solve the aforementioned equations are expressed as follows:

$$\Psi(\tilde{\gamma},\breve{\xi}) = \sum_{n=1}^{\infty}\Psi_n(\tilde{\gamma})\breve{\xi}^n;\ \Psi(0,\breve{\xi}) = \sum_{n=1}^{\infty}\Psi_n(0)\breve{\xi}^n;\ \Phi_i(\tilde{\gamma},\breve{\xi}) = \sum_{n=0}^{\infty}\Phi_{i,n}(\tilde{\gamma})\breve{\xi}^n;$$

$$\Phi_i(0,\breve{\xi}) = \sum_{n=0}^{\infty}\Phi_{i,0}(\tilde{\gamma})\breve{\xi}^n;\ \Upsilon_b(\tilde{\gamma},\breve{\xi}) = \sum_{n=0}^{\infty}\Upsilon_{b,n}(\tilde{\gamma})\breve{\xi}^n;\ \Upsilon_b(0,\breve{\xi}) = \sum_{n=0}^{\infty}\Upsilon_{b,n}(0)\breve{\xi}^n;$$

Multiply $\breve{\xi}^n$ and summing over n, ($n = 0, 1, 2, \ldots$) from (2) to (14),

$$\frac{\partial}{\partial\tilde{\gamma}}\Psi(\tilde{\gamma},\breve{\xi}) + (\vartheta + \varphi(\tilde{\gamma}))\Psi(\tilde{\gamma},\breve{\xi}) = 0 \tag{16}$$

$$\frac{\partial}{\partial \tilde{\gamma}}\Phi_1(\tilde{\gamma},\breve{\xi}) + (\vartheta(1-\breve{\xi}) + \vartheta\beta\breve{\xi} + \sigma_1 + \zeta_1(\tilde{\gamma}))\Phi_1(\tilde{\gamma},\breve{\xi}) = 0 \tag{17}$$

$$\frac{\partial}{\partial \tilde{\gamma}}\Phi_2(\tilde{\gamma},\breve{\xi}) + (\vartheta(1-\breve{\xi}) + \sigma_2 + \zeta_2(\tilde{\gamma}))\Phi_2(\tilde{\gamma},\breve{\xi}) = 0 \tag{18}$$

$$\frac{\partial}{\partial \tilde{\gamma}}\Upsilon_b(\tilde{\gamma},\breve{\xi}) + (\vartheta(1-\breve{\xi}) + \alpha + \phi + \eta_b(\tilde{\gamma}))\Upsilon_b(\tilde{\gamma},\breve{\xi}) = 0 \tag{19}$$

$$\Psi(0,\breve{\xi}) = \bar{\theta}\int_0^\infty \Phi_1(\tilde{\gamma},\breve{\xi})\zeta_1(\tilde{\gamma})d\tilde{\gamma} + \int_0^\infty \Phi_2(\tilde{\gamma},\breve{\xi})\zeta_2(\tilde{\gamma})d\tilde{\gamma} + \int_0^\infty \Upsilon_b(\tilde{\gamma},\breve{\xi})\eta_b(\tilde{\gamma})d\tilde{\gamma} - (\vartheta+\phi)\Omega_0 \tag{20}$$

$$\Phi_1(0,\breve{\xi}) = \frac{1}{\breve{\xi}}\int_0^\infty \Psi(\tilde{\gamma},\breve{\xi})\varphi(\tilde{\gamma})d\tilde{\gamma} + \vartheta\int_0^\infty \Psi(\tilde{\gamma},\breve{\xi})d\tilde{\gamma} + (\alpha+\phi)\int_0^\infty \Upsilon_b(\tilde{\gamma},\breve{\xi})d\tilde{\gamma} + (\sigma_1+\beta\vartheta)\int_0^\infty \Phi_1(\tilde{\gamma},\breve{\xi})d\tilde{\gamma} \tag{21}$$

$$\Phi_2(0,\breve{\xi}) = \theta\int_0^\infty \Phi_1(\tilde{\gamma},\breve{\xi})\zeta_1(\tilde{\gamma})d\tilde{\gamma} \tag{22}$$

$$\Upsilon_b(0,\breve{\xi}) = \vartheta\Omega_0 \tag{23}$$

After some mathematics calculation, finally we obtain

$$\Phi_1(0,\breve{\xi}) = \frac{\Psi(0,\breve{\xi})}{\breve{\xi}}\{S^*(\vartheta) + \breve{\xi}[1-S^*(\vartheta)]\} + \vartheta\Omega_0 B(\breve{\xi}) + \vartheta\Psi_0 + \mathscr{H}_1(\breve{\xi}) \tag{24}$$

where $B(\breve{\xi}) = \frac{\alpha+\phi}{\alpha+\phi+\vartheta(1-\breve{\xi})}(1-\mathscr{G}_b^*(\mathscr{A}_b(\breve{\xi})))$, $\mathscr{H}_1(\breve{\xi}) = \frac{\sigma_1+\beta\vartheta}{\sigma_1+\beta\vartheta\breve{\xi}+\vartheta(1-\breve{\xi})}(1-\mathscr{H}_1^*(\mathscr{A}_1(\breve{\xi})))$

$$\Phi_2(0,\breve{\xi}) = \theta\Phi_1(0,\breve{\xi})\mathscr{H}_1^*(\mathscr{A}_1(\breve{\xi})) \tag{25}$$

$$\Psi(0,\breve{\xi})\{\breve{\xi} - [\bar{\theta}\mathscr{H}_1^*(\mathscr{A}_1(\breve{\xi})) + \theta\mathscr{H}_1^*(\mathscr{A}_1(\breve{\xi}))\mathscr{H}_2^*(\mathscr{A}_2(\breve{\xi}))](S^*(\vartheta) + \breve{\xi}[1-S^*(\vartheta)])\}$$
$$= \breve{\xi}\left\{\Omega_0[(\bar{\theta}\mathscr{H}_1^*(\mathscr{A}_1(\breve{\xi})) + \theta\mathscr{H}_1^*(\mathscr{A}_1(\breve{\xi}))\mathscr{H}_2^*(\mathscr{A}_2(\breve{\xi})))(\phi + \vartheta M(\breve{\xi}))\right.$$

$$+\vartheta(\mathscr{G}_b^*(\mathscr{A}_b(\check{\xi}))-1)-\phi]+\mathscr{H}_1(\check{\xi})[\mathscr{H}_1^*(\mathscr{A}_1(\check{\xi}))+(\bar{\theta}+\theta\mathscr{H}_1^*(\mathscr{A}_2(\check{\xi})))] \tag{26}$$

Utilising Eq. (26) in (24), we obtain

$$\Phi_1(0,\check{\xi})=\frac{Ne(\check{\xi})}{\check{\xi}De(\check{\xi})}(S^*(\vartheta)+\check{\xi}[1-S^*(\vartheta)])+\Omega_0[\phi+\vartheta B(\check{\xi})]+\mathscr{H}_1(\check{\xi}) \tag{27}$$

Utilising Eq. (27) in (25), we obtain

$$\Phi_2(0,\check{\xi})$$
$$=\{\theta\mathscr{H}_1^*(\mathscr{A}_1(\check{\xi}))\}\left\{\frac{Ne(\check{\xi})}{\check{\xi}De(\check{\xi})}(S^*(\vartheta)+\check{\xi}[1-S^*(\vartheta)])+\Omega_0[\phi+\vartheta B(\check{\xi})]+\mathscr{H}_1(\check{\xi})\right\} \tag{28}$$

where $\mathscr{A}_1(\check{\xi})=\sigma_1+\beta\vartheta\check{\xi}+\vartheta(1-\check{\xi})$, $\mathscr{A}_2(\check{\xi})=\sigma_2+\vartheta(1-\check{\xi})$ and $\mathscr{A}_b(\check{\xi})=\alpha+\phi+\vartheta(1-\check{\xi})$

3.3 The Steady-State Solution

The PGFs are as follows if the system is in SS constraint $\tau<1$:

(i) The no. of consumers in the orbit while the server is available:

$$\Psi(\check{\xi})=\frac{Ne(\check{\xi})}{De(\check{\xi})} \tag{29}$$

$$Ne(\check{\xi})=\check{\xi}\Big\{\Omega_0[(\bar{\theta}\mathscr{H}_1^*(\mathscr{A}_1(\check{\xi}))+\theta\mathscr{H}_1^*(\mathscr{A}_1(\check{\xi}))\mathscr{H}_2^*(\mathscr{A}_2(\check{\xi})))(\phi+\vartheta M(\check{\xi}))$$
$$+\vartheta(\mathscr{G}_b^*(\mathscr{A}_b(\check{\xi}))-1)-\phi]$$
$$+\mathscr{H}_1(\check{\xi})[\mathscr{H}_1^*(\mathscr{A}_1(\check{\xi}))+(\bar{\theta}+\theta\mathscr{H}_1^*(\mathscr{A}_2(\check{\xi})))]$$

$$De(\check{\xi})=\{\check{\xi}-[\bar{\theta}\mathscr{H}_1^*(\mathscr{A}_1(\check{\xi}))+\theta\mathscr{H}_1^*(\mathscr{A}_1(\check{\xi}))\mathscr{H}_2^*(\mathscr{A}_2(\check{\xi}))](S^*(\vartheta)+\check{\xi}[1-S^*(\vartheta)])\}$$

(ii) When the service is frequently busy, the no. of consumers in orbit is

$$\Phi_1(\check{\xi})=(1-\mathscr{H}_1^*(\mathscr{A}_1(\check{\xi}))\{\Omega_0[(\phi+\vartheta B(\check{\xi}))(\check{\xi}+\mathscr{H}_1^*(\mathscr{A}_1(\check{\xi}))[\bar{\theta}+\theta\mathscr{H}_2^*(\mathscr{A}_2(\check{\xi}))]$$
$$[1-(S^*(\vartheta)+\check{\xi}(1-S^*(\vartheta)))])+\vartheta[\mathscr{G}_b^*(\mathscr{A}_b(\check{\xi}))-1]-\phi]$$
$$+\check{\xi}\mathscr{H}_1(\check{\xi})\}/\mathscr{A}_1(\check{\xi})De(\check{\xi}) \tag{30}$$

$$\Phi_2(\check{\xi}) = \frac{\theta \mathscr{H}_1^*(\mathscr{A}_1(\check{\xi}))Ne(\check{\xi})}{\mathscr{A}_2(\check{\xi})De(\check{\xi})}\{1 - \mathscr{H}_2^*(\mathscr{A}_2(\check{\xi}))\} \tag{31}$$

(iii) The no. of consumers in the orbit when the server is at a slow speed service:

$$\Upsilon_b(\check{\xi}) = \frac{\vartheta \Omega_0 B(\check{\xi})}{\alpha + \phi} \tag{32}$$

Integrating the Eqs. (26)–(28) with respect to $\tilde{\gamma}$ and compute the partial PG $\Psi(\check{\xi}) = \int_0^\infty \Psi(\tilde{\gamma}, \check{\xi})d\tilde{\gamma}$, $\Phi_1(\check{\xi}) = \int_0^\infty \Phi_1(\tilde{\gamma}, \check{\xi})d\tilde{\gamma}$, $\Phi_2(\check{\xi}) = \int_0^\infty \Phi_2(\tilde{\gamma}, \check{\xi})d\tilde{\gamma}$, $\Upsilon_b(\check{\xi}) = \int_0^\infty \Upsilon_b(\tilde{\gamma}, \check{\xi})d\tilde{\gamma}$. We calculate the probability that the server is available using the normalisation condition (Ψ_0) by establishing functions as follows: when there is no consumer in the orbit $\check{\xi} = 1$ in (29) – (32) and whenever the L'Hospital rule is required, we produce $\Psi_0 + \Omega_0 + \Psi(1) + \Phi_1(1) + \Phi_2(1) + \Upsilon_b(1) = 1$.

Corollary 1 *Utilising the PGF, the number of consumers in the system* $(Ks(\check{\xi}))$ *and orbit* $(K0(\check{\xi}))$*are computed by*

$$K_s(z) = \Psi_0 + \Omega_0 + \Psi(\check{\xi}) + z(\Phi_1(\check{\xi}) + \Phi_2(\check{\xi}) + \Upsilon_b(\check{\xi})) \tag{33}$$

$$K_0(z) = \Psi_0 + \Omega_0 + \Psi(\check{\xi}) + \Phi_1(\check{\xi}) + \Phi_2(\check{\xi}) + \Upsilon_b(\check{\xi}) \tag{34}$$

4 System Performance Measures

When the system is on SS, (i) With respect to $\check{\xi}$, (34) and providing $\check{\xi} = 1$ represents the mean no. of consumers in the orbit (L_q)

$$L_q = K_0^{'}(1) = \lim_{\check{\xi}\to 1} \frac{d}{d\check{\xi}} K_0(\check{\xi}) = \Omega_0 \left[\frac{Ne_q^{''}(1)De_q^{'}(1) - De_q^{''}(1)Ne_q^{'}(1)}{2(De_q^{'}(1))^2} \right] \tag{35}$$

(ii) With respect to $\check{\xi}$, (33) and providing $\check{\xi} = 1$ represents the mean no. of consumers in the system (L_s)

$$L_s = K_s^{'}(1) = \lim_{\check{\xi}\to 1} \frac{d}{d\check{\xi}} K_s(\check{\xi}) = \Omega_0 \left[\frac{Ne_s^{''}(1)De_q^{'}(1) - De_q^{''}(1)Ne_q^{'}(1)}{2(De_q^{'}(1))^2} \right] \tag{36}$$

(iii) Little's approach is used to estimate the mean time for a consumer in the system (W_s) and the mean period for a consumer in the queue (W_q). $W_s = \frac{L_s}{\vartheta}$ and $W_q = \frac{L_q}{\vartheta}$, respectively.

5 Special Cases

Case (i) Let $\theta = 0$ and $S^*(\vartheta) \to 1$. Here the framework may be lowered to a $M/G/1$ RQ with standard retrial times. These results are consistent with Kalidass and Kasturi [7].

Case (ii) Let $\theta = 0, \sigma_1 = \sigma_2 = \beta = 0$ and $S^*(\vartheta) \to 1$. Our system can be reduced to a $M/G/1$ RQ with WVs [6].

6 Numerical Analysis

In this part, we use the MATLAB to show the various options for measuring system behaviour. We examine retrial times, two-stage service times, slower service times, vacation times, and exponentially distributed repair times. In order to satisfy the stability condition, numerical measures are chosen at random.

With the effect of the parameters σ_1, σ_2, ϕ, ζ_1, ζ_2, η_b, and α, Fig. 1a–c displays the 3-D graph depicting the system performance measures. The surface of Fig. 1a shows that as the arrival rate (ϑ) increases, (L_q) and (W_q) diminish. In Fig. 1b, we found that (L_q) and (W_q) diminish while the first stage of service rate ζ_1 increases. Figure 1c shows that as the second stage of service rate ζ_2 escalates, (L_q) and (W_q) diminish. Figure 1d, e displays that as ϑ andζ_1 escalates, (L_q) and (Ψ_0) diminish as well.

7 Conclusion

In this article, we explored a single-server RQ with two-phase service and preemptive resume under WV and WB. If the required and appropriate criteria are viable, the system can be stabilised. When the system is ideal, normally busy, or operating at a slower speed, the PGF of the no. of consumers in the system and its orbit are calculated utilising the PGF approach and the SVT. This model combines the $FCFS$ and $LCFS$ preemptive resume service disciplines (for $\beta = 0$ and $\beta = 1$, respectively). The impact of system parameters is finally investigated using a variety of numerical outcomes. This concept was inspired by a range of real-time systems, such as computer and communication networks, in which signals are received by a single-server retrial queue under a variety of working vacation and working failure conditions. Hopefully, system managers will be able to utilise this study to create a system that is cost-effective to administer and create well-informed judgments about system size and other concerns.

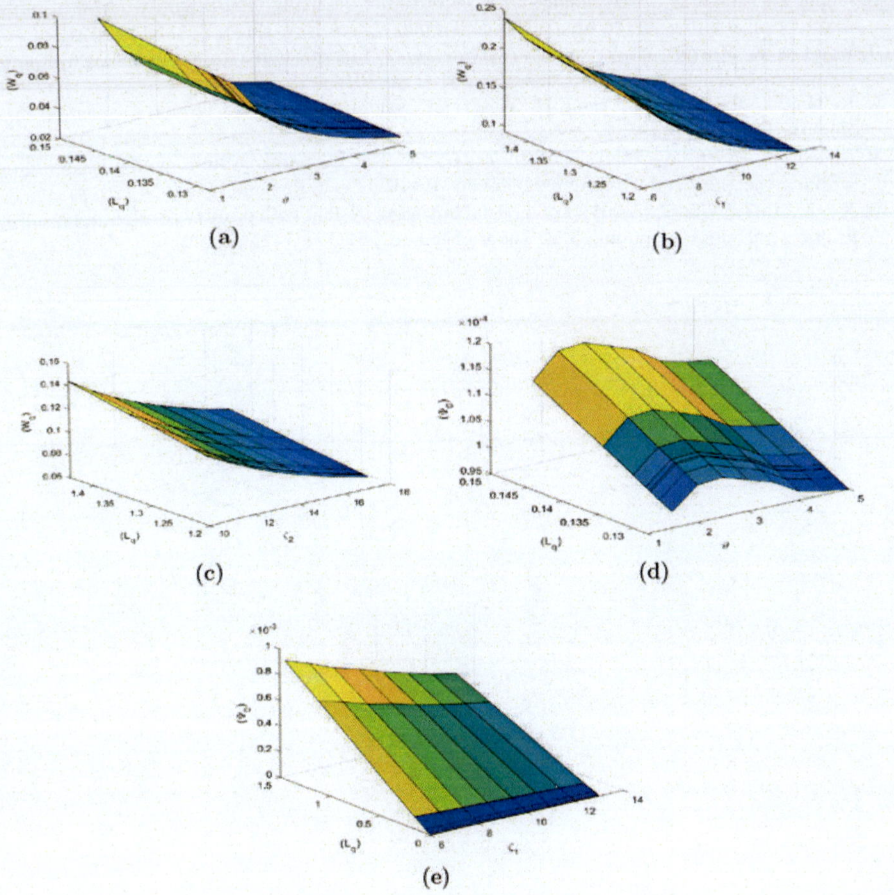

Fig. 1 **(a)** L_q, W_q verses arrival rate ϑ. **(b)** L_q, W_q verses FRS ζ_1. **(c)** L_q, W_q verses SOS ζ_2. **(d)** L_q, Ψ_0 verses arrival rate ϑ. **(e)** L_q, Ψ_0 verses FRS ζ_1

References

1. Artalejo, J.R.: Accessible bibliography on retrial queues. Math. Comput. Model. **51**, 1071–1081 (2010)
2. Artalejo, J.R., Gomez-Corral, A.: Retrial Queueing Systems. Springer, Berlin (2008)
3. Bong Dae, C., Kwang kyu, P.: The $M/G/1$ retrial queue with Bernoulli schedule. Queueing Syst. **7**(2), 219–228 (1990)
4. Chandrasekaran, V.M., Indhira, K., Saravanarajan, M.C., Rajadurai, P.: A survey on working vacation queueing models. Int. J. Pure Appl. Math. **106**, 33–41 (2016)
5. GnanaSekar, M.M.N., Kandaiyan, I.: Analysis of an $M/G/1$ retrial queue with delayed repair and feedback under working vacation policy with impatient customers. Symmetry **14**(10), 2024 (2022)

6. Gupta, P., Kumar, N.: Performance analysis of retrial queueing model with working vacation, interruption, waiting server, breakdown and repair. J. Sci. Res. **13**(3), 833–844 (2021)
7. Kalidass, K., Kasturi, R.: A two phase service $M/G/1$ queue with a finite number of immediate Bernoulli feedbacks. OPSEARCH **51**(2), 201–218 (2014)
8. Keilson, J., Cozzolino J., Yang, H.: A service system with unfilled requests repeated. Oper. Res. **16**, 1126–1137 (1968)
9. Rajadurai: Sensitivity analysis of an $M/G/1$ retrial queueing system with disaster under working vacations and working breakdowns. RAIRO Oper. Res. **52**(1), 35–54 (2018)
10. Xu, J., Liu, L., Wu, K.: Analysis of a retrial queueing system with priority service and modified multiple vacations. Commun. Stat. Theory Methods, **52**(17), 1–25 (2022)

Models of Goal Programming and R Programming to Earmark Acreage

K. Shalini and Sridevi Polasi

1 Introduction

In Karnataka, more than 70% of the population directly or indirectly depends on agriculture, especially rice crops. In the latest report, the number of farmers' suicides reduced in Karnataka compared to the last few years. The main key points of agriculture are finance, water resources, land availability, crop allocation, accessibility of labor, machine power, etc. The fast growth in the Indian population and industrialization makes us think about the optimization of land allocation in agriculture. Dependency upon the condition of climate, soil attributes, rain, socioeconomic status, and so on are important for increasing the yield of the crops. Using suitable mathematical models, the decision-makers may be able to achieve more profit and growth in the production of crops.

In agricultural planning, several operational methods were implemented. Goal programming (GP) is one of the advanced methods of operations research. Biswas et al. [1] used a fuzzy goal programming technique for land use planning in agriculture. Tiwari et al. [19] gave an implemented model for fuzzy goal programming. Ramadan [20] explained the relationship between goal programming and fuzzy programming with numerical examples. Singha et al. [9] applied a linear programming model for cropping patterns in the cannel command area. Bijay Baran and Bhola Nath [5] contributed to a long-range production planning in agricultural systems using fuzzy goal programming. Sahoo et al. [4] implemented optimal planning for the Land, Water, and Crop system using fuzzy and linear programming models. Dinesh et al. [10] applied fuzzy goal programming for agricultural land

K. Shalini (✉) · S. Polasi
Department of Mathematics, REVA University, Bangalore, Karnataka, India

F. Kamalov et al. (eds.), *Advances in Mathematical Modeling and Scientific Computing*, Trends in Mathematics, https://doi.org/10.1007/978-3-031-41420-6_47

allocation problem. Baran Pal [6] developed a genetic algorithm–based hybrid goal programming approach to land allocation in the agricultural system. Sharma et al. [8] applied a GP problem for land use planning in an agricultural system. Baran Pal et al. [7] developed an interval-valued GP for land allocation problem. Vashistha [11] used goal programming in farm agricultural planning. Mirkarimi [21] developed cropping pattern management of selected crops using fuzzy GP. Patel et al. [16] applied linear programming problems for agricultural land allocation to the major crops. Mishra and Singh [3] solved optimal allocation of land through fuzzy goal programming. Tamta [17] gave optimum farm planning methods under different farming systems using linear programming. Anuradha [2] developed a goal programming model for solving land allocation problems for five different crops in the Anuradha district. Srivastava et al. [18] applied a multi-objective fuzzy goal programming problem in agricultural land allocation for crop planning. Mosleh [25] introduced a sustainable distribution of agricultural lands and water resources using the goal programming method. Sasank et al. [22] has used the goal programming technique for agricultural land allotment. Zenis et al. [12] solved linear programming models for the optimization of agricultural farmlands. Kagita et al. [15] developed a fuzzy goal programming for solving land allocation problems in Krishna Delta. Farshad et al. [13] contributed with an optimal cultivation crops' model using fuzzy goal programming. Tirupathi et al. [23] gave a simple and multiple GP in optimal land distribution problems for agricultural income maximization. Usha Rani et al. [24] proposed fuzzy multi-objective linear programming for agriculture production planning.

In most real-world problems, we have to target so many goals with multiple objective functions. Goal programming is one of the best procedures to solve the problem of having several goals with disagreeing objectives. The GP attains each goal successively with an acceptable level. In 1961 Channes and Cooper introduced a goal programming model. In 1965 Ijiri implemented different priority levels for inconsistent goals and assigned different weights for that goal with the same priority level. In the agricultural land allocation problem, there are several goals to attain like increasing the profit with the maximum utilization of land, water, labor, and machine power. The main objective here is to find the best optimal model for agricultural land allocation using the goal programming model in the Karnataka district. The assigned goals are maximum utilization of land (depending upon the duration and restriction of the land allotted to crops), water (amount of water required for the crops) and manpower. Here, the main aim is to control the production of various crops so that the decision-maker can get the maximum benefit and also can look back on the required water resources.

2 Data from Various Crops

In Karnataka, the total available land is approximately 1.92 lakh sq. km, out of which 1,23,100 sq. km is used for agricultural purposes. Agricultural land of around

Fig. 1 Rainfall in Kharif season

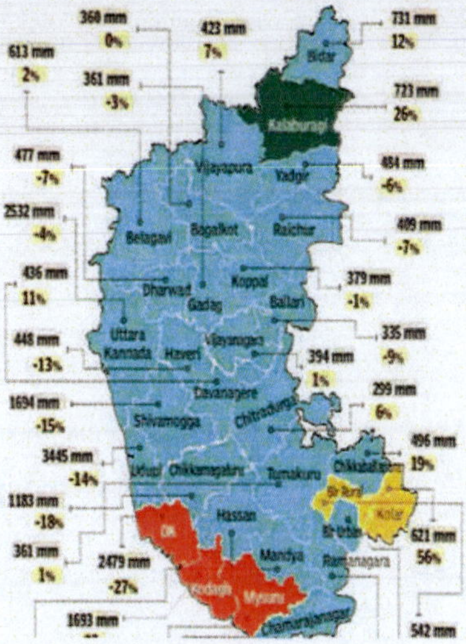

30,900 sq. km covers an irrigation area. In the year 2021, the population of the state was 66.17 million (as per Aadhar statistics 2021/2022). In the Kharif season, a total of 77,000 sq. km area is dedicated for sowing under various crops; around 20,000 sq. km is used as irrigated area; and 57,000 sq. km is under rainfed conditions. The rainfall (see Fig. 1) of the state in the Kharif season is around 115 mm (pre-monsoon from March-1 to May-31), 852 mm (southwestern monsoon from June-1 to September-30), and 7 mm (September). The land of the state has different types of soils black soil, red soil, red loam soils, and mixed red and black soil. Around 80 lakh farmers are there in the state. Here we took some crops of the state (see Fig. 2) rice, sugarcane, sunflower, cotton, millets, and chilies. In this paper, we considered the Kharif season. The data was collected (see Tables 1 and 2) by the farmers of the state and from the Karnataka government web-site [14].

3 Model Formulation

Goal programming is a technique for achieving (possibly competing) goals by ranking them in order of importance. The following is a general goal programming model:

$$\text{Minimize} \quad Z = \sum_{i=0}^{m} P_i W_i \left(d_i^+ + d_i^-\right) \quad \text{where } i = 1, 2, \ldots, \text{m}$$

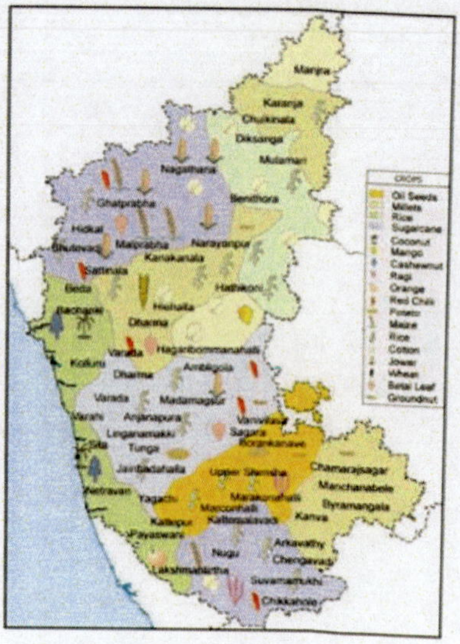

Fig. 2 Water resources and crops

Table 1 Information for crops, water requirement, yield, and manpower

Crops	Water requirement (cubic meter/hectare)	Crop duration (days)	Man power requirement (hour/hectare)
Oryza sativa (rice)	13,500	120	150
Arachis hypogaea (groundnut)	6000	130	80
Saccharum officinarum (sugarcane)	25,000	435	600
Capsicum frutescens (chilies)	9000	85	200
Gossypium sps (cotton)	10,000	150	200
Pennisetum glaucum (millets)	6500	120	60
Helianthus annuus (sunflower)	8000	90	50

$$\text{Subject to } \sum_{j=1}^{n} a_{ij}x_j + d_i^- - d_i^+ = b_i$$

Where $j = 1, 2, \ldots. n$ and $x_j, d_i^+, d_i^- \geq 0$, for all i and j.

The goals are expressed by using the m-component column b_i in relation to variables m. The coefficient for the jth decision variable in the ith constraint is represented as a_{ij}. Decision variable x_j. W_i stands for each goal's weights. P_i represents the priority of the goal-assigned rank order, which is $P_1 > P_2 > P_3 > P_4 \ldots \ldots P_n$.

Table 2 Information for crops, cost of production, yield (quintal/ha), price, income, and profit

Crops	Cost of production (Rs. per hectare)	Selling price (Rs. per quintal)	Yield (quintal per hectare)	Income (Rs. per hectare)	Profit (Rs. per ha)
Oryza sativa (rice)	80,000	3200	60	1,92,000	1,12,000
Arachis hypogaea (groundnut)	37,000	4875	16	78,000	41,000
Saccharum officinarum (sugarcane)	75,000	265	800	2,12,000	1,37,000
Capsicum frutescens (chilies)	75,000	4300	25	1,07,500	32,500
Gossypium sps (cotton)	55,000	1425	57	81,225	26,225
Pennisetum glaucum (millets)	50,000	6500	18	1,17,000	67,000
Helianthus annuus (sunflower)	25,000	3700	14	51,800	26,800

The overachievement deviational variable is represented by d_i^+ (surplus variable in LP). The underachievement deviational variable is represented by d_i^- (slack variable in LP). There are many inconsistent goals to satisfy the agricultural department within the state. We construct the following goals and objective functions using the general GP model.

Priority 1: Utilizing maximum availability of land.
Priority 2: Utilizing maximum availability of water.
Priority 3: Available land of *Oryza sativa* and *Saccharum officinarum* is restricted to 18,60,000 hectares.
Priority 4: Available land of *Arachis hypogaea* and *Gossypium* is more than 10,73,940 hectares.
Priority 5: Utilizing maximum land that takes less than or equal to 120 days.
Priority 6: Utilizing maximum availability of agriculture laborers.
Priority 7: More than 18,00,000 hectares amounts of *Pennisetum glaucum* should be cultivated.

Construction of Goals
Let the various crops be cultivated in $x_1, x_2, x_3 \ldots\ldots x_7$ quantity of hectares.

Goal 1: Utilization of land.

The first goal is to utilize the maximum available land of 77,00,000 hectares.

$$\sum_{i=1}^{7} x_i + U_1 - O_1 = 7700000$$

x_i is the ith crop in hectares.
U_1 is the underutilization of land.
O_1 is the overutilization of land.

Goal 2: Utilization of water.

The rainfall noted in the Kharif season amounts to 905 mm in various districts of Karnataka. 1 mm = 0.001 m^3 for 1 square meter, 1 hectare = 10,000 square meters; available land for agriculture is 77 lakh hectares, and 3075 tmc of water is used from various water resources like dams, borewells, and water wells. 1 tmc = 28,000,000 m^3. A total of 86,106,968,500 m^3 of water should be fully utilized.

$$\sum_{i=1}^{7} a_i x_i + U_2 - O_2 = 86106968500$$

a_i is the requirement of water per hectare for the ith crop in m^3,
U_2 is the underutilization of water, and
O_2 is the overutilization of water.

Goal 3: *Oryza sativa* (rice) and *Saccharum officinarum* production restriction (sugarcane).

Oryza sativa and *Saccharum officinarum* crops require more water, so we restrict the underutilization of land of 18,62,000 hectares for both crops.

$$x_1 + x_3 + U_3 - O_3 = 1862000$$

Goal 4: *Arachis hypogaea* and *Gossypium officinarum* production restriction.

We restrict the underutilization of *Arachis hypogaea* and *Gossypium* to more than 1,073,940 hectares.

$$x_5 + x_2 + U_4 - O_4 = 1073940$$

Goal 5: Duration of crops less than or equal to 120 days.

We have to overutilize the less-duration crops that can yield in 120 or lesser days.

$$x_1 + x_4 + x_6 + x_7 + U_5 - O_5 = 4764060$$

Goal 6: Labor.

The total availability of agriculture laborers is 80 lakh, and their total working hours are 64,000,000 (8 h/day).

$$\sum_{i=1}^{7} b_i x_i + U_6 - O_6 = 64000000$$

b_i is the labor required for the *i*th crop (hours/hectare)
U_6 is the underutilization of laborers
O_6 is the overutilization of laborers

Goal 7: *Pennisetum glaucum* (millets).

Pennisetum glaucum crop should be cultivated on more than 1,800,000 hectares.

$$x_6 + U_7 - O_7 = 1800000$$

The main objective of the function is to minimize the deviation from a number of goals by assigning appropriate priorities. By formulating the above constraints, the objective function is given below as:

$$\text{Min } Z = P_1 U_1 + P_2 U_2 + P_3 U_3 + P_4 O_4 + P_5 O_5 + P_6 U_6 + P_7 O_7$$

Implementation of R Programming.
Goal programming problem is:

$$\text{Min } Z = P_1 U_1 + P_2 U_2 + P_3 U_3 + P_4 O_4 + P_5 O_5 + P_6 U_6 + P_7 O_7$$

Subjected to the constraints:

$$\sum_{i=1}^{7} x_i + U_1 - O_1 = 7700000$$

$$\sum_{i=1}^{7} a_i x_i + U_2 - O_2 = 86106968500$$

$$x_1 + x_3 + U_3 - O_3 = 1862000$$

$$x_5 + x_2 + U_4 - O_4 = 1073940$$

$$x_1 + x_4 + x_6 + x_7 + U_5 - O_5 = 4764060$$

$$\sum_{i=1}^{7} b_i x_i + U_6 - O_6 = 64000000$$

$$x_6 + U_7 - O_7 = 1800000$$

Where $x_1, x_2, x_3, x_4, x_5, x_6, x_7, U_1, U_2, U_3, U_4, U_5, U_6, U_7, O_1, O_2, O_3, O_4, O_5, O_6, O_7 \geq 0$.

R programming code for the problem is:

```
library(lpSolve)
library(goalprog)
coefficient <- matrix (c (1, 1, 1, 1, 1, 1, 1, 13500, 6000, 25000,
 9000, 1000, 6500, 8000, 1, 0, 1, 0, 0,     0,0,
0, 1, 0, 0, 1, 0,0,1, 0, 0, 1,0, 1, 1, 150, 80, 600, 200, 200, 60,
 50, 1, 0, 0, 0, 0, 0, 0), nrow = 7, ncol = 7, byrow = TRUE)
coefficient
targets <- c (7700000, 86106968500, 1862000, 1073940, 4764060,
  64000000, 1800000)
targets
achievements <- data. Frame (matrix (c (1, 1, 0, 1, 2, 2, 0, 1, 3,
  3, 0, 1, 4, 4, 1, 0, 5, 5, 1, 0, 6, 6, 0, 1, 7, 7, 1, 0),
  nrow = 7, byrow=TRUE))
names (achievements) <- c ("objective", "priority", "p", "n")
achievements
soln <- llgp (coefficient, targets, achievements)
soln$out
print (soln$out)
```

4 Results

Figure 3 (goal programming) shows the over and underachievement of goals. Except for P3, P4, P6, and P7, the goals are fully achieved (Table 3). By using LiPS v1.9.4 software, this GP model was solved. The agricultural land is not distributed for the cultivation of *Oryza sativa* (rice), *Gossypium officinarum* (groundnut), *Capsicum frutescens* (Chilies) and *Pennisetum glaucum* (millets) because x_1, x_2, x_4, and x_6 are zeros. From this R programming, we achieved all the goals except G6 and G7 (Table 3). Figure 3 (R programming) shows the under- and overachievement of goals by using R programming.

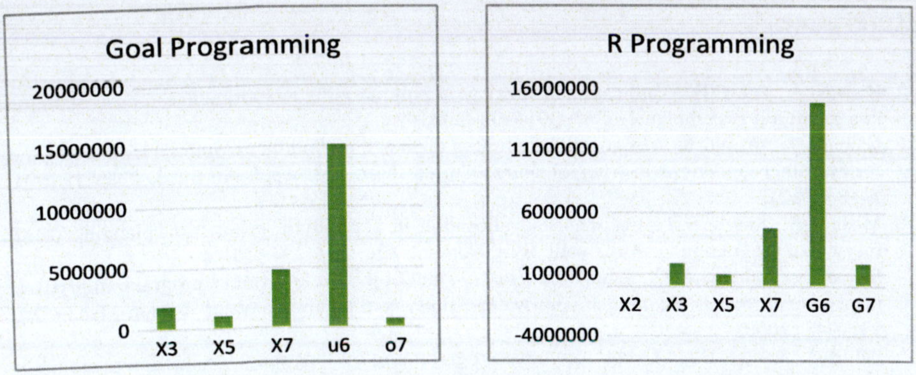

Fig. 3 Graph showing the result using linear programming solver and R programming solver

Table 3 Result for the allocation problem

Priorities	Goal achieved	Decision variable using LiPS	Decision variable using R program	Deviation variable using LiPS	Deviation variable using R program
1	YES	$x_1 = 0$	$x_1 = 0$	$U_1 = 0$	$G_1 = 0$
				$O_1 = 0$	$O_1 = 0$
2	YES	$x_2 = 0$	$x_2 = 74109$	$U_2 = 0$	$G_2 = 0$
				$O_2 = 0$	$O_2 = 0$
3	NO	$x_3 = 1877439.5$	$x_3 = 1862000$	$U_3 = 15439.520$	$G_3 = 15439.520$
				$O_3 = 0$	$O_3 = 0$
4	NO	$x_4 = 0$	$x_4 = 0$	$U_4 = 0$	$G_4 = 0$
				$O_4 = 15439.520$	$O_4 = 15439.520$
5	YES	$x_5 = 1058500.48$	$x_5 = 999830.3$	$U_5 = 0$	$G_5 = 0$
				$O_5 = 0$	$O_5 = 0$
6	NO	$x_6 = 0$	$x_6 = 0$	$U_6 = 151236681$	$G_6 = 149729800$
				$O_6 = 0$	$O_6 = 0$
7	NO	$x_7 = 4764048$	$x_7 = 4764060$	$U_7 = 0$	$G_7 = 0$
				$O_7 = 741499.52$	$O_7 = 741499.52$

5 Conclusion

The goal programming problem is constructed with decision variables (7), priority goals (7), and deviational variables (14). The underachievement of manpower (G_6 in R programming and U_6 in Goal programming) shows that the state needs more people to attract towards farming. This chapter helps to increase the cultivation of land as it visibly increases the production, also fulfills the needs of agricultural resources, and it helps the government to plan the proper allocation of farmland and help to grow the farm economy of the Karnataka state.

References

1. Biswas, A., Pal, B.B.: Application of fuzzy goal programming technique to land use planning in agricultural system. Omega. **33**, 391–398 (2005)
2. Gamage, A.: Application of goal programming approach on finding an optimal land allocation for five other field crops in Anuradhapura District. Oper. Res. Appl. An Int. J. (ORAJ). **4**(2), 1–13 (2017)
3. Mishra, B., Singh, S.R.: Optimal land allocation in agricultural production planning using fuzzy goal programming. Adv. Intell. Syst. Comput. **436**, 287–298 (2016)
4. Sahoo, B., Lohani, A.K., Sahu, R.K.: Fuzzy multiobjective and linear programming based management models for optimal land-water-crop system planning. Water Resour. Manag. **20**, 931–948 (2006)
5. Pal, B.B., Moitra, B.N.: Using fuzzy goal programming for long range production planning in agricultural systems. Indian J. Agric. Econ. **59**(1), 75–90 (2004)
6. Pal, B.B., Chakraborti, D., Biswas, P.: A genetic algorithm based hybrid goal programming approach to land allocation problem for optimal cropping plan in agricultural system. In: Fourth International Conference on Industrial and Information Systems, pp. 28–31. IEEE (2009)
7. Pal, B.B., Kumar, M., Sen, S.: A Priority Based Interval-Valued Goal Programming Approach for Land Utilization Planning in Agricultural System: a Case Study, pp. 1–9. Communication and Networking Technologies, IEEE (2010)
8. Sharma, D.K., Gaur, A., Ghosh, D.: Goal programming model for agricultural land allocation problems. Int. J. Model. Simul. **28**(1), 43–48 (2008)
9. Singha, D.K., Jaiswalb, C.S., Reddya, K.S., Singha, R.M., Bhandarkarb, D.M.: Optimal cropping pattern in a canal command area. Agric. Water Manag. **50**, 1–8 (2001)
10. Dinesh, K.S., JANA, R.K., Avinash, G.: Fuzzy goal programming for agricultural land allocation problems. Yugoslav J. Oper. Res. **17**(1), 31–42 (2007)
11. Vashistha, P.K.: Application of goal programming in farm agricultural planning. Int. J. Res. Sci. Technol. **1**(1), 1–10 (2011)
12. Zenis, F.M., Supian, S., Lesmana, E.: Optimization of land use of agricultural farms in Sumedang regency by using linear programming models. Mater. Sci. Eng. **332**, 1–6 (2018)
13. Mohammadian, F., Heydari, M.: Application of fuzzy goal programming to determine the optimal cultivation crops model. Int. J. Stud. Res. Technol. Manag. **7**(1), 01–08 (2019)
14. https://raitamitra.karnataka.gov.in./info-4/ANNUAL++REPORT/en
15. Kagita, M.K., Tata, H.K., Mannava, A.C.: A fuzzy environment strategy for optimal agricultural land allocation in Krishna Delta. Int. Res. J. Comput. Sci. **05**(02), 57–64 (2018)
16. Patel, N., Thaker, M., Chaudhari, C.: Agricultural land allocation to the major crops through linear programming model. Int. J. Sci. Res. **6**(4), 519–552 (2015)
17. Tamta, P., Singh, H.N., Singh, J., Ruhela, A., Pal, V.K., Pant, M.: Optimum farm planning under different farming systems on small and marginal farms in Rampur district, Uttar Pradesh A linear programming approach. Int. J. Res. Manag. **6**(4), 42–53 (2016)
18. Srivastava, P., Singh, R.M.: Agricultural land allocation for crop planning in a canal command area using fuzzy multiobjective goal programming. J. Irrig. Drain. Eng. **143**(6), 1–9 (2017)
19. Tiwari, R.N., Dharmar, S., Rao, J.R.: Fuzzy goal programming- an additive model. Fuzzy Sets Syst. **24**, 27–34 (1987)
20. Mohamed, R.H.: The relationship between goal programming and fuzzy programming. Fuzzy Sets Syst. **89**, 215–222 (1997)
21. Mirkarimi, S.H., Joolaie, R., Eshraghi, F., Abadi, F.S.B.: Application of fuzzy goal programming in cropping pattern management of selected crops in Mazandaran Province (case study Amol township). Int. J. Agric. Crop Sci. **6**(15), 1062–1067 (2013)
22. Kommerce, S.M., Reddy, R., Babu, J.R.: Agricultural land apportionment through goal programming. Int. J. Sci. Res. Sci. Technol. **3**(8), 1011–1015 (2017)

23. Padi, T.R., Paidipat, K.K., Oram, M.: Optimal land allocation programming problems for agricultural revenue maximization in Andhra Pradesh. Agric. Sci. Dig. Res. Article, 1–5 (2020)
24. Basumatary, U.R., Dipak Kr. Mitra: A study on optimal land allocation through fuzzy multi-objective linear programming for agriculture production planning in Kokrajhar District, BTAD, Assam, India. Int. J. Appl. Eng. Res. **15**(1), 94–100 (2020)
25. Mosleh, Z., Salehi, M.H., Fasakhodi, A.A., Jafari, A., Mehnatkesh, A., Borujeni, I.E.: Sustainable allocation of agricultural lands and water resources using suitability analysis and mathematical multi-objective programming. Geoderma. **303**, 52–59 (2017)

Neutral-Bipolar Fuzzy Sets and Its Applications

G. Aruna and J. Jesintha Rosline

1 Introduction

The notion of ambiguity and haziness in real life can be simply dealt with the flexibility of fuzzy set theory presented by Professor Lotfi. A. Zadeh in the year 1965 [9]. Further, the theory of Fuzzy numbers was established by various authors like Krassimir T. Atanassov [1], Dubois, R. Yager (1993), and Mizumoto, J. Buckley [9]. Wen–Ran Zhang initiated the concept bipolar fuzzy sets in 1998 [8]. The bipolar fuzzy set theory is the blending of both polarization and fuzziness [8]. It delivers a theoretical basis for clustering and conflict determinations. Bui Cong Cuong [2] introduced the concept called Picture fuzzy sets (PFSs). The idea of neutrality invoked the PFS as a dynamic area in fuzzy sets. Muhammad Riaz [6, 7] presented the hybrid idea of bipolar picture fuzzy operators. Fuzzy arithmetic models are useful innovations because of their differences between the ancient numerical models in sciences and symbolic models used in expert systems. Basic description, examples, properties of fuzzy sets defined by authors like Kaufmann, Dubois, Prade, Yager, Garg and Jesintha Rosline et al. [3, 4, 5] are stepping stones for Neutral-Bipolar fuzzy sets (NBFSs). The features of bipolarity and its applications for organic sciences, Psychology, neural networking, decision-making, business, and applied mathematics motivate to frame the conception Neutral-Bipolar fuzzy sets (NBFSs). In human life, the existence of uncertainty involves in decision-making of each possession. Typically, human beings have the nature of neutrality and bipolarity in making decisions. The exploration of neutrality and bipolarity with the impact of fuzzy set paves the way of new notion Neutral-Bipolar fuzzy sets. The Neutral-Bipolar fuzzy set is a fuzzy arithmetic tool, which has a far greater

G. Aruna · J. Jesintha Rosline (✉)
Department of Mathematics, Auxilium College (Autonomous), Thiruvalluvar University, Serkadu, Vellore, Tamil Nadu, India

F. Kamalov et al. (eds.), *Advances in Mathematical Modeling and Scientific Computing*, Trends in Mathematics, https://doi.org/10.1007/978-3-031-41420-6_48

expressive power than interval arithmetic by virtue of gradation of membership in a fuzzy set.

The main objective of this chapter is to present the distinctive conception termed as Neutral-Bipolar fuzzy sets (NBFSs). The neutrality degree of membership ranges from $[-1, 1]$ in NBFS. The section two comprises of basic definitions. The Neutral-Bipolar fuzzy set is well defined in the section three. The properties of Neutral-Bipolar fuzzy sets, its operations, and theorems are discussed in section four and five. Finally, the applications of NBFS in MCDM technique of decision-making are illustrated.

2 Preliminaries

Picture Fuzzy Sets [2] A picture fuzzy set (PFS) on the universe U is defined as $P = \{(x, \mu(x), \eta(x), \nu(x))\}$ where $\mu(x) \epsilon\ [0, 1]$ is the positive membership degree of x in P, $\eta(x) \epsilon\ [0, 1]$ is the neutral membership degree and $\nu(x) \epsilon\ [0, 1]$ is the negative membership degree.

Bipolar Fuzzy Sets [8] Let X be the nonempty fuzzy set. The bipolar fuzzy set $B = \{B^-, B^+\}$ is defined in X. $B^+ \in [0, 1]$ is the satisfaction degree of x in B, $B^- \in [-1, 0]$ is the satisfaction degree of the implicit counter property of x in B.

Operations on Picture Fuzzy Sets [2] Let $K_1, K_2 \in PFS$: (i) $K_1 \subseteq K_2$ if $\mu_{K_1}(x) \le \mu_{K_2}(x)$, $\eta_{K_1}(x) \le \eta_{K_2}(x)$ and $\nu_{K_1}(x) \ge \nu_{K_2}(x) \forall x \in X$. (ii) $K_1 = K_2$ if $K_1 \subseteq K_2$ and $K_1 \supseteq K_2$, (iii) $K_1 \cup K_2 = \Big\{ < x, \max\left(\mu_{K_1}(x), \mu_{K_2}(x)\right), \min\left(\eta_{K_1}(x), \eta_{K_2}(x)\right), \min(\nu_{K_1}(x), \nu_{K_2}(x)) > \} \forall x \in X$. (iv) $K_1 \cap K_2 = \Big\{ < x, \min\left(\mu_{K_1}(x), \mu_{K_2}(x)\right), \min\left(\eta_{K_1}(x), \eta_{K_2}(x)\right), \max(\nu_{K_1}(x), \nu_{K_2}(x)) > \} \forall x \in X$. (v) ${K_1}^C = \left\{ < x, \ \nu_{K_1}(x), \eta_{K_1}(x), \mu_{K_1}(x) > \right\} \forall x \in X$.

3 Neutral-Bipolar Fuzzy Sets

The spatial relation is expressed by human mind in the alternate concept of "left" and "right" as opposite direction is as equivalent polarization. But this does not mean that one of them is the negation of the other. The symmetric of 'opposite' captures a notion of semantics rather than a complexity. We are familiar with positive and negative membership degree. In Neutral-Bipolar fuzzy set, the neutral membership refers to the rate of chances between positive and negative side, which exists in the desirable and counter desirable rate. A chance is a reaction, which exists between the stimulation and the responses. Here, the stimulation may come from both positive and negative sides. Hence, the level of indefiniteness in desirable and counter desirable intensifies the neutral level. A human brain not only thinks about the positive and negative poles but also about the neutrality level due to circumstance. The level of neutral need not be zero at all stages. Positive stimulators intensify the neutrality over positive side. Negative stimulators intensify the neutrality over negative side. The intensification range of positive stimulators may result in the diminished rate of negative stimulators and the intensification range of negative stimulators may result in the diminished rate of positive stimulators. The circumstance plays a major role in NBFS. Circumstance sometimes refers to self-motivation, workouts, maintaining good health, people around us, biological diseases, etc. The circumstance constitutes the region with neutrality in positive range and negative range. Hence, we consider the range of neutrality lies in the interval $[-1, 1]$.

Neutral-Bipolar fuzzy set Let X be a nonempty set. A Neutral-Bipolar Fuzzy set (NBFS) is defined as

$$NB = \{< x,\ \ \mu(x),\ \ \eta(x),\ \ \gamma(x), \lambda\ (x) \mid x \epsilon X\},$$

where positive degree of membership $\mu(x) \in [0, 1]$ refers to the desirable rate of x in NB, negative degree of membership $\gamma(x) \in [-1, 0]$ is the counter desirable rate of x in NB, the neutral degree $\eta(x) \in [-1, 1]$ is the rate of satisfaction of the property and counter property to some extent of x in NB, and $\lambda\ (x) \in [0, 1]$ is the undesirable value of x in NB with the conditions $0 \leq \mu(x) + \lambda\ (x) \leq 1$ and $0 \leq |\mu(x)| + |\gamma(x)| + |\eta(x)|\ \leq 2\ (\forall x \epsilon X)$. Now $2 - (|\mu(x)| + |\gamma(x)| + |\eta(x)|)$ is the refusal degree of x in NB. Let NBFS(x) denote the set of all Neutral-Bipolar fuzzy sets on the universe U.

Neutral-Bipolar fuzzy set can be applied in the circumstances where the neutrality exists. Suppose a person is undergoing a treatment for a particular disease, then

- The rate of curable is referred to as desirable rate.
- The rate of deterioration is referred to as counter desirable rate.
- The rate of chances being positive and negative due to health issues like high immune power system, blood pressure, and diabetes is referred to as neutrality rate.
- If the person is not taking any treatment, it is referred to as undesirable rate.

Degree of NBFS Let $NB_1 \in NB(x)$, then

1. The degree of neutral membership is defined as

$$d\,(\eta(x)) = 2 - [|\mu(x)| + |\gamma(x)|]\ \ (\forall x \epsilon X)$$

2. The degree of positive desirable membership is defined as

$$d_f\,(\mu(x)) = \mu(x) + \mu(x).\ |\ \eta(x)\ |\ (\forall x \epsilon X)$$

3. The degree of negative desirable membership is defined as

$$d_f\ (\gamma(x)) = \mid \gamma(x) \mid + \mid \gamma(x) \mid . \mid \eta(x) \mid \ (\forall x \in X)$$

For example, Let A be the Neutral-Bipolar fuzzy set with $\mu(x) = 0.6$, $|\gamma(x)| = 0.5$ then $d(\eta(x)) = 0.9$, $d_f(\mu(x)) = 1.14$ and $d_f(\gamma(x)) = 0.95$. Suppose if we take the positive response for the product is 0.6, the negative response for the product is 0.5, the neutral response is 0.9, then by using the degree of NBFS we can find the favorablc positive response is 1.14 and the favorable negative response is 0.95.

Remark

1. If $\mu(x) \neq 0$, $\gamma(x) = \eta(x) = \lambda(x) = 0$, then NBFS reduces to FS.
2. If $\mu(x) \neq 0$, $\lambda(x) \neq 0$, $\gamma(x) = \eta(x) = 0$,then NBFS reduces to IFS.

4 Properties of Neutral-Bipolar Fuzzy Sets (NBFSs)

Subset Let X be a nonempty set and NB_1, $NB_2 \in NBFS(x)$. Then NB_1 is a subset of NB_2 denoted by$NB_1 \subseteq NB_2$ if for each $x \in X$; $\mu_{NB_1}(x) \leq \mu_{NB_2}(x)$, $\mid \eta_{NB_1}(x) \mid \leq \mid \eta_{NB_2}(x) \mid$ and $\gamma_{NB_1}(x) \geq \gamma_{NB_2}(x)$.

Complement of NBFS The complement of the set $NB_1 \in NB(x)$ is symbolized as ${NB_1}^C$ and it is indicated by ${NB_1}^C = \{1 - \mu_{NB_1}(x), 2 - |\eta_{NB_1}(x)|\ , -1 - \gamma_{NB_1}(x)\}$ $(\forall x \in X)$.

Equal Set The two nonempty set NB_1, $NB_2 \in NB(x)$ are said to be equal set if for each $x \in X$; $\mu_{NB_1}(x) = \mu_{NB_2}(x)$, $\mid \eta_{NB_1}(x) \mid = \mid \eta_{NB_2}(x) \mid$ and $\gamma_{NB_1}(x) = \gamma_{NB_2}(x)$.

Equivalent The two NBFS set NB_1, $NB_2 \in NB(x)$ are alleged to be equivalent to each other. $NB_1 \sim NB_2$. If there exists function $f : \mu_{NB_1}(x) \rightarrow \mu_{NB_2}(x)$, $f : \mid \eta_{NB_1}(x) \mid \rightarrow \mid \eta_{NB_2}(x) \mid$ and $f : \gamma_{NB_1}(x) \rightarrow \gamma_{NB_2}(x)$, which are bijective. Then the function has a 1–1 correspondence between domain and codomain.

Proper Subset If $\mu_{NB_1}(x) \leq \mu_{NB_2}(x)$, $\mid \eta_{NB_1}(x) \mid \leq \mid \eta_{NB_2}(x) \mid$, $\gamma_{NB_1}(x) \geq \gamma_{NB_2}(x) \forall x \in X$ and $NB_1 \subseteq NB_2$, $NB_1 \neq NB_2$. Then $NB_1$is the proper subset of NB_2.

Definition 1 If NB_1, NB_2, $NB_3 \in NB(x)$. Then we have the following:

1. If $NB_1 \precsim NB_1$, then NB_1 is reflexive.
2. If $NB_1 \precsim NB_2$, $NB_2 \precsim NB_1$,then it is symmetric.
3. If $NB_1 \precsim NB_2$, $NB_2 \precsim NB_3$ implies $NB_1 \precsim NB_3$, then it is transitive relation.

5 Operations on Neutral-Bipolar Fuzzy Sets

5.1 Operations and Theorems on NBFS

Operations on NBFS Let $NB_1 = (\mu_1(x), \ \eta_1(x), \gamma_1(x))$ and $NB_2 = (\mu_2(x), \eta_2(x), \gamma_2(x))$ be two NBFS. Then

(i) Union: $NB_1 \cup NB_2 = \{ < x, \max(\mu_{NB_1}(x), \mu_{NB_2}(x)), \max(|\eta_{NB_1}(x)|, |\eta_{NB_2}(x)|), \min(\gamma_{NB_1}(x), \gamma_{NB_2}(x)) >; \forall x \in X \}$

(ii) Intersection: $NB_1 \cap NB_2 = \{ < x, \min(\mu_{NB_1}(x), \mu_{NB_2}(x)), \min(|\eta_{NB_1}(x)|, |\eta_{NB_2}(x)|), \max(\gamma_{NB_1}(x), \gamma_{NB_2}(x)) >; \forall x \in X \}$

Proposition 1 Bipolar fuzzy set is the special instance of Neutral-Bipolar fuzzy set (NBFS).

Proof For any $x \in X$, consider an NBFS set $NB_1 = (\mu_1(x), \ \eta_1(x), \gamma_1(x))$. Suppose the value of $\eta_1(x)$ is zero, then NBFS diminishes to Bipolar fuzzy set.

Proposition 2 The NBFS is not the generalization of Picture Fuzzy set.

Proof For any $x \in X$, consider an NBFS set $NB_1 = (\mu_1(x), \ \eta_1(x), \gamma_1(x))$.

Although NBFS encompasses the neutrality property, it contrasts in ranges from PFS. Moreover, the length of neutrality is functional in NBFS. So, NBFS is not the generalization of PFS.

Theorem 1 Let $NB_1, NB_2, NB_3 \in NB(x)$ in the universe U. Then, we have various operational rules:

(i) Idempotent Rule: $NB_1 \cup NB_1 = NB_1$ and $NB_1 \cap NB_1 = NB_1$
(ii) Commutative Rule: $NB_1 \cup NB_2 = NB_2 \cup NB_1$ and $NB_1 \cap NB_2 = NB_2 \cap NB_1$.
(iii) Complementary Rule: $({NB_1}^C)^C = NB_1$
(iv) Associative Rule: $(NB_1 \cup NB_2) \cup NB_3 = NB_1 \cup (NB_2 \cup NB_3)$ and

$$(NB_1 \cap NB_2) \cap NB_3 = NB_1 \cap (NB_2 \cap NB_3)$$

(v) Distributive Rule: $NB_1 \cup (NB_2 \cap NB_3) = (NB_1 \cup NB_2) \cap (NB_1 \cup NB_3)$ and $NB_1 \cap (NB_2 \cup NB_3) = (NB_1 \cap NB_2) \cup (NB_1 \cap NB_3)$
(vi) De Morgan's Rule: $(NB_1 \cup NB_2)^C = {NB_1}^C \cap {NB_2}^C$

$$(NB_1 \cap NB_2)^C = {NB_1}^C \cup {NB_2}^C$$

Proof

(i)

$$NB_1 \cup NB_1 = \left\{<x, \max\left(\mu_{NB_1}(x), \mu_{NB_1}(x)\right), \max\left(|\eta_{NB_1}(x)|, |\eta_{NB_1}(x)|\right), \min\left(\gamma_{NB_1}(x), \gamma_{NB_1}(x)\right) >; \forall x \in X\right\}$$

$$= NB_1. i.e, NB_1 \cup NB_1 = NB_1.$$

Similarly, $NB_1 \cap NB_1 = NB_1$.

(ii)

$$NB_1 \cup NB_2 = \left\{< x, \max\left(\mu_{NB_1}(x), \mu_{NB_2}(x)\right), \max\left(|\eta_{NB_1}(x)|, |\eta_{NB_2}(x)|\right), \min\left(\gamma_{NB_1}(x), \gamma_{NB_2}(x)\right) >; \forall x \in X \right\}$$

$=\{<x, \max\left(\mu_{NB_2}(x), \mu_{NB_1}(x)\right), \max\left(|\eta_{NB_2}(x)|, |\eta_{NB_1}(x)|\right), \min(\gamma_{NB_2}(x), \gamma_{NB_1}(x))>; \forall x \in X\} = NB_2 \cup NB_1$.

Similarly, $NB_1 \cap NB_2 = NB_2 \cap NB_1$.

(iii)

$$NB_1{}^C = \left\{< x, 1 - \mu_{NB_1}(x), 2 - |\eta_{NB_1}(x)|, -1 - \gamma_{NB_1}(x) >; \forall x \in X\right\}.$$

$$\left(NB_1{}^C\right)^C = \left\{< x, 1 - \mu_{NB_1{}^C}(x), 2 - |\eta_{NB_1{}^C}(x)|, -1 - \gamma_{NB_1{}^C}(x) >; \forall x \in X\right\}.$$

Hence, $\left(NB_1{}^C\right)^C = NB_1$.

Proof is obvious for (iv) and (v).

Example

Let NB_1, $NB_2 \in NB(x)$ be two NBFS sets in the universe U. If $NB_1 = (0.8, 0.5, -0.7)$ and $NB_2 = (0.7, 0.3, -0.6)$. Then the arithmetic results follow:

$$NB_1 \cup NB_2 = (0.8, 0.5, -0.7), NB_1 \cap NB_2 = (0.7, 0.3, -0.6), \left(NB_1{}^C\right) = (0.2, 1.5, -0.3), \left(NB_1{}^C\right)^C = (0.8, 0.5, -0.7) \text{ Hence, } \left(NB_1{}^C\right)^C = NB_1.$$

Definition 2 Let $NB_1 = (\mu_1(x), \eta_1(x), \gamma_1(x))$ and $NB_2 = (\mu_2(x), \eta_2(x), \gamma_2(x))$ be two NBFS. Then

(i) $NB_1 \bigvee NB_2 = \Big\{ \max\left(\mu_{NB_1}(x), \mu_{NB_2}(x)\right), \max\left(|\eta_{NB_1}(x)|, |\eta_{NB_2}(x)|\right), \min(\gamma_{NB_1}(x), \gamma_{NB_2}(x)\Big\}$

(ii) $NB_1 \wedge NB_2 = \Big\{ \min\left(\mu_{NB_1}(x), \mu_{NB_2}(x)\right), \min\left(|\eta_{NB_1}(x)|, |\eta_{NB_2}(x)|\right), \max(\gamma_{NB_1}(x), \gamma_{NB_2}(x)\Big\}$

(iii) For $\lambda > 0$, $NB_1{}^{\lambda} = \left(\left(\mu_{NB_1}(x)\right)\right)^{\lambda}, \left(|\eta_{NB_1}(x)|\right)^{\lambda}, \left(1 - \left(1 - \left(-\gamma_{NB_1}(x)\right)^{\lambda}\right)\right)$

(iv) For $\lambda > 0$, $NB_1 \otimes NB_2 = \{\mu_{NB_1}(x).\mu_{NB_2}(x)\ ,\ |\eta_{NB_1}(x)|.|\eta_{NB_2}(x)|,$

$$\{1 - (1 - (-\gamma_{NB_1}(x))\} + \{1 - (1 - (-\gamma_{NB_2}(x)))$$
$$- \{1 - (1 - \gamma_{NB_1}(x))\}\{1 - (1 - \gamma_{NB_2}(x))\}\}$$

5.2 *Certain Results on Neutral-Bipolar Fuzzy Operations*

Definition 3 Let $NB_1 = (\mu_1(x),\ \eta_1(x), \gamma_1(x))$ be the NBFS. Then score valued function $\delta(NB_1)$ and the accuracy valued function $\alpha(NB_1)$ of NBFS is defined as

(i) $\delta\,(NB_1) = \frac{1}{2}\,(\mu_1 + |\eta_1| + \gamma_1)$where $\delta(NB_1) \in [-1, 1]$
(ii) $\alpha\,(NB_1) = \frac{1}{2}\,(\mu_1 + |\eta_1| - \gamma_1)$where $\delta(NB_1) \in [0, 1]$.

Definition 4 Let $NB_1, NB_2 \in NBFS$, the results as follows:

- If $\delta(NB_1) < \delta(NB_2)$ then $NB_1 < NB_2$.
- If $\delta(NB_1) > \delta(NB_2)$ then $NB_1 > NB_2$.
- If $\delta(NB_1) = \delta(NB_2)$,then we can find the grades using accuracy function. Case 1: If $\alpha(NB_1) < \alpha(NB_2)$ then $NB_1 < NB_2$. Case 2: If $\alpha(NB_1) > \alpha(NB_2)$ then $NB_1 > NB_2$. Case 3: If $\alpha(NB_1) = \alpha(NB_2)$ then $NB_1 = NB_2$.

Neutral-Bipolar Weighted Geometric Mean (NBWGM) The Neutral-Bipolar Weighted Geometric Mean (NBWGM) is defined as NBWGM $(NB_1, NB_2, \ldots..NB_n) = NB_1{}^{q_1} \otimes NB_2{}^{q_2} \otimes \cdots \otimes NB_n{}^{q_n}$, where q $= (q_1, q_2, \ldots., q_n)$, $q_k \in [0, 1]$ is the weight vector of NB_k and $\sum_{k=1}^{n} q_k = 1$.

Theorem 3 The Neutral-Bipolar Weighted Geometric Mean of NBFS $NB_i(i = 1, 2, \ldots, n)$ of NBWGM operator is

$$\text{NBWGM}\,(NB_1, NB_2, \ldots..NB_n) = \Big\{\prod_{k=1}^{n} (\mu_{NB_k})^{q_k}, \prod_{k=1}^{n} |(\eta_{NB_k})^{q_k}|\ ,\ \Big(-\prod_{k=1}^{n} (-\ (-\gamma_{NB_k})^{q_k}\Big)\Big\}$$

Proof The proof follows by the method of mathematical induction. For k = 2

$$NB_1{}^{q_1} \bigotimes NB_2{}^{q_2} = \Big((\mu_{NB_1})^{q_1} \cdot (\mu_{NB_2})^{q_2},\ \big|(\eta_{NB_1})^{q_1}\big|.\big|(\eta_{NB_2})^{q_2}\big|,$$
$$\Big(1-\Big(1-\Big(-\gamma_{NB_1}^{q_1}\Big)\Big)\Big)+\Big(1-\Big(1-\Big(-\gamma_{NB_2}^{q_2}\Big)\Big)\Big)$$
$$-\Big(\Big(1-\Big(1-\Big(-\gamma_{NB_1}^{q_1}\Big)\Big).\Big(1-\Big(1-\Big(-\gamma_{NB_2}^{q_2}\Big)\Big)\Big)\Big)$$

$$=\textstyle\prod_{k=1}^{2} (\mu_{NB_k})^{q_k}, \prod_{k=1}^{2} \mid (\eta_{NB_k})^{q_k} \mid, \Big(1-\Big(1-(-\gamma)_{NB_1}^{q_1}\Big)+1$$
$$-\Big(1-(-\gamma)_{NB_2}^{q_2}\Big)-1+\Big(1-(-\gamma)_{NB_1}^{q_1}\Big)+\Big(1-(-\gamma)_{NB_2}^{q_2}\Big)-\Big(1-(-\gamma)_{NB_1}^{q_1}\Big)$$
$$\Big(1-(-\gamma)_{NB_2}^{q_2}\Big)\Big)$$

$$=\left\{\prod_{k=1}^{2} (\mu_{NB_k})^{q_k}, \prod_{k=1}^{2} |(\eta_{NB_k})^{q_k}|\ , \left(1-\prod_{k=1}^{2}(1-\ (-\gamma_{NB_k})^{q_k}\right)\right\}$$

Let us assume it holds for k = n.$NB_1{}^{q_1} \bigotimes NB_2{}^{q_2} \bigotimes \cdots \bigotimes NB_n{}^{q_n} = \left\{\prod_{k=1}^{n} (\mu_{NB_k})^{q_k}, \prod_{k=1}^{n} |(\eta_{NB_k})^{q_k}|\ , \left(-\prod_{k=1}^{n}(-\ (-\gamma_{NB_k})^{q_k}\right)\right\}$

We prove for k = n + 1,

$$NB_1{}^{q_1} \bigotimes NB_2{}^{q_2} \bigotimes \cdots \bigotimes NB_n{}^{q_n} \bigotimes NB_{n+1}{}^{q_{n+1}} = \bigotimes_{k=1}^{n} NB_k{}^{q_k} \bigotimes NB_{k+1}{}^{q_{k+1}}$$

Consider, $\bigotimes_{k=1}^{n} NB_k{}^{q_k} \bigotimes NB_{k+1}{}^{q_{k+1}} = \Big(\bigotimes_{k=1}^{n} (\mu_{NB_k})^{q_k} \cdot (\mu_{NB_{k+1}})^{q_{k+1}},$
$\bigotimes_{k=1}^{n} \mid (\eta_{NB_k})^{q_k} \mid . \mid (\eta_{NB_{k+1}})^{q_{k+1}} \mid, \bigotimes_{k=1}^{n} \Big(1-\Big(1-\Big(-\gamma_{NB_k}^{q_k}\Big)\Big)\Big)$

$$+\Big(1-\Big(1-\Big(-\gamma_{NB_{k+1}}^{q_{k+1}}\Big)\Big)\Big)-\Big(\bigotimes_{k=1}^{n}\Big(1-\Big(1-\Big(-\gamma_{NB_k}^{q_k}\Big)\Big).$$
$$\Big(1-\Big(1-\Big(-\gamma_{NB_{k+1}}^{q_{k+1}}\Big)\Big)\Big)\Big)$$
$$=\left\{\prod_{k=1}^{n+1} (\mu_{NB_k})^{q_k}, \prod_{k=1}^{n+1} |(\eta_{NB_k})^{q_k}|\ , \left(1-\prod_{k=1}^{n+1}(1-\ (-\gamma_{NB_k})^{q_k}\right)\right\}$$

Hence, the theorem is proved.

Theorem on Idem-potency Let $NB_i (i = 1, 2, \ldots, n)$is the set of all NBFS sets. If $NB_k = NB$ for each k, then NBFSWA($NB_1, NB_2, \ldots . NB_n$) = NB.

Proof We know that NBWGM $(NB_1, NB_2, \ldots .NB_n) = \left\{\prod_{k=1}^{n} (\mu_{NB_k})^{q_k},\right.$ $\left.\prod_{k=1}^{n} |(\eta_{NB_k})^{q_k}|\ , \left(1-\prod_{k=1}^{n}(1-\ (-\gamma_{NB_k})^{q_k}\right)\right\}$

If $NB_k = NB$ (for each k) $= \left\{ (\mu_{NB})^{\sum_{k=1}^{n} q_k}, \ |(\eta_{NB})^{\sum_{k=1}^{n} q_k}|, \left(1 - \left(1 - (-\gamma_{NB})^{\sum_{k=1}^{n} q_k}\right)\right\}\right.$. NBFSWA$(NB_1, NB_2, \ldots . NB_n) = \{\mu_{NB}, \ |(\eta_{NB})|, 1 - (1 - (-\gamma_{NB}))\ \} = $ NB.

Theorem on Monotonicity Let NB_k and $NB_{k'}$ be the pair of sets in NBFS. If $NB_k < NB_{k'}$ for each k, then NBFSWA$(NB_1, NB_2, \ldots . NB_n) < (NB_{1'}, NB_{2'}, \ldots . NB_{n'})$.

Proof If $NB_k, NB_{k'} \in$ NBFS and $NB_k < NB_{k'}$ for each value of k, then

$$\left\{ (\mu_{NB})^{q_k}, \ |(\eta_{NB})|^{q_k}|, \left(1 - \left(1 - (-\gamma_{NB})^{q_k}\right)\right\} < \left\{ (\mu_{NB'})^{q_k}, \ |(\eta_{NB}), |^{q_k}, \left(1 - \left(1 - (-\gamma_{NB'})^{q_k}\right)\right\}\right.\right.$$

Hence, NBFSWA $(NB_1, NB_2, \ldots . NB_n) < (NB_{1'}, NB_{2'}, \ldots . NB_{n'})$.

Theorem on Boundedness Let $NB_i(i = 1, 2, \ldots, n)$is the set of all NBFS sets. Let$(NB)^- = \min(NB_1, NB_2, \ldots . NB_n)$ and $(NB)^+ = \max(NB_1, NB_2, \ldots . NB_n)$. Then $(NB)^- \leq (NB_1, NB_2, \ldots . NB_n) \leq (NB)^+$.

Proof The proof is obvious.

6 Multicriteria Decision-Making in NBFS

In this section, Multicriteria decision-making method using the geometric operators defined for NBFS is presented. Presume that $R = \{R_1, R_2, R_3, \ldots, R_q\}$is the collection of alternatives and $C = \{C_1, C_2, C_3, \ldots, C_r\}$ is the collection of criteria. $\sum_{k=1}^{n} q_k = 1\ (k = 1, 2, 3, \ldots n)$ and it represents the weight of C_k. The decision-makers examined the alternatives with the mentioned conditions and presented the details in the form of Neutral-Bipolar fuzzy sets.

We demonstrate the application of NBFS in bipolar disorder. Manic-depressive illness is a psychological health condition that grounds extreme moods swings. It is a lasting condition one can handle our emotional instability and other indicators by a proper health care. Bipolar type 1, bipolar type 2 and cyclothemic are the major types of bipolar disorder. It refers the three different behaviors manic behavior, depression, hypomanic with depression combinations. Manic behavior refers elevated emotions, highly energetic and active. Depression is loss of attention and persistent hopelessness. Cyclothemic is the combination of hypomanic and depression. We can easily compare this with NBFS like manic behavior as positive value of membership, depression as negative value of membership, and the combination level as neutral value. Experts believed that this biological difference arises due to the substance variations in the brain. The main reasons of bipolar disorder are hereditary factor, environmental influence, physical complaint, and medication.

Table 1 The alternatives with age factor

Alternatives	Age factors
R_1	below 25
R_2	around 25–45
R_3	above 45

Table 2 Decision matrix of NBFS

C_1	C_2	C_3	C_4
R_1(0.7, 0.2, −0.8)	(0.8, 0.3, −0.6)	(0.7, 0.3, −0.5)	(0.7, 0.7, −0.2)
R_2(0.8, 0.1, −0.7)	(0.3, 0.7, −0.4)	(0.9, 0.7, −0.1)	(0.5, 1.0, −0.5)
R_3 (0.4, 0.3, −0.6)	(0.8, 0.3, −0.7)	(0.5, 0.4,-0.6)	(0.6, 0.8, −0.6)

Table 3 NBFS normalized matrix

C_1	C_2	C_3	C_4
R_1(0.7, 0.2, −0.8)	(0.8, 0.3, −0.6)	(0.7, 0.3, −0.5)	(0.3, 1.3, −0.8)
R_2(0.8, 0.1, −0.7)	(0.3, 0.7, −0.4)	(0.9, 0.7, −0.1)	(0.5, 1.0, −0.5)
R_3 (0.4, 0.3, −0.6)	(0.8, 0.3, −0.7)	(0.5, 0.4,-0.6)	(0.4, 1.2, −0.4)

Assume that the main reasons of bipolar disorder as criteria $C = \{C_1, C_2, C_3, C_4\}$ (C_1= hereditary factor, C_2= environmental influence,C_3= physical complaint, C_4= medication). The age factors as the set of alternatives$R = \{R_1, R_2, R_3\}$ (R_1= below 25, R_2= around 25–45, R_3= above 45) are presented in Table 1.

Reasons for Alternatives

- Below 25: The person below 25 ages mostly belongs to the student's community. They might be taking care by their parents, surrounded by the friends. And those peoples are on the safe side of life.
- Around 25–45: The person around 25–45 ages mostly belong to those who just start their career. They are in the urge to settle down in life. They should take care of their self and their surroundings. They are in the struggling to get a progression and enduring in the career. This stage is the learning process so they are at risk side of life.
- Above 45: The persons around 45 age belongs to those who settled down in a career and those who are in retired life. This is the stage with full of experience, so they can easily manage the complications in life. These persons are on the safe side of life. Now by using the NBFS, we can find the most affected age group people for the mentioned criteria.

Step 1. Construct the NBFS decision matrix $(NB_{ij})_{q\times r}$ for the given condition relates to the alternatives. The NBFS decision matrix is given in Table 2.

Step 2. Normalized$(NB_{ij})_{q\times r}$ matrix is given in Table 3. By taking the complement of cost value and benefit value as presented. Here, C_4= medication is the cost value.

Case 1 The weight vector for this particular case is $q_1 = (0.4, 0.2, 0.2, 0.2)$

Step 3. Compute the value of N_i for the given data using NBWGM for each alternative. NBWGM(N_1) = {0.6068775, 0.342018,-0.6562456}, NBWGM(N_2) = {0.6127774, 0.3451749, 0.524532}, NBWGM(N_3) = {0.4804497, 0.4192962, −0.590466}.

Step 4. Estimate$\delta(N_i)$,$\delta(N_1) = 0.146325$, $\delta(N_2) = 0.216710$, $\delta(N_3) = 0.154640$.

Step 5. Arrange the N_i according to δ value to choose the appropriate assessment.$\delta(N_2) > \delta(N_3) > \delta(N_1)$.

Result. The persons with age group between 25 and 40 were affected due to high hereditary factor.

Case 2 The weight vector for this particular case is $q_2 = (0.2, 0.4, 0.2, 0.2)$. We can have the same data as given in step 1 and step 2. By arranging $\delta(N_2) > \delta(N_1) > \delta(N_3)$. *Result.* The persons with age group between 25 and 40 were highly influenced by the environment.

Case 3 The weight vector for this particular case is $q_2 = (0.2, 0.2, 0.4, 0.2)$. We can have the same data as given in step 1 and step 2. We get $\delta(N_2) > \delta(N_1) > \delta(N_3)$. The persons with age group between 25 and 40 were highly affected with physical complaints.

Case 4 The weight vector for this particular case is $q_4 = (0.2, 0.2, 0.2, 0.4)$. We can have the same data as given in step 1 and step 2. By arranging the $\delta(N_i)$, we get $\delta(N_2) > \delta(N_3) > \delta(N_1)$. In this case, the persons with age group between 25 and 40 were highly pretentious with medicines.

Conclusion Evidently the persons between 25 and 45 were affected with all the mentioned criteria. Since, they are at the risk side of life. Fuzziness on Neutral-Bipolar fuzzy set (NBFS) supports us to find the people who are to be cared first.

7 Conclusion

The new notion of Neutral-Bipolar fuzzy sets (NBFSs) has been introduced and it is defined with basic properties. Also, we have discussed the NBFS operations, score function value, accuracy value, and aggregated values. In addition, the demonstration of NBFS in bipolar disorder using the operators of NBFS was provided. In MCDM problem, Neutral-Bipolar fuzzy set supports us to find the desirable alternative. NBFS can be effective in the areas such as the neutral thought of human brains in certain situation, neutral results in medicines due to some health factor, neutral financial status in business, and neutral thoughts in decision-making. In forthcoming work, we can elongate the applications in various fields like neural networking and artificial intelligence.

References

1. Atanassov, K.T.: Intuitionistic Fuzzy Sets. Fuzzy Sets Syst. **20**(1), 87–96 (1986)
2. Cuong, B.C..: Picture fuzzy sets. (February 2015) (2016)
3. Garg, H.: Some picture fuzzy aggregation operators and their applications to multicriteria decision-making. Arab. J. Sci. Eng. **42**(12), 5275–5290 (2017)
4. Jesintha Rosline, J., Mike Dison, E.: Symmetric pentagonal fuzzy numbers. Int. J. Pure Appl. Math. **119**(9), 245–253 (2018)
5. Kirupa, A., Jesintha Rosline, J.: Domination in double layered fuzzy graphs. AIP Conference Proceedings. **020117**, 1–7 (2019)
6. Riaz, Muhammad, Harish Garg, Hafiz Muhammad, et al. 2021. Multi-criteria decision making based on bipolar picture fuzzy operators and new distance measures multi-criteria decision making based on bipolar picture fuzzy. (April).
7. Riaz, M., Garg, H., Farid, H.M.A., Chinram, R.: Multi-criteria decision making based on bipolar picture fuzzy operators and new distance measures. Comput. Model. Eng. Sci. **127**(2), 771–800 (2021)
8. Zhang, W-R.: YinYang bipolar fuzzy sets. (June 1998). (2016)
9. Zwick, R., Zimmermann, H.-J.: 106 The American Journal of Psychology. Fuzzy Set Theory and Its Applications. (1993)

Critical Path in an Intuitionistic Triangular Fuzzy Number for Time Cost Trade-Off in Project Network by the Modified Traditional Method

S. Priyadharshini and G. Deepa

1 Introduction

In the modern environment, it is difficult for project managers and decision-makers to influence the projects using effective project management strategies. However, because of the complexity of the few initiatives that were undertaken in the late 1950s, it was necessary to develop a new strategy that would be more appropriate and effective. Operations Research has adopted the Project Evaluation and Review Technique (PERT) and the Critical Path Method (CPM). While Morgan R. Walker and James E. Kelley were in the height of their success in 1957, the US Naval Forces were created [15]. In essence, PERT and CPM are time-oriented methodologies. The main distinction among CPM and PERT is as follows: how the time estimations were applied? Although they were deterministic in CPM, PERT assigns a probabilistic value to time. It was well known for being a vital tool for large-scale project planning and programming. The decision-maker may retain and exert control over the project's budget, schedule, and level of work with the use of the critical path method. This method is often employed in many industries to assess and improve a project's effectiveness [17]. Decision-makers and administrators require efficient project management tools to tackle initiatives in the competitive domain. The tactics that will enable projects to be finished and carried out within a given time range must be chosen by project managers [11]. Because of the unpredictable nature of the data and the inconsistent leveling for the desired task, Lofti Asker Zadeh created the fuzzy set principles, which are crucial in this type of decision-making environment [26]. The fuzzy critical path problem (FCP) had many solutions documented in the literature. Fuzzy Program Evaluation and

S. Priyadharshini · G. Deepa (✉)
Department of Mathematics, SAS, VIT, Vellore, Tamil Nadu, India
e-mail: deepa.g@vit.ac.in

F. Kamalov et al. (eds.), *Advances in Mathematical Modeling and Scientific Computing*, Trends in Mathematics, https://doi.org/10.1007/978-3-031-41420-6_49

Review Technique (FPERT), developed by Chanas and Kamburowski, was the first technique to identify an efficient path [7]. Instead of using FPERT, where the time to discover the critical path, decision-makers may use the Fuzzy Critical Path Method (FCPM) when they have deterministic data to locate a critical path. By utilizing octagonal fuzzy numbers, Stephen Dinakar and Rameshan [22] presented a method for analyzing the critical path. By using intuitionistic triangular fuzzy integers rather than precise values, Balaganesan and Ganesan [3] proposed a novel method for identifying the critical path for network diagrams.

Reference [14] proposed the novel JOSE Method, which has 13 parameters and a ranking mechanism known as the Euclidean ranking method, to find FCP. It was shown that triskaidecagonal and fuzzy triskaidecagonal numbers may be used to define the dynamic encoding recursion of the critical path. Reference [19] presented a unique analytical method for identifying critical path. They implemented, tabulated, and computed the findings for their new trapezoidal fuzzy number defuzzification method on the float time for each operation in the project network. Using table data, they were able to identify the crucial route [19]. They discovered the intuitionistic fuzzy critical path (IFCP) in an acyclic network and suggested the "Maximum edge distance" method [18] in terms of intuitionistic triangular fuzzy numbers. As a result, fuzzy critical path issues have been the subject of several papers. A few of them are [1, 10, 21].

As new firms enter the market and existing organizations increase their employment opportunities, the construction industry is growing more difficult every day. To gain an advantage over rivals, construction companies work to keep costs as low as feasible. To do this, however, precise scheduling and planning of construction projects are required. To shorten the project's duration, critical activities could be hastened for an additional fee. Crashing a critical activity causes the project to take longer and costs to increase. The indirect cost goes down as the project's duration gets shorter. The lowest sum of direct and indirect project costs is sought while solving Time Cost Trade-Off (TCT) problems. The relevance of the TCT problem was originally recognized by scholars Fulkerson (1961) [12] and Kelly (1961) [16] more than 50 years ago, almost concurrent with the development of project analytic tools. Numerous heuristic algorithms have been developed in an effort to discover the optimal solution to TCT problems. The heuristic method created by Siemens (1971) [20], which represented the first important attempt to address the TCT issue, was further refined by Siemens and Gooding, Goyal (1975) [13], and other researchers. The techniques provide the lowest project cost for a certain completion time. Additionally, Barber and Boardman (1988) [4] and Chiu and Chiu (2005) [8] proposed heuristic algorithms for the calculation of TCT using linear cost curves. For continuous crashing functions, the aforementioned algorithms provide the optimal solution, but they are unable to guarantee convergence to the global optimum for nonlinear or discrete crashing solutions. After considering the hazards connected to each activity, Siadat et al. provided a way to identify the critical route of the project network (2020) [9]. They used intervalued intuitionistic fuzzy. The method they proposed was tested for both its pace of convergence and its capacity to locate the global optimum. According to test findings (2017) [5], the algorithm visibly converges to optimum or very optimal solutions. In the realm of construction engineering and management, it was discovered that the MAWA-TLBO algorithm was effective for the TCTP (2018) [24]. To make the suggested model practical, they take into account a range of project coefficients. Last but not least, there are several approaches to add uncertainty to the problem (2018) [23]. They examined time resource-cost trade-off models for building projects in uncertain contexts (2021) using uncertain theory [25].

2 Preliminaries

2.1 *Fuzzy Set [FS] [26]*

Fuzzy sets are those whose bounds represent a degree of membership in [0, 1], then $X = \{(p, \alpha_A(p)): p \in P\}$, where X is a FS with ordered pair, P is a set, the membership function is defined as $\alpha_x: P \rightarrow [0, 1]$ and the degree of membership is indicated by $\alpha_x(p)$, for each element $p \in P$.

2.2 *Intuitionistic Fuzzy Set (IFS) [2]*

IFS X of the type $X = \{(p, \alpha_x(p), \gamma_x(p)): p \in P\}$, then the degree of membership is $\alpha_x: P \rightarrow [0, 1]$ and the degree of nonmembership is $\beta_x: P \rightarrow [0, 1]$, $0 \leq \alpha_x(p) + \beta_x(p) \leq 1$, of the element $p \in P$ to the set X, $\forall p \in P$.

2.3 *Fuzzy Number [26]*

If it satisfies, Normal, Piecewise continues and Convex then it is Fuzzy Number.

2.4 *Triangular Fuzzy Number [TFN] [26]*

A triplet (p, q, r; 1), where $p < q < r$; $p, q, r \in \Re$ used to construct a triangular fuzzy number, X. The membership function $\alpha_x(p)$ is (Fig. 1)

Fig. 1 TFN

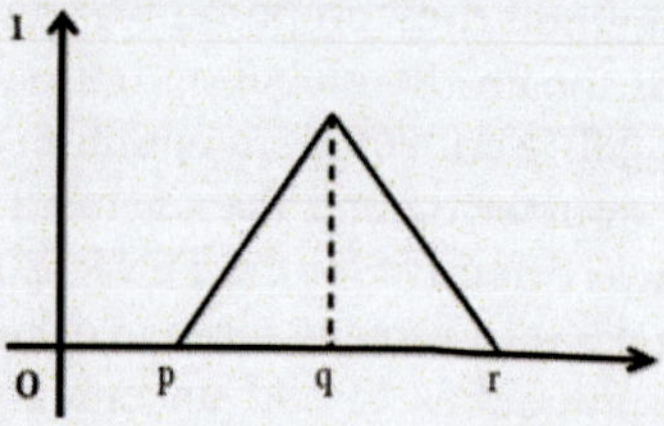

$$\alpha_x(p) = \begin{cases} \frac{x-p}{q-p}, & p \le x < q \\ 1, & x = q \\ \frac{r-x}{r-q}, & q < x \le r \\ 0, & otherwise. \end{cases}$$

2.5 Intuitionistic Triangular Fuzzy Number (ITFN) [2]

A triplet (p', q, r'; 1), which is used to define an ITFN, $p' < p < q < r < r'$; p', q, r' $\in \Re$. The following is the nonmembership function $\beta_x(p)$ (Fig. 2):

$$\beta_x(p) = \begin{cases} \frac{x-p'}{q-p'}, & p' \le x < q \\ 1, & x = q \\ \frac{r'-x}{r'-q}, & q < x \le r' \\ 0, & otherwise \end{cases}$$

2.6 Addition Operation on ITFN [2]

Consider $X = (p_1, q_1, r_1)\,(p'_1, q_1, r'_1)$ and $Y = (p_2, q_2, r_2)\,(p'_2, q_2, r'_2)$ be two ITFN, then

$$X \bigoplus Y = (p_1 + p_2, q_1 + q_2, r_1 + r_2)\left(p'_1 + p'_2, q_1 + q_2, r'_1 + r'_2\right).$$

Fig. 2 ITFN

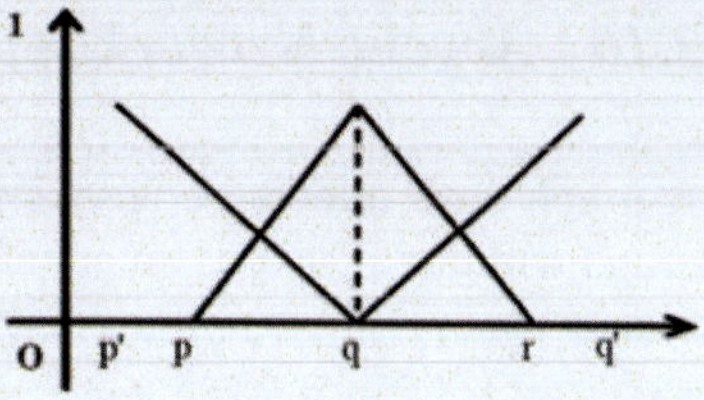

2.7 *Subtraction Operation on ITFN [2]*

Consider $X = (p_1, q_1, r_1)(p'_1, q_1, r'_1)$ and $Y = (p_2, q_2, r_2)(p'_2, q_2, r'_2)$ be two ITFN, then

$$X \ominus Y = (p_1 - r_2, q_1 - q_2, r_1 - p_2)\left(p'_1 - r'_2, q_1 - q_2, r'_1 - p'_2\right).$$

2.8 *Maximum Operation on ITFN [2]*

Consider $X = (p_1, q_1, r_1)(p'_1, q_1, r'_1)$ and $Y = (p_2, q_2, r_2)(p'_2, q_2, r'_2)$ be two ITFN, then

$$M_{max} = \max\ (X, Y)$$

$$= \left(\max\ (p_1, p_2), \max\ (q_1, q_2), \max\ (r_1, r_2)\right), \left(\max\ \left(p'_1, p'_2\right), \max\ (q_1, q_2), \max\ \left(r'_1, r'_2\right)\right).$$

2.9 *Minimum Operation on ITFN [2]*

Consider $X = (p_1, q_1, r_1)(p'_1, q_1, r'_1)$ and $Y = (p_2, q_2, r_2)(p'_2, q_2, r'_2)$ be two ITFN, then

$$M_{min} = \min\ (X, Y)$$

$$= \left(\min\ (p_1, p_2), \min\ (q_1, q_2), \min\ (r_1, r_2)\right), \left(\min\ \left(p'_1, p'_2\right), \min\ (q_1, q_2), \min\ \left(r'_1, r'_2\right)\right).$$

2.10 Acyclic Network [6]

A graph with directed edges is referred to as a digraph. Consequently, a directed graph without a cycle is an acyclic digraph.

3 Methodology

3.1 General Technique in ITFN for IFCP

Step 1:
Create an acyclic network called G(V, E), where V stands for vertices and E for edges. In a real context, each arc length or edge weight refers to the expense, duration, etc.
Step 2:
Find the suitable path lengths L_k, $k = 1$ to n and all viable pathways P_k, $k = 1$ to n from source I to destination "j." Let $L_k = (p_k, q_k, r_k)$.
Step 3:
Calculate the length of each potential route, $L_P = \sum_{\substack{ij \in p \\ p \in P_k}} d_{ij}$, where $k = 1$ to n. The path with the longest path length is found as the IFCP.

Notations
D_{ij} – Duration of the Activity from i to j
IE_{ij} – Intuitionistic Fuzzy Earliest Time
IL_{ij} – Intuitionistic Fuzzy Latest Time
IL_P – Intuitionistic Fuzzy Path Length
ITF_{ij} – Intuitionistic Fuzzy Total Float
IFP_P – Intuitionistic Fuzzy Float of the path
P_C – Project Cost
D_C – Direct Cost
I_C – Indirect Cost
N_T – Normal Time
C_T – Crash Time
N_C – Normal Cost
C_C – Crash Cost

Intuitionistic Fuzzy Earliest Time (IFET)
The time at the earliest in which an activity can begin is IFET, which is IE_{ij}. It is formulated by

$$IE_{ij} = \max_{p\in P_{1-k}} \{IL_P\} = \max_{p\in P_{1-k}} \sum_{ij \in p} d_{ij} \quad (1)$$

Intuitionistic Fuzzy Latest Time (IFLT)

The IFLT, which is determined by comparing the lengths of the paths from 1 to n and k to n, is the latest time at which a task may be finished. The Symbol for it is IL_{ij}. It is determined by

$$IL_{ij} = \max_{p\epsilon\ P_{1-n}}\{IL_P\} - \max_{p\epsilon\ P_{k-n}}\{IL_P\}$$

$$IL_{ij} = \max_{p\epsilon\ P_{1-n}} \sum_{ij\ \epsilon\ p} d_{ij} - \max_{p\epsilon\ P_{k-n}} \sum_{ij\ \epsilon\ p} d_{ij} \tag{2}$$

Intuitionistic Fuzzy Total Float (IFTF)

The variation between IFLT and IFET is the amount of time that an activity can be delayed without having an impact on the project's new deadline, which is denoted by ITF_{ij}. It is formulated as

$$ITF_{ij} = IL_{ij} \ominus IE_{ij}$$

$$ITF_{ij} = \max_{p\epsilon\ P_{1-n}} \sum_{ij\ \epsilon\ p} d_{ij} - \max_{p\epsilon\ P_{k-n}} \sum_{ij\ \epsilon\ p} d_{ij} - \max_{p\epsilon\ P_{1-k}} \sum_{ij\ \epsilon\ p} d_{ij} \tag{3}$$

Intuitionistic Fuzzy Path Float (IFPF)

The difference between the longest path ($L_P{}^*$) and the path's length, or IFPF, determines how long an activity can be postponed after starting early without affecting how long it will take to complete the project (L_P). The designation is IPFP. It is stated as follows:

$$IPF_p = L_P^* - L_P = \max_{p\epsilon\ P_{1-n}} \sum_{ij\ \epsilon\ p} d_{ij} - \sum_{\substack{ij\ \epsilon p \\ p\epsilon P_{1-n}}} d_{ij} \tag{4}$$

3.2 *Magnitude Measure for ITfn*

If (p,q,r)(p',q,r') be the intuitionistic triangular fuzzy number, then magnitude measure (Mag) is defined as

$$\text{Magnitude Measure} = \left\{\frac{p + 7q + r}{12}, \frac{2p' + 5q + 2r'}{12}\right\} \tag{5}$$

Theorem
In an Intuitionistic fuzzy environment, the Intuitionistic Fuzzy Total Float for a network is calculated as follows:

$$\mathrm{ITF}_{\mathrm{i,j}} = \min_{p\in P_{1-n}} \left\{ IPF_p \mid (i, j) \in p \right\} \tag{6}$$

Proof
Consider the R.H.S,

$$\Rightarrow \min_{p\in P_{1-n}} \left\{ IPF_p \mid (i, j) \in p \right\}$$

$$\Rightarrow \min_{p\in P_{1-n}} \left\{ \max_{p\epsilon\ P_{1-n}} \sum_{i,j\ \epsilon\ p} d_{i,j} - \sum_{\substack{i,\ j\ \epsilon p \\ p\epsilon P_{1-n}}} d_{i,j} \right\}$$

$$\Rightarrow \min_{p\in P_{1-n}} \left\{ \max_{p\epsilon\ P_{1-n}} \sum_{i,j\ \epsilon\ p} d_{i,j} + \left(- \sum_{\substack{i,\ j\ \epsilon p \\ p\epsilon P_{1-n}}} d_{i,j} \right) \right\}$$

$$\Rightarrow \min_{p\in P_{1-n}} \left\{ \max_{p\epsilon\ P_{1-n}} \sum_{i,\ j\ \epsilon\ p} d_{i,j} \right\} + \min_{p\in P_{1-n}} \left\{ - \sum_{\substack{i,\ j\ \epsilon p \\ p\epsilon P_{1-n}}} d_{i,j} \right\}$$

$$\Rightarrow \left\{ \max_{p\epsilon\ P_{1-n}} \sum_{i,\ j\ \epsilon\ p} d_{i,j} \right\} - \max_{p\in P_{1-n}} \left\{ \sum_{\substack{i,\ j\ \epsilon p \\ p\epsilon P_{1-n}}} d_{i,j} \right\}$$

$$\Rightarrow \left\{ \max_{p\epsilon\ P_{1-n}} \sum_{i,\ j\epsilon\ p} d_{i,j} \right\}$$

$$- \max_{p\in P_{1-k}} \left\{ \sum_{\substack{a,\ b\ \epsilon p \\ p\epsilon P_{1-k}}} d_{a,b} \right\} - \max_{p\in P_{k-n}} \left\{ \sum_{\substack{i,\ j\ \epsilon p \\ p\epsilon P_{k-n}}} d_{i,j} \right\}$$

$$\Rightarrow \left\{ \max_{p\epsilon\ P_{1-n}} \sum_{i,\ j\ \epsilon\ p} d_{i,j} \right\} - \max_{p\in P_{k-n}} \left\{ \sum_{\substack{i,\ j\ \epsilon p \\ p\epsilon P_{i-n}}} d_{i,j} \right\}$$

$$- \max_{p \in P_{1-k}} \left\{ \sum_{\substack{i,\, j\, \epsilon p \\ p \epsilon P_{1-k}}} d_{i,j} \right\}$$

$$\Rightarrow ITF_{i,j} \forall \text{i, j } \varepsilon \text{ p, p } \varepsilon \text{ P.}$$

$$\Rightarrow \text{L.H.S.}$$

Hence proved.

3.3 Algorithm

Step 1: Create the network. Determine the length of each path and all potential path by
$L_P = \sum_{ij\epsilon\ p} d_{ij}$
Step 2: The maximum path length (L_P^*) is calculated. $L_P^* = \max_{p \in P} \sum_{ij\epsilon\ p} d_{ij}$
Step 3: In the network for the all the path IPF_p is found. (Using Eq. (4))
Step 4: Using IFPF, calculate the ITF_{ij} for all the activity. (Using Eq. (6))
Step 5: Defuzzify the IFTF value using the magnitude measure formula. The defuzzified values that reach zero determine the critical activity and associated critical path.

3.4 Proposed Method to Find Time Cost Trade-Off Problem

Step 1: Using the aforementioned algorithm (3.2), identify the critical activity and the critical path.
Step 2: Determine the total project cost and total critical path duration. $P_C = D_c + (I_c * d_{ij})$
Step 3: Calculate the crash cost (ΔC) and the crash time (ΔT) to find the Cost Slope (C_s).
$C_s = \frac{\Delta T}{\Delta C}$; $\Delta T = N_T - C_T$; $\Delta C = C_C - N_C$
Step 4: Find the least C_s in the critical activities and crash that activity.
Step 5: Find the new Critical path and estimate the new project's cost using the formula.
$P_C = D_c + C_C$ of $C_A + (I_c * d_{ab})$

Step 6: Simultaneously, in the critical path, all the activities are crashed until the P_C increases.

Numerical Illustration (Table 1)
Construct an eight-way road from Salem to Chennai in Fig. 3, using FPCM with optimum cost (Fig. 4; Tables 2, 3 and 4).

Table 1 Description of Node in road map

Node	Description
1	Salem
2	Vellore
3	Polur
4	Thiruvanamalai
5	Cheyyar
6	Villupuram
7	Chennai

Fig. 3 Road map from Chennai to Salem

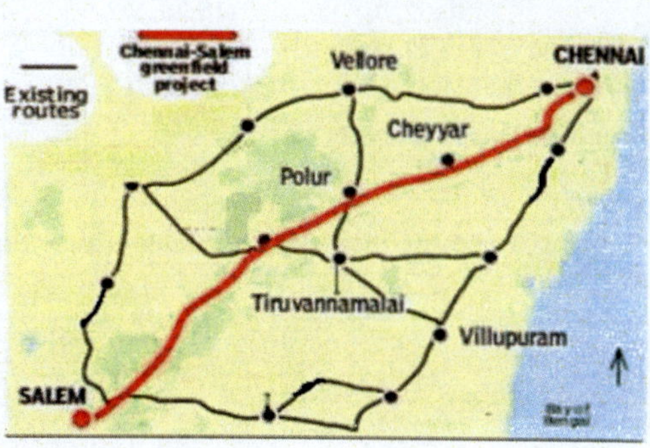

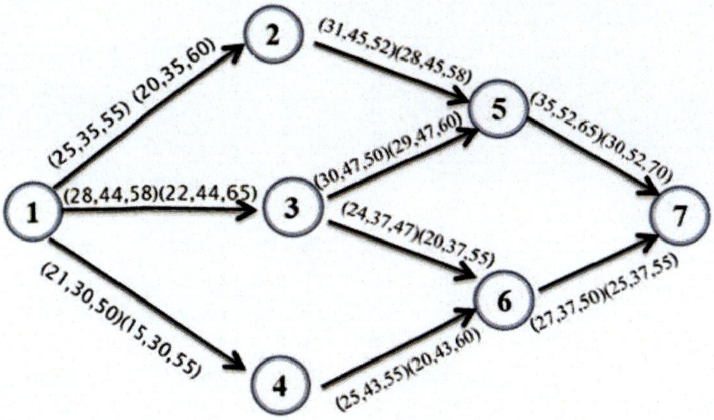

Fig. 4 Network diagram

Table 2 Duration and cost for construction

Activity	Duration	Crash time	Crash cost	Normal cost
1–2	(25,35,55) (20,35,60)	(13,14,15) (12,14,16)	(760,800,900) (750,800,1000)	(400,500,600) (300,500,700)
1–3	(28,44,58) (22,44,65)	(13,15,17) (10,15,20)	(600,700,900) (500, 700, 950)	(410,420,430) (400,420,450)
1–4	(21,30,50) (15,30,55)	(8,10,12) (5,10,15)	(800,900,950) (700,900,1000)	(600,650,680) (500,650,690)
2–5	(31,45,52) (28,45,58)	(7,10,13) (5,10,18)	(600,700,800) (500,700,900)	(200,300,400) (100,300,450)
3–5	(30,47,50) (29,47,60)	(10,15,20) (5,15,25)	(800,850,950) (700,850,1000)	(400,550,600) (350,550,650)
3–6	(24,37,47) (20,37,55)	(8,10,12) (5,10,15)	(900,950,1000) (800,950,1050)	(700,750,780) (650,750,790)
4–6	(25,43,55) (20,43,60)	(15,16,17) (12,16,18)	(600, 700, 800) (500,700,900)	(300,400,450) (200,400,480)
5–7	(35,52,65) (30,52,70)	(18,20,22) (10,20,24)	(700,800,900) (600, 800, 950)	(400,500,550) (350,500,580)
6–7	(27,37,50) (25,37,55)	(7,10,12) (5,10,15)	(1000,1100,1200) (900,1100,1500)	(700,800,850) (500,800,880)

Solution: Construct the network

Table 3 The defuzzified values for Table 2 using Eq. (5)

Activity	Duration (d_{ab})	Crash time	Crash cost	Normal cost
1–2	(27.083, 27.9167)	(10.5, 10.5)	(605, 610)	(375, 375)
1–3	(32.833, 32.67)	(11.25,11.25)	(533, 542)	(315, 315)
1–4	(23.416, 24.33)	(7.5, 7.5)	(671, 667)	(486, 484)
2–5	(33.167, 32.583)	(7.5, 7.5)	(525, 525)	(225, 225)
3–5	(34.0833,32.9167)	(11.25,11.25)	(642, 646)	(404, 396)
3–6	(27.5, 27.25)	(7.5, 7.5)	(713, 713)	(561, 559)
4–6	(31.75, 31.25)	(12, 12)	(525, 525)	(296, 292)
5–7	(38.67, 38.33)	(15, 15)	(600, 600)	(371, 367)
6–7	(28, 28.25)	(7.417, 7.333)	(825, 825)	(596, 592)

Table 4 IFPF

Path (Pp)	Path length (Lp)	$IPF_p = L_P^* - L_P$
P_1: 1-2-5-7	(98.920, 98.829)	(6.667, 5.087)
P_2: 1-3-5-7	**(105.586, 103.917)L_P^***	**(0, 0)**
P_3: 1-3-6-7	(88.333, 88.170)	(17.253, 15.747)
P_4: 1-4-6-7	(83.166, 83.830)	(22.420, 20.087)

Table 5 Critical path

Activity	IFTF
1–2	(6.667, 5.087)
1–3	**(0, 0)**
1–4	(22.420, 20.087)
2–5	(6.667, 5.087)
3–5	**(0, 0)**
3–6	(17.253, 15.747)
4–6	(22.420, 20.087)
5–7	**(0, 0)**
6–7	(17.253, 15.747)

Table 6 Cost slope

Activity	Δc	ΔT	Cost slope
1–2	(230, 235)	(16.583, 17.416)	(13.87, 13.49)
1–3	**(218, 227)**	**(21.583, 21.42)**	**(10.101, 10,598)**
1–4	(185, 183)	(15.916, 16.83)	(11.624, 10.873)
2–5	(300, 300)	(25.667, 25.083)	(11.688, 11.960)
3–5	**(238, 250)**	**(22.833, 21.667)**	**(10.424, 11.538)**
3–6	(152, 154)	(20, 19.75)	(7.6, 7.798)
4–6	(229, 233)	(19.75, 19.25)	(11.595, 12.104)
5–7	**(229, 233)**	**(23.67, 23.33)**	**(9.675, 9.987)**
6–7	(229,233)	(20.583, 20.92)	(11.126, 11.138)

Table 7 Crashing

Iteration	Crash	Crash duration	Crash cost
1	5–7 → (15, 15)	(90.586, 88.917)	(12,684, 12,497)
2	1–3 → (11.25, 11.25)	(79.336, 77.667)	(11,559, 11,372)
3	3–5 → (11.25, 11.25)	(68.086, 66.417)	(10,434, 10,247)

From Table 5, we conclude that the critical path is 1-3-5-7 that is Salem– Polur–Cheyyar–Chennai and path length that is total duration is (93, 143, 173) (81, 143, 195) (Tables 6 and 7).

Therefore, for constructing a road from Salem to Chennai is (105.586, 103.917) and crashed to (68.086, 66.417) and the total direct cost (14,184, 13,997), which is crashed to (10,434, 10,247).

Simulation

The suggested algorithm is developed in MATLAB. Figure 5a shows a network, Fig. 5b represents the IFCP, Fig. 5c demonstrates crashing and Fig. 5d for cost estimating.

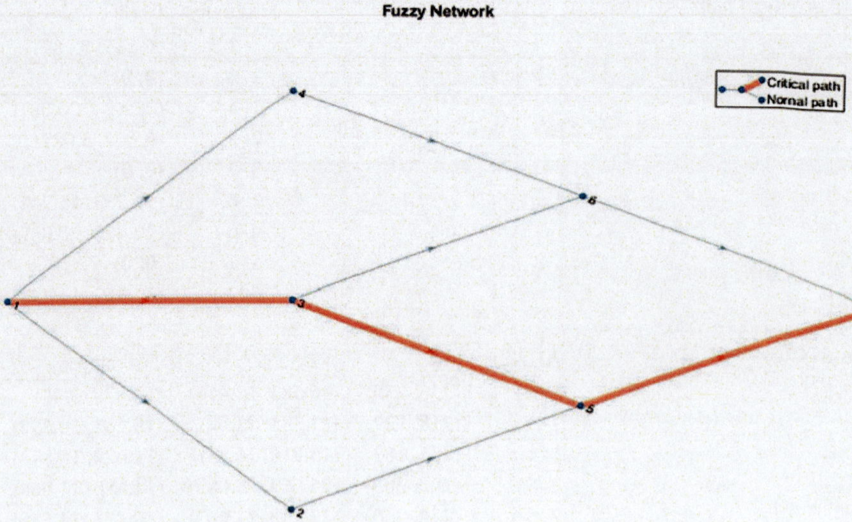

Fig. 5a A Network

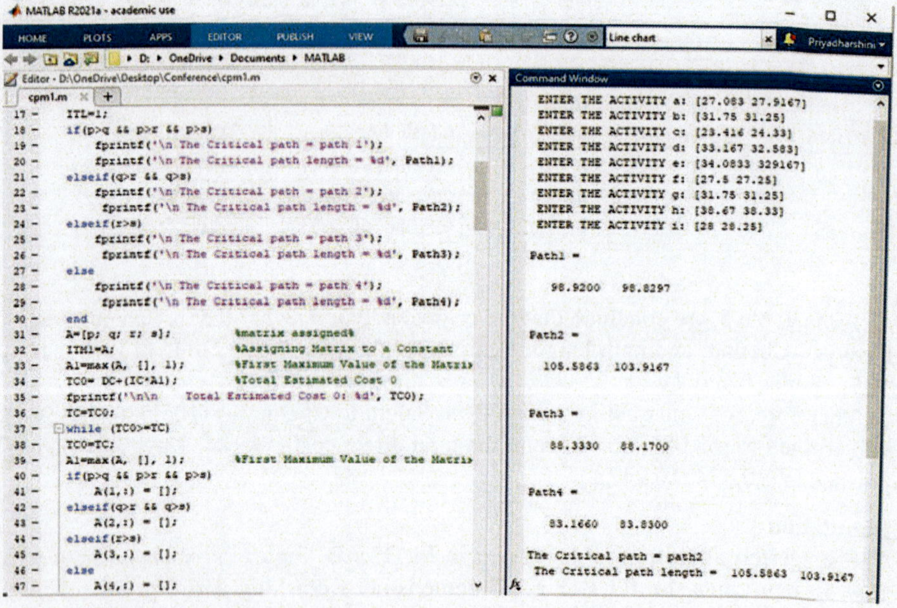

Fig. 5b IFCP representation

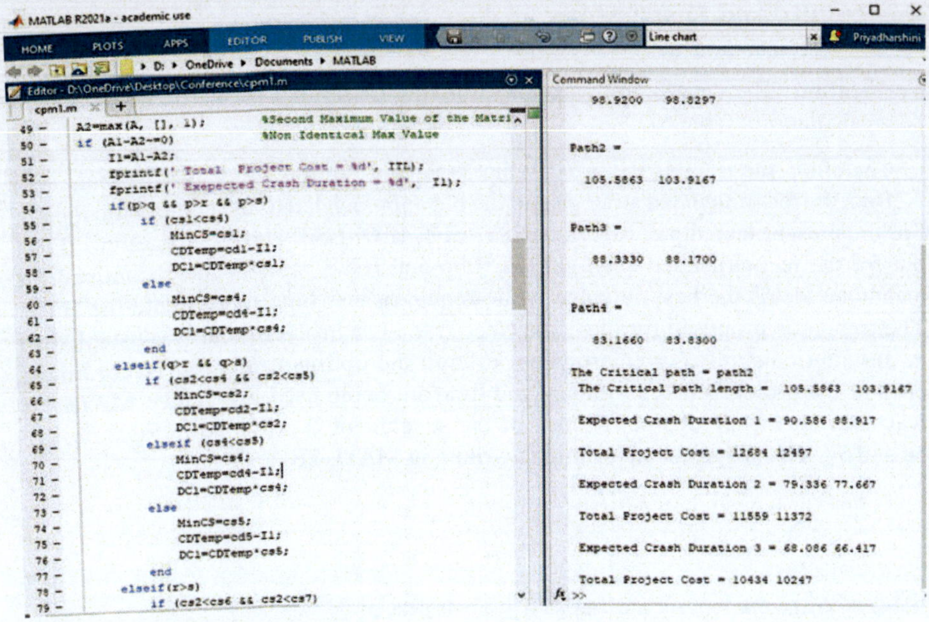

Fig. 5c Crashing demonstration

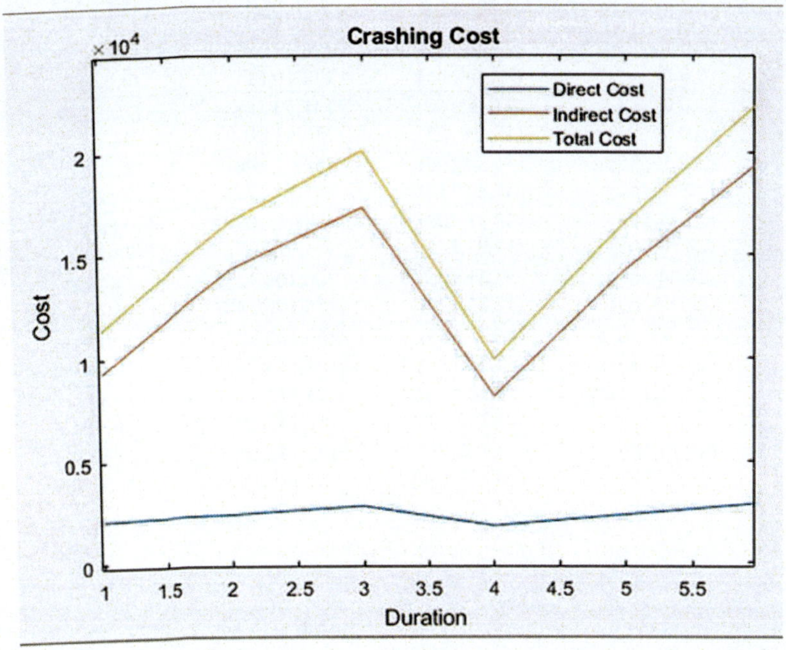

Fig. 5d Cost estimation

4 Results and Discussions

Verification The verification was done utilizing traditional forward and backward pass calculation (Table 8).

Therefore, the IFCP is P_2: 1-3-5-7 that is Salem–Polur–Cheyyar– Chennai.

Both the recommended strategy and the old approach led to the same critical path. For the present technique, calculate EST, EFT, LFT, LST, and TF to get the IFCP, but for the recommended way, get the IPFp and IFTF values. The recommended technique yields the best outcome while requiring less time to compute the IFCP. The technique proposed for the Time Cost Trade-Off Problem that was established in an intuitionistic fuzzy environment offered the optimum answer. Accordingly, 60% of the project's overall money and time are being used efficiently. MATLAB was used for the simulation. A copy of the attachment is provided in Figs. 5a, 5b, 5c and 5d. The suggested algorithm is written in MATLAB.

Table 8 Traditional forward and backward pass calculation

Activity	Earliest start Time (EST)	Earliest finish Time (EFT)	Latest finish Time (LFT)	Total Float (TF)
1–2	(0,0,0) (0,0,0)	(25, 35, 55) (20, 35, 60)	(27,46,56) (23,46,67)	(2,11,1) (3,11,7)
1–3	(0,0,0) (0,0,0)	(28,44,58) (22, 44, 65)	(28,44,58) (22,44,65)	**(0,0,0)** **(0,0,0)**
1–4	(0,0,0) (0,0,0)	(21, 30, 50) (15, 30, 55)	(41,63,58) (36,63,80)	(20,33,8) (21,33,25)
2–5	(25, 35, 55) (20, 35, 60)	(56,80,107) (48,80,118)	(58,91,108) (51,91,125)	(2,11,1) (3,11,7)
3–5	(28,44,58) (22, 44, 65)	(58,91,108) (51,91,125)	(58,91,108) (51,91,125)	**(0,0,0)** **(0,0,0)**
3–6	(28,44,58) (22, 44, 65)	(52,81,105) (42,81,120)	(66,106,123) (56,106,140)	(14,25,18) (14,25,20)
4–6	(21, 30, 50) (15, 30, 55)	(46,73,105) (35,73,110)	(66,106,123) (56,106,140)	(20,33,18) (21,33,30)
5–7	(58,91,108) (51,91,125)	(93,143,163) (81,143,195)	(93,143,163) (81,143,195)	**(0,0,0)** **(0,0,0)**
6–7	(46,73,105) (35,73,115)	(73,110,155) (60,110,170)	(93,143,163) (81,143,195)	(20,30,58) (20,30,25)

5 Conclusion

To plan and manage challenging projects, decision-makers might use intuitionistic fuzzy critical path and cost information. The decision-maker can model their project and define words like "maybe," "in between," "nearly," and other language variables for activity time, as opposed to clear models. This study offers a brand new method for crashing networks that makes use of intuitionistic fuzzy triangular numbers. Simulation result done using MATLAB reveals the same outcome. As a result, the methodology described in this study offers a means of obtaining the ideal cost and time in an intuitionistic context.

References

1. Anushya, V., Balasowandari, P.: Fuzzy critical path with area measure. Int. J. Pure Appl. Math. **118**(6), 167–173 (2018)
2. Atanassov, K.: Intuitionistic fuzzy sets. Fuzzy Sets Syst. **20**(1), 87–96 (1986). https://doi.org/10.1016/s0165-0114(86)80034-3
3. Balaganesan, M., Ganesan, K.: An approach to find critical path in intuitionistic fuzzy environment. Int. J. Pure Appl. Math. **119**(9), 395–403 (2018)
4. Barber, J.T., Boardman, J.T.: Knowledge-based project control employing heuristic optimisation. IEEE Proceedings A (Physical Science, Measurement and Instrumentation, Management and Education, Reviews). **135**(8), 529–538 (1988)
5. Bettemir, Ö.H., Talat Birgönül, M.: Network analysis algorithm for the solution of discrete time-cost trade-off problem. KSCE J. Civ. Eng. **21**(4), 1047–1058 (2017)
6. Blue, M., Bush, B., Puckett, J.: Unified approach to fuzzy graph problems. Fuzzy Sets Syst. **125**, 355–368 (2002)
7. Chanas, S., Kamburowski, J.: The use of fuzzy variables in PERT. Fuzzy Sets Syst. **5**, 11–19 (1981)
8. Chiu, Y.-S.P., Chiu, S.W.: Incorporating expedited time and cost of the end product into the product structure diagram. Int. J. Mach. Tools Manuf. **45**(7–8), 987–991 (2005)
9. Dorfeshan, Y., Siadat, A.: Determining project characteristics and critical path by a new approach based on modified NWRT method and risk assessment under an interval type-2 fuzzy environment. Sci. Iran. **26**(4) (2018)
10. Elizabeth, S., Sujatha, L.: Project scheduling method using triangular intuitionistic fuzzy numbers and triangular fuzzy numbers. Appl. Math. Sci. **9**(4), 185–198 (2015)
11. Elizabeth, S., Sujatha, L., Bharathi, K.: On searching critical path in a project network under fuzzy environment. In: National Conference on recent approaches of Mathematics to Science and Technology (NCRAM 2015)., ISBN:978-81-929362-1-5 (2015)
12. Fulkerson, D.R.: A network flow computation for project cost curves. Manag. Sci. **7**(2), 167–178 (1961)
13. Goyal, S.K.: Note—a note on "a simple cpm time-cost tradeoff algorithm". Manag. Sci. **21**(6), 718–722 (1975)
14. Jose Parvin Praveena, N., Sagaya Nathan Stalin, C., Rajkumar, A.: Critical path problem through intuitionistic triskaidecagonal fuzzy number using two different algorithms. Advances in Artificial Intelligence and Data Engineering. **1133**, 159–167 (2020)
15. Kelley, J., Walker, M.: Critical-Path Planning and Scheduling. In: *1959 Proceedings of the Eastern Joint Computer Conference* (1959)

16. Kelley, J.R., James, E.: Critical-path planning and scheduling: mathematical basis. Oper. Res. **9**(3), 296–320 (1961)
17. Newell, M., Grashina, M.: The project management question and answer book. Am. Manag. Assoc. **98** (2003)
18. Priyadharshini, S., Deepa, G.: Critical path interms of intuitionistic triangular fuzzy numbers using maximum edge distance method. Reliab. Theory Appl. **17.1**(67), 382–390 (2022)
19. Ravi Shankar, N., Sireesha, V., Phani Bushan Rao, P.: An analytical method for finding critical path in a fuzzy project network. Int. J. Contemp. Math. Sci. **5**(20), 953–962 (2010)
20. Siemens, N.: A simple CPM time-cost trade off algorithm. Manag. Sci. **17**(6), B-354 (1971)
21. Sireesha, V., Ravishankar, N.: A new approach to find total float time and critical path in a fuzzy project network. Int. J. Eng. Sci. Technol. **2**(4), 600–609 (2010)
22. Stephen Dinagar, D., Rameshan, N.: A new approach for fuzzy critical path method using octogonal fuzzy numbers. Int. J. Pure Appl. Math. **119**(13), 357–364 (2018)
23. Amiri, T., Javad, M.: Multi-project time-cost optimization in critical chain with resource constraints. KSCE J. Civ. Eng. **22**(10), 3738–3752 (2018)
24. Toğan, V., Azim Eirgash, M.: Time-cost trade-off optimization of construction projects using teaching learning based optimization. KSCE J. Civ. Eng. **23**(1), 10–20 (2019)
25. Toğan, V., Berberoğlu, N., Dede, T.: Optimizing of discrete time-cost in construction projects using new adaptive weight formulations. KSCE J. Civ. Eng. **26**(2), 511–521 (2022)
26. Zadeh, L.A.: Fuzzy sets. Inf. Control. **8**, 338–353 (1965)

Operations on Alternate Quadra–Submerging Polar (AQSP) Fuzzy Graphs and Its Applications

A. Anthoni Amali and J. Jesintha Rosline

1 Introduction

Lottfi A. Zadeh [13] has innovated Fuzzy set theory in 1965 as a tool to study uncertainty. Imprecision rose due to submerge conflict evolution, which has an impact in determining the fuzzy level fixed membership grades alternatively. The methods of fuzzy graphs and its logic allow for qualification of qualitative data modeling of complex fuzzy graph relationship and systems involving nonrandom uncertainty are introduced by Rosenfeld (2016) in 1975. Dissimilar categories of fuzzy graph correspondents and concepts were acquaint by Rosenfeld (Akram 2017), J.N. Mordeson (1993), P.S. Nair (Of and In 2010), D.S. Malik (Wang et al. 2009), K.R. Bhutani (Of and In 2010), M. Akram [1, 2, 3] T. Al-Hawary (2017) [4], A.N. Gani [6], M.S. Sunitha (2020), S. Mathew (1972), M.G. Karunambigai et al. [9]. Edification on fuzzy graph operations and possessions were discussed in a specific way by mathematicians A. Somasunaram (Jun 2014). Bai [5], Javaid [7], Jesintha Rosline [8], Pathinathan [10], Lin, J.L [11] and Mankena [12] have acquaint with global domination and fuzzy independent dominating set in a fuzzy graph.

2 Preliminaries

Definition of AQSP Fuzzy Graphs AQSP fuzzy graph is defined with crisp graph $G = (\sigma_{AQSP}, \mu_{AQSP})$,$G^* = \left(\sigma^*_{AQSP},\ \mu^*_{AQSP}\right)$ is as, V = $\{\sigma^P(x), \sigma^N(x), \rho^P(x), \rho^N(x)\}$ and E = $\{\mu^P(x, y), \mu^N(x, y), \gamma^P(x, y), \gamma^N(x, y)\}$ vertex, edge set. $\sigma_P = V \rightarrow [0, 1]$,

A. Anthoni Amali · J. Jesintha Rosline (✉)
Department of Mathematics, Auxilium College (Autonomous), Thiruvalluvar University, Serkadu, Vellore, Tamil Nadu, India

F. Kamalov et al. (eds.), *Advances in Mathematical Modeling and Scientific Computing*, Trends in Mathematics, https://doi.org/10.1007/978-3-031-41420-6_50

$\sigma_N = V \rightarrow [-1, 0]$, $\rho_P = d \mid 0.5, \sigma^P(x) \mid$ and $\rho_N = -d \mid -0.5, \sigma^N(x) \mid$. By Condition,

(i) $\mu^P(x, y) \leq \min(\sigma^P(x), \sigma^P(y))$,
(ii) $\mu^N(x, y) \geq \max(\sigma^N(x), \sigma^N(y))$.
(iii) $\gamma^P(x, y) \leq \min(\rho^P(x), \rho^P(y))$,
(iv) $\gamma^N(x, y) \geq \max(\rho^N(x), \rho^N(y))$.

It implies the result such as $-1 \leq \sigma^P(x) + \sigma^N(x) \leq 1$ and $|\rho^P(x) + \rho^N(x)| \leq 1$ with condition $0 \leq \sigma^P(x) + \sigma^N(x) + |\rho^P(x) + \rho^N(x)| \leq 2$.

3 Interface on Level of Fixation in AQSP Coordinates

Consider a situation in which the value is uncertain. Suppose, the mid-values $\{[0, 0], [0, 0.5], [0.5, 0.5], [0.5, 0]\}$ and $\{[0, 0], [-0.5, 0], [-0.5, -0.5], [0, -0.5]\}$ is positive and negative membership of conflicts in human thoughts processes, the Fuzzy set of $\{[0, 0], [0, 1], [1, 1], [1, 0]\}$ and $\{[0, 0], [-1, 0], [-1, -1], [0, -1]\}$. AQSP Fuzzy is submerging with fixation $[-0.5, 0] \subseteq [-1, 0]$, $[-0.5, 0.5] \subseteq [-1, 1]$ and $[0, 0.5] \subseteq [0, 1]$, $[0.5, 0.5] \subseteq [1, 1]$.$[-0.5, 0.5] \subseteq [-1, 1]$ are finite. AQSP Fuzzy graph is useful, to consider that of maximum values [−1, 1]. By using the max – min values as] − 0.5, 0.5] or (−0.5, 0.5]: [−0.5, 0.5[or [−0.5, 0.5) and]−0.5, 0.5[or (−0.5, 0.5), which is open at the left and at the right. Note that the membership values $x \in [0.5, 0.5] \subseteq [1, 1]$, which denotes that "*x*" is an uncertain value.

4 Operations on AQSP Fuzzy Graphs

4.1 Cartesian Product of AQSP Fuzzy Graphs

Let $\sigma_1 = \{\sigma_1^P(x), \sigma_1^N(x), \rho_1^P(x), \rho_1^N(x)\}$ and $\sigma_2 = \{\sigma_2^P(x), \sigma_2^N(x), \rho_2^P(x), \rho_2^N(x)\}$ be two AQSP Fuzzy vertex sets V_1 *and* V_2. Then the Fuzzy edge sets of (E_1, E_2) is given as$\mu_1 = \{\mu_1^P(x, y), \mu_1^N(x, y), \gamma_1^P(x, y), \gamma_1^N(x, y)\}$ and the other pair is denoted as$\mu_2 = \{\mu_2^P(x, y), \mu_2^N(x, y), \gamma_2^P(x, y), \gamma_2^N(x, y)\}$. We define the Cartesian product of AQSP, $G_1 \times G_2 = \{(\sigma_1, \rho_1) \times (\sigma_2, \rho_2), (\mu_1, \gamma_1) \times (\mu_2, \gamma_2)\}$ as the product of two AQSP, where $\mu_1 \subseteq ((\sigma_1, \rho_1) \times (\sigma_1, \rho_1))$ and $\mu_2 \subseteq ((\sigma_2, \rho_2) \times (\sigma_2, \rho_2))$.

4.2 Proposition

The Cartesian product of AQSP $G_1 = (\sigma_{AQSP}, \mu_{AQSP})$, $G_2 = (\sigma_{AQSP}, \mu_{AQSP})$ is of the crisp fuzzy graphs $G_1{}^* = \left(\sigma^*_{AQSP}, \mu^*_{AQSP}\right)$ and $G_2{}^* = \left(\sigma^*_{AQSP}, \mu^*_{AQSP}\right)$. We prove that $G_1 \times G_2$, such as $\rho^P(x) = d \mid 0.5, \sigma^P(x) \mid$ and $\rho^P(y) = d \mid 0.5, \sigma^P(y) \mid$.

Proof

(i) $\sigma_1^P(x_1) \times \sigma_2^P(x_2) = (\sigma_1^P(x_1) \bigwedge \sigma_2^P(x_2))$,

$$\sigma_1^N(x_1) \times \sigma_2^N(x_2) = (\sigma_1^N(x_1) \bigvee \sigma_2^N(x_2))$$

$$\rho_1^P(x_1) \times \rho_2^P(x_2) = (\rho_1^P(x_1) \bigwedge \rho_2^P(x_2)),$$
$$\rho_1^N(x_1) \times \rho_2^N(x_2) = (\rho_1^N(x_1) \bigvee \rho_2^N(x_2))$$

for all values of $(x_1, x_2) \in (\sigma_{AQSP}, \rho_{AQSP})$.

(ii) $\mu_1^P(x, x_2) \times \mu_2^P(x, y_2) = \left(\sigma_1^P(x) \wedge \mu_2^P(x_2, y_2)\right)$,

$$\mu_1^N(x, x_2) \times \mu_2^N(x, y_2) = \left(\sigma_1^N(x) \bigvee \mu_2^N(x_2, y_2)\right),$$

$$\gamma_1^P(x, x_2) \times \gamma_2^P(x, y_2) = \left(\rho_1^P(x) \bigwedge \gamma_2^P(x_2, y_2)\right),$$

$$\gamma_1^N(x, x_2) \times \gamma_2^N(x, y_2) = \left(\gamma_1^N(x) \bigvee \gamma_2^N(x_2, y_2)\right),$$

for all values of $x \in (\sigma_1, \rho_1)$ *and* $(x_2, y_2) \in (\mu_2, \gamma_2)$.

(iii) $\mu_1^P(x_1, z) \times \mu_2^P(x, y_2) = \left(\sigma_1^P(x_1, y_1) \wedge \sigma_2^P(z)\right)$,

$$\mu_1^N(x_1, z) \times \mu_2^N(y_1, z) = \left(\sigma_1^N(x_1, y_1) \bigvee \sigma_2^N(z)\right),$$

$$\forall z \in (\sigma_2, \rho_2), (x_1, y_1) \in (\mu_1, \gamma_1).$$

Fig. 1 $(G_1 \times G_2) = (\sigma_{AQSP}, \mu_{AQSP})$

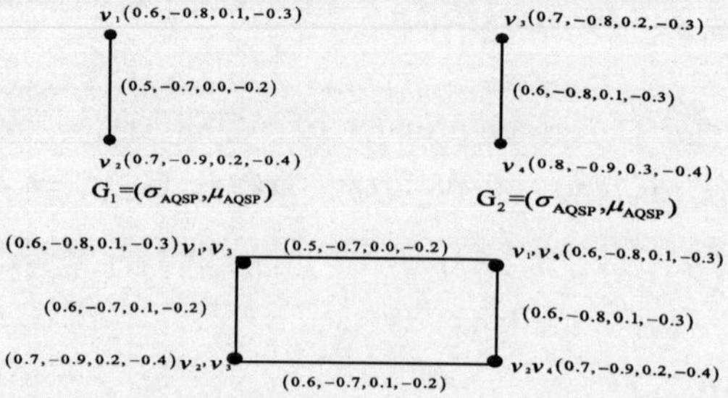

Similarly, we get results for γ_1^P, γ_1^N. Hence, the proof is completed.

4.3 Theorem

The Cartesian product of two AQSP fuzzy graphs is also an AQSP fuzzy graph is given with the result such as, $G_1 \times G_2 = \{(\sigma_1, \rho_1) \times (\sigma_2, \rho_2), (\mu_1, \gamma_1) \times (\mu_2, \gamma_2)\}$.

Proof Let $x \in (\sigma_1., \rho_1)$ *and* $(x_2, y_2) \in (\mu_2, \gamma_2)$ then we have the following:

(i) $(\mu_1^P \times \mu_2^P)((x,x_2),(x,y_2)) = \wedge\left\{\left(\sigma_1^P \times \sigma_2^P\right)(x,x_2),\ \sigma_1^P \times \sigma_2^P\right)(x,y_2)$

(ii) $(\mu_1^N \times \mu_2^N)((x,x_2),(x,y_2)) = \vee\left\{\left(\sigma_1^N \times \sigma_2^N\right)(x,x_2),\ \sigma_1^N \times \sigma_2^N\right)(x,y_2)$,

similarly we get same results for γ_1^P, γ_1^N.

Let $z \in (\sigma_2, \rho_2)\ (x_1, y_1) \in (\mu_1, \gamma_1)$ then we define the following,

(iii) $(\mu_1^P \times \mu_2^P)((x_1,z),(y_1,z)) = \wedge\left\{\left(\sigma_1^P \times \sigma_2^P\right)(x_1,z),\ \sigma_1^P \times \sigma_2^P\right)(y_1,z)$

(iv) $(\mu_1^N \times \mu_2^N)((x_1,z),(y_1,z)) = \vee\left\{\left(\sigma_1^N \times \sigma_2^N\right)(x_1,z),\ \sigma_1^N \times \sigma_2^N\right)(y_1,z)$

Similarly, we get results for γ_1^P, γ_1^N from the Fig. 1, hence the above results complete the proof.

4.4 Composition of AQSP Fuzzy Graphs

Let $\sigma_1 = \left\{\sigma_1^P(x), \sigma_1^N(x), \rho_1^P(x), \rho_1^N(x)\right\}, \sigma_2\left\{\sigma_2^P(x), \sigma_2^N(x), \rho_2^P(x), \rho_2^N(x)\right\}$ be AQSP Fuzzy vertex sets of V_1 *and* V_2. Then the Fuzzy edge sets of (E_1, E_2) is given as follows: $\mu_1 = \left\{\mu_1^P(x,y), \mu_1^N(x,y), \gamma_1^P(x,y), \gamma_1^N(x,y)\right\}$,

$$\mu_2 = \left\{\mu_2^P(x,y), \mu_2^N(x,y), \gamma_2^P(x,y), \gamma_2^N(x,y)\right. .$$

We define $G_1[\ G_2] = \{[(\sigma_1, \rho_1) \circ (\sigma_2, \rho_2)],\ [(\mu_1, \gamma_1) \circ (\mu_2, \gamma_2)]\ \}$ where we have denoted that $\mu_1 \subseteq ((\sigma_1, \rho_1) \circ (\sigma_1, \rho_1))$ and $\mu_2 \subseteq ((\sigma_2, \rho_2) \circ (\sigma_2, \rho_2))$.

4.5 Proposition

The composition of AQSP $G_1 = (\sigma_{AQSP}, \mu_{AQSP})$, $G_2 = (\sigma_{AQSP}, \mu_{AQSP})$, is of the crisp graphs $G_1{}^* = \left(\sigma^*_{AQPFG}, \mu^*_{AQPFG}\right)$, $G_2{}^* = \left(\sigma^*_{AQPFG}, \mu^*_{AQPFG}\right)$. Then we prove that $(G_1 \circ G_2)$ with $\rho^P(x) = d$ ı 0.5, $\sigma^P(x)$ ı and $\rho^P(y) = d$ ı 0.5, $\sigma^P(y)$ ı .

Proof

(i) $\sigma_1^P(x_1) \circ \sigma_2^P(x_2) = (\sigma_1^P(x_1) \bigwedge \sigma_2^P(x_2))$,

$$\sigma_1^N(x_1) \circ \sigma_2^N(x_2) = (\sigma_1^N(x_1) \bigvee \sigma_2^N(x_2)),$$

$$\rho_1^P(x_1) \circ \rho_2^P(x_2) = (\rho_1^P(x_1) \bigwedge \rho_2^P(x_2)),$$

$$\rho_1^N(x_1) \circ \rho_2^N(x_2) = (\rho_1^N(x_1) \bigvee \rho_2^N(x_2)),$$

for all values of $(x_1, x_2) \in (\sigma_{AQSP}, \rho_{AQSP})$.

(ii) $\mu_1^P(x, x_2) \circ \mu_2^P(x, y_2) = \left(\sigma_1^P(x) \bigwedge \mu_2^P(x_2, y_2)\right)$,

$$\mu_1^N(x, x_2) \circ \mu_2^N(x, y_2) = \left(\sigma_1^N(x) \bigvee \mu_2^N(x_2, y_2)\right),$$

for all values of $x \in (\sigma_1, \rho_1)$ *and* $(x_2, y_2) \in (\mu_2, \gamma_2)$.

(iii) $\mu_1^P(x_1, z) \circ \mu_2^P(x, y_2) = \left(\sigma_1^P(x_1, y_1) \bigwedge \sigma_2^P(z)\right)$,

$$\mu_1^N(x_1, z) \circ \mu_2^N(y_1, z) = \left(\sigma_1^N(x_1, y_1) \bigvee \sigma_2^N(z)\right),$$

for all values of $z \in (\sigma_2, \rho_2)$ *and* $(x_1, y_1) \in (\mu_1, \gamma_1)$.

(iv) $\mu_1^P(x_1, x_2) \circ \mu_2^P(y_1, y_2) = \bigwedge\left(\sigma_2^P(x_2), \sigma_2^P(y_2), \mu_1^P(x_1, y_1)\right)$

Fig. 2 $G_1[G_2]$

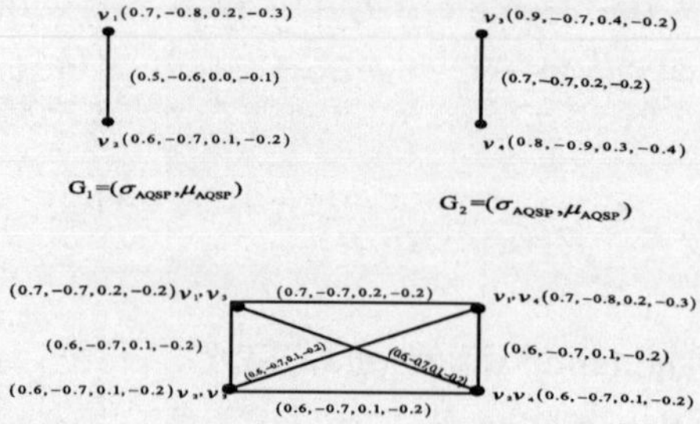

$$\mu_1^N(x_1, x_2) \circ \mu_2^N(y_1, y_2) = \bigvee\left(\sigma_2^N(x_2), \sigma_2^N(y_2), \mu_1^N(x_1, y_1)\right),$$

similarly we get γ_1^P, γ_1^N. for all values of $(x_1, x_2)(y_1, y_2) \in (\mu^0, \gamma^0) - (\mu, \gamma)$.

Where $\mu^0 = \mu \cup \{((x_1, x_2)(y_1, y_2))|\ (x_1, x_2) \in \mu_1,\quad x_2 \neq y_2\}$ and $\gamma^0 = \gamma \cup \{((x_1, x_2)(y_1, y_2))|\ (x_1, x_2) \in \gamma_1,\quad x_2 \neq y_2\}$. Hence $G_1 \circ G_2 = \{(\sigma_1, \rho_1) \circ (\sigma_2, \rho_2),\ (\mu_1, \gamma_1) \circ (\mu_2, \gamma_2)\}$ is proved.

4.6 Theorem

The composition of two AQSP fuzzy graphs proves that the following *function* $G_1[G_2] = \{[(\sigma_1, \rho_1) \circ (\sigma_2, \rho_2)],\ [(\mu_1, \gamma_1) \circ (\mu_2, \gamma_2)]\}$ is also AQSP fuzzy graphs. And we have denoted as $\mu_1 \subseteq ((\sigma_1, \rho_1) \circ (\sigma_1, \rho_1))$ and $\mu_2 \subseteq ((\sigma_2, \rho_2) \circ (\sigma_2, \rho_2))$ respectively.

Proof Let $x \in (\sigma_1, \rho_1)$ *and* $(x_2, y_2) \in (\mu_2, \gamma_2)$ then we have,

(i) $\mu_1^P \circ \mu_2^P)((x, x_2), (x, y_2)) = \bigwedge\left\{\left(\sigma_1^P \circ \sigma_2^P\right)(x, x_2),\ \left(\sigma_1^P \circ \sigma_2^P\right)(x, y_2)\right\}$

(ii) $(\mu_1^N \circ \mu_2^N)((x, x_2), (x, y_2)) = \bigvee\left\{\left(\sigma_1^N \circ \sigma_2^N\right)(x, x_2),\ \left(\sigma_1^N \circ \sigma_2^N\right)(x, y_2)\right\}$,

similarly we get results for γ_1^P, γ_1^N. Let $z \in (\sigma_2, \rho_2)\ (x_1, y_1) \in (\mu_1, \gamma_1)$,

(iii) $(\mu_1^P \circ \mu_2^P)((x_1, z), (y_1, z)) = \bigwedge\left\{\left(\sigma_1^P \circ \sigma_2^P\right)(x_1, z),\ \left(\sigma_1^P \circ \sigma_2^P\right)(y_1, z)\right\}$

(iv) $(\mu_1^N \circ \mu_2^N)((x_1, z), (y_1, z)) = \bigvee\left\{\left(\sigma_1^N \circ \sigma_2^N\right)(x_1, z),\ \left(\sigma_1^N \circ \sigma_2^N\right)(y_1, z)\right\}$,

Similarly, we get results for γ_1^P, γ_1^N; hence, the proof is finalized (Fig. 2).

4.7 Union of Two AQSP Fuzzy Graphs

Let $\sigma_1=\left\{\sigma_1^P(x),\sigma_1^N(x),\rho_1^P(x),\rho_1^N(x)\right\}$, $\sigma_2=\left\{\sigma_2^P(x),\sigma_2^N(x),\rho_2^P(x),\rho_2^N(x)\right\}$ be two AQSP Fuzzy vertex sets of V_1 *and* V_2. Then the Fuzzy edge sets of (E_1,E_2) is given as $\mu_1=\left\{\mu_1^P(x,y),\mu_1^N(x,y),\gamma_1^P(x,y),\gamma_1^N(x,y)\right\}$ and
$\mu_2=\left\{\mu_2^P(x,y),\mu_2^N(x,y),\gamma_2^P(x,y),\gamma_2^N(x,y)\right\}$. We define as,
$G_1\cup G_2=\{[(\sigma_1,\rho_1)\cup(\sigma_2,\rho_2)],\ [(\mu_1,\gamma_1)\cup(\mu_2,\gamma_2)]\}$ where we have denoted that $\mu_1\subseteq((\sigma_1,\rho_1)\cup(\sigma_1,\rho_1))$ and $\mu_2\subseteq((\sigma_2,\rho_2)\cup(\sigma_2,\rho_2))$, respectively.

Proposition The Union of two AQSP fuzzy graphs $G_1=(\sigma_{AQSP},\mu_{AQSP})$ and $G_2=(\sigma_{AQSP},\mu_{AQSP})$, crisp graphs$G_1{}^*=\left(\sigma^*_{AQSP},\ \mu^*_{AQSP}\right)$, $G_2{}^*=\left(\sigma^*_{AQSP},\ \mu^*_{AQSP}\right)$. Then we prove that AQSP Fuzzy graph $G_1\cup G_2$ with the sufficient condition

$$\rho^P(x)=d\mid 0.5,\sigma^P(x)\mid\quad\text{and}\ \rho^P(y)=d\mid 0.5,\sigma^P(y)\mid.$$

Proof

(i) $\left(\sigma_1^P\cup\sigma_2^P\right)(x)=\sigma_1^P(x)\ if\ x\in\sigma_1\cap\sigma_2'$ and

$$\left(\sigma_1^P\cup\sigma_2^P\right)(x)=\sigma_2^P(x)\ if\ x\in\sigma_2\cap\sigma_1'$$

$$\left(\sigma_1^P\cup\sigma_2^P\right)(x)=\bigvee\left(\sigma_1^P(x),\sigma_2^P(x),\ \ if\ x\in\sigma_1\cap\sigma_2\right.$$

(ii) $\left(\sigma_1^N\cup\sigma_2^N\right)(x)=\sigma_1^N(x)\ if\ x\in\sigma_1\cap\sigma_2'$ and

$$\left(\sigma_1^N\cup\sigma_2^N\right)(x)=\sigma_2^N(x)\ if\ x\in\sigma_2\cap\sigma_1'$$

$$\left(\sigma_1^N\cup\sigma_2^N\right)(x)=\bigwedge\left(\sigma_1^N(x),\sigma_2^N(x),\ \ if\ x\in\sigma_1\cap\sigma_2\right.$$

Similarly, we get results for ρ_1^P, ρ_1^N. For the edge set membership function is as

(iii) $\left(\mu_1^P\cup\mu_2^P\right)(x,y)=\mu_1^P\ (x,y)\ if\ \ (x,y)\in\mu_1\cap\mu_2'$

$$\left(\mu_1^P\cup\mu_2^P\right)(x,y)=\mu_2^P\ (x,y)\ if\ \ (x,y)\in\mu_2\cap\mu_1'$$

$$\left(\mu_1^P \cup \mu_2^P\right)(x, y) = \bigvee \left(\mu_1^P\ (x, y), \mu_2^P\ (x, y),\ if\ (x, y) \in \mu_1 \cap \mu_2\right)$$

(iv) $\left(\mu_1^N \cup \mu_2^N\right)(x, y) = \mu_1^N\ (x, y)\ if\ (x, y) \in \mu_1 \cap \mu_2'$

$$\left(\mu_1^N \cup \mu_2^N\right)(x, y) = \mu_2^N\ (x, y)\ if\ (x, y) \in \mu_2 \cap \mu_1'$$

$$\left(\mu_1^N \cup \mu_2^N\right)(x, y) = \bigwedge \left(\mu_1^N\ (x, y), \mu_2^N\ (x, y)\right),\ if\ (x, y) \in \mu_1 \cap \mu_2,$$

Similarly, we get results for γ_1^P, γ_1^N.

4.8 Theorem

The Union of two AQSP fuzzy graphs is an AQSP fuzzy graphs is given with the result, $G_1 \cup G_2 = \{(\sigma_1, \rho_1) \cup (\sigma_2, \rho_2),\ (\mu_1, \gamma_1) \cup (\mu_2, \gamma_2)\}$.

Proof Let$(x, y) \in \mu_1 \cap \mu_2$, then we have,

(i) $\left(\mu_1^P \cup \mu_2^P\right)(x, y) = \bigwedge\ \ \sigma_1^P \cup \sigma_2^P)(x),\ \sigma_1^P \cup \sigma_2^P)(y))$

(ii) $\left(\mu_1^N \cup \mu_2^N\right)(x, y) = \bigvee ((\sigma_1^N \cup \sigma_2^N)(x),\ \sigma_1^N \cup \sigma_2^N)(y))$

Similarly, we get results for γ_1^P, γ_1^N. And we prove that if, $(x, y) \in \mu_1 \cap \overline{\mu}_2$

(iii) $(\mu_1^P \cup \mu_2^P)(x, y) = \mu_1^P\ (x, y)) \leq (\bigwedge (\sigma_1^P\ (x), \sigma_1^P\ (y)),$

$$= \bigwedge \left(\left(\sigma_1^P \cup \sigma_2^P\right)(x),\ \left(\sigma_1^P(x) \cup \sigma_2^P(y)\right)\right), If\ (x, y) \in (\sigma_1 \cap \sigma_2).$$

(iv) $(\mu_1^N \cup \mu_2^N)(x, y) = \mu_1^N\ (x, y)) \geq \bigvee (\sigma_1^N\ (x), \sigma_1^N\ (y)))$

$= \bigvee ((\sigma_1^N \cup \sigma_2^N)(x),\ \ \sigma_1^N \cup \sigma_2^N)(x, y)),\ if\ x \in (\sigma_1 \cap \overline{\sigma}_2), y \in (\sigma_1 \cap \sigma_2)$, this implies the following, $\geq \bigvee (\bigwedge (\sigma_1^N(x), \sigma_2^N(x))\ \bigwedge (\sigma_1^N(y), \sigma_2^N)$ (y)), $= \bigvee \left(\ (\sigma_1^N \cup \sigma_2^N)(x),\ \ (\sigma_1^N(x) \cup \sigma_2^N(y)\right)$, *If* $(x, y) \in (\sigma_1 \cap \sigma_2)$, similarly we get results for γ_1^P, γ_1^N. Again if we consider $(x, y) \in \mu_1' \cap \mu_2$ then we prove the results (Fig. 3).

Fig. 3 $(G_1 \cup G_2)$

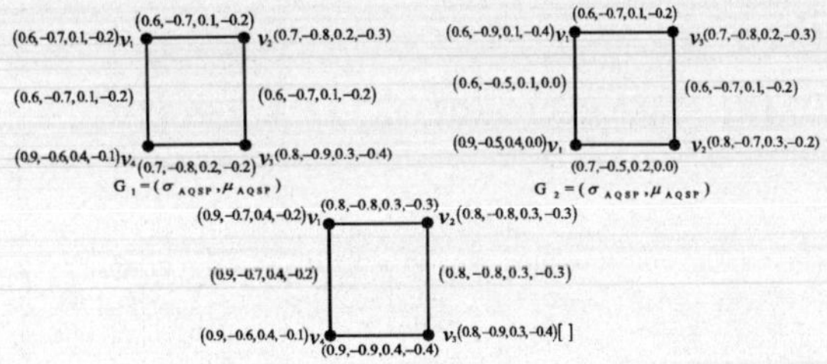

$$\left(\mu_1^P \cup \mu_2^P\right)(x, y) = \bigwedge\left(\left(\sigma_1^P \cup \sigma_2^P\right)(x),\ \left(\sigma_1^P \cup \sigma_2^P\right)(y)\right)$$

$$\left(\mu_1^N \cup \mu_2^N\right)(x, y) = \bigvee\left(\left(\sigma_1^N \cup \sigma_2^N\right)(x), \left(\sigma_1^N \cup \sigma_2^N\right)(y)\right)$$

$$\left(\gamma_1^P \cup \gamma_2^P\right)(x, y) = \bigwedge\left(\left(\rho_1^P \cup \rho_2^P\right)(x), \left(\rho_1^P \cup \rho_2^P\right)(y)\right)$$

$$\left(\gamma_1^N \cup \gamma_2^N\right)(x, y) = \bigvee\left(\left(\rho_1^N \cup \rho_2^N\right)(x), \left(\rho_1^N \cup \rho_2^N\right)(y)\right)$$

4.9 Join of Two AQSP Fuzzy Graphs

Let $\sigma_1 .= \left\{ \sigma_1^P(x), \sigma_1^N(x), \rho_1^P(x), \rho_1^N(x)\right\}$, $\sigma_2 = \left\{ \sigma_2^P(x), \sigma_2^N(x), \rho_2^P(x), \rho_2^N(x)\right\}$ be two AQSP Fuzzy vertex sets of V_1 *and* V_2. Then the Fuzzy edge sets of (E_1, E_2) is given as $\mu_1 = \left\{ \mu_1^P(x, y), \mu_1^N(x, y), \gamma_1^P(x, y), \gamma_1^N(x, y)\right\}$

$$\mu_2 = \left\{ \mu_2^P(x, y), \mu_2^N(x, y), \gamma_2^P(x, y), \gamma_2^N(x, y)\right\}.$$

We define $G_1 + G_2 = \{[(\sigma_1, \rho_1) + (\sigma_2, \rho_2)],\ [(\mu_1, \gamma_1) + (\mu_2, \gamma_2)]\}$.

4.10 Proposition

Join of two AQSP fuzzy graphs $G_1 = (\sigma_{AQSP}, \mu_{AQSP})$, $G_2 = (\sigma_{AQSP}, \mu_{AQSP})$ is of the crisp graphs $G_1{}^* = \left(\sigma^*_{AQSP},\ \mu^*_{AQSP}\right)$ and$G_2{}^* = \left(\sigma^*_{AQSP},\ \mu^*_{AQSP}\right)$. We prove $G_1 + G_2$ with condition, $\rho^P(x) = d \mid 0.5, \sigma^P(x) \mid$, $\rho^P(y) = d \mid 0.5, \sigma^P(y) \mid$.

Proof

(i) $\left(\sigma_1^P+\sigma_2^P\right)(x)=\sigma_1^P\ (x)\ if\ x\in\sigma_1\cap\sigma_2'$

$$\left(\sigma_1^P+\sigma_2^P\right)(x)=\sigma_2^P\ (x)\ if\ x\in\sigma_2\cap\sigma_1'$$

$$\left(\sigma_1^P+\sigma_2^P\right)(x)=\bigvee\left(\sigma_1^P\ (x),\sigma_2^P\ (x),\ \ if\ x\in\sigma_1\cap\sigma_2\right.$$

(ii) $\left(\sigma_1^N+\sigma_2^N\right)(x)=\sigma_1^N\ (x)\ if\ x\in\sigma_1\cap\sigma_2'$

$$\left(\sigma_1^N+\sigma_2^N\right)(x)=\sigma_2^N\ (x)\ if\ x\in\sigma_2\cap\sigma_1'$$

$$\left(\sigma_1^N+\sigma_2^N\right)(x)=\bigwedge\left(\sigma_1^N\ (x),\sigma_2^N\ (x),\ \ if\ x\in\sigma_1\cap\sigma_2\right.$$

Similarly, we get results for ρ_1^P, ρ_1^N. And the edge set membership is given as

(iii) $\left(\mu_1^P+\ \mu_2^P\right)(x,y)=\mu_1^P\ \ (x,y)\ if\ \ (x,y)\in\mu_1\cap\mu_2'$

$$\left(\mu_1^P+\ \mu_2^P\right)(x,y)=\mu_2^P\ \ (x,y)\ if\ \ (x,y)\in\mu_2\cap\mu_1'$$

$$\left(\mu_1^P+\ \mu_2^P\right)(x,y)=\bigvee\left(\mu_1^P\ \ (x,y)\,,\mu_2^P\ \ (x,y)\,,\ \ if\ \ (x,y)\in\mu_1\cap\mu_2\right.$$

(iv) $\left(\mu_1^N+\mu_2^N\right)(x,y)=\mu_1^N\ \ (x,y)\ if\ \ (x,y)\in\mu_1\cap\mu_2'$

$$\left(\mu_1^N+\ \mu_2^N\right)(x,y)=\mu_2^N\ \ (x,y)\ if\ \ (x,y)\in\mu_2\cap\mu_1'$$

$$\left(\mu_1^N+\ \mu_2^N\right)(x,y)=\bigwedge\left(\mu_1^N\ \ (x,y)\,,\mu_2^N\ \ (x,y)\,,\ \ if\ \ (x,y)\in\mu_1\cap\mu_2\right.$$

Similarly, we get results for γ_1^P, γ_1^N.

(v) $\left(\mu_1^P + \mu_2^P\right)(x, y) = \bigvee\left(\mu_1^P(x), \mu_2^P(y), (\mu_1^N + \mu_2^N)(x, y)\right.$
$= \bigwedge\left(\mu_1^N(x), \mu_2^N(y)\right.$

$$\left(\gamma_1^P + \gamma_2^P\right)(x, y) = \bigvee\left(\gamma_1^P(x), \gamma_2^P(y), \left(\gamma_1^N + \gamma_2^N\right)(x, y)\right.$$
$$= \bigwedge\left(\gamma_1^N(x), \gamma_2^N(y)\right.$$

if $(x, y) \in (\mu', \gamma')$ where (μ', γ') represents the set of all edges $(\sigma_1, \rho_1), (\sigma_1, \rho_1)$.

4.11 Theorem

Join of two AQSP fuzzy graphs is an AQSP fuzzy graphs is proved with the result, which is defined as, $(G_1 + G_2) = \{(\sigma_1, \rho_1) + (\sigma_2, \rho_2), (\mu_1, \gamma_1) + (\mu_2, \gamma_2)\}$.

Proof Let $(x, y) \in (\mu^0, \gamma^0)$, then we prove the following with the result

(i) $\left(\mu_1^P + \mu_2^P\right)(x, y) = \bigvee\left(\sigma_1^P(x), \sigma_2^P(y), = \bigvee\left(\left(\sigma_1^P + \sigma_2^P\right)(x), \left(\sigma_1^P + \sigma_2^P\right)(y)\right.\right.$

$$\left(\mu_1^N + \mu_2^N\right)(x, y) = \bigwedge\left(\sigma_1^N(x), \sigma_2^N(y), = \bigwedge\left(\left(\sigma_1^N + \sigma_2^N\right)(x), \left(\sigma_1^N + \sigma_2^N\right)(y)\right.\right.$$

(ii) $\left(\mu_1^P + \mu_2^P\right)(x, y) = \bigwedge\left(\left(\sigma_1^P + \sigma_2^P\right)(x), \left(\sigma_1^P + \sigma_2^P\right)(y)\right)$

$$\left(\mu_1^N + \mu_2^N\right)(x, y) = \bigvee\left(\left(\sigma_1^N + \sigma_2^N\right)(x), \left(\sigma_1^N + \sigma_2^N\right)(y)\right)$$

Similarly, we get results for γ_1^P, γ_1^N. And we prove that if$(x, y) \in \mu_1 \cap \overline{\mu}_2$,

(iii) $\left(\mu_1^P + \mu_2^P\right)(x, y) = \bigwedge\left(\left(\sigma_1^P + \sigma_2^P\right)(x), \left(\sigma_1^P + \sigma_2^P\right)(y)\right)$,

(iv) $\left(\mu_1^N + \mu_2^N\right)(x, y) = \bigvee\left(\left(\sigma_1^N + \sigma_2^N\right)(x), \left(\sigma_1^N + \sigma_2^N\right)(y)\right)$,

Similarly, we get results for γ_1^P, γ_1^N.Again if we consider $(x, y) \in \mu_1' \cap \mu_2$,

$$\left(\mu_1^P + \mu_2^P\right)(x, y) = \bigwedge\left(\left(\sigma_1^P + \sigma_2^P\right)(x), \left(\sigma_1^P + \sigma_2^P\right)(y)\right)$$

Fig. 4 $(G_1 + G_2)$

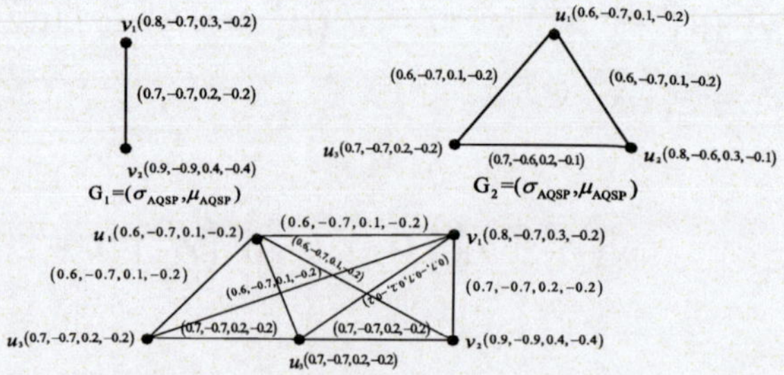

$$\left(\mu_1^N + \mu_2^N\right)(x, y) = \bigvee \left(\left(\sigma_1^N + \sigma_2^N\right)(x), \left(\sigma_1^N + \sigma_2^N\right)(y)\right)$$

Similarly, we get results for γ_1^P, γ_1^N. Hence, the proof is completed.

Remark A strong AQSP Fuzzy graph $G = (\sigma_{AQSP}, \mu_{AQSP})$ is called as Self-Complementary and not always complete if $\overline{\overline{G}} \asymp G$ (Fig. 4).

4.12 *Complement of a Strong AQSP Fuzzy Graph*

The Complement of a strong AQSP $G = (\sigma_{AQSP}, \mu_{AQSP})$ $G^* = \left(\sigma^*_{AQSP}, \ \mu^*_{AQSP}\right)$ is a strong AQSP of $\overline{G} = \left(\overline{\sigma}_{AQSP}, \overline{\mu}_{AQSP}\right)$ *on* $\overline{G^*}$. Where $\overline{\sigma}_{AQSP} = \overline{\sigma}^P, \ \overline{\rho}^P, \overline{\sigma}^N, \overline{\rho}^N$), $\overline{\mu}_{AQSP} = (\overline{\mu}^P, \ \overline{\gamma}^P, \overline{\mu}^N, \overline{\gamma}^N$) is the notations of the edges.

(i) $\overline{\sigma}_{AQSP} = \sigma_{AQSP}, \overline{\rho}_{AQSP} = \rho_{AQSP}$
(ii) $\overline{\sigma}^P(x) = \sigma^P(x), \overline{\sigma}^N(x) = \sigma^N(x), \overline{\rho}^P(x) = \rho^P(x)$ and $\overline{\rho}^N(x) = \rho^N(x)$

$$\forall \ x \in (\sigma, \rho)$$

(iii) $\overline{\mu}^P(x, y) = \left\{ 0\, if \ \ \mu^P(x, y) > 0, \bigwedge \left(\mu^P(x), \mu^P(y)\right) \ if \ \mu^P(x, y) = 0, \right.$

$$\overline{\mu}^N(x, y) = \left\{ 0\, if \ \ \mu^N(x, y) > 0, \bigvee \left(\mu^N(x), \mu^N(y)\right) \ if \ \mu^N(x, y) = 0, \forall (\mu, \gamma) \right.$$

Similarly, we get results for γ_1^P, γ_1^N (Fig. 5).

Fig. 5 $\overline{\overline{G}}$ AQSP fuzzygraph

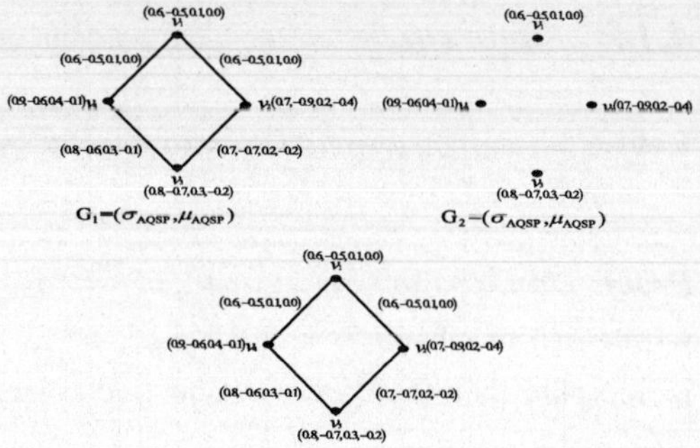

4.13 Proposition

If $G = (\sigma_{AQSP}, \mu_{AQSP})$ is a Self-complementary strong AQSP fuzzy graph, then we prove,

(i) $\sum_{x \neq y} \mu^P(x, y) = \sum_{x \neq y} \bigwedge \left(\mu^P(x), \mu^P(y)\right)$,
(ii) $\sum_{x \neq y} \mu^N(x, y) = \bigvee \sum_{x \neq y} \left(\mu^N(x), \mu^N(y)\right)$
(iii) $\sum_{x \neq y} \gamma^P(x, y) = \sum_{x \neq y} \bigwedge \left(\gamma^P(x), \gamma^P(y)\right)$,
(iv) $\sum_{x \neq y} \gamma^N(x, y) = \bigvee \sum_{x \neq y} \left(\gamma^N(x), \gamma^N(y)\right)$

Proof Let $G = (\sigma_{AQSP}, \mu_{AQSP})$ be a self-complementary AQSP fuzzy graph, then there exist an automorphism $\varphi : \sigma \to \sigma$ such that $\overline{\mu}^P(x) = \mu^P(x)$, $\overline{\mu}^N(x) = \mu^N(x) \forall \sigma \in \sigma_{AQSP}$ then for $\overline{\mu}^P(x, y) = \mu^P(x, y)$ and $\overline{\mu}^N(x, y) = \mu^N(x, y) \forall \ (x, y) \in \sigma_{AQSP}$, by definition we have the following:

(i) $\overline{\mu}^P(\varphi(x), \ \varphi(y)) = \bigwedge \left(\overline{\mu}^P\left(\varphi(x), \overline{\mu}^P \varphi(y)\right)\right)$,

$$\overline{\mu}^P(x, y) = \bigwedge \left(\mu^P(x), \mu^P(y) \Longrightarrow \sum_{x \neq y} \mu^P(x, y) = \sum_{x \neq y} \bigwedge \left(\mu^P(x), \mu^P(y)\right).\right.$$

(ii) $\overline{\mu}^N(\varphi(x), \ \varphi(y)) = \bigvee \left(\overline{\mu}^N\left(\varphi(x), \overline{\mu}^N \varphi(y)\right)\right)$,

$$\overline{\mu}^N(x, y) = \bigvee \left(\mu^N(x), \mu^N(y) \Longrightarrow \sum_{x \neq y} \mu^N(x, y) = \sum_{x \neq y} \bigvee \left(\mu^N(x), \mu^N(y)\right).\right.$$

Similarly, we get results for $\overline{\gamma}^P$, $\overline{\gamma}^N$. Hence the proof is completed.

4.14 Theorem

Let $G_1 = (\sigma_{AQSP}, \mu_{AQSP})$,$G_2 = (\sigma_{AQSP}, \mu_{AQSP})$ be the strong AQSP fuzzy graphs, then $G_1 \cong G_2 \iff \overline{G_1} \cong \overline{G_2}$.

Proof Let $G_1 = (\sigma_{AQSP}, \mu_{AQSP})$, $G_2 = (\sigma_{AQSP}, \mu_{AQSP})$ be AQSP of fuzzy graphs with Crisp graphs $G_1{}^* = \left(\sigma^*_{AQSP}, \mu^*_{AQSP}\right)$, $G_2{}^* = \left(\sigma^*_{AQSP}, \mu^*_{AQSP}\right)$ are isomorphic then there exist a bijective corresponding mapping $\varphi : \sigma_1 \to \sigma_2$ and $\varphi : \rho_1 \to \rho_2$,

$\sigma_1^P(x) = \sigma_2^P(\varphi(x)), \sigma_1^N(x) = \sigma_2^N(\varphi(x))\ \forall\ x \in \sigma_1$ for the vertices set and edges,

$$\mu_1^P(x, y) = \mu_2^P(\varphi(x), \varphi(y)), \mu_1^N(x, y) = \mu_2^N(\varphi(x), \varphi(y))\ \forall\ (x, y) \in \mu_1.$$

By definition of self-complement we have the following results such that,

(i)

$$\mu_1^P(x, y) = \wedge\left(\mu_1^P(x), \mu_1^P(y), = \wedge\left(\mu_2^P\left(\varphi\left(x\right), \mu_2^P\left(\varphi\left(y\right)\right)\right) = \overline{\mu}_2^P(\varphi(x), \varphi(y))\right.\right.$$

(ii)

$$\mu_1^N(x, y) = \vee\left(\mu_1^N(x), \mu_1^N(y), = \vee\left(\mu_2^N\left(\varphi\left(x\right), \mu_2^N\left(\varphi\left(y\right)\right)\right) = \overline{\mu}_2^N(\varphi(x), \varphi(y))\right.\right.$$

$$\forall (x, y) \in \mu_1.$$

Similarly we have results for the submerging polar mapping as follows:

$\rho_1^P(x) = \rho_2^P(\varphi(x)), \rho_1^N(x) = \rho_2^N(\varphi(x))\ \forall\ x \in \rho_1$ for the vertices set and edges,

$$\gamma_1^P(x, y) = \gamma_2^P(\varphi(x), \varphi(y)), \gamma_1^N(x, y) = \gamma_2^N(\varphi(x), \varphi(y))\ \forall\ (x, y) \in \gamma_1.$$

By definition of self-complement we have the following results such that,

(iii)

$$\gamma_1^P(x, y) = \wedge\left(\gamma_1^P(x), \gamma_1^P(y) = \wedge\left(\gamma_2^P\left(\varphi\left(x\right), \gamma_2^P\left(\varphi\left(y\right)\right)\right) = \overline{\gamma}_2^P(\varphi(x), \varphi(y))\right.\right.$$

(iv)

$$\gamma_1^N(x, y) = \vee\left(\gamma_1^N(x), \gamma_1^N(y) = \vee\left(\gamma_2^N\left(\varphi\left(x\right), \gamma_2^N\left(\varphi\left(y\right)\right)\right) = \overline{\gamma}_2^N(\varphi(x), \varphi(y))\right.\right.$$

$\forall (x, y) \in \gamma_1$. Hence, the result $\overline{G_1} \cong \overline{G_2}$ is completed. And the converse is obvious.

5 Arithmetic Conflicts Thought Analysis on Board of Syllabus (Tables 1, 2, 3, 4 and 5)

The invention of AQSP fuzzy functions and is applications is useful tool to find the level of distention area in using Cartesian Alternate quadrants $[-1, 1]$.

Table 1 Five schools in Chennai, distinct board of syllabus taken as vertices

Vertices	Board of syllabus	Schools
v_1	CBSE	School – I
v_2	ICSE	School – II
v_3	State Board	School – III
v_4	International	School – IV
v_5	Other boards	School – V

Table 2 The stress and conflicts of the person is given as fuzzy edges

Edges	Parent feelings	Edges	Parent feelings
v_1v_2	Depression	v_2v_4	Being overwhelmed
v_1v_3	Anxiety	v_2v_5	Cognitive difficulties
v_1v_4	Discouragement	v_3v_4	Curiosity
v_1v_5	Irritability	v_3v_5	Excitement
v_2v_3	pessimism	v_4v_5	Contentment

Table 3 Fuzzy vertices score of membership/nonmembership degree

V. Score	v_1	v_2	v_3
	$(0.7, -0.8, 0.2, -0.3)$ $(0.8, -0.9, 0.3, -0.4)$ $=0.550$	$(0.5, -0.6, 0.0, -0.1)$ $(0.9, -0.9, 0.4, -0.4)$ $=0.475$	$(0.9, -0.6, 0.4, -0.1)$ $(0.6, -0.9, 0.1, -0.4)$ $=0.450$
	v_4	v_5	
	$(0.6, -0.6, 0.1, -0.1)$ $(0.9, -0.9, 0.4, -0.4)$ $=0.500$	$(0.8, -0.5, 0.3, 0.0)$ $(0.6, -0.5, 0.1, 0.0)$ $=0.350$	

Table 4 Fuzzy edges score values of membership/nonmembership degree

(μ_s, γ_s)*Score*	v_1v_2	v_1v_3	v_1v_4
	$(0.6, -0.6, 0.1, -0.1)$ $(0.8, -0.9, 0.3, -0.4)$ $=0.475$	$(0.7, -0.6, 0.2, -0.1)$ $(0.6, -0.7, 0.1, -0.2)$ $=0.400$	$(0.6, -0.6, 0.1, -0.1)$ $(0.8, -0.8, 0.3, -0.3)$ $=0.450$
	v_2v_4	v_2v_5	v_3v_4
	$(0.5, -0.6, 0.0, -0.1)$ $(0.9, -0.9, 0.4, -0.4)$ $=0.470$	$(0.5, -0.5, 0.0, 0.0)$ $(0.6, -0.5, 0.1, 0.0)$ $=0.345$	$(0.6, -0.6, 0.1, -0.1)$ $(0.6, -0.9, 0.1, -0.4)$ $=0.425$
v_1v_5	v_2v_3	v_3v_5	v_4v_5
$(0.7, -0.5, 0.2, 0.0)$ $(0.6, -0.5, 0.0, 0.0)$ $=0.325$	$(0.6, -0.6, 0.1, -0.1)$ $(0.8, -0.9, 0.3, -0.4)$ $=0.350$	$(0.8, -0.5, 0.3, 0.0)$ $(0.6, -0.5, 0.1, 0.0)$ $=0.475$	$(0.6, -0.5, 0.1, 0.0)$ $(0.6, -0.5, 0.1, 0.0)$ $=0.300$

Table 5 Convolution energy level membership/nonmembership degree

Convolution energy level	Membership / Non membership value	Convolution energy level	Membership/ non membership
$\varphi_C(0)$	0/0.2	$\varphi_C(8)$	8/0.7
$\varphi_C(1)$	1/0.9	$\varphi_C(9)$	9/0.7
$\varphi_C(2)$	2/1	$\varphi_C(10)$	10/0.8
$\varphi_C(3)$	3/0.9	$\varphi_C(11)$	11/0.7
$\varphi_C(4)$	4/0.7	$\varphi_C(12)$	12/0.6
$\varphi_C(5)$	5/0.8	$\varphi_C(13)$	13/0.7
$\varphi_C(6)$	6/0.8	$\varphi_C(14)$	14/0.5
$\varphi_C(7)$	7/0.8	$\varphi_C(15)$	15/0.3

6 Method on AQSP Fuzzy Graphs

The AQSP satisfies (a) $\mu_S^P(x, y) \leq \sigma_S^P(x) \wedge \sigma_S^P(y)$ and (b) $\mu_S^N(x, y) \geq \sigma_S^N(x) \vee \sigma_S^N(y)$, score value such that, $\frac{1}{2}\left(\frac{1}{d^P}\sum \varphi_x^P - \frac{1}{d^N}\sum \varphi_x^N\right)$ is a formula to find the dominating thought, this leads the person to choose the final decision.

The computation of convolution is defined with the score formula as,

$$\varphi_{(\sigma_s,\rho_s)+(\mu_s,\gamma_s)}(x+y) = \bigvee\left\{\varphi_{\sigma_s}(x) \bigwedge \varphi_{\mu_s}(y)\right. \quad \text{we denote as } (\sigma_s + \mu_s) = C$$

We can conclude the result of (σ_s, ρ_s) *Score* and (μ_s, γ_s) *Score* is defined as follows:

$(v_1 > v_4 > v_2 > v_3 > v_5)$ and $(v_1v_2 \geq v_3v_5 > v_2v_4 > v_1v_4 > v_3v_4 > v_1v_3 > v_2v_3 > v_2v_5 > v_1v_5 > v_4v_5)$. The dominated feeling leads the person to choose CBSE board of syllabus in the conflict process. Here the convolution computation result shows vividly the most dominated feeling is defined as v_1v_2= Depression which has full membership degree 1. The weightage of the other mixed feelings are denoted in the above table.

7 Conclusion

In AQSP fuzzy graph, we show through convolution computation that the energy level of an individual is affected by conflicts and stress, which are submerged by dominant feelings. The frustration of the person is the result of two or more incompatible feelings. Obviously, the convolution result shows that the dominating feelings must be moderated to avoid the massive attack, high BP, sugar, suicidal tendency, and other body ailments. AQSP fuzzy graph is a model for a psychologist to precisely pinpoint the human mind's dominant feeling, which is uncertain, and at the same time, it is the leading point for their final decision. Using the new invention of AQSP fuzzy graph tool, we can maintain an emotionally balanced life through medication and psychological guidance.

References

1. Akram, M.: Bipolar fuzzy graphs. Inf. Sci. **181**(24), 5548–5564 (2011). https://doi.org/10.1016/j.ins.2011.07.037
2. Akram, M., Dudek, W.A.: "Interval-Valued Fuzzy Graphs." Computers and Mathematics with Applications 61(2): 289–99. http://dx.doi.org/10.1016/j.camwa.2010.11.004. 2012. "Regular Bipolar Fuzzy Graphs". Neural Comput. Appl. **21**(SUPPL. 1), 197–205 (2011)
3. Akram, M., Zafar, F.: 380 Studies in Fuzziness and Soft Computing. Intuitionistic Fuzzy Rough Graphs (2020)
4. Al-Hawary, T.: "Complete Fuzzy Graphs." Int. J. Math. Comb. 2011(4): 26–34.2017. "Certain Classes of Fuzzy Graphs". Eur. J. Pure Appl. Math. **10**(3), 552–560 (2011)
5. Bai, W., Ding, J., Zhang, C.: Dual hesitant fuzzy graphs with applications to multi-attribute decision making. Int. J. Cogn. Comput. Eng. **1**(July), 18–26 (2020). https://doi.org/10.1016/j.ijcce.2020.09.002
6. Gani, A.N., Muruganantham, P., Nafiunisha, A.: A new type of dominating fuzzy graphs. **20**(6), 1085–1091 (2021)
7. Javaid, M., Kashif, A., Rashid, T.: Hesitant fuzzy graphs and their products. Fuzzy Inf. Eng. **12**(2), 238–252 (2020). https://doi.org/10.1080/16168658.2020.1817658
8. Jesintha Rosline, J., Pathinathan, T.: Different types of products on Hesitancy Fuzzy Graphs. Int. J. Pure Appl. Math. **119**(9), 255–265 (2018)
9. Karunambigai, M.G., Parvathi, R., Kalaivani, O.K.: A study on Atanassov's intuitionistic fuzzy graphs. IEEE Int. Conf. Fuzzy Syst. **2**(ii), 649–655 (2011)
10. Pathinathan, T., Jesintha Rosline, J.: Double layered fuzzy graph. Ann. Pure Appl. Math. **8**(1), 135–143 (2014)
11. Lin, J.L., Khomnotai, L., Liu, H.C.: Oscillation of repeated max-weighted power mean compositions of fuzzy matrices. Axioms. **10**(4), 1–13 (2021)
12. Mankena, Ramakrishna, .V. Pradeep Kumar, CH. Ramprasad, and J. Vijaya Kumar. 2021. "Density of an m-Bipolar fuzzy graph." Malaya J. Matematik 9(1): 551–556
13. Zadeh, L.A.: A fuzzy-set-theoretic interpretation of linguistic hedges. J. Cybernetics. **2**(3), 4–34 (1972)

References

[illegible]

[illegible]

[illegible]

[illegible]

[illegible]

[illegible]

[illegible]

[illegible]

[illegible]

[illegible]

[illegible]

[illegible]

Optimal Solution for Transportation Problems Using Trapezoidal Fuzzy Numbers

J. Boobalan and P. Raja

1 Introduction

In today's highly competitive global market, transportation models can maximize profit and minimize costs. Transportation problems (TPs) involve the transportation of items from different origins to different destinations. The classical TP developed by Hitchcock [9] and the transportation costs (TCs), demand, and supply are fixed numbers.

In real-life situations, fixed numbers do not come about naturally. In uncertain environments, Zimmermann [19] elevates a TP to a new model, which he labels as a fuzzy transportation problem (FTP). One FTP was formulated on the basis of the integer–value problem of Chanas and Kuchta [3]. Bellman and Zadeh [18] also aimed to solve FTPs with their technique. Pandian and Natarajan [12] aimed to find the optimal solution to the TCs, demand, and supply, which first requires using trapezoidal fuzzy numbers (TFNs). The optimal solution to any problem depends on the design function.

In the literature, a few initial basic feasible solutions are available, including the northwest corner method (NWCM) [10], the least cost method (LCM) [10], Vogel's approximation method (VAM) [13], Goyal's version of VAM (GVAM) [8], and the initial basic feasible solution (IBFS) proposed to the TP [1, 4, 5]. Also, many researchers have modified VAM to obtain an IBFS to the TP. In general, TPs are solved in fixed environments, but fixed environments do not appear in nature.

J. Boobalan (✉)
Department of Mathematics, Annamalai University, Deputed to Manbumigu Dr. Puratchithalaivar, M.G.R., Government Arts and Science College, Kattumannarkoil, Tamil Nadu, India

P. Raja
Department of Mathematics, Periyar University College of Arts and Science, Pappireddipatti, Tamil Nadu, India

F. Kamalov et al. (eds.), *Advances in Mathematical Modeling and Scientific Computing*, Trends in Mathematics, https://doi.org/10.1007/978-3-031-41420-6_51

In the literature [6, 7, 11, 14–17], TFNs have been converted into fixed values, and researchers have used the robust ranking technique to solve FTPs. In this chapter, a TFN is converted into a fixed number by using the median method of ranking functions to solve FTPs. In this chapter, we present an SFZM and achieve the optimal solution. Our latest results are then compared with those from [10, 12–14].

2 Groundwork

In this section, we present a basic definition of TFN and a basic definition of the median ranking method.

2.1 *Definition [12]*

A fuzzy set $\widetilde{\rho}$ serves as a fuzzy number whose membership function has the following characteristics:

- $\nu_{\widetilde{\rho}}(\zeta) := \rho \rightarrow [0, 1]$ is continuous.
- $\nu_{\widetilde{\rho}}(\zeta) = 0$ for all $\zeta \in (-\infty, \zeta_1] \cup [\zeta_4, \infty)$.
- $\nu_{\widetilde{\rho}}(\zeta)\ is \rightleftarrows$ strictly increasing for $[\zeta_1, \zeta_2]$ and strictly decreasing for $[\zeta_3, \zeta_4]$.
- $\nu_{\widetilde{\rho}}(\zeta) = 01$ for all $\zeta \in [\zeta_2, \zeta_3]$, where $\zeta_1 < \zeta_2 < \zeta_3 < \zeta_4$.

2.2 *Definition [12]*

A fuzzy number, $\widetilde{\rho} = (\zeta_1, \zeta_2, \zeta_3, \zeta_4)$, is a TFN if its membership function is given by the following:

$$\nu_{\widetilde{\rho}}(\zeta) = \begin{cases} \frac{(\zeta-\zeta_1)}{(\zeta_2-\zeta_1)}, \zeta_1 \le \zeta \le \zeta_2, \\ 1, \zeta_2 \le \zeta \le \zeta_3, \\ \frac{(\zeta-\zeta_4)}{(\zeta_3-\zeta_4)}, \zeta_3 \le \zeta \le \zeta_4, \\ 0, \text{otherwise}. \end{cases}$$

2.3 *Ranking Function [2]*

Let $\widetilde{\rho} = (\zeta_1, \zeta_2, \zeta_3, \zeta_4)$ be the TFN, in which case the median ranking method yields $R\left(\widetilde{\rho}\right) = \left(\frac{\zeta_2+\zeta_3}{2}\right)$

3 Proposed Algorithm

In this section, the steps for the SFZM are presented, as follows:

Step 1.

An unbalanced FTP refers to a balanced FTP that has zero fuzzy costs (ZFCs), adding minimum fuzzy cost.

Step 2.

The total supply ≠ the total demand, so ZFCs are added.

Step 3.

3(a) Select the lowest fuzzy cost (FC) value in each row, and then remove that from the FC values in that row.

3(b) Select the lowest FC value in each column, and then remove that from the FC values in that column. At minimum, one ZFC should now be available in each row and each column.

Step 4.

4(a) Confirm that the total fuzzy supply is lower than or equal to the total fuzzy demand.

4(b) Confirm that the total fuzzy demand is now lower than or equal to the total fuzzy supply.

4(c) If Step 4(a) and Step 4(b) have been completed, then jump ahead to Step 7; otherwise, proceed to Step 5.

Step 5.

The lowest-level lines are drawn parallel and perpendicular to the ZFCs in Step 3.

Step 6.

6(a) Starting from any cell not enclosed by lines, select the lowest FC.

6(b) Remove each FC in a cell not enclosed by lines.

6(c) Add to each (FC) in a cell enclosed by two lines.

6(d) If an FC in a cell enclosed by lines remains unchanged, then return to Step 4.

Step 7.

7(a) Select the lowest fuzzy TC value from Step 4.
7(b) Select a row single fuzzy-zero cell (RSFZC) and a column single fuzzy-zero cell (CSFZC) for allocation.
7(c) Allocate the maximum possible, and cross off in the usual manner.
7(d) If an RSFZC or a CSFZC does not appear in both rows and columns, then repeat the process from Step 7(a) to Step 7(c) until this requirement has been satisfied.

4 Numerical Examples

4.1 Fuzzy Data for Example 1 [12, 14] (Table 1)

The chosen fuzzy data for Example 1 are balanced. We used the median ranking method for the defuzzification process (Tables 2 and 3).

The minimum TC value according to the NWCM fixed cost is **₹138.5** (Table 4).
The minimum TC value according to the LCM fixed cost is **₹134.5** (Table 5).
The minimum TC value according to VAM's fixed cost is **₹123.5** (Table 6).

Table 1 Supply Vs Demand

	$Demand_1$	$Demand_2$	$Demand_3$	$Demand_4$	Supply
$Supply_1$	(1,2,3,4)	(1,3,4,6)	(9,11,12,14)	(5,7,8,11)	(1,6,7,12)
$Supply_2$	(0,1,2,4)	(−1,0,1,2)	(5,6,7,8)	(0,1,2,3)	(0,1,2,3)
$Supply_3$	(3,5,6,8)	(5,8,9,12)	(12,15,16,17)	(7,9,10,12)	(5,10,12,17)
Demand	(5,7,8,10)	(1,5,6,10)	(1,3,4,6)	(1,2,3,4)	

Table 2 Defuzzied data for Example 1

	$Demand_1$	$Demand_2$	$Demand_3$	$Demand_4$	Supply
$Supply_1$	2.5	3.5	11.5	7.5	6.5
$Supply_2$	1.5	0.5	6.5	1.5	1.5
$Supply_3$	5.5	8.5	15.5	9.5	11
Demand	7.5	5.5	3.5	2.5	

Table 3 Northwest corner method (NWCM) [10]

	$Demand_1$	$Demand_2$	$Demand_3$	$Demand_4$	Supply
$Supply_1$	6.5×2.5	3.5	11.5	7.5	6.5
$Supply_2$	1×1.5	0.5×0.5	6.5	1.5	1.5
$Supply_3$	5.5	5×8.5	3.5×15.5	2.5×9.5	11
Demand	7.5	5.5	3.5	2.5	

Table 4 Least cost method (LCM) [10]

	$Demand_1$	$Demand_2$	$Demand_3$	$Demand_4$	Supply
$Supply_1$	6.5 × 2.5	3.5	11.5	7.5	6.5
$Supply_2$	1.5	1.5 × 0.5	6.5	1.5	1.5
$Supply_3$	1 × 5.5	4 × 8.5	3.5 × 15.5	2.5 × 9.5	11
Demand	7.5	5.5	3.5	2.5	

Table 5 Vogel's approximation method (VAM) [13]

	$Demand_1$	$Demand_2$	$Demand_3$	$Demand_4$	Supply
$Supply_1$	1 × 2.5	5.5 × 3.5	11.5	7.5	6.5
$Supply_2$	1.5	0.5	6.5	1.5 × 1.5	1.5
$Supply_3$	6.5 × 5	8.5	3.5 × 15.5	1 × 9.5	11
Demand	7.5	5.5	3.5	2.5	

Table 6 Modified distribution method (MODI) [10]

	$Demand_1$	$Demand_2$	$Demand_3$	$Demand_4$	Supply
$Supply_1$	2.5	5.5 × 3.5	1 × 11.5	7.5	6.5
$Supply_2$	1.5	0.5	1.5 × 6.5	1.5	1.5
$Supply_3$	7.5 × 5.5	8.5	1 × 15.5	2.5 × 9.5	11
Demand	7.5	5.5	3.5	2.5	

Table 7 SFZM

	$Demand_1$	$Demand_2$	$Demand_3$	$Demand_4$	Supply
$Supply_1$	2.5	5.5 × 3.5	1 × 11.5	7.5	6.5
$Supply_2$	1.5	0.5	1.5 × 6.5	1.5	1.5
$Supply_3$	7.5 × 5.5	8.5	1 × 15.5	2.5 × 9.5	11
Demand	7.5	5.5	3.5	2.5	

The minimum TC value according to the modified distribution method's fixed cost is **₹121** (Table 7).

The minimum TC value according to the SFZM's fixed cost is **₹121.**

4.2 *SFZM with [10, 12–14]*

According to a comparison of the SFZM with the methods from [10, 12–14], as shown in Table 8, SFZM clearly provides lowest cost and thus the optimal result (Table 9; Figs. 1 and 2).

Table 8 Comparison of SFZM with other methods, including those from [10, 12–14]

S. No	NWCM	LCM	VAM	[12]	[14]	MODI	SFZM
1	**₹138.5**	**₹134.5**	**₹123.5**	**₹132.17**	**₹122.5**	**₹121**	**₹121**

Table 9 The deviation percentage of various methods

S. No	NWCM	LCM	VAM	[12]	[14]	MODI	SFZM
1	**12.6**	**10**	**2**	**8.5**	**1.2**	**0**	**0**

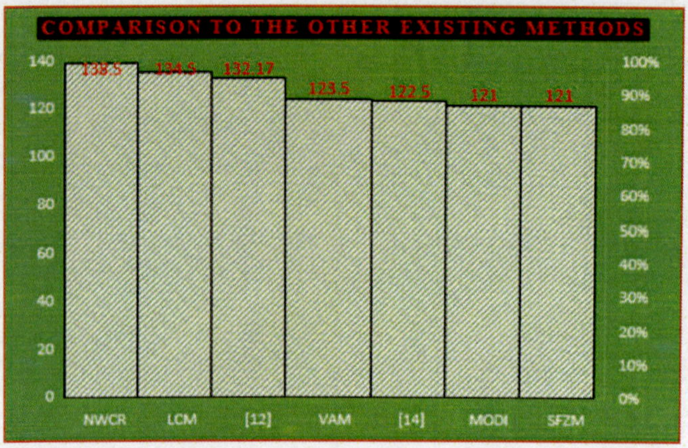

Fig. 1 Comparison to the other existing methods (bar chart)

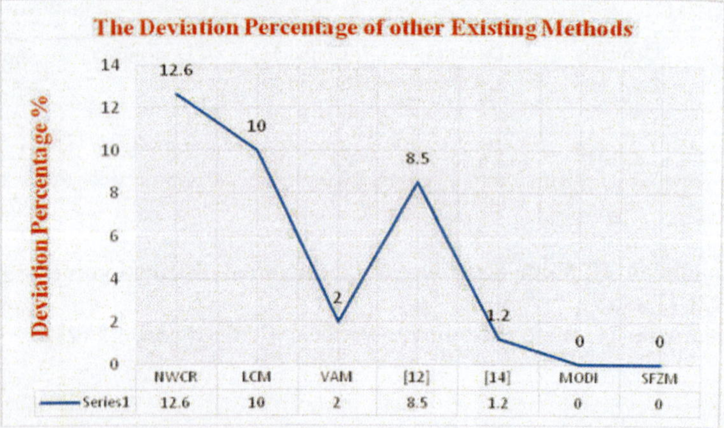

Fig. 2 The deviation percentage of other existing methods (up and down graph)

Table 10 Fuzzy Supply Vs Fuzzy Demand

	Fuzzy Demand$_1$	Fuzzy Demand$_2$	Fuzzy Demand$_3$	Fuzzy Demand$_4$	Fuzzy Demand$_5$	Fuzzy Supply
Fuzzy Supply$_1$	(5,7,8,11)	(1,6,7,12)	(2,4,5,7)	(2,5,6,9)	(7,9,10,12)	(20,35,45,60)
Fuzzy Supply$_2$	(5,8,9,12)	(2,5,6,9)	(1,6,7,12)	(5,7,8,11)	(5,8,9,12)	(15,25,35,45)
Fuzzy Supply$_3$	(1,6,7,12)	(5,8,9,12)	(7,9,10,12)	(1,6,7,12)	(2,5,6,9)	(10,15,25,30)
Fuzzy Supply$_4$	(2,5,6,9)	(5,7,8,11)	(5,7,8,11)	(5,8,9,12)	(1,6,7,12)	(5,8,12,15)
Fuzzy Demand	(15,25,35,45)	(15,25,35,45)	(8,14,16,22)	(10,15,25,30)	(2,4,6,8)	

Table 11 The deviation Value (₹) of various methods

S. No	NWCM	LCM	VAM	[14]	MODI	SFZM
1	**₹685**	**₹610**	**₹560**	**₹577.5**	**₹560**	**₹560**

Table 12 The deviation percentage of various methods

S. No	NWCM	LCM	VAM	[14]	MODI	SFZM
1	**18.2**	**8.2**	**0**	**3.1**	**0**	**0**

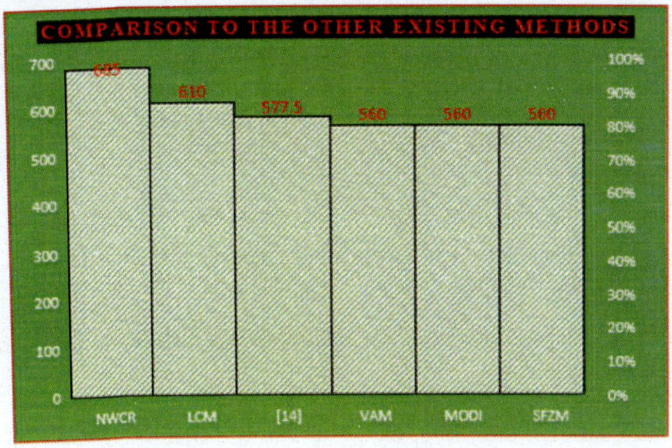

Fig. 3 The deviation percentage of other existing methods (bar chart)

4.3 Fuzzy Data for Example 2 [14] (Tables 10, 11 and 12; Figs. 3 and 4)

4.4 Merits of Using the SFZM

In this chapter, the SFZM converted TFNs into fixed numbers, and we compared our results with those from [10, 12–14], which shows that the SFZM offers better

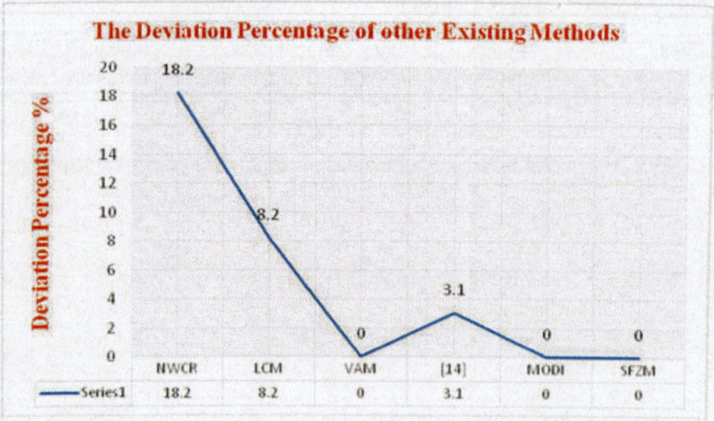

Fig. 4 The deviation percentage of other existing methods (up and down graph)

solutions than those offered in [10, 12–14]. The SFZM also has the following merits:

1. The SFZM is easy to understand.
2. The SFZM achieves the optimal solution.
3. The SFZM is not needed to find the initial basic feasible solution.
4. SFZM solutions can be compared with [10, 12–14] by using fixed numbers.

5 Conclusion

In this chapter, the SFZM used the mean ranking method to convert the TFNs of transportation problems into fixed numbers.

- For the total cost of Example 1, the results of the SFZM solution, in Table 8, show that value to be ₹121, which is better than the cost of [12] and that of [14], which were ₹132.17 and ₹122.5, respectively.
- For the total cost of Example 2, the results of the SFZM solution, in Table 11, show that value to be ₹560, which is better than the cost of [14], which was ₹577.5.

References

1. Akpana, T., Ugbeb, J.U., Ajahd, O.: A modified Vogel approximation method for solving balanced transportation problems. Am. Sci. Res. J. Eng. Technol. Sci. **14**, 289–302 (2015)
2. Ashale, T.R.: Newly Proposed Matrix Reduction Technique under Mean Ranking Method for Solving Trapezoidal Fuzzy Transportation Problems Under Fuzzy Environment (2021). https://doi.org/10.20944/preprints202106.0573

3. Charnes, A., Cooper, W.W., Henderson, A.: An Introduction to Linear Programming. Wiley, New York (1953)
4. Deshmukh, N.M.: An innovative method for solving transportation problem. Int. J. Phys. Math. Sci. **2**, 86–91 (2012)
5. Ekanayake, E.M.U.S.B., Perera, S.P.C., Daundasekara, W.B., Juman, Z.A.M.S.: A modified ant colony optimization algorithm for solving a transportation problem. J. Adv. Math. Comput. Sci. **35**, 83–101 (2020)
6. Fegade, M.R., Jadhav, V.A., Muley, A.A.: Solving Fuzzy transportation problem using zero suffix and robust ranking methodology. IOSR J. Eng. **2**, 36–39 (2012)
7. Allung, F.M., Blegur, N.K.F., Dethan.: Modified Hungarian method for solving balanced fuzzy transportation problems. J. Varian. **5**, 161–170 (2022)
8. Goyal, S.K.: Improving VAM for unbalanced transportation problems. J. Oper. Res. Soc. **35**, 1113–1114 (1984)
9. Hitchcock, F.L.: The distribution of a product from several sources to numerous localities. J. Phys. Sci. **20**, 224–230 (1941)
10. Hamdy, A.T.: Operations Research: An Introduction, 8th edn, p. 2007. Pearson Prentice Hall, Upper Saddle River (2007)
11. Karthy, T., Ganesan, K.: Revised improved zero point method for the trapezoidal fuzzy transportation problems. AIP Conference Proceedings. (2019)
12. Pandian, P., Natarajan, G.: A new algorithm for finding a fuzzy optimal solution for fuzzy transportation problems. Appl. Math. Sci. **4**, 79–90 (2010)
13. Reinfeld, N.V., Vogel, W.R.: Mathematical Programming. Prentice-Hall, Englewood Cliffs, NJ (1958)
14. Dhouib, S.: Solving the trapezoidal fuzzy transportation problems via new heuristic: The Dhouib-Matrix-TP1. Int. J. Oper. Res. Inf. Syst. **12**, 1–16 (2022)
15. Soomro, A.S., Junaid, M., Tularam, G.A.: Modified Vogel's approximation method for solving transportation problems. Math. Theory Model. **5** (2015)
16. Ngastiti, P.T.B., Bayu Surarso, S.: Zero point and zero suffix methods with robust ranking for solving fully fuzzy transportation problems. IOP Conference Series: Journal of Physics: Conference Series. (2018)
17. Ngastiti, P.T.B., Surarso, B., Sutimin.: Comparison between zero point and zero suffix methods in fuzzy transportation problems. J. Mat. Mantik. **6**, 38–46 (2020)
18. Zadeh, L.A.: Fuzzy sets as a basis for a theory of possibility. Fuzzy Sets Syst. **1**, 3–28 (1978)
19. Zimmermann, H.J.: Fuzzy programming and linear programming with several objective functions. Fuzzy Sets Syst. **1**, 45–55 (1978)

Heterogeneous Queueing Model with Intermittently Obtainable Server with Feedback

K. Divya, K. Indhira, and M. Seenivasan

1 Introduction

Queueing systems with service feedback have several uses in manufacturing, computing, and telecommunications systems. In the queueing literature, one of the most important elements of queueing systems is the feedback phenomenon. Customers wish to receive service again when they receive substandard service. Customers can keep using the service until they're satisfied. If there is a free server when a customer arrives, the server begins to be served immediately, and the customer leaves the system after the service is completed. In two heterogeneous servers, server 1 is always available and server 2 is only accessible intermittently. Server 2 performs some peculiar and unusual jobs. Interference in service on server 2 occurs for a period of time but only after the service in hand is performed. This service is referred to as intermittently obtainable.

The MGM is used for the analysis of quasi-birth-death phenomena; it is a continuous-time Markov chain whose transition rate matrices have a repetitive block structure. It necessitates the use of a transition rate matrix with a tridiagonal block structure. Neuts [11] discussed several matrix geometric stochastic model solutions. MGM is used to get a computable explicit formula for probability distributions and some other system efficiencies. Agarwal [1] introduced the idea of an intermittently available server. Sharda [15] presented a queueing model in which arrivals and departures occur in variable-size batches where the server is intermittently available. A multi-server queueing system with heterogeneous servers was investigated by

K. Divya · K. Indhira (✉)
Department of Mathematics, School of Advanced Sciences, Vellore Institute of Technology, Vellore, Tamil Nadu, India
e-mail: kindhira@vit.ac.in

M. Seenivasan
Mathematics Wing-DDE, Annamalai University, Chidambaram, Tamil Nadu, India

F. Kamalov et al. (eds.), *Advances in Mathematical Modeling and Scientific Computing*, Trends in Mathematics, https://doi.org/10.1007/978-3-031-41420-6_52

Krishnakumar et al. [8]. Moreover, Krishnamoorthy and Sreenivasan [9] developed a $M/M/2$ queueing system with heterogeneous servers, where one server remains inactive while the other goes on vacation when no customers are waiting.

Kalyanaraman and Seenivasan investigated a multi-server retrial queue with breakdown and geometric loss [7]. Kumar and Madheswari [10] studied a Markovian queue with two heterogeneous servers and server-dependent vacations. Ayyappan et al. [3] investigated a $M/M/1$ retrial queue with feedback in the context of preemptive priority service. Montazer Haghighi and Mishev [5]–[4] investigated delayed and network queues. Some other notable MGM works are [6, 13, 14, 16]. MGM was employed in a recent paper by Seenivasan and Indumathi [12] to model a queueing model of two heterogeneous servers where one is intermittently obtainable. However, there is a chance that customers will be dissatisfied for some reason. In order to overcome this issue, we have added the feedback concept to the paper [12]. Therefore, we have proposed a heterogeneous feedback queueing model with intermittently obtainable server, and thus the unsatisfied customers are also considered.

The following is an overview of the remaining portions of this paper: Our model's construction has been presented in Sect. 2. The geometric structure of the matrix has been presented in Sect. 3, along with the solution to the matrix. In addition, we have derived the model's stability criterion and various assessment formulas in Sect. 3.2. Section 4 covers the numerical study. The system performance measurements have been stated in Sect. 5, and the conclusion is provided in the last part.

2 Construction of the Model

Our queueing model takes feedback into account for heterogeneous and intermittently obtainable servers (servers 1 and 2). The arrival of the customer follows a Poisson process with rate λ. Customers enter the system in accordance with the state of the servers, namely, their intermittently obtainable state, working state, or idle state. Server 1 and server 2 follow different rates for their services: μ_1 and μ_2, respectively, where both follow an exponential distribution. Server 2 is either idle or intermittently accessible, while server 1 is always accessible. When the line length seems greater than or equal to zero, server 2 begins doing some unusual and critical jobs. Before that, server 2 must execute the job at hand. In addition, the retrieval capacity time on server 2 follows an exponential distribution with the rate κ. Idle server 2 either enters the working period to provide the customer with service at the rate η_0 or enters the intermittently obtainable period to provide the customer with service at the rate η_1. After receiving a service, an unfulfilled or dissatisfied customer will elect to enter for another service with rate $\bar{\nu}$, which is known as "feedback," or he may decide to exit the system permanently after satisfied service completion with the rate ν $(= 1 - \bar{\nu})$. The model's transition diagram is depicted in the below figure (Fig. 1). We omit server 1's workout from the transition diagram since it is always working.

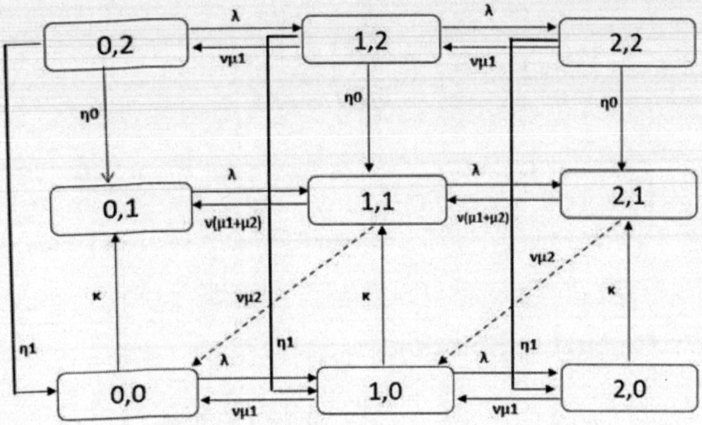

Fig. 1 Transition diagram of the model

Let $X(t) = \{(N(t), S(t)); t \geq 0\}$ be a Markov process (MP) with a state space at time t, where $N(t)$ represents the no. of customer in the system and

$$S(t) = \begin{cases} 0, & \text{if server 2 is } \textit{Intermittently obtainable} \\ 1, & \text{if server 2 is } \textit{Working} \\ 2, & \text{if server 2 is } \textit{Idle} \end{cases}$$

The system state is symbolized by $N(t)$ and $S(t)$. Let $\{(N(t), S(t)); t \geq 0\}$, with the state space organized in lexicographical manner as follows:

$\Omega = \bigcup (i, j); i \geq 0, j = 0, 1, 2$. Infinitesimal generator matrix G is expressed as follows

$$G = \begin{pmatrix} D_0 & E_0 & 0 & 0 & 0 & 0 & 0 \cdots \\ E_2 & E_1 & E_0 & 0 & 0 & 0 & 0 \cdots \\ 0 & E_2 & E_1 & E_0 & 0 & 0 & 0 \cdots \\ 0 & 0 & E_2 & E_1 & E_0 & 0 & 0 \cdots \\ 0 & 0 & 0 & E_2 & E_1 & E_0 & 0 \cdots \\ \vdots & \vdots & \vdots & \vdots & \vdots & \vdots & \vdots \ldots \end{pmatrix}$$

where

$$D_0 = \begin{pmatrix} -(\lambda+\kappa) & \kappa & 0 \\ 0 & -\lambda & 0 \\ \eta_0 & \eta_1 & -(\lambda+\eta_0+\eta_1) \end{pmatrix}; \quad E_0 = \begin{pmatrix} \lambda & 0 & 0 \\ 0 & \lambda & 0 \\ 0 & 0 & \lambda \end{pmatrix};$$

$$E_2 = \begin{pmatrix} \nu\mu_1 & 0 & 0 \\ \nu\mu_2 & \nu(\mu_1+\mu_2) & 0 \\ 0 & 0 & \nu\mu_1 \end{pmatrix}$$

$$\text{and} \quad E_1 = \begin{pmatrix} -(\lambda+\nu\mu_1+\kappa) & \kappa & 0 \\ 0 & -(\lambda+\nu\mu_1+2\nu\mu_2) & 0 \\ \eta_0 & \eta_1 & -(\lambda+\nu\mu_1+\eta_0+\eta_1) \end{pmatrix}.$$

3 Matrix Geometric Solution

We define $P_{ij} = \{(N(t) = i, S(t) = j)\}$, where i indicates the total number of customers in the queue and j reflects the server state. The prob. vector is described as $T = (\tau_0, \tau_1, \tau_2, \ldots)$, where $\tau_i = (\tau_{i0}, \tau_{i1}, \tau_{i2})$, $i = 0, 1, 2, \ldots$.

The static prob. matrix T is solved by using the condition $TG = 0$.

$$\tau_0 D_0 + \tau_1 E_2 + \ldots = 0 \tag{1}$$

$$\tau_0 E_0 + \tau_1 E_1 + \tau_2 E_2 + \ldots = 0 \tag{2}$$

$$\tau_1 E_0 + \tau_2 E_1 + \tau_3 E_2 + \ldots = 0 \tag{3}$$

$$\vdots$$

$$\tau_i E_0 + \tau_{i+1} E_1 + \tau_{i+2} E_2 + \ldots = 0 \tag{4}$$

$$\tau_k = \tau_0 R^k, \ \text{where } k \geq 1 \tag{5}$$

Let matrix R be a rate matrix. By substituting Eq. (5) into Eqs. (1)–(4), we obtain

$$\tau_0(D_0 + RE_2) = 0 \tag{6}$$

$$\tau_0(E_0 + RE_1 + R^2 E_2) = 0 \tag{7}$$

$$\tau_i E_0 + \tau_{i+1} E_1 + \tau_{i+2} E_2 = 0. \tag{8}$$

$$\tau_0 R^i (E_0 + RE_1 + R^2 E_2) = 0, \quad i \geq 1. \tag{9}$$

The normalizing equation is given by

$$\tau_0 [I - R]^{-1} e = 1. \tag{10}$$

Here e is a column vector in which all elements are 1s in the corresponding column. R is a minimal nonnegative solution of the matrix quadratic equations.

$$E_0 + RE_1 + R^2 E_2 = 0 \tag{11}$$

$$R = -E_1^{-1}[E_0 + R^2 E_2] \tag{12}$$

where $R \geq 0$ and it's an irreducible nonnegative matrix of radius smaller than 1. Matrix R can be calculated using an iterative approach as shown below.

$$R_0 = 0 \tag{13}$$

$$R_{n+1} = -E_0E_1^{-1} - R_n^2E_2E_1^{-1} \quad n \geq 1 \tag{14}$$

All values of R will expand monotonically, and R is converging to $[E_0 + R_n^2E_2^{-1}]$ and $-E_1^{-1}$ where both are nonnegative. By considering matrix geometric procedure, the steady state is obtained.

$$\tau E_0e \leq \tau E_2e \tag{15}$$

where row vector $\tau = (\tau_0, \tau_1, \tau_2)$ is obtained from the generator matrix $E = E_0 + E_1 + E_2$ which is given below.

$$E = \begin{pmatrix} -\kappa & \kappa & 0 \\ \nu\mu_2 & -\nu\mu_2 & 0 \\ \eta_0 & \eta_1 & -(\eta_0 + \eta_1) \end{pmatrix}. \tag{16}$$

Matrix E is irreducible and is the least generator in a Markov process. Unique equations are also satisfied by T

$$TE = 0; \quad Te = 1 \tag{17}$$

From Eq. (17), we obtain

$$\tau_2 = 0; \tau_1 = \left(\frac{\kappa}{\nu\mu_2}\right)\tau_0; \tau_0 = \left[1 + \frac{\kappa}{\nu\mu_2}\right]^{-1} \tag{18}$$

When we apply the stability condition (15) for the Quasi Birth-Death (QBD) procedure in matrix G, we get

$$\lambda[\tau_0 + \tau_1 + \tau_2] \leq \nu\mu_1[\tau_0 + \tau_1 + \tau_2] + 2\nu\mu_2\tau_1 \tag{19}$$

3.1 Special Cases

Case (i) No feedback

Let $\nu = 0$. Our model is reduced to a two heterogeneous server queueing model with intermittently obtainable servers which coincides with [12].

Case (ii) No intermittently obtainable server

Our model is reduced to an instantaneous feedback queue with heterogeneous servers, which coincides with [2].

3.2 Evaluation of the Result

The probability of the server status is determined by the result evaluation given below:

1. The prob. of server 2 when it is in intermittently obtainable state is $PS_{2IO} = \sum_{n=0}^{\infty} \tau_{n,0}$
2. The prob. of server 2 in a working state is $PS_{2W} = \sum_{n=0}^{\infty} \tau_{n,1}$
3. The prob. of server 2 in an idle state is $PS_{2Id} = \sum_{n=0}^{\infty} \tau_{n,2}$
4. If server 2 is in intermittently obtainable state, then the average no. of customers present in the system is $E[N_1] = \sum_{n=0}^{\infty} n\tau_{n,0}$
5. If server 2 is in a working state, then the average no. of customers present in the system is $E[N_2] = \sum_{n=0}^{\infty} n\tau_{n,1}$
6. If server 2 is in an idle state, then the average no. of customers present in the system is $E[N_3] = \sum_{n=0}^{\infty} n\tau_{n,2}$
7. The average no. of customers present in the system is $E[N] = E[N_1] + E[N_2] + E[N_3]$

4 Numerical Study

In this part, a numerical study has been presented by utilizing the MGM to learn more about and comprehend the implementation of the λ parameter in the system's unique probabilistic description characteristics. We have established the probability vectors of the system by altering λ values while keeping the other variables constant.

Illustration 5 Assume that $\lambda = 0.3$, $\mu_1 = 1.6$, $\mu_2 = 2.8$, $\kappa = 0.5$, $\eta_0 = 0.1$, $\eta_1 = 0.2$, $\nu = 0.3$, $\bar{\nu} = 0.7$, and

$$R = \begin{pmatrix} 0.2958 & 0.0919 & 0 \\ 0.0067 & 0.1316 & 0 \\ 0.0982 & 0.0992 & 0.3246 \end{pmatrix}$$

The probability vectors in Table 1 have been calculated by substituting the R matrix in Eqs. (6) and (10). Therefore, we get the vector $\tau_0 = (0.1335, 0.6705, 0.0053)$. This τ_0 can be used to compute the rest of the vectors $\tau_i' s$ using the equation $\tau_k = \tau_0 R^k$. Lastly, we verified that the cumulative probability is $0.9996 \approx 1$.

Table 1 Probability vector for $\lambda = 0.3$

τ_i	τ_{i0}	τ_{i1}	τ_{i2}	Total
τ_0	0.1335	0.6705	0.0053	0.8093
τ_1	0.0445	0.1010	0.0017	0.1472
τ_2	0.0140	0.0176	0.0006	0.0322
τ_3	0.0043	0.0037	0.0002	0.0082
τ_4	0.0013	0.0009	0.0001	0.0023
τ_5	0.0004	0.0001	0.0000	0.0004
Total				0.9996

Table 2 Probability vector for $\lambda = 0.35$

τ_i	τ_{i0}	τ_{i1}	τ_{i2}	Total
τ_0	0.1414	0.6240	0.0033	0.7687
τ_1	0.0549	0.1114	0.0012	0.1675
τ_2	0.0201	0.0233	0.0004	0.0438
τ_3	0.0072	0.0059	0.0002	0.0133
τ_4	0.0026	0.0017	0.0001	0.0044
τ_5	0.0009	0.0005	0.0000	0.0014
τ_6	0.0003	0.0001	0.0000	0.0004
Total				0.9995

Illustration 6 Assume that $\lambda = 0.35$, $\mu_1 = 1.6$, $\mu_2 = 2.8$, $\kappa = 0.5$, $\eta_0 = 0.1$, $\eta_1 = 0.2$, $\nu = 0.3$, $\bar{\nu} = 0.7$, and

$$R = \begin{pmatrix} 0.3455 & 0.1130 & 0 \\ 0.0089 & 0.1521 & 0 \\ 0.1423 & 0.1415 & 0.3669 \end{pmatrix}$$

The probability vectors in Table 2 were calculated by substituting the R matrix in Eqs. (6) and (10). Therefore, we get the vector τ_0= (0.1414, 0.6240, 0.0033). This τ_0 can be used to compute the rest of the vectors $\tau_i' s$ using the equation $\tau_k = \tau_0 R^k$. Lastly, we verified that the cumulative probability is 0.9995 $\approx$1.

Illustration 7 Assume that $\lambda = 0.4$, $\mu_1 = 1.6$, $\mu_2 = 2.8$, $\kappa = 0.5$, $\eta_0 = 0.1$, $\eta_1 = 0.2$, $\nu = 0.3$, $\bar{\nu} = 0.7$, and

$$R = \begin{pmatrix} 0.3969 & 0.1371 & 0 \\ 0.0115 & 0.1724 & 0 \\ 0.2122 & 0.2079 & 0.4061 \end{pmatrix}$$

The probability vectors in Table 3 were calculated by substituting the R matrix in Eqs. (6) and (10). Therefore, we get the vector $\tau_0 = (0.1466, 0.5792, 0.0001)$. This τ_0 can be used to compute the rest of the vectors $\tau_i' s$ using the equation $\tau_k = \tau_0 R^k$. Lastly, we verified that the cumulative probability is 0.9994 $\approx$1.

Table 3 Probability vector for $\lambda = 0.4$

τ_i	τ_{i0}	τ_{i1}	τ_{i2}	Total
τ_0	0.1466	0.5792	0.0001	0.7259
τ_1	0.0649	0.1200	0.0000	0.1849
τ_2	0.0271	0.0296	0.0000	0.0567
τ_3	0.0111	0.0088	0.0000	0.0199
τ_4	0.0045	0.0030	0.0000	0.0075
τ_5	0.0018	0.0013	0.0000	0.0031
τ_6	0.0007	0.0004	0.0000	0.0011
τ_7	0.0002	0.0001	0.0000	0.0003
Total				0.9994

Table 4 Probability vector for $\lambda = 0.45$

τ_i	τ_{i0}	τ_{i1}	τ_{i2}	Total
τ_0	0.1489	0.5251	0.0000	0.6740
τ_1	0.0748	0.1167	0.0000	0.1915
τ_2	0.0355	0.0304	0.0000	0.0659
τ_3	0.0165	0.0096	0.0000	0.0261
τ_4	0.0076	0.0036	0.0000	0.0112
τ_5	0.0035	0.0015	0.0000	0.0050
τ_6	0.0016	0.0007	0.0000	0.0023
τ_7	0.0007	0.0003	0.0000	0.0010
τ_8	0.0003	0.0001	0.0000	0.0004
τ_9	0.0002	0.0000	0.0000	0.0002
τ_{10}	0.0001	0.0000	0.0000	0.0001
Total				0.9777

Illustration 8 Assume that $\lambda = 0.45, \mu_1 = 1.6, \mu_2 = 2.8, \kappa = 0.5, \eta_0 = 0.1, \eta_1 = 0.2, \nu = 0.3, \bar{\nu} = 0.7$, and

$$R = \begin{pmatrix} 0.4518 & 0.1659 & 0 \\ 0.0144 & 0.1923 & 0 \\ 0.3432 & 0.3316 & 0.4421 \end{pmatrix}$$

The probability vectors in Table 4 were calculated by substituting the R matrix in Eqs. (6) and (10). Therefore, we get the vector $\tau_0 = (0.1489, 0.5251, 0.0000)$. This τ_0 can be used to compute the rest of the vectors $\tau_i's$ using the equation $\tau_k = \tau_0 R^k$. Lastly, we verified that the cumulative probability is 0.9777 $\approx$1.

Illustration 9 Assume that $\lambda = 0.5, \mu_1 = 1.6, \mu_2 = 2.8, \kappa = 0.5, \eta_0 = 0.1, \eta_1 = 0.2, \nu = 0.3, \bar{\nu} = 0.7$, and

$$R = \begin{pmatrix} 0.5153 & 0.2039 & 0 \\ 0.0177 & 0.2121 & 0 \\ 0.7332 & 0.6984 & 0.4754 \end{pmatrix}$$

Table 5 Probability vector for $\lambda = 0.5$

τ_i	τ_{i0}	τ_{i1}	τ_{i2}	Total
τ_0	0.1480	0.4080	0.0000	0.5560
τ_1	0.0844	0.1279	0.0000	0.2119
τ_2	0.0458	0.0444	0.0000	0.0902
τ_3	0.0244	0.0187	0.0000	0.0431
τ_4	0.0129	0.0089	0.0000	0.0218
τ_5	0.0068	0.0045	0.0000	0.0113
τ_6	0.0036	0.0023	0.0000	0.0059
τ_7	0.0019	0.0012	0.0000	0.0031
τ_8	0.0009	0.0003	0.0000	0.0012
τ_9	0.0005	0.0003	0.0000	0.0008
τ_{10}	0.0003	0.0002	0.0000	0.0005
τ_{11}	0.0002	0.0001	0.0000	0.0003
Total				0.9461

Table 6 Performance measures for λ

λ	0.30	0.35	0.40	0.45	0.50
PS_{2IO}	0.1980	0.2274	0.2569	0.2897	0.3297
PS_{2W}	0.7938	0.7669	0.7424	0.6880	0.6168
PS_{2Id}	0.0079	0.0052	0.0002	0.0000	0.0000
$E[N1]$	0.0926	0.1334	0.1850	0.2629	0.3866
$E[N2]$	0.1514	0.1856	0.2272	0.2353	0.3613
$E[N3]$	0.0039	0.0030	0.0001	0.0000	0.0000
$E[N]$	0.2479	0.3220	0.4123	0.4982	0.7479

The probability vectors in Table 5 were calculated by substituting the R matrix in Eqs. (6) and (10). Therefore, we get the vector $\tau_0 = (0.1480, 0.4080, 0.0000)$. This τ_0 can be used to compute the rest of the vectors $\tau_i' s$ using the equation $\tau_k = \tau_0 R^k$. Lastly, we verified that the cumulative probability is 0.9461 $\approx$1.

5 System Performance Measures

Numerical measurements are chosen at random to satisfy the stability condition. The following Table 6 shows the estimated values for our model when the parameter λ varies from 0.4 to 0.6:

Figure 2 reveals that if the arrival rate λ rises, the prob. of server 2 when it is in an intermittently obtainable state (PS_{2IO}) and the average number of customers in the system when server 2 is in intermittently available state ($E[N1]$), in a working state ($E[N2]$), and in a system ($E[N]$) increase, whereas if the arrival rate λ increases, the prob. of server 2 when it is in working state (PS_{2W}), the prob. of server 2 when it is in idle state (PS_{2Id}), and the average number of customers in the system when server 2 is in idle state ($E[N3]$) subside.

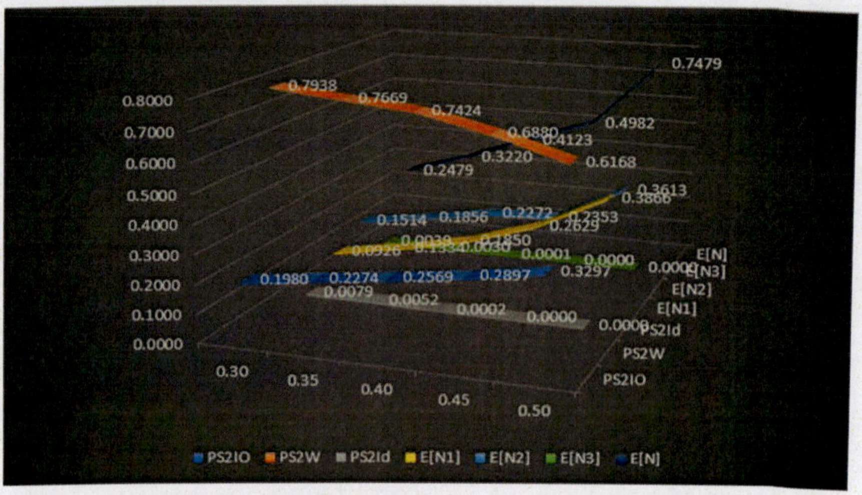

Fig. 2 Arrival rate λ vs PS_{2IO}, PS_{2W}, PS_{2Id}, $E[N1]$, $E[N2]$, $E[N3]$, $E[N]$

6 Conclusion

This study focused on heterogeneous servers with feedback, where server 1 is constantly available and server 2 is intermittently obtainable. An approach based on MGM is used to produce steady-state probability vectors. Finally, we addressed the consequences of performance measures such as prob. of server 2 in an intermittently available state; prob. of server 2 in a working state; prob. of server 2 being idle; the average number of customers in the system when server 2 is intermittently available, working, and in an idle state; and the average number of customers in the system with different arrival rate values.

References

1. Agarwal, N.N.: Some problems in the theory of reliability and queues. Ph.D. Thesis, Kurukshetra university, Kurukshetra (1965)
2. Agassi, M., Aliyeva, S., Sztrik, J.: Analysis of instantaneous feedback queue with heterogeneous servers. Mathematics **8**, 2186 (2020)
3. Ayyappan, G., Subramanian, A.M.G, Sekar, G.: M/M/1 retrial queueing system with preemptive priority service. Int. J. Comput. Appl. **5**(4), 487–500 (2010)
4. Haghighi, A.M., Mishev, D.P.: Queuing models in industry and business. Nova Publishers (2008)
5. Haghighi, A.M., Mishev, D.P.: Delayed and Network Queues. John Wiley and Sons Inc., New Jersey (2016)
6. Indra, Rajan, V.: Queuing analysis of markovian queue having two heterogeneous servers with catastrophes using matrix geometric technique. Int. J. Stat. Appl. **12**, 205–212 (2017)

7. Kalyanaraman, R., Seenivasan, M.: Multi-server retrial queuing system with unreliable server. Int. J. Comput Cog. **7**(3), 113–120 (2010)
8. Krishnakumar, B.K., Madheswari, S.P., Vijayakumar, A.: The M/G/1 retrial queue with feedback and starting failures. Appl. Math. Model., **26**(11), 1057–1075 (2002)
9. Krishnamoorthy, A., Sreenivasan, C.: An M/M/2 queueing system with heterogeneous servers including one with working vacation. Int. J. Stoch. Anal., **2012**, (2012)
10. Kumar, B.K., Madheswari S.P.: An M/M/2 queueing system with heterogeneous servers and multiple vacations. Math. Comput. Model. **13**(41), 1415–1429 (2005)
11. Neuts, M.F.: Matrix-Geometric Solutions in Stochastic Models, vol. 2. Johns Hopkins University, Baltimore (1981)
12. Seenivasan, M., Indumathi, M.: Performance analysis of two heterogeneous server queuing model with intermittently obtainable server using matrix geometric method. J. Phys. Conf. Ser. **1724**(1), IOP Publishing (2021)
13. Seenivasan, M., Chakravarthy, V.J., Abinaya, R.: Markovian Queueing Model with Server Breakdown, Single Working Vacation, and Catastrophe. Advances in Electrical and Computer Engineering, pp. 409–421. Springer, Singapore (2022)
14. Seenivasan, M., Manikandan, H., Subasri, K.S.: Analysis of Heterogeneous Queueing Model with Unreliable Server and Working Vacation. Advances in Electrical and Computer Engineering, pp. 331–345. Springer, Singapore (2022)
15. Sharda: A queueing problem with intermittantly available server and arrivals and departures in batches of variable size. J. Appl. Math. Mech./Z. Angew. Math. Mech. **48**(7), 471–476 (1968)
16. Singh, V.P.: Two-server Markovian queues with balking: heterogeneous vs. homogeneous servers. Oper. Res. **1**(18), 145–159 (1970)

An Economic Order Quantity Inventory Model for the Food Supply Chain with Waste Minimization based on a Circular Economy

S. Vennila and K. Karthikeyan

1 Introduction

"The circular economy is based on the principles of designing out waste and pollution, keeping products and materials in use and regenerating natural systems." The circular economy is followed by the linear economy, the linear economy is defined as take-make-use-dispose, and the circular economy is take–make–use and recycle. Boualem Rabta [1] introduced an EOQ model with a circular economy to make a profit maximization and assert the level of circularity. Sanni et al. [2]. in their paper aimed to develop a mathematical model for the system, including an optimized reordering policy for the model, and identify the sufficient and necessary conditions for the global optimum. Dr. Chittaranjan Mallic [3] determined the optimal order level to minimize the inventory due to the demand rate varying across production cycles, with a finite production rate that no shortages. Enes Ünal et al.'s [4] study contributed to the theoretical understanding of the circular economy implementation and managers with a road map to achieve higher levels of circularity. Anne P.M. Velenturf et al. [5] discussed the important of circular economy to reduce the zero-waste residue technologies and aims to inspire that promotes the transition to a sustainable circularity. According to Abishek Narasimmalu et al. [6]., the major principles of circular economy, which focus on energy and resource recovery from food waste for a sustainable environment, are recycling and reuse. Their review focuses on technological advancements in food waste management, including processing methods and product recovery.

The model by Sebatjane M [7], aims to reduce the inventory cost throughout the supply chain for food preservation technology. Amir Hossein Nobil [8], In their

S. Vennila · K. Karthikeyan (✉)
Department of Mathematics, SAS, VIT, Vellore, Tamil Nadu, India
e-mail: k.karthikeyan@vit.ac.in

F. Kamalov et al. (eds.), *Advances in Mathematical Modeling and Scientific Computing*, Trends in Mathematics, https://doi.org/10.1007/978-3-031-41420-6_53

study, Salameh and Jaber [12] calculated the optimal reorder point for an inventory model to obtain the reordering point and control decisions to manage the volatility of the system. Ata Allah Taleizadeh et al. [9]. in their study developed an EOQ model for partial back ordering and cost increase to find the concavity of optimal profit. Ata Allah Taleizadeh.et.al [10]. proposed an EOQ model for a special quantity greater than the EOQ policy to find maximizing of concavity. Haolan Liao et al. [11]. in their study discussed the EOQ model with uncertain demand under the trade process and carbon cap to solve the multioptimizations results. Imre Dobos et al. [12]. studied the quality thought of the production/recycling model and determined the optimal quality control by the manufacturer and repurchased all units by reducing inventory holding costs. Hovelaque et al. [13]. proposed a unique model that allows a corporation to maximize profit while limiting emissions are identified, as are processes that allow a company to maximize profit while reducing carbon emissions. Andrea Urbinati et al. [14]. in their study investigated the difference between the network value of circular economy and customer value proposition and interface. Khunt Ritchter [15] modifies the variable order cost m and n for determining the minimum cost shown convex and reduced the waste disposal rates to be concave are discussed.

The proposed "circular" form of the "EOQ" inventory model thinks about how conceivable it is for an organization to settle fair and the square of circularity index of the item as estimated by a record. We assume that product circularity influences demand, costs, and prices due to consumer behavior, material choice, supplier selection, technical processes, and perhaps government taxes, subsidies, and restrictions. Thus, the organization must make the best selection based on the factors listed above. Section 2 discusses the analysis of EOQ with CE and the relationship between the demand and unit profit function on the one hand side and the circularity index level on the other hand by using the linear and nonlinear forms with profit maximization.

2 Notations

θ: Food product of circularity index.
Q: Quantity of order (positive).
R: Length of the period.
$\lambda(\theta) > 0$: Rate of demand (units /day) $\lambda(\theta)=\lambda_0$.
$P(\theta) > 0$: Gross profit, $P(\theta)=P_0$.
h: cost of holding $h \neq 0$
K: Fixed ordering cost

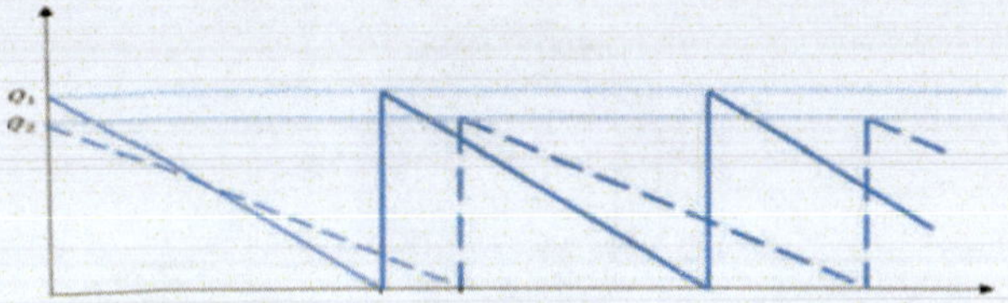

Fig. 1 Variation of circularity in inventory model

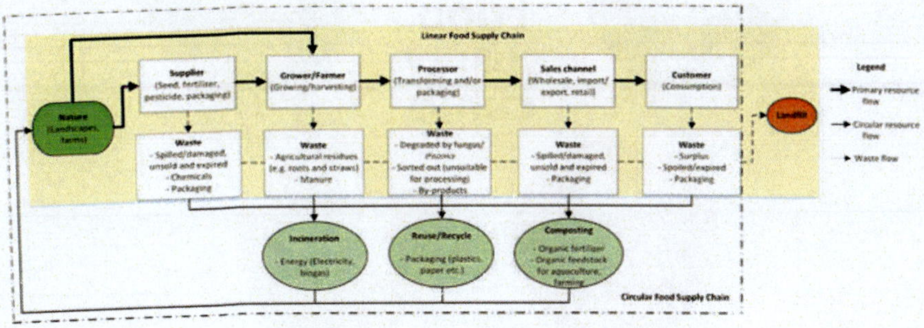

Fig. 2 Structure of circular food supply

3 Analysis of EOQ with CE

According to the EOQ model, demand is deterministic and all expenses are linearly related. Let us assume the model of the product is calculated by circularity index. The manager orders a definite amount Q at the start of each period to meet the exact demand during the period. A circular version of the product can be offered or a standard version, with the circularity based on an index. Ensure that food waste is reduced: prevent food waste from being generated.

(i) Reuse – redistribution for human consumption tends to animal feed.
(ii) Recycle/recover – Extraction of compounds of interest → industrial uses → anaeorabic digestion → composting → thermal treatment with energy recovery → land spreading.

3.1 *Model's Assumption*

1. A Rate of demand is a function of quadratic with CI of an item.
2. For a fixed θ,the interest rate is known, constant, and deterministic.
3. Gross profit per unit, purchase cost are functions of θ.
4. Fixed set-up cost and carrying cost is an independent form of θ.
5. Since the demand is deterministic and known, shortages are not allowed.

Fig. 3 The process of the food supply chain

3.2 *Unit Profit Function and Demand*

The complex approach to represent the impact of the CE naming on the unit profit and demand is to expect the quadratic function's $\lambda(\theta) = a\theta^2 + b\theta + \lambda_0$ and P $(\theta)= a\theta^2 + b\theta + P_0$, in which a and b are constants. As a result, the linearity of their two segments in our model may be simplified and unrealistic. A nonlinear quadratic relationship might be a more likely way to express the effect of Circular Economy Index on demand and cost. We investigated the quadratic linear and quadratic nonlinear relationships between cost and unit profit and a quadratic demand function can also be described.

$$\mathrm{P}\ (\theta) = a\left(e^{\delta(\theta-1)}\right)^2 + b\left(e^{\delta(\theta-1)}\right) + P_0 \tag{1}$$

where P_0, a, and b are constant. Likewise, the logistic function

$$\mathrm{P}\ (\theta) = \mathrm{a}\left(\frac{1}{1+e^{-\delta(\theta-1)}}\right)^2 + \mathrm{b}\left(\frac{1}{1+e^{-\delta(\theta-1)}}\right) + P_0 \tag{2}$$

Here, the assumed quadratic demand function can be expressed in a nonlinear relationship and we proposed the following logarithmic quadratic function:

$$\lambda = a(\log(1+\gamma\theta))^2 + \mathrm{b}\ \log\ (1+\gamma\theta) + \lambda_0 \tag{3}$$

In this equation, λ_0,a, b, γ, and θ are constants. Logic functions are defined as follows:

$$\lambda = \lambda_0 + \left(\frac{a}{1 + e^{-\gamma(\theta-\theta_0)}}\right)^2 + \left(\frac{b}{1 + e^{-\gamma(\theta-\theta_0)}}\right) \tag{4}$$

where λ_0, a, and γ are constants.

3.3 Profit Functions

In contrast to the classical EOQ model, we will maximize profit instead of minimizing cost to account for the effect of circularity on selling prices and on-demand. This work involves measuring the lot size quantity Q in]0, + ∞[as well as the profit order Circular Economy Index $\theta*$ in [0,1] that maximizes the profit. The two functions $\lambda(\theta)$ and P(θ) decrease in θ and the other increases.

To find the trivial solutions, assumed the period profit we may write:

$$\mathrm{P}\ (\theta) = \mathrm{Q} - \mathrm{K} - \frac{\mathrm{hRQ}}{2}. \tag{5}$$

To construct the average profit is:

$$\pi\ (Q, \theta) = \lambda\ (\theta)\ P\ (\theta) - \frac{\lambda\ (\theta)\ K}{Q} - \frac{hQ}{2}. \tag{6}$$

We aim to maximize the average profit $\pi(Q, \theta)$ gives to $0 \leq \theta \leq 1$ for $(Q, \theta) \epsilon$ (0, +∞). Hence we write:

$$\begin{cases} \max \pi\ (Q, \theta) \\ s.t, \\ \theta \leq 1, \\ Q, \theta \geq 0. \end{cases}$$

Subject to constraints $\mathrm{ay} + \mathrm{s}^2 = \mathrm{b}$ and

$$-\mathrm{y} + \mathrm{t}^2 = 0.$$

the Lagrangian equation is

$$\mathrm{L}(\mathrm{Q}, \theta, \lambda, \mu) = \lambda\ (\theta)\ P\ (\theta) - \frac{\lambda\ (\theta)\ K}{Q} - \frac{hQ}{2} - \lambda \left(\mathrm{ay} + \mathrm{s}^2 - \mathrm{b}\right) \theta - \mu \left(-\mathrm{y} + t^2\right) \theta. \tag{7}$$

Karush-kuhn-tucker quadratic programming conditions are given by

$$\frac{\partial \mathrm{L}\ (\mathrm{Q}, \theta, \lambda, \mu)}{\partial Q} = \frac{\lambda\,(\theta)\,K}{Q^2} - \frac{h}{2} = 0. \tag{8}$$

$$\frac{\lambda\,(\theta)\,K}{Q^2} = \frac{h}{2} \tag{9}$$

$$\mathrm{Q}^2 = \frac{2\lambda\,(\theta)\,K}{h} \tag{10}$$

$$\mathrm{Q}{*} = \sqrt{\frac{2\lambda\,(\theta)\,K}{h}} \tag{11}$$

$$\frac{\partial L}{\partial \theta} = \lambda\,(\theta)\,P'\ (\theta) + \lambda'\,(\theta)\,P\,(\theta) - \frac{\lambda'\,(\theta)\,K}{Q} - \lambda\left(ay + s^2 - b\right) - \mu\left(-y + t^2\right) = 0. \tag{12}$$

$$\text{if } \mathrm{ay} + \mathrm{s}^2 - \mathrm{b} = 0;\ \mu y - t^2 = 0$$

3.4 Optimal Profit

$$\pi^* = \pi\left(Q^*, \theta^*\right) = \left[\lambda\left(\theta^*\right) P\left(\theta^*\right) - \frac{\lambda\,(\theta^*)\,K}{Q^*} - \frac{hQ^*}{2}\right] \tag{13}$$

Substitute the value $\mathrm{Q}{*} = \theta^*$, we obtain

$$\pi\left(Q^*, \theta^*\right) = \pi\left(\theta^*\right) = \lambda\,(\theta)\,P\,(\theta) - \sqrt{2h\lambda\,(\theta)\,K}. \tag{14}$$

3.5 Linear Relationship

Consider the Hessian Matrix to find the Concavity of Objective function:

$$\frac{\partial f}{\partial Q} = -\frac{\left[\lambda_0 + a\theta^2 + b\theta\right] K}{Q^2} - \frac{h}{2}. \tag{15}$$

$$\frac{\partial^2 f}{\partial Q^2} = -\frac{2\left(a\theta^2 + b\theta + \lambda_0\right)K}{Q^3}$$

$$\frac{\partial^2 f}{\partial Q \partial \theta} = \frac{(2a\theta + b)\,K}{Q^2} \tag{16}$$

$$\frac{\partial^2 f}{\partial \theta^2} = 2(2a\theta + b)^2 + \left[4^2 a^2\theta^2 + 4ab\theta + 2aP_0 + 2a\lambda_0 - \frac{2aK}{Q}\right] \tag{17}$$

Therefore, the hessian matrix is

$$\mathrm{H} = \begin{bmatrix} -2\frac{\left(a\theta^2+b\theta+\lambda_0\right)K}{Q^3} & \frac{(2a\theta+b)K}{Q^2} \\ \frac{(2a\theta+b)K}{Q^2} & 2(2a\theta + b)^2 + \left[4a^2\theta^2 + 4ab\theta + 2aP_0 + 2a\lambda_0 - \frac{2aK}{Q}\right] \end{bmatrix} \tag{18}$$

Hessian Matrix solution should be Negative semidefinite to secure the concavity. Note that

$$-2\frac{\left(a\theta^2 + b\theta + \lambda_{0K}\right)}{Q^3} < 0 \text{ and } 2(2a\theta + b)^2$$

$$+\left[4a^2\theta^2 + 4ab\theta + 2aP_0 + 2a\lambda_0 - \tfrac{2aK}{Q}\right] < 0$$

Also, $|H|>0$,

$$|H| = 2K\left[4a^2\theta^2 + 4ab\theta + 2a\left(\lambda_0 + P_0\right) - \frac{2aK}{Q}\right] > -\frac{(2a\theta + b)}{Q}\mathrm{K} \tag{19}$$

Therefore, the result H is concavity because the solution is in the negative semi-definitive.

3.6 *Nonlinear Relationships with Exponential Unit Profit Function*

To find the nonlinear by using the efforts of Lagrange and to note the local maximum and local minimum in the objective function, consider the quadratic functions:

$$\lambda(\theta) = \lambda_0 + a\theta^2 + b\theta > 0,$$

$$P\ (\theta) = \left[P_0 + ae^{\delta(\theta-1)^2} + be^{\delta(\theta-1)}\right] > 0 \tag{20}$$

Then the Profit function is

$$\pi\ (Q, \theta) = \left(a\theta^2 + b\theta + \lambda_0\right)\left(P_0 + a\left(e^{\delta(\theta-1)^2}\right) + \mathrm{b}e^{\delta(\theta-1)}\right) \tag{21}$$

$$-\left(\frac{(a\theta^2+b\theta+\lambda_0)K}{Q}\right) - \frac{hQ}{2}.$$

$$\frac{\partial^2 L}{\partial Q^2} = \frac{-2\left(a\theta^2 + b\theta + \lambda_0\right)K}{Q^3}$$

$$\frac{\partial^2 L}{\partial Q \partial\theta} = \frac{(2a\theta + b)\ K}{Q^2}$$

$$\frac{\partial^2 L}{\partial\theta^2} = 2aP_0 + \left[\left((2a\delta + b\delta)\ a\theta^2 + 2b\theta + 2a\theta + \lambda_0\right)\right.$$

$$\left. + \left(\left(2a^2e^{\delta(\theta-1)}\right) + 2ab + 4a^2\delta\theta + 2ab\delta\theta + 2ab\delta + b^2\delta\right)\right]e^{\delta(\theta-1)} - \frac{2aK}{Q} \tag{22}$$

Therefore, the hessian matrix is

$$H = \begin{bmatrix} \frac{-2\left(a\theta^2+b\theta+\lambda_0\right)K}{Q^3} & \frac{(2a\theta+b)K}{Q^2} \\ \frac{(2a\theta+b)K}{Q^2} & \begin{matrix} 2aP_0 + (2a\delta + b\delta)\left(a\theta^2 + 2b\theta + 2a\theta + \lambda_0\right) + \\ \left(2a^2e^{\delta(\theta-1)} + 2ab + 4a^2\delta\theta + 2ab\delta\theta + 2ab\delta + b^2\delta\right) \\ e^{\delta(\theta-1)} - \frac{2aK}{Q} \end{matrix} \end{bmatrix}$$

It should be negative semidefinitive to execute the concave. The parameters of equations are a, b, and δ are must be a > 0, $b < 0$ and $\delta > 0$. Then determinant term is negative and

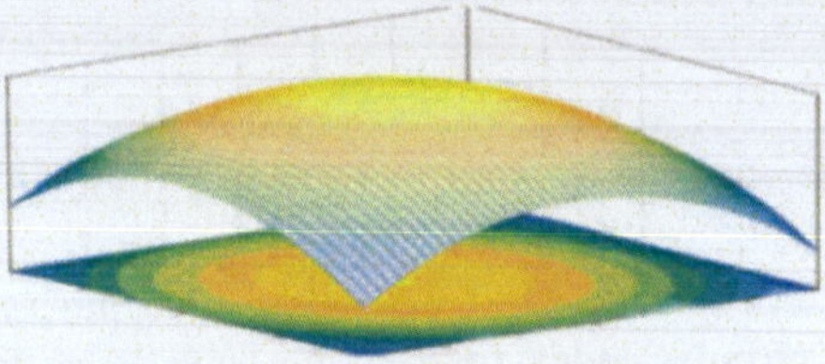

Fig. 4 A concavity plot of profit functions

$$|H| = \left[\left(\frac{-2(a\theta^2+b\theta+\lambda_0)K}{Q^3}\right)\left[\left[2aP_0 + (2a\delta + b\delta)\left(a\theta^2 + 2b\theta + 2a\theta + \lambda_0\right)\right.\right.\right.$$
$$\left.\left. + \left(2a^2e^{\delta(\theta-1)} + 2ab + 4a^2\delta\theta + 2ab\delta\theta + 2ab\delta + b^2\delta\right)e^{\delta(\theta-1)}\right] - \frac{2aK}{Q}\right]$$
$$\left. -\left[\frac{(2a\theta+b)K}{Q^2}\right]^2\right] > 0 \tag{23}$$

$$\text{Det H} = \left[2\left(a\theta^2 + b\theta + \lambda_0\right)\left[\left[\left(2aP_0 + \left(2a\delta + b\delta\right)\left(a\theta^2 + 2a\theta + 2b\theta + \lambda_0\right)\right.\right.\right.\right.$$
$$\left. + \left(\left(2a^2e^{\delta(\theta-1)}\right) + 2\text{ab} + 4a^2\delta\theta + 2ab\delta\theta + 2ab\delta + b^2\delta\right)e^{\delta(\theta-1)}\right]$$
$$\left.\left. -2\text{aK}Q^2\right]\right] - \frac{(2a\theta+b)^2K^2}{Q} > 0. \tag{24}$$

Hence, H is negative semidefinitive to find the concavity (Fig. 4).

3.7 Illustration

Let's us have an example to demonstrate the findings of the study. Examine the impact of the demand's shape and size in the model. Consider the cost of ordering is set as K = "\$200, h = \$ 0.4 as a holding cost. Assume demand rate θ = 70,000 and P_0="\$ 3" is the basic unit profit. A product is in quadratic demand function, circularity indicator $\lambda(\theta)$=$\lambda_0 + a\theta^2 + b\theta$. And demand a = 8600 units and also P (θ)= $a\theta^2 + b\theta + P_0$. Accordingly, we assume a maximum 13.5% increase in demand and a 12% decrease in unit gross profit as the circularity level of the product increases. The optimal solution is defined as Q* = 13,860 and θ^*= 0.56, Annual profit π^* = "\$ 236,120"

4 Conclusion

In this Chapter, we introduced an EOQ inventory model for food supply chain with waste minimization. In the circular economy, food waste is a major issue, to overcome the problem we recycle food waste. The purpose of this research is to determine the best circular index for reusing food so as to maximize profit by protecting the environment. This chapter explained that recycled food product can be purchased at any time; a customer can decide what they want to buy according to the decision variable. This work measures the product by circularity index with linear and non-linear relationships related to the Karush-Kuhn Tucker quadratic programming problem to maximize profit.

References

1. Rabta, B.: An economic order quantity inventory model for a circular economy indicator. Comput. Ind. Eng. **140**, 106215 (2020)
2. S.S Sanni, W.I.E. Chukwu. An EOQ model for items with three parameter Weibull distribution deterioration ramp- type demand and shortages. Appl. Math. Model. 37,9698–9706 (2013)
3. Mallic, C.: Formulation of optimal economic order quantity under different inventory models. IJERT. **2**(10), 2278–0181 (2013)
4. Ünal, E., Shao, J.: A taxonomy of circular economy implementation strategies for manufacturing firms: analysis of 391 cradle-to-cradle products. J. Clean. Prod. **212**, 754e765 (2019)
5. Velenturf, A.P.M., Archer, S.A., Gomes, H.I., Christgen, B., Lag-Brotons, A.J., Purnell, P.: Circular economy and the matter of integrated resources. Sci. Total Environ. **689**, 963–969 (2019)
6. Narasimmalu, A., Ramasamy, R.: Food processing industry waste and circular economy. IOP Conf. Series: Mater. Sci. Eng. **955 955**, 012089 (2020)
7. Sebatjane, M.: The impact of preservation technology investments on lot-sizing and shipment strategies in a three- echelon food supply chain involving growing and deteriorating items. Oper. Res. Perspect. **100241** (2022)
8. Nobil, A.H., Sedigh, A.H.A., Cárdenas-Barrón, L.E.: Reorder point for the EOQ inventory model with imperfect quality items. Ain Shams Eng. J. **11**, 1339–1343 (2020)
9. Ata Allah Taleizadeh, A., Pentico, D.W.: An economic order quantity model with a known price increase and partial backordering. Eur. J. Oper. Res. **228**, 516–525 (2013)
10. Taleizadeh, A.A., Pentico, D.W., Aryanezhad, M., Ghoreyshi, S.M.: An economic order quantity model with partial backordering and a special sale price. Eur. J. Oper. Res. **221**, 571–583 (2012)
11. Liao, H., Deng, Q.: A carbon constrained EOQ model with uncertain demand for remanufactured products. J. Clean. Prod. **199**, 334–347 (2018)
12. Dobos, I.: Knut Richer: a production/ recycling model with quality consideration. Int. J. Prod. Econ. **104**, 571–579 (2006)
13. Hovelaque, V., Bironneau, L.: The carbon-constrained EOQ model with carbon emission dependent demand. Int. J. Prod. Econ. **164**, 285–291 (2015)
14. Urbinati, A., Chiaroni, D., Chiesa, V.: An extended production/recycling model with stationary demand and return rates. J. Clean. Prod. **168**, 487–498 (2017)
15. Richter, K.: The EOQ repair and waste disposal model with variable setup numbers. Eur. J. Oper. Res. **96**, 313–324 (1996)

Achievement Estimations of Priority Queue System in Fuzzy Environment

R. Ramesh, M. Kannan, and M. Seenivasan

1 Introduction

For the time being, we confront countless priority-based issues in the queueing ambient such as hospitals, banks, medical shops, petrol stations, ration stores, and dialing calls in mobile networks. Underscoring the importance of time management is the uttermost concentration of the researcher. At this juncture, the queueing model employs a massive emphatic part. The basic definitions, theorems [3, 5, 9] and models of queueing theory are very important for our research intentions. In our perpetual period, most of the time we touch the fuzzy idea and utilizations [10, 23]. We implement the priorities [15] in the queueing environments very often. Occasionally, either the priorities are gained instantly or these get hold of a long duration. These achievements can also be measured by fuzzy sanity [7, 13].

Altogether, class 1 customers (with priority) and class 2 customers (without priority) are serviced by same server. But the cost measures are entirely different for these customers [20]. In this everlasting situation, the preemptive priority-based customers are basically intended as special class 1 customers. This is an added privilege of the abovementioned customers. But the non-preemptive priority-based customers [14, 16, 17] are always normal in consideration. However, practically, these customers are far better than the class 2 customers. More researchers have contributed various ranking techniques [1, 2, 4, 6, 12].

R. Ramesh
Department of Mathematics, Arignar Anna Government Arts College, Musiri, Tamil Nadu, India

M. Kannan
Department of Mathematics, Government Arts College, Ariyalur, Tamil Nadu, India

M. Seenivasan (✉)
Mathematics Wing – DDE, Annamalai University, Annamalainagar, Chidambaram, Tamil Nadu, India

F. Kamalov et al. (eds.), *Advances in Mathematical Modeling and Scientific Computing*, Trends in Mathematics, https://doi.org/10.1007/978-3-031-41420-6_54

Many authors contributed more ordering procedures for detecting the achievements of fuzzy system like Robust ranking method [11]. But the distance-based ordering procedures are very popular. In particular, the area-based [19, 22], centroid of centroids [16, 21], and circumcenter of centroids [18] ordering methods are very noteworthy.

Here we utilize the in-center-based [8] ordering procedure for finding the achievements of fuzzy queueing models of non-preemptive priority.

2 Preludes

2.1 Fuzzy Set

A set $\tilde{A}$ is determined as a fuzzy set if it is accomplished by a membership function $\phi_{\tilde{A}}$, i.e., $\tilde{A}$ = {(x, $\phi_{\tilde{A}}$ (x)); xU}. Here, the function mapping from members of the universal set U to [0,1] and $\phi_{\tilde{A}}$ (x) is interpreted as the membership grade of xU in $\tilde{A}$.

2.2 Triangular Fuzzy Number

If a number $\tilde{A}$ $(a_1, a_2, a_3; 1)$ is called triangular fuzzy number, it is accomplished by a membership function $\phi_{\tilde{A}}(x) = \begin{cases} \frac{x-a_1}{a_2-a_1}, & a_1 \le x \le a_2 \\ 1, & x = a_2 \\ \frac{x-a_3}{a_2-a_3}, & a_2 \le x \le a_3 \\ 0, & otherwise \end{cases}$.

2.3 Trapezoidal Fuzzy Number

If a number $\tilde{A}$ $(a_1, a_2, a_3, a_4; 1)$ is called trapezoidal fuzzy number, it is accomplished by a membership function $\phi_{\tilde{A}}(x) = \begin{cases} \frac{x-a_1}{a_2-a_1}, & a_1 \le x \le a_2 \\ 1, & a_2 \le x \le a_3 \\ \frac{x-a_4}{a_3-a_4}, & a_3 \le x \le a_4 \\ 0, & otherwise \end{cases}$.

3 Non-Preemptive Priority Fuzzy Queues

Consider the Kth priority clients occurring at solo queue (Fig. 1) by a Poisson stream with the fuzzy rate $\tilde{\lambda}_k$, (k = 1, 2, 3, ..., r) and waiting for service, which has the fuzzy rate $\tilde{\mu}$ in FIFO mode. Regardless of priorities, the service distribution for those clients keeps exponentially with mean $1/\tilde{\mu}_k$ and completes the service within next client is entered. They got service by their individual precedence.

We start with $\rho_k = \frac{\lambda_k}{\mu_k} < 1, (1 \le k < r), \sigma_k = \sum_{i=1}^{k} \rho_i\ (\sigma_0 \approx 0, \sigma_r \approx \rho)$.

The steady-state condition for $\sigma_r = \rho < 1$. Let $\varphi_{\tilde{\lambda}}(s), \varphi_{\tilde{\mu}}(t)$ are the membership functions of fuzzy sets $\widetilde{\lambda}_i = \left\{ \left(s, \varphi_{\widetilde{\lambda}}(s)\right) \middle| s \in S\right\}$ and $\widetilde{\mu} = \left\{ \left(t, \varphi_{\widetilde{\mu}}(t)\right) \middle| t \in T\right\}$.

By queuing theory postulations,

The average number of customers in the Queue $L_q = \sum_{i=1}^{r} L_q^{(i)} = \sum_{i=1}^{r} \frac{\lambda_i \sum_{k=1}^{r} \frac{\rho_k}{\mu_k}}{(1-\sigma_{i-1})(1-\sigma_i)}$.

The waiting time in queue $W_q = \sum_{i=1}^{r} \frac{\lambda_i w_q^{(i)}}{\lambda}$, where $w_q^{(i)} = \frac{\sum_{k=1}^{r} \frac{\rho_k}{\mu_k}}{(1-\sigma_{i-1})(1-\sigma_i)}$.

We consider M/M/1 queueing system, which will be 3-priorities with same service rates.

$$\tilde{\mu}_1 = \tilde{\mu}_2 = \tilde{\mu}_3 = \tilde{\mu}.$$

Now $\tilde{\rho}_1 = \frac{\tilde{\lambda}_1}{\tilde{\mu}_1}, \tilde{\rho}_2 = \frac{\tilde{\lambda}_2}{\tilde{\mu}_2}$, and $\tilde{\rho}_3 = \frac{\tilde{\lambda}_3}{\tilde{\mu}_3}$. Since $\tilde{\rho} = \tilde{\rho}_1+\tilde{\rho}_2+\tilde{\rho}_3, \tilde{\lambda} = \tilde{\lambda}_1+\tilde{\lambda}_2+\tilde{\lambda}_3\ \tilde{\rho} = \frac{\tilde{\lambda}_1+\tilde{\lambda}_2+\tilde{\lambda}_3}{\tilde{\mu}}$.

Also $w_q^{(i)} = \frac{\frac{\tilde{\rho}_1+\tilde{\rho}_2+\tilde{\rho}_3}{\tilde{\mu}}}{(1-\sigma_{i-1})(1-\sigma_i)}$ and $L_q^{(i)} = \frac{(\tilde{\rho}_1+\tilde{\rho}_2+\tilde{\rho}_3)\tilde{\rho}_i}{(1-\sigma_{i-1})(1-\sigma_i)}$.

The mean clients in the system $L_s = \frac{L_q}{\rho} = \frac{\sum_{i=1}^{r} L_q^{(i)}}{\sum_{i=1}^{r} \rho_i}$.

The waiting time in the system $W_s = \frac{L_S}{\lambda} = \frac{\sum_{i=1}^{r} L_s^{(i)}}{\sum_{i=1}^{r} \lambda_i}$.

From which we deduce that

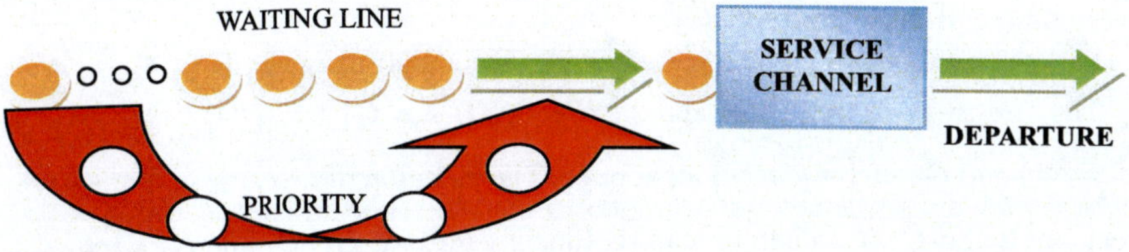

Fig. 1 Priority queue

$$w_q^{(1)} = \frac{\tilde{\rho}}{\tilde{\mu} - \tilde{\lambda}_1}$$

$$w_q^{(2)} = \frac{\tilde{\rho}}{(1 - \tilde{\rho}_1 - \tilde{\rho}_2)\left(\tilde{\mu} - \tilde{\lambda}_1\right)}$$

$$w_q^{(3)} = \frac{\tilde{\rho}}{(1 - \tilde{\rho}_1 - \tilde{\rho}_2 - \tilde{\rho}_3)\left(\tilde{\mu} - \tilde{\lambda}_1 - \tilde{\lambda}_2\right)}$$

$$L_q^{(1)} = \frac{\tilde{\rho}\tilde{\rho}_1}{(1 - \tilde{\rho}_1)},$$

$$L_q^{(2)} = \frac{\tilde{\rho}\tilde{\rho}_2}{(1 - \tilde{\rho}_1 - \tilde{\rho}_2)(1 - \tilde{\rho}_1)}$$

$$L_q^{(3)} = \frac{\tilde{\rho}\tilde{\rho}_3}{(1 - \tilde{\rho}_1 - \tilde{\rho}_2 - \tilde{\rho}_3)(1 - \tilde{\rho}_1 - \tilde{\rho}_2)}$$

$$L_S^{(1)} = \frac{L_q^{(1)}}{\tilde{\rho}_1}\; L_S^{(2)} = \frac{L_q^{(2)}}{\tilde{\rho}_2}\; L_S^{(3)} = \frac{L_q^{(3)}}{\tilde{\rho}_3}\; W_S^{(1)} = \frac{L_S^{(1)}}{\tilde{\lambda}_1}\; W_S^{(2)} = \frac{L_S^{(2)}}{\tilde{\lambda}_2}\; W_S^{(3)} = \frac{L_S^{(3)}}{\tilde{\lambda}_3}$$

4 Circumcenter of Centroids Ranking Method

Consider a trapezium (Fig. 2) APQD and extract the trapezium into one rectangle (*BPQC*), and two triangles (*APB&CQD*). The circumcenter of the centroids of these rectangle and triangles is observed as the reference point for ordering the generalized fuzzy numerals. The circumcenter of these centroids is equidistant from every tip (centroids) of $\Delta G_1G_2G_3$.

Let $\tilde{A} = (a, b, c, d; w)$ be a generalized trapezoidal fuzzy number (Fig. 2).

The centroids of (*BPQC*), and (*APB&CQD*) are $G_1 = \left(\frac{a+2b}{3}, \frac{w}{3}\right)$, $G_2 = \left(\frac{b+c}{2}, \frac{w}{2}\right)$ and $G_3 = \left(\frac{2c+d}{3}, \frac{w}{3}\right)$ all at once. The equation of $\overleftrightarrow{G_1G_3}$ is $y = \frac{w}{3}$ and G_2 does not lie on $\overleftrightarrow{G_1G_3}$. Then the non-collinear centroids G_1, G_2 and G_3 are from a triangle.

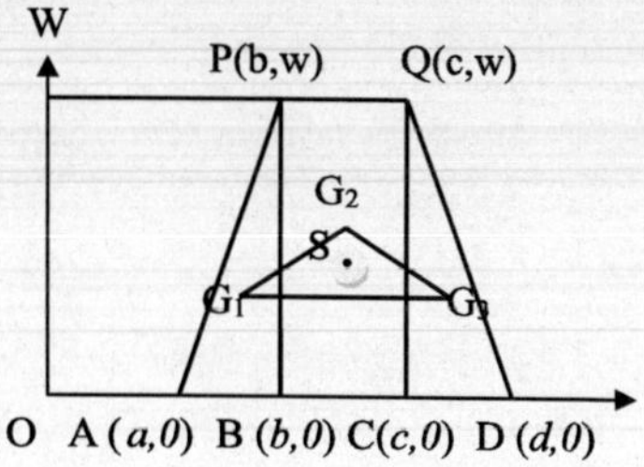

Fig. 2 Circumcenter of centroids

We detect the circumcenter $S_{\tilde{A}} = (\overline{x}_0, \overline{y}_0)$ of the $\Delta G_1G_2G_3$ for $\tilde{A} = (a, b, c, d; w)$ as

$$S_{\tilde{A}}(\overline{x}_0, \overline{y}_0) = \left(\frac{a + 2b + 2c + d}{6}, \frac{(2a + b - 3c)(2d + c - 3b) + 5w^2}{12w}\right) \tag{1}$$

If c = b in $\tilde{A} = (a, b, c, d; w)$, we get the circumcenter of $\tilde{A} = (a, b, d; w)$ is

$$S_{\tilde{A}}(\overline{x}_0, \overline{y}_0) = \left(\frac{a + 4b + d}{6}, \frac{4(a - b)(d - b) + 5w^2}{12w}\right) \tag{2}$$

Finally the ranking index is $R(\tilde{A}) = \sqrt{\overline{x}_0^{\,2} + \overline{y}_0^{\,2}}$, which is the Euclidean distance from the circumcenter of the centroids.

5 Illustration

Imagine an abnormal situation where mysterious disease cases are engaged in Chennai Kaveri Hospital, and some patients are waiting there to occupy beds with oxygen ventilators. They all want to get medical treatment. At once, some patients need immediate treatment, particularly for low-oxygen-level critical patients. The situation is very critical to handle when the hospital admits serious patients. We can detect the extended waiting time and the queue length for this present situation.

Illustration (i): (Trapezoidal Fuzzy Number)

Consider the arrival rates of 1st, 2nd, and 3rd priority units being $\tilde{\lambda}_1 = [1,2,4,5:1]$, $\tilde{\lambda}_2 = [2,3,5,6:1]$, $\tilde{\lambda}_3 = [3,4,6,7:1]$, and the same service rate $\tilde{\mu}_1 = [53,54,56,57:1]$/h all at once.

The membership function of [1,2,4,5:1] is

$$\varphi_{\tilde{\lambda}}(x) = \begin{cases} \frac{(x-1)}{(2-1)}, & 1 \le x \le 2 \\ 1, & 2 \le x \le 4 \\ \frac{(x-5)}{(4-5)}, & 4 \le x \le 5 \\ 0, & otherwise \end{cases}$$

$$\text{And } \mathrm{R}\left(\tilde{\lambda}_1\right) = \mathrm{R}\,(1, 2, 4, 5 : 1)$$
$$= \left(\frac{1+2(2)+2(4)+5}{6}, \frac{(2(1)+2-3(4))(2(5)+4-3(2))+5(1)^2}{12(1)}\right)$$
$$= \sqrt{(3)^2 + (-4.92)^2} = 5.76$$

$$\mathrm{R}\left(\tilde{\lambda}_2\right) = \mathrm{R}\,(2, 3, 5, 6 : 1) = 6.34, \mathrm{R}\left(\tilde{\lambda}_3\right) = \mathrm{R}\,(3, 4, 6, 7 : 1) = 7.01, \mathrm{R}\,(\tilde{\mu}_1)$$
$$= \mathrm{R}\,(53, 54, 56, 57 : 1) = 55.22$$

$$\text{And } \mathrm{R}\left(\tilde{\lambda}\right) = 19.11, \mathrm{R}\,(\tilde{\rho}_1) = 0.10, \mathrm{R}\,(\tilde{\rho}_2) = 0.12, \mathrm{R}\,(\tilde{\rho}_3) = 0.13, \mathrm{R}\,(\tilde{\rho}) = 0.35$$

Then

Mean waiting span of 1st priority in the queue is

$$w_q^{(1)} = \frac{\tilde{\rho}}{\tilde{\mu}-\tilde{\lambda}_1}$$
$$= \frac{0.35}{55.22-5.76} = 0.007$$

Mean waiting span of 2nd priority in the queue is

$$w_q^{(2)} = \frac{\tilde{\rho}}{(1-\tilde{\rho}_1-\tilde{\rho}_2)\left(\tilde{\mu}-\tilde{\lambda}_1\right)}$$
$$= \frac{0.35}{(1-0.10-0.12)(55.22-5.76)} = 0.009$$

Mean waiting span of 3rd priority in the queue is

$$w_q^{(3)} = \frac{\tilde{\rho}}{(1-\tilde{\rho}_1-\tilde{\rho}_2-\tilde{\rho}_3)\left(\tilde{\mu}-\tilde{\lambda}_1-\tilde{\lambda}_2\right)}$$
$$= \frac{0.35}{(1-0.10-0.12-0.13)(55.22-5.76-6.34)} = 0.012$$

The average number of customers in the queue of 1st priority is

$$L_q^{(1)} = \frac{\tilde{\rho}\tilde{\rho}_1}{(1-\tilde{\rho}_1)}$$
$$= \frac{0.035}{(1-0.10)} = 0.04$$

The average number of customers in the queue of 2nd priority is

$$L_q^{(2)} = \frac{\tilde{\rho}\,\tilde{\rho}_2}{(1-\tilde{\rho}_1-\tilde{\rho}_2)(1-\tilde{\rho}_1)}$$
$$= \frac{0.042}{(1-0.10-0.12)(1-0.10)} = 0.06$$

The average number of customers in the queue of 3rd priority is

$$L_q^{(3)} = \frac{\tilde{\rho}\,\tilde{\rho}_3}{(1-\tilde{\rho}_1-\tilde{\rho}_2-\tilde{\rho}_3)(1-\tilde{\rho}_1-\tilde{\rho}_2)}$$
$$= \frac{0.0455}{(1-0.10-0.12-0.13)(1-0.10-0.12)} = 0.09$$

The average number of customers in the system of 1st priority is

$$L_S^{(1)} = \frac{L_q^{(1)}}{\tilde{\rho}_1} = 0.4$$

The average number of customers in the system of 2nd priority is

$$L_S^{(2)} = \frac{L_q^{(2)}}{\tilde{\rho}_2} = 0.5$$

The average number of customers in the system of 3rd priority is

$$L_S^{(3)} = \frac{L_q^{(3)}}{\tilde{\rho}_3} = 0.7$$

Mean waiting span of 1st priority in the system is

$$W_S^{(1)} = \frac{L_S^{(1)}}{\tilde{\lambda}_1} = 0.07$$

Mean waiting span of 2nd priority in the system is

$$W_S^{(2)} = \frac{L_S^{(2)}}{\tilde{\lambda}_2} = 0.08$$

Mean waiting span of 3rd priority in the system is

$$W_S^{(3)} = \frac{L_S^{(3)}}{\tilde{\lambda}_3} = 0.10$$

Similarly, we can proceed for some illustrations we got.

Table 1 Performance measures of $w_q^{(1)}$

$w_q^{(1)}$ (Units)	$\widetilde{\mu}_1 = 55.22$	$\widetilde{\mu}_2 = 56.22$	$\widetilde{\mu}_3 = 57.21$	$\widetilde{\mu}_4 = 58.21$	$\widetilde{\mu}_5 = 59.20$
$\lambda = 19.11$	0.007	0.006	0.006	0.005	0.005
$\lambda = 25.71$	0.010	0.010	0.009	0.009	0.008
$\lambda = 33.45$	0.014	0.013	0.012	0.012	0.011
$\lambda = 41.71$	0.018	0.017	0.017	0.016	0.015
$\lambda = 50.23$	0.023	0.022	0.021	0.020	0.020

Table 2 Performance measures of $w_q^{(2)}$

$w_q^{(2)}$ (Units)	$\widetilde{\mu}_1 = 55.22$	$\widetilde{\mu}_2 = 56.22$	$\widetilde{\mu}_3 = 57.21$	$\widetilde{\mu}_4 = 58.21$	$\widetilde{\mu}_5 = 59.20$
$\lambda = 19.11$	0.009	0.008	0.006	0.005	0.005
$\lambda = 25.71$	0.015	0.014	0.013	0.012	0.011
$\lambda = 33.45$	0.022	0.021	0.020	0.019	0.018
$\lambda = 41.71$	0.034	0.033	0.031	0.029	0.028
$\lambda = 50.23$	0.056	0.053	0.049	0.046	0.044

Table 3 Performance measures of $w_q^{(3)}$

$w_q^{(3)}$ (Units)	$\widetilde{\mu}_1 = 55.22$	$\widetilde{\mu}_2 = 56.22$	$\widetilde{\mu}_3 = 57.21$	$\widetilde{\mu}_4 = 58.21$	$\widetilde{\mu}_5 = 59.20$
$\lambda = 19.11$	0.012	0.012	0.011	0.011	0.010
$\lambda = 25.71$	0.016	0.022	0.020	0.019	0.017
$\lambda = 33.45$	0.046	0.041	0.039	0.038	0.034
$\lambda = 41.71$	0.106	0.097	0.089	0.081	0.072
$\lambda = 50.23$	0.446	0.380	0.297	0.239	0.212

Table 4 Performance measures of $L_q^{(1)}$

$L_q^{(1)}$ (Units)	$\widetilde{\mu}_1 = 55.22$	$\widetilde{\mu}_2 = 56.22$	$\widetilde{\mu}_3 = 57.21$	$\widetilde{\mu}_4 = 58.21$	$\widetilde{\mu}_5 = 59.20$
$\lambda = 19.11$	0.04	0.04	0.03	0.03	0.02
$\lambda = 25.71$	0.09	0.08	0.07	0.07	0.06
$\lambda = 33.45$	0.14	0.13	0.13	0.12	0.11
$\lambda = 41.71$	0.22	0.22	0.21	0.20	0.20
$\lambda = 50.23$	0.37	0.35	0.34	0.32	0.31

Here, Tables 1, 2, 3, 4, 5, 6, 7, 8, 9, 10, 11, and 12 and the Figs. 3, 4, 5, 6, 7, 8, 9, 10, 11, 12, 13, and 14 show that if we allow the priorities, then the performance measures will increase with respect to the arrival and service rates.

Illustration (ii): (Triangular Fuzzy Number)

Conceive the arrival rates of 1st, 2nd, and 3rd priority units are $\widetilde{\lambda}_1 = [1,4,5:1]$, $\widetilde{\lambda}_2 = [2,5,6:1]$, $\widetilde{\lambda}_3 = [3,6,7:1]$ and the same service rate $\widetilde{\mu}_1 = [54,57,58:1]$/h all at once.

The membership function of [1,4,5:1] is

Table 5 Performance measures of $L_q^{(2)}$

$L_q^{(2)}$ (Units)	$\widetilde{\mu}_1 = 55.22$	$\widetilde{\mu}_2 = 56.22$	$\widetilde{\mu}_3 = 57.21$	$\widetilde{\mu}_4 = 58.21$	$\widetilde{\mu}_5 = 59.20$
$\lambda = 19.11$	0.06	0.06	0.05	0.05	0.04
$\lambda = 25.71$	0.12	0.12	0.11	0.11	0.09
$\lambda = 33.45$	0.25	0.23	0.21	0.21	0.20
$\lambda = 41.71$	0.47	0.46	0.43	0.40	0.38
$\lambda = 50.23$	0.94	0.89	0.82	0.78	0.73

Table 6 Performance measures of $L_q^{(3)}$

$L_q^{(3)}$ (Units)	$\widetilde{\mu}_1 = 55.22$	$\widetilde{\mu}_2 = 56.22$	$\widetilde{\mu}_3 = 57.21$	$\widetilde{\mu}_4 = 58.21$	$\widetilde{\mu}_5 = 59.20$
$\lambda = 19.11$	0.09	0.08	0.07	0.07	0.06
$\lambda = 25.71$	0.22	0.20	0.18	0.17	0.17
$\lambda = 33.45$	0.56	0.49	0.46	0.46	0.40
$\lambda = 41.71$	1.56	1.42	1.33	1.13	1.06
$\lambda = 50.23$	7.89	6.86	5.29	4.19	3.53

Table 7 Performance measures of $L_S^{(1)}$

$L_S^{(1)}$ (Units)	$\widetilde{\mu}_1 = 55.22$	$\widetilde{\mu}_2 = 56.22$	$\widetilde{\mu}_3 = 57.21$	$\widetilde{\mu}_4 = 58.21$	$\widetilde{\mu}_5 = 59.20$
$\lambda = 19.11$	0.40	0.40	0.39	0.37	0.35
$\lambda = 25.71$	0.55	0.55	0.54	0.54	0.46
$\lambda = 33.45$	0.74	0.72	0.70	0.68	0.65
$\lambda = 41.71$	0.96	0.96	0.94	0.91	0.90
$\lambda = 50.23$	1.28	1.25	1.22	1.18	1.16

Table 8 Performance measures of $L_S^{(2)}$

$L_S^{(2)}$ (Units)	$\widetilde{\mu}_1 = 55.22$	$\widetilde{\mu}_2 = 56.22$	$\widetilde{\mu}_3 = 57.21$	$\widetilde{\mu}_4 = 58.21$	$\widetilde{\mu}_5 = 59.20$
$\lambda = 19.11$	0.5	0.45	0.45	0.44	0.44
$\lambda = 25.71$	0.78	0.77	0.73	0.73	0.64
$\lambda = 33.45$	1.25	1.15	1.11	1.11	1.05
$\lambda = 41.71$	1.88	1.84	1.79	1.67	1.63
$\lambda = 50.23$	3.13	2.98	2.84	2.68	2.59

Table 9 Performance measures of $L_S^{(3)}$

$L_S^{(3)}$ (Units)	$\widetilde{\mu}_1 = 55.22$	$\widetilde{\mu}_2 = 56.22$	$\widetilde{\mu}_3 = 57.21$	$\widetilde{\mu}_4 = 58.21$	$\widetilde{\mu}_5 = 59.20$
$\lambda = 19.11$	0.69	0.62	0.58	0.58	0.53
$\lambda = 25.71$	1.26	1.26	1.13	1.06	1.06
$\lambda = 33.45$	2.55	2.33	2.19	2.19	2.00
$\lambda = 41.71$	5.78	5.46	5.12	4.52	4.24
$\lambda = 50.23$	24.66	21.43	17.06	13.96	11.75

$$\phi_{\tilde{\lambda}}(x) = \begin{cases} \frac{(x-1)}{(4-1)}, & 1 \leq x \leq 4 \\ 1, & x = 4 \\ \frac{(x-5)}{(4-5)}, & 4 \leq x \leq 5 \\ 0, & otherwise \end{cases}$$

Table 10 Performance measures of $W_S^{(1)}$

$W_S^{(1)}$ (Units)	$\widetilde{\mu}_1 = 55.22$	$\widetilde{\mu}_2 = 56.22$	$\widetilde{\mu}_3 = 57.21$	$\widetilde{\mu}_4 = 58.21$	$\widetilde{\mu}_5 = 59.20$
$\lambda = 19.11$	0.071	0.070	0.069	0.069	0.068
$\lambda = 25.71$	0.074	0.072	0.070	0.069	0.067
$\lambda = 33.45$	0.076	0.074	0.073	0.070	0.069
$\lambda = 41.71$	0.077	0.075	0.074	0.071	0.070
$\lambda = 50.23$	0.084	0.082	0.081	0.080	0.070

Table 11 Performance measures of $W_S^{(2)}$

$W_S^{(2)}$ (Units)	$\widetilde{\mu}_1 = 55.22$	$\widetilde{\mu}_2 = 56.22$	$\widetilde{\mu}_3 = 57.21$	$\widetilde{\mu}_4 = 58.21$	$\widetilde{\mu}_5 = 59.20$
$\lambda = 19.11$	0.080	0.079	0.077	0.075	0.071
$\lambda = 25.71$	0.096	0.095	0.095	0.094	0.071
$\lambda = 33.45$	0.11	0.108	0.106	0.104	0.09
$\lambda = 41.71$	0.14	0.136	0.132	0.124	0.121
$\lambda = 50.23$	0.19	0.18	0.17	0.16	0.15

Table 12 Performance measures of $W_S^{(3)}$

$W_S^{(3)}$ (Units)	$\widetilde{\mu}_1 = 55.22$	$\widetilde{\mu}_2 = 56.22$	$\widetilde{\mu}_3 = 57.21$	$\widetilde{\mu}_4 = 58.21$	$\widetilde{\mu}_5 = 59.20$
$\lambda = 19.11$	0.10	0.09	0.088	0.086	0.083
$\lambda = 25.71$	0.13	0.128	0.122	0.114	0.111
$\lambda = 33.45$	0.21	0.19	0.188	0.183	0.17
$\lambda = 41.71$	0.39	0.37	0.35	0.30	0.29
$\lambda = 50.23$	1.39	1.21	0.96	0.79	0.66

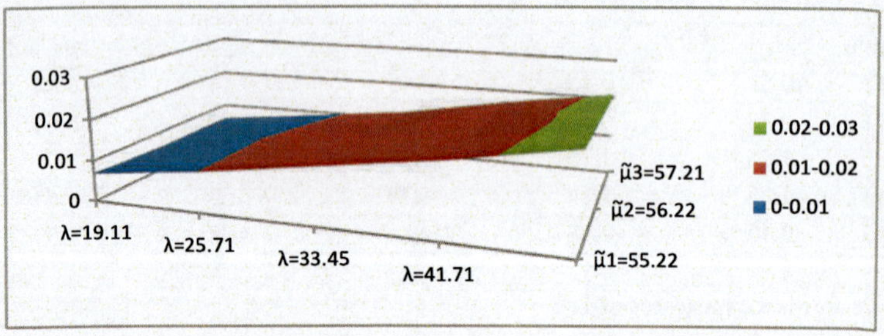

Fig. 3 Performance measures of $w_q^{(1)}$

$$\text{And } R\left(\tilde{\lambda}_1\right) = R\,(1, 4, 5 : 1)$$
$$= \left(\frac{1+4(4)+5}{6}, \frac{4(1-4)(5-4)+5(1)}{12(1)}\right)$$
$$= \sqrt{(3.67)^2 + (-0.58)^2} = 3.72$$

$$R\left(\tilde{\lambda}_2\right) = R\,(2, 5, 6 : 1) = 4.71$$

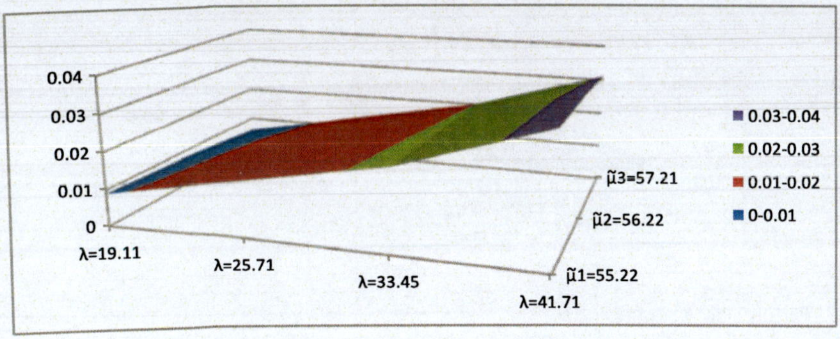

Fig. 4 Performance measures of $w_q^{(2)}$

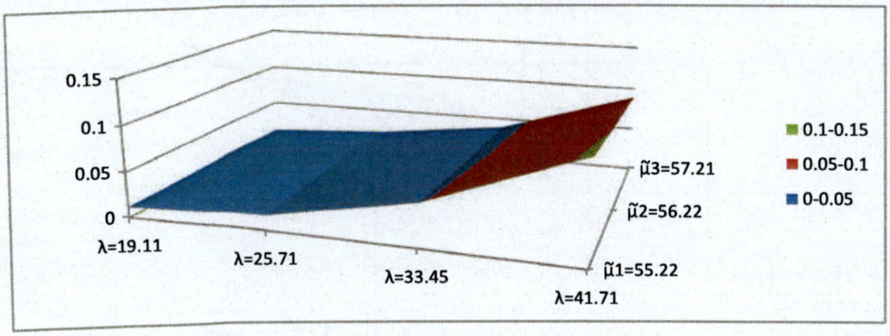

Fig. 5 Performance measures of $w_q^{(3)}$

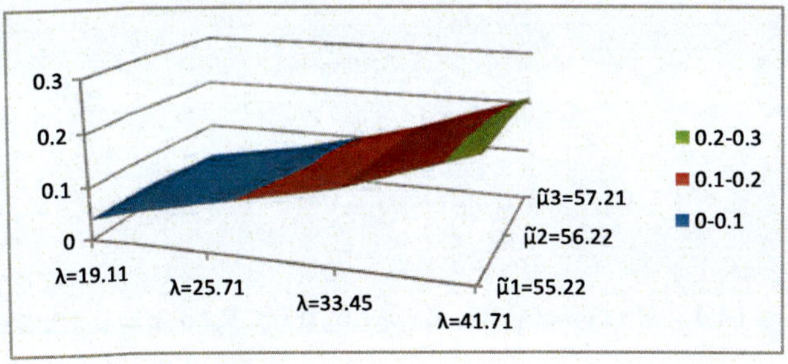

Fig. 6 Performance measures of $L_q^{(1)}$

$$\mathrm{R}\left(\tilde{\lambda}_3\right) = \mathrm{R}\,(3, 6, 7 : 1) = 5.70$$

$$\mathrm{R}\,(\tilde{\mu}_1) = \mathrm{R}\,(54, 57, 58 : 1) = 56.67$$

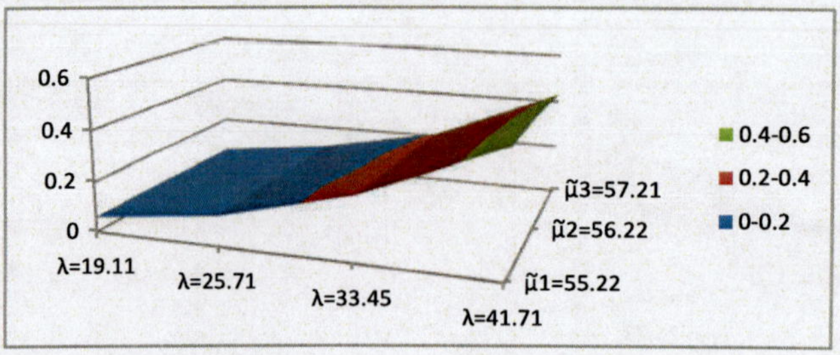

Fig. 7 Performance measures of $L_q^{(2)}$

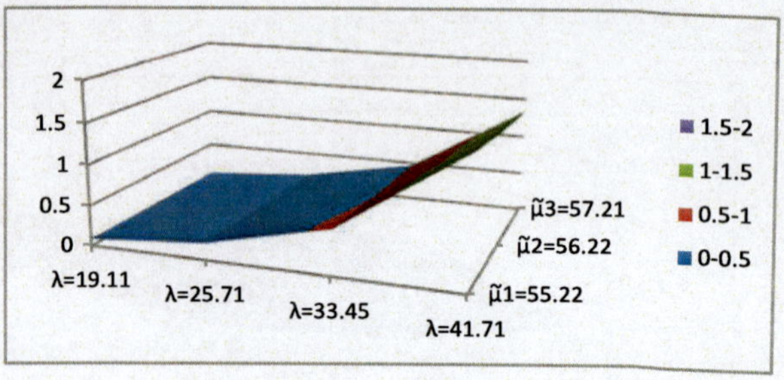

Fig. 8 Performance measures of $L_q^{(3)}$

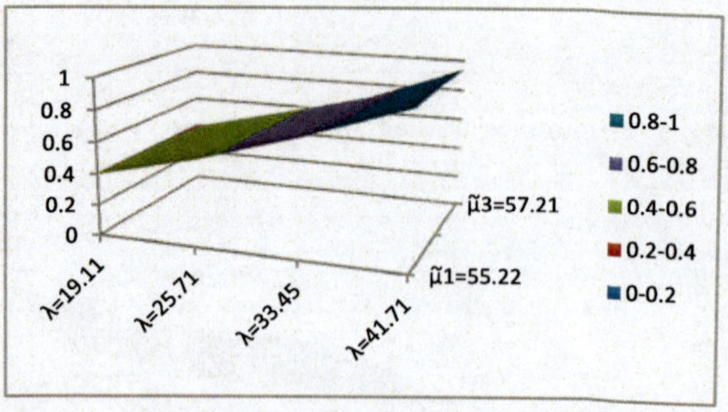

Fig. 9 Performance measures of $L_S^{(1)}$

And $\mathrm{R}\left(\tilde{\lambda}\right) = 14.13, \mathrm{R}\left(\tilde{\rho}_1\right) = 0.07, \mathrm{R}\left(\tilde{\rho}_2\right) = 0.08, \mathrm{R}\left(\tilde{\rho}_3\right) = 0.10, \mathrm{R}\left(\tilde{\rho}\right) = 0.25$

Then

Mean waiting span of 1st priority in the queue is

$$w_q^{(1)} = \frac{\tilde{\rho}}{\tilde{\mu}-\tilde{\lambda}_1}$$
$$= \frac{0.25}{(56.67-3.72)} = 0.005$$

Fig. 10 Performance measures of $L_S^{(2)}$

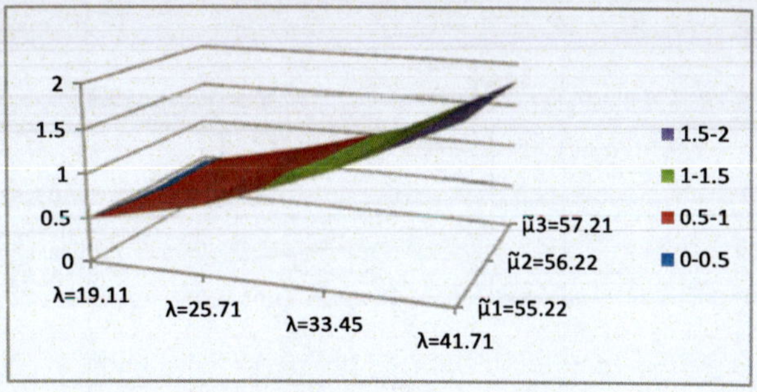

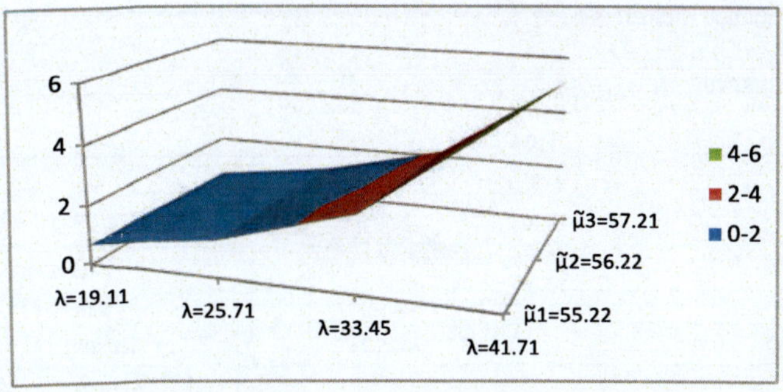

Fig. 11 Performance measures of $L_S^{(3)}$

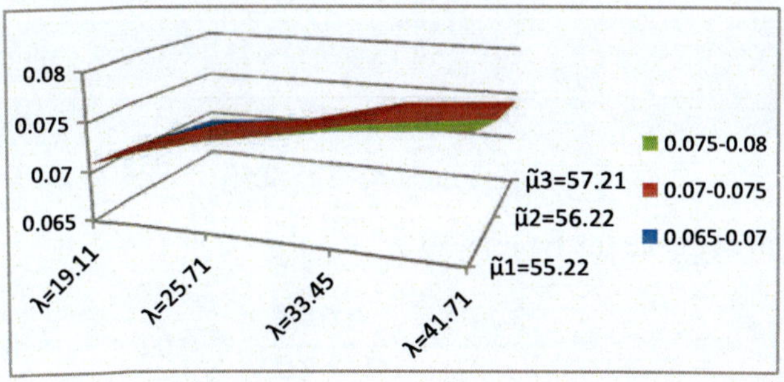

Fig. 12 Performance measures of $W_S^{(1)}$

Mean waiting span of 2nd priority in the queue is

$$w_q^{(2)} = \frac{\tilde{\rho}}{(1-\tilde{\rho}_1-\tilde{\rho}_2)\left(\tilde{\mu}-\tilde{\lambda}_1\right)}$$

$$= \frac{0.25}{(1-0.07-0.08)(56.67-3.72)} = 0.006$$

Mean waiting span of 3rd priority in the queue is

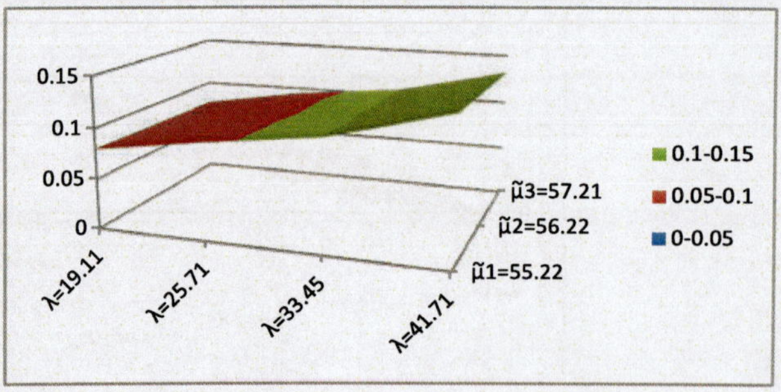

Fig. 13 Performance measures of $W_S^{(2)}$

Fig. 14 Performance measures of $W_S^{(3)}$

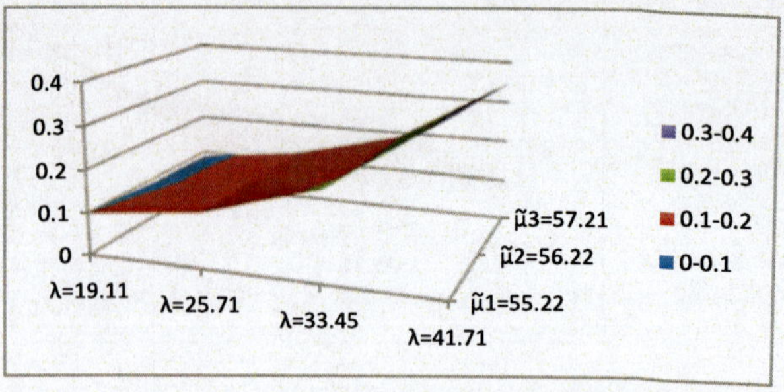

$$w_q^{(3)} = \frac{\tilde{\rho}}{(1-\tilde{\rho}_1-\tilde{\rho}_2-\tilde{\rho}_3)\left(\tilde{\mu}-\tilde{\lambda}_1-\tilde{\lambda}_2\right)}$$
$$= \frac{0.25}{(1-0.07-0.08-0.10)(56.67-3.72-4.71)} = 0.007$$

Mean length of queue for 1st priority is

$$L_q^{(1)} = \frac{\tilde{\rho}\ \tilde{\rho}_1}{(1-\tilde{\rho}_1)} = \frac{0.0175}{(1-0.07)} = 0.02$$

Mean length of queue for 2nd priority is

$$L_q^{(2)} = \frac{\tilde{\rho}\ \tilde{\rho}_2}{(1-\tilde{\rho}_1-\tilde{\rho}_2)\,(1-\tilde{\rho}_1)} = \frac{0.02}{(1-0.07-0.08)\,(1-0.07)} = 0.03$$

Mean length of queue for 3rd priority is

$$L_q^{(3)} = \frac{\tilde{\rho}\tilde{\rho}_3}{(1-\tilde{\rho}_1-\tilde{\rho}_2-\tilde{\rho}_3)(1-\tilde{\rho}_1-\tilde{\rho}_2)}$$
$$= \frac{0.025}{(1-0.07-0.08-0.10)(1-0.07-0.08)} = 0.04$$

The average number of customers in the system of 1st priority is

$$L_S^{(1)} = \frac{L_q^{(1)}}{\tilde{\rho}_1} = 0.29$$

The average number of customers in the system of 2nd priority is

$$L_S^{(2)} = \frac{L_q^{(2)}}{\tilde{\rho}_2} = 0.38$$

The average number of customers in the system of 3rd priority is

$$L_S^{(3)} = \frac{L_q^{(3)}}{\tilde{\rho}_3} = 0.4$$

Mean waiting span of 1st priority in the system is

$$W_S^{(1)} = \frac{L_S^{(1)}}{\tilde{\lambda}_1} = 0.08$$

Mean waiting span of 2nd priority in the system is

$$W_S^{(2)} = \frac{L_S^{(2)}}{\tilde{\lambda}_2} = 0.08$$

Mean waiting span of 3rd priority in the system is

$$W_S^{(3)} = \frac{L_S^{(3)}}{\tilde{\lambda}_3} = 0.07$$

Similarly we can proceed for some illustrations we got

Here, Tables 13, 14, 15, 16, 17, 18, 19, 20, 21, 22, 23, and 24 and the Figs. 15, 16, 17, 18, 19, 20, 21, 22, 23, 24, 25, and 26 show that if we allow the priorities, then the performance measures will increase with respect to the arrival and service rates.

6 Upshot

This chapter diagnoses with a strange manner for obtaining the achievement estimations of Priority Queuing system with fuzzy parameters, which is very fashionable. This method is used to get concrete solutions and furnish more legitimate verdicts. In order to reinforce these suggested tactics, a real case study will be in the offing for analysis.

Table 13 Performance measures of $w_q^{(1)}$

$w_q^{(1)}$ (Units)	$\tilde{\mu}_1 = 56.67$	$\tilde{\mu}_2 = 57.67$	$\tilde{\mu}_3 = 58.67$	$\tilde{\mu}_4 = 59.67$	$\tilde{\mu}_5 = 60.67$
$\lambda = 14.13$	0.005	0.005	0.004	0.004	0.003
$\lambda = 23.08$	0.009	0.008	0.008	0.007	0.006
$\lambda = 32.06$	0.013	0.012	0.011	0.010	0.010
$\lambda = 41.04$	0.03	0.02	0.02	0.01	0.01
$\lambda = 50.04$	0.03	0.03	0.02	0.02	0.01

Table 14 Performance measures of $w_q^{(2)}$

$w_q^{(2)}$ (Units)	$\tilde{\mu}_1 = 56.67$	$\tilde{\mu}_2 = 57.67$	$\tilde{\mu}_3 = 58.67$	$\tilde{\mu}_4 = 59.67$	$\tilde{\mu}_5 = 60.67$
$\lambda = 14.13$	0.006	0.005	0.005	0.004	0.003
$\lambda = 23.08$	0.011	0.010	0.010	0.009	0.009
$\lambda = 32.06$	0.019	0.018	0.017	0.017	0.016
$\lambda = 41.04$	0.04	0.04	0.03	0.03	0.02
$\lambda = 50.04$	0.06	0.05	0.05	0.04	0.03

Table 15 Performance measures of $w_q^{(3)}$

$w_q^{(3)}$ (Units)	$\tilde{\mu}_1 = 56.67$	$\tilde{\mu}_2 = 57.67$	$\tilde{\mu}_3 = 58.67$	$\tilde{\mu}_4 = 59.67$	$\tilde{\mu}_5 = 60.67$
$\lambda = 14.13$	0.007	0.006	0.006	0.005	0.004
$\lambda = 23.08$	0.016	0.015	0.014	0.014	0.013
$\lambda = 32.06$	0.037	0.034	0.032	0.031	0.030
$\lambda = 41.04$	0.08	0.08	0.07	0.07	0.06
$\lambda = 50.04$	0.30	0.26	0.20	0.19	0.16

Table 16 Performance measures of $L_q^{(1)}$

$L_q^{(1)}$ (Units)	$\tilde{\mu}_1 = 56.67$	$\tilde{\mu}_2 = 57.67$	$\tilde{\mu}_3 = 58.67$	$\tilde{\mu}_4 = 59.67$	$\tilde{\mu}_5 = 60.67$
$\lambda = 14.13$	0.04	0.04	0.03	0.03	0.02
$\lambda = 23.08$	0.07	0.06	0.05	0.05	0.04
$\lambda = 32.06$	0.12	0.12	0.11	0.10	0.09
$\lambda = 41.04$	0.21	0.20	0.19	0.18	0.17
$\lambda = 50.04$	0.34	0.32	0.31	0.30	0.29

Table 17 Performance measures of $L_q^{(2)}$

$L_q^{(2)}$ (Units)	$\tilde{\mu}_1 = 56.67$	$\tilde{\mu}_2 = 57.67$	$\tilde{\mu}_3 = 58.67$	$\tilde{\mu}_4 = 59.67$	$\tilde{\mu}_5 = 60.67$
$\lambda = 14.13$	0.04	0.03	0.03	0.02	0.01
$\lambda = 23.08$	0.09	0.08	0.08	0.07	0.06
$\lambda = 32.08$	0.20	0.19	0.18	0.17	0.16
$\lambda = 41.04$	0.41	0.40	0.38	0.36	0.35
$\lambda = 50.04$	0.82	0.79	0.72	0.69	0.64

Table 18 Performance measures of $L_q^{(3)}$

$L_q^{(3)}$ (Units)	$\widetilde{\mu}_1 = 56.67$	$\widetilde{\mu}_2 = 57.67$	$\widetilde{\mu}_3 = 58.67$	$\widetilde{\mu}_4 = 59.67$	$\widetilde{\mu}_5 = 60.67$
$\lambda = 14.13$	0.04	0.04	0.03	0.03	0.02
$\lambda = 23.08$	0.14	0.13	0.13	0.12	0.11
$\lambda = 32.08$	0.43	0.40	0.38	0.36	0.32
$\lambda = 41.04$	1.24	1.13	1.06	0.99	0.91
$\lambda = 50.04$	5.29	4.72	3.78	3.42	2.81

Table 19 Performance measures of $L_S^{(1)}$

$L_S^{(1)}$ (Units)	$\widetilde{\mu}_1 = 56.67$	$\widetilde{\mu}_2 = 57.67$	$\widetilde{\mu}_3 = 58.67$	$\widetilde{\mu}_4 = 59.67$	$\widetilde{\mu}_5 = 60.67$
$\lambda = 14.13$	0.29	0.30	0.31	0.32	0.33
$\lambda = 23.08$	0.47	0.46	0.44	0.44	0.43
$\lambda = 32.08$	0.72	0.71	0.65	0.64	0.63
$\lambda = 41.04$	0.92	0.91	0.89	0.86	0.85
$\lambda = 50.04$	1.21	1.19	1.17	1.15	1.12

Table 20 Performance measures of $L_S^{(2)}$

$L_S^{(2)}$ (Units)	$\widetilde{\mu}_1 = 56.67$	$\widetilde{\mu}_2 = 57.67$	$\widetilde{\mu}_3 = 58.67$	$\widetilde{\mu}_4 = 59.67$	$\widetilde{\mu}_5 = 60.67$
$\lambda = 14.13$	0.38	0.35	0.30	0.26	0.26
$\lambda = 23.08$	0.63	0.61	0.58	0.58	0.56
$\lambda = 32.08$	1.05	1.05	1.00	0.97	0.94
$\lambda = 41.04$	1.71	1.67	1.65	1.57	1.52
$\lambda = 50.04$	2.83	2.72	2.57	2.46	2.37

Table 21 Performance measures of $L_S^{(3)}$

$L_S^{(3)}$ (Units)	$\widetilde{\mu}_1 = 56.67$	$\widetilde{\mu}_2 = 57.67$	$\widetilde{\mu}_3 = 58.67$	$\widetilde{\mu}_4 = 59.67$	$\widetilde{\mu}_5 = 60.67$
$\lambda = 14.13$	0.41	0.40	0.37	0.32	0.31
$\lambda = 23.08$	0.94	0.89	0.84	0.83	0.81
$\lambda = 32.08$	2.05	2.00	1.90	1.78	1.68
$\lambda = 41.04$	4.77	4.52	4.24	3.96	3.79
$\lambda = 50.04$	17.06	15.23	12.60	11.40	9.69

Table 22 Performance measures of $W_S^{(1)}$

$W_S^{(1)}$ (Units)	$\widetilde{\mu}_1 = 56.67$	$\widetilde{\mu}_2 = 57.67$	$\widetilde{\mu}_3 = 58.67$	$\widetilde{\mu}_4 = 59.67$	$\widetilde{\mu}_5 = 60.67$
$\lambda = 14.13$	0.092	0.090	0.088	0.086	0.083
$\lambda = 23.08$	0.08	0.08	0.07	0.07	0.06
$\lambda = 32.08$	0.09	0.08	0.08	0.07	0.07
$\lambda = 41.04$	0.08	0.07	0.07	0.06	0.06
$\lambda = 50.04$	0.08	0.08	0.07	0.06	0.06

Table 23 Performance measures of $W_S^{(2)}$

$W_S^{(2)}$ (Units)	$\tilde{\mu}_1 = 56.67$	$\tilde{\mu}_2 = 57.67$	$\tilde{\mu}_3 = 58.67$	$\tilde{\mu}_4 = 59.67$	$\tilde{\mu}_5 = 60.67$
$\lambda = 14.13$	0.085	0.082	0.070	0.069	0.067
$\lambda = 23.08$	0.089	0.086	0.083	0.080	0.075
$\lambda = 32.08$	0.11	0.10	0.09	0.09	0.08
$\lambda = 41.04$	0.13	0.12	0.12	0.11	0.10
$\lambda = 50.04$	0.17	0.16	0.15	0.15	0.14

Table 24 Performance measures of $W_S^{(3)}$

$W_S^{(3)}$ (Units)	$\tilde{\mu}_1 = 56.67$	$\tilde{\mu}_2 = 57.67$	$\tilde{\mu}_3 = 58.67$	$\tilde{\mu}_4 = 59.67$	$\tilde{\mu}_5 = 60.67$
$\lambda = 14.13$	0.073	0.072	0.071	0.06	0.05
$\lambda = 23.08$	0.12	0.11	0.10	0.10	0.09
$\lambda = 32.08$	0.18	0.17	0.16	0.15	0.14
$\lambda = 41.04$	0.33	0.31	0.29	0.27	0.26
$\lambda = 50.04$	0.96	0.86	0.71	0.64	0.55

Fig. 15 Performance measures of $w_q^{(1)}$

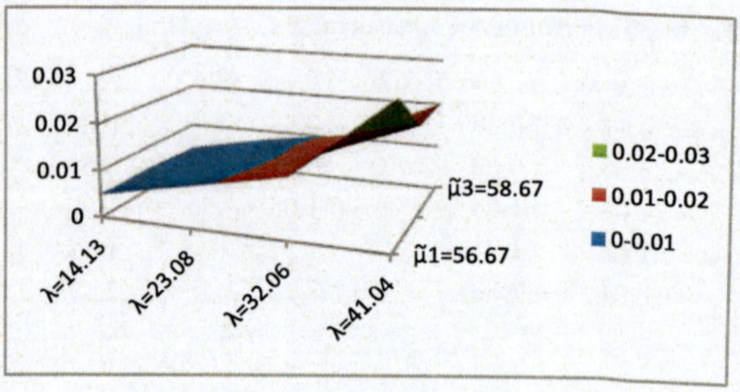

Fig. 16 Performance measures of $w_q^{(2)}$

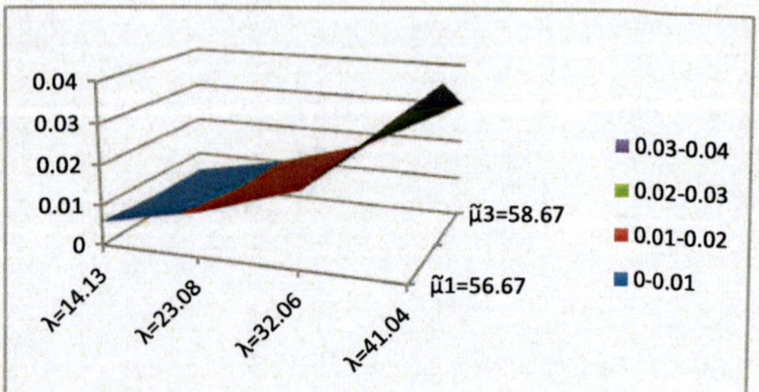

Fig. 17 Performance measures of $w_q^{(3)}$

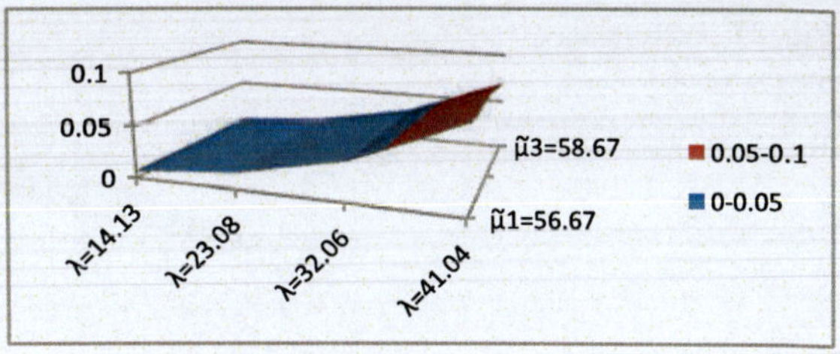

Fig. 18 Performance measures of $L_q^{(1)}$

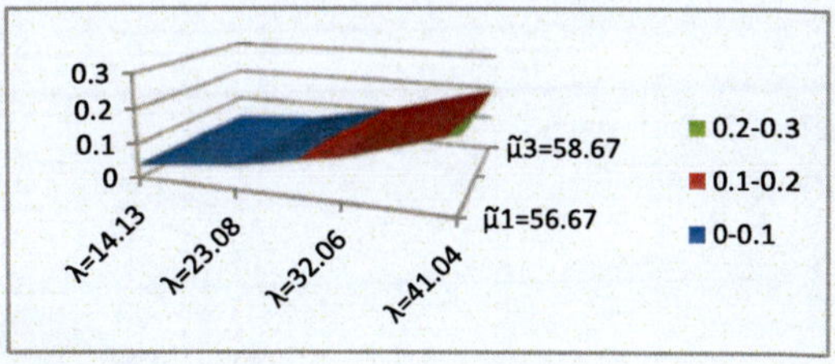

Fig. 19 Performance measures of $L_q^{(2)}$

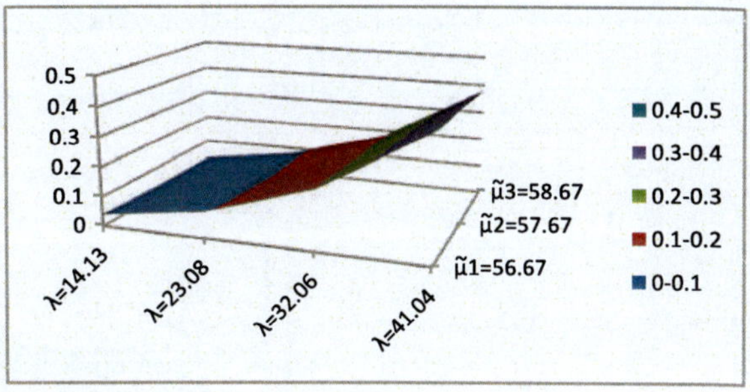

Fig. 20 Performance measures of $L_q^{(3)}$

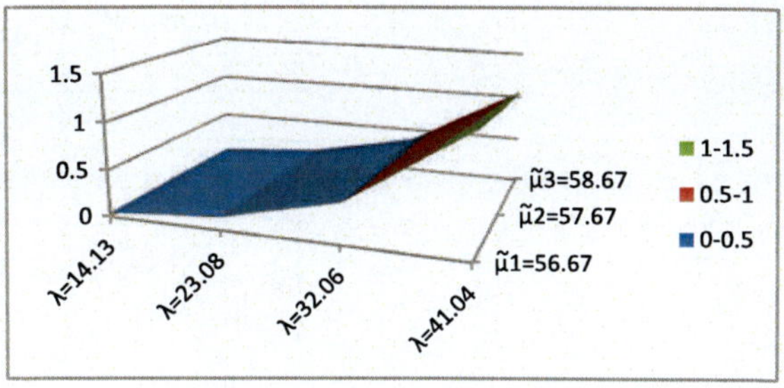

Fig. 21 Performance measures of $L_S^{(1)}$

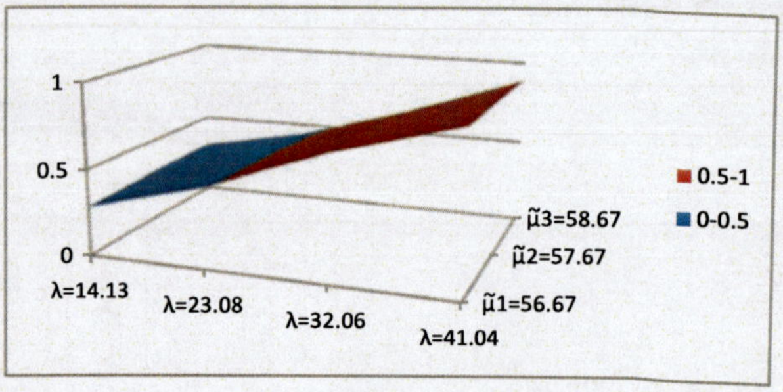

Fig. 22 Performance measures of $L_S^{(2)}$

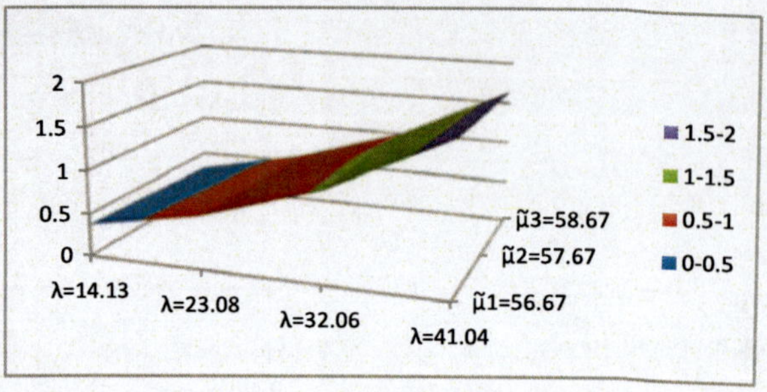

Fig. 23 Performance measures of $L_S^{(3)}$

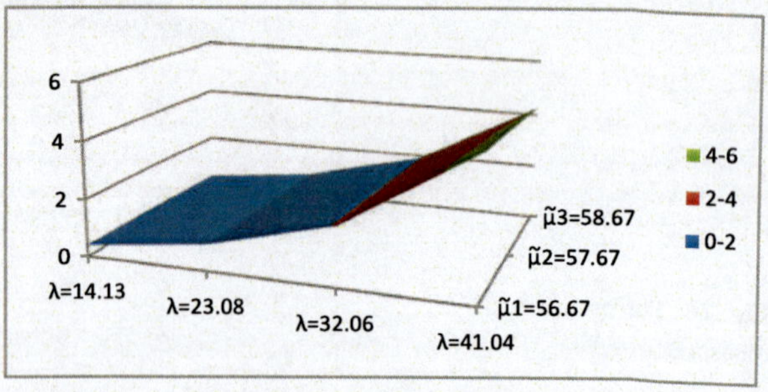

Fig. 24 Performance measures of $W_S^{(1)}$

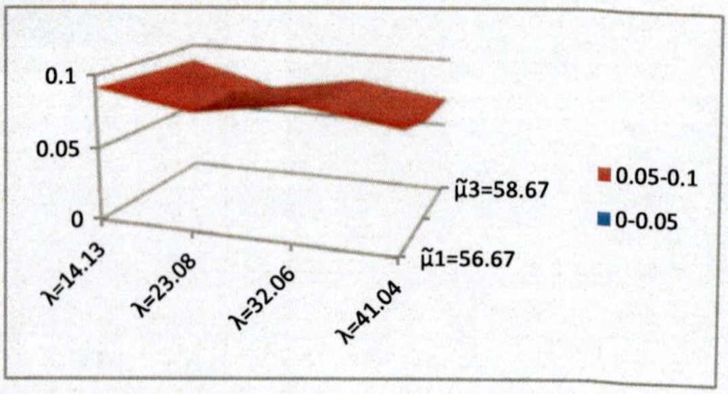

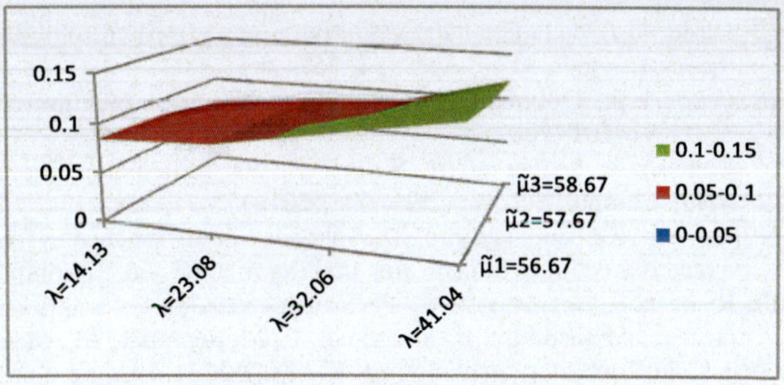

Fig. 25 Performance measures of $W_S^{(2)}$

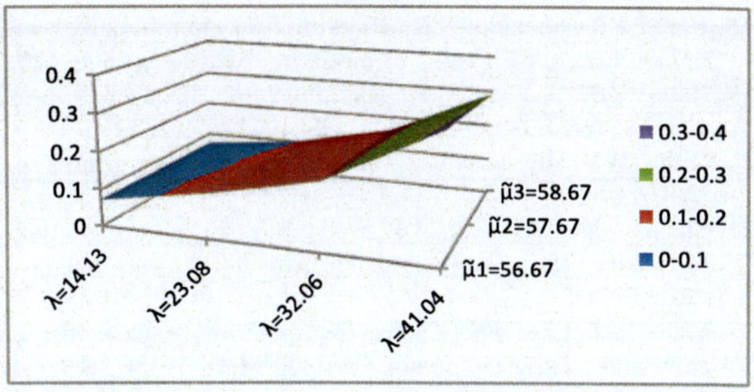

Fig. 26 Performance measures of $W_S^{(3)}$

References

1. Allaviranloo, T., Jahantigh, M.A., Hajighasemi, S.: A new distance measure and ranking method for generalized trapezoidal fuzzy numbers. Math. Probl. Eng. **2013**, 623757, 6 pages (2013)
2. Azman, F.N., Abdullah, L.: Ranking fuzzy numbers by centroid method. Malays. J. Fundam. Appl. Sci. **8**(3), 117–121 (2012)
3. Bose, S.: An Introduction to Queueing Systems. Kluvar Academic/Plenum Publishers, New York (2008)
4. Chu, T.C., Tsao, C.T.: Ranking fuzzy numbers with an area between the centroid point and original point. Comput. Math. Appl. **43**(1–2), 111–117 (2002)
5. Cooper, R.: Introduction to Queueing Theory, 3rd edn. CEE Press, Washington, DC (1990)
6. Dat, L.Q., Yu, V.F., Chou, S.Y.: An improved ranking method for fuzzy numbers based on the centoid index. Int. J. Fuzzy Syst. **14**(3), 413–419 (2012)
7. Devaraj, J., Jayalakshmi, D.: A fuzzy approach to non-preemptive priority queues. Int. J. Math. Arch. **3**, 2704–2712 (2012)
8. Hari Ganesh, A., Helen Shobana, A., Ramesh, R.: Identification of critical path for the analysis of bituminous road transport network using integrated FAHP – FTOPSIS method. Mater. Today Proc. **37**(Part 2), 193–206 (2021)
9. Janos, S.: Basic Queueing Theory. Globe Edit Publishers, Omniscriptum GMBH (2016)
10. Klir, G.J., Yuvan, B.: Fuzzy Sets and Fuzzy Logic Theory and Applications. Prentice Hall of India (2005)
11. Palpandi, B., Geetharamani, G.: Computing performance measures of fuzzy non-preemptive priority queues using robust ranking technique. Appl. Math. Sci. **7**(102), 5095–5102 (2013)
12. Parandin, N., Araghi, M.A.F.: Ranking of fuzzy numbers by distance method. J. Appl. Math. **15**(19), 47–55 (2008)

13. Pardo, M.J., de la Fluente, D.: Optimizing a priority discipline queueing model using fuzzy set theory. Comput. Math. Appl. **54**, 267–281 (2007)
14. Pekoz, E.A.: Optimal policies for multi-server non-preemptive priority queues. Queueing Syst. **42**(1), 91–101 (2002)
15. Ramesh, R., Kumaraghuru, S.: Priority disciplined queueing models with fuzzy parameters. J. Math. Comput. Sci. **4**(3), 594–602 (2014)
16. Ramesh, R., Kumaraghuru, S.: Analysis of performance in four non-preemptive priority fuzzy queues by centroid of centroids ranking method. Int. J. Comput. Tech. **4**(1), 12–20 (2017)
17. Ramesh, R., Seenivasan, M.: Performance analysis of single server low-priority queue based on electronic transmitter. In: Sengodan, T., Murugappan, M., Misra, S. (eds.) Advances in Electrical and Computer Technologies. ICAECT 2021. Lecture Notes in Electrical Engineering, vol. 881. Springer, Singapore (2022)
18. Rao, P.P.B., Shankar, N.R.: Ranking fuzzy numbers with a distance method using circumcenter of centroids and an index of modality. Adv. Fuzzy Syst. **2011**, 178308 (2011)
19. Rao, P.P.B., Shankar, N.R.: Ranking generalized fuzzy numbers using area, mode, spreads and weights. Int. J. Appl. Sci. Eng. **10**(1), 41–57 (2012)
20. Ritha, W., Lilly, R.: Fuzzy queues with priority discipline. Appl. Math. Sci. **4**(12), 575–582 (2010)
21. Shankar, N.R., Sarathi, B.P., Babu, S.S.: Fuzzy critical path method based on a new approach of ranking fuzzy numbers using centroid of centroids. Int. J. Fuzzy Syst. Appl. **3**(2), 16–31 (2013)
22. Wang, Y.J., Lee, H.S.: The revised method of ranking fuzzy numbers with an area between the centroid and original points. Comput. Math. Appl. **55**(9), 2033–2042 (2008)
23. Zadeh, L.A.: Fuzzy sets as a basis for a theory of possibility, fuzzy sets and systems. Fuzzy Sets Syst. **1**, 3–28 (1978)

Successive Approximation of Neutral Stochastic Partial Integrodifferential Systems with State-Dependent Delay and Poisson Jumps

R. Pradeepa and R. Jayaraman

2010 Mathematics Subject Classification 65C30, 26A33, 34A08, 34H05

1 Introduction

In this article, we discuss the successive approximation of SNPID systems with SD delay and PJ [6, 10]. We contemplate the subsequent of the following SNPID system with SD delay and PJ [1, 2]:

$$
\begin{aligned}
&d\Big[y(\ell)+\mathscr{G}_1\Big(\ell, y_{\varrho(\ell,y_\ell)}, \int_0^\ell a_1(\ell,\hbar, y_{\varrho(\hbar,y_\hbar)})d\hbar\Big)\Big]\\
&=\Big[\mathscr{A}\Big[y(\ell)+\mathscr{G}_1\Big(\ell, y_{\varrho(\ell,y_\ell)}, \int_0^\ell a_1(\ell,\hbar, y_{\varrho(\hbar,y_\hbar)})d\hbar\Big)\Big]\\
&+\int_0^\ell R(\ell-\hbar)\Big[y(\hbar)+\mathscr{G}_1\Big(\hbar, y_{\varrho(\hbar,y_\hbar)}, \int_0^\hbar a_1(\hbar,\ell, y_{\varrho(\ell,y_\ell)})d\ell\Big)\Big]d\hbar\Big]dt\\
&+\mathscr{G}_2\Big(\ell, y_{\varrho(\ell,y_\ell)}, \int_0^\ell a_2(\ell,\hbar, y_{\varrho(\hbar,y_\hbar)})d\hbar\Big)dt\\
&+\mathscr{G}_3\Big(\ell, y_{\varrho(\ell,y_\ell)}, \int_0^\ell a_3(\ell,\hbar, y_{\varrho(\hbar,y_\hbar)})d\hbar\Big)dW(\ell)+\int_Z g(\ell, y(\ell-), z)\widetilde{N}(dt,dz),\\
&\quad \ell\in\mathscr{I}:=[0,\Upsilon], \qquad (1)
\end{aligned}
$$

R. Pradeepa (✉)
Department of Mathematics, Periyar University, Salem, Tamil Nadu, India

R. Jayaraman
Department of Mathematics, Periyar University College of Arts and Science, Dharmapuri, Tamil Nadu, India

F. Kamalov et al. (eds.), *Advances in Mathematical Modeling and Scientific Computing*, Trends in Mathematics, https://doi.org/10.1007/978-3-031-41420-6_55

$$y_0 = \varphi \in \mathscr{B}. \tag{2}$$

The strongly continuous semigroup of bounded linear operators (LOs) $\Upsilon, \ell \geq 0$ on ζ has an infinitesimal generator called $\mathscr{A}$, where $D(\mathscr{A})$ is a subset of ζ and $R(\ell), \ell \in \mathscr{I}$ is a closed LO on ζ. The history of y_ℓ is $(-\infty, 0] \to \zeta, y_\ell(\eta) = y(\ell + \eta)$ for $\ell \geq 0$.

As stated axiomatically in Sect. 2, the state $y(\cdot)$ takes values in a separable real Hilbert space (HS) ζ with inner product (IP) $(\cdot, \cdot)$ and norm $\|\cdot\|$ and is a member of an abstract phase space $\mathscr{B}$. Let $\mathscr{K}$ be another separable HS with the IP $\langle \cdot, \cdot \rangle_{\mathscr{K}}$ and the norm $\|\cdot\|_{\mathscr{K}}$. Let's say you have $\{W(\hbar)$: A given Wiener process with increment covariance set by the nuclear operators is denoted as $\{W(\hbar) : 0 \leq \hbar \leq \ell\}$. Poisson point process (PPP) $\hat{K}(\cdot)$ induces a compensatory martingale measure that is unrelated to the $\widetilde{N}(d\hbar, dz)$ defined on a full probability space $(\Omega, \mathscr{F}, \{\mathscr{F}_\ell\}_{\ell \geq 0}, \mathscr{P})$ with a normal filtration $\{\mathscr{F}_\ell\}_{\ell \geq 0}$. This filtration is created by the Wiener process W.

The functions $\mathscr{G}_1, \mathscr{G}_2, \mathscr{G}_3, g$ and $\varrho : \mathscr{I}\ \ell \mathscr{B} \to (-\infty, \Upsilon]$ are approximate versions of the functions mentioned subsequently. In this instance, the space of all $Q-$Hilbert-Schmidt operators from $\mathscr{K}_0$ to ζ, as specified in Sect. 2, is denoted by $\mathscr{L}_2^0 = \mathscr{L}_2(\mathscr{K}_0; \zeta)$. An $\mathscr{F}_0-$adapted, $\mathscr{B}-$valued random variable with a second moment of finite that is independent of the WP W is the starting data $\varphi = \{\varphi(\ell) : -\infty < \ell \leq 0\}$.

2 Preliminaries

Wiener Process The separable HS is defined as $(\zeta, |\cdot|_\zeta, \langle \cdot, \cdot \rangle_\zeta)$ and $(\zeta, |\cdot|_{\mathscr{K}}, \langle \cdot, \cdot \rangle_{\mathscr{K}})$. The term $\mathscr{L}(\mathscr{K}, \zeta)$ refers to the set of all bounded LOs from K to ζ with the norm $\|\cdot\|$. For further information on these concepts, please see [3] and [4]. Let $(\Omega, \mathscr{F}, \mathscr{P}, \mathscr{F})$ be a thoroughly filtered probability space that meets the standard requirement with $(\mathscr{F} = \{\mathscr{F}_\ell\}_{\ell \geq 0})$, demonstrating accurate and continuous expanding family filtering, and $\mathscr{F}_0$ includes all fsF-null sets of $\mathscr{P}$.

Assume that $\{p(\ell), \ell \geq 0\}$ is true. A stationary σ-finite PPP accepts values in the measurable space $(Z, R(Z))$ and is $\mathscr{F}_\ell$-adapted. According to the formula $N_p((\ell_1, \ell_2]\ \ell \Lambda)$, the random measure N_p is defined as $N_p((\ell_1, \ell_2]\ \ell \Lambda) := \sum_{\hbar \in (\ell_1, \ell_2]} 1_\Lambda(p(\hbar))$. It is called Poisson random measure $\Lambda \in \mathscr{B}(Z - 0)$ induced by $p(\cdot)$.

Since v is the characteristic measure of N_p, also known as the compensated Poisson random measure, we can define the measure $\widetilde{N}$ as $\widetilde{N}(dt, dz) = N_p(dt, dz) - v(dz)dt$,. Let $W = (W_\ell)_{\ell \geq 0}$, be a K-valued WP defined on $(\Omega, \mathscr{F}, \mathscr{P}, \mathscr{F})$ with covariance operator Q. This WP is independent of the PPP, i.e.,

$$\mathbb{E} \langle W(\ell), y \rangle_{\mathscr{K}} \langle W(\ell), z \rangle_{\mathscr{K}} = (\ell \wedge \hbar) \langle Qy, z \rangle_{\mathscr{K}}, \quad y, z \in \mathscr{K}.$$

2.1 Equations with Partial Integrodifferentials in Banach Spaces

If ζ is a BS, $\mathscr{A}$ and $R(\ell)$ are closed LOs on ζ. Y represents the BS $D(\mathscr{A})$ equipped with the graph norm defined by $|y|_Y := |\mathscr{A}y| + |y|$, for $y \in Y$.

Think about the Cauchy problem's consequences.

$$x'(\ell) = \mathscr{A}x(\ell) + \int_0^\ell R(\ell - \hbar)x(\hbar)d\hbar, \quad \text{for} \quad \ell \geq 0, \tag{3}$$

$$x(0) = x_0 \in \zeta. \tag{4}$$

Theorem 2.1 *Let us consider the circumstances as being* $(\chi 1)$ *and* $(\chi 2)$ *hold. The Cauchy problem equation then has a specific resolvent operator that is unique of 3–4.*

Now, we give some findings on the underlying integrodifferential equation's existence of solutions.

$$x'(\ell) = \mathscr{A}x(\ell) + \int_0^\ell R(\ell - \hbar)x(\hbar)d\hbar + q(\ell), \quad \text{for} \quad \ell \geq 0, \tag{5}$$

$$x(0) = x_0 \in \zeta, \tag{6}$$

where $q : [0, +\infty) \to \zeta$ is a continuous function (CF).

Definition 2.1 ([4]) A CF $x : [0, +\infty) \to \zeta$ is said to be an equation's rigorous solution (Se) (2.3)–(2.4) if

1. $x \in C^1([0, +\infty); \zeta) \cap C([0, +\infty); Y)$
2. x satisfies equations 5–6 for $\ell \geq 0$

Theorem 2.2 ([4]) *Suppose that* $(\chi 1)$ *and* $(\chi 2)$ *hold. If* x *is a strict Se 5–6, then we have*

$$x(\ell) = \Upsilon x_0 + \int_0^\ell \Upsilon(\ell - \hbar)q(\hbar)d\hbar, \quad \textit{for} \quad \ell \geq 0. \tag{7}$$

Definition 2.2 ([4]) For $x_0 \in \zeta$, a MS of 5–6 is a function $x : [0, +\infty) \to \zeta$, if x satisfies equation 7.

Theorem 2.3 ([4]) *Let* $q \in C^1([0, +\infty); \zeta)$ *and* x *be defined by (2.5). If* $x_0 \in D(\mathscr{A})$, *then* x *is a strict Se 5–6.*

Definition 2.3 ([5]) The set of $\mathscr{F}_0$-measurable functions from $(-\infty, 0]$ to ζ furnished with a seminorm $\|\cdot\|_{\mathscr{B}}$ is known as the phase space $\mathscr{B}((-\infty, 0], \zeta)$. The following are the axioms for $\mathscr{B}$:

(B1) If $y : (-\infty, \Upsilon] \to \zeta, \Upsilon > 0$, is such that $y_0 \in \mathscr{B}$, then, for every $\ell \in \mathscr{I} := [0, \Upsilon]$, if these circumstances are true:

1. $y_\ell \in \mathscr{B}$;
2. $\|y(\ell)\| \leq C\|y_\ell\|_{\mathscr{B}}$;
3. $\|y_\ell\|_{\mathscr{B}} \leq \overline{M}(\ell) \sup_{0\leq\hbar\leq\ell} |y(\hbar)| + \overline{N}(\ell)\|y_0\|_{\mathscr{B}}$,

(B2) $\mathscr{B}$ is complete space.

Lemma 2.1 *Let $y : (-\infty, \Upsilon] \to \zeta$ be an $\mathscr{F}_\ell$-adapted measurable process such that the $\mathscr{F}_0$-adapted process $y_0 = \varphi(\ell) \in \mathscr{L}_2(\Omega, \mathscr{B})$. Then $\|y_\hbar\|_{\mathscr{B}} \leq N_\Upsilon \mathbb{E}\|\varphi\|_{\mathscr{B}} + K_\Upsilon \sup_{0\leq\hbar\leq\Upsilon} \mathbb{E}\|x(\hbar)\|$, where $K_\Upsilon = \sup\{K(\ell) : 0 \leq \ell \leq \Upsilon\}$ and $N_\Upsilon = \sup\{N(\ell) : 0 \leq \ell \leq \Upsilon\}$.*

Definition 2.4 An $\mathscr{F}_\ell$-adapted, ζ-valued stochastic process (VSP) $y(\ell)$ defined on $-\infty < \ell \leq \Upsilon, \Upsilon > 0$ is called a MS for (1)–(2) if

(a1) $y(\ell)$ is cadlag and $\{y_\ell : 0 \leq \ell \leq \Upsilon\}$ is a $\mathscr{B}$-VSP

(a2) $\int_0^{\Upsilon} |y(\theta)|^2 d\theta < \infty$ almost surely

(a3) For arbitrary $0 \leq \ell \leq \Upsilon$, $y(\ell)$ satisfies the following integral equation:

$$
\begin{aligned}
y(\ell) =& \Upsilon[\varphi(0) + \mathscr{G}_1(0, \varphi, 0)] - \mathscr{G}_1\left(\ell, y_{\varrho(\ell, y_\ell)}, \int_0^\ell a_1(\ell, \hbar, y_{\varrho(\hbar, y_\hbar)})d\hbar\right) \\
&+ \int_0^\ell \Upsilon(\ell-\hbar)\mathscr{G}_2\left(\hbar, y_{\varrho(\hbar, y_\hbar)}, \int_0^\hbar a_2(\hbar, \ell, y_{\varrho(\ell, y_\ell)})d\ell\right)d\hbar \\
&+ \int_0^\ell \Upsilon(\ell-\hbar)\hbar\left(\hbar, y_{\varrho(\hbar, y_\hbar)}, \int_0^\hbar a_3(\hbar, \ell, y_{\varrho(\ell, y_\ell)})d\ell\right)dW(\hbar) \\
&+ \int_0^\ell \Upsilon(\ell-\hbar)\int_Z g(\hbar, y(\hbar-), z)\widetilde{N}(d\hbar, dz),
\end{aligned}
\tag{8}
$$

(a4) $y_0 = \varphi \in \mathscr{B}$.

We denote by $\mathscr{M}^2((-\infty, \Upsilon], \zeta)$ the space of all ζ-valued $\mathscr{F}_\ell$-adapted process $y = \{y(\ell), -\infty < \ell \leq \Upsilon\}$ such that

(b1) $y_0 = \varphi \in \mathscr{B}$ and $y(\ell)$ is cadlag on $[0, \Upsilon]$

(b2) For all $y \in \mathscr{M}^2((-\infty, \Upsilon], \zeta)$

$$\|y\|^2_{\mathcal{M}^2} := \mathbb{E}\|\varphi\|^2_{\mathcal{B}} + \mathbb{E}\int_0^{\Upsilon} |y(\ell)|^2 dt < \infty. \tag{9}$$

Lemma 2.2 *The space $\mathcal{M}^2((-\infty, \Upsilon], \zeta)$ is a BS with the norm defined by (9).*

Proof We skip the evidence here because it is normal. □

By using the following assumptions, we establish the existence and uniqueness findings in this section:

(χ1) There exists a constant $M > 0$ such that $|\Upsilon|^2 \leq M$, for all $\ell \in \mathcal{I}$.

(χ2) The function $\ell \to \varphi_\ell$ is continuous from $\widetilde{\mathcal{L}}(\varrho^-) = \{\varrho(\hbar, \psi) \leq 0, (\hbar, \psi) \in \mathcal{I}\ \ell \mathcal{B}\}$ to $\mathcal{B}$, and there exists a continuous and bounded function $\mathcal{W}^{\varphi}$: $\widetilde{\mathcal{L}}(\varrho^-) \to (0, \infty)$ such that $\|\varphi_\ell\|_{\mathcal{B}} \leq \widetilde{W}^{\varphi}(\ell)\mathbb{E}|\varphi|_{\mathcal{B}}$ for each $\ell \in \widetilde{\mathcal{L}}(\varrho^-)$.

(χ3) The function $\mathcal{G}_1 : \mathcal{I} \times \mathcal{B} \times \zeta \to \chi$ is continuous and there exists constant $K^2_{\mathcal{G}_1} > 0$ such that for any $x_1, x_2 \in \mathcal{B}$ and $y_1, y_2 \in \zeta, \ell \geq 0$

$$|\mathcal{G}_1(\ell, x_1, y_1) - \mathcal{G}_1(\ell, x_2, y_2|^2 \leq K^2_{\mathcal{G}_1}(\|x_1 - x_2\|^2_{\mathcal{B}} + \|y_1 - y_2\|^2_{\zeta}),$$

and we further assume that $\mathcal{G}_1(\ell, 0, 0) \equiv 0$ for all $\ell \geq 0$.

(χ4) The function $a_i : \Delta\ \ell \mathcal{B} \to \zeta, i = 1, 2, 3$, where $\Delta = \{(\ell, \hbar) \in \mathcal{I}\ \ell \mathcal{I}, 0 \leq \hbar \leq \Upsilon\}$ satisfies the following:

(1) The function $a_i(\ell, \hbar, 0) : \mathcal{B} \to \zeta$ is continuous for each $(\ell, \hbar) \in \Delta$ and the function $a_i(\cdot, \cdot, x) : \Delta \to \zeta$ is strongly measurable for each $x \in \mathcal{B}$.

(2) There exist constants $K_{a_1}, K_{a_2} > 0$, and $K_{a_3} > 0$ such that for all $\ell, \hbar \in \mathcal{I}$ and $x \in \mathcal{B}$ we have

$$\|a_i(\ell, \hbar, x)\|^2 \leq K_{a_i}(1 + \|x\|^2_{\mathcal{B}}), \quad \text{for} \quad i = 1, 2, 3.$$

(χ5) The measurable mappings $\mathcal{G}_2(\cdot), \mathcal{G}_3(\cdot)$ and $g(\cdot)$ convince the subsequence of situation:

(1) For all $\ell \in \mathcal{I}, x_1, x_2 \in \mathcal{B}, y_1, y_2 \in \zeta$

$$|\mathcal{G}_2(\ell, x_1, y_1) - \mathcal{G}_2(\ell, x_2, y_2)|^2 \leq K(\|x_1 - x_2\|^2_{\mathcal{B}} + \|y_1 - y_2\|^2_{\zeta}),$$

$$|\mathcal{G}_3(\ell, x_1, y_1) - \mathcal{G}_3(x_2, y_2)|^2_{\mathcal{L}_2^0} \leq K(\|x_1 - x_2\|^2_{\mathcal{B}} + \|y_1 - y_2\|^2_{\zeta}).$$

(2) For any ζ-valued process $x(\hbar), y(\hbar), \ell \in \mathcal{I}$

$$\int_0^{\ell}\int_Z |g(\hbar, x(\hbar-), z) - g(\hbar, y(\hbar-), z)|^2 v(dz)d\hbar$$

$$\leq \int_0^\ell K(|x(\hbar) - y(\hbar)|^2)d\hbar,$$

$$\left(\int_0^\ell \int_Z |g(\hbar, x(\hbar-), z) - g(\hbar, y(\hbar-), z)|^4 v(dz)d\hbar\right)^{\frac{1}{2}}$$
$$\leq \int_0^\ell K(|x(\hbar) - y(\hbar)|^2)d\hbar,$$

$$\left(\int_0^\ell \int_Z |g(\hbar, x(\hbar-), z)|^4 v(dz)d\hbar\right)^2$$
$$\leq \int_0^\ell K(|x(\hbar)|^2)d\hbar.$$

where $K(\cdot) : \mathbb{R}_+ \to \mathbb{R}_+$ is a concave nondecreasing function such that $K(0) = 0$, $K(u) > 0$ for $u > 0$ and $\int_{0+} \frac{du}{K(u)} = \infty$.

($\chi 6$) There exists a constant $M_1 > 0$, $\forall\, \ell \in \mathscr{I}$ such that

$$|\mathscr{G}_2(\ell, 0, 0)|^2 \vee |\mathscr{G}_3(\ell, 0, 0)|^2 \vee \int_Z |g(\ell, 0, z)|^2 v(dz) \leq M_1.$$

Lemma 2.3 *Let $y : (-\infty, \Upsilon] \to \zeta$ such that $y_0 = \varphi$. If ($\chi 4$) holds, then*

$$\|y_\hbar\|_{\mathscr{B}} \leq (N_\Upsilon + \mathscr{W}_0^\varphi)\|\varphi\|_{\mathscr{B}} + K_\Upsilon \sup\{\|y(\theta); \theta \in [0, \max\{0, \hbar\}]\}, \ \hbar \in \widetilde{\mathscr{L}}(\varrho^-) \cup \mathscr{I},$$

where $\mathscr{W}_0^\varphi = \sup_{\ell \in \widetilde{\mathscr{L}}(\varrho^-)} \mathscr{W}^\varphi(\ell)$.

3 Main Results

We demonstrate the existence and uniqueness theorem in this section. We now provide the series of subsequent approximations, which is defined as follows:

$$
\begin{aligned}
y^0(\ell) &= \Upsilon\varphi(0), \quad t \in \mathscr{I}, \\
y^q(\ell) &= \varphi(\ell), \quad \ell \in (-\infty, 0], \quad q = 1, 2, \cdots, \\
y^q(\ell) &= \iota(\ell)[\varphi(0) + \mathscr{G}_1(0, \varphi, 0)] - \mathscr{G}_1\Big(\ell, y^q_{\varrho(\ell, y^q_\ell)}, \int_0^\ell a_1(\ell, \hbar, y^q_{\varrho(\hbar, y^q_\hbar)})d\hbar\Big) \\
&+ \int_0^\ell \iota(\ell-\hbar)\mathscr{G}_2\Big(\hbar, y^{q-1}_{\varrho(\hbar, y^{q-1}_\hbar)}, \int_0^\hbar a_2(\hbar, \ell, y^{q-1}_{\varrho(\ell, y^{q-1}_\ell)})d\ell\Big)d\hbar \\
&+ \int_0^\ell \iota(\ell-\hbar)\mathscr{G}_3\Big(\hbar, y^{q-1}_{\varrho(\hbar, y^{q-1}_\hbar)}, \int_0^\hbar a_3(\hbar, \ell, y^{q-1}_{\varrho(\ell, y^{q-1}_\ell)})d\ell\Big)dW(s) \\
&+ \int_0^\ell \iota(\ell-\hbar)\int_Z g(\hbar, y^{q-1}(\hbar-), z)\widetilde{N}(d\hbar, dz), \quad \ell \in \mathscr{I}, \quad q = 1, 2, \cdots.
\end{aligned} \tag{10}
$$

The following Bihari inequality (BI), which is described in [2], is required to create the uniqueness of MSs of the systems (1)–(2).

Lemma 3.4 (BI [2]) *Let $\Upsilon > 0$ and $u_0 \geq 0, u(\ell), x(\ell)$ be CF on $\mathscr{I}$. Let $K : \mathbb{R}^+ \to \mathbb{R}^+$ be a concave and nondecreasing CF such that $K(r) > 0$ for $r > 0$. If*

$$u(\ell) \leq u_0 + \int_0^\ell x(\ell)K(u(\hbar))d\hbar,$$

then

$$u(\ell) \leq G^{-1}\Big(G(u_0) + \int_0^\ell x(\hbar)d\hbar\Big),$$

$\forall \ell \in \mathscr{I}$ and $G(u_0) + \int_0^\ell x(\hbar)d\hbar \in Dom(G^{-1})$, where $G(r) = \int_1^r \frac{d\hbar}{K(\hbar)}(r \geq 0)$ and G^{-1} is the inverse function of the G. In particular, if moreover, $u_0 = 0$ and $\int_{0+} \frac{d\hbar}{K(\hbar)} = +\infty$, then $u(\ell) = 0, \forall \ell \in \mathscr{I}$.

Lemma 3.5 *You can suppose that $(\chi 1) - (\chi 6)$ hold and together with $K^2M^2 < \frac{1}{2}$ hold; then $\forall \ell \in (-\infty, \Upsilon], q \geq 0$; there exists a PC K_1 such that $\|y^q\|^2_{\mathscr{M}^2} \leq K_1$.*

Proof It is obvious that $y^0 \in \mathscr{M}^2((-\infty, 0], \Upsilon)$. For $q \geq 1$, we have

$$\mathbb{E}\sup_{0\le\hbar\le\ell}|y^q(\ell)|^2 \le 5\mathbb{E}\sup_{0\le\hbar\le\ell}|\iota(\hbar)[\varphi(0)+\mathscr{G}_1(0,\varphi,0)]|^2$$
$$+5\mathbb{E}\sup_{0\le\hbar\le\ell}\left|\mathscr{G}_1\left(\hbar, y^q_{\varrho(\hbar,y^q_\hbar)}, \int_0^\hbar a_1(\hbar,\ell,y^q_{\varrho(\ell,y^q_\ell)})d\ell\right)\right|$$
$$+5\mathbb{E}\sup_{0\le\hbar\le\ell}\left|\int_0^\hbar \iota(\hbar-r)\mathscr{G}_2\left(r, y^{q-1}_{\varrho(r,y^{q-1}_r)}, \int_0^r a_2(r,\ell,y^{q-1}_{\varrho(\ell,y^{q-1}_\ell)})d\ell\right)dr\right|^2$$
$$+5\mathbb{E}\sup_{0\le\hbar\le\ell}\left|\int_0^\hbar \iota(\hbar-r)\mathscr{G}_3\left(r, y^{q-1}_{\varrho(r,y^{q-1}_r)}, \int_0^r a_3(r,\ell,y^{q-1}_{\varrho(\ell,y^{q-1}_\ell)})d\ell\right)dW(r)\right|^2$$
$$+5\mathbb{E}\sup_{0\le\hbar\le\ell}\left|\int_0^\hbar\int_Z \iota(\hbar-r)g(r,y^{q-1}(r-),z)\widetilde{N}(dr,dz)\right|^2 = 5\sum_{i=1}^5 J_i. \tag{11}$$

Now, $J_i, i=1,2,\cdots,5$. By conditions $(\chi 3)$ and $(\chi 5)$, we have

$$J_1 \le 2\left[\mathbb{E}\sup_{0\le\hbar\le\ell}|\iota(\hbar)\varphi(0)|^2 + \mathbb{E}\sup_{0\le\hbar\le\ell}|\iota(\hbar)\mathscr{G}_1(0,\varphi,0)|^2\right] \le 2M(1+K^2_{\mathscr{G}_1})\mathbb{E}\|\varphi\|^2_{\mathscr{B}}.$$

By using Holder's inequality (HI), Lemma 2.1, and the conditions $(\chi 5)$ and $(\chi 6)$, we get

$$J_2 \le \mathbb{E}\sup_{0\le\hbar\le\ell}\left|\mathscr{G}_1\left(\hbar, y^q_{\varrho(\hbar,y^q_\hbar)}, \int_0^\hbar a_1(\hbar,\ell,y^q_{\varrho(\ell,y^q_\ell)})d\ell\right)\right|^2$$
$$\le K^2_{\mathscr{G}_1}(1+\Upsilon K_{a_1})\mathbb{E}\sup_{0\le\hbar\le\ell}\|y^q_{\varrho(\hbar,y^q_\hbar)}\|^2_{\mathscr{B}}.$$

The statement for applying here $(\chi 1)$, $(\chi 6)$ and HI, we have

$$J_3 \le \Upsilon\mathbb{E}\sup_{0\le\hbar\le\ell}\int_0^\hbar\left\|\iota(\hbar-r)\mathscr{G}_2\left(r, y^{q-1}_{\varrho(r,y^{q-1}_r)}, \int_0^r a_2(r,\ell,y^{q-1}_{\varrho(\ell,y^{q-1}_\ell)})d\ell\right)\right\|^2 dr$$
$$\le 2M\Upsilon\left[M_1\Upsilon + \int_0^\ell \mathbb{E}K\big((1+\Upsilon K_{a_2})\|y^{q-1}_{\varrho(\hbar,y^{q-1}_\hbar)}\|^2_{\mathscr{B}}\big)d\hbar\right].$$

With our assumptions $(\chi 4)$, $(\chi 5)$ and Burkholder-type inequality [3], it follows that there exist some PC D_1 and D_2 such that

$$J_4 \le MD_1Tr(Q)\int_0^\ell \mathbb{E}\left\|\mathscr{G}_3\left(r, y^{q-1}_{\varrho(r,y^{q-1}_r)}, \int_0^r a_3(r,\ell,y^{q-1}_{\varrho(\ell,y^{q-1}_\ell)})d\ell\right)\right\|^2_{\mathscr{L}^0_2} d\hbar$$

$$\leq 2MD_1Tr(Q)\Big[M_1\Upsilon + \int_0^{\ell} K((1+\Upsilon K_{a_3})\mathbb{E}\|y^{q-1}_{\varrho(\hbar, y^{q-1}_{\hbar})}\|^2_{\mathscr{B}})d\hbar\Big].$$

And

$$J_5 \leq MD_2\Big[\mathbb{E}\int_0^{\ell} |g(\hbar, y^{q-1}(\hbar-), z)|^2 v(dz)d\hbar$$
$$+\mathbb{E}\Big(\int_0^{\ell}\int_Z |g(\hbar, y^{q-1}(\hbar-), z)|^4 v(dz)d\hbar\Big)^{\frac{1}{2}}\Big]$$
$$\leq MD_2\Big[\mathbb{E}\int_0^{\ell}\int_Z |g(\hbar, y^{q-1}(\hbar-), z) - g(\hbar, 0, z) + g(\hbar, 0, z)|^2 v(dz)d\hbar$$
$$+\mathbb{E}\Big(\int_0^{\ell}\int_Z |g(\hbar, y^{q-1}(\hbar-), z)|^4 v(dz)d\hbar\Big)^{\frac{1}{2}}\Big]$$
$$\leq 2MD_2\Big[M_1\Upsilon + \int_0^{\ell} \mathbb{E}K(|y^{q-1}(\hbar)|^2)d\hbar\Big] + MD_2\int_0^{\ell} \mathbb{E}K(|y^{q-1}(\hbar)|^2)d\hbar$$
$$= 2MD_2M_1\Upsilon + 3MD_2\int_0^{\ell} \mathbb{E}K(|y^{q-1}(\hbar)|^2)d\hbar.$$

Substituting the estimates $(J_1)-(J_5)$ together with (3.1), we have

$$\mathbb{E}\sup_{0\leq\hbar\leq\ell} |y^q(\hbar)|^2 \leq Q_1 + 5K^2_{\mathscr{G}_1}(1+\Upsilon K_{a_1})\mathbb{E}\sup_{0\leq\hbar\leq\ell} \|y^q_{\varrho(\hbar, y^q_{\hbar})}\|^2_{\mathscr{B}}$$
$$+ 10M\Upsilon(1+\Upsilon K_{a_2})\int_0^{\ell} \mathbb{E}K(\|y^{q-1}_{\varrho(\hbar, y^{q-1}_{\hbar})}\|^2_{\mathscr{B}})d\hbar$$
$$+ 10MD_1(1+K_{a_3})\int_0^{\ell} \mathbb{E}K(\|y^{q-1}_{\varrho(\hbar, y^{q-1}_{\hbar})}\|^2_{\mathscr{B}})d\hbar + 15MD_2\int_0^{\ell} \mathbb{E}K(|y^{q-1}(\hbar)|^2_{\mathscr{B}})d\hbar$$

where $Q_1 = 10M(1+K^2_{\mathscr{G}_1})\mathbb{E}\|\varphi\|^2_{\mathscr{B}} + 10MM_1\Upsilon^2 + 10MM_1Tr(Q)D_1\Upsilon + 10MM_1D_2\Upsilon$. We have

$$\mathbb{E}\sup_{0\leq\hbar\leq\ell} |y^q(\hbar)|^2 \leq Q_2 + 5K^2_{\mathscr{G}_1}(1+\Upsilon K_{a_1})\sup_{0\leq\hbar\leq\ell} \|y^q_{\varrho(\hbar, y^q_{\hbar})}\|^2_{\mathscr{B}} + 10Md\Big(\Upsilon(1+\Upsilon K_{a_2})$$
$$+MD_1(1+\Upsilon K_{a_3})\Big)\int_0^{\ell} K\big(\mathbb{E}\|y^{q-1}_{\varrho(\hbar, y^{q-1}_{\hbar})}\|^2_{\mathscr{B}}\big)d\hbar + 15MdD_2\int_0^{\ell} \mathbb{E}|y^{q-1}(\hbar)|^2_{\mathscr{B}}d\hbar,$$

where $Q_2 = Q_1 + 10cM\Upsilon^2(1+\Upsilon K_{a_2}) + 10cMTD_1Tr(Q)(1+\Upsilon K_{a_3}) + 15cMTD_2$. Then

$$\|y^{q-1}_{\varrho(\hbar, y^{q-1}_{\hbar})}\|^2_{\mathscr{B}} \leq 2\left[(N_\Upsilon + \mathscr{W}^{\varphi}_0)\mathbb{E}\|\varphi\|_{\mathscr{B}}\right]^2 + 2K^2_\Upsilon \sup_{0\leq\hbar\leq\Upsilon} \mathbb{E}\|y(\hbar)\|,$$

we get

$$\mathbb{E} \sup_{0\leq\hbar\leq\ell} |y^q(\hbar)|^2 \leq \frac{1}{1 - 10K^2_{\mathscr{G}_1} K^2_\Upsilon(1 + \Upsilon K_{a_1})}$$

$$\left[Q_3 + 5d(4TK^2_\Upsilon(1 + \Upsilon K_{a_2}) + 4D_1 MK^2_\Upsilon(1 + \Upsilon K_{a_3}) + 3D_2)(\ell) \int_0^\ell \mathbb{E} \sup_{0\leq r\leq\hbar} |y^{q-1}(r)|^2 d\hbar\right],$$

where $Q_3 = Q_2 + [10K^2_{\mathscr{G}_1}(1 + \Upsilon K_{a_1})\mathbb{E} \sup_{0\leq\hbar\leq\ell} \left[(N_\Upsilon + \mathscr{W}^{\varphi}_0)\mathbb{E}\|\varphi\|^2_{\mathscr{B}}\right]^2 + 20Md\Upsilon\big(\Upsilon(1+\Upsilon K_{a_2}) + D_1 Tr(Q)(1+\Upsilon K_{a_3})\big)\left[(N_\Upsilon + \mathscr{W}^{\varphi}_0)\mathbb{E}\|\varphi\|^2_{\mathscr{B}}\right]^2$. It follows that for some constants Q_4, Q_5

$$\max_{1\leq q\leq K} \mathbb{E} \sup_{0\leq\hbar\leq\ell} |y^q(\hbar)|^2 \leq Q_4 + Q_5 \int_0^\ell \max_{1\leq q\leq K} \mathbb{E} \sup_{0\leq r\leq\hbar} \left|y^q(\hbar)\right|^2 d\hbar,$$

where

$$Q_4 = \frac{1}{1 - K^2_{\mathscr{G}_1} K^2_\Upsilon(1 + \Upsilon K_{a_1})}$$

$$\left[Q_3 + 5Md(4TK^2_\Upsilon(1 + \Upsilon K_{a_2}) + 4K^2_\Upsilon D_1 Tr(Q)(1 + \Upsilon K_{a_3}) + 3D_2)\Upsilon\mathbb{E}\|\varphi\|^2_{\mathscr{B}}\right].$$

$$Q_5 = \frac{1}{1 - K^2_{\mathscr{G}_1} K^2_\Upsilon(1 + \Upsilon K_{a_1})}\left[Q_3 + 5Md(4TK^2_\Upsilon(1+\Upsilon K_{a_2}) + 4K^2_\Upsilon D_1 Tr(Q)(1+\Upsilon K_{a_3}) + 3D_2)\right].$$

$$\mathbb{E} \sup_{0\leq\hbar\leq\ell} |y^q(\hbar)|^2 \leq Q_4 e^{Q_5\Upsilon}.$$

Furthermore,

$$\|y^q\|^2_{\mathscr{M}^2} = \mathbb{E}\|y^q_0\|^2_{\mathscr{B}} + \mathbb{E}\int_0^\Upsilon |y^q(\hbar)|^2 d\hbar \leq \mathbb{E}\|\varphi\|^2_{\mathscr{B}} + TQ_4 e^{Q_5\Upsilon} < \infty,$$

which implies $y^q(\cdot) \in \mathscr{M}^2((\infty, \Upsilon], \zeta)$ and $\|y^q\|^2_{\mathscr{M}^2} \leq K_1$ with $K_1 = \mathbb{E}\|\varphi\|^2_{\mathscr{B}} + TQ_4 e^{Q_5\Upsilon}$. □

Lemma 3.6 *The premises of Lemma 3.6 are true. Further, we presume that*

$$4K^2\widetilde{M}^2 + 5K^2M^2\widetilde{M}^2 < 1. \tag{12}$$

Positive constants do exist for K_2, K_3 *such that for all* $\ell \in \mathscr{I}$, $p, q \geq 1$,

$$\mathbb{E}\bigg(\sup_{0<\hbar\leq\ell}|y^{p+q}(\hbar)-y^{q}(\hbar)|^2\bigg)$$
$$\leq K_2\int_0^\ell K\bigg(\widetilde{M}^2\mathbb{E}\sup_{0\leq r\leq\hbar}|y^{p+q-1}(r)-y^{q-1}(r)|^2\bigg)d\hbar \tag{13}$$

and

$$\mathbb{E}\bigg(\sup_{0\leq\hbar\leq\ell}|y^{p+q}(\hbar)-y^{q}(\hbar)|^2\bigg)\leq K_3\ell. \tag{14}$$

Proof By using as our definition of y^q we can determine that any $p, q \geq 1$ and $\ell \in \mathscr{I}$

$$\mathbb{E}\bigg(\sup_{0\leq\hbar\leq\ell}|y^{p+q}(\hbar)-y^{q}(\hbar)|^2\bigg)$$
$$\leq 4\mathbb{E}\bigg(\sup_{0\leq\hbar\leq\ell}\bigg|\mathscr{G}_1\bigg(\hbar, y^{p+q}_{\varrho(\hbar,y^{p+q}_\hbar)}, \int_0^\hbar a_1(\hbar,\ell,y^{p+q}_{\varrho(\ell,y^{p+q}_\ell)})d\ell\bigg)$$
$$-\mathscr{G}_1\bigg(\hbar, y^{q}_{\varrho(\hbar,y^{q}_\hbar)}, \int_0^\hbar a_1(\hbar,\ell,y^{q}_{\varrho(\ell,y^{q}_\ell)})d\ell\bigg)\bigg|^2$$
$$+4\mathbb{E}\sup_{0\leq\hbar\leq\ell}\bigg|\int_0^\hbar \iota(\hbar-r)\bigg[\mathscr{G}_2\bigg(r, y^{p+q-1}_{\varrho(r,y^{p+q-1}_r)}, \int_0^r a_2(r,\ell,y^{p+q-1}_{\varrho(\ell,y^{p+q-1}_\ell)})d\ell\bigg)$$
$$-\mathscr{G}_2\bigg(r, y^{q-1}_{\varrho(r,y^{q-1}_r)}, \int_0^r a_2(r,\ell,y^{q-1}_{\varrho(\ell,y^{q-1}_\ell)})d\ell\bigg)\bigg]dr\bigg|^2$$
$$+4\mathbb{E}\sup_{0\leq\hbar\leq\ell}\bigg|\int_0^\hbar \iota(\hbar-r)\bigg[\mathscr{G}_3\bigg(r, y^{p+q-1}_{\varrho(r,y^{p+q-1}_r)}, \int_0^r a_3(r,\ell,y^{p+q-1}_{\varrho(\ell,y^{p+q-1}_\ell)})d\ell\bigg)$$
$$-\mathscr{G}_3\bigg(r, y^{q-1}_{\varrho(r,y^{q-1}_r)}, \int_0^r a_3(r,\ell,y^{q-1}_{\varrho(\ell,y^{q-1}_\ell)})d\ell\bigg)\bigg]dW(r)\bigg|^2$$
$$+4\mathbb{E}\sup_{0\leq\hbar\leq\ell}\bigg|\int_0^\hbar\int_Z \iota(\hbar-r)\bigg[g(r,y^{p+q-1}(r--),z)-g(r,y^{q-1}(r-),z)\bigg]$$
$$\widetilde{N}(dr,dz)\bigg|^2 = 4\sum_{i=6}^{9} J_i. \tag{15}$$

Now, we estimate every phrase J_i, remembering that

$$\|y^q_\hbar - y^p_\hbar\|_{\mathscr{B}} \leq \widetilde{M}\sup_{0\leq r\leq\hbar}|y^q(r)-y^p(r)|,$$

we have

$$J_6 \leq K^2 \mathbb{E} \sup_{0\leq \hbar \leq \ell} (1+\Upsilon K_{a_1}) \| y_{\hbar}^{p+q}$$

$$- y_{\hbar}^{p}\|_{\mathscr{B}}^2 \leq K^2 \widetilde{M}^2 \mathbb{E} \sup_{0\leq \hbar \leq \ell} (1+\Upsilon K_{a_1}) |y^{p+q}(\hbar) - y^p(\hbar)|^2.$$

Continuing in this manner, we obtain

$$J_7 \leq \Upsilon \int_0^\ell K \widetilde{M}^2 \mathbb{E} \sup_{0\leq r\leq \hbar} (1+\Upsilon K_{a_2}) |y^{p+q-1}(r) - y^{q-1}(r)|^2 d\hbar$$

$$J_8 \leq D_1 \int_0^\ell K \widetilde{M}^2 \mathbb{E} \sup_{0\leq r\leq \hbar} (1+\Upsilon K_{a_3}) |y^{p+q-1}(r) - y^{q-1}(r)|^2 d\hbar$$

$$J_9 \leq 2D_2 \int_0^\ell K \widetilde{M}^2 \mathbb{E} \sup_{0\leq r\leq \hbar} |y^{p+q-1}(r) - y^{q-1}(r)|^2 d\hbar.$$

The estimations are combined $(J_6) - (J_9)$ together with (3.5). There are

$$\mathbb{E}\left(\sup_{0\leq \hbar \leq \ell} |y^{p+q}(\hbar) - y^p(\hbar)|^2 \right)$$

$$\leq K_2 \int_0^\ell K\left(\widetilde{M}^2 \mathbb{E} \sup_{0\leq r\leq \hbar} |y^{p+q-1}(r) - y^{q-1}(r)|^2 \right) d\hbar$$

$$\leq K_2 \int_0^\ell K(2\widetilde{M}^2 K_1) d\hbar = K_3 \ell,$$

where $K_3 = K_2 K(2\widetilde{M}^2 K_1)$, $K_2 = \frac{Q_6}{1-4K^2\widetilde{M}^2(1+\Upsilon K_{a_1})}$, $Q_6 = 4\Upsilon(1+\Upsilon K_{a_1}) + 4D_1(1+\Upsilon K_{a_2}) + 8D_2$. It is referred to as the inequality (14). □

Theorem 3.4 *Assuming Lemmas 3.5 and 3.6 are true, the systems (1)–(2) satisfy a unique MS* $y(\ell) \in \mathscr{M}^2((-\infty, \Upsilon], \zeta)$.

Proof This theorem's proof consists of two phases.

Phase 1: Let's illustrate that $\{y^q(\ell), \ell \in \mathscr{I}\}$ is a Cauchy sequence. Let $K_1(u) = K_2 K(u)$. Choose $\Upsilon_1 \in \mathscr{I}$ such that $K_1(\widetilde{M}^2 K_3 u) \leq K_3$ for $u \in [0, \Upsilon_1]$.

In the beginning, we define two function sequences. $\{\delta_q(\ell)\}_{q\in \mathbb{N}}$ and $\{\delta_{p,q}(\ell)\}_{p,q\in\mathbb{N}}$ by $\delta_1(\ell) = K_3\ell$, $\delta_{q+1}(\ell) = \int_0^\ell K_1(\widetilde{M}^2\delta_q(u))du$. $\delta_{p,q}(\ell) = \mathbb{E}\left(\sup_{0\leq u\leq \ell} |y^{p+q}(u) - y^q(u)|^2 \right)$.

Then $\{\delta_q(\ell)\}_{q\in\mathbb{N}}$ is monotonically decreasing when $q \to \infty$ and $0 \leq \delta_{p,q}(\ell) \leq \delta_q(\ell)$ for all $p, q \geq 1$, $\ell \in [0, \Upsilon_1]$. In fact, it is obvious that $\delta_{1,p}(\ell) \leq \delta_1(\ell)$ and

$$\begin{aligned}\delta_{2,p}(\ell) &\leq \mathbb{E}\Big(\sup_{0\leq u\leq \ell}|y^{p+2}(u)-y^{2}(u)|^{2}\Big)\\ &\leq \int_0^{\ell} K_1\Big(\widetilde{M}^2\mathbb{E}\sup_{0\leq u\leq \ell}|y^{p+1}(u)-y^{1}(u)|^{2}\Big)d\hbar\\ &\leq \int_0^{\ell} K_1(\widetilde{M}^2\delta_1(\hbar))d\hbar \leq \int_0^{\ell} K_1(\widetilde{M}^2K_3\hbar)d\hbar = \delta_2(\ell)\leq K_3\ell=\delta_1(\ell),\end{aligned}$$

which implies that $\delta_{2,p}(\ell) \leq \delta_2(\ell) \leq \delta_1(\ell)$. Now consider the result holds for q; then

$$\begin{aligned}\delta_{q+1,p}(\ell) =&\mathbb{E}\Big(\sup_{0\leq u\leq \ell}|y^{p+q+1}(u)-y^{q+1}(u)|^{2}\Big)\leq \int_0^{\ell} K_1\big(\widetilde{M}^2\delta_{p,q}(\hbar)\big)d\hbar\\ &\leq \int_0^{\ell} K_1(\widetilde{M}^2\delta_q(\hbar))d\hbar = \delta_{q+1}(\ell)\leq \int_0^{\ell} K_1(\widetilde{M}^2\delta_{q-1}(\hbar))d\hbar=\delta_q(\ell).\end{aligned}$$

This demonstrates $\delta_q(\ell)$ is a nonnegative and decreasing CF on $[0, \Upsilon_1]$ by induction on q, so we can assume a function $\delta(\ell)$ by $\delta_k(\ell) \downarrow \delta(\ell)$, and it is easy to verify that $\delta(0) = 0$ is CF on $[0, \Upsilon_1]$. Consequently $\delta(\ell) = \lim_{q\to\infty}, \delta_q(\ell) = \lim_{q\to\infty}\int_0^{\ell} K_1(\delta_{q-1}(\hbar))d\hbar = \int_0^{\ell} K_1(\delta(\hbar))d\hbar$. From $\delta(0) = 0$, $\int_{0+}\frac{du}{K_1(u)} = +\infty$ together with BI, we have $\delta(\ell) \equiv 0$. Thus, $0 \leq \delta_{p,q}(\ell) \leq \delta_q(\Upsilon_1) \to 0$ as $q \to \infty$.

This demonstrates $\{y^q(\ell), \ell \in [0, \Upsilon_1]\}$ is a Cauchy sequence in $\mathscr{M}^2((-\infty, \Upsilon], \zeta)$, so there exists an $y(\ell) \in \mathscr{M}^2((-\infty, \Upsilon], \zeta)$ such that $y^q(\ell)$ converge to $y(\ell)$ in $\mathscr{M}^2((-\infty, \Upsilon], \zeta)$ as $q \to \infty$. By Borel-Cantelli lemma, there exists a subsequence $\{y^{q_k}(\ell)\} \subseteq \{y^q(\ell)\}$ such that $y^{q_k}(\ell) \to y(\ell)$ uniformly on $[0, \Upsilon_1]$ as $q_k \to \infty$. Using standard method, it shows that $y(\ell)$ satisfies the systems (1)–(2).

Phase 2: We are going to show how unique it is. Let $y(\ell)$ and $z(\ell)$ be two MSs of the systems (1.1)–(1.2). For $\ell \in \mathscr{I}$, by the same calculations in Lemma 3.1, we can simplify that

$$\mathbb{E}\sup_{0\leq\hbar\leq\ell}|y(\ell)-z(\ell)|^2 \leq K_2\int_0^{\ell} K\Big(\widetilde{M}^2\mathbb{E}\sup_{0\leq r\leq\hbar}|y(r)-z(r)|^2\Big)d\hbar.$$

It may be expressed as follows using Lemma 3.1: $\mathbb{E}\Big(\sup_{0\leq\hbar\leq\ell}|y(\hbar)-z(\hbar)|^2\Big) = 0$, on $\ell \in \mathscr{I}$. Nevertheless, it is clear that $y(\ell) = z(\ell)$ for $\ell \in (-\infty, 0]$; thus, $y(\ell) = z(\ell)$ almost surely for $\ell \in (-\infty, 0]$. □

4 Conclusion

In this paper we have present inquiry demonstrates the moderate solution for SNPID systems under the circumstance of state-dependent delay and Poisson jumps. To overcome the Lipschitz criteria which is unable to fit the present equation, we developed a resolvent operator and successive approximation of neutral stochastic partial integrodifferential systems.

References

1. Benchohra, M., Litimein, S., Guerekata, G.N.: On fractional integro-differential inclusions with state-dependent delay in Banach spaces. Appl. Anal. **92** , 335–350 (2013)
2. Bihari, I.: A generalization of a Lemma of Bellman and its application to uniqueness problem of differential equations. Acta Math. Acad. Sci. Hungar **7**, 71–94 (1956)
3. Da Prato, G., Zabczyk, J.: Stochastic Equations in Infinite Dimensions. Cambridge University Press, Cambridge (1992)
4. Grimmer, R.: Resolvent operators for integral equations in a Banach space. Trans. Am. Math. Soc. **273**(1), 333–349 (1982)
5. Hale, J., Kato, J.: Phase space for retarded equations with infinite delay. Funkcialaj Ekvacioj **21**, 11–41 (1978)
6. Hernandez, E., Sakthivel, R., Tanaka, S.: Existence results for impulsive evolution differential equations with state-dependent. Electron. J. Differ. Equ. **28**, 1–11 (2008)
7. Hino, Y., Murakami, S., Naito, T.: Functional Differential Equations with Unbounded Delay. Springer, Berlin (1991)
8. Pruss, J.: Evolutionary Integral Equations and Applications. Birkhauser, Basel (1993)
9. Varun Bose, C.S., Udhayakumar, R., Elshenhab, A.M., Sathish Kumar, M., Ro, J.S.: Discussion on the approximate controllability of Hilfer fractional neutral integro-differential inclusions via almost sectorial operators. Fractal and Fractional **6**(10), 607 (2022)
10. Yan, Z., Zhang, H.: Asymptotic stability of fractional impulsive stochastic partial integro-differential equations with state-dependent delay. Electron. J. Differ. Equ. **206**, 1–29 (2013)

A Hybrid Genetic Algorithm-Based Linear Programming Model to Optimize Feed Cost for Indian Ruminants: With Stochastic Model in Comparison

Ravinder Singh Kuntal, Vishal Patil, Radha Gupta, and Rajendran Duraisamy

1 Introduction

In India, ruminant production contributes about 4.8% and 3.4% of total meat and milk production, it also adds on to family income about 15–27%. Goat contributes 3% in milk production and 14.25% in meat production. As per 19th livestock census report, there are 65 million goats and 135 million goat population in India [3]. Milk production and meat production are crucial in India as annual milk growth rate is 6.3% and meat production is also increased to 0.4% in 2016–2017 compared to 2015–2016 [3, 14]. From the census report, it has been confirmed that goat population has increased a lot in recent years. As per the census report, goats have more capacity as compared to other ruminants to convert some feeds to milk and meat. Naturally, goats are very sturdy and can tolerate weather change highly. They can even survive for longer duration without water in heavy drought. In addition, during the natural disasters and food crises, goats can even consume less quality food, and this type of survival with low quality of food is not possible for other animals [13]. As per report, the quality of goat feed ensures the quality of meat or milk production. Hence, providing adequate feed according to the demand of goat

R. S. Kuntal (✉)
Department of Mathematics, Nitte Meenakshi Institute of Technology, Bangalore, Karnataka, India

V. Patil
Department of Mathematics, FET-Jain (Deemed-to-be) University, Bangalore, Karnataka, India

R. Gupta
Department of Mathematics, Dayananda Sagar College of Engineering, Bangalore, Karnataka, India

R. Duraisamy
NIANP (National Institutes of Animal Nutrient and Physiology), Bangalore, Karnataka, India

F. Kamalov et al. (eds.), *Advances in Mathematical Modeling and Scientific Computing*, Trends in Mathematics, https://doi.org/10.1007/978-3-031-41420-6_56

helps them to grow faster and produce more milk and meat [3, 10, 14]. Naturally, goats has tendency to eat all types of food, but they do not like to eat same food always; hence, there should be some diversity in their feeding habits. Goats also do not like to eat leftovers of another goat. Even though they can eat such food and consume the food that other animals do not eat, but for commercial purpose, farmers should be very careful about goat feeding, as it is necessary to feed the goat with quality of feedstuffs with proper nutrition and energy [11]. Unlike other dairy animals like cattle and buffalos, goat feed also should contain proper amount of protein, fat, minerals, and vitamins. Goats drink very less water compared to other livestock animals. Goat milk and its products like yogurt and cheese contribute about threefold significance in human nutrition. Milk protein in goat milk is 1% more than cow milk, and goat milk has more fatty acids as compared to cow milk, which are important for human health, especially in cardiovascular conditions [15]. Goats are multipurpose livestock ruminant, because it not only provides milk and meat but also fiber and skin [16]. Compared to cow farming, goat farming requires less space and money [5]. As discussed above, goats are contributing to socioeconomic growth of India in all the possible ways, but production of goat can improved only by improving the feeding standards. Feeding of goat should contain all nutrients, vitamins, and minerals and other ingredients in proper ratio and should be cheap as possible. To satisfy this requirement, an Excel-based nutrient calculator is developed, which calculates the nutrient requirement of goat based on body weight (BW) and daily gain, as per Indian Council of Agricultural Research [5] standards [10, 11]. Once the nutrient requirement is calculated, then a linear programming model can be developed to formulate the least-cost ration for goat. LP model is solved using LP simplex, GRG Nonlinear, and results have been compared with real-coded genetic algorithm.

2 Introduction

2.1 Nutrient Requirement

This study is based on primary data, in which a least cost formulation for ruminant animal is formulated based on body weight and average daily gain (ADG). Ruminant animals like goat and sheep have more capacity of converting feed to milk and meat, which provide an important source of protein and other nutrients for human consumption. In India, ruminant animals always play an important role in agriculture economy as they provide livelihood to many small and large farmers. Feeding of goat should contain all nutrients, vitamins, and minerals and other ingredients in proper ratio and should be cheap as possible. To satisfy this requirement, an Excel-based nutrient calculator is developed which calculates the nutrient requirements like dry matter (DM), crude protein (CP), total digestible nutrients (TDN), calcium (Ca), phosphorus (P), and metabolizable energy (ME) based on body weight and daily gain of goat as per Indian Council of Agricultural Research standards [5]. Once the nutrient requirement is calculated, then a linear programming model has been developed to formulate the least-cost ration for goats. LP model is solved using Excel solver-based LP simplex, GRG Nonlinear, and results are compared with hybrid real-coded genetic algorithm [7, 8]. Using local feeds, and ICAR Standards, Excel-based nutrient calculator is developed. Goats have unique choice of shrubs and trees leaves particularly woody plants compared to other domestic animals. Their pattern of diet selection compares closely with that of deer. Because of this unique choice, they are sometimes exploited as weapons against encroaching brush species. So, in many countries like the USA and South Africa, goats have been used in the control of intensive grazing for short period, and this method only works when the invading plant is palatable to the goat. If the balanced ration has been given to goats, they can grow well, produce maximum milk, and meat for which goat farmers need to feed 12–20% protein-containing feeds with all types of necessary vitamins and minerals in proper ratio. Small-scale goat farmers have lack of knowledge about modern feeding practices, which results in low productivity and more expense. Goats will eat from 6% to 11% of their body weight daily (based on dry matter: DM) of forage [12]. The exact amount they will eat varies depending on climate, temperature, types of browse available (palatability), and the age of the goat. Among the ruminants, goats seem to do the best at digesting high quantities of roughage because of its unique ability to digest roughage. Hence, if growing goat does not get required nutrients at reasonable price it will affect in milk yield and meat production. From Table 1, we can see that feed composition comprises major factors like dry matter, energy, protein, calcium, and phosphorus. Dry matter intake is very important as far as nutrition is concerned. Dry matter intake is dependent upon many factors like fodder quality and quantity, climate condition, and nutrient requirement of goat. Total DM must contain the required amount of energy and protein, which is digestible. Energy is expressed as TDN "Total Digestible Nutrients" and measured in terms of "Kilo Calories" (Kcal) or kilo joules (KJ). TDN represents the percentage of energy present in feedstuff. Protein is expressed as crude protein (CP). CP represents the percentage of protein present in feedstuff; CP is very essential for the maintenance, as it increases the forage intake. CP level will differ at different stages of a goat's life. As per Patil et al. [10, 11] while producing balanced ration, these three requirements should satisfy compulsorily with least cost, then the production will be more and farmers will be benefitted.

2.2 *Least-Cost Ration Formulation*

This study is based on primary data, in which a least cost formulation for ruminant animal was formulated. In brief, feeding of goat should contain all nutrients, vitamins, minerals, and other ingredients in proper ratio and should be cheap as possible. To satisfy this requirement, an Excel-based nutrient calculator is

Table 1 Composition of feed ingredients

Feedstuff	Cos/kg	DM (%)	CP (%)	TDN (%)	Ca (%)	P (%)
Wheat straw	11	92.7	4.8	47.5	0.31	0.1
Corn silage, mature	5.25	44.2	8.5	65.4	0.26	0.25
Sorghum, Sudan hay	16.5	86.5	9.4	54.4	0.54	0.2
Barley silage, headed	5	35.5	12	60.2	0.48	0.3
Grass hay	18	88.1	10.6	56.3	0.58	0.23
Corn grain, ground, dry	11.5	88.1	9.4	88.7	0.04	0.3
Sorghum, grain dry rolled	30	88.6	11.6	80.6	0.07	0.35
Barley grain rolled	18	91	12.4	82.7	0.06	0.39
Oats grain rolled	65	90	13.2	78.5	0.11	0.4
Wheat grain rolled	45	89.4	14.2	86.6	0.05	0.43
Rice bran	15	90.6	15.5	84.8	0.07	1.78
Wheat bran	25.5	89.1	17.3	71.5	0.13	1.18
Peanut meal	30	92.3	51.8	74.8	0.2	0.64
Soybean meal	32	89.6	46.3	88.5	0.36	0.66
Sunflower meal	19	92.2	28.4	59.9	0.48	1
Cotton seed meal	30	90.5	44.9	66.4	0.2	1.15
Linseed meal	40	90.3	32.6	65.4	0.4	0.83
De-oiled canola meal	15	90.3	37.8	69.9	0.75	1.1
Salt (sodium chloride) NaCl	4.1	100	0	0	0	0
Calcite (calcium carbonate)	2.1	100	0	0	30.71	0
DCP (dicalcium phosphate)	28	97	0	0	22	0

developed, which calculates the nutrient requirement of goat based on body weight (BW) and daily gain, as per Indian Council of Agricultural Research (ICAR) standards. The ration has been formulated using following steps: Step-1: Calculation of nutrient requirement of goat as per Indian Council of Agricultural Research [5] standards. Step-2: Selection of feedstuffs. Step-3: Fixing the constraints. Step-4: Formulating mathematical modeling. Step-5: Finding results. Starting from Step-1, an Excel-based nutrient calculator is developed as per ICAR [5] standards, which provide nutrient requirements like DM, CP, TDN, Ca, and P for goat based on body weight and average daily gain. In this study, we have selected 21 feedstuffs in which 5 are forages, 13 are concentrates, and 3 are minerals. When a goat farmer has to feed a balanced diet to a goat, all the nutrients should be mixed with proper ratio; hence, we have kept 15–30% forage and 80–90% concentrates on dry matter intake; see Table 2 for nutrient requirement. To obtain the least cost feed, we have five constraints for each nutrient, and it is unique for each type of goat. The minimum and maximum levels of crude protein, total digestible nutrient, calcium, and phosphorus for goat-1 and goat-2 are given in Table 3. Forage (15–30%) and concentrates (80–90%) have been calculated on total dry matter intake for each type of goats. The linear programming model has been formulated below.

LP model for goat-1: Min $z = \sum_{i=1}^{21} cixj$	LP model for goat-2:
S.to: $\sum_{i=1}^{21}$ DM = 0.8288 kg	S.to $\sum_{i=1}^{21}$ DM = 0.7018 kg
$0.120951 \le \sum_{i=1}^{21} CP \le 0.145142$ kg	$0.098755 \le \sum_{i=1}^{21} CP \le 0.118506$ kg
$0.545968 \le \sum_{i=1}^{21} TDN \le 0.655162$ kg	$0.45664 \le \sum_{i=1}^{21} TDN \le 0.547968$ kg
$0.003353 \le \sum_{i=1}^{21} Ca \le 0.004024$ kg	$0.002792 \le \sum_{i=1}^{21} Ca \le 0.003351$ kg
$0.002479 \le \sum_{i=1}^{21} P \le 0.002975$ kg	$0.002075 \le \sum_{i=1}^{21} P \le 0.00249$ kg
$0.12432 \le \sum_{i=1}^{5} Forage \le 0.261072$ kg	$0.10527 \le \sum_{i=1}^{5} Forage \le 0.221067$ kg
$0.66304 \le \sum_{i=6}^{21} Conc \le 1.392384$ kg	$0.56144 \le \sum_{i=6}^{21} Conc \le 1.179024$ kg;
Min $z = \sum_{i=1}^{21} cixj$	$0 \le xj \le cmax$, j = 1, 2, 3 ... n.

Table 2 Nutrient requirement

	Animal details	Nutrient requirement
Goat-1	Weight (kg): 30 ADG (g/d): 100	DM(kg): 0.8288, CP(g): 120.9514, TDN(g): 545.9683, Ca(g): 3.3533, P(g): 2.4789
Goat-2	Weight (kg): 20 ADG (g/d): 75	DM(kg): 0.7018, CP(g): 98.7549, TDN(g): 456.640034, Ca(g): 2.7923, P(g): 2.0754

Table 3 Min-max range of constraints

	Goat-1		Goat-2	
Constraints	Min (kg)	Max (kg)	Min (kg)	Max (kg)
CP	0.120951	0.145142	0.098755	0.118506
TDN	0.545968	0.655162	0.45664	0.547968
Ca	0.003353	0.004024	0.002792	0.003351
P	0.002479	0.002975	0.002075	0.00249
Forage	0.12432	0.261072	0.10527	0.221067
Concentrate	0.66304	1.392384	0.56144	1.179024

Where *ci* is the cost of each feed ingredient, *xj* represents number of feeds, *cmax* is the maximum limit for each feed where the upper and lower bounds have been calculated on DM basis. LPP has been solved using real-coded genetic algorithm, and the results are compared with conventional techniques like simplex method and GRG nonlinear.

3 Result

The results obtained by various techniques, namely, LP simplex, GRG nonlinear for least-cost ration are presented in Table 4 for goat-1 and goat-2, respectively. Linear programming model is also solved using real-coded genetic algorithm, which has been developed by Ravinder Singh Kuntal et al. to formulate least-cost ration for dairy cattle during pregnancy [8]. The results obtained by various techniques, namely, LP simplex, GRG nonlinear, evolutionary algorithm (EA), and real-coded genetic algorithm (RGA) for least-cost ration are presented in Tables 4 and 5 for goat-1 and goat-2, respectively. The linear programming model 1,

Table 4 Results for goat-1

Feedstuff	LP-simplex	GRG NL	EA	RGA
Wheat straw	0.249305	0.1815	0.2452992	0.1138
Corn silage, mature	0	0	0.0001443	0.0519
Sorgham, Sudan hay	0	0.073341	0.0000566	0
Barley silage, headed	0	0	0.0000994	0
Grass hay	0	0	0.0000403	0
Corn grain, ground, dry	0.16576	0.068608	0.1657441	0.1658
Sorghum, grain dry rolled	0.16576	0.144685	0.0153729	0.1658
Barley grain, rolled	0.16576	0.130521	0.1656859	0.1658
Oats grain, rolled	0	0.139072	0.1654134	0
Wheat grain, rolled	0	0.090549	0.0000794	0
Rice bran	0	0	0.0000770	0
Wheat bran	0	0	0.0000507	0
Peanut meal	0.146682	0.074937	0.1354271	0
Soybean meal	0	0	0.0000418	0.1526
Sunflower meal	0.00545	0	0.0000520	0
Cotton seed meal	0	0	0.0000272	0
Linseed meal	0	0	0.0000240	0
De-oiled canola meal	0.000746	0	0.0000155	0
Salt (sodium chloride) NaCl	0.004144	0.004144	0.0041438	0.0041
Calcite (calcium carbonate)	0.008288	0.002236	0.0026366	0.0083
DCP (dicalcium phosphate)	0.00045	0.008288	0.0082471	0.0007

LPP: Obj fun: 17.1673, DM: 0.8288, CP: 0.14514, TDN: 0.64964, Ca: 0.00402, P: 0.00297 Forage: 0.24931

GRG NL: Obj fun: 26.3018, DM: 0.8288, CP: 0.12506, TDN: 0.65516, Ca: 0.00402, P: 0.00297, Forage: 0.25484 Concentrate: 0.66304

EA: Obj fun: 23.1293, DM: 0.82697, CP: 0.14179, TDN: 0.64458, Ca: 0.00402, P: 0.00298 Forage: 0.24564 Concentrate 0.66304

Hybrid RGA: Obj fun: 16.3256, DM: 0.8288, CP: 0.135905, TDN: 0.640865, Ca: 0.004022, P: 0.002975 Forage: 0.1657, Concentrate 0.6631

2 for each goat has 21 variables, which are too complex for finding an optimal solution graphically; hence, it was used LP simplex method, which provides an iterative algorithm to locate the corner points systematically until it was get an optimal solution. In 2016, Kuntal et al. [7] have developed a real-coded genetic algorithm to formulate the ration for dairy cattle during pregnancy. They tested the results with simplex LP, GRG nonlinear, and evolutionary algorithm (EA) and found that there was no significant difference ($p > 0.05$) between various techniques adopted [8]. They concluded that real-coded genetic algorithm (RGA) can also be used for ration formulation to find least cost feed stuffs in dairy cattle without any nutrient's deficiency. In addition, in 2016, Kuntal et al. [7] developed real-coded genetic algorithm with hybrid function for formulating the least-cost ration for dairy buffaloes. LPP for ruminants has very rigid constraints. Due to this rigidity after specifying RGA for LP model, it returns a local minimum instead

Table 5 Results for goat-2

Feedstuff	LP-simplex	GRG NL	EA	RGA
Wheat straw	0.211977	0.188612	0.207990	0.117
Corn silage, mature	0	0	0.000048	0.0234
Sorgham, Sudan hay	0	0.024193	0.000010	0
Barley silage, headed	0	0	0.000080	0
Grass hay	0	0	0.000112	0
Corn grain, ground, dry	0.140360	0.032633	0.140289	0.1404
Sorghum, grain dry rolled	0.140360	0.135658	0.023298	0.1404
Barley grain, rolled	0.140360	0.118288	0.140357	0.1404
Oats grain, rolled	0.031915	0.132428	0.140054	0
Wheat grain, rolled	0	0.067512	0.000051	0
Rice bran	0	0	0.000135	0
Wheat bran	0	0	0.000009	0
Peanut meal	0.074379	0.062413	0.104613	0
Soybean meal	0	0	0.000003	0.1295
Sunflower meal	0.021507	0	0.000023	0
Cotton seed meal	0	0	0.000002	0
Linseed meal	0	0	0.000003	0
De-oiled canola meal	0	0	0.000002	0
Salt (sodium chloride) NaCl	0.003509	0.003509	0.003507	0.0035
Calcite (calcium carbonate)	0.002032	0.001981	0.002078	0.007
DCP (dicalcium phosphate)	0.007018	0.007018	0.007015	0.0003

LPP: Obj fun: 15.6128, DM: 0.7018, CP: 0.105904, TDN: 0.547968, Ca: 0.003351, P: 0.002490 Forage: 0.211977, Concentrate 0.56144
GRG NL: Obj fun: 22.78142, DM: 0.7018, CP: 0.104196, TDN: 0.547968, Ca: 0.063351, P: 0.002490 Forage: 0.212804, Concentrate 0.56144
EA: Obj fun: 19.5917, DM: 0.700127, CP: 0.116021, TDN: 0.546612, Ca: 0.003348, P: 0.002491 Forage: 0.208241, Concentrate 0.561439
Hybrid RGA: Obj fun: 13.9433, DM: 0.7019, CP: 0.114457, TDN: 0.539294, Ca: 0.003344, P: 0.00249 Forage: 0.1404, Concentrate 0.5615

of global minimum; hence, to overcome this deficiency of RGA, they introduced gradient-based search technique inside genetic algorithm where gradient-based search technique uses sequential quadratic programming (SQP) subproblem where SQP methods are representation of nonlinear programming methods. It has been concluded that RGA with hybrid function could be implemented to find least-cost ration when there is rigidity in constraints [8]. In the present study, similar hybrid RGA is used to find least-cost ration for ruminant cattle and the result is compared with Excel-based LP simplex, GRG nonlinear, evolutionary algorithm (EA). From Table 4, we can see that evolutionary algorithm and GRG nonlinear fail to provide least-cost ration for goat-1(GRG: Rs 26.3018, EA: Rs 23.1293) because the algorithm has selected costly feed stuffs compared to LP simplex and Hybrid GA (LP: Rs 17.1673, Hybrid RGA: Rs 16.3256). For goat-1, all the constraints have been satisfied by LP, GRG nonlinear, and Hybrid RGA except Excel-based

EA fails to satisfy DM (see graph 1). Similarly, from Table 5, we can observe that evolutionary algorithm and GRG nonlinear fail to provide least-cost ration for goat-2 (GRG: Rs 22.78, EA: Rs 19.5917) because the algorithm has selected costly feedstuffs compared to LP simplex and Hybrid GA (LP: Rs 15.6128, Hybrid RGA: Rs 13.9433). LP, GRG, is satisfying all the constraints and Hybrid RGA except EA fails to satisfy DM (see graph 2). From the result, it is seen that Hybrid RGA had tried to neglect costly feedstuffs in LP model 1 and 2 where all the constraints including the most important constraint (dry matter intake) have been satisfied. From all four techniques, we can conclude that for goat-1 with body weight 30 kg and ADG of 100 g needs total mixed ration about 0.98 kg on as such basis, where CP: 0.1359 kg, TDN: 0.6408 kg, Ca: 0.04 kg, and P: 0.0029 kg amounting Rs 16.3256 per goat on as dry matter basis is required. For goat-2 with body weight 25 kg and average daily gain of 75 g needs total mixed ration about 0.81 kg on as such basis, in which CP: 0.11445 kg, TDN: 0.53929 kg, Ca: 0.00334 kg, and P: 0.00249 kg amounting Rs 13.943 per goat on as dry matter basis. Dry matter (DM basis) is the value of nutrient after removing moisture, for example, if wheat straw has 92.7% DM, it means there is 7.3% moisture present in it. Goat farmers need total mixed ration to feed the ruminant on "as such basis," whereas such basis includes the moisture. Moisture content varies considerably, between feeds contributing to the as such basis weight. Therefore, we have converted the least cost and total mixed ration obtained by all techniques to "as such basis" and the result shows that goat-1 needs 0.98 kg TMR on as such basis amounting Rs 18.52 and goat-2 needs TMR of 0.81 kg on as such basis amounting Rs 15.69.

4 Discussion

According to the foregoing investigations and assessment, the estimated values for TDN, CP, and DMI are almost identical to the values provided by Mandal et al. [9] and Patil et al. [12] and are shown in Table 6. ICAR [5] values have been taken into account while comparing DMI. For goat-1, TDN values are similar, but found to be 12% higher than the values provided by Mandal et al. [9] 15.9% higher than the values provided by Patil et al. [12], and 17.2% higher than the values provided by ICAR [5] standards to that of compared values. The evaluated CP values are reported 13.3% higher than the values provided by Mandal et al. [9] and Patil et al. [12], whereas 12.3% higher than ICAR [5] standards. The DMI estimation was similar to that of ICAR [5] but showed a lower value compared with the NRC 1981 for about 0.2–1.4%. For goat-2, TDN is found to be 8.5% higher than the values provided by Patil et al. [12] and 18% higher than the values provided by ICAR [5]. CP is recorded 15.6% higher than the values provided by Patil et al. [12] and 3.9% higher than the values provided by ICAR [5]. The DMI estimation showed a lower value of 1.4–3.8% than the values provided by Patil et al. [12] and ICAR [5]. Nutrient needs are known to be genotype, environmental condition, and dietary scenario dependent [2]. The dietary requirements of goats are influenced by age, gender,

Table 6 Comparison of present study requirements TDN, CP, and DMI with those prescribed by existing feeding standards[a]

	TDN (g)	CP (g)	DMI (kg)
BW-30 kg, ADG-100 g			
Present	640.86	135.9	**0.83**
Mandal et al. [9]	572(+12)	120(+13.3)	NS
Kearl [6]	552(+16.1)	84(+61.8)	NS
NRC [17]	562(+14)	79(+72)	1.2(−31)
Patil et al. [12]	553(+15.9)	120(+13.3)	0.83(−0.2)
ICAR [5]	547(+17.2)	121(+12.3)	0.84(−1.4)
BW-25 kg ADG-75 g			
Present	539.29	114.4	0.7019
Mandal et al. [9]	406(+32.8)	89(+28.5)	NS
Kearl [6]	NS	NS	NS
NRC [17]	NS	NS	NS
Patil et al. [12]	497.25(+8.5)	99(+15.6)	0.71(−1.4)
ICAR [5]	457(+18)	110.08(+3.9)	0.73(−3.8)

[a]NS: not stated in the feeding standards. Values in parentheses indicate that percentage higher (+) or lower (−) is the present estimate as compared to the estimated values of the present study

Table 7 Paired t-test of present and stochastic model [12] values of TDN, CP, and DMI[a] at 5% significance level

Parameter	TDN	CP	DMI
p-value	0.2159	0.01017	0.5
t value	2.835	62.6	−1
Sample SD	32.4	0.3536	0.005728

[a]In case of TDN, rejecting null hypothesis is too high (21.59%). In case of CP, rejecting null hypothesis is small (1.02%). In case of DM I, rejecting null hypothesis is too high (50%)

climate, genetics, body composition, parasites, metabolic cycle, and activities [4]. According to Mandal et al. [9] and Kearl [6] the information provided was based on a few publications on Indian animals and used protein requirement estimations derived from nitrogen balancing tests or studies with nonproducing animals, which could explain the disparity between the two databases [1] provided in Table 7.

A paired t-test is done between estimated values and the values provided by Patil et al. [12] (see Table 7) revealing that there is a non-significant huge difference for TDN between the study of Patil ($M = 525.1$, $SD = 39.4$) and the present study ($M = 590.1$, $SD = 71.8$), $t(1) = 2.8$, $p = 0.216$. H0 cannot be rejected since the p-value is greater than the average of the study provided by Patil et al. [12]. Population is regarded equivalent to the average of the present study population. The paired-t test findings for CP demonstrated a statistically huge difference between the study of Patil et al. [12] ($M = 109.5$, $SD = 14.8$) and the present study ($M = 125.2$, $SD = 15.2$), $t(1) = 62.6$, $p = 0.010$. H0 has been rejected because of the p-value. The average of the study of Patil et al. [12] population is not equivalent to the average of the present study population. The paired-t test results for DMI revealed a non-significant medium difference between the study of Patil et al. [12] (M = 0.8, SD = 0.08) and the present study (M = 0.8, SD = 0.09), t (1) = 1, p = 0.500. H0 cannot be rejected since the p-value is greater then. The average of the study of Patil et al. [12] population is regarded equivalent to the average of the present study population. In other words, the difference between the study of Patil et al. [12] and the present study averages is not statistically significant. A non-significant finding does not establish that H0 is right; it only shows that the null assumption cannot be rejected. The p-value is 0.5, with (P (x <= −1) = 0.25). It implies that the probability of a type I mistake, which rejects a valid H0, is excessively high: 0.5 (50%). The higher the p-value, the stronger the support for H0. LP helps in the addition of a new feed to achieve a livestock ration that meets particular nutritional values as well as other requirements at the lowest possible cost [12]. Stochastic Programming methods have been utilized to compensate for the variability of nutrients present in various feed ingredients [10]. Hence, it is essential to compare the solutions of LP and the stochastic model. The comparisons of solution on a dry matter basis are given in Table 6.

5 Conclusion

The study optimizes the least-cost ration of Indian goats using BW and ADG for several feeding tests done throughout India in a wide range of dietary and climatic conditions. The prediction equations constructed in this study were true and sufficiently accurate for estimating needs under various situations based on linear programming. The proposed diet for Indian goats utilizing data has the lowest cost and meets the DMI, CP, and TDN nutritional values. ANOVA test at 5% level of significance has been performed on all four different techniques, LP simplex, GRG nonlinear, evolutionary algorithm (EA), and hybrid real-coded genetic algorithm (RGA), found there is no significance difference between the methods as $p > 0.05$, and concluded that hybrid real code genetic algorithm (RGA) method can be also be used, which provides least cost. However, fixing of constrains and use of code for making software are also considered while choosing the techniques for making least cost feed formulation. Further detailed study with various species of animals and with different physiological needs may be required to fine-tune the techniques for farmer use. We are able to economize the TMR cost such that the feed requirements of the animals have been met without any nutrient deficiency.

References

1. Cannas, A., Tedeschi, L.O., Atzori, A.S., Lunesu, M.F.: How can nutrition models increase the production efficiency of sheep and goat operations? Anim. Front. **9**, 33–44 (2019)

2. Ferket, P.R., van Heugten, E., van Kempen, T.A.T.G., Angel, R.: Nutritional strategies to reduce environmental emissions from non-ruminants. J. Anim. Sci. **80**, E168–E182 (2002)
3. Goat Farming In India: Full business guide for the beginners (2022). https://www.roysfarm.com/goat-farming-in-india/
4. Gurung, N.: Nutritional requirements of different classes of meat goats. Prof. Agric. Work. J. **6**, 10 (2020)
5. ICAR: Nutrient Requirements of Sheep, Goat and Rabbit. ICAR, New Delhi (2013)
6. Kearl, L.C.: Nutrient Requirements of Ruminants in Developing Countries. International Feedstuffs Institute, Utah State University, Logan (1982) https://www.scirp.org/(S(czeh2tfqyw2orz553k1w0r45))/reference/ReferencesPapers.aspx?ReferenceID=1438897
7. Kuntal, R.S., Gupta, R., Rajendran, D., Patil, V.: Application of real coded genetic algorithm (RGA) to find least cost feedstuffs for dairy cattle during pregnancy. Asian J. Anim. Vet. Adv. **11**, 594–607 (2016)
8. Kuntal, R.S., Gupta, R., Rajendran, D., Patil, V.: Study of real-coded hybrid genetic algorithm (RGA) to find least-cost ration for non-pregnant dairy buffaloes. In: Soft Computing for Problem Solving. Advances in Intelligent Systems and Computing, vol. 817, pp. 369–389. Springer Verlag, Singapore (2019)
9. Mandal, A.B., Paul, S.S., Mandal, G.P., Kannan, A., Pathak, N.N.: Deriving nutrient requirements of growing Indian goats under tropical condition. Small Rumin. Res. **58**, 201–217 (2005)
10. Patil, V., Gupta, R., Rajendran, D., Singh Kuntal, R.: Comparative study on feed formulation software: a short review. Int. J. Res. Granthaalayah. **5**, 105–115 (2017)
11. Patil, V., Gupta, R., Rajendran, D., Kuntal, R.S.: Formation and designing of "least-cost ration formulation application of cattle" using Excel VBA. In: Logistics, Supply Chain and Financial Predictive Analytics. Asset Analytics, pp. 101–110. Springer, Singapore (2019)
12. Patil, V., Gupta, R., Duraisamy, R., Kuntal, R.S.: Nutrient requirement equations for Indian goat by multiple regression analysis and least cost ration formulation using a linear and non-linear stochastic model. J. Anim. Physiol. Anim. Nutr. (Berl). **106**, 968–977 (2021). https://doi.org/10.1111/JPN.13653
13. Pradip Kumar, D.E.: 20th livestock census-2019: all India report (2022). https://ruralindiaonline.org/en/library/resource/20th-livestock-census-2019-all-india-report/
14. Reports | Department of Animal Husbandry & Dairying (2022). https://www.dahd.nic.in/documents/reports
15. Singh, A.: Livestock production statistics of India – 2018. Vet Extension (2019). www.vetextension.com. https://www.vetextension.com/livestock-animal-production-statistics-of-india-2018/
16. Singh, A.: Livestock production statistics of India – 2019. Vet Extension (2020). www.vetextension.com. https://www.vetextension.com/livestock-animal-production-statistics-of-india-2019/
17. NRC: Nutrient Requirements of Goats: Angora, Dairy, and Meat Goats in Temperate and Tropical Countries. National Academy of Sciences, National Academic Press, Washington, DC (1981)

Part IV
Statistics

Outlier Detection Using the Range Distribution

Dania Dallah and Hana Sulieman

1 Introduction

Outliers are anomalous observations found in a various kind of data. According to Schwertman et al. [5], outliers are defined as "*observations which are presumed to come from a different distribution than the majority of the data and they have a profound influence on the data analysis often leading to erroneous conclusions*." Generally, they may appear to be a nuisance, impeding the inference process and lowering the quality of data prediction, making the model less accurate and ineffective. Many researchers believe that outliers should be detected and eliminated as soon as possible so that data analysis can progress. However, because outliers include important information, this approach is frequently too restricted. Furthermore, outliers may be caused from experiments that may not have been run correctly or data that have been coded wrongly. In other circumstances, determining whether these outliers are poor data might be difficult because they could be due to integral random variation and represent a scientifically relevant discovery. As a result, it's important to understand these data and how they function as a separate variable.

2 Background

Many existing outlier identification techniques are found in statistics literature. In this section, we briefly describe some common techniques. Section 2.1 provides a

D. Dallah (✉) · H. Sulieman
American University of Sharjah, Sharjah, SH, UAE
e-mail: g00078476@alumni.aus.edu; hsulieman@aus.edu

F. Kamalov et al. (eds.), *Advances in Mathematical Modeling and Scientific Computing*, Trends in Mathematics, https://doi.org/10.1007/978-3-031-41420-6_57

description of techniques appropriate for univariate data, while Sect. 2.2 discusses outlier detection for multivariate data.

2.1 *Outlier Detection for Univariate Data*

Boxplot Tukey [6] proposed the concept of a *boxplot*, a method to detect outliers using graphical representation of the data based on their quantiles. It uses the concept of Interquartile range (IQR) technique to visualize outliers in a simple but effective way. Outliers are defined as the observations that are below the *boxplot lower whisker* defined by ($Q_3 - 1.5x\ IQR$) or above the *boxplot upper whisker* defined by ($Q_3 + 1.5x\ IQR$)).

Further enhancement and modification to the rule have been also included in the literature. Schwertman et al. [5] suggested an extension of Tukey's box-plot method termed the "box-plot rule" in the literature. They proposed a simple, more general fencing technique that provides for greater flexibility in determining the "outside rate," or the likelihood that an instance from a noncontaminated normal population falls beyond a certain limit or boundary. Their generalized fence technique consists of a confidence interval when variance is known and for a sample of size one, however, using the expected value of the IQR sample size *n*. Assuming normal distribution of the data, the generalized fences by Schwertman et al. [5] are:

$$F_n = q_2 \pm \frac{IQR}{k_n} Z_\alpha$$

where q_i is the ith quartile, IQR $=q_3 - q_1$, k_n is an adjustment factor relating the mean value of the IQR to the standard deviation of the data with various sample sizes, and Z_α is determined such that $P(Z \geq Z_\alpha) = \alpha$ for the standard normal variable Z and the total outside rate or tail area is 2α.

Extreme Studentized Deviate Rosner [4] proposed the extreme studentized deviate (ESD) procedure that detects multiple outliers in a univariate data set where an upper bound for the assumed number of outliers is specified. This technique follows an approximately normal distribution.

For a given upper bound, *b*, the generalized ESD method performs *b* different tests for testing the existence of one outlier, two outliers, and so on.

The hypotheses of the ESD test are:

H_0: no outlier exists in the data set.
H_a: at most b outliers exist in the data set.

The test statistic is defined by $t_i = \frac{\max_i |x_i - \overline{x}|}{s}$ where $\overline{x}$ and s refer to the sample mean and sample standard deviation, respectively. The process starts out by removing the observation that maximizes $|x_i - \overline{x}|$, computes again t_i, and repeats the process again until it reaches the assumed number of outliers, *b*. For a significance level α, the critical regions corresponding to the b test statistics are constructed by computing the b critical values.

$$\lambda_i = \frac{(n-i)\ t_{p,n-i-1}}{\sqrt{\left(n-i-1+t_{p,n-\mathrm{i}-1}^2\right)(n-\mathrm{i}+1)}}, \mathrm{i} = 1, 2, \ldots, b.$$

where $t_{p,\,v}$ is the $100p$-percentile point from the t distribution with v degrees of freedom and $p = 1 - \frac{\alpha}{2(n-i+1)}$.

The largest i for which $t_i > \lambda_i$ is used to determine the number of existing outliers in the sample data.

Rosner [4] has done numerous simulation studies to show that the critical value is accurate for $n \geq 25$ and reasonably accurate for $n \geq 15$.

ESD and boxplots are common outlier detection methods for univariate data. Multivariate outlier detection techniques have been also developed and utilized in many application domains.

2.2 *Outlier Detection for Multivariate Data*

Ramaswamy et al. [3] suggested a distance-based outlier formulation based on the distance of points from the nearest neighbors. This outlier detection algorithm starts with clustering the data points. Then, it calculates both the lower and upper bounds on $D^k$for data points in each partition where for a k and a point, $D^K(p)$ denote the distance of the k^{th}nearest neighbor of p and measures how much outlying a point can be. Furthermore, points that have larger values of $D^K(p)$ are more thinly distributed neighborhoods. This makes them strong outliers comparing to denser cluster points that have a lower value of $D^K(p)$. This technique then takes this information of the partitions that cannot contain the top percent of outliers and removes them from the dataset. The remaining partitions of points are then used to compute the outliers.

Kamalov and Leung [2] proposed a novel outlier detection algorithm that used dimensionality reduction and density estimation. This technique starts with reducing the dimension of the data using Principle Component Analysis (PCA) and then estimates the density distribution of the data using the Kernel Density Estimation method (KDE) to calculate the outlier points. In their paper, they make a comparison with other known techniques such as the LOF method, to demonstrate how their algorithms have shown to perform favorably in comparison.

In this chapter, we explore the use of the range statistic and its distribution in order to identify outliers in univariate data. We propose to standardize the range statistic by the interquartile range used in the construction of the boxplot discussed above to create a more robust tool to detect outliers. The proposed approach is applied to real datasets well known in machine learning literature and the findings are compared to those obtained by using the conventional standardization using the standard deviation, σ.

This chapter is structured as follows: in Sect. 3, we will construct the probability distributions of the range and interquartile range, IQR, and evaluate their characteristics through empirical simulations using the kernel density function; in Sect. 4, we will implement standardized range statistic to detect outlier in real datasets, and in Sect. 5, we wrap up with some conclusions and insights for future work.

3 Methodology

3.1 The Range Statistic

In statistics, the **range** is used to measure the spread of univariate data. It is given by

$$Range\,\{X_1, \ldots, X_n\} = R = X_n - X_1$$

where X_n is the maximum value and X_1 is the minimum value in the distribution.

The ***range statistics*** can be used to diagnose for the existence of outliers. It is a more reliable indicator of data variability when the distribution has no extreme values. However, the range can be misleading when outliers exist in the data set. A single outlier in the data will result in an entirely different range. As a result, the range can be used to spot these endpoints.

For a normal population, let's assume $X_1, X_2, \ldots, X_n$ be a sample of randomly selected n observations from the population. R is calculated from this sample, having the following properties:

1. R is nonnegative.
2. R is such that the distribution function of $W = \frac{R}{\sigma}$ is independent of σ, where σ is the population standard deviation.

Harter [1] produced a table of percentile values of the range distribution for normally distributed data with different sample sizes up to sample size $n = 100$. The authors proposed two approaches to contaminate the data with outliers: *mean value contamination* and *standard deviation contamination.*

The standardization of R by σ makes the statistic highly sensitive to unusual values in the data because the standard deviation itself is influenced by unusual and outlier values. Consequently, using W as a tool to identify outliers can be misleading.

In this chapter, we propose to standardize the range statistic R by the interquartile range IQR discussed in the boxplot method above.

Define the statistic:

$$K = \frac{R}{IQR}$$

Given that IQR corresponds to the middle 50% of the data points, it is influenced by the existence of outliers. As a result, the statistic ***K*** is more robust than ***W*** and so using ***K*** to deduct outliers provides a more precise identification of such anomalous observations.

3.2 *Order Statistic*

In statistics, the ith order statistic of a statistical sample is equal to its ith ordered value, where $i - 1$ of the data fall below and $n - i$ fall above:

$$X_1 < X_2 < \cdots < X_i < X_{i+1} < \cdots < X_n,$$

when $i = 1$, we obtain the smallest order statistic, $X_1 = \min\{X_1, \ldots, X_n\}$

Similarly, when $i = n$, we obtain the nth order (largest) statistic, $X_n = \max\{X_1, \ldots, X_n\}$

The **probability density function (PDF)** for the ith order statistics can be found as

$$f_{X(i)}(x) = \frac{n!}{(i-1)!\,(n-i)!} f_x(x)[F_x(x)]^{i-1}[1 - F_x(x)]^{n-i} \tag{1}$$

The **cumulative distribution function (CDF)** for the ith order statistics $\boldsymbol{F_x(x)}$ can be given by the following:

$$F_{X(i)}(x) = \sum_{i=r}^{n} \binom{n}{i} [F_x(x)]^i [1 - F_x(x)]^{n-i} \tag{2}$$

It also can be shown that the **joint distribution of two order** $X_{i:n}$, *and* $X_{j:n}$ can be equal to

$$f_{i,j:n}(x_i, x_j) = \frac{n!}{(i-1)!\,(j-i-1)!\,(n-j)!} \cdot [F_x(x_i)]^{i-1}\left[F_x(x_j) - F_x(x_i)\right]^{j-i-1}$$

$$\cdot \left[1 - F_x(x_j)\right]^{n-j} f_x(x_i)\, f_x(x_j) \tag{3}$$

$$-\infty < x_i < x_j < \infty$$

$$1 \le i \le j \le n$$

Based on Eq. (3), one can deduce that the CDF and PDF of the range are as follows:

CDF of R

$$F_R(x, r) = n \int_{-\infty}^{\infty} [F(x + r) - F(x)]^{n-1} f(x) dx \tag{4}$$

PDF of R

$$f_R(r) = \int_{-\infty}^{\infty} n(n-1)\{F(x_1 + r) - F(x_1)\}^{n-2} f(x_1) f(x_1 + r)\, dx_1 \tag{5}$$

Now specifically, when $i = \frac{n+1}{4}$ and $j = \frac{3(n+1)}{4}$, we obtain the joint distribution of the two-order statistic $X_i = Q_1,$ *and* $X_j = Q_3$ following Eq. 3. From the resulting equation, we can obtain the PDF of the interquartile range IQR=Q_3- Q_1as follows:

PDF of IQR

$$f_Q(q) = \frac{n!}{\left(\frac{n}{4}-\frac{3}{4}\right)!\left(\frac{n}{2}-\frac{1}{2}\right)!\left(\frac{n}{4}-\frac{1}{4}\right)!} \int_{-\infty}^{\infty} [F_x(x_i)]^{\frac{n}{4}-\frac{3}{4}} [F_x(q+x_i) - F_x(x_i)]^{\frac{n}{2}-\frac{1}{2}}$$

$$\cdot [1 - F_x(q + x_i)]^{\frac{n}{4}-\frac{1}{4}} \cdot f_x(x_i) f_x(q + x_i)\, dx_i \tag{6}$$

where Q = IQR

3.3 *Empirical Evaluation of the probability distributions of W and K*

In this section, we examine the characteristics of the probability distributions of the standardized range statistic by σ, W and the standardized range by IQR statistic, K. For convenience we assume continuous data values coming from a normal distribution. The KDE method used by Firuz et al. (2019) is utilized here to approximate the PDF functions of W and K. Given that the PDF of R constructed in Eq. (5) depends on the sample size, n, the simulation work considers several values of n. For each run a random sample of size n is generated and the values of the two statistics W and K are calculated. The process is repeated 10,000 times.

The resulting empirical PDFs of W and K for the different n*'s* are plotted in Fig. 1.

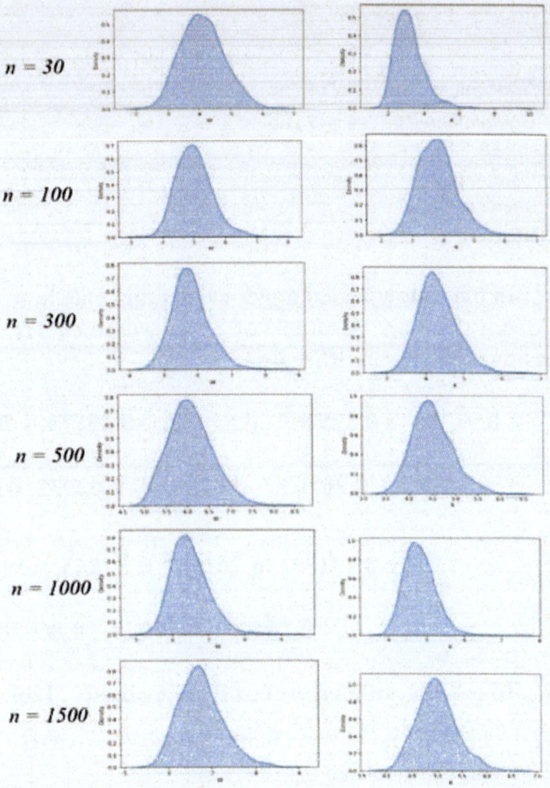

Fig. 1 Probability density function of W and K with different n's

Table 1 Percentile values for different n's

	W		K	
	95 percentile	97.5 percentile	95 percentile	97.5 percentile
$n = 30$	5.3082	5.4943	4.6413	5.1678
$n = 100$	5.9915	6.3022	4.8821	5.1887
$n = 300$	6.7361	6.9343	5.1944	5.4625
$n = 500$	6.9629	7.1801	5.3080	5.4983
$n = 1000$	7.3798	7.5829	5.5358	5.7073
$n = 1500$	7.6292	7.8419	5.7321	5.8672

Figure 1 demonstrates that overall, the PDFs for W and K agree in behavior for each sample size n with K showing slightly more skewness to the right than W which implies that K can assume larger values than W. Based on these constructed PDFs, we calculated the 95 and 97.5 percentiles for each n to be used as threshold values for determining the outlier values. Table 1 below shows the resulting percentile values. As expected, the values of the 95 and 97.5 percentiles change with different *sample sizes*.

Table 2 Example of n = 25

	W	95 percentile	97.5 percentile	K	95 percentile	97.5 percentile
Noncontaminated	3.224	no	no	3.583	no	no
Contaminated (mean = 3.5)	4.545	no	no	5.052	yes	no
Contaminated (mean = 6)	7.872	yes	yes	8.750	yes	yes

In the following example, we demonstrate how the range statistic can be used to identify outliers. A sample of 25 measurements is randomly generated from *N(0, 1)* using the software RStudio:

– 0.91281, –0.82526, –0.81438, –0.58179, –0.50219, –0.43809, –0.38885,

– 0.35986, –0.20163, –0.07892, –0.02932, 0.089886, 0.096274, 0.116971,

0.12338, 0.131531, 0.261961, 0.31863, 0.510856, 0.714533, 0.739841,

0.764061, 0.773405, 0.886785, 2.310297

To test for outliers, we test the hypothesis (Table 2):

H_0: There are no outliers (sample came from *N(0,1)*)
H_a: There are outlier(s)

For the noncontaminated data, at 5% significance level, we fail to reject H_0 and conclude that there are no outliers found in the sample, since W is less than percentile. Similarly, we obtain K and conclude there are no outliers as per the 95th percentile in the above table. Another sample is now generated by replacing the maximum value with 3.6315, an observation drawn from $N(\mu = 3.5, \sigma = 1)$ to the above sample. The objective is to test the null hypothesis that the observations belong to the same normal population with $\mu = 0$ and $\sigma = 1$. Hence, according to the given percentile, we conclude that H_0 *is not rejected* for W but is *rejected* for K. Therefore, only K was able to suspect the presence of an outlier in the sample data. Now, we try another sample generated by replacing the maximum with 6.95908, an observation drawn from $N(\mu = 6, \sigma = 1)$ to the above sample. Hence, according to the given percentile at 5% significance level, we conclude that H_0 is *rejected* for both W and K. Therefore, an outlier is suspected to be present in the sample data.

In conclusion, we can infer that K is not very sensitive for moderate (mild) outliers as W is. K is more robust and capable of detecting outliers.

4 Numerical Experiments

For numerical experiments with real data sets, we will use the ***Iris*** dataset that is found in the library of Rstudio and the three datasets obtain from Firuz's paper: ***Ecoli*** dataset found in Kaggle platform, ***Yeast*** dataset, and ***Spectrometer*** dataset found in UCI (Firuz et al. [2]) (Table 3).

The parent distribution of the variables of interest selected from these data sets will be assumed Normal. The observations will be first standardized and the range of standardized scores R will be used for the range statistic test described in Sect. 3.

Looking at the boxplots of the selected variables, we can see the possibility of finding outliers in each of the selected datasets (Figs. 2 and 3).

Therefore, we construct a table summary to decide if there exist any outliers (Table 4).

The above table compares the calculated W and K with the 95 and 97.5 percentiles we found from the randomly samples data for different n*'s*. For each variable, we compared the values of W and K with the closest n that is in Table 1. For example, for the first variable, AAC that is in the *Ecoli* dataset, we calculated W to be 7.1909. From Table 2, we see that the 95 percentile for $n = 300$ is 6.7361,

Table 3 Summary of all the datasets

	No. of observations	No. of variables	Variable(s) used
Iris	150	5	Sepal.Width
E. coli	336	9	AAC, GVH
Yeast	1484	9	alm, gvh
Spectrometer	531	103	bluebandflux8, Red_base_1, red.band.flux_20

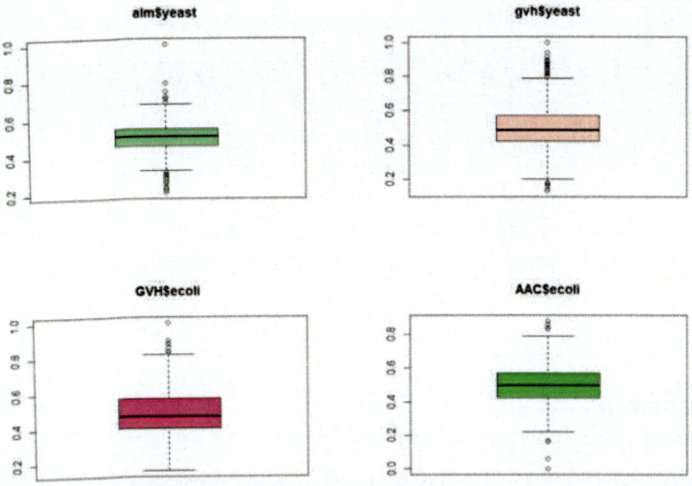

Fig. 2 Boxplots of variables from Yeast and *Ecoli* datasets

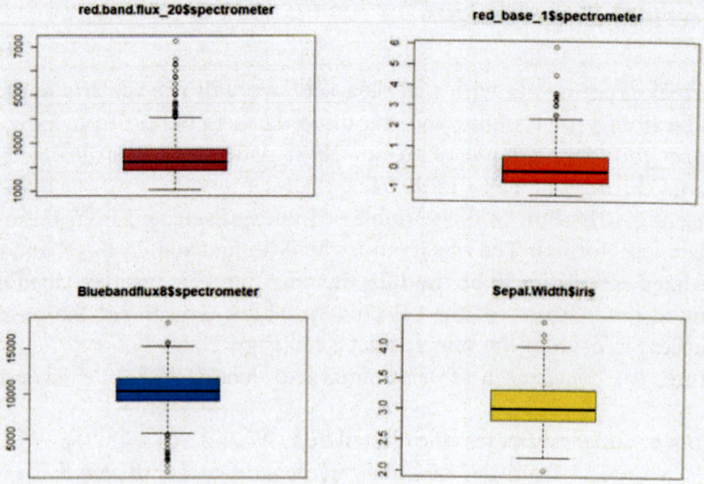

Fig. 3 Boxplots of variables from Iris and Spectrometer datasets

Table 4 Summary of all the datasets

	n	W	95 percentile	97.5 percentile	K	95 percentile	97.5 percentile
Ecoli$AAC	336	7.1909	yes	yes	5.8666	yes	yes
Spectrometer$ blue.band.flux_8	531	7.0530	yes	no	6.6871	yes	yes
Yeast$alm	1484	9.1150	yes	yes	8.7777	yes	yes
Iris$Sepa.Width	150	5.5602	no	no	4.800	no	no
Ecoli$GVH	336	5.6696	no	no	4.9411	no	no
Spectrometer$ RedBand1	531	7.1620	no	no	5.6715	no	no
Spectrometer$ res.band.flux_20	531	6.7569	no	no	6.9434	yes	yes
Yeast$gvh	1484	7.0204	no	no	5.800	yes	no

which is less than *W*; thus, we conclude that we suspect the presence of an outlier. The table continues and compares the values obtained in Table 2.

In conclusion, we can infer that *K* is less sensitive toward outliers compared to *W* because *IQR* is more robust. The robustness of the *K* statistic makes it more capable of detecting outliers than *W*.

5 Conclusion

Outlier detection has been an active research field utilized in many application domains. In this work, we explore the use of the range statistic standardized by the standard deviation and IQR to identify outliers in univariate data. This study is considered the first attempt to use the range statistic for the intent of identifying potential outliers. The characteristics of the probability distributions for the W and K statistics have been evaluated through simulation studies using data generated from normal distribution. Using the kernel density estimation procedure, the values of the 95 and 97.5 percentiles of the distributions of W and K have been estimated and used as threshold values to determine outliers. The numerical experiments have shown that the performance of the K statistic in detecting outliers leads to better and more precise identification of such anomalous observations. Future work would consider the development of a sequential outlier detection strategy using the range statistic standardized by the interquartile range.

References

1. Harter, H.L.: Use of tables of percentage points of range and studentized range. Technometrics. **3**(3), 407–411 (1961)
2. Kamalov, F., Leung, H.H.: Outlier detection in high dimensional data. J. Inf. Knowl. Manag. **19**(01), 2040013 (2020)
3. Ramaswamy, S., Rastogi, R., Shim, K.: Efficient algorithms for mining outliers from large data sets. In: Proceedings of the 2000 ACM SIGMOD international conference on Management of data, pp. 427–438 (2000, May)
4. Rosner, B.: Percentage points for a generalized ESD many-outlier procedure. Technometrics. **25**(2), 165–172 (1983)
5. Schwertman, N.C., de Silva, R.: Identifying outliers with sequential fences. Comput. Stat. Data Anal. **51**(8), 3800–3810 (2007)
6. Tukey, J.W.: Exploratory data analysis. **2**, 131–160 (1977)

Presenting a Flexible Class of INAR(p) Models to Analyze the COVID-19 Series in Mauritius

Noha Youssef, Naushad Mamode Khan, Ashwinee Devi Soobhug, Azmi Chutoo, and Shakil Ameerudden

1 Introduction

Time series of counts are commonly encountered in disciplines ranging from the epidemiological, environmental, meteorological, biopharmaceutical, economics and financial, transport to last but not least other interdisciplinary studies. Some examples of such counting time series include the intraday transaction series of stocks [1, 2], the daily frequency of claims in insurance [3], the weekly number of fatal accidents [4], the monthly number of domestic violence cases [5], and, the most recent instance, the daily number of new COVID-19 infected cases in a country [6, 7]. The COVID-19 series analysis is primordial at this moment due to the high death tolls and the rapid weakening of the social, economic, and health system. We lay focus on the COVID-19 analysis in an economically vulnerable state like

---

Authors Naushad Mamode Khan, Ashwinee Devi Soobhug, Azmi Chutoo, and Shakil Ameerudden have contributed equally to this work

N. Youssef (✉)
Department of Mathematics and Actuarial Science, The American University in Cairo, Cairo, Egypt
e-mail: nayoussef@aucegypt.edu

N. M. Khan · A. Chutoo
Faculty of Social Science and Humanities, University of Mauritius, Réduit, Mauritius
e-mail: n.mamodekhan@uom.ac.mu

A. D. Soobhug
Statistics Mauritius, Ministry of Finance, Economic Development and Planning, Port-Louis, Mauritius

S. Ameerudden
Ministry of Health and Wellness, Port-Louis, Mauritius

F. Kamalov et al. (eds.), *Advances in Mathematical Modeling and Scientific Computing*, Trends in Mathematics, https://doi.org/10.1007/978-3-031-41420-6_58

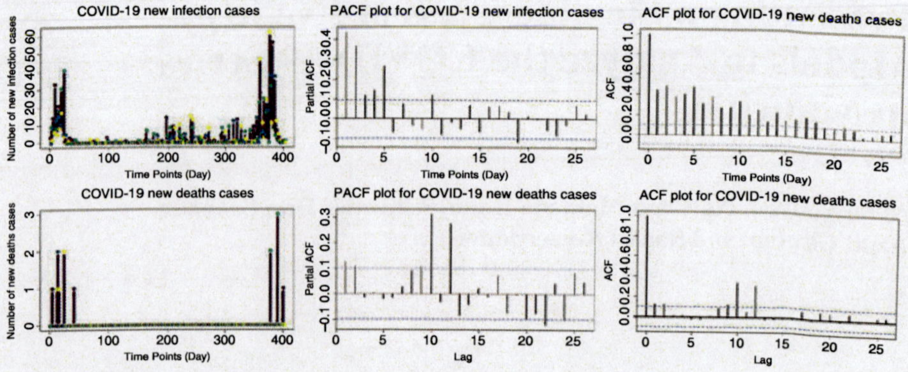

Fig. 1 The time, PACF, and ACF plots of new COVID-19 infection and death cases in Mauritius, 18 March 2020 to 25 April 2021

Mauritius since unlike other developed countries Mauritius was able to overcome the first wrath of COVID-19 pandemic in 2020 via strict sanitary measures. However, due to high dependence on international trade, its economy faced an unprecedented crisis, causing an economic contraction of around 15%. Figure 1 represents 404 observations each under the COVID-19 new infection and death cases, with respective partial autocorrelation function (PACF) and autocorrelation function (ACF) plots. The ACF plots for the COVID-19 new infection and death cases, as shown in Fig. 1, clearly depict non-stationarity due to slow decay, and the respective PACF plots even confirm that the COVID-19 new infection and death series in Mauritius for that specified period are high-ordered ($p = 7$), qualifying for high-ordered integer-valued auto-regressive time series structure. Based on the *qcc.overdispersion.test* in R statistical software via the *qcc* package, it is confirmed that the series is over-dispersed (variance > mean), and the Van den Broek test was also significant, proving that the series is zero-inflated. In fact, from June 2020 until October 2020, a long sequence of zero COVID-19 new infection cases was reported. Further, the Ljung-Box test ascertains the existence of serial correlation in the series. The presence of trend via the Cox-Stuart test was also significant. In fact, via the Cox-Stuart test results, it was found that the COVID-19 new case series has a decreasing trend. So, the presence of trend did induce non-stationarity in the series [8]. Based on the above graphs, the COVID-19 new infected series for Mauritius described numerous features that include over-dispersion phenomenon, excess zeros, non-stationarity due to time-varying covariates, and among other features. Thus, it requires a high-ordered integer-valued time series process to model such series. The paper is organized as follows: Sect. 1 highlights the gap in the current literature via graphical representation. The aim of this study has been described as well. Section 2 reviews the existing literature. The proposed INAR(p) models and its properties are in Sect. 3. Section 4 eventually summarizes the research findings and shows some short-term results. Finally Sect. 5 concludes the study with some useful recommendations for future research.

2 Background

The commonest model in the literature to model count time series is the simple INAR(1) process based on the binomial thinning operation [9]. The INAR equation consists of a survival term based on the thinning and an innovation term that is responsible to handle any error effects. From [10], the INAR(1) is expressed as

$$Y_t = \rho * Y_{t-1} + R_t \tag{1}$$

where $*$ is the thinning operator. The ρ is treated as a fixed nonnegative constant that lies in the interval [0,1) and

$$\rho * Y_{t-1} = \sum_{j=1}^{Y_{t-1}} B_j,$$

with B_j which is identically and independently Bernoulli-distributed r.v. with fixed probability ρ, that is, $\rho * Y_{t-1} \mid Y_{t-1} \sim \text{Binomial}(Y_{t-1}, \rho)$. The sequence of $\{R_t\}$ is assumed to be independently distributed with mean μ_t and variance σ_t^2. Under these assumptions, [11] confirms that the INAR process in 1 is Markovian. Additionally, the covariance between the previous lagged observations, Y_{t-k}, for $k \geq 1$ and the current innovation term R_t is assumed to be zero, that is, $Cov(Y_{t-k}, R_t)$=0, while $Cov(Y_t, R_t) = V(R_t)$. The conditional probability generating function of $Y_t \mid Y_{t-1}$, denoted by $G_{Y_t|Y_{t-1}}(s) = E[s^{Y_t|Y_{t-1}}]$, is given as:

$$G_{Y_t|Y_{t-1}}(s) = G_{\rho * Y_t|Y_{t-1}}(s) \times G_{R_t}(s), \tag{2}$$

where for the binomial thinning,

$$G_{Y_t|Y_{t-1}}(s) = (1 - \rho + \rho s)^{Y_{t-1}} \times G_{R_t}(s), \tag{3}$$

In the existing INAR(1), see Eq. 1; several types of thinning operators and innovation distributions have been applied in an attempt to model the over-dispersion and the other features more efficiently. These changes include mainly the binomial and negative binomial thinnings [12], while the Poisson distributions were commonly referred to represent the innovation part. However, the simple INAR(1) is not enough adequate and flexible to accommodate for series of multiple lags and severe over-dispersion with non-stationarity and zero inflation like the case of Mauritius. Hence, we require a high-ordered INAR process (INAR(p)) model with a robust thinning and flexible innovation.

In the existing literature, the INAR(p) model has been studied to some extent with Poisson and negative binomial innovations [13, 14]. We can notice as well that the non-stationarity in [14] was induced by time-dependent covariates in the innovation functions, but there is yet no research on the INAR(p) with zero-inflated innovations together with non-stationarity.

Thus, this paper proposes a wider family of INAR(p) models under generalized binomial and other thinning mechanism with some new Poisson mixed innovations and the associated zero-inflated innovations. In terms of the inferential methods, the conditional maximum likelihood (CML) approach is used. Note that by altering the probability properties of $\rho * Y_{t-1}$ [15], different $G_{Y_t|Y_{t-1}}(s)$ expressions are obtained, leading to different conditional likelihoods.

3 Model Formulation and Properties

From Eq. 1, the INAR(p) process is obtained by extending the previous lagged observations to obtain

$$Y_t = \rho_1 * Y_{t-1} + \rho_2 * Y_{t-2} + \ldots + \rho_p * Y_{t-p} + R_t \tag{4}$$

with the same assumptions of the previous lagged terms, Y_{t-k}, $1 \leq k \leq p$ on R_t, that is, $Cov(Y_{t-k}, R_t) = 0$. Further to the specification on R_t defined previously, it is also assumed that $\sum_{k=1}^{p} \rho_k < 1$ and R_t follows a probability distribution D, that is, $R_t \sim D(Y_{t,1}, Y_{t,2}, \ldots, Y_{t,s}, \nu)$, with its link or mean function θ_t given by:

$$log\theta_t = \beta_0 + \beta_1 Y_{t,1} + \beta_2 Y_{t,2} + \ldots + \beta_s Y_{t,s} \tag{5}$$

and ν is the extra-Poisson or over-dispersion parameter. The vector $\tilde{Y}_t = (Y_{t,1}, Y_{t,2}, \ldots, Y_{t,s})'$ represents the coefficient for the s covariates. Thus, with this specification on R_t, the expectation and variance of Y_t, that is, $E(Y_t)$ and $V(Y_t)$, vary with time.

The moments under the generalized binomial and negative binomial thinnings are computed as follows:

1. Under the binomial thinning assumption,

$$\rho * Y_t \mid Y_t \sim Bin(Y_t, \rho)$$

$$E(Y_t) = \sum_{k=1}^{p} \rho_k E(Y_{t-k}) + E(R_t)$$

where $E(R_t) = \mu_t$ and

$$Var(Y_t) = E(Y_t) + \sum_{k=1}^{p} \rho_k^2 [Var(Y_{t-k}) - E(Y_{t-k})] +$$

$$2 \sum_{k=1}^{p} \sum_{k<k'}^{p} \rho_k \rho_{k'} Cov(Y_{t-k}, Y_{t-k'}) + [Var(R_t) - E(R_t)].$$

2. Under the generalized binomial (GB) thinning,

$$E(Y_t) = \sum_{k=1}^{p} \rho_k E(Y_{t-k}) + E(R_t)$$

$$V(Y_t) = \sum_{k=1}^{p} \left(1 + \frac{2\gamma}{(1-\gamma)}\right) \rho_k (1-\rho_k) E(Y_{t-k})$$

$$\sum_{k=1}^{p} \rho_k^2 Var(Y_{t-k}) + Var(R_t) +$$

$$2 \sum_{k=1}^{p} \sum_{k<k'}^{p} \rho_k \rho_{k'} Cov(Y_{t-k}, Y_{t-k'}).$$

For $\gamma = 0$, we obtain the properties of the binomial thinning.

3. For the negative binomial (NB) thinning,

$$E(Y_t) = \sum_{k=1}^{p} \left(\frac{1-\rho_k}{\rho_k}\right) E(Y_{t-k}) + E(R_t)$$

$$V(Y_t) = \sum_{k=1}^{p} \left(\frac{1-\rho_k}{\rho_k}\right) E(Y_{t-k})$$

$$+ \sum_{k=1}^{p} \left(\frac{1-\rho_k}{\rho_k}\right)^2 Var(Y_{t-k}) + Var(R_t)$$

$$+2 \sum_{k=1}^{p} \sum_{k<k'}^{p} \left(\frac{1-\rho_k}{\rho_k}\right) \left(\frac{1-\rho_{k'}}{\rho_{k'}}\right) Cov(Y_{t-k}, Y_{t-k'})$$

Overall, we can see that, under the different thinnings, the INAR(p) process is capable of capturing any forms of over-dispersion with over-dispersed innovations. Let η be the vector of unknown parameters and the conditional likelihood is then constructed from:

$$L(\eta) = log\left[\prod_{t=2}^{T} f(y_t \mid y_{t-1}, y_{t-2}, \ldots, y_{t-p})\right]$$

where $f(y_t \mid y_{t-1}, y_{t-2}, \ldots, y_{t-p})$ is obtained by inverting the generating function in Davies [16]. The solution vector is obtained by the *optim* function in R. Bu et al. [17] established the asymptotic properties of the CML approach.

4 Results and Findings

This section presents the fitting of the different INAR models to the COVID-19 series in Mauritius from 18 March 2020 to 25 April 2021. We consider the following distributions with their respective generating functions $G_{R_t}(s)$ for the innovation distributions in the INAR process:

1. Geometric: $G_{R_t}(s) = \frac{\lambda_t^* s}{1-(1-\lambda_t^*)s}$ where $\lambda_t^* \in (0, 1)$ and $\lambda_t^* = \frac{\lambda_t}{\lambda_t+1}, \lambda_t > 0$
2. Negative Binomial (NB): $G_{R_t}(s) = [1 + \nu\lambda_t(1 - s)]^{-\nu^{-1}}$ where $\lambda_t > 0$ and $\nu > 0$.

And their zero-inflated (ZI) versions are obtained as:

$$G_{\bar{R_t}}(s) = \pi + (1 - \pi)G_{R_t}(s), \tag{6}$$

where π indicates the probability of the zero occurrence. The "ZIM" package was used for running these ZI models. The following time-dependent covariates are described:

- The COVID-19 Stringency Index (SI): This index gives an indication of the strictness of government policies to the COVID-19 pandemic. The score is between 0 and 100 where an index nearing 100 indicates strict response and less strict response otherwise. In this study, the logarithm of the nominal value of the SI (log(SI)) was used for analysis purposes.
- The event of vaccination (Vaccine): The event of vaccination was categorized into two possibilities (binary) where 1 indicates that the event for vaccination is being done and 0 (ref) that the event of vaccination is not being done.
- The reproduction rate (ReR): This variable is expected to change over time depending on the sanitary restrictions, the speed of the contact tracing exercise, the behavior of every person, and the resilience of the human immunity. The data on ReR was obtained from[1] and the logarithm of the nominal value of ReR (log(ReR)) was considered.
- The Relative COVID-19 Risk due to Weather and Air Pollution (CRW): This variable represents some environmental factors and explains their impact on COVID-19 transmission. The CRW was categorized into 0 and 1 (binary) where 0 (ref) is when an index is below 1 referring to relatively lower impact of weather factors on spread of COVID-19 and 1 is when an index is above 1, indicating otherwise. Data has been extracted from[2] and imputation based on observations was done for missing values.

The estimates of the above variables are shown in Table 1:

[1] https://ourworldindata.org/covid-cases.

[2] https://projects.iq.harvard.edu/covid19.

Table 1 Results under binomial, generalized binomial (GB), and negative binomial (NB) thinnings for selected innovation distributions

Innovation	Parameter	Thinning	ρ_1	ρ_2	ρ_3	ρ_4	ρ_5	ρ_6	ρ_7	Intercept	ReR	SI	Vaccine	CRW	η_0	η_1	η_2	η_3	η_4	ν	γ	AIC
ZI geometric	Estimates	Binomial	0.186	0.094	0.089	0.068	0.030	0.036	0.031	0.565	0.163	0.504	−0.748	0.122	−0.014	0.503	−0.193	−0.100	0.324			2507.8
	Std error		0.019	0.006	0.035	0.081	0.060	0.016	0.007	0.008	0.001	0.011	0.011	0.039	0.006	0.009	0.016	0.018	0.013			
	P-value		0.000	0.000	0.011	0.402	0.620	0.026	0.000	0.000	0.000	0.000	0.000	0.000	0.033	0.000	0.000	0.000	0.000			
	Estimates	GB	0.202	0.095	0.113	0.051	0.038	0.051	0.045	0.278	0.164	−0.459	−1.043	0.167	−0.086	0.184	0.075	0.163	0.121		0.842	1469.8
	Std error		0.034	0.008	0.008	0.012	0.004	0.008	0.004	0.000	0.000	0.000	0.000	0.301	0.003	0.000	0.000	0.003	0.901		0.014	
	P-value		0.000	0.000	0.000	0.000	0.000	0.000	0.000	0.000	0.000	0.000	0.000	0.578	0.000	0.000	0.000	0.000	0.893		0.000	
	Estimates	NB	0.194	0.093	0.063	0.044	0.035	0.025	0.069	0.836	0.184	−0.407	−1.094	0.581	−0.143	0.534	0.485	0.180	0.556			2014.1
	Std error		0.000	0.000	0.000	0.000	0.000	0.000	0.000	0.000	0.000	0.000	0.001	0.411	0.000	0.000	0.000	0.000	0.801			
	P-value		0.000	0.000	0.000	0.000	0.000	0.000	0.000	0.000	0.000	0.000	0.000	0.157	0.000	0.000	0.000	0.000	0.488			
ZI-NB	Estimates	Binomial	0.188	0.114	0.111	0.056	0.042	0.041	0.041	0.095	0.063	−0.098	−0.107	0.034	0.028	0.17	−0.369	0.149	0.232			2348.8
	Std error		0.000	0.000	0.000	0.000	0.000	0.000	0.000	0.001	0.000	0.000	0.001	0.128	0.001	0.000	0.000	0.001	0.125			
	P-value		0.000	0.000	0.000	0.000	0.000	0.000	0.000	0.000	0.000	0.000	0.000	0.788	0.000	0.000	0.000	0.000	0.063			
	Estimates	GB	0.319	0.144	0.171	0.140	0.021	0.113	0.055	0.330	0.452	−0.308	−0.467	0.266	−0.846	0.625	0.505	0.100	0.180	0.735	0.852	1693.1
	Std error		0.003	0.015	0.025	0.046	0.006	0.001	0.001	0.001	0.040	0.038	0.000	0.194	1.656	0.149	0.357	0.009	0.694	0.087	0.023	
	P-value		0.000	0.000	0.000	0.003	0.000	0.000	0.000	0.000	0.000	0.000	0.000	0.170	0.609	0.000	0.157	0.000	0.795	0.000	0.000	
	Estimates	NB	0.206	0.089	0.077	0.054	0.044	0.046	0.055	0.262	0.158	−0.308	−0.023	0.253	−0.189	0.390	0.401	0.752	0.122	0.844		2185.4
	Std error		0.001	0.002	0.002	0.002	0.002	0.002	0.001	0.619	0.001	0.002	0.006	0.149	1.026	0.001	0.001	0.026	0.276	0.001		
	P-value		0.000	0.000	0.000	0.000	0.000	0.000	0.000	0.672	0.000	0.000	0.000	0.089	0.854	0.000	0.000	0.000	0.660	0.000		

The Akaike Information Criterion (AIC) is satisfactory under both INAR(7) with ZI geometric and ZI negative binomial (ZI-NB) innovations and the different thinnings. In fact, the estimates under the generalized binomial thinning are at lower standard errors and the ZI-NB gives slightly better fitting. Based on the respective p-values, the variables "ReR," "SI," and "Vaccine" were highly significant in reducing the number of COVID-19 infection in Mauritius, as compared to "CRW."

The ReR is directly associated with the number of new active cases [18]. This is because by observing the evolution of the series in Mauritius, it can be deduced that in October 2020 when there was an increase in international mobility [19], an exponential increase in the number of active COVID-19 cases was reported. At that point, a worrisome "ReR" of above 1 was being reported, indicating high risk of getting infected. Based on the reversed estimates of "Vaccine," it can be deduced that as the vaccination campaigns take place, this is reflected positively in the share of Mauritian population which has already received at least one dose of the vaccines, and likewise, the risk of getting infected with acute symptoms is expected to decrease considerably. Next, timely imposition of immediate sanitary measures during the peak COVID-19 phases also plays an important role in curbing the spread of the virus. In fact, the quicker and earlier the sanitary measures are imposed, the more rapidly is the SARS-CoV-2 contained in the local community. Conversely, unlike other European regions, Mauritius reported its highest cases of COVID-19 in both warmer and colder regions, and in both weather conditions, summer and winter, "CRW" was proven to be insignificant in curbing the number of active COVID-19 cases. In fact, given the constant mutation of the SARS-CoV-2 in different regions and Mauritius having a comparatively restricted regional disparity, possibly a larger dataset on "CRW" will allow better exploration of its association with the number of infection [20, 21]. The estimates of the over-dispersion parameters in the ZI models are also significant.

Based on these results, an attempt was also made to forecast the number of COVID-19 cases for the next 10 days. Results were that the ZI-NB model had the relatively lower root mean square errors (RMSEs) of 3.61, compared to the ZI geometric which had RMSE of 4.47. It can also be seen that the 95% confidence interval lies between 0 and 1 which means that during the next 10 days, that is, from 26 April 2021 till 05 May 2021, there was 95% chance that the new COVID-19 infection case will lie between 0 and 1. Through this forecasting exercise, it was found that due to some unmeasurable and unpredictable latent effects, the forecasted number of locally acquired new COVID-19 cases may not be easily estimated, especially in the long run. In fact, in the rise of a sudden shock or spike, the forecasted values are naturally disrupted, since the predictor functions in the innovation distribution may not include a new physical or latent effect. Such a situation may be circumvented by updating the list of covariates on a daily basis. These short-term forecasts in this study, however, do help policy makers to understand the severity of the current situation and take timely preventive measures to contain the spread of the coronavirus at that point in time (Fig. 2).

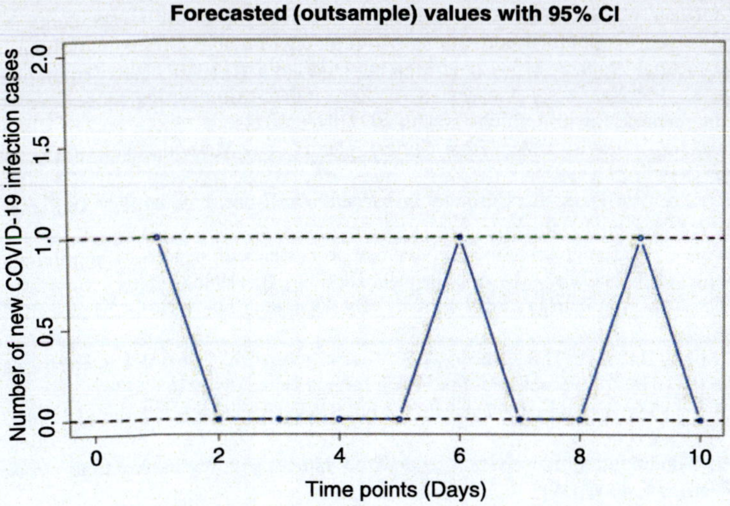

Fig. 2 Forecasted values (out-sample) with 95% confidence interval

5 Conclusion and Future Scope

Up to this extent, this work has focused on the INAR(p) processes with different Poisson-mixture innovations and their corresponding zero-inflated parts under different thinning mechanisms, taking into account the effect of time-varying covariate specifications and non-stationarity. As discussed earlier, such processes are highly suitable to model over-dispersed time series with higher orders. The INAR(p) with ZI geometric and ZI-NB innovations was implemented. As a future work, the INAR(p) can be extended to multivariate INAR(p) where we can consider more response variables such as the number of COVID-19 infected and death cases.

References

1. Brannas, K., Quoreshi, A.M.M.S.: Integer-valued moving average modelling of the number of transactions in stocks. Appl. Financ. Econ. **20**, 1429–1440 (2010)
2. Quoreshi, S., Uddin, R., Mamode Khan, N.: Quasi-maximum likelihood estimation for long memory stock transaction data–under conditional heteroskedasticity framework. J. Risk Financ. Manage. **12**, 1–13 (2020)
3. Gouri'eroux, C., Jasiak, J.: Heterogeneous inar (1) model with application to car insurance. Insur. Math. Econ. **34**, 177–192 (2004)
4. Karlis, D., Sermaidis, G.., Brijs, T.: Discrete valued time series models for examining weather effects in daily accident counts. In: De Castillo et al. (eds.) Proceedings of the 22th International Workshop on Statistical Modelling, Barcelona, pp. 368–373 (2007). ISBN: 978-84-690-5943-2
5. Bakouch, H.S., Ristic, M.M.: Zero truncated poisson integer valued ar(1) model. Metrika **72**, 265–280 (2010)

6. MamodeKhan, N., Soobhug, A.D., Jannoo, Z.: Modeling road traffic accidents in mauritius using clustered longitudinal com-poisson with gamma random effects. Commun. Stat. Case Stud. Data Anal. Appl. **7**, 113–127 (2021). https://doi.org/10.1080/23737484.2020.1842269
7. Tawiah, K., Iddrisu, W.A., Asosega, K.A.: Zero-inflated time series modelling of covid-19 deaths in ghana. J. Environ. Public Health **2021**, 1–9 (2021)
8. Brockwell, P.J., Davis, R.A.: Time Series: Theory and Methods, vol. 82(5), pp. 1–580. Springer, New York (1991)
9. Steutel, F.W., Van Harn, K.: Discrete analogues of self-decomposability and stability. Ann. Probab. **7**, 893–899 (1979)
10. McKenzie, E.: Autoregressive moving-average processes with negative binomial and geometric marginal distributions. Adv. App. Probab. **18**, 679–705 (1986)
11. Weiß, C.: A poisson inar(1) model with serially dependent innovations. Metrika **78**, 829–851 (2015)
12. Ristic, M.M., Bakouch, H.S., Nastic, A.S.: A new geometric first-order integervalued autoregressive (NGINAR(1)) process. J. Stat. Plan. Infer. **136**, 2218–2226 (2009)
13. Du, J., Li, Y.: The integer-valued autoregressive (INAR(p)) model. J. Time Ser. Anal. **12**, 129–142 (1991)
14. Joe, H.: Likelihood inference for generalized integer autoregressive time series models. Econometrics **7**, 43 (2019)
15. Latour, A.: Existence and stochastic structure of a non-negative integer-valued autoregressive process. J. Time Ser. Anal. **19**, 439–455 (1998)
16. Davies, R.B.: Numerical inversion of a characteristic function. Biometrika **60**, 415–17 (1973)
17. Bu, R., McCabe, B.P.M., Hadri, K.: Maximum likelihood estimation of higherorder integer-valued autoregressive processes. J. Time Ser. Anal. **29**, 973–994 (2008)
18. Billah, M.A., Miah, M.M., Khan, M.A., Khan, N.: Reproductive number of coronavirus: a systematic review and meta-analysis based on global level evidence. PLOS ONE **15**, 1–17 (2020)
19. Linka, K., Peirlinck, M., Kuhl, E.: The reproduction number of covid-19 and its correlation with public health interventions. medRxiv **3**, 1–16 (2020)
20. Dobricic, S., Pisoni, E., Pozzoli, L., Van-Dingenen, R., Lettieri, T., Wilson, J., Vignati, E.: Do environmental factors such as weather conditions and air pollution influence covid-19 outbreaks? Science for Policy report by the Joint Research Centre (JRC), The European Commission (2020)
21. Bukhari, Q., Massaro, J.M., D'Agostino, R.B., Khan, S.: Effects of weather on coronavirus pandemic. Int. J. Environ. Res. Public Health **17**(15), 5399 (2020). https://doi.org/10.3390/ijerph17155399

Prediction of Social Status on Depression by Using Logistic Regression

K. Karthikeyan, Rashi Khubnani, Ishika Ahuja, and M. Seenivasan

1 Introduction

Current society is highly vulnerable to psychological disorders, and the term depression has been around for quite a while now. There have also been many papers examining the causes and factors that could lead to depression. The consensus from these papers points to a positive correlation between socioeconomic status and depression, but most of these studies were done on people living in cities or students. A substantial part of papers on this topic found anxiety and self-esteem to be more affected by the above factors. While anxiety and self-esteem issues are less grave compared to depression, but these make a person highly vulnerable to depression. Thus, in this chapter, we concentrate on cases of depression among farmers, taking into consideration their age, education level, saved asset, and living expenses, as significant factors to determine socioeconomic status. For this study, we only consider cases of depression and considerable anxiety or self-esteem issues.

In this chapter, depression status is the value predicted. Education level, saved assets, and living expenses are the primary measures used to scale the socioeconomic status of a person. Age is one of the most common predictors for depression in various studies, since people of different age groups see the same situation in

K. Karthikeyan (✉)
Department of Maths, School of Advanced Sciences, VIT, Vellore, Tamil Nadu, India
e-mail: k.karthikeyan@vit.ac.in

R. Khubnani
New Horizon College, Bangalore, India

I. Ahuja
VIT, Vellore, Tamil Nadu, India

M. Seenivasan
Mathematics wing, Annamalai University, Annamalainagar, Tamil Nadu, India

F. Kamalov et al. (eds.), *Advances in Mathematical Modeling and Scientific Computing*, Trends in Mathematics, https://doi.org/10.1007/978-3-031-41420-6_59

different ways and likewise have a different view of economic and social status. Although there are limitations to this study, the ultimate objective is to determine a clear relationship between the lifestyle of people and depression in people of different ages living in small towns where the primary source of income is farming. It is essential to do so because most of the time, studies on mental illnesses take place with the main focus toward people in cities, and people who might not even be aware of psychological conditions and have no help, are overlooked. The hope is to encourage greater government's involvement and gain attention toward these people, and this might help decrease the silent suffering of these people.

2 Literature Survey

Numerous studies focused on the relationship among depression, economy, and age. Most of them have found a positive correlation among depression, economy, and age. Bjelland et al. [1] introduced weather there is higher education is safe from depression and anxiety. Sancakoglu et al. [2] published a paper showing a relation between socioeconomic status and depression, anxiety, and self-esteem in early adolescents. The article concluded that lower socioeconomic status was associated with higher rates of depression and trait anxiety. Domenech-Abella et al. [3] studied about depression and socio economic status about adults in various countries. Freeman et al. [4] developed a relation between depression and SES in different countries. Lorant et al. [5] presented the study about the depression due to inequalities in socio economies. Mezuk et al. [6] analyzed about variation in depressions by using genetically informed designed (GID). Murphy et al. [7] discussed about social status by using depressions and anxiety. Peyrot et al. [8] explained about how lower educational affects depressions. Turner et al. [9] discussed about dispersion by stress and socio-economic status. Age is another factor considered in many theses. Brandyn M. Road, in his review, attempted to find a connection among age and depression, calling it a self-viability model. The chapter featured the meaning of the review. This investigation discovered that from the social-mental viewpoint, side effects of depression connected to individuals' convictions that they can't direct or control their working. The motivation behind their review was to test social mental hypothesis and its cases about self-viability by looking at whether age and sex contrasts in depression are a reason for feeling guideline, profound self-adequacy, and reaction styles to depression. Results showed that more established associates endured fundamentally lower depression and rumination scores than school grown-ups and were more productive in overseeing gloomy feelings. Just profound self-viability for gloomy feelings, rumination, and interruption made sense of remarkable difference in burdensome side effects.

Aislinne Freeman, Stefanos Tyrovolas, in collaboration with many researchers in their study about the role of socioeconomic status in depression, showed prominent relation or impact of social economic degree of statue, SES, on depression across all countries in Europe. Subsequent to adapting to confounders, the chances of depression essentially diminished for each unit expansion in the SES score. Also, an interesting point in this study was, advanced education altogether diminished the chances of depression in every nation, except monthly inflow of money didn't have a consistent effect.

The HUNT study by Bjelland I, Krokstad S, which focused on higher education's contribution to protecting people from depression, also showed a significant relationship among age, education level on depression in their longitudinal analysis. Josje ten Kate, Willem de Koster &Jeroen van der Waal Broke down broadly agent review information gathered in the USA in 2014 (n = 1932) to find the reason why less taught individuals were more inclined to burdensome side effects. They figured out and tried an original clarification for the instructive slope in burdensome side effects. They speculated that status as social capital drives this slope notwithstanding deep rooted monetary and social variables, through the sensations of social privilege it rouses.

3 Data Sources and Graphical Representations

The data collected from a study about the living conditions of people who live in rural zones. Because all the columns were not considered, we delete them or ignore them. The final features or columns were the following:

Age, education, level, save asset, living expenses, depressed: [Zero: No depressed] or [One: depressed] (Binary for target class) the main objective is to show statistics analysis to find a correlation between age, education level, and socioeconomic status. For the initial level of the study, only save asset and living expenses were considered to measure socioeconomic status. The dataset has six columns and a total of 1432 people who took part in the survey. The pair plot of all attributes used is given below.

(i) Age versus education level: The below graph demonstrates the variation in education level with age. It can be observed that people in their twenties paid great attention to education and showed higher education levels compared aged people (Fig. 1).

(ii) Age versus living expense: Even though the expenses of youth is distributed in a huge range, trends of higher living expenses have only been noted mainly in young people even though older people tend to have higher living expenses due to usual medical and extensive care factors (Fig. 2).

(iii) Age versus save asset: Graphs of saved asset and living expenses against age showed quite a lot of similarity. With increasing expenses the young youth have also showed good savings (Fig. 3).

(iv) Age versus depression: Age is considered as one of the major causes of depression but alone in village areas, it is evident that young people show approximately equivalent depression cases as aged people (Fig. 4).

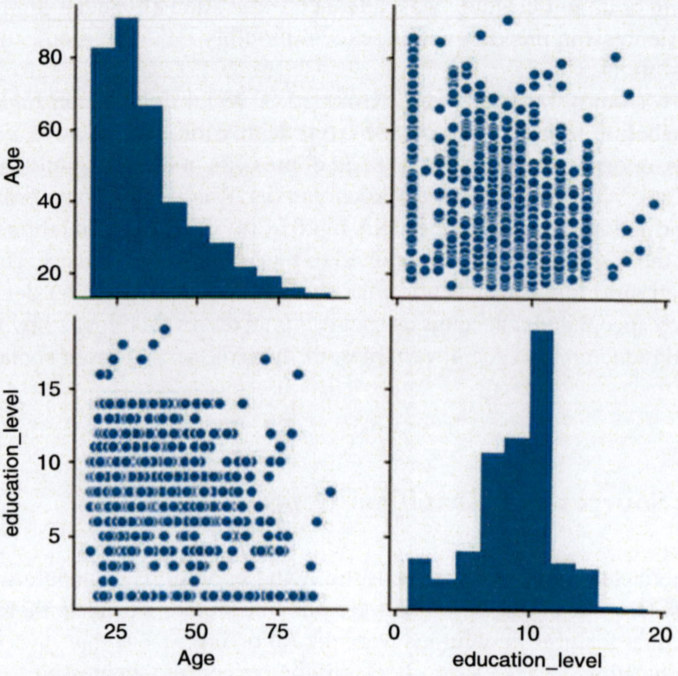

Fig. 1 Age versus education Level

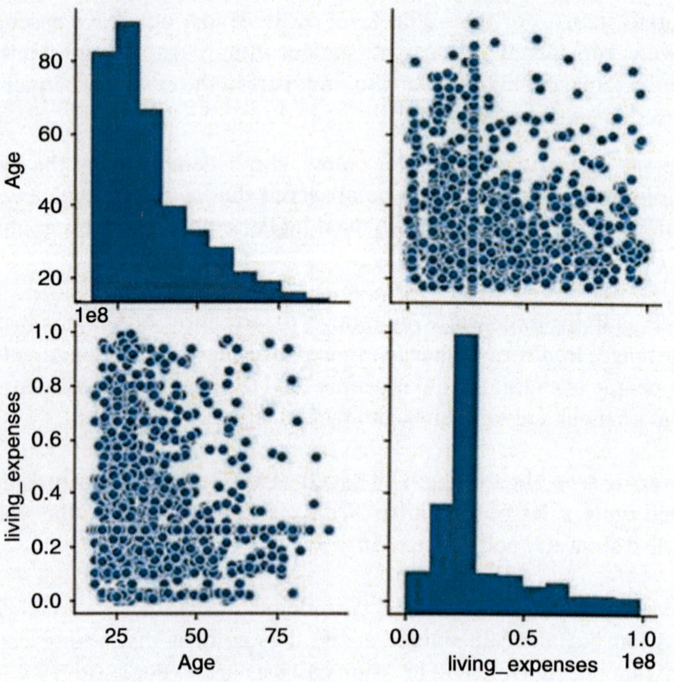

Fig. 2 Age vs living expenses

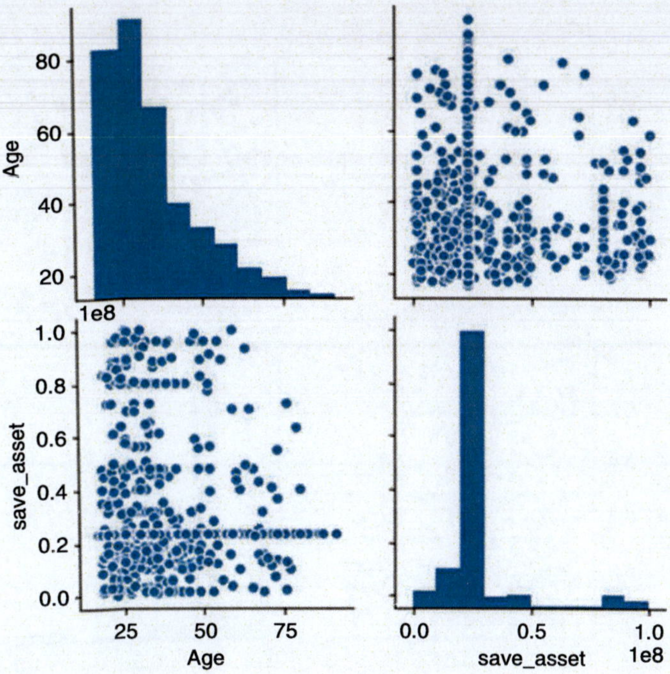

Fig. 3 Age versus saved asset

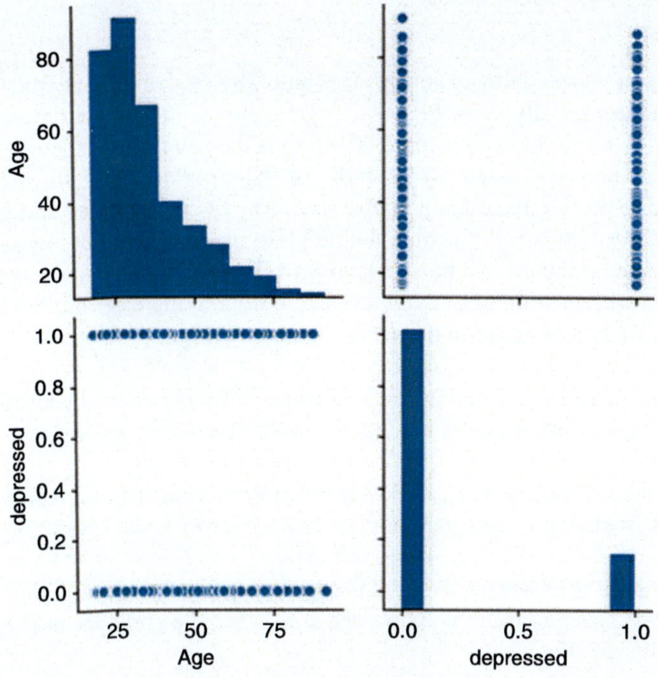

Fig. 4 Age versus depression

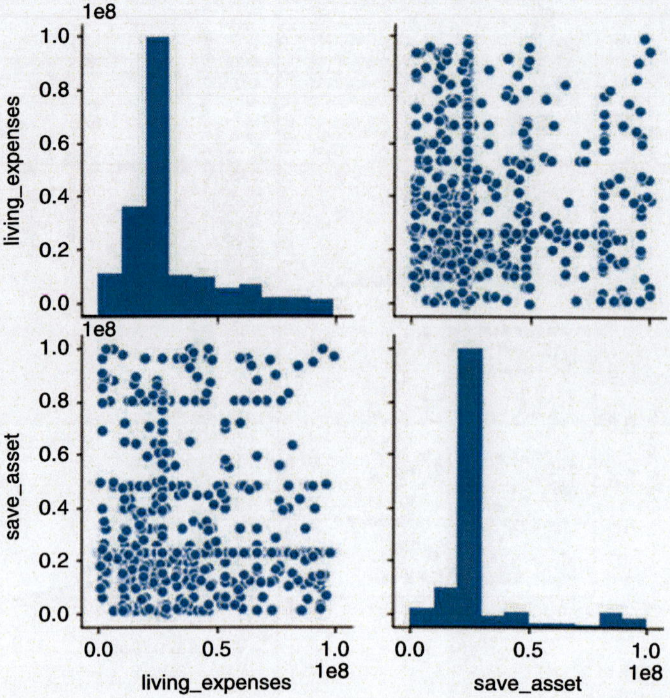

Fig. 5 Living expenses versus saved asset

(v) Save asset versus living expenses: It is naturally evident that people with lower saved assets usually show higher living expenses. The whole dataset could be divided into four kinds of people. The first is the people who showed low living expenses and low saved assets, most of the people belong to this category. Second is the group of people who have low living expenses and high saved assets. Third are a set of people with high living expenses yet low saved asset, they make up the second most majority in the set of people surveyed. The last set is of people with high saved assets and high living expense; these kinds of people are in a minority in the surveyed dataset (Fig. 5).

(vi) Save asset versus education level: Unexpectedly increase in education didn't lead to a significant increase in higher saved assets (Fig. 6).

(vii) Save asset versus depressed: As for the graphs, it can be said that most of the people with depression were noticed to have lower saved assets (Fig. 7).

(viii) Education level versus living expenses: Education level and living expenses showed no significant trend between each other, the graph seems to be equally distributed (Fig. 8).

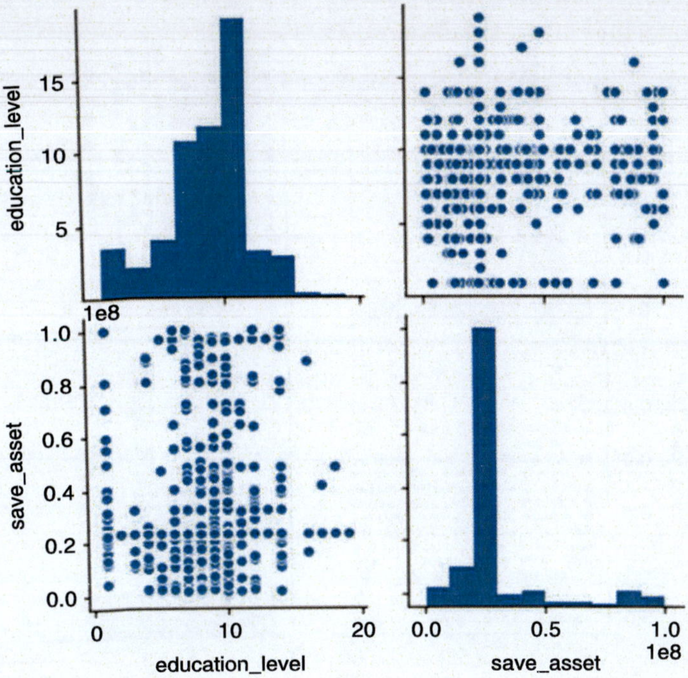

Fig. 6 Education level versus saved asset

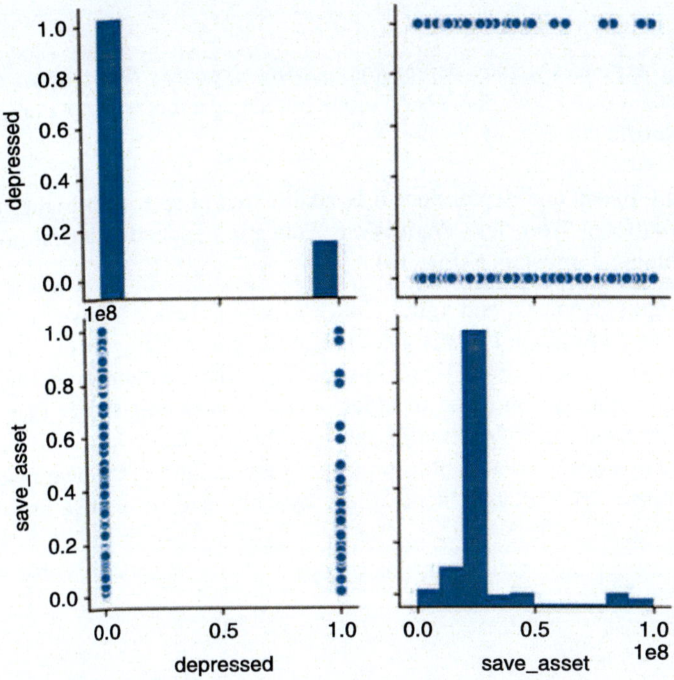

Fig. 7 Depressed versus saved asset

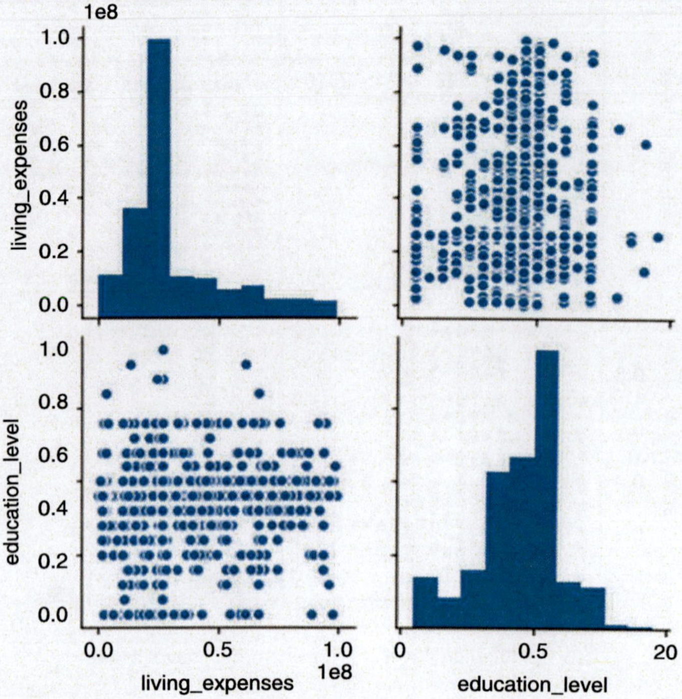

Fig. 8 Living Expenses versus Education Level

(ix) Living expenses versus depression: Living expenses don't seem to show a significant trend with depression alone but act as a factor when combined with other attributes (Fig. 9).

(x) Education level and depression: It is visible that people with higher education levels suffered from less depression. Education indeed acted as a shield to psychological imbalance (Fig. 10).

From graphs iv, vii, ix, and x, our choice of using Binary Logistic regression is justified. Also from graph I it was great to see the youth being more educated than the previous generation. Graph ii and iii show high living expenses yet high savings of younger people as compared to older people. Graph viii shows that educated people tend to have controlled living expenses. The correlation heat map was used to find any highly correlated variables to prevent overfitting and the heat map showed no highly correlated variables hence there was no need to remove any variables (Fig. 11).

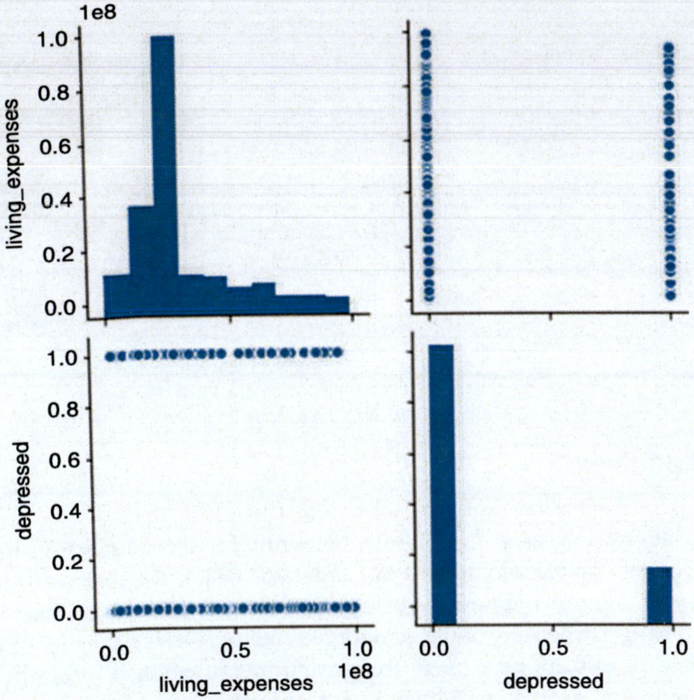

Fig. 9 Living expenses versus depression

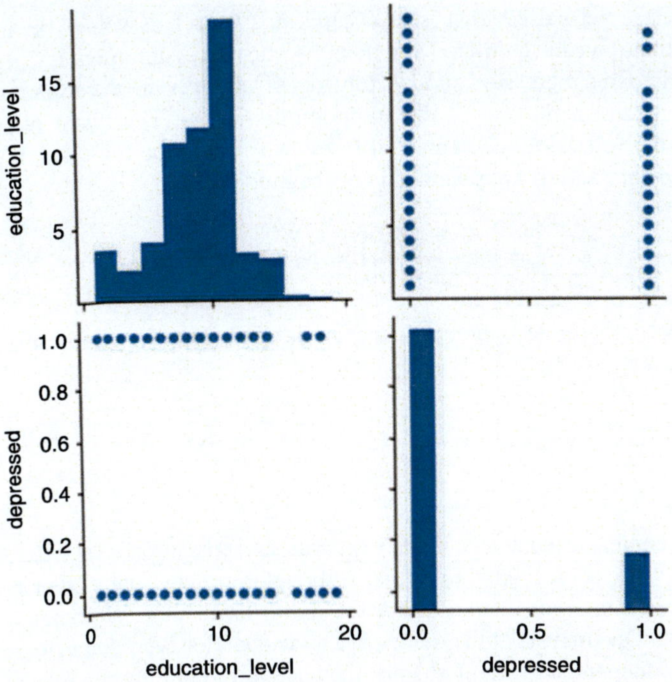

Fig. 10 Education level versus depression

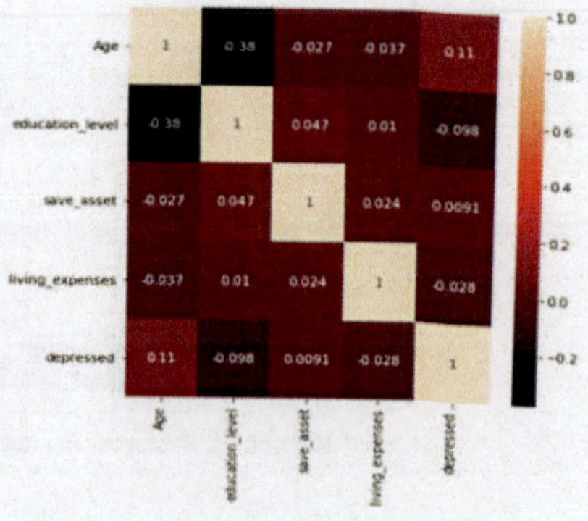

Fig. 11 Correlation heat map

4 Methodology

Regression analysis is used to define a blueprint for the connection between a solitary reaction variable and at least one indicator factor. Regression examination assists one with figuring out how the value of the reliant variable (or "standard variable' or Y) changes when any of the autonomous factors (X) is changed. Conversely, the other free factors are kept fixed. Regression investigation is generally utilized for expectation and estimating, which is our objective here. After a dry synopsis of the information and looking at the pair plots of the information, it recommends nonstraight connections as a rule. It is likewise seen that the anticipated variable isn't ceaseless yet paired, that is, it accepts just 0 or 1 as values. In this way, to represent them, we have utilized a double logistic regression model.

In the logistic regression model, the log of changes in the reliant variable is displayed as a straight mix of the autonomous factors. The linear relationship is written in the following mathematical form (where ℓ is the log-odds, b is the base of the logarithm and 'β'are parameters of the model):

$$l = \log\left(\frac{p}{1-p}\right) = \beta_0 + \beta_1 x_1 + \beta_2 x_2$$

By simple algebraic operations, the probability that Y = 1 (or the person is depressed) is.

$$p = \frac{e^l}{e^l + 1}$$

In this chapter, saved assets, living expenses, and education level were the factors used to measure socioeconomic status. Although there are many other factors that could be taken into consideration, for the initial level, these attributes or features were picked. On analysis of the model, age and education were found to be the most significant independent variables, since they had minimum |p| or p esteem value.

5 Result Analysis

In the outline of the model made with save assets, living expenses, training, and age as independent factors, we discovered that living expenses had p esteem >0.5 this inferred it had the least importance in the discovery of depression. Age had a p-value closest to zero. Hence, it had maximum significance in the prediction of depression (Fig. 12).

The anticipated likelihood of depression, when estimated against age utilizing the model, made is given underneath. Graphs were made by education level equal to 8.69, save asset set to 27424707.84, and living expenses set to 32482565.52 (Figs. 13 and 14).

An expansion in age has prompted an increment in instances of depression, which is similar to the pattern in urban communities and is statistically most significant. Subsequently, we can accept age to be one of the primary considerations to predict depression.

The predicted probability of depression, when measured against living expenses using the regression model made, is given below. Education level adjusted to 8.69, save asset was set to 27424707.84 and age was set to 34.78 (Figs. 15 and 16).

We identified that the ascent in living expenses had diminished the probability of depression; however, the thing that matters is not huge contrasted with the impact of education level, Age and saved assets. The anticipated likelihood of depression, when estimated against save assets utilizing the model, made is given underneath.

```
> summary(model)

Call:
glm(formula = depressed ~ Age + education_level + save_asset +
    living_expenses, family = binomial(link = "logit"), data = data)

Deviance Residuals:
    Min       1Q   Median       3Q      Max
-0.9608  -0.6083  -0.5595  -0.5168   2.1571

Coefficients:
                  Estimate Std. Error z value Pr(>|z|)
(Intercept)     -1.571e+00  3.811e-01  -4.122 3.75e-05 ***
Age              1.344e-02  5.194e-03   2.588  0.00965 **
education_level -5.687e-02  2.533e-02  -2.245  0.02476 *
save_asset       2.287e-09  3.960e-09   0.578  0.56354
living_expenses -3.286e-09  3.518e-09  -0.934  0.35033
---
Signif. codes:  0 '***' 0.001 '**' 0.01 '*' 0.05 '.' 0.1 ' ' 1

(Dispersion parameter for binomial family taken to be 1)

    Null deviance: 1287.2  on 1428  degrees of freedom
Residual deviance: 1266.1  on 1424  degrees of freedom
AIC: 1276.1

Number of Fisher Scoring iterations: 4
```

Fig. 12 Summary of the model built

Fig. 13 Prediction results with the major variable as age

Predicted values of depressed
x = Age

x	Predicted	SE	95% CI
10	0.12	0.15	[0.09, 0.16]
20	0.14	0.11	[0.11, 0.16]
30	0.15	0.08	[0.13, 0.17]
40	0.17	0.08	[0.15, 0.19]
60	0.21	0.14	[0.17, 0.26]
70	0.24	0.19	[0.18, 0.31]
80	0.26	0.24	[0.18, 0.36]
100	0.32	0.34	[0.19, 0.48]

Adjusted for:
* education_level = 8.69
* save_asset = 27424707.84
* living_expenses = 32482565.52

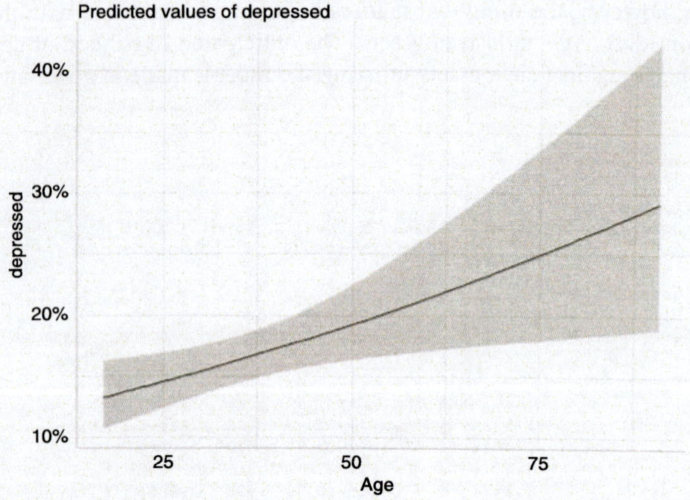

Fig. 14 Graph for prediction results with the major variable as age

Graphs were made by education level equal to 8.69; Age set to 34.78, and living expenses set to 32482565.52 (Figs. 17 and 18).

We saw expansion in saved assets had prompted an expansion in the likelihood of depression, which is in opposition to what we saw on account of individuals from urban communities. Even though the expansion is not measurably massive, this part

```
> ggpredict(model,terms="living_expenses")
Data were 'prettified'. Consider using `terms="living_expenses [all]"` to get smooth plots.

# Predicted values of depressed
# x = living_expenses

       x | Predicted |   SE |       95% CI
---------------------------------------------
       0 |      0.18 | 0.13 | [0.14, 0.22]
 1.3e+07 |      0.17 | 0.10 | [0.15, 0.20]
 2.5e+07 |      0.17 | 0.08 | [0.15, 0.19]
 3.7e+07 |      0.16 | 0.08 | [0.14, 0.18]
   5e+07 |      0.15 | 0.10 | [0.13, 0.18]
 6.3e+07 |      0.15 | 0.13 | [0.12, 0.19]
 7.5e+07 |      0.14 | 0.17 | [0.11, 0.19]
   1e+08 |      0.13 | 0.25 | [0.09, 0.20]

Adjusted for:
*             Age =       34.78
* education_level =        8.69
*      save_asset = 27424707.84
```

Fig. 15 Prediction results with the major variable as living expenses

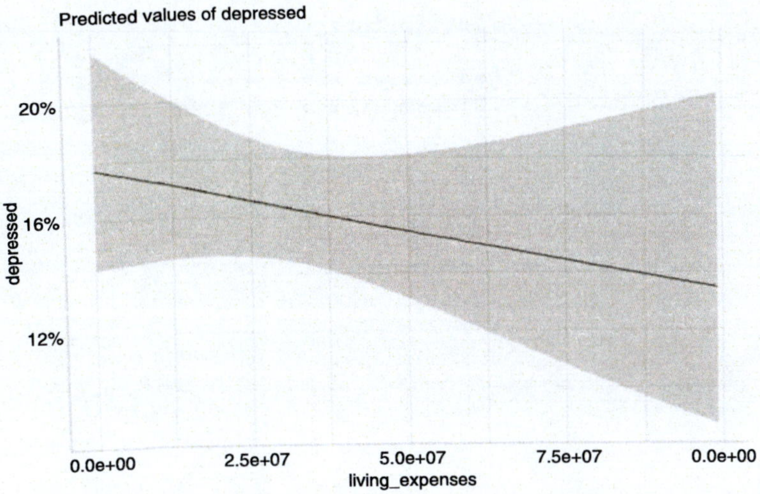

Fig. 16 Graph for prediction results with the major variable as Living Expenses

of the result is more shocking since we believe that the primary cause of suicides and depression in farmers is financial pressure.

The predicted probability of depression, when measured against education using the regression model made, is given below. Saved asset adjusted to 27424707.84, living_expenses set to 32482565.52 and age set to 34.78 (Figs. 19 and 20).

As noted on account of metropolitan urban areas, ascend in education level shields individuals from depression better compared to saved resources. Thus, monitoring oneself and one's general public helps an individual protect himself/herself from psychological issues, and results can be more assuring than financial status.

Fig. 17 Prediction results with the major variable as saved assets

```
# Predicted values of depressed
# x = save_asset

       x | Predicted |   SE |       95% CI
---------------------------------------------
       0 |      0.15 | 0.13 | [0.12, 0.19]
1.3e+07 |      0.16 | 0.09 | [0.13, 0.18]
2.5e+07 |      0.16 | 0.07 | [0.14, 0.18]
3.7e+07 |      0.17 | 0.08 | [0.14, 0.19]
   5e+07 |      0.17 | 0.11 | [0.14, 0.20]
6.3e+07 |      0.17 | 0.16 | [0.13, 0.22]
7.5e+07 |      0.18 | 0.20 | [0.13, 0.24]
   1e+08 |      0.19 | 0.29 | [0.11, 0.29]

Adjusted for:
*                Age =       34.78
* education_level =        8.69
* living_expenses = 32482565.52
```

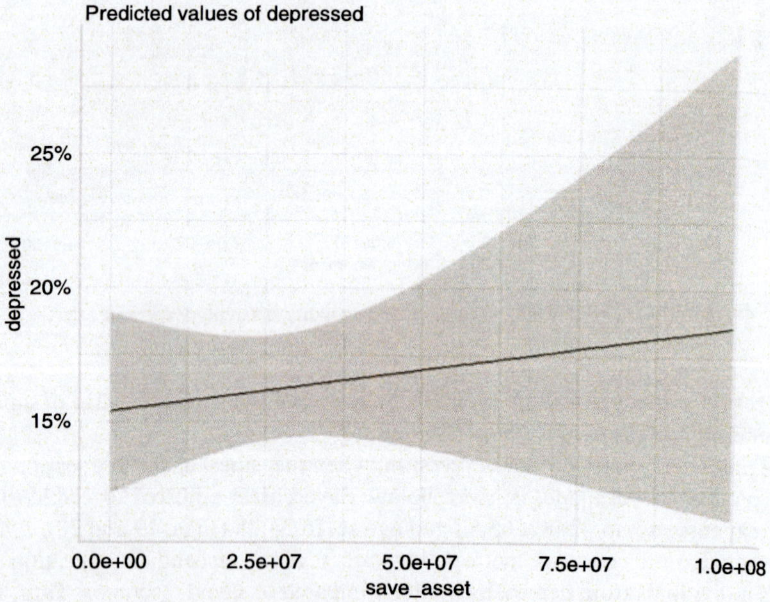

Fig. 18 Graph for prediction results with the major variable as saved assets

Fig. 19 Prediction results with the major variable as education level

```
> ggpredict(model,terms="education_level")

# Predicted values of depressed
# x = education_level

 x | Predicted |   SE |        95% CI
--------------------------------------
 1 |      0.23 | 0.20 | [0.17, 0.31]
 3 |      0.21 | 0.16 | [0.16, 0.27]
 5 |      0.19 | 0.11 | [0.16, 0.23]
 7 |      0.18 | 0.08 | [0.15, 0.20]
10 |      0.15 | 0.08 | [0.13, 0.17]
12 |      0.14 | 0.12 | [0.11, 0.17]
14 |      0.13 | 0.16 | [0.09, 0.16]
19 |      0.10 | 0.28 | [0.06, 0.16]

Adjusted for:
*              Age =       34.78
*       save_asset = 27424707.84
* living_expenses = 32482565.52
```

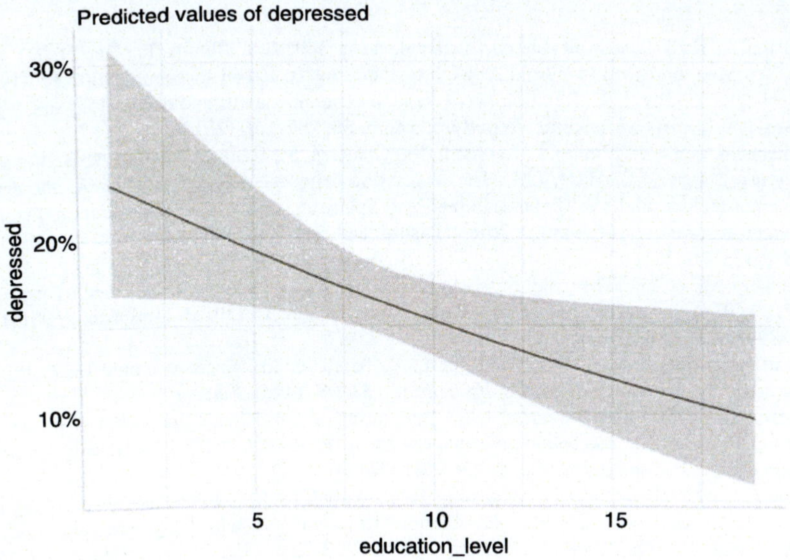

Fig. 20 Graph for prediction results with the major variable as education level

6 Conclusion

In this chapter, we analyzed and studied various factors that could contribute to the vulnerability of a person toward depression. We considered age, education level, living expenses, and saved assets together against depression to find factors among these that are most significant and also to create a model to detect depression. The results showed that age was the most contributing factor, which we expected since we noticed a similar pattern in urban communities. However, contrary to what we expected, living expenses and saved assets showed very little significance compared to education level and age. On comparing with the results of urban cities, we noticed that financial status has less significance as a driving factor to depression in the case of farmers. In farmers with an increase in education level, there was a steady decrease in depression cases.

Living expenses also drove the probability of depression on the negative side. Increasing age, on the other hand, led to an increased probability of depression. Saved assets also promoted depression.

References

1. Bjelland, I., Krokstad, S., Mykletun, A., Dahl, A.A., Tell, G.S., Tambs, K.: The HUNT study. Soc. Sci. Med. **66**(6), 1334–1345 (2008). https://doi.org/10.1016/j.socscimed.2007.12.019
2. Sancakoglu, S., Sayar, K.: Relation between socioeconomic status and depression, anxiety, and self-esteem in early adolescents. Yeni Symposium. **50**, 207–220 (2012)
3. Domenech-Abella, J., Mundo, J., Leonardi, M., Chatterji, S., Tobiasz- Adamczyk, B., Koskinen, S., Ayuso-Mateos, J.L., Haro, J.M.: The association between socioeconomic status and depression among older adults in Finland, Poland and Spain: A comparative cross-sectional study of distinct measures and pathways. J. Affect. Disord. **241**, 311–318 (2018). https://doi.org/10.1016/j.jad.2018.08.077
4. Freeman, A., et al.: The role of socio-economic status in depression: Results from the COURAGE (aging survey in Europe). BMC Public Health. **16** (2016). https://doi.org/10.1186/s12889-016-3638-0
5. Lorant, V., Deliege, D., Eaton, W., Robert, A., Ansseau, M.: Socioeconomic inequalities in depression: A meta-analysis. Am. J. Epidemiol. **157**, 98–112 (2003)
6. Mezuk, B., Myers, J.M., Kendler, K.S.: Integrating social science and behavioral genetics: Testing the origin of socioeconomic disparities in depression using a genetically informed design. Am. J. Public Health. **103**, S145–S151 (2013)
7. Murphy, J.M., Olivier, D.C., Monson, R.R., Sobol, A.M., Federman, E.B., Leighton, A.H.: Depression and anxiety in relation to social status: A prospective epidemiologic study. Arch. Gen. Psychiatry. **48**(3), 223–229 (1991). https://doi.org/10.1001/archpsyc.1991.01810270035004
8. Peyrot, W.J., et al.: The association between lower educational attainment and depression owing to shared genetic effects? Results in ~25,000 subjects. Mol. Psychiatry. **20**, 735–743 (2015). https://doi.org/10.1038/mp.2015.50
9. Turner, R.J., Lloyd, D.A.: The stress process and the social distribution of depression. J. Health Soc. Behav. **40**, 374–404 (1999)

Markovian Queueing Model with Single Working Vacation, Breakdown with Backup Server

M. Seenivasan and R. Abinaya

AMS Subject Classification Number 60K25,60K30, 90B22

1 Introduction

Queueing frameworks with SWV and perfect assistance have obtained significance over several dozens due to enormous degree utilizes chiefly producing framework, administration framework, media communications, and PC framework. In various certified queueing conditions, after help satisfaction, on the off chance that no client in the line, a server goes on a vacation. This type is known as vacation queue. Utilizing overview paper by Doshi [3], numerous scientists presented queueing model with vacation. In this article, we analysed SWV and breakdown with backup server in M/M/1 queueing model. This queueing system has been examined using MGM. The arrival follows FCFS schedule. The remaining model follows as given. In the second section, a brief model description is given. In the third section, performance measures are presented. The fourth section contains numerical study. The final section gives short conclusion about the model.

Queueing model with server vacations has been introduced by many researchers referred to the survey paper by Doshi [3]. An M/M/1 queue with WV has been developed by Servi and Finn [8]. Wu and Takagi [9] analysed an M/G/1 queue with multiple working vacations. The service times while a service interval, service times while a vacation interval and vacation times are all exponentially assigned. Analysis of GI/M/1 queue with multiple working vacations was studied by Baba [1]. The

Authors M. Seenivasan and R. Abinaya have contributed equally to this work

M. Seenivasan (✉)
Mathematics Wing - DDE, Annamalai University, Chidambaram, Tamil Nadu, India

R. Abinaya
Department of Mathematics, Annamalai University, Chidambaram, Tamil Nadu, India

F. Kamalov et al. (eds.), *Advances in Mathematical Modeling and Scientific Computing*, Trends in Mathematics, https://doi.org/10.1007/978-3-031-41420-6_60

server will only take one WV when there is no customer in the system. So, if the system has no customers when the server comes back from SWV, it will idle on the system and wait for the customers to arrive instead of picking up another WV. A multi-server system with an SWV were proposed by Lin and Ke [4]. The notion that a disaster is happening randomly leads to the destruction of all the customers there, and it is not uncommon for a service facility to temporarily disable until a new costumer arrives. Disasters can come from outside the organization or from another service centre, for example, an event taking place in the computer system, production system, and service system.

Di Crescenzo et al. (2003) introduced the M/M/1 queue with catastrophes and its continuous approximation. Avi-Itzhak and Naor (1963) explained about some queueing problems with the service station subject to breakdown. In the same year, Amsterdam and Thiruvengadam (1963) surveyed queuing with breakdowns. A Markovian queue with N servers subject to breakdowns and repairs was demonstrated by Neuts and Lucantoni (1979). A queueing model with server breakdowns, repairs, vacations, and backup server was analysed by Srinivas R. Chakravarthy et al. [2].

A Matrix Geometric Method (MGM) is the best tool for finding the more difficult queueing problems. Neuts [5] deliberated Markov chains with applications queueing theory, which have a matrix geometric invariant probability vector. Neuts [6] derived matrix geometric solution in stochastic models. Kalyanaraman et al. (2019) described a Markovian working vacation queue with server state-dependent arrival rate and with partial breakdown. Seenivasan et al. (2020) studied performance analysis of two heterogeneous servers queuing model with an intermittently obtainable server using Matrix Geometric Method. A retrial queueing model with two heterogeneous servers using the Matrix Geometric Method was investigated by Seenivasan et al. [7].

2 Model Description

In this section, we studied a single-server queueing model with single working vacation and breakdown with backup server. The clients appear in line according to Poisson process with rate λ. They make a line subject to her/his request for appearance. At a normal working period, influx customers were served at a service rate μ_b, following an exponential distribution. The server starts a single vacation of arbitrary length if there is no customer in the system with parameter η following an exponential distribution. During a SWV period, influx customers get service with rate μ_v, following an exponential distribution. If the queue forms, then the server chops and shifts its rate from μ_v to μ_b, and the normal working interval starts. For next vacation server waits idle to serve influx customers. This type of vacation is called SWV. Then the server starts a normal working period when a customer arrival occurs.

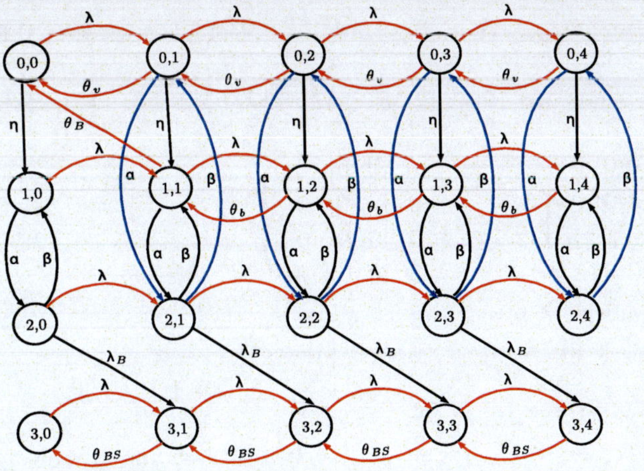

Fig. 1 The rate transition diagram

When the server is subject to sudden breakdown with rate α, it should be repaired for it to go back to normal with rate β. Whenever the server undergoes breakdown, a backup server will appear and serve the customer with rate μ_{BS}, exponentially distributed. After breakdown the customer in the queue will be sent to the backup server with rate λ_B. The rate transition diagram is given in Fig. 1.

Let $\{(k(t), n(t)) : t \geq 0\}$ be a Markov process (MP), where $k(t)$ and $n(t)$ denote state of process at time t.

$$k(t) = \begin{cases} 0, & \text{when server is single working vacation} \\ 1, & \text{when server is normal working period} \\ 2, & \text{when server is breakdown} \\ 3, & \text{Backup server state} \end{cases}$$

n(t) denotes the total number of customers in the system.

The quasi-birth-death process along with the state space Ω is as follows:

$\Omega = \{(0, 0), (1, 0), (2, 0)\}U\{(i, j); i = 0, 1, j = 1, 2, \ldots., n \geq 1\}$

Consider a quasi-birth-death process with infinitesimal generator matrix Q presented below:

$$Q = \begin{pmatrix} L_{00} & M_0 & 0 & 0 & 0 & 0 & \cdots\cdots \\ N_{00} & L_1 & M_0 & 0 & 0 & 0 & \cdots\cdots \\ 0 & N_1 & L_1 & M_0 & 0 & 0 & \cdots\cdots \\ 0 & 0 & N_1 & L_1 & M_0 & 0 & \cdots\cdots \\ 0 & 0 & 0 & N_1 & L_1 & M_0 & \cdots\cdots \\ \cdots & \cdots & \cdots & \cdots & \cdots & \cdots & \cdots\cdots \end{pmatrix}$$

where

$$L_{00} = \begin{pmatrix} -(\lambda+\eta) & \eta & 0 & 0 \\ 0 & -(\lambda+\alpha) & \alpha & 0 \\ 0 & \beta & -(\beta+\lambda+\lambda_B) & 0 \\ 0 & 0 & 0 & \lambda \end{pmatrix}; N_{00} = \begin{pmatrix} \mu_v & 0 & 0 & 0 \\ \mu_b & 0 & 0 & 0 \\ 0 & 0 & 0 & 0 \\ 0 & 0 & 0 & \mu_{BS} \end{pmatrix};$$

$$M_0 = \begin{pmatrix} \lambda & 0 & 0 & 0 \\ 0 & \lambda & 0 & 0 \\ 0 & 0 & \lambda & \lambda_B \\ 0 & 0 & 0 & \lambda \end{pmatrix}; N_1 = \begin{pmatrix} \mu_v & 0 & 0 & 0 \\ 0 & \mu_b & 0 & 0 \\ 0 & 0 & 0 & 0 \\ 0 & 0 & 0 & \mu_{BS} \end{pmatrix};$$

$$L_1 = \begin{pmatrix} -(\lambda+\mu_v+\alpha+\eta) & \eta & \alpha & 0 \\ 0 & -(\lambda+\mu_b+\alpha) & \alpha & 0 \\ \beta & \beta & -(2\beta+\lambda+\lambda_B) & 0 \\ 0 & 0 & 0 & -(\mu_{BS}+\lambda) \end{pmatrix}.$$

We define $p_{ij} = \{k = i, n = j\}$, where j denotes the number of customers in the system and i denotes the server state. Probability vector is defined as $P = (P_0, P_1, P_2, \ldots)$ where $P_j = (p_{0j}, p_{1j}, p_{2j}, p_{3j})$, $j = 0, 1, 2, 3, \ldots$

The static probability row matrix is represented by using $PQ = O$.

$$P_0 L_{00} + P_1 N_{00} = 0 \tag{1}$$

$$P_0 M_0 + P_1 L_1 + P_2 N_1 + \cdots = 0 \tag{2}$$

$$P_1 M_0 + P_2 L_1 + P_3 N_1 + \cdots = 0 \tag{3}$$

$$P_i M_0 + P_{i+1} L_1 + P_{i+2} N_1 + \cdots = 0 \tag{4}$$

Given the geometric relation

$$P_i = P_0 R^i, i \geq 1 \tag{5}$$

From Eq. (1), we get

$$P_0[L_{00} + RN_{00}] = 0 \tag{6}$$

Using geometric relation in Eq. (4), we have

$$P_0 R^i (M_0 + RL_1 + R^2 N_1) = 0, i \geq 1 \tag{7}$$

The normalizing equation is defined by

$$P_0 [I - R]^{-1} e = 1 \tag{8}$$

where 'e' is the column unit vector with all its element equal to one.

The static condition of such a QBD (see Neuts [6]) can be obtained by the drift condition

$$PM_0e < PN_1e \tag{9}$$

where the row vector $P = (P_0, P_1, P_2, P_3)$ is obtained from the infinitesimal generator $W = M_0 + L_1 + N_1$.

W is given by

$$W = \begin{pmatrix} -(\alpha+\eta) & \eta & \alpha & 0 \\ 0 & -\alpha & \alpha & 0 \\ \beta & \beta & 2\beta & 0 \\ 0 & 0 & 0 & 0 \end{pmatrix} \tag{10}$$

W is irreducible and the row vector P can be shown to be unique such that

$$PW = 0, Pe = 1 \tag{11}$$

From Eq. (10), we have

$$P_1 = (\frac{2\eta+\alpha}{\alpha})P_0$$

$$P_2 = (\frac{\eta+\alpha}{\beta})P_0$$

$$P_3 = 0$$

$$P_0 = [1 + (\frac{2\eta+\alpha}{\alpha}) + (\frac{\eta+\alpha}{\beta})]^{-1} \tag{12}$$

The static condition takes format

$$\lambda[P_0 + P_1 + P_2 + P_3] + \lambda_B P_2 < \mu_{BS} P_3 + \mu_b P_1 + \mu_v P_0 \tag{13}$$

Equation (13) gives the static probability of W.

Once the rate matrix R is obtained, the probability vectors P_js $(j \geq 1)$ are obtained from Eqs. (6) and (8).

3 Performance Measures

Performance measures have been found using steady-state probabilities as given below:

When the server is idle, the mean no. of customers present is $E(I) = P_0$ (14)

When the server is on SWV, the mean no. of customers present is

$$E(SWV) = \sum_{j=0}^{\infty} jp_{0j} \tag{15}$$

When the server is in normal busy period, the mean no. of customers present is

$$E(BP) = \sum_{j=1}^{\infty} jp_{1j} \tag{16}$$

When the server is on breakdown, the mean no. of customers present is

$$E(BD) = \sum_{j=1}^{\infty} jp_{2j} \tag{17}$$

When the backup server is activated, then the mean no. of customers present is

$$E(BS) = \sum_{j=1}^{\infty} jp_{3j} \tag{18}$$

The mean no. of customers in the system is

$$E(N) = E(WV) + E(BP) + E(BD) \tag{19}$$

The total outcome obtained from the system is $\Delta = \mu_v \sum_{j=0}^{\infty} jp_{0j} +$

$$\mu_b \sum_{j=1}^{\infty} jp_{1j} \tag{20}$$

4 Numerical Study

Here, we make mathematical calculation for model given by the segment above. Our goals are to show effect of a parameter on system features. By modifying λ four illustrations are presented in these sections.

The parameter λ value varies and all other argument values are fixed. 4.1. Illustration to 4.4. Illustration is presented below.

4.1 Illustration

We take $\lambda = \lambda_B = 0.2, \mu_b = 1.5, \mu_v = 1.2, \mu_{BS} = 1.0, \alpha = 0.2, \eta = 0.1, \beta = 0.3$ and the rate matrix is $R = \begin{pmatrix} 0.1453 & 0.0312 & 0.0266 & 0.0794 \\ 0.0125 & 0.1333 & 0.0226 & 0.0629 \\ 0.0863 & 0.1115 & 0.2148 & 0.4382 \\ 0 & 0 & 0 & 0.2000 \end{pmatrix}$.

By substituting R matrix in Equation (9.6) vector P_0 are obtained and normalization equation $P_0 [I - R]^{-1} e = 1$ for the mathematical argument selected previously, row vector P_1 is granted by $P_0 = (0.4286, 0.2000, 0.0571, 0.0000)$. Moreover, the balance vectors P_js are gained from $P_j = P_0 R^j$, $j = 1, 2, 3, \ldots$ and are shown in Table 1. Columns 2, 3, and 4 contain the three elements of

Table 1 Probability vectors

P_i	p_{0j}	p_{1j}	p_{2j}	p_{3j}	Total
P_0	0.4286	0.2000	0.0571	0.0000	0.6857
P_1	0.0697	0.0464	0.0282	0.0716	0.2159
P_2	0.0131	0.0115	0.0090	0.0351	0.0687
P_3	0.0028	0.0029	0.0025	0.0127	0.0209
P_4	0.0007	0.0008	0.0007	0.0041	0.0063
P_5	0.0002	0.0002	0.0002	0.0012	0.0018
P_6	0.0000	0.0001	0.0000	0.0003	0.0004
P_7	0.0000	0.0000	0.0000	0.0001	0.0001
Total					0.9998

Table 2 Probability vectors

P_i	p_{0j}	p_{1j}	p_{2j}	p_{3j}	Total
P_0	0.4432	0.1023	0.0227	0.0000	0.5682
P_1	0.0995	0.0480	0.0224	0.0786	0.2485
P_2	0.0248	0.0188	0.0102	0.0544	0.1082
P_3	0.0069	0.0069	0.0040	0.0276	0.0454
P_4	0.0021	0.0024	0.0014	0.0123	0.0182
P_5	0.0006	0.0008	0.0005	0.0051	0.0070
P_6	0.0002	0.0003	0.0002	0.0020	0.0027
P_7	0.0001	0.0001	0.0001	0.0008	0.0011
P_8	0.0000	0.0000	0.0000	0.0003	0.0003
P_9	0.0000	0.0000	0.0000	0.0001	0.0001
Total					0.9997

$P_j, j = 0, 1, 2, \ldots$. The final column constitutes the total of two elements. Total probability was confirmed to be $0.9998 \approx 1$.

4.2 Illustration

We take $\lambda = \lambda_B = 0.3, \mu_b = 1.5, \mu_v = 1.2, \mu_{BS} = 1.0, \alpha = 0.2, \eta = 0.1, \beta = 0.3$ and the rate matrix is $R = \begin{pmatrix} 0.2144 & 0.0530 & 0.0308 & 0.1233 \\ 0.0181 & 0.2017 & 0.0264 & 0.0958 \\ 0.1175 & 0.1711 & 0.2643 & 0.6239 \\ 0 & 0 & 0 & 0.3000 \end{pmatrix}$.

By substituting R matrix in Equation (9.6) vector P_0 are obtained and normalization equation $P_0\,[I - R]^{-1}\,e = 1$ for the mathematical argument selected previously, row vector P_1 is granted by $P_0 = (0.4432, 0.1023, 0.0227, 0.0000)$. Moreover, the balance vectors P_js are gained from $P_j = P_0 R^j, j = 1, 2, 3, \ldots$ and are shown in Table 2. Columns 2, 3, and 4 contain the three elements of $P_j, j = 0, 1, 2, \ldots$. The final column constitutes the total of two elements. Total probability was confirmed to be $0.9997 \approx 1$.

Table 3 Probability vectors

P_i	p_{0j}	p_{1j}	p_{2j}	p_{3j}	Total
P_0	0.3796	0.0696	0.0126	0.0000	0.4618
P_1	0.1092	0.0493	0.0183	0.0802	0.2570
P_2	0.0340	0.0252	0.0105	0.0700	0.1397
P_3	0.0115	0.0115	0.0050	0.0446	0.0726
P_4	0.0041	0.0050	0.0022	0.0249	0.0362
P_5	0.0015	0.0021	0.0009	0.0129	0.0174
P_6	0.0006	0.0009	0.0004	0.0064	0.0083
P_7	0.0002	0.0004	0.0002	0.0031	0.0039
P_8	0.0001	0.0001	0.0001	0.0014	0.0017
P_9	0.0000	0.0001	0.0000	0.0006	0.0007
P_{10}	0.0000	0.0000	0.0000	0.0003	0.0003
P_{11}	0.0000	0.0000	0.0000	0.0001	0.0001
Total					0.9997

4.3 Illustration

We take $\lambda = \lambda_B = 0.4, \mu_b = 1.5, \mu_v = 1.2, \mu_{BS} = 1.0, \alpha = 0.2, \eta = 0.1, \beta = 0.3$ and the rate matrix is $R = \begin{pmatrix} 0.2790 & 0.0739 & 0.0330 & 0.1632 \\ 0.0220 & 0.2673 & 0.0285 & 0.1241 \\ 0.1358 & 0.2132 & 0.2989 & 0.7628 \\ 0 & 0 & 0 & 0.4000 \end{pmatrix}$.

By substituting R matrix in Equation (9.6) vector P_0 are obtained and normalization equation $P_0 [I - R]^{-1} e = 1$ for the mathematical argument selected previously, row vector P_1 is granted by $P_0 = (0.3796, 0.0696, 0.0126, 0.0000)$. Moreover, the balance vectors P_js are gained from $P_j = P_0 R^j, j = 1, 2, 3, \ldots$ and are shown in Table 3. Columns 2, 3, and 4 contain the three elements of $P_j, j = 0, 1, 2, \ldots$. The final column constitutes the total of two elements. Total probability was confirmed to be $0.9997 \approx 1$.

4.4 Illustration

We take $\lambda = \lambda_B = 0.5, \mu_b = 1.5, \mu_v = 1.2, \mu_{BS} = 1.0, \alpha = 0.2, \eta = 0.1, \beta = 0.3$ and the rate matrix is $R = \begin{pmatrix} 0.3387 & 0.0918 & 0.0339 & 0.1o973 \\ 0.0242 & 0.3284 & 0.0295 & 0.1472 \\ 0.1438 & 0.2348 & 0.3244 & 0.8590 \\ 0 & 0 & 0 & 0.5000 \end{pmatrix}$.

By substituting R matrix in Equation (9.6) vector P_0 are obtained and normalization equation $P_0 [I - R]^{-1} e = 1$ for the mathematical argument selected

Table 4 Probability vectors

P_i	p_{0j}	p_{1j}	p_{2j}	p_{3j}	Total
P_0	0.3249	0.0398	0.0061	0.0000	0.3708
P_1	0.1119	0.0443	0.0142	0.0752	0.2456
P_2	0.0410	0.0282	0.0097	0.0784	0.1573
P_3	0.0160	0.0153	0.0054	0.0597	0.0964
P_4	0.0065	0.0077	0.0027	0.0399	0.0568
P_5	0.0028	0.0038	0.0013	0.0247	0.0326
P_6	0.0012	0.0018	0.0006	0.0146	0.0182
P_7	0.0005	0.0008	0.0003	0.0084	0.0100
P_8	0.0002	0.0004	0.0001	0.0047	0.0054
P_9	0.0001	0.0002	0.0001	0.0026	0.0030
P_{10}	0.0000	0.0001	0.0000	0.0014	0.0015
P_{11}	0.0000	0.0000	0.0000	0.0007	0.0007
P_{12}	0.0000	0.0000	0.0000	0.0004	0.0004
P_{13}	0.0000	0.0000	0.0000	0.0002	0.0002
P_{14}	0.0000	0.0000	0.0000	0.0001	0.0001
Total					0.9990

Table 5 Performance measures with the effect of λ

λ	$E(I)$	$E(SWV)$	$E(BP)$	$E(BD)$	$E(BS)$	$E(N)$	δ
0.2	0.6857	0.1081	0.0829	0.0575	0.2048	1.1390	0.4589
0.3	0.5682	0.1831	0.1224	0.0648	0.3658	1.3043	0.7691
0.4	0.4618	0.2414	0.1746	0.0722	0.5989	1.5489	1.1505
0.5	0.3708	0.2951	0.2188	0.0745	0.9321	1.8913	1.6144

previously, row vector P_1 is granted by $P_0 = (0.3249, 0.0398, 0.0061, 0.0000)$. Moreover, the balance vectors P_js are gained from $P_j = P_0 R^j, j = 1, 2, 3, \ldots$ and are shown in Table 4. Columns 2, 3, and 4 contain the three elements of $P_j, j = 0, 1, 2, \ldots$. The final column constitutes the total of two elements. Total probability was confirmed to be $0.9990 \approx 1$.

From the probability vectors, we find the performance measure using Equations (14), (15), (16), (17), (18), (19), and (20), respectively. By analysing the mean no. of customers if server is idle and on single working vacation, busy period, and breakdown, mean no. of customers throughout system, and total output gained from the system in Table 5. 5.

Figure 2 shows that if arrival λ rate increases, the mean number of customers if server is idle decreases, whereas the mean number of customers if server is on SWV, mean number of customers if server is on busy period, total output gained, mean number of customers if server is on breakdown and if backup server gets activated, and mean number of customers in the system increase.

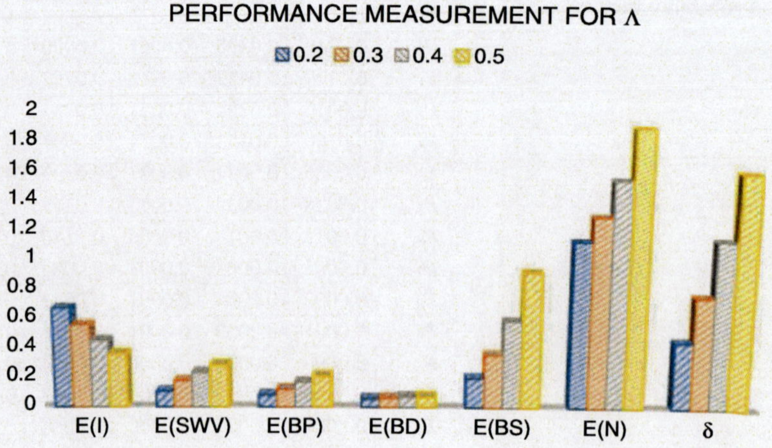

Fig. 2 Vacation interruption rate versus E(I), E(SWV), E(BP), E(BD), E(BS), E(N), and δ

5 Conclusion

In this article, we have studied a single-server queueing model along with single working vacation and breakdown with backup server. We have obtained the static probability row vector by MGM, and also we derived performance measures with the effect of λ such as the mean number of customers if the server is idle, mean number of customers if server is on single working vacation, mean number of customers if server is on busy period, mean number of customers if server is on breakdown and if backup server is activated, mean number of customers in the system, total output gained from the system, and delay time, respectively.

References

1. Baba, Y.: Analysis of a GI/M/1 queue with multiple working vacations. Oper. Res. Lett. **33**, 201–209 (2005)
2. Chakravarthy, S.R., Rakhee, S., Kulshrestha: A queueing model with server breakdowns, repairs, vacations, and backup server. Oper. Res. Perspect. **7** (2020)
3. Doshi, B.T.: Single - server queue with vacation a servey. Queueing Syst. **1**, 29–66 (1986)
4. Lin, C.H., Ke, J.C.: Multi-server system with single working vacation. Appl. Math. Model. **33**, 2967–2977 (2009)
5. Neuts, M.F.: Markov chains with applications queueing theory, which have a matrix geometric invariant probability vector. Adv. Appl. Prob. **10**, 185–212 (1978)
6. Neuts, M.F.: Matrix-Geometric Solutions in Stochastic Models, vol. 2. Johns Hopkins series in the mathematical series. Johns Hopkins University Press, Baltimore (1981)
7. Seenivasan, M., Rajarajan, G., Indumathi, M.: Retrial queueing model with two heterogeneous servers using the Matrix Geometric Method. Mater. Sci. Eng. **1070** (2021)
8. Servi, L.D., Finn, S.G.: M/M/1 queues with working vacation (M/M/1/WV). Perform. Eval. **50**, 41–52 (2002)
9. Wu, D., Takagi, H.: M/G/1 queue with multiple working vacation. In: Proceedings of the Queueing Symposium, Stochastic Models and Their Application, Kakegowa, pp. 51–60 (2003)

Exploring ARIMA Models with Interacted Lagged Variables for Forecasting

Baskaran Thangarajan, Nagaraja M. S., and B. V. Dhandra

1 Introduction

Time series forecasting has been a subject of constant research. Box and Jenkins propounded a coherent and iterative approach – popularly known as the Box-Jenkins approach – for time series identification, estimation, and verification [1]. This had made enormous impact on the modern theory and practice of time series forecasting and the use of Autoregressive Integrated Moving Average (ARIMA) models and their extensions in many areas of science [2]. Forecasting requires the knowledge of the factors or variables of influence in the data generating process and also the right data. Further, giving cognizance to the genuine fluctuations and ignoring random ones is essential in forecasting.

The objective of time series forecasting is to know how the process will move further into the future. However, the forecasting performance depends on inclusion of the right set of contributing variables in the forecasting model [3]. The objective of the present article is to explore this aspect and to bring the interactions among the lagged variables into the ARIMA model for improving the forecast performance.

2 Background and Literature Review

Many references to the inclusion of interactions among variables in regression models exist [4, 5]. Inclusion of interaction of two or more factors in the regression model improves prediction efficiency [6] and [7]. However, at most care needs to

B. Thangarajan · Nagaraja M. S. · B. V. Dhandra (✉)
CHRIST (Deemed to be University), Bengaluru, India
e-mail: t.baskaran@res.christuniversity.in; nagaraja.ms@christuniversity.in; basanna.veeranna@christuniversity.in

F. Kamalov et al. (eds.), *Advances in Mathematical Modeling and Scientific Computing*, Trends in Mathematics, https://doi.org/10.1007/978-3-031-41420-6_61

be taken [5] while doing so. In an additive linear regression model, inclusion of a product term as an interaction term is justified [8].

In spite of several references to including interaction in a linear regression model, to the best of our knowledge, no such attempt has been made to take interaction into consideration in time series forecasting analysis, especially in ARIMA modeling. This has motivated us to explore inclusion of interacted lagged variables in ARIMA modeling. The paper lays emphasis on comparing models – with and without interaction terms – rather than identifying ARIMA models for forecasting.

2.1 Box-Jenkins Approach to ARIMA Modeling

Though several approaches exist in the literature for time series forecasting, Box-Jenkins ARIMA modeling approach is the most popular one, due to its widespread real-time applications. It assumes linearity and stationarity of the data generating process. In this approach, the plausible model is first identified by comparing the theoretical and observed autocorrelations, and then the parameters are estimated. Through diagnostic check of the residuals, the goodness of fit of the model is established and then the model is used for forecasting. If the model is not adequate, a new model needs to be identified, parameters are to be estimated and goodness of fit is to be checked, and the procedure may be repeated till the goodness of fit is established.

In an ARIMA model, the future value of a variable is assumed to be a linear function of its past observations, and the present and past random errors. Thus, it has the following form:

$$y_t = \phi_0+\phi_1 y_{t-1}+\phi_2 y_{t-2}+\cdots+\phi_p y_{t-p}+\epsilon_t+\theta_1\epsilon_{t-1}+\theta_2\epsilon_{t-2}+\cdots+\theta_q\epsilon_{t-q} \quad (1)$$

where y_t is the actual value of the response variable at time t. ϵ_t is the random error at time t and distributed as i.i.d. with mean 0 and variance σ_ϵ^2. $\phi_i (i = 0, 1, 2, \ldots, p)$ and $\theta_j (j = 1, 2, \ldots, q)$ are the model parameters. p and q are the number of lags in the autoregressive and moving average parts, respectively. These are integers and are called orders of the model. The process $\{y_t\}$ is assumed linear and stationary. If it is not stationary, it is made stationary with suitable transformations and differencing before implementing the Box-Jenkins approach.

3 Interactions in ARIMA Modeling

The Autoregressive (AR) part of the ARIMA model in Eq. (1) does have the regressors as the lagged variables in y, namely, $y_{t-1}, y_{t-2}, \ldots, y_{t-p}$, while the Moving Average (MA) part has the lagged variables in ϵ. Each coefficient in the AR part tells the influence of the specific lagged variable on y_t, while that of the MA part tells the influence of the error lagged variable on y_t. As the moving average part consists only of random error terms, interaction is not contemplated at this stage. However, interaction is envisaged in the autoregressive part. Let us consider a simple ARIMA(2,0,0) model as in Eq. (2) for easy interpretation.

$$y_t = \phi_1 y_{t-1} + \phi_2 y_{t-2} + \epsilon_t \tag{2}$$

It is appropriate to ask if only the lagged variables y_{t-1} *and* y_{t-2} influence the response y_t. If so, it may shorten the scope of the model and the model may not characterize the true data generating process. For example, in the case of stock price series, the price of yesterday (y_{t-1}) definitely affects the price of today (y_t). The price of day before yesterday (y_{t-2}) also affects the price of today (y_t). However, the prices of yesterday and day before yesterday alone cannot explain the entire variation in the price of today (y_t). Therefore, it is prudent to expect that the prices of yesterday as well as day before yesterday can *jointly* also influence the price of today. The model in Eq. (2) does not have the scope to include such a joint influence. In this paper, an attempt is made to include this joint influence in the form of interaction between y_{t-1} and y_{t-2} while fitting ARIMA model.

In the AR(2,0,0) model, there are only two lagged variables and therefore there can be only one interaction term, namely, $y_{t-1} * y_{t-2}$. The interaction term here is conceived as a product of the two lagged variables, mainly because including addition or subtraction of the variables will only modify the parameters in the model and will not introduce the effect of interaction. If the interaction is in any other functional form, the model will become complicated and the coefficients may become uninterpretable. On introduction of the interaction term, the model takes the form

$$y_t = \phi_1 y_{t-1} + \phi_2 y_{t-2} + \phi_{12}(y_{t-1} * y_{t-2}) + \epsilon_t \tag{3}$$

where ϕ_1, ϕ_2 and ϕ_{12} are the parameters giving the individual influences of y_{t-1} and y_{t-2} and the joint influence of the interaction $y_{t-1} * y_{t-2}$, respectively.

3.1 Proposed Methodology for Including Interacted Lagged Variables in ARIMA

While considering inclusion of interacted lagged variables in an ARIMA model is required, we should first know whether interaction exists and, if it exists, how many such interactions need to be included and how to include. Including interactions is a step that comes after identifying the order of the model or in other words after knowing the number of lags. For each lag we have a lagged variable.

In order to know if interactions exist, it is suggested that the correlations between the response variable and the different interactions may be computed along with their significance. Existence of significant correlations can suggest existence of influence of the interacted lagged variables on the response variable y_t. If any of the correlations is found to be significant, it may be assumed that interactions exist. The next issue is which interactions are to be included. Ideally all the interactions with significant correlations with y_t should be included. However, this becomes very cumbersome, and therefore, as a practical solution, it is suggested that only the pairwise interactions may be included in the modeling, and in fact higher-order interactions are likely to have weak correlation with response variable by intuition.

Further, we also need to evolve a methodology to include interactions in the model so as to enhance its performance. In spite of the interaction term being the product of the regressors, it is suggested to include it as another exogenous variable. Thus, in ARIMA modeling, it is proposed to use the interactions as external regressors and increase the number of columns of the design matrix by the number of interactions and proceed with the modeling.

After fitting the model, if any of the estimates of the interaction terms is not significant, the term should be ignored and the model rebuilt without the specific interaction term, since a nonsignificant coefficient for an interaction term indicates the absence of interaction [9]. The model can be specified as INTARIMA(p,d,q)[a,m], where p, d, and q represent the usual model parameters in the ARIMA model, while *a* represents the number of interactions included in the model with reference to the AR lagged variables and *m* with reference to the MA variables. For example, INTARIMA(4,0,2)[6,0] represents the ARIMA model with p = 4, d = 0, and q = 2 and a = 6 pairwise autoregressive interactions and m = 0 moving average interactions.

Though inclusion of interaction term in the model actually increases forecast accuracy, it may lead to the problem of multicollinearity. However, in interaction models, it does not pose much problem [10] and [7].

In the present paper, two error metrics, namely, mean absolute error, $MAE = mean(|ei|)$, and root mean squared error, $RMSE = \sqrt{mean(e_i^2)}$, where y_i is the observation, $\hat{y}_i$ is the forecast, and $ei = y_i - \hat{y}_i$ [3] where e_i is the error, are used.

3.2 Estimation of Parameters

An INTARIMA(p,0,0)[a,0] model with *a* pairwise interactions in the autoregressive part can be formulated as

$$
\begin{aligned}
y_t = &\phi_0 + \phi_1 y_{t-1} + \phi_2 y_{t-2} + \cdots + \phi_p y_{t-p} \\
&+ \phi_{12} y_{t-1} y_{t-2} + \phi_{23} y_{t-2} y_{t-3} + \cdots + \phi_{(p-1),p} y_{t-(p-1)} y_{t-p} \\
&+ \phi_{13} y_{t-1} y_{t-3} + \phi_{24} y_{t-2} y_{t-4} + \cdots + \phi_{(p-2)p} y_{t-(p-2)} y_{t-p} \\
&\vdots \\
&+ \phi_{1,p} y_{t-1} y_p + \epsilon_t^*
\end{aligned} \tag{4}
$$

The model can be written in matrix form as

$$
y = X^* \phi^* + \epsilon^* \tag{5}
$$

where

$y = \left[y_{p+1} \; y_{p+2} \; \cdots \; y_{n-1} \; y_n \right]^T_{(n-p)\times 1}$,

$\phi^* = \left[\phi_0 \; \phi_1 \; \ldots \; \phi_p \; \phi_{12} \; \phi_{23} \; \ldots \; \phi_{1,p} \right]^T_{(p+a+1)\times 1}$,

$\epsilon^* = \left[\epsilon^*_{p+1} \; \epsilon^*_{p+2} \; \cdots \; \epsilon^*_n \right]^T_{(n-p)\times 1}$ and

$$
X^* = \begin{bmatrix}
1 & y_p & y_{p-1} & \cdots & y_1 & y_p y_{p-1} & \ldots & y_p y_1 \\
1 & y_{p+1} & y_p & \cdots & y_2 & y_{p+1} y_p & \cdots & y_{p+1} y_2 \\
\vdots & \vdots & \vdots & \cdots & \vdots & \vdots & \cdots & \vdots \\
1 & y_{n-1} & y_{n-2} & \cdots & y_{n-p} & y_{n-1} y_{n-2} & \cdots & y_{n-1} y_{n-p}
\end{bmatrix}
$$

Here, the X^* matrix has $n - p$ rows and $p + a + 1$ columns, where $a = \binom{p}{2}$ is the total number of pairwise interactions included in the model. Therefore, the least squares estimate of parameters ϕ^* is given by

$$
\hat{\phi^*} = (X^{*\prime} X^*)^{-1} X^{*\prime} y \tag{6}
$$

4 Empirical Analysis

4.1 Data Sets

In order to demonstrate the effectiveness of including interaction term in the ARIMA modeling, two simulated time series, two well-known data sets – Wolf's sunspot data and the Canadian lynx data – and a financial time series, Cipla series, are used in the present paper. These data sets are selected to demonstrate the inclusion of interactions in different situations. Each time series is divided into training and test sets for the purpose of fitting the model and forecasting. The error metrics are then examined to understand the performance of the model. The training set is exclusively used for training the model, while the test set is used for forecasting. As the objective of this paper is to demonstrate the effect of including interaction, the ARIMA models used by other researchers have been used, when available, without fresh identification.

4.2 *Analysis and Results*

Simulated ARIMA(2,0,0) Time Series with Interaction

In order to demonstrate the importance of the interaction in the specification of the model, we simulate 100, 200, 500, and 1000 observations for an ARIMA(2,0,0) process with the AR coefficients $\phi_1 = 0.88$ and $\phi_2 = -0.48$ and with an interaction coefficient $\phi_{12} = 0.08$. The AR coefficients have been selected following the stationarity conditions for an AR(2) process. The simulation model is

$$y_t = 0.88 * y_{t-1} - 0.48 * y_{t-2} + 0.08 * (y_{t-1} * y_{t-2}) + \epsilon_t \tag{7}$$

For this purpose, we need to first generate data for the error term ϵ_t and then the variable y_t. But this y_t contains interactions. Therefore, we first generate data for an interaction-free variable z_t and then later add interactions. This is required as otherwise the series explodes in view of the inclusion of interaction terms multiple times. Taking $z_0 = 0$ and $z_{-1} = 0$, we simulate the data for z_t with $\phi_1 = 0.88$ and $\phi_2 = -0.48$.

$$\begin{aligned} z_1 &= \phi_1 * z_0 + \phi_2 * z_{-1} + \epsilon_1 = \epsilon_1 \\ z_2 &= \phi_1 * z_1 + \phi_2 * z_0 + \epsilon_2 = \phi_1 * z_1 + \epsilon_2 \\ z_3 &= \phi_1 * z_2 + \phi_2 * z_1 + \epsilon_3 \\ z_i &= \phi_1 * z_i + \phi_2 * z_{i-1} + \epsilon_i \quad i = 4, 5, \ldots, n \end{aligned} \tag{8}$$

Once we generate n observations for z_t, we proceed to add the interaction term with $\phi_{12} = 0.08$ to have y_t as

$$\begin{aligned} y_1 &= z_1 + \phi_{12} * (z_0 * z_{-1}) = z_1 \quad \text{since } z_0 = 0 \textit{ and } z_{-1} = 0 \\ y_2 &= z_2 + \phi_{12} * (z_1 * z_0) = z_2 \quad \text{since } z_0 = 0 \\ y_3 &= z_3 + \phi_{12} * (z_2 * z_1) \\ y_j &= z_j + \phi_{12} * (z_{j-1} * z_{j-2}) \quad j = 4, 5, \ldots, n \end{aligned} \tag{9}$$

The error terms are not added as they had already been present in the $z_t^{\prime}s$. Once the data for y_t are generated, the lagged variables, namely, y_{t-1} and y_{t-2}, are generated so as to fit any model. As there are only two lags in the AR part, there are only two lagged variables y_{t-1} and y_{t-2} and one interaction $y_{t-1} * y_{t-2}$. After

Table 1 Performance of the simulated ARIMA(2,0,0) series with interactions

Sample size	Model	RMSE	MAE
100	Without interaction	1.211202	1.023228
100	With one interaction	1.167893	1.001355
200	Without interaction	1.536906	1.09778
200	With one interaction	1.532218	1.09759
500	Without interaction	1.259966	1.056963
500	With one interaction	1.259912	1.062001
1000	Without interaction	1.336828	1.102149
1000	With one interaction	1.327218	1.091410

Table 2 Performance of the simulated ARIMA(2,0,2) series without interactions

Model	RMSE	MAE
Without interaction	**1.267284**	**1.051181**
With one interaction	1.308095	1.089415

constructing the lagged and interaction variables, the correlation of the interaction variable $(y_{t-1} * y_{t-2})$ with y_t is checked for its significance. Two ARIMA(2,0,0) models are fitted – one without interaction and the other with one interaction. As the data are simulated with interactions, the model with interaction should show better performance. The results are displayed in Table 1.

From Table 1, one may find that both RMSE and MAE decrease on inclusion of the interaction term in the model. Thus, the model with interaction has performed better with lower errors than the model without interaction.

Simulated ARIMA(2,0,2) Time Series Without Interaction

This series is simulated using *arima.sim*() function of the *stats* package of the R software for an ARIMA(2,0,2) model. In this data set, interactions are not added. As such the data do not have interactions. 100 observations of ARIMA(2,0,2) are generated.

As the model has two lags in the AR part and two lags in the MA part, we should have one interaction in the AR part and one in the MA part. However, we consider and propose to include one interaction, only with respect to the AR part. In order to fit the model with interactions, lagged variables are compiled, and the correlation between the interaction term and the response variable is calculated, and it is found that the correlation is not significant. One reason for the insignificance of the correlation is that the data generating process did not have interaction. In this situation, one may expect and rightly so that the model without interaction is the appropriate model. For the purpose of clarity, two models are constructed – one without interactions and the other with interaction. The performance of the two models for the data set is given in Table 2.

From Table 2, one may find that the errors RMSE and MAE are more in the model with interaction than in the model without interaction. Thus, we conclude that if the data set does not contain any interaction, introduction of interactions in the model increases the error.

Wolf's Sunspot Dataset

The sunspot dataset studied here is a set of annual number of sunspots from the year 1700 to 1987 and has 288 observations. The series has a historical interest for statisticians, as it is used by [11]. The data set is considered nonlinear and non-Gaussian and is widely used [12]; in view of the implications in a number of fields including climatology and environmental sciences, the sunspot data has continuously been a subject of research.

The ARIMA(9,0,0) model is fitted without interactions as this is the model fitted by earlier researchers [12] and [13] for this dataset. The same model is fitted with interactions too. The model has 9 lagged variables resulting in 36 pairwise interactions, of which 27 have been empirically found to have significant correlations with y_t.

Taking the top one, two, three, and other interactions, different AR(9,0,0) models are constructed along with a model without interaction. This is done for two forecasting horizons h = 35 and h = 67 as done by the previous researchers [12]. The performance of the models in these two horizons is presented in Table 3.

From Table 3 for the sunspot data, it is seen that the errors increase when we add more and more interactions. We may conclude that the effect of interactions in the sunspot data series is not significant enough to reduce the forecasting errors. However, appropriate model can be decided by examining the residuals and other criteria.

The Canadian Lynx Data Series

The Canadian lynx data series is about the number of lynx trapped in the Mackenzie River district of Northern Canada. It has a total of 114 observations. This data set has also been extensively analyzed in the time series literature [12]. In order to

Table 3 Performance of the sunspot series

	Horizon = 35		Horizon = 67	
Model	RMSE	MAE	RMSE	MAE
No interaction	**34.42397**	**22.86284**	**50.38593**	**34.66572**
Top 1 interactions	39.91899	29.71007	52.24681	38.15206
Top 2 interactions	40.74129	30.45259	51.12665	37.57086
Top 3 interactions	41.84142	31.18101	53.65753	39.72602
Top 4 interactions	42.71343	31.78311	54.82468	40.82654
Top 5 interactions	50.51314	50.51314	59.42757	59.42757

Table 4 Performance of the lynx series

Model	RMSE	MAE
No interaction	0.35772761	0.28601436
Top 1 interaction	0.27132435	0.20392396
Top 2 interaction	**0.07397827**	**0.07288690**
Top 3 interaction	0.08936157	0.08465978
Top 4 interaction	0.86740714	0.77284263
Top 5 interaction	0.93660789	0.93660789

have compatibility with other studies [12, 13] and [14], logarithm to the base 10 have been taken for the data and used for analysis. ARIMA(12,0,0) has been fitted. This model has 12 lags and therefore 12 lagged variables. These give rise to 66 pairwise interactions. The correlations of these interactions with the y_t variable were examined for their significance, and it was found that 47 out of the 66 interactions were significant.

The performance of the interaction models with top one to five interactions and of the model without interaction is presented in Table 4.

From Table 4, it is clear that there is a reduction in the errors while including the first two interactions when compared to the model without interactions. As we increase the number of interactions, the errors increase. Therefore, it is concluded that the interaction model with the top two interactions will be the appropriate model.

Financial Series (Cipla Series)

The time series, here, is the historical data of closing stock price of the Cipla Limited, a pharmaceutical company in India, from the National Stock Exchange (NSE), India. It has daily closing price from 3 January 2000 to 30 April 2021 for all working days. In all, it has 5306 observations. The data series is normalized using min-max normalization technique to have values between *zero* and *one*.

The series is also found to be non-stationary and by differencing once, the data become stationary. On identification, ARIMA(3,1,3) model is found to be appropriate to the Cipla series. As the order of the model is $AR = 3$ and $MA = 3$, there will be three lagged variables in the AR part and three lagged variables in the MA part. However, we consider the interactions only in the AR part. Thus, we need to include three pairwise interaction terms in the model. All the three correlations of the interactions with the y_t variable are found to be significant. Different ARIMA(3,1,3) models were fitted with the top one, two, and three interactions, along with a model without interaction. The performance of these models is presented in Table 5.

From Table 5, we find that the model with the top one interaction has less error, and therefore, it is concluded that the model with top one interaction is the appropriate model.

Table 5 Performance of the financial (Cipla) series

Model	RMSE	MAE
No interaction	0.2728865	0.2438373
Top 1 interaction	**0.1988333**	**0.1832178**
Top 2 interactions	0.2071377	0.1899316
Top 3 interactions	0.2020295	0.1859046

5 Conclusions and Future Scope

The objective of the present paper is to understand the effect of including the interacted lagged variables in ARIMA modeling. For this purpose, five data sets have been used. In the simulated ARIMA(2,0,0) series, wherein interactions were purposely introduced, the experimental results prove that the model with interaction has less errors conforming to the data generating process. In the simulated ARIMA(2,0,2) series, wherein interactions were absent, the results show that the model with interaction is not appropriate as it gives higher errors as compared to model without interaction. With reference to the sunspot data series, we find that the series may not have interaction effect as the model without interaction performs better. On the other hand, for the lynx data series, the performance of the model with interaction is better than the model without interaction. The Cipla data series also shows better results for interaction model as compared to model without interaction. Overall, it is concluded from the results obtained from all the five data sets that if the data series actually contains interactions, then including appropriate interactions improves the performance of the ARIMA model. However, more research is needed to examine the different facets of including interactions in ARIMA model.

Acknowledgments The authors are thankful to the authorities of the CHRIST (Deemed to be University) for their encouragement and support for carrying out this research.

References

1. Box, G., Jenkins, G.: Time Series Analysis: Forecasting and Control. Holden-Day (1976)
2. De Gooijer, J.G., Hyndman, R.J.: 25 years of time series forecasting. Int. J. Forecast. **22**, 443–473 (2006)
3. Hyndman, R.J., Athanasopoulos, G.: Forecasting: Principles and Practice, 2nd edn. OTexts, Melbourne (2018)
4. Vatcheva, K.P., Lee, M.: The effect of ignoring statistical interactions in regression analyses conducted in epidemiologic studies: an example with survival analysis using Cox proportional hazards regression model. Epidemiology (2016)
5. Balli, H.O., Sørensen, B.E.: Interaction effects in econometrics. Empir. Econ. **45**, 583–603 (2013)
6. Box, G.: Do interactions matter? J. Quality Eng. **2**, 365–369 (1990)
7. Friedrich, R.J.: In defense of multiplicative terms in multiple regression equations. Am. J. Pol. Sci. **26**, 797–833 (1982)

8. Allison, P.D.: Testing for interaction in multiple regression. Am. J. Sociol. **83**(1), 144–153 (1977). https://www.jstor.org/stable/2777767
9. Kasim, R.M.: Interaction effect. In: Encyclopedia of Survey Research Methods, pp. 340–342. Sage Publications (2015). https://doi.org/10.4135/9781412983907.n962
10. Brambor, T., Clark, W.R., Golder, M.: Understanding interaction models: Improving empirical analyses. Polit. Anal. **14**, 63–82 (2006)
11. Udny Yule, G.: On a method of investigating periodicities in disturbed series, with special reference to Wolfer's sunspot. Philos. Trans. R. Soc. Lond. Ser. A, Contain. Pap. A Math. or Phys. Character **226**, 267–298 (1927)
12. Zhang, G.P.: Time series forecasting using a hybrid ARIMA and neural network model. Neurocomputing **50**, 159–175 (2003)
13. Rao, T.S., Gabr, M.M.: An Introduction to Bispectral Analysis and Bilinear Time Series Models (1984)
14. Campbell, M.J., Walker, A.M.: A survey of statistical work on the Mackenzie river series of annual Canadian lynx trappings for the years 1821–1934 and a new analysis. Source J. R. Stat. Soc. Ser. A **140**, 411–431 (1977)

A Novel Hybrid Model for Time Series Forecasting Using Artificial Neural Network and Autoregressive Integrated Moving Average Models

Baskaran Thangarajan, Nagaraja M. S., and B. V. Dhandra

1 Introduction

Time series forecasting is attracting considerable interest of the researchers due to its applications in different sectors, including information and technology (IT) sector, and also due to its forecasting power, despite its simplicity. When the data generating process of the time series is unknown – usually the case – linear Gaussian stochastic process is assumed, mainly due to several of its advantages [1]. The alternative is to assume the existence of a nonlinear process. However, a real-world data generating process consists neither only of the linear process nor only of the nonlinear process. It is generally a combination of both the processes. Therefore, substantial research has been devoted to combine the forecasts of different models so as to use the capabilities of each component model in modeling the time series. The idea of combining forecasts emerged even in 1906, when Francis Galton observed, in a fair guessing the weight of an ox (1198 pounds) was close to the average (1197 pounds) of 787 estimates of its weight [2]. However, Bates and Granger in 1969 [3] established a sound footing to the idea of combining forecasts and popularized it into a scientific method for enhancing forecast accuracy. When the data generating process is unknown, obtaining an estimate from a single model and treating it as the best estimate may be deceptive, due to the uncertainties in the data, or in parameters, or in the model, to mention a few. In view of the potentiality for improving the forecast accuracy, Clemen in his review article encourages combination forecasts [4]. Reinforcing this, Adhikari et al. assert that no single model can provide best forecasts, unlike the combined forecasts from multiple models [5].

B. Thangarajan (✉) · Nagaraja M. S. · B. V. Dhandra
CHRIST (Deemed to be University), Bengaluru, India
e-mail: t.baskaran@res.christuniversity.in; nagaraja.ms@christuniversity.in; basanna.veeranna@christuniversity.in

F. Kamalov et al. (eds.), *Advances in Mathematical Modeling and Scientific Computing*, Trends in Mathematics, https://doi.org/10.1007/978-3-031-41420-6_62

Zhang in 2003 proposed a hybrid model by adding the forecasts of linear and nonlinear components in the data generating process [6]. He suggests to compute the linear forecast with ARIMA model and the nonlinear forecast with ANN model and to combine them through simple addition. He also exhibits the superiority of the proposed model over the component models through empirical results. Zou et al. in 2007 built up a combined forecasting model using a linear combination of forecasts of ARIMA and ANN models, giving equal weights to the forecasts [7]. The proposed model generally outperforms ARIMA and ANN models at individual level. As against the assumption of additive relationship by Zhang [6], a multiplicative relationship is assumed by Wang et al., and the linear and nonlinear forecasts are multiplied to produce the combined forecast with enhanced accuracy over the forecasts from the component models [8].

Several researchers use weights constructed in different methods to combine forecasts. Adhikari et al. propose a weighted ensemble approach that combines several ANN algorithms to increase the forecast accuracy [9]. They propose a homogeneous ensemble technique of several different algorithms of ANN. The technique applies a scientific methodology for constructing weights and uses them appropriately to enhance the forecast accuracy. In 2012, Chee Yin et al. formulated a new procedure called Forecast Weights Averaging (FWA) using weights derived from a number of averaging methods to improve the forecast accuracy [10]. Nagahamulla et al. investigate the application of ensemble neural networks (ENN) consisting of several algorithms and use a weighted average method to combine the forecasts [11]. Adhikari et al. combine forecasts of different ANN models with selected weights inversely proportional to the mean squared error (MSE) from the concerned model [5].

There is a clear distinction between the hybrid approaches using weights and those using other methods like data pre-processing [12]. Attempts have been made to first decompose the data into linear part and nonlinear part and then combine their estimates. Khandelwal et al. decompose the data into linear component and nonlinear component using discrete wavelet transformation (DWT) and combine the ARIMA and ANN forecasts of the two components, respectively, by adding them [13]. While forecasting the stock prices, Yu and Li use a moving average filter to decompose the data into linear and nonlinear components and then combine the ARIMA and recurrent neural network (RNN) forecasts of the respective components by adding them [14]. In a similar attempt, Madan and Mangipudi also use DWT decomposition but apply ANN instead of RNN [15]. Wang et al. present a vivid review of the forecast combinations emerged during the last five decades [16].

The above literature review has motivated us to introduce a new method to combine forecasts of two models of different classes using weights proportional to the inverse of the error.

The remaining paper has five sections. While Sect. 2 briefly describes the ARIMA and ANN models, Sect. 3 introduces the proposed methodology for combining forecasts. Section. 4 compares empirical results from two data sets, while Sect. 5 presents the concluding remarks.

2 Component Models

2.1 *Autoregressive Integrated Moving Average (ARIMA) Model*

The ARIMA model has the characteristic that the future value of a variable depends linearly on the past observations and random errors. The general form of the ARIMA model is

$$y_t = \phi_1 y_{t-1} + \phi_2 y_{t-2} + \cdots + \phi_p y_{t-p} + \epsilon_t + \theta_1 \epsilon_{t-1} + \theta_2 \epsilon_{t-2} + \cdots + \theta_q \epsilon_{t-q} \quad (1)$$

where y_t is the value of a variable y at time t. $\phi_1, \phi_2, \ldots, \phi_p, \theta_1, \theta_2, \ldots, \theta_q$ are the parameters to be estimated. p and q are integral values representing the number of lags in the autoregressive observations and errors, respectively. They are referred to as the orders of the model. p and q may not have any relationship between them. These are decided by examining the autocorrelation structure of the variable y and by considering the minimum value of Akaike's Information Criterion (AIC) or the corrected AIC or the Bayesian Information Criterion (BIC) from different plausible models. ϵ_t is the random error at time t and is i.i.d. with mean 0 and variance σ_ϵ^2. Box and Jenkins developed an iterative approach [17] for identifying the appropriate model for a given time series by comparing the properties of the theoretical and empirical autocorrelations and partial autocorrelations. After identifying the model, estimates of parameters with least errors are obtained. The appropriateness of the model is examined through diagnostic checking of the residuals. If the identified model is not appropriate, the procedure of identification, estimation, and diagnostic check is repeated for ensuring the goodness of fit. Once the adequate model is finalized, it is used for forecasting purposes. The approach works for a stationary process only. If the process is not stationary, it is made stationary with suitable transformations and differencing before proceeding with Box-Jenkins approach.

2.2 *Artificial Neural Network (ANN) Model*

Artificial neural network is a popular and widely used artificial intelligence-based algorithm for time series forecasting. The unique features of the ANN like self-adaptive data-driven nature, nonlinearity, easy generalization, and universal functional approximation characteristic have made the algorithm to draw overwhelming attention [18]. It has one input layer, one or more hidden layers, and one output layer with an activation (transfer) function to transform the inputs to the output. Each layer has a number of nodes for processing. Every node in a layer is linked to one or more nodes in the next layer through weights, in such a manner that they are non-revertible (feed forward). The input layer has as many nodes as the predictors required in the problem. As far as the nodes in the hidden layers are concerned, they depend on the complexity of the input-output relationship. Too many hidden layers may lead to over fitting while too few to under fitting. Usually one or two hidden layers should be sufficient for most practical situations. The count of nodes in the output layer is based on the forecasting horizon or the number of forecasts that are to be generated. Logistic function is the most commonly used activation function. Among the different structures of the ANN, the forward network with one hidden layer feed is the one that is well recognized for time series forecasting [6]. The model ANN(p,q,1) has three layers – with p nodes in the input layer corresponding to the p inputs, q nodes in the single hidden layer, and one node in the output. The above model has the following form:

$$y_t = \omega_0 + \sum_{j=1}^{q} \omega_j g(\beta_{0j} + \sum_{i=1}^{p} \beta_{ij} y_{t-i}) + \epsilon_t \tag{2}$$

where ω_j, $j = 1, 2, \cdots, q$ and βij, $i = 1, 2, \cdots, p$ & $j = 1, 2, \cdots, q$ are the connection weights, ω_0, β_{0j} are the bias terms, and g is the logistic activation function given by

$$g(x) = \frac{1}{1 + exp(-x)} \tag{3}$$

By minimizing the errors, the parameters are estimated [19].

2.3 *Forecast-Error Metrics*

In the present paper, the model performance is proposed to be measured using three error metrics (i) the mean absolute error (MAE), which is the mean of the absolute differences between the actual observations and the estimated values, (ii) the mean squared error (MSE), which is the mean of the squared deviations, and (iii) the mean absolute percentage error (MAPE).

3 Proposed Hybrid Model

The objective is to build a new model by combining ARIMA forecasts and ANN forecasts to have better forecast accuracy. Let $y_1, y_2, \cdots, y_n$ be the time series of actual observations on a variable y at different points of time. Let us divide this time series of observations into training, validation, and test sets as

$$y_{tr} = y_1, y_2, \cdots, y_r \quad \text{where } r < n$$

$$y_{tv} = y_{r+1}, y_{r+2}, \cdots, y_{r+s} \quad \text{where } r + s < n,\ s < r \text{ and}$$

$$y_{ts} = y_{r+s+1}, y_{r+s+2}, \cdots, y_n$$

As noted earlier, the time series generally consists of linear as well as nonlinear components [6]. The linear component is modeled using the ARIMA while the nonlinear portion using the ANN. In the proposed approach, an appropriate ARIMA model is conceived using y_{tr} and y_{tv}, and three errors, namely, MAE_1, MSE_1 and $MAPE_1$ are calculated. Using these errors, the weight with reference to the ARIMA model is calculated as

$$w_1 = exp\left(\frac{1}{MAE_1 + MSE_1 + MAPE_1}\right) \tag{4}$$

The training is again performed using the combined set $y_{tr} + y_{tv}$ and an appropriate model is arrived at using the test set y_{ts}. With this model, the forecast F_1 is obtained using ARIMA model. This forecast pertains to the linear part of the data generating process.

In order to model the nonlinear part, we use the ANN in a similar manner and obtain the three errors, namely, MAE_2, MSE_2 and $MAPE_2$, the weight with reference to the ANN model

$$w_2 = exp\left(\frac{1}{MAE_2 + MSE_2 + MAPE_2}\right) \tag{5}$$

and also the forecast F_2. The two forecasts F_1 and F_2 are combined together to produce the combined forecast F_c as the weighted average of the forecasts with weights w_1 and w_2. Thus

$$F_c = \frac{w_1 * F_1 + w_2 * F_2}{w_1 + w_2} \tag{6}$$

In this approach, two nonhomogeneous forecasts, i.e., one from the linear model ARIMA and the other from the nonlinear model ANN, are combined. Further, the sum of the errors in the denominator in Eqs. (4) and (5) helps to have the weights inversely proportional to the combined error. It is possible that the three errors may be in different magnitudes and may affect the weights differently. The exponential function in the weight computation smoothens this effect [9].

4 Empirical Analysis

In order to demonstrate the efficiency of the proposed method of combining forecasts, we use two time series – the lynx series and the series of closing stock prices of Indian Oil Corporation (IOC). The lynx time series is a stationary series. It contains the number of lynx trapped in each year in Northwest Canada during 1821–1934. This time series has a total of 114 observations. In view of the extremely large

Fig. 1 Original lynx series

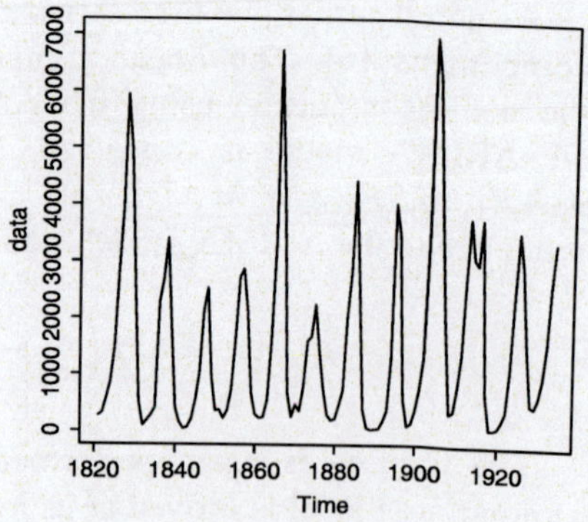

Fig. 2 Closing stock of IOC

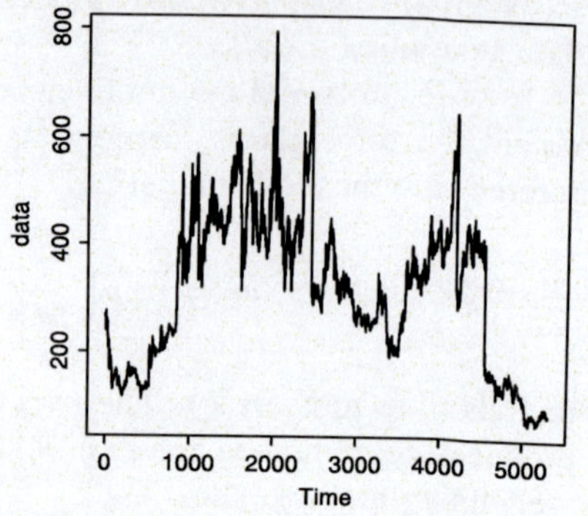

values at each peak of this time series, a logarithmic transformation is applied [20–22]. Figure 1 gives the plot of the Lynx time series.

The second time series is the closing prices of stocks of the Indian Oil Corporation (IOC) Limited from the National Stock Exchange, India, for the period from 3 January 2000 to 30 April 2021 (a total of 5306 observations). This is a nonstationary time series. The plot of the IOC time series is given in Fig. 2.

For assessing the model performance, we divide both the time series into training (70%), validation (20%), and test (10%) sets. Using the training and the validation data sets, appropriate ARIMA and ANN models are conceived. Using these models, separate errors and weights are calculated for each time series. This step is done exclusively for the purpose of building the weights using the forecasting errors arising out of these models. The training and validation data sets are then pooled to form a new training set and this is used for training again. Using this training and test data sets, appropriate ARIMA and ANN models are decided. At this stage, the forecasts are obtained for each time series using the two models. The forecasts obtained from these models are combined as a weighted arithmetic mean using the weights obtained earlier, to have the combined forecast. The performance of the models with respect to the two time series is given in Table 1.

The table clearly shows that the errors from the combined forecasts are less than the errors from the ARIMA as well as the ANN models. Thus, the proposed

Table 1 Performance of the models for lynx and IOC time series

	Lynx			IOC		
Model	MAE	MSE	MAPE	MAE	MSE	MAPE
ARIMA	0.345	0.166	10.712	232.992	55082.474	229.782
ANN	0.190	0.053	6.354	47.047	2977.695	51.380
Combined	0.189	0.048	6.211	45.917	2853.620	50.223

methodology of combining forecasts gives better forecast than the individual component-model forecasts.

5 Conclusion

Improving the forecast accuracy in time series is an interesting area of research. Several models including the very popular ARIMA and ANN models exist for forecasting. Combining the forecasts using weights constructed in a specific manner enhances the forecast accuracy. The present paper proposes a novel methodology to combine the forecasts of the ARIMA and ANN models in terms of weights obtained using different error metrics. The empirical analysis using the lynx and IOC time series proves that the proposed methodology gives less forecast error than the forecast errors from the component models.

Acknowledgments The authors are thankful to the authorities of the CHRIST (Deemed to be University) for their encouragement and support for carrying out this research.

References

1. Medeiros, M.C., Veiga, Á.: A hybrid linear-neural model for time series forecasting. IEEE Trans. Neural Netw. **11**(6), 1402–1412 (2000)
2. Surowiecki, J.: The Wisdom of Crowds (2005)
3. Bates, J.M., Granger, C.W.J.: The combination of forecasts. J. Oper. Res. Soc. **20**, 451–468 (1969)
4. Clemen, R.T.: Combining forecasts: a review and annotated bibliography. Int. J. Forecast. **5**, 559–583 (1989)
5. Adhikari, R., Verma, G., Khandelwal, I.: A model ranking based selective ensemble approach for time series forecasting. Proc. Comput. Sci. **48**, 14–21 (2015)
6. Zhang, G.P.: Time series forecasting using a hybrid ARIMA and neural network model. Neurocomputing **50**, 159–175 (2003)
7. Zou, H.F., Xia, G.P., Yang, F.T., Wang, H.Y.: An investigation and comparison of artificial neural network and time series models for Chinese food grain price forecasting. Neurocomputing **70**, 2913–2923 (2007)
8. Wang, L., Zou, H., Su, J., Li, L., Chaudhry, S.: An ARIMA-ANN hybrid model for time series forecasting. Syst. Res. Behav. Sci. **30**, 244–259 (2013)

9. Adhikari, R., Agrawal, R.K.: A homogeneous ensemble of artificial neural networks for time series forecasting. Int. J. Comput. Appl. **32**, 1–8 (2011)
10. Chee Yin, Y., Kok-Haur, N., Hock-Eam, L.: Combining forecast weights: why and how? In: International Conference on Fundamental and Applied Sciences 2012 AIP Conf. Proc. 1482, pp. 351–356 (2012). https://doi.org/10.1063/1.4757493
11. Nagahamulla, H.R.K., Ratnayake, U.R., Ratnaweera, A.: An ensemble of artificial neural networks in rainfall forecasting. In: The International Conference on Advances in ICT for Emerging Regions, pp. 176–181 (2012)
12. Tascikaraoglu, A., Uzunoglu, M.: A review of combined approaches for prediction of short-term wind speed and power. Renew. Sustain. Energy Rev. **34**, 243–254 (2014)
13. Khandelwal, I., Satija, U., Adhikari, R.: Efficient financial time series forecasting model using DWT decomposition. In: 2015 IEEE International Conference on Electronics, Computing and Communication Technologies (CONECCT) (2015). https://doi.org/10.1109/CONECCT.2015.7383917
14. Yu, S.L., Li, Z.: Stock price prediction based on ARIMA-RNN combined model. In: 4th International Conference on Social Science (2017)
15. Madan, R., Mangipudi, P.S.: Predicting computer network traffic: a time series forecasting approach using DWT, ARIMA and RNN. In: Eleventh International Conference on Contemporary Computing (2018)
16. Wang, X., Hyndman, R.J., Li, F., Kang, Y.: Forecast combinations: an over 50-year review (2022)
17. Box, G., Jenkins, G.: Time series analysis: forecasting and control. Holden-Day (1976)
18. Khashei, M., Bijari, M.: An artificial neural network (p, d, q) model for timeseries forecasting. Expert Syst. Appl. **37**, 479–489 (2010)
19. Khashei, M., Bijari, M.: A new class of hybrid models for time series forecasting. Expert Syst. Appl. **39**, 4344–4357 (2012)
20. Moran, P.A.P.: The statistical analysis of the sunspot and lynx cycles. J. Anim. Ecol. **18**, 115–116 (1949)
21. Campbell, M.J., Walker, A.M.: A Survey of statistical work on the Mackenzie river series of annual Canadian lynx trappings for the years 1821–1934 and a new analysis. Source J. R. Stat. Soc. Ser. A **140**, 411–431 (1977)
22. Rao, T.S., Gabr, M.M.: An Introduction to Bispectral Analysis and Bilinear Time Series Models (1984)

Part V
Graph Theory

Sigma Chromatic Number of Mycielski Transformation of Graphs

C. Yogalakshmi and B. J. Balamurugan

1 Introduction

The vertex coloring of a graph $G(V, E)$ with vertex set $V(G)$ and an edge set $E(G)$ is the process of assigning colors to the vertices so that no two adjacent vertices share the same color. The chromatic number, denoted by $\chi(G)$, is the minimum number of colors required for a vertex coloring of the graph G. The open neighborhood of the vertex v, denoted by $N(v)$, is the set of vertices that are adjacent to v.

In 2010 [3], Chartrand et al. developed a new type of neighbor distinguishing coloring known as a sigma coloring of G by adding few conditions to the vertex coloring $c : V(G) \to N$ as follows:

(i) The adjacent vertices of G may receive the same color.
(ii) The color sum of a vertex $v \in V$ is $\sigma(v) = \sum_{u \in N(v)} c(u)$, where $N(v)$ is the open neighborhood of the vertex v.
(iii) $\sigma(u) \neq \sigma(v)$ for any two adjacent vertices u and v in G.

The minimum number of colors used for such a coloring is called the sigma chromatic number of G, and it is denoted by $\sigma(G)$. The key feature of sigma coloring is that $\sigma(G) \leq \chi(G)$. The sigma chromatic numbers of some standard graphs, such as path, cycle, complete graph, etc., were discussed in [3]. The behavior of sigma colorings of some graphs using graph operations such as join operation [1, 4, 6, 7], corona operation [5], and middle graph [8] were studied by the researchers. In this manuscript, an unary operation called Mycielski transformation

C. Yogalakshmi · B. J. Balamurugan (✉)
Division of Mathematics, School of Advanced Sciences, Vellore Institute of Technology, Chennai, India
e-mail: balamurugan.bj@vit.ac.in

F. Kamalov et al. (eds.), *Advances in Mathematical Modeling and Scientific Computing*, Trends in Mathematics, https://doi.org/10.1007/978-3-031-41420-6_63

of graphs is considered, and the sigma colorings of the graphs obtained through the Mycielski transformation have been investigated with appropriate illustration.

Duplication of a vertex v of G generates a new graph G' by adding a new vertex v' such that the vertices adjacent to v are adjacent to v'. The clique S of a graph G is a subset of V such that every two distinct vertices are adjacent to each other. In a graph G, a maximum clique is one in which no clique can be formed by adding vertices to an existing clique. The clique number $\omega(G)$ is defined as the number of vertices in a maximum clique on the graph G. To construct a graph with small clique number and high chromatic number, Mycielski [9] introduced a new method of transforming a graph into another graph known as Mycielski transformation of graphs. The Mycielski graph, denoted by $\mu(G)$, is the graph obtained by duplicating each vertex u_i of the graph G, and these duplicate vertices are adjacent to an additional vertex w. If the graph G is a triangle-free graph, then $\mu(G)$ is also a triangle-free graph.

In this article, the sigma colorings of the Mycielski transformation of path graph P_n, cycle graph C_n, complete graph K_n, complete bipartite graph $K_{m,n}$, wheel graph W_n, and helm graph H_n are determined with appropriate examples. The terminologies and notions of a graph G used in this article have been cited from [2].

2 Mycielski Transformation of a Graph and Its Sigma Coloring

In this section, the definition of the Mycielski transformation of graph G is recalled with an appropriate example, and the sigma chromatic numbers of Mycielski transformation of path, complete bipartite, and complete graph have been determined.

Definition 1 Let G be a graph of order n whose vertices are labeled as $u_1, u_2, \ldots u_n$. The Mycielski transformation of a graph G, denoted by $\mu(G)$, is the graph obtained by duplicating each vertex of the graph G, labeled as $u'_1, u'_2, \ldots u'_n$, and these duplicate vertices are adjacent to one additional vertex w. The Mycielski transformation of graph has $2n + 1$ vertices.

Example 1 The graph G and its Mycielski transformation are shown in Fig. 1.

2.1 *Sigma Coloring of Mycielski Transformation of Path and Complete Bipartite Graph*

In [3], Chartrand et al. have proven that the sigma chromatic number of the path graph P_n for $n \geq 2$ is

Fig. 1 **(a)** A graph G. **(b)** The Mycielski transformation of the graph G

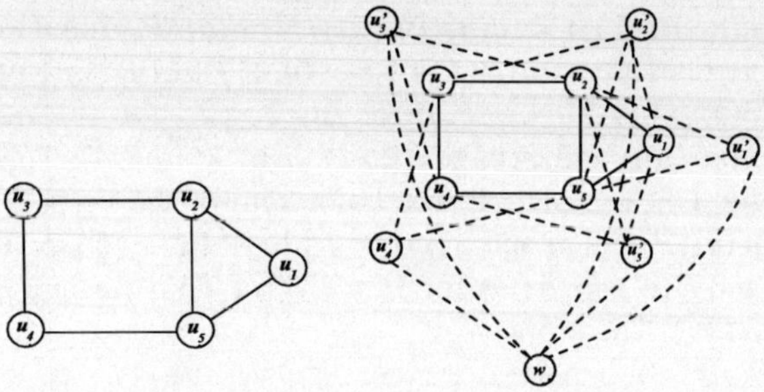

(a) (b)

$$\begin{cases} 1, & for\ n = 3, \\ 2, & otherwise. \end{cases}$$

In this result, the sigma coloring c and the color sum of the vertices of the graph P_n are determined as follows:

$$c(u_i(P_n)) = \begin{cases} a, & if\ i \equiv 1 (mod\ 4), \\ b, & if\ i \not\equiv 1 (mod\ 4). \end{cases}$$

and

$$\sigma(u_i(P_n)) = \begin{cases} a+b, & for\ i\ is\ even\ and\ i \neq n, \\ 2b, & for\ i\ is\ odd\ and\ i \neq 1, \\ a, & for\ i = n\ when\ n \equiv 2(mod\ 4), \\ b, & for\ i = 1, n\ except\ for\ n \equiv 2(mod\ 4). \end{cases}$$

The following theorem shows the sigma chromatic number of Mycielski transformation of the path graph.

Theorem 1 *The sigma chromatic number of the Mycielski transformation of the path graph* P_n *for* $n \geq 2$ *is*

$$\sigma(\mu(P_n)) = \begin{cases} 3, & for\ n = 2, \\ 2, & for\ n \geq 3. \end{cases}$$

Proof When $n = 2$, the graph $\mu(P_2) \cong C_5$; thus $\sigma(\mu(P_2)) = \sigma(C_5) = 3$. The vertices of the subgraph P_n in $\mu(P_n)$ are labeled as u_i for any $1 \leq i \leq n$, and the duplicate vertices of u_i are labeled as u_i', and the vertex adjacent to each duplicate vertex is labeled as w. Since the adjacent vertices u_i and u_{i+1} for $2 \leq i \leq n-2$ are of the same degree, $\sigma(\mu(P_n))$ must be greater than or equal to 2. Let $c : V(\mu(P_n)) \rightarrow \{a, b\}$ be the vertex coloring of the graph $\mu(P_n)$. For $1 \leq i \leq n$,

$c(u_i(\mu(P_n))) = c(u'_i(\mu(P_n))) = c(u_i(P_n))$ and $c(w) = a$. The color sum of the vertices are $\sigma(u_i(\mu(P_n))) = c(u_{i-1})+c(u'_{i-1})+c(u_{i+1})+c(u'_{i+1}) = 2(c(u_{i-1})+c(u_{i+1}) = 2(\sigma(u_i(P_n))$ for any $2 \leq i \leq n-1$, $\sigma(u_1) = 2b$, $\sigma(u_n) = 2a$ *if* $n \equiv 2(mod\ 4)$ and $\sigma(u_n) = 2b$ *if* $n \not\equiv 2(mod\ 4)$. This implies that $\sigma(u_i) \neq \sigma(u_{i+1})$ for $1 \leq i \leq n-1$. The color sum of the vertices u'_i and w are $\sigma(u'_i(\mu(P_n))) = \sigma(u_i(P_n)) + a$ and $\sigma(w) = \lceil \frac{n}{4} \rceil a + (n - \lceil \frac{n}{4} \rceil)b$. This implies that $\sigma(u_i) \neq \sigma(u'_i)$ and $\sigma(u'_i) \neq \sigma(w)$ for $1 \leq i \leq n$. Thus for any $n \geq 2$, c is a sigma coloring of $\mu(P_n)$. Therefore

$$\sigma(\mu(P_n)) = \begin{cases} 3, & for\ n = 2, \\ 2, & for\ n \geq 3. \end{cases}$$

An example of sigma coloring of Mycielski transformation of the path graph P_7 is shown in Fig. 2a.

Let G be a complete bipartite graph $K_{m,n}$. The sigma chromatic number of $K_{m,n}$ is

$$\sigma(K_{m,n}) = \begin{cases} 1, & if\ m \neq n, \\ 2, & if\ m = n. \end{cases}$$

The following theorem shows the sigma chromatic number of Mycielski transformation of a complete bipartite graph.

Theorem 2 *The sigma chromatic number of Mycielski transformation of complete bipartite graph is*

$$\sigma(\mu(K_{m,n})) = \begin{cases} 1, & when\ m \neq n,\ except\ for\ n = 2m-1, \\ & m = 2n-1\ and\ either\ n = 1\ or\ m = 1 \\ 2, & otherwise. \end{cases}$$

Proof Let $|V_1| = m$ and $|V_2| = n$ be the two partite vertex sets of $K_{m,n}$, and the degrees of the vertices in V_1 and V_2 are n and m, respectively. The Mycielski transformation of a complete bipartite graph has $2(m+n)+1$ vertices. The vertices of the subgraph $K_{m,n}$ of $\mu(K_{m,n})$ are labeled as u_i for $1 \leq i \leq |V_1|$ and v_j for $1 \leq j \leq |V_2|$, and the duplicate vertices of u_i and v_j are labeled as u'_i for $1 \leq i \leq |V_1|$ and v'_j for $1 \leq j \leq |V_2|$, and an additional vertex which is adjacent to each duplicate vertex is labeled as w. We prove the result by two cases.

Case 1 When $m \neq n$

The degrees of the vertices of the graph $\mu(K_{m,n})$ for any $1 \leq i \leq m$ and $1 \leq j \leq n$ are $d(u_i) = 2n$, $d(u'_i) = n+1$, $d(v_j) = 2m$, $d(v'_j) = m+1$ and $d(w) = m+n$. Since the degrees of the adjacent vertices are different, consider the vertex coloring of the graph $\mu(K_{m,n})$ to be $c : V(\mu(K_{m,n})) \rightarrow \{a\}$. In this case, $\sigma(x) = d(x)$ for any $x \in V(\mu(K_{m,n}))$. Now we have to prove that c is a sigma coloring of the graph $K_{m,n}$ for any m, n where $m \neq n$. Since $m \neq n$, $\sigma(u_i) \neq \sigma(v_j)$. Assume that $\sigma(w) \neq \sigma(u_i')$, that is, $m + n \neq n + 1$. This implies that $m \neq 1$. Similarly when considering $\sigma(w) \neq \sigma(v_j')$, we get $n \neq 1$. This infers that the coloring c does not satisfy the condition of sigma coloring if either $m = 1$ or $n = 1$. Now consider $\sigma(u_i) \neq \sigma(v_j')$, that is, $2n \neq m + 1$. We get $n \neq (m + 1)/2$. Similarly, while considering $\sigma(v_j) \neq \sigma(u_j')$, we get $m \neq \frac{n+1}{2}$. This implies that c satisfies the condition of sigma coloring if $n \neq \frac{m+1}{2}$ and $m \neq \frac{n+1}{2}$. Hence $\sigma(\mu(K_{m,n})) = 1$ if $m \neq n$ provided that $n \neq \frac{m+1}{2}$, $m \neq \frac{n+1}{2}$ and either $m \neq 1$ or $n \neq 1$.

Case 2 When $m = n$

When $m = n$, $d(u_i) = d(v_j)$, therefore $\sigma(\mu(K_{m,n})) \geq 2$. Let $c : V(\mu(K_{m,n})) \rightarrow \{a, b\}$ be the vertex coloring of the graph $\mu(K_{m,n})$ for any $m = n$ which are defined as follows: $c(u_i) = c(u_i') = c(w) = a$ and $c(v_i) = c(v_i') = b$. Then the color sums of the vertices are $\sigma(u_i) = 2nb, \sigma(v_i) = 2ma, \sigma(u_i') = a + nb, \sigma(v_i') = (m + 1)a, \sigma(w) = ma + nb$. Since $m = n$ and $a \neq b$, every two adjacent vertices has distinct color sum. Therefore $\sigma(\mu(K_{m,n})) = 2$ for any $m = n$.

Similar method of coloring as in Case 2 satisfies the sigma coloring of $\mu(K_{m,n})$ for $n = 2m - 1, m = 2n - 1$ and for either $m = 1$ or $n = 1$ when $m \neq n$.

Example 2 The sigma coloring of Mycielski transformation of the path graph P_7 and the complete bipartite graph $K_{3,3}$ is shown in Fig. 2.

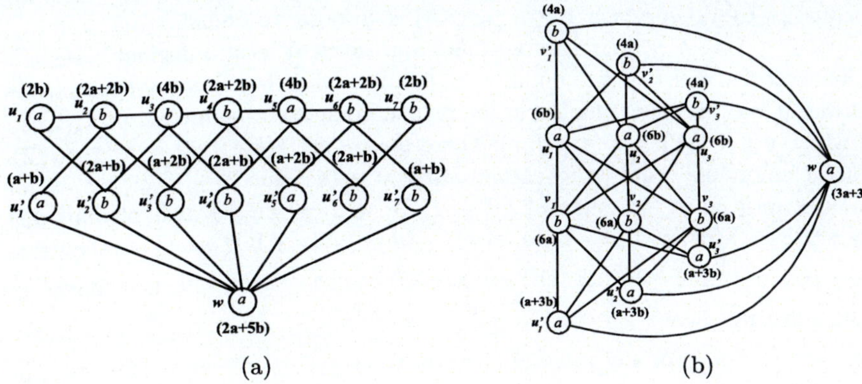

Fig. 2 **(a)** The sigma coloring of Mycielski transformation of P_7. **(b)** The sigma coloring of Mycielski transformation of $K_{3,3}$

2.2 Sigma Coloring of Mycielski Transformation of Complete Graph

For the integers $l \geq 2$ and $r = 2$, let $g = g(l, r) = \binom{l+r-1}{l-1}$ be the number of distinct r-element multisubsets using l colors. For any n, the $2n$ vertices $u_1, u_2, \ldots u_n, u_1^{'}, u_2^{'}, \ldots u_n^{'}$ of $\mu(K_n)$ are partitioned into n subsets of 2-element multisubset as $A_i = (u_i, u_i^{'})$ for $1 \leq i \leq n$ where u_i and $u_i^{'}$ are nonadjacent vertices. Let $A = \{a_1, a_2\}$ be the 2-element multiset of positive integer. Define $\sigma(A) = \sum_{i=1}^{2} a_i$. In [3], it has been proven that for any l, r there exists $g(l, r)$ number of distinct multisubset $A_1, A_2, \ldots, A_g$ for which $\sigma(A_i) \neq \sigma(A_j)$ for any i, j except for $i = j$.

In [3], it has been determined that $\sigma(K_n) = n$. The Mycielski transformation of complete graph K_n is denoted as $\mu(K_n)$. The graph $\mu(K_n)$ has $2n + 1$ vertices. Now the sigma chromatic number of Mycielski transformation of complete graph is determined in the following theorem.

Theorem 3 *For any positive integer $n \geq 3$, $\sigma(\mu(K_n)) = l$ for which $g(l-1, 2) < n \leq g(l, 2)$.*

Proof Let $G = \mu(K_n)$ be the Mycielski transformation of complete graph K_n. The vertices of the subgraph K_n in $\mu(K_n)$ are labeled as u_i for $1 \leq i \leq n$, and the duplicate vertices of u_i are labeled as u_i' for $1 \leq i \leq n$, and one additional vertex that is adjacent to all the duplicate vertices is labeled as w. Let $A_i = (u_i, u_i^{'})$ for $1 \leq i \leq n$ be the partite set of $\mu(K_n)$ where u_i and $u_i^{'}$ are nonadjacent. Since the vertex u_i is adjacent to each u_j and $u_j^{'}$ for $1 \leq i < j \leq n$, $\sigma(A_i)$ must not be equal to $\sigma(A_j)$ for each $1 \leq i < j \leq n$ in order to obtain $\sigma(u_i) \neq \sigma(u_k)$ for any $1 \leq i < k \leq n$. Therefore there must exist at least n distinct 2-element multisubset that satisfies the condition $\sigma(A_i) \neq \sigma(A_j)$. In [3], it has been proven that the sigma chromatic number of the regular complete k-partite graph $K_{k(r)}$ is l for which $g(l-1, r) < k \leq g(l, r)$. Therefore we get $g(l-1, 2) < n \leq g(l, 2)$. Hence l colors are required to obtain $\sigma(u_i) \neq \sigma(u_j)$ for $1 \leq i < j \leq n$ for which n must be in the range of $g(l-1, 2) < n \leq g(l, 2)$. It is obvious that $\sigma(u_i) \neq \sigma(u_j^{'})$ for any $i \neq j$ since $N[u_j'] - \{w\} \subset N[u_i]$. It is enough to show that $\sigma(w) \neq \sigma(u_j^{'})$ for $1 \leq i \leq n$. The vertex $c(w)$ can be colored with any color from l colors except when $\sum_{i=1}^{n} c(u_i) = \sum_{j=1}^{n} c(u_j^{'})$. If $\sum_{i=1}^{n} c(u_i) = \sum_{j=1}^{n} c(u_j^{'})$, we get $\sigma(w) = \sigma(u_j^{'})$ for any value of $c(w)$ for some j, $1 \leq j \leq n$. For this case, in order to obtain distinct color sum, interchange the color of one multisubset $(u_i, u_i^{'})$ for $1 \leq i \leq n$ provided that the interchanged vertex color $c(u_i)$ is not equal to $c(u_i')$. Thus $\sigma(\mu(K_n)) = l$ for which the value of n ranges between $g(l-1, 2) < n \leq g(l, 2)$.

Example 3 The sigma coloring of Mycielski transformation of complete graph $\mu(K_4)$ is shown in Fig. 3.

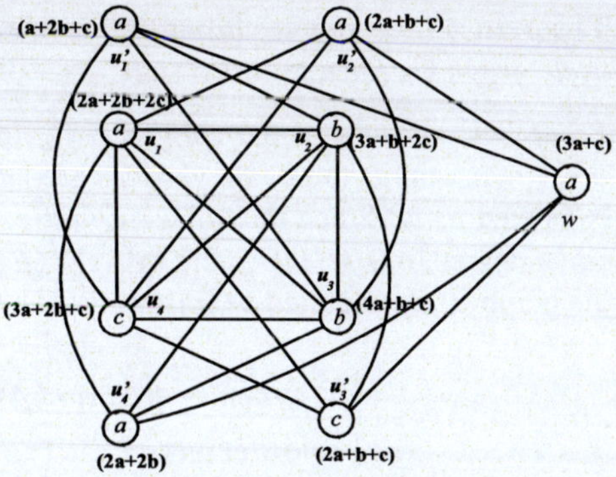

Fig. 3 The sigma coloring of Mycielski transformation of complete graph $\mu(K_4)$

3 Sigma Coloring of Mycielski Transformation of Cycle, Wheel, and Helm Graph

In this section, the sigma chromatic numbers of Mycielski transformation of cycle, wheel, and helm graph are determined. In [3], Chartrand et al. have proven that the sigma chromatic number of the cycle graph C_n for $n \geq 3$ is

$$\begin{cases} 2, & when\ n\ is\ even, \\ 3, & when\ n\ is\ odd. \end{cases}$$

In this result, the sigma coloring c and the color sum of the vertices of the graph C_n were determined as follows:

$$c(u_i(C_n)) = \begin{cases} a, & if\ i \equiv 1\ (mod\ 4), \\ d, & if\ i = n\ when\ n\ is\ odd, \\ b, & otherwise. \end{cases}$$

and

$$\sigma(u_i(C_n)) = \begin{cases} a+b, & for\ i\ is\ even\ but\ except\ for\ i = n-1\ when\ n\ is\ odd\ and \\ & \quad except\ for\ i = n\ when\ n \equiv 2 (mod\ 4) \\ 2b, & for\ i\ is\ odd\ except\ for\ i = 1\ when\ n\ is\ odd, \\ b+c, & for\ i = 1,\ n-1,\ when\ n\ is\ odd, \\ 2a, & for\ i = n\ when\ n \equiv 2 (mod 4). \end{cases} \quad (1)$$

The following theorem determines the sigma chromatic number of Mycielski transformation of the cycle graph C_n.

Theorem 4 *The sigma chromatic number of the Mycielski transformation of the cycle graph C_n for $n \geq 4$ is $\sigma(\mu(C_n)) = 2$.*

Proof The vertices of the subgraph C_n in $\mu(C_n)$ for $n \geq 4$ are labeled as u_i for $1 \leq i \leq n$, and its duplicate vertices are labeled as u'_i, and the one additional vertex is labeled as w. Since the adjacent vertices u_i and u_{i+1} for $1 \leq i \leq n-1$ are of equal degree, $\sigma(\mu(C_n)) \geq 2$. Let $c : V(\mu(C_n)) \to \{a, b\}$ be the vertex coloring of the graph $\mu(C_n)$ defined as follows:

$$c(u_i(\mu(C_n))) = \begin{cases} a, \ when\ i\ \equiv 1(mod\ 4) \\ b, \ otherwise \end{cases} \tag{2}$$

$$c(u'_i(\mu(C_n))) = \begin{cases} a, \ when\ i \equiv 1(mod\ 4),\ except\ for\ i = n\ when\ n\ is\ odd, \\ b, \ when\ i \not\equiv 1(mod\ 4)\ except\ for\ i = n\ when\ n\ is\ odd. \end{cases} \tag{3}$$

$$c(u'_n(\mu(C_n))) = \begin{cases} a, \ if\ c(u_n) = b\ when\ n\ is\ odd, \\ b, \ if\ c(u_n) = a\ when\ n\ is\ odd. \end{cases} \tag{4}$$

$$c(w(\mu(C_n))) = a \tag{5}$$

We now show that c is a sigma coloring of $\mu(C_n)$. The color sums of the vertices are as follows: When n is even, $c(u_i(\mu(C_n)) = c(u'_i(\mu(C_n))) = c(u_i(C_n))$. Therefore $\sigma(u_i(\mu(C_n))) = 2(\sigma(u_i(C_n)))$, $\sigma(u'_i(C_n)) = \sigma(u_i(C_n)) + c(w) = \sigma(u_i(C_n)) + a$ and $\sigma(w) = \lceil \frac{n}{4} \rceil a + (n - \lceil \frac{n}{4} \rceil)b$. It is clear that no two adjacent vertices have same color sum. When n is odd, the color sums of the vertices are as follows:

$$\sigma(u_i) = \begin{cases} a + 3b, \ for\ i = 1\ and\ i = n-1\ when\ n \equiv 1(mod\ 4) \\ 2a + 2b, \ when\ i\ is\ even,\ i = n\ and\ i \neq n-1 \\ 4b, \quad when\ i\ is\ odd,\ for\ 3 \leq i \leq n-2 \\ 3a + b, \ for\ i = n-1\ when\ n \equiv 3(mod\ 4) \end{cases} \tag{6}$$

$$\sigma(u'_i) = \begin{cases} a + 2b, \ when\ i\ is\ odd\ and\ for \\ \qquad i \neq 1, n\ when\ i \equiv 1(mod\ 4) \\ \qquad i \neq n\ when\ i \equiv 3(mod\ 4) \\ 2a + b, \ when\ i\ is\ even\ and\ for \\ \qquad i = 1, n\ when\ i \equiv 1(mod\ 4) \\ \qquad i = n\ when\ i \equiv 3(mod\ 4) \end{cases} \tag{7}$$

$$\sigma(w) = \begin{cases} \lfloor \frac{n}{4} \rfloor a + (n - \lfloor \frac{n}{4} \rfloor)b \ \ when\ n \equiv 1(mod\ 4) \\ (\lceil \frac{n}{4} \rceil + 1)a + (n - (\lceil \frac{n}{4} \rceil + 1)) \ \ when\ n \equiv 3(mod\ 4) \end{cases} \tag{8}$$

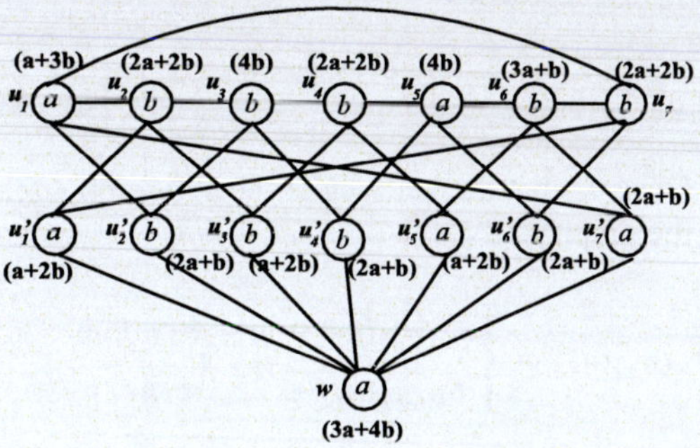

Fig. 4 Sigma coloring of Mycielski transformation of cycle graph C_7

It is observed that the obtained color sums of u_i and $u_i^{'}$ are distinct. In the case of vertex w, $d(w) > d(u_i^{'})$ for $n \geq 4$. Hence $\sigma(w) \neq \sigma(u_i^{'})$ for any $n \geq 4$. Therefore c is a sigma coloring of $\mu(C_n)$ and hence $\sigma(\mu(C_n)) = 2$.

Example 4 The sigma coloring of Mycielski transformation of cycle graph $\mu(C_7)$ is shown in Fig. 4.

The wheel graph W_n is the graph obtained by the join of C_n and K_1. It has been proved in [10] that

$$\sigma(W_n) = \begin{cases} 4, if n = 3 \\ 3, if n \geq 4 \text{ } where \text{ } n \text{ } is \text{ } odd \\ 2, if n \geq 4 \text{ } where \text{ } n \text{ } is \text{ } even. \end{cases}$$

The following theorem shows the sigma chromatic number of $\mu(W_n)$.

Theorem 5 *Let W_n be a wheel graph of order $n \geq 4$. Then $\sigma(\mu(W_n)) = 2$.*

Proof A subgraph of the Mycielski transformation of W_n is the Mycielski transformation of C_n since the cycle graph C_n is a subgraph of the wheel graph W_n. The vertex set of $\mu(W_n)$ is $V(\mu(W_n)) = V(\mu(C_n)) \cup \{v, v'\}$ and the edge set of $\mu(W_n)$ is $E(\mu(W_n)) = E(\mu(C_n)) \cup \{vu_i, 1 \leq i \leq n\} \cup \{v^{'}u_i, 1 \leq i \leq n\} \cup \{wv^{'}\} \cup \{u_i^{'}v, 1 \leq i \leq n\}$. Let c_1 be the vertex coloring of the graph $\mu(W_n)$. Since $\mu(C_n) \subset \mu(W_n)$, the vertex coloring of $\mu(W_n)$ is $c_1(u_i(\mu(W_n))) = c(u_i(\mu(C_n)))$, $c_1(u_i'(\mu(W_n))) = c(u_i'(\mu(C_n)))$, $c_1(v(\mu(W_n))) = b$,

$$c_1(v'(\mu(W_n))) = \begin{cases} a, \text{ } when \text{ } n \equiv 3 (mod \text{ } 4) \\ b, \text{ } otherwise \end{cases}$$

and

$$c_1(w(\mu(W_n))) = \begin{cases} b, & if\ n \equiv 3(mod\ 4), \\ a, & otherwise. \end{cases}$$

Observe that the vertices of the subgraph $\mu(C_n)$ in $\mu(W_n)$ are colored as defined for the graph $\mu(C_n)$.

$$\sigma_{\mu(W_n)}(u_i) = \begin{cases} \sigma_{\mu(C_n)}(u_i) + a + b, \ when\ n \equiv 3(mod\ 4), \\ \sigma_{\mu(C_n)}(u_i) + 2b, \ \ otherwise \end{cases} \tag{9}$$

$$\sigma_{\mu(W_n)}(u_i') = \sigma_{\mu(C_n)}(u_i') + b \tag{10}$$

$$\sigma_{\mu(W_n)}(w) = \begin{cases} \sigma_{\mu(C_n)}(w) + a, \ when\ n \equiv 3(mod\ 4) \\ \sigma_{\mu(C_n)}(w) + b, \ \ otherwise \end{cases} \tag{11}$$

$$= \begin{cases} \lceil \frac{n}{4} \rceil a + (n - \lceil \frac{n}{4} \rceil + 1)b, \ \ when\ n\ is\ even, \\ \lfloor \frac{n}{4} \rfloor a + (n - \lfloor \frac{n}{4} \rfloor + 1)b, \ \ when\ n \equiv 1(mod\ 4), \\ (\lceil \frac{n}{4} \rceil + 2)a + (n - (\lceil \frac{n}{4} \rceil + 1))b, \ \ when\ n \equiv \ 3(mod\ 4). \end{cases} \tag{12}$$

$$\sigma_{\mu(W_n)}(v) = \begin{cases} 2(\lceil \frac{n}{4} \rceil a + (n - \lceil \frac{n}{4} \rceil b), \ \ when\ n\ is\ even, \\ (2(\lceil \frac{n}{4} \rceil) - 1)a + 2nb - (2\lceil \frac{n}{4} \rceil - 1)b, \ \ when\ n \equiv 1(mod\ 4), \\ (2(\lceil \frac{n}{4} \rceil) + 1)a + 2nb - (2\lceil \frac{n}{4} \rceil + 1)b, \ \ when\ n \equiv 3(mod\ 4), \end{cases} \tag{13}$$

$$\sigma_{\mu(W_n)}(v') = \begin{cases} \lceil \frac{n}{4} \rceil a + (n - \lceil \frac{n}{4} \rceil + 1)b, \ \ when\ n \equiv 3(mod\ 4), \\ (\lceil \frac{n}{4} \rceil + 1)a + (n - \lceil \frac{n}{4} \rceil)b, \ \ otherwise \end{cases} \tag{14}$$

Now we need to show that the above obtained color sums of any two adjacent vertices are distinct. Theorem 4 proves that the color sum of every vertex of $\mu(C_n)$ are distinct. Hence the color sums of the vertex $u_i'(\mu(W_n))$ and $w(\mu(W_n))$ are distinct and $\sigma_\mu(W_n)(u_i') \neq \sigma_\mu(W_n)(u_i)$. Observe from (9) that each vertex u_i is incremented with same color, therefore $\sigma(u_i) \neq \sigma(u_{i+1}))$ for $1 \leq i \leq n-1$ and $\sigma(u_1) \neq \sigma(u_n)$, and since $\sigma(v) = \sum_{i=1}^{n} c(u_i) + \sum_{i=1}^{n} c(u_i')$, it is obvious that $\sigma(v) \neq \sigma(u_i)$ and $\sigma(v) \neq \sigma(u_i')$. From (12) and (14), it can be observed that $\sigma(v) \neq \sigma(w)$. Since every two adjacent vertices has distinct color sum, c_1 is the sigma coloring of $\mu(W_n)$ and hence $\sigma(\mu(W_n)) = 2$.

The wheel graph W_n is the subgraph of H_n. A helm graph, denoted by H_n is a graph obtained by joining a pendent vertex to each vertex of C_n in the wheel graph W_n. In [10], it has been proven that $\sigma(H_n) = 2$. The following theorem shows the sigma chromatic number of $\mu(H_n)$.

Theorem 6 *If H_n is a helm graph of order $n \geq 4$, then $\sigma(\mu(H_n)) = 2$.*

Proof The wheel graph W_n is a subgraph of helm graph H_n. Therefore the Mycielski transformation of wheel graph is a subgraph of Mycielski transformation of helm graph. The vertex set of $\mu(H_n)$ is $V(\mu(H_n)) = V(\mu(W_n)) \cup \{x_i, x_i^{'}\}$, and the edge set of $\mu(H_n)$ is $E(\mu(H_n)) = E(\mu(W_n)) \cup \{u_i x_i, 1 \leq i \leq n\} \cup \{x_i^{'} u_i, 1 \leq i \leq n\} \cup \{u_i^{'} x_i\} \cup \{w x_i^{'}\}$. The vertices of the subgraph $\mu(W_n)$ in the graph $\mu(H_n)$ are colored as in Theorem 5. The remaining vertices x_i and $x_i^{'}$ for $1 \leq i \leq n$ are colored with the color a, that is, $c(x_i) = c(x_i^{'}) = a$. Hence the color sums of the vertices of the graph are as follows: $\sigma_{\mu(H_n)}(w) = \sigma_{\mu(W_n)}(w) + na$, $\sigma_{\mu(H_n)}(u_i^{'}) = \sigma_{\mu(W_n)}(u_i^{'}) + a$, $\sigma_{\mu(H_n)}(u_i) = \sigma_{\mu(W_n)}(u_i) + 2a$, $\sigma_{\mu(H_n)}(v) = \sigma_{\mu(W_n)}(v)$, $\sigma_{\mu(H_n)}(v') = \sigma_{\mu(W_n)}(v')$. It can be observed that no two adjacent vertices have same color sum. Therefore $\sigma(\mu(H_n)) = 2$.

4 Conclusion

In this article, the sigma chromatic numbers of Mycielski transformation of path graph, complete bipartite graph, complete graph, cycle graph, wheel graph, and helm graph were determined. It is interesting to note that the graphs considered in this article have a special characteristic that $\sigma(\mu(G)) \leq \sigma(G)$. As a future work, one can verify whether $\sigma(\mu(G)) \leq \sigma(G)$ for any graph G and also verify whether $\sigma(\mu(G)) = \sigma(G)$ whenever $\sigma(G) = 2$.

References

1. Bullay-og, M.C.A., Garciano, A.D., Marcelo, R.M.: On the sigma value and sigma range of the join of a finite number of even cycles of the same order. J. Phys.: Conf. Series **1836**, 012015 (2021)
2. Bondy, J.A., Murty, U.S.R.: Graph Theory with Applications, vol. 290. Macmillan, London (1976)
3. Chartrand, G., Okamoto, F., Zhang, P.: The sigma chromatic number of a graph. Graphs Comb. **26**, 755–733 (2010)
4. Garciano, A., Lagura, M., Marcelo, R.: On the sigma chromatic number of the join of a finite number of paths and cycles. Asian-Eur. J. Math. **14**(2), 2150019 (2019)
5. Garciano, A.D., Lagura, M., Marcelo, R.: Sigma chromatic number of the graph coronas involving complete graphs. J. Phys.: Conf. Series **1538**, 012003 (2020)
6. Garciano, A.D., Lagura, M.C.T., Marcelo, R.M.: Join of paths and cycles of equal order with sigma chromatic number 2. Asian-Eur. J. Math. **15**(1), 2250006 (2022)
7. Lagura, M.C.: On the sigma chromatic number of the join of paths, cycles, and complete multipartite graphs Master's Thesis (Ateneo de Manila University), 9 (2016)
8. Manamtam, J.A., Garciano, A.D., Tolentino, M.A.C.: Sigma chromatic number of the middle graph of some families of graphs. J. Phys.: Conf. Series **2157**, 012001(2022)
9. Mycielski, J.: Sur le coloriage des graphes. Colloq. Math. **3**(2), 161–162 (1955)
10. Suresh Kumar, J., Pillai, P.K.: Sigma coloring of some cycle related graphs. J. Emer. Technol. Innovative Res. **7**, 284–289 (2022)

Bipartite Decomposition of Graphs Using Chromatic Number

M. Yamuna and K. Karthika

1 Introduction

A general task in mathematics is to decompose complex object into smaller objects retaining back a certain structure. Many problems in graph theory like chromatic partition and domatic partition can be phrased as graph decomposition problems. Many researchers have been interested in these kinds of problems. A decomposition of a graph G = (V, E) aims to partition the given graph G = (V, E) into a collection of subgraphs $\Im_1, \Im_2, \ldots, \Im_k$, such that each edge $e \in E(G)$ belongs to exactly one subgraph $\Im_i$, $1 \leq i \leq k$. A set X contained in V (G) is said to be an independent set if no two vertices in X are adjacent. Minimum number of independent sets that includes all the vertices of G is called a chromatic partition of G [1]. Throughout this chapter, we consider only simple, finite, and connected graphs with n vertices. For interesting results on graph decomposition, one may refer to [2–6].

Graph coloring has the unique property of partitioning a vertex set into color classes. This is defined as a chromatic partition in the literature of graph theory. The chromatic partition, partitions V (G) into independent sets. We also know that the cardinality of the chromatic partition equals to the chromatic number of G. Each set in the chromatic partition equals to the chromatic number of G. Each set in the chromatic partition is an independent set. In this chapter, we target to determine the decomposition of G into bipartite graphs. We know that a bipartite graph partitions V(G) into two sets V_1 and V_2 such that each V_i is independent, where i = 1, 2. We focused to use the independence property of a chromatic partition to determine decomposition into bipartite graphs. This is possible, since any given graph has at least one chromatic partition.

M. Yamuna · K. Karthika (✉)
SAS, VIT, Vellore, Tamil Nadu, India
e-mail: myamuna@vit.ac.in; karthika.k@vit.ac.in

F. Kamalov et al. (eds.), *Advances in Mathematical Modeling and Scientific Computing*, Trends in Mathematics, https://doi.org/10.1007/978-3-031-41420-6_64

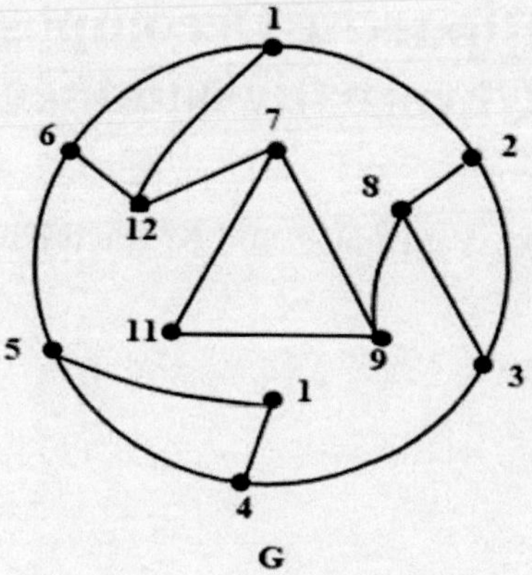

Fig. 1 Graph for Sect. 2 discussion

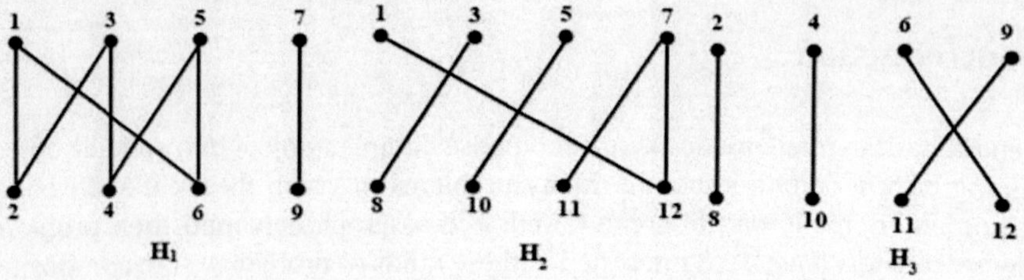

Fig. 2 Construction of bipartite graphs

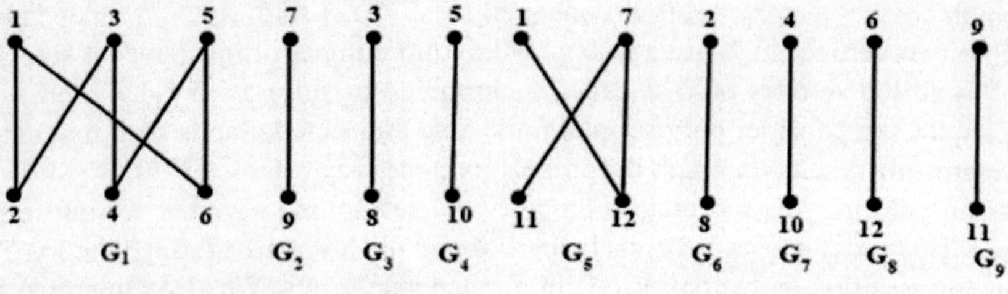

Fig. 3 Graph decomposition

As an illustration $P = \{X_1, X_2, X_3\} = \{\{1, 3, 5, 7\}, \{2, 4, 6, 9\}, \{8, 10, 11, 12\}\}$ is a chromatic partition for G in Fig. 1. We need to determine bipartite graphs including all edges in G. Since each X_i is independent, edges in G are between (X_i, X_j), $i \neq j$. We shall determine all possible subsets of size two for P, namely, $\{\{X_1, X_2\}, \{X_1, X_3\}, \{X_2, X_3\}\}$. We shall now construct bipartite graphs for each of the subsets of size two. The resulting graphs are seen in Fig. 2.

Note that each graph is a bipartite graph. We also note that for our example, all three graphs are disconnected graphs. Since $\langle X_i \rangle$ is independent all components are bipartite graphs. The given graph G is decomposed into nine bipartite graphs $G_1, G_2, \ldots G_9$ (Fig. 3).

2 Results and Discussion

In this section, we aim to develop an algorithm for decomposing a graph G into bipartite graphs. For this purpose, let us develop two algorithms: one for determining independent sets of a graph G and another for vertex partition for G. Each algorithm shall be supported with examples. Based on this algorithm, we shall develop a MATLAB code for determining the decomposition.

2.1 Algorithm for Independent Sets of G

"A dominating set for a graph G is a set D that is adjacent to every vertex in V – D. Let G be any graph with n – vertices. Let A denote the adjacency matrix of G. Let N denote a n × n matrix, where

$$N = [n_{ij}]n \times n = \begin{cases} 1 & \text{if } i = j \\ a^{ij} & \text{the } (i, j)^{th} \text{ entry in the adjacency matrix.} \end{cases}$$

Let $x = \langle x(v_1), x(v_2), \ldots, x(v_n) \rangle^T$ be a {0, 1} vector. If x represents any dominating set, then $Nx \geq 1$, that is in the resulting matrix Nx, all the entry values are nonzero (Fig. 4).

We use this technique for determining an independent set. In matrix Ax, if for every nonzero entry in x the corresponding entry in Ax is 0, then x represents an independent set. This is true, since the vertices in x do not dominate each other, which means that they are not adjacent to each other. In other words, the induced subgraph of the vertices in x is a null graph.

Example
Consider {1, 3, 6} is one of the independent set for G in Fig. 5 and its corresponding vector x = ⟨ **1**, 0, **1**, 0, 0, **1** ⟩. Note that every nonzero entry in x, the corresponding entry in Ax = ⟨ **0**, 3, **0**, 1, 1, **0**, 2 ⟩ is 0, implies x represents an independent set of G.

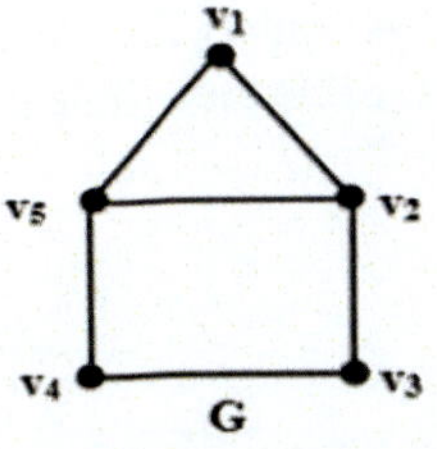

$$\underset{N}{\begin{bmatrix} 1 & 1 & 0 & 0 & 1 \\ 1 & 1 & 1 & 0 & 1 \\ 0 & 1 & 1 & 1 & 0 \\ 0 & 0 & 1 & 1 & 1 \\ 1 & 1 & 0 & 1 & 1 \end{bmatrix}} \underset{x}{\begin{bmatrix} 1 \\ 0 \\ 0 \\ 1 \\ 0 \end{bmatrix}} = \underset{Nx}{\begin{bmatrix} 1 \\ 1 \\ 1 \\ 1 \\ 2 \end{bmatrix}} \geq \begin{bmatrix} 1 \\ 1 \\ 1 \\ 1 \\ 1 \end{bmatrix};$$

Fig. 4 $\{v_1, v_4\}$ is a dominating set for G [7]"

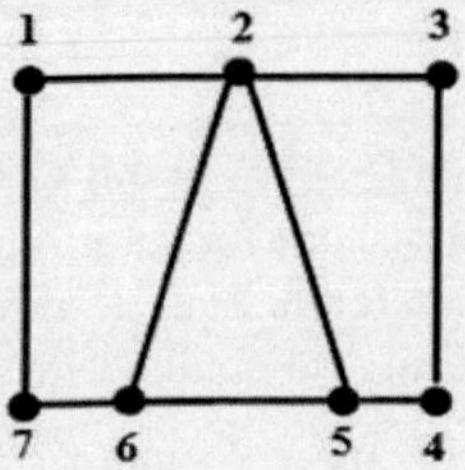

G

$$A=\begin{bmatrix}0&1&0&0&0&0&1\\1&0&1&0&1&1&0\\0&1&0&1&0&0&0\\0&0&1&0&1&0&0\\0&1&0&1&0&1&0\\0&1&0&0&1&0&1\\1&0&0&0&0&1&0\end{bmatrix};\ x=\begin{bmatrix}1\\0\\1\\0\\0\\1\\0\end{bmatrix}\ \text{and}\ Ax=\begin{bmatrix}0\\3\\0\\1\\1\\0\\2\end{bmatrix}.$$

Fig. 5 Graph G, adjacency matrix A and the product matrix Ax

This matrix method of finding a dominating set can be extended to determine all possible independent sets of a graph G.

1. "Let G be a graph with n – vertices $\{v_1, v_2, \ldots, v_n\}$. Consider all possible subsets of size k, $1 \le k \le n$. Label them as $S_1, S_2, \ldots, S_p$, where, $p = \rho(n) - 1$, where $\rho(n)$ denotes a power set of all vertices. Let $x = \{x_1, x_2, \ldots, x_p\}$ be a set of {0, 1} vectors defined by $x_i = \langle x(v_1), x(v_2), \ldots, x(v_n) \rangle^T$, where

$$x\ (v_i) = \begin{cases} 1, & if\ v_i \in S_i, \\ 0, & otherwise. \end{cases}$$

Following the above notation if k = 3, then $S_1 = \{v_1\}$, $S_2 = \{v_2\}$, $S_3 = \{v_3\}$, $S_4 = \{v_1, v_2\}$, $S_5 = \{v_1, v_3\}$, $S_6 = \{v_2, v_3\}$, $S_7 = \{v_1, v_2, v_3\}$. $x_1 = \langle 1, 0, 0 \rangle^T$, $x_2 = \langle 0, 1, 0 \rangle^T$, $x_3 = \langle 0, 0, 1 \rangle^T$, $x_4 = \langle 1, 1, 0 \rangle^T$, $x_5 = \langle 1, 0, 1 \rangle^T$, $x_6 = \langle 0, 1, 1 \rangle^T$, $x_7 = \langle 1, 1, 1 \rangle^T$".

Let $X = [x_1\ x_2\ \ldots\ x_p]$, that is X is a matrix of vectors where each column represent vector x_i.

2. Determine Ax. Ax is a $n \times p$ matrix. Let us denote each column of Ax as ax_i, $1 \le i \le p$.
3. Consider the ith column of Ax and the ith vector x_i. We know that "$ax_i = \langle ax_i(v_1), ax_i(v_2), \ldots, ax_i(v_n) \rangle^T$ "and "$x_i = \langle x(v_1), x(v_2), \ldots, x(v_n) \rangle^T$". If for every nonzero entry in vector x_i, the corresponding entry in vector ax_i is 0, then x_i is an independent set with respect to G.
4. Let $F \subseteq X$ be the set of all vectors such that x_i and ax_i satisfy Step – 2. Note that F is a submatrix of X which is a collection of all possible independent sets of G. Let $|F| = k$, $k < p$.

Example

$$AX = \begin{bmatrix} 0 & 1 & 1 & 1 \\ 1 & 0 & 1 & 0 \\ 1 & 1 & 0 & 1 \\ 1 & 0 & 1 & 0 \end{bmatrix} \begin{bmatrix} 0&0&0&1&0&0&0&1&1&1&0&1&1&1&1 \\ 0&0&1&0&0&1&1&0&0&1&1&0&1&1&1 \\ 0&1&0&0&1&0&1&0&1&0&1&1&0&1&1 \\ 1&0&0&0&1&1&0&1&0&0&1&1&1&0&1 \end{bmatrix}$$

$$= \begin{bmatrix} 1&1&1&0&2&2&2&1&1&1&3&2&2&2&3 \\ 0&1&0&1&1&0&1&1&2&1&1&2&1&2&2 \\ 1&0&1&1&1&2&1&2&1&2&2&2&3&3&3 \\ 0&1&0&1&1&0&1&1&2&1&1&2&1&2&2 \end{bmatrix}$$

$$F = \begin{bmatrix} 0&0&0&1&0 \\ 0&0&1&0&1 \\ 0&1&0&0&0 \\ 1&0&0&0&1 \end{bmatrix}$$

$F = \{x_1, x_2, x_3, x_4, x_6\}$ are independent sets of G, since the nonzero entry in F and the corresponding entry in AX is 0.

2.2 *Algorithm for Vertex Partition of G*

From Algorithm – 1, we have determined all possible independent sets for any graph G. We now shall continue to determine a partition for G. Consider the independent sets generated from Algorithm – 1.

Example
For the graph G in Fig. 6, let us choose the vectors $F_1 = \{4\}$, $F_4 = \{1\}$. Add the corresponding vectors f_1, f_4 in matrix F. $f_1 + f_4 = [1\ 0\ 0\ 1]^T$. We observe that $F_1 \cup F_4$ is not a partition for V (G).

Consider the vectors $F_1 = \{4\}$, $F_2 = \{3\}$, $F_4 = \{1\}$, $F_5 = \{2, 4\}$. Add the corresponding vectors f_1, f_2, f_4, and f_5 in matrix F. $f_1 + f_2 + f_4 + f_5 = [1\ 1\ 1\ 2]^T$. We note that $\{F_1, F_2, F_4, F_5\}$ is not a partition for V (G).

Consider the vectors $F_2 = \{3\}$, $F_4 = \{1\}$, $F_5 = \{2, 4\}$. Add the corresponding vectors f_2, f_4 and f_5 in matrix F. $f_2 + f_4 + f_5 = [1\ 1\ 1\ 1]^T$. We note that $\{F_2, F_4, F_5\}$

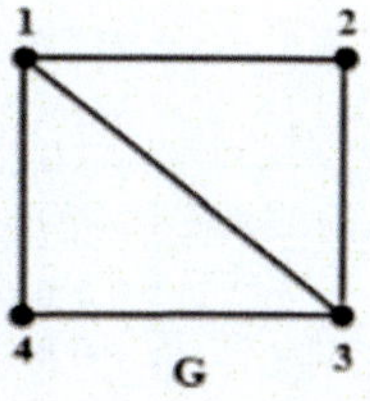

Fig. 6 Graph for Sect. 2.1 discussion

forms a partition for V (G), since $F_2 \cup F_4 \cup F_5 = V(G)$ and $F_2 \cap F_4 \cap F_5 = \phi$. We use this observation for developing an algorithm to determine the partition for G.

Consider a matrix F. let $F = [f_1\ f_2\ \ldots\ f_k]$.

1. Determine the power set of F, say $\rho(F) = \{H_1, H_2, \ldots, H_q\}$, where $q = 2\ k - 1$ (leaving the empty set). Note that each H_i represents set of column vector of matrix F.
2. Add the vectors in each H_i to generate a set of vectors $Y = \{y_1, y_2, \ldots, y_q\}$. Note that each y_i is a vector of order $n \times 1$.
3. y_i is $\begin{cases} a\ partition\ for\ G & if\ every\ entry\ of\ y_i\ is\ 1. \\ not\ a\ partition\ for\ G & otherwise. \end{cases}$

Example

For the graph in Fig. 6, $\rho(F) = \{H_1, H_2, \ldots, H_{32}\}$, where

$$H_1 = \begin{bmatrix} 0 \\ 0 \\ 0 \\ 1 \end{bmatrix}; H_2 = \begin{bmatrix} 0 \\ 0 \\ 1 \\ 0 \end{bmatrix}; H_3 = \begin{bmatrix} 0 \\ 1 \\ 0 \\ 0 \end{bmatrix}; H_4 = \begin{bmatrix} 1 \\ 0 \\ 0 \\ 0 \end{bmatrix}; H_5 = \begin{bmatrix} 0 \\ 1 \\ 0 \\ 1 \end{bmatrix};$$

$$H_7 = \begin{bmatrix} 0\ 0 \\ 0\ 0 \\ 0\ 1 \\ 1\ 0 \end{bmatrix}; H_8 = \begin{bmatrix} 0\ 1 \\ 0\ 0 \\ 0\ 0 \\ 1\ 0 \end{bmatrix}; H_9 = \begin{bmatrix} 0\ 0 \\ 0\ 1 \\ 0\ 0 \\ 1\ 1 \end{bmatrix};$$

$$H_{10} = \begin{bmatrix} 0\ 0 \\ 0\ 1 \\ 1\ 0 \\ 0\ 1 \end{bmatrix}; H_{11} = \begin{bmatrix} 0\ 1 \\ 0\ 0 \\ 1\ 0 \\ 0\ 0 \end{bmatrix}; H_{12} = \begin{bmatrix} 0\ 0 \\ 0\ 1 \\ 1\ 0 \\ 0\ 1 \end{bmatrix}; H_{13} = \begin{bmatrix} 0\ 1 \\ 1\ 0 \\ 0\ 0 \\ 0\ 0 \end{bmatrix}; H_{14} = \begin{bmatrix} 0\ 0 \\ 1\ 1 \\ 0\ 0 \\ 0\ 1 \end{bmatrix};$$

$$H_{14} = \begin{bmatrix} 0\ 0 \\ 1\ 1 \\ 0\ 0 \\ 0\ 0 \end{bmatrix}; H_{15} = \begin{bmatrix} 1\ 0 \\ 0\ 1 \\ 0\ 0 \\ 0\ 1 \end{bmatrix};$$

$$H_{16} = \begin{bmatrix} 0\ 0\ 0 \\ 0\ 0\ 1 \\ 0\ 1\ 0 \\ 1\ 0\ 0 \end{bmatrix}; H_{17} = \begin{bmatrix} 0\ 0\ 1 \\ 0\ 0\ 0 \\ 0\ 1\ 0 \\ 1\ 0\ 0 \end{bmatrix}; H_{18} = \begin{bmatrix} 0\ 0\ 0 \\ 0\ 0\ 1 \\ 0\ 1\ 0 \\ 1\ 0\ 1 \end{bmatrix}; H_{19} = \begin{bmatrix} 0\ 0\ 1 \\ 0\ 1\ 0 \\ 0\ 0\ 0 \\ 1\ 0\ 0 \end{bmatrix};$$

$$H_{20} = \begin{bmatrix} 0\ 0\ 0 \\ 0\ 1\ 1 \\ 0\ 0\ 0 \\ 1\ 0\ 1 \end{bmatrix}; H_{21} = \begin{bmatrix} 0\ 1\ 0 \\ 0\ 0\ 1 \\ 0\ 0\ 0 \\ 1\ 0\ 1 \end{bmatrix}; H_{22} = \begin{bmatrix} 0\ 0\ 1 \\ 0\ 1\ 0 \\ 1\ 0\ 0 \\ 0\ 0\ 0 \end{bmatrix};$$

$$H_{23} = \begin{bmatrix} 0\ 0\ 0 \\ 0\ 1\ 1 \\ 1\ 0\ 0 \\ 0\ 0\ 1 \end{bmatrix}; H_{24} = \begin{bmatrix} 0\ 1\ 0 \\ 0\ 0\ 1 \\ 1\ 0\ 0 \\ 0\ 0\ 1 \end{bmatrix}; H_{25} = \begin{bmatrix} 0\ 1\ 0 \\ 1\ 0\ 1 \\ 0\ 0\ 0 \\ 0\ 0\ 1 \end{bmatrix}; H_{26} = \begin{bmatrix} 0\ 0\ 0\ 1 \\ 0\ 0\ 1\ 0 \\ 0\ 1\ 0\ 0 \\ 1\ 0\ 0\ 0 \end{bmatrix};$$

$$H_{27} = \begin{bmatrix} 0\ 0\ 0\ 0 \\ 0\ 0\ 1\ 1 \\ 0\ 1\ 0\ 0 \\ 1\ 0\ 0\ 1 \end{bmatrix}; H_{28} = \begin{bmatrix} 0\ 0\ 1\ 0 \\ 0\ 0\ 0\ 1 \\ 0\ 1\ 0\ 0 \\ 1\ 0\ 0\ 1 \end{bmatrix}; H_{29} = \begin{bmatrix} 0\ 0\ 1\ 0 \\ 0\ 1\ 0\ 1 \\ 0\ 0\ 0\ 0 \\ 1\ 0\ 0\ 1 \end{bmatrix}; H_{30} = \begin{bmatrix} 0\ 0\ 1\ 0 \\ 0\ 1\ 0\ 1 \\ 1\ 0\ 0\ 0 \\ 0\ 0\ 0\ 1 \end{bmatrix}; H_{31} = F$$

All possible partitions of G are $Y = \{y_1, y_2\}$, where

$$y_1 = \begin{bmatrix} 0\,1\,0 \\ 0\,0\,1 \\ 1\,0\,0 \\ 0\,0\,1 \end{bmatrix};\quad y_2 = \begin{bmatrix} 0\,0\,0\,1 \\ 0\,0\,1\,0 \\ 0\,1\,0\,0 \\ 1\,0\,0\,0 \end{bmatrix}$$

2.3 *Algorithm for Graph Decomposition of G*

1. Let $Z = \{z_1, z_2, \ldots, z_t\} \subseteq Y$ be a collection of vectors such that each z_i is a partition for G, $1 \le i \le t$, $t < q$.
2. Choose any random z_i from Z and the corresponding sets from ρ (F). Let z_i be the union of b sets in ρ (F), say $W = \{W_1, W_2, \ldots, W_b\}$. Note that $W \subseteq \rho$ (F).
3. Determine the bC_2 subsets of W. Let $B_{ij} = \{W_i, W_j\}$ $i < j$.
4. Construct graphs B_{ij}, where
   (i) $V(B_{ij}) = V(W_i) \cup V(W_j)$.
   (ii) $E(B_{ij})$ = All possible edges between $V(W_i)$ and $V(W_j)$ from G.
5. Note that
   (i) Each B_{ij} is a bipartite graph, since $\langle W_i \rangle \cup \langle W_j \rangle$ is a null graph.
   (ii) $\langle B_{ij} \rangle$ is either a connected or a disconnected graph (may be a null graph).

   Hence, each $\langle B_{ij} \rangle$ is the union of bipartite graphs.
6. $B = \bigcup_{\substack{i \neq j \\ i < j}} B_{ij}$, *where* $1 \le i \le b$ *and* $1 \le j \le b$ is a decomposition for G such that each B_{ij} is a bipartite graph, $1 \le B_{ij} \le m$.

Example
For the graph in Fig. 6, $Z = \{z_1. z_2\}$, where $z_1 = y_1$ and $z_2 = y_2$. Choose $z_1 = \{W_1, W_2, W_3\}$, $W_1 = [0\,0\,1\,0]^T$, $W_2 = [1\,0\,0\,0]^T$, $W_3 = [0\,1\,0\,1]^T$. The $3C_2$ subsets of W are $B_{12} = \{W_1, W_2\}$, $B_{13} = \{W_1, W_3\}$ and $B_{23} = \{W_2, W_3\}$. The graphs of B_{12}, B_{13} and B_{23} are seen in Fig. 6.

3 MATLAB Program for Graph Decomposition

Based on the discussion in Sect. 2, we determined a MATLAB code for determining the graph decomposition of any graph G. The output for the matrix representation for graph decomposition of Fig. 6 is as seen in Snapshot 1 and graphs in Fig. 7 are same as the graphs in the Snapshot 2.

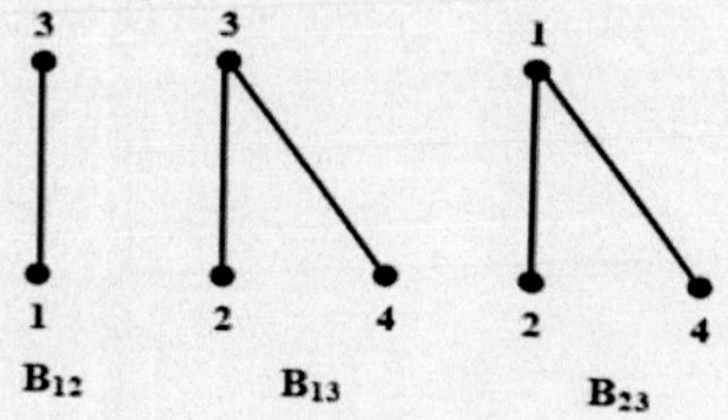

Fig. 7 Bipartite graph decomposition of graph G in Fig. 6

4 Conclusion

The proposed method determines

- All possible independent sets for G
- All possible partitions for G
- A decomposition for G into bipartite graphs

This algorithm to determine a decomposition helps in determining the independent sets, and hence, it can be used to determine the independence number of G. It also determines all possible partitions of the vertex set of G and hence can be used to determine a chromatic partition for G. This means that the proposed algorithm can also be used for determining the chromatic number of any graph G. This method of decomposing a given graph G into bipartite graphs can be extended to determine a decomposition of G into multipartite graphs by considering all possible subsets of sizes three, four, etc. This chapter would be of interest for researchers working in graph coloring and graph decomposition.

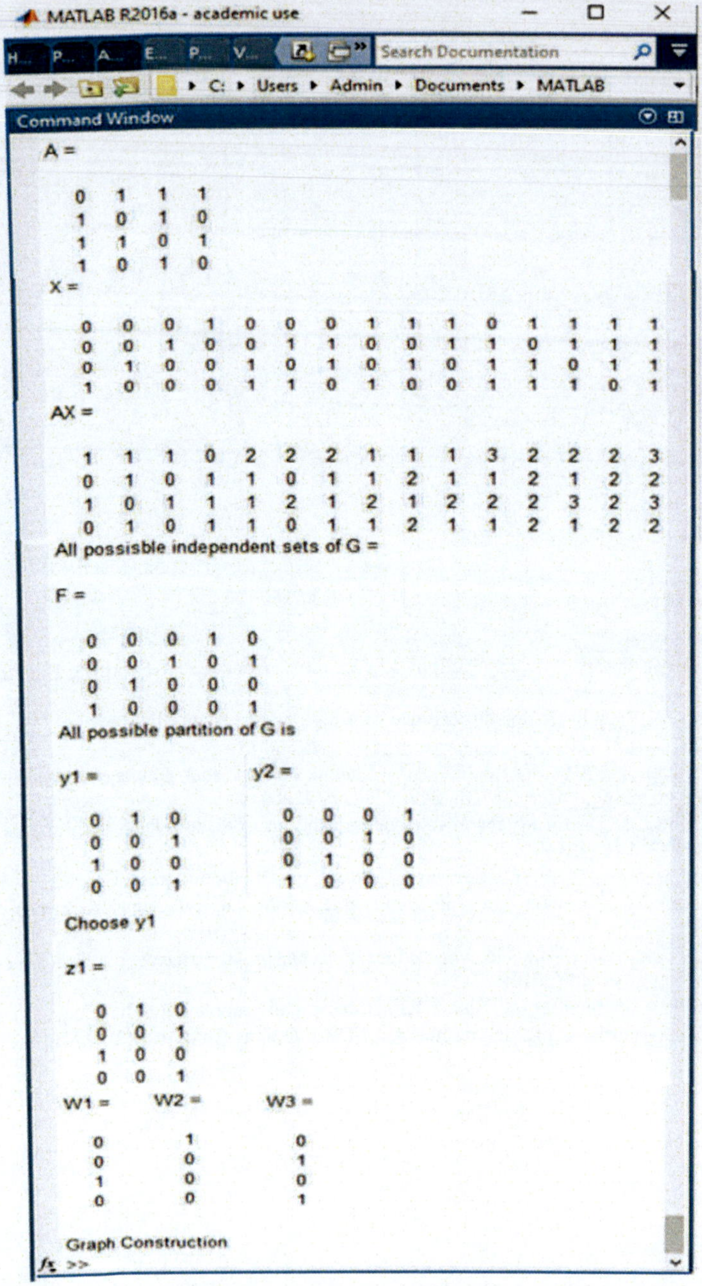
MATLAB R2016a - academic use
Search Documentation
C: ▸ Users ▸ Admin ▸ Documents ▸ MATLAB
Command Window
A =
0 1 1 1
1 0 1 0
1 1 0 1
1 0 1 0
X =
0 0 0 1 0 0 0 1 1 1 0 1 1 1 1
0 0 1 0 0 1 1 0 0 1 1 0 1 1 1
0 1 0 0 1 0 1 0 1 0 1 1 0 1 1
1 0 0 0 1 1 0 1 0 0 1 1 1 0 1
AX =
1 1 1 0 2 2 2 1 1 1 3 2 2 2 3
0 1 0 1 1 0 1 1 2 1 1 2 1 2 2
1 0 1 1 1 2 1 2 1 2 2 2 3 2 3
0 1 0 1 1 0 1 1 2 1 1 2 1 2 2
All possisble independent sets of G =
F =
0 0 0 1 0
0 0 1 0 1
0 1 0 0 0
1 0 0 0 1
All possible partition of G is
y1 =
0 1 0
0 0 1
1 0 0
0 0 1
y2 =
0 0 0 1
0 0 1 0
0 1 0 0
1 0 0 0
Choose y1
z1 =
0 1 0
0 0 1
1 0 0
0 0 1
W1 =
0
0
1
0
W2 =
1
0
0
0
W3 =
0
1
0
1
Graph Construction
fx >>

Snapshot 1

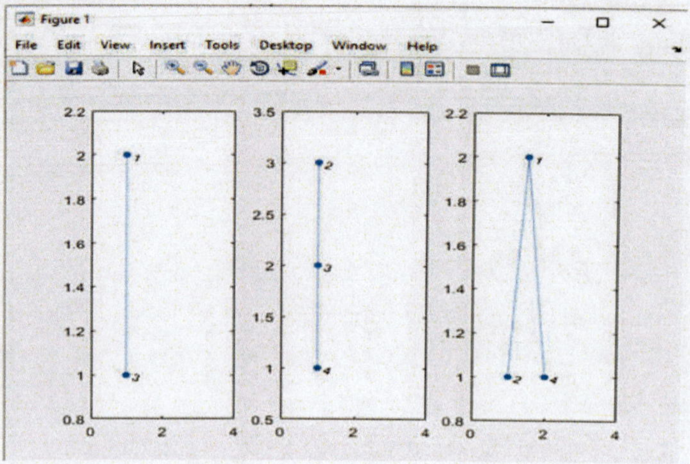

Snapshot 2

References

1. Deo, N.: Graph theory with applications to computer science. PHI Learning Private Limited (2013)
2. Markstrom, K.: Even cycle decompositions of 4-regular graphs and line graphs. Discret. Math. **312**(17), 2676–2681 (2012)
3. Oleg, V.B., Ivanova, A.O., Kostochka, A.V., Sheikh, N.N.: Planar graphs decomposable into a forest and a matching. Discret. Math. **309**, 277–279 (2009)
4. Milici, S., Tuza, Z.: Uniformly resolvable decompositions of K_v into P_3 and K_3 graphs. Discret. Math. **331**, 137–141 (2014)
5. Gioan, E.: Christophe Paul: Split decomposition and graph-labelled trees: Characterizations and fully dynamic algorithms for totally decomposable graphs. Discret. Appl. Math. **160**, 708–733 (2012)
6. Boode, A.H.: Hajo Broersma: Decompositions of graphs based on a new graph product. Discret. Appl. Math. **259**, 31–40 (2019)
7. Rubalcaba, R.R., Schneider, A., Slater, P.J.: A survey on graphs which have equal domination and closed neighborhood packing numbers. AKCE J. Graphs Combin. **3**(2), 93–114 (2006)

Degree-Based Topological Indices and QSPR Analysis of Some Drugs Used in the Treatment of Dengue

W. Tamilarasi and B. J. Balamurugan

1 Introduction

A chemical graph theory is an important area of research in mathematical chemistry where scientists use graph theory techniques to study the behavior and properties of chemical structures. A chemical molecule is transformed into a real number, known as a topological index, and with the use of chemical graph theory tools, the structure of a molecule is then analyzed using these indices. A chemical structure is converted into a molecular graph by taking atoms as vertices, bonds between atoms as edges, and double bonds as parallel edges. There are numerous uses of topological index in QSPR and QSAR studies, where physicochemical properties and biological activities of molecules are correlated with the topological indices, which can aid in identifying the most reliable index that can correlate with all physical characteristics of chemical compound.

Dengue is a tropical disease spread by Aedes mosquitoes infected by a dengue virus, and the disease is more prevalent in urban areas and more than 400 million people get infected annually by this disease. There exists no specific treatment or drug for dengue fever. However, some drugs like Acetaminophen, Chloroquine, Ranitidine, and Remdesivir can control dengue fever. In this study, molecular graphs of above aforesaid medications are subjected to some well-known degree-based topological indices such as Reduced Reciprocal Randić index $RRR(G)$, Arithmetic Geometri c_1 index $AG_1(G)$, SK index $SK(G)$, SK_1 index SK_1 (G), SK_2 index SK_2 (G), General Sum Connectivity index $SCI(G)$, and Forgotten index $F(G)$. In

W. Tamilarasi · B. J. Balamurugan (✉)
Division of Mathematics, School of Advanced Sciences, Vellore Institute of Technology, Chennai, Tamil Nadu, India
e-mail: tamilarasi.w2019@vitstudent.ac.in; balamurugan.bj@vit.ac.in

F. Kamalov et al. (eds.), *Advances in Mathematical Modeling and Scientific Computing*, Trends in Mathematics, https://doi.org/10.1007/978-3-031-41420-6_65

chemical graph theory, these indices are used frequently to describe various physical and chemical properties of molecular structures.

2 Terminologies and Notations

In this section, the definition of degree-based topological indices, M-polynomial, and some notations used to deduce these indices from M-polynomial are stated here.

Definition 1 The Reduced Reciprocal Randić index $RRR(G)$, proposed by Gutman et al. in [1], is defined as

$$RRR(G) = \sum_{ij \in E(G)} \sqrt{(d_i - 1)\left(d_j - 1\right)}.$$

Definition 2 The Arithmetic Geometri c_1 index $AG_1(G)$, proposed by Shigehalli and Kanabur in [2], is defined as

$$AG_1(G) = \sum_{ij \in E(G)} \frac{d_i + d_j}{2\sqrt{d_i d_j}}.$$

Definition 3 SK index $SK(G)$, proposed by Shigehalli and Kanabur in [3], is defined as

$$SK(G) = \sum_{ij \in E(G)} \frac{d_i + d_j}{2}.$$

Definition 4 SK_1 index $SK_1\ (G)$, proposed by Shigehalli and Kanabur in [3], is defined as

$$SK_1\ (G) = \sum_{ij \in E(G)} \frac{d_i . d_j}{2}.$$

Definition 5 SK_2 index $SK_2\ (G)$, proposed by Shigehalli and Kanabur in [3], is defined as

$$SK_2(G) = \sum_{ij \in E(G)} \left(\frac{d_i + d_j}{2}\right)^2.$$

Definition 6 The general sum connectivity index $SCI(G)$, proposed by Du et al. [4], is defined as

$$SCI(G) = \sum_{ij \in E(G)} \frac{1}{\sqrt{d_i + d_j}}.$$

Definition 7 The Forgotten index (*G*), proposed by Gutman and Furtula [5], is defined as

$$F(G) = \sum_{ij \in E(G)} \left(d_i^2 + d_j^2\right).$$

Definition 8 Deutsch and Klavzar in [6] proposed M-polynomial of a graph *G*, which is used to deduce degree-based topological indices and is defined as follows:

$M\ (G; x, y) = \sum_{\delta \le i \le j \le \Delta} e_{i,j} x^i y^j$ where $e_{i,j}$, $i, j \ge 1$ denote the number of edges $uv \in E$ (G) and $(d_u, d_v) = (i,j)$, d_u and d_v represent the number of vertices incident at *u* and *v*, respectively, and $(\delta, \Delta) = (\min d_v, \max d_v)$ where $u, v \in V$ (G) [7].

The topological indices and its derivative formulae from M-polynomial are given in Table 1.

Topological descriptor plays a significant role in mathematical chemistry, especially in QSPR/QSAR studies. Mondal S. et al. in [8] and Kirmani et al. in [9] have retrieved some degree-based and neighborhood degree sum-based topological indices by

M-polynomial and NM-polynomial approach for some antiviral drugs used for the treatment of COVID-19 and carried out QSPR analysis. M. C. Shanmukha et al. have investigated some degree-based topological indices on anticancer drugs and also established QSPR analysis [10]. Havareh in [11] designed QSPR model using degree-based, Mostar type, and distance-based topological indices for drugs that are used in the treatment of COVID-19. QSPR analysis using well-known degree-based topological indices for drugs used in the treatment of breast cancer

Table 1 Formulae for topological indices deduced from $M(G, x, y)$

Index	Derivation from $M(G, x, y) = h(x, y)$
Reduced reciprocal Randić Index $RRR(G)$	$D_x^{1/2} D_y^{1/2} Q_{y(-1)} Q_{x(-1)} [h\ (x, y)]_{x=y=1}$
Arithmetic Geometric c_1 index $AG_1(G)$	$\frac{1}{2} D_x J S_x^{1/2} S_y^{1/2} [h\ (x, y)]_{x=1}$
SK index $SK(G)$	$\frac{1}{2}\left(D_x + D_y\right) [h\ (x, y)]_{x=y=1}$
SK_1 index $SK_1\ (G)$	$\frac{1}{2} D_x D_y [h\ (x, y)]_{x=y=1}$
SK_2 index $SK_2\ (G)$	$\frac{1}{4} D_x^2 J [h\ (x, y)]_{x=y=1}$
General sum connectivity index $SCI(G)$	$S_x^{1/2} J [h\ (x, y)]_{x=1}$
Forgotten index $F(G)$	$\left(D_x^2 + D_y^2\right)\ [h\ (x, y)]_{x=1}$

was investigated in [12] and was revealed that some of the indices possess high correlation value with the physicochemical properties of drugs. Zhong et al. in [13] have investigated the QSPR analysis of valency-based topological indices of COVID-19 drugs and found that these indices serve a good predictive means in QSPR investigations. Recently, in [7], the ADMET properties and QSPR analysis of drugs against COVID-19 Omicron variant with some degree based topological indices via M-polynomial are investigated. By the motivation of the above works, in this chapter, some degree-based indices are computed from molecular graphs of aforesaid drugs used to treat dengue fever using M-polynomial. Further QSPR analysis was established between these indices and physicochemical properties by using cubic regression model. where the operators are defined as follows: $D_x h(x, y) = \frac{x\partial(h(x,y))}{\partial x}$, $D_y h(x, y) = \frac{y\partial(h(x,y))}{\partial y}$, $Jh(x, y) = h(x, x)$, $Q_{x(\alpha)}h(x, y) = x^{\alpha}h(x, y)$, $D_x^{1/2} h(x, y) = \sqrt{\frac{x\partial(h(x,y))}{\partial x}}.\sqrt{h(x, y)}$, $D_y^{1/2} h(x, y) = \sqrt{\frac{y\partial(h(x,y))}{\partial y}}.\sqrt{h(x, y)}$ $S_x^{\frac{1}{2}} h(x, y) = \sqrt{\int_0^x \frac{h(t,y)}{t}dt}.\sqrt{h(x, y)}$, $S_y^{\frac{1}{2}} h(x, y) = \sqrt{\int_0^y \frac{h(x,t)}{t}dt}.\sqrt{h(x, y)}$.

3 Computation of Topological Indices

The edge partition approach is used to derive the M-polynomial of the molecular graphs of Acetaminophen (A), Chloroquine (C), Ranitidine (RA), and Remdesivir (RE). The drug's chemical composition is referred from https://www.chemspider.com.

The formulae in Table 1 are used to derive the indices from their M-polynomial.

Theorem 1 The M-polynomial of the chemical graph A of Acetaminophen is

$$M(A; x, y) = xy^2 + xy^3 + xy^4 + 3x^2y^4 + 3x^3y^3 + 8x^3y^4.$$

Proof The chemical structure and chemical graph of Acetaminophen are shown in Fig. 1a, b respectively. Let A be the chemical graph of Acetaminophen with 13 vertices and 17 edges. Let $E_{i,j}$ be defined as $E_{i,j} = \{uv \in E(G) : d_u = i, d_v = j\}$ and $|E_{i,j}| = e_{i,j}$. From Fig. 1, $e_{1,2} = 1$, $e_{1,3} = 1$, $e_{1,4} = 1$, $e_{2,4} = 3$, $e_{3,3} = 3$, $e_{3,4} = 8$. Therefore, by definition of M-polynomial, we have

$$M(P; x, y) = \sum_{\delta \le i \le j \le \Delta} e_{i,j} x^i y^j$$

$$= e_{1,2}xy^2 + e_{1,3}xy^3 + e_{1,4}xy^4 + e_{2,4}x^2y^2 + e_{3,3}x^2y^3 + e_{3,4}x^2y^4.$$

$$= xy^2 + xy^3 + xy^4 + 3x^2y^4 + 3x^3y^3 + 8x^3y^4.$$

(a) (b)

Fig. 1 (**a**) Chemical structure and (**b**) Chemical graph of Acetaminophen

Theorem 2 Let A be the chemical graph of Acetaminophen and $M(A; x, y) = xy^2 + xy^3 + xy^4 + 3x^2y^4 + 3x^3y^3 + 8x^3y^4$, then

(i) $RRR(A) = 34.9383$, (ii) $AG_1(A) = 17.7302$, (iii) $SK(A) = 52$, (iv) $SK_1(A) = 78$, (v) $SK_2(A) = 201.5$, (vi) $SCI(A) = 6.9977$, (vii) $F(A) = 346$.

Proof Let $M(A; x, y) = xy^2 + xy^3 + xy^4 + 3x^2y^4 + 3x^3y^3 + 8x^3y^4 = h(x, y)$

$$Q_{x(-1)}h(x, y) = y^2 + y^3 + y^4 + 3xy^4 + 3x^2y^3 + 8x^2y^4$$

$$Q_{y(-1)}Q_{x(-1)}h(x, y) = y + y^2 + y^3 + 3xy^3 + 3x^2y^2 + 8x^2y^3$$

$$D_y^{1/2}Q_{y(-1)}Q_{x(-1)}h(x, y) = y+\sqrt{2}y+\sqrt{3}y+3\sqrt{3}xy^3+3\sqrt{2}x^2y^2+8\sqrt{3}x^2y^3$$

$$D_x^{1/2}D_y^{1/2}Q_{y(-1)}Q_{x(-1)}h(x, y) = y+\sqrt{2}y+\sqrt{3}y+3\sqrt{3}xy^3+6x^2y^2+8\sqrt{6}x^2y^3$$

$$S_y^{1/2}h(x, y) = \frac{1}{\sqrt{2}}xy^2 + \frac{1}{\sqrt{3}}xy^3 + \frac{1}{2}xy^4 + \frac{3}{2}x^2y^4 + \sqrt{3}x^3y^3 + 4x^3y^4$$

$$S_x^{\frac{1}{2}}S_y^{\frac{1}{2}}h(x, y) = \frac{1}{\sqrt{2}}xy^2 + \frac{1}{\sqrt{3}}xy^3 + \frac{1}{2}xy^4 + \frac{3}{2\sqrt{2}}x^2y^4 + x^3y^3 + \frac{4}{\sqrt{3}}x^3y^4$$

$$JS_x^{1/2}S_y^{1/2}h(x, y) = \frac{1}{\sqrt{2}}x^3 + \frac{1}{\sqrt{3}}x^4 + \frac{1}{2}x^5 + \frac{3}{2\sqrt{2}}x^6 + x^6 + \frac{4}{\sqrt{3}}x^7$$

$$D_xJS_x^{1/2}S_y^{1/2}h(x, y) = \frac{3}{\sqrt{2}}x^3 + \frac{4}{\sqrt{3}}x^4 + \frac{5}{2}x^5 + \frac{18}{2\sqrt{2}}x^6 + 6x^6 + \frac{28}{\sqrt{3}}x^7$$

$$\frac{1}{2}D_x J S_x^{1/2} S_y^{1/2} h(x, y) = \frac{3}{2\sqrt{2}}x^3 + \frac{2}{\sqrt{3}}x^4 + \frac{5}{4}x^5 + \frac{9}{2\sqrt{2}}x^6 + 3x^6 + \frac{14}{\sqrt{3}}x^7$$

$$D_x h(x, y) = xy^2 + xy^3 + xy^4 + 6x^2y^4 + 9x^3y^3 + 24x^3y^4$$

$$D_y h(x, y) = 2xy^2 + 3xy^3 + 4xy^4 + 12x^2y^4 + 9x^3y^3 + 32x^3y^4$$

$$\left(D_x + D_y\right) h(x, y) = 3xy^2 + 4xy^3 + 5xy^4 + 18x^2y^4 + 18x^3y^3 + 56x^3y^4$$

$$\frac{1}{2}\left(D_x + D_y\right) h(x, y) = \frac{3}{2}xy^2 + 2xy^3 + \frac{5}{2}xy^4 + 9x^2y^4 + 9x^3y^3 + 28x^3y^4$$

$$D_x D_y h(x, y) = 2xy^2 + 3xy^3 + 4xy^4 + 24x^2y^4 + 27x^3y^3 + 96x^3y^4$$

$$\frac{1}{2}\left(D_x D_y\right) h(x, y) = xy^2 + \frac{3}{2}xy^3 + 2xy^4 + 12x^2y^4 + \frac{27}{2}x^3y^3 + 48x^3y^4$$

$$J h(x, y) = x^3 + x^4 + x^5 + 3x^6 + 3x^6 + 8x^7$$

$$D_x^2 J h(x, y) = 9x^3 + 64x^4 + 125x^5 + 108x^6 + 108x^6 + 392x^7$$

$$\frac{1}{4}D_x^2 J h(x, y) = \frac{9}{4}x^3 + \frac{64}{4}x^4 + \frac{125}{4}x^5 + \frac{108}{4}x^6 + \frac{108}{4}x^6 + \frac{392}{4}x^7$$

$$S_x^{1/2} J h(x, y) = \frac{1}{\sqrt{3}}x^3 + \frac{1}{2}x^4 + \frac{1}{\sqrt{5}}x^5 + \frac{3}{\sqrt{6}}x^6 + \frac{3}{\sqrt{6}}x^6 + \frac{8}{\sqrt{7}}x^7$$

$$D_x^2 h(x, y) = xy^2 + xy^3 + xy^4 + 12x^2y^4 + 27x^3y^3 + 72x^3y^4$$

$$D_y^2 h(x, y) = 4xy^2 + 9xy^3 + 16xy^4 + 48x^2y^4 + 27x^3y^3 + 128x^3y^4$$

$$\left(D_x^2 + D_y^2\right) h(x, y) = 5xy^2 + 10xy^3 + 17xy^4 + 60x^2y^4 + 54x^3y^3 + 200x^3y^4$$

Fig. 2 (**a**) Chemical structure and (**b**) Chemical graph of Chloroquine

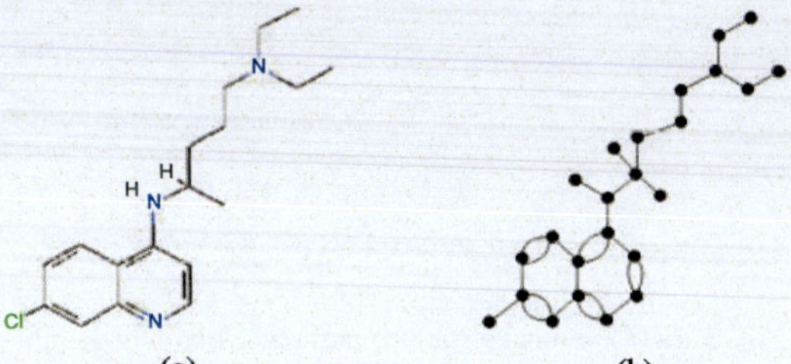

(a) (b)

(i) $RRR(A) = D_x^{1/2} D_y^{1/2} Q_{y(-1)} Q_{x(-1)} [h\,(x,\,y)]_{x=y=1} = 34.9383$
(ii) $AG_1(A) = \frac{1}{2} D_x J S_x^{1/2} S_y^{1/2} [h\,(x,\,y)]_{x=1} = 17.7302$
(iii) $SK(A) = \frac{1}{2}\left(D_x + D_y\right)[h\,(x,\,y)]_{x=y=1} = 52$
(iv) $SK_1\,(A) = \frac{1}{2} D_x D_y [h\,(x,\,y)]_{x=y=1} = 78$
(v) $SK_2\,(A) = \frac{1}{4} D_x^2 J [h\,(x,\,y)]_{x=y=1} = 201.5$
(vi) $SCI(A) = S_x^{1/2} J [h\,(x,\,y)]_{x=1} 6.9977$
(vii) $F(A) = \left(D_x^2 + D_y^2\right)[h\,(x,\,y)]_{x=1} = 346$

Theorem 3 Let C be the chemical graph of Chloroquine, then the M-polynomial of C is $M(A; x, y) = 2xy^2 + xy^3 + 3xy^4 + 2x^2y^2 + 3x^2y^3 + x^2y^4 + 5x^3y^3 + 10x^3y^4 + 3x^4y^4$.

Proof The chemical structure and graph of Chloroquine are shown in Fig. 2a, b respectively. Let C be the molecular graph of Chloroquine with 24 vertices and 30 edges. Let $E_{i,j}$ be defined as $E_{i,j} = \{uv \in E(G) : d_u = i, d_v = j\}$ and $|E_{i,j}| = e_{i,j}$. From Fig. 2, we get $e_{1,2} = 2$, $e_{1,3} = 1$, $e_{1,3} = 3$, $e_{2,2} = 2$, $e_{2,3} = 3$, $e_{2,4} = 1$, $e_{3,3} = 5$, $e_{3,4} = 10$, $e_{4,4} = 3$. Therefore, by definition of M-polynomial, we have

$$M\,(C; x,\,y) = \sum_{\delta \le i \le j \le \Delta} e_{i,j} x^i y^j$$

$$= e_{1,2}\,xy^2 + e_{1,3}xy^3 + e_{1,3}xy^4 + e_{2,2}x^2y^2 + e_{2,3}x^2y^3 + e_{2,4}x^2y^4 + e_{3,3}x^3y^3 + e_{3,4}x^3y^4 + e_{4,4}x^4y^4.$$

$$= 2xy^2 + xy^3 + 3xy^4 + 2x^2y^2 + 3x^2y^3 + x^2y^4 + 5x^3y^3 + 10x^3y^4 + 3x^4y^4.$$

(a) (b)

Fig. 3 (**a**) Chemical structure and (**b**) Chemical graph of Ranitidine

The method applied in Theorem 2 is used to get the various degree-based topological indices for the chemical graph of Chloroquine in the subsequent theorem.

Theorem 4 Let C be the chemical graph of Chloroquine and $M(C;x,y) = 2xy^2 + xy^3 + 3xy^4 + 2x^2y^2 + 3x^2y^3 + x^2y^4 + 5x^3y^3 + 10x^3y^4 + 3x^4y^4$, then

(i) $RRR(C) = 60.0799$, (ii) $AG_1(C) = 31.2521$, (iii) $SK(C) = 89$, (iv) $SK_1\ (C) = 133$, (v) $SK_2\ (C) = 278.5$, (vi) $SCI(C) = 12.6277$, (vii) $F(C) = 582$.

The M-polynomial of the molecular graph of Ranitidine is established in the next theorem.

Theorem 5 The M-polynomial of the chemical graph RA of Ranitidine is $M(A;x,y) = 5xy^3 + 2xy^4 + 3x^2y^2 + 2x^2y^3 + 6x^2y^4 + x^3y^3 + 6x^3y^4 + 3x^4y^4$.

Proof The chemical structure and graph of Ranitidine are shown in Fig. 3a, b, respectively. Let RA be the chemical graph of Ranitidine with 24 vertices and 28 edges. Let $E_{i,j}$ be defined as $E_{i,j} = \{uv \in E(G) : d_u = i, d_v = j\}$ and $|E_{i,j}| = e_{i,j}$. From Fig. 3, $e_{1,3} = 5$, $e_{1,4} = 2$, $e_{2,2} = 3$, $e_{2,3} = 2$, $e_{2,4} = 6$, $e_{3,3} = 1$, $e_{3,4} = 6$, $e_{4,4} = 3$. Therefore, by definition of M-polynomial, we have

$$M\,(RA;\,x,\,y) = \sum_{\delta \le i \le j \le \Delta} e_{i,j}x^i y^j$$

$$= e_{1,3}xy^3 + e_{1,4}xy^4 + e_{2,2}x^2y^2 + e_{2,3}x^2y^3 + e_{2,4}x^2y^4 + e_{3,3}x^3y^3 + e_{3,4}x^3y^4 + e_{4,4}x^4y^4.$$

$$= 5xy^3 + 2xy^4 + 3x^2y^2 + 2x^2y^3 + 6x^2y^4 + x^3y^3 + 6x^3y^4 + 3x^4y^4.$$

By using the M-polynomial of chemical graph of Ranitidine, the indices values are computed in Theorem 6 by following same procedure as in Theorem 2.

Theorem 6 The M-polynomial of the chemical graph RA of Ranitidine is $M(RA;x,y) = 5xy^3 + 2xy^4 + 3x^2y^2 + 2x^2y^3 + 6x^2y^4 + x^3y^3 + 6x^3y^4 + 3x^4y^4$, then

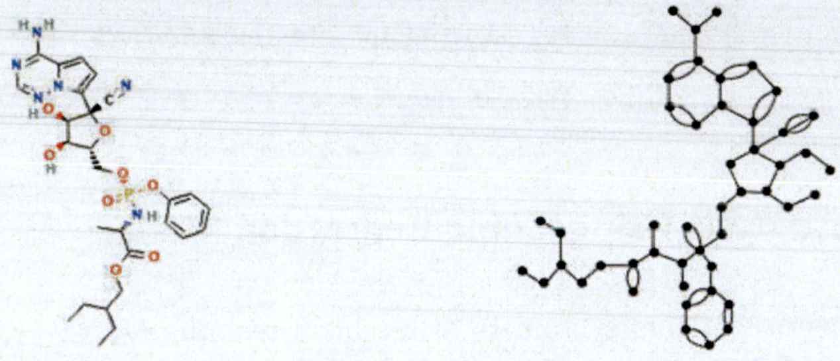

(a) (b)

Fig. 4 (a) Chemical structure and (b) Chemical graph of Remdesivir

(i) $RRR(RA) = 51.4528$, (ii) $AG_1(RA) = 28.8781$, (iii) $SK(RA) = 80$, (iv) SK_1 $(RA) = 112$, (v) SK_2 $(RA) = 241.5$, (iv) $SCI(RA) = 12.0806$, (vii) $F(RA) = 518$.

Theorem 7 The M-polynomial of the chemical graph *RE* of Remdesivir is

$$M(RE; x, y) = 4xy^2 + 4xy^3 + 2x^2y^2 + 7x^2y^3 + 5x^2y^4 + 4x^2y^5 + 14x^3y^3 + 17x^3y^4 + x^3y^5 + 3x^4y^4.$$

Proof The chemical structure and chemical graph of Remdesivir are shown in Fig. 4a, b respectively. Let *RE* be the molecular graph of Remdesivir with 47 vertices and 61 edges. Let $E_{i,j}$ be defined as $E_{i,j} = \{uv \in E(G) : d_u = i, d_v = j\}$ and $|E_{i,j}| = e_{i,j}$. From Fig. 4, $e_{1,2} = 4$, $e_{1,3} = 4$, $e_{2,2} = 2$, $e_{2,3} = 7$, $e_{2,4} = 5$, $e_{2,5} = 4$, $e_{3,3} = 14$, $e_{3,4} = 17$, $e_{3,5} = 1$, $e_{4,4} = 3$. Therefore, by definition of M-polynomial we have, $M(RE; x, y) = \sum_{\delta \le i \le j \le \Delta} e_{i,j} x^i y^j$

$$= e_{1,2}xy^2 + e_{1,3}xy^3 + e_{2,2}x^2y^2 + e_{2,3}x^2y^3 + e_{2,4}x^2y^4 + e_{2,5}x^2y^5 + e_{3,3}x^3y^3 + e_{3,4}x^3y^4 + e_{3,5}\,x^3y^5 + e_{4,4}x^4y^4.$$

$$= 4xy^2 + 4xy^3 + 2x^2y^2 + 7x^2y^3 + 5x^2y^4 + 4x^2y^5 + 14x^3y^3 + 17x^3y^4 + x^3y^5 + 3x^4y^4.$$

By using the M-polynomial of chemical graph of Remdesivir, the indices values are computed in Theorem 8 by following same procedure as in Theorem 2.

Theorem 8 Let $M(RE; x, y) = 4xy^2 + 4xy^3 + 2x^2y^2 + 7x^2y^3 + 5x^2y^4 + 4x^2y^5 + 14x^3y^3 + 17x^3y^4 + x^3y^5 + 3x^4y^4$ be the M-polynomial of the chemical graph *RE* of Remdesivir then (i) $RRR(RE) = 119.6863$, (ii) $AG_1(RE) = 62.9452$, (iii) $SK(RE)$

= 182, (iv) SK_1 (*RE*) = 271.5, (v) SK_2 (*RE*) = 569, (vi) *SCI*(*RE*) = 25.5480, (vii) *F*(*RE*) = 1190.

4 QSPR Analysis Using Cubic Regression Model

The main objective of this section is to establish quantitative structure-property relationship (QSPR) between the above defined degree based topological indices and physicochemical properties of the aforesaid drugs in order to access the effectiveness of the topological indices. The properties that are taken into consideration in this chapter are boiling point, enthalpy of vaporization, flash point, molar refraction, polar surface area, polarizability, surface tension and molar volume of drugs that are used to treat dengue. The values of physical and chemical properties of dengue therapies in Table 3 are obtained from https://www.chemspider.com, an online chemical structure database. In Table 2 the values of the topological indices obtained from Theorems 2, 4, 6 and 8 are enlisted. The cubic regression model was used for QSPR analyses. The following is the cubic regression equation.

$$P = a + b(TI) + c(TI)^2 + d(TI)^3 \tag{1}$$

In Eq. (1), *P* – physicochemical characteristics, *a* -constant, {*b*, *c*, *d*}- regression coefficient and *TI* – topological indices. Using SPSS software, regression analyses were performed. Table 4, represent the correlation coefficient value (R^2) determined between the indices and the physicochemical characteristics of dengue medication compounds. The statistical parameters for the cubic regression model having highest correlation coefficient value (R^2) are listed in Tables 5, 6, 7, 8, 9, 10, and 11. For

Table 2 The values of topological indices for the molecular graph of dengue drugs

Antidengue drugs	*RRR*	AG_1	*SK*	SK_1	SK_2	*SCI*	*F*
Acetaminophen	34.9383	17.7302	52	78	201.5	6.9977	346
Chloroquine	60.0799	31.2521	89	133	278.5	12.6277	582
Ranitidine	51.4528	28.8781	80	112	241.5	12.0806	518
Remdesivir	119.686	62.9452	182	271.5	569	25.5480	1190

Table 3 Physicochemical characteristics value of dengue therapies

Antidengue drugs	*BP*	*EV*	*FP*	*MR*	*PSA*	*P*	*ST*	*MV*
Acetaminophen	387.8	66.2	188.4	42.4	49	16.8	52.8	120.9
Chloroquine	460.6	72.1	232.3	97.4	28	38.6	44	287.9
Ranitidine	437.1	69.4	218.2	85.6	112	34	45	265.4
Remdesivir	–	–	–	149.5	213	59.3	62.3	409.0

Table 4 The correlation coefficient value (R^2) determined between indices and the physicochemical characteristics of dengue medication compounds

Topological indices	*BP*	*EV*	*FP*	*MR*	*PSA*	*P*	*ST*	*MV*
RRR	**1.000**	0.999	**1.000**	0.998	0.805	0.997	0.995	0.989
AG_1	0.999	0.999	0.999	0.999	0.811	0.999	**1.000**	**1.000**
SK	0.999	0.999	0.999	**1.000**	0.808	**1.000**	0.999	0.998
SK_1	**1.000**	**1.000**	**1.000**	0.995	0.807	0.995	0.992	0.984
SK_2	**1.000**	**1.000**	**1.000**	0.982	0.818	0.982	0.977	0.960
SCI	0.999	0.999	0.998	0.995	0.818	0.996	**1.000**	0.999
F	**1.000**	0.999	**1.000**	**1.000**	0.808	**1.000**	0.998	0.996

Max (R^2) value are bolded for each physicochemical property

Table 5 Parameters of the cubic QSPR model for *RRR* index statistically

Cubic regression equation	R	F	S. E	P
$BP = 4.47E2 - 7.8\,(RRR) + 0.24\,(RRR)^2 - 1.68E{-}3\,(RRR)^3$	1	3407.285	4.540	0.012
$FP = 2.08E2 - 3.74\,(RRR) + 0.1\,(RRR)^2 - 8.77E{-}4\,(RRR)^3$	1	3784.835	2.151	0.011

Table 6 Parameters of the cubic QSPR model for AG_1 index statistically

Cubic regression equation	R	F	S. E	P
$ST = 72.66 - 1.28\,(AG_1) + 5.57E{-}3\,(AG_1)^2 + 1.93E{-}3\,(AG_1)^3$	1	94962.634	0.034	0.002
$MV = -2.73E2 + 28.96\,(AG_1) - 0.42\,(AG_1)^2 + 2.03E{-}3\,(AG_1)^3$	1	11310.091	1.361	0.007

Table 7 Parameters of the cubic QSPR model for *SK* index statistically

Cubic regression equation	R	F	S. E	P
$MR = -60.76 + 2.24\,(SK) - 4.4E{-}3\,(SK)^2 - 8.45E{-}6\,(SK)^3$	1	106571.312	0.165	0.002
$P = -26.63 + 0.98\,(SK) - 2.72E{-}3\,(SK)^2 - 3.09E{-}7\,(SK)^3$	1	12504026.34	0.006	0.000

Table 8 Parameters of the cubic QSPR model for SK_1 index statistically

Cubic regression equation	R	F	S. E	P
$BP = 3.53E2 - 1.25\,(SK_1) + 0.03\,(SK_1)^2 - 1.13\,E{-}4\,(SK_1)^3$	1	24674.488	1.687	0.010
$EV = 92.53 - 0.85\,(SK_1) + 8.47E{-}3\,(SK_1)^2 - 2.43\,E{-}5\,(SK_1)^3$	1	1363.319	1.151	0.019
$FP = 1.54E2 - 0.35\,(SK_1) + 0.01\,(SK_1)^2 - 5.68E{-}5\,(SK_1)^3$	1	78980.303	0.471	0.003

Table 9 Parameters of the cubic QSPR model for SK_2 index statistically

Cubic regression equation	R	F	S. E	P
$BP = -2.74E2 + 5.13\,(SK_2) - 9.68E{-}3\,(SK_2)^2 + 2.66\,E{-}6\,(SK_2)^3$	1	5328.810	3.630	0.005
$EV = 78.44 - 0.3\,(SK_2) + 1.65E{-}3\,(SK_2)^2 - 2.42\,E{-}6\,(SK_2)^3$	1	14362.408	0.355	0.006
$FP = -2.33E2 + 3.36\,(SK_2) - 6.93E{-}3\,(SK_2)^2 + 3.05\,E{-}6\,(SK_2)^3$	1	2592.425	2.599	0.014

Table 10 Parameters of the cubic QSPR model for *SCI* index statistically

Cubic regression equation	R	F	S. E	P
$ST = 42.27 + 4.6\,(SCI) - 0.55\,(SCI)^2 + 0.02\,(SCI)^3$	1	2330.436	0.215	0.015

Table 11 Parameters of the cubic QSPR model for *F* index statistically

Cubic regression equation	R	F	S. E	P
$BP = 5.99E2 - 1.64\,(F) + 3.79E{-3}\,(F)^2 - 2.39\,E{-6}\,(F)^3$	1	1327.465	7.272	0.019
$FP = 3E2 - 0.88\,(F) + 2.08E{-3}\,(F)^2 - 1.3\,E{-6}\,(F)^3$	1	1133.040	3.930	0.021
$MR = -1.12E2 + 0.61\,(F) - 5.16E{-4}\,(F)^2 + 1.61\,E{-7}\,(F)^3$	1	3468.176	0.916	0.012
$P = -46.93 + 0.25\,(F) - 2.27E{-4}\,(F)^2 + 7.45E{-8}\,(F)^3$	1	2558.915	0.423	0.014

goodness of fit, maximum correlation coefficient (R) value, minimum standard error (S. E), maximum F – value and p value less than 0.05 are considered.

5 Conclusion

The study of topological indices is used to determine the properties of molecular structure in pharmaceutical sciences, which is important for the creation of a new drug. In this chapter, some drugs for dengue fever are considered and from the expression of M-polynomial, the degree-based topological indices are derived. In the QSPR study, cubic regression model is analyzed using eight descriptors and seven topological indices. It is apparent that the majority of the topological indices derived in this chapter have demonstrated good prediction abilities for the physicochemical properties of drugs used to treat dengue. In particular, the best predictive topological indices are as follows:

1. *RRR* index for Boiling point (*BP*) and Flash point (*FP*)
2. AG_1 index for Surface tension (*ST*) and Molar volume (*MV*)
3. *SK* index for Molar refraction (*MR*) and Polarizability (*P*)
4. SK_1 and SK_2 index for Boiling point (*BP*), Enthalpy of vaporization (*EV*), and Flash point (*FP*)
5. *SCI* index for Surface tension (*ST*)
6. *F* index for Boiling point (*BP*), Flash point (*FP*), Molar refraction (*MR*) and Polarizability (*P*)

The correlation coefficients for various topological indices were established in this work, and it will assist chemists in creating novel drugs by combining those with high correlations.

References

1. Gutman, I., Furtula, B., Elphick, C.: Three new/old vertex–degree–based topological indices. MATCH Commun. Math. Comput. Chem. **72**, 617–632 (2014)
2. Shigehalli, V.S., Kanabur, R.: Arithmetic-geometric indices of path graph. J. Comp. Math. Sci. **6**(4), 194–199 (2015)
3. Shigehalli, V., Kanabur, R.: Computing degree based topological indices of Polyhex Nanotube. J. Math. Nanosci. **6**(1–2), 47–55 (2016)
4. Du, Z., Zhou, B., Trinajstić, N.: On the general sum-connectivity index of trees. Appl. Math. Lett. **24**(3), 402–405 (2011)
5. Gutman, I., Furtula, B.: A forgotten topological index. J. Math. Chem. **53**, 1184–1190 (2015)
6. Emeric, D., Klavzar, S.: M-polynomial and degree-based topological indices. Iran. J. Math. Chem. **6**, 93–102 (2015)
7. Tamilarasi, W., Balamurugan, B.J.: ADMET and quantitative structure property relationship analysis of anti-Covid drugs against omicron variant with some degree-based topological indices. Int. J. Quantum Chem., e26967 (2022). https://doi.org/10.1002/qua.26967
8. Mondal, S., De, N., Pal, A.: Topological indices of some chemical structures applied for the treatment of COVID-19 patients. Polycycl. Aromat. Compd. (2020). https://doi.org/10.1080/10406638.2020.1770306
9. Kirmani, S.A.K., Ali, P., Azam, F.: Topological indices and QSPR/QSAR analysis of some antiviral drugs investigated for the treatment of COVID-19 patients. J. Quantum Chem. **121**(9), e26594 (2021)
10. Shanmukha, M.C., Basavarajappa, N.S., Shilpa, K.C., Usha, A.: Degree-based topological indices in anticancer drugs with QSPR analysis. Heliyon. **6**, e04235 (2020)
11. Havare, O.C.: Quantitative structure analysis of some molecules in drugs used in the treatment of COVID-19 with topological indices. Polycycl. Aromat. Compd. https://doi.org/10.1080/10406638.2021.1934045
12. Bokhary, S.A.U.I.H., Adnan, Siddiqui, M.K., Cancan, M.: On topological indices and QSPR analysis of drugs used for the treatment of breast cancer. Polycycl. Aromat. Compd. (2021). https://doi.org/10.1080/10406638.2021.1977353
13. Zhong, J.-F., Rauf, A., Naeem, M., Rahman, J., Aslam, A.: Quantitative structure-property relationship (QSPR) of valency based topological indices with Covid-19 drugs and application. Arab. J. Chem. **14**, 103240 (2021)

Orientation Number of Two Complete Bipartite Graphs with Linkages

G. Rajasekaran and R. Sampathkumar

1 Introduction

In this article, a graph G is a tuple (V, E), V is a non-empty vertex set, and $E \subseteq V$. The elements of E are called edges. Let $d_G(u)$ be the *degree* of a vertex u in G (i.e., the number of edges incident with u), and the maximum degree of G is Δ. The length of a shortest (a, b)-path in G is denoted by $d_G(a, b)$. The value $e_G(a) = max\{d_G(a, b) \mid b \in V\}$ is called *eccentricity of* a. Let (D, V, A) be a directed graph and $A \subseteq V$ called arc set. The notions $e_D(v)$, for $v \in V(D)$, and $d(D)$ are defined as in the undirected graph. An *orientation* D is derived from G by giving a direction to each edges of G. If there is a $x \sim y$ directed path in D, then a vertex x is called *reachable* from y. An oriented graph D is *strong* if there are directed $u \sim v$ and $v \sim u$ paths that exist for all $u, v \in V(D)$. Robbins' one-way street theorem [7] states that "an orientation D of G is strong if and only if G is 2-edge-connected." Let $\mathscr{D}$ denote the set of all strong orientations of bridge-less G. The number $\vec{d}\,(G) = \min\{d(D) \mid D \in \mathscr{D}(G)\}$ is called *orientation number* (*minimum oriented diameter*) of G. An orientation D is called an *optimal orientation* of G if for any $D \in \mathscr{D}(G)$ with $d(D) = \vec{d}\,(G)$.

For a set of disjoint graphs, Koh and Ng [3] investigated the orientation number and designed an optimal orientation for a resulting graph obtained by connecting the

---

G. Rajasekaran (✉)
Department of Mathematics, School of Advanced Sciences, Vellore Institute of Technology, Vellore, India
e-mail: rajasekaran.ganapathy@vit.ac.in

R. Sampathkumar
Department of Mathematics, Annamalai University, Annamalainagar, India

F. Kamalov et al. (eds.), *Advances in Mathematical Modeling and Scientific Computing*, Trends in Mathematics, https://doi.org/10.1007/978-3-031-41420-6_66

given graphs with a set of additional edges. Koh and Ng [3] discussed the following: for $r = 2$,

- Let n_1 and n_2 be the orders of bridge-less graphs, respectively, and H_1 and H_2, and $\mathscr{H}_2^*$ be a class of graphs formed by linking two edges between H_1 and H_2. If $\Delta_1 = n_1 - 1$ and $\Delta_2 = n_2 - 1$, then $\vec{d}\ (\mathscr{H}_2^*) = 4$.
- $min\{m : \vec{d}\ (2; m) = 3\} = 4$.
- $\vec{d}\ (\mathscr{H}(s, s; 2s)) = \vec{d}\ (\mathscr{H}(s, s+1; 2s)) = \vec{d}\ (\mathscr{H}(s, s+2; 2s+1)) = \vec{d}\ (\mathscr{H}(s, s+3; 2s+2)) = 2, \forall s \geq 5$.

Also, Ng [6] examined the following:

- $\vec{d}\ (\mathscr{H}(s, s+4; 2s+3)) = 2$.
- $\vec{d}\ (\mathscr{H}(s, t; 2s+4)) = 2, \forall t \geq s+5$.

For a family of graphs $\mathscr{H}$, $\vec{d}\ (\mathscr{H}) = \min\{\vec{d}\ (H) : H \in \mathscr{H}\}$. Hence, $\vec{d}\ (r; m) = \vec{d}\ (\mathscr{H}_m^r)$. Very recently, Rajasekaran and Sampathkumar [8] found some results on three complete graphs with linkages. For more details refer to [8]. Also, they [8] posed the following problems:

- Examine $min\{\ell : \vec{d}\ (3; \ell) = 3\}$.
- Verify $min\{\ell : \vec{d}\ (\mathscr{H}(n, n, n; \ell)) = 2\}$.
- Determine $min\{\ell : \vec{d}\ (3; \ell) = 2\}$.

We focus on the minimum oriented diameter and form a corresponding optimal orientation for two complete bipartite graphs with additional edges in this article.

For $c, f \in V$, use $c \rightarrow f$ or $f \leftarrow c$ if $(c, f) \in A$. Generally, for $C, F \subseteq V$ with $C \cap F = \phi$, write, $C \rightarrow F$ if for every $c \in C$ and $f \in F$, $c \rightarrow f$. Write $c \rightarrow F$ for $\{c\} \rightarrow F$ and $C \rightarrow f$ for $C \rightarrow \{f\}$. The *converse* of D, denoted as $\widetilde{D}$, is the digraph derived from D by reversing each of its arcs. Hence, $d(D) = d(\widetilde{D})$. For notions and terminologies not completely covered here, see [1]. For wide number of results on minimum oriented diameter, see Koh and Tay's survey [4] and also refer to [2] and [5].

2 Linkages of Complete Bipartite Networks

This section starts with minimum oriented diameter of links between two complete bipartite networks.

For *four* fixed integers p_1, p_2, q_1, q_2 with $p_2 \geq p_1 \geq 2$ and $q_2 \geq q_1 \geq 2$, an integer m with $2 \leq m \leq (p_1 + p_2)(q_1 + q_2)$, and the number of edges with one end in K_{p_1,p_2} and other end in K_{q_1,q_2}, let $\mathscr{B}(p_1, p_2; q_1, q_2; m)$ be the class of graphs without cut edges and that are formed from the disjoint union of K_{p_1,p_2} and K_{q_1,q_2} by adding m edges between the vertices of K_{p_1,p_2} and K_{q_1,q_2}.

Let $X = \{x_1, x_2, \ldots, x_{p_1}\}$ and $Y = \{y_1, y_2, \ldots, y_{p_2}\}$ be the partite sets of K_{p_1,p_2} and $U = \{u_1, u_2, \ldots, u_{q_1}\}$ and $V = \{v_1, v_2, \ldots, v_{q_2}\}$ be that of K_{q_1,q_2}.

Theorem 1 $min\{m : \vec{d}\ (\mathscr{B}(2, 2; 2, 2; m)) = 5\} = 2.$

Proof Consider the digraph D with arcs $x_1 \to u_1$, $x_2 \to y_2$, $y_1 \to \{x_1, x_2\}$, $y_2 \to x_1$, $u_1 \to \{v_1, v_2\}$, $u_2 \to v_1$, $v_1 \to y_1$, $v_2 \to u_2$. Clearly, $\vec{d}\ (D) = 5$. □

Theorem 2 $min\{m : \vec{d}\ (\mathscr{B}(2, 2; 2, 2; m)) = 4\} \le 6.$

Proof Consider the digraph D with arcs $x_1 \to \{y_1, u_1\}$, $x_2 \to \{y_2, u_2\}$, $y_1 \to x_2$, $y_2 \to \{x_1, v_1\}$, $u_1 \to \{v_1, x_2\}$, $u_2 \to v_2$, $v_1 \to \{u_2, y_1\}$, $v_2 \to \{u_1, y_2\}$. Clearly, $\vec{d}\ (D) = 4$. □

Theorem 3 *For* $p \ge 2$, $q \ge 2$, *and* $p + q \ge 5$,

$$min\{m : \vec{d}\ (\mathscr{B}(2, p; 2, q; m)) = 6\} = 2.$$

Proof Let $G = K_{2,p} \cup K_{2,q} \cup \{x_1u_1, x_2u_2\}$. Then $G \in \mathscr{B}(2, p; 2, q; 2)$. Orient G as $x_1 \to u_1 \to V \to u_2 \to x_2 \to Y \to x_1$. Let's call the resulting directed graph D. Prove that $d(D) \le 6$. It follows from the dipath $x_1 \to u_1 \to \{v_i, v_j\} \to u_2 \to x_2 \to \{y_r, y_s\} \to x_1$. Consequently, $\vec{d}\ (\mathscr{B}(2, p; 2, q; 2)) \le 6$. □

Theorem 4 *For* $p \ge 2$, $q \ge 2$, *and* $p + q \ge 5$,

$$min\{m : \vec{d}\ (\mathscr{B}(2, p; 2, q; m)) \le 5\} \ge 3.$$

Proof If $G_0 \in \mathscr{B}(2, p; 2, q; 2)$ and D_0 is an orientation of G_0 such that $d(D_0) \le 5$. One of the two linked arcs in D_0 must have tail in $X \cup Y$ and head in $U \cup V$, and so, if necessary after relabeling, it must be one of the following: $x_1 \to u_1$, $x_1 \to v_1$, $y_1 \to u_1$, $y_1 \to v_1$. So we consider *four* cases. The other linked arc in D_0 must have tail in $U \cup V$ and head in $X \cup Y$.

Case 1. $x_1 \to u_1$.

The other linked arc in D_0 must be one of the following: $u_1 \to x_2$, $u_1 \to y_1$, $u_2 \to x_1$, $u_2 \to x_2$, $u_2 \to y_1$, $v_1 \to x_1$, $v_1 \to x_2$, $v_1 \to y_1$.

- $u_1 \to x_2$ is an arc in D_0 : Then $d_{D_0}(u_2, x_1) \le 5$ implies that $u_2 \to v_i \to u_1 \to x_2 \to y_j \to x_1$ for some i and j, and hence, $d_{D_0}(x_2, v_i) > 5$, a contradiction.
- $u_1 \to y_1$ is an arc in D_0 : Then $d_{D_0}(u_2, y_2) \le 5$ implies that $u_2 \to v_i \to u_1 \to y_1 \to x_j \to y_2$ for some i and j, and hence, $d_{D_0}(x_2, v_i) > 5$, a contradiction.
- $u_2 \to x_1$ is an arc in D_0 : Then $d_{D_0}(x_2, u_2) \le 5$ implies that $x_2 \to y_i \to x_1 \to u_1 \to v_j \to u_2$ for some i and j, and hence $d_{D_0}(u_1, y_i) > 5$, a contradiction.

- $u_2 \to x_2$ is an arc in D_0 : Then, for every $i \in \{1, 2, \ldots, p\}$, $d_{D_0}(u_1, y_i) \leq 5$ implies that $u_1 \to v_j \to u_2 \to x_2 \to y_i$ for some j. $d_{D_0}(y_1, y_2) \leq 5$ implies that $y_1 \to x_1 \to y_2$, and hence $d_{D_0}(y_2, y_1) > 5$, a contradiction.
- $u_2 \to y_1$ is an arc in D_0 : Then, for every $i \in \{2, 3, \ldots, p\}$, $d_{D_0}(u_1, y_i) \leq 5$ implies that $u_1 \to v_j \to u_2 \to y_1 \to x_k \to y_i$ for some $j \in \{1, 2, \ldots, q\}$ and $k \in \{1, 2\}$. If $k = 1$ for some $i \in \{2, 3, \ldots, p\}$, then $d_{D_0}(y_i, u_2) > 5$, a contradiction. Hence, for every $i \in \{2, 3, \ldots, p\}$, $k = 2$. $d_{D_0}(y_2, y_1) \leq 5$ implies that $y_2 \to x_1 \to y_1$, and hence $d_{D_0}(y_1, u_2) > 5$, a contradiction.
- $v_1 \to x_1$ is an arc in D_0 : Then $d_{D_0}(x_2, u_2) \leq 5$ implies that $x_2 \to y_i \to x_1 \to u_1 \to v_j \to u_2$ for some $i \in \{1, 2, \ldots, p\}$ and $j \in \{1, 2, \ldots, q\}$, and hence $d_{D_0}(v_2, y_i) > 5$, a contradiction.
- $v_1 \to x_2$ is an arc in D_0 : Then, for every $i \in \{2, 3, \ldots, q\}$, $d_{D_0}(v_i, x_1) \leq 5$ implies that $v_i \to u_k \to v_1 \to x_2 \to y_j \to x_1$ for some $k \in \{1, 2\}$ and $j \in \{1, 2, \ldots, p\}$. If $k = 1$ for some $i \in \{2, 3, \ldots, q\}$, then $d_{D_0}(x_2, v_i) > 5$, a contradiction. Hence, for every $i \in \{2, 3, \ldots, q\}$, $k = 2$. $d_{D_0}(v_1, u_2) \leq 5$ implies that $v_1 \to u_1$ and hence, $d_{D_0}(u_1, x_1) > 5$, a contradiction.
- $v_1 \to y_1$ is an arc in D_0 : Then for every $i \in \{2, 3, \ldots, q\}$ and $j \in \{2, 3, \ldots, p\}$, $d_{D_0}(v_i, y_j) \leq 5$ implies that $v_i \to u_k \to v_1 \to y_1 \to x_\ell \to y_j$ for some $k, \ell \in \{1, 2\}$. If $\ell = 1$ for some $i \in \{2, 3, \ldots, q\}$ and $j \in \{2, 3, \ldots, p\}$, then $d_{D_0}(y_j, u_2) > 5$, a contradiction. Hence, for every $i \in \{2, 3, \ldots, q\}$ and $j \in \{2, 3, \ldots, p\}$, $\ell = 2$. If $k = 1$ for some $i \in \{2, 3, \ldots, q\}$ and $j \in \{2, 3, \ldots, p\}$, then $d_{D_0}(x_2, v_i) > 5$, a contradiction. Hence, for every $i \in \{2, 3, \ldots, q\}$ and $j \in \{2, 3, \ldots, p\}$, $k = 2$. As $p + q \geq 5$, either $p \geq 3$ or $q \geq 3$. If $p \geq 3$, then for $j_1, j_2 \in \{2, 3, \ldots, p\}$ with $j_1 \neq j_2$, $d_{D_0}(y_{j_1}, y_{j_2}) \leq 5$ implies that $y_{j_1} \to x_1$, and hence $\{y_2, y_3, \ldots, y_p\} \to x_1$; now $d_{D_0}(y_2, y_3) \leq 5$ implies that $x_1 \to y_1$; consequently, $d_{D_0}(y_1, u_2) > 5$, a contradiction. Similarly, if $q \geq 3$, then for $i_1, i_2 \in \{2, 3, \ldots, q\}$ with $i_1 \neq i_2$, $d_{D_0}(v_{i_1}, v_{i_2}) \leq 5$ implies that $u_1 \to v_{i_2}$, and hence $u_1 \to \{v_2, v_3, \ldots, v_q\}$; now $d_{D_0}(v_2, v_3) \leq 5$ implies that $v_1 \to u_1$; consequently, $d_{D_0}(u_1, y_2) > 5$, a contradiction.

Case 2. $x_1 \to v_1$.

The other linked arc in D_0 must be one of the following: $v_1 \to x_2$, $v_1 \to y_1$, $u_1 \to x_1$, $u_1 \to x_2$, $u_1 \to y_1$, $v_2 \to x_1$, $v_2 \to x_2$, $v_2 \to y_1$.

- $v_1 \to x_2$ is an arc in D_0 : Then, for every $i \in \{2, 3, \ldots, q\}$, $d_{D_0}(v_i, x_1) \leq 5$ implies that $v_i \to u_k \to v_1 \to x_2 \to y_j \to x_1$ for some $k \in \{1, 2\}$ and $j \in \{1, 2, \ldots, p\}$, and hence $d_{D_0}(x_2, u_k) > 5$, a contradiction.
- $v_1 \to y_1$ is an arc in D_0 : Then, for every $i \in \{2, 3, \ldots, q\}$, $d_{D_0}(x_2, v_i) \leq 5$ implies that $x_2 \to y_j \to x_1 \to v_1 \to u_k \to v_i$ for some $j \in \{1, 2, \ldots, p\}$ and $k \in \{1, 2\}$, and hence $d_{D_0}(u_k, y_2) > 5$, a contradiction.
- $u_1 \to x_1$ is an arc in D_0 : Then $d_{D_0}(u_2, x_2) \leq 5$ implies that $u_2 \to v_i \to u_1 \to x_1 \to y_j \to x_2$ for some $i \in \{1, 2, \ldots, q\}$ and $j \in \{1, 2, \ldots, p\}$. If $i \neq 1$, then $d_{D_0}(y_j, v_i) > 5$, a contradiction. Hence $i = 1$, but then $d_{D_0}(y_j, v_2) > 5$, again a contradiction.

- $u_1 \to x_2$ is an arc in D_0 : Then, for every $i \in \{2, 3, \ldots, q\}$, $d_{D_0}(x_2, v_i) \le 5$ implies that $x_2 \to y_j \to x_1 \to v_1 \to u_k \to v_i$ for some $j \in \{1, 2, \ldots, p\}$ and $k \in \{1, 2\}$. If $k = 1$ for some $i \in \{2, 3, \ldots, q\}$, then $d_{D_0}(v_i, x_1) > 5$, a contradiction. Hence, for every $i \in \{2, 3, \ldots, q\}$, $k = 2$. $d_{D_0}(u_2, v_1) \le 5$ implies that $u_1 \to v_1$, and hence $d_{D_0}(x_2, u_1) > 5$, a contradiction.
- $u_1 \to y_1$ is an arc in D_0 : Then, for every $i \in \{2, 3, \ldots, p\}$, $d_{D_0}(u_2, y_i) \le 5$ implies that $u_2 \to v_j \to u_1 \to y_1 \to x_k \to y_i$ for some $j \in \{1, 2, \ldots, q\}$ and $k \in \{1, 2\}$. If $j = 1$ for some $i \in \{2, 3, \ldots, p\}$, then $d_{D_0}(x_2, u_2) > 5$, a contradiction, and hence $j \ne 1$. If $k = 1$ for some $i \in \{2, 3, \ldots, p\}$, then $d_{D_0}(y_i, v_2) > 5$, a contradiction, and hence $k = 2$. For every $\ell \in \{2, 3, \ldots, q\}$, $d_{D_0}(x_2, v_\ell) \le 5$ implies that $x_2 \to y_r \to x_1 \to v_1 \to u_s \to v_\ell$ for some $r \in \{2, 3, \ldots, p\}$ and $s \in \{1, 2\}$. If $s = 1$ for some $\ell \in \{2, 3, \ldots, q\}$, then $d_{D_0}(v_\ell, y_r) > 5$, a contradiction, and hence $s = 2$. As $p + q \ge 5$, either $p \ge 3$ or $q \ge 3$. If $p \ge 3$, then for $i_1, i_2 \in \{2, 3, \ldots, p\}$ with $i_1 \ne i_2$, $d_{D_0}(y_{i_1}, y_{i_2}) \le 5$ implies that $y_{i_1} \to x_1$, and hence $\{y_2, y_3, \ldots, y_p\} \to x_1$; now $d_{D_0}(y_2, y_3) \le 5$ implies that $x_1 \to y_1$; consequently, $d_{D_0}(y_1, v_2) > 5$, a contradiction. Similarly, if $q \ge 3$, then for $\ell_1, \ell_2 \in \{2, 3, \ldots, q\}$ with $\ell_1 \ne \ell_2$, $d_{D_0}(v_{\ell_1}, v_{\ell_2}) \le 5$ implies that $v_{\ell_1} \to u_1$, and hence $\{v_2, v_3, \ldots, v_q\} \to u_1$; now $d_{D_0}(v_2, v_3) \le 5$ implies that $u_1 \to v_1$; consequently, $d_{D_0}(v_1, y_2) > 5$, a contradiction.
- $v_2 \to x_1$ is an arc in D_0 : Then, for every $i \in \{1, 2, \ldots, p\}$, $d_{D_0}(y_i, v_2) \le 5$ implies that $y_i \to x_1 \to v_1 \to u_k \to v_2$ for some $k \in \{1, 2\}$, and hence $d_{D_0}(v_2, y_i) > 5$, a contradiction.
- $v_2 \to x_2$ is an arc in D_0 : Then, for every $i \in \{1, 2, \ldots, p\}$, $d_{D_0}(y_i, v_2) \le 5$ implies that $y_i \to x_1 \to v_1 \to u_k \to v_2$ for some $k \in \{1, 2\}$. $d_{D_0}(y_1, y_2) \le 5$ implies that $y_1 \to x_2 \to y_2$, and hence $d_{D_0}(y_2, y_1) > 5$, a contradiction.
- $v_2 \to y_1$ is an arc in D_0 : Then, for every $i \in \{1, 2, \ldots, p\}$, $d_{D_0}(y_i, v_2) \le 5$ implies that $y_i \to x_1 \to v_1 \to u_k \to v_2$ for some $k \in \{1, 2\}$. For every $j \in \{2, 3, \ldots, p\}$, $d_{D_0}(y_1, y_j) \le 5$ implies that $y_1 \to x_2 \to y_j$. Hence, $d_{D_0}(x_1, y_2) > 5$, a contradiction.

Case 3. $y_1 \to u_1$.

The other linked arc in D_0 must be one of the following: $u_1 \to x_1$, $u_1 \to y_2$, $u_2 \to x_1$, $u_2 \to y_1$, $u_2 \to y_2$, $v_1 \to x_1$, $v_1 \to y_1$, $v_1 \to y_2$.

- $u_1 \to x_1$ is an arc in D_0 : This is the converse digraph of the subcase $x_1 \to u_1 \to y_1$.
- $u_1 \to y_2$ is an arc in D_0 : This is similar to the subcase $v_2 \to x_1 \to v_1$.
- $u_2 \to x_1$ is an arc in D_0 : This is similar to the subcases $x_1 \to v_1$ and $u_1 \to x_2$. To see this, apply the permutation $(x_i, u_i)(y_i, v_i)$ and consider the converse digraph.
- $u_2 \to y_1$ is an arc in D_0 : This is similar to the subcase $x_1 \to v_1 \to x_2$.
- $u_2 \to y_2$ is an arc in D_0 : This is similar to the subcases $x_1 \to v_1$ and $v_2 \to x_2$.
- $v_1 \to x_1$ is an arc in D_0 : This is the converse digraph of the subcases $x_1 \to v_1$ and $u_1 \to y_1$.

- $v_1 \to y_1$ is an arc in D_0 : This is similar to the subcase $x_1 \to v_1 \to y_1$. To see this, apply the permutation $(x_i, u_i)(y_i, v_i)$ and consider the converse digraph.
- $v_1 \to y_2$ is an arc in D_0 : This is similar to the subcases $x_1 \to v_1$ and $v_2 \to y_1$. To see this, apply the permutation $(x_i, u_i)(y_i, v_i)$ and consider the converse digraph.

Case 4. $y_1 \to v_1$.

The other linked arc in D_0 must be one of the following: $u_1 \to x_1$, $u_1 \to y_1$, $u_1 \to y_2$, $v_1 \to x_1$, $v_1 \to y_2$, $v_2 \to x_1$, $v_2 \to y_1$, $v_2 \to y_2$.

- $u_1 \to x_1$ is an arc in D_0 : This is the converse digraph of the subcases $x_1 \to u_1$ and $v_1 \to y_1$.
- $u_1 \to y_1$ is an arc in D_0 : This is similar to the subcase $x_1 \to v_1 \to y_1$.
- $u_1 \to y_2$ is an arc in D_0 : This is similar to the subcases $x_1 \to v_1$ and $v_2 \to y_1$.
- $v_1 \to x_1$ is an arc in D_0 : This is the converse digraph of the subcase $x_1 \to v_1 \to y_1$.
- $v_1 \to y_2$ is an arc in D_0 : Then, for every $i \in \{2, 3, \ldots, p\}$ and $j \in \{2, 3, \ldots, q\}$, $d_{D_0}(y_i, v_j) \leq 5$ implies that $y_i \to x_k \to y_1 \to v_1 \to u_\ell \to v_j$ for some $k, \ell \in \{1, 2\}$, and hence $d_{D_0}(u_\ell, y_1) > 5$, a contradiction.
- $v_2 \to x_1$ is an arc in D_0 : This is similar to the subcases $x_1 \to v_1$ and $v_2 \to y_1$. To see this, apply the permutation (v_1, v_2) and consider the converse digraph.
- $v_2 \to y_1$ is an arc in D_0 : This is similar to the subcase $y_1 \to v_1 \to y_2$.
- $v_2 \to y_2$ is an arc in D_0 : Then, $d_{D_0}(v_1, y_1) \leq 5$ implies that $v_1 \to u_k \to v_2 \to y_2 \to x_\ell \to y_1$ for some $k, \ell \in \{1, 2\}$. As $p + q \geq 5$, either $p \geq 3$ or $q \geq 3$. If $p \geq 3$, then $d_{D_0}(y_1, y_3) \leq 5$ implies that $y_1 \to x_{\ell'}$, where $\{\ell, \ell'\} = \{1, 2\}$, and hence $d_{D_0}(x_{\ell'}, v_2) > 5$, a contradiction. If $q \geq 3$, then $d_{D_0}(v_2, v_3) \leq 5$ implies that $v_2 \to u_{k'}$, where $\{k, k'\} = \{1, 2\}$, and hence $d_{D_0}(u_{k'}, y_1) > 5$, a contradiction.

This completes the proof. □

Theorem 5 *For* $p_2 \geq p_1 \geq 3$ *and* $q_2 \geq q_1 \geq 3$,

$$min\{m : \vec{d}\,(\mathscr{B}(p_1, p_2; q_1, q_2; m)) = 5\} = 2.$$

Proof Define $G = K_{p_1,p_2} \cup K_{q_1,q_2} \cup \{x_1u_1, y_1v_1\}$. Then $G \in \mathscr{B}(p_1, p_2; q_1, q_2; 2)$. Orient G as follows:

(i) $x_1 \to y_1 \to v_1 \to u_1 \to x_1$;
(ii) $x_1 \to \{y_2, y_3, \ldots, y_{p_2}\}$, $\{x_2, x_3, \ldots, x_{p_1}\} \to y_1$, $\{v_2, v_3, \ldots, v_{q_2}\} \to u_1$, $v_1 \to \{u_2, u_3, \ldots, u_{q_1}\}$; and
(iii) Orient G_1 as $G[\{x_2, x_3, \ldots, x_{p_1}, y_2, y_3, \ldots, y_{p_2}\}]$ and $G_2 = G[\{u_2, u_3, \ldots, u_{q_1}, v_2, v_3, \ldots, v_{q_2}\}]$ such that $\vec{d}\,(G_1) \leq 4$ and $\vec{d}\,(G_2) \leq 4$.

Let D be the resulting directed one. Claim $d(D) \leq 5$. To prove the claim, compute eccentricities of the vertices of G.

Observe that, as $\vec{d}(G_1) \leq 4$, for every $i \in \{2, 3, \ldots, p_1\}$, there exists $j \in \{2, 3, \ldots, p_2\}$ such that $y_j \rightarrow x_i$; as $\vec{d}(G_2) \leq 4$, for every $k \in \{2, 3, \ldots, q_2\}$, there exists $\ell \in \{2, 3, \ldots, q_1\}$ such that $u_\ell \rightarrow v_k$.

Existence of the dipath: $x_1 \rightarrow y_1 \rightarrow v_1 \rightarrow \{u_1, u_2, \ldots, u_{q_1}\}$ and $x_1 \rightarrow \{y_2, y_3, \ldots, y_{p_2}\}$, in D, together with the above observation, we have $e_D(x_1) \leq 4$.

Let $i \in \{2, 3, \ldots, p_1\}$. Existence of the dipaths: $x_i \rightarrow y_1 \rightarrow v_1 \rightarrow u_1 \rightarrow x_1$ and $x_i \rightarrow y_1 \rightarrow v_1 \rightarrow \{u_2, u_3, \ldots, u_{q_1}\}$, in D, together with the above observation and $\vec{d}(G_1) \leq 4$, we have $e_D(x_i) \leq 4$.

Existence of the dipath: $y_1 \rightarrow v_1 \rightarrow u_1 \rightarrow x_1 \rightarrow \{y_2, y_3, \ldots, y_{p_2}\}$ and $y_1 \rightarrow v_1 \rightarrow \{u_2, u_3, \ldots, u_{q_1}\}$, in D, together with the above observation, we have $d_D(y_1) \leq 5$.

Let $i \in \{2, 3, \ldots, p_2\}$. Then there exists $r \in \{2, 3, \ldots, p_1\}$ such that $y_i \rightarrow x_r$. The dipaths exist from: $x_r \rightarrow y_1 \rightarrow v_1 \rightarrow u_1 \rightarrow x_1$ and $x_r \rightarrow y_1 \rightarrow v_1 \rightarrow \{u_2, u_3, \ldots, u_{q_1}\}$, in D, together with the above observation and $\vec{d}(G_1) \leq 4$, we have $e_D(y_i) \leq 5$.

The dipaths exist from: $u_1 \rightarrow x_1 \rightarrow y_1 \rightarrow v_1 \rightarrow \{u_2, u_3, \ldots, u_{q_1}\}$ and $u_1 \rightarrow x_1 \rightarrow \{y_2, y_3, \ldots, y_{p_2}\}$, in D, together with the above observation, we have $e_D(u_1) \leq 5$.

Let $i \in \{2, 3, \ldots, q_1\}$. Then there exists $r \in \{2, 3, \ldots, q_2\}$ such that $u_i \rightarrow v_r$. The dipaths exist from: $v_r \rightarrow u_1 \rightarrow x_1 \rightarrow y_1 \rightarrow v_1$ and $v_r \rightarrow u_1 \rightarrow x_1 \rightarrow \{y_2, y_3, \ldots, y_{p_2}\}$, in D, together with the above observation and $\vec{d}(G_2) \leq 4$, we have $e_D(u_i) \leq 5$.

The dipaths exist from: $v_1 \rightarrow \{u_2, u_3, \ldots, u_{q_1}\}$ and $v_1 \rightarrow u_1 \rightarrow x_1 \rightarrow \{y_1, y_2, \ldots, y_{p_2}\}$, in D, together with the above observation, we have $e_D(v_1) \leq 4$.

Let $i \in \{2, 3, \ldots, q_2\}$. The dipaths exist from: $v_i \rightarrow u_1 \rightarrow x_1 \rightarrow \{y_1, y_2, \ldots, y_{p_2}\}$ and $v_i \rightarrow u_1 \rightarrow x_1 \rightarrow y_1 \rightarrow v_1$, in D, together with the above observation and $\vec{d}(G_2) \leq 4$, we have $e_D(v_i) \leq 4$.

Hence, $d(D) \leq 5$. This ends the proof. □

Theorem 6 *For* $p_2 \geq p_1 \geq 3$ *and* $q_2 \geq q_1 \geq 3$, $min\{m : \vec{d}(\mathscr{B}(p_1, p_2; q_1, q_2; m)) \leq 4\} \geq 3$.

Proof If $G_0 \in \mathscr{B}(p_1, p_2; q_1, q_2; 2)$ and D_0 is an orientation of G_0 such that $d(D_0) \leq 4$. One of the two linked arcs in D_0 must have tail in $X \cup Y$ and head in $U \cup V$, and so, if necessary after relabeling, it must be one of the following: $x_1 \rightarrow u_1$, $x_1 \rightarrow v_1$, $y_1 \rightarrow u_1$, $y_1 \rightarrow v_1$. So we consider *four* cases. In each of the *four* cases, the other linked arc in D_0 must have tail in $U \cup V$ and head in $X \cup Y$.

Case 1. $x_1 \rightarrow u_1$.

The other linked arc in D_0 must be one of the following: $u_1 \rightarrow x_2$, $u_1 \rightarrow y_1$, $u_2 \rightarrow x_1$, $u_2 \rightarrow x_2$, $u_2 \rightarrow y_1$, $v_1 \rightarrow x_1$, $v_1 \rightarrow x_2$, $v_1 \rightarrow y_1$. But then $d_{D_0}(x_2, u_2) > 4$, a contradiction.

Case 2. $x_1 \to v_1$.

The other linked arc in D_0 must be one of the following: $v_1 \to x_2$, $v_1 \to y_1$, $u_1 \to x_1$, $u_1 \to x_2$, $u_1 \to y_1$, $v_2 \to x_1$, $v_2 \to x_2$, $v_2 \to y_1$. But then $d_{D_0}(x_2, v_2) > 4$, a contradiction.

Case 3. $y_1 \to u_1$.

The other linked arc in D_0 must be one of the following: $u_1 \to x_1$, $u_1 \to y_2$, $u_2 \to x_1$, $u_2 \to y_1$, $u_2 \to y_2$, $v_1 \to x_1$, $v_1 \to y_1$, $v_1 \to y_2$. But then $d_{D_0}(y_2, u_2) > 4$, a contradiction.

Case 4. $y_1 \to v_1$.

The other linked arc in D_0 must be one of the following: $u_1 \to x_1$, $u_1 \to y_1$, $u_1 \to y_2$, $v_1 \to x_1$, $v_1 \to y_2$, $v_2 \to x_1$, $v_2 \to y_1$, $v_2 \to y_2$. But then $d_{D_0}(y_2, v_2) > 4$, a contradiction.

This ends the proof. □

Theorem 7 *For $p_2 \geq p_1 \geq 4$ and $q_2 \geq q_1 \geq 4$,* $min\{m : \vec{d}(\mathscr{B}(p_1, p_2; q_1, q_2; m)) = 4\} \leq 8$.

Proof Let $G = K_{p_1,p_2} \cup K_{q_1,q_2} \cup \{x_1u_1, x_2u_2, y_1v_1, y_2v_2, x_1v_2, x_2v_1, y_1u_2, y_2u_1\}$. Then $G \in \mathscr{B}(p_1, p_2; q_1, q_2; 8)$. Orient G as follows:

(i) $y_2 \to \{u_1, v_2\} \to x_1$, $u_2 \to \{x_2, y_1\} \to v_1$;
(ii) $x_1 \to y_2 \to x_2 \to y_1 \to x_1$, $u_1 \to v_1 \to u_2 \to v_2 \to u_1$;
(iii) $x_1 \to \{y_3, y_4, \ldots, y_{p_2}\} \to x_2$, $y_1 \to \{x_3, x_4, \ldots, x_{p_1}\} \to y_2$, $u_1 \to \{v_3, v_4, \ldots, v_{q_2}\} \to u_2$, $v_1 \to \{u_3, u_4, \ldots, u_{q_1}\} \to v_2$;
(iv) Orient G_1 as $G[\{x_3, x_4, \ldots, x_{p_1}, y_3, y_4, \ldots, y_{p_2}\}]$ and $G_2 = [\{u_3, u_4, \ldots, u_{q_1}, v_3, v_4, \ldots, v_{q_2}\}]$ such that $\vec{d}(G_1) \leq 4$ and $\vec{d}(G_2) \leq 4$.

Let D be the oriented resulting one. Claim $d(D) \leq 4$. By computing eccentricities for the vertices of G, demonstrate the claim.

Observe that, as $\vec{d}(G_1) \leq 4$, for every $i \in \{3, 4, \ldots, p_1\}$, there exists $j \in \{3, 4, \ldots, p_2\}$ such that $y_j \to x_i$ and for every $k \in \{3, 4, \ldots, p_2\}$, there exists $\ell \in \{3, 4, \ldots, p_1\}$ such that $x_\ell \to y_k$; as $\vec{d}(G_2) \leq 4$, for every $i \in \{3, 4, \ldots, q_2\}$, there exists $j \in \{3, 4, \ldots, q_1\}$ such that $u_j \to v_i$ and for every $k \in \{3, 4, \ldots, q_1\}$, there exists $\ell \in \{3, 4, \ldots, q_2\}$ such that $v_\ell \to u_k$.

Existence of the dipaths: $x_1 \to y_2 \to x_2 \to y_1$, $x_1 \to \{y_3, y_4, \ldots, y_{p_2}\}$, $x_1 \to y_2 \to u_1 \to \{v_3, v_4, \ldots, v_{q_2}\}$, $x_1 \to y_2 \to u_1 \to v_1 \to u_2$, and $x_1 \to y_2 \to v_2$, in D, together with the above observation, we have $e_D(x_1) \leq 4$.

From these existence of dipaths: $x_2 \to y_1 \to x_1 \to y_2 \to u_1$, $x_2 \to y_1 \to x_1 \to \{y_3, y_4, \ldots, y_{p_2}\}$, $x_2 \to y_1 \to v_1 \to u_2 \to v_2$, and $x_2 \to y_1 \to v_1 \to \{u_3, u_4, \ldots, u_{q_1}\}$, in D, together with the above observation, we have $e_D(x_2) \leq 4$.

Let $i \in \{3, 4, \ldots, p_1\}$. The dipaths exist from: $x_i \to y_2 \to x_2 \to y_1 \to x_1$, $x_i \to y_2 \to u_1 \to \{v_3, v_4, \ldots, v_{q_2}\} \to u_2$, $x_i \to y_2 \to u_1 \to v_1$, and

$x_i \to y_2 \to v_2$, in D, together with the above observation and $\vec{d}\ (G_1) \leq 4$, we have $e_D(x_i) \leq 4$.

From these existence of dipaths: $y_1 \to x_1 \to y_2 \to x_2$, $y_1 \to \{x_3, x_4, \ldots, x_{p_1}\}$, $y_1 \to v_1 \to u_2$, $y_1 \to v_1 \to \{u_3, u_4, \ldots, u_{q_1}\}$, and $y_1 \to x_1 \to y_2 \to \{u_1, v_2\}$, in D, together with the above observation, we have $e_D(y_1) \leq 4$.

From these existence of dipaths: $y_2 \to x_2 \to y_1 \to \{x_1\} \cup \{x_3, x_4, \ldots, x_{p_1}\}$, $y_2 \to u_1 \to \{v_3, v_4, \ldots, v_{q_2}\}$, $y_2 \to u_1 \to v_1 \to u_2$, and $y_2 \to v_2$, in D, together with the above observation, we have $e_D(y_2) \leq 4$.

Let $i \in \{3, 4, \ldots, p_2\}$. The dipaths exist from: $y_i \to x_2 \to y_1 \to x_1 \to y_2$, $y_i \to x_2 \to v_1 \to \{u_2\} \cup \{u_3, u_4, \ldots, u_{q_1}\}$, and $y_i \to x_j \to y_2 \to \{u_1, v_2\}$ for some $j \in \{3, 4, \ldots, p_1\}$, in D, together with the above observation and $\vec{d}$ $(G_1) \leq 4$, we have $e_D(y_i) \leq 4$.

From these existence of dipaths: $u_1 \to x_1 \to y_2 \to x_2 \to y_1$, $u_1 \to x_1 \to \{y_3, y_4, \ldots, y_{p_2}\}$, $u_1 \to v_1 \to \{u_3, u_4, \ldots, u_{q_1}\}$, and $u_1 \to v_1 \to u_2 \to v_2$, in D, together with the above observation, we have $e_D(u_1) \leq 4$.

From these existence of dipaths: $u_2 \to y_1 \to x_1 \to y_2 \to x_2$, $u_2 \to y_1 \to x_1 \to \{y_3, y_4, \ldots, y_{p_2}\}$, and $u_2 \to v_2 \to u_1 \to \{v_1\} \cup \{v_3, v_4, \ldots, v_{q_2}\}$, in D, together with the above observation, we have $e_D(u_2) \leq 4$.

Let $i \in \{3, 4, \ldots, q_1\}$. The dipaths exist from: $u_i \to v_2 \to u_1 \to v_1 \to u_2$, $u_i \to v_2 \to x_1 \to \{y_3, y_4, \ldots, y_{p_2}\}$, $u_i \to v_2 \to x_1 \to y_2 \to x_2$, and $u_i \to v_j \to u_2 \to y_1$ for some $j \in \{3, 4, \ldots, q_2\}$, in D, together with the above observation and $\vec{d}\ (G_2) \leq 4$, we have $e_D(u_i) \leq 4$.

From these existence of dipaths: $v_1 \to \{u_3, u_4, \ldots, u_{p_1}\}$, $v_1 \to u_2 \to v_2 \to u_1$, $v_1 \to u_2 \to y_1 \to \{x_3, x_4, \ldots, x_{p_1}\}$, $v_1 \to u_2 \to y_1 \to x_1 \to y_2$, and $v_1 \to u_2 \to x_2$, in D, together with the above observation, we have $e_D(v_1) \leq 4$.

From these existence of dipaths: $v_2 \to x_1 \to \{y_3, y_4, \ldots, y_{p_2}\}$, $v_2 \to x_1 \to y_2 \to x_2 \to y_1$, $v_2 \to u_1 \to \{v_3, v_4, \ldots, v_{q_2}\}$, and $v_2 \to u_1 \to v_1 \to u_2$, in D, together with the above observation, we have $e_D(v_2) \leq 4$.

Let $i \in \{3, 4, \ldots, q_2\}$. From these existence of dipaths: $v_i \to u_2 \to v_2 \to u_1 \to v_1$, $v_i \to u_2 \to y_1 \to \{x_3, x_4, \ldots, x_{p_1}\}$, $v_i \to u_2 \to y_1 \to x_1 \to y_2$, and $v_i \to u_2 \to x_2$, in D, together with the above observation and $\vec{d}\ (G_2) \leq 4$, we have $e_D(v_i) \leq 4$. Hence, $d(D) \leq 4$. This completes the proof. □

3 Conclusion

We end this article by the following problems:

Problem 1 Find $min\{m : \vec{d}\ (\mathscr{B}(2, 2; 2, 2; m)) = 4\}$.

Problem 2 For $p \geq 2$, $q \geq 2$, and $p + q \geq 5$, find: *(i)* $min\{m : \vec{d}\ (\mathscr{B}(2, p; 2, q; m)) = 5\}$, and

(ii) $min\{m : \overrightarrow{d}\ (\mathscr{B}(2, p; 2, q; m)) = 4\}$.

Problem 3 For $p_2 \geq p_1 \geq 3$ and $q_2 \geq q_1 \geq 3$, find: $min\{m : \overrightarrow{d}\ (\mathscr{B}(p_1, p_2; q_1, q_2; m)) = 4\}$.

Problem 4 For $p_2 \geq p_1 \geq 4$ and $q_2 \geq q_1 \geq 4$, find: $min\{m : \overrightarrow{d}\ (\mathscr{B}(p_1, p_2; q_1, q_2; m)) = 4\}$.

Notes and Comments This article presents some new findings regarding the minimum oriented diameter of two complete bipartite networks with connecting edges. This article also encourages future research on the same parameter.

References

1. Bang-Jensen, J., Gutin, G.: Digraphs: Theory, Algorithms and Applications, 2nd edn. Springer, London (2009)
2. Boesch, F., Tindell, R.: Robbin's theorem for mixed multigraphs. Am. Math. Monthly **87**, 716–719 (1980)
3. Koh, K.M., Ng, K.L.: The orientation number of two complete graphs with linkages. Discrete Math. **295**, 91–106 (2005)
4. Koh, K.M., Tay, E.G.: Optimal orientations of graphs and digraphs: a survey. Graphs Combin. **18**, 745–756 (2002)
5. Maurer, S.B.: The king chicken theorems. Math. Mag. **53**, 67–80 (1980)
6. Ng, K.L.: On a conjecture concerning the orientation number of a graph. Discrete Math. **309**, 1603–1610 (2009)
7. Robbins, H.: A theorem on graphs with an application to a problem of traffic control. Am. Math. Monthly **46**, 281–283 (1939)
8. Rajasekaran, G., Sampathkumar, R.: The orientation number of three complete graphs with linkages. TWMS J. Appl. Eng. Math. **11**, 13–24 (2021)

Packing Chromatic Number of Windmill Related Graphs and Chain Silicate Networks

Tony Augustine and Roy Santiago

1 Introduction

Graph coloring has many applications in the field of frequency assignments. To prevent interference, radio and television transmitters use different frequencies to operate. But given the limited spectrum and growing number of broadcast stations, some transmitters might use the same frequency. Transmitters using the same frequency must be placed far enough apart from one another to minimize interference. Goddard et al. [1] introduced the idea of broadcast coloring, which Brešar et al. [2] later referred to as packing chromatic number, as a mathematical solution to this issue. After studying a few transformation graphs, Durgun and Dörtok expanded on their packing chromatic numbers [3]. By demonstrating that the packing chromatic number is more than 13 in the class of subcubic graphs, Brešar et al. [4] addressed an unresolved issue from Gastineau and Togni [5]. A sharp upper bound and a sharp lower bound were provided for the packing chromatic number of the lexicographic product of graphs by Božović and Peterin [6]. The packing chromatic number for circular fans with two and four chords and a mesh of trees was provided by William et al. [7]. A known upper bound on the packing chromatic number of the square lattice was improved by Soukal and Holub [8]. A few generalizations of the packing chromatic number of transformation graphs were provided by Aruldoss and Murugarajan [9]. Some general characteristics of the packing chromatic numbers of finite super subdivisions of graphs were described by Lemdania et al. [10]. They also calculated the packing chromatic numbers of various neighborhood corona graphs, cycles, and finite super subdivisions of complete

T. Augustine · R. Santiago (✉)
Department of Mathematics, School of Advanced Sciences, Vellore Institute of Technology, Vellore, India
e-mail: roy.santiago@vit.ac.in

F. Kamalov et al. (eds.), *Advances in Mathematical Modeling and Scientific Computing*, Trends in Mathematics, https://doi.org/10.1007/978-3-031-41420-6_67

graphs. For graphs with a diameter of at least 3, determining the packing chromatic number is NP-complete [11]. Argiroffo et al. demonstrated that this problem may be solved in polynomial time for various nontrivial graph classes [12]. For numerous graph families, the packing chromatic number has been investigated [10–20]. For a given graph G, a mapping $\varphi : V(G) \longrightarrow \{1, 2, \ldots, r\}$, such that $\varphi(x) = \varphi(y) = k$ implies $d(x, y) > k$, is termed a packing coloring of G. The packing chromatic number of G is the least number of colors in a packing coloring of G, denoted by $\chi_\rho(G)$. This article provides the best lower bound for packing chromatic number of windmill related graphs and chain silicate networks.

2 Results

Definition 1 ([21]) The windmill graph $W(k, r)$ is an undirected graph constructed for $k \geq 2$ and $r \geq 2$ by joining n copies of the complete graph K_k at a shared vertex. It has $(k-1)r + 1$ vertices.

Definition 2 ([21]) An n-uniform windmill star graph $WS_n(k, r)$ includes a star S_{n+1} along with hub at p so that the deletion of the n edges of S_{n+1} partitions the graph into n independent windmills $W^i(k, r), i \in [1, n]$ and an isolated vertex. See Fig. 1.

In the following theorems, for $c(v)$, $d(u, v)$, and $diam(G)$, we mean color of a vertex v, distance between vertices u and v, and diameter of graph G, respectively.

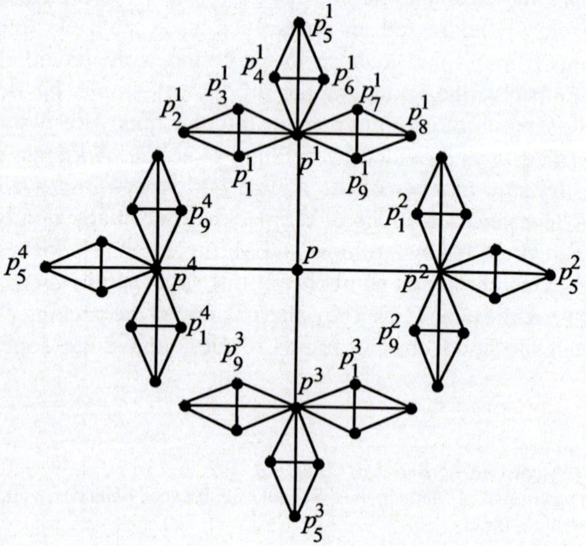

Fig. 1 A 4-uniform windmill star graph $WS_4(4, 3)$

Theorem 1 *For n-uniform windmill star graph* $WS_n(k,r)$, $\chi_\rho(WS_n(k,r)) = 3 + n(kr - 2r - 1)$, $n \geq 4, k \geq 3, r \geq 2$.

Proof Let $W^i(k,r), i \in [1,n], k \geq 3, r \geq 2$ be the windmills of $WS_n(k,r)$. Let $V(W^i(k,r)) = \{p^i_j, p^i, p : i \in [1,n], j \in [1,(k-1)r]\}$, where p^i is the hub of $W^i(k,r)$ and p is the hub of S_{n+1}. See Fig. 1. Since $\text{diam}(WS_n(k,r)) = 4$, no vertex of $WS_n(k,r)$ accepts color greater than 3.

Case 1: Suppose $c(p) = 3$. Since $d(p, p^i_j) = 2$ and $d(p, p^i) = 1, i \in [1,n], j \in [1,(k-1)r]$, no other vertices of $WS_n(k,r)$ accept 3. Choose any one vertex p^i of $W^i(k,r)$ and suppose $c(p^i) = 3$. Since all other vertices of $WS_n(k,r)$ are at distance ≤ 3 to chosen vertex p^i, no other vertices of $WS_n(k,r)$ can accept 3. Similarly, choose any one vertex p^i_j of $W^i(k,r)$ and suppose $c(p^i_j) = 3$. Since $\text{diam}(W^i(k,r)) = 2$ and $d(p^i_j, p^l_m) = 4, i \neq l, i,l \in [1,n], j,m \in [1,(k-1)r]$, no more than n vertices can accept 3.

Case 2: Following the same arguments of case 1 for vertices p, p^i, p^i_j for color 2 and since $d(p^i_j, p^l_m) = 4$ and $d(p^i, p^l_m) = 3, i \neq l, i,l \in [1,n], j,m \in [1,(k-1)r]$, allocating 2 to p^i_j or to p^i, no more than n vertices can accept 2.

Case 3: If 1 is set to any p^i, to the highest $(n-1)r + 1$ vertices can have color 1. If $c(p) = 1$ and color of one vertex in each $W^i(k,r)$ is 1, no more than $nr + 1$ vertices can accept 1.

There are $n(kr - r + 1) + 1$ vertices in $WS_n(k,r)$ and $nr + 2n + 1$ vertices received colors 1, 2, and 3. Thus, the remaining $n(kr - 2r - 1)$ vertices could be given different colors from 4. Hence, $\chi_\rho(WS_n(k,r)) \geq 3 + n(kr - 2r - 1)$.

The following is the procedure to color the n-uniform windmill star graph $WS_n(k,r)$ and the sharpness of the bound is proved as well.

Agenda: Packing Coloring $WS_n(k,r)$,$n \geq 4, k \geq 3, r \geq 2$
Input: A uniform n-fan split graph $WS_n(k,r)$
Algorithm:
Rule 1: Assign 1 to p and $p^i_{(k-1)j-(k-2)}$,$i \in [1,n], j \in [1,r]$.
Rule 2: Assign 2 and 3 to p^i_2 and $p^i_{(k-1)r}$ respectively.
Rule 3: Assign 4 to $3 + n(kr - 2r - 1)$ to the remaining vertices of $WS_n(k,r)$.
Output: A packing $3 + n(kr - 2r - 1)$ coloring of $WS_n(k,r)$

Proof of Correctness: The $\text{diam}(W^i(k,r)) = 2$. Since $d(p^i_{(k-1)j-(k-2)}, p^l_{(k-1)m-(k-2)}) \leq 2, j \neq m, i,l \in [1,n], j,m \in [1,r]$ and $d(p^i_{(k-1)j-(k-2)}, p) = 2$, coloring vertices $p^i_{(k-1)j-(k-2)}$,$i \in [1,n], j \in [1,r]$ and p by 1, $nr + 1$ vertices accept 1. Since $\text{diam}(WS_n(k,r)) = 4$ and $d(p^i_j, p^l_m) = 4$, $i \neq l$, $i,l \in [1,n], j,m \in [1,(k-1)r]$, by giving 2 and 3 to vertices p^i_2 and $p^i_{(k-1)r}$ respectively, $2n$ vertices accept 2 and 3 as there are n independent windmills. Now, if we color remaining $n(kr - 2r - 1)$ vertices with different colors starting from 4, we get a packing $3 + n(kr - 2r - 1)$ coloring of $WS_n(k,r)$. Hence, $\chi_\rho(WS_n(k,r)) \leq 3 + n(kr - 2r - 1)$. □

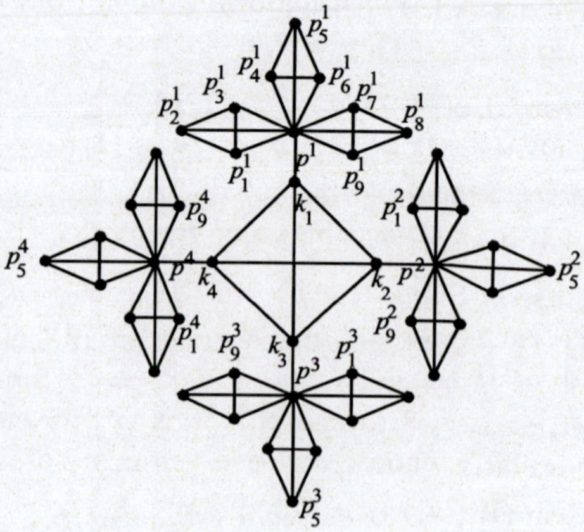

Fig. 2 A 4-uniform windmill complete graph $WK_4(4, 3)$

Definition 3 ([21]) Let $k_i, i \in [1, n]$ be the vertices of the complete graph K_n. Let $W^i(k, r)$ be the windmills with hubs $p^i, i \in [1, n]$. Let $k_i p^i, i \in [1, n]$ be an edge. The graph constructed is called n-uniform windmill complete graph and denoted by $WK_n(k, r)$.

Theorem 2 *For n-uniform windmill complete graph* $WK_n(k, r)$, $\chi_\rho(WK_n(k, r)) \geq 3 + n(kr - 2r - 1), n \geq 4, k \geq 3, r \geq 3$.

Proof Let $W^i(k, r), i \in [1, n], k \geq 3, r \geq 2$ be the windmills of $WK_n(k, r)$. Let $W^i(k, r)) = \{p^i_j, p^i : i \in [1, n], j \in [1, (k-1)r]\}$, where p^i is the hub of $W^i(k, r)$. See Fig. 2. Since diam$(WK_n(k, r)) = 5$, no vertex of $WK_n(k, r)$ accepts color greater than 4 more than once.

Case 1: Suppose $c(k_i) = 4$ or $c(p^i) = 4$ for any $i \in [1, n]$. Since $d(k_i, p^i_j) = 2$ and $d(k_i, p^i) = 1, i \in [1, n], j \in [1, (k-1)r]$, no other vertices of $KW(n, r)$ can receive 4. Now if we choose any one vertex p^i_j of any $W^i(k, r)$, then $c(p^i_j) = 4$. Since $d(p^i_j, p^l_m) = 5, i \neq l, i, l \in [1, n], j, m \in [1, (k-1)r]$ and there are n windmills, no more than n vertices can accept 4.

Case 2: Suppose $c(k_i) = 3$ for any $i \in [1, n]$. Since $d(k_i, p^i_j) = 2$ and $d(k_i, p^i) = 1, i \in [1, n], j \in [1, (k-1)r]$, no other vertices of $WK_n(k, r)$ can accept 3. Similarly, if 3 is set to p^i or to p^i_j, no more than n vertices of $WK_n(k, r)$ can accept 3.

Case 3: Suppose $c(k_i) = 2$ or $c(p^i) = 2$ or $c(p^i_j) = 2$ for any $i \in [1, n]$. Since $d(p^i_j, p^l_m) = 5$, $d(p^i, p^l_m) = 4$ and $d(u_i, w^l_m) = 3, i \neq l, i, l \in [1, n], j, m \in [1, (k-1)r]$, no more than n vertices of $WK_n(k, r)$ can accept 2.

Case 4: If 1 is allocated to p^i for any $i \in [1, n]$, r vertices in each $W^i(k, r)$ and one vertex in K_n can be colored as there are n windmills. Thus, $nr + 1$ vertices can accept 1.

Fig. 3 SL_3

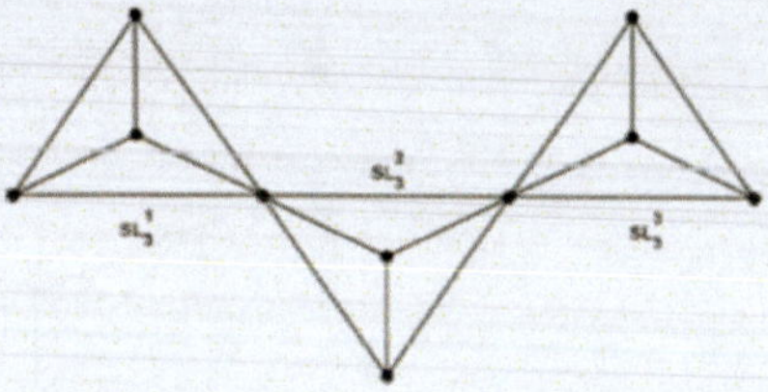

Fig. 4 Packing coloring of SL_3

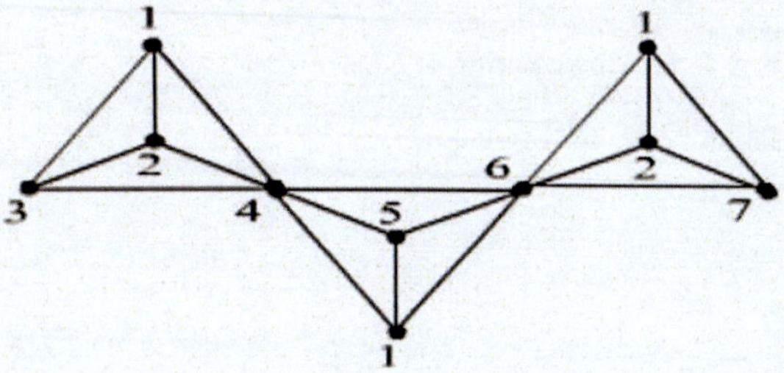

There are $n(kr - r + 2)$ vertices in $WK_n(k, r)$ and $nr + 3n + 1$ vertices received 1, 2, 3, and 4. Thus the remaining $nr(k-2)-(n+1)$ vertices should accept different colors from 5. Hence, $\chi_\rho(WS_n(k, r)) \geq 4 + nr(k - 2) - (n + 1)$. □

Definition 4 ([22]) Consider a single tetrahedron (i.e., a pyramid having a triangular base). The silicon atom is joined to evenly spaced oxygen atoms when oxygen atoms are placed at each of a tetrahedron's four corners. It produces a silicate tetrahedron, and when it unites with other tetrahedra linearly, a single-row silicate chain is created. Each tetrahedron is a complete graph with four vertices (Fig. 3).

Lemma 1 $\chi_\rho(SL_n) \geq \begin{cases} 7 \ for n = 3 \\ 8 \ for n = 4 \\ 9 \ for n = 5. \end{cases}$

Proof

Case 1: When $n = 3$

Consider SL_3; since $diam(SL_3) = 3$ and SL_3 consists of 3 copies of K_4 (Fig. 4), maximum of two vertices can receive 2, and one vertex in each K_4 can receive 1. The remaining 5 vertices should receive a distinct color starting from 3. Thus $\chi_\rho(SL_3) \geq 7$.

Case 2: When $n = 4$

Since SL_4 consists of 4 copies of K_4 and $diam(SL_4) = 4$ and SL_3, maximum of two pairs of vertices can receive colors 2 and 3 (Fig. 5) and one vertex in each K_4 can receive color 1. The remaining 5 vertices should receive a distinct color starting from 4. Then $\chi_\rho(SL_4) \geq 8$.

Case 3: When $n = 5$

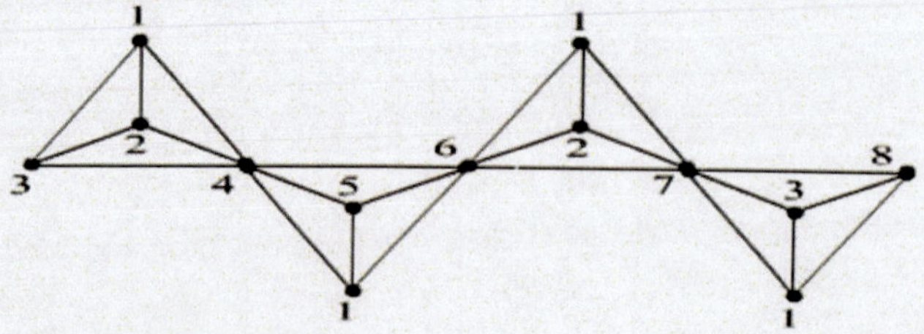

Fig. 5 Packing coloring of SL_4

Fig. 6 Packing coloring of SL_5

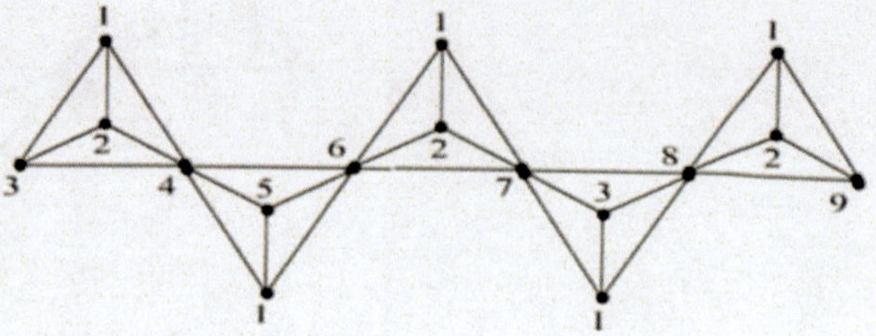

Consider SL_5; it contains five copies of K_4 and number of vertices is 16. Since $diam(SL_3) = 5$, we can give color 1 to each vertex of K_4. Also maximum three vertices can receive color 2 and two vertices can receive color 3; then the remaining 6 vertices should receive distinct colors starting from 4 (Fig. 6). Then $\chi_\rho(SL_5) \geq 9$. □

Lemma 2 $\chi_\rho(SL_n) \leq \begin{cases} 7 \; forn = 3 \\ 8 \; forn = 4 \\ 9 \; forn = 5. \end{cases}$

Proof From Figs. 4, 5, and 6, we get the result. □

Theorem 3 $\chi_\rho(SL_n) = \begin{cases} 7 \; forn = 3 \\ 8 \; forn = 4 \\ 9 \; forn = 5. \end{cases}$

Proof From Lemmas 1 and 2, we get the result. □

3 Conclusion

This article has calculated the packing chromatic number of n-uniform windmill star graph, n-uniform windmill complete graph, and chain silicate networks for $n = 3, 4, 5$. The problem of packing coloring of subdivision of n-uniform windmill star graph and n-uniform windmill complete graph and general chain silicate networks are under investigation.

Acknowledgments The authors would like to express their sincere gratitude to the anonymous referees for their suggestions for the improvement of this article.

References

1. Goddard, W., Hedetniemi, S.M., Hedetniemi, S.T., Harris, J.M., Rall, D.F.: Broadcast chromatic numbers of graphs. Ars Combin. **86**, 33–49 (2008)
2. Brešar, B., Klavžar, S., Rall, D.: On the packing chromatic number of Cartesian products, hexagonal lattice, and trees. Discrete Appl. Math. **155**, 2303–2311 (2007)
3. Durgun, D., Dörtok, B.: Packing chromatic number of transformation graphs. Therm. Sci. **23**, 1991–1995 (2019)
4. Brešar, B., Klavar, S., Rall, D., Wash, K.: Packing chromatic number under local changes in a graph. Discrete Math. **340**, 1110–1115 (2017)
5. Gastineau, N., Togni, O.: S-packing colorings of cubic graphs. Discrete Math. **339**, 2461–2470 (2016)
6. Božović, D., Peterin, I.: A note on the packing chromatic number of lexicographic products. Discrete Appl. Math. **293**, 34–37 (2021)
7. William, A., Roy, S., Rajasingh, I.: Packing chromatic number of circular fans and mesh of trees. Int. J. Soft Comput. **4**(2), 145–152 (2014)
8. Holub, P., Soukal, R.: A note on packing chromatic number of the square lattice. Electron. J. Comb. **17**, N17–N17 (2010)
9. Aruldoss, R., Murugarajan, P.: A generalized packing chromatic number of families of unicycle hairy graphs. J. Comput. Sci. **10**(5), 518–523 (2020)
10. Lemdania, R., Abbasa, M., Fermec, J.: Packing chromatic numbers of finite super subdivisions of graphs. Combinatorics **34**(10), 3275–3286 (2020)
11. Kim, M., Lidick, B., Masak, T., Pfender, F.: Notes on complexity of packing coloring. Inform. Process. Lett. **137**, 6–10 (2018)
12. Argiroffo, G., Nasini, G., Torres, P.: The packing coloring problem for lobsters and partner limited graphs. Discrete Appl. Math. **164**, 373–382 (2014)
13. Brešar, B., Klavžar, S., Rall, D.F.: On the packing chromatic number of Cartesian products, hexagonal lattice, and trees. Discrete Appl. Math. **155**, 2303–2311 (2007)
14. Fres'an-Figueroa, J., Gonz'alez-Moreno, D., Olsen, M.: On the packing chromatic number of Moore graphs. Discret. Appl. Math. **289**, 185–193 (2021)
15. Balogh, J., Kostochka, A., Liu, X.: Packing chromatic number of cubic graphs. Discrete Math. **341(2)**, 474–483 (2018)
16. Brešar, B., Klavžar, S., Rall, D.F.: Packing chromatic number of base-3 Sierpi nski graphs. Graphs Combin. **32**, 1313–1327 (2016)
17. Finbow, A.S., Rall, D.F.: On the packing chromatic number of some lattices. Discrete Appl. Math. **158**, 1224–1228 (2010)
18. Gastineau, N., Holub, P., Togni, O.: On the packing chromatic number of subcubic outerplanar graphs. Discrete Appl. Math. **255**, 209–221 (2019)
19. Martin, B., Raimondi, F., Chen, T., Martin, J.: The packing chromatic number of the infinite square lattice is between 13 and 15. Discrete Appl. Math. **225**, 136–142 (2017)
20. Torres, P., Valencia-Pabon, M.: The packing chromatic number of hypercubes. Discrete Appl. Math. **190191**, 127–140 (2015)
21. Roy, S.: Packing chromatic number of certain fan and wheel related graph. AKCE Int. J. Graphs Comb. **14(1)**, 63–69 (2017)
22. Liu, J.B., Shafiq, M., Ali, H., Naseem, A., Asghar, S.S.: Topological indices of mth chain silicate graphs. Mathematics **7**, 42 (2019)

Complementary Triple Connected Total Domination Number of a Graph

K. Priya, G. Mahadevan, and C. Sivagnanam

1 Introduction

Defining G as (V(G),E(G)), we get V(G) for vertex sets and E(G) for edge sets. Here, undirected, finite, and non-empty sets. Examine [2, 3]. The idea of total domination in graphs was first proposed by Cockayne, Dawes, and Hedetniemi [1] and has been extensively studied by several academics over the past year. G. Mahadevan et al. [4] were the first to present triple connected domination numbers. In this paper we introduce the new parameter by just adding a triple connected condition on the complement of total dominating set and the parameter to be complementary triple connected total domination number of a graph.

Helm graph H_a is obtained by attaching pendant vertices to each vertices of wheel graph $W_{1,a}$. The middle graph $M(G)$ is created by subdividing each edge of G exactly once and joining all of the newly added middle vertices of adjacent edges of G. The central graph of G, denoted by $C(G)$, is obtained by subdividing each edge of G exactly once and joining all the nonadjacent vertices of G. The a-barbell graph is obtained by connecting a copies of a complete graph K_a by a bridge, and it is denoted by $K_a \cup K_a + e$. The banana tree $B_{r,k}$ is a tree that is obtained from union of r copies of a k star (a star graph with k vertices) and K_1, where $V(K_1) = \{v\}$, by joining a leaf of every k star with v by an edge. $F_{a,b}$ is a firecracker graph which is obtained from a path P_a by attaching a pendant vertex of b star graph to every vertex of P_a. The total graph [5] $T(G)$ of a graph G is the graph whose vertex set is $V(G) \cup E(G)$ and two vertices are adjacent if they are either neighboring or

K. Priya · G. Mahadevan (✉)
Department of Mathematics, The Gandhigram Rural Institute - Deemed to be University, Gandhigram, India

C. Sivagnanam
Department of General Requirements, University of Technology and Applied Sciences, Sur, Oman

F. Kamalov et al. (eds.), *Advances in Mathematical Modeling and Scientific Computing*, Trends in Mathematics, https://doi.org/10.1007/978-3-031-41420-6_68

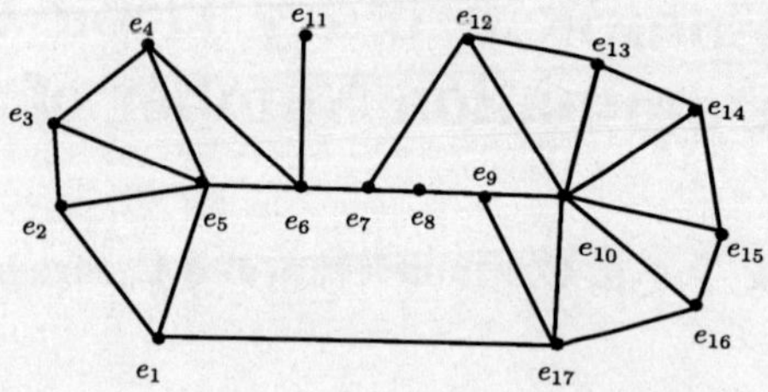

Fig. 1 A graph with $\gamma_t \leq CTCTD$

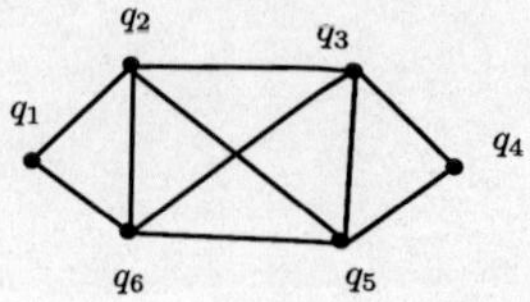

Fig. 2 $\gamma = \gamma_t = CTCTD$

incident in G. A lollipop graph $L_{b,a}$ is a graph obtained from $P_a \cup K_b$ by joining an end vertex of P_a and a vertex K_b by an edge. Let G_1 and G_2 be two copies of a connected graph G. The shadow graph $D_2(G)$ is obtained from $G_1 \cup G_2$ by joining each vertex $u \in V(G_1)$ to all the vertices in $N(u') \cap V(G_2)$ where $u' \in V(G_2)$ corresponding to the vertex u. The splitting graph $S'(G)$ of a graph G is created by adding a new vertex v' matching to each vertex v in G and joining v' and all the vertices in $N_G(v)$. The umbrella graph $U_{b,a}$ is created by joining a path P_a to the central vertex of a fan F_b. The book graph $B_a = K_{1,a} \times P_2$. Let G' be a copy of G. The mirror graph of a graph G is denoted by $M_r(G)$ which is obtained from $G \cup G'$ by joining the vertex $v \in V(G)$ and $v' \in V(G')$ where v' is a vertex corresponding to v.

1.1 Complementary Triple Connected Total Domination Number

Observation For any graph G in Fig. 1, $\gamma_t(G) \leq CTCTD(G)$.

Observation There exists a graph G in Fig. 2, with $\gamma(G) = \gamma_t(G) = CTCTD(G)$.

Observation CTCTD-number does not exist for path, cycle, star, bistar, helm graph, barbell graph, friendship graph, middle graph of path and cycle, central graph of path and cycle, windmill graph, splitting graph, and book graph.

Observation If G is a complete graph or fan graph or wheel graph or complete bipartite graph which is not a star, then $CTCTD(G) = 2$.

Observation

1. $CTCTD(L_{b,a}) = a + 1 \ if \ a, b \geq 2$.

2. $CTCTD(U_{b,a}) = a.$
3. $CTCTD(B_{r,k}) = rk - 2\ if\ r, k \geq 3.$
4. $CTCTD(F_{a,b}) = a(b-1)\ if\ a \geq 3, b \geq 2.$

1.2 CTCTD-Number for Some Specific Type of Graphs

Theorem 1 *For a path P_a, $a \geq 3$*

$$CTCTD\ (T(P_a)) = \begin{cases} 4\lceil \frac{a}{7} \rceil & if\ a\ \equiv 0\ or\ 6\ (mod\ 7) \\ 4\lfloor \frac{a}{7} \rfloor + 1 & if\ a\ \equiv 1\ (mod\ 7) \\ 4\lfloor \frac{a}{7} \rfloor + 2 & if\ a\ \equiv 2\ or\ 3\ or\ 4\ (mod\ 7) \\ 4\lceil \frac{a}{7} \rceil - 1 & if\ a\ \equiv 5\ (mod\ 7). \end{cases}$$

Proof Let $V(T(P_a)) = \{v_1, v_2, \ldots, v_a, u_1, u_2, \ldots, u_{a-1}\}$, and let
$E(T(P_a)) = \{v_i u_i, v_i v_{i+1}, u_i v_{i+1}, u_j u_{j+1}; 1 \leq i \leq a-1; 1 \leq j \leq a-2\}$.
$B_1 = \{v_i, u_j\ :\ i\ \equiv 2\ or\ 3\ (mod\ 7);\ j\ \equiv 5\ or\ 6\ (mod\ 7)\}$.

$$\text{Assume } S = \begin{cases} B_1 & if\ a\ \equiv 0\ or\ 3\ or\ 4\ (mod\ 7) \\ B_1 \cup \{u_{a-1}\} & if\ a\ \equiv 1\ or\ 2\ (mod\ 7) \\ B_1 \cup \{v_{a-1}\} & if\ a\ \equiv 5\ (mod\ 7) \\ B_1 \cup \{v_a\} & if\ a\ \equiv 7a+6. \end{cases}$$

Then S is a CTCTD-set of $T(P_a)$ and hence

$$CTCTD\ (T(P_a)) \leq |S| = \begin{cases} 4\lceil \frac{a}{7} \rceil & if\ a\ \equiv 0\ or\ 6\ (mod\ 7) \\ 4\lfloor \frac{a}{7} \rfloor + 1 & if\ a\ \equiv 1\ (mod\ 7) \\ 4\lfloor \frac{a}{7} \rfloor + 2 & if\ a\ \equiv 2\ or\ 3\ or\ 4\ (mod\ 7) \\ 4\lceil \frac{a}{7} \rceil - 1 & if\ a\ \equiv 5\ (mod\ 7). \end{cases}$$

Let S' be a CTCTD-set of $T(P_a)$. Since $< V - D >$ is disconnected for any dominating set D of cardinality less than or equal to k, where

$$k = \begin{cases} 4\lceil \frac{a}{7} \rceil - 1 & if\ a\ \equiv 0\ or\ 6\ (mod\ 7) \\ 4\lfloor \frac{a}{7} \rfloor & if\ a\ \equiv 1\ (mod\ 7) \\ 4\lfloor \frac{a}{7} \rfloor + 1 & if\ a\ \equiv 2\ or\ 3\ or\ 4\ (mod\ 7) \\ 4\lceil \frac{a}{7} \rceil - 2 & if\ a\ \equiv 5\ (mod\ 7), \end{cases}$$

we have

$$|S'| \geq k+1 = \begin{cases} 4\lceil \frac{a}{7} \rceil & if\ a\ \equiv 0\ or\ 6\ (mod\ 7) \\ 4\lfloor \frac{a}{7} \rfloor + 1 & if\ a\ \equiv 1\ (mod\ 7) \\ 4\lfloor \frac{a}{7} \rfloor + 2 & if\ a\ \equiv 2\ or\ 3\ or\ 4\ (mod\ 7) \\ 4\lceil \frac{a}{7} \rceil - 1 & if\ a\ \equiv 5\ (mod\ 7). \end{cases}$$

Hence the result follows. □

Example 1

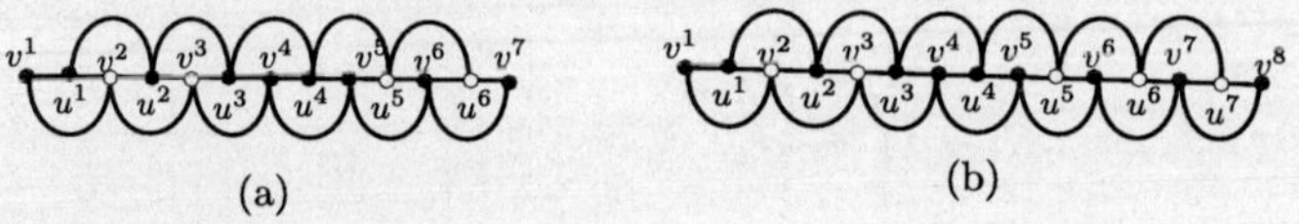

Fig. 3 (**a**) Total graph of $T(P_r)$ (**b**) Total graph of $T(P_s)$

1.3 Illustration

Here, lightened vertices found a CTCTD-set and its denoted by S.

In Fig. 3a $|S| = 4$. Since $a = 7$, $CTCTD(T(P_a)) = 4\lceil \frac{a}{7} \rceil$ implies $CTCTD(T(P_7)) = 4\lceil \frac{7}{7} \rceil = 4$.

In Fig. 3b $|S| = 5$. Since $a = 8$, $CTCTD(T(P_a)) = 4\lceil \frac{a}{7} \rceil - 1$ implies $CTCTD(T(P_8)) = 4\lceil \frac{8}{7} \rceil - 1 = 5$.

Theorem 2 *For a cycle C_a, $a \geq 3$*

$$CTCTD\,(T(C_a)) = \begin{cases} 4\lceil \frac{a}{7} \rceil & if\ a\ \equiv 7q\ or\ 7q+5\ or\ 7q+6 \\ 4\lfloor \frac{a}{7} \rfloor + 1 & if\ a\ \equiv 7q+1 \\ 4\lfloor \frac{a}{7} \rfloor + 2 & if\ a\ \equiv 7q+2\ or\ 7q+3 \\ 4\lceil \frac{a}{7} \rceil - 1 & if\ a\ \equiv 7q+4\,. \end{cases}$$

Proof Let $V(T(C_a)) = \{v_1, v_2, \ldots, v_a, u_1, u_2, \ldots, u_a\}$, and let $E(T(C_a)) = \{v_iu_i, v_iv_{i+1}, u_iv_{i+1}, u_iu_{i+1}, v_av_1, u_av_1, u_au_1;\ 1 \leq i \leq a-1\}$.
Let $S = \{v_i\ :\ i\ \equiv 7a\ or\ 7a+2;\ u_j\ :\ j\ \equiv 7a+4\ or\ 7a+5\ \}$.
Then S is a CTCTD-set of $T(C_a)$ and hence

$$CTCTD\,(T(C_a)) \leq |S| = \begin{cases} 4\lceil \frac{a}{7} \rceil & if\ a\ \equiv 7q\ or\ 7q+5\ or\ 7q+6 \\ 4\lfloor \frac{a}{7} \rfloor + 1 & if\ a\ \equiv 7q+1 \\ 4\lfloor \frac{a}{7} \rfloor + 2 & if\ a\ \equiv 7q+2\ or\ 7q+3 \\ 4\lceil \frac{a}{7} \rceil - 1 & if\ a\ \equiv 7q+4. \end{cases}$$

Let S' be a CTCTD-set of $T(C_a)$. Since $< V - D >$ is disconnected for any dominating set D of cardinality less than or equal to k, where

$$k = \begin{cases} 4\lceil \frac{a}{7} \rceil - 1 & if\ a\ \equiv 7q\ or\ 7q+5\ or\ 7q+6 \\ 4\lfloor \frac{a}{7} \rfloor & if\ a\ \equiv 7q+1 \\ 4\lfloor \frac{a}{7} \rfloor + 1 & if\ a\ \equiv 7q+2\ or\ 7q+3 \\ 4\lceil \frac{a}{7} \rceil - 2 & if\ a\ \equiv 7q+4, \end{cases}$$

we have

$$|S'| \geq k+1 = \begin{cases} 4\lceil \frac{a}{7} \rceil & if\ a\ \equiv 7q\ or\ 7q+5\ or\ 7q+6 \\ 4\lfloor \frac{a}{7} \rfloor + 1 & if\ a\ \equiv 7q+1 \\ 4\lfloor \frac{a}{7} \rfloor + 2 & if\ a\ \equiv 7q+2\ or\ 7q+3 \\ 4\lceil \frac{a}{7} \rceil - 1 & if\ a\ \equiv 7q+4. \end{cases}$$

Hence the result follows. □

Theorem 3 *For a path P_a, $a \geq 3$*

$$CTCTD\,(M_r(P_a)) = \begin{cases} 4\lceil \frac{a}{6} \rceil + 1 & if\ a\ \equiv 0\ (mod\ 6) \\ 4\lfloor \frac{a}{6} \rfloor + 2 & if\ a\ \equiv 1\ or\ 2\ (mod\ 6) \\ 4\lceil \frac{a}{6} \rceil - 1 & if\ a\ \equiv 3\ (mod\ 6) \\ 4\lceil \frac{a}{6} \rceil & otherwise. \end{cases}$$

Proof Let $V(M_r(P_a)) = \{v_1, v_2, \ldots, v_a, u_1, u_2, \ldots, u_a\}$ and let $E(M_r(P_a)) = \{u_i v_i, u_j u_{j+1}, v_j v_{j+1} : 1 \leq i \leq a, 1 \leq j \leq a-1\}$. Let $B_1 = \{v_i, u_j\ :\ i\ \equiv 1\ or\ 2\ (mod\ 6);\ j\ \equiv 4\ or\ 5\ (mod\ 6)\}$.

Assume $S = \begin{cases} B_1 \cup \{u_a\} & if\ a\ \equiv 0\ or\ 1\ (mod\ 6) \\ B_1 & if\ a\ \equiv 2\ or\ 5\ (mod\ 6) \\ B_1 \cup \{v_a\} & if\ a\ \equiv 3\ or\ 4\ (mod\ 6). \end{cases}$

Then S is a CTCTD-set of $M_r(P_a)$ and hence

$$CTCTD\,(M_r(P_a)) \leq |S| = \begin{cases} 4\lceil \frac{a}{6} \rceil + 1 & if\ a\ \equiv 0\ (mod\ 6) \\ 4\lfloor \frac{a}{6} \rfloor + 2 & if\ a\ \equiv 1\ or\ 2\ (mod\ 6) \\ 4\lceil \frac{a}{6} \rceil - 1 & if\ a\ \equiv 3\ (mod\ 6) \\ 4\lceil \frac{a}{6} \rceil & otherwise. \end{cases}$$

Let S' be a CTCTD-set of $M_r(P_a)$. Since $< V - D >$ is disconnected for any dominating set D of cardinality less than or equal to k, where

$$k = \begin{cases} 4\lceil \frac{a}{6} \rceil & if\ a\ \equiv 0\ (mod\ 6) \\ 4\lfloor \frac{a}{6} \rfloor + 1 & if\ a\ \equiv 1\ or\ 2\ (mod\ 6) \\ 4\lceil \frac{a}{6} \rceil - 2 & if\ a\ \equiv 3\ (mod\ 6) \\ 4\lceil \frac{a}{6} \rceil - 1 & otherwise, \end{cases}$$

we have

$$|S'| \geq k + 1 = \begin{cases} 4\lceil \frac{a}{6} \rceil + 1 & if\ a\ \equiv 0\ (mod\ 6) \\ 4\lfloor \frac{a}{6} \rfloor + 2 & if\ a\ \equiv 1\ or\ 2\ (mod\ 6) \\ 4\lceil \frac{a}{6} \rceil - 1 & if\ a\ \equiv 3\ (mod\ 6) \\ 4\lceil \frac{a}{6} \rceil & otherwise. \end{cases}$$

Hence the result follows. □

Observation

1. $CTCTD\,(M_r(K_a)) = 2$.
2. $CTCTD\,(M_r(W_{1,a})) = 2$.
3. $CTCTD\,(M_r(\bar{P}_a)) = 4, a \geq 4$.

Theorem 4 *For a cycle C_a, $a \geq 3$*

$$CTCTD\,(M_r(C_a)) = \begin{cases} 4\lceil \frac{a}{6} \rceil & if\ a\ \equiv 0\ or\ 4\ or\ 5\ (mod\ 6) \\ 4\lfloor \frac{a}{6} \rfloor + 1 & if\ a\ \equiv 1\ (mod\ 6) \\ 4\lfloor \frac{a}{6} \rfloor + 2 & if\ a\ \equiv\ 2\ (mod\ 6) \\ 4\lceil \frac{a}{6} \rceil - 1 & if\ a\ \equiv 3\ (mod\ 6). \end{cases}$$

Proof Let $V(M_r(C_a)) = \{v_1, v_2, \ldots, v_a, u_1, u_2, \ldots, u_a\}$ and let
$E(M_r(C_a)) = \{u_i v_i, u_j u_{j+1}, v_j v_{j+1}, v_a v_1, u_a u_1 : 1 \leq i \leq a, 1 \leq j \leq a-1\}$.
Let $B_1 = \{v_i, u_j\ :\ i\ \equiv 1\ or\ 2\ (mod\ 6);\ j\ \equiv 4\ or\ 5\ (mod\ 6)\}$.

Assume $S = \begin{cases} B_1 \cup \{v_a\} & if\ a \equiv 3\ or\ 4\ (mod\ 6) \\ B_1 & otherwise. \end{cases}$

Then S is a CTCTD-set of $M_r(C_a)$ and hence

$$CTCTD\ (M_r(C_a)) \leq |S| = \begin{cases} 4\lceil \frac{a}{6} \rceil & if\ a \equiv 0\ or\ 4\ or\ 5\ (mod\ 6) \\ 4\lfloor \frac{a}{6} \rfloor + 1 & if\ a \equiv 1\ (mod\ 6) \\ 4\lfloor \frac{a}{6} \rfloor + 2 & if\ a \equiv 2 (mod\ 6) \\ 4\lceil \frac{a}{6} \rceil - 1 & if\ a \equiv 3\ (mod\ 6). \end{cases}$$

Let S' be a CTCTD-set of $M_r(C_a)$. Since $< V - D >$ is disconnected for any dominating set D of cardinality less than or equal to k, where

$$k = \begin{cases} 4\lceil \frac{a}{6} \rceil - 1 & if\ a \equiv 0\ or\ 4\ or\ 5\ (mod\ 6) \\ 4\lfloor \frac{a}{6} \rfloor & if\ a \equiv 1\ (mod\ 6) \\ 4\lfloor \frac{a}{6} \rfloor + 1 & if\ a \equiv 2 (mod\ 6) \\ 4\lceil \frac{a}{6} \rceil - 2 & if\ a \equiv 3\ (mod\ 6), \end{cases}$$

we have

$$|S'| \geq k + 1 = \begin{cases} 4\lceil \frac{a}{6} \rceil & if\ a \equiv 0\ or\ 4\ or\ 5\ (mod\ 6) \\ 4\lfloor \frac{a}{6} \rfloor + 1 & if\ a \equiv 1\ (mod\ 6) \\ 4\lfloor \frac{a}{6} \rfloor + 2 & if\ a \equiv 2 (mod\ 6) \\ 4\lceil \frac{a}{6} \rceil - 1 & if\ a \equiv 3\ (mod\ 6). \end{cases}$$

Hence the result follows. □

Theorem 5 *For a path P_a, $a \geq 4$*

$$CTCTD\ (D_2(P_a)) = \begin{cases} (\frac{a}{2}) + 2 & if\ a \equiv 0\ (mod\ 4) \\ 2\lceil \frac{a}{4} \rceil + 1 & if\ a \equiv 4q + 1\ or\ 4q + 3 \\ 2\lceil \frac{a}{4} \rceil & if\ a \equiv 4q + 2\,. \end{cases}$$

Proof Let $V(D_2(P_a)) = \{v_1, v_2, \ldots, v_a, u_1, u_2, \ldots, u_a\}$ and let $E(D_2(P_a)) = \{v_i v_{i+1}, u_i u_{i+1}, u_i v_{i+1}, v_i u_{i+1} : 1 \leq i \leq a - 1\}$.

Let $B_1 = \{v_i\ :\ i\ \equiv 4a\ or\ 4a + 2\}$.

$$\text{Assume } S = \begin{cases} B_1 \cup \{v_{a-1}, v_a\} & if\ a \equiv 0\ (mod\ 4) \\ B_1 \cup \{v_{a-2}, v_{a-1}\} & if\ a \equiv 4q + 1 \\ B_1 & if\ a \equiv 4q + 2 \\ B_1 \cup \{v_a\} & if\ a \equiv 4q + 3. \end{cases}$$

Then S is a CTCTD-set of $D_2(P_a)$ and hence

$$CTCTD\ (D_2(P_a)) \leq |S| = \begin{cases} (\frac{a}{2}) + 2 & if\ a \equiv 0\ (mod\ 4) \\ 2\lceil \frac{a}{4} \rceil + 1 & if\ a \equiv 4q + 1\ or\ 4q + 3 \\ 2\lceil \frac{a}{4} \rceil & if\ a \equiv 4q + 2\,. \end{cases}$$

Let S' be a CTCTD-set of $D_2(P_a)$. Since $< V - D >$ is disconnected for any dominating set D of cardinality less than or equal to k, where

$$k = \begin{cases} (\frac{a}{2}) + 1 & if\ a \equiv 0\ (mod\ 4) \\ 2\lceil \frac{a}{4} \rceil & if\ a \equiv 4q + 1\ or\ 4q + 3 \\ 2\lceil \frac{a}{4} \rceil - 1 & if\ a \equiv 4q + 2, \end{cases}$$

we have

$$|S'| \geq k+1 = \begin{cases} \lceil \frac{a}{2} \rceil + 2 & if\ a \equiv 0\ (mod\ 4) \\ 2\lceil \frac{a}{4} \rceil + 1 & if\ a \equiv 4q+1\ or\ 4q+3 \\ 2\lceil \frac{a}{4} \rceil & if\ a \equiv 4q+2. \end{cases}$$

Hence the result follows. □

Theorem 6 *For a cycle C_a, $a \geq 4$*

$$CTCTD\ (D_2(C_a)) = \begin{cases} 2\lceil \frac{a}{4} \rceil - 1 & if\ a \equiv 1\ (mod\ 4) \\ 2\lceil \frac{a}{4} \rceil & otherwise. \end{cases}$$

Proof Let $V(D_2(C_a)) = \{v_1, v_2, \ldots, v_a, u_1, u_2, \ldots, u_a\}$ and let $E(D_2(C_a)) = \{v_i v_{i+1}, u_i u_{i+1}, u_i v_{i+1}, v_i u_{i+1}, v_a v_1, u_a u_1, u_a v_1, v_a u_1 : 1 \leq i \leq a-1\}$. Let $S = \{v_i : i \equiv 4q\ or\ 4q+2\}$.

Then S is a CTCTD-set of $D_2(C_a)$ and hence

$$CTCTD\ (D_2(C_a)) \leq |S| = \begin{cases} 2\lceil \frac{a}{4} \rceil - 1 & if\ a \equiv 1\ (mod\ 4) \\ 2\lceil \frac{a}{4} \rceil & otherwise. \end{cases}$$

Let S' be a CTCTD-set of $D_2(C_a)$. Since $< V - D >$ is disconnected for any dominating set D of cardinality less than or equal to k, where

$$k = \begin{cases} 2\lceil \frac{a}{4} \rceil - 2 & if\ a \equiv 1\ (mod\ 4) \\ 2\lceil \frac{a}{4} \rceil - 1 & otherwise, \end{cases}$$

we have $|S'| \geq k+1 = \begin{cases} 2\lceil \frac{a}{4} \rceil - 1 & if\ a \equiv 1\ (mod\ 4) \\ 2\lceil \frac{a}{4} \rceil & otherwise. \end{cases}$

Hence the result follows. □

Notes and Comments In this paper, we have introduced and analyzed CTCTD-number and found the exact values for some standard graphs and specific types of graphs.

References

1. Cockayne, E.J., Dawes, R.M., Hedetniemi, S.T.: Total dominations in graphs. Network **10**, 211–219 (1980)
2. Harary, F.: Graph Theory. Addison Wesley, Reading (1972)
3. Haynes, T.W., Hedetniemi, S.T., Slater, P.J.: Fundamentals of Dominations in Graphs. Marcel Dekker, New York (1998)
4. Mahadevan, G., Selvam,A.,Paulraj Joseph, J., Subramanian,T.: Triple connected domination number of a graph. Int. J. Math. Comb. **3**, 93–104 (2012)
5. Sivagnanam C.: Neighborhood connected domination number of total graphs. Gen. Math. Notes **25**(1), 27–32 (2014)

Pebbling Number and 2-Pebbling Property for the Middle Graphs of the Graph Obtained from Fan Graph by Deleting f Independent Edges

Arokiam Lourdusamy, Irudayaraj Dhivviyanandam, and Susaimanikam Kither Iammal

1 Introduction

Lagarias and Saks were the first ones to introduce the concept of pebbling, and F.R.K. Chung [1] used the concept of pebbling to solve a number theoretic conjecture. Glenn Hulbert published a survey of pebbling variants [2]. The subject of graph pebbling has seen massive growth after Hulbert's survey. In the past 30 years, so many new variants in graph pebbling have been developed which can be applied to the field of transportation, computer memory allocation, game theory and the installation of mobile towers.

Let G be a connected graph. Consider a graph with a distribution of pebbles over the vertices of G. Let us define the pebbling number of a vertex v, $\pi(G, v)$, in a graph G as follows: It is the smallest number with the property that one pebble can be moved to v using pebbling moves regardless of how these pebbles are distributed on the vertices of G. Then, $\pi(G)$, the pebbling number of G, is defined as the maximum of $\pi(G, v)$ over all v.

Consider a configuration of pebbles with property $p \geq 2\pi(G) - q + 1$ where q is the number of vertices having at least one pebble and p is the total number of pebbles placed on G. If we can move two pebbles to a specified vertex, then we say G satisfies the 2-pebbling property.

Lourdusamy et al. [3–6] showed that n-cube, the complete, even cycle, complete r-partite, star, wheel and fan graphs satisfy the $2t$-pebbling property. In this paper,

A. Lourdusamy
Department of Mathematics, St. Xavier's College (Autonomous), Palayamkottai, India

I. Dhivviyanandam (✉) · S. Kither Iammal
Department of Mathematics, St. Xavier's College (Autonomous), Manonmaniam Sundaranar University, Palayamkottai, India
Manonmaniam Sundaranar University, Abisekapatti, India

F. Kamalov et al. (eds.), *Advances in Mathematical Modeling and Scientific Computing*, Trends in Mathematics, https://doi.org/10.1007/978-3-031-41420-6_69

we discuss the pebbling number and the 2-pebbling property for the middle graph of F_n^{-f}.

2 Preliminaries

Definition 1 Let G be a graph. We construct a new graph from G whose vertex set is the set of vertices in G and new vertices which are inserted into every edge of G and whose edge set is the set of all edges in G plus edges obtained by joining those pairs of new vertices which lie on adjacent edges of G. The graph constructed is a middle graph of G and denoted by $M(G)$.

Definition 2 Consider the path $P_{n-1} : x_1, \cdots, x_{n-1}$. Then add an extra vertex x_0 to the path P_{n-1} and connect it to all of the path's vertices. The graph obtained by this process is called a fan graph and it is denoted by F_n. Considering the fan graph F_n, we construct the F_n^{-f} graph by deleting f independent edges. For convenience, we always assume that $V(F_n^{-f}) = x_1, x_2, \cdots, x_{n-1}, x_0$ and the independent edges of F_n^{-f} are $x_1x_2, x_3x_4, \cdots, x_{n-2}x_{n-1}$ if n is odd and $x_1x_2, x_3x_4, \cdots, x_{n-3}x_{n-2}$ if n is even.

Note 1

If n is even, then the graph F_n^{-f} will have $\left\lfloor \frac{n-1}{2} \right\rfloor$ edges and one singleton vertex x_1 on P_{n-1} path of the fan graph after deleting the f independent edges. All the edges and the singleton vertex will have a common vertex x_0. Hence, the edge set is $E(F_n^{-f}) = \{x_0x_j, x_1, x_2x_3, \cdots, x_{n-2}x_{n-1}\}$ where $j = 1, 2, \cdots, n-1$ (Fig. 1)

- If n is odd, then the graph F_n^{-f} will have $\left\lceil \frac{n-1}{3} \right\rceil$ edges and two singleton vertices x_1 and x_{n-1} on P_{n-1} path of the fan graph after deleting the f independent

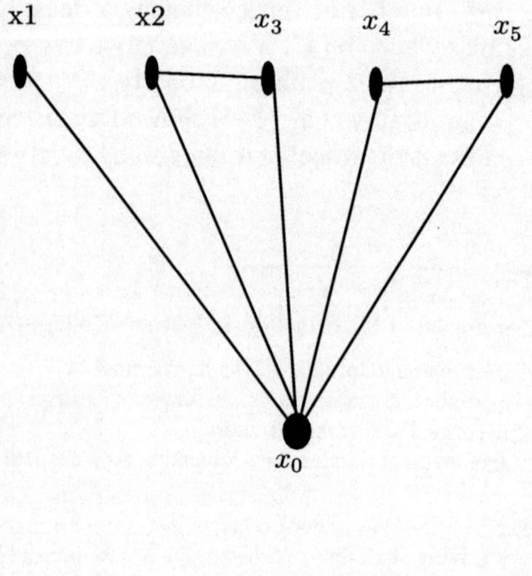

Fig. 1 n is even

Fig. 2 *n* is odd

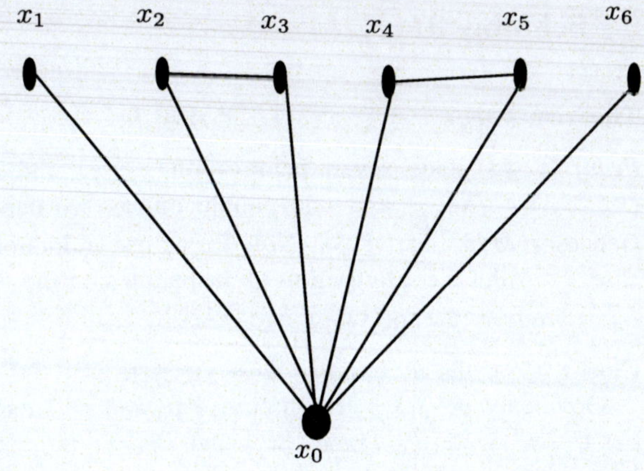

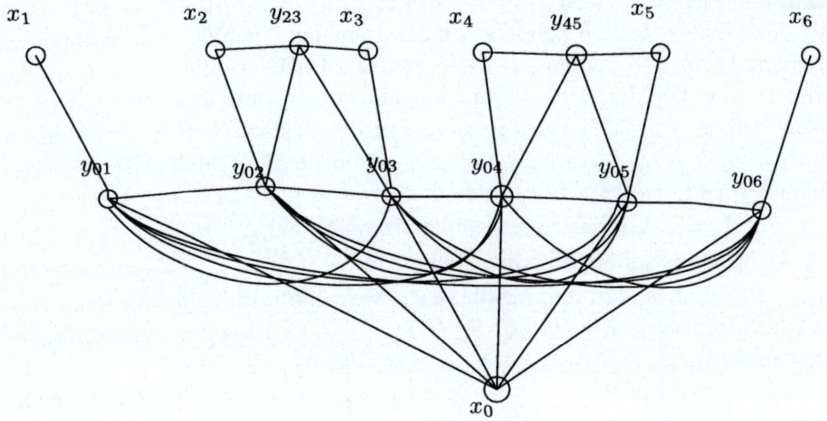

Fig. 3 Middle graph of F_7^{-3} where n is odd

edges. The edge set is $E(F_n^{-f}) = \{x_0x_j, x_1,\ x_2x_3, \cdots, x_{n-3}x_{n-2},\ x_{n-1}\}$ where $j = 1,\ 2, \cdots,\ n-1$ (Fig. 2)

The following theorems and definitions are used to prove the theorems in Sect. 3 (Fig. 3).

Theorem 1 ([3]) *The pebbling number of a path P_n on n vertices $f(P_n) = 2^{n-1}$.*

Notation The symbol $p(v)$ stands for the number of pebbles on the vertex v. Let $p^{\sim}(v)$ be the number of pebbles remaining on v after the sequence of pebbling moves. The notation $x \underset{\rightarrow}{i} y$ stands for the number of pebbles transferred from a vertex x to a vertex y where $1 \leq i \leq \lfloor \frac{\pi(G)}{2} \rfloor$.

Let $S \subseteq V(G)$. The number of pebbles on S after the operation of pebbling moves is denoted as $p^{\sim}(S)$. The target vertex is denoted as w.

3 Pebbling $M(F_n^{-f})$

Theorem 2 *For* $n \geq 4$, $\pi(M(F_n^{-f})) = 3n$.

Proof If we place seven pebbles on $x_{(n-1)}$, three pebbles on each vertex of $\{x_2, x_3, \cdots, x_{n-2}\}$ and one pebble on x_0, we cannot reach the destination x_1. Hence, $\pi(M(F_n^{-f})) \geq 3n$. Now let us use induction on n to prove the sufficient case. Consider a configuration of $3n$ pebbles on the vertices of the middle graph of F_n^{-f}. We have the following five cases:

Case 1 Consider $w = x_1$ $(or\ w = x_{(n-1)})$.

Obviously $p(y_{01}) \leq 1$. Otherwise $p(y_{01}) > 1$ can bring a pebble to the target. Let $C_i = \{y_{i(i+1)}\}$, $D_j = \{x_j\}$ and $E_{0k} = \{x_0, y_{01}, y_{02}, \cdots, y_{0(n-1)}\}$, where $1 \leq j \leq n-1$, $1 \leq k \leq n-1$ and $i = 2, 4, 6, \cdots, n-3$ when n is odd. When n is even, $i = 2, 4, 6, \cdots, n-2$. Let $p(w) = 0$, $p(E_{0k}) = 0$. If a vertex of D_j or C_i has at least 8 pebbles, we can transmit 1 pebble to w without using the remaining $3n - 8$ pebbles. If there exist 2 vertices x_l and x_m of D_j $(l, m \neq j)$ with $4 \leq p(x_l), p(x_m) \leq 7$, we can transmit 1 pebble to w using at most 14 pebbles. If there exist two vertices $y_{l(l+1)}$ and $y_{s(s+1)}$ of C_i $(l, s \neq i)$ with $4 \leq p(y_{l(l+1)}), p(y_{s(s+1)}) \leq 7$, we can transmit a pebble to w. If there exist one vertex of D_j with $4 \leq p(x_l) \leq 7$ and one vertex of C_i with $2 \leq p(y_{l(l+1)}) \leq 3$, then $x_l \xrightarrow{3} y_{0l}$ and $y_{l(l+1)} \xrightarrow{1} y_{0l}$. After shifting pebbles from D_j and C_i, $p^{\sim}(y_{0l}) \geq 4$. Clearly, $y_{0l} \xrightarrow{2} y_{01}$ which is sufficient to put a pebble on w. If there exist a vertex of D_j and a vertex of C_i with at least four pebbles each, we can transmit a pebble to w. If there exist 4 vertices x_l, x_m x_s and x_t of D_j $(l, m, s, t \neq j)$ using $2 \leq p(x_l), p(x_m) \leq 3$ and 2 vertices $y_{l(l+1)}$ and $y_{s(s+1)}$ of C_i $(l, s \neq i)$ using $2 \leq p(y_{l(l+1)}), p(y_{s(s+1)}) \leq 3$, we can transmit 1 pebble to w with at most 18 pebbles. (Note that l and s are to be even where $m = l+1$ and $t = s+1$.) If there exist a vertex of D_j with $6 \leq p(x_l) \leq 7$ where $l \neq 1, n-1$ and a vertex of C_i adjacent to x_l with at least two pebbles, then $x_l \xrightarrow{3} y_{0l}$ and $y_{l(l+1)} \xrightarrow{1} y_{0l}$. Now $p^{\sim}(y_{0l}) \geq 4$. Thus we are done.

Subcase 1.1 Let $p(x_1) = 0$, $p(y_{01}) = 1$. If there exists any one of the vertices of E_{0k} $(k \neq 1)$ with at least two pebbles or any one of the vertices of D_j or C_i with at least four pebbles, we could have $V(E_{0k}) \xrightarrow{1} y_{0l}$ or $V(D_j) \xrightarrow{2} V(E_{0K}) \xrightarrow{1} y_{0l}$. Similarly from $V(C_i)$ we can transform a pebble to y_{0l}. Hence, $p^{\sim}(y_{0l}) \geq 2$. Then could reach w. If $p(x_0) \geq 2$, then we have $x_0 \xrightarrow{1} y_{0l}$ which is sufficient to reach w.

Subcase 1.2 Let $p(x_1) = 0$, $p(y_{01}) = 1$, $p(x_0) \leq 1$. If there exists a vertex of E_{0k} with four pebbles, we are done. If there exist one pebble each on all the vertices of E_{0k} and at least two pebbles on any one of the vertices of C_i or D_j, we could move a pebble to y_{01}. Hence, we are done. If there exist two pebbles on any one of the vertices of C_i or D_j and one pebble on the vertex of E_{0k} adjacent to some vertex of C_i and D_j, we could move a pebble to y_{01}. Hence we can put a pebble to w. If there exists any vertex x_l of D_j or $y_{l(l+1)}$ of C_i with $6 \leq p(x_l) \leq 7$ or $6 \leq p(y_{l(l+1)}) \leq 7$, we are done. If there exist a vertex x_l of D_j and $y_{l(l+1)}$ of C_i with $2 \leq p(x_l)$, $p(y_{l(l+1)}) \leq 3$ or two vertices $x_l, x_{(l+1)}$ of D_j with $2 \leq p(x_l), p(x_{(l+1)}) \leq 3$ and $p(y_{l(l+1)}) = 1$, we could put two pebbles to a vertex of E_{0k} adjacent to either of the vertices x_l, $x_{l+1)}$. Hence, we can shift a pebble from a vertex of E_{0k} to y_{0l} which is sufficient to reach w. The following conditions satisfy the abovementioned cases:

Similarly, the result follows for x_{n-1}.

Case 2 Consider $w \in D_j$ $(j \neq 1)$.

Let $p(x_0) \leq 1$, $p(V(E_{0k})) \leq 1$, $p(V(C_i)) \leq 1$.

Subcase 2.1 Consider $p(x_0) = p(V(E_{0K})) = p(V(C_i)) = 0$. Without loss of generality, let $w = x_2$. If $p(x_3) \geq 4$ or there exists a vertex in D_j where $p(V(D_j)) \geq 8(j \neq 2, 3)$, then w receives a pebble. If there exist two vertices in D_j with $4 \leq P(V(D_j)) \leq 7$, we could put at most three pebbles each on two vertices of E_{0k} which is sufficient to reach w.

If $p(y_{23}) = 1$ or $p(y_{02}) = 1$ and $p(x_3) \geq 2$ or $p(y_{03}) \geq 2$, then w receives a pebble. By symmetry using Case 1, we can reach the destination for all x_j where $j = 3, 4, \cdots, n-2$.

Case 3 Consider $w = y_{23}$ or $w = y_{(n-2)(n-1)}$. It is obvious that $p(y_{01}) \leq 1$ and $p(x_j) \leq 1$ $(j = 2, 3)$. The proof is similar to Case 1.

Case 4 Let $w = y_{0k}$ where $k = 1, 2, \ldots, n-1$. Without loss of generality, let $p(w) = P(y_{01}) = 0$. If $p(x_0) \geq 2$ or $p(V(E_{0k})) \geq 2$, $(k \neq 1)$ or $p(x_j) \geq 4$ where $j = 2, 3, \cdots, n-1$ or $p(x_1) \geq 2$, we are done. Let $p(y_{0m}) = 1$ where $m = 2, 3, \cdot, n-1$. If $2 \leq p(x_m) \leq 3$, we could put a pebble on w. By symmetry, we can prove for all the vertices of E_{0k}.

Case 5 Let $w = x_0$. If there is a vertex of D_j or C_i or E_{0k} with four pebbles, we are done. By using Case 4, we can transform a pebble into a destination. Hence $\pi(M(F_n^{-f})) = 3n$, $(n \geq 4)$.

Theorem 3 *The graph $M(F_n^{-f})$ satisfies the 2-pebbling property.*

Proof The graph $M(F_n^{-f})$ has at least $2\pi(M(F_n^{-f})) - q + 1 \geq 6n - q + 1$ pebbles on it. Let $A = \{x_1\ x_2\ \cdots\ x_{n-1}\}$, $B = \{y_{23}, y_{45}, y_{67} \cdots y_{(n-2)(n-1)}\}$ and $C = \{x_0, y_{01}, y_{02}, \cdots, y_{0(n-1)}\}$. By notation $q_a, q_b,$ *and* q_c stand for the number of vertices with at least one pebble in A, B and C, respectively. Hence $q = q_a + q_b + q_c = n - 1 + \lceil \frac{n}{3} \rceil + n = 2n + \lceil \frac{n}{3} \rceil - 1$.

Case 1 Let $w = x_k$ be the target vertex where $k \in \{1, 2, 3 \cdots, n-1\}$

Without loss of generality, let $w = x_1$ and $p(y_{01}) \leq 1$. The total number of vertices occupied is $q = q_a + q_b + q_c - 1$. If we place $p(x_1) = 0$, $p(y_{01}) = 1$, then we have $p(V(M(F_n^{-f}))) \geq 6n - q$. If $q = 2n + \lceil \frac{n}{3} \rceil$, then we have $p(V(M(F_n^{-f}))) \geq 6n - (2n + \lceil \frac{n}{3} \rceil)$. If we use $2n + \lceil \frac{n}{3} \rceil + 11$ pebbles on $2n + \lceil \frac{n}{3} \rceil$ vertices, we could shift 3 pebbles to y_{01} and 2 pebbles to w.

Let $q \leq 2(n-1) + \lceil \frac{n}{3} \rceil$. Assume $q = 2(n-1) + \lceil \frac{n}{3} \rceil$. Then $p(V(M(F_n^{-f}))) \geq 6n - (2(n-1) + \lceil \frac{n}{3} \rceil + 1) = 4n - \lceil \frac{n}{3} \rceil + 3$. Having only $2(n-1) + \lceil \frac{n}{3} \rceil + 12$ pebbles, we could move 2 pebbles to w. Then we are left with $2n - 2\lceil \frac{n}{3} \rceil + 7$ pebbles.

Let $p(y_{01}) = 0$, $q \leq 2(n-1) + \lceil \frac{n}{3} \rceil - 1$. Suppose $q = 2(n-1) + \lceil \frac{n}{3} \rceil - 1$; then we have $p(V(M(F_n^{-f}))) = 4n - \lceil \frac{n}{3} \rceil + 4$. Using $2(n-1) + \lceil \frac{n}{3} \rceil + 10$ pebbles we can shift 4 pebbles to y_{01}. Hence we are done. Then there will be $2n - 2\lceil \frac{n}{3} \rceil - 4$ pebbles remaining on the vertices of $M(F_n^{-f})$.

Let $q < 2(n-1) + \lceil \frac{n}{3} \rceil - 1$; then $p(V(M(F_n^{-f}))) \leq 4n - \lceil \frac{n}{3} \rceil + 5$. Using $2(n-1) + \lceil \frac{n}{3} \rceil - 14$ pebbles we can shift 4 pebbles to y_{01}. Hence we are done. Then we are left with $2n - 2\lceil \frac{n}{3} \rceil - 7$ pebbles. Similarly, we can prove the vertices of A and B.

Case 2 Let $w \in C - x_0$. Without loss of generality, let $w = y_{01}$ and $p(x_1) \leq 1$ and $p(V(C_i)) \leq 1$. Consider $p(x_1) = 1$ and $p(V(C_i)) = 1$; then $q \leq 2(n-1) + \lceil \frac{n}{3} \rceil$. Let $q = 2(n-1) + \lceil \frac{n}{3} \rceil$. If there exist two pebbles each on any two vertices, then we could move two pebbles to w. Hence the total number of pebbles used is $2(n-1) + \lceil \frac{n}{3} \rceil + 2$.

When $q < 2(n-1) + \lceil \frac{n}{3} \rceil$, then $p(y_{0(n-1)}) = 0$ and $p(y_{01}) = 0$. We can shift two pebbles, if there exist two pebbles each on some vertices of A except on $x_{(n-1)}$ or eight pebbles on $x_{(n-1)}$. Similarly, we can prove for the rest of vertices of $C - x_0$.

Case 3 Let $w = x_0$.

Let $p(x_0) = 0$ and $q \leq 2(n-1) + \lceil \frac{n}{3} \rceil$. Consider $q = 2(n-1) + \lceil \frac{n}{3} \rceil$. Then we will have $4n - \lceil \frac{n}{3} \rceil + 3$ pebbles. If there exist two pebbles each on two vertices adjacent to x_0 or two pebbles each on two vertices of A or B, we could shift two pebbles to w. Let all the neighbourhood vertices of x_0 contain 0 pebbles; then we will have $q \leq (n-1) + \lceil \frac{n}{3} \rceil$. Then $p(V(M(F_n^{-f}))) = 5n - \lceil \frac{n}{3} \rceil + 2$. If there exist eight pebbles on any one of the vertices of A or two vertices of A containing four pebbles each, we can shift two pebbles to w. Similarly, if there exist eight pebbles on any one of the vertices of B or two vertices of B containing four pebbles each, we can shift two pebbles to w. In both cases, we have used only $(n-1) + \lceil \frac{n}{3} \rceil + 7$ pebbles. Hence, the number of pebbles remaining on the vertices of $M(F_n^{-f})$ is at least $4n - 2\lceil \frac{n}{3} \rceil - 6$. Hence $M(F_n^{-f})$ satisfies the 2-pebbling property.

References

1. Chung, F.R.K.: Pebbling in hypercubes. SIAMJ. Disc. Math. **2**(4), 467–472 (1989)
2. Hurlbert, G.: A survey of graph pebbling. Congressus Numerantium **139**, 41–64 (1999)
3. Lourdusamy, A.: *t*-pebbling the product of graphs. Acta Ciencia Indica **XXXII**(M.No.1), 171–176 (2006)
4. Lourdusamy, A., Jeyaseelan, S.S., Tharani, A.P.: t-pebbling the product of fan graphs and the product of wheel graphs. Int. Math. Forum **32**, 1573–1585 (2009)
5. Lourdusamy, A., Tharani, A.P.: On *t*-pebbling graphs. Utilitas Math. **87**, 331–342 (2012)
6. Lourdusamy, A., Tharani, A.P.: The *t*-pebbling conjecture on products of complete r-partite graphs. Ars Comb. **102**, 201–212 (2011)

The de Bruijn Graph of Sequential Repetition of Patterns in DNA Strings

Wan Heng Fong and Ahmed Ildrussi

1 Introduction

The important computational and challenging problem in molecular biology is to assemble a genome from huge amount of reads of different DNA strings. Therefore, many assembly algorithms have been introduced to conquer this problem (see [1–3]). Most of these assembly algorithms are based on three steps: firstly, an algorithm finds the overlaps among the reads, then it lays out these reads, and the last step is to find the consensus genome.

In the overlap step, all the assembly algorithms construct a graph to illustrate the data, called the overlap graph, where every vertex corresponds to a read and two reads are linked by an arc if and only if they have an overlap of enough length [4–6]. However, this strategy faced a lot of difficulties in terms of the huge amount of data from the next-generation sequencers and in comparing all the pairs of reads. To solve these problems, many modern assembler algorithms use a graph called the de Bruijn graph in the overlap step [7–12], instead of the overlap graph.

A de Bruijn graph is a graph in which every vertex corresponds to a $(k-1)$-mer existing in a DNA string and every two vertices are linked by an arc if and only if there exists a k-mer in which its $(k-1)$ suffix and $(k-1)$ prefix correspond to these two vertices. The de Bruijn graph was first introduced by Nicolas De Bruijn in 1946 [13]. De Bruijn graphs make it possible to represent the overlap of all words of a given length by a directed graph.

Again the de Bruijn graph faced another problem because its storing requires a huge amount of memory. Therefore, two data structures have been proposed to

W. H. Fong (✉) · A. Ildrussi
Department of Mathematical Sciences, Faculty of Science, Universiti Teknologi Malaysia, Johor, Malaysia
e-mail: fwh@utm.my; hussin.mohamed.ahmed@graduate.utm.my

F. Kamalov et al. (eds.), *Advances in Mathematical Modeling and Scientific Computing*, Trends in Mathematics, https://doi.org/10.1007/978-3-031-41420-6_70

reduce the size of memory for the de Bruijn graph [14, 15]. Again in 2012, a new succinct representation of a de Bruijn graph has been proposed to reduce its storing size [16]. Recently, new research on the properties of de Bruijn graph has been done. For instance, Mudaber et al. [17] used a splicing system model to generate the DNA splicing languages based on cutting sites of initial strings, crossing, and context factors of restriction enzymes, while Ruslim et al. [18] have given an exhaustive review on the application of graph in DNA splicing systems. On the other hand, Rizzi et al. [19] discussed the most recent advances in the problem of constructing, representing, and navigating assembly graphs, focusing on very large data sets, while Ben-Ari et al. [20] improved the efficiency of de Bruijn graph construction using compact universal hitting sets.

In this article, some properties of the de Bruijn graph of two times and three times repetition patterns in a DNA string in sequence are presented. Furthermore, a generalization of these properties for m times repetition patterns in a DNA string in sequence is obtained. This research is a theoretical work which can be developed into an algorithm in order to deduce the number of loops and multiple edges in the de Bruijn graph, thus reducing the storage size in terms of computations.

The research in this article is organized as follows: in Sect. 2, the de Bruijn graphs of two times and three times repetition patterns in a DNA string in sequence are presented. In Sect. 3, a biological example is provided to illustrate the results. In Sect. 4, a generalization of the de Bruijn graphs of m times repetition patterns in a DNA string in sequence is obtained.

2 The de Bruijn Graph of Two Times and Three Times Repetition Patterns in Sequence

In this section, the de Bruijn graphs of a DNA string containing n copies of the same element a occurring two times and three times in sequence and n distinct elements occurring two times and three times in sequence are found.

In the following lemma, the de Bruijn graph of two times repetition pattern $\underbrace{aa\ldots a}_{n}$ in sequence, i.e., $(\underbrace{aa\ldots a}_{n})^2$ where $n \geq 2$, is presented.

Lemma 1 *Let R be a strand of a DNA string and let $(\underbrace{aa\ldots a}_{n})^2$ be a substring of R with length n, where $n \geq 2$. Then the number of 3-mers in the string R is equal to $(2n-1)$ and the de Bruijn graph of R has $(2n-2)$ self-loops.*

Next, Lemma 2 obtains the de Bruijn graph of two times repetition of a pattern $a_1a_2\ldots a_n$ in sequence, i.e., $(a_1a_2\ldots a_n)^2$ where $n \geq 3$.

Lemma 2 *Let R be a strand of a DNA string and let $(a_1a_2\ldots a_n)^2$ be a substring of R where $n \geq 3$. Then the number of 3-mers in the string R which are repeated* *two times is equal to* $(n-2)$*, and the de Bruijn graph of R contains* $(n-2)$ *pairs of vertices linked by two directed edges.*

Note that the proofs of Lemmas 1 and 2 are omitted because they are easy and similar to the proofs of Lemmas 3 and 4, respectively.

In the following lemma, the de Bruijn graph of three times repetition pattern $\underbrace{aa\ldots a}_{n}$ in sequence, i.e., $(\underbrace{aa\ldots a}_{n})^3$ where $n \geq 2$, is presented.

Lemma 3 *Let R be a strand of a DNA string and let* $(\underbrace{aa\ldots a}_{n})^3$ *be a substring of R with length n, where* $n \geq 2$*. Then the number of* 3*-mers in the string R is equal to* $(3n-1)$ *and the de Bruijn graph of R has* $(3n-2)$ *self-loops.*

Proof This lemma is proven by using induction on n. If $n = 2$, then the pattern is given as $aaaaaax$, where x is an arbitrary sequence of letters. Therefore, all the 3-mers of this pattern are given as aaa, aaa, aaa, aaa, and aax. Now, split these 3-mers into left and right 2-mers, and connect each left 2-mer to its corresponding right 2-mer by a directed edge, presented as

$$aa \xrightarrow{aaa} aa \xrightarrow{aaa} aa \xrightarrow{aaa} aa \xrightarrow{aaa} aa \xrightarrow{aax} ax.$$

The number of 3-mers is given by $3(2) - 1 = 5$ in the string R which are aaa, aaa, aaa, aaa, and aax. These 3-mers give 2-mer of the form aa and form $3(2) - 2 = 4$ self-loops on the vertex aa in the de Bruijn graph. Hence, the result is true for $n = 2$.

Assume that the result is true for $n = k$, which means that for

$$\underbrace{aa\ldots a}_{k}\underbrace{aa\ldots a}_{k}\underbrace{aa\ldots a}_{k}x,$$

there are $(3k-1)$ 3-mers in the string R, which give 2-mer of the form aa and form $(3k-2)$ self-loops on the vertex aa in the de Bruijn graph.

Now, for $n = k + 1$, the pattern is given as

$$\underbrace{aa\ldots a}_{k+1}\underbrace{aa\ldots a}_{k+1}\underbrace{aa\ldots a}_{k+1}x = \underbrace{aa\ldots a}_{k}\underbrace{aa\ldots a}_{k}\underbrace{aa\ldots a}_{k}aaax,$$

for any arbitrary sequence of letters x. Clearly, this pattern increases by three letters aaa as compared to the case of $n = k$. Thus, there are three additional 3-mers in this case of the form aaa. Therefore, the number of 3-mers in this case is given by

$$3k - 1 + 3 = 3(k + 1) - 1.$$

Also, the number of self-loops increases by three, shown as follows:

$$3k - 2 + 3 = 3(k + 1) - 2.$$

Thus, the result is true for $n = k + 1$ and hence the proof is completed. □

In the following lemma, the de Bruijn graph of three times repetition pattern $a_1a_2 \ldots a_n$ in sequence, i.e., $(a_1a_2 \ldots a_n)^3$ where $n \geq 3$, is obtained.

Lemma 4 *Let R be a strand of a DNA string and let $(a_1a_2 \ldots a_n)^3$ be a substring of R with length n, where $n \geq 3$. Then the string R contains $(n - 2)$ 3-mers being repeated three times and two 3-mers being repeated two times, and the de Bruijn graph of R contains $(n - 2)$ pairs of vertices linked by three directed edges and two pairs of vertices linked by two directed edges.*

Proof This lemma is proven by using induction on n. If $n = 3$, then the pattern is written as $a_1a_2a_3a_1a_2a_3a_1a_2a_3x$, where x is an arbitrary sequence of letters. The 3-mers of this pattern are given by $a_1a_2a_3, a_2a_3a_1, a_3a_1a_2, a_1a_2a_3, a_2a_3a_1, a_3a_1a_2,$ $a_1a_2a_3$, and a_2a_3x. By splitting these 3-mers into left and right 2-mers, the de Bruijn graph of R is shown in

$$a_1a_2 \xrightarrow{a_1a_2a_3} a_2a_3 \xrightarrow{a_2a_3a_1} a_3a_1 \xrightarrow{a_3a_1a_2} a_1a_2 \xrightarrow{a_1a_2a_3} a_2a_3 \xrightarrow{a_2a_3a_1}$$
$$a_3a_1 \xrightarrow{a_3a_1a_2} a_1a_2 \xrightarrow{a_1a_2a_3} a_2a_3 \xrightarrow{a_2a_3x} a_3x,$$

and illustrated in Fig. 1.

Thus, the number of 3-mers that are repeated three times in the string R is given by $3 - 2 = 1$, namely, $a_1a_2a_3$, and there are two 3-mers in the string R, repeated two times which are $a_2a_3a_1$ and $a_3a_1a_2$. The 3-mer $a_1a_2a_3$ forms $3 - 2 = 1$ pair of vertices, namely, (a_1a_2, a_2a_3), linked by three directed edges, and the 3-mers $a_2a_3a_1$ and $a_3a_1a_2$ form two pairs of vertices, namely, (a_2a_3, a_3a_1) and (a_3a_1, a_1a_2), linked by two directed edges in the de Bruijn graph. Hence, the result is true for $n = 3$.

Suppose the result is true for $n = k$. Then, for

$$a_1a_2 \ldots a_ka_1a_2 \ldots a_ka_1a_2 \ldots a_kx,$$

there are $(k - 2)$ 3-mers repeated three times in the string R, which are $a_{i-2}a_{i-1}a_i$, $i = 3, 4, \ldots, k$, and two 3-mers, namely, $a_{k-1}a_ka_1$ and $a_ka_1a_2$, repeated two times. The 3-mers $a_{i-2}a_{i-1}a_i$, $i = 3, 4, \ldots, k$ form $(k - 2)$ pairs of vertices of the form

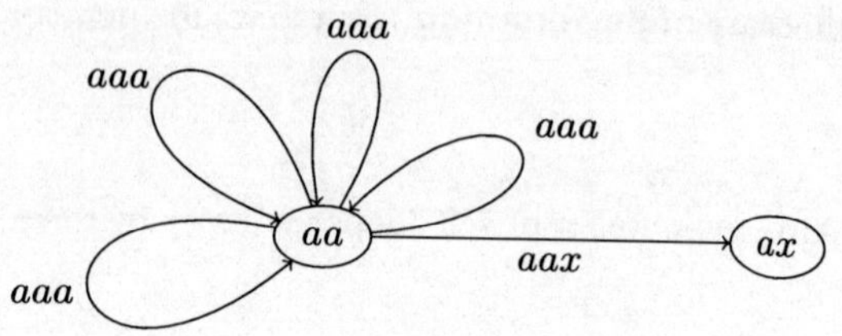

Fig. 1 A de Bruijn graph with substring $a_1a_2a_3$ occurring three times in sequence

$(a_{i-2}a_{i-1}, a_{i-1}a_i)$, $i = 3, 4, \ldots, k$, which are linked by three directed edges, and the 3-mers $a_{k-1}a_ka_1$ and $a_ka_1a_2$ form two pairs of vertices, namely, $(a_{k-1}a_k, a_ka_1)$ and (a_ka_1, a_1a_2), linked by two directed edges in the de Bruijn graph.

Now, assume that $n = k + 1$. Then, the pattern is given as follows:

$$a_1 \ldots a_{k+1}a_1 \ldots a_{k+1}a_1 \ldots a_{k+1}x = a_1 \ldots a_ka_{k+1}a_1 \ldots a_ka_{k+1}a_1 \ldots a_ka_{k+1}x,$$

for any arbitrary sequence of letters x. The number of elements of the string R increases by three as compared to the case of $n = k$, which is the element a_{k+1} being repeated three times. Thus, there are five additional 3-mers in this case, three of the form $a_{k-1}a_ka_{k+1}$ and the others $a_ka_{k+1}a_1$ and $a_{k+1}a_1a_2$ in place of the 3-mers $a_{k-1}a_ka_1$ and $a_ka_1a_2$, respectively. Then, the string R has one more element being repeated three times. Therefore, the number of 3-mers in the string R which are repeated three times in this case is given by

$$k - 2 + 1 = (k + 1) - 2.$$

Hence, the set of pairs of vertices linked by three directed edges in the de Bruijn graph of R increases by one, which is $(a_{k-1}a_k, a_ka_{k+1})$, and the pairs of vertices that are linked by two directed edges are $(a_ka_{k+1}, a_{k+1}a_1)$ and $(a_{k+1}a_1, a_1a_2)$. Thus, the number of pairs of vertices linked by three directed edges in the de Bruijn graph of R in this case is found as follows:

$$k - 2 + 1 = (k + 1) - 2.$$

Hence, the result is true for $n = k + 1$ and the proof is completed. □

In the next section, a biological example is provided to illustrate the results in Sect. 2.

3 A Biological Example of the de Bruijn Graph of Sequential Repetition of Patterns

In this section, the result in Sect. 2 is illustrated by a biological example.

Example 1 Let $AGTCAGTCAGTCTACA$ be a DNA string with length 16. It is clear that the substring $AGTC$ occurs three times in sequence. Therefore, all the 3-mers of this string are given by

$$AGT, GTC, TCA, CAG, AGT, GTC, TCA, CAG, AGT, GTC,$$
$$TCT, CTA, TAC, ACA.$$

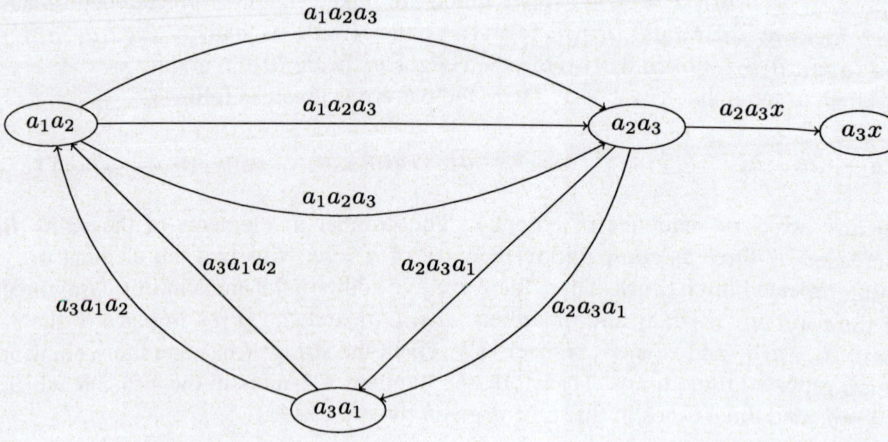

Fig. 2 The de Bruijn graph of $AGTCAGTCAGTCTACA$ with substring $AGTC$ occurring three times in sequence

By splitting these 3-mers into left and right 2-mers and connecting each left 2-mer to its corresponding right 2-mer by a directed edge, the de Bruijn graph of the DNA string is given by

$$AG \xrightarrow{AGT} GT \xrightarrow{GTC} TC \xrightarrow{TCA} CA \xrightarrow{CAG} AG \xrightarrow{AGT} GT \xrightarrow{GTC} TC \xrightarrow{TCA}$$

$$CA \xrightarrow{CAG} AG \xrightarrow{AGT} GT \xrightarrow{GTC} TC \xrightarrow{TCT} CT \xrightarrow{CTA} TA \xrightarrow{TAC} AC \xrightarrow{ACA} CA,$$

and shown in Fig. 2.

From Fig. 2, the number of 3-mers that are repeated three times in the DNA string is equal to two, which are AGT and GTC, and the 3-mers TCA and CAG are repeated two times. The 3-mers AGT and GTC form two pairs of vertices, namely, (AG, GT) and (GT, TC), linked by three directed edges, and the 3-mers TCA and CAG form two pairs of vertices, namely, (TC, CA) and (CA, AG), linked by two directed edges in the de Bruijn graph.

By using Lemma 4, the same result is obtained where the number of 3-mers that are repeated three times in the DNA string is given by $4 - 2 = 2$, and these 3-mers form $4 - 2 = 2$ pairs of vertices linked by three directed edges in the de Bruijn graph.

In the next section, the de Bruijn graphs of m times repetition patterns in sequence are obtained.

4 The de Bruijn Graph of m Times Repetition Pattern in Sequence

In this section, the de Bruijn graphs of a DNA string containing n copies of the same element a occurring m times in sequence and n distinct elements occurring m times in sequence are obtained.

In the following theorem, the de Bruijn graph of m times repetition pattern $\underbrace{aa\ldots a}_{n}$ in sequence, i.e., $(\underbrace{aa\ldots a}_{n})^m$ where $n \geq 2$, is presented.

Theorem 1 *Let R be a strand of a DNA string and let* $(\underbrace{aa\ldots a}_{n})^m$ *be a substring of R with length n, where* $n \geq 2$. *Then the number of* 3*-mers in the string R is equal to* $(mn-1)$ *and the de Bruijn graph of R has* $(mn-2)$ *self-loops.*

Proof By assumption, the pattern of m times repetition of the substring $\underbrace{aa\ldots a}_{n}$ in sequence, i.e., $(\underbrace{aa\ldots a}_{n})^m$, is written as

$$\underbrace{\underbrace{aa\ldots a}_{n}\underbrace{aa\ldots a}_{n}\ldots\underbrace{aa\ldots a}_{n}}_{m}x,$$

where x is an arbitrary sequence of letters.

This theorem is proven by using induction on m. If $m = 2$, then by Lemma 1, the result is true.

Suppose the result is true for $m = k$, which means that for

$$\underbrace{\underbrace{aa\ldots a}_{n}\underbrace{aa\ldots a}_{n}\ldots\underbrace{aa\ldots a}_{n}}_{k}x,$$

there are $(kn-1)$ 3-mers in the string R, which give 2-mer of the form aa and form $(kn-2)$ self-loops on the vertex aa in the de Bruijn graph.

Now, for $m = k+1$, the pattern is written as

$$\underbrace{\underbrace{aa\ldots a}_{n}\ldots\underbrace{aa\ldots a}_{n}}_{k+1}x = \underbrace{\underbrace{aa\ldots a}_{n}\ldots\underbrace{aa\ldots a}_{n}}_{k}\underbrace{aa\ldots a}_{n}x,$$

for any arbitrary sequence of letters x. It is clear that this pattern increases by n letters $\underbrace{aa\ldots a}_{n}$ as compared to the pattern in the case of $m = k$. Therefore, the number of 3-mers increases by n, i.e., the number of 3-mers in this case is

$$kn - 1 + n = (k + 1)n - 1.$$

Also, the number of self-loops increases by n, shown as follows:

$$kn - 2 + n = (k + 1)n - 2.$$

Thus, the result is true for $m = k + 1$ and hence the proof is completed. □

In the following theorem, the de Bruijn graph of m times repetition pattern $a_1a_2 \ldots a_n$ in sequence, i.e., $(a_1a_2 \ldots a_n)^m$ where $n \geq 3$, is obtained.

Theorem 2 *Let R be a strand of a DNA string and let $(a_1a_2 \ldots a_n)^m$ be a substring of R with length n, where $n \geq 3$. Then the string R contains $(n-2)$ 3-mers being repeated m times and two 3-mers being repeated $m-1$ times, and the de Bruijn graph of R contains $(n-2)$ pairs of vertices linked by m directed edges and two pairs of vertices linked by $m-1$ directed edges.*

Proof By assumption, the pattern of m times repetition of the substring $a_1a_2 \ldots a_n$ in sequence, i.e., $(a_1a_2 \ldots a_n)^m$, is given by

$$\underbrace{(a_1a_2 \ldots a_n)(a_1a_2 \ldots a_n) \ldots (a_1a_2 \ldots a_n)}_{m} x,$$

where x is an arbitrary sequence of letters.

This theorem is proven by using induction on m. If $m = 3$, then by Lemma 4, the result is true.

Suppose the result is true for $m = k$. Then for

$$\underbrace{(a_1a_2 \ldots a_n)(a_1a_2 \ldots a_n) \ldots (a_1a_2 \ldots a_n)}_{k} x,$$

there are $(n-2)$ 3-mers in the string R that are repeated k times, namely, $a_{i-2}a_{i-1}a_i$, $i = 3, 4, \ldots, n$, and two 3-mers, namely, $a_{n-1}a_na_1$ and $a_na_1a_2$, repeated $k-1$ times. The 3-mers $a_{i-2}a_{i-1}a_i$, $i = 3, 4, \ldots, n$ form $(n-2)$ pairs of vertices of the form $(a_{i-2}a_{i-1}, a_{i-1}a_i)$, $i = 3, 4, \ldots, n$, which are linked by k directed edges, and the 3-mers $a_{n-1}a_na_1$ and $a_na_1a_2$ form two pairs of vertices, namely, $(a_{n-1}a_n, a_na_1)$ and (a_na_1, a_1a_2), linked by $k-1$ directed edges in the de Bruijn graph.

Now, assume that $m = k + 1$. Then the pattern is given as follows:

$$\underbrace{(a_1a_2 \ldots a_n) \ldots (a_1a_2 \ldots a_n)}_{k+1} x = \underbrace{(a_1a_2 \ldots a_n) \ldots (a_1a_2 \ldots a_n)}_{k}(a_1a_2 \ldots a_n)x,$$

for any arbitrary sequence of letters x. The substring $(a_1a_2 \ldots a_n)$ is repeated one more time in this case as compared to the case of $m = k$. This implies that all the 3-mers in the case $m = k$ must increase by one. Therefore, the string R contains $(n-2)$ 3-mers being repeated $k+1$ times, namely, $a_{i-2}a_{i-1}a_i$, $i = 3, 4, \ldots, n$, and two 3-mers, namely, $a_{n-1}a_na_1$ and $a_na_1a_2$, being repeated k times. Also, the 3-mers $a_{i-2}a_{i-1}a_i$, $i = 3, 4, \ldots, n$ form $(n-2)$ pairs of vertices of the form $(a_{i-2}a_{i-1}, a_{i-1}a_i)$, $i = 3, 4, \ldots, n$, which are linked by $k+1$ directed edges, and the 3-mers $a_{n-1}a_na_1$ and $a_na_1a_2$ form two pairs of vertices, namely, $(a_{n-1}a_n, a_na_1)$ and (a_na_1, a_1a_2), linked by k directed edges in the de Bruijn graph. Hence, the result is true for $m = k+1$ and the proof is completed. □

5 Conclusion

In this article, the de Bruijn graphs of two times and three times repetitions of n copies of the same element a where $n \geq 2$, and n distinct elements where $n \geq 3$ in sequence in DNA strings, are obtained in Sect. 2. In addition, the results are illustrated by a biological example in Sect. 3. Furthermore, these results are generalized for m times repetition in Sect. 4 as follows: if the DNA string contains a pattern $(\underbrace{aa\ldots a}_{n})^m$ where $n \geq 2$, then the string has $(mn-1)$ 3-mers, and its de Bruijn graph contains $mn-2$ directed self-loops, while in case of a pattern of the form $(a_1a_2\ldots a_n)^m$ where $n \geq 3$, the string contains $(n-2)$ 3-mers being repeated m times and two 3-mers being repeated $m-1$ times, and the de Bruijn graph contains $(n-2)$ pairs of vertices linked by m directed edges and two pairs linked by $m-1$ directed edges.

Acknowledgments The authors would like to acknowledge Universiti Teknologi Malaysia (UTM) and Research Management Centre for the financial support through Universiti Teknologi Malaysia Research Research (UTMFR) Vote Number Q.J130000.3854.22H45 and UTMShine Grant Vote Number Q.J130000.2454.09G89.

References

1. Kasahara, M., Morishita, S.: Large-Scale Genome Sequence Processing, Illustrated edn. Imperial College Press, London (2006)
2. Miller, J.R., Koren, S., Sutton, G.: Assembly algorithms for next-generation sequencing data. Genomics **95**(6), 315–327 (2010)
3. Pop, M.: Genome assembly reborn: recent computational challenges. Briefings Bioinf. **10**(4), 354–366 (2009)
4. Batzoglou, S., Jaffe, D.B., Stanley, K., Butler, J., Gnerre, S., Mauceli, E., Berger, B., Mesirov, J.P., Lander, E.S.: Arachne: a whole-genome shotgun assembler. Genome Res. **12**(1), 177–189 (2002)
5. Huang, X., Yang, S.P.: Generating a genome assembly with PCAP. Current Protocols in Bioinformatics Ch. 11, Unit 11.3 (2005)
6. Myers, E.W., Sutton, G.G., Delcher, A.L., Dew, I.M., Fasulo, D.P., Flanigan, M.J., Kravitz, S.A., Mobarry, C.M., Reinert, K.H.J., Remington, K.A., Anson, E.L., Bolanos, R.A., Chou, H., Jordan, C.M., Halpern, A.L., Lonardi, S., Beasley, E.M., Brandon, R.C., Chen, L., Dunn, P.J., Lai, Z., Liang, Y., Nusskern, D.R., Zhan, M., Zhang, Q., Zheng, X., Rubin, G.M., Adams,

M.D., Venter, J.C.: A whole-genome assembly of drosophila. Science **287**(5461), 2196–2204 (2000)
7. Li, R., Zhu, H., Ruan, J., Qjan, W., Fang, X., Shi, Z., Li, Y., Li, S., Shan, G., Kristiansen, K., Yang, H., Wang, J.: De novo assembly of human genomes with massively parallel short read sequencing. Genome Res. **20**(2), 265–272 (2009)
8. MacCallum, I., Przybylski, D., Gnerre, S., Burton, J., Shlyakhter, I., Gnirke, A., Malek, J., McKernan, K., Ranade, S., Shea, T.P., Williams, L., Young, S., Nusbaum, C., Jaffe, D.B.: Allpaths 2: small genomes assembled accurately and with high continuity from short paired reads. Genome Biol. **10**(10), R103 (2009)
9. Pevzner, P.A., Tang, H., Waterman, M.S.: An eulerian path approach to dna fragment assembly. Proc. Natl. Acad. Sci. **98**(17), 9748–9753 (2001)
10. Sahli, M., Shibuya, T.: Arapan-S: a fast and highly accurate whole-genome assembly software for viruses and small genomes. BMC Res. Notes **5**(1), 243 (2012)
11. Simpson, J.T., Wong, K., Jackman, S.D., Schein, J.E., Jones, S.J.: Abyss: a parallel assembler for short read sequence data. Genome Res. **19**(6), 1117–1123 (2009)
12. Zerbino, D.R., Birney, E.: Velvet: algorithms for de novo short read assembly using de Bruijn graphs. Genome Res. **18**(5), 821–829 (2008)
13. De Bruijn, N.G.: A combinatorial problem. Proc. Sect. Sci. Koninklijke Nederlandse Akademie van Wetenschappen te Amsterdam **49**(7), 758–764 (1946)
14. Conway, T.C., Bromage, A.J.: Succinct data structures for assembling large genomes. Bioinformatics **27**(4), 479–486 (2011)
15. Ye, C., Ma, Z.S., Cannon, C.H., Pop, M., Yu, D.W.: Exploiting sparseness in de novo genome assembly. BMC Bioinf. **13**(6), S1 (2012)
16. Bowe, A., Onodera, T., Sadakane, K., Shibuya, T.: Succinct de Bruijn graphs. In: Raphael, B., Tang, J. (eds.) Algorithms in Bioinformatics, WABI 2012. Lecture Notes in Computer Science, vol. 7534, pp. 225–235. Springer, Berlin (2012)
17. Mudaber, M.H., Yusof, Y., Mohamad, M.S., Ramli, A.N.M., Lim, W.L.: Modelling of two stages DNA splicing languages on de Bruijn graph. Jurnal Teknologi **78**(1), 73–78 (2016)
18. Ruslim, N.M., Elizabeth, M., Yusof, Y., Mohamad, M.S., Adzhar, N.: Deoxyribonucleic acid (DNA) splicing system from graph theoretic perspective. J. Phys. Conf. Ser. **1988**(1), 012081 (2021)
19. Rizzi, R., Beretta, S., Patterson, M., Della Vedova, G., Bonizzoni, P.: Overlap graphs and De Bruijn graphs: data structures for de novo genome assembly in the big data era. Quant. Biol. **7**(4), 278–292 (2019)
20. Ben-Ari, Y., Flomin, D., Pu, L., Orenstein, Y., Shamir, R.: Improving the efficiency of De Bruijn graph construction using compact universal hitting sets. In: Jiang, H., Huang, X., Zhang, J. (eds.) Proceedings of the 12th ACM International Conference on Bioinformatics, Computational Biology and Health Informatics, Gainesville, FL, pp. 1–9 ACM, New York (2021)

Independent Domination Number of Cyclic and Acyclic Graphs

S. Thilsath Parveen and B. J. Balamurugan

1 Introduction

Let $G = (V, E)$ be the graph with V as the vertex set and E as the edge set. The number of edges incident to a vertex v in G is called the degree of the vertex v and is denoted as $d_G(v)$. The maximum and minimum degrees of the graph G are denoted as $\Delta(G)$, $\delta(G)$, respectively. A vertex of degree one is called a pendant vertex and an edge incident with the pendant vertex is called a pendant edge.

Domination in graphs is a rapidly emerging subject of graph theory that attracted a large number of researchers due to its variety and relevant real-world problems. There are many variations in this area, which are available in the literature. One such emerging variation is independent domination in graphs [1]. Though independent domination [2] has its origin in late 1950, it is still a predominant area of research in graph theory. The independent domination concept was defined by Berge [2] and Ore [3] in 1962. Cockayne and Hedetniemi [4] introduced the notation $i(G)$ as independent domination number.

The first equality result, $\gamma(G) = i(G)$, using claw free graphs [5], was proved by Allan and Laskar in 1978. Later this result was further extended by Jerzy Topp and Volkmann by considering a class of graphs whose domination and independent domination number are equal [6]. Purnima Gupta et al. [7] found a sufficient condition for a graph to satisfy $\gamma(G) = i(G)$ and also constructed several families of graphs satisfying $\gamma(G) = i(G)$ using this sufficient condition. S. K. Vaidya and R.M. Pandit [8] have proved that $\gamma(G) = i(G)$ for middle graphs and complete bipartite graphs. Favaron [9] found independent domination number of general graphs having

S. Thilsath Parveen · B. J. Balamurugan (✉)
Division of Mathematics, School of Advanced Sciences, Vellore Institute of Technology, Chennai, Tamil Nadu, India
e-mail: balamurugan.bj@vit.ac.in

F. Kamalov et al. (eds.), *Advances in Mathematical Modeling and Scientific Computing*, Trends in Mathematics, https://doi.org/10.1007/978-3-031-41420-6_71

Table 1 Notations used in the chapter

S.No.	Notations	Symbols
1.	Open neighborhood of $v \in V$ of G	$N_G(v)$
2	Closed neighborhood of $v \in V$ of G	$N_G[v]$
3	Degree of $v \in V$ of G	$d_G(v)$
4	Maximum degree of G	$\Delta(G)$
5	Minimum degree of G	$\delta(G)$
6	Ceiling function of x	$\lceil x \rceil$
7	Floor function of x	$\lfloor x \rfloor$
8	Path graph with n vertices	P_n
9	Cycle graph with n vertices	C_n
10	Star graph on n vertices	S_n
11	Null graph on n vertices	$\overline{K_n}$
12	Corona product of paths P_m and P_n	$P_m \circ P_n$
13	Domination number of G	$\gamma(G)$
14	Independent domination number of G	$i(G)$

sharp upper bounds as a function of n and δ, where n and δ are referred as the order and minimum degree of the graph. Haviland [10] extended this work for larger values of δ. The independent domination number of wheel related graph [11] was found by Vaidya and Pandit. Senthurpriya and Meenakshi [12] investigated the independent domination number of triangular and quadrilateral snake graph. J. Anitha et al. [13] computed the independent domination number of various necklace graphs such as path, cycle, complete, and star necklace graph and proved that they have $i(G) = \gamma(G)$. The corona product on graphs G_1 and G_2 was introduced by Frechet and Harary [14]. The terminologies of graph G and domination concepts used in this chapter have been cited from D.B. West [15] and Haynes et al. [16], respectively. The independent domination has various applications in many fields such as interconnection networks, army camps location problem, wireless network, biological networks, IP addresses in computer networks, and many more. In this chapter, we find independent domination number of various cyclic and acyclic graphs. The notations followed in this chapter are summarized in the Table 1.

Remark Throughout the chapter, we call independent dominating set as *ID* set and independent domination number as *IDN* for simplicity.

2 Basic Terminologies

In this section, we recall the concepts and definitions associated with domination in graphs.

Definition 1 [17]: A set $S \subseteq V$ is called a dominating set if every vertex u in V is either in S or adjacent to a vertex in S. The minimum cardinality of the dominating set of G is the domination number $\gamma(G)$.

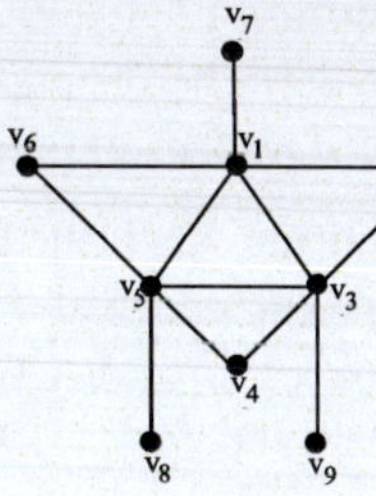

Fig. 1 The graph G with $\gamma(G) = 3$ and $i(G) = 4$

Definition 2 [15]: An independent set in a graph G is a set of pairwise nonadjacent vertices of G. The independence number is the maximum cardinality of an independent set in G.

Definition 3 [1]: Let $G = (V, E)$ be a graph. A subset S of V, which is both independent and dominating in G, is called the independent dominating set of G.The minimum size of an independent dominating set is the independent domination number $i(G)$.

Example 1 The graph G with dominating set $\{v_1, v_3, v_5\}$ and independent dominating set $\{v_1, v_4, v_8, v_9\}$ are shown in Fig. 1.

Definition 4 A fan graph denoted by $F_{m,\,n}$ is a graph obtained from the null graph $\overline{K_m}$ on m vertices and the path P_n on n vertices using the join graph operator.

Definition 5 The firecracker graph denoted by $F\ (m, n)$ is obtained by the concatenation of n-stars S_n by linking one leaf from each star.

Definition 6 A tadpole graph $T_{m,\,n}$ is obtained by connecting a cycle graph C_m to a path graph P_n using a bridge. The $T_{m,\,n}$ tadpole graph is also called as dragon graph (Truszczyński 1984) or kite graph (Kim and Park 2006).

Definition 7 Let the family of disjoint stars be $G_1, G_2, \ldots, G_n$. Let v be a new vertex. The banana tree $B(m, n)$ is obtained by joining v to one pendant vertex of each star graph.

Definition 8 A bistar $B_{m,\,n}$ is the graph obtained by connecting the center vertex of $K_{1,\,m}$ and $K_{1,\,n}$ by an edge.

Definition 9 [18]: A coconut tree $CT(m, n)$ is constructed from the path P_n by appending m new pendant edges at an end vertex of P_n.

Definition 10 The diamond snake graph D_n is a graph consists of n cycles C_4 and these cycles are connected such that any two adjacent cycles share a common vertex.

Definition 11 [14]: The corona product $G_1 \circ G_2$ of two graphs G_1 and G_2 is the graph formed from one copy of G_1 and $|V(G_1)|$ copies of G_2 such that the ith vertex of G_1 is adjacent to every vertex in the ith copy of G_2.

Definition 12 Let $G = (V, E)$ be a bipartite Levi graph associated with an incidence structure. The Pappus graph is a bipartite cubic graph with 18 vertices and 27 edges formed as the Levi graph of Pappus configuration.

Definition 13 [19]: Let $G = (V, E)$ be a graph with $V = S_1 \cup S_2 \cup \cdots \cup S_t \cup T$ where each S_i, $1 \le i \le t$ is a set having at least two vertices of same degree and $T = V - \bigcup_{i=1}^{t} S_i$. The degree splitting graph $DS(G)$ is obtained from G by adding vertices $w_1, w_2, \cdots, w_t$ and joining w_i to each vertex of S_i $(1 \le i \le t)$. If $V(G) = \bigcup_{i=1}^{t} S_i$ then $T = \varnothing$.

3 Independent Domination Number of Cyclic Graphs

The *IDN* of cyclic graphs such as tadpole graph, fan graph and corona product of two path graphs have been computed in this section.

Theorem 1 The *IDN* of the tadpole graph $T_{m,n}$ $m, n \ge 1$ is

$$i\left(T_{m,n}\right) = \begin{cases} \frac{m+n}{3}, \text{ if } m \equiv 1 \bmod 3 \textit{ and } n \equiv 2 \bmod 3 \\ \left\lceil \frac{m}{3} \right\rceil + \left\lceil \frac{n-1}{3} \right\rceil, \textit{ otherwise} \end{cases}$$

Proof Let $T_{m,n}$ be a tadpole graph with $m + n$ vertices and $m + n$ edges. Let the vertices of the subgraph C_m of $T_{m,n}$ be $\{v_1, v_2, \ldots, v_m\}$ and the vertices of the subgraph P_n of $T_{m,n}$ be $\{v_{m+1}, v_{m+2}, \cdots, v_{m+n}\}$, where the vertex v_m of the cycle and v_{m+1} of the path are connected by a bridge in $T_{m,n}$.

Case (i) When $m \equiv 1$ *mod* 3 and $n \equiv 2$ *mod* 3

Let the $3k$ vertices be $v_1, v_2, \ldots, v_{m-1}$ and v_m be a vertex in cycle C_m. Let v_{m+3}, $v_{m+4}, \ldots, v_{m+n}$ be the $3j$ vertices and v_{m+1}, v_{m+2} be two vertices in path P_n. Let $S = V_1 \cup V_2 \cup V_3$ where V_1 contains $k = \left(\frac{m-1}{3}\right)$ vertices $v_2, v_5, \cdots, v_{m-2}$, V_2 contains the $j = \left(\frac{n-2}{3}\right)$ vertices $v_{m+4}, v_{m+7}, \cdots, v_{m+n-1}$ and $V_3 = \{v_{m+1}\}$. Observe that the k vertices in V_1 dominates the $3k$ vertices in C_m, the j vertices in V_2 dominates the $3j$ vertices in P_n and the vertex in V_3 dominates the three vertices v_m, v_{m+1}, v_{m+2}. Hence S dominates all the vertices in T_{m+n} and also S contains the smallest number of vertices. It is evidently clear that the set S is independent, implying that S is a *ID* set having $\left(\frac{m-1}{3}\right) + \left(\frac{n-2}{3}\right) + 1 = \frac{m+n}{3}$ vertices.

Case (ii) The remaining possibilities are discussed in two subcases.

Subcase (a) When $m \not\equiv 1$ *mod* 3 and $n \equiv 2$ *mod* 3

By assuming the labeling used in Case (i), even though $j + 1 = \left\lceil \frac{n-1}{3} \right\rceil$, vertices in $V_2 \cup V_3$ dominate all vertices in P_n and also the vertex v_m in C_m, the remaining $m - 1$ vertices in C_m, which is of the form either $3k - 1$ or $3k - 2$ can only be

Fig. 2 The graph $T_{3,1}$

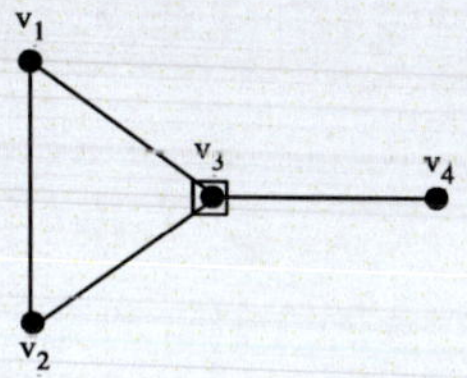

Fig. 3 The graph $T_{5,2}$

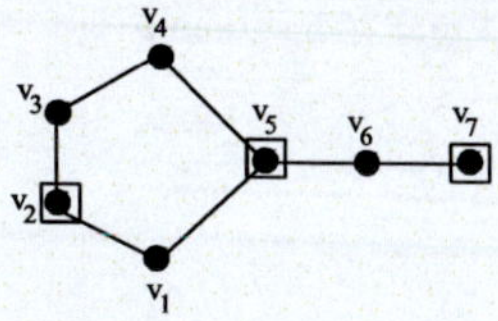

dominated by $\lceil \frac{m}{3} \rceil$ vertices. Hence, the minimum number of vertices in the *ID* set is $\lceil \frac{m}{3} \rceil + \lceil \frac{n-1}{3} \rceil$.

Subcase (b) In all other cases, select a *ID* set in C_m having $\lceil \frac{m}{3} \rceil$ number of vertices. Without loss of generality, let us assume that the vertex v_n of the cycle is in the chosen dominating set of C_m, which is connected with the vertex v_{m+1} of the path P_n. For the remaining subgraph with $(n - 1)$ vertices of the path P_n, the *IDN* is $\lceil \frac{n-1}{3} \rceil$. Therefore, the $\lceil \frac{m}{3} \rceil$ vertices of C_m and $\lceil \frac{n-1}{3} \rceil$ vertices of the path form an *ID* set for $T_{m,n}$. Hence, in this case,$i\left(T_{m,n}\right) = \lceil \frac{m}{3} \rceil + \lceil \frac{n-1}{3} \rceil$.

Example 2 The tadpole graphs $T_{3,1}$ and $T_{5,2}$ with $i(T_{3,1}) = 1$ and $i(T_{5,2}) = 3$ are given in Figs. 2 and 3 respectively.

Theorem 2 The *IDN* of the fan graph $F_{m,n}$ $m \geq 1$ $n \geq 2$ is $i\left(F_{m,n}\right) = \lceil \frac{n}{3} \rceil$.

Proof Let $F_{m,n}$ be a fan graph. The graph $F_{m,n}$ is obtained by joining an empty graph $\overline{K_m}$ on m vertices with each of the vertices $v_1, v_2, \ldots, v_n$ of the path graph P_n. Let the vertices of the null graph be $u_1, u_2, \ldots, u_m$. The sum of vertices and edges of the graph are $m + n$, $2m + 1$ respectively.

The vertices of P_n dominates each and every vertex of the null graph $\overline{K_m}$ and it is independent. Hence the *IDN* of $F_{m,n}$, $m \geq 1$, $n \geq 2$ is same as the *IDN* of P_n, $n \geq 2$. Therefore, the *IDN* of the fan graph $F_{m,n}$, $m \geq 1$, $n \geq 2$ is $\lceil \frac{n}{3} \rceil$.

Example 3 The fan graphs $F_{4,2}$ and $F_{3,3}$ with $i(F_{4,2}) = i(F_{3,3}) = 1$ are shown in Figs. 4 and 5 respectively.

Theorem 3 Let P_m and P_n be two path graphs on m $(m \geq 1)$ and n $(n \geq 1)$ vertices respectively. Then the corona product of $P_m \circ P_n$ is $i\left(P_m \circ P_n\right) = \lceil \frac{m}{2} \rceil + \lfloor \frac{m}{2} \rfloor \cdot \lceil \frac{n}{3} \rceil$

Fig. 4 The graph $F_{4,2}$

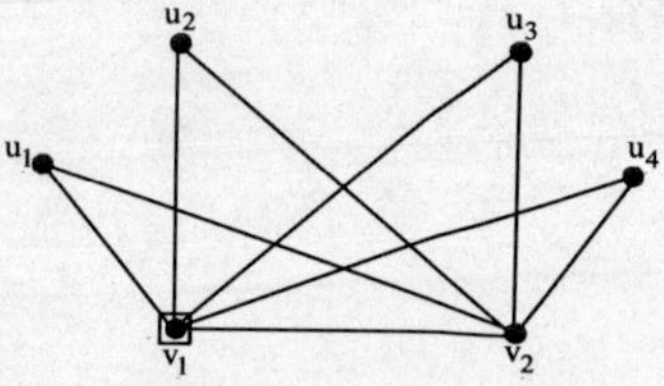

Fig. 5 The graph $F_{3,3}$

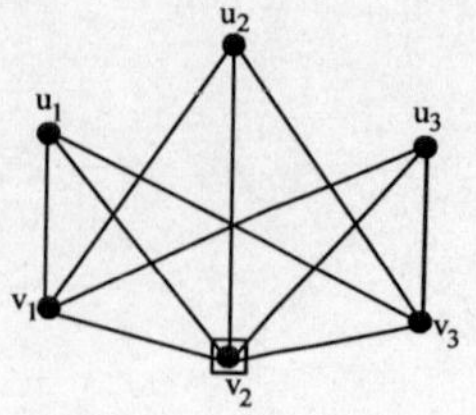

Proof Let $P_m = (v_1, v_2, \ldots, v_m)$ and $P_n = (u_1, u_2, \ldots, u_n)$ be the path graphs on m and n vertices respectively. In order to obtain a minimum independent dominating set S for $P_m \circ P_n$ we proceed as follows:

Let the vertices of the ith copy of $P_m \circ P_n$ be $u_1^i, u_2^i, \ldots, u_m^i$. We construct S as follows:

Step 1: Set, $S = \varnothing$ an empty set.
2. Add vertices $\{v_1, v_2, \ldots, v_k\}$ where $k = m$ if m is not of multiple of 2 or $k = m - 1$ if m is multiple of 2 to S.
3. For $j = 2$ to m in steps of 2. Add a minimum *ID* set of $\lceil \frac{n}{3} \rceil$ vertices of the jth copy of P_n to S.
4. Output S.

We claim that S is the minimum *ID* set for $P_m \circ P_n$. The $\lceil \frac{n}{2} \rceil$ vertices (the maximum number of independent vertices) added to S in step 2, dominates the maximum number of vertices in the copies of P_n. Therefore, reducing these vertices would imply lesser number of dominating vertices of the copies of P_n implying that the collection of vertices will not be minimum. In step3, the minimum number of $\lceil \frac{n}{3} \rceil$ vertices from each not dominated copies of P_n is added to S. This implies that the set is independent and dominates all vertices of $P_m \circ P_n$ and contains the minimum number of vertices. Therefore, $i\ (P_m \circ P_n) = \lceil \frac{m}{2} \rceil + \lfloor \frac{m}{2} \rfloor \bullet \lceil \frac{n}{3} \rceil$.

4 Independent Domination Number of Acyclic Graphs

The independent domination numbers of acyclic graphs such as banana tree, bistar, and firecracker graph are obtained in the following section.

Theorem 4 The *IDN* of banana tree $B(m, n)$ with $m \geq 1, n \geq 1$ is $m + 1$.

Fig. 6 The graph $B(4,2)$

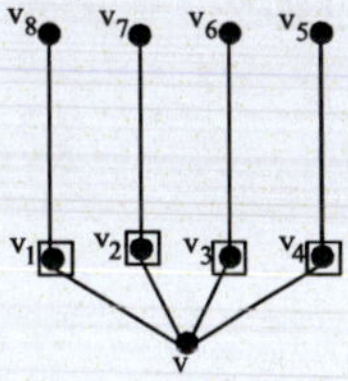

Fig. 7 The graph $B(4,4)$

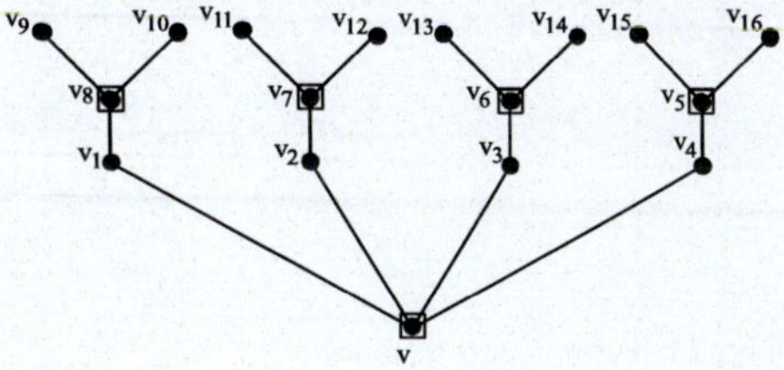

Fig. 8 The graph $(B_{5,6})$

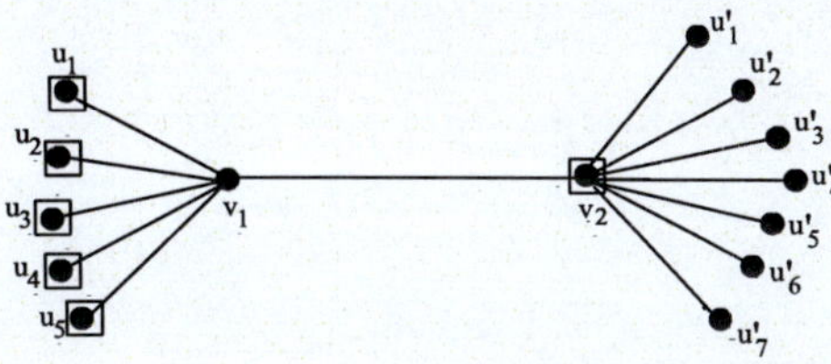

Proof Let $B(m,n)$ be a banana tree with $mn+1$ vertices and mn edges. Banana tree graph consists of m $(m \geq 1)$ copies of n stars $(n \geq 1)$, by linking one vertex of each copy of star to a new single root vertex $'v'$. The set of all m central vertices and the root vertex v forms the *ID* set for $B(m,n)$. Hence, the *IDN* of the banana tree is $m+1$.

Example 4 The Banana tree graphs $B(4,2)$ and $B(4,4)$ with $i(B(4,2)) = 4$ and $i(B(4,4)) = 5$ are given in Figs. 6 and 7 respectively.

Theorem 5 For the bistar graph $B_{r,\,s}$, $(r, s \geq 1)$, the *IDN* is $i(B_{r,\,s}) = \min(r, s) + 1$.

Proof Bistar graph $B_{r,s}$ is obtained from K_2 by joining r leaves to one end of K_2 and s leaves to the other end of K_2. Let v_1 be an apex vertex with r pendent vertices $\{u_1, u_2, \ldots, u_r\}$ and v_2 be another apex vertex with s pendent vertices $\{u'_1, u'_2, \cdots, u'_s\}$ with $r+s+2$ vertices and $r+s+1$ edge in $B_{r,\,s}$.

Without loss of generality, let us assume that $s > r$ and v_1 is adjacent to v_2. The minimum dominating set for $B_{r,s}$ is $\{v_1, v_2\}$ but it is not an independent set. The minimum *ID* set for $B_{r,\,s}$ is $\{u_1, u_2, \cdots, u_r, v_2\}$. Hence $i(B_{r,\,s}) = r + 1 = \min(r, s) + 1$. Therefore, *IDN* of bistar graph $B_{r,\,s}$ is $\min(r, s) + 1$.

Example 5 The bistar graph $B_{5,\,6}$ with $i(B_{5,\,6}) = 6$ is shown in Fig. 8.

Theorem 6 The *IDN* of a firecracker graph $F(m, n)$ $m \geq 1, n \geq 2$ is $i(F(m, n)) = m$.

Fig. 9 The graph $F(2, 4)$

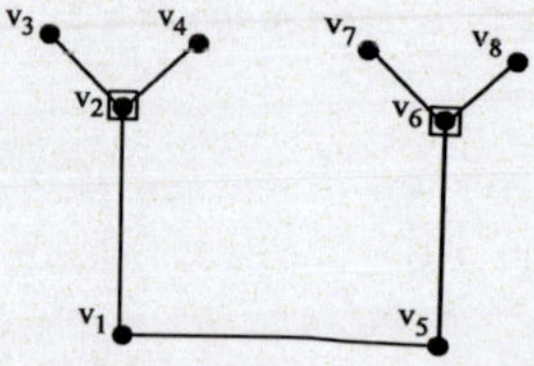

Fig. 10 The graph $F(4, 3)$

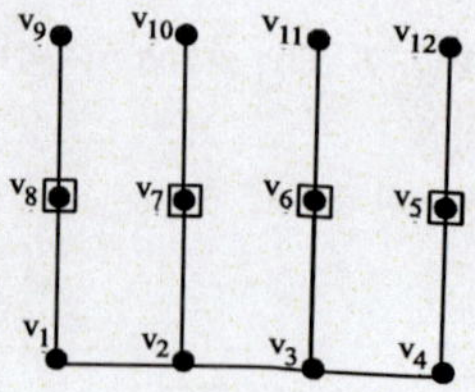

Fig. 11 Pappus graph having $i(G) = 5$

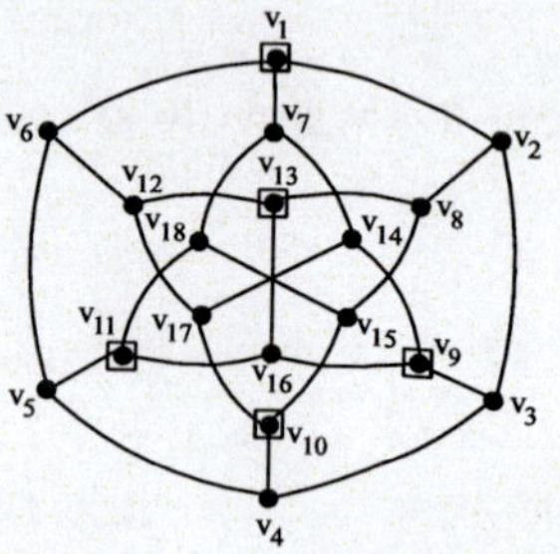

Proof Let G be the firecracker graph $F\ (m, n)\ m \geq 1, n \geq 2$. Let the total number of vertices and edges of the graph be mn and $(mn) - 1$, respectively. The central vertex of the star graph is linked to all other remaining vertices of the graph. Let the collection of all m central vertices in $F(m, n)$ be S. It is clearly evident that the vertices in S are independent and dominate remaining vertices of $F(m, n)$. Hence, the *IDN* of $F(m, n)$ is m. That is, $i(F(m, n)) = m$.

Example 6 The firecracker graphs $F(2, 4)$ and $F(4, 3)$ with $i(F(2, 4)) = 2$ and $i(F(4, 3)) = 4$ are shown in Figs. 9 and 10 respectively.

5 Independent Domination Number of Other Graphs

The *IDN* of pappus graph, diamond snake graph, coconut tree, degree splitting graph of path and degree splitting graph of wheel have been computed.

1. For a pappus graph both *IDN* and domination number is five (Fig. 11).

2. The dominating set of a diamond snake D_n with $n + 1$ vertices is independent. Hence, the *IDN* for the diamond snake graph is $i(D_n) = n + 1$.

3. The dominating set of a coconut tree $CT(m, n)$ with $m + 1$ vertices and $m + n - 1$ edges is independent and the *IDN* is $i\ (CT\ (m, n)) = \left\lceil \frac{m-2}{3} \right\rceil + 1 = \left\lceil \frac{m+1}{3} \right\rceil$.
4. The dominating set of degree splitting graph of path graph is independent and the *IDN* is $i(DS(P_n) = 2, n \geq 3$.
5. The dominating set of degree splitting graph of wheel graph is independent and the *IDN* is $i(DS(W_n)) = 2, n \geq 5$.

6 Conclusion

In this chapter, the *IDN* of graphs such as fan graph $F_{m,\,n}$, firecracker graph $F(m, n)$, tadpole graph $T_{m,\,n}$,bistar graph $B_{r,\,s}$, diamond snake graph D_n, banana tree $B(m, n)$, coconut tree $CT(m, n)$, papus graph, degree splitting of path P_n, degree splitting of wheel W_n, and the corona product of $P_m \circ P_n$ have been computed. Finding *IDN* of various graphs is potential area of research and future scope too. Domination in graphs has a lot of application in science and technology and also in day-to-day real-life problems.

References

1. Goddard, W., Henning, M.: Independent domination in graphs: a survey and recent results. Discret. Math. **313**, 839–854 (2013)
2. Berge, C.: Theory of Graphs and Its Applications. Methuen, London (1962)
3. Ore, O.: Theory of graphs. Am. Math. Soc. Transl. **38**, 206–212 (1962)
4. Cockayne, E.J., Hedetniemi, S.T.: Towards a theory of domination in graphs. Networks. **7**, 247–261 (1977)
5. Allan, R.B., Laskar, R.: On domination and independent domination numbers of a graph. Discret. Math. **23**(2), 73–76 (1978)
6. Topp, J., Volkmann, L.: On graphs with equal domination and independent domination numbers. Discret. Math. **96**(1), 75–80 (1991)
7. Gupta, P., Singh, R., Arumugam, S.: Graphs with equal domination and independent domination numbers. AKCE Int. J. Graphs Comb. **17**(2), 691–696 (2020)
8. Vaidya, S.K., Pandit, R.M.: Graphs with equal domination and independent domination number. TWMS J. Appl. Eng. Math. **5**(1), 74–79 (2015)
9. Favaron, O.: Two relations between the parameters of independence and irredundance. Discret. Math. **70**, 17–20 (1988)
10. Haviland, J.: Independent domination in triangle-free graphs. Discret. Math. **308**(16), 3545–3550 (2008)
11. Vaidya, S.K., Pandit, R.M.: Independent domination in some wheel related graphs. Appl. Appl. Math.: Int. J. (AAM). **11**(1), 397–407 (2016)
12. Senthurpriya, N., Meenakshi, S.: Independent domination number in triangular and quadrilateral snake graph. Int. J. Recent Technol. Eng. **8**(45), 20–23 (2019)
13. Anitha, J., Suganya, V., Jayaraman, G.: Independent domination number in necklace graph. J. Int. Pharm. Res. **46**(1), 388–392 (2019)
14. Frucht, R., Harary, F.: On the corona of two graphs. Aequationes Math. **4**(3), 322–325 (1970)
15. West, D.B.: Introduction to Graph Theory, 2nd edn. Prentice-Hall of India, New Delhi (2009)

16. Haynes, T.W., Hedetniemi, S.T., Slater, P.J.: Fundamentals of Domination in Graphs. Marcel Dekker Inc., New York (1998)
17. Haynes, T.W., Hedetniemi, S.T., Henning, M.A.: Topics in Domination in Graphs, vol. 64. Springer Nature, Cham (2020)
18. Anuthiya, S., Mahadeven, G.: Some more results on at most twin extendable separated domination number of graphs. J. Phys. Conf. Ser. **1724**, 012018 (2021)
19. Basavanagoud, B., Patil, P.V., Sunilkumar, M.H.: Domination in degree splitting graphs. J. Anal. Comput. **8**(1), 1–8 (2012)

Computation of Complete Partite k-Zumkeller Graphs

M. Kalaimathi and B. J. Balamurugan

1 Introduction

Let $G = (V, E)$ be a finite, simple, connected graph where V is the vertex set and E is the edge set of G. A labeled graph G is a graph that results from labeling the vertices of V or the edges of E or both through appropriate functions defined on V and/or E and/or both. The labels may be the numbers or alphabets. In the past 60 years, more than a thousand research articles have been published based on various labeling techniques. The labeled graphs were first discussed by Rosa [1] in the year 1967. The recent updates and developments in the field of graph labeling can be found in the survey article by Gallian [2]. The labeled graphs with numbers have been studied due to its wide range of applications in the software engineering, radar detection, network engineering, circuit design, communication network, and database management. In 1977, Bloom and Golomb [3] studied some graph labeling applications in terms of coding theory, circuit design, and network design. In cryptography [4, 5], the labeled graphs play a crucial part in the process of encryption and decryption. Nowadays, research in graph labeling is increasingly expanding and it has become a significant branch of multidisciplinary research between graph theory and scientific fields.

In 2003, Zumkeller introduced the Zumkeller numbers by generalizing the concept of practical numbers [6]. These numbers can be seen in the A083207 edition of the encyclopedia of integer sequences. One can find the Zumkeller numbers and their partitions in [7]. The Zumkeller numbers and their properties are referred to [8]. In the year 2015, Balamurugan et al. [9] first proposed the notion of k-

M. Kalaimathi · B. J. Balamurugan (✉)
Division of Mathematics, School of Advanced Sciences, Vellore Institute of Technology, Chennai, Tamil Nadu, India
e-mail: balamurugan.bj@vit.ac.in

F. Kamalov et al. (eds.), *Advances in Mathematical Modeling and Scientific Computing*, Trends in Mathematics, https://doi.org/10.1007/978-3-031-41420-6_72

Zumkeller graphs by limiting the number of distinct Zumkeller numbers in optimal way through appropriate functions on G. Balamurugan et al. [10] expanded the k-Zumkeller labeling to a few more graphs in the year 2018. Recently, in the year 2021, Basher in [11, 12] showed the existence of k-Zumkeller labeling of super subdivision of some graphs, cartesian products of cycles and paths. They have also proved that the tensor products of cycles and paths admit k-Zumkeller labeling. The study in this chapter examines whether the complete bipartite, tripartite, and multipartite graphs admit the k- Zumkeller labeling. For the concepts and terminologies of graphs, Harary [13] and West [14] have been cited.

2 Terminologies and Definitions

The Zumkeller numbers [8] and complete partite graphs are discussed with appropriate examples in this section. The concept of k-Zumkeller labeling with suitable example is also provided.

Definition 1

Let n be a positive integer. Let $P_1 = \{m_1, m_2, m_3, \ldots, m_{p_1}\}$ and $P_1 = \{n_1, n_2, n_3, \ldots, n_{p_2}\}$ be two disjoint sets of positive factors n. If $\sum_{\mathrm{i}=1}^{\mathrm{p}_1} \mathrm{m_i} = \sum_{j=1}^{p_2} n_i$, then n is called a Zumkeller number.

For example, 56 can be partitioned as $P_1 = \{1, 2, 7, 8, 14, 28\}$ and $P_2 = \{4, 56\}$ such that $\sum_{\mathrm{i}=1}^{\mathrm{p}_1} \mathrm{m_i} = \sum_{j=1}^{p_2} n_i = 60$. Hence, 56 is a Zumkeller number.

Definition 2

A m – partite is a graph whose vertex set is partitioned into m disjoint subsets with cardinality $n_1, n_2, \ldots, n_m$ such that every edge might connect vertices in separate subsets and that no edge has both of its endpoints in the same subset. The complete m – partite graph is a graph such that every vertex of a subset of m – partite graph is connected to every vertex of other subsets. It is denoted by $K_{n_1,n_2,\ldots,n_m}$, where n_1, $n_2, \ldots, n_m$ are the cardinality of the disjoint subsets of the vertex set of $K_{n_1,n_2,\ldots,n_m}$. The graph is complete bipartite and complete tripartite in case of $m = 2$ and $m = 3$, respectively.

Definition 3

Let $G = (V, E)$ be a simple graph with vertex set V and edge set E. A $1-1$ function $f : V \to N$ is said to induce a k-Zumkeller graph G if the induced edge function $f^* : E \to N$ defined by $f^*(xy) = f(x)f(y)$ satisfies the following conditions:

(i) $f^*(xy)$ is a Zumkeller number for every $xy \in E$.
(ii) The distinct Zumkeller numbers on the edges of G are k.

Definition 4

A simple graph $G(V, E)$ is said to be a k-Zumkeller graph if G admits a k-Zumkeller labeling.

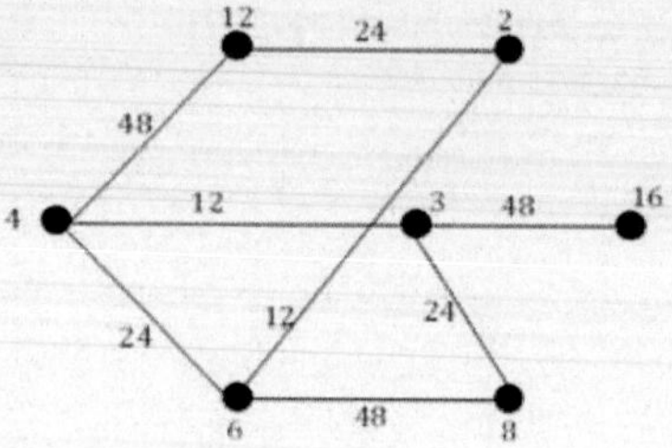

Fig. 1 A 3-Zumkeller graph

Example 1
A 3-Zumkeller graph is given in Fig. 1.

3 *k*- Zumkeller Labeling of Complete Partite Graph

This section examines, with suitable examples, the *k*-Zumkeller labeling of complete bipartite, complete tripartite, and complete *m* partite graphs.

Theorem 1
The complete bipartite graph $K_{n_1,n_2}, n_1, n_2 \geq 2$ is a $[n_1 + n_2 - 1]$-Zumkeller graph.

Proof Let $V = \{A_i : 1 \leq i \leq n_1\} \cup \{B_j : 1 \leq j \leq n_2\}$ and $E = \{A_iB_j : 1 \leq i \leq n_1, 1 \leq j \leq n_2\}$ be the vertex set and edge set of K_{n_1,n_2}. The graph has $n_1 + n_2$ vertices and n_1n_2 edges.

Define an one to one function $f : V \to N$ such that

$$f(A_i) = 2^i, 1 \leq i \leq n_1$$

$$f(B_j) = p_1 2^j, 1 \leq j \leq n_2$$

where p_1 is a prime number such that $2 < p_1 < 10$.

Define an induced edge function $f^* : E \to N$ such that $f^*(xy) = f(x)f(y)$ for all $xy \in E$. Then the function provides the Zumkeller numbers on edges as follows:

$$f^*(A_iB_j) = p_1 2^{i+j}, 1 \leq i \leq n_1, 1 \leq j \leq n_2$$

The edges of the complete bipartite graph K_{n_1,n_2} receive the Zumkeller numbers viz.,

$p_1 2^2, p_1 2^3, \ldots, p_1 2^{n_1+n_2}$, which are all distinct and the total count of different Zumkeller numbers is $[n_1 + n_2 - 1]$. Hence, the complete bipartite graph K_{n_1,n_2} is a $[n_1 + n_2 - 1]$ Zumkeller graph where $n_1, n_2 \geq 2$.

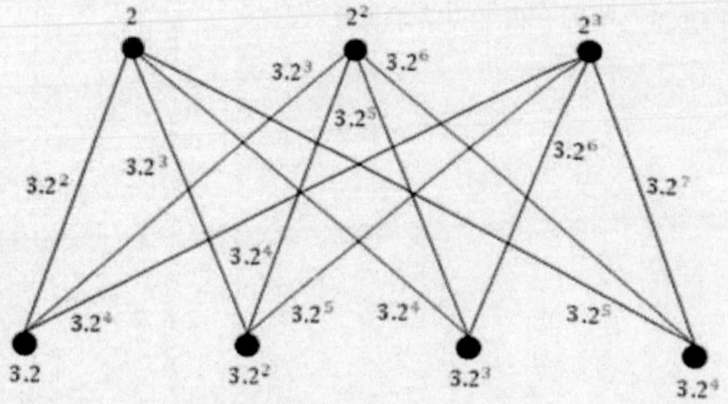

Fig. 2 6-Zumkeller graph of $K_{3,4}$

Example 2
A 6-Zumkeller graph of complete bipartite $K_{3,4}$ is shown in Fig. 2.

Theorem 2
The complete tripartite graph K_{n_1,n_2,n_3}, $n_1, n_2, n_3 \geq 2$ is a $[2(n_1 + n_2 + n_3) - 3]$-Zumkeller graph.

Proof Let $V = \{A_i : 1 \leq i \leq n_1\} \cup \{B_j : 1 \leq j \leq n_2\} \cup \{C_l : 1 \leq l \leq n_3\}$ and $E = \{A_iB_j : 1 \leq i \leq n_1, 1 \leq j \leq n_2\} \cup \{A_iC_l : 1 \leq i \leq n_1, 1 \leq l \leq n_3\} \cup \{B_jC_l : 1 \leq j \leq n_2, 1 \leq l \leq n_3\}$ be the vertex and edge set of K_{n_1,n_2,n_3}. The graph has $n_1 + n_2 + n_3$ vertices and $n_1(n_2 + n_3) + n_2n_3$ edges.

Define a one-to-one function $f : V \to N$ such that

$$f(A_i) = 2^i, 1 \leq i \leq n_1$$

$$f(B_j) = p_1\, 2^j, 1 \leq j \leq n_2$$

$$f(C_l) = p_2\, 2^l, 1 \leq l \leq n_3$$

where p_1, p_2 are distinct prime numbers such that $2 < p_1, p_2 < 10$.

Define an induced edge function $f^* : E \to N$ such that $f^*(xy) = f(x)f(y)$ for all $xy \in E$. Then the function provides the Zumkeller numbers on edges as follows:

$$f^*(A_iB_j) = p_1\, 2^{i+j}, 1 \leq i \leq n_1, 1 \leq j \leq n_2$$

$$f^*(A_iC_l) = p_2\, 2^{i+l}, 1 \leq i \leq n_1, 1 \leq l \leq n_3$$

$$f^*(B_jC_l) = p_2p_1\, 2^{j+l}, 1 \leq j \leq n_2, 1 \leq l \leq n_3$$

The edges of the complete tripartite graph K_{n_1,n_2,n_3} receive the Zumkeller numbers viz., $p_1\, 2^2, p_1\, 2^3, \ldots, p_1\, 2^{n_1+n_2}, p_2\, p_1\, 2^2, p_2\, p_1\, 2^3, \ldots, p_2\, p_1\, 2^{n_2+n_3}$,

Fig. 3 15-Zumkeller graph of $K_{3,3,3}$

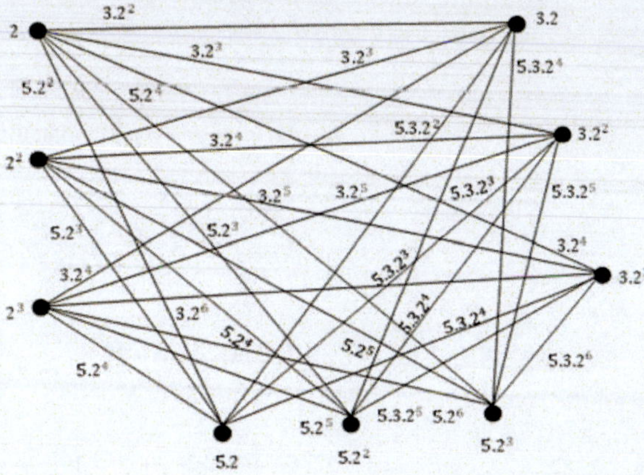

$p_2\,2^2, p_2\,2^3, \ldots, p_2\,2^{n_1+n_3}$, which are all distinct and the total count of different Zumkeller number is $[2(n_1+n_2+n_3)-3]$. Hence, the complete tripartite graph K_{n_1,n_2,n_3} is a $[2(n_1+n_2+n_3)-3]$ - Zumkeller graph when $n_1, n_2, n_3 \geq 2$.

Example 3
The 15- Zumkeller graph of $K_{3,3,3}$ is shown in Fig. 3.

Theorem 3
The complete *m* partite graph $K_{n_1,n_2,\ldots,n_m}$, $4 \leq m \leq 8$ is a *k*-Zumkeller graph.

Proof Let $V = \{A_i : 1 \leq i \leq n_1\} \cup \{B_j : 1 \leq j \leq n_2\} \cup \{C_l : 1 \leq l \leq n_3\} \cup \{D_n : 1 \leq n \leq n_4\} \cup \{E_p : 1 \leq p \leq n_5\} \cup \{F_q : 1 \leq q \leq n_6\} \cup \{G_r : 1 \leq r \leq n_7\} \cup \{H_s : 1 \leq s \leq n_8\}$ be the vertex set of $K_{n_1,n_2,\ldots,n_8}$.

Let $E = \{A_iB_j : 1 \leq i \leq n_1, 1 \leq j \leq n_2\} \cup \{A_iC_l : 1 \leq i \leq n_1, 1 \leq l \leq n_3\} \cup \{A_iD_n : 1 \leq i \leq n_1, 1 \leq n \leq n_4\} \cup \ldots \cup \{A_iH_s : 1 \leq i \leq n_1, 1 \leq s \leq n_8\} \cup \{B_jC_l : 1 \leq j \leq n_2, 1 \leq l \leq n_3\} \cup \{B_jD_n : 1 \leq j \leq n_2, 1 \leq n \leq n_4\} \cup \ldots \cup \{B_jH_s : 1 \leq j \leq n_2, 1 \leq s \leq n_8\} \cup \{C_lD_n : 1 \leq l \leq n_3, 1 \leq n \leq n_4\} \cup \ldots \cup \{C_lH_s : 1 \leq l \leq n_3, 1 \leq s \leq n_8\} \cup \{D_nE_p : 1 \leq n \leq n_4, 1 \leq p \leq n_5\} \cup \ldots \cup \{D_nH_s : 1 \leq n \leq n_4, 1 \leq s \leq n_8\} \cup \{E_pF_q : 1 \leq p \leq n_5, 1 \leq q \leq n_6\} \cup \ldots \cup \{E_pH_s : 1 \leq p \leq n_5, 1 \leq s \leq n_8\} \cup \{F_qG_r : 1 \leq q \leq n_6, 1 \leq r \leq n_7\} \cup \ldots \cup \{F_qH_s : 1 \leq q \leq n_6, 1 \leq s \leq n_8\} \cup \{G_rH_s : 1 \leq r \leq n_7, 1 \leq s \leq n_8\}$ be the edge set of $K_{n_1,n_2,\ldots,n_8}$.

The graph has $n_1 + n_2 + n_3 + \ldots + n_m$ vertices and $n_1(n_2 + n_3 + \ldots + n_m) + n_2(n_3 + n_4 + \ldots + n_m) + n_3(n_4 + n_5 + \ldots + n_m) + n_5(n_6 + n_7 + \ldots + n_m) + \ldots + n_{m-1}n_m$ edges.

Define a one-to-one function $f : V \to N$ such that

$$f(A_i) = 2^i, 1 \leq i \leq n_1$$

$$f(B_j) = p_1\,2^j, 1 \leq j \leq n_2$$

$$f\,(C_l) = p_2\, 2^l, 1 \le l \le n_3$$

$$f\,(D_n) = p_2\; p_1\, 2^n, 1 \le n \le n_4$$

$$f\,(E_p) = p_3\, 2^p, 1 \le p \le n_5$$

$$f\,(F_q) = p_3\; p_1\, 2^q, 1 \le q \le n_6$$

$$f\,(G_r) = p_3\; p_2\, 2^r, 1 \le r \le n_7$$

$$f\,(H_s) = p_3\; p_2\; p_1\, 2^s, 1 \le s \le n_8$$

where p_1, p_2, p_3 are distinct prime numbers such that $2 < p_1, p_2, p_3 < 10$.

Define an induced edge function $f^* : E \rightarrow N$ such that $f^*(xy) = f(x)f(y)$ for all $xy \in E$. Then the function provides the Zumkeller numbers on edges as follows:

$f^*(A_iB_j) = p_1\, 2^{i+j}, 1 \le i \le n_1, 1 \le j \le n_2$
$f^*(A_iC_l) = p_2\, 2^{i+l}, 1 \le i \le n_1, 1 \le l \le n_3$
$f^*(A_iD_n) = p_2p_1\, 2^{i+n}, 1 \le i \le n_1, 1 \le n \le n_4$
$f^*(A_iE_p) = p_3\, 2^{i+p}, 1 \le i \le n_1, 1 \le p \le n_5$
$f^*(A_iF_q) = p_3p_1\, 2^{i+q}, 1 \le i \le n_1, 1 \le q \le n_6$
$f^*(A_iG_r) = p_3p_2\, 2^{i+r}, 1 \le i \le n_1, 1 \le r \le n_7$
$f^*(A_iH_s) = p_3p_2p_1\, 2^{i+s}, 1 \le i \le n_1, 1 \le s \le n_8$

$f^*(B_jC_l) = p_2p_1\, 2^{j+l}, 1 \le j \le n_2, 1 \le l \le n_3$
$f^*\left(B_jD_n\right) = p_2p_1^2\, 2^{j+n}, 1 \le j \le n_2, 1 \le n \le n_4$
$f^*(B_jE_p) = p_3\, p_1 2^{j+p}, 1 \le j \le n_2, 1 \le p \le n_5$
$f^*\left(B_jF_q\right) = p_3p_1^2\, 2^{j+q}, 1 \le j \le n_2, 1 \le q \le n_6$
$f^*(B_jG_r) = p_3p_2p_1\, 2^{j+r}, 1 \le j \le n_2, 1 \le r \le n_7$
$f^*\left(B_jH_s\right) = p_3p_2p_1^2\, 2^{j+s}, 1 \le j \le n_2, 1 \le s \le n_8$

$f^*\,(C_lD_n) = p_2^2p_1\, 2^{l+n}, 1 \le l \le n_3, 1 \le n \le n_4$
$f^*(C_lE_p) = p_3p_2\, 2^{l+p}, 1 \le l \le n_3, 1 \le p \le n_5$
$f^*(C_lF_q) = p_3p_2p_1\, 2^{l+q}, 1 \le l \le n_3, 1 \le q \le n_6$
$f^*\,(C_lG_r) = p_3p_2^2\, 2^{l+r}, 1 \le l \le n_3, 1 \le r \le n_7$
$f^*\,(C_lH_s) = p_3p_2^2p_1\, 2^{l+s}, 1 \le l \le n_3, 1 \le s \le n_8$

$f^*(D_nE_p) = p_3p_2p_1\, 2^{n+p}, 1 \le n \le n_4, 1 \le p \le n_5$
$f^*\left(D_nF_q\right) = p_3p_2p_1^2\, 2^{n+q}, 1 \le n \le n_4, 1 \le q \le n_6$

$f^*(D_nG_r) = p_3p_2^2p_1\ 2^{n+r}, 1 \leq n \leq n_4, 1 \leq r \leq n_7$
$f^*(D_nH_s) = p_3p_2^2p_1^2\ 2^{n+s}, 1 \leq n \leq n_4, 1 \leq s \leq n_8$

$f^*(E_pF_q) = p_3^2p_1\ 2^{p+q}, 1 \leq p \leq n_5, 1 \leq q \leq n_6$
$f^*(E_pG_r) = p_3^2p_2\ 2^{p+r}, 1 \leq p \leq n_5, 1 \leq r \leq n_7$
$f^*(E_pH_s) = p_3^2p_2p_1\ 2^{p+s}, 1 \leq p \leq n_5, 1 \leq s \leq n_8$

$f^*(F_qG_r) = p_3^2p_2p_1\ 2^{q+r}, 1 \leq q \leq n_6, 1 \leq r \leq n_7$
$f^*(F_qH_s) = p_3^2p_2p_1^2\ 2^{q+s}, 1 \leq q \leq n_6, 1 \leq s \leq n_8$

$f^*(G_rH_s) = p_3^2p_2^2p_1\ 2^{r+s}, 1 \leq r \leq n_7, 1 \leq s \leq n_8$

Case 1 When $m = 4$

The edges of the complete 4-partite graph K_{n_1,n_2,n_3,n_4} receive the Zumkeller numbers viz.,

$p_1\,2^2, p_1\,2^3, \ldots, p_1\,2^{n_1+n_2}, p_2\,2^2, p_2\,2^3, \ldots, p_2\,2^{n_1+n_3}, p_2p_1\,2^2, p_2p_1\,2^3, \ldots, p_2p_1\,2^{n_1+n_4}, p_2p_1\,2^2, p_2p_1\,2^3, \ldots, p_2p_1\,2^{n_2+n_3}, p_2p_1^2\,2^2, p_2p_1^2\,2^3, \ldots, p_2p_1^2\,2^{n_2+n_4}, p_2^2p_1\,2^2, p_2^2p_1\,2^3, \ldots, p_2^2p_1\,2^{n_3+n_4}$

Therefore, the total count of different Zumkeller numbers on the edges is given by

$$[n_2 + (n_1 - 1) + n_3 + (n_1 - 1) + \mathbf{n_4} + (\mathbf{n_1} - \mathbf{1}) + n_3 + (n_2 - 1) + n_4 + (n_2 - 1) + n_4 + (n_3 - 1)]$$

where the highlighted terms $[n_4 + (n_1 - 1)]$ can be replaced as follows:

$$n_4 + (n_1 - 1) = \begin{cases} n_4 + (n_1 - 1) - (n_3 + (n_2 - 1)), & n_4 + (n_1 - 1) > (n_3 + (n_2 - 1)) \\ 0, & n_4 + (n_1 - 1) \leq (n_3 + (n_2 - 1)) \end{cases}$$

Hence, the total count of different Zumkeller numbers on the edges of the complete 4-partite graph K_{n_1,n_2,n_3,n_4} is as follows:

(i) $3n_4 + 2n_3 + 2n_2 + 3n_1 - 5$, when $n_4 + (n_1 - 1) > (n_3 + (n_2 - 1.))$
(ii) $2n_4 + 3n_3 + 3n_2 + 2n_1 - 5$, when $n_4 + (n_1 - 1) \leq (n_3 + (n_2 - 1))$

Case 2 When $m = 5$

The edges of the complete 5-partite graph K_{n_1,n_2,n_3,n_4,n_5} receive the following Zumkeller numbers:

$p_1\,2^2, p_1\,2^3, \ldots, p_1\,2^{n_1+n_2}, p_2\,2^2, p_2\,2^3, \ldots, p_2\,2^{n_1+n_3}, p_2p_1\,2^2, p_2p_1\,2^3, \ldots, p_2p_1\,2^{n_1+n_4}, p_3\,2^2, p_3\,2^3, \ldots, p_3\,2^{n_1+n_5}, p_2p_1\,2^2, p_2p_1\,2^3, \ldots, p_2p_1\,2^{n_2+n_3}, p_2p_1^2\,2^2, p_2p_1^2\,2^3, \ldots, p_2p_1^2\,2^{n_2+n_4}, p_3\ p_1\,2^2, p_3\ p_1\,2^3, \ldots, p_3\ p_1\,2^{n_2+n_5},$

$p_2^2 p_1 \; 2^2, p_2^2 p_1 \; 2^3, \ldots, p_2^2 p_1 \; 2^{n_3+n_4}, \; p_3 p_2 \; 2^2, p_3 p_2 \; 2^3, \ldots, p_3 p_2 \; 2^{n_3+n_5}$, $p_3 p_2 p_1 \, 2^2, p_3 p_2 p_1 \, 2^3, \ldots, p_3 p_2 p_1 \, 2^{n_4+n_5}$

Therefore, edges of the complete 5-partite graph give the following total count of different Zumkeller numbers:

$$\begin{aligned}&[n_2 + (n_1 - 1) + n_3 + (n_1 - 1) + \boldsymbol{n_4} + (\boldsymbol{n_1} - \boldsymbol{1}) + n_5 + (n_1 - 1) + n_3 + (n_2 - 1)\\ &+ n_4 + (n_2 - 1) + n_5 + (n_2 - 1) + n_4 + (n_3 - 1) + n_5 + (n_3 - 1) + n_5 + (n_4 - 1)]\end{aligned}$$

where the highlighted terms can be replaced as follows:

$$\begin{aligned}&n_4 + (n_1 - 1)\\ &= \begin{cases} n_4 + (n_1 - 1) - (n_3 + (n_2 - 1)), & n_4 + (n_1 - 1) > (n_3 + (n_2 - 1))\\ 0, & n_4 + (n_1 - 1) \le (n_3 + (n_2 - 1)) \end{cases}\end{aligned}$$

Case 3 When $m = 6$

The edges of the complete 6-partite graph $K_{n_1,n_2,n_3,n_4,n_5,n_6}$ receive the following Zumkeller numbers:

$p_1 \, 2^2, p_1 \, 2^3, \ldots, p_1 \, 2^{n_1+n_2}, p_2 \, 2^2, p_2 \, 2^3, \ldots, p_2 \, 2^{n_1+n_3}, p_2 p_1 \, 2^2, p_2 p_1 \, 2^3, \ldots,$ $p_2 p_1 \; 2^{n_1+n_4}, \; p_3 \; 2^2, p_3 \; 2^3, \ldots, p_3 \; 2^{n_1+n_5}, p_3 p_1 \; 2^2, p_3 p_1 \; 2^3, \ldots, p_3 p_1 \; 2^{n_1+n_6}$, $p_2 p_1 \; 2^2, p_2 p_1 \; 2^3, \ldots, p_2 p_1 \; 2^{n_2+n_3}, \; p_2 p_1^2 \; 2^2, p_2 p_1^2 \; 2^3, \ldots, p_2 p_1^2 \; 2^{n_2+n_4}$, $p_3 \; p_1 \; 2^2, p_3 \; p_1 \; 2^3, \ldots, p_3 \; p_1 \; 2^{n_2+n_5}, \; p_3 p_1^2 \; 2^2, p_3 p_1^2 \; 2^3, \ldots, p_3 p_1^2 \; 2^{n_2+n_6}$, $p_2^2 p_1 \; 2^2, p_2^2 p_1 \; 2^3, \ldots, p_2^2 p_1 \; 2^{n_3+n_4}, \; p_3 p_2 \; 2^2, p_3 p_2 \; 2^3, \ldots, p_3 p_2 \; 2^{n_3+n_5}$, $p_3 p_2 \, 2^2, p_3 p_2 \, 2^3, \ldots, p_3 p_2 \, 2^{n_3+n_6}, p_3 p_2 p_1 \, 2^2, p_3 p_2 p_1 \, 2^3, \ldots, p_3 p_2 p_1 \, 2^{n_4+n_5}$, $p_3 p_2 p_1^2 \, 2^2, p_3 p_2 p_1^2 \, 2^3, \ldots, p_3 p_2 p_1^2 \, 2^{n_4+n_6}, p_3^2 p_1 \, 2^2, p_3^2 p_1 \, 2^3, \ldots, p_3^2 p_1 \, 2^{n_5+n_6}$

Therefore, edges of the complete 6-partite graph give the following total count of different Zumkeller numbers:

$$\begin{aligned}&[\, n_2 + (n_1 - 1) + n_3 + (n_1 - 1) + \boldsymbol{n_4} + (\boldsymbol{n_1} - \boldsymbol{1}) + n_5 + (n_1 - 1) + \boldsymbol{n_6}\\ &+ (\boldsymbol{n_1} - \boldsymbol{1}) + n_3 + (n_2 - 1) + n_4 + (n_2 - 1) + n_5 + (n_2 - 1) + n_6 + (n_2 - 1)\\ &+ n_4 + (n_3 - 1) + n_5 + (n_3 - 1) + \boldsymbol{n_6} + (\boldsymbol{n_3} - \boldsymbol{1}) + n_5 + (n_4 - 1) + n_6\\ &+ (n_4 - 1) + n_6 + (n_5 - 1)]\end{aligned}$$

where the highlighted terms can be replaced as follows:

$$\begin{aligned}&n_4 + (n_1 - 1)\\ &= \begin{cases} n_4 + (n_1 - 1) - (n_3 + (n_2 - 1)), & n_4 + (n_1 - 1) > (n_3 + (n_2 - 1))\\ 0, & n_4 + (n_1 - 1) \le (n_3 + (n_2 - 1)) \end{cases}\end{aligned}$$

$$n_6 + (n_1 - 1) = \begin{cases} n_6 + (n_1 - 1) - (n_5 + (n_2 - 1)), & n_6 + (n_1 - 1) > (n_5 + (n_2 - 1)) \\ 0, & n_6 + (n_1 - 1) \leq (n_5 + (n_2 - 1)) \end{cases}$$

$$n_6 + (n_3 - 1) = \begin{cases} n_6 + (n_3 - 1) - (n_5 + (n_4 - 1)), & n_6 + (n_3 - 1) > (n_5 + (n_4 - 1)) \\ 0, & n_6 + (n_3 - 1) \leq (n_5 + (n_4 - 1)) \end{cases}$$

Case 4 When $m = 7$

The edges of the complete 7-partite graph $K_{n_1,n_2,n_3,n_4,n_5,n_6,n_7}$ receive the following Zumkeller numbers:

$p_1\,2^2, p_1\,2^3, \ldots, p_1\,2^{n_1+n_2}, p_2\,2^2, p_2\,2^3, \ldots, p_2\,2^{n_1+n_3}, p_2p_1\,2^2, p_2p_1\,2^3, \ldots, p_2p_1\,2^{n_1+n_4}, p_3\,2^2, p_3\,2^3, \ldots, p_3\,2^{n_1+n_5}, p_3p_1\,2^2, p_3p_1\,2^3, \ldots, p_3p_1\,2^{n_1+n_6}, p_3p_2\,2^2, p_3p_2\,2^3, \ldots, p_3p_2\,2^{n_1+n_7}, p_2p_1\,2^2, p_2p_1\,2^3, \ldots, p_2p_1\,2^{n_2+n_3}, p_2p_1^2\,2^2, p_2p_1^2\,2^3, \ldots, p_2p_1^2\,2^{n_2+n_4}, p_3\,p_1\,2^2, p_3\,p_1\,2^3, \ldots, p_3\,p_1\,2^{n_2+n_5}, p_3p_1^2\,2^2, p_3p_1^2\,2^3, \ldots, p_3p_1^2\,2^{n_2+n_6}, p_3p_2p_1\,2^2, p_3p_2p_1\,2^3, \ldots, p_3p_2p_1\,2^{n_2+n_7}, p_2^2p_1\,2^2, p_2^2p_1\,2^3, \ldots, p_2^2p_1\,2^{n_3+n_4}, p_3p_2\,2^2, p_3p_2\,2^3, \ldots, p_3p_2\,2^{n_3+n_5}, p_3p_2\,2^2, p_3p_2\,2^3, \ldots, p_3p_2\,2^{n_3+n_6}, p_3p_2^2\,2^2, p_3p_2^2\,2^3, \ldots, p_3p_2^2\,2^{n_3+n_7}, p_3p_2p_1\,2^2, p_3p_2p_1\,2^3, \ldots, p_3p_2p_1\,2^{n_4+n_5}, p_3p_2p_1^2\,2^2, p_3p_2p_1^2\,2^3, \ldots, p_3p_2p_1^2\,2^{n_4+n_6}, p_3p_2^2p_1\,2^2, p_3p_2^2p_1\,2^3, \ldots, p_3p_2^2p_1\,2^{n_4+n_7}, p_3^2p_1\,2^2, p_3^2p_1\,2^3, \ldots, p_3^2p_1\,2^{n_5+n_6}, p_3^2p_2\,2^2, p_3^2p_2\,2^3, \ldots, p_3^2p_2\,2^{n_5+n_7}, p_3^2p_2p_1\,2^2, p_3^2p_2p_1\,2^3, \ldots, p_3^2p_2p_1\,2^{n_6+n_7}$

Therefore, edges of the complete 7-partite graph give the following total number of distinct Zumkeller numbers:

$$\begin{aligned} [\, & n_2 + (n_1 - 1) + n_3 + (n_1 - 1) + \boldsymbol{n_4} + \boldsymbol{(n_1 - 1)} + n_5 + (n_1 - 1) + \boldsymbol{n_6} + \boldsymbol{(n_1 - 1)} \\ & + \boldsymbol{n_7} + \boldsymbol{(n_1 - 1)} + n_3 + (n_2 - 1) + n_4 + (n_2 - 1) + n_5 + (n_2 - 1) + n_6 \\ & + (n_2 - 1) + \boldsymbol{n_7} + \boldsymbol{(n_2 - 1)} + n_4 + (n_3 - 1) + n_5 + (n_3 - 1) + \boldsymbol{n_6} + \boldsymbol{(n_3 - 1)} \\ & + n_7 + (n_3 - 1) + n_5 + (n_4 - 1) + n_6 + (n_4 - 1) + n_7 + (n_4 - 1) + n_6 \\ & + (n_5 - 1) + n_7 + (n_5 - 1) + n_7 + (n_6 - 1)\,] \end{aligned}$$

where the highlighted terms can be replaced as follows:

$$n_4 + (n_1 - 1) = \begin{cases} n_4 + (n_1 - 1) - (n_3 + (n_2 - 1)), & n_4 + (n_1 - 1) > (n_3 + (n_2 - 1)) \\ 0, & n_4 + (n_1 - 1) \leq (n_3 + (n_2 - 1)) \end{cases}$$

$$n_6 + (n_1 - 1) = \begin{cases} n_6 + (n_1 - 1) - (n_5 + (n_2 - 1)), & n_6 + (n_1 - 1) > (n_5 + (n_2 - 1)) \\ 0, & n_6 + (n_1 - 1) \le (n_5 + (n_2 - 1)) \end{cases}$$

$$n_6 + (n_3 - 1) = \begin{cases} n_6 + (n_3 - 1) - (n_5 + (n_4 - 1)), & n_6 + (n_3 - 1) > (n_5 + (n_4 - 1)) \\ 0, & n_6 + (n_3 - 1) \le (n_5 + (n_4 - 1)) \end{cases}$$

$$n_7 + (n_1 - 1) = \begin{cases} n_7 + (n_1 - 1) - (n_5 + (n_3 - 1)), & n_7 + (n_1 - 1) > (n_5 + (n_3 - 1)) \\ 0, & n_7 + (n_1 - 1) \le (n_5 + (n_3 - 1)) \end{cases}$$

$$n_7 + (n_2 - 1) = n_7 + (n_2 - 1) - [n_5 + (n_4 - 1) + (n_6 + (n_3 - 1) - (n_5 + (n_4 - 1)))],$$

if $n_7 + (n_2 - 1) > [n_5 + (n_4 - 1) + (n_6 + (n_3 - 1) - (n_5 + (n_4 - 1)))]$
otherwise 0.

Therefore, the complete *m*- partite graph is a *k*- Zumkeller graph, where the value of *k* differs as per the above cases. Hence the proof.

4 Conclusion

In this chapter, the *k*- Zumkeller labeling of complete *m* partite graphs where $2 \le m \le 8$ has been computed. Computing *k*-Zumkeller labeling of graphs with minimum value of *k* is challenging and fascinating one. Eventually, this can be extended to *n* partite graph as well in future, which is an interesting area of research.

References

1. Rosa, A.: On certain valuations of the vertices of a graph. In: Theory of Graphs (Internat. Sympos. Rome. 1966), pp. 349–359. Gordan and Breach, Newyork, Dunod, Paris (1967)
2. Gallian, J.A.: A dynamic survey of graph labeling. Electron. J. Comb., 17 # DS6 (2021)
3. Bloom, G.S., Golomb, S.W.: Applications of numbered undirected graphs. Proc. IEEE. **165**(4), 526–570 (1977)

4. Gurjar, D.K., Krishnaa, A.: Lexicographic labeled graphs in cryptography. Adv. Appl. Discrete Math. **27**(2), 209–232 (2021)
5. Kuppan, R., Shobana, L., Cangul, I.N.: Encryption and decryption algorithms using strong face graph of a tree. Int. J. Comput. Math.: Comput. Syst. Theory. **5**, 225–233 (2020)
6. Srinivasan, A.K.: Practical numbers. Curr. Sci. **17**, 179–180 (1948)
7. Buss, F.: Zumkeller numbers and partitions. http://www.luschny.de/math/seq/FrankBussZumkellerNumbers.html
8. Peng, Y., Bhaskara Rao, K.P.S.: On Zumkeller numbers. J. Number Theory. **133**(4), 1135–1155 (2013)
9. Balamurugan, B.J., Thirusangu, K., Thomas, D.G.: k-Zumkeller labeling for twig graphs. Electron Notes Discrete Math. **48**, 119–126 (2015)
10. Balamurugan, B.J., Thirusangu, K., Thomas, D.G., Murali, B.J.: k-Zumkeller labeling of graphs. Int. J. Eng. Technol. **7**, 460–463 (2018)
11. Basher, M.: *k*-Zumkeller labeling of super subdivision of some graphs. J. Egypt Math. Soc. **29**(12) (2021). https://doi.org/10.1186/s42787-021-00121-y
12. Basher, M.: *k*-Zumkeller labeling of the Cartesian and tensor product of paths and cycles. J. Intell. Fuzzy Syst. **40**, 5061–5070 (2021)
13. Harary, F.: Graph Theory. CRC Press, Philippines (2019)
14. West, D.B.: Introduction to Graph Theory, 2nd edn. PHI Learning Private Limited, New Delhi (2009)

Characterizations of (γ_i, γ_{DDS}, γ_{DSNS}) – Trees

K. Karthika and M. Yamuna

1 Introduction

"A graph G = (V, E) consists of a set of V = {v_1, v_2, ..., v_n}, whose elements are called vertices and a set E = {e_1, e_2, ..., e_m}, whose elements are called edges, such that each edge e_k is identified with an unordered pair (v_i v_j) of vertices. The number of vertices is called the order of graph and denoted by |V| = n. The number of edges is called the size of the graph and denoted by |E| = m. The degree of a vertex v in graph G, denoted deg (v), is the number of edges incident with v. For any set S of vertices of G, the induced subgraph ⟨S⟩ is the maximal subgraph of G with vertex set S. Two vertices of S are adjacent in ⟨S⟩ if and only if they are adjacent in G. The vertex identification of a pair of vertices v_1 and v_2 of a graph produces a graph in which the vertices v_1 and v_2 are replaced with a single vertex v such that v is adjacent to the union of the vertices to which v_1 and v_2 were originally adjacent. An edge contraction is an operation, which removes an edge from a graph while simultaneously identifying the two vertices that it previously joined. The subdivision of some edge e with endpoints {u, v} yields a graph containing one new vertex w, with an edge set replacing e by two new edges, (u w) and (w v), respectively. A subdivision of a graph G is a graph resulting from the subdivision of edges in G. P_n and K_n denote the path and complete graph with n vertices [1].

A dominating set D of G is a set of vertices of G such that every vertex in V – D is adjacent to a vertex in D. If D has the smallest possible cardinality of any dominating set of G, then D is called a minimum dominating set. The cardinality of any minimum dominating set for G is called the domination number of G, and

K. Karthika · M. Yamuna (✉)
SAS, VIT, Vellore, Tamilnadu, India
e-mail: karthika.k@vit.ac.in; myamuna@vit.ac.in

F. Kamalov et al. (eds.), *Advances in Mathematical Modeling and Scientific Computing*, Trends in Mathematics, https://doi.org/10.1007/978-3-031-41420-6_73

it is denoted by γ (G). A γ – set denotes a dominating set for G with minimum cardinality. The private neighborhood of v ∈ D, denoted by pn (v, D), is defined by pn (v, D) = N (v) − N (D − {v}). A vertex in V – D is k – dominated if it is dominated by at least 2 – vertices in D, that is, | N (v) ∩ D | ≥ 2. A vertex v is said to be a, down vertex if γ (G − v) < γ (G), level vertex if γ (G − v) = γ (G), up vertex if γ (G − v) > γ (G). A subset D of V is an independent dominating set if D is an independent set and every vertex of Vis adjacent to some vertex in D. Independent domination number denoted i (G) is the minimum cardinality of an independent dominating set. An independent dominating set of cardinality i (G) is called an i (G) – set [2]. Comparing different types of domination parameters and characterized tress are well studied by various authors [3–10].

A graph G is said to be domination dot stable (DDS) if γ (G·uv) = γ (G) for all u, v ∈ V (G), u adjacent to v. A graph G is said to be a domination non-subdivision stable (DSNS) if γ (G_{sd} uv) = γ (G) + 1 for all u, v ∈ V (G), u adjacent to v [11, 12]. An example of DDS and DSNS is seen in Fig. 1a, b. For the graph in Fig. 1a, D = {7, 8} is one possible γ – set for G, pn (7, D) = {1, 6} and a vertex 2 is 2 – dominated, since it is dominated by 7 and 8. Also γ (G) = γ (G·28) = 2. This is true for all for all u, v ∈ V (G), u adjacent to v ⇒ G is a DDS graph. For the graph in Fig. 1b, γ (G) = 3, γ (G_{sd} 9 10) = 4. This is true for all u, v ∈ V (G), u adjacent to v ⇒ G is a DSNS graph".

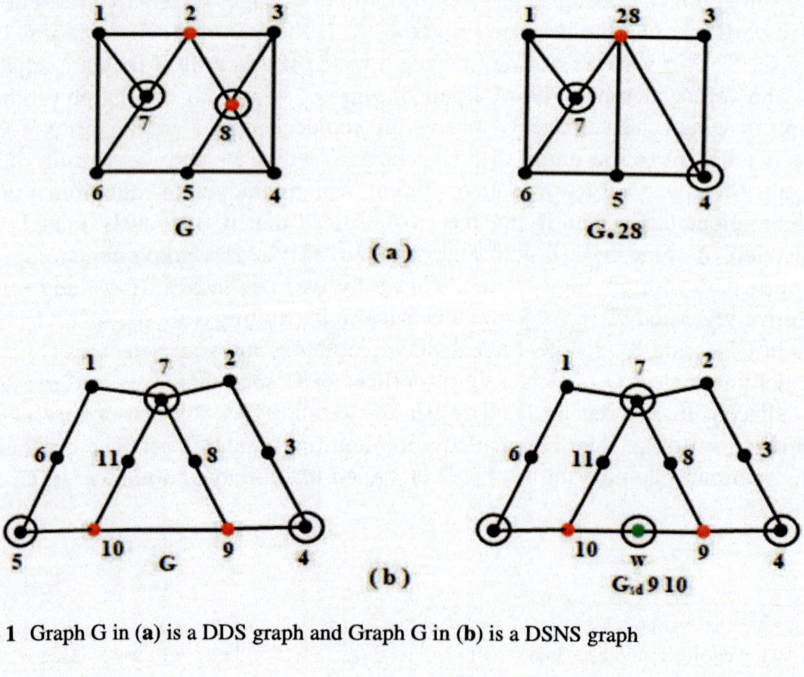

Fig. 1 Graph G in (**a**) is a DDS graph and Graph G in (**b**) is a DSNS graph

2 A Characterization of (γ_i, γ_{DDS}, γ_{DSNS}) – Trees

Graph operations often have an impact on various graph parameters. It also has its influence in changing the domination number by applying these graph operations. The domination number either increases, decreases, or remains unchanged. Sometimes two unrelated graph operations share a common property. A DDS graph targets to retain the same domination number, while a DSNS graph aims to increase the domination number. Surprisingly, both are possible when the γ – set of G is independent. A DSNS graph requires the additional property of having no 2 – dominated vertices. One of the focuses of research in graph theory that has received attention is characterizing graphs, which have equal parameters or nonequal parameters. Characterizing trees with equal domination numbers has required attention from audiences interested in domination theory. The tree in Fig. 2a is both DDS and DSNS. Consider the tree in Fig. 2b, the tree G_2 is DDS, since $\gamma(G_2 \cdot uv) = \gamma(G_2)$, as seen in Fig. 2c. The graph G_2 is not DSNS, since $\gamma(G_2) = \gamma(G_{2\,sd}\, uv) = 3 \neq \gamma(G_2) + 1$ as seen in Fig. 2d. So, a DDS tree need not be DSNS.

We recollect the following results.

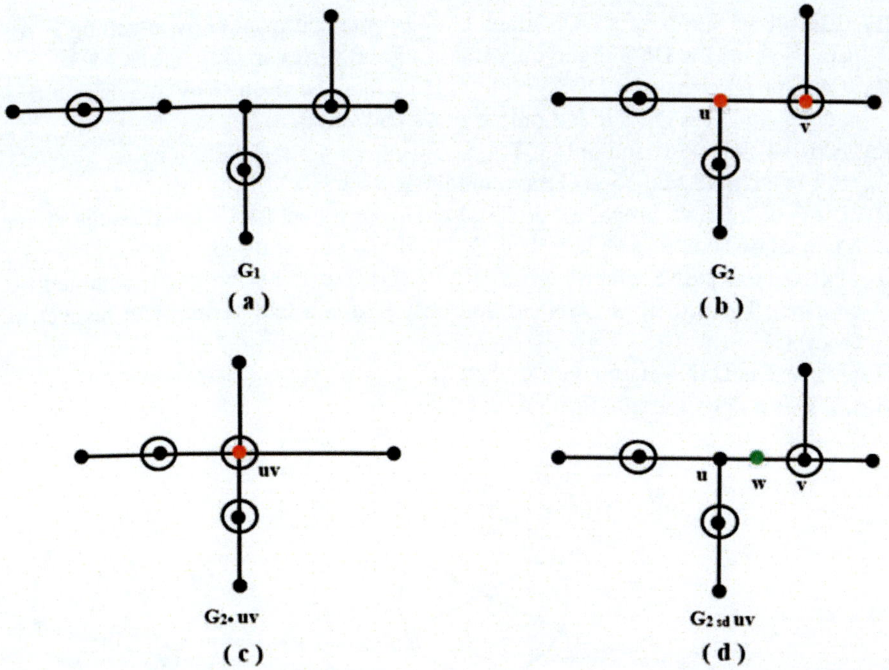

Fig. 2 Graph G_1 in (**a**) is both DDS and DSNS, whereas Graph G_2 in (**b**) is only DDS. The graph (**c**) denotes the graph obtained by merging vertex u and v, while (**d**) represents the graph obtained by subdividing u and v in G_2

R_1. "A graph G is DDS if and only if every γ – set of G is an independent dominating set."

R_2. "A graph G is a DSNS graph if and only if

(i) every γ – set of G is independent, and
(ii) G has no two dominated vertices."

For these two theorems, we understand that both the dominating sets are independent. "A γ – set D is said to be independent if $\langle D \rangle$ is independent."

For the graph in Fig. 3a, γ (G) = i (G) = 2 and in Fig. 3b, γ (G) = 2 and i (G) = 3. So, we understand that i (G) need not be equal to γ (G) always. The scope of this research work is limited to graphs with γ (G) = i (G). We shall denote the γ – set of these graphs as γ_i (G).

From Sect. 1, we observe that traditionally, tree characterization techniques focus on trees with two domination parameters of equal size. In this chapter, we extend this to possible tree characterizations with three equal domination parameters, namely, independence domination number, DDS number, and DSNS number. We shall denote such trees as (γ_i, γ_{DDS}, γ_{DSNS}) – trees. We proceed to characterize the same expecting that this is possible, since γ_{DSNS} (T) – set and γ_{DDS} (T) – set are independent. We recollect the following results which may be of use for characterizing (γ_i, γ_{DDS}, γ_{DSNS}) – trees [11, 12].

R_3. "Let u be a vertex of a DDS graph G. H is obtained from G by attaching a tree path P_2 at u. H is DDS if and only if u is a good vertex in G.

R_4. Let u be a vertex of a DDS graph G. H is obtained from G by attaching a tree path P_3 at u. H is DDS if and only if u is a bad vertex in G.

R_5. A tree T is DDS if and only if T $\in \tau$.

R_6. If T is a DSNS tree, then G has a unique γ – set.

R_7. Let G be a DSNS graph, u $\in$ V (G). Let D be a γ – set for G. Let H be the graph obtained by attaching P_2 to u. H is DSNS if and only if u$\in$ D.

R_8. Let G be a DSNS graph and let u $\in$ V (G). Let H be the graph obtained by attaching P_4 to u. H is DSNS if and only if u is a bad vertex with respect to (w.r.t) G.

R_9. A tree T is DSNS if and only if T $\in \tau$

R_{10}. If G is a DSNS graph, then pn (u, D) $\geq$ 2.

Fig. 3 γ (G) = i (G) in (**a**) and γ (G) $\neq$ i (G) in (**b**)

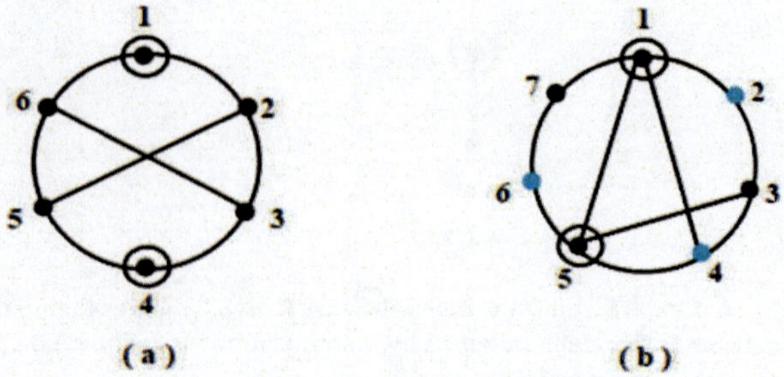

R_{11}. If G is a DSNS graph, then a pendant vertex is always a level vertex".

Let T_1 and T_2 denote the following two operations on any tree T. By attaching P_n to a vertex of T, we mean adding the path P_n and joining v to a pendant of P_n.

Operation T_1: Attach a tree path P_2 to a vertex of tree T that belong to any γ_{DSNS} (T) – set.

Operation T_2: Attach a tree path P_4 to a vertex v of tree T s.t. (such that) $v \notin \gamma_{DSNS}$ (T) – set.

As usual, let τ be a family defined by "τ = {T / T is obtained from P_3 by a finite sequence of operations T_1 or T_2}". We will now prove that any tree T ∈ τ has equal γ_i, γ_{DDS} and γ_{DSNS} numbers.

Theorem 1
If T ∈ τ, then T is a (γ_i, γ_{DDS}, γ_{DSNS}) – tree.

Proof Let S (T) = {operations needed to construct T}. We shall use induction on the set, when T = P_3, that is, when S (T) = 0, T is a (γ_i, γ_{DDS}, γ_{DSNS}) – tree. So, the induction hypothesis is true for the trivial case.

Assume that the result is true for any tree T′ ∈ τ s.t. S (T′) < p for some integer p ≥ 1, that is, T′ is a (γ_i, γ_{DDS}, γ_{DSNS}) – tree. Let T be a tree in the family τ s.t. T is obtained from T′ by applying the Operation T_1 or T_2 p times that is, S (T) = p. By induction hypothesis T′ is a (γ_i, γ_{DDS}, γ_{DSNS}) – tree. Let us now consider the following two cases, which depends on whether T is obtained from T′ by applying Type – 1 or Type – 2.

Case – 1 T is obtained from T′ by applying Operation – T_1. Suppose T is obtained from T′ by adding a vertex u and the edge uv, where v ∈ V (T′), then we know that v is in some γ_{DSNS} (T′) – set. Let D′ be γ_{DSNS} (T′) – set. By result R_6, we know that D′ is the only γ – set for T′. D′ is also a γ – set for T′, since v dominates u. By R_9, we know that T is a DSNS tree and hence, D′ is unique with respect to (w.r.t) T. D′ is a γ – set for T s.t.

(a) D′ is independent.
(b) There are no 2 – dominated vertices w.r.t. D′.

From R_1 and R_2, γ_i (T) = γ_{DSNS} (T) = γ_{DDS} (T) ⇒ D′ is a (γ_i, γ_{DDS}, γ_{DSNS}) – set and hence T is a (γ_i, γ_{DDS}, γ_{DSNS}) – tree.

Case – 2 T is obtained from T′ by operation T_2. Suppose T is obtained from T′ by adding a path $u_1u_2u_3$ and the edge u_1v, where v ∈ V (T′). Let D′ be a γ_{DSNS} (T′) – set. This γ_{DSNS} (T′) can be extended to a γ_{DSNS} (T′) – set by adding to it the vertex u_2. Hence, γ_{DSNS} (T) = γ_{DSNS} (T′) + 1. Let D = D′ ∪ {u_2}.

A. **D is unique.**
Suppose the γ_{DSNS} (T) is not unique, then D′ is not unique. This is true, since v is a bad vertex w.r.t D′. u_2 dominates u_1 and u_3. Hence if D is not unique, then there is at least two γ_{DSNS} (T′) – sets, say D_1, D_2 s.t. D_1 ∪ {u_2} and D_2 ∪ {u_2}

are γ_{DSNS} (T′) – set, a contradiction to our assumption that T′ is a DSNS – tree, by the result R_6.

B. **There are no two dominated vertices w.r.t D.**

By (A), we know that:

(i) D is unique w.r.t. T.
(ii) D′ is unique w.r.t. T′.
(iii) $D = D' \cup \{u_2\}$.
(iv) $v \notin D', D$.
(v) There are no two dominated vertices w.r.t. D, since T is DSNS by induction hypothesis.
(vi) $\{u_1, u_3\} \in pn(u_2, D) \Rightarrow$ there are no 2 – dominated vertices w.r.t. D.

C. **⟨D⟩ is independent.**

Since ⟨D, D′⟩ are unique and $D = D' \cup \{u_2\}$, u_2 is not adjacent to any vertex in T. Since D is γ_{DSNS} (T′) – set, ⟨D′⟩ is independent ⇒ ⟨D⟩ is independent.

From A, B, C, R_1, and R_2, γ_i (T) = γ_{DSNS} (T) = γ_{DDS} (T) ⇒ D′ is a (γ_i, γ_{DDS}, γ_{DSNS}) – set and hence T is a (γ_i, γ_{DDS}, γ_{DSNS}) – tree. Hence, if $T \in \tau$, T is a (γ_i, γ_{DDS}, γ_{DSNS}) – tree.

We shall now prove the converse part of Theorem 1.

Theorem 2
Any tree T is a (γ_i, γ_{DDS}, γ_{DSNS}) – tree, then T is a member of the family τ.

Proof We shall use induction on the number of vertices of a (γ_i, γ_{DDS}, γ_{DSNS}) – tree, $n \geq 3$. Let us consider T be a tree rooted at a leaf r in a longest possible path T. Let P be an r – u path as seen in Fig. 4.

By T_a, we mean a subtree, that is, induced along with a vertex "a" and its descendants in T. By R_{10}, since T is a (γ_i, γ_{DDS}, γ_{DSNS}) – tree it contains a γ_{DSNS} (T) – set D that contains no – pendant vertices of T. In particular, u is not contained in D. Since D contains no pendant vertices, it is obvious that all the support vertices are in D ⇒ v is included in D.

Let $T' = \{T - T_u\}$. T is a (γ_i, γ_{DDS}, γ_{DSNS}) – tree. Suppose deg (v) ≥ 3. Then v is adjacent to at least one pendant vertex w.r.t. T′. Since u is a level vertex w.r.t. T (by the result R_{11}), any γ_{DSNS} (T) – set is a γ – set for T′ also. T′ has a unique γ – set. Suppose T′ has two distinct γ – sets, say D and D_1. Let u_1 be adjacent to v, a pendant vertex w.r.t. T′. Either v or u_1 belongs to D_1. If $v \in D_1$, then D and D_1 are two distinct γ – sets for T′ containing v. If $u_1 \in D_1$, then $D_2 = D_1 - \{u_1\} \cup \{u\}$ is a γ – set for T′.

- D and D_2 are two distinct γ – sets for T′ containing v.

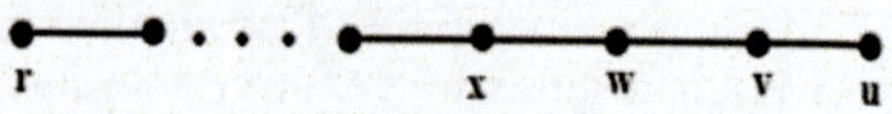

Fig. 4 r – u path

- Either D and D_1 or D and D_2 are two distinct γ – sets for T, a contradiction to the fact that T is DSNS by the result R_6.
- D is the only possible γ – sets for T, T′. $\Rightarrow$ D is unique.
- D is a γ_{DSNS} – set for T′ $\Rightarrow$ D is independent and T′ has no 2 – dominated vertices. Hence, T can be obtained from T′ by operation T_1. Thus, we may assume that deg (v) = 2 w.r.t. T.

If deg (v) = 2, then $T'' = T'_w - \{w\}$ is either K_1 or K_2. We know that $w \notin D$ as D is independent. If $T'' = K_1$, say w_1, then $w_1 \in D$, a contradiction as w is 2 – dominated w.r.t. T. If $T'' = K_2$, say w_1, w_2, then either $w_1 \in D$ or $w_2 \in D$. If $w_1 \in D$, then w is 2 – dominated w.r.t. T, a contradiction to the fact that T is DSNS. If $w_2 \in D$, then w is 2 – dominated, then $D_3 = D - \{w_2\} \cup \{w_1\}$ is a γ – set for T s.t. w is two dominated, a contradiction to the fact that T is DSNS $\Rightarrow$ deg (w) = 2. If x is a good vertex w.r.t. T″, then there is a γ – set for T″, say D_4 containing x. D and $D_4 \cup \{v\}$ are two distinct γ – sets for T, which is not possible as T is DSNS $\Rightarrow$ x is a bad vertex w.r.t. T″. Hence, we state that T can be generated from T″ by applying Type – 2 operation.

Let $T' = \{T - T_w\}$. If x is a good vertex w.r.t. T′, then there is a γ – set for T′, say D_4 containing x. D and $D_4 \cup \{v\}$ are two distinct γ – sets for T, which is not possible as T is DSNS $\Rightarrow$ x is a bad vertex w.r.t. T′. Hence, we state that T can be generated from T′ by applying Type – 2 operation.

From the above two theorems, we provide the following characterization.

Theorem 3
Any tree T is a $(\gamma_i, \gamma_{DDS}, \gamma_{DSNS})$ – tree if and only if T is a member of the family τ.

3 Conclusion

This chapter provided an iterative technique for constructing and characterizing trees with three equal domination parameters, which is a new approach away from the traditional method of characterizing trees with two equal domination parameters. We believe that this characterization will provide scope to improve the tree characterization techniques for focusing on characterizing trees with equal n – domination parameters.

References

1. Deo, N.: Graph Theory with Applications to Computer Science. PHI Learning Private Limited. Englewood Cliffs, NJ (2013)
2. Haynes, T.W., Hedetniemi, S.T., Slater, P.J.: Fundamentals of Domination in Graphs. Marcel Dekker, Inc., New York (1998)
3. Zepeng, L., Jin, X.: A characterization of trees with equal independent domination and secure domination numbers. Inf. Process. Lett. **119**, 14–18 (2017)

4. Chellali, M., Volkmann, L.: Characterization of trees with equal 2-domination number and domination number plus two. Discuss. Math. Graph Theory. **31**, 687–697 (2011)
5. Chellali, M., Meddah, N.: Trees with equal 2-domination and 2-independence numbers. Discuss. Math. Graph Theory. **32**, 263–270 (2012)
6. Brause, C., Henning, M.A., Krzywkowski, M.: A characterization of trees with equal 2-domination and 2-independence numbers. Discret. Math. Theor. Comput. Sci. **19**, 1–14 (2017)
7. Larson, C.E., Pepper, R.: Graphs with equal independence and annihilation numbers. Electron. J. Combin. **18**, 180 (2011)
8. Krzywkowski, M.: On trees with equal 2-domination and 2-outer-independent domination numbers. Indian J. Pure Appl. Math. **46**, 191–195 (2015)
9. Shan, E., Kang, L.: A characterization of trees with equal total domination and paired-domination numbers. Australas. J. Comb. **30**, 31–39 (2004)
10. Dorflinga, M., Goddard, W., Henning, M.A., Mynhardt, C.M.: Construction of trees and graphs with equal domination parameters. Discret. Math. **306**, 2647–2654 (2006)
11. Yamuna, M., Karthika, K.: Domination dot stable trees. Int. J. Pharm. Technol. **8**(2), 13564–13568 (2016)
12. Yamuna, M., Elakkiya, A.: Non – domination subdivision stable graphs. IOP Conf. Ser.: Mater. Sci. Eng. **263**, 042128 (2017)

Mobius Cordial Labeling of Graphs

A. AshaRani, K. Thirusangu, and B. J. Balamurugan

1 Introduction

The study of labeling of the vertices (edges) of graphs with real numbers or subsets of sets, which satisfies certain constraints, has been motivated by a number of real-world practical issues. In 1967, Rosa [7] introduced the concept of graph labeling. Gallian [3] provides a comprehensive and dynamic overview of graph labeling methods. In 1987, Cahit [2] introduced the notion of cordial labeling of graphs.

In number theory [1], it is interesting to study about the integers with specified properties. It is established by the fundamental theorem of arithmetic that primes are the building blocks of the set of natural numbers. Multiplicative functions play an important role in the study of the prime factorization of an integer.

The Mobius function [8] is a multiplicative function, which is an interesting function to investigate. This function was introduced by Mobius in 1832 [6] and is defined by

$$\mu(n) \equiv \begin{cases} 0 & \text{if } n \text{ has one or more repeated prime factors} \\ 1 & \text{if } n = 1 \\ (-1)^k & \text{if } n \text{ is a product of } k \text{ distinct primes} \end{cases}$$

and so that $\mu(n) \neq 0$ indicates that n is square free [5].

A. AshaRani
Department of Mathematics with Computer Applications, PSG College of Arts & Science, Coimbatore, Tamil Nadu, India

K. Thirusangu
Department of Mathematics, SIVET College, Gowrivakkam, Chennai, Tamil Nadu, India

B. J. Balamurugan (✉)
Division of Mathematics, School of Advanced Sciences, Vellore Institute of Technology, Chennai, Tamil Nadu, India
e-mail: balamurugan.bj@vit.ac.in

F. Kamalov et al. (eds.), *Advances in Mathematical Modeling and Scientific Computing*, Trends in Mathematics, https://doi.org/10.1007/978-3-031-41420-6_74

It is interesting to note that $\mu(1) = 1$, $\mu(p) = -1$ for prime p and $\mu(p^r) = 0$ for prime powers p^r with $r > 1$.

Some applications [9] of this function can be found in traditional graph coloring problems and in network flows [5]. Here, we present the concept of new graph labeling known as Mobius cordial labeling, and it has been investigated to certain family of graphs. To understand graph notations, concepts and terminology, we refer to Harary [4].

2 Main Result

Definition 1
Let V be the vertex set and E be the edge set for the graph G = (V,E). A 1–1 function $f : V \to N$ is said to be a Mobius cordial labeling of the graph G if the induced edge function $f^* : E \to \{-1, 0, 1\}$ defined by

$$f^*(uv) = |f\ (u) - f\ (v)|$$

$$= n = \begin{cases} 1 & if\ n = 1 \\ (-1)^k & if\ n\ is\ a\ product\ of\ k\ distinct\ primes \\ 0 & if\ n\ has\ one\ or\ more\ repeated\ prime\ factors \end{cases}$$

where $uv \in E$ satisfies the following conditions:

(i)

$$\left|e_{f^*}(0) - e_{f^*}(1)\right| \le 1$$

(ii)

$$\left|e_{f^*}(0) - e_{f^*}(-1)\right| \le 1$$

(iii)

$$\left|e_{f^*}(-1) - e_{f^*}(1)\right| \le 1.$$

Definition 2
Mobius cordial graph G is a graph that admits the Mobius cordial labeling.

Example 1
The Mobius cordial labeling of a graph G is shown in Fig. 1.

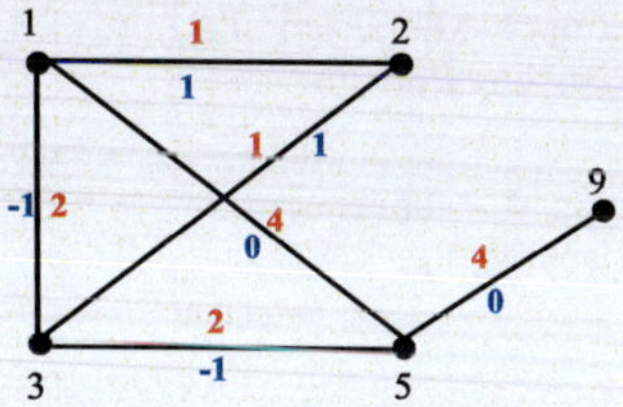

Fig. 1 Mobius cordial labeling of G

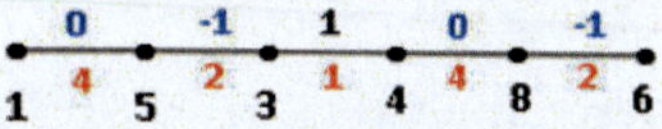

Fig. 2 Mobius cordial labeling P_6

Theorem 1 The path P_n, $n \geq 1$ is a Mobius cordial graph.

Proof Let P_n be a path graph with the vertex set $V = \{u_1, u_2, u_3 \ldots u_n\}$ and the edge set $E = \{u_1u_2, u_2u_3, \ldots u_n - {}_1u_n\}$.

The vertices of the graph P_n are labeled through the 1–1 function

$$f(u_l) = \begin{cases} 1 & for\ l = 1 \\ 2^2 + u_{l-1} & for\ l \equiv 2\,(\text{mod}3)\,;\ 1 < l \leq n \\ l & otherwise \end{cases} \tag{1}$$

By the definition of the induced edge function of Mobius cordial labeling, for every $u_lu_l - {}_1 \in E$,

$$f^*(u_lu_{l-1}) = \left|f(u_l) - f(u_{l-1})\right| \ for\ 1 < l \leq n$$

$$f^*(u_lu_{l-1}) = \begin{cases} 4 & for\ l \equiv 2(\text{mod } 3) \\ 2 & for\ l \equiv 0(\text{mod } 3) \\ 1 & for\ l \equiv 1(\text{mod } 3) \end{cases}$$

$$\text{(i.e)} \qquad f^*(u_lu_{l-1}) = \begin{cases} 0\ forl \equiv 2(\text{mod } 3) \\ -1\ forl \equiv 0(\text{mod } 3) \\ 1\,forl \equiv 1(\text{mod } 3) \end{cases}$$

Case 1 When n≡0 (*mod* 3)

Let the vertices of P_n be labeled as in Eq. (1). By the definition of the induced edge function f^*, the number of edges with label -1, 0, 1 are $\left\lceil \frac{n-1}{3} \right\rceil$, $\left\lceil \frac{n-1}{3} \right\rceil$, and $\left\lfloor \frac{n-1}{3} \right\rfloor$ respectively (Fig. 2).

Case 2 When n≡1 (*mod* 3)

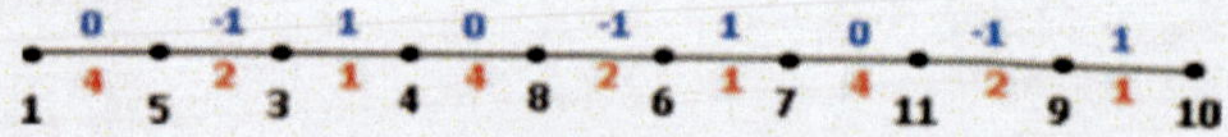

Fig. 3 Mobius cordial labeling P_{10}

Fig. 4 Mobius cordial labeling P_8

Let the vertices of P_n be labeled as in Eq. (1).By the definition of the induced edge function f^*, the number of edges with label -1, 0, 1 are $\frac{n-1}{3}$, $\frac{n-1}{3}$, and $\frac{n-1}{3}$, respectively (Fig. 3).

Case 3 When $n \equiv 2(mod\ 3)$

Let the vertices of P_n be labeled as in Eq. (1). By the definition of the induced edge function f^*, the number of edges with label -1, 0, 1 are $\left\lfloor \frac{n-1}{3} \right\rfloor$, $\left\lceil \frac{n-1}{3} \right\rceil$, and $\left\lfloor \frac{n-1}{3} \right\rfloor$, respectively (Fig. 4).

In all the above cases, we find that edges satisfy the conditions

$$\left|e_{f^*}(0) - e_{f^*}(1)\right| \leq 1, \left|e_{f^*}(0) - e_{f^*}(-1)\right| \leq 1 \text{ and } \left|e_{f^*}(-1) - e_{f^*}(1)\right| \leq 1.$$

Thus, the path P_n admits a Mobius cordial labeling.

Theorem 2 The cycle C_n, $n \geq 3$ is a Mobius cordial graph.

Proof Let $V = \{u_1, u_2, u_3 \ldots u_n\}$ be the vertex set and $E = \{u_1u_2, u_2u_3, \ldots u_{n-1}u_n, u_nu_1\}$ be the edge set of the cycle C_n.

The vertices of cycle C_3 are labeled as $f(u_1) = 1$, $f(u_2) = 2$, $f(u_3) = 5$. By the definition of the induced edge function f^*,

$$f^*(u_1u_2) = \left|f(u_1) - f(u_2)\right| = 1;$$

$f^*(u_2u_3) = |f(u_2) - f(u_3)| = -1; f^*(u_3u_1) = |f(u_3) - f(u_1)| = 0$. Thus, the cycle C_3 admits a Mobius cordial labeling.

When $n > 3$, we have the following cases.

Case 1 For $n \equiv 1\ (mod\ 3)$

Let the vertices of C_n be labeled as (Fig. 5)

$$f(u_l) = \begin{cases} 1 & for\ l = 1 \\ 2^2 + u_{l-1} & for\ l \equiv 2\,(\text{mod}3)\,;\, 1 < l \leq n \\ l & otherwise \end{cases} \quad (2)$$

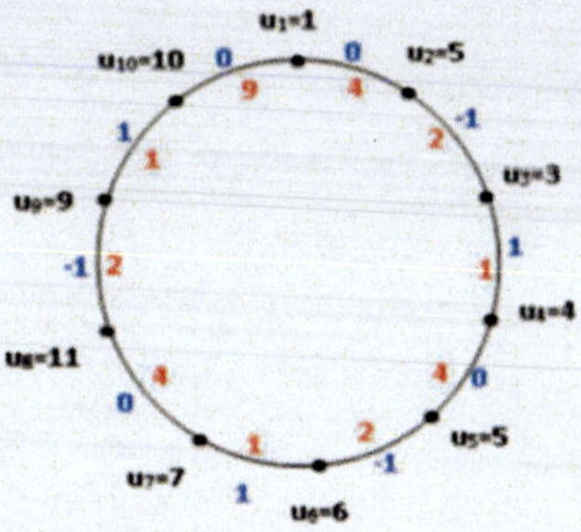

Fig. 5 Mobius cordial labeling of C_{10}

By the definition of the induced edge function, for every $u_l u_{l-1} \in E$

$$f^*(u_l u_{l-1}) = |f(u_l) - f(u_{l-1})| \; for \; 1 < l \leq n-2$$

$$f^*(u_l u_{l-1}) = \begin{cases} 4 & for \; l \equiv 2 \,(\text{mod } 3) \\ 2 & for \; l \equiv 0 \,(\text{mod } 3) \\ 1 & for \; l \equiv 1 \,(\text{mod } 3) \end{cases}$$

(i.e) $$f^*(u_l u_{l-1}) = \begin{cases} 0 & for \; l \equiv 2 \,(\text{mod } 3) \\ -1 & for \; l \equiv 0 \,(\text{mod } 3) \\ 1 & for \; l \equiv 1 \,(\text{mod } 3) \end{cases}$$

In this case, we find that the number of edges with label −1, 0, 1 are $\left\lfloor \frac{n-1}{3} \right\rfloor$, $\left\lceil \frac{n-1}{3} \right\rceil$, and $\left\lfloor \frac{n-1}{3} \right\rfloor$, respectively.

Case 2 For $n \equiv 0(mod\ 3)$

In this case, first n − 2 vertices of C_n are labeled as given by the Eq. (2) and nth vertex is labeled as $f(u_n) = n$.

We compute $f^*(u_n u_1) = |f(u_n) - f(u_1)| = k\ (say)$.

The vertex u_{n-1} is labelled as follows:

Subcase (i) *If k* is a product of odd number of distinct primes, then

$$f(u_{n-1}) = f(u_n) + 4$$

By Definition 1, we find that

$$f^{*}(u_{n-2}u_{n-1}) = \left|f(u_{n-2}) - f(u_{n-1})\right| = 1$$

$$f^{*}(u_{n-1}u_{n}) = \left|f(u_{n-1}) - f(u_{n})\right| = 0$$

$$f^{*}(u_{n}u_{1}) = |f(u_{n}) - f(u_{1})| = -1$$

Subcase (ii) If k is a product of even number of distinct primes, then

$$f(u_{n-1}) = f(u_{n-2}) + 4$$

By Definition 1,

$$f^{*}(u_{n-2}u_{n-1}) = \left|f(u_{n-2}) - f(u_{n-1})\right| = 0$$

$$f^{*}(u_{n-1}u_{n}) = \left|f(u_{n-1}) - f(u_{n})\right| = -1$$

$$f^{*}(u_{n}u_{1}) = |f(u_{n}) - f(u_{1})| = 1$$

Subcase (iii) If k has one or more repeated prime factors, then

$$f(u_{n-1}) = f(u_{n}) + 1$$

By Definition 1,

$$f^{*}(u_{n-2}u_{n-1}) = \left|f(u_{n-2}) - f(u_{n-1})\right| = -1$$

$$f^{*}(u_{n-1}u_{n}) = \left|f(u_{n-1}) - f(u_{n})\right| = 1$$

$$f^{*}(u_{n}u_{1}) = |f(u_{n}) - f(u_{1})| = 0$$

The Mobius cordial labeling of C_9, C_{12}, and C_{15} are shown in the following figures (Figs. 6, 7, and 8).

By the definition of the induced edge function f^*, the number of edges with label $-1, 0, 1$ are $\frac{n-1}{3}$, $\frac{n-1}{3}$, and $\frac{n-1}{3}$, respectively.

Case 3 For $n \equiv 2(mod\ 3)$

In this case first n $-$ 2 vertices are labeled as given by Eq. (2), and nth vertex is labeled as $f(u_n) = n + 3$.We compute $f^*(u_n u_1) = |f(u_n) - f(u_1)| = k\ (say)$ .The vertex u_{n-1} is labeled as follows:

Subcase (i) If k is a product of odd number of distinct primes, then

$$f\ (u_{n-1}) = f\ (u_n) - 1$$

By Definition 1,

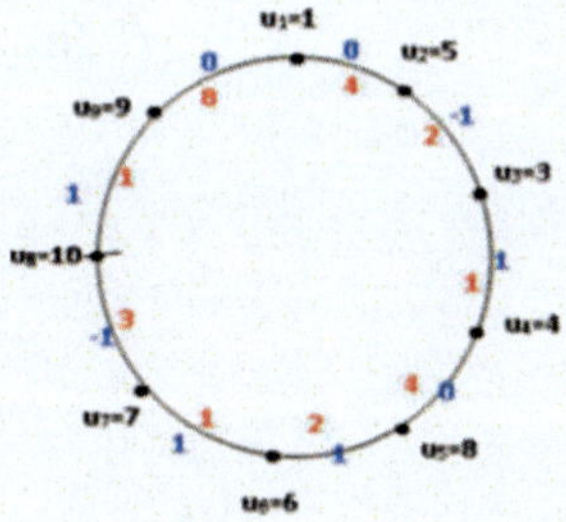

Fig. 6 Mobius cordial labeling of C_9

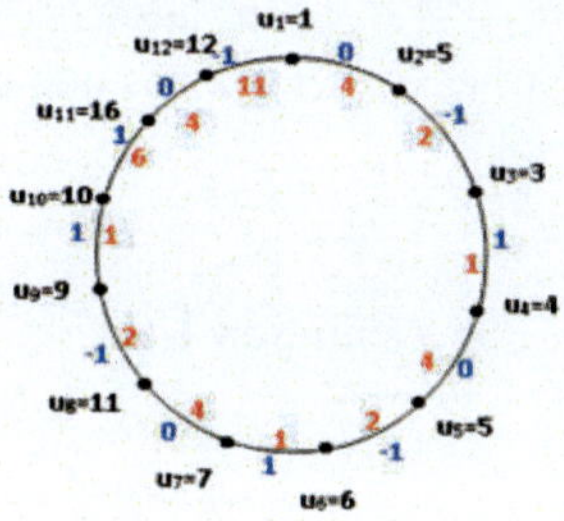

Fig. 7 Mobius cordial labeling of C_{12}

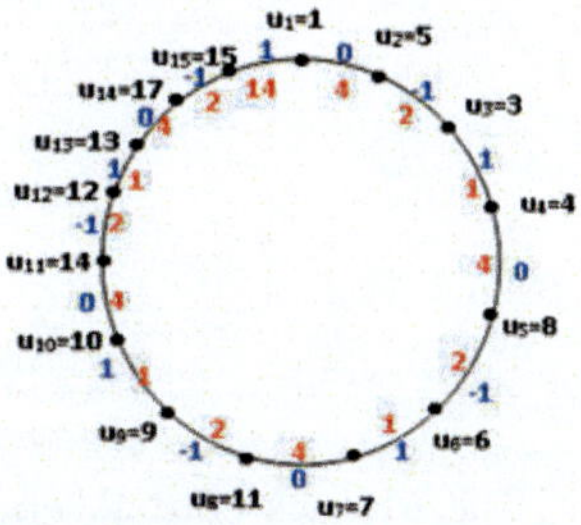

Fig. 8 Mobius cordial labeling of C_{15}

$$f^*(u_{n-2}u_{n-1}) = \left|f(u_{n-2}) - f(u_{n-1)}\right| = 0$$

$$f^*(u_{n-1}u_n) = \left|f(u_{n-1}) - f(u_{n)}\right| = 1$$

$$f^*(u_n u_1) = |f(u_n) - f(u_1)| = -1$$

Subcase (ii) If k is a product of even number of distinct primes, then

$$f(u_{n-1}) = f(u_n) + 3$$

By Definition 1,

$$f^*(u_{n-2}u_{n-1}) = \left|f(u_{n-2}) - f(u_{n-1)}\right| = 0$$

$$f^*(u_{n-1}u_n) = \left|f(u_{n-1}) - f(u_{n)}\right| = -1$$

$$f^*(u_n u_1) = |f(u_n) - f(u_1)| = 1$$

Subcase (iii) If k has one or more repeated prime factors, then

$$f(u_{n-1}) = f(u_n) + 6$$

By Definition 1,

$$f^*(u_{n-2}u_{n-1}) = \left|f(u_{n-2}) - f(u_{n-1)}\right| = -1$$

$$f^*(u_{n-1}u_n) = \left|f(u_{n-1}) - f(u_{n)}\right| = 1$$

$$f^*(u_n u_1) = |f(u_n) - f(u_1)| = 0$$

The Mobius cordial labeling of C_{14}, C_{17}, and C_{20} are shown in the following figures (Figs. 9, 10, and 11).

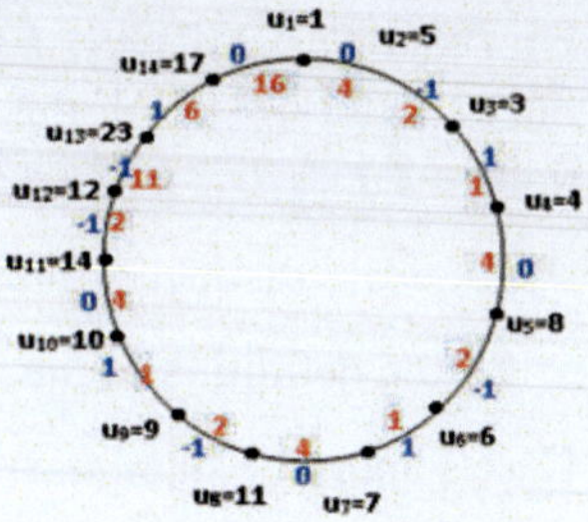

Fig. 9 Mobius cordial labeling of C_{14}

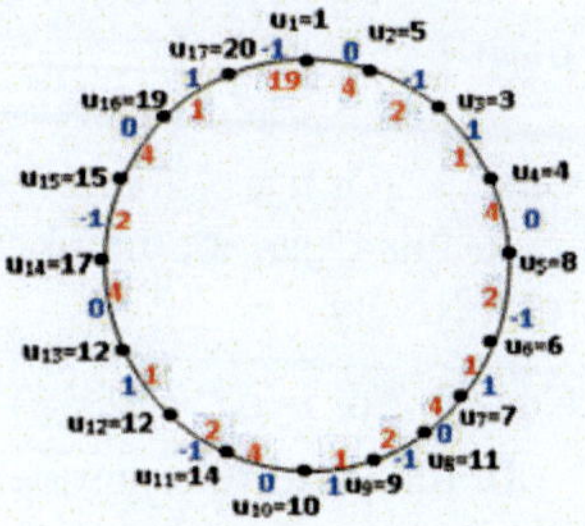

Fig. 10 Mobius cordial labeling of C_{17}

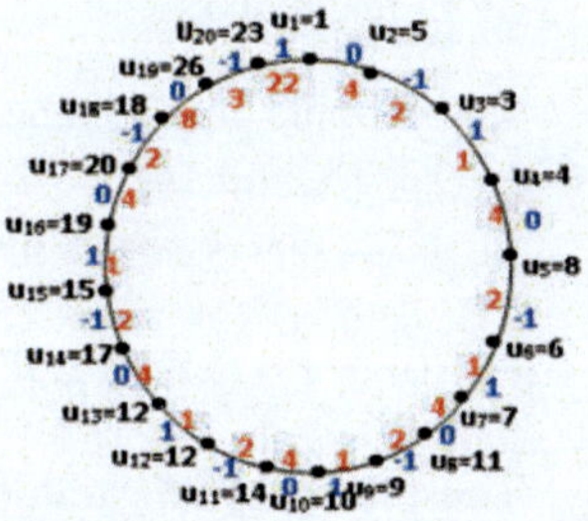

Fig. 11 Mobius cordial labeling of C_{20}

Thus, in this case, the number of edges with label -1, 0, 1 are $\left\lceil \frac{n-1}{3} \right\rceil$, $\left\lceil \frac{n-1}{3} \right\rceil$ and $\left\lfloor \frac{n-1}{3} \right\rfloor$ respectively.

In all the cases, we find that the labeled edges satisfy the conditions

$$\left|e_{f^*}(0) - e_{f^*}(1)\right| \leq 1, \left|e_{f^*}(0) - e_{f^*}(-1)\right| \leq 1 \text{ and } \left|e_{f^*}(-1) - e_{f^*}(1)\right| \leq 1.$$

Thus, the path C_n admits a Mobius cordial labeling.

Theorem 3 The star graph S_n is a Mobius cordial graph.

Proof The star graph S_n is a graph with the vertex set $V = \{u_1, u_2, u_3 \ldots u_n\}$ and the edge set $E = \{u_1u_2, u_1u_3, \ldots u_1u_n\}$. Let $\deg(u_1) = n - 1$ and $\deg(u_i) = 1$ for each $i = 2, 3, \ldots, n$

Let the vertices of S_n be labeled as follows:

$$f(u_1) = 1$$

For $l = 1, 2, 3, \ldots \lfloor \frac{n}{3} \rfloor$, $f\,(u_{l+1}) = \left(\prod_{i=1}^{n} p_i\right) + 1$, n is an even integer
For $l = \lfloor \frac{n}{3} \rfloor + 1, \lfloor \frac{n}{3} \rfloor + 2, \ldots 2\lfloor \frac{n}{3} \rfloor$, $f\,(u_{l+1}) = (p_i^2) + 1$.
For $l = 2\lfloor \frac{n}{3} \rfloor + 1, 2\lfloor \frac{n}{3} \rfloor + 2, \ldots n\text{-}1$, $f\,(u_{l+1}) = \left(\prod_{i=1}^{m} p_i\right) + 1$, m is an odd integer

where p_i are distinct primes.

By the definition of the induced edge function f^*,

Case 1 For $n \equiv 0\ (mod\ 3)$

Among the n − 1 edges, the number of edges with label −1, 0, 1 are $\lceil \frac{n-1}{3} \rceil$, $\lceil \frac{n-1}{3} \rceil$, and $\lfloor \frac{n-1}{3} \rfloor$, respectively.

Case 2 For $n \equiv 1\ (mod\ 3)$

In this case, the number of edges with label −1, 0, 1 are $\frac{n-1}{3}$, $\frac{n-1}{3}$, and $\frac{n-1}{3}$, respectively.

Case 3 For $n \equiv 2\ (mod\ 3)$

In this case, the number of edges with label −1, 0, 1 are $\lfloor \frac{n-1}{3} \rfloor$, $\lfloor \frac{n-1}{3} \rfloor$, and $\lceil \frac{n-1}{3} \rceil$, respectively.

The Mobius cordial labeling of S_9, S_7, and S_8 are shown in the following figures (Figs. 12, 13, and 14).

In all the above cases, the labeled edges satisfy the conditions

$$\left|e_{f^*}(0) - e_{f^*}(1)\right| \le 1,\ \left|e_{f^*}(0) - e_{f^*}(-1)\right| \le 1 \text{ and } \left|e_{f^*}(-1) - e_{f^*}(1)\right| \le 1.$$

Thus, the graph S_n admits a Mobius cordial labeling.

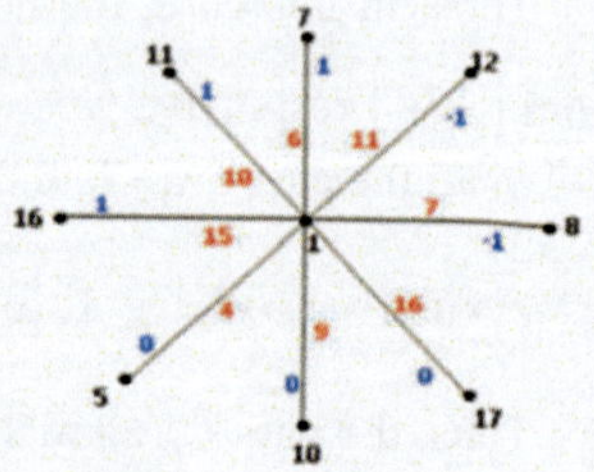

Fig. 12 Mobius cordial labeling of S_9

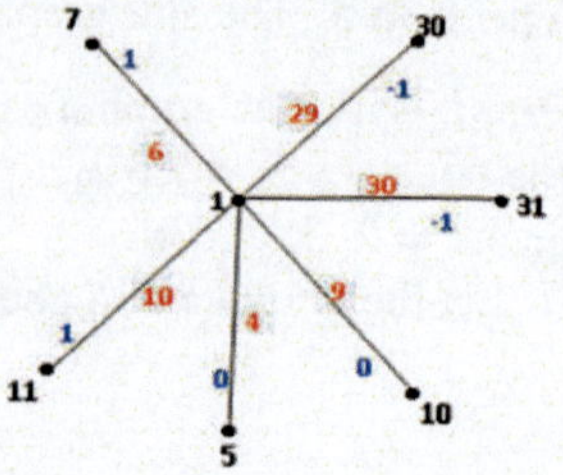

Fig. 13 Mobius cordial labeling of S_7

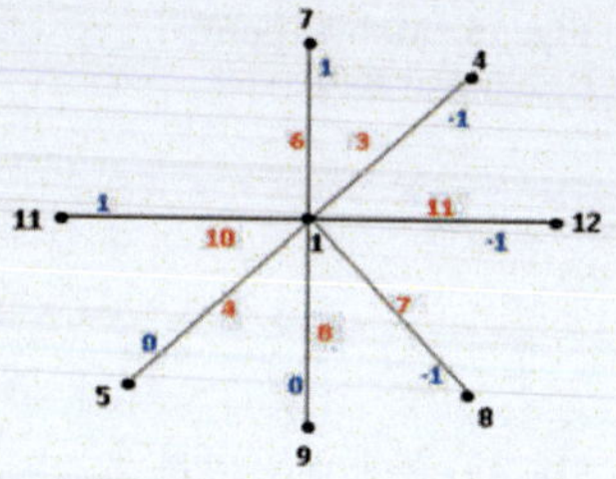

Fig. 14 Mobius cordial labeling of S_8

Theorem 4 The bi-star graph $B_{n,\,n}(n > 3)$ admits a Mobius cordial labeling.

Proof The bi-star graph $B_{n,\,n}$ is the graph obtained by joining the center vertices of two copies of star graph S_n by an edge. The vertex set of $B_{n,\,n}$ is $V = \{u_i, v_i; 1 \le i \le n\}$, where u_1, v_1 are the center vertices (vertices of degree n − 1) and remaining vertices are pendent vertices (vertices of degree 1). The edge set of $B_{n,\,n}$ is $E = \{u_1u_i, v_1v_i, u_1v_1, 1 < i \le n\}$. $B_{n,\,n}$ has 2n vertices and 2n − 1 edges.

The vertices of $B_{n,\,n}$ are labeled as follows:

$$f(u_1) = 1;\ f(v_1) = 2$$

For $l = 1, 2, 3, \ldots \lfloor \frac{n}{3} \rfloor$, $f(u_{l+1}) = \left(\prod_{i=1}^{n_1} p_i\right) + 1$, n_1 is an even integer

$$f(v_{l+1}) = \left(\prod_{i=1}^{n_2} p_i\right) + 2,\ n_2 \text{ is an odd integer}$$

For $l = \lfloor \frac{n}{3} \rfloor + 1, \lfloor \frac{n}{3} \rfloor + 2, \ldots 2\lfloor \frac{n}{3} \rfloor$, $f(u_{l+1}) = \left(p_i^2\right) + 1$

$$f(v_{l+1}) = \left(p_i^2\right) + 2$$

For$l = 2\lfloor \frac{n}{3} \rfloor + 1, 2\lfloor \frac{n}{3} \rfloor + 2, \ldots n - 1,$

$f(u_{l+1}) = \left(\prod_{i=1}^{m_1} p_i\right) + 1$, $m_1$is an odd integer
$f(v_{l+1}) = \left(\prod_{i=1}^{m_2} p_i\right) + 2$, $m_2$is an even integer

where p_i are distinct primes.

Case 1 When $n \equiv 0\ (mod\ 3)$

By the definition of the induced edge function f^*, the number of edges with label −1, 0, 1 are $\left\lfloor \frac{2n-1}{3} \right\rfloor$, $\left\lceil \frac{2n-1}{3} \right\rceil$, and $\left\lceil \frac{2n-1}{3} \right\rceil$, respectively (Fig. 15).

Case 2 When $n \equiv 1\ (mod\ 3)$

By the definition of the induced edge function f^*, the number of edges with label −1, 0, 1 are $\left\lfloor \frac{2n-1}{3} \right\rfloor$, $\left\lfloor \frac{2n-1}{3} \right\rfloor$, and $\left\lceil \frac{2n-1}{3} \right\rceil$, respectively (Figs. 16 and 17).

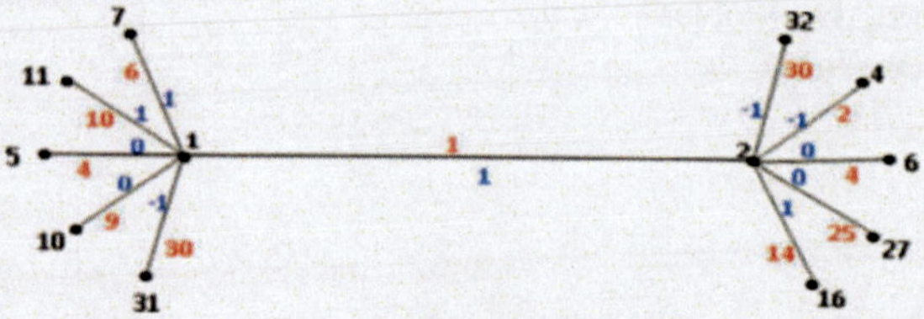

Fig. 15 Mobius cordial labeling of $B_{6,6}$

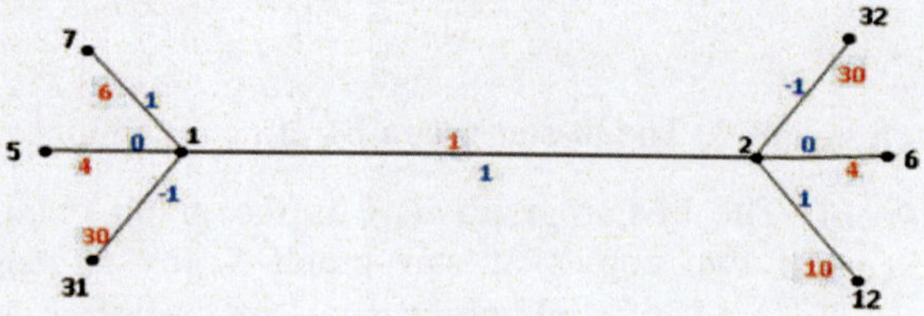

Fig. 16 Mobius cordial labeling of $B_{4,4}$

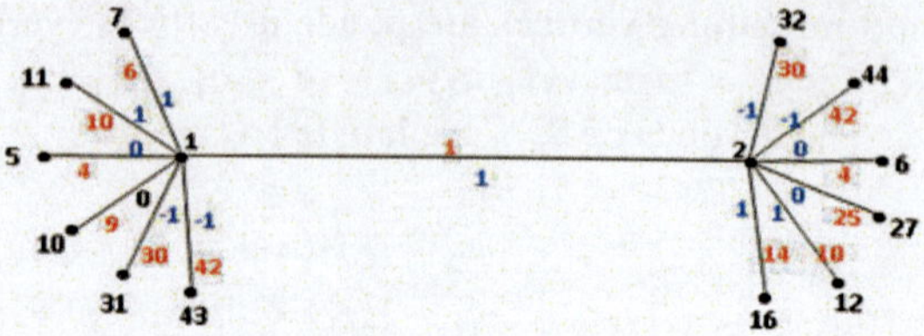

Fig. 17 Mobius cordial labeling of $B_{7,7}$

Case 3 When $n \equiv 2\ (mod\ 3)$

From the definition of the induced edge function f^*, the number of edges with label $-1, 0, 1$ are $\left\lfloor \frac{2n-1}{3} \right\rfloor$, $\left\lfloor \frac{2n-1}{3} \right\rfloor$, and $\left\lceil \frac{2n-1}{3} \right\rceil$, respectively.

Thus, the labeled edges satisfy the conditions

$$\left|e_{f^*}(0) - e_{f^*}(1)\right| \leq 1, \left|e_{f^*}(0) - e_{f^*}(-1)\right| \leq 1 \text{ and } \left|e_{f^*}(-1) - e_{f^*}(1)\right| \leq 1.$$

Thus, the graph $B_{n,\,n}$ admits a Mobius cordial labeling.

3 Conclusion

In this chapter, the Mobius cordial labeling has been investigated for the graphs path, cycle, star graph, and bi-star graph with appropriate illustrations. Finding the classes of graphs that permit this labeling is a potentially fruitful and intriguing field of research.

References

1. Apostol, T.M.: Introduction to Analytic Number Theory. Springer Science & Business Media, Berlin (1998)
2. Cahit, I.: Cordial graphs: a weaker version of graceful and harmonious graphs. Ars Comb. **23**, 201–207 (1987)
3. Gallian, J.A.: A dynamic survey of graph labeling. Electron. J. Comb., 17 # DS6 (2021)
4. Harary, F.: Graph Theory. CRC Press, Boca Raton (2019)
5. Havil, J.: Gamma: Exploring Euler's Constant. Princeton University Press, Princeton (2003)
6. Möbius, A.F.: Über eine besondere Art von Umkehrung der Reihen. J. für die Reine und Angew. Math. **9**, 105–123 (1832)
7. Rosa, A.: On certain valuations of the vertices of a graph. In: Theory of Graphs (Internat. Sympos. Rome. 1966), pp. 349–359. Gordan and Breach, Newyork, Dunod, Paris (1967)
8. Rota, G.C.: On the foundations of combinatorial theory: I. Theory of Möbius functions. Classic Papers in Combinatorics. Boston, MA: Birkhauser Boston, 332–360 (1964)
9. Koshy, T.: Elementary Number Theory with Applications. Academic Press, New Delhi (2005)

Selfipendant and Extremal Pendant Graphs

Jomon Kottarathil and Sudev Naduvath

MSC 2010 05C70, 05C38, 05C40

1 Introduction

For terms and definitions in graph theory, we refer to [1], and for more on graph classes, see [2, 3]. In this discussion, we consider only undirected, simple, finite, and connected graphs. Unless mentioned otherwise, the order and size of the graphs under consideration are denoted by n and m. Let u be a vertex of a given graph G. Then, $d(u)$ represents the degree of the vertex u. The join $G + H$ of the graphs G and H is the graph with the vertex set $V(G + H) = V(G) \cup V(H)$ and the edge set $E(G + H) = E(G) \cup E(H) \cup uv : u \in V(G), v \in V(H)$. $G + H$ thus consists of a copy of G and a copy of H together with all possible edges between them.

The *decomposition* of a graph G is defined as a partition of G into edge-disjoint subgraphs. The *path decomposition* is the partitioning of its edges into subgraphs, where each of the subgraphs is a path in G. Path decomposition can be considered a special case of acyclic decomposition. The *pendant number* of a graph G, denoted by $\Pi_p(G)$, is the least number of vertices in the graph G such that they are the end vertices of a path in a given path decomposition (see [7]). An end vertex of a path in a path decomposition of G is called a *terminus* or a *terminal vertex* of G, concerning the path decomposition concerned (see [5]). The collection of all terminal vertices of G is denoted by $V_p(G)$. Thus, $\Pi_p(G) = \min\{|V_p(G)|\}$. We use the convention of representing terminal vertices by white nodes whenever we draw figures.

More studies on pendant number can be seen in [4, 6, 8, 9, 11]. A similar study on star graphs, namely, *star number of graphs*, is introduced in [10].
The following results are helpful while we move on to the major sections.

J. Kottarathil (✉)
Department of Mathematics, St. Joseph's College, Moolamattom, Arakulam, Kerala, India

S. Naduvath
Department of Mathematics, CHRIST (Deemed to be University), Bangalore, Karnataka, India

F. Kamalov et al. (eds.), *Advances in Mathematical Modeling and Scientific Computing*, Trends in Mathematics, https://doi.org/10.1007/978-3-031-41420-6_75

Theorem 1 (**[7]**) *Let G be a connected graph with n vertices. If G has l odd-degree vertices, then* $l \leq \Pi_p(G) \leq n$.

Theorem 2 (**[7]**) *If G is a connected graph with* $n \geq 2$ *vertices, then* $2 \leq \Pi_p(G) \leq n$.

Theorem 3 (**[6]**) *Let G be a graph on n vertices. Then,* $\Pi_p(G) = n$ *if and only if every vertex of G is of odd degree.*

Theorem 4 (**[9]**) *If both the graphs G and its complement* $\overline{G}$ *are connected, then* $4 \leq \Pi_p(G) + \Pi_p(\overline{G}) \leq 2(n-1)$.

Motivated by the studies mentioned above, in this paper, we introduce the parameters such as selfipendant number, equipendant number, and extremal pendant number of a graph and characterise these parameters.

2 Results and Discussions

Proposition 1 *Not all vertices of an odd-order graph are its terminal vertices.*

Proof Let G be a graph of odd order n. If possible, assume that $\Pi_p(G) = n$. Then, by Theorem 3, every vertex of G has an odd degree. It implies that G has no even-degree vertex, which contradicts the fact that the number of odd-degree vertices in a graph is even, hence the result.

Proposition 2 *The pendant number of any tree* T *is an even number.*

Proof Let T be a tree on n vertices, of which k vertices are of even degree. Then, $\Pi_p(T) = n - k$ (see [7]). It is proved in [7] that the pendant number of a tree T equals the number of odd-degree vertices in it. Since the number of odd-degree vertices in a graph is even, the pendant number of the given tree is an even number.

Proposition 3 *Among all types of trees of order n, the star graph* $K_{1,n-1}$ *has the maximum pendant number.*

Proof We know that the pendant number of any tree T on n vertices is even (by Proposition 2). Moreover, if all the vertices of a graph are of odd degrees, then its pendant number is n (by Theorem 1). Thus, a tree that attains a maximum pendant number implies either of the following two cases happens:

(i) If n is even, then the pendant number is n itself.
(ii) If n is odd, then the pendant number is $n - 1$.

As it is proved in [7], for a star graph $K_{1,n-1}$ on n vertices,

$$\Pi_p(K_{1,n-1}) = \begin{cases} n-1 & \text{if } n \text{ is odd;} \\ n & \text{if } n \text{ is even.} \end{cases}$$

In other words, for $K_{1,n-1}$, n is even implies pendant number is n itself, and n is odd implies the pendant number is $n - 1$. Hence, $K_{1,n-1}$ always has a maximum possible number of terminal vertices.

Definition 1 ([9]) A *Π_p-realisation* of a positive integer $k \geq 2$ is a minimal connected graph G, whose pendant number is k.

For every positive integer $k \geq 2$, there is a Π_p-realisation for k (see [9]). That is, corresponding to any integer $k \in \mathbb{N}$, we can draw a simple graph G with $\Pi_p(G) = k$. Another result says that there exists a connected graph G of order n with $\delta(G) \geq 2$ corresponding to any natural number $k \geq 2$ and corresponding to any order $n \geq 3$ with desired pendant number (see [9]).

Consider the path P_n; $n \geq 2$ and $n \in \mathbb{N}$; then its pendant number is 2 (see [7]). This fact can be expressed as Π_p-realisation for the lower bound.

Proposition 4 *There exists graphical realisation of order n such that $\Pi_p(G) = 2$ for any positive integer $n \geq 2$.*

The following result deals with Π_p-realisation satisfying the minimum criteria for edges and vertices.

Theorem 5 *There exists a minimal graphical realisation (in the sense that the number of both vertices and edges is minimum) with $\Pi_p(G) = k$ for any positive integer $k \geq 2$.*

Proof Note that trees are minimally connected graphs and contain a minimum number of edges for a minimum number of vertices. Among them, the star graph $K_{1,n}$ contains a maximum number of leaf vertices. By Proposition 2, the pendant number of any tree is an even number. Thus, we shall take the star graphs $K_{1,2r-1}$; $r \in \mathbb{N}$ to get the pendant number of the graphical realisations of even natural numbers. Recall that the pendant number of the star graph $K_{1,n-1}$; $n \geq 2$ is n, when n is an even number (see [7]).

Consider the graphical realisation of G corresponding to any odd nonnegative integer r. By Theorem 1, to get a graph G with the pendant number r, we need at least $r + 1$ vertices. Thus, the required graph contains at least one cycle. Now, consider the pineapple graph $K_3^{(p)}$; p is an odd natural number, which is obtained by appending p pendant vertices to a vertex of the complete graph K_3. It is the same as the one-point union of a C_3 and a star graph $K_{1,2r-1}$, where the one-point union is taken on the root of the star graph. The pendant number of $K_3^{(p)}$ is always $p + 2$ which is an odd number. Hence, if we take the graphs $K_{1,2r-1}$; $r \in \mathbb{N}$ and $K_3^{(p)}$; $p = 1, 3, 5, \ldots$ alternatively, we get the required result (see Fig. 1).

Next, we define the equipendant graphs and selfipendant graphs and find their relationship with self-complementary graphs.

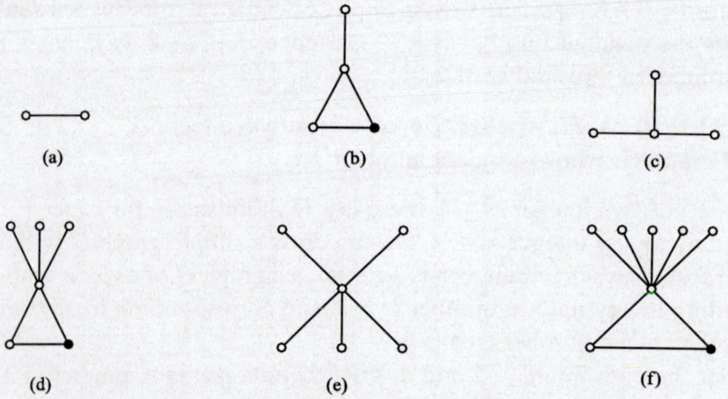

Fig. 1 The Π_p-realisation of graphs with minimum n and m. **(a)** $K_{1,1}$ with pendant number 2. **(b)** $K_3^{(1)}$ with pendant number 3. **(c)** $K_{1,3}$ with pendant number 4. **(d)** $K_3^{(3)}$ with pendant number 5. **(e)** $K_{1,5}$ with pendant number 6. **(f)** $K_3^{(5)}$ with pendant number 7

3 Selfipendant Graphs

Definition 2 Let G and H be two graphs such that $\Pi_p(G) = \Pi_p(H)$; then G and H are called *equipendant* graphs. The pendant number of equipendant graphs is called *equipendancy*.

For example, P_n and C_n are equipendant graphs with equipendancy 2.

Definition 3 Let G be a graph and $\overline{G}$ be its complement. If $\Pi_p(G) = \Pi_p(\overline{G})$, then G and $\overline{G}$ are called *selfipendant graphs*.

Theorem 6 *Let G be a graph. Let $v \in V(G)$. If G is selfipendant, then:*

(i) $1 \leq d(v) \leq n-2$.
(ii) $n-1 \leq m \leq \frac{n^2-3n+2}{2}$.

Proof

(i) Suppose $d(v) = 0$ for some $v \in V(G)$. It implies that v is an isolated vertex of G and so G is disconnected. Hence, $d(v) \geq 1$. Similarly, if $d(v) = n-1$, then the corresponding vertex $v \in \overline{G}$ will be disconnected. Thus, $d(v) \leq n-2$.
(ii) A connected graph G contains at least $n-1$ edges. Then, $m(G) + m(\overline{G}) = \frac{n(n-1)}{2}$. Suppose G contains $n-1$ edges. Then $\overline{G}$ has $\frac{n(n-1)}{2} - (n-1) = \frac{n^2-3n+2}{2}$ edges and vice versa. Therefore, $n-1 \leq m \leq \frac{n^2-3n+2}{2}$.

The bounds in the above inequalities are sharp. For example, Fig. 2 shows the kite graph and its complement on $n = 5$; both are selfipendant and equipendant with pendant number 4 and $\delta(G) = 1$ and $\Delta(G) = 3$. For kite, $m = 6$ and for *cokite*, $m = 4$. Hence, both graphs attain either of the bounds. In Theorem 4, we

Fig. 2 Two selfipendant graphs which attain the bounds mentioned in Theorem 6. **(a)** *Kite*. **(b)** Cokite

have $4 \le \Pi_p(G) + \Pi_p(\overline{G}) \le 2(n-1)$. It can also be noted that this example attains the upper bound for Theorem 4. The lower bound for Theorem 4 is attained by P_4 and its complement.

Now, we discuss a few properties of equipendant and selfipendant graphs. Recall the pendant number of an r-regular graph G given in [6];

$$\Pi_p(G) = \begin{cases} 2 & \text{if } r \text{ is even;} \\ n & \text{if } r \text{ is odd.} \end{cases}$$

Using this result we can prove the following:

Proposition 5 *All even regular graphs are equipendant with equipendancy* 2.

A graph G and its complement $\overline{G}$ are said to be self-complementary if $G \cong \overline{G}$. If $G \cong \overline{G}$, then it is trivial that $\Pi_p(G) = \Pi_p(\overline{G})$. Hence, we get the following results on the relations between self-complementary, selfipendant, and equipendant graphs. Since the proof is clear from the definitions themselves, we omit it.

Proposition 6 *For any graph G, we have the following:*

(i) If G is self-complementary, then G is both selfipendant and equipendant.
(ii) If G and $\overline{G}$ are selfipendant, then they are equipendant having the same equipendancy number.

But the converse of the above proposition need not be true. That is:

(i) A selfipendant graph or an equipendant graph may not be self-complementary. For example, the graphs given in Fig. 2 are both selfipendant and equipendant but not self-complementary.
(ii) An equipendant graph need not be a selfipendant. For example, P_3 and C_3 are equipendant with equipendancy 2, but they are not selfipendant graphs.

Proposition 7 *There are no selfipendant graphs of orders* 2 *and* 3. *But there exist a pair each of selfipendant graphs of orders* 4 *and* 6 *and five pairs of order* 5.

Proof The path K_2 is the only connected graph of order 2, and two connected graphs of order 3 are P_3 and C_3. Their complements are disconnected graphs. The only pair of selfipendant graphs of order 4 is P_4 and $\overline{P_4}$ with pendant number 2. There are five pairs of selfipendant graphs, namely, kite graph (with $\Pi_p = 4$), P_5 (with $\Pi_p = 2$), P or $4 - pan$ or banner (with $\Pi_p = 3$), bull (with $\Pi_p = 4$), and

C_5 (with $\Pi_p = 2$) along with their complements, whereas there is only one pair of graphs of order 6 ($K_{3,3} - e$ and $\overline{K_{3,3} - e}$) as selfipendant (with $\Pi_p = 4$).

Though there are *three* self-complementary graphs of orders up to 5, namely, P_4, C_5, and the bull graph, we have *seven* selfipendant graphs up to the order 6.

Proposition 8 *There does not exist a selfipendant graph G on n vertices such that* $\Pi_p(G) = n$.

Proof Let G be a graph. If G is selfipendant, then $\Pi_p(G) = \Pi_p(\overline{G})$. Assume that $\Pi_p(G) = n$. Then, $\Pi_p(\overline{G}) = n$. Then, $\Pi_p(G) + \Pi_p(\overline{G}) = n + n = 2n$, contradicting the Theorem 4. Hence, $\Pi_p(G) \neq n$, completing the proof.

4 Extremal Pendant Graphs

When we discuss the pendant number of graphs, it can be observed that the complete graphs K_n; $n \in \mathbb{N}$ attain the lower bounds and the upper bounds alternatively. For instance, the pendant number of $K_3, K_4, K_5, K_6, \ldots$ is $2, 4, 2, 6, \ldots$, respectively. This property is exhibited by many other classes of graphs too. In view of this fact, we introduce the notion of extremal pendant graphs as follows:

Definition 4 A graph G on n vertices is called an *extremal pendant graph*, if the graph $G + K_1$ satisfies any one of the following:

(i) $\Pi_p(G) = 2$ and $\Pi_p(G + K_1) = n + 1$.
(ii) $\Pi_p(G) = n$ and $\Pi_p(G + K_1) = 2$.

For the sake of convenience, we denote an extremal pendant graph as an EP-graph.

In other words, let G be a graph satisfying either of the bounds 2 or n. If by adding a new vertex and joining it to all the existing vertices, the pendant number of G attains the other extreme, then G is called an EP-graph. The smallest EP-graph with $\Pi_p(G) = 2$ is K_3 and with $\Pi_p(G) = n$ is $K_{1,3}$.

Proposition 9 *If G is an EP-graph with* $\Pi_p(G) = 2$*, then it is Eulerian.*

Proof Let G be an EP-graph with $\Pi_p(G) = 2$. It implies that either there are two odd-degree vertices or two even degree vertices which act as end vertices of every path in G. Suppose u and v be the two odd-degree vertices of G, which act as the pendant vertices for every path. Take $G+K_1$. Then, except u and v, all other vertices of G become odd. We know that if G is a graph of order at least four and has at least two even-degree vertices, then $\Pi_p(G) \leq n-2$ (see [9]). Thus, it is possible to make a path decomposition of this graph so that these two vertices are internal vertices of every path. Thus $\Pi_p(G + K_1) \neq n + 1$. By Definition 4, G is not EP-graph. Hence, G has no odd-degree vertex, and so it is Eulerian.

Proposition 10 *Let G be an EP-graph on n vertices with* $\Pi_p(G) = 2$*. Then, n must be an odd number.*

Proof Let G be an EP-graph on n vertices with $\Pi_p(G) = 2$. Since G is EP-graph, we have $\Pi_p(G + K_1) = n + 1$. That is, every vertex of $G + K_1$ is of odd degree. Since the number of odd-degree vertices is even, $n + 1$ must be an even number. Hence, n must be an odd number.

Proposition 11 *An Eulerian graph G is EP-graph if and only if $\Pi_p(G) = 2$.*

Proof Let G be an Eulerian graph on n vertices. Assume that G is an EP-graph. Then, either $\Pi_p(G) = n$ or $\Pi_p(G) = 2$. The pendant number of G is n which implies that every vertex of G is of odd degree, a contradiction. Hence, $\Pi_p(G) = 2$.

Now assume that G is an Eulerian graph with $\Pi_p(G) = 2$. Take $G + K_1$. Then, the degree of each vertex of G is increased by one and becomes an odd number. By the Proposition 10, the number of vertices of G is odd. Hence, the new vertex added too must be of an odd degree. Thus, $\Pi_p(G+K_1) = n+1$. Hence, G is an EP-graph.

Proposition 12 *Let G be graph on n vertices such that $\Pi_p(G) = 2$. The graph G is an EP-graph if and only if G is Eulerian.*

Proof Let G be graph on n vertices such that $\Pi_p(G) = 2$. Assume G is an EP-graph. Then, by the Proposition 9, every vertex of G is even, and hence G is Eulerian. Now, assume the converse. That is, G is Eulerian, and the pendant number of G is 2. Since every vertex of $G + K_1$ is of odd degree, G is an EP-graph.

Proposition 13 *Let G be an EP-graph on n vertices with $\Pi_p(G) = n$. Then n must be an even number.*

Proof Let G be an EP-graph on n vertices with $\Pi_p(G) = n$. It implies that the degree of every vertex of G is an odd number. But the number of odd-degree vertices is even. Hence n is an even number.

Proposition 14 *If G is an EP-graph, then*

$$\Pi_p(G + K_2) = \begin{cases} 2 & \text{if } \Pi_p(G) = 2; \\ n + 2 & \text{if } \Pi_p(G) = n. \end{cases}$$

Proof Let G be an EP-graph. Then, there are two cases:

Case 1: Let $\Pi_p(G) = 2$. Then, G is an Eulerian graph such that $\Pi_p(G + K_1) = n + 1$ having all the vertices are of odd degree. Then $G + K_2$, where u, v are the vertices of K_2, is again Eulerian such that $d(u) = d(v) = n + 1$. Being an Eulerian graph, it is sufficient for $G + K_2$ to have path decomposition using u and v as end vertices. Hence, $\Pi_p(G + K_2) = 2$.

Case 2: Let $\Pi_p(G) = n$. Then, by Proposition 1 and Proposition13, n is even and all the vertices of G are of odd degree. It implies that the vertices of $G + K_2$ are also of odd degrees. Thus, $\Pi_p(G + K_2) = n + 2$.

The converse need not be true. That is, $\Pi_p(G) = \Pi_p(G + K_2) = 2$ need not imply that G is an EP-graph. For instance, consider the cycle C_4. $\Pi_p(C_4) = 2$ and $\Pi_p(C_4 + K_2) = 2$. But $C_4 + K_1 = W_5$, the wheel graph with 5 vertices

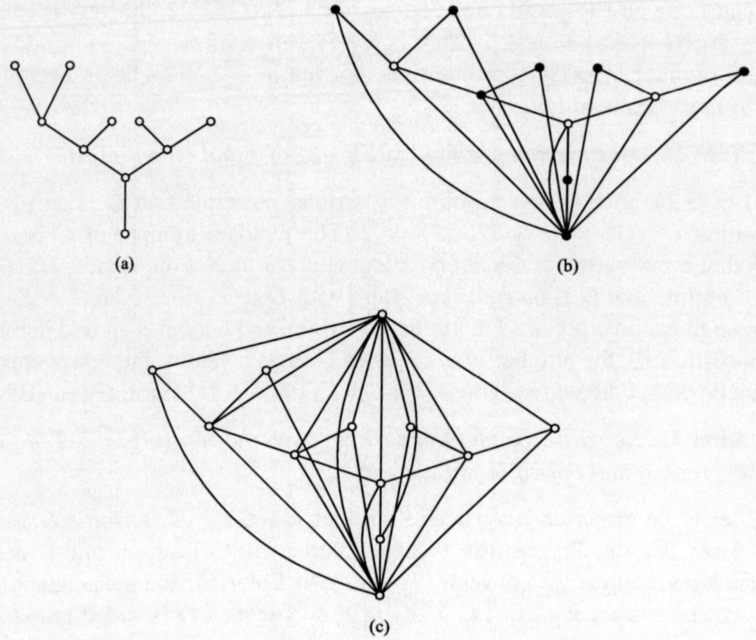

Fig. 3 Examples for $\Pi_p(G)$ and $\Pi_p(G + K_2)$ are same as total number of vertices, but G is not EP-graph. (a) A graph G with $\Pi_p = n$. (b) The graph $G + K_1$ with $\Pi_p \neq 2$. (c) The graph $G + K_2$ with $\Pi_p = n + 2$

such that $\Pi_p(W_5) = 4$ which is not an EP-graph by definition. Again, $\Pi_p(G) = \Pi_p(G + K_2) = n + 2$ need not imply that G is an EP-graph (see Fig. 3). Moreover, these examples indicate that a graph G on n vertices with either $\Pi_p(G) = 2$ or $\Pi_p(G) = n$ need not be an EP-graph.

Proposition 15 *Let G be a graph on n vertices with $\Pi_p(G) = n$. If $\Delta(G) = n - 1$, then G is an EP-graph.*

Proof Let G be a graph on n vertices with $\Pi_p(G) = n$. It implies that every vertex of G is of an odd degree. Let $\Delta(G) = n - 1$. Then, there is at least one vertex u of G such that $d(u) = n - 1$. Now, consider $G + K_1$ on $n + 1$ vertices. Let v be the newly added vertex. Then, $G + K_1$ is an Eulerian graph with $d(u) = d(v) = n$. Clearly, u and v can be the endpoints of every path in $G + K_1$. Thus, $\Pi_p(G + K_1) = 2$. Hence, G is an EP-graph.

Theorem 7 *Let G be an EP-graph. Then, $\Pi_p(G) = n$ if and only if $G + K_1$ is an Eulerian graph.*

Proof Let G be an EP-graph. Assume that $\Pi_p(G) = n$. Then, every vertex of G is odd and the order of G is even. Thus, every vertex of $G + K_1$ is even, hence the result.

Conversely, let $G + K_1$ be an Eulerian graph. Suppose $\Pi_p(G) \neq n$. Since G is an EP-graph, $\Pi_p(G) = 2$ and $\Pi_p(G + K_1) = n$, which contradicts the fact that $G + K_1$ is Eulerian. Hence $\Pi_p(G) = n$.

Invoking Theorem 12 and Theorem 7, we get the following:

Theorem 8 *If a graph G is an EP-graph, then either G or* $G + K_1$ *is an Eulerian graph.*

Theorem 9 *An even regular graph is EP-graph if and only if the number of vertices is odd.*

Proof Let G be both an even regular and EP-graph on n vertices. Since every even regular graph has pendant number 2, we have $\Pi_p(G) = 2$ and $\Pi_p(G+K_1) = n+1$. It shows that n is odd.

Conversely, let n be an odd number and $\Pi_p(G) = 2$. Then, the vertex degree of every vertex of $G + K_1$ is an odd number. Thus, $\Pi_p(G + K_1) = n + 1$. Hence, it is an EP-graph.

5 Conclusion

In this paper, we defined the equipendant graphs, selfipendant graphs, and extremal pendant (in short, EP-graph) graphs. We discussed the pendant number of these classes of graphs together with that of self-complementary graphs. We identified certain characteristic properties of these graphs. The Π_p-realisation of graphs having the minimum number of vertices and edges was also identified. We have initiated the study of the pendant number of the join of graphs also. The various properties and characteristic properties of pendant numbers of graphs such as join of graphs, Eulerian graphs, etc. are promising and give scope for further studies.

Acknowledgments The authors would like to acknowledge the motivation and support from Prof. Dr. Joseph Varghese Kureethara, Director, Centre for Research, CHRIST (Deemed to be University), Bangalore, India.

References

1. Chartrand, G., Zhang, P.: Introduction to Graph Theory. McGraw Hill Education (India), Chennai (2017)
2. De Ridder, H.N.: Information system on graph classes and their inclusions (ISGCI) (2015). www.graphclasses.org
3. Harary, F.: Graph Theory. Narosa Publishing House, New Delhi (2001)
4. Kottarathil, J.: On pseudo-balancing of path-induced signed graphs. Jordan J. Math. Stat. (2022). To appear
5. Kottarathil, J., Naduvath, S.: On path-induced signed graphs. Communicated (2022)

6. Kottarathil, J., Naduvath, S., Kureethara, J.V.: The pendant number of line graphs and total graphs. In: Data Science and Security, pp.457–468. Springer, Berlin (2022)
7. Sebastian, J.K., Kureethara, J.V.: Pendant number of graphs. Int. J. Appl. Math. **31**(5), 679–689 (2018)
8. Sebastian, J.K., Kureethara, J.V., Naduvath, S., Dominic, C.: On the pendant number of some new graph classes. Res. Rev. Discrete Math. Struct. **6**(1), 15–21 (2019)
9. Sebastian, J.K., Kureethara, J.V., Naduvath, S., Dominic, C.: A study on the pendant number of graph products. Acta Universi. Sap. Inform. **11**(1), 24–40 (2019)
10. Sebastian, J.K., Kureethara, JV., Naduvath, S., Dominic, C.: On star decomposition and star number of some graph classes. Int. J. Sci. Res. Math. Stat. Sci. **5**(6), 81–85 (2018)
11. Sebastian, J.K., Naduvath, S., Kureethara, J.V.: The pendant number of some graph products. In: Recent Advancements in Graph Theory, pp. 221–231. CRC Press, Boca Raton (2020)

Part VI
Mathematical Education

Computing the CD-Number of Strong Product of Graphs

L. Praveenkumar, G. Mahadevan, C. Sivagnanam, S. Anuthiya, and S. Kaviya

1 Introduction

All the graphs $G = (V(G), E(G))$ considered here are finite, simple, undirected, and without isolated vertex. The neighbor of a vertex $v \in V(G)$ is denoted by $N(v)$, where $N(v) = \{u \in V(G) : d(u, v) = 1\}$ and $deg(v) = |N(v)|$ denotes the degree of the vertex v. Let P_r denote a path of length r, C_r denote a cycle of length r, and G^c denote the complement of G. Let G be a graph; an *induced subgraph* H is obtained from G by deleting some vertices of G. If G is said to be a *bipartite* [3], then its vertex set can be separated into two nonempty subsets Y_1 and Y_2 such that each edge of G has one end in Y_1 and other in Y_2 and is denoted by $G(Y_1, Y_2)$. A bipartite graph $G(Y_1, Y_2)$ is said to be *complete bipartite* [3] if each vertex of Y_1 is adjacent to all the vertices of Y_2. If $G(Y_1, Y_2)$ is complete with $|Y_1| = r$ and $|Y_2| = s$, then $G(Y_1, Y_2)$ is denoted by $K_{r,s}$, where $K_{1,s}$ is a *star graph* [3]. A wheel graph $W_{1,r}, r \geq 3$ is obtained by joining a single vertex to all the vertices of a cycle C_r. A helm graph H_r is obtained by joining a pendant edge to every vertex of the outer cycle of $W_{1,r}$. Let $L(G)$ be the *line graph* of G where $V(L(G)) = E(G)$ and two vertices in $L(G)$ are adjacent if and only if the corresponding edges have an end vertex in common. It is clear that $L(P_r) = P_{r-1}$ and $L(C_r) = C_r$. A *dominating set* S is a set of vertices of G with the condition that every vertex $v \in V - S$, $d(v, S) = 1$. If dominating set D of G with minimum cardinality, then the domination number $\gamma(G) = |D|$. The *strong product* [2] of graphs H_1 *and* H_2 is the graph $H_1 \boxtimes H_2$ with vertex set $V(H_1) \times V(H_2)$, and any two of its vertices (v_1, u_1) *and* (v_2, u_2)

L. Praveenkumar (✉) · G. Mahadevan (✉) · S. Anuthiya · S. Kaviya
Department of Mathematics, Gandhigram Rural Institute - Deemed to be University, Gandhigram, Tamil Nadu, India

C. Sivagnanam
Department of General Requirements, University of Technology and Applied Sciences - Sur, Sur, Sultanate of Oman

F. Kamalov et al. (eds.), *Advances in Mathematical Modeling and Scientific Computing*, Trends in Mathematics, https://doi.org/10.1007/978-3-031-41420-6_76

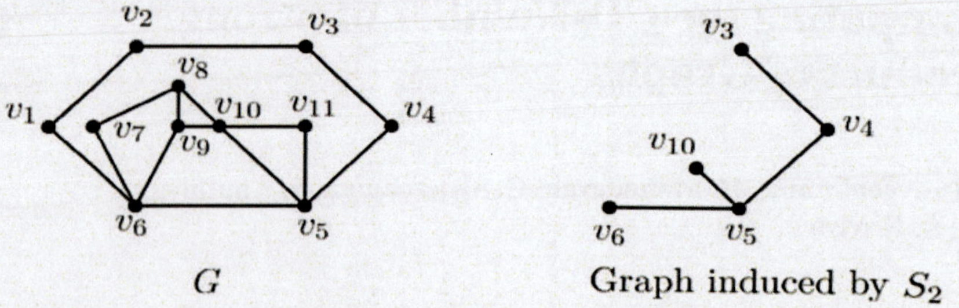

Fig. 1 Example

are adjacent whenever $v_1v_2 \in E(H_1)$ *and* $u_1 = u_2$ or $u_1u_2 \in E(H_2)$ *and* $v_1 = v_2$ or $v_1v_2 \in E(H_1)$ *and* $u_1u_2 \in E(H_2)$. The concept of corona domination number was introduced by G. Mahadevan et al. [4]. The corona domination number (CD-number) γ_{CD} is a minimum cardinality of the dominating set S, with the subgraph induced by S having either pendant or support vertices only. $S_1 = \{v_3, v_6, v_{10}\}$ and $S_2 = \{v_3, v_4, v_5, v_6, v_{10}\}$ are the dominating set and corona dominating set, respectively (Fig. 1).

Graph domination and associated concepts have been analyzed for many years; among them, many authors studied the domination number of the various product graphs, especially for the product involving paths, cycles, the domination number for the cross product of paths and cycles [1], etc. Motivated by the above, we determined the exact values of $\gamma_{CD}(P_r \boxtimes P_s)$, $\gamma_{CD}(P_r \boxtimes P_s^c)$, $\gamma_{CD}(P_r \boxtimes C_s)$, $\gamma_{CD}(P_r \boxtimes K_s)$, *etc*. In the future, we will find the exact value of $\gamma_{CD}(P_r \boxtimes G)$ for any given graph G. Also, we see the relationship between the $\gamma_{CD}(P_r)$ with $\gamma_{CD}(P_r \boxtimes G)$ and $\gamma_{CD}(G)$ with $\gamma_{CD}(P_r \boxtimes G)$.

Here, we have given the CD-number for some trivial graphs. $\gamma_{CD}(P_r^c) = 2,\ r \geq 4$, $\gamma_{CD}(C_4^c) = 4$, $\gamma_{CD}(C_5^c) = 3$, $\gamma_{CD}(C_r^c) = 2,\ r \geq 6$, $\gamma_{CD}(P_2 \boxtimes K_s) = \gamma_{CD}(P_3 \boxtimes K_s) = 2$.

2 CD-Number of Strong Product for Some Standard Graphs

We begin this section with some observations. The CD-number of the strong product of the paths P_1, P_2, and P_3 with P_r is same as the CD-number of P_r, and the remaining cases are given below.

2.1 *Examining the* γ_{CD} *for* $P_r \boxtimes P_s$ *and* $P_r \boxtimes P_s^c$

Let $P_r = (v_1, v_2, \ldots, v_r)$ and $P_s = (u_1, u_2, \ldots, u_s)$.

Then $V(P_r \boxtimes P_s) = \{(v_i, u_j) : 1 \le i \le r, 1 \le j \le s\}$ and $E(P_r \boxtimes P_s) = \{(v_i, u_j)(v_i, u_{j+1}) : 1 \le i \le r, 1 \le j \le s-1\} \cup \{(v_i, u_j)(v_{i+1}, u_j) : 1 \le i \le r-1, 1 \le j \le s\} \cup \{(v_i, u_j)(v_{i+1}, u_{j+1}) : 1 \le i \le r-1, 1 \le j \le s-1\} \cup \{(v_i, u_j)(v_{i+1}, u_{j-1}) : 1 \le i \le r-1, 2 \le j \le s\}$.

Theorem 1 *If $s \ge 4$, then $\gamma(P_4 \boxtimes P_s) = 2\lceil \frac{s}{3} \rceil$.*

Proof Let $P_4 = (v_1, v_2, v_3, v_4)$ and $P_s = (u_1, u_2, \ldots, u_s)$.

Then $V(P_4 \boxtimes P_s) = \{(v_i, u_j) : 1 \le i \le 4, 1 \le j \le s\}$ and $E(P_4 \boxtimes P_s) = \{(v_i, u_j)(v_i, u_{j+1}) : 1 \le i \le 4, 1 \le j \le s-1\} \cup \{(v_i, u_j)(v_{i+1}, u_j) : 1 \le i \le 3, 1 \le j \le s\} \cup \{(v_i, u_j)(v_{i+1}, u_{j+1}) : 1 \le i \le 3, 1 \le j \le s-1\} \cup \{(v_i, u_j)(v_{i+1}, u_{j-1}) : 1 \le i \le 3, 2 \le j \le s\}$. Let $S_1 = \{(v_1, u_j), (v_4, u_j) : j = 3l+2\}$. Then

$$S = \begin{cases} S_1 & \text{if } s = 3k+3 \text{ or } 3k+2, \\ S_1 \cup \{(v_1, u_s), (v_4, u_s)\} & \text{if } s = 3k+1, \end{cases}$$

is a dominating set of $P_4 \boxtimes P_s$. Thus $\gamma(P_4 \boxtimes P_s) \le |S| = 2\lceil \frac{s}{3} \rceil$. Assume that D is a dominating set of $P_4 \boxtimes P_s$. Since $\gamma(P_s) = \lceil \frac{s}{3} \rceil$ and any dominating set of ith row of $P_4 \boxtimes P_s$ dominates all the vertices in $(i-1)^{th}$ row and $(i+1)^{th}$ row, any dominating set of a row dominates at most three rows. Hence $|D| \ge 2\lceil \frac{s}{3} \rceil$. Then $\gamma(P_4 \boxtimes P_s) \ge 2\lceil \frac{s}{3} \rceil$ and hence the result follows.

Theorem 2 *If $s \ge 4$, then $\gamma_{CD}(P_4 \boxtimes P_s) = 2\lceil \frac{s}{3} \rceil$.*

Proof Let $P_4 = (v_1, v_2, v_3, v_4)$ and $P_s = (u_1, u_2, \ldots, u_s)$. Then $V(P_4 \boxtimes P_s) = \{(v_i, u_j) : 1 \le i \le 4, 1 \le j \le s\}$ and $E(P_4 \boxtimes P_s) = \{(v_i, u_j)(v_i, u_{j+1}) : 1 \le i \le 4, 1 \le j \le s-1\} \cup \{(v_i, u_j)(v_{i+1}, u_j) : 1 \le i \le 3, 1 \le j \le s\} \cup \{(v_i, u_j)(v_{i+1}, u_{j+1}) : 1 \le i \le 3, 1 \le j \le s-1\} \cup \{(v_i, u_j)(v_{i+1}, u_{j-1}) : 1 \le i \le 3, 2 \le j \le s\}$. Let $S_1 = \{(v_i, u_j) : i = 4l+2 \text{ or } 4l+3, j = 3l+2\}$.

$$\text{Then } S = \begin{cases} S_1 & \text{if } s = 3k+3 \text{ or } 3lk+2, \\ S_1 \cup \{(v_i, u_s) : i = 4l+2 \text{ or } 4l+3\} & \text{if } s = 3k+1, \end{cases}$$

is a $CD-set$ of $P_4 \boxtimes P_S$. Thus $\gamma_{CD}(P_4 \boxtimes P_s) \le |S| = 2\lceil \frac{s}{3} \rceil$. Since $\gamma(P_4 \boxtimes P_s) = 2\lceil \frac{s}{3} \rceil$, the result follows.

Theorem 3 *If $k \ge 1$, $r = 4k+4$ or $4k+1$, $r \ge 5$ and $s \ge r$, then $\gamma_{CD}(P_r \boxtimes P_s) = \lceil \frac{r}{2} \rceil \lceil \frac{s}{3} \rceil$.*

Proof *The proof this theorem has been subdivided into two cases based on r.*

Case 1 $r = 4l+4$. *Let $S_1 = \{(v_i, u_j) : i = 4l+2$ or $4l+3, j = 3l+2\}$.*

$$\text{Then } S = \begin{cases} S_1 & \text{if } s = 3k+3 \text{ or } 3k+2, \\ S_1 \cup \{(v_i, u_s) : i = 4l+2 \text{ or } 4l+3\} & \text{if } s = 3k+1, \end{cases} \text{ is}$$

a $CD-set$ of $P_r \boxtimes P_s$. Thus $\gamma_{CD}(P_r \boxtimes P_s) \le |S| = \lceil \frac{r}{2} \rceil \lceil \frac{s}{3} \rceil$. Suppose there exists a dominating set $D_1 \subseteq V$ of cardinality at most $d_1 = \lceil \frac{r}{2} \rceil \lceil \frac{s}{3} \rceil - 1$; then $< D_1 >$ contains an isolated vertex. Hence $|D_1| \ge d_1 + 1 = \lceil \frac{r}{2} \rceil \lceil \frac{s}{3} \rceil$.

Case 2 $r = 4l+1$. *Let $S_2 = \{(v_i, u_j) : i = 4l+2$ or $4l+3, j = 3l+2\} \cup \{(v_{r-1}, u_j) : j = 3l+2\}$. Then*

$$S = \begin{cases} S_2 & \textit{if } s = 3k+3 \textit{ or } 3k+2, \\ S_2 \cup \{(v_i, u_s) : i = 4l+2 \textit{ or } 4l+3\} \cup \{(v_{r-1}, u_s)\} & \textit{if } s = 3k+1, \end{cases}$$

is a $CD-set$ of $P_r \boxtimes P_s$. Thus $\gamma_{CD}(P_r \boxtimes P_s) \leq |S| = \lceil \frac{r}{2} \rceil \lceil \frac{s}{3} \rceil$. Suppose there exists a dominating set $D_2 \subseteq V$ of cardinality at most $d_2 = \lceil \frac{r}{2} \rceil \lceil \frac{s}{3} \rceil - 1$; then $< D_2 >$ contains an isolated vertex. Thus $|D_2| \geq d_2 + 1 = \lceil \frac{r}{2} \rceil \lceil \frac{s}{3} \rceil$, therefore the proof.

Theorem 4 *If $k \geq 2$, $r = 4k + 2$, $r \geq 10$ and $s \geq r$, then*

$$\gamma_{CD}(P_r \boxtimes P_s) = \begin{cases} (\frac{r}{2} + 1)\lceil \frac{s}{3} \rceil & \textit{if } s = 3m + 3 \textit{ or } 3m + 2 \\ (\frac{r}{2} + 1)\lceil \frac{s}{3} \rceil - 4 & \textit{if } s = 3m + 1. \end{cases}$$

Proof Let $S_1 = \{(v_i, u_j) : i = 4l+1 \textit{ or } 4l+2,\ j = 3l+2\}$, $S_2 = \{(v_i, u_j) : i = 4l+2 \textit{ or } 4l+3,\ j = 3l+2\} \cup \{(v_{r-1}, u_j) : j = 3l+2\} - \{(v_r, u_{s-2}), (v_{r-3}, u_{s-2})\}$ *and* $A = (\{(v_i, u_s) : i = 4l + 2 \textit{ or } 4l + 3\} \cup \{(v_{r-1}, u_{s-1})(v_{r-4}, u_{s-1})\}) - \{(v_r, u_s), (v_{r-3}, u_s), (v_{r-4}, u_s)\}$. Then $S = \begin{cases} S_1 & \text{if } s = 3l + 2 \textit{ or } 3l + 3, \\ S_2 \cup A & \text{if } s = 3l + 1, \end{cases}$ is a $CD-set$ of $P_r \boxtimes P_s$. Thus

$$\gamma_{CD}(P_r \boxtimes P_s) \leq |S| = \begin{cases} (\frac{r}{2} + 1)\lceil \frac{s}{3} \rceil & \text{if } s = 3m + 2 \textit{ or } 3m + 3, \\ (\frac{r}{2} + 1)\lceil \frac{s}{3} \rceil - 4 & \text{if } s = 3m + 1. \end{cases}$$

Suppose there exists a dominating set $D \subseteq V$ of cardinality at most

$$d = \begin{cases} (\frac{r}{2} + 1)\lceil \frac{s}{3} \rceil - 1 & \text{if } s = 3m + 2 \textit{ or } 3m + 3, \\ (\frac{r}{2} + 1)\lceil \frac{s}{3} \rceil - 5 & \text{if } s = 3m + 1. \end{cases}$$

Then $< D >$ contains an isolated vertex. Thus $|D| \geq d+1 = \begin{cases} (\frac{r}{2} + 1)\lceil \frac{s}{3} \rceil & \text{if } s = 3m + 2 \textit{ or } 3m + 3, \\ (\frac{r}{2} + 1)\lceil \frac{s}{3} \rceil - 4 & \text{if } s = 3m + 1. \end{cases}$

Hence the theorem follows.

Theorem 5 *If $s \geq 6$, then* $\gamma_{CD}(P_6 \boxtimes P_s) = \begin{cases} s & \textit{if } s = 4l + 4, \\ s + 1 & \textit{if } s = 4l + 1 \textit{ or } 4l + 3, \\ s + 2 & \textit{if } s = 4l + 3. \end{cases}$

Proof Let $P_6 = (v_1, v_2, \ldots, v_6)$ and $P_s = (u_1, u_2, \ldots, u_s)$.

Then $V(P_6 \boxtimes P_s) = \{(v_i, u_j) : 1 \leq i \leq 6, 1 \leq j \leq s\}$ and $E(P_6 \boxtimes P_s) = \{(v_i, u_j)(v_i, u_{j+1}) : 1 \leq i \leq 6, 1 \leq j \leq s - 1\} \cup \{(v_i, u_j)(v_{i+1}, u_j) : 1 \leq i \leq 5, 1 \leq j \leq s\} \cup \{(v_i, u_j)(v_{i+1}, u_{j+1}) : 1 \leq i \leq 5, 1 \leq j \leq s - 1\} \cup \{(v_i, u_j)(v_{i+1}, u_{j-1}) : 1 \leq i \leq 5, 2 \leq j \leq s\}$. Let $S_1 = \{(v_i, u_j) : i = 3l+2, j = 4l + 2 \textit{ or } 4l + 3\}$.

Then $S = \begin{cases} S_1 & \text{if } s = 4l + 4 \textit{ or } 4l + 3, \\ S_1 \cup \{(v_i, u_{s-1}) : i = 3l + 2\} & \text{if } s = 4l + 1 \textit{ or } 4l + 2, \end{cases}$ is a $CD-set$ of $P_6 \boxtimes P_s$. Thus $\gamma_{CD}(P_6 \boxtimes P_s) \leq |S| = \begin{cases} s & \text{if } s = 4l + 4, \\ s + 1 & \text{if } s = 4l + 1 \textit{ or } 4l + 3, \\ s + 2 & \text{if } s = 4l + 3. \end{cases}$

Suppose there exists a dominating set $D \subseteq V$ of cardinality at most

$d = \begin{cases} s-1 \text{ if } s = 4l+4, \\ s \quad\ \text{if } s = 4l+1 \text{ or } 4l+3, \\ s+1 \text{ if } s = 4l+3. \end{cases}$ Then $< D >$ has an isolated vertex.

Thus $|D| \geq d+1 = \begin{cases} s \quad\ \text{if } s = 4l+4, \\ s+1 \text{ if } s = 4l+1 \text{ or } 4l+3, \\ s+2 \text{ if } s = 4l+3. \end{cases}$ hence the result.

Theorem 6 *If $k \geq 1$, $r = 4k+3$, $r \geq 7$, and $s \geq r$, then*

$$\gamma_{CD}(P_r \boxtimes P_s) = \begin{cases} \lceil \frac{r}{2} \rceil \lceil \frac{s}{3} \rceil & \text{if } s = 3m+3 \text{ or } 3m+2, \\ \lceil \frac{r}{2} \rceil \lceil \frac{s}{3} \rceil - 2 & \text{if } s = 3m+1. \end{cases}$$

Proof *Let* $S_1 = \{(v_i, u_j) : i = 4l+2 \text{ or } 4l+3,\ j = 3l+2\}$ *and* *let* $A = (\{(v_i, u_s) : i = 4l+2 \text{ or } 4l+3\} \cup \{(v_{r-1}, u_{s-1})\}) - \{(v_r, u_s), (v_{r-1}, u_s)\}$

Then $S = \begin{cases} S_1 & \text{if } s = 3l+3 \text{ or } 3l+2, \\ (S_1 \cup A) - \{(v_r, u_{s-2})\} & \text{if } s = 3l+1, \end{cases}$ is a $CD-set$ of $P_r \boxtimes P_s$.

Thus $\gamma_{CD}(P_r \boxtimes P_s) \leq |S| = \begin{cases} \lceil \frac{r}{2} \rceil \lceil \frac{s}{3} \rceil & \text{if } s = 3m+3 \text{ or } 3m+2, \\ \lceil \frac{r}{2} \rceil \lceil \frac{s}{3} \rceil - 2 & \text{if } s = 3m+1. \end{cases}$

Suppose there exists a dominating set $D \subseteq V$ of cardinality at most

$d = \begin{cases} \lceil \frac{r}{2} \rceil \lceil \frac{s}{3} \rceil - 1 & \text{if } s = 3m+3 \text{ or } 3m+2, \\ \lceil \frac{r}{2} \rceil \lceil \frac{s}{3} \rceil - 3 & \text{if } s = 3m+1. \end{cases}$ Then $< D >$ contains an isolated vertex in $\{(v_i, u_j) : i = 4l+2 \text{ or } 4l+3,\ j = 3l+2\} \cup \{(v_{r-1}, u_j) : j = 3l+1\}$.

Thus $|D| \geq d+1 = \begin{cases} \lceil \frac{r}{2} \rceil \lceil \frac{s}{3} \rceil & \text{if } s = 3m+3 \text{ or } 3m+2, \\ \lceil \frac{r}{2} \rceil \lceil \frac{s}{3} \rceil - 2 & \text{if } s = 3m+1. \end{cases}$ Hence, the statement of the theorem is established.

Theorem 7 *If $l \geq 0$, $r \geq 3$ and $2 \leq s \leq 3$, then*

$$\gamma_{CD}(P_r \boxtimes P_s^c) = \begin{cases} r+2 & \text{if } r = 4l+2, \\ 2\lceil \frac{r}{2} \rceil & \text{otherwise.} \end{cases}$$

Proof $V(P_r \boxtimes P_s^c) = \{(v_i, u_j) : 1 \leq i \leq r, 1 \leq j \leq s\}$ and $E(P_r \boxtimes P_s^c) = \{(v_i, u_j)(v_{i+1}, u_j) : 1 \leq i \leq r-1, 1 \leq j \leq s\} \cup \{(v_i, u_j)(v_{i+1}, u_k) : 1 \leq i \leq r-1, 1 \leq j \leq s-2, j+2 \leq k \leq s\} \cup \{(v_i, u_j)(v_{i+1}, u_l) : 1 \leq i \leq r-1, 3 \leq j \leq s, 1 \leq l \leq j-2\} \cup \{(v_i, u_j)(v_i, u_k) : 1 \leq i \leq r, 1 \leq j \leq s-2, j+2 \leq k \leq s\}$.
Let $S_1 = \{(v_i, u_j) : i = 4l+2 \text{ or } 4l+3,\ j = 1, 2\}$. Then

$S = \begin{cases} S_1 \cup \{(v_{r-1}, u_j) : j = 1, 2\} & \text{if } r = 4l+1 \text{ or } 4l+2, \\ S_1 & \text{otherwise,} \end{cases}$

is a CD-set of $(P_r \boxtimes P_s^c)$. Thus $\gamma_{CD}(P_r \boxtimes P_s^c) \leq |S| = \begin{cases} r+2 & \text{if } r = 4l+2, \\ 2\lceil \frac{r}{2} \rceil & \text{otherwise.} \end{cases}$

Suppose there exists a dominating D of cardinality at most

$d = \begin{cases} r+1 & \text{if } r = 4l+2, \\ 2\lceil \frac{r}{2} \rceil - 1 & \text{otherwise,} \end{cases}$ Then $< D >$ has an isolated vertex.

Thus $|D| \geq d+1 = \begin{cases} r+2 \text{ if } r = 4l+2, \\ 2\lceil \frac{r}{2} \rceil \text{ otherwise.} \end{cases}$ Hence the theorem follows.

Theorem 8 *If $r \geq 4, s \geq 4$ and $s \geq r$, then $\gamma_{CD}(P_r \boxtimes P_s^c) = 2\lceil \frac{r}{3} \rceil$.*

Proof Let $P_r = (v_1, v_2, \ldots, v_r)$ and $P_s = (u_1, u_2, \ldots, u_s)$.

Then $V(P_r \boxtimes P_s^c) = \{(v_i, u_j) : 1 \leq i \leq r, 1 \leq j \leq s\}$ and $E(P_r \boxtimes P_s^c) = \{(v_i, u_j)(v_{i+1}, u_j) : 1 \leq i \leq r-1, 1 \leq j \leq s\} \cup \{(v_i, u_j)(v_{i+1}, u_k) : 1 \leq i \leq r-1, 1 \leq j \leq s-2, j+2 \leq k \leq s\} \cup \{(v_i, u_j)(v_{i+1}, u_l) : 1 \leq i \leq r-1, 3 \leq j \leq s, 1 \leq lj-2\} \cup \{(v_i, u_j)(v_i, u_k) : 1 \leq i \leq r, 1 \leq j \leq s-2, j+2 \leq k \leq s\}$. Let $S_1 = \{(v_i, u_1), (v_i, u_{s-1}) : i = 3l+2\}$. Then

$S = \begin{cases} S_1 & \text{if } r = 3l+2 \text{ or } 3l+3. \\ S_1 \cup \{(v_r, u_1), (v_r, u_{s-1})\} & \text{if } r = 3l+1, \end{cases}$

is a $CD-set$ of $P_r \boxtimes P_s^c$. Thus $\gamma_{CD}(P_r \boxtimes P_s^c) \leq |S| = 2\lceil \frac{r}{3} \rceil$.

Suppose there exists a dominating set $D \subseteq V$ of cardinality at most $d = 2\lceil \frac{r}{3} \rceil - 1$; then $< D >$ contains an isolated vertex. Thus $|D| \geq d+1 = 2\lceil \frac{r}{3} \rceil$. Hence the theorem follows.

Observation 2.1.

1. If $r \geq 3$ *and* $s \geq r$, then $\gamma_{CD}(C_r \boxtimes K_s) = \gamma_{CD}(P_r \boxtimes K_s)$.
2. $\gamma_{CD}(P_1 \boxtimes K_s) = \gamma_{CD}(P_1^c \boxtimes K_s) = 2$.
3. If $2 \leq r \leq 3$, *then* $\gamma_{CD}(P_r^c \boxtimes K_s) = 4$.
4. If $r \leq 2$, then $\gamma_{CD}(P_r^c \boxtimes P_s^c)$ does not exists.
5. $\gamma_{CD}(P_3^c \boxtimes P_3^c)$ does not exists.
6. If $r \geq 3$ and $s \geq 4$, then $\gamma_{CD}(P_r^c \boxtimes P_s^c) = 4$.

2.2 Exact γ_{CD} for $P_r \boxtimes K_s$ and $P_r \boxtimes K_{1,s}$

Theorem 9 *If $l \geq 0$, $r \geq 4$ and $s \geq r$, then*

$$\gamma_{CD}(P_r \boxtimes K_s) = \begin{cases} \frac{r}{2} + 1 & \text{if } r = 4l+2, \\ \lceil \frac{r}{2} \rceil & \text{otherwise.} \end{cases}$$

Proof Let $P_r = (v_1, v_2, \ldots, v_r)$ and $V(K_s) = \{u_1, u_2, \ldots, u_s\}$.

Then $V(P_r \boxtimes K_s) = \{(v_i, u_j) : 1 \leq i \leq r, 1 \leq j \leq s\}$ and $E(P_r \boxtimes K_s) = \{(v_i, u_j)(v_i, u_{j+1}) : 1 \leq i \leq r, 1 \leq j \leq j \leq s-1\} \cup \{(v_i, u_j)(v_i, u_k) : 1 \leq i \leq r, 1 \leq j \leq s-2, j+2 \leq k \leq s\} \cup \{(v_i, u_j)(v_{i+1}, u_k) : 1 \leq i \leq r-1, j+1 \leq k \leq s, 1 \leq j \leq s-1\} \cup \{(v_i, u_j)(v_{i-1}, u_k) : 2 \leq i \leq r, 1 \leq j \leq s-1, j+1 \leq k \leq s\} \cup \{(v_i, u_j)(v_{i+1}, u_j) : 1 \leq i \leq r-1, 1 \leq j \leq s\}$. Let $S_1 = \{(v_i, u_1) :$

$i = 4l+2$ *or* $4l+3\big\}$. Then $S = \begin{cases} S_1 \cup \{(v_{r-1}, u_1)\} \text{ if } r = 4l+1 \text{ or } 4l+2, \\ S_1 \qquad\qquad\qquad\qquad \text{otherwise,} \end{cases}$ is a $CD-$set of $P_r \boxtimes K_s$. Thus $\gamma_{CD}(P_r \boxtimes K_s) \le |S| = \begin{cases} \frac{r}{2}+1 \text{ if } r = 4l+2, \\ \lceil \frac{r}{2} \rceil \quad \text{otherwise.} \end{cases}$

Suppose there exists a dominating set $D \subseteq V$ of cardinality at most $d = \begin{cases} \frac{r}{2} & \text{if } r = 4l+2, \\ \lceil \frac{r}{2} \rceil - 1 & \text{otherwise,} \end{cases}$ Then $< D >$ contains an isolated vertex. Thus $|D| \ge d+1 = \begin{cases} \frac{r}{2}+1 \text{ if } r = 4l+2, \\ \lceil \frac{r}{2} \rceil \quad \text{otherwise.} \end{cases}$ therefore the proof.

Theorem 10 *For any nontrivial path* P_r *and* $K_{t,s}, t \ge 1$, $\gamma_{CD}(P_r \boxtimes K_{t,s}) = 2\lceil \frac{r}{3} \rceil$.

Proof Let $P_r = (v_1, v_2, v_3, \ldots, v_r)$ and $V(K_{t,s}) = (V_1, V_2)$, where $V_1 = (x_1, x_2, x_3, \ldots, x_t)$ and $V_2 = (y_1, y_2, y_3, \ldots, y_s)$. Then $V(P_r \boxtimes K_{t,s}) = \{(v_i, x_j), (v_i, y_k) : 1 \le i \le r, 1 \le j \le t, 1 \le k \le s\}$ and $E(P_r \boxtimes K_{t,s}) = \{(v_i, x_j)(v, y_k) : 1 \le i \le r, i \le k \le s, 1 \le j \le t\} \cup \{(v_i, x_j)(v_{i+1}, x_j), (v_i, y_k)(v_{i+1}, y_k) : 1 \le i \le r-1, i \le k \le s, 1 \le j \le t\} \cup \{(v_i, x_j)(v_{i+1}, y_k) : 1 \le i \le r-1, i \le k \le s, 1 \le j \le t\} \cup \{(v_i, x_j)(v_{i-1}, y_k) : 2 \le i \le r, i \le k \le s, 1 \le j \le t\}$. Let $S_1 = \big\{(v_i, u_1), (v_i, u_{s-1}) : i = 3l+2\big\}$. Then

$$S = \begin{cases} S_1 & \text{if } r = 3l+2 \text{ or } 3l+3, \\ S_2 \cup \{(v_r, u_1), (v_r, u_{s-1})\} & \text{if } r = 3l+1, \end{cases}$$

is a $CD-set$ of $P_r \boxtimes K_{t,s}$. Thus $\gamma_{CD}(P_r \boxtimes P_s^c) \le |S| = 2\lceil \frac{r}{3} \rceil$. Suppose there exists a dominating set $D \subseteq V$ of cardinality at most $d = 2\lceil \frac{r}{3} \rceil - 1$; then $< D >$ contains an isolated vertex. Thus $|D| \ge d+1 = 2\lceil \frac{r}{3} \rceil$. Hence the theorem follows.

2.3 *Identifying the* γ_{CD} *for* $P_r \boxtimes C_s$, $C_r \boxtimes C_s$, *and* $P_r \boxtimes H_s$

Theorem 11 $\gamma_{CD}(C_r \boxtimes C_s) = \gamma_{CD}(P_r \boxtimes P_s)$, $s \ge r$.

Proof Let $C_r = (v_1, v_2, \ldots, v_r, v_1)$ and $C_s = (u_1, u_2, \ldots, u_s, u_1)$ be the cycle on r and s vertices, respectively. The proof of this theorem has been subdivided into many cases, and each case will be explained in an identical way as in Theorems 3, 4, 5, and 6.

Observation 2.2.

1. If $2 \le r \le 3$ and $s \ge r$, then $\gamma_{CD}(P_r \boxtimes P_s^c) = 4$.
2. If $s \ge 3$, then $\gamma_{CD}(C_3 \boxtimes P_s^c) = 4$.
3. If $r, s \ge 4$, then $\gamma_{CD}(C_r \boxtimes P_s^c) = 2\lceil \frac{r}{3} \rceil$.
4. If $2 \le r \le 3$ and $s \ge 4$, then $\gamma_{CD}(P_r \boxtimes C_s^c) = 4$.
5. If $r, s \ge 4$, then $\gamma_{CD}(P_r \boxtimes C_s^c) = 2\lceil \frac{r}{3} \rceil$.
6. If $s \ge 3$, then $\gamma_{CD}(C_3 \boxtimes C_s^c) = 4$.
7. If $r, s \ge 4$, then $\gamma_{CD}(C_r \boxtimes C_s^c) = 2\lceil \frac{r}{3} \rceil$.

8. If $2 \leq r \leq 3$ and $s \geq r$, then $\gamma_{CD}(P_r \boxtimes P_s^c) = 4$.
9. Let G be a totally disconnected graph of order s; then

$$\gamma_{CD}(P_r \boxtimes G) = \begin{cases} \frac{s(r+2)}{2} & \text{if } s \equiv 2 \ (mod\, 4), \\ s\lceil \frac{r}{2} \rceil & \text{otherwise.} \end{cases}$$

Theorem 12 *If $l \geq 0$, $r \leq 3$, and $s \geq 3$, then*

$$\gamma_{CD}(P_r \boxtimes C_s) = \begin{cases} \frac{s}{2}+1 & \text{if } s = 4l+2, \\ \lceil \frac{s}{2} \rceil & \text{otherwise.} \end{cases}$$

Proof $V(P_r \boxtimes C_s) = \{(v_i, u_j) : 1 \leq i \leq r, 1 \leq j \leq s\}$ and $E(P_r \boxtimes C_s) = \{(v_i, u_j)(v_i, u_{j+1}) : 1 \leq i \leq r, 1 \leq j \leq s-1\} \cup \{(v_i, u_j)(v_{i+1}, u_j) : 1 \leq i \leq r-1, 1 \leq j \leq s\} \cup \{(v_i, u_j)(v_{i+1}, u_{j+1}) : 1 \leq i \leq r-1, 1 \leq j \leq s-1\} \cup \{(v_i, u_j)(v_{i+1}, u_{j-1}) : 1 \leq i \leq r-1, 2 \leq j \leq s\} \cup \{(v_i, u_1)(v_{i+1}, u_s) : 1 \leq i \leq r-1\} \cup \{(v_i, u_1)(v_{i-1}, u_s) : 2 \leq i \leq r\} \cup \{(v_i, u_1)(v_i, u_s) : 1 \leq i \leq r\}$. Let $S_1 = (v_2, u_i) : i = 4l+2$ *or* $4l+3$. Then $S = \begin{cases} S_1 \cup \{(v_2, u_{s-1})\} & \text{if } s = 4l+2, \\ S_1 & \text{otherwise,} \end{cases}$ is a CD-set of $P_r \boxtimes C_s$. Thus $\gamma_{CD}(P_r \boxtimes C_s) \leq |S| = \begin{cases} \frac{s}{2}+1 & \text{if } s = 4l+2, \\ \lceil \frac{s}{2} \rceil & \text{otherwise.} \end{cases}$ Suppose there exists a dominating set $D \subseteq V$ of cardinality at most $d = \begin{cases} \frac{s}{2} & \text{if } s = 4l+2, \\ \lceil \frac{s}{2} \rceil - 1 & \text{otherwise,} \end{cases}$ then $< D >$ contains an isolated vertex. Thus $|D| \geq d+1 = \begin{cases} \frac{s}{2}+1 & \text{if } s = 4l+2, \\ \lceil \frac{s}{2} \rceil & \text{otherwise,} \end{cases}$ therefore the proof.

Theorem 13 *If $r, s \geq 4$, then $\gamma_{CD}(P_r \boxtimes C_s) = \gamma_{CD}(P_r \boxtimes P_s)$.*

Proof Let $P_r = (v_1, v_2, \ldots, v_r)$ and $C_s = (u_1, u_2, \ldots, u_s, u_1)$. Then $V(P_r \boxtimes C_s) = \{(v_i, u_j) : 1 \leq i \leq r, 1 \leq j \leq s\}$ and $E(P_r \boxtimes C_s) = \{(v_i, u_j)(v_i, u_{j+1}) : 1 \leq i \leq r, 1 \leq j \leq s-1\} \cup \{(v_i, u_j)(v_{i+1}, u_j) : 1 \leq i \leq r-1, 1 \leq j \leq s\} \cup \{(v_i, u_j)(v_{i+1}, u_{j+1}) : 1 \leq i \leq r-1, 1 \leq j \leq s-1\} \cup \{(v_i, u_j)(v_{i+1}, u_{j-1}) : 1 \leq i \leq r-1, 2 \leq j \leq s\} \cup \{(v_i, u_1)(v_{i+1}, u_s) : 1 \leq i \leq r-1\} \cup \{(v_i, u_1)(v_{i-1}, u_s) : 2 \leq i \leq r\} \cup \{(v_i, u_1)(v_i, u_s) : 1 \leq i \leq r\}$. The proof of this theorem has been subdivided into many cases, and each case will be explained in an identical way as in Theorems 3, 4, 5, and 6.

Theorem 14 *If $r \geq 2$ and $s \geq 3$, then*

$$\gamma_{CD}(P_r \boxtimes H_s) = \begin{cases} \frac{(s+1)(r+2)}{2} & \text{if } s \equiv 2 \ (mod\, 4), \\ (s+1)\lceil \frac{r}{2} \rceil & \text{otherwise.} \end{cases}$$

Proof Let $P_r = (v_1, v_2, v_3, \ldots, v_r)$ and $V(H_s) = \{u_1, u_2, u_3, \ldots, u_s, u'_1, u'_2, u'_3, \ldots, u'_s, u\}$, where $\Delta(H_s) = u$, $deg(u'_i) = 1$ and the vertices $u_1, u_2, u_3, \ldots, u_s$ form a cycle. Then $V(P_r \boxtimes H_s) = \{(v_i, u_j)(v_i, u'_j)(v_i, u) : 1 \leq i \leq r, 1 \leq j \leq s\}$ and $E(P_r \boxtimes H_s) = \{(v_i, u_j)(v_{i+1}, u_j), (v_i, u'_j)(v_{i+1}, u'_j), (v_i, u)(v_{i+1}, u) : 1 \leq$

$i \le r-1, 1 \le j \le s\} \cup \{(v_i,u_j)(v_{i+1},u_j), (v_i,u_1)(v_i,u_s) : 1 \le i \le r, 1 \le j \le s-1\} \cup \{(v_i,u_j)(v_i,u'_j), (v_i,u_j)(v_i,u) : 1 \le i \le r, 1 \le j \le s\} \cup \{(v_i,u_j)(v_{i+1},u_{j+1}) : 1 \le i \le r-1, 1 \le j \le s-1\} \cup \{(v_i,u_j)(v_{i+1},u'_j) : 1 \le i \le r-1, 1 \le j \le s\} \cup \{(v_i,u_j)(v_{i+1},u) : 1 \le i \le r-1, 1 \le j \le s\} \cup \{(v_{i+1},u_j)(v_i,u_{j+1}) : 1 \le i \le r-1, 2 \le j \le s-1\} \cup \{(v_{i+1},u_j)(v_i,u'_j) : 1 \le i \le r-1, 1 \le j \le s\} \cup \{(v_{i+1},u_j)(v_i,u) : 1 \le i \le r-1, 1 \le j \le s\}$. Let $S_1 = \{(v_i,u'_j), (v_i,u) : i \equiv 2 \; or \; 3 \; (mod \; 4), 1 \le j \le s\}$. Then $S = \begin{cases} S_1 \cup \{(v_{r-1},u'_j), (v_{r-1},u) : 1 \le j \le s\} & \text{if } r \equiv 1 \; or \; 2 \; (mod \; 4), \\ S_1 & \text{otherwise,} \end{cases}$ is a CD-set of $P_r \boxtimes H_s$. Thus $\gamma_{CD}(P_r \boxtimes H_s) \le |S| = \gamma_{CD}(P_r \boxtimes H_s) = \begin{cases} \frac{(s+1)(r+2)}{2} & \text{if } s \equiv 2 \; (mod \; 4), \\ (s+1)\lceil \frac{r}{2} \rceil & \text{otherwise.} \end{cases}$ Suppose there exists a D dominating of cardinality at most $d = \begin{cases} \frac{(s+1)(r+2)}{2} - 1 & \text{if } s \equiv 2 \; (mod \; 4), \\ (s+1)\lceil \frac{r}{2} \rceil - 1 & \text{otherwise,} \end{cases}$ Then $< D >$ has an isolated vertex. Thus $|D| \ge d+1 = \begin{cases} \frac{(s+1)(r+2)}{2} & \text{if } s \equiv 2 \; (mod \; 4), \\ (s+1)\lceil \frac{r}{2} \rceil & \text{otherwise,} \end{cases}$ hence the theorem.

Observation 2.3.

1. $\gamma_{CD}(P_r \boxtimes W_{1,s}) = \gamma_{CD}(P_r) = \begin{cases} \frac{r}{2} + 1 & \text{if } s \equiv 2 \; (mod \; 4), \\ \lceil \frac{r}{2} \rceil & \text{otherwise.} \end{cases}$
2. $\gamma_{CD}(L(P_r) \boxtimes L(P_s)) = \gamma_{CD}(P_{r-1} \boxtimes P_{s-1})$.
3. $\gamma_{CD}(L(P_r) \boxtimes P_s) = \gamma_{CD}(P_{r-1} \boxtimes P_s)$.
4. $\gamma_{CD}(L(C_r) \boxtimes L(C_s)) = \gamma_{CD}(C_r \boxtimes C_s)$.

Conclusion

Finding the CD-number for a general graph is an NP-complete problem. In this paper we find out the exact value of $\gamma_{CD}(P_r \boxtimes P_s)$, $\gamma_{CD}(P_r \boxtimes P_s^c)$, $\gamma_{CD}(P_r \boxtimes C_s)$, $\gamma_{CD}(P_r \boxtimes K_s)$, $\gamma_{CD}(P_r \boxtimes W_{1,s})$, $\gamma_{CD}(L(P_r) \boxtimes P_s)$, *etc*. Also we have given the minimum CD-set for the above noticed graph, and the comparison of this parameter with other dominating parameters will be described in the successive paper.

References

1. Gravier, S., Khelladi, A.: On the domination number of cross products of graphs. Discrete Math. **145**(1–3), 273–277 (1995)
2. Hammack, R.H., Imrich, W., Klavžar, S., Imrich, W., Klavžar, S.: Handbook of Product Graphs, vol. 2. Boca Raton, CRC Press (2011)
3. Harary, F.: Graph Theory. Addison Wesley, Reading (1972)
4. Mahadevan, G.., Vimala Suganthi, M., Sivagnanam, C.: Corona domination of graphs. Mathematical Modelling and Computational Intelligence Techniques, Spinger Proceedings in Mathematics and Statistic, vol. 376, pp. 255–265 (2021)

Detection of TCC-Domination Number for Some Product Related Graphs

G. Mahadevan, S. Kaviya, C. Sivagnanam, L. Praveenkumar, and S. Anuthiya

1 Introduction

The graph G taken throughout this paper was supposed to be finite, nontrivial, simple and undirected. For any undefined term, refer to [2, 6, 7]. Let $V(G)$ and $E(G)$ be the vertex set and edge set. The triple connected graph was initiated by Paulraj Joseph et al. [5], subsequently, triple connected domination number of a graph was introduced by Mahadevan et al. [3], and the idea of certified domination number was initiated by Magda Dettlaff et al. [1]. Inspired by the above papers, we introduced a fresh parameter called triple connected certified domination number [4]. The purpose concerning this product graph is from the parameter certified domination, which already has the product graphs. The strong product [6] is the union of the Cartesian product and tensor product. The octagonal grid is the strong product of two paths. The results of this attempt are interesting because they can lead us to a better understanding of product related graphs. In Sect. 2 we discuss triple connected certified domination for Cartesian [7], lexicographic [6] and corona product graphs [6] and related work reviews. Section 3 gives triple connected certified domination for strong product graphs on path and cycle (Fig. 1).

Definition A dominating set S is said to be a triple connected certified dominating set (TCCD set), if for every v∈ S $|N(v) \cap (V - S)| = 0$ or $k, k \geq 2$ and any three vertices of S lie on a path in $< S >$. The minimum cardinality of a TCCD set is called the triple connected certified domination number (TCCD-number) and is denoted by $\gamma_{TCC}(G)$.

G. Mahadevan (✉) · S. Kaviya (✉) L. Praveenkumar · S. Anuthiya
The Gandhigram Rural Institute - Deemed to be University, Gandhigram, Tamil Nadu, India

C. Sivagnanam
Department of General Requirements, University of Technology and Applied Sciences - Sur, Sultanate of Oman

F. Kamalov et al. (eds.), *Advances in Mathematical Modeling and Scientific Computing*, Trends in Mathematics, https://doi.org/10.1007/978-3-031-41420-6_77

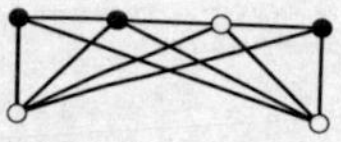

Fig. 1 A graph with $\gamma_{TCC} = 3$

2 Triple Connected Certified Domination for Cartesian, Lexicographic and Corona Product Graphs

Here, we discussed some results on the Cartesian, lexicographic and corona products and obtained the entire values of γ_{TCC} on path and cycle.

1. If $r \geq 3$, then $\gamma_{TCC}(P_r \square P_3) = \gamma_{TCC}(P_r \square C_3) = \gamma_{TCC}(C_r \square C_3) = r$.
2. If $r = 2$, then $\gamma_{TCC}(P_2 \odot P_s) = \gamma_{TCC}(P_2 \odot C_s) = 3$.
3. If $r \geq 3$ and $s \geq 2$, then $\gamma_{TCC}(P_r \odot P_s) = r$.
4. If $r, s \geq 3$, then $\gamma_{TCC}(P_r \odot C_s) = \gamma_{TCC}(C_r \odot C_s) = r$.
5. If $r = 2$ and $s = 2$, then $\gamma_{TCC}(P_r \circ P_s) = 4$
6. If $s \geq 3$, then $\gamma_{TCC}(P_r \circ P_s) = \gamma_{TCC}(P_r \circ C_s) = \gamma_{TCC}(C_r \circ C_s) = \begin{cases} 3 & if\ r \leq 4, \\ r-2 & if\ r \geq 5. \end{cases}$

3 TCCD-Number for Strong Product of Graphs

Throughout this section, we consider $r \geq s$. Also we have extracted some results on the strong product of graphs involving paths and obtained the exact γ_{TCC} values. Here the graph taken contains the vertex and edge set

$$V(P_r \boxtimes P_s) = \{v_{ij} : 1 \leq i \leq r, 1 \leq j \leq s\}\ and$$
$$E(P_r \boxtimes P_s) = \{v_{ij}v_{i,j+1} : 1 \leq i \leq r,\ 1 \leq j \leq s-1\}$$
$$\cup \{v_{ij}v_{i+1,j} : 1 \leq i \leq r-1,\ 1 \leq j \leq s\}$$
$$\cup \{v_{ij}v_{i+1,j+1} : 1 \leq i \leq r-1,\ 1 \leq j \leq s-1\}$$
$$\cup \{v_{ij}v_{i+1,j-1} : 1 \leq i \leq r-1,\ 2 \leq j \leq s-1\}.$$

Observation:

1. For $\gamma_{TCC}(P_2 \boxtimes P_2) = 4$, and $\gamma_{TCC}(P_2 \boxtimes P_1)$ does not exist.
2. If $3 \leq r \leq 4\ and\ 2 \leq s \leq 3$, then $\gamma_{TCC}(P_r \boxtimes P_s) = 3$.
3. If $r \geq 5\ and\ 2 \leq s \leq 3$, then $\gamma_{TCC}(P_r \boxtimes P_s) = r - 2$.
4. If $r \geq 4$, then $\gamma_{TCC}(P_r \boxtimes P_4) = r$.
5. $\gamma_{TCC}(P_{12} \boxtimes P_{11}) = 37$.
6. If $5 \leq r \leq 9$, then $\gamma_{TCC}(P_r \boxtimes P_5) = \begin{cases} (r-4)+(s+1) & if\ r = 5 \\ 2(r-3)+(s-3) & if\ otherwise. \end{cases}$
7. If $r \geq 10$, then $\gamma_{TCC}(P_r \boxtimes P_5) = \begin{cases} \lceil \frac{r}{4} \rceil \lfloor \frac{s}{3} \rfloor + 3\frac{r}{2} - 2 & if\ r \equiv 0\ or\ 2\ (mod\ 4), \\ \lceil \frac{r}{4} \rceil \lfloor \frac{s}{3} \rfloor + 3\lfloor \frac{r}{2} \rfloor & if\ r \equiv 1\ (mod\ 4), \\ \lceil \frac{r}{4} \rceil \lfloor \frac{s}{3} \rfloor + 3\lfloor \frac{r}{2} \rfloor - 1 & if\ r \equiv 3\ (mod\ 4). \end{cases}$

Theorem 1 *If* $r \geq 6$, *then*

$$\gamma_{TCC}(P_r \boxtimes P_6) = \begin{cases} 6\lfloor \frac{r}{4} \rfloor + \lceil \frac{r}{2} \rceil + 2 \; if \; r \equiv 3 \; (mod \; 4), \\ 2(r-1) \quad otherwise. \end{cases}$$

Proof Take $S_1 = \{v_{2j} : 2 \le j \le 5\} \cup \{v_{r-1,j} : 2 \le j \le 5\}$, $S_2 = \{v_{i2} : 3 \le i \le r-2\} \cup \{v_{i4} : 5 \le i \le r-2\}$, $S_3 = \{v_{ij} : i \equiv 2 \, (mod \, 4),\ j \equiv 2 \, (mod \, 3),\ 2 \le i \le r-2\}$, $S_4 = \{v_{ij} : i \equiv 3 \; (mod \; 4),\ j \equiv \; 0 \, or \, 1 \, (mod \; 3),\ 2 \le i \le r-1,\ 2 \le j \le 5\}$, $S_5 = \{v_{i5} : i \equiv 4 \, or \, 5 \, (mod \; 8),\ 2 \le i \le r-1\} \cup \{v_{i5} : i \equiv 3 \, (mod \, 4),\ 2 \le i \le r-1\}$, $S_6 = \{v_{i2} : i \equiv 0 \, or \, 1 \, (mod \; 8),\ 2 \le i \le r-1\} \cup \{v_{i2} : i \equiv 3 \; (mod \; 4),\ 2 \le i \le r-1\}$, $S_7 = \{v_{r-1,j} : 2 \le j \le 5\}$.

Then $S = \begin{cases} S_3 \cup S_4 \cup S_5 \cup S_6 & if \; r \equiv \; 0 \; (mod \; 4), \\ S_1 \cup S_2 & if \; r \equiv \; 1 \; or \; 2 \; (mod \; 4), \\ S_3 \cup S_4 \cup S_5 \cup S_6 \cup S_7 & if \; r \equiv \; 3 \; (mod \; 4), \end{cases}$

is a TCCD set of $P_r \boxtimes P_6$ and hence $\gamma_{TCC}(P_r \boxtimes P_6) \le |S| =$ $\begin{cases} 6\lfloor \frac{r}{4} \rfloor + \lceil \frac{r}{2} \rceil + 2 \; if \; r \equiv 3 \; (mod \; 4), \\ 2(r-1) \quad otherwise. \end{cases}$

Let S' be a TCCD set of $P_r \boxtimes P_6$. Since any dominating set D of cardinality at most $k = \begin{cases} 6\lfloor \frac{r}{4} \rfloor + \lceil \frac{r}{2} \rceil + 1 \; if \; r \equiv 3 \; (mod \; 4), \\ 2r-3 \quad otherwise, \end{cases}$ whose induced subgraph $< D >$ is not triple connected, then we have

$|S'| \ge k+1 = \begin{cases} 6\lfloor \frac{r}{4} \rfloor + \lceil \frac{r}{2} \rceil + 2 \; if \; r \equiv 3 \; (mod \; 4), \\ 2(r-1) \quad otherwise. \end{cases}$

Hence the result follows. □

Theorem 2 *If* $r \ge 8, r \equiv \; 0 \; (mod \; 4)$ *and* $s \equiv \; 0 \; (mod \; 4)$, *then* $\gamma_{TCC}(P_r \boxtimes P_s) =$ $\begin{cases} \frac{s}{2}(\frac{r}{2}+1) - 2 & if \; r \equiv 0 \; (mod \; 3), \\ 3\frac{s}{4}\lceil \frac{r}{3} \rceil - 2 & if \; r \equiv 1 \; (mod \; 3), \\ 3\frac{s}{4}\lceil \frac{r}{3} \rceil + \frac{s}{4} - 2 & if \; r \equiv 2 \; (mod \; 3). \end{cases}$

Proof Take $S_1 = \{v_{ij} \; : i \equiv 2 \; (mod \; 3),\ j \equiv 2 \; (mod \; 4),\ 2 \le i \le r-1,\ 2 \le j \le s-1\}$, $S_2 = \{v_{ij} \; : i \equiv 0 \; or \; 1 \; (mod \; 3),\ j \equiv 3 \; (mod \; 4),\ 2 \le i \le r-1,\ 2 \le j \le s-1\}$, $S_3 = \{v_{r-1,j} \; : j \equiv \; 4 \; or \; 5 \; (mod \; 8),\ 2 \le j \le s-1\} \cup \{v_{r-1,j} \; : j \equiv 2 \; (mod \; 4),\ 2 \le j \le s-1\}$, $S_4 = \{v_{2j} : j \equiv \; 0 \; or \; 1 \; (mod \; 8),\ 2 \le j \le s-1\} \cup \{v_{2j} : j \equiv \; 3 \; (mod \; 4),\ 2 \le j \le s-1\}$, $S_5 = \{v_{r-1,j} \; : j \equiv \; 4 \; or \; 5 \; (mod \; 8),\ 2 \le j \le s-1\} \cup \{v_{r-1,j} \; : j \equiv 3 \; (mod \; 4),\ 2 \le j \le s-1\}$.

Then $S = \begin{cases} S_1 \cup S_2 \cup S_4 \cup S_5 \text{ if } r \equiv \; 0 \; (mod \; 3), \\ S_1 \cup S_2 \cup S_3 \cup S_4 \text{ if } r \equiv \; 1 \; or \; 2 \; (mod \; 3), \end{cases}$ is a TCCD set of $P_r \boxtimes$ P_s and hence $\gamma_{TCC}(P_r \boxtimes P_s) \le |S| = \begin{cases} \frac{s}{2}(\frac{r}{2}+1) - 2 & if \; r \equiv 0 \; (mod \; 3), \\ 3\frac{s}{4}\lceil \frac{r}{3} \rceil - 2 & if \; r \equiv 1 \; (mod \; 3), \\ 3\frac{s}{4}\lceil \frac{r}{3} \rceil + \frac{s}{4} - 2 & if \; r \equiv 2 \; (mod \; 3). \end{cases}$

Let S' be a TCCD set of $P_r \boxtimes P_s$. Since any dominating set D of cardinality at most $k = \begin{cases} \frac{s}{2}(\frac{r}{2}+1) - 3 & if \; r \equiv 0 \; (mod \; 3), \\ 3\frac{s}{4}\lceil \frac{r}{3} \rceil - 3 & if \; r \equiv 1 \; (mod \; 3), \\ 3\frac{s}{4}\lceil \frac{r}{3} \rceil + \frac{s}{4} - 3 & if \; r \equiv 2 \; (mod \; 3), \end{cases}$ whose induced

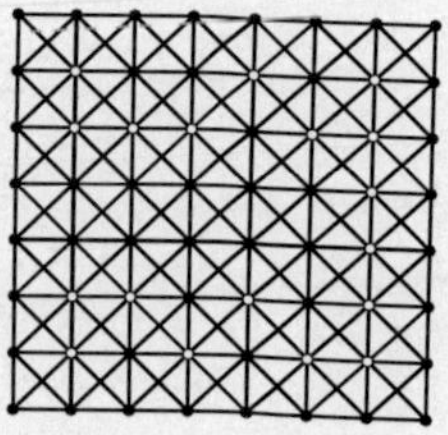

Fig. 2 A graph with $\gamma_{TCC}(P_8 \boxtimes P_8) = 18$

subgraph $< D >$ is not triple connected, then we have $|S'| \geq k + 1 = \begin{cases} \frac{s}{2}(\frac{r}{2}+1) - 2 & if\ r \equiv 0\ (mod\ 3), \\ 3\frac{s}{4}\lceil\frac{r}{3}\rceil - 2 & if\ r \equiv 1\ (mod\ 3), \\ 3\frac{s}{4}\lceil\frac{r}{3}\rceil + \frac{s}{4} - 2 & if\ r \equiv 2\ (mod\ 3). \end{cases}$ Hence the result follows. □

Theorem 3 *If* $r \geq 8,\ r \equiv 0\ (mod\ 4)$ *and either* $s \equiv 1$ *or* $2\ (mod\ 4)$ *or* $s \equiv 3\ (mod\ 4)$ *when* $s \geq 15$, *then* $\gamma_{TCC}(P_r \boxtimes P_s) = \begin{cases} \frac{3r}{4}\lceil\frac{s}{3}\rceil + \frac{r}{2} - 2 & if\ s \equiv 0\ (mod\ 3), \\ \frac{3r}{4}\lceil\frac{s}{3}\rceil - 2 & if\ s \equiv 1\ (mod\ 3), \\ \frac{3r}{4}\lceil\frac{s}{3}\rceil + \frac{r}{4} - 2 & if\ s \equiv 2\ (mod\ 3). \end{cases}$

Proof Take $S_1 = \{v_{ij}\ : i \equiv 2\ (mod\ 4),\ j \equiv 2\ (mod\ 3),\ 2 \leq i \leq r-1,\ 2 \leq j \leq s-1\}$, $S_2 = \{v_{ij}\ : i \equiv 3\ (mod\ 4);\ j \equiv 0\ or\ 1\ (mod\ 3),\ 2 \leq i \leq r-1,\ 2 \leq j \leq s-1\}$, $S_3 = \{v_{i,s-1}\ : i \equiv 4\ or\ 5\ (mod\ 8),\ 2 \leq i \leq r-1\} \cup \{v_{i,s-1}\ : i \equiv 2\ (mod\ 4),\ 2 \leq i \leq r-1\}$, $S_4 = \{v_{i2} : i \equiv 0\ or\ 1\ (mod\ 8),\ 2 \leq i \leq r-1\} \cup \{v_{i2} : i \equiv 3\ (mod\ 4),\ 2 \leq i \leq r-1\}$, $S_5 = \{v_{i,s-1}\ : i \equiv 4\ or\ 5\ (mod\ 8),\ 2 \leq i \leq r-1\} \cup \{v_{i,s-1}\ : i \equiv 3\ (mod\ 4),\ 2 \leq i \leq r-1\}$.

Then $S = \begin{cases} S_1 \cup S_2 \cup S_4 \cup S_5 \text{ if } r \equiv 0\ (mod\ 3), \\ S_1 \cup S_2 \cup S_3 \cup S_4 \text{ if } r \equiv 1\ or\ 2\ (mod\ 3), \end{cases}$ is a TCCD set of $P_r \boxtimes P_s$ and hence

$$\gamma_{TCC}(P_r \boxtimes P_s) \leq |S| = \begin{cases} \frac{3r}{4}\lceil\frac{s}{3}\rceil + \frac{r}{2} - 2 & if\ s \equiv 0\ (mod\ 3), \\ \frac{3r}{4}\lceil\frac{s}{3}\rceil - 2 & if\ s \equiv 1\ (mod\ 3), \\ \frac{3r}{4}\lceil\frac{s}{3}\rceil + \frac{r}{4} - 2 & if\ s \equiv 2\ (mod\ 3). \end{cases}$$

Let S' be a TCCD set of $P_r \boxtimes P_s$. Since any dominating set D of cardinality at most $k = \begin{cases} \frac{3r}{4}\lceil\frac{s}{3}\rceil + \frac{r}{2} - 3 & if\ s \equiv 0\ (mod\ 3), \\ \frac{3r}{4}\lceil\frac{s}{3}\rceil - 3 & if\ s \equiv 1\ (mod\ 3), \\ \frac{3r}{4}\lceil\frac{s}{3}\rceil + \frac{r}{4} - 3 & if\ s \equiv 2\ (mod\ 3), \end{cases}$ whose induced subgraph $< D >$ is not triple connected, then we have $|S'| \geq k + 1 = \begin{cases} \frac{3r}{4}\lceil\frac{s}{3}\rceil + \frac{r}{2} - 2 & if\ s \equiv 0\ (mod\ 3), \\ \frac{3r}{4}\lceil\frac{s}{3}\rceil - 2 & if\ s \equiv 1\ (mod\ 3), \\ \frac{3r}{4}\lceil\frac{s}{3}\rceil + \frac{r}{4} - 2 & if\ s \equiv 2\ (mod\ 3). \end{cases}$ Hence the result follows. □

See Fig. 2.

Theorem 4 *If* $r \geq 8, s \leq 11, r \neq 12, r \equiv 0 \ (mod\ 4)$ *and* $s \equiv 3 \ (mod\ 4)$, *then*

$$\gamma_{TCC}(P_r \boxtimes P_s) = \begin{cases} 3\lceil\frac{r}{3}\rceil\lfloor\frac{s}{4}\rfloor + \lceil\frac{s}{2}\rceil + r - 4 & if\ r \equiv 0\ (mod\ 3), \\ 3\lceil\frac{r}{3}\rceil\lfloor\frac{s}{4}\rfloor - 2\lfloor\frac{s}{4}\rfloor + \lceil\frac{s}{2}\rceil + r - 4 & if\ r \equiv 1\ (mod\ 3), \\ 3\lceil\frac{r}{3}\rceil\lfloor\frac{s}{4}\rfloor - \lfloor\frac{s}{4}\rfloor + \lceil\frac{s}{2}\rceil + r - 4 & if\ r \equiv 2\ (mod\ 3). \end{cases}$$

Proof Take $S_1 = \{v_{ij} : i \equiv 2\ (mod\ 3),\ j \equiv 2\ (mod\ 4),\ 2 \leq i \leq r-1,\ 2 \leq j \leq s-2\}$, $S_2 = \{v_{ij} : i \equiv 0\ or\ 1\ (mod\ 3),\ j \equiv 3\ (mod\ 4),\ 2 \leq i \leq r-1,\ 2 \leq j \leq s-2\}$, $S_3 = \{v_{r-1,j} : j \equiv 4\ or\ 5\ (mod\ 8),\ 2 \leq j \leq s-2\} \cup \{v_{r-1,j} : j \equiv 3\ (mod\ 4),\ 2 \leq j \leq s-2\}$, $S_4 = \{v_{2j} : j \equiv 0\ or\ 1\ (mod\ 8),\ 2 \leq j \leq s-1\} \cup \{v_{2j} : j \equiv 3\ (mod\ 4),\ 2 \leq j \leq s-1\}$, $S_5 = \{v_{i,s-1} : 2 \leq i \leq r-1\}$, $S_6 = \{v_{r-1,j} : j \equiv 4\ or\ 5\ (mod\ 8),\ 2 \leq j \leq s-2\} \cup \{v_{r-1,j} : j \equiv 2\ (mod\ 4),\ 2 \leq j \leq s-2\}$. Then $S = \begin{cases} S_1 \cup S_2 \cup S_3 \cup S_4 \cup S_5 & if\ r \equiv 0\ (mod\ 3), \\ S_1 \cup S_2 \cup S_4 \cup S_5 \cup S_6 & if\ r \equiv 1\ or\ 2\ (mod\ 3), \end{cases}$
is a TCCD set of $P_r \boxtimes P_s$ and hence

$$\gamma_{TCC}(P_r \boxtimes P_s) \leq |S| = \begin{cases} 3\lceil\frac{r}{3}\rceil\lfloor\frac{s}{4}\rfloor + \lceil\frac{s}{2}\rceil + r - 4 & if\ r \equiv 0\ (mod\ 3), \\ 3\lceil\frac{r}{3}\rceil\lfloor\frac{s}{4}\rfloor - 2\lfloor\frac{s}{4}\rfloor + \lceil\frac{s}{2}\rceil + r - 4 & if\ r \equiv 1\ (mod\ 3), \\ 3\lceil\frac{r}{3}\rceil\lfloor\frac{s}{4}\rfloor - \lfloor\frac{s}{4}\rfloor + \lceil\frac{s}{2}\rceil + r - 4 & if\ r \equiv 2\ (mod\ 3). \end{cases}$$

Let S' be a TCCD set of $P_r \boxtimes P_s$. Since any dominating set D of cardinality at most $k = \begin{cases} 3\lceil\frac{r}{3}\rceil\lfloor\frac{s}{4}\rfloor + \lceil\frac{s}{2}\rceil + r - 5 & if\ r \equiv 0\ (mod\ 3), \\ 3\lceil\frac{r}{3}\rceil\lfloor\frac{s}{4}\rfloor - 2\lfloor\frac{s}{4}\rfloor + \lceil\frac{s}{2}\rceil + r - 5 & if\ r \equiv 1\ (mod\ 3), \\ 3\lceil\frac{r}{3}\rceil\lfloor\frac{s}{4}\rfloor - \lfloor\frac{s}{4}\rfloor + \lceil\frac{s}{2}\rceil + r - 5 & if\ r \equiv 2\ (mod\ 3), \end{cases}$ whose induced subgraph $< D >$ is not triple connected, then we have

$$|S'| \geq k + 1 = \begin{cases} 3\lceil\frac{r}{3}\rceil\lfloor\frac{s}{4}\rfloor + \lceil\frac{s}{2}\rceil + r - 4 & if\ r \equiv 0\ (mod\ 3), \\ 3\lceil\frac{r}{3}\rceil\lfloor\frac{s}{4}\rfloor - 2\lfloor\frac{s}{4}\rfloor + \lceil\frac{s}{2}\rceil + r - 4 & if\ r \equiv 1\ (mod\ 3), \\ 3\lceil\frac{r}{3}\rceil\lfloor\frac{s}{4}\rfloor - \lfloor\frac{s}{4}\rfloor + \lceil\frac{s}{2}\rceil + r - 4 & if\ r \equiv 2\ (mod\ 3). \end{cases}$$

Hence the result follows. □

Theorem 5 *If* $r \geq 9, r \equiv 1\ (mod\ 4)$ *and* $s \equiv 0\ (mod\ 4)$, *then* $\gamma_{TCC}(P_r \boxtimes P_s) = \begin{cases} 3\lceil\frac{r}{3}\rceil\frac{s}{4} + \frac{s}{2} - 2 & if\ r \equiv 0\ (mod\ 3), \\ 3\lceil\frac{r}{3}\rceil\frac{s}{4} - 2 & if\ n \equiv 1\ (mod\ 3), \\ 3\lceil\frac{r}{3}\rceil\frac{s}{4} + \frac{s}{4} - 2 & if\ r \equiv 2\ (mod\ 3). \end{cases}$

Proof Take $S_1 = \{v_{ij} : i \equiv 2\ (mod\ 3),\ j \equiv 2\ (mod\ 4),\ 2 \leq i \leq r-1,\ 2 \leq j \leq s-1\}$, $S_2 = \{v_{ij} : i \equiv 0\ or\ 1\ (mod\ 3),\ j \equiv 3\ (mod\ 4),\ 2 \leq i \leq r-1,\ 2 \leq j \leq s-1\}$, $S_3 = \{v_{r-1,j} : j \equiv 4\ or\ 5\ (mod\ 8),\ 2 \leq j \leq s-1\} \cup \{v_{r-1,j} : j \equiv 2\ (mod\ 4),\ 2 \leq j \leq s-1\}$, $S_4 = \{v_{2j} : j \equiv 0\ or\ 1\ (mod\ 8), 2 \leq j \leq s-1\} \cup \{v_{2j} : j \equiv 3\ (mod\ 4),\ 2 \leq j \leq s-1\}$, $S_5 = \{v_{r-1,j} : j \equiv 4\ or\ 5\ (mod\ 8),\ 2 \leq j \leq s-1\} \cup \{v_{r-1,j} : j \equiv 3\ (mod\ 4),\ 2 \leq j \leq s-1\}$.

Then $S = \begin{cases} S_1 \cup S_2 \cup S_4 \cup S_5 \text{ if } r \equiv 0\ (mod\ 3), \\ S_1 \cup S_2 \cup S_3 \cup S_4 \text{ if } r \equiv 1\ or\ 2\ (mod\ 3), \end{cases}$ is a TCCD set of $P_r \boxtimes P_s$ and hence $\gamma_{TCC}(P_r \boxtimes P_s) \leq |S| = \begin{cases} 3\lceil\frac{r}{3}\rceil\frac{s}{4} + \frac{s}{2} - 2 & if\ r \equiv 0\ (mod\ 3), \\ 3\lceil\frac{r}{3}\rceil\frac{s}{4} - 2 & if\ r \equiv 1\ (mod\ 3), \\ 3\lceil\frac{r}{3}\rceil\frac{s}{4} + \frac{s}{4} - 2 & if\ r \equiv 2\ (mod\ 3). \end{cases}$

Let $S^{'}$ be a TCCD set of $P_r \boxtimes P_s$. Since any dominating set D of cardinality at most $k = \begin{cases} 3\lceil\frac{r}{3}\rceil\frac{s}{4} + \frac{s}{2} - 3 & if\ r \equiv 0\ (mod\ 3), \\ 3\lceil\frac{r}{3}\rceil\frac{s}{4} - 3 & if\ r \equiv 1\ (mod\ 3), \\ 3\lceil\frac{r}{3}\rceil\frac{s}{4} + \frac{s}{4} - 3 & if\ r \equiv 2\ (mod\ 3), \end{cases}$ whose induced subgraph $< D >$ is not triple connected, then we have $|S^{'}| \geq k + 1 = \begin{cases} 3\lceil\frac{r}{3}\rceil\frac{s}{4} + \frac{s}{2} - 2 & if\ r \equiv 0\ (mod\ 3), \\ 3\lceil\frac{r}{3}\rceil\frac{s}{4} - 2 & if\ r \equiv 1\ (mod\ 3), \\ 3\lceil\frac{r}{3}\rceil\frac{s}{4} + \frac{s}{4} - 2 & if\ r \equiv 2\ (mod\ 3). \end{cases}$

Hence the result follows. □

Theorem 6 *If $r \geq 9, r \equiv 1\ (mod\ 4)$ and $s \equiv 1 (mod\ 4)$, then $\gamma_{TCC}(P_r \boxtimes P_s) =$*

$$\begin{cases} 3\lfloor\frac{r}{4}\rfloor\lceil\frac{s}{3}\rceil + \lfloor\frac{r}{2}\rfloor + s - 4 & if\ s \equiv 0\ (mod\ 3), \\ 3\lfloor\frac{r}{4}\rfloor\lceil\frac{s}{3}\rceil - 2\lfloor\frac{r}{4}\rfloor + \lfloor\frac{r}{2}\rfloor + s - 4 & if\ s \equiv 1\ (mod\ 3), \\ 3\lfloor\frac{r}{4}\rfloor\lceil\frac{s}{3}\rceil - \lfloor\frac{r}{4}\rfloor + \lfloor\frac{r}{2}\rfloor + s - 4 & if\ s \equiv 2\ (mod\ 3). \end{cases}$$

Proof Take $S_1 = \{v_{ij}\ : i \equiv 2\ (mod\ 4),\ j \equiv 2\ (mod\ 3),\ 2 \leq i \leq r-1,\ 2 \leq j \leq s-1\}$, $S_2 = \{v_{ij}\ : i \equiv 3\ (mod\ 4),\ j \equiv 0\ or\ 1\ (mod\ 3),\ 2 \leq i \leq r-1,\ 2 \leq j \leq s-1\}$, $S_3 = \{v_{i,s-1}\ : i \equiv 4\ or\ 5\ (mod\ 8),\ 2 \leq i \leq r-1\} \cup \{v_{i,s-1}\ : i \equiv 3\ (mod\ 4),\ 2 \leq i \leq r-1\}$, $S_4 = \{v_{i2}\ : i \equiv 0\ or\ 1\ (mod\ 8), 2 \leq i \leq r-1\} \cup \{v_{i2}\ : i \equiv 3\ (mod\ 4),\ 2 \leq i \leq r-1\}$, $S_5 = \{v_{r-1,j} : 2 \leq j \leq s-1\}$, $S_6 = \{v_{i,s-1}\ : i \equiv 4\ or\ 5\ (mod\ 8),\ 2 \leq i \leq r-1\} \cup \{v_{i,s-1}\ : i \equiv 2\ (mod\ 4),\ 2 \leq i \leq r-1\}$. Then $S = \begin{cases} S_1 \cup S_2 \cup S_3 \cup S_4 \cup S_5 & if\ s \equiv 0\ (mod\ 3), \\ S_1 \cup S_2 \cup S_4 \cup S_5 \cup S_6 & if\ s \equiv 1\ or\ 2\ (mod\ 3), \end{cases}$ is a TCCD set of $P_r \boxtimes P_s$ and

hence $\gamma_{TCC}(P_r \boxtimes P_s) \leq |S| = \begin{cases} 3\lfloor\frac{r}{4}\rfloor\lceil\frac{s}{3}\rceil + \lfloor\frac{r}{2}\rfloor + s - 4 & if\ s \equiv 0\ (mod\ 3), \\ 3\lfloor\frac{r}{4}\rfloor\lceil\frac{s}{3}\rceil - 2\lfloor\frac{r}{4}\rfloor + \lfloor\frac{r}{2}\rfloor + s - 4 & if\ s \equiv 1\ (mod\ 3), \\ 3\lfloor\frac{r}{4}\rfloor\lceil\frac{s}{3}\rceil - \lfloor\frac{r}{4}\rfloor + \lfloor\frac{r}{2}\rfloor + s - 4 & if\ s \equiv 2\ (mod\ 3). \end{cases}$

Let $S^{'}$ be a TCCD set of $P_r \boxtimes P_s$. Since any dominating set D of cardinality at most $k = \begin{cases} 3\lfloor\frac{r}{4}\rfloor\lceil\frac{s}{3}\rceil + \lfloor\frac{r}{2}\rfloor + s - 5 & if\ s \equiv 0\ (mod\ 3), \\ 3\lfloor\frac{r}{4}\rfloor\lceil\frac{s}{3}\rceil - 2\lfloor\frac{r}{4}\rfloor + \lfloor\frac{r}{2}\rfloor + s - 5 & if\ s \equiv 1\ (mod\ 3), \\ 3\lfloor\frac{r}{4}\rfloor\lceil\frac{s}{3}\rceil - \lfloor\frac{r}{4}\rfloor + \lfloor\frac{r}{2}\rfloor + s - 5 & if\ s \equiv 2\ (mod\ 3), \end{cases}$ whose induced subgraph $< D >$ is not triple connected, then we have

$$|S^{'}| \geq k + 1 = \begin{cases} 3\lfloor\frac{r}{4}\rfloor\lceil\frac{s}{3}\rceil + \lfloor\frac{r}{2}\rfloor + s - 4 & if\ s \equiv 0\ (mod\ 3), \\ 3\lfloor\frac{r}{4}\rfloor\lceil\frac{s}{3}\rceil - 2\lfloor\frac{r}{4}\rfloor + \lfloor\frac{r}{2}\rfloor + s - 4 & if\ s \equiv 1\ (mod\ 3), \\ 3\lfloor\frac{r}{4}\rfloor\lceil\frac{s}{3}\rceil - \lfloor\frac{r}{4}\rfloor + \lfloor\frac{r}{2}\rfloor + s - 4 & if\ s \equiv 2\ (mod\ 3). \end{cases}$$

Hence the result follows. □

Theorem 7 *If $r \geq 9, r \equiv 1\ (mod\ 4)$ and $s \equiv 2\ or\ 3\ (mod\ 4)$, then $\gamma_{TCC}(P_r \boxtimes P_s) = \begin{cases} 3\lceil\frac{r}{3}\rceil\lfloor\frac{s}{4}\rfloor + \lceil\frac{s}{2}\rceil + r - 4 & if\ r \equiv 0\ (mod\ 3), \\ 3\lceil\frac{r}{3}\rceil\lfloor\frac{s}{4}\rfloor - 2\lfloor\frac{s}{4}\rfloor + \lfloor\frac{s}{2}\rfloor + r - 4 & if\ r \equiv 1\ (mod\ 3), \\ 3\lceil\frac{r}{3}\rceil\lfloor\frac{s}{4}\rfloor - \lfloor\frac{s}{4}\rfloor + \lfloor\frac{s}{2}\rfloor + r - 4 & if\ r \equiv 2\ (mod\ 3). \end{cases}$*

Proof Take $S_1 = \{v_{ij}\ : i \equiv 2\ (mod\ 3),\ j \equiv 2\ (mod\ 4),\ 2 \leq i \leq r-1,\ 2 \leq j \leq s-2\}$, $S_2 = \{v_{ij}\ : i \equiv 0\ or\ 1\ (mod\ 3),\ j \equiv 3\ (mod\ 4),\ 2 \leq i \leq r-1,\ 2 \leq$

$j \leq s-1\}$, $S_3 = \{v_{r-1,j} : j \equiv 4\ or\ 5\ (mod\ 8),\ 2 \leq j \leq s-2\} \cup \{v_{r-1,j} : j \equiv 3\ (mod\ 4),\ 2 \leq j \leq s-2\}$, $S_4 = \{v_{2j} : j \equiv 0\ or\ 1\ (mod\ 8),\ 2 \leq j \leq s-2\} \cup \{v_{2j} : j \equiv 3\ (mod\ 4),\ 2 \leq j \leq s-2\}$, $S_5 = \{v_{i,s-1} : 2 \leq i \leq r-1\}$, $S_6 = \{v_{r-1,j} : j \equiv 4\ or\ 5\ (mod\ 8),\ 2 \leq j \leq s-2\} \cup \{v_{r-1,j} : j \equiv 2\ (mod\ 4),\ 2 \leq j \leq s-2\}$.

Then $S = \begin{cases} S_1 \cup S_2 \cup S_3 \cup S_4 \cup S_5 & if\ r \equiv 0\ (mod\ 3), \\ S_1 \cup S_2 \cup S_4 \cup S_5 \cup S_6 & if\ r \equiv 1\ or\ 2\ (mod\ 3), \end{cases}$ is a TCCD set of $P_r \boxtimes P_s$ and hence $\gamma_{TCC}(P_r \boxtimes P_s) \leq |S| =$
$\begin{cases} 3\lceil \frac{r}{3} \rceil \lfloor \frac{s}{4} \rfloor + \lceil \frac{s}{2} \rceil + r - 4 & if\ r \equiv 0\ (mod\ 3), \\ 3\lceil \frac{r}{3} \rceil \lfloor \frac{s}{4} \rfloor - 2\lfloor \frac{s}{4} \rfloor + \lfloor \frac{s}{2} \rfloor + r - 4 & if\ r \equiv 1\ (mod\ 3), \\ 3\lceil \frac{r}{3} \rceil \lfloor \frac{s}{4} \rfloor - \lfloor \frac{s}{4} \rfloor + \lfloor \frac{s}{2} \rfloor + r - 4 & if\ r \equiv 2\ (mod\ 3). \end{cases}$

Let S' be a TCCD set of $P_r \boxtimes P_s$. Since any dominating set D of cardinality at most $k = \begin{cases} 3\lceil \frac{r}{3} \rceil \lfloor \frac{s}{4} \rfloor + \lceil \frac{s}{2} \rceil + r - 5 & if\ r \equiv 0\ (mod\ 3), \\ 3\lceil \frac{r}{3} \rceil \lfloor \frac{s}{4} \rfloor - 2\lfloor \frac{s}{4} \rfloor + \lfloor \frac{s}{2} \rfloor + r - 5 & if\ r \equiv 1\ (mod\ 3), \\ 3\lceil \frac{r}{3} \rceil \lfloor \frac{s}{4} \rfloor - \lfloor \frac{s}{4} \rfloor + \lfloor \frac{s}{2} \rfloor + r - 5 & if\ r \equiv 2\ (mod\ 3), \end{cases}$ whose induced subgraph $< D >$ is not triple connected, then we have

$$|S'| \geq k+1 = \begin{cases} 3\lceil \frac{r}{3} \rceil \lfloor \frac{s}{4} \rfloor + \lceil \frac{s}{2} \rceil + r - 4 & if\ r \equiv 0\ (mod\ 3), \\ 3\lceil \frac{r}{3} \rceil \lfloor \frac{s}{4} \rfloor - 2\lfloor \frac{s}{4} \rfloor + \lfloor \frac{s}{2} \rfloor + r - 4 & if\ r \equiv 1\ (mod\ 3), \\ 3\lceil \frac{r}{3} \rceil \lfloor \frac{s}{4} \rfloor - \lfloor \frac{s}{4} \rfloor + \lfloor \frac{s}{2} \rfloor + r - 4 & if\ r \equiv 2\ (mod\ 3). \end{cases}$$

Hence the result follows. □

Theorem 8 *If* $r \geq 10$, $r \equiv 2\ (mod\ 4)$ *and* $s \equiv 0\ (mod\ 4)$, *then* $\gamma_{TCC}(P_r \boxtimes P_s) =$
$\begin{cases} 3\lceil \frac{r}{3} \rceil \frac{s}{4} + \frac{s}{2} - 2 & if\ r \equiv 0\ (mod\ 3), \\ 3\lceil \frac{r}{3} \rceil \frac{s}{4} - 2 & if\ r \equiv 1\ (mod\ 3), \\ 3\lceil \frac{r}{3} \rceil \frac{s}{4} + \frac{s}{4} - 2 & if\ r \equiv 2\ (mod\ 3). \end{cases}$

Proof Take $S_1 = \{v_{ij} : i \equiv 2\ (mod\ 3),\ j \equiv 2\ (mod\ 4),\ 2 \leq i \leq r-1,\ 2 \leq j \leq s-1\}$, $S_2 = \{v_{ij} : i \equiv 0\ or\ 1\ (mod\ 3),\ j \equiv 3\ (mod\ 4),\ 2 \leq i \leq r-1,\ 2 \leq j \leq s-1\}$, $S_3 = \{v_{r-1,j} : j \equiv 4\ or\ 5\ (mod\ 8),\ 2 \leq j \leq s-1\} \cup \{v_{r-1,j} : j \equiv 3\ (mod\ 4),\ 2 \leq j \leq s-1\}$, $S_4 = \{v_{2j} : j \equiv 0\ or\ 1\ (mod\ 8),\ 2 \leq j \leq s-1\} \cup \{v_{2j} : j \equiv 3\ (mod\ 4),\ 2 \leq j \leq s-1\}$, $S_5 = \{v_{r-1,j} : j \equiv 4\ or\ 5\ (mod\ 8),\ 2 \leq j \leq s-1\} \cup \{v_{r-1,j} : j \equiv 2\ (mod\ 4),\ 2 \leq j \leq s-1\}$.

Then $S = \begin{cases} S_1 \cup S_2 \cup S_3 \cup S_4 & if\ r \equiv 0\ (mod\ 3), \\ S_1 \cup S_2 \cup S_4 \cup S_5 & if\ r \equiv 1\ or\ 2\ (mod\ 3), \end{cases}$ is a TCCD set of $P_r \boxtimes P_s$ and hence $\gamma_{TCC}(P_r \boxtimes P_s) \leq |S| = \begin{cases} 3\lceil \frac{r}{3} \rceil \frac{s}{4} + \frac{s}{2} - 2 & if\ r \equiv 0\ (mod\ 3), \\ 3\lceil \frac{r}{3} \rceil \frac{s}{4} - 2 & if\ r \equiv 1\ (mod\ 3), \\ 3\lceil \frac{r}{3} \rceil \frac{s}{4} + \frac{s}{4} - 2 & if\ r \equiv 2\ (mod\ 3). \end{cases}$

Let S' be a TCCD set of $P_r \boxtimes P_s$. Since any dominating set D of cardinality at most $k = \begin{cases} 3\lceil \frac{r}{3} \rceil \frac{s}{4} + \frac{s}{2} - 3 & if\ r \equiv 0\ (mod\ 3), \\ 3\lceil \frac{r}{3} \rceil \frac{s}{4} - 3 & if\ r \equiv 1\ (mod\ 3), \\ 3\lceil \frac{r}{3} \rceil \frac{s}{4} + \frac{s}{4} - 3 & if\ r \equiv 2\ (mod\ 3), \end{cases}$ whose induced subgraph $< D >$ is not triple connected, then we have $|S'| \geq k + 1 = \begin{cases} 3\lceil \frac{r}{3} \rceil \frac{s}{4} + \frac{s}{2} - 2 \; if \; r \equiv 0 \; (mod \; 3), \\ 3\lceil \frac{r}{3} \rceil \frac{s}{4} - 2 \quad if \; r \equiv 1 \; (mod \; 3), \\ 3\lceil \frac{r}{3} \rceil \frac{s}{4} + \frac{s}{4} - 2 \; if \; r \equiv 2 \; (mod \; 3). \end{cases}$

Hence the result follows. □

Theorem 9 *If* $r \geq 10, r \equiv 2 \; (mod \; 4)$ *and* $s \equiv 1 (mod \; 4)$, *then* $\gamma_{TCC}(P_r \boxtimes P_s) = \begin{cases} 3\lfloor \frac{r}{4} \rfloor \lceil \frac{s}{3} \rceil + \frac{r}{2} + s - 4 & if \; s \equiv 0 \; (mod \; 3), \\ 3\lfloor \frac{r}{4} \rfloor \lceil \frac{s}{3} \rceil - 2\lfloor \frac{r}{4} \rfloor + \frac{r}{2} + s - 4 & if \; s \equiv 1 \; (mod \; 3), \\ 3\lfloor \frac{r}{4} \rfloor \lceil \frac{s}{3} \rceil - \lfloor \frac{r}{4} \rfloor + \frac{r}{2} + s - 4 & if \; s \equiv 2 \; (mod \; 3). \end{cases}$

Proof Take $S_1 = \{v_{ij} : i \equiv 2 \; (mod \; 4), \; j \equiv 2 \; (mod \; 3), \; 2 \leq i \leq r-1, \; 2 \leq j \leq s-1\}$, $S_2 = \{v_{ij} : i \equiv 3 \; (mod \; 4), \; j \equiv 0 \; or \; 1 \; (mod \; 3), \; 2 \leq i \leq r-1, \; 2 \leq j \leq s-1\}$, $S_3 = \{v_{i,s-1} : i \equiv 4 \; or \; 5 \; (mod \; 8), \; 2 \leq i \leq r-2\} \cup \{v_{i,s-1} : i \equiv 2 \; (mod \; 4), \; 2 \leq i \leq r-2\}$, $S_4 = \{v_{i2} : i \equiv 0 \; or \; 1 \; (mod \; 8), \; 2 \leq i \leq r-2\} \cup \{v_{i2} : i \equiv 3 \; (mod \; 4), \; 2 \leq i \leq r-1\}$, $S_5 = \{v_{r-1,j} : 2 \leq j \leq s-1\}$, $S_6 = \{v_{i,s-1} : i \equiv 4 \; or \; 5 \; (mod \; 8), \; 2 \leq i \leq r-2\} \cup \{v_{i,s-1} : i \equiv 3 \; (mod \; 4), \; 2 \leq i \leq r-2\}$.

Then $S = \begin{cases} S_1 \cup S_2 \cup S_4 \cup S_5 \cup S_6 & if \; s \equiv 0 \; (mod \; 3), \\ S_1 \cup S_2 \cup S_3 \cup S_4 \cup S_5 & if \; s \equiv 1 \; or \; 2 \; (mod \; 3), \end{cases}$

is a TCCD set of $P_r \boxtimes P_s$ and hence $\gamma_{TCC}(P_r \boxtimes P_s) \leq |S| = \begin{cases} 3\lfloor \frac{r}{4} \rfloor \lceil \frac{s}{3} \rceil + \frac{r}{2} + s - 4 & if \; s \equiv 0 \; (mod \; 3), \\ 3\lfloor \frac{r}{4} \rfloor \lceil \frac{s}{3} \rceil - 2\lfloor \frac{r}{4} \rfloor + \frac{r}{2} + s - 4 & if \; s \equiv 1 \; (mod \; 3), \\ 3\lfloor \frac{r}{4} \rfloor \lceil \frac{s}{3} \rceil - \lfloor \frac{r}{4} \rfloor + \frac{r}{2} + s - 4 & if \; s \equiv 2 \; (mod \; 3). \end{cases}$

Let S' be a TCCD set of $P_r \boxtimes P_s$. Since any dominating set D of cardinality at most $k = \begin{cases} 3\lfloor \frac{r}{4} \rfloor \lceil \frac{s}{3} \rceil + \frac{r}{2} + s - 5 & if \; s \equiv 0 \; (mod \; 3), \\ 3\lfloor \frac{r}{4} \rfloor \lceil \frac{s}{3} \rceil - 2\lfloor \frac{r}{4} \rfloor + \frac{r}{2} + s - 5 & if \; s \equiv 1 \; (mod \; 3), \\ 3\lfloor \frac{r}{4} \rfloor \lceil \frac{s}{3} \rceil - \lfloor \frac{r}{4} \rfloor + \frac{r}{2} + s - 5 & if \; s \equiv 2 \; (mod \; 3), \end{cases}$ whose induced subgraph $< D >$ is not triple connected, then we have

$$|S'| \geq k + 1 = \begin{cases} 3\lfloor \frac{r}{4} \rfloor \lceil \frac{s}{3} \rceil + \frac{r}{2} + s - 4 & if \; s \equiv 0 \; (mod \; 3), \\ 3\lfloor \frac{r}{4} \rfloor \lceil \frac{s}{3} \rceil - 2\lfloor \frac{r}{4} \rfloor + \frac{r}{2} + s - 4 & if \; s \equiv 1 \; (mod \; 3), \\ 3\lfloor \frac{r}{4} \rfloor \lceil \frac{s}{3} \rceil - \lfloor \frac{r}{4} \rfloor + \frac{r}{2} + s - 4 & if \; s \equiv 2 \; (mod \; 3). \end{cases}$$

Hence the result follows. □

Theorem 10 *If* $r \geq 10, r \equiv 2 \; (mod \; 4)$ *and* $s \equiv 2 \; or \; 3 \; (mod \; 4)$, *then* $\gamma_{TCC}(P_r \boxtimes P_s) = \begin{cases} 3\lceil \frac{r}{3} \rceil \lfloor \frac{s}{4} \rfloor + \lceil \frac{s}{2} \rceil + r - 4 & if \; r \equiv 0 \; (mod \; 3), \\ 3\lceil \frac{r}{3} \rceil \lfloor \frac{s}{4} \rfloor - 2\lfloor \frac{s}{4} \rfloor + \lceil \frac{s}{2} \rceil + r - 4 & if \; r \equiv 1 \; (mod \; 3), \\ 3\lceil \frac{r}{3} \rceil \lfloor \frac{s}{4} \rfloor - \lfloor \frac{s}{4} \rfloor + \lceil \frac{s}{2} \rceil + r - 4 & if \; r \equiv 2 \; (mod \; 3). \end{cases}$

Proof **Case 1:** $r \equiv 2 \; (mod \; 4)$ and $s \equiv 2 \; (mod \; 4)$

Take $S_1 = \{v_{ij} : i \equiv 2 \; (mod \; 3), \; j \equiv 2 \; (mod \; 4), \; 2 \leq i \leq r-1, \; 2 \leq j \leq s-2\}$, $S_2 = \{v_{ij} : i \equiv 0 \; or \; 1 \; (mod \; 3), \; j \equiv 3 \; (mod \; 4), \; 2 \leq i \leq r-1, \; 2 \leq j \leq s-1\}$, $S_3 = \{v_{r-1,j} : j \equiv 4 \; or \; 5 \; (mod \; 8), \; 2 \leq j \leq s-1\} \cup \{v_{r-1,j} : j \equiv 3 \; (mod \; 4), \; 2 \leq j \leq s-1\}$, $S_4 = \{v_{2j} : j \equiv 0 \; or \; 1 \; (mod \; 8), \; 2 \leq j \leq s-2\} \cup \{v_{2j} : j \equiv 3 \; (mod \; 4), \; 2 \leq j \leq s-1\}$, $S_5 = \{v_{i,s-1} : 2 \leq i \leq r-1\}$, $S_6 = \{v_{r-1,j} : j \equiv 4 \; or \; 5 \; (mod \; 8), \; 2 \leq j \leq s-2\} \cup \{v_{r-1,j} : j \equiv 2 \; (mod \; 4), \; 2 \leq j \leq s-1\}$.

Then $S = \begin{cases} S_1 \cup S_2 \cup S_3 \cup S_4 \cup S_5 & if\ r \equiv 0\ (mod\ 3), \\ S_1 \cup S_2 \cup S_4 \cup S_5 \cup S_6 & if\ r \equiv 1\ or\ 2\ (mod\ 3), \end{cases}$

Case 2: $r \equiv 2\ (mod\ 4)$ and $s \equiv 3\ (mod\ 4)$

Take $S_1 = \{v_{ij} : i \equiv 2\ (mod\ 3),\ j \equiv 2\ (mod\ 4),\ 2 \le i \le r-1,\ 2 \le j \le s-2\}$, $S_2 = \{v_{ij} : i \equiv 0\ or\ 1\ (mod\ 3),\ j \equiv 3\ (mod\ 4),\ 2 \le i \le r-1,\ 2 \le j \le s-1\}$, $S_3 = \{v_{r-1,j} : j \equiv 4\ or\ 5\ (mod\ 8),\ 2 \le j \le s-2\} \cup \{v_{r-1,j} : j \equiv 3\ (mod\ 4),\ 2 \le j \le s-2\}$, $S_4 = \{v_{2j} : j \equiv 0\ or\ 1\ (mod\ 8),\ 2 \le j \le s-2\} \cup \{v_{2j} : j \equiv 3\ (mod\ 4),\ 2 \le j \le s-2\}$, $S_5 = \{v_{i,s-1} : 2 \le i \le r-1\}$, $S_6 = \{v_{r-1,j} : j \equiv 4\ or\ 5\ (mod\ 8),\ 2 \le j \le s-2\} \cup \{v_{r-1,j} : j \equiv 2\ (mod\ 4),\ 2 \le j \le s-2\}$.

Then $S = \begin{cases} S_1 \cup S_2 \cup S_3 \cup S_4 \cup S_5 & if\ r \equiv 0\ (mod\ 3), \\ S_1 \cup S_2 \cup S_4 \cup S_5 \cup S_6 & if\ r \equiv 1\ or\ 2\ (mod\ 3), \end{cases}$ are a TCCD set of $P_r \boxtimes P_s$ and hence

$$\gamma_{TCC}(P_n \boxtimes P_m) \le |S| = \begin{cases} 3\lceil \frac{r}{3} \rceil \lfloor \frac{s}{4} \rfloor + \lceil \frac{s}{2} \rceil + r - 4 & if\ r \equiv 0\ (mod\ 3), \\ 3\lceil \frac{r}{3} \rceil \lfloor \frac{s}{4} \rfloor - 2\lfloor \frac{s}{4} \rfloor + \lceil \frac{s}{2} \rceil + r - 4 & if\ r \equiv 1\ (mod\ 3), \\ 3\lceil \frac{r}{3} \rceil \lfloor \frac{s}{4} \rfloor - \lfloor \frac{s}{4} \rfloor + \lceil \frac{s}{2} \rceil + r - 4 & if\ r \equiv 2\ (mod\ 3). \end{cases}$$

Let S' be a TCCD set of $P_r \boxtimes P_s$. Since any dominating set D of cardinality at most $k = \begin{cases} 3\lceil \frac{r}{3} \rceil \lfloor \frac{s}{4} \rfloor + \lceil \frac{s}{2} \rceil + r - 5 & if\ r \equiv 0\ (mod\ 3), \\ 3\lceil \frac{r}{3} \rceil \lfloor \frac{s}{4} \rfloor - 2\lfloor \frac{s}{4} \rfloor + \lceil \frac{s}{2} \rceil + r - 5 & if\ r \equiv 1\ (mod\ 3), \\ 3\lceil \frac{r}{3} \rceil \lfloor \frac{s}{4} \rfloor - \lfloor \frac{s}{4} \rfloor + \lceil \frac{s}{2} \rceil + r - 5 & if\ r \equiv 2\ (mod\ 3), \end{cases}$ whose induced subgraph $< D >$ is not triple connected, then we have

$$|S'| \ge k + 1 = \begin{cases} 3\lceil \frac{r}{3} \rceil \lfloor \frac{s}{4} \rfloor + \lceil \frac{s}{2} \rceil + r - 4 & if\ r \equiv 0\ (mod\ 3), \\ 3\lceil \frac{r}{3} \rceil \lfloor \frac{s}{4} \rfloor - 2\lfloor \frac{s}{4} \rfloor + \lceil \frac{s}{2} \rceil + r - 4 & if\ r \equiv 1\ (mod\ 3), \\ 3\lceil \frac{r}{3} \rceil \lfloor \frac{s}{4} \rfloor - \lfloor \frac{s}{4} \rfloor + \lceil \frac{s}{2} \rceil + r - 4 & if\ r \equiv 2\ (mod\ 3). \end{cases}$$

Hence the result follows. □

Theorem 11 *If* $r \ge 7, r \equiv 3\ (mod\ 4)$ *and* $s \equiv 0\ (mod\ 4)$, *then* $\gamma_{TCC}(P_r \boxtimes P_s) = \begin{cases} 3\lceil \frac{r}{3} \rceil \frac{s}{4} + \frac{s}{2} - 2 & if\ r \equiv 0\ (mod\ 3), \\ 3\lceil \frac{r}{3} \rceil \frac{s}{4} - 2 & if\ r \equiv 1\ (mod\ 3), \\ 3\lceil \frac{r}{3} \rceil \frac{s}{4} + \frac{s}{4} - 2 & if\ r \equiv 2\ (mod\ 3). \end{cases}$

Proof Take $S_1 = \{v_{ij} : i \equiv 2\ (mod\ 3),\ j \equiv 2\ (mod\ 4),\ 2 \le i \le r-1,\ 2 \le j \le s-1\}$, $S_2 = \{v_{ij} : i \equiv 0\ or\ 1\ (mod\ 3),\ j \equiv 3\ (mod\ 4),\ 2 \le i \le r-1,\ 2 \le j \le s-1\}$, $S_3 = \{v_{r-1,j} : j \equiv 4\ or\ 5\ (mod\ 8),\ 2 \le j \le s-1\} \cup \{v_{r-1,j} : j \equiv 3\ (mod\ 4),\ 2 \le j \le s-1\}$, $S_4 = \{v_{2j} : j \equiv 0\ or\ 1\ (mod\ 8),\ 2 \le j \le s-1\} \cup \{v_{2j} : j \equiv 3\ (mod\ 4),\ 2 \le j \le s-1\}$, $S_5 = \{v_{r-1,j} : j \equiv 4\ or\ 5\ (mod\ 8),\ 2 \le j \le s-1\} \cup \{v_{r-1,j} : j \equiv 2\ (mod\ 4),\ 2 \le j \le s-1\}$.

Then $S = \begin{cases} S_1 \cup S_2 \cup S_3 \cup S_4 & if\ r \equiv 0\ (mod\ 3), \\ S_1 \cup S_2 \cup S_4 \cup S_5 & if\ r \equiv 1\ or\ 2\ (mod\ 3), \end{cases}$

is a TCCD set of $P_r \boxtimes P_s$ and hence

$$\gamma_{TCC}(P_n \boxtimes P_m) \le |S| = \begin{cases} 3\lceil \frac{r}{3} \rceil \frac{s}{4} + \frac{s}{2} - 2 & if\ r \equiv 0\ (mod\ 3), \\ 3\lceil \frac{r}{3} \rceil \frac{s}{4} - 2 & if\ r \equiv 1\ (mod\ 3), \\ 3\lceil \frac{r}{3} \rceil \frac{s}{4} + \frac{s}{4} - 2 & if\ r \equiv 2\ (mod\ 3). \end{cases}$$

Let S' be a TCCD set of $P_r \boxtimes P_s$. Since any dominating set D of cardinality at most $k = \begin{cases} 3\lceil\frac{r}{3}\rceil\frac{s}{4} + \frac{s}{2} - 3 & if\ r \equiv 0\ (mod\ 3), \\ 3\lceil\frac{r}{3}\rceil\frac{s}{4} - 3 & if\ r \equiv 1\ (mod\ 3), \\ 3\lceil\frac{r}{3}\rceil\frac{s}{4} + \frac{s}{4} - 3 & if\ r \equiv 2\ (mod\ 3), \end{cases}$ whose induced subgraph $< D >$ is not triple connected, then we have

$$|S'| \geq k+1 = \begin{cases} 3\lceil\frac{r}{3}\rceil\frac{s}{4} + \frac{s}{2} - 2 & if\ r \equiv 0\ (mod\ 3), \\ 3\lceil\frac{r}{3}\rceil\frac{s}{4} - 2 & if\ r \equiv 1\ (mod\ 3), \\ 3\lceil\frac{r}{3}\rceil\frac{s}{4} + \frac{s}{4} - 2 & if\ r \equiv 2\ (mod\ 3). \end{cases}$$

Hence the result follows. □

Theorem 12 *If $r \geq 7, r \equiv 3\ (mod\ 4)$ and $s \equiv 1$ or $2\ (mod\ 4)$, then* $\gamma_{TCC}(P_r \boxtimes P_s) = \begin{cases} 3\lfloor\frac{r}{4}\rfloor\lceil\frac{s}{3}\rceil + \lceil\frac{r}{2}\rceil + s - 4 & if\ s \equiv 0\ (mod\ 3), \\ 3\lfloor\frac{r}{4}\rfloor\lceil\frac{s}{3}\rceil - 2\lfloor\frac{r}{4}\rfloor + \lceil\frac{r}{2}\rceil + s - 4 & if\ s \equiv 1\ (mod\ 3), \\ 3\lfloor\frac{r}{4}\rfloor\lceil\frac{s}{3}\rceil - \lfloor\frac{r}{4}\rfloor + \lceil\frac{r}{2}\rceil + s - 4 & if\ s \equiv 2\ (mod\ 3). \end{cases}$

Proof Take $S_1 = \{v_{ij}\ : i \equiv 2\ (mod\ 4),\ j \equiv 2\ (mod\ 3),\ 2 \leq i \leq r-2,\ 2 \leq j \leq s-1\}$, $S_2 = \{v_{ij}\ : i \equiv 3\ (mod\ 4),\ j \equiv 0$ or $1\ (mod\ 3),\ 2 \leq i \leq r-1,\ 2 \leq j \leq s-1\}$, $S_3 = \{v_{i,s-1}\ : i \equiv 4$ or $5\ (mod\ 8),\ 2 \leq i \leq r-2\} \cup \{v_{i,s-1}\ : i \equiv 2\ (mod\ 4),\ 2 \leq i \leq r-2\}$, $S_4 = \{v_{i2} : i \equiv 0$ or $1\ (mod\ 8),\ 2 \leq i \leq r-1\} \cup \{v_{i2} : i \equiv 3\ (mod\ 4),\ 2 \leq i \leq r-1\}$, $S_5 = \{v_{r-1,j} : 2 \leq j \leq s-1\}$, $S_6 = \{v_{i,s-1} : i \equiv 4$ or $5\ (mod\ 8),\ 2 \leq i \leq r-2\} \cup \{v_{i,s-1}\ : i \equiv 3\ (mod\ 4),\ 2 \leq i \leq r-2\}$.

Then $S = \begin{cases} S_1 \cup S_2 \cup S_4 \cup S_5 \cup S_6 & \text{if } r \equiv 0\ (mod\ 3), \\ S_1 \cup S_2 \cup S_3 \cup S_4 \cup S_5 & \text{if } r \equiv 1 \text{ or } 2\ (mod\ 3), \end{cases}$

is a TCCD set of $P_r \boxtimes P_s$ and hence

$$\gamma_{TCC}(P_n \boxtimes P_m) \leq |S| = \begin{cases} 3\lfloor\frac{r}{4}\rfloor\lceil\frac{s}{3}\rceil + \lceil\frac{r}{2}\rceil + s - 4 & if\ s \equiv 0\ (mod\ 3), \\ 3\lfloor\frac{r}{4}\rfloor\lceil\frac{s}{3}\rceil - 2\lfloor\frac{r}{4}\rfloor + \lceil\frac{r}{2}\rceil + s - 4 & if\ s \equiv 1\ (mod\ 3), \\ 3\lfloor\frac{r}{4}\rfloor\lceil\frac{s}{3}\rceil - \lfloor\frac{r}{4}\rfloor + \lceil\frac{r}{2}\rceil + s - 4 & if\ s \equiv 2\ (mod\ 3). \end{cases}$$

Let S' be a TCCD set of $P_r \boxtimes P_s$. Since any dominating set D of cardinality at most $k = \begin{cases} 3\lfloor\frac{r}{4}\rfloor\lceil\frac{s}{3}\rceil + \lceil\frac{r}{2}\rceil + s - 5 & if\ s \equiv 0\ (mod\ 3), \\ 3\lfloor\frac{r}{4}\rfloor\lceil\frac{s}{3}\rceil - 2\lfloor\frac{r}{4}\rfloor + \lceil\frac{r}{2}\rceil + s - 5 & if\ s \equiv 1\ (mod\ 3), \\ 3\lfloor\frac{r}{4}\rfloor\lceil\frac{s}{3}\rceil - \lfloor\frac{r}{4}\rfloor + \lceil\frac{r}{2}\rceil + s - 5 & if\ s \equiv 2\ (mod\ 3), \end{cases}$ whose induced subgraph $< D >$ is not triple connected, then we have

$$|S'| \geq k+1 = \begin{cases} 3\lfloor\frac{r}{4}\rfloor\lceil\frac{s}{3}\rceil + \lceil\frac{r}{2}\rceil + s - 4 & if\ s \equiv 0\ (mod\ 3), \\ 3\lfloor\frac{r}{4}\rfloor\lceil\frac{s}{3}\rceil - 2\lfloor\frac{r}{4}\rfloor + \lceil\frac{r}{2}\rceil + s - 4 & if\ s \equiv 1\ (mod\ 3), \\ 3\lfloor\frac{r}{4}\rfloor\lceil\frac{s}{3}\rceil - \lfloor\frac{r}{4}\rfloor + \lceil\frac{r}{2}\rceil + s - 4 & if\ s \equiv 2\ (mod\ 3). \end{cases}$$

Hence the result follows. □

Theorem 13 *If $r \geq 7, r \equiv 3\ (mod\ 4)$ and $s \equiv 3\ (mod\ 4)$, then* $\gamma_{TCC}(P_r \boxtimes P_s) = \begin{cases} 3\lceil\frac{r}{3}\rceil\lfloor\frac{s}{4}\rfloor + \lceil\frac{s}{2}\rceil + r - 4 & if\ r \equiv 0\ (mod\ 3), \\ 3\lceil\frac{r}{3}\rceil\lfloor\frac{s}{4}\rfloor - 2\lfloor\frac{s}{4}\rfloor + \lceil\frac{s}{2}\rceil + r - 4 & if\ r \equiv 1\ (mod\ 3), \\ 3\lceil\frac{r}{3}\rceil\lfloor\frac{s}{4}\rfloor - \lfloor\frac{s}{4}\rfloor + \lceil\frac{s}{2}\rceil + r - 4 & if\ r \equiv 2\ (mod\ 3). \end{cases}$

Proof Take $S_1 = \{v_{ij}\ : i \equiv 2\ (mod\ 3),\ j \equiv 2\ (mod\ 4),\ 2 \leq i \leq r-1,\ 2 \leq j \leq s-2\}$, $S_2 = \{v_{ij}\ : i \equiv 0$ or $1\ (mod\ 3),\ j \equiv 3\ (mod\ 4),\ 2 \leq i \leq r-1,\ 2 \leq j \leq s-1\}$, $S_3 = \{v_{r-1,j}\ : j \equiv 4$ or $5\ (mod\ 8),\ 2 \leq j \leq s-2\} \cup \{v_{r-1,j}\ : j \equiv 3\ (mod\ 4),\ 2 \leq j \leq s-2\}$, $S_4 = \{v_{2j} : j \equiv 0$ or $1\ (mod\ 8),\ 2 \leq j \leq s-1\} \cup \{v_{2j} :$

$j \equiv 3\ (mod\ 4),\ 2 \leq j \leq s-1\},\ S_5 = \{v_{i,s-1} : 2 \leq i \leq r-1\},\ S_6 = \{v_{r-1,j} : j \equiv 4\ or\ 5\ (mod\ 8),\ 2 \leq j \leq s-2\} \cup \{v_{r-1,j} : j \equiv 2\ (mod\ 4),\ 2 \leq j \leq s-2\}$.

Then $S = \begin{cases} S_1 \cup S_2 \cup S_3 \cup S_4 \cup S_5 & if\ r \equiv 0\ (mod\ 3), \\ S_1 \cup S_2 \cup S_4 \cup S_5 \cup S_6 & if\ r \equiv 1\ or\ 2\ (mod\ 3), \end{cases}$ is a TCCD set of $P_r \boxtimes P_s$ and hence

$$\gamma_{TCC}(P_n \boxtimes P_m) \leq |S| = \begin{cases} 3\lceil \frac{r}{3} \rceil \lfloor \frac{s}{4} \rfloor + \lceil \frac{s}{2} \rceil + r - 4 & if\ r \equiv 0\ (mod\ 3), \\ 3\lceil \frac{r}{3} \rceil \lfloor \frac{s}{4} \rfloor - 2\lfloor \frac{s}{4} \rfloor + \lceil \frac{s}{2} \rceil + r - 4 & if\ r \equiv 1\ (mod\ 3), \\ 3\lceil \frac{r}{3} \rceil \lfloor \frac{s}{4} \rfloor - \lfloor \frac{s}{4} \rfloor + \lceil \frac{s}{2} \rceil + r - 4 & if\ r \equiv 2\ (mod\ 3). \end{cases}$$

Let S' be a TCCD set of $P_r \boxtimes P_s$. Since any dominating set D of cardinality at most $k = \begin{cases} 3\lceil \frac{r}{3} \rceil \lfloor \frac{s}{4} \rfloor + \lceil \frac{s}{2} \rceil + r - 5 & if\ r \equiv 0\ (mod\ 3), \\ 3\lceil \frac{r}{3} \rceil \lfloor \frac{s}{4} \rfloor - 2\lfloor \frac{s}{4} \rfloor + \lceil \frac{s}{2} \rceil + r - 5 & if\ r \equiv 1\ (mod\ 3), \\ 3\lceil \frac{r}{3} \rceil \lfloor \frac{s}{4} \rfloor - \lfloor \frac{s}{4} \rfloor + \lceil \frac{s}{2} \rceil + r - 5 & if\ r \equiv 2\ (mod\ 3), \end{cases}$ whose induced subgraph $< D >$ is not triple connected, then we have

$$|S'| \geq k + 1 = \begin{cases} 3\lceil \frac{r}{3} \rceil \lfloor \frac{s}{4} \rfloor + \lceil \frac{s}{2} \rceil + r - 4 & if\ r \equiv 0\ (mod\ 3), \\ 3\lceil \frac{r}{3} \rceil \lfloor \frac{s}{4} \rfloor - 2\lfloor \frac{s}{4} \rfloor + \lceil \frac{s}{2} \rceil + r - 4 & if\ r \equiv 1\ (mod\ 3), \\ 3\lceil \frac{r}{3} \rceil \lfloor \frac{s}{4} \rfloor - \lfloor \frac{s}{4} \rfloor + \lceil \frac{s}{2} \rceil + r - 4 & if\ r \equiv 2\ (mod\ 3). \end{cases}$$

Hence the result follows. □

4 Conclusion

After discussing the concept of triple connected certified domination here, we have generalized some results on product related graphs for paths and cycles. In future, we will discuss this parameter for some fractal graphs.

References

1. Detlaff, M., Lemanska, M., Topp, J.: Certified domination. AKCE Int. J. Graphs Comb. **17**, 86–97 (2020)
2. Haynes, T.W., Hedetniemi, S.T., Slater, P.J.: Fundamentals of Domination in Graphs. Marcel Dekker Inc., Newyork (1998)
3. Mahadevan, G., Avadayappan Selvam, J., Joseph, P., Subraaniam, T. : Triple connected domination number of a graph. Int. J. Math. Comb. **3**, 93–104 (2012)
4. Mahadevan, G., Kaviya, S., Sivagnanam, C.: Triple Connected Certified Domination in Graphs. Indian J. of Natural Sci. (Accepted).
5. Paulraj Joseph, J., Angel Jebitha, M.K., Chitra Devi, P., Sudhana, G.: Triple connected graphs. Indian J. Math. Math. Sci. **8**(1), 61–75 (2012)
6. Sabidussi, G.: Graph multiplication. Math. Z **72**, 446–457 (1960)
7. Vizing, V.G.: The Cartesian product of Graphs. Vycisl. Sistemy **9**, 30–43 (1963)

Learners' Mental Constructions in Learning Circle Geometry

F. Abakah and D. Brijlall

1 Introduction

Circle geometry has been identified as a difficult-to-teach and a difficult-to-learn Grade 11 mathematical concept in South African schools [5]. Teachers are required to guide Grade 11 mathematics learners to depend on the knowledge of geometry they learnt in earlier grades, to guide their conception of this new mathematical knowledge, that is, circle geometry.

From APOS theory, the ACE teaching cycle is based on two general hypotheses/assumptions: (1) assumption on mathematical knowledge: finding solutions to mathematical problems by applying appropriate mental structures and by approaching them from a social context; and (2) hypothesis on learning: mathematical concepts are not learnt directly but by applying appropriate mental structures to make meaning from it.

From the above postulates, the ACE teaching cycle seeks to find appropriate and effective strategies, which can guide an individual to develop appropriate mental structures, which may be applicable in any given problem-solving situation. This is aimed at guiding students to develop detailed understanding of mathematical concepts by applying their known mental structures to seek for new mathematical knowledge. This study will guide teacher-trainers and teachers to find an efficacious instructional approach for circle geometry – a difficult and problematic mathematical content area for both teachers and learners in South Africa. In fulfillment of the above aim, the following critical research questions were formulated: (1) *What mental constructions do Grade 11 learners demonstrate when solving circle*

F. Abakah · D. Brijlall (✉)
Department of Mathematics, Durban University of Technology, KwaZulu-Natal, South Africa
e-mail: deonarainb@dut.ac.za

F. Kamalov et al. (eds.), *Advances in Mathematical Modeling and Scientific Computing*, Trends in Mathematics, https://doi.org/10.1007/978-3-031-41420-6_78

geometry problems? (2) *How can the mental constructions demonstrated by these learners guide the revision of the preliminary genetic decomposition?*

2 Theoretical Framework

APOS theory was adopted as the theoretical framework for this study. In doing so, how mathematical knowledge, in this case, circle geometry, was formulated through mental constructions: actions, processes, objects, and organizing these in schemas, are sequentially elaborated below, with reference to Fig. 1 below.

In the diagram below, two circles intersect at K and Y. The larger circle passes through O, the centre of the smaller circle. T is a point on the smaller circle such that KT is a tangent to the larger circle. TY produced meets the larger circle at W.
WO produced meets KT at E.

Let $\hat{W}_1 = x$

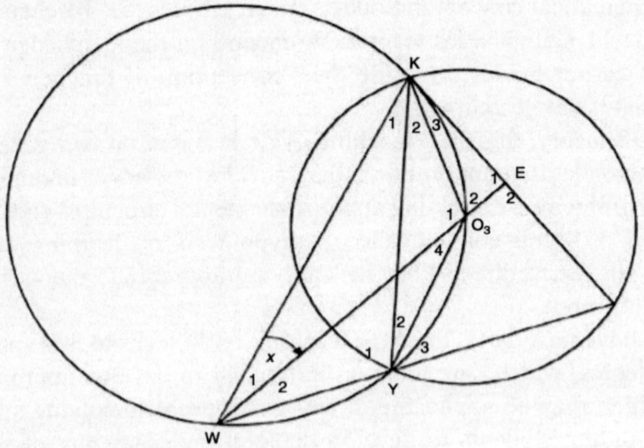

12.1	Determine FOUR other angles, each equal to *x*.	(8)
12.2	Prove that $\hat{T} = 90° - x$	(3)
12.3	Prove that KE = ET.	(3)
12.4	Prove that $KE^2 = OE.WE$	(6) [20]

Fig. 1 Exemplar-mental construction task

2.1 Action-Conception Stage

Question 12.1 (see Fig. 1) typifies action stage of geometry mental construction. Students are required to give four other angles equal to x. Students' ability to provide four other angles depends on them being able to develop explicit expression/stimuli to mainly identify and recognize appropriate relevant geometric properties, axioms, and theorems. Students are required to recall the circle geometry theorems, so that they will be able to identify and recognize the specific theorem(s) which will be applicable, to obtain an appropriate solution to the question under consideration.

2.2 Process-Conception Stage

Question 12.2 (see Fig. 1) is the typical example of the process stage of geometry mental construction. Learners who can prove that $\hat{T} = 90° - x$ may be at the process stage of geometry conception, since they are expected to apply their knowledge of geometry, to interpret the geometric diagram without necessarily following any step-by-step procedures. Learners are required to reflect on the action process of "finding other angles equal to x, which is interiorized mentally by the students, and this can enable them to reverse the action processes (converse of theorems), when necessary, which are mental processes [6].

2.3 Object-Conception Stage

For an example of this level instance (see Fig. 1), learners who can prove that KE = ET, geometrically, can be judged to be at the object stage of solving circle geometry problems. In this case learners will be expected to perform an action (midpoint theorem) on a process conception of finding the angle in a semi-circle. Learners are required to find on their own, appropriate techniques and skills in solving geometry problems intuitively based on his/her imaginations. Learners are required to have encapsulated geometry theorems, converses of geometry theorems, and their effective applications into a cognitive object. The action process of giving four other angles equal to x and the process stage of proving that $\hat{T} = 90° - x$ are expected to be applied, cognitively.

2.4 Schema-Conception Stage

From Fig. 1 above, learners who are capable of providing appropriate solution to Question 12.4 are required to connect the action process of finding equal angles in

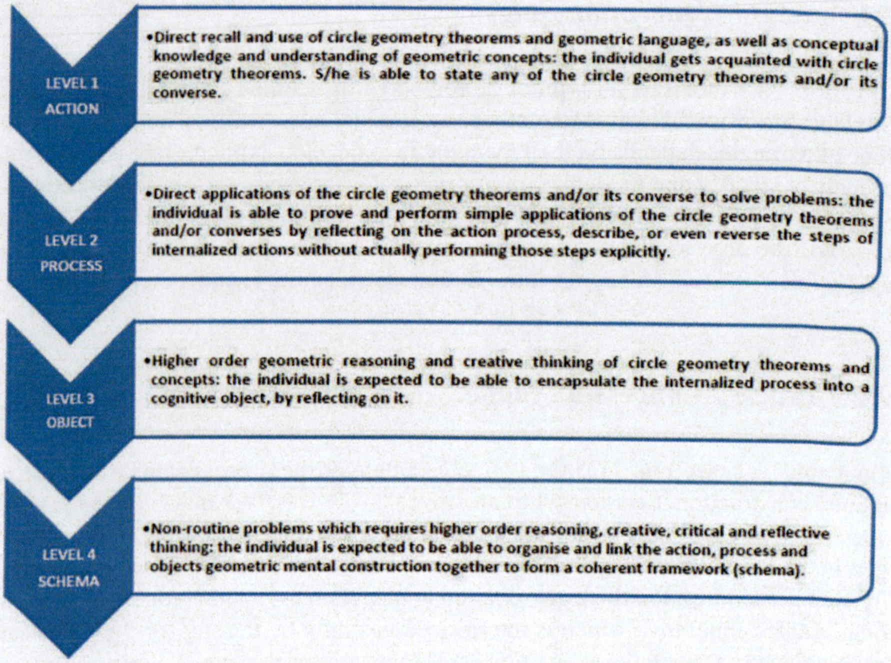

Fig. 2 Preliminary genetic decomposition of circle geometry

two triangles, then demonstrate a process conception by extracting the relevant sides of the two triangles, and, thereafter, act on the sides to provide the expected outcome as desired by the question. Making effective connections across the geometry concepts to find a meaningful solution path to form a schema, to solve more complex problems is what is required to be able to prove that $KE^2 = OE.\ WE$, as shown in Question 12.4 [6] (Fig. 2).

3 Data Analysis and Discussion

3.1 Procedure for Data Analysis

Participants' responses to the standardized tests were thoroughly analyzed to ascertain the mental constructions they demonstrated. In this regard, varied mental constructions portrayed by participants for each of the four standardized tests were analyzed. This was done by implementing the following rating scale: (1) level 0 – preaction mental construction level; (2) level 1 – action mental construction level; (3) level 2 – process mental construction level; (4) level 3 – object mental

Table 1 Overall summary of APOS categorization of students' mental construction on each of the four standardized test items on circle geometry

Participants	APOS categorization of students' mental constructions							
	Action		Process		Object		Schema	
	(ST 1)		(ST 2)		(ST 3)		(ST 4)	
	A	NA	A	NA	A	NA	A	NA
Participants	27	3	27	3	10	20	10	20

construction level; (5) level 4 – schema mental construction level. This rating scale was a modification of the evaluation codes by [2].

3.2 *Analysis of Participants' Responses in View of Each PGD*

The overall summary of the categorization of participants' mental constructions, according to APOS, on each of the four standardized test (ST) items on circle geometry is presented in Table 1.

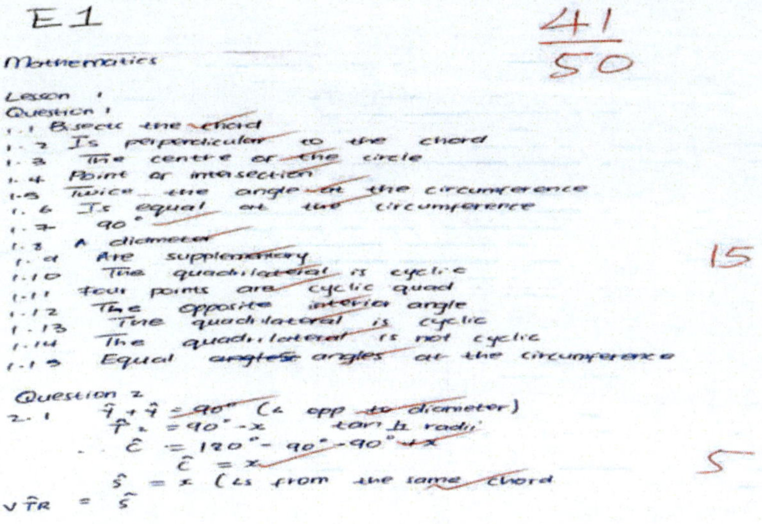
E1
41/50
Mathematics
Lesson 1
Question 1
1.1 Bisect the chord
1.2 Is perpendicular to the chord
1.3 The centre of the circle
1.4 Point of intersection
1.5 Twice the angle at the circumference
1.6 Is equal at the circumference
1.7 90°
1.8 A diameter
1.9 Are supplementary
1.10 The quadrilateral is cyclic
1.11 four points are cyclic quad
1.12 The opposite interior angle
1.13 The quadrilateral is cyclic
1.14 The quadrilateral is not cyclic
1.15 Equal angles at the circumference
15
Question 2
2.1 $\hat{T} + \hat{T} = 90°$ (∠ opp diameter)
$\hat{T}_2 = 90° - x$ tan ⊥ radii
$\therefore \hat{E} = 180° - 90° - 90° + x$
$\hat{E} = x$
$\hat{S} = x$ (∠s from the same chord)
$\hat{VTR} = \hat{S}$
5

Scan 1 Written response of E_1

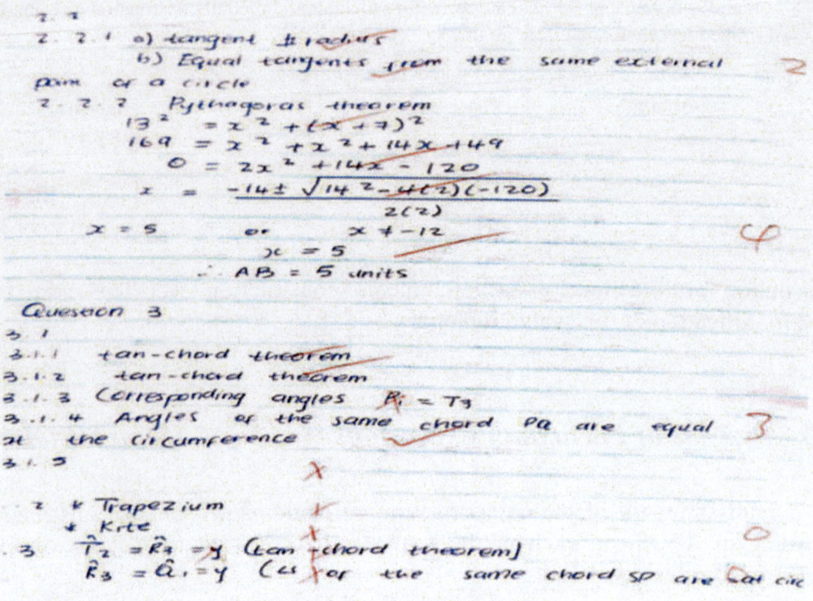

Scan 2 Written response of E_1

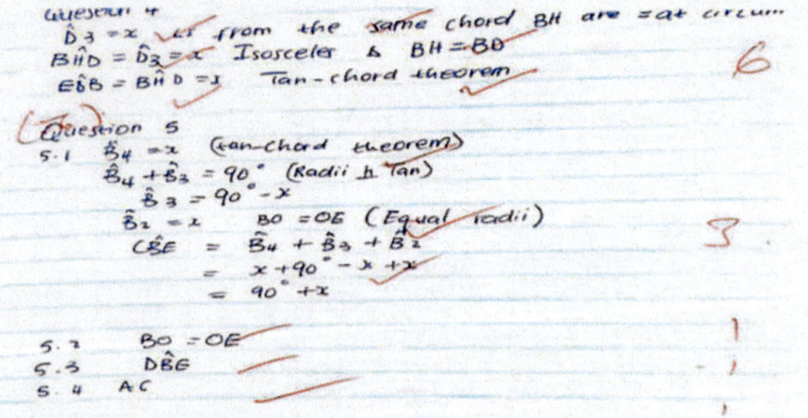

Scan 3 Written response of E_1

Participants' Responses at Action Level

Vignette 1: Exemplar of participants' written responses at action level

The written responses to question 1 and part of question 2 are presented above (see Scan 1). Questions 1 & 2 assessed participants' knowledge on circle geometry theorems and converse of theorems. It can be observed that s/he could provide all the correct responses- an indicator of substantial understanding of circle geometry theorems and/or converse of theorems. The rest of the written response to question 2, is presented on Scan 2.

Scan 2 presents the rest of the written responses for question 2-all written responses were correct. Scan 2 also presents the written responses to question 3- written responses were partly correct (inappropriate responses on geometric terminologies were given by the participant). The other part of question 2 assessed participants' knowledge on circle geometry theorems while question 3 assessed participants' conceptual knowledge and understanding of geometric concepts. The next presentation (Scan 3) follows.

The written responses for Questions 4 & 5 are presented in Scan 3. These questions assessed participants' knowledge on circle geometry theorems and geometric terminologies. Impressively, s/he got all responses correct. The written responses (see Scans 1, 2, and 3) were analyzed and discussed below.

Analysis and Discussion of Participant's Written Responses

As said earlier, 27 out of the 30 (90%) of participants were able to attain this level. The written response (see Scans 1, 2, and 3) is an exhibit of the work done by these participants at the action level to highlight their mental constructions at this level. From question 1 (see Scan 1), it could be observed that 15 out of the expected 15 responses were correct. This question tested participants' knowledge and understanding of circle geometry theorems and converse of theorems. These 27 participants were able to directly recall from memory, the appropriate expected answers on circle geometry theorems. However, some participants had challenges with giving appropriate responses to questions on converse of theorems. This is an indication that they might have just memorized the theorems without conceptually understanding the geometric concepts involved [4].

A thorough perusal of the written responses by E_1 inform that in question 1 (see Scan 1), 15 out of the expected 15 responses were correct. E_1 could recall all the appropriate expected answers from memory. This testified that these participants conceptually, understood the circle geometry concepts, terminologies and geometric language well [4].

With reference to Scans 1 and 2, the unedited written responses by E_1 to question 2 can be found. This testified that the participant was able to recall from memory the appropriate expected responses to question 2. S/he provided correct statements and reasons, used geometric language and terminologies correctly, and could also perform basic computations correctly by applying the Pythagoras theorem effectively (see sub-question 2.2.2). They were as well able to recall from memory the circle geometry theorems and they identified and recognized the specific theorems which were applicable to provide a meaningful solution to the question under consideration. They understood every detail of what they were expected to do and they were able to demonstrate that well, as required. The correct substitutions were, as well, made when applying the Pythagoras theorem, that is:

$13^2 = AB^2 + (x + 7)^2$. This was another indication that the participant was operating at the action level. In addition, the participant was able to expand, simplify, and used the quadratic formula effectively, to solve for the value of x. Interestingly, the participant was able to recognise that x is a distance; since negative distances do not exist, the answer to this subquestion cannot be negative. It can be observed that based on this, s/he rejected the negative solution and accepted the positive value of x as the final solution to this sub-question. This testifies that s/he was able to follow step-by-step procedures efficiently. These are indications to ascertain that the participant was operating at the action level of circle geometry mental construction [2, 3].

Again, the researcher examined the participant's written responses to question 3 (see Scan 2). It was observed that this participant also had challenges, interpreting the diagram well. In addition, the participant's written responses to questions 4 and 5 were examined (see Scan 3). S/he was able to recall from memory the appropriate expected answers. This was evident in the written response as s/he provided correct geometric statements and corresponding reasons. It can be observed from sub-question 5.1 above that E_1 correctly stated $\hat{B}_2 = x$, with the appropriate reason "Tan-chord theorem." S/he was also able to recognise that the "tangent is perpendicular to radius." Based on this, s/he correctly stated that $\hat{B}_4 + \hat{B}_3 = 90^0$. Furthermore, the participant demonstrated his/her conceptual understanding of circle geometry concepts when s/he indicated that $C\hat{B}E = \hat{B}_4 + \hat{B}_3 + \hat{B}_2$. S/he went ahead to correctly substitute the values of angles $\hat{B}_4$, $\hat{B}_3$, $\hat{B}_2$ into the above equation. S/he then simplified to obtain the final answer of $C\hat{B}E = 90° - x$. All these provided enough evidence for the researchers to ascertain that this participant had attained the action level of mental construction, in accordance with the proposed PGD at this level [2].

Even though their responses to the questions were not perfect, these 27 participants were able to recall the majority of the appropriate expected answers from memory. Hence, the researchers asserted that these participants demonstrated enough conception of circle geometry concepts at the action mental construction level in line with the PGD. On this basis, the researchers rated these 27 participants as having attained this level. However, 3 participants were unable to recall the majority of the appropriate expected answers from memory. Thus, the researchers asserted that they demonstrated little or no conception at this level. To this end, the researchers rated them as "not attained," based on their written responses [4]. In view of the above, they were judged as performing at the "Pre-action mental construction level." They either provided irrelevant responses mostly, or they provided no responses at all [1].

Modified Genetic Decomposition (MGD)

From the analysis and findings, we altered the PGD to form the MGD (see Fig. 3).

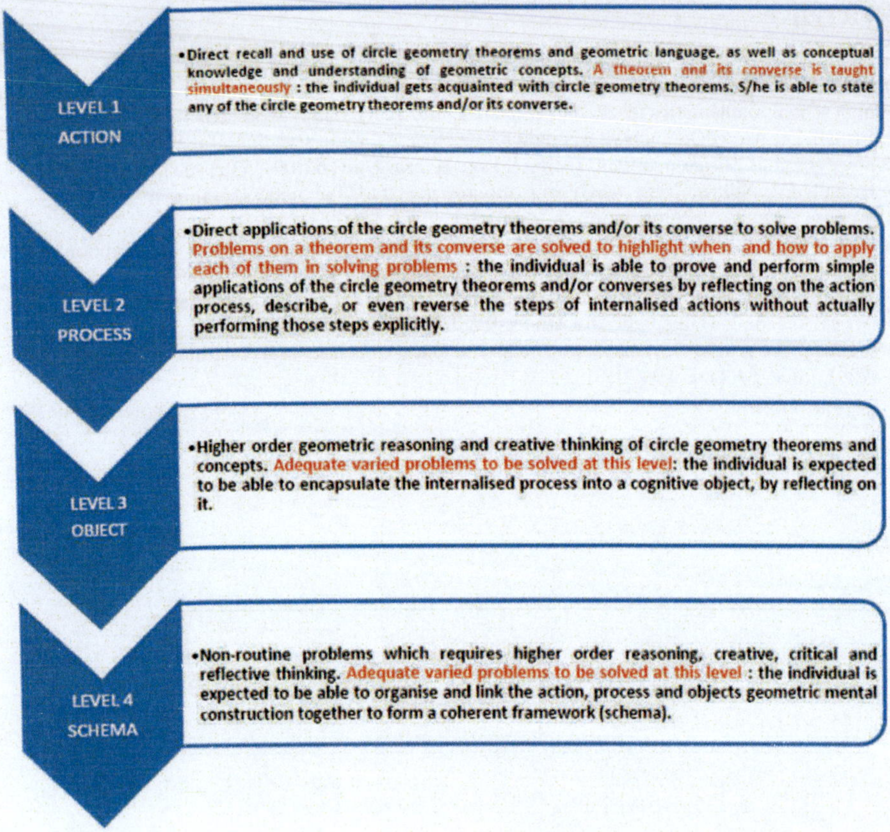

Fig. 3 The modified genetic decomposition of circle geometry concepts

4 Conclusion

The content analysis performed on participants' written responses to four standardized test instruments to demonstrate each level of circle geometry mental conception was carried out. This showed that 27 (90%) of participants operated at the action level of the APOS theory as expected, while 3 (10%) did not. Also, 27 (90%) of participants operated at the process level, while 3 (10%) of participants did not attain this level. For the object level of conception, 10 (33%) of participants operated at this level, while 20 (67%) did not. For the schema level of conception, 10 (33%) of participants operated at the schema level, while 20 (67%) did not. The analysis of data helped alter the PGD.

References

1. Bansilal, S., Brijlall, D., Mkhwanazi, T.: An exploration of the common content knowledge of high school mathematics teachers. Perspect. Educ. **32**(1), 0258–2236 (2014)
2. Brijlall, D.: An APOS analysis of engineering students' understanding of the nature of solutions of a system of linear equations. Technol. Rep. Kansai Univ. **62**(09), 5361–5374 (2020)
3. Brijlall, D., Ndlazi, N.J.: Analysing engineering students' understanding of integration to propose a genetic decomposition. J. Math. Behav. **55**, 100690 (2019)
4. Chagwiza, C.J., Maharaj, A., Brijlall, D.: University students' mental construction when learning the Convergence of a Series concept. Pythagoras. **41**(1), a567 (2020)
5. Department of Basic Education: Mathematics Teaching and Learning Framework for South Africa: Teaching Mathematics for Understanding. DBE, Pretoria (2018)
6. Maharaj, A.: An APOS analysis of natural science students' understanding of derivatives. S. Afr. J. Educ. **33**(1), 1–19 (2013)

N_α- Separation Axioms in Topological Spaces

Nadia M. Ali Abbas and Shuker Khalil

1 Introduction

A topological space (TS) is a fairly broad concept. More specificity is frequently desirable. Some studies in TS and their extensions in nonclassical TS are shown by many mathematicians [1–13]. One method is to define topological spaces with more constrained attributes using the separation axioms. In general, it is not true that a sequence in a topological space has only one limit. However, using the separation axioms, a type of space may be established in which the limit, if it exists at all, is unique. In 2015, N_a- open sets were initially examined by N.A Dawood and N.M Ali, see [14]; by using these sets, we study some classes of N_α- separation axioms and N_α- Ti separation for each $i = 0, 1, 2$, and look into some of their characteristics. Separation axioms have also been generalized to other generic topological spaces such as ordered topological spaces [15]. Ibrahim [16] presented and explored the features of a strong variant of α-open sets termed α_γ-open via operation in 2013. Khalaf and Ibrahim [17] extended their investigation of the features of operations defined on the collection of α-open sets introduced by Ibrahim [16], defining and discussing numerous properties of α_γ-regular, α-β-compact, and α_γ-connected spaces, as well as α-(γ, β)-continuous functions.

In this chapter, we use N_α- open sets in topological spaces to create new types of N_α-separation axioms and study some of their properties. There are also some definitions and theorems offered. Here in this work, all spaces X and Y are topological spaces, also the closure (interior resp.) of a subset A of X is denoted by cl (A) (int (A) resp.)

N. M. Ali Abbas
Ministry of Education, Directorate General of Education, Baghdad, Al-Kark, Iraq

S. Khalil (✉)
Department of Mathematics, College of Science, University of Basrah, Basrah, Iraq

F. Kamalov et al. (eds.), *Advances in Mathematical Modeling and Scientific Computing*, Trends in Mathematics, https://doi.org/10.1007/978-3-031-41420-6_79

2 Fundamental Concepts

We will cover some fundamental principles that will be useful in our work.

Definition 1 [15]
Assume that X is a topological space (TS), a set A is named N_a- open set ($N_\alpha - OS$) if for some α-open set $B \neq \varnothing$ satisfies $cl\ (B) \subseteq A$. Also, its complement is named a N_a- closed set ($N_\alpha - CS$). The collection of all N_α- open sets is referred as $N_\alpha O(X)$, and its complement by $N_\alpha C(X)$.

Remark 1 [15]
A set A is a ($N_\alpha - CS$) if for some α-closed set $\varnothing \neq B \neq X$ satisfies $\subseteq int\ (B)$.

Remark 2 [15]
(i) X and $\varnothing$ are N_α- open sets ($N_\alpha - OS$ s) in any (TS) X.
(ii) Any clopen set is ($N_\alpha - OS$).
(iii) Any set in discrete space is ($N_\alpha - OS$).

Theorem 1 [15]
Let X_1, X_2 be topological spaces (TSs). Then A_1 and A_2 are ($N_\alpha - OS$ s) in X_1, X_2 resp., if and only if $A_1 \times A_2$ is ($N_\alpha - OS$) in $X_1 \times X_2$.

Proposition 1 [15]
Let X be (TS). Then:

(i) Finite union of ($N_\alpha - OS$ s) is ($N_\alpha - OS$) also.
(ii) Finite intersection of ($N_\alpha - CS$ s) is ($N_\alpha - CS$) also.

Definition 2 [15]
The union of all N_a- open set of X contained in A is named N_a- interior of A and is denoted ${N_\alpha}^{int}(A)$, and the intersection of all ($N_\alpha - CS$) containing A is called N_α-closure of A, refereed by ${N_\alpha}^{cl}(A)$.

Definition 3 [18]
We say A is generalized N_α – closed set ($gN_\alpha - CS$) of a space X, if ${N_\alpha}^{cl}(A) \subseteq B$ whenever $A \subseteq B$ and B is ($N_\alpha - OS$).

The complement of ($gN_\alpha - CS$) is generalized N_α- open set ($gN_\alpha - OS$) in X.

Theorem 2 [18]
(i) If A is ($N_\alpha - CS$) in X, then it is ($gN_\alpha - CS$).
(ii) If A is ($N_\alpha - OS$) in X, then it is ($gN_\alpha - OS$).

Proposition 2 [15]
Suppose that (Y, t_y) is a subspace of a (TS) X with $A \subseteq Y \subseteq X$. Then:

(i) If $A \in N_a O(X)$, then $A \in N_\alpha O(Y)$.
(ii) If $A \in N_a O(Y)$, then $A \in N_\alpha O(X)$, where Y is clopen set in X

Definition 4 [19]
Let X_1, X_2 be (TSs) where, $f : X_1 \to X_2$ is a mapping, then f is named:

(i) $N_\alpha(N_{\alpha*}$- continues) resp. if f^{-1} (A) is $(N_\alpha - OS)$ in X_1 for each A open set $((N_\alpha - OS))$in X_2 resp.
(ii) $N_\alpha(N_{\alpha*}$- open) mapping if $f(A)$ is $(N_\alpha - OS)$ in X_2 for each open set $((N_\alpha - OS))$ A in X_1 resp.
(iii) gN_α-continuous ($gN_{\alpha*}$ -continuous) resp. mapping if for each open set $((gN_\alpha - OS))$ set A in Y respectively then $f^{-1}(A)$ is $(gN_\alpha - OS)$ in X.

3 Some Characteristics of N_α-Separation N_α- Axioms

In this section, we study ${N_\alpha}^{Ti}$- space X for each $i = 0, 1, 2$ and we discuss some of these spaces' characteristics and remarks. We will prove certain theorems in the following cases when X is a finite space.

Definition 5
Assume that X is a (TS). We say X is a ${N_\alpha}^{TO}$-space if for any $x \neq y$ in X, there exists $(N_\alpha - OS)$ A containing one of them but not other.

Theorem 3
Let X be a (TS). Then X is ${N_\alpha}^{To}$-space if and only if ${N_\alpha}^{Cl\{x\}} \neq {N_\alpha}^{Cl\{y\}}$.

Proof Let ${N_\alpha}^{Cl\{x\}} \neq {N_\alpha}^{Cl\{y\}}$, $\forall x \neq y$ in X. This implies ${N_\alpha}^{Cl\{x\}} \nsubseteq {N_\alpha}^{Cl\{y\}}$ *or* ${N_\alpha}^{Cl\{y\}}$ $\nsubseteq {N_\alpha}^{Cl\{x\}}$. Suppose ${N_\alpha}^{Cl\{x\}} \nsubseteq {N_\alpha}^{Cl\{y\}}$, hence $X \notin {N_\alpha}^{Cl\{y\}}$, thus $x \in ({N_\alpha}^{Cl\{y\}})^c$, which is $(N_\alpha - OS)$ and $y \notin ({N_\alpha}^{Cl\{y\}})^c$. Thus, X is ${N_\alpha}^{To}$-space, assume X is ${N_\alpha}^{To}$-space; hence, for each $x \neq y$ in X, there exists $(N_\alpha - OS)$ G such that $x \in G$, $y \notin G$ *or* $y \in G$, $x \notin G$. Hence, G^c is $(Na - CS)$. $x \notin G^c$, $y \in G^c y$; hence, $x \notin {N_\alpha}^{Cl\{y\}}$, $x \in {N_\alpha}^{Cl\{x\}}$; this means $x \notin {N_\alpha}^{Cl\{y\}}$. Thus, ${N_\alpha}^{Cl\{x\}} \neq {N_\alpha}^{\{y\}}$.

Definition 6
Let X be a (TS). Then X is named ${N_\alpha}^{T1}$-space if each pair of distinct points x and y of X, there exist two N_α- open sets A, B containing x and y, respectively, such that $y \notin A$, $x \notin B$.

Proposition 3
Let X be a (TS). Then X is ${N_\alpha}^{T1}$ -space if and only if $\{x\}$ is $(N_\alpha - CS)$ $\forall x \in X$.

Proof Assume that X is ${N_\alpha}^{T1}$-space, to show that each $\{x\}$ is $(N_\alpha - CS)$, this means we must show that $X/\{x\}$ is $(N_\alpha - OS)$ for each singleton set $\{x\}$ in X.

Let $y \in X/\{x\}$, then $y \neq x$ in X, since X is ${N_\alpha}^{T1}$ space, then there exists $(N_\alpha - OS)$ G with $y \in G$ and $x \notin G$. This implies that $y \in G \subseteq \frac{X}{\{x\}}$; this implies $X/\{x\}$ is $(N_\alpha - OS)$. Hence, $\{x\}$ is $(N_\alpha - CS)$.

Conversely: Let $\{x\}$ be $(N_\alpha - CS)$, $\forall x \in X$, to prove X is ${N_\alpha}^{T1}$ -space. Let $x \neq y$ in X, hence $\{x\}$, $\{y\}$ are $(N_\alpha - CSs)$ hence $\{x\}^c$, $\{y\}^c$ are $(N_\alpha - OSs)$ and $y \in \{x\}^c$, $x \notin \{x\}^c$, $x \in \{y\}^c$, $y \notin \{y\}^c$. Therefore, X is ${Na}^{T1}$-space.

Definition 7
Let X be a (TS). Then X is named $N_\alpha{}^{T2}$-space if for any two distinct points x, y in X there exists two ($N_\alpha - OSs$) X satisfy $x \in A_1$, $y \in A_2$ and $A_1 \cap A_2 = \varnothing$

Proposition 4
If X is Na^{T2}- space, then $A = \{(x, y) : x = y, x, y \in X\}$ is $(Na - CS)$.

Proof Assume that X is $N_\alpha{}^{T2}$- space, to prove A is $(N_\alpha - CS)$, let $(x, y) \in A^c \subseteq X \times X/A$, this mean x and y are two distinct points in X, where X is $N_\alpha{}^{T1}$-space then for some $A_1, A_2 \in N_\alpha O(X)$ satisfy $x \in A_1$, $y \in A_2$ and A_1, A_2 are disjoint sets, hence $(x, y) \in A_1 \times A_2 \subseteq A^c$, but $A_1 \times A_2 \in N_\alpha O(X \times X)$ (see Theorem 1), hence A^c is N_α- open set, thus A is $(N_\alpha - CS)$.

Proposition 5
If f, $g : x \rightarrow y$ are $N_{\alpha *}$- continuous and Y is $N_\alpha{}^{T2}$ space, then the set $A = \{x : x \in X f(x) = g(x)\}$ is $(N_\alpha - CS)$.

Proof If $x \notin A$, then $x \in A^c$ this mean that $f(x) \neq g(x)$ in Y, since Y is $N_\alpha{}^{T2}$-space, then there exist $B_1, B_2 \in N_\alpha O(Y)$ such that $f(x) \in B_1$, $g(x) \in B_2$ and $B_1 \cap B_2 = \varnothing$, but $f^{-1}(B_1)$, $g^{-1}(B_2) \in N_\alpha O(X)$ since f, g are $N_{\alpha *}$-continuous, hence $x \in f^{-1}(B_1)$, $x \in g^{-1}(B_2)$ hence $x \in f^{-1}(B_1) \cap g^{-1}(B_2)$, let $B = f^{-1}(B_1) \cap g^{-1}(B_2)$, where B is $(N_\alpha - OS)$. Now we shall prove $B \subseteq A^c$, i.e $B \cap A = \varnothing$. Suppose that $B \cap A \neq \varnothing$ this mean $y \in B \cap A$; thus, $y \in A$, $y \in B$. Hence, $y \in f^{-1}(B_1)$, $y \in g^{-1}(B_2)$, hence $f(y) \in B_1$, $g(y) \in B_2$, $y \in A$. Thus, $f(y) = g(y)$, since $y \in A$, hence $B_1 \cap B_2 \neq \varnothing$, which is a contradiction, thus $B \subseteq A^c$, thus $A^c \in N_\alpha O(X)$,, hence $A \in N_\alpha C(x)$.

Proposition 6
If X and Y are $N_\alpha{}^{Ti}$- space, then $X \times Y$ is $N_\alpha{}^{Ti}$- space $\forall i = 0, 1, 2$

Proof Assume that X and Y are $N_\alpha{}^{Ti}$- space. Put $i = 0$ and take $(x_1, y_1) \neq (x_2, y_2)$ in $X \times Y$, then for any two distinct points x_1 and x_2 in X, there exists $A_1 \in N_\alpha O(X)$ such that $x_1 \in A_1$, $x_2 \notin A_1$ *or* $x_1 \notin A_1$, $x_2 \in A_1$, also $y_1 \neq y_2$, then there exists $A_2 \in N_\alpha O(Y)$ such that $y_1 \in A_2$, $y_2 \notin A_2$ or $y_1 \notin A_2$, $y_2 \in A_2$ then $(x_1, y_1) \in A_1 \times A_2$ $(x_2, y_2) \notin A_1 \times A_2$ *or* $(x_1, y_1) \notin A_1 \times A_2 (x_2, y_2) \in A_1 \times A_2$ but $A_1 \times A_2$ is $(N_\alpha - OS)$ in $X \times Y$ (see Theorem 1). Hence $X \times Y\, N_\alpha{}^{Ti}$- space. Similarly, we can prove other states for $i = 1, 2$.

Proposition 7
If X is $N_\alpha{}^{Ti}$, then it is $N_\alpha{}^{Ti-1}$ –space, where $i = 2, 1$.

Proof The proof is consider from Definitions 5, 6, and 7.

Theorem 4
The inverse image of $N_\alpha{}^{Ti}$ -space under injective $N_{\alpha *}$- continuous mapping is also $N_\alpha{}^{Ti}$ space, where $i = 0, 1, 2$

We shall prove only when $i = 2$ and the other cases are similarly.

Proof Let $f : X \rightarrow Y$ be injective, $N_{\alpha *}$- continuous mapping and $x_1 \neq x_2$ in X, since f is injective then $y_1 = f(x_1) \neq f(x_2) = y_2$ in Y where Y is $N_\alpha{}^{T2}$ then there exist two disjoint N_α- open set A_1, A_2 in Y satisfy $y_1 \in A_1$, $y_2 \in A_2$, since f is $N_{\alpha *}$- continuous

Diagram 1 The relationship between $N_\alpha{}^{Ti}$-spaces and $gN_\alpha{}^{Ti}$ spaces

$$\begin{array}{ccccc} N_\alpha T2 & \rightarrow & N_\alpha T1 & \rightarrow & N_\alpha T0 \\ \downarrow & & & & \downarrow \\ gN_\alpha T2 & \rightarrow & gN_\alpha T1 & \rightarrow & gN_\alpha T0 \end{array}$$

than $f^{-1}(A_1), f^{-1}(A_2)$ are ($N_\alpha - OSs$) in X such that $x_1 \in f^{-1}(A_1)$, $x_2 \in f^{-1}(A_2)$ and $f^{-1}(A_1) \cap f^{-1}(A_2) = \varnothing$. Therefore, X is $N_\alpha{}^{T2}$-space.

Theorem 5
If $f : X \rightarrow Y$ is injective N_α- continuous and Y is T_2 space, then X is $N_\alpha{}^{T2}$ – space.

Proof Similar to the proof of Theorem 4.

Definition 8
Let X be a (TS). Then X is called $gN_\alpha{}^{Ti}$-space, where $i = 0, 1, 2$ if:

(i) $i = 0$ if for any $x \neq y$ in X, there exists ($gN_\alpha - OS$) A containing one of them but not other.
(ii) $i = 1$ if for any $x \neq y$ in X, there exist two ($gN_\alpha - OSs$) A, B containing x and y, respectively, satisfy $y \notin A$, $x \notin B$.
(iii) $i = 2$ if for each pair of distinct point x, y in X there exist disjoint ($gN_\alpha - OSs$) A, B such that $x \in A$, $y \in B$.

Proposition 8
Every $N_\alpha{}^{Ti}$- space is $gN_\alpha{}^{Ti}$ space.

Proof The proof is in hand, from Theorem 2 where every N_α- open set is gN_α- open set.

By Propositions 7 and 8 we have the following Diagram 1

Theorem 6
If $f : X \rightarrow Y$ is injective gN_α- continuous and Y is T_2 – space than X is $gN_\alpha - T_2$ space.

Proof Assume that $x \neq y$ in X, since f is injective, thus $f(x) \neq f(y)$ in Y where Y is T_2 space, then there exists disjoint open sets A, B satisfy $f(x) \in A$, $f(y) \in B$ and $A \cap B = \varnothing$, since f is gN_α- continuous, then $f^{-1}(A), f^{-1}(B)$ *are* ($gN_\alpha - OSs$) in X see (Definition 4(iii)) where $x \in f^{-1}(A)$, $y \in f^{-1}(B)$ and $f^{-1}(A) \cap f^{-1}(B) = \varnothing$. Hence, X is $gN_\alpha - T_2$ space.

4 Conclusion and Future Work

We use N_α-open sets in topological spaces to generate new sorts of N_α-separation axioms and investigate some of their features in this research. Some theorems are also provided. In future work, we will discuss in nonclassical (TS) such as neutrosophic/fuzzy/soft topological spaces.

References

1. Jafari, S., Noiri, T.: Contra – α – continuous mappings between topological space. Iran. Int. J. Sci. **2**, 153–167 (2001)
2. Damodharan, K., Vigneshwaran, M., Khalil, S.M.: $N_{\delta * g\alpha}$-continuous and irresolute functions in neutrosophic topological spaces. Neutrosophic Sets Syst. **38**(1), 439–452 (2020)
3. Khalil, S.M.: On neurosophic delta generated per-continuous functions in neutrosophic topological spaces. Neutrosophic Sets Syst. **48**, 122–141 (2022)
4. Khalil, S.M., Suleiman, E., Ali Abbas, N.M.: New technical to generate permutation measurable spaces. In: 2021 1st Babylon International Conference on Information Technology and Science (BICITS), pp. 160–163 (2021). https://doi.org/10.1109/BICITS51482.2021.9509892
5. Hasan, M.A., Khalil, S.M., Abbas, N.M.A.: Characteristics of the soft-(1, 2)-gprw–closed sets in soft bi-topological spaces. In: Conf., IT-ELA 2020, vol. 9253110, pp. 103–108 (2020)
6. Abbas, N.M.A., Khalil, S.M., Hamza, A.A.: On α^*- continuous and contra α^*- continuous mappings in topological spaces with soft setting. Int. J. Nonlinear Anal. Appl. **12**, 1107–1113 (2021)
7. Hasan, M.A., Ali Abbas, N.M., Khalil, S.M.: On soft α^*- open sets and soft contra α^*- continuous mappings in soft topological spaces. J. Interdiscip. Math. **24**, 729–734 (2021)
8. Khalil, S.M., Hameed, F.: An algorithm for generating permutations in symmetric groups using soft spaces with general study and basic properties of permutations spaces. J. Theor. Appl. Inf. Technol. **96**, 2445–2457 (2018)
9. Khalil, S.M., Hameed, F.: An algorithm for generating permutation algebras using soft spaces. J. Taibah Univ. Sci. **12**(3), 299–308 (2018)
10. Ali Abbas, N.M., Khalil, S.M.: On new classes of neutrosophic continuous and contra mappings in neutrosophic topological spaces. Int. J. Nonlinear Anal. Appl. **12**(1), 718–725 (2021)
11. Nivetha, A.R., Vigneshwaran, M., Abbas, N.M.A., Khalil, S.M.: On $N_{*g\alpha}$- continuous in topological spaces of neutrosophy. J. Interdiscip. Math. **24**(3), 677–685 (2021)
12. Ali Abbas, N.M., Khalil, S.M., Vigneshwaran, M.: The neutrosophic strongly open maps in neutrosophic bi-topological spaces. J. Interdiscip. Math. **24**(3), 667–675 (2021)
13. Jafari, S., Noiri, T.: On contra pre- continuous mapping. Bull. Malays. Math. Soc. **25**, 115–128 (2002)
14. Dawood, N.A., Ali, N.M.: N_α- open set and N_α- regularity in topological space. Int. J. Adv. Sci. Tech. Res. **5**(3), 87–96 (2015)
15. Al-shami, T.M., Abo-Elhamayel, M.: Novel class of ordered separation axioms using limit points. Appl. Math. Inf. Sci. **14**(6), 1103–1111 (2020)
16. Ibrahim, H.Z.: On a class of $\alpha\gamma$-open sets in a topological space. Acta Sci. Technol. **35**(3), 539–545 (2013)
17. Khalaf, A.B., Jafari, S., Ibrahim, H.Z.: Bioperations on α-open sets in topological spaces. Int. J. Pure Appl. Math. **103**(4), 653–666 (2015)
18. Ali, N.M.: Applications of Na and generalized Na- closed sets in topological space. AL – Bahir J. Nat. Eng. Sci. **8**(15), 55–63 (2018)
19. Ali, N.M.: N_α- continuous and contra N_α- continuous mappings in topological space. AL – Bahir J. Nat. Eng. Sci. **3**(5), 67–76 (2016)

Investigating How the Activity, Classroom Discussion, and Exercise (ACE) Teaching Cycle Influences Learners' Problem-Solving and Achievement in Circle Geometry

F. Abakah and D. Brijlall

1 Introduction

Finding effective problem-solving instructional approaches for teaching and learning of mathematics has taken center stage for researchers for more than a century. Such instructional approaches are expected to serve modern demands of applying mathematical ideas to problem-solving not only in mathematics but also in real life situations. Problem-solving has become a focal point of most contemporary mathematics curricula, in which the world is perceived to be gearing toward the fourth industrial revolution; hence, a curriculum, which will teach students reasoning and thinking skills, is essential. This will require students to engage in explicit organized thinking about mathematical contents and enable them to apply their reflective understanding of such thinking. This approach will afford students the opportunity to be engaged in more practice by employing the thinking skills they have acquired in the mathematics classroom in diverse content [7].

The inception of APOS theory as a constructivist approach has being either wholly or partially adopted in a plethora of contemporary mathematics education research studies [7]. The theory has been adopted and implemented effectively as a teaching and learning approach for different mathematical concepts, although only a handful of these studies have focused on geometry. The researchers for this study, therefore, believe that the theory is relevant for this research study, aimed at improving the teaching and learning of circle geometry. This belief is due to the positive acknowledgements these instructional approaches had received from studies that had employed APOS theory and its associated ACE teaching cycle. This study aimed to demonstrate the same level of efficacy in adopting the ACE

F. Abakah · D. Brijlall (✉)
Department of Mathematics, Durban University of Technology, KwaZulu-Natal, South Africa
e-mail: deonarainb@dut.ac.za

F. Kamalov et al. (eds.), *Advances in Mathematical Modeling and Scientific Computing*, Trends in Mathematics, https://doi.org/10.1007/978-3-031-41420-6_80

teaching cycle for this study, so that participants' problem-solving competences will be optimized [6]. With this aim in mind, we formulated the main research question as: *How can the ACE teaching cycle influence Grade 11 learners' achievements in circle geometry?* In partly answering this research question, the hypothesis and the null hypothesis of the study were formulated as follows:

Hypothesis (H_A): *There is a statistically significant improvement in the circle geometry test scores of the learners who used the ACE teaching cycle as a problem-solving instructional approach.*

Null hypothesis (H_0): *There is no statistically significant improvement in the circle geometry test scores of the learners who used the ACE teaching cycle as a problem-solving instructional approach.*

2 Theoretical Framework

The constructivism perspective of teaching and learning has proven to be sacrosanct to modern demands of finding an effective problem-solving instructional approach to serve modern curricula needs [7]. This emanated from a series of research studies, which might have commenced, a century ago. Observations, experimentations, individual and corporative learning, learners being given the opportunity to make their own mathematical constructions and to justify their mathematical constructions, and brainstorming, among others, are key features of the constructivism epistemology. These also serve as inherent techniques, which are necessary for teaching and learning of problem solving. Furthermore, Maharaj [5] provided details about the constructivists' approach by giving the following characteristics of the constructivist mathematics curriculum: It guarantees meaningful learning and understanding; it promotes creative thinking, reasoning, or exploration, by providing students an experimenting learning environment, which enables them to guess, conjecture, and test hypotheses; independent learning, which accords students the freedom to meaningfully construct their mathematical ideas and for them to also ponder on their mathematical constructions; social interaction, which promotes effective communications between pedagogues and students, when they are alone, in groups, or in class.

3 Methodology

This study implemented the mixed-method research design. Creswell [3] defined mixed methods research as "a procedure for collecting, analysing, and 'mixing' both quantitative and qualitative research and methods in a single study to understand a research problem." To establish the effects the ACE teaching approach under investigation in this study had on the study participants (experimental group), both numerical and interpretive narrations were required. Hence, the need to employ a mixed-method research design for this study was felt. The quantitative part was required to provide numerical presentations of the study participants' achievements at the research field. From the same reasoning, the qualitative part was required so that more detailed narrations and interpretations of events and activities at the research field can be obtained. The researchers realized that both quantitative and qualitative research techniques were required to adequately answer the research question of this study, which had inherent characteristics of both quantitative and qualitative research designs. In this study, two participant groups were involved: the control group and the experimental group. The control group had 30 members, while the experimental group had 32 members. Participants in both groups were grade 11 mathematics learners in a South African high school, who were targeted to serve as participants for this study.

3.1 How the ACE Teaching Cycle Was Implemented in this Study

Activities, Classroom discussions, and Exercises are the three components of the ACE teaching approach, which is the pedagogical component of the APOS theory; these three components are repeated in a circular pattern. In this study, these three components were partitioned into four: circle geometry problems (activities), which were presented to learners to enable them to develop appropriate mental structures. Each group was thereupon given the opportunity to present their conjectured solutions on the board. Thereafter, the class discussion stage followed where learners thoroughly discussed how they obtained solutions to the activities in stage 1. Lastly, exercises in relation to the learnt concepts at stages 1, 2, & 3, were given to participants. This stage was characterized by transfer and application of knowledge. These sequential processes were followed to guide learners to develop appropriate mental structures, for solving mathematical problems, that is, circle geometry [5, 7].

3.2 Data Analysis and Discussion

Data analysis was carried out in twofold – quantitative data analysis and qualitative data analysis. The results from these two procedures were then integrated – from which concise and empirically proven findings, conclusion, and recommendations were made. First, quantitative data analysis was done – descriptive and inferential statistics techniques were used. The inferential statistics data analysis procedure is delineated below.

Part 1: Quantitative Data Analysis

Inferential Statistics Data Analysis Results

The composite result for all the four standardized tests was used for the hypothesis test. This was done by adding the marks each participant in both the control group and the experimental group, obtained in each of the four standardized tests. The researcher realized that using the composite scores for the hypothesis test might be better as larger set of data will afford a degree of reliability to the findings, in juxtaposition to using the scores for each of the four standardized tests.

Hypothesis Test

Our main aim in this section is to apply statitsical methods to test the effectiveness of the new teaching method. Hence, we seek to test the hypothesis that:

- H_0: The differences between the mean marks of treatment and controlled group is zero.
- H_1: The differences between the mean marks of treatment and controlled group is not zero.

In practice, for a large data set, due to the central limit theorem, marks are supposed to be approximately normally distributed. However, due to the aggregate nature of the marks obtained from participants' composite scores from this study, the assumption may not hold. In addition, the marks were recorded as count variables. The study utilized an independent measured controlled experiment. The result of this type of design is prone to differences between participants as a result of variations in learning abilities and barriers; thus, it is necessary to account for the effects of such nuisance factors (random effects) in the appropriate model. A statistical methodology, which is able to test for the differences between two population means from a count distribution characterized by random variations, is the generalized mixed effect model. What, then, are generalized mixed effect models?

Generalized Mixed Effect Models

According to Bono, Alarcón, and Blanca [1], skewness and kurtosis values from educational research data generally deviate from the normal distribution. This is when the generalized linear mixed models become essential. This statistical model is efficient and conducive for analyzing nonnormally distributed data (continuous or discrete). This includes categorical data, count data, proportional response data, etc.. Factors to be considered when using this model are distribution of the data, link function and nature of random effects, estimating the parameters, and testing significance. Although this model may be complex in its implementation, it gives more detailed analysis [1].

Justification for Using Negative Binomial Mixed Effect Model

This is an educational research study involving learners in their natural classroom settings. Hence, it produced real data (quantitative), which are not normally distributed – standardized tests data, which were used for this analysis, are count data, which belong to this category. Integral to count data is "overdispersion," which generates incorrect standard errors. This implies that there exist greater variance in the data this model will be applied on [1].

To this end, the researchers opted for generalized mixed effect model- an appropriate statistical data analysis technique for analyzing data that is not normally distributed. This contrasted with exploring other avenues that are not directly applicable to the nature of the collected data which will result in the data being transformed to obtain normality or using nonparametric analyses, so that classical approaches may definitely have to be used [1]. The normality test was conducted on the count data to ascertain why it is not a normally distributed data, hence the mixed effect model was ideal for modeling this data.

Normality Test

The generalized linear mixed effect model for count data assume that the error distribution comes from a Poisson or the negative Binomial distribution; thus, to determine the appropriateness of the data, it is necessary that the possibility of the data coming from the normal distribution is ruled out as this will invalidate the final results.

The normal quantile plot, the Jarque-Bera test, and the Shapiro-Wilk test were, therefore, used in testing the normality or the nonnormality of the data. From all these tests, the null hypothesis of normality is not rejected when the *p-values* < 0.005. The results of the tests were presented in Table 1, while the normal quantile plot was displayed in Fig. 1.

In Table 1, the results of the two tests converge and support the presence of nonnormality of the group marks. The Shapiro test and the Jarque-Bera test reject the null of normality (p-value < 0.005). This result is confirmed by the clear deviations of the left sample quantile points from the theoretical quantile as

Table 1 Normality test

Test name	Test statistic	*P*-value
Shapiro-Wilk	0.909	0.000224
Jarque-Bera	60.512	0.000698

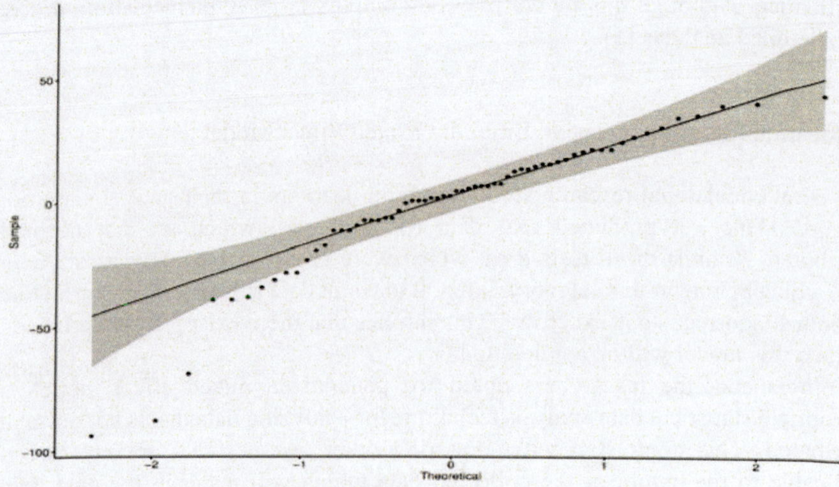

Fig. 1 qq plot for control and treatment groups

indicated on Fig. 1. It was, therefore, concluded that the marks are not normally distributed suggesting that a generalized linear model may be appropriate.

Count data can be modeled by either using the Poisson mixed effect model or negative binomial mixed effect model. These two mixed effect models share many similarities. However, unlike the Poisson, the negative binomial distribution has the scale parameter, which permits the variance to be larger/smaller than the mean. This may decrease or solve overdispersion [1]. Since count data in practice are usually overdispersed, the overdispersion test was conducted to decide on whether to use Poisson mixed effect or negative binomial mixed effect model.

Overdispersion Test

Sequence Identification numbers assigned to each of the participants was used to proxy the random effects. Johnson [4] reconciled classical and Bayesian methods of significance testing for many papers published in psychology journals and found out that *p-values* of 0.005 and 0.001 correspond to strong and very strong evidence against the null hypothesis, while the *p-values* in the neighborhood of 0.05 and 0.01 reflect only modest evidence. Therefore, the level of significance in this study was set at 0.005 (0.5%).

As mentioned earlier, we first test for the nature of the overdispersion in the data before we can decided to use either the mixed effect Poisson or the mixed effect negative Binomial model in testing for the effectiveness of the new teaching method. If the data is oversidpersed, the dispersion parameter should be greater than 1; hence, we are interested in testing the following hypothesis:

Table 2 Overdispersion test

Alternative hypothesis	True dispersion is greater than 1
Dispersion parameter	6.125867
Z -statistic	3.0873
P-value	0.00101

H_0: dispersion = 1
H_1: dispersion > 1

We fail to reject the null hypothesis if the *p-value* > 0.005. These results were presented in Table 2.

It is evident from Table 2 that the dispersion parameter is greater than 1. The *p*-value of the test confirms this observation (*P*-value < 0.005); thus, we reject the null hypothesis and conclude that the data is overdispersed. This conclusion indicates that the negative Binomial mixed effect model is appropriate for modeling the data. Hence, the Negative Binomial mixed effect model was used to test if the teaching method was effective using the MASS package of R programming language to estimate the parameters of the model.

4 Results

From the box plots in Fig. 2, there is a difference between the mean marks of control and treatment groups. The mean mark of the treatment group is clearly bigger than the mean mark of the control group. Also, the violin plot illustrated in Fig. 3 indicates that the width of the treatment group and its mean is wider and bigger than those of the control group. These observations provide a possible indication that the new teaching method may be more effective.

4.1 Model Outputs

After fitting the data to the Negative Binomial mixed effect model, the estimated parameters obtained were summarized in Table 3. Valid conclusions and generalizations cannot be made based on an inadequate estimated model; thus, before interpreting the estimated output and make inferences, it is imperative to assess the adequacy of the estimated model. The Chi-square test was used in this regard and the test result was presented in Table 4. This test reported the residual deviance, the degrees of freedom (df), and the associated *p-value*.

The residual deviance is the difference between the deviance of the current model and the maximum deviance of the ideal model where the predicted values are identical to the observed. Therefore, if the residual difference is small enough (*p-value* > 0.002) and the goodness of fit test is not significant, it indicates that the

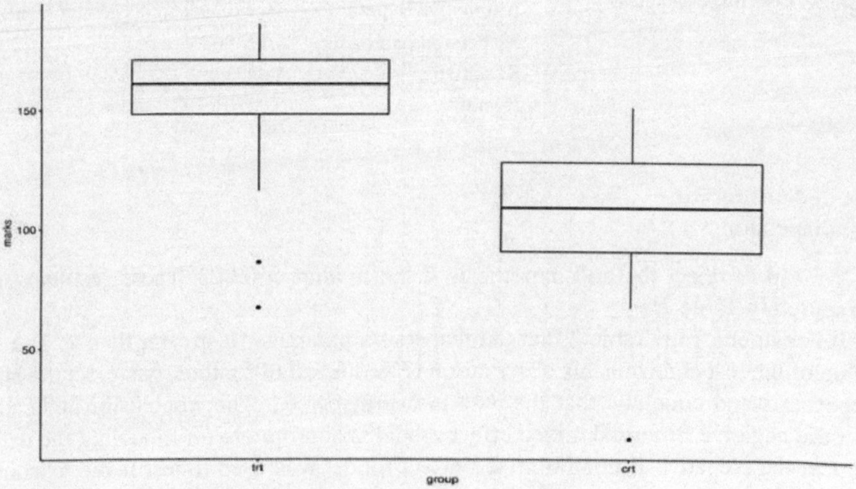

Fig. 2 Box plot of aggregated marks

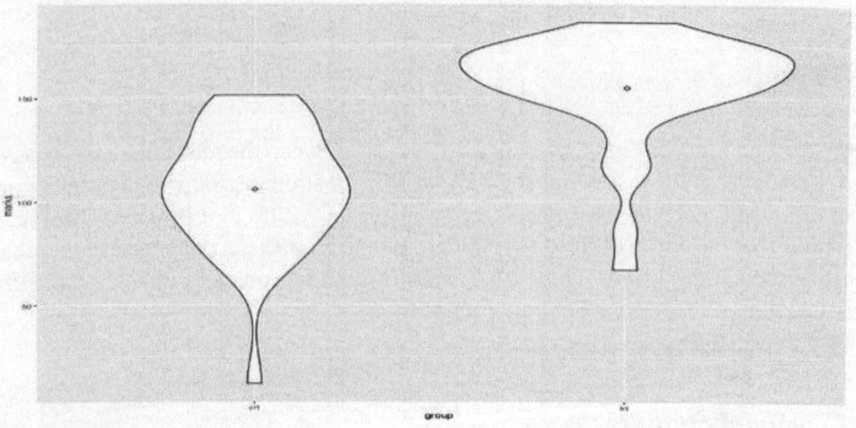

Fig. 3 Violin plot of aggregated marks

Table 3 Estimated parameters

Parameter	Estimate	Std. Error	Pr(>|z|)
Intercept	4.610054	0.185542	2e−16
Treatment	0.414302	0.134926	0.0021
Random effect	0.001097	0.003766	0.7709

Table 4 Chi-square test

Residual deviance	Df	Pr(>|z|)
67.34626	59	0.2131441

model fits the data well. Since the *p-value* > 0.002, we conclude that the model fits reasonably and it is adequate for making inferences about the two populations.

The associated standard errors for the estimates are fairly small, suggesting that the estimates are not far from the true or population estimates. The difference between the means of the control group and the treatment group is 0.414302 and the corresponding *p-value* is 0.0021, which is less than the 0.5% level of significance; thus, there is a significant difference between the mean marks of the control and the treatment groups.

4.2 Part 2: Qualitative Data Analysis

As said earlier, content analysis was carried out on the qualitative component of the collected data from the standardized tests. The researchers supported this data with observations from group discussions and group presentations. This guided the researchers ascertain "how" participants conjectured solutions to each of the standardized tests. From this, the researchers deduced the implementation of the ACE teaching cycle on the experimental group improved participants' problem-solving competence.

4.3 Findings

The findings that emanated from conducting this study were presented and discussed in-line with the research question – H*ow can the ACE teaching cycle influence Grade 11 learners' achievements in circle geometry?* This study empirically established that the ACE teaching instructional approach significantly guided and enhanced participants' cognitive engagements and development, which, ultimately, optimized their problem-solving competence [7]. This is in relation to the assertion that mathematical processes are purely cognitive processes, characterized by reasoning and thinking skills [6]. Circle geometry, as explained, involves modeling and problem-solving abilities, in association with logical, inductive, and deductive reasoning, which are all cognitive processes. These instructional objectives are best accomplished, by implementing the constructivist perspective on teaching and learning.

The performance testified that the experimental group who were taught by the ACE teaching cycle demonstrated better understanding, better thinking, and better problem-solving competence in circle geometry in comparison to their counterparts in the control group who were taught by the traditional approach.

This study also established that the implementation of the ACE teaching cycle placed learners at the center of the learning process instead of the teacher; thus, the emphasis shifted from teacher-centered instructional approach (in the control group) to learner-centered instructional approach (in the experimental group). The latter group were encouraged to think, question, solve problems, and discuss their ideas, strategies, and solutions. The ACE teaching cycle, therefore, created an intellectual environment where advanced mathematical thinking was nurtured and developed [2].

5 Conclusion

In generic terms, the mean marks recorded from the experimental group's scores for each of the four standardized tests, as well as, from the composite scores were greater than the mean marks, respectively, recorded from the control group's scores. This established that the experimental group participants performed better than the control group participants. Also, the hypothesis test carried out justified that there was a significant improvement in the marks of the experimental group and this was achieved, primarily, from the implementation of the intervention approach – the ACE teaching cycle; the ACE teaching cycle, hence, enhanced learners' conceptual understanding and geometric reasoning. The researchers, thus, deduced from the conduct of this study that the ACE teaching cycle, as an intervention approach for teaching cycle geometry, elicited and evoked participants' thinking and problem-solving competence, which, in turn, optimized participants' achievements in circle geometry. The researchers, hence, recommend that the ACE teaching cycle should be used, either wholly or partly, as an instructional approach for circle geometry.

References

1. Bono, R., Alarcón, R., Blanca, M.J.: Report quality of generalized linear mixed models in psychology: a systematic review. Front. Psychol. **12**, 666182 (2021)
2. Brijlall, D., Ndlazi, N.J.: Analysing engineering students' understanding of integration to propose a genetic decomposition. J. Math. Behav. **55**, 100690 (2019)
3. Creswell, J.W.: Educational Research: Planning, Conducting, and Evaluating Quantitative and Qualitative Research, 4th edn. Pearson Education, Upper Saddle River (2012)
4. Johnson, V.E.: Revised standards for statistical evidence. Proc. Natl. Acad. Sci. U. S. A. **110**(48), 19313–19317 (2013)
5. Maharaj, A.: An APOS analysis of students' understanding of the concept of a limit of a function. Pythagoras. **71**, 41–52 (2010)
6. Mudrikah, A.: Problem-based learning associated by action-process-object-schema (APOS) theory to enhance students' high order mathematical thinking ability. Int. J. Res. Educ. Sci. **2**(1), 125–135 (2016)
7. Syarifuddin, H., Atweh, B.: The use of activity, classroom discussion, and exercise (ACE) teaching cycle for improving students' engagement in learning elementary linear algebra. Eur. J. Sci. Math. Educ. **10**(1), 104–138 (2022)

Exploring Possible Teacher and Learner Support Structures to Improve Learner Mathematics Performance

S. J. Ivasen and D. Brijlall

1 Introduction

The competence of learners, in Mathematics and the Sciences, is important for their social and economic improvement [1], which have a direct impact on the country's scientific and technological standing [3]. In a teacher training program, teachers simply do not learn everything they need to know about Mathematics, such as how students learn Mathematics and how to teach it effectively. Therefore, teacher education must be seen as an ongoing process and teachers need a basis for ongoing learning [1]. Kilpatrick et al. add that teachers need to adapt to evolving curriculum, new materials, technological changes, improvements in student thinking and teaching practice, and changes in the way education is delivered. Hence, the researchers concluded that sustained continuous support structures for professional development of Mathematics teachers will impact on the performance of their learners.

2 Literature Review

In its Curriculum Assessment and Policy Statement (CAPS), the Department of Basic Education of South Africa [4] defines Mathematics as a language that uses symbols and notations to describe numerical, geometric, and graphical relationships.

S. J. Ivasen
Department of Mathematics Durban, University of Technology, South Africa
e-mail: Selvami@nect.org.za

D. Brijlall (✉)
Department of Mathematics, Durban University of Technology, KwaZulu-Natal, South Africa
e-mail: deonarainb@dut.ac.za

F. Kamalov et al. (eds.), *Advances in Mathematical Modeling and Scientific Computing*, Trends in Mathematics, https://doi.org/10.1007/978-3-031-41420-6_81

According to Hersh [5], Mathematics is about human-created or invented ideas, that, the teacher's perception of Mathematics, her content knowledge, pedagogical practices, and preknowledge of the teacher's recipients all influence the teaching of Mathematics in the classroom. Most public primary school teachers in South Africa do not have Mathematics as a specialist subject in their education qualifications. As a result, teachers' motivation of students in the early grades is limited. The teacher's conception of Mathematics, the way Mathematics is to be taught, and his or her content knowledge are critical in motivating students to improve their Mathematics performance. Ramanujam et al. [6] claim that there is very little systemic communication between primary and secondary school Mathematics teachers, and none between secondary school Mathematics teachers and college/university teachers. In the Phambela Circuit, there exists one combined (Grades R to 12) and two senior primary (Grades R to 9) schools. Learners in Grades 3 and 6 are enrolled in the circuit's primary schools. Most Grade 9 learners are enrolled in the circuit's secondary schools. According to Ivasen [7], there is a significant disparity in primary school Grades 3 and 6 Mathematics performance when compared to the performance of the same learners in Grade 9 students in the secondary schools. This is evidence of possible compartmentalization, as there is no correlation between Grades 3 and 6 pass rates when compared to the Grade 9 pass rate. The lack of a Mathematics qualification among most primary school Mathematics teachers aggregates this disparity [2]. In this regard, district and circuit officials play a special role in supporting teaching and learning, to eliminate this disparity. They also use credible data to identify which schools and students are consistently underperforming across grades and subjects; which teachers and principals have these students in common; and what other characteristics of these schools may help explain these results.

3 Research Design and Methodology

3.1 Population and Sample

The *population* for this study included 7 principals, 8 deputy principals, 7 department heads who supervise Mathematics teaching and learning, 30 Mathematics teachers, and 3565 Grade 8 to 12 students in the Phambela Circuit's seven high schools. The researcher used deliberate (or purposeful) sampling. The criteria of the chosen sample in this study are participation in management, supervision, teaching, and/or learning of Mathematics in high schools in the Phambela Circuit. The *sample* included seven principals or deputy principals, seven departmental heads who supervise Mathematics teaching and learning, 14 Mathematics teachers (one who teaches Grades 8 and/or 9 and one who teaches Grades 10–12), and 35 learners (one for each grade from 8 to 12) from each of the seven high schools in the Phambela circuit.

3.2 *Choice of Approach*

According to Johnson and Onwuegbuzie [8], the purpose of the mixed methods approach is not to replace qualitative or quantitative approaches, but rather to build on the strengths and minimize the shortcomings of both in single and cross-study research studies. The researchers would be unable to observe the natural settings of individual schools and draw conclusions about how participants respond verbally to the questions, if only a quantitative method was used. The converse is also true: If the researchers used only a qualitative method, they would be unable to use statistics to quantify the relationships between variables (barriers) and how they affect learner Mathematics performance. To make recommendations on policy decisions, a combined qualitative and quantitative approach was adopted [9]. Hence a mixed-method approach was chosen. The researchers chose to adopt a sequential transformative strategy [10] as the questions in the interview schedules and questionnaires were guided by the research question.

The key methods used in this mixed-method study were a combination of closed-ended and open-ended questionnaires, structured interviews, and a literature review. Prior to the interviews, the researchers explained the study's goal, that their and the school's identities were kept strictly confidential, and that their participation in the study was entirely voluntary. For each category of participants, namely, principals/deputies, departmental heads, Mathematics teachers, and learners offering Mathematics within each school, four separate structured interview schedules of open-ended questions were used. IsiZulu was the mother tongue of most of the participants. Hence the interview schedules were presented in both English and IsiZulu. This bilingualism helped to clarify issues. The researchers verbally clarified questions in English (the language of learning and teaching) and used probing to make it easier for participants to respond. Each interview was recorded on a smartphone with a digital audio recorder application and saved in folders labeled A to G for each of the seven schools: AP (for principal/deputy principal), ADH (for Departmental Head), AFET (for a teacher offering Mathematics in Grades 10–12), and AGET (for a teacher offering Mathematics in Grades 8 and 9). For each interview with learners offering Mathematics in Grades 8–12, the labels A8, A9, A10, A11, and A12 were used. Ethical issues were adhered to when interviewing participants, and parental permission was requested when interviewing learners.

4 Data Analysis

Data analysis was an ongoing process for the researchers, beginning with the initial stages of reading the literature on the study's subject, determining the suitability of theoretical frameworks and methodology that would suit this study [11]. To extract categories, themes, patterns, and symmetries from the data from interviews, the focus was on the participants' context in relation to the phenomenon. The researchers adopted a six-phase method of qualitative and quantitative analysis of data, adapted from the seven-step mixed method. The researchers presented a qualitative analysis of data from the interviews and thereafter a quantitative analysis of the questionnaires, through the phases illustrated in Table 1.

The researchers made use of the qualitative analysis application, ATLAS.ti version 8. Each data reduction of interviews was uploaded, and codes were assigned to pieces of responses (quotations) from each participant. All the data reduction of interviews for principals/deputies and teachers of Mathematics and learners, one from each of Grades 8–12 was coded, and these codes were examined in ways to see if they could be meaningfully grouped together. The codes were compared and analyzed to search for patterns in the data and codes with similar themes was organized in Code Groups. The ATLAS.ti application allowed the researchers to keep track of a large amount of information as well as create a complex coding system that translated into frequency tables from which themes emerged. The researchers used closed ended (structured) questionnaires for the Principals/deputies, departmental heads, and teachers of Mathematics. The Rickert Scale type of questions was used because the intention was to measure degrees and intensity of the responses. The SPSS Data editor was used to create variables for each question and value-coded all options to the questions for each questionnaire, as quantitative analysis is associated with numbers. The researchers, thereafter, analyzed the different questionnaire categories – principals/deputy principals, departmental heads, teachers of mathematics, and learners of Mathematics. Statistics of various variables in all categories of the

Table 1 Six phases in the mixed-method data analysis process

Phase in the mixed methods analysis of data process	Application in qualitative data analysis using Atlas.ti Version 8	Application in the quantitative data analysis using SPSS
Phase 1 (data reduction, capture, coding, and display)	1(a) Qualitative data reduction, data capture, coding, charts, and diagrams	1(b) Data capture, data display (descriptive stats, tables, and graphs)
Phase 2 (data transformation)	2(a) Qualitative data is represented statistically	2(b) Quantitative data is translated into narrative data that can be analyzed qualitatively
Phase 3 (Correlation of data)	Correlation between qualitative with quantitative data	
Phase 4 (Consolidation of data)	Qualitative and quantitative data are combined	
Phase 5 (Comparison of data)	Qualitative and quantitative data are compared	
Phase 6 (Integration of data)	Qualitative and quantitative data are integrated into a coherent whole (final findings and discussions)	

questionnaires and a frequency descriptive statistic of all variables relating to the research questions was presented. These frequency statistics were converted into narrative data that could be qualitatively analyzed.

The researchers examined the interview data and questionnaire data to search for evidence of correlation between them. In addition to the aspects that indicated correlation, the researcher examined the remaining interview and questionnaire data to search for themes that were closely connected to the research questions of this study. With SPSS, there exists an opportunity to check for correlation between responses from the questionnaires, while interviews do not provide this, statistically. After transforming, correlating, consolidating, and comparing the interview and questionnaire data, the researcher arranged the themes discussed so that they could address the research aims/questions.

5 Findings and Discussion

5.1 Systemic Challenges

Class sizes and staffing in South African public schools have been inherently problematic in post-Apartheid South Africa. While the learner–teacher ratio in ex-model C and private schools was low, most public schools are overcrowded. Despite the post provisioning requirements and standards for staffing of schools and funding policies by the Department of Basic Education to resolve the issues of equity and redress (overcrowded and poor schools), most public schools in the Phambela Circuit remain overcrowded. The learner–teacher ratio for Mathematics in the Phambela Circuit was 71:1. This clearly indicates that the postprovisioning model implemented by the Provincial Education Department is not working. At the time of writing this chapter, South Africa was in a Covid-19 lockdown. The difficulties in respect of learner–teacher ratio in public schools were exposed. In fact, all imbalances in public schools were exposed. Most public-school principals and SMTs were tasked to adjust their timetables to accommodate classes of 20:1, in keeping with the Department of Health's, social distancing COVID-19 protocols. In view of the overcrowding many Phambela schools adopted the alternate day, alternate grade option to comply with COVID – 19 protocols. This resulted in lesser notional time. The Annual Teaching Plans (in Grades 8–11) and School Based Assessments (in Grades 8–12) were adjusted in each subject to accommodate the lesser notional time. This differed from school to school, as the problem was further compounded by teachers, who cited comorbidities, applied to work from home. In addition, some learners, who citied anxiety and fear of the COVID-19 virus, did not return to school. The National Senior Certificate timetable was revisited to accommodate the reduced notational time and social distancing policies. This had a ripple effect on the internal examination timetables.

The Annual Teaching Plans (ATPs) for Mathematics are tight and cannot be successfully completed in the allocated time. Mathematics teachers were compelled to make use of additional classes to complete the ATPs. The Mathematics teacher is not allowed freedom to adapt the ATPs to suit her students' level and the unique challenges experienced in her school. The topics tested in Grades 10, 11, and 12; provincial common tests; and National Mathematics papers prove this. The COVID-19 lockdown proved that ATPs could be adjusted to accommodate challenges faced in the different schools.

Noninclined Mathematics Departmental Heads form 81.8% of school management teams. The absence of this Mathematical background provides no motivation in improving learner Mathematics performances. To improve learner Mathematics performances, the Departmental Heads must make a meaningful contribution and play a supportive role in content creation, interactions, discussions, encourage teamwork, and provide alternate solutions to problems.

The promotion and progression requirements for the GET phase vastly differ from those requirements in the FET Phase. Every year there is an adjustment with the promotion requirements, in the GET phase. In 2019, learners in the GET phase, who met all other requirements, but failed Mathematics, were allowed to progress to the next grade. This further allowed learners (and in some cases, teachers) not to work harder in Mathematics in the GET phase because learners are guaranteed the option of progression to the next grade despite a failure in Mathematics. The negative attitude in Mathematics by learners is carried over in Grades 10 and 11, where they are allowed to fail Mathematics and pass all other subjects and still be promoted to the next grade. There is an exodus of learners from Mathematics to Mathematical Literacy in the FET phase. The permitted combination of Mathematical Literacy with Accounting in the FET phase further lends impetus to this exodus.

The shortage of Mathematics LTSM ensures that teachers and learners are not actively engaged in any investigative work due to the lack of not only Mathematics textbooks but Mathematical instruments like calculators, compasses, dividers, protractors, rulers and set squares. The nonreturn of LTSM, especially textbooks on loan and calculators, by learners also poses a problem for the recipient learners in the following year.

5.2 Societal Challenges

Socioeconomic status of learners' families impacts on leaners Mathematics performance. Over 50% of learners interviewed come from families that rely on state pension, social grants or care giver sponsors and they form the lowest income group of South African society. The affordability of additional references books, guides and stationery like scientific calculators, Mathematics instruments like compasses, dividers, and set squares posed challenges to learners who come from these backgrounds.

There exist serious *language barrier* problems with learners. This hinders the translation of definitions of Mathematical concepts. Most of the learners have isiZulu as their mother tongue. Further, there is the element of embarrassment for learners to participate in group/class discussions because as they are not eloquent speakers of English. The wording and phrasing of questions in the common tests and examinations differ from those used in daily classwork and homework exercises.

Most learners have an inborn *negative perception and attitude for Mathematics* from an early age. This is compounded by comments made by parents, peers, and some teachers that Mathematics is difficult.

Teachers of Mathematics cite *poor parental involvement* as a serious challenge. They receive little or no cooperation from parents with regard to their children's Mathematics performances. Open days are attended by very few interested parents of learners who display a positive attitude to Mathematics.

5.3 Pedagogical Challenges

Teachers of Mathematics complain about learners having *poor Mathematics backgrounding*. Teachers cannot complete the respective grade ATPs, as they spend additional time reteaching basic Mathematical fundamentals like addition, subtraction, multiplication, and division. Learners learn Mathematics in water-tight compartments. They believe that what is taught in the previous grades is not applicable in the next grade.

Poor learner behavior poses the highest challenge to teachers and interested learners in the Mathematics classroom. Only a few learners in class are interested in the lesson, while the majority are rowdy and disrupt the lesson.

Two thirds of learners are *unaware of the ATPs and the SBAs*. These documents provide scope, give direction, and furnishes a plan of action towards completing the syllabus for the year. They provide vital information about weighting (in terms of marks) for each section in Mathematics, which impacts on the time learners and teachers need to spend on that section. Furthermore, learners must be aware of how SBAs and their weightings feature during their progress with the Mathematics syllabus.

Over 50% of the SMT and teachers of Mathematics were unaware of the *Framework for Teaching Mathematics for Understanding*. This framework adopts a conceptual fields approach as an approach that enables learners to see Mathematics as a linked network of concepts and relationships and connections between the different concepts and related processes. The goal is to improve teaching awareness and teaching skills.

Teacher preparedness for the modern classroom poses another challenge to learners.

Although teachers have the necessary Mathematics content knowledge, they lack the ability to present Mathematics lessons to modern children who come from diverse backgrounds and who face a variety of social challenges. There is a need for the didactics of the presentation of lessons to be revised continuously according to the current classroom situation. Presentation of lessons that worked previously may not work in the present or in the future. The teacher's bag of pedagogic techniques must be revamped to meet modern trends in Mathematics teaching and learning.

Although Phambela high schools Mathematics teachers are adequately qualified, they lack a range of *Mathematics teaching strategies*. Most learners complained that they struggled to complete classwork and homework, without the assistance of the teacher. This lack of using alternatives to the "chalk and talk" technique revealed itself during the COVID-19 pandemic. Due to closure of schools, teachers were tasked to communicate with learners and assign tasks via social media like WhatsApp chat groups through text and voice notes, use Google classroom and other electronic media. The Department of Education and most teachers were not adequately prepared to adapt to this situation as they were not trained or workshopped in using internet-based learning.

6 Recommendations

First, the researchers recommend the establishment of Mathematics Professional Learning Communities (PLCs) between high schools and their feeder primary schools, in the Phambela Circuit. When high school and feeder primary school Mathematics teachers meet in PLCs, most learner and teacher challenges, strengths and weaknesses, and gaps in Mathematics teaching and learning can be identified and addressed. The topics covered/not covered in the Annual Teaching Plans (ATPs) of the previous year are handed over. This allows the teacher in Grade 8 (high school) to revisit/adjust the Grade 8 Mathematics ATPs to cover work that is prerequisite for a topic in Grade 8, likewise amongst other grades. There must exist PLCs within the school, which must address the issues of handing over, adjustment of ATPs, collaborative, cooperative, cross curricula, and rotational teaching. The researchers also recommend the formation of PLCs between the Phambela High Schools and mathematicians and researchers from post school institutions of higher learning. High school and post school institutions of higher learning PLCs will address gaps in high school Mathematics teaching; identify and place more emphasis on specific topics; understand the nature of assessments, teaching, and learning styles; and address the issues of accountability for results.

Second, to address the issue of language barriers, the researchers suggest a strategy called "Teaching Mathematics across the Curriculum" be implemented. Every subject offered in each grade level throughout the child's education must include a mandatory Mathematics module. For example, Mathematics vocabulary on concepts may be taught in English, and a translation or interpretation of English Mathematics concepts may be taught in the vernacular language. A child's education must be viewed as a collective responsibility of all subject teachers. This will provide a support base for Mathematics teachers to eliminate, or minimize, the teaching of preknowledge and Mathematics vocabulary. The researchers recommend that all education associations, teacher labor unions, nongovernmental organizations with interest in the improvement of learner Mathematics performance, and the Department of Basic Education form a task team to develop Mathematics materials in the vernacular, in phases. The researchers propose a pilot project, beginning with the Foundation Phase (Grades 1–3), that permits Mathematics to be taught and answered in the local vernacular to see if there is a positive improvement in learner Mathematics performance. The results could be compared with a control group of second language learners being taught Mathematics through the medium of English. This project must be phased gradually into the Intermediate Phase (Grades 4–6), the GET Phase (Grades 7–9), and, finally, the FET Phase (Grades 10–12). Ultimately, schools will have the option of assessing the Mathematics curriculum in English, the local vernacular, or the student's mother tongue.

Third, the prescribed Mathematics syllabi must be changed to a modular/credit approach. In all grades, learners (supported by parents/teachers) must have the choice to choose modules and that will acquire credits, upon passing. The attainment of credits must determine a pass/fail in the subject. In this way, the choice of modules will assist with career choices.

Fourth, to encourage parents to participate in the school's programs, the researchers suggest a "Partnership Model". According to this model, teachers are regarded as educational experts, while parents are regarded as experts on their children. The goal is to establish a relationship in which teachers and parents share knowledge and influence to provide the best possible education for the children. This relationship is based on mutual respect, which includes respect for the opinions of others as well as respect for the integrity of others. To address barriers to parental engagement, the school must have a supportive environment, with school staff being respectful and sensitive to parents, and vice versa. While most schools use students to communicate meeting notices to parents, other methods of communication must be explored. Each parent/guardian may be given the school calendar of activities, programs, and dates for parental meetings at the start of each year. Learners should distribute a bi-lingual (English/IsiZulu) newsletter to families on a monthly/quarterly basis. Photographs of parents at meetings may be used in the newsletter or on the Facebook page with their permission. This will foster a sense of community and belonging.

Last, the fourth industrial revolution's demands in South Africa are increasing by the day. Teachers are at the forefront of preparing students to be technologically literate so that teaching and learning activities can benefit from scientific and technological advancements. There is no technology that can compensate for poor teaching. As a result, it is critical that teachers learn how to use technology while focusing on improved learning. Teachers must work hard to become familiar with the latest technology and how to incorporate it into Mathematics teaching and learning. Teachers are expected to create Mathematics programs that are relevant to the school's Mathematical activities. Technology is not the solution; rather, it is a facilitator.

7 Conclusion

This study also suggests that, regardless of how many barriers exist in a learning situation, the principal and her staff can make use various support structures to improve the Mathematics academic performance of learners. All sciences and technology, and, thus, all human progress, are built on Mathematics. Teachers of Mathematics must tailor their lessons to their students' needs and learning styles. Technology has brought the outside world into the classroom as well as taken the classroom into the outside world, allowing teaching, and learning to occur at any time and in any location. Teachers must plan e-learning activities with the future in mind, because the students they teach will be using technologies that are far beyond what is currently available. Policymakers must come on board and provide the necessary training before acquiring the relevant technological hardware and software to develop adaptable and technologically competent Mathematics teachers and learners. Teachers must also understand that no amount of new technology will ever be able to compensate for a poor Mathematics teacher. As a result, to prepare teaching materials and assessments for software-based e-learning, the Mathematics teacher must consult learning and teaching theories (Vygotsky's sociocultural and conceptual development, ZPD, and Mathematics proficiency, via scaffolding in all its forms). To implement the recommendations and strategies outlined in this study, we must refocus our priorities and resources, because passing high school Mathematics opens the door for our children to be accepted into tertiary institutions of higher learning, resulting in better career opportunities.

References

1. Kilpatrick, J., Swafford, J., Findell, B.: Adding It Up: Helping Children Learn Mathematics. The National Academies Press, Washington, DC (2001)
2. Reddy, V., et al.: Beyond Benchmarks: What 20 Years of TIMSS Data Tell Us About South African Education. HSRC Press, Cape Town (2015)
3. Othman, M., Yusof, Y., Mahmood, A.: A conceptual framework for mathematical ability analysis through the lens of cultural neuroscience. Proc.: Soc. Behav. Sci. **56**, 175–182 (2012)
4. Department of Basic Education.: Curriculum and Assessment Policy Statement (2011)
5. Hersh, R.: Some proposals for reviving the philosophy of mathematics. Adv. Math. **31**, 31–50 (1979)
6. Ramanujam, R., et al.: Teaching of Mathematics. National Council of Educational Research and Training, New Delhi (2006)
7. Ivasen, S.J.: A Brief Report on Status/Challenges Identified in Grade 3/6/9, Mathematics, in the Phambela Circuit. Phambela Mathematics Committee, KwaDukuza (2018)
8. Johnson, R., Onwuegbuzie, A.: Mixed methods research: a research paradigm whose time has come. Educ. Res. **33**(7), 14–26 (2004)
9. Hart, L., Smith, S., Swars, S.L., Smith, M.: An examination of research methods in mathematics education. J. Mixed Methods Res. **3**(1), 26–41 (2009)
10. Creswell, J.W.: Research design, qualitative, quantitative & mixed method approaches. In: Third Edition: Research Design, pp. 203–223. Sage Journals, California (2009)
11. Naicker, P.: Integrating Learner Drifters from Rural and Township Schools to Suburban Schools. University of KwaZulu Natal, Durban (2014)

A Study of Mathematical Epidemiology Model of Dengue Spread with Fractional Properties

Sonal Jain, Ho-Hon Leung, and Firuz Kamalov

1 Introduction

In tropical areas of the world, dengue disease is a typical arboviral illness. The *Aedes* mosquito bite is how humans are exposed to it. There are four different types of viruses, classified as I, II, III, and IV. The bites of *Aedes* mosquitoes transmit these viruses to people. A person who contracts one of the four viral serotypes once loses immunity to the other three serotypes, according to Gubler [8], but never again contracts that particular serotype. To comprehend the dynamics of infectious illness transmission better, numerous mathematical models have been created and tested. The investigation of dengue illness dynamics and the identification of the variables affecting the disease's spread in order to support control strategies have both benefited greatly in recent years from modeling. Many researchers have proposed epidemic model [5–8, 10, 12, 13, 15, 17] to research the kinetics of the dengue virus.

The dengue virus cannot be treated with a specific medication. Programs that raise public awareness may be useful in lowering the disease's prevalence. Various epidemic models have been put out to investigate how awareness affects dengue disease management. One method of preventing mosquito bites is dengue disease prevention. When the lights are on, both day and night, mosquitoes bite people. Therefore, people can use mosquito nets and repellents to prevent mosquito bites.

S. Jain (✉)
School of Technology, Woxsen University, Hyderabad, Telangana, India

H.-H. Leung
Department of Mathematical Sciences, UAE University, Al Ain, UAE
e-mail: hohon.leung@uaeu.ac.ae

F. Kamalov
Department of Electrical Engineering, Canadian University Dubai, Dubai, UAE
e-mail: firuz@cud.ac.ae

F. Kamalov et al. (eds.), *Advances in Mathematical Modeling and Scientific Computing*, Trends in Mathematics, https://doi.org/10.1007/978-3-031-41420-6_82

Infected hosts can recover quickly if they recognize disease signs and visit the doctor in time for supportive care. This level of awareness can aid in disease management. Another way of controlling dengue is destroying larval breeding sites of mosquitoes and killing them. Spray of insecticides may be applied to control larvae or adult mosquitoes which can transmit dengue viruses. This type of biological model have two properties as we observed Markovian and Non-Markovian. In dengue spread model does not really follow the Markovian process therefore does have memory effect, thus can well be described using the concept of nonlocal differential operators with non-local and non-singular kernel as these operators have a crossover from exponential decay law to power law as waiting time distribution.

Numerous areas, including mathematical biology, physics, fluid dynamics, electrochemistry, signal processing, viscoelasticity, mathematical biology, economics, and many more, employ fractional calculus in various ways. Fractional calculus derivatives are significant concepts in the discipline of fractional calculus. Many scientists and researchers have recently examined problems in this particular branch; see [3, 9]. Atangana-Baleanu fractional derivatives feature both Markovian and non-Markovian qualities, offering applied mathematics new methods to more correctly describe complicated real-world issues. The well-known Caputo-Fabrizio derivative is not Markovian, whereas the Riemann-Liouville derivative is exclusively Markovian. In contrast to the simple power law of the Riemann-Liouville derivative and the simple exponential decay of the Caputo-Fabrizio derivative, the derivative waiting time exhibits power law, stretched exponential, and Brownian motion simultaneously. The derivative mean square displacement distinguishes sub-diffusion from conventional diffusion, whereas Riemann-Liouville is simply a power law and scale-invariant. This shows that the Atangana-Baleanu fractional derivative can be used to describe multiscale or real-world phenomena whose properties change across time and space, such as the spread of disease, the transport of toxins inside fractured aquifers, and many others. This crossover tendency is present in many empirical systems.

The main goal of this research is to apply the fixed point theorem to a fractional Caputo-Fabrizio derivative mathematical system in order to uncover memory and crossover effects and impart information about the precision and uniqueness of the solution. The progression of this article is as follows: In Sect. 2, the Caputo-Fabrizio derivative and Atangana-Baleanu fractional derivative are discussed. In Sect. 3 and Sect. 4 the fractional dengue spread model and approximate solution pertaining to novel CF fractional derivative are discussed. Discussion of numerical solutions is provided in Sect. 5. The final section, Sect. 6, contains the concluding remarks.

2 Preliminaries

In this part, we give a few definitions and traits of the fractional derivative.

Definition 1 When p is a function that is not necessarily differentiable and λ is a real number $\ni 0 \leq \lambda \leq 1$, the Riemann-Liouville derivative with λ order and power law is given as citesamko.

$$^{RL}D_t{}^{\lambda}[p(t)] = \frac{1}{\Gamma(1-\lambda)} \frac{d}{dt} \int_0^t (t-x)^{-\lambda} p(x) dy. \quad (1)$$

Definition 2 ([4]) The Caputo-Fabrizio derivative of order λ for $p \in H^1(a, b), b > a, \lambda \in [0, 1]$ is defined by:

$$D_t^{\lambda}(p(t)) = \frac{M(\lambda)}{(1-\lambda)} \int_a^t p'(x) \exp\left[-\lambda \frac{t-x}{1-\lambda}\right] dx. \quad (2)$$

where $M(\lambda)$ is a normalization function $\ni M(0) = M(1) = 1$. However, the derivative can be recast as follows if the function does not belong to $H_1(a, b)$.

$$D_t^{\lambda}(p(t)) = \frac{M(\lambda)}{(1-\lambda)} \int_a^t (p(t) - p(x)) \exp\left[-\lambda \frac{t-x}{1-\lambda}\right] dx. \quad (3)$$

Definition 3 ([11]) Consider $0 < \lambda < 1$ for a function p is fractional integral of order λ which is defined by

$$I_{\lambda}^t(p(t)) = \frac{2(1-\lambda)}{(2-\lambda)M(\lambda)} f(t) + \frac{2\lambda}{(2-\lambda)M(\lambda)} \int_0^t p(s) ds, t \geq 0. \quad (4)$$

Definition 4 The function f with $f \in H^1(x, y), y > x, \lambda \in [0, 1]$ is differentiable; then, the definition of Atangana-Baleanu derivative is as follows:

$${}_a^{ABC}D_t^{\lambda}[p(t)] = \frac{B(\lambda)}{1-\lambda} \int_a^t \frac{d}{dt} p(x) E_{\lambda}\left[-\lambda \frac{(t-x)^{\lambda}}{1-\lambda}\right] dx. \quad (5)$$

with $B(0) = B(1) = 1$

Definition 5 The fractional integral associate with Atangana-Baleanu fractional integral is as follows [1, 2]:

$${}_a^{ABC}I_t^{\lambda}[p(t)] = \frac{1-\lambda}{B(\lambda)} p(t) + \frac{\lambda}{B(\lambda)\Gamma(\lambda)} \int_a^t p(j)(t-j)^{\lambda-1} dj. \quad (6)$$

When lambda is zero, the original function is recovered; if lambda is also 1, the ordinary integral is obtained.

3 Model Description

The model divides the whole host population (Mh) into four classes: Three classes are defined by the variables R_h (recovered), I_h (infected), E_h (exposed), and S_h (susceptible), and the total vector (mosquito) population N_v is divided into three classes—I_v (infectious), E_v (exposed), and S_v (susceptible). We presume that w_1 of susceptible hosts use insect repellents to avoid being bitten by mosquitoes. In this manner, infectious mosquitoes engage with the fraction $(1 - w_1)$ of susceptible hosts. The proportion of infectious hosts $w2$ seeks out prompt supportive care and recovers quickly at the rate $r\gamma h(r > 1)$. Mosquito repellents are used by the proportion $s1w2$ of infectious hosts (s_1 is the proportionality constant). An insecticide spraying attempt is represented by the control variable w_3. The recruitment rate of the mosquito population is consequently decreased by a factor of $1 - w_3$. Additionally, it is assumed that mosquito mortality rates rise at a rate of s_2w_3 (s_2 is the proportionality constant).

The dynamics of dengue fever are discussed in this section, along with prevention strategies. The current SEIR-SEI vector host model's system of differential equations is discussed by [12].

$$
\begin{aligned}
\frac{dS_h}{dt} &= \lambda_h M_h - (1 - w_1)\frac{b\zeta_h}{M_h} S_h I_\phi - \lambda_h S_h \\
\frac{dE_h}{dt} &= (1 - w_1)\frac{b\zeta_h}{M_h} S_h I_\phi - (\phi_h + \lambda_h) E_h \\
\frac{dI_h}{dt} &= \phi_h E_h - [r w_2 \gamma h + (1 - w_2)\gamma h + \lambda_h] I_h \\
\frac{dR_h}{dt} &= [r w_2 \gamma_h + (1 - w_2)\gamma_h] I_h - \lambda_h R_h \\
\frac{dS_v}{dt} &= (1 - w_3)\pi_\phi - (1 - s_1 w_2)\frac{b\zeta_h\phi}{M_h} S_\phi I_h - (s_2 w_3 + \lambda_h\phi) S_\phi \\
\frac{dE_\phi}{dt} &= (1 - s_1 w_2)\frac{b\zeta_h\phi}{M_h} S_\phi I_h - (s_2 w_3 + \phi_\phi + \lambda_h\phi) E_\phi \\
\frac{dI_\phi}{dt} &= \phi_\phi E_\phi - (s_2 w_3 + \mu\phi) I_\phi
\end{aligned}
\tag{7}
$$

The model parameters are given as follows: λ_h is the death rate of host population, ϕ_h is given as host's incubation rate, γ_h represents recovery rate of host population, ζ_h is defined as transmission probability from vector to host, π_ϕ represents recruitment rate of vector population, μ_ϕ gives death rate of vector population, ϕ_ϕ represents vector's incubation rate, β_ϕ represents transmission probability from host to vector, and b is defined as biting rate of vector. Total host population, $M_h = S_h + E_h + I_h + R_h$, total vector population, $N_\phi = S_\phi + E_\phi + I_\phi$.

$$\frac{dM_h}{dt} = 0 \text{ and } \frac{dN_\phi}{dt} = (1 - w_3)\pi_\phi - (s_2 w_3 + \mu_\phi)N_\phi.$$

So, M_h remains constant and N_ϕ approaches the equilibrium $(1 - w_3)\pi_\phi(s_2 w_3 + \mu_\phi \phi)$ as $t \to \infty$. Introducing the proportions

$$A_h = \frac{S_h}{M_h},\ B_h = \frac{E_h}{M_h},\ C_h = \frac{I_h}{M_h},\ D_h = \frac{R_h}{M_h},\ E_\phi = \frac{S_\phi}{(1 - w_3)\pi\phi/(s_2 w_3 + \mu_\phi)}$$

$$e_\phi = \frac{E_\phi}{(1 - w_3)\pi_\phi \phi/(s_2 w_3 + \mu_\phi \phi)},\ i_\phi = \frac{I_\phi}{(1 - w_3)\pi_\phi/(s_2 w_3 + \mu_\phi \phi)}$$

with $r_h = 1 - A_h - B_h - C_h$ and $E_\phi = 1 - e_\phi - i_\phi$, the corresponding five-dimensional nonlinear system of ODEs can be used to represent the equation system (7) as follows:

$$\begin{aligned}
\frac{dA_h}{dt} &= \lambda_h(1 - A_h) - \zeta A_h i_\phi \\
\frac{dB_h}{dt} &= \zeta A_h i_\phi - \beta B_h \\
\frac{dC_h}{dt} &= \phi_h B_h - \gamma C_h \\
\frac{de_\phi}{dt} &= \delta E_\phi C_h - (\epsilon + \phi_\phi)e_\phi \\
\frac{di_\phi}{dt} &= \phi_\phi e_\phi - \epsilon i_\phi
\end{aligned} \tag{8}$$

Here,

$$\zeta = \frac{b\zeta_h \pi_\phi(1 - w_1)(1 - w_3)}{M_h(s_2 w_3 + \mu\phi)},\quad \beta = \phi_h + \lambda_h,\quad \gamma = r w_2 \gamma_h + (1 - w_2)\gamma_h + \lambda_h,$$

$$\delta = (1 - s_1 w_2)b\beta_\phi,$$

$\epsilon = s_2 w_3 + \mu_\phi$.

This system has no memory and is exponentially stable due to the Markovian process. Therefore, we regulate the system by a non-Markovian process using the recently proposed to incorporate the Caputo-Fabrizio arbitrarily ordered derivative as offered by the memory effect.

$$\begin{aligned}
{}_0^{CF}D_t^{\lambda}A_h &= \lambda_h(1-A_h) - \zeta A_h i_\phi \\
{}_0^{CF}D_t^{\lambda}B_h &= \zeta A_h i_\phi - \beta B_h \\
{}_0^{CF}D_t^{\lambda}C_h &= \phi_h B_h - \gamma C_h \\
{}_0^{CF}D_t^{\lambda}e_\phi &= \delta E_\phi C_h - (\epsilon + \phi_\phi)e_\phi \\
{}_0^{CF}D_t^{\lambda}i_\phi &= \phi_\phi e_\phi - \epsilon i_\phi
\end{aligned} \tag{9}$$

The initial conditions are:

$$A_h(0) = \delta_1, \quad B_h(0) = \delta_2, \quad C_h(0) = \delta_3, \quad e_\phi(0) = \delta_4, \quad i_\phi(0) = \delta_5. \tag{10}$$

4 Numerical Solution

In this section, we provide a numerical framework for the fractional model based on the CF derivative. This takes into account the following nonlinear fractional ordinary equation.

$$\begin{cases} {}_0^{CF}D_t^{\lambda}u(p) = g(t, u(p)) \\ u(0) = u_0 \end{cases} \tag{11}$$

The equation above can be converted into a fractional integral equation by applying the fundamental theorem of fractional calculus:

$$u(p) - u(0) = \frac{2(1-\zeta)}{(2-\zeta)M(\zeta)} g(p, u(p)) + \frac{2\zeta}{(2-\zeta)M(\zeta)} \int_0^p g(\xi, u(\xi))d\xi, \tag{12}$$

The preceding equation is rewritten as $n = 0, 1, 2, \ldots$ at a given position p_{n+1}.

$$\begin{aligned}
u(p_{n+1}) - u(p_0) = &\frac{2(1-\zeta)}{(2-\zeta)M(\zeta)}[g(p_{n+1}) - g(p_n)] \\
&+ \frac{2\zeta}{(2-\zeta)M(\zeta)} \int_{p_n}^{p_{n+1}} g(\xi, u(\xi))d\xi
\end{aligned} \tag{13}$$

The second step of our numerical scheme consists of approximation of the function $g(p, u(p))$. We therefore approximate $f(p, u(p))$ within the range $[p_n, p_n + 1]$ using the well-known Lagrange interpolation polynomial which yields:

$$P(\xi)(\approx g(\xi, u(\xi))) = \left\{\frac{(\xi - p_{n-1})}{(p_n - p_{n-1})}\right\} g(p_n, u_n) + \left\{\frac{(\xi - p_n)}{(p_{n-1} - p_n)}\right\} g(p_{n-1}, u_{n-1}) \tag{14}$$

$$P(\xi)(\approx g(\xi, u(\xi))) = \left\{\frac{(\xi - p_{n-1})}{(p_n - p_{n-1})}\right\} g_n + \left\{\frac{(\xi - p_n)}{(p_{n-1} - p_n)}\right\} g_{n-1} \tag{15}$$

Consequently, the aforementioned approximation can be added to Eq. (13) to get

$$\begin{aligned} u(p_{n+1}) - u(t_0) = &\frac{2(1-\zeta)}{(2-\zeta)M(\zeta)}[g(p_{n+1}) - g(p_n)] \\ &+ \frac{2\zeta}{(2-\zeta)M(\zeta)} \int_{p_n}^{p_{n+1}} \left[\left\{\frac{(\xi - p_{n-1})}{(p_n - p_{n-1})}\right\} g_n \right. \\ &\left. + \left\{\frac{(\xi - p_n)}{(p_{n-1} - p_n)}\right\} g_{n-1}\right] d\xi \end{aligned} \tag{16}$$

As a result, after several simplifications and integration, the equation shown below is created:

$$u_{n+1} - u_n = \frac{2(1-\zeta)}{(2-\zeta)M(\zeta)}[g_{n+1} - g_n] + \frac{2\zeta}{(2-\zeta)M(\zeta)} h \left[\frac{3}{2} g_n - \frac{1}{2} g_{n-1}\right] \tag{17}$$

Now for finding the numerical solution of fractional model based on the CF derivative. We arrive at a solution to Eq. (9).

$$s_{h_{n+1}} - s_{h_n} = \frac{2(1-\zeta)}{(2-\zeta)M(\zeta)} \left[\lambda_h(1 - s_{h_{n+1}}) - \zeta s_{h_{n+1}} i_{\phi_{n+1}} - \lambda_h(1 - s_{h_n}) + \zeta s_{h_n} i_{\phi_n}\right]$$

$$+ \frac{2\zeta h}{(2-\zeta)M(\zeta)} \left[\frac{3}{2}[\lambda_h(1 - s_{h_n}) - \zeta s_{h_n} i_{\phi_n}] - \frac{1}{2}[\lambda_h(1 - s_{h_{n-1}}) - \zeta s_{h_{n-1}} i_{\phi_{n-1}}]\right]$$

$$e_{h_{n+1}} - e_{h_n} = \frac{2(1-\zeta)}{(2-\zeta)M(\zeta)} \left[\zeta s_{h_{n+1}} i_{\phi_{n+1}} - \beta e_{h_{n+1}} - \zeta s_{h_n} i_{\phi_n} + \beta e_{h_n}\right]$$

$$+ \frac{2\zeta h}{(2-\zeta)M(\zeta)} \left[\frac{3}{2}\left[\zeta s_{h_n} i_{\phi_n} - \beta e_{h_n}\right] - \frac{1}{2}\left[\zeta s_{h_{n-1}} i_{\phi_{n-1}} - \beta e_{h_{n-1}}\right]\right]$$

$$i_{h_{n+1}} - i_{h_n} = \frac{2(1-\zeta)}{(2-\zeta)M(\zeta)} \left[\phi_{h_{n+1}} e_{h_{n+1}} - \gamma i_{h_{n+1}} - \phi_{h_n} e_{h_n} + \gamma i_{h_n}\right]$$

$$+ \frac{2\zeta h}{(2-\zeta)M(\zeta)} \left[\frac{3}{2}\left[\phi_{h_n} e_{h_n} - \gamma i_{h_n}\right] - \frac{1}{2}[\phi_{h_{n-1}} e_{h_{n-1}} - \gamma i_{h_{n-1}}\right]$$

$$e_{\phi_{n+1}} - e_{\phi_n} = \frac{2(1-\zeta)}{(2-\zeta)M(\zeta)}$$

$$\left[\delta s_{\phi_{n+1}} i_{h_{n+1}} - (\epsilon + \phi_{\phi_{n+1}}) e_{\phi_{n+1}} - \delta s_{\phi_n} i_{h_n} + (\epsilon + \phi_{\phi_n}) e_{\phi_n}\right]$$

$$+ \frac{2\zeta h}{(2-\zeta)M(\zeta)} \left[\frac{3}{2}\left[\delta s_{\phi_n} i_{h_n} - (\epsilon + \phi_{\phi_n}) e_{\phi_n}\right] - \frac{1}{2}\left[\delta s_{\phi_{n-1}} i_{h_{n-1}} - (\epsilon + \phi_{\phi_{n-1}}) e_{\phi_{n-1}}\right]\right]$$

$$i_{\phi_{n+1}} - i_{\phi_n} = \frac{2(1-\zeta)}{(2-\zeta)M(\zeta)}\ \phi_{\phi_{n+1}} e_{\phi_{n+1}} - \epsilon i_{\phi_{n+1}} - \phi_{\phi_n} e_{\phi_n} + \epsilon i_{\phi_n}]$$

$$+ \frac{2\zeta h}{(2-\zeta)M(\zeta)}\left[\frac{3}{2}\ \phi_{\phi_n} e_{\phi_n} - \epsilon i_{\phi_n}] - \frac{1}{2}\ \phi_{\phi_{n-1}} e_{\phi_{n-1}} - \epsilon i_{\phi_{n-1}}]\right] \tag{18}$$

5 Numerical Simulation

In this part, we show how the model can be numerically replicated using the recommended numerical scheme for different values of fractional order.

Figures 1, 2, 3, 4, and 5 display the results of the numerical simulations. Figures 1 and 2 are each given a λ value of 1, λ values of 0.75 and 0.55 in Fig. 3, λ values of 0.35 in Fig. 4, and λ values of 0.15 in Fig. 5, respectively.

The aforementioned initial circumstances and settings were used to produce our numerical simulation [14].

$M_h = 5071126,\ \pi_\phi = 2500000,\ \phi_h = 0.1667,\ \lambda_h = 0.0045,\ \mu_\phi = 0.02941,\ \gamma_h = 0.328833,\ b\zeta_h = 0.75,\ b\beta_\phi = 0.375,\ \phi_\phi = 0.1428.$

Fig. 1 For $\lambda = 1$

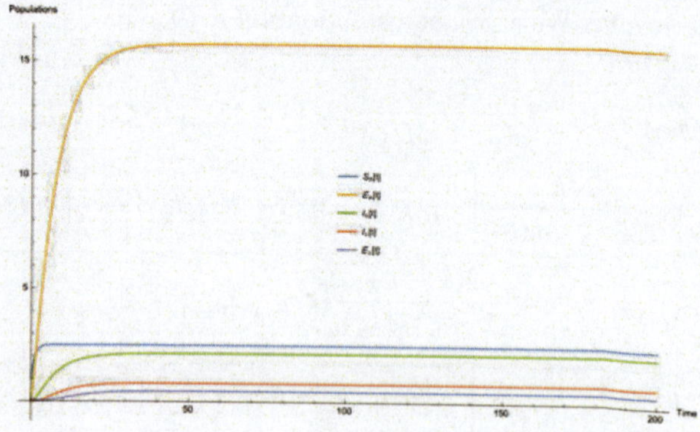

Fig. 2 For $\lambda = 0.75$

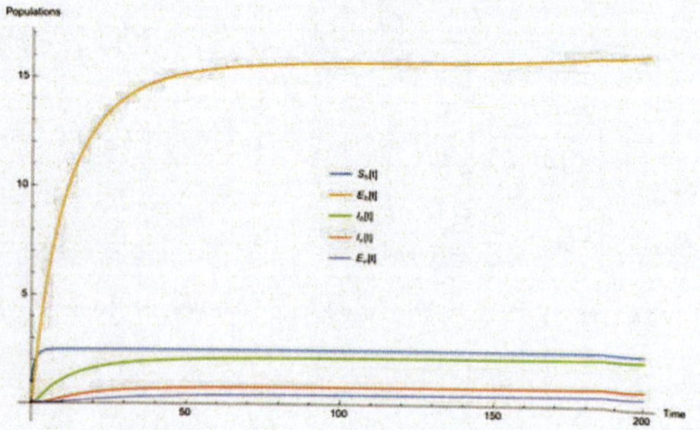

Fig. 3 For $\lambda = 0.55$

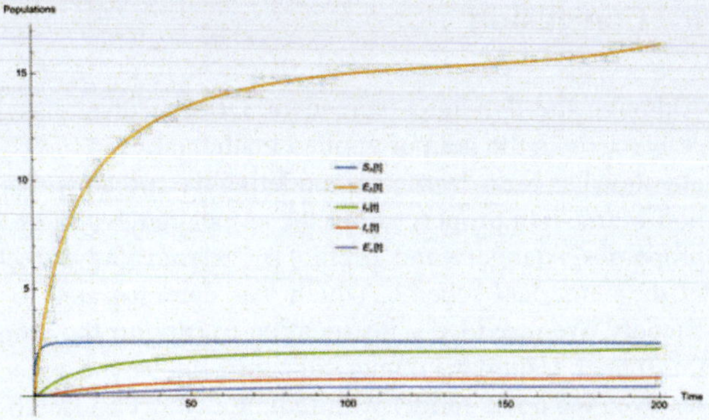

Fig. 4 For $\lambda = 0.35$

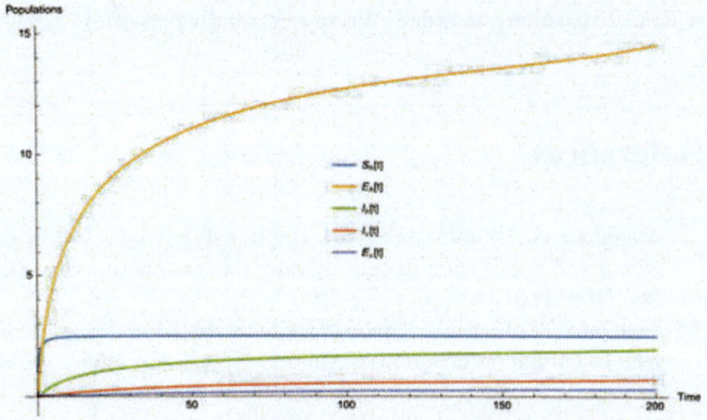

Fig. 5 For $\lambda = 0.15$

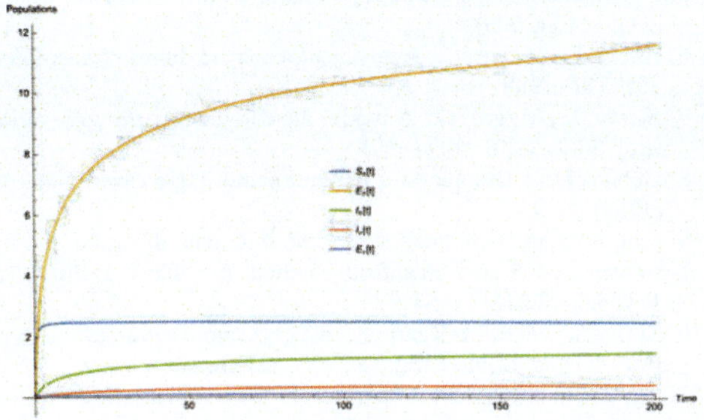

6 Conclusion

Although the dynamics spread of Dengue fever has attracted attention of many researchers in the field of applied mathematics in biology, it is worth noting that, no attention has been devoted to modeling the spread with a differential operator having non-Markovian properties but the associated evolution equation having Markovian properties. Atangana and Toufik [16] offered a novel technique to raise the precision of the numerical scheme, which was demonstrated to be very accurate and user-friendly. We provided a justification to support the employment of this operator in this model. Using this numerical approach, we introduced numerical simulation and resolved the novel model with fading memory caused by the exponential kernel.

Acknowledgments The United Arab Emirates' Ministry of Education (MOE) provides financial support for the project. The assistance is based on the 2019 Collaborative Research Program Grant (CRPG 2019) being awarded. We appreciate the opportunity the MOE has provided for us.

References

1. Atangana, A., Dumitru, B.: New fractional derivatives with nonlocal and non-singular kernel: theory and application to heat transfer model. Therma Sci. **18** (2016). https://doi.org/10.2298/TSCI160111018A
2. Atangana, A., Koca, I.: Chaos in a simple nonlinear system with Atangana-Baleanu derivatives with fractional order. Chaos Soliton Fract. (2016). https://doi.org/10.1016/j.chaos.2016.02.012
3. Atangana, A., Jain S.: A new numerical approximation of the fractal ordinary differential equation. Eur. Phys. J. Plus **133**, 37 (2018). https://doi.org/10.1140/epjp/i2018-11895-1
4. Caputo, M., Fabrizio, M.: A new definition of fractional derivative without singular kernel. Prog. Fract. Differ. Appl. **1**, 73–85 (2015)
5. Derouich, M., Boutayeb, A., Twizell, E.H.: A model of dengue fever. BioMed. J. Line Cent. **2**(1), 1–10 (2003)
6. Esteva, L., Vargas, C.: Analysis of a dengue disease transmission model. Math. Biosci. **150**(2), 131–151 (1988)
7. Esteva, L., Vargas, C.: A model for dengue disease with variable human population. J. Math. Biol. **38**(3), 220–240 (1999)
8. Gubler, D.J.: Dengue and dengue hemorhagic fever. Clin. Microbiol. Rev. **11**(3), 480–496 (1988)
9. Jain, S.: Numerical analysis for the fractional diffusion and fractional Buckmaster's equation by two step Adam- Bashforth method. Eur. Phys. J. Plus **133**, 19 (2018). https://doi.org/10.1140/epjp/i2018-11854-x
10. Kermack, W.O., McKendrick, A.G.: A contribution to the mathematical theory fo epidemics. Proc. R. Soc. Lond. **115**(772), 700–721 (1927)
11. Losada, J., Nieto, J.J.: Properties of the new fractional derivative without singular kernel. Prog. Fract. Differ. Appl. **1**, 87–92 (2015)
12. Phaijoo, G.R., Gurung, D.B.: Mathematical model of dengue disease transmission dynamics with control measures. J. Adv. Math. Comput. Sci. **23**(3), 1–12 (2017)
13. Sardar, T., Rana, S., Chattopadhyay: A mathematical model of dengue transmission with memory. Commun. Nonlin. Simmul. **22**(1), 511–525 (2015)
14. Side, S., Noorani, M.S.M.: SEIR model for transmission of dengue fever. Int. J. Adv. Sci. Eng. Inf. Technol. **2**(5), 380–389 (2012)

15. Soewono, E., Supriatna, A.K.: A two-dimensional model for the transmission of dengue fever disease. Bull. Malays. Math. Sc. Soc. **24**(1), 49–57 (2001)
16. Toufik, M., Atangana: New numerical approximation of fractional derivative with non-local and non-singular kernel: application to chaotic models. A. Eur. Phys. J. Plus **132**, 444 (2017). https://doi.org/10.1140/epjp/i2017--11717-0
17. Zizhang, Z., Jain, S.: Mathematical model of Ebola and Covid-19 with fractional differential operators: non-Markovian process and class for virus pathogen in the environment. Chaos Soliton Fract. 140, 110175 (2020)